Bernhard Welz
Michael Sperling

Atomic Absorption Spectrometry

 WILEY-VCH

Bernhard Welz
Michael Sperling

Atomic Absorption Spectrometry

Third, Completely Revised Edition

 **WILEY-VCH**

Weinheim · New York · Chichester · Toronto · Brisbane · Singapore

Dr. Bernhard Welz
In den Auen 20
D-88690 Uhldingen, Germany

Dr. Michael Sperling
Bodenseewerk
Perkin-Elmer GmbH
Alte Nußdorfer Straße
D-88662 Überlingen

Library of Congress Card No.: Applied for

British Library Cataloguing-in-Publication Data: A catalogue record for this book is available from the Britsh Library.

© WILEY-VCH Verlag GmbH, D-69469 Weinheim (Federal Republic of Germany), 1999

Printed an acid-free and chlorine-free paper.

Layout: Inge Bertsch
Composition: Kühn & Weyh, D-79111 Freiburg
Printing: Strauss Offsetdruck, D-69509 Mörlenbach
Bookbinding: Wilhelm Osswald & Co, D-67433 Neustadt
Printed in the Federal Republic of Germany

Preface to the third Edition

Atomic absorption spectrometry (AAS) is today, more than 40 years after it was proposed by Walsh as an analytical procedure, well established in numerous fields of instrumental analysis. Due to its high specificity and selectivity, as well as the fact that operation is relatively simple, AAS has gained its place alongside ICP OES and ICP-MS. It is used to perform numerous routine tasks in the laboratory ranging from the determination of trace contents through to major constituents. The fact that more than 1000 original papers dealing with AAS are published every year is a clear indication that there are numerous new developments over and above the routine applications.

These include new knowledge on atomization and other reaction mechanisms, improved analytical methods, especially in the fields of trace and ultratrace analysis, as well as in solids analysis, particularly using slurries. A major contribution has been the developments in instrumentation, such as transversely-heated graphite furnaces, integrated platforms, the application of solid-state detectors, or simultaneous multielement AAS, as well as new sample introduction and on-line pretreatment techniques such as flow injection. New areas of application including the analysis of 'high tech' materials and speciation analysis must also be mentioned.

To do justice to these manifold developments, this monograph has been rearranged, completely revised, and correspondingly extended. Thus, Chapter 1 on the historical development of AAS is new; in the first instance it demonstrates the maturity of the technique and provides the historically interested reader with the background information, and in the second instance it frees the technical chapters from historical ballast, leaving them free for a discussion of the current state of knowledge. Chapter 2 on the physical principles of AAS is also largely new; many of the topics discussed in this chapter are only inadequately treated in standard textbooks or not at all, and other topics have only been thoroughly developed in recent years.

Chapters 5, 6, and 7 are likewise new. Chapter 5 deals with procedures of measurement and calibration, the principles of quality control and assessment, and the basics of the statistical evaluation of the analytical results. Chapters 6 and 7 deal with automation and species analysis, and present a short review of the developments in these areas during recent years. On the other hand, the comparative chapter on other analytical procedures is no longer included since the recent developments in ICP OES and ICP-MS are well beyond the scope of this monograph.

Compared to the last edition, Chapter 8 includes significantly improved knowledge on the mechanisms of atomization and interferences, particularly for GF AAS and HG AAS. Chapter 9, dealing with the individual elements, now includes information on the stability and storage of test sample and calibration solutions, as well as the determination of species. In the treatment of applications in Chapter 10, methods no longer relevant, such as the determination of volatile elements by GF AAS using atomization from the tube wall with peak height evaluation, have been eliminated. On the other hand, all relevant procedures for speciation analysis are newly included.

Throughout this edition the terminology proposed by ISO and IUPAC has been used consistently. Thus, instead of ambiguous units such as ppm, ppb, etc., we have consistently used the ISO units mg/L, μg/L, ng/L and mg/g, μg/g, ng/g, etc. We have also attempted to avoid concentrations quoted in percent, which nevertheless was not possible in all cases since it was not clear from the original papers whether the concentration of acids, for example, were quoted in weight or volume percent.

To establish the bibliography we applied a relational databank (PELIDAS, © M. Sperling) to evaluate more than 55 000 entries from the field of atomic spectroscopy using plausibility checks to guarantee the quality of the citations. For the selection of the 6500 or so citations in this monograph, next to their information content, their topicality and availability also played a role. It is clear that for such a selection subjectivity comes into play, even though we have always attempted to be objective; we therefore ask for our readers' understanding if any paper that they deem to be important is not cited.

To maintain the topicality of this monograph we have also departed from traditional methods of production. Since the entire work, including the layout, was produced on the authors' PCs, we were able to update the contents until shortly before publication. We feel sure our readers will excuse the inadequacies of the word processor used in producing the layout since these are more than compensated by the advantage of topicality.

Überlingen, September 1998   Bernhard Welz
                                                                    Michael Sperling

Preface to the second Edition

In the nine years since the publication of the first edition of this monograph, atomic absorption spectrometry has undergone a remarkable development. This is perhaps no entirely true for flame AAS, which nowadays is established as a routine procedure in all branches of elemental analysis, but it is certainly the case with all other techniques of AAS. Even though flame AAS had already found acceptance in many standard methods due to its reliability in the mg/L range, it is only a few years ago that considerable doubt was cast on the ability of graphite furnace and hydride generation AAS to provide correct results at all in the μg/L and ng/L ranges.

The difficulties observed by many analysts using these techniques were due in part to the shortcomings of the instruments employed and in part to non-optimum application, since the significance of a number of parameters had not been recognized. In addition, the general problems of trace and nanotrace analysis had to be taken into consideration, since these newer techniques opened this concentration range to AAS.

These days, the causes of the majority of interferences and also the possibilities for their elimination are known. Even if all technical problems have not been completely solved, the way to their solution has been shown.

Thus, as well as the flame technique, the graphite furnace, hydride generation, and cold vapor techniques are nowadays of equal significance. The major field of application of these newer techniques is in trace, nanotrace and ultratrace analysis. Each of these techniques has its own atomizer, its own specific mechanisms of atomization and interference, and of course its own preferred field of application. In this second edition, these three techniques are thus treated separately whenever this appears expedient.

This made it necessary to substantially revise numerous chapters. Chapter 3 now deals only with atomizers, their historical development, and their specific characteristics for each technique. A new chapter 8 has been introduced in which the mechanisms of atomization and the interferences for each technique are discussed in detail. Additionally, typical interferences and their elimination are mentioned. A general discussion and classification of interferences is presented in chapter 7. Application of the Zeeman effect for background correcttion is also treated in detail in this chapter. This treatment includes the theoretical aspects of the method, the various configurations, and their advantages and disadvantages. In the chapters on individual elements and specific applications, the various techniques are, wherever applicable, weighed against each other.

A discussion on trace and nanotrace analysis has also been newly introduced, since the newer techniques of AAS are among the most sensitive methods for elemental analysis. Solids analysis is also treated since this has become possible with the graphite furnace technique. A section on environmental analysis has been included in the chapter on specific applications, and topical questions on the analysis of air, waste water and sewage sludge are addressed.

Among associated analytical methods, atomic emission spectrometry employing an inductively coupled argon plasma is discussed especially, since it is frequently regarded as a competitive technique to flame AAS. However, a broad treatment of this theme is outside the scope of this book.

Graphite furnace atomic emission spectrometry has also received attention even though, like atomic fluorescence spectrometry, it is rarely used in practice.

Finally, terms, nomenclature and units of measurement have been brought into line with the latest international standards – a fact reflected in the changed title of this monograph. Of particular help in this respect was my work on the committee of material testing within the German Institute of Standardization. This committee was chaired by Dr. Hans Massmann, who, until his death, worked on the completion of DIN 51 401 and who also made valuable suggestions for the second edition of this book – a fact greatly appreciated.

I should also like to thank those readers who wrote to me pointing out errors in the first edition; they have made valuable contributions to improving this work. I should particularly like to thank Sir Alan Walsh who drew my attention to a number of errors and who proposed numerous improvements and more precise definitions.

The numerous new diagrams were prepared with the customary care by Mr. E. Klebsattel who receives my grateful thanks. I should also like to thank Mr. J. Storz for designing the cover.

This book is the English-language version of its German forerunner "Atomabsorptionsspektrometrie" (formerly "Atom-Absorptions-Spektroskopie") which is now in its third edition. As for the first edition the translation has been very capably carried out by Christopher Skegg to whom I extend my thanks.

Meersburg, May 1985   Bernhard Welz

Preface to the first Edition

It was very convenient that the translation of my book into the English languagee was undertaken just as I was completing the second German edition. Therefore, all the latest developments and publications could be incorporated directly in the English edition

Usually, years go by between the publication of the original book and the completion of a translation which, therefore, typically does not represent the latest state. Here, however, the translation could be published about a year after the original German edition. This is of special importance for the rapidly growing field of furnace atomic absorption which was hardly known a few years ago when the first edition of my book was published. In the meantime it has found worldwide acceptance among analysts.

So, to all my friends and colleagues who have been involved in the translation and completion of this book, I would like to express my thanks for the time that they have spent and for all the effort that they have put into it so that it could be published so early. Last not least, I want to express my pleasure that my book on Atomic Absorption Spectroscopy has been accepted for translation into English. I hope that it will prove a stimulus to atomic absorption spectroscopy and will help analysts and spectroscopists in their daily work.

Meersburg, March 1976   Bernhard Welz

Contents

Abbreviations and Acronyms

The following abbreviations and acronyms are used in this monograph:

AA	acetylacetone
AAS	atomic absorption spectrometry
A/D (conversion)	analog-to-digital
AES	Auger electron spectroscopy
AFM	atomic force microscopy
ANOVA	analysis of variance
APDC	ammonium pyrrolidine dithiocarbamate
AsB	arsenobetaine
ASV	anode stripping voltammetry
BC	background correction; background corrector
BCR	Bureau Commun de Référence, Belgium
BERM	International Symposium for Biological and Environmental Reference Materials
BG	borosilicate glass
BOC	baseline offset correction
BPTH	1,5-bis[phenyl-(2-pyridyl)methylene]thiocarbohydrazide
CARS	coherent anti-Stokes Raman spectroscopy
CCD	charge coupled device
CE	concentration efficiency
CF	continuous flow
CGC	capillary gas chromatography
CI	consumptive index
CID	charge injection device
CPG	controlled pore glass
CRA	carbon rod atomizer
CRM	certified reference material
CSIRO	Commonwealth Scientific and Industrial Research Organization (Australia)
CT	cryotrapping
CV AAS	cold vapor AAS
DAL	dialkyl-lead
DBT	dibutyltin
DCTA	1,2-diaminocyclohexane-N,N,N',N'-tetraacetic acid
DDAB	didodecyl-dimethylammonium-bromide
DDC	diethyldithiocarbamate
DDTC	diethyldithiocarbamate
DDTP	dimethoxydithiophosphate
DESe	diethylselenium
DIBK	di-isobutyl ketone
DIN	*Deutsches Institut für Normung* (German Standards Institute)

DMA	dimethylarsonate
DMF	dimethylformamide
DMSe	dimethylselenium
DMSO	dimethylsulfoxide
DPTH	1,5-bis(di-2-pyridylmethylene)thiocarbonhydrazone
DTC	dithiocarbamate
DTP	dithiophosphoric acid
EDL	electrode discharge lamp
EDTA	ethylendiamine-tetraactic acid
EDX	energy-dispersive X-ray spectrometry
EF	enrichment factor
EG	polycrystalline electrographite (also *Erfassungsgrenze* in Section 5.2.3)
EMP	electron microprobe
EPA	Environmental Protection Agency (U.S.A.)
ESCA	electron spectroscopy for chemical analysis
ET AAS	electrothermal AAS
ETV	electrothermal vaporization
ETV-ICP-MS	electrothermal vaporization inductively coupled plasma mass spectrometry
F AAS	flame AAS
FANES	furnace atomic non-thermal excitation spectrometry
FEP	fluorinated engineering polymers (perfluoro-ethylene-propylene)
FG	flint glass
FI	flow injection
FIA	flow injection analysis
FID	flame ionization detector
FIMS	flow injection mercury system (Perkin-Elmer)
FIT	flame-in-tube (non-heated quartz tube atomizer with flame)
FWHM	full width at half maximum
GAP	good analytical practice
GC	gas chromatography (also glassy carbon in Chapter 1)
GF AAS	graphite furnace AAS
GLP	good laboratory practice
GLS	gas-liquid separator
HCL	hollow cathode lamp
HG AAS	hydride-generation AAS
HGA	longitudinally-heated graphite atomizer (Perkin-Elmer)
HMA-HMDTC	hexamethyleneammonium-hexamethylenedithiocarbamate
HMDTC	hexamethylenedithiocarbamate
HPLC	high performance liquid chromatography
IAEA	International Atomic Energy Authority (Austria)
IBMK	isobutyl methyl ketone
ICP	inductively coupled plasma
ICP-MS	inductively coupled plasma mass spectrometry

ILOD	instrument limit of detection
INAA	instrumental neutron activation analysis
IR	infrared (wavelength range > 800 nm)
ISO	International Organization for Standardization
IUPAC	International Union of Pure and Applied Chemistry
KR	knotted reactor
LC	liquid chromatography
LD-TOF-MS	laser-dispersion time-of-flight mass spectrometry
LEAFS	laser-enhanced atomic fluorescence spectrometry
LOD	limit of detection
LOQ	limit of quantitation
MBT	monobutyltin
MeHg	methylmercury
MeT	methyltin
MIP	microwave induced plasma
MMA	monomethylarsonate
MMT	monomethyltin
MS	mass spectrometry
NAA	neutron activation analysis
NaHEDC	bis(2-hydroxyethyl)dithiocarbamate sodium salt
NIST	National Institute of Standards and Technology (U.S.A.)
NL	non-linearity
NTA	nitrilotriacetic acid
OES	optical emission spectroscopy
PAN	1-(2-pyridylazo)-2-naphthol
PAR	4-(2-pyridylazo)resorcinol
PC	polycarbonate
Pd-Mg	mixed modifier of palladium nitrate and magnesium nitrate
PE	polyethylene
PET	poly(ethyleneterephthalate)
PFA	perfluoroalkoxy plastics
PG	pyrolytic graphite
PI	polyimides
PIXE	particle [proton] induced X-ray emission spectroscopy
PMPB	1-phenyl-3-methyl-4-benzoyl-5-pyrazolone
PMT	photomultiplier tube
PP	polypropylene
PSF	polysulfone
PTFE	polytetrafluoroethylene
PU	polyurethane
PVC	polyvinyl chloride
PVD	physical vapor deposition
QF	quartz tube furnace with a small flame burning in it
QTA	quartz tube atomizer
RBS	Rutherford backscattering spectroscopy

REE	rare earth elements
RF	radio frequency
RNAA	radiochemical neutron activation analysis
ROC (model)	reduction of oxides by carbon
S	silica
S-H	Smith-Hieftje high current pulsing background correction
S/N	signal-to-noise ratio
SEM	scanning electron microscopy
SeMet	selenomethionine
SIMS	secondary ion mass spectrometry
SRM	standard reference material
SS	solid sampling
SSD	solid state detector
SSF	spectral shadow filming
STM	scanning tunneling microscopy
STPF	stabilized temperature platform furnace (a concept for quasi isothermal atomization)
TAL	tetraalkyl-lead
TAR	4-(2-thiazolylazo)resorcinol
TBT	tributyltin
TEL	tetraethyl-lead
TEM	transmission electron microscopy
TGL	temperature gradient lamp
THF	tetrahydrofuran
THGA	transversely-heated graphite atomizer with integrated platform (Perkin-Elmer)
TMDTC	tetramethylenedithiocarbamate
TML	tetramethyl-lead
TMSe	trimethylselenonium
TOMA	tri-N-octylmethylammonium
TOPO	trioctylphosphine oxide
TPG	total pyrolytic graphite
TPN (patients)	total parenteral nutrition
TPP	triphenylphosphine
TTFA	thenoyltrifluoroacetone
UV	ultraviolet (wavelength range < 400 nm)
WHO	World Health Organization
XAD	ion exchange resin
XPS	X-ray photoelectron spectroscopy
XRD	X-ray diffraction analysis
ZBC	Zeeman-effect background correction

'Atomic absorption spectrometry (AAS) is a spectroanalytical procedure for the qualitative detection and quantitative determination of elements employing the absorption of optical radiation by free atoms in the gaseous state' [1501].

1 The Historical Development of Atomic Absorption Spectrometry

1.1 The Early History

The beginning of optical spectroscopy is generally attributed to Sir ISAAC NEWTON [3205] who, in a letter to the Royal Society in 1672, described the observation that sunlight is split into various colors when it is passed through a prism. Albeit JOANNES MARCUS MARCI VON KRONLAND (1595–1667), professor of medicine at the University of Prague (Figure 1-1), had already explained the origin of the rainbow on the basis of the diffraction and scattering of light in water droplets in his book *Thaumantias. Liber de arcu coelesti deque colorum apparentium natura ortu et causis* published in 1648; he can thus be looked upon as the first spectroscopist.

The history of absorption spectrometry is closely connected with the observation of sunlight [5934]. In 1802 Wollaston discovered the black lines in the sun's spectrum. These were later investigated in detail by Fraunhofer, who assigned letters to the

Figure 1-1. 'Joannes Marcus Marci, Doctor of Philosophy and Medicine, and Professor, born in Kronland in Bohemia, 17 June 1595.'

strongest lines, starting at the red end of the spectrum with the letter A. Even nowadays it is common to refer to the 'sodium D line', a designation originated by Fraunhofer.

In 1820 Brewster expressed the view that these Fraunhofer lines were caused by absorption processes in the sun's atmosphere. The underlying principles of this absorption were established by KIRCHHOFF and BUNSEN [3121–3124] during their systematic examination of the line reversal in the spectra of alkali and alkaline-earth elements. They conclusively demonstrated that the typical yellow line emitted by sodium salts in a flame is identical to the black D line of the sun's spectrum. The classical experimental arrangement is shown in Figure 1-2.

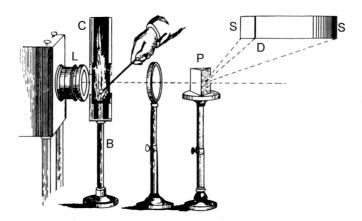

Figure 1-2. Experimental setup of KIRCHHOFF and BUNSEN for investigating the line reversal in the sodium spectrum (according to [5934]). Radiation from a lamp is focused by lens **L** through the flame of a Bunsen burner **B** into which sodium chloride is introduced with a spatula. The radiation beam is dispersed by prism **P** and observed on screen **S**. The sodium D line appears as a black discontinuity in the otherwise continuous spectrum.

The relationship between emission and absorption was formulated by Kirchhoff in his law, which is generally valid and states that any material that can emit radiation at a given wavelength will also absorb radiation of that wavelength.

The connection between atomic structure and the interaction of atoms with radiation was established by Planck (1900) in the quantum law of absorption and emission of radiation, according to which an atom can only absorb radiation of well defined wavelength λ or frequency v, i.e. it can only take up and release definite amounts of energy ε:

$$\varepsilon = hv = \frac{hc}{\lambda},$$

(1.1)

where h is Planck's constant and c is the speed of light. Characteristic values of ε and v exist for each atomic species.

On the basis of this and many other spectroscopic observations, Bohr proposed his atomic model in 1913, the fundamental principle of which is that atoms do not exist in random energy states, but only in certain fixed states which differ from each other by integral quantum numbers. Upon absorbing a quantum of energy, an atom is transformed into a particular, energy-enriched state 'containing' the radiation energy which has been taken up. After a period of around 10^{-9} s to 10^{-8} s the atom can re-emit this energy and thus return to the ground state.

Although Kirchhoff had already recognized the principle of atomic absorption in 1860 and the theoretical basis was steadily extended during the following decades, the practical significance of this technique was not recognized for a long time. Since the work of Kirchhoff, the principle of atomic absorption was mainly used by astronomers to determine the composition and concentration of metals in the atmospheres of stars. Chemical analyses were only carried out very sporadically by this technique; the determination of mercury vapor did, however, acquire a degree of importance [4225, 6380] (see Section 1.8.1).

The actual year of birth of modern AAS was 1955. In that year, publications authored independently by WALSH [6135] and ALKEMADE and MILATZ [125, 126] recommended AAS as a generally applicable analytical procedure.

1.2 Sir Alan Walsh and the Period 1952–1962

Even though the publications by ALKEMADE and MILATZ [125, 126] in The Netherlands and WALSH [6135] in Australia appeared in the same year, making it difficult to answer the question as to who actually rediscovered AAS, Alan Walsh (Figure 1-3) is generally recognized as the 'father' of modern AAS. This privilege is his just due since he campaigned with untiring energy against the resistance to this new idea for more than a decade, spending much time to overcome the disinterest and misunderstanding. Best of all, let Alan Walsh himself describe the events and developments of the period 1952 to 1962 [6137].

'My initial interest in atomic absorption spectroscopy was a result of two interacting experiences: one of the spectrochemical analysis of metals over the period 1939–1946; the other of molecular spectroscopy over the period from 1946–1952. The interaction occurred early in 1952, when I began to wonder why, as in my experience,

Figure 1-3. Sir Alan Walsh.

molecular spectra were usually obtained in absorption and atomic spectra in emission. The result of this musing was quite astonishing: there appeared to be no good reasons for neglecting atomic absorption spectra; on the contrary, they appeared to offer many vital advantages over atomic emission spectra as far as spectrochemical analysis was concerned. There was the attraction that absorption is, at least for atomic vapours produced thermally, virtually independent of the temperature of the atomic vapour and of excitation potential. In addition, atomic absorption methods offered the possibility of avoiding excitation interference, which at that time was thought by many to be responsible for some of the interelement interferences experienced in emission spectroscopy when using an electrical discharge as light source. In addition, one could avoid problems due to self-absorption and self-reversal which often make it difficult to use the most sensitive lines in emission spectroscopy.'

'As far as possible experimental problems were concerned, I was particularly fortunate in one respect. For several years prior to these first thoughts on atomic absorption, I had been regularly using a commercial i.r. spectrophotometer employing a modulated light source and synchronously tuned detection system. A feature of this system is that any radiation emitted by the sample produces no signal at the output of the detection system. This experience had no doubt prevented the formation of any possible mental block associated with absorption measurements on luminous atomic vapours.'

In an internal report for the period February–March 1952 Walsh suggested that the same type of modulated system should be considered for recording atomic absorption spectra. 'Assuming that the sample is vaporised by the usual methods, e.g. flame, arc, or spark, then the emission spectrum is "removed" by means of the chopper principle. Thus the emission spectrum produces no output signal and only the absorption spectrum is recorded.'

In the same report he continued: 'For analytical work it is proposed that the sample is dissolved and then vaporised in a Lundegardh flame. Such flames have a low temperature (2000 K) compared to arcs and sparks (5000 K) and have the advantage that few atoms would be excited, the great majority being in the ground state. Thus absorption will be restricted to a small number of transitions and a simple spectrum would result. In addition, the method is expected to be sensitive since transitions will be mainly confined to those from the ground level to the first excited state.'

The next report for the period April–May 1952 included the diagram shown in Figure 1-4 and described the first experiment as follows: 'The sodium lamp was operated from 50 cycles/s and thus had an alternating output so that it was not necessary to use a chopper. The D lines from the lamp were isolated—but not resolved from each other—by means of a direct vision spectroscope and their intensities were measured by means of a photomultiplier tube, the output from which was recorded on a cathode ray oscillograph. Amplification of the signal was achieved by the a.c. amplifiers in the oscillograph. With the slit-width used the signal gave full-scale deflection on the oscillograph screen. A Meker flame was interposed between the sodium lamp and the entrance slit of the spectroscope. When a solution of sodium chloride was atomised into the air supply of the flame the signal at the oscillograph was reduced to zero. The principle of the method is therefore established.'

Figure 1-4. The first outline by Walsh for the measurement of atomic absorption from his report for April–May 1952.

In retrospect Walsh admitted to 'optimistic naivety'; he nevertheless recalls: 'This simple experiment gave me a great thrill, and I excitedly called John Willis, who at the time was working on infrared spectroscopy and was later to make important contributions to the development of atomic absorption methods of chemical analysis. "Look," I said, "that's atomic absorption." "So what?" was his reply, which was the precursor of many disinterested reactions to our atomic absorption project over the next few years.'

In his report for June–July 1952 Walsh discussed the problem of recording atomic absorption spectra using a continuum source and came to the conclusion that a resolution of approximately 2 pm would be required. This was far beyond the capabilities of the best spectrometer available in his laboratory at that time. The report concluded: 'One of the main difficulties is due to the fact that the relations between absorption and concentration depend on the resolution of the spectrograph, and on whether one measures peak absorption or total absorption as given by the area under the absorption/wavelength curve.'

This realization led him to conclude that the measurement of atomic absorption requires line radiation sources with the sharpest possible emission lines. The task of the monochromator is then merely to separate the line used for the measurement from all the other lines emitted by the source. The high resolution demanded for atomic absorption measurements is in end effect provided by the line source.

At this point Walsh had already recognized the salient points of AAS: The use of line radiation sources, which make a high-resolution monochromator unnecessary, the principle of modulation, which makes the technique selective and eliminates the radiation emitted by the atomizer, and the use of laminar flames of relatively low temperature to atomize the sample. A patent for the technique was lodged at the end of 1953 and an atomic absorption spectrometer was publicly demonstrated in Melbourne in March 1954. However, during the exhibition the instrument aroused little interest.

The last changes to the specifications for the patent were submitted in October 1954 and immediately thereafter WALSH sent his first manuscript about AAS to Spectrochimica Acta, which was published at the beginning of 1955 [6135]. This paper was followed by others from Walsh's group [5010] and also from ALLAN [135] in New Zealand and DAVID [1426] in Australia. Nevertheless the technique was still looked upon as a 'scientific curiosity' rather than a practical analytical technique.

In the meantime, Hilger and Watts had built an instrument, but since the radiation source was not modulated it could not exploit the technique. Other manufacturers later made the same mistake. WALSH [6137] recalls that 'by 1958 there was no sign of any instrument manufacturer willing to produce the type of instrument which we thought desirable.'

WALSH then decided to arrange for the production of appropriate equipment. The necessary items were manufactured by three small companies in Melbourne and then assembled by the user. 'As it transpired,' he wrote [6137], 'for the next few years the members of our research group were increasingly involved in supporting the commercial production in Australia of atomic absorption equipment. That a new type of Australian industry was eventually created was, of course, cause for much satisfaction, but it was inevitable that there was a substantial reduction in our research effort over a period of several years.'

Figure 1-5. The Perkin-Elmer Model 303, the first spectrometer built exclusively for AAS.

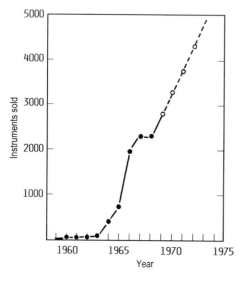

Figure 1-6. Sales figures for atomic absorption spectrometers during the first 15 years (from [6138]).

'For me, one of the "great moments" was in 1962 when I described to various members of staff of the Perkin-Elmer Corp. in Norwalk the impressive results which were being obtained by the laboratories in Australia which were by that time successfully using the technique. It was during these discussions that Chester Nimitz (the then current General Manager of the Instrument Division of Perkin-Elmer) asked, rather tersely: "If this goddam technique is as useful as you say it is, why isn't it being used right here in the USA?" My reply, which my friends in Norwalk have never allowed me to forget, was to the effect that the USA was an underdeveloped country! The Perkin-Elmer decision to embark on a large-scale project relating to the production of atomic absorption equipment was made shortly afterward.' The very first instrument built to Walsh's ideas, the Model 303, is depicted in Figure 1-5. From this time point there was a remarkable growth in AAS, which can be demonstrated by the number of spectrometers produced (Figure 1-6).

1.3 The Development of Spectral Radiation Sources

One of the decisive steps during the rediscovery of AAS was the realization that the high resolution required for measurements by this technique could be provided by the sharp emission lines of line radiation sources. For his experiments, WALSH used *hollow cathode lamps* (HCLs). This type of lamp was first described by PASCHEN [4555] in 1916. However, early lamps of this type were unstable and not at all easy to operate since argon under a pressure of approximately 1 kPa had to be continuously pumped through them. Only after the introduction of sealed lamps with fused-in electrodes [1524], which were modified by WALSH et al. [2945, 5011] for use in AAS, did this type of source become applicable for routine use. The construction of one of these early HCLs is depicted in Figure 1-7.

Despite the clear advantages of sealed sources for routine operation, demountable HCLs with exchangeable cathodes have been reported from time to time [258, 2174, 3129, 4968, 5593]. However, these sources have only acquired a limited significance for research purposes, since their operation is more complicated and requires experience of high vacuum techniques.

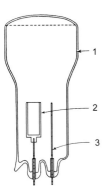

Figure 1-7. Schematic construction of a simple hollow cathode lamp as used by JONES and WALSH [2945]. **1** – glass cylinder; **2** – hollow cathode; **3** – nickel anode.

At the beginning of the 1960s a development phase set in during which the HCL was made into an extremely reliable spectral source that can be used for virtually all elements determinable by AAS. SLAVIN [5439] recalls in particular the very fruitful cooperation between Jack Sullivan at CSIRO and Carl Sebens and John Vollmer from Perkin-Elmer. A further impulse came from WHITE [6297] who showed that after several hundred operating hours a normal cylindrical cathode changes its geometry due to the transport of sputtered metal from the hot inner side to the surface. Thereby a cathode in the form of a hollow sphere with a relatively small opening is produced. When the ratio of the diameter of the opening to the internal diameter of the sphere is 1:4, virtually no more metal atoms can escape from inside the cathode, thus largely eliminating the problems of self-absorption and self-reversal.

After years of research, MANNING and VOLLMER [3849] introduced an HCL that was marketed under the tradename Intensitron®. Careful isolation of the anode and cathode with mica disks and ceramic materials contributed markedly to the stabilization and localization of the glow discharge, leading to a more intense emission for the same lamp current, to a better ratio of the emission of the atomic lines in comparison to the ion lines, to less self-absorption, and to a longer lifetime (see Section 3.1.1). For a number of elements, particularly stable and intense emission could be obtained when the metal was in the molten state in the cathode. VOLLMER [6100] reported an HCL containing a mixture of bismuth and lithium that had a significantly improved radiation emission compared to a lamp with a cathode made of bismuth only. The operating current was chosen in such a way that a thin film of molten bismuth formed on the inside of the cathode. Similar lamps, in which the analyte is molten during operation, have also been described for a number of other elements [3850, 6101].

Numerous working groups turned their attention to the construction of *multielement* lamps in order to avoid a too frequent change of lamp. JONES and WALSH [2945] constructed a lamp containing several cathodes in the same glass cylinder, each of a different metal. MASSMANN [3834], and BUTLER and STRASHEIM [982], constructed multielement lamps whose cathodes consisted of rings of various metals pressed together. The concept used nowadays for the construction of multielement lamps was developed by SEBENS *et al.* [5207], in which various metals in powder form are mixed, pressed, and sintered. Using this method, virtually any metal combination can be produced [1882, 2496, 3842].

As well as the work on the development of reliable radiation sources for routine use, considerable efforts were directed to increasing the radiant intensity while at the same time decreasing the self-absorption. SULLIVAN and WALSH [5681] constructed a special type of hollow cathode in which the glow discharge on the cathode was used principally to sputter as much cathode material as possible. The cloud of atoms, which in normal HCLs is responsible for the self-absorption, was then excited by a second discharge produced in front of the cathode by a supplementary electrode isolated from the first discharge. The authors reported a hundred-fold increase in the intensity of these lamps compared to conventional HCLs. The half-width and profile of the resonance lines were said not to have altered and, above all, practically no more self-absorption or self-reversal was to be observed. Accordingly, the calibration curves published by SULLIVAN and WALSH showed good linearity up to high absorbance values. These findings were verified through work by CARTWRIGHT *et al.* [1092–1094], in which attention was also drawn to the considerable improvements for such difficultly vaporized elements as silicon, titanium, and vanadium. At that time only lamps of low radiant intensity were available for these elements. However, despite occasional reports about their use [120], these high-intensity lamps did not attain lasting significance in AAS. Review articles on the development of radiation sources for AAS were published by BUTLER and BRINK [984] and by SULLIVAN [5687].

Recently, various HCLs have been described in which the glow discharge is enhanced by microwaves [1079, 1080] or by radio waves [245, 1848]. Compared to conventional HCLs, an improved signal-to-noise (S/N) ratio and lower limits of detection were achieved with these lamps [4830, 6173]. Nevertheless, since only a two- to threefold improvement is obtained it is questionable whether the increased operating complexity is worthwhile. A monograph edited by CAROLI [1078] provides a good overview of the various attempts to construct improved HCLs for AAS.

Electrodeless discharge lamps (EDLs) are among the radiation sources exhibiting the highest radiant intensity and the narrowest line widths. A thorough investigation of this lamp type was made by BLOCH and BLOCH [670] in 1935 and in subsequent years they were used in a variety of high resolution studies. At the end of the 1960s EDLs found increasing interest in atomic fluorescence spectrometry [301, 887, 1400–1402, 1405, 3867]. The more intense radiation of EDLs by at least a power of ten compared to normal HCLs was of considerable interest for this technique since the fluorescence signal is, in the first approximation, directly proportional to the radiant power. At about the same time the first applications in AAS were described [888, 1400, 1404, 1405].

The early papers published on EDLs were largely concerned with the design of these lamps and a number of authors gave detailed instructions on their construction [1404, 1405, 3867, 6381]. Although their design is really very simple—a few micrograms to a few milligrams of the analyte, either as the element or the halide, are fused into a quartz tube under an argon pressure of about 1 kPa—their operation gives rise to a number of problems. Various factors, such as the purity and dimensions of the quartz tube, the pressure of the inert fill gas, and the quantity and chemical form of the analyte, play an important role. The excitation frequency and the location of the quartz tube within the radio frequency coil are further important factors. SNEDDON *et al.* [5480] have published a review article which provides the interested reader with detailed information on the development, manufacture, and operation of this type of lamp.

In an entertaining manner, HERB KAHN [3000] described how, through a conversation with L'vov, the enormous difficulties in the manufacture of EDLs could be virtually eliminated in a single stroke. At that time L'vov was routinely using EDLs without having observed any difficulties. During the conversation it became clear that two factors were essential for the stable operation of EDLs: firstly the metal content in the lamp should be in the microgram and not in the milligram range, and secondly the lamp should be operated at a frequency of 100 MHz or less and not at a frequency of 2450 MHz which was usual up to that time. As Kahn said, within three hours of reporting this conversation to his engineers he saw the 'best cadmium EDL that had ever been seen in the western world.'

Very volatile metals, such as mercury, thallium, zinc, and the alkali metals, earlier were determined using *low pressure vapor discharge lamps* since these were offered commercially at low prices and gave a high radiant intensity. Various papers were published in which these lamps were compared to HCLs [3841, 5010, 5409]. The disadvantage of these lamps is that—under normal operating conditions—they emit strongly broadened lines due to self-absorption and self-reversal of the high internal atom concentration and are therefore little suited to atomic absorption measurements [5891]. For use in AAS, vapor discharge lamps had to be operated at highly reduced currents to avoid excessive self-reversal which in turn resulted in increased instability. Since excellent EDLs for the volatile elements are nowadays commercially available, vapor discharge lamps are no longer of significance in AAS.

The *temperature gradient lamps* (TGLs) introduced by GOUGH and SULLIVAN [2185], which are in principle also vapor discharge lamps, gained a degree of importance. The section of the lamp containing the analyte element (as the pure metal) is heated. Two electrodes are located in a separate section of the lamp; a relatively high current (approx. 500 mA) at low voltage (approx. 30 V) is applied to the electrodes. The quartz window is located in the discharge zone where it is heated to prevent condensation and self-reversal of the metal vapor. It was reported that these lamps provided a greater radiant power for the same line width as EDLs [2185]. Naturally a TGL can only be manufactured for those elements that have a sufficiently high vapor pressure at the operating temperature, such as As, Cd, K, Na, P, S, Se, Te, and Zn.

In 1955 ALKEMADE and MILLATZ [126] proposed using a *flame*, into which high metal salt concentrations were sprayed, as a radiation source for AAS; various authors followed this stimulus [5395]. A flame as the primary radiation source has the advantage that it is cheap, universal, and very flexible. Especially for multielement analyses [982] it admits to practically any desired element combination. As further variants, electric sparks [5594, 5596] and arcs [2721, 3280] were investigated as radiation sources by a number or authors. A disadvantage of these radiation sources—compared to lamps—is their low stability and intensity. However, the greatest disadvantage is that, depending on the temperature of the source, the half-widths of the emission lines are at least as wide and mostly wider than those of the absorption lines, leading to non-linear calibration curves (see Section 2.5.3). These disadvantages have prevented the wider application of this type of source.

Despite the unfavorable prognosis that Walsh made for the application of *continuum sources* in AAS, time and again experiments have been made with such sources. Continuum sources demand a spectral apparatus with a very high power of resolution. In his first paper about AAS, WALSH [6135] showed that a monochromator with a resolution of at least 2 pm was necessary to measure the line profiles of absorption lines. Walsh was also doubtful whether there would be enough radiant energy in this small interval to provide a usable S/N ratio. This was confirmed in the earliest papers published on the use of continuum sources since the sensitivity obtained was around two powers of ten lower than with HCLs [3840]. A little later, however, FASSEL et al. [1851], by using a 150 W xenon arc lamp or a 650 W halogen lamp, were able to attain sensitivities for a number of elements that were comparable to those obtained using HCLs. Similar results were also reported by MCGEE and WINEFORDNER [4016], and KELIHER and WOHLERS [3070]. However, the calibration curves in many cases were significantly more non-linear than when using HCLs. More recently the application of continuum lamps as the primary source has been judged increasingly positive. In a review article, MARSHALL et *al.* [3903] particularly emphasize the possibility of performing multielement analyses in AAS.

The first really convincing application of continuum sources in AAS come from ZANDER *et al.* [6515, 6516], O'HAVER *et al.* [4409], and HARNLY and O'HAVER [2386]. These authors used xenon arc lamps with a power of 150–300 W, an Echelle polychromator, and wavelength modulation with a quartz refractor plate. This permitted measurements to be performed at various positions along the line profile. HARNLY [2389] and also O'HAVER and MESSMAN [4410] have published review articles on their experience of using continuum sources for multielement AAS.

The greatest problem for the application of continuum sources is that the radiant intensity of the xenon arc lamps normally employed falls off drastically below 280 nm [4409]. SMITH *et al.* [5460] found that at the zinc line at 213.9 nm a line radiation source emits 500 times more photons than a continuum source. To overcome this problem, experiments have been performed more recently to pulse continuum sources [4209, 4210, 5160] or to use flash lamps [459] since the radiant intensity is higher over a short time interval.

Lasers, whose properties, such as intensity and spectral bandwidth, can be optimized for practically every spectroscopic experiment, have been proposed as radiation sources. *Tunable dye lasers* can be set to virtually any atomic line between 213 nm and 900 nm with a bandwidth corresponding to the natural line width of an atomic line. These properties would seem to make them suitable for use in analytical atomic spectroscopy [1807]. Despite the extremely good spectroscopic properties of tunable dye lasers there are a number of practical and economic reasons why they have not found their way into common use: these laser systems are expensive, frequently unreliable, and difficult to operate.

In contrast to dye lasers, *diode lasers* would seem to be more suitable to one day replace HCLs and EDLs. Diode lasers are cheap, reliable, easy to operate, and have long lifetimes [2519, 4362]. HERGENRÖDER and NIEMAX [2518] presented a laser AAS instrument in which the HCL had been replaced by a diode laser tuned to the absorption wavelength. The monochromator with photomultiplier could then be replaced by a simple semiconductor photodiode [3470]. These authors point to a number of advantages of laser AAS which can be ascribed to the fact that, in contrast to HCLs and EDLs, lasers are sources with selectable wavelengths. It is possible to set the wavelength to the flank of an absorption line, thus allowing higher analyte concentrations to be determined. The background in the direct neighborhood of the absorption line can be measured and corrected, or a reference element can be measured. Similar suggestions for the use of diode lasers as sources in AAS have also come from the group around WINEFORDNER [377, 378, 4339–4341] and from AXNER [3562].

As attractive as laser AAS may appear, it is unlikely that it will attain practical significance in this century. The reason lies in the wavelength range available to diode lasers. Currently the lowest reachable wavelengths are around 660–670 nm [377, 378, 3470]. By cooling to 0 °C or with liquid nitrogen, respectively, it has been possible to come down to 620 nm or 585 nm, but this technique is unsuitable for routine use. Even with frequency doubling in crystals it is likely to take years until the important wavelength range for AAS of 190–325 nm can be attained with diode lasers [3470].

1.4 Single-beam, Double-beam, Single-channel, Multi-channel

The first AAS instrument presented by WALSH and co-workers in Melbourne in 1954 was a double-beam spectrometer. WALSH recalls that 'after our initial experiments we regarded this as absolutely necessary because of the poor stability of many of our HCLs' [6137]. Resulting from the close contact that Walsh had with Perkin-Elmer (compare Section 1.2), the first AAS instrument developed by that company, the Model 303, was a double-beam spectrometer (Figure 1-8).

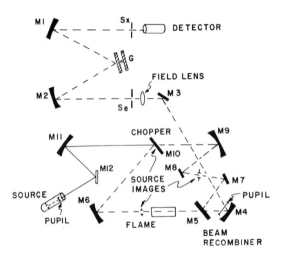

Figure 1-8. Beam path of the Perkin-Elmer Model 303 double-beam spectrometer (from [2990]).

The 'do-it-yourself' instruments produced by Walsh *et al.* from 1958 obviously had to be of simple design and thus employed the single-beam principle. From these beginnings the Techtron company later developed a complete atomic absorption spectrometer that was also a single-beam instrument. Likewise, the early instruments manufactured by Hilger and Watts and several other companies were also single-beam spectrometers, since they were built from existing emission spectrometers or their components.

The question as to the better principle—single-beam or double-beam—split AAS theorists and users into two often hostile camps for more than two decades. The greatest advantage ascribed to single-beam instruments was the use of a minimum number of optical components and thus a lower loss of radiant energy. Long term stability was seen as the major advantage of double-beam spectrometers. A number of authors showed that the precision attainable with a single-beam spectrometer was in fact better [5559], especially at low absorbance values near to the limit of detection [3518]. Under these conditions noise is mainly due to photon noise, lamp flicker noise, and flame noise.

Even at an early stage a number of authors recognized that measurements with a single-beam spectrometer were virtually independent of larger fluctuations in the operating conditions, provided that a sufficiently large number of reference measurements were performed [3418]. Electrical circuits for drift correction were also described that provided drift-free signals for several hours [5716].

Nowadays there is more or less general agreement that the advantages of double-beam spectrometers frequently have been overvalued. The double-beam principle can merely compensate for changes in intensity of the radiation source and changes in sensitivity of the detector, but not drift phenomena in the atomizer, such as the warmup of a burner. Further, it has been shown that during the warmup phase of a radiation source, not only the radiant intensity, but also the line profile, and thus the sensitivity, change [1357]. This means that even with a double-beam instrument a certain warmup time is required. But since the quality of radiation sources nowadays is significantly better than in the early days of AAS and their radiant intensity changes only very slowly after a relatively short warmup time, increasing numbers of single-beam instruments are coming into use that have double-beam characteristics by automatically measuring and correcting the baseline directly prior to the measurement.

As well as classical monochromators, *resonance detectors* were investigated in the early days of AAS. SULLIVAN and WALSH [5682, 5683, 5686] reported the use of resonance detectors as 'monochromators' in atomic absorption spectrometers. The mode of operation of a resonance detector is depicted schematically in Figure 1-9. After passing through the atomizer, the spectral radiation emitted from the primary source falls on the resonance detector. The detector consists of a device for producing a cloud of neutral analyte atoms, similar to an HCL. If the radiation passing through the atomizer again falls on analyte atoms in the ground state, the radiation of the resonance lines is absorbed, while the remaining (non-absorbable) radiation passes through the atom cloud. The absorbed resonance radiation excites the metal atoms to fluoresce. The fluorescent radiation can be measured, for example, by a detector at right angles to the incident radiation beam. The fluorescence signal is proportional to the absorbed radiation.

Since the spectrum of the fluorescence radiation consists only of spectral lines that were absorbed by the cloud of atoms, resonance detectors can take over the function of the monochromator. It was shown experimentally that, due to the absence of non-absorbable background radiation, the calibration curves obtained with resonance detectors were linear over a wider range than those obtained when the radiation of the primary source falls directly on the detector. The major disadvantages of this type of detector are that only a small portion of the fluorescence radiation emitted in all directions falls on the radiation detector and that an individual resonance detector is required for each element. Resonance detectors were thus used only rarely in practice [746] and are nowadays merely of academic interest.

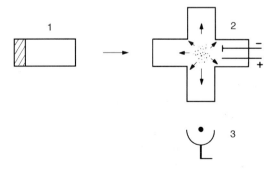

Figure 1-9. Schematic design of a resonance detector system. **1** – radiation source; **2** – resonance detector; **3** – radiation detector (from [5190]). Details see text.

SULLIVAN and WALSH [5684–5686] used resonance detectors to construct *simultaneous multielement instruments*. They employed several HCLs that irradiated a flame in different positions and the radiation was intercepted by resonance detectors. Likewise, MAVRODINEANU and HUGHES [3984] described a multichannel spectrometer that operated with several HCLs and detectors. The design of this spectrometer is depicted in Figure 1-10. BUTLER and STRASHEIM [982] developed a simultaneous instrument with several HCLs and a movable detector. MITCHELL *et al.* [4138] used a Vidicon detector to construct a multichannel atomic absorption spectrometer and were able to determine ten elements simultaneously. BUSCH and MORRISON [976] published a review on the early attempts to construct multielement instruments.

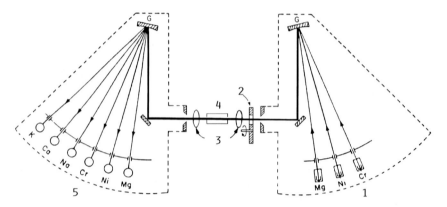

Figure 1-10. Multichannel spectrometer for flame AAS designed by MAVRODINEANU and HUGHES [3984]. **1** – radiation sources; **2** – chopper; **3** – lenses; **4** – flame; **5** – detectors; **G** – gratings.

In subsequent years, numerous attempts were made to construct multielement atomic absorption spectrometers. In most cases two to four elements could be determined either simultaneously or in rapid sequence. Normally, classical components such as HCLs or EDLs, mono- or polychromators, and conventional detectors were employed [108, 414, 1622, 3343, 3471, 4275, 4854]. Such instruments were characterized by insufficient sensitivity or poor S/N ratio [414, 4275, 4854], or by a lack of flexibility [3343, 3471]. Typical for this is the suggestion by LUNDBERG and JOHANSSON [3658] to use a multi-element HCL and a rotating sector disk that allowed the radiation of three elements to fall in rapid sequence on the detector. This instrument is naturally restricted to the three elements for which it was built.

A much more flexible concept was developed by the group around O'HAVER [6515, 6516] on the basis of a continuum source. The principle problems for the use of continuum sources, which had already been recognized by Walsh, were overcome in a very elegant manner. O'Haver and co-workers used a very intense xenon arc lamp, an Echelle polychromator of high resolution, and wavelength modulation through an oscillating quartz plate. An advantage of this system is that the resolution of the Echelle polychromator is roughly in the same magnitude as the width of an atomic line [3071], so that spectral overlap is hardly to be expected. At the same time the non-absorbable radiation is effectively masked and thus a sensitivity comparable to normal AAS attained.

Nevertheless, the problem of the simultaneous determination of several analytes by AAS has only been satisfactorily solved in recent years after the combination of Echelle polychromators and matrix detectors [4783, 4784, 5332] (see Section 3.7).

1.5 The Non-specific Attenuation of Radiation

As already recognized by Walsh (compare Section 1.2), the selectivity and specificity of AAS are based on the use of element-specific line sources, modulation of this radiation, and selective amplification. Nevertheless, within the atomizer the radiation can be attenuated not only by atomic absorption but also by a number of other effects that have been known collectively as 'non-specific' or 'background absorption'. The most frequent causes are radiation absorption by gaseous molecules and radiation scattering on particles within the absorption volume.

These effects were recognized in principle at an early stage and described by a number of authors. As early as 1962, WILLIS [6338] showed that radiation losses, caused by scattering on particles or absorption by molecules not dissociated in the flame, led to errors that could not be corrected by the analyte addition technique. In 1965 BILLINGS [636] measured the radiation scattering caused by various salts at higher concentrations in an air-propane flame. Through the use of an air-acetylene flame and a premix burner, background attenuation is not a serious problem, at least not for solutions whose concentrations are not too high. Thus in his monograph on AAS published in 1968, SLAVIN [5413] devoted only one page to interferences caused by radiation scattering.

In graphite furnaces however, and especially in the longitudinally-heated version proposed by Massmann, very high non-specific attenuation of radiation can be observed. GÜÇER and MASSMANN [2256] made very detailed investigations on molecular spectra in graphite furnaces and recorded the dissociation continua of a number of alkali metal halides (see Section 5.4.1.1).

The major problem in the 1960s was that there were no suitable means for eliminating non-specific attenuation. The frequent recommendation to use a 'matrix blank sample', which exhibits the same background absorption as the true sample but does not contain the analyte, could rarely be performed in practice. Such matrix blank samples are seldom available and attempts to prepare them synthetically mostly fail due to the high purity requirements placed on each constituent. Furthermore, the composition of the concomitants often varies from sample to sample and their true content is frequently unknown.

Most procedures originally proposed for the elimination of background attenuation were based on the fact that while atomic absorption occurs in a very narrow spectral range, molecular absorption and radiation scattering are both broad band phenomena. One such procedure is the *reference element technique* in which the absorption of another element at a neighboring wavelength is measured and subtracted from the absorption at the analytical wavelength. A prerequisite for this technique is naturally that the reference element is not contained in the sample. Further, 'broad band' does not mean that the background attenuation remains constant over a range of several tens of nanometers. Thus for measurement of the background attenuation using the reference element technique the 'next best' element must be chosen [5530], i.e., the element whose wavelength lies closest to the analytical wavelength. Frequently, however, the next-lying absorption line is so far removed that only a qualitative estimation of the background attenuation at the analytical line can be made.

WILLIS [6339] and SLAVIN [5406] therefore proposed correcting background attenuation by *measuring the effect at a non-absorbable line*. Although such lines do not exhibit element-specific absorption, they can be attenuated by scattering or molecular absorption, i.e., exhibit background attenuation. The advantage of using such lines for correction is that any radiation attenuation measured can only be non-specific, and that many more non-absorbable lines are available than absorption lines. Nevertheless, even this technique is not sufficiently accurate and requires two sequential measurements at different lines.

In 1965, KOIRTYOHANN and PICKETT [3196] proposed using a continuum source in addition to the line source for quasi-simultaneous measurement of the total attenuation and the background attenuation. The atomic absorption signal is obtained by subtraction. The first commercial system for automatic background correction (BC) based on the principle proposed by KOIRTYOHANN and PICKETT [3196] was introduced in 1968 [2994]. Among other things this principle was chosen because it could be incorporated without notable changes into the then most widely used double-beam spectrometer, the Perkin-Elmer Model 303. As depicted in Figure 1-11, the sector mirror which originally served to split the beam was used to send radiation from the HCL and a deuterium lamp in rapid sequence through the flame. The radiation from the deuterium lamp was processed as the reference signal by the electrical measuring system (for details see Section 3.4.1).

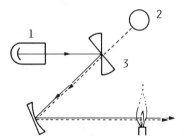

Figure 1-11. Schematic of the background corrector used in the Model 303. **1** – HCL; **2** – continuum source; **3** – rotating sector mirror (from [2994]).

Although this system of BC represented a considerable improvement compared to the usual practice at that time, nevertheless its limitations soon became apparent. MASSMANN [3938] pointed out that continuum sources were not capable of correcting the background attenuation caused by electron excitation spectra since these consist of many narrow lines. These spectra are based on electronic transitions in the molecules. The structure of the bands is derived from the transitions of the rotational and the various vibrational levels of the one electronic state to the rotational and vibrational levels of the other electronic state, i.e., between discrete energy states of the molecule.

DE GALAN [2043] and MASSMANN [3938, 3941] in particular repeatedly stressed that the uncontrolled use of a background corrector cannot guarantee correct results; in fact it may even introduce errors. It is important when analyzing unknown samples to work both with and without BC. In this way important information about the appearance and magnitude of background attenuation can be obtained that would otherwise be lost when working solely with BC. Among other things this is important because the accuracy of BC decreases with increasing background attenuation.

MARKS *et al.* [3891] pointed out a further source of error that can occur when using continuum sources for BC; radiation from the continuum source can be absorbed by atoms of concomitant elements within the spectral bandwidth passed by the monochromator. The result here is also overcompensation. MANNING [3857] reported a spectral interference of this sort in the determination of selenium in the presence of iron, and VAJDA [5973] found a whole series of similar examples.

Although these limitations of continuum source BC have been known for years, many analysts have had a blind faith in the infallibility of background correction. Especially in graphite furnace AAS (GF AAS) this led to considerable errors. For example, virtually all determinations of chromium in biological materials performed prior to 1978 must be viewed with skepsis owing to incomplete BC [2295, 6014]. The same is true for the determination of selenium in blood due to the presence of iron [2963, 3497, 3857, 5039]. In a review article, SLAVIN and CARNRICK [5434] presented a multitude of similar examples.

In work published in 1966 and 1967, BARRINGER [395] and LING [3533] found that improved BC could be achieved by using radiation close to the analytical line rather than with continuum sources. They utilized the principle of self-reversal of resonance lines where the line is strongly broadened and thus the original wavelength virtually disappears. In 1983 SMITH and HIEFTJE [5474] took up the proposals of Barringer and Ling for BC using *high current pulsing*. The total absorption is measured while operating the HCL under normal conditions, then the lamp is pulsed with a high current, which leads to self-reversal (refer to Sections 2.3.3 and 3.4.3), so that the background is measured with the strongly broadened line close to the analytical line. This technique has the principal disadvantage that it can only be applied to volatile elements since only these exhibit sufficient self-reversal. Additionally, a considerable loss in sensitivity can occur.

Compared to continuum source BC, high current pulsing has the advantage of requiring only one lamp. Further, higher background attenuation can be corrected. Compared to Zeeman-effect BC (discussed below), high current pulsing has the major disadvantage that the signal dissipates only slowly so that a lower modulation frequency is required which, together with the above mentioned loss in sensitivity, makes it unattractive for GF AAS. For flame AAS, on the other hand, continuum source BC is usually adequate.

A by far more efficient system for BC, especially for GF AAS, is obtained by utilizing the *Zeeman effect* (refer to Sections 2.6 and 3.4.2). As early as 1969, PRUGGER and TORGE [4721] applied for a patent for the *application of the Zeeman effect for background correction*. Shortly thereafter HADEISHI and MCLAUGHLIN [2306] worked on the same subject. In the following years, especially in Japan, the application of the Zeeman effect for BC was studied intensely [2307, 3211, 3212]. However, the first really effective system for BC in GF AAS was not introduced until 1981 [1892]. The reason for this is to be found in the multitude of ways in which the Zeeman effect can be applied in AAS. It took a relatively long period until the optimum way had been found.

The Zeeman effect—the splitting of spectral lines in a magnetic field—is treated in detail in Section 2.6. However, the technical application to BC is mentioned here very briefly. As depicted in Figure 1-12, the magnet can be mounted at either the radiation source or the atomizer; the magnetic field can be orientated either parallel or perpendicular to the radiation beam, and a constant or an alternating magnetic field can be applied. These possibilities give eight different configurations (refer to Section 3.4.2). Nevertheless they differ in part considerably in their realization, application and, above all, in their performance.

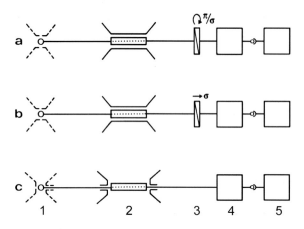

Figure 1-12. Schematic of the possible configurations of the magnet for utilizing the Zeeman effect for the measurement and correction of background attenuation. **a** – transverse constant field; **b** – transverse alternating field; **c** – longitudinal alternating field; **1** – radiation source; **2** – atomizer; **3** – polarizer; **4** –monochromator; **5** – detector and signal processing (from [2042]).

PRUGGER and TORGE [4721] used a permanent magnet mounted at the radiation source with the magnetic field at right angles to the radiation beam path. The same configuration was also examined later by KOIZUMI and YASUDA [3208], and STEPHENS and RYAN [5567, 5568]. From 1971, HADEISHI et al. [2306, 2307] made extensive studies of various Zeeman systems. They started by examining a longitudinal constant field at the radiation source. As will be shown later (Section 3.4.2), no atomic absorption can be measured using this configuration since the radiation is split into two components—one component (σ^+) is shifted to shorter wavelengths while the other (σ^-) is shifted to longer—and the actual analytical line disappears from the spectrum. Hadeishi used an EDL filled with the pure isotope ^{199}Hg (or in later work with the isotope ^{198}Hg) as the radiation source. In a suitable magnetic field (about 0.7 Tesla), the isotope line splits in such a way that the σ^- component coincides with the center of the absorption line for natural mercury, while the σ^+ component is situated at the far end of the absorption profile. The atomic absorption and background attenuation are measured with the σ^- component, while the background attenuation alone is measured with the σ^+ component. A rotating polarizer is used to discriminate between the polarized σ components. This procedure can naturally be used only for the given element and can thus be regarded as a scientific curiosity, rather than a generally applicable technique.

The configuration of the magnet at the radiation source, which was used almost exclusively in earlier studies, has a number of principal problems. The greatest is that conventional HCLs cannot be operated in a magnetic field and EDLs only with reduced performance. Mercury is the only element that has a sufficiently high vapor pressure to emit stably under these conditions. It is therefore hardly surprising that the initial investigations on the Zeeman effect were concerned with this element almost exclusively [2306–2308, 3208]. With other elements, even when they form highly volatile halides, the EDL must be heated to between 250 °C and 300 °C to obtain a vapor pressure of around 1 kPa. This vapor pressure is required if the radiation source should emit stably when run at 2 W power at 100 MHz [3209, 6465]. After evaluating more than sixty publications on EDLs HADEISHI [2311] came to the conclusion that this type of source can be used routinely only for elements of high vapor pressure. For other elements it would be necessary to develop special sources that emit stably in a strong magnetic field.

For a number of elements, STEPHENS [5569] constructed capacitively coupled sources, run at a relatively low frequency of 2 MHz, that were magnetically stable. KOIZUMI and YASUDA [3210] designed a lamp in which the anode and the cathode were in the form of two parallel plates. This source was operated at 500 V under a high frequency field of 100 MHz. MURPHY and STEVENS [4247] attempted to run conventional HCLs at 2.5 MHz, but found that the cathodes were destroyed rather quickly. Other attempts to operate HCLs in strong magnetic fields were also not particularly successful [5570, 5572].

HADEISHI [2310] developed a 'magnetically concentrated' lamp for low-volatile elements in which the cathode could be exchanged to permit a change of element. To reduce self-absorption to a minimum, operating conditions had to be maintained by continuous evacuation and supply of argon. The magnetic field localized the discharge in this lamp so that a high radiant power could be attained [6361]. Nevertheless, HADEISHI [2311] regarded the high technical requirements with special lamps as the greatest disadvantage of the direct Zeeman effect with the magnetic field at the radiation source.

Resulting from these difficulties with sources located in the magnetic field, from the middle of the 1970s investigations were increasingly made with the magnetic field at the atomizer. The investigations by KOIZUMI and YASUDA [3211, 3212], DAWSON et al. [1443], and FERNANDEZ et al. [1890] are mentioned in particular. These authors used a constant magnetic field, which did not however prove to be optimum. Constant magnetic fields, regardless of whether they are applied at the source or the atomizer, often lead to a loss in sensitivity, and to strong curvature and 'rollover' of the calibration curves.

The ideal system, a longitudinal alternating field at the atomizer, was first described in 1975 by UCHIDA and HATTORI [5946] and a year later by OTRUBA et al. [4499]. These publications received little interest, however, probably because in both cases a flame was used for atomization, and continuum source BC is normally fully adequate for this type of atomizer.

In retrospect it can be seen that the Zeeman effect took so long to become established as a technique for BC because all work published earlier used unsuitable configurations. Time and again it can be observed that a system which offers principal advantages cannot establish itself if there are technical problems (design of new lamps), or analytical disadvantages (loss in sensitivity, rollover of the calibration curves), or badly selected applications.

From 1978 DE LOOS-VOLLEBREGT and DE GALAN [2042, 3585] made detailed studies on the various configurations for utilizing the Zeeman effect for the measurement and correction of background attenuation in GF AAS. They came to the conclusion that in principle a longitudinal alternating field at the atomizer would be the best, but that this might require too many compromises with respect to the dimensions of the graphite tube. This configuration was in fact introduced commercially but more than a decade later in combination with a totally new atomizer concept [5154]. Since a transverse alternating field at the atomizer was easily feasible, DE LOOS-VOLLEBREGT and DE GALAN [2042, 3585] proposed this as an optimum compromise between the efficiency of BC and the tube dimensions. At the same time this configuration also demanded the lowest analytical compromises since the loss in sensitivity and rollover of the calibration curves were relatively low [3587]. FERNANDEZ *et al.* [1892] investigated the efficiency of this system for the correction of higher background attenuation in GF AAS and were able to confirm the findings of de Loos-Vollebregt and de Galan. This system was without doubt the most efficient and successful concept for GF AAS until the introduction of transversely-heated atomizers with a longitudinal magnetic field for Zeeman-effect BC [5154].

1.6 Burners and Flames

The history of atomic spectroscopy is closely linked with the development of burners and flames. The experiments performed by KIRCHHOFF and BUNSEN [3121–3124] on the line reversal in the spectra of alkali and alkaline earth metals could only have been performed using the burner developed by Bunsen a few years previously, since for the first time this allowed spectral observations to be made with a *non-luminous, virtually transparent flame*. The breakthrough for flame emission spectrometry ('flame photometry') for the quantitative spectral analysis of elements came in 1929 when LUNDEGARDH [3668] introduced a *premix burner with a pneumatic nebulizer*. This burner, which is depicted in Figure 1-13, does not differ in principle from modern AAS burners.

It was without doubt a decisive contribution to the success of AAS that Walsh used exactly this type of burner in his first instrument and that it was also used in the Perkin-Elmer Model 303. Nonetheless, a very heated dispute broke out in this area since a group of spectroscopists and analysts from the field of flame emission spectrometry preferred the *direct injection or total consumption burner* that was commonly used in that discipline. The major advantages of this type of burner, which is depicted in Figure 1-14, are the simple design, enabling it to be manufactured cheaply, safety (a flashback of the flame is impossible), and the possibility of using virtually all imaginable gas mixtures. This latter aspect was of some importance particularly during the experimental phase of AAS in the 1960s.

The major disadvantage of the direct injection burner is that the aerosol and the flame gases are not premixed and thus a turbulent flame with a diffuse or disrupted flame front is produced. With a premix burner, on the other hand, a laminar, transparent flame with a smooth flame front is produced. As depicted in Figure 1-15, the dramatic differences in the transparency of these flame types, which is so important in AAS, can be excellently demonstrated with schlieren photographs [699, 5412]. In addition, with a premix burner there is a distinct zonal structure of the concentration of the analyte atoms, since the more uniform droplet size allows atomization within a restricted zone of the flame [1340, 2342, 4801]. The concentration distribution of analyte atoms for two elements in a laminar air-acetylene flame is shown in Figure 1-16.

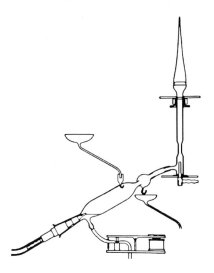

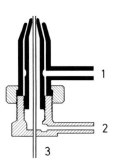

Figure 1-13. Premix burner by LUNDEGARDH.

Figure 1-14. Direct injection burner. **1** – fuel gas line; **2** – oxidant line; **3** – sample capillary.

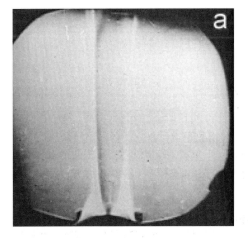

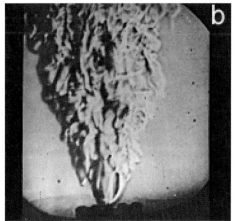

Figure 1-15. Schlieren photographs of an air-acetylene flame. **a** – premix burner with a laminar flame; **b** – direct injection burner with a turbulent flame (from [5412]).

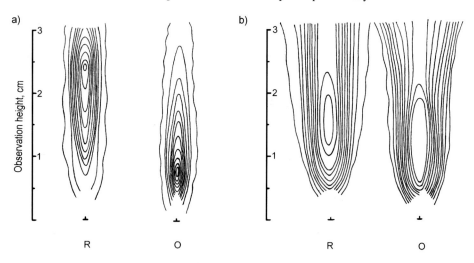

Figure 1-16. Concentration distribution of analyte atoms in a laminar air-acetylene flame. **a** – calcium; **b** – silver; **R** – fuel-rich reducing flame; **O** – fuel-lean oxidizing flame (from [4801]).

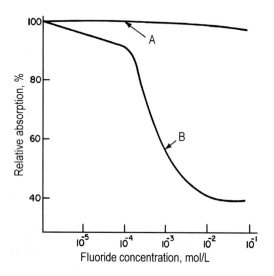

Figure 1-17. The influence of fluoride on the determination of barium. **A** – laminar air-acetylene flame; **B** – turbulent oxygen-hydrogen flame (from [33]).

A number of authors compared the advantages and disadvantages of both burner types [3197, 5412]. The occurrence of non-spectral interferences in particular was investigated [33, 6532]. As an example, the influence of fluoride on the determination of barium in a laminar air-acetylene flame and in a turbulent oxygen-hydrogen flame is shown in Figure 1-17. The most common cause for the occurrence of interferences can be traced to the influence of concomitants in the sample on the degree of conversion of the aerosol droplets into free analyte atoms [128]. Various authors found that physical factors in particular influence the rate of vaporization and atomization.

KERBYSON and RATZKOWSKI [3082] carried out interesting comparative measurements in which the aerosol produced by a direct injection burner was examined directly and then with a spray chamber interposed. The maximum droplet volume measured at 5 cm above the burner (no flame) was found to be approximately 7 nL for the direct injection burner and about 0.05 nL for the burner with spray chamber. GIBSON et al. [2105] found that with direct injection burners even water droplets could pass through the whole flame without being completely vaporized. This finding was verified by other authors [1447, 4543].

Decreasing droplet size of the aerosol, higher flame temperatures, and a longer residence time in the flame reduce or eliminate numerous interferences [2543, 3200, 6341]. This has also been verified by the finding that interferences are often less pronounced in higher flame zones than directly above the burner slot [127, 2013]. Detailed investigations have shown that, for example, the influence of phosphate [5470] and similar anions on calcium can be more or less totally eliminated by the choice of suitable flames and burners [1908, 4205, 4640].

The final argument brought against the premix burner at that time was that solely air could be used as the oxidant. It was thus limited to the relatively cool air-propane, air-hydrogen, and air-acetylene flames with which only 30–35 elements could be determined satisfactorily. Attempts were thus repeatedly made to determine those elements that form stable oxides by using the hot, turbulent oxygen-acetylene flame [1604, 1850, 3169, 4203], frequently with the addition of organic solvents [139, 1133, 5405], or with modified burners [165, 3839], or even with an oxygen-cyanogen flame [2031, 4894]. However, since not only the temperature, but also more importantly the composition of the flame, i.e. the partial pressures of oxygen, hydrogen, etc., play an important role in the atomization of these elements, these endeavors met with only limited success.

Without doubt the most important development in flames was the introduction of the *nitrous oxide-acetylene flame* by WILLIS [6340, 6347] in 1965. As a result of its low burning velocity, this hot flame offers a favorable chemical, thermal, and optical environment for virtually all metals that are problematical in the air-acetylene flame due to the latter's lower temperature and relatively high partial pressure of oxygen. The nitrous oxide-acetylene flame can be run without notable difficulties in a premix burner and thus was responsible for displacing the direct injection burner once and for all from AAS. Nowadays air-acetylene and nitrous oxide-acetylene flames are used almost exclusively in AAS.

Of historical interest are a number of flames with hydrogen as the fuel gas since they led to conclusions about atomization mechanisms. ALLAN [140] found that the sensitivity for tin was markedly better in an *air-hydrogen flame* compared to air-acetylene or nitrous oxide-acetylene. Other authors confirmed this finding [1054, 1056]. A thorough investigation of the air-hydrogen flame [5131] showed that it offered advantages for a number of other elements, for example due to the lower degree of ionization of alkali metals compared to the air-acetylene flame. In the wavelength range 200–230 nm it also exhibits lower radiation absorption. Nevertheless, a number of interferences must be expected and the use of organic solvents, although possible, is more critical [5786].

During his work on atomic fluorescence spectrometry, WINEFORDNER examined an *argon-hydrogen-entrained air flame* [6009, 6507], and DAGNALL reported the low radiation absorption of a *nitrogen-hydrogen-entrained air flame* [1403, 1404]. In both of these flame types, hydrogen serves as the fuel gas while argon or nitrogen is used to nebulize the sample into the spray chamber. When the flame is ignited, the hydrogen, diluted with the inert gas, burns in the surrounding air. This results in a flame with a unique profile, which is particularly noticeable if a three-slot burner head is used [698]. On the outside, where mixing with the surrounding air takes place, it has a temperature of about 850 °C, while in the middle the temperature reaches only 300–500 °C, depending on the flame height [1403]. This type of flame is termed a *diffusion flame*.

KAHN [2993] investigated the applicability of the argon-hydrogen diffusion flame in AAS and found that for arsenic, selenium, cadmium, and tin there were considerable improvements in the sensitivity and limits of detection. In all probability atomization takes place with the active participation of hydrogen [4269, 4987, 4988], although the reaction mechanism is not completely clear. All authors, however, draw attention to the considerable spectral and non-spectral interferences that may occur in a flame of such low temperature. This flame shows its greatest advantage at the start of the vacuum UV range because of its high transparency compared to other flame types. It is therefore particularly suitable for the determination of arsenic and selenium. For use in the hydride-generation technique the usual disadvantages are of no great importance since in this technique the analyte element is separated from the remaining matrix and conducted to the burner as a gas [3854]. Nowadays these flames have been replaced as atomizers in the hydride-generation technique by heated quartz tube atomizers.

Various groups investigated the design of *burner heads* for the premix burner. BUTLER [983] proposed a wide flame burner in which the burner head had a row of holes instead of a slot. The most important property of this burner was that the flame could not flash back. Also, the burner provided somewhat better sensitivity and a 'softer' flame. WEST *et al.* [113, 2598] constructed a series of burners by welding or bonding a number of steel capillaries together. Safety was one of the most important aspects of this type of burner. Additionally, virtually any geometric form could be constructed, for example adapted to the shape of the radiation beam [4797]. Further, because of their safety against a flashback of the flame, burners with capillaries or holes were particularly suited for investigations on combustion and atomization processes using unusual gas combinations [4797, 3310]. However, the narrow holes or capillaries in these burner heads have a tendency to clog and the necessity of regular cleaning prevented their general acceptance.

In contrast, the *three-slot burner head* (Figure 1-18) proposed by BOLING [698, 699] was used widely for routine analyses for many years. The flame is so wide that it reliably encompasses the widest beam cross-section. In addition, atmospheric oxygen can only enter the edge of the flame so that optimum reducing conditions exist in the center of the flame. Elements whose oxides have a high dissociation energy exhibit improved sensitivity in the three-slot burner [47, 5529, 6088]. The absorbance is also less dependent on the height of observation so that it is easier to adjust the flame [698, 5529]. Three-slot burners exhibit less noise [699] and can tolerate a much higher total solids content without clogging [698].

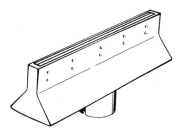

Figure 1-18. Three-slot burner head by BOLING (from [5529]).

Particularly during the second half of the 1960s efforts were increasingly turned to improving the sensitivity of AAS. HELL [2485] and VENGHIATTIS [6021] independently attempted to improve on the low use of the aspirated solution caused by separation of the larger droplets by *heating the spray chamber*. For a large number of elements an increase in the sensitivity by a factor of around ten was found. However, this procedure can only be used with dilute aqueous solutions. During the following years this idea was taken up time and again [69, 2294, 2485, 4396, 4826], but up to the present without noticeable success.

To obtain a more suitable droplet size and thus higher utilization of the aerosol, experiments were also performed with *ultrasonic nebulizers*. However, in practice these nebulizers were not easy to use and the effective improvement obtained in the sensitivity due to the higher efficiency was offset by the slow nebulization rate [2597, 5526, 5615, 6284]. ISAAQ and MORGENTHALER [2812–2814] described a design in which a thermostatted heating chamber and a cooler to predry the sample aerosol were interposed between the ultrasonic nebulizer and the burner. With this system the sensitivity was improved by a power of ten, but it was too complicated for routine use.

In 1963, FUWA and VALLEE [2029, 2031] introduced a *long tube burner* in which the flame of a direct injection burner was directed at the opening of a tube of ceramic material mounted in the beam path of a spectrometer. In this way the atoms were forced to remain longer in the radiation beam; the absorbance should thus increase proportionally to the length of the tube. This was also confirmed to a limited degree [3195], but the increase in sensitivity was restricted by the lifetime of the atoms since these recombine under the decreasing temperature along the tube [192]. A number of difficulties, such as carryover and strong interferences, also occur due to the lower temperatures prevailing in the tube [5614]; these could be partly eliminated by heating the tube [4161, 4983, 4985]. Various fields of application have been treated by AGAZZI [46], CHAKRABARTI [1134], FUWA [2029, 2030], KOIRTHYOHANN [3195], RAMAKRISHNA [4793], RUBEŠKA [4983, 4986], and ŠTUPAR [5613, 5614].

Working on a similar principle, but much simpler, is the use of a slotted quartz tube ('atom trap') mounted in the radiation beam above a premix burner [3066, 3461, 3462, 4119, 6170, 6171]. The flame is thus forced to spread out horizontally so that the atoms remain for a longer period in the absorption volume. The improvement in the sensitivity is only by a factor of 1.1–5, however [871, 872, 967–969, 1002, 2397, 4009, 6290].

As a forerunner of electrothermal atomization, the *sampling boat technique* [2995] acquired a degree of importance. In this technique, a sample volume of 0.1–1 mL is pipetted into a small tantalum boat, the solution is dried, and then the boat is introduced into the flame. The sample in the boat is thereby atomized within a few seconds, and a transient signal is obtained. Since the temperatures that can be attained with the boat technique are clearly lower than the flame temperature, this technique was suitable only for easily atomized elements such as arsenic, bismuth, cadmium, lead, mercury, selenium, silver, tellurium, and zinc. For these elements improvements in the limits of detection by factors of 20–50 times were obtained. This technique found rapid acceptance [970, 1251, 1743, 2033, 2632, 3403, 3643] because it was exactly those elements of toxicological and environmental interest that could be determined with enhanced sensitivity, and biological samples such as blood and urine could be analyzed directly [2996].

A short period later, DELVES [1474] proposed a modification of the boat technique that he had developed especially for the determination of lead in blood. The tantalum boat was replaced by a small, round nickel cup, and to increase the sensitivity an open nickel or quartz tube, through which the radiation beam passed, was mounted above the cup (Figure 1-19). This tube increased the residence time of the atoms in the absorption volume. The blood samples in the nickel cups were dried on a hotplate and pretreated before they were introduced into the flame. Other authors checked [1883] and modified [404, 1673, 1886, 4960] this technique for the determination of lead in the smallest blood samples and found it particularly suited for screening. This technique, like the boat technique, was subsequently extended to other elements and materials [1121, 1123, 1124, 1674, 2314, 2498, 2686, 2853, 2958–2960, 3519, 5318]. However, during the 1970s these techniques were replaced steadily by electrothermal AAS (GF AAS), which is much more sensitive, so that they will not be discussed further.

Figure 1-19. Delves system. The sample is introduced in a nickel cup into the flame of a three-slot burner. The atoms pass through a hole into the quartz tube which is mounted above the burner directly in the beam path.

1.7 Electrothermal Atomization

1.7.1 Boris L'vov and the Graphite Cuvette

BORIS L'VOV (Figure 1-20) was without doubt one of the earliest supporters of Walsh and his idea; and he became one of its keenest pioneers. His contact with AAS came about by fortunate accident. After completing his studies at the Leningrad State University he began work in 1955 at the Isotope Laboratory of the State Institute for Applied Chemistry on the spectrochemical analysis of samples which were marked with stable radioactive isotopes. For this work he employed a demountable HCL.

As L'VOV [3720] wrote: 'Another circumstance which proved to be lucky for me is that my chief, Dr Kibisov, was at that time one of the editors of the Russian abstract journal Khimiya. That is how in the winter of 1956 I came across a paper by Alan Walsh in Spectrochimica Acta which had been sent to Kibisov for review, the paper which was fated to become a turning point in the history of analytical atomic spectroscopy. The impression produced on me by this paper was so strong that I made up my mind to check the validity of the author's ideas. Since using my official working hours to perform seemingly dubious experiments was unthinkable, I sacrificed the next summer vacation for this purpose. As an atomizer, I decided to use a tubular graphite furnace installed in a commercial stand for the preconcentration of volatile impurities from refractory materials prior to emission spectrochemical analysis. The equipment stood unused in the corner of the room which was another lucky circumstance for me. Having obtained sufficiently stable sodium D line emission from a hollow cathode tube, which I observed visually through a glass prism monochromator, I started heating slowly the graphite furnace placed between the source and the monochromator, having put a pinch of NaCl in the furnace preliminarily. Imagine my wonder when the bright sodium lines from the hollow cathode tube started to weaken and then disappeared completely ...'

Figure 1-20. Boris L'vov.

Like Walsh he, too, received the same lack of understanding for his enthusiasm—but, likewise, he remained true to his idea. In 1958 he held his first lecture on the 'spectrochemical determination of impurities in radioactive samples' and his first paper on 'the investigation of atomic absorption spectra through total vaporization of the sample in a graphite cuvette' [3679] was published in 1959. This first paper was translated into English 25 years later [3721]. L'vov's work initially became known through his first publication in English in Spectrochimica Acta in 1961 [3680]; however, this does not mean that it was acknowledged and evaluated by a larger audience. This took place after the First International Conference on Atomic Absorption Spectroscopy in Prague in 1967 during which L'vov had been able to discuss his ideas with other participants.

L'vov in fact did not invent the 'graphite cuvette' (as he termed his atomizer) just as Walsh did not discover AAS. KING [3106] had used a heated graphite tube for emission measurements at the start of this century and this type of furnace was subsequently used by various groups for spectroanalytical measurements. MARSHALL et al. [3903] were able to trace this furnace back into the nineteenth century; LOCKYER had used the apparatus depicted in Figure 1-21 to study the absorption spectra of metal vapors at increased temperature.

However, L'vov was without doubt the first person after the rediscovery of AAS to use an electrically heated furnace as an atomizer. Not only did L'vov develop this atomizer, but he also established the theoretical principles, which he presented in a book published in Russian in 1966 [3683]; the English translation appeared in 1970 [3690]. L'vov has earned the greatest credit for his persistent support of GF AAS and the new impulses he has given to it time and again. The success of this technique is due to his untiring efforts.

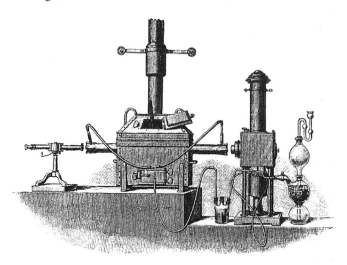

Figure 1-21. Apparatus used by LOCKYER for measuring atomic absorption (from [3570]). The atomizer was an iron tube mounted in a coal-fired furnace. Hydrogen was generated in a Kipp's apparatus to provide a non-oxidizing atmosphere in the tube. The electric radiation source is to the right and a spectroscope is to the left.

Before he started to design his first graphite cuvette, L'vov concluded that the analyte must be volatilized in such a short time (τ_1) that losses due to diffusion could be neglected. 'A calculation of the residence time of the atoms in the furnace (τ_2) showed that this would require the sample to be vaporized in a fraction of a second. Considering that the furnace heating time is much longer, I decided to give up sample vaporization on the furnace wall in favor of introducing the sample on a carbon electrode into a preheated furnace. To accelerate the vaporization process still more, the electrode was additionally heated by a direct current arc and later by making use of ohmic resistance of the electrode-furnace contact' [3720]. L'vov's first graphite cuvette is depicted in Figure 1-22.

'In the subsequent few months I managed to assemble a simple single-beam spectrometer which I used to verify the expected advantage of the principle of the complete sample vaporization of the sample into an isothermal furnace under the condition:

$$\tau_1 \ll \tau_2 .\tag{1.2}$$

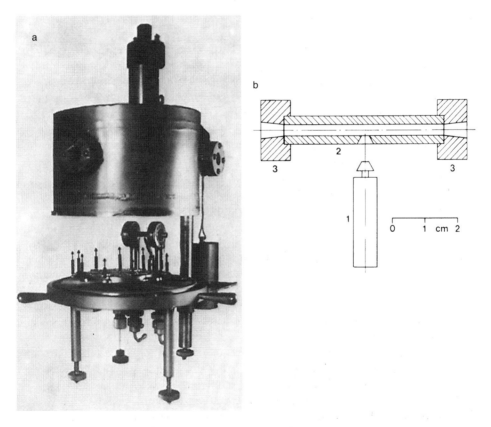

Figure 1-22. Graphite cuvette by L'vov. **a** – argon chamber with atomizer; **b** – schematic cross-section. **1** – graphite electrode; **2** – graphite tube; **3** – contacts.

It was much later [3683] that I concluded that condition 1.2 is superfluous with an isothermal furnace when measuring the pulse area since irrespective of pulse shape the integrated absorbance, Q_A , is related to the amount of analyte in the sample, m_0, through the expression:

$$Q_A \equiv \int_0^\infty A_t \, dt = S \, m_0 , \qquad (1.3)$$

where S is the sensitivity $(dQ/dm)'$ [3720]. Exactly this very important realization was completely ignored for the next fifteen years and only came slowly into the awareness of analysts during the 1980s.

Even during the 1960s L'vov and co-workers were able to determine more than 40 elements with absolute limits of detection of between 0.01 pg and 10 pg using his graphite cuvette. At the same time he was able to show that matrix influences were much lower than in the flame technique. L'vov was even able to determine sulfur, phosphorus, and iodine directly at their resonance lines in the vacuum UV, and phosphorus and iodine at the non-resonance lines at 213.6 nm and 206.2 nm, respectively, with limits of detection of 0.2 ng and 2 ng absolute. Graphite furnace AAS was used for the analysis of water, high purity reagents, metals, alloys, semiconductors, radioactive substances, rock samples, etc. A review article on this early work was published in 1969 [3684].

Apart from L'vov and his group, very few analysts took an interest in the graphite furnace as an atomizer. NIKOLAEV and ALESKOVSKII [4366] confirmed the advantages of the graphite cuvette and published new metallurgical applications. In 1968 WOODRIFF *et al.* [6377-6379] developed a furnace that only differed from that of L'vov in its length. It was an isothermally heated tube furnace into which the sample was introduced in a capsule. Later Woodriff used a nebulizer to introduce the sample; a technique that did not prove to be worthwhile. KOIRTYOHANN [3203] found that the furnace described by Woodriff was relatively free of spectral and non-spectral interferences, and that it provided a stable platform for performing basic studies; due to its size and complicated design it was however unsuitable for routine measurements.

WEST [6288] proposed a very simplified graphite atomizer that is depicted in Figure 1-23. A carbon rod, 1–2 mm in diameter and about 20 mm long, is clamped between two electrodes in a glass jacket purged with argon. The sample (e.g. 1 μL solution) is placed onto the carbon rod by means of a micropipet or syringe. The rod is then brought to incandescence by resistance heating (100 A, 5 V). A measurement took 5–10 s and the next sample could be introduced after about two minutes. In a later version the glass jacket was abandoned and the carbon rod was merely flushed with argon from beneath. Because of its ease of access, this open version was particularly simple to operate.

With this system West found limits of detection of 0.1 ng for silver and magnesium. Nevertheless, sample introduction was rather critical and required a degree of experience. In addition, a large number of interferences were found that could be attributed to the steep temperature gradient between the carbon rod and the environment. This was confirmed by experiments in which the radiation beam was focused at different heights above the incandescent carbon rod; a clear increase in the interferences and a decrease in the sensitivity were observed with increasing distance from the rod.

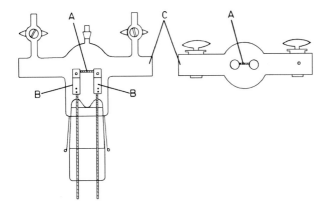

Figure 1-23. Carbon rod atomizer designed by WEST [6288]. **A** – Carbon rod; **B** – contacts; **C** – glass jacket.

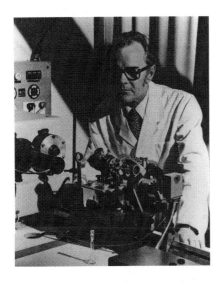

Figure 1-24. Hans Massmann.

A similar principle was proposed by MONTASER *et al.* [4168], who used a braid of 1.5–2 mm diameter plaited from graphite fibers. Advantages shown by this system included a low current consumption and the ability to handle somewhat larger sample volumes, since the braid absorbed liquids better. However, it was not possible to obtain graphite braids of particularly high purity, and the lifetime was relatively short, even at lower temperatures. The limits of detection attained were hardly better than with the flame technique [4167], so that this system was more suitable for microanalyses rather than trace determinations.

Without doubt the most important contribution to the further development of GF AAS came from MASSMANN (Figure 1-24), who reported his first results in 1965 [3935]. Massmann simplified the furnace design drastically; rather than introducing the sample into a preheated tube, it was atomized from the tube wall by rapid heating.

A

B

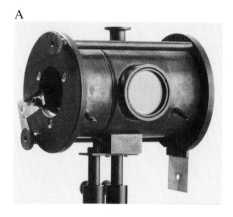
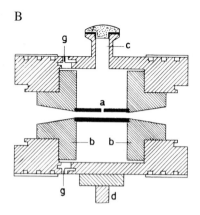

Figure 1-25. Graphite furnace designed by MASSMANN [3937]. **A** – Massmann's original furnace. **B** – cross-sectional view; **a** – graphite tube; **b** –steel flanges; **c** – sample introduction port; **d** – mount; **g** – plastic insulator (from [3937]).

Massmann's graphite furnace (depicted in Figure 1-25) comprised a 5 cm long graphite tube which was heated by passing a high current (500 A) at low voltage (10 V) along the tube. Massmann's furnace was not mounted in a closed argon chamber like that of L'vov and therefore had to be purged continuously with an inert gas to prevent the ingress of atmospheric oxygen. The sample (maximum 0.5 mg solid or 50 µL liquid) was introduced into the graphite tube through a small hole in the tube wall.

With his system Massmann achieved limits of detection that were typically ten times poorer than those quoted by L'vov. The reason for this can be found in the forced inert gas stream through the tube which reduces the residence time of the atoms in the absorption volume. Also, the heating rate of resistance heating is slower than that of atomization in an arc, so that the condition in equation 1.2 is not fulfilled. Massmann [3935] observed background attenuation to varying degrees in his furnace, which he eliminated by using a dual-channel spectrometer, and also considerable dependence of the signal height on the nature and quantity of the matrix [3936].

The cause of these difficulties was to be found almost entirely in the altered design of the furnace, i.e. heating of the tube and volatilization of the sample from the tube wall were interdependent and thus isothermal conditions were not given. In his book [3690], L'vov clearly expressed his views on Massmann's attempt to simplify the graphite furnace technique:

1. 'The sample atomizes from the cuvette wall much more slowly than from an additional electrode. In the first place, therefore, there is fractional atomization of elements with different volatilities, and, in the second place, the peaks corresponding to maximum absorption are lower. The correctness of using the integration method of measuring absorption is also dubious in this case, since the temperature of the cuvette is steadily rising while the measurements are being made, and τ_2, the time spent by the atoms in the cuvette, is not constant.'

2. 'The maximum temperature to which the sample is heated is governed by the temperature of the inner wall of the cuvette. In order, therefore, to atomize even elements

of average volatility, such as iron and copper, a very powerful source of current must be used for heating the cuvette. It is also not satisfactory to raise the temperature of the cuvette owing to the increase in the continuous background from the cuvette walls.'

3. 'When samples are introduced into the cuvette by this method, analysis is slower than with the normal variation of the method, since the sequence followed by all stages of the analysis for each sample definitely takes longer than analysing by measuring series of several samples.'

'Definitely, therefore, this method of simplification is not advisable.'

Despite these weaknesses (which at that time were known only to a few people) virtually all commercial electrothermal atomizers built in the 1970s and 1980s were based on Massmann's principle. The first commercial graphite furnace, the HGA-70, introduced by Perkin-Elmer in 1970, is shown in Figure 1-26.

The first applications performed with this furnace confirmed the improved sensitivity of two to three powers of magnitude compared to the flame technique [3853, 6204], and the applicability to micro- and trace-analysis of natural waters, biological samples, and direct solids analysis [6204]. At the same time attention was drawn to the presence of considerable non-specific attenuation and strong matrix effects 'not to be observed to this degree in the nebulizer-flame technique' [3853]. Despite numerous instrumental improvements, among which automatic sample dispensing must especially be mentioned [6206], these disadvantages became increasingly apparent during the following years as the technique found increasing application. Despite the generally recognized high sensitivity of the technique, there was hardly a paper published that did not lament spectral interferences, matrix effects, and variable sensitivity. These publications will not be mentioned here further since they no longer represent the current state of the art. A good overview of this work can be found in a monograph published by SLAVIN [5428].

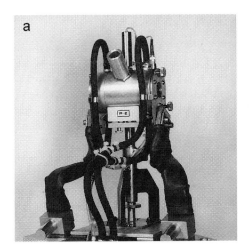
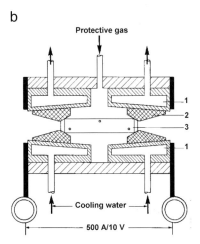

Figure 1-26. The Perkin-Elmer HGA-70, the first commercial graphite furnace (HGA = Heated Graphite Atomizer). **a** – atomizer; **b** – schematic cross-section; **1** – cooling chambers; **2** – graphite contacts; **3** – graphite tube.

It is characteristic that the solutions to these problems were again proposed by L'VOV [3704]. In his view, the considerable difficulties in the graphite furnace technique arose mostly from the non-isothermal absorption zone in the atomizers (in terms of both time and space), from the use of peak height for signal evaluation, and the formation of gaseous monohalides of the analyte. To eliminate the main problem, the temporally non-isothermal state of the absorption volume, three solutions could be considered: atomization by capacitive discharge [3868], probe atomization [3706], and platform atomization [3701].

Capacitive discharge was examined thoroughly by CHAKRABARTI *et al.* [1139–1142, 1188] at the beginning of the 1980s, and heating rates up to 100 K/ms were reported. The relatively high voltages of 150–250 V required by this technique demanded graphite material of very high resistivity to attain an effective conversion into heat energy. For these experiments total pyrolytic graphite (TPG) was used exclusively such that the voltage drop was at right angles to the layers in the graphite. Chakrabarti reported that this technique was virtually independent of the concomitants, that interferences were markedly reduced, and that BC was not required. However, McCAFFREY and MICHEL [4007, 4008] later showed that this technique was not as free of problems as reported by Chakrabarti. They found that the tubes used exhibited substantial temperature gradients across their circumference and thus did not provide an isothermal environment for the atoms.

The idea of introducing the sample into the preheated furnace on a *probe* was taken up quickly, since it brings the behavior of a Massmann furnace very close to that of an isothermal furnace. SLAVIN *et al.* [3860, 5423] made a thorough investigation of probe atomization (Figure 1-27) and showed that matrix effects in comparison to wall atomization were substantially lower. Other groups also successfully applied probe atomization [1390, 2135]. Nevertheless, the changes required to the furnace for the implementation of this technique delayed its commercial introduction for many years [877].

It is hardly surprising that the simplest solution, *platform atomization*, has found the widest application [1137, 1891, 3006, 3862, 5415, 5416, 5419, 5421], since no changes are required to either the furnace or the technique. It suffices to place a small graphite platform in the tube under the sample introduction hole and to dispense the sample onto it. Since the platform is largely heated by radiation from the walls of the tube it attains the atomization temperature when the tube and the inert gas have already reached

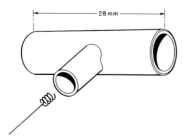

Figure 1-27. Initial experiments on probe atomization by SLAVIN *et al.* The measurement solution is introduced on a tungsten spiral into a modified graphite furnace (from [3861]).

their final temperature. Should a local thermal equilibrium exist during atomization, variations in the volatilization of the analyte and thus matrix effects can be eliminated by integration of the peak area (integrated absorbance) [6264], as L'vov recognized in 1966 [3683]. In effect this platform made of anisotropic TPG proposed by L'vov started an evolution that led to the general acceptance of the graphite furnace technique as a reliable analytical procedure.

Nevertheless, the form of the platform has undergone considerable change over the years. While in the early days fragments or sections of graphite tubes were placed loosely as platforms into the graphite tube [2555, 3006, 5860], complicated constructions in TPG were later used. It was found to be a problem that although the platform should be firmly fixed in the tube, it should nevertheless have minimum contact [5325–5327]. This problem was solved with the introduction of the 'integrated platform' in which the tube and platform are made of a single piece of graphite and have only one point of connection (see Section 4.2.2.3).

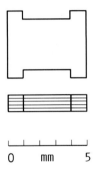

Figure 1-28. L'vov platform of anisotropic pyrolytic graphite to retard atomization; the hatching shows the arrangement of the layers (from [3704]).

1.7.2 Walter Slavin and the STPF Concept

Based on L'vov's findings [3704], WALTER SLAVIN (Figure 1-29) introduced the concept of the 'Stabilized Temperature Platform Furnace' (STPF) in 1981 [5419]; in principle this was the translation of L'vov's ideas into practice. This concept comprised a 'package' of conditions that should bring the temporally and spatially non-isothermal Massmann furnace as close as possible to the ideal of the L'vov isothermal furnace. The visible 'heart' of this concept is the L'vov platform. Nevertheless, Slavin emphasized repeatedly that the use of the platform alone would not bring the desired effect. It is absolutely necessary to maintain **all** conditions of the concept to achieve the corresponding improvement in the performance.

In addition to the L'vov platform, SLAVIN [5419] included the following elements as important components of the STPF concept: a *fast heating rate* for the graphite tube (1500–2000 °C/s), since, as depicted schematically in Figure 1-30, with increasing heating rate the platform effect (i.e. the retardation of sample atomization) is enhanced; *shutdown of the forced gas stream* through the graphite tube during atomization (*stopped flow)*; *fast-reacting electronics* to register the fast signals free of distortion; and integration of the peak area (*integrated absorbance*).

Figure 1-29. Walter Slavin.

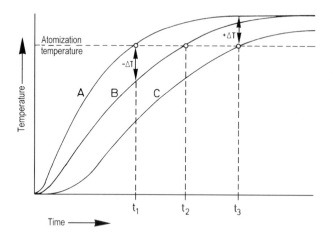

Figure 1-30. Schematic heating profiles for **A** – the graphite tube wall, **B** – the gas phase, and **C** – the L'vov platform. For atomization from the tube wall the sample is volatilized very early (t_1) while the inert gas atmosphere is still cooler than the tube wall ($-\Delta T$). Atomization from the L'vov platform is substantially retarded (t_3) so that the tube and the inert gas are at a higher temperature ($+\Delta T$) and have stabilized in temperature.

How dependent all elements of the STPF concept are upon each other can particularly be seen for integrated absorbance. At an early stage, L'VOV [3690] emphasized that integration of the signal area can only eliminate interferences resulting from different volatilization behavior when the sample is atomized under isothermal conditions. As further important elements of the STPF concept, SLAVIN [5419] mentioned the use of graphite tubes with a good *pyrolytic graphite coating*, the use of *modifiers,* and a powerful *background corrector* (e.g. utilizing the Zeeman effect). The effectiveness of the STPF concept can be guaranteed only when *all* of these conditions are used, even if success is occasionally achieved by employing a part of them. The introduction of the STPF concept started a development during the 1980s that made GF AAS into a highly sensitive, reliable, and versatile technique for trace and ultratrace analyses in a wide range of matrices.

The technique of chemical modification to reduce spectral and non-spectral interferences played an important role in the success of the STPF concept. This technique was initially examined by EDIGER *et al.* [1677]; they also introduced the frequently used, but controversial, term *matrix modification*. This technique uses chemical additives to obtain a controlled alteration and matching of the physical and chemical properties of the sample. Thus the formation of thermally stable compounds of the analyte or of concomitants can be enhanced or depressed.

Even during the first analytical work with the graphite furnace it became clear that to avoid or reduce interferences for many samples it would be necessary to separate the analyte as completely as possible from concomitants (the matrix) prior to atomization. In a Massmann furnace it is possible to use a temperature program that typically comprises the steps drying, pyrolysis, atomization, and cleaning for this purpose. Solvents are vaporized during the drying step, easily vaporized concomitants are removed during the pyrolysis step, and difficultly vaporized substances are removed in the cleaning step. The analyte must be volatilized solely during the atomization step, but under no circumstances during pyrolysis. However, since the volatility of the analyte depends not only on its bonding form but also on the concomitants, optimization of the pyrolysis temperature with changing sample composition is often extremely difficult.

Even at an early stage various authors described the use of chemical additives to better control the behavior of the analyte and concomitants during the pyrolysis and atomization steps. In 1973, MACHATA and BINDER [3769] used lanthanum, strontium, aluminium, and cesium salts for the determination of lead and thallium in biological materials. BRODIE and MATOUŠEK [852] added phosphoric acid to their samples to stabilize the analyte. EDIGER *et al.* [1676] used additions of ammonium nitrate for the determination of copper in sea-water to largely remove the interfering sodium chloride during pyrolysis according to:

$$NaCl + NH_4NO_3 \rightarrow NaNO_3 + NH_4Cl. \tag{1.4}$$

The non-volatile sodium chloride (melting point 801 °C; boiling point 1413 °C) is converted to sodium nitrate, which decomposes at 380 °C, and ammonium chloride, which sublimes at 335 °C. Excess ammonium nitrate decomposes above 210 °C, so that all concomitants can either be removed at relatively low temperature or converted into compounds that do not interfere.

EDIGER *et al.* [1677] carried out the first systematic investigations in 1975; they proposed a whole series of reagents and reactions to stabilize the analytes and vaporize the concomitants. Among the modifiers proposed by Ediger are nickel to stabilize arsenic, selenium and tellurium, ammonium fluoride, ammonium sulfate and ammonium hydrogen phosphate for the determination of cadmium, and ammonium sulfide to stabilize mercury. Ediger further investigated the influence of oxidizing reagents on the determination of germanium and gallium and found that perchloric acid or hydrogen peroxide in nitric acid solution brought the best results.

In the years that followed the number of proposed chemical additives increased, for example molybdenum [2499], potassium dichromate [3134, 5273], and also magnesium nitrate [5421], which can be used relatively universally. FRECH and CEDERGREN also mentioned that the introduction of gases such as hydrogen [1960] or the use of strong acids [1961] aided in the separation of concomitants prior to atomization. The number of proposed chemical additives, most of which had been determined empirically, had grown so much within a few years that SLAVIN in his article in 1981 on the STPF concept could only quote a few of them. A comprehensive and systematic overview of modifiers was published by TSALEV et al. [5904].

However, the enormous variety of chemical additives finally turned out to be an obstacle for the introduction of the STPF concept since there were in practice no rules of selection. From the middle of the 1980s systematic investigations were then undertaken with the aim of simplifying and unifying the use of modifiers. Based on investigations by SHAN et al. [1216, 5263, 5264, 5266–5268], SCHLEMMER and WELZ [5150] in 1986 proposed a mixture of palladium nitrate and magnesium nitrate as a possible universal modifier, initially for nine elements. It was later shown by WELZ et al. [6265] that this mixture could be used successfully for more than 20 elements and could thus replace virtually all previously suggested modifiers. Because this mixed modifier allowed determinations to be performed better than with previously proposed modifiers, these authors concluded that the search for further improvement was hardly justified. Every further attempt to optimize the conditions for a given analyte or application would probably only introduce complications for routine analyses. Problems that could not be solved with the universal modifier could probably be solved by using other techniques (e.g. separation of the analyte) rather than searching for new additives.

1.7.3 The Two-step Atomizer

The two earliest designs of furnace for GF AAS, those of L'VOV [3690] and WOODRIFF [6377–6379], came very close to the ideal embodiment of an isothermal atomizer in which vaporization and atomization of the sample are independent. Nevertheless, in both atomizers electrical contact took place at the ends of the tube so that there was a temperature profile along the length of the tube. The biggest disadvantage of these atomizers, however, was their size and difficult operation.

The extreme opposite to these atomizers was the 'carbon rod atomizer' introduced by Varian at the beginning of the 1970s, which is illustrated in Figure 1-31. Resulting from the electrical contact *transverse to the tube*, a much more constant temperature was obtained along the length of the tube. Nevertheless, since the sample was atomized from the wall of the tube and the dimensions of the tube were very small, various authors reported major interferences due to concomitants [3203] and poor repeatability of the measurements [4200]. From this it is clear that a *spatially isothermal atomizer* is alone not a guarantee for better performance.

FRECH and JONSSON [1964] utilized this concept of a transversely-heated graphite tube to develop a quasi *temporally and spatially isothermal atomizer*, which allowed largely independent control of the vaporization and atomization of the sample. This

Figure 1-31. CRA-90 carbon rod atomizer in the 'Mini-Massmann' version (with kind permission of Varian Associates).

atomizer consisted of a transversely-heated graphite tube, underneath which a graphite cup was mounted that could be heated independently. The graphite tube could be preheated to a preselected temperature before the sample was vaporized electrothermally from the graphite cup. The authors termed this atomizer a *two-step atomizer*.

FRECH and JONSSON [1964] were able to show in their first paper that by vaporizing the sample into an absorption volume at temporally and spatially constant temperature, numerous known interferences from concomitants could be eliminated or at least markedly reduced. With this system, evaluation of the integrated absorbance signal was a major prerequirement for freedom from interferences. This is plausible since in principle the atomizer was very similar to L'vov's graphite cuvette; nevertheless sample vaporization (τ_1) was definitely slower and the residence time (τ_2), due to the smaller tube dimensions, was shorter, so that the stipulation $\tau_1 \ll \tau_2$ was more poorly met than in L'vov's graphite cuvette and thus integration of the signal area was essential.

In the following years a number of other two-step atomizers were proposed and developed [4905, 4906, 5344, 5507], however these will not be discussed here further. FRECH *et al.* were also able to improve their two-step atomizer, for example by using longer atomization tubes which improved the sensitivity and precision [3664]. The major weakness of this system with respect to its reliability was the electrical contact of the tube in the hot zone, as can be seen for the atomizer illustrated in Figure 1-31. Frequently an arc formed between the contacts and the tube, making the heating non-reproducible and increasing the burn-up of the tube.

FRECH *et al.* [1970] were finally able to solve this problem by making the tube and the contacts from a single piece of graphite; electrical contact was then made in the cold zone. LUNDBERG *et al.* [3667] developed an improved two-step atomizer based on this principle which is depicted schematically in Figure 1-32. In this furnace the graphite tube and the graphite cup were both made from a single piece with integrated contacts.

In several publications, FRECH *et al.* [1966, 1967, 3664] compared the performance of the two-step atomizer with that of a conventional atomizer of the Massmann type with L'vov platform. As to be expected, these authors found that the isothermal furnace was

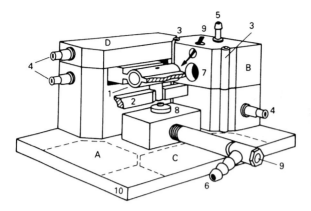

Figure 1-32. Two-step atomizer designed by LUNDBERG *et al.* [3667]. **1** – graphite tube with integrated contacts; **2** – graphite cup with integrated contacts; **3** – gaskets; **4** – cooling water; **5** and **6** – gas ports; **7** – radiation beam aperture; **8** – block for inert gas and optical sensor; **9** – optical sensor; **10** – baseplate; **A**, **B** – blocks for cup heating; **C**, **D** – blocks for tube heating; the system is covered with a quartz plate.

less subject to interferences due to concomitants, since higher atomization temperatures could be used which in addition could be selected independent of the volatility of the analyte. Nevertheless the important recognition was made that even for atomization under isothermal conditions and at higher temperatures, not all interferences due to concomitants could be eliminated. This finding was also confirmed through high temperature equilibrium calculations [1967]; this is discussed in detail in Section 8.2.2.4, part 2.

FRECH *et al.* also found that the two-step atomizer was ideal for performing investigations on reaction mechanisms [434, 1972, 1974] and also for the direct analysis of solid samples [3667]. In several further publications these authors compared the two-step atomizer with the transversely-heated tube with integrated contacts (depicted in Figure 1-33) in which the sample was atomized from a platform [553, 1970, 3666]. They found that the spatial thermal equilibrium offered by both furnace types contributed the most to the performance. Principally this was low interferences due to the concomitants and low memory effects compared to furnaces with electrical contact at the ends of the tube. Although the two-step atomizer offered a number of advantages, the authors considered that, compared to the transversely-heated tube with platform, the complexity would not justify routine use [3667]. A transversely-heated graphite tube with integrated

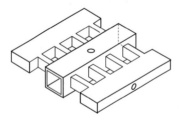

Figure 1-33. Graphite tube with integrated contacts (from [1970]). Electrical contact is at the cold zone of the graphite wings.

contacts which was developed from this design is the heart of the current generation of graphite furnaces; these are characterized by very high performance and freedom from interferences (see Sections 4.2.2.5 and 8.2.4).

1.7.4 Graphite Tubes and Other Atomizers

The preferred material for the manufacture of graphite tubes and other graphite parts is polycrystalline electrographite (EG). This material can be easily mechanically worked and has favorable thermal and electrical properties. However, at higher temperatures polycrystalline EG becomes increasingly porous.

Through experiments with radioactive tracers, L'VOV and KHARTSYZOV [3689] were able to show that the diffusion of metal atoms through the walls of the graphite tube (in dependence on the temperature) could be greater than from the open tube ends. In the initial experiments they covered the inner surface of the graphite tube with tantalum foil which immediately prevented these losses, and also enhanced atomization of carbide-forming elements [3703, 3707]. Later, L'vov found that tubes made of pyrolytic graphite (produced by pyrolyzing hydrocarbons, e.g. methane, at about 2000 °C), or tubes that were coated with pyrolytic graphite (PG), had the same properties. With these 'pyrocoated' tubes, doubled sensitivity could be obtained for a number of elements because of the prevention of diffusion losses.

CLYBURN et al. [1305] were the first to perform PG-coating directly in the graphite furnace by passing a methane-inert gas mixture through the graphite tube at a temperature slightly above 2000 °C. A dense, hard, impermeable and oxidation-resistant PG layer is deposited on the tube surface. Initially, considerable problems were experienced with the durability of the PG layer [358], so MORROW and MCELHANEY [4201] mixed 10% methane with the inert purge gas to obtain a continuous renewal of the PG layer and thereby a marked increase in the tube lifetime.

MANNING and SLAVIN [3858, 3859] treated PG-coated tubes with molybdenum and were able to obtain reproducible signals, which they attributed to a sealing of cracks and flaws in the PG layer. ORTNER and KANTUSCHER [4480] were the first to propose the impregnating of the graphite tubes with metal salts. They found that tubes impregnated with sodium tungstate gave the highest sensitivity and also the most reproducible tube lifetimes. RUNNELS et al. [5009] used a layer of either lanthanum or zirconium to prevent contact of the sample with the graphite and thereby the formation of carbides. THOMPSON et al. [5832] likewise proposed coating with lanthanum, while ZATKA [6520] used tantalum in hydrofluoric acid solution, and HAVEZOV et al. [2448] used zirconium. The generally proposed mechanism is that at increased temperature the metal salts react with the carbon to form carbides and that these have a similar density and inertness as a layer of pyrolytically deposited graphite.

NORVAL et al. [4387] first coated graphite tubes with tungsten using physical vapor deposition (PVD) and then with a layer of PG. They reported that these tubes had a very long lifetime. The PVD technique for coating tubes with metals such as tungsten and tantalum or with metal carbides was later investigated by other groups [4071, 4483, 4736]. A disadvantage of this technique was that although the layer on the outer tube surface was even and dense, it was not on the inner surface, where it would have been far more important [6251]. For this reason, and because PG-coating had in the meantime been further perfected, coatings of metals or metal carbides did not come into general use except for special applications.

The surfaces of an uncoated polycrystalline EG tube and one coated with PG are depicted in Figure 1-34. The significant differences in the properties of the surfaces are logically to be found in varying reactivities and the occurrence of matrix influences. One of the most frequent causes for the occurrence of interferences in uncoated EG tubes is the penetration of sample constituents into the graphite; these are then vaporized together with the analyte during the atomization step. SLAVIN *et al.* [5418, 5422] have repeatedly shown that interferences caused by perchloric acid, for example, cannot be observed with tubes having a good PG coating.

At an early stage L'VOV used tubes and, especially, platforms made of total pyrolytic graphite (TPG) [3689, 3701, 3704]. This material has a number of properties, such as low gas permeability and high chemical resistance, that make it ideal for use in GF AAS. At the same time the mass of the tube can be reduced using this material and thus a higher heating rate can be attained. A number of authors investigated tubes made of

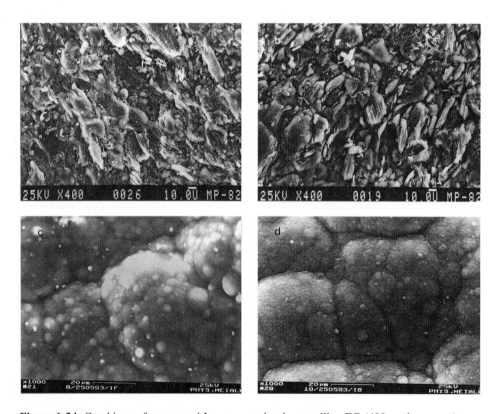

Figure 1-34. Graphite surfaces. **a** and **b** – uncoated polycrystalline EG (400x enlargement): **a** – unused; **b** – after 250 atomization cycles; **c** and **d** – PG coating (1000x enlargement): **c** – unused; **d** – after 320 atomization cycles (from [4482, 4937]).

TPG and reported that this could be a promising alternative to tubes made of polycrystalline EG [875, 1221, 1643, 3543, 3545, 4716]. One of the reasons for the high expectations placed on this material was the hope that, even though graphite vaporizes slowly, the surface of the tube would not change during its lifetime, since under every PG layer was more PG [3543]. During a detailed investigation, however, DE LOOS-VOLLEBREGT *et al.* [3596] came to the conclusion that although the increased heating rate led to higher sensitivity in peak height, the surface of TPG tubes was at the best comparable to polycrystalline EG tubes coated with PG.

In a series of long-term tests, WELZ *et al.* [6251, 6252] showed that the surface of TPG can be strongly attacked and changed due to pitting and intercalation. These effects are shown in Figure 1-35 on the example of two scanning electron microscope photographs. As can be seen from Figure 1-36, such corrosion effects can cause dramatic changes to the magnitude and shape of the measured signals. In a number of cases it was not possible to obtain reproducible signals at any time during the lifetime of the tubes. Since the expectations placed on TPG could not be met in practice, this material did not come into general use.

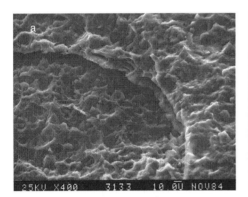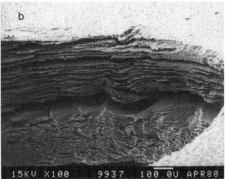

Figure 1-35. Corrosion of a TPG tube. **a** – pitting on the inner surface after 560 determinations of vanadium; **b** – flaking and delamination of the layers at the sample introduction hole after 550 measurements with 2 g/L La (from [6252]).

Another material that attracted the interest of analysts during the 1980s is glassy carbon (GC). This is a monolithic carbon with a high degree of structural disorder and exhibiting a glassy fracture. Glassy carbon is manufactured by the pyrolysis of polymer resins, which accounts for its structure, as depicted schematically in Figure 1-37. In the L_C direction the graphite layers are not longer than 5 nm and the strata are folded and completely disordered. The low gas permeability and high resistance to oxidation of GC is due to a skin effect that prevents diffusion into defects in the lattice. As soon as this surface is damaged or removed, however, the microstructure is extremely sensitive to oxidation [2706].

YANAGISAWA and TAKEUCHI [6449] and KITEGAWA *et al.* [3139] were the first to report the use of GC tubes for GF AAS, without however going into detail about the properties of this material. A number of authors then later compared GC with other materials for the manufacture of graphite tubes [2044, 3248, 3590, 3591, 4619, 5151, 6238, 6562]. Initially the expectations placed on GC were also high, since this material offers longer lifetimes for the tubes and a good long-term consistency of the atomization signals can be obtained [2044, 3590]. Nevertheless it became later clear that GC could not meet these expectations and that the low reactivity of the surface could even be a disadvantage [5628]. Clear indications were also found that at the atomization temperatures used in GF AAS GC has a reactivity similar to uncoated EG [3663, 3754, 5151, 6238, 6246, 6248].

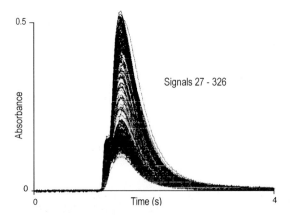

Figure 1-36. Three hundred overlaid atomization signals for 0.5 ng Cu in 0.5 mol/L perchloric acid in a TPG tube (from [6252]).

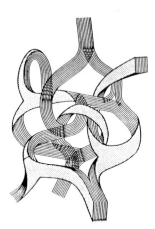

Figure 1-37. Glassy carbon; schematic structure model (from [2902]).

This reactivity of GC was confirmed through detailed studies using scanning electron microscopy [6251, 6255]. As shown in Figure 1-38, a new tube has, with the exception of a few flaws, a completely smooth surface. In continuous analytical use, however, the surface can change dramatically due to pitting and catalytic graphitization, as shown in Figure 1-39. This change from amorphous carbon into crystalline graphite under the catalytic influence of metals has been described in the literature [3897, 3898, 4508].

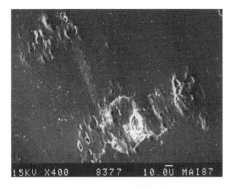

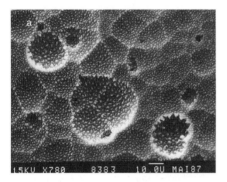

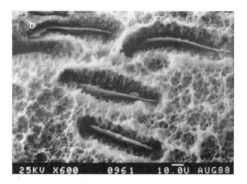

Figure 1-38. Inner surface of a GC tube with flaws (from [6255]).

Figure 1-39. Surface of a GC tube after continuous testing. **a** – pitting after 600 determinations of vanadium; **b** – catalytic graphitization and pitting after 550 determinations in the presence of 2 g/L La. The lamina in the deeper cavities are crystalline graphite (from [6255]).

It is hardly surprising that such changes to the surface structure also lead to changes in the analytical signal. For example, the overlaid signals for molybdenum in an iron matrix with a new GC tube and after 400–500 determinations are shown in Figure 1-40. Although the tube had not been mechanically destroyed, an analytical evaluation is practically impossible. Further, the signals often displayed tailing from the beginning so that integration of the signal area (integrated absorbance) is not meaningful.

The most radical way of eliminating reactions of the analyte or concomitants with graphite is to use metal atomizers. L'VOV and KHARTSYZOV [3689] lined the graphite tube with tantalum foil, and later L'VOV et al. [3729, 3732] reported considerable improvements in the determination of the alkaline-earth and rare-earth metals using this technique and with a tantalum platform. SUZUKI and OHTA [5713] and SYCHRA et al. [5721] went a step further by doing away with the graphite substrate altogether and constructed atomizers of molybdenum or tungsten. These metal furnaces were characterized by very fast heating rates that improved the peak height sensitivity considerably [4736, 6112]. At the same time however the signals were extremely narrow, leading to difficulties in their evaluation [5276, 5280].

Compared to graphite, metal atomizers can only offer improvements for those few elements that form refractory oxides and carbides, such as the rare earths. On the other hand, metal atomizers also have specific reactions that can lead to their premature destruction. Among these are the formation of alloys with metallic concomitants and the formation of carbides with carbon from biological samples, etc. Nevertheless, even without such reactions, the lifetime of metal atomizers is limited by recrystallization, which leads to increased brittleness. As shown in Figure 1-41, this recrystallization leads finally to fracture of the material.

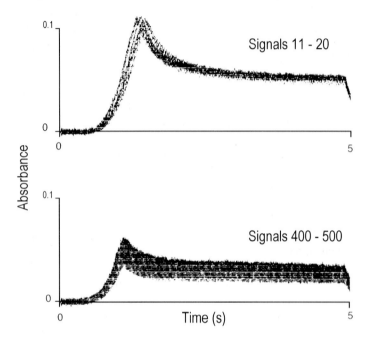

Figure 1-40. Atomization signals for 1 ng Mo in the presence of 20 µg Fe in a GC tube; signals 11–20 and 400–500 are overlaid (from [6255]).

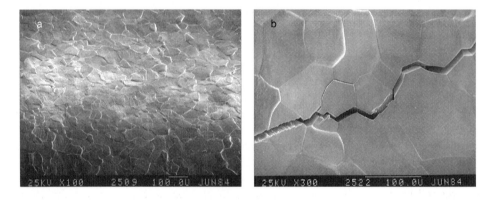

Figure 1-41. Tungsten atomizer after 200 atomization cycles for manganese in sea-water. **a** – crystalline surface; **b** – intercrystalline fracture (from [4483]).

1.8 Chemical Vapor Generation

1.8.1 The Cold Vapor Technique

Mercury is the only metallic element that has a vapor pressure as high as 0.0016 mbar at 20 °C, which corresponds to a concentration of approximately 14 mg/m^3 of atomic mercury in the vapor phase. The possibility thus exists of determining mercury directly by AAS without an atomizer. The element must merely be reduced to the metal from its compounds and transferred to the vapor phase. This procedure is termed the *cold vapor technique*. This unique property led many workers to attempt the determination of mercury even at an early stage in the development of elemental analysis. Another reason is that the toxicity of mercury has been known for a long time and there was thus great interest in developing a sensitive technique for its determination. Therefore, the determination of mercury by atomic absorption was mentioned variously in the literature even before the rediscovery of AAS by Walsh. The first paper, published in 1930, was authored by MÜLLER and PRINGSHEIM [4225]. Nine years later WOODSON [6380] described an apparatus for the determination of mercury in air. In modified form, this apparatus was used by several other workers.

The most successful technique for the determination of mercury was discovered by POLUEKTOV and VITKUN [4682, 4683]. During their investigations on the determination of mercury by flame AAS, they discovered an unusually large increase in the absorbance, by one or two orders of magnitude, if tin(II) chloride was added to the sample being aspirated. This effect could be attributed to the reducing action of this reagent, which ensured that virtually all of the mercury being aspirated passed into the flame in the atomic state. They thereupon eliminated the nebulizer and flame, passed air through the sample after tin(II) chloride had been added, and then conducted the air through a 30 cm quartz cell mounted in the radiation beam of the atomic absorption spectrometer. The limit of detection (LOD) obtained by this technique amounted to 0.5 ng Hg.

Poluektov and co-workers were certainly not the first to describe the reduction of mercury salts to metallic mercury with tin(II) chloride, but they were the first to use this reaction in combination with AAS. HATCH and OTT [2429] extended this technique and employed it for the analysis of metals, rock and soil samples. The first commercial system for the determination of mercury by the cold vapor technique was introduced in 1971 [2998]; it is depicted schematically in Figure 1-42. Air is circulated in a closed system until it is saturated with mercury vapor.

The necessity of determining mercury in the lowest concentrations led to the development of preconcentration and separation techniques. Many of these techniques were based on the fact that mercury is an extremely noble metal and can therefore be easily deposited either chemically or electrolytically from solution, for example onto copper. BRANDENBERGER and BADER [776, 777] developed a procedure in 1967 for the electrolytic deposition of mercury on a copper spiral with subsequent vaporization into a quartz cell by electrically heating the spiral. Later, next to this dynamic technique in which the mercury was transported through the cell by a gentle gas stream, they

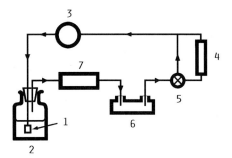

Figure 1-42. Closed circulatory system for the determination of mercury by the cold vapor technique. Free metallic mercury, released by reduction with tin(II) chloride in the reaction flask (**2**), is circulated by a pump (**3**) through the absorption cell (**6**). **1** – bubbler; **4** – absorbant; **5** – three-way valve; **7** – drying agent (from [2998]).

developed a static procedure that was simpler to operate [778, 779]. In both cell types absolute quantities of 0.2 ng Hg could be detected. HINKLE and LEARNED [2568] used a similar procedure except that they deposited mercury by chemical means onto a wire gauze which was then subsequently heated.

KAISER et al. [3001] established that when the concentration of mercury is at or below 10 µg/L, the yield from these 'static procedures' is poor, and cannot be improved through long electrolysis time of ten hours or more, or through stirring or ultrasonics. They therefore used a small column packed with copper gauze as the cathode and pumped the solution repeatedly through in a closed circuit. The copper gauze column acted rather like an ion exchange column. With this system, 50 pg Hg could be deposited quantitatively from 10 mL nitric acid solution in five minutes.

KAISER et al. [3001] further reported on digestion equipment in which biological or non-volatile inorganic samples were digested in a stream of oxygen and the vaporized mercury was collected by amalgamation on a gold gauze. Mercury vapor in air or stack gases, etc., could also be collected and preconcentrated using this procedure.

Apart from these techniques to separate and preconcentrate mercury from solution or from solid or gaseous samples, the combination of reduction and vaporization of metallic mercury with subsequent amalgamation from the gas phase has been described much more often in the literature. Frequently, the deposition of mercury vapor on tin [3012], or better on silver [3001, 3961, 4248] or gold [3001, 3002, 3950, 5958, 6004, 6223, 6436] was used to avoid kinetic effects or to improve the sensitivity. Graphite tubes coated with palladium [6440] or lined with platinum [435] were also used to preconcentrate mercury. SCHROEDER et al. [5190] published a review on the preconcentration of mercury on noble metals. Nowadays, gold or gold-platinum gauzes are used almost exclusively for the preconcentration of mercury because of their effectiveness, durability, and ease of regeneration (refer to Section 4.3.2.1).

SANZ-MEDEL et al. [5094, 5095] were the first to describe the determination of *cadmium* by the cold vapor technique. When sodium tetrahydroborate is added, in all probability the unstable cadmium hydride is formed, which then decomposes to atomic cadmium, permitting measurement by AAS.

1.8.2 The Hydride-Generation Technique

The fact that arsenic and a number of the other representative elements of Groups IV, V and VI of the periodic table form volatile, covalent hydrides with 'nascent' hydrogen has been known and utilized for more than 100 years (for example, in Marsh's and Gutzeit's Tests for arsenic). In the early 1950s a number of techniques were introduced for the determination of arsenic and other hydride-forming elements using colorimetric methods. The hydride was generated with zinc in acidic solution and the gaseous reaction products were conducted into solutions containing ammonium molybdate or hydrazine sulfate, for example, which form characteristic, colored complexes with the hydride. Some of these techniques are still in use even today.

HOLAK [2630], in 1969, was the first to apply hydride generation for the determination of arsenic using AAS. He generated hydrogen by adding zinc to the test sample solution acidified with hydrochloric acid and collected the arsine in a trap cooled in liquid nitrogen. At the end of the reaction he warmed the trap and conducted the arsine with a stream of nitrogen into an argon-hydrogen diffusion flame to measure the atomic absorption. The apparatus developed by Holak is depicted schematically in Figure 1-43. The advantage of volatilization as a gaseous hydride lies clearly in the separation and preconcentration of the analyte element and thus a reduction or even complete elimination of interferences.

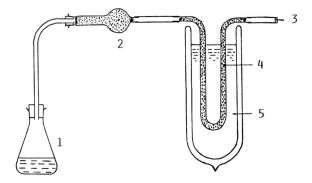

Figure 1-43. Classical apparatus used by HOLAK for the generation and collection of arsine for subsequent determination by AAS. **1** – arsine generator; **2** – calcium chloride; **3** – connection to the AA instrument; **4** – glass beads; **5** – liquid nitrogen (from [2630]).

The most commonly used method in former times for the generation of nascent hydrogen, and thereby hydrides such as arsine, was the reaction of metals such as zinc with hydrochloric acid. It is therefore hardly surprising that this technique was used initially for AAS as well. The reaction vessels frequently comprised flasks fitted with dropping funnels that permitted zinc to be added to the acidified test sample solution without having to open the system [3854]. LICHTE and SKOGERBOE [3516] used a column packed with granulated zinc through which the test sample solution was run.

GOULDEN and BROOKSBANK [2187], in an automated system, employed a suspension of aluminium powder in water as the reductant. In this system, however, the hydride had to be driven from a heated packed column by a stream of inert gas. Other authors albeit were unable to obtain satisfactory results using this reductant [4651]. For the determination of selenium, LANDSFORD *et al.* [3445] proposed the addition of tin(II) chloride to the test sample solution acidified with hydrochloric acid; the selenium hydride had to be driven from solution by a stream of inert gas. POLLOCK and WEST [4679, 4680] employed a mixture of magnesium and titanium trichloride to generate the hydride by adding it to the test sample solution acidified with sulfuric and hydrochloric acids.

Metal-acid reactions have a number of disadvantages, however, that played no small part in preventing the wider acceptance of the hydride-generation technique. When zinc is used as the reductant, only antimony, arsenic and selenium can be determined. Also, granulated metals often cannot be obtained with the required degree of purity, so that it is necessary to work with significant, frequently varying, blank values. McDANIEL *et al.* [4012] established that using this reaction, only about 8% of the hydride is released, while around 90% is trapped in the precipitated zinc sludge or does not react. A yield as low as this is certainly unsatisfactory for trace determinations.

The introduction of *sodium tetrahydroborate* ($NaBH_4$; earlier frequently called sodium borohydride) as the reductant brought about a marked change in the hydride-generation technique [771]. Using this reductant SCHMIDT and ROYER [5158] determined antimony, arsenic, bismuth and selenium, POLLOCK and WEST [4680] determined germanium, and FERNANDEZ [1887] optimized the conditions for these elements and also for tellurium and tin. THOMPSON and THOMERSON [5830] reported the successful determination of lead using this reductant and thus increased to eight the number of metals that can be determined by this technique. In recent years the vaporization of cadmium [1023], thallium [1663], and copper [5645] has also been reported.

Initially this reductant was used in a similar manner to zinc, i.e. sodium tetrahydroborate pellets were dropped into the reaction flask containing the acidified measurement solution. This mode of operation proved unsatisfactory, since poorly reproducible results were often obtained and contamination was also a problem, as with zinc. The reaction was also difficult to control because an alkaline zone formed around the tetrahydroborate pellet in which the processes were completely different from those in an acidic environment. McDANIEL *et al.* [4012] found that when using tetra-hydroborate pellets in measurement solutions acidified with 0.6 mol/L hydrochloric acid solution, only about 10% of the hydride (for selenium) was released, while in 6 mol/L hydrochloric acid solution under additional mixing with a stream of nitrogen, the yield could be increased to 50–60%.

The necessary reproducibility and control over the reaction were first achieved with the introduction of sodium tetrahydroborate in solution. Such solutions can be stabilized by preparing them in sodium hydroxide solution. The technique is also easier to automate since only solutions are involved, so that a higher sample throughput can be achieved. Depending on the type of reaction vessel used, to obtain good mixing and to drive out the hydride, the solution is either stirred with a magnetic stirrer [3221] or a stream of inert gas is passed through it [1325, 5341, 5471]. JACKWERTH *et al.* [2861] showed that by using a reaction vessel with a conical bottom and by introducing the tetrahydroborate solution through a capillary into the bottom of the vessel, stirring is no longer required.

Through the violent reaction of the alkaline reductant solution with the acidified test sample solution and the conical shape of the vessel, thorough, turbulent mixing takes place, guaranteeing a rapid and complete reaction.

As well as sodium tetrahydroborate, sodium cyanotrihydroborate ($NaBH_3CN$) was proposed as a reductant for the hydride-generation technique [883]. The authors reported a substantial improvement in yield of the hydride and greater freedom from interferences in the presence of higher concentrations of cations such as nickel. The biggest disadvantage of this reductant was the very slow reaction, taking several minutes, which made collection of the hydride, for example in a trap cooled with liquid nitrogen, necessary.

RIGIN et al. [4864–4866] proposed an electrochemical technique for the reduction of arsenic and tin to their hydrides on a platinum electrode in alkaline medium. This idea was later taken up independently in Germany and China for the determination of arsenic, antimony and selenium. Both groups employed the flow injection technique with an electrolytic flowcell in which the hydride was separated from the test sample solution by a membrane. BROCKMAN et al. [849, 850] found the very low reagent consumption and the freedom from interferences an advantage, while LIN et al. [3527] wanted to avoid the use of sodium tetrahydroborate at all costs. Compared with more recent work [6256] this appears to be the only advantage, compared to the disadvantage of reduced sensitivity by one to two orders of magnitude.

In his initial experiments, HOLAK [2630] first collected arsine in a cold trap cooled in liquid nitrogen prior to warming to vaporize it for measurement. This procedure was also later used by a number of other workers since it was highly sensitive (in peak height) and a high degree of freedom from interferences could be achieved [2222, 4475]. This was especially the case when an interference was caused by the hydride developing faster or slower from test sample solutions than from the calibration solution. Nowadays, collection of the hydride in a cold trap is used only for speciation analysis [1582] or for research applications [6216, 6234, 6254].

Since freezing of the hydride in a cold trap with subsequent vaporization by warming was a rather time-consuming procedure, FERNANDEZ and MANNING [1884] developed a system in which the hydride was collected in a balloon. After 15–30 s collection time, the hydride and the hydrogen were conducted to the atomizer by a stream of inert gas. The system is depicted schematically in Figure 1-44. A disadvantage of this procedure was that the collection time for a number of elements, for example bismuth and selenium, had to be maintained very exactly since their hydrides decompose very easily [1455, 2008]. The procedure was also relatively complicated, so that it did not come into general use.

In 1980, DRASCH et al. [1606] described a procedure in which arsine was conducted via a glass tube into a graphite furnace preheated to 370 °C, as depicted schematically in Figure 1-45. The hydride decomposes and arsenic is preconcentrated in the graphite tube and can subsequently be determined by GF AAS.

LEE [3482] determined bismuth using a similar apparatus. Using radiotracers, he found that 72% of the bismuth could be collected reproducibly in the tube independent of the temperature in the range 25–350 °C. BROVKO et al. [865–868] in Russia and, especially, STURGEON et al. [5629, 5631, 5632, 5634–5636, 6332] in Canada

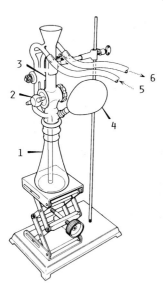

Figure 1-44. Apparatus for the generation of hydride and collection in a balloon. **1** – reaction vessel; **2** – stopcock for NaBH$_4$ pellet; **3** – dispensing column; **4** – balloon; **5** – inert gas supply; **6** – gas line to burner (from [1884]).

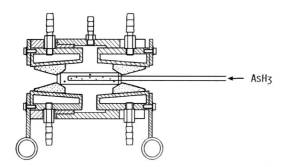

AsH₃ → (AsH$_3$)

Figure 1-45. Graphite furnace with side inlet tube for arsine (from [1606]).

investigated this form of hydride preconcentration during the following years. It was found that next to the temperature the condition of the tube surface played an important role [264, 5629, 5631].

This problem was brought under control when, independently from each other, ZHANG *et al.* [6539, 6540, 6542] in China and STURGEON and co-workers [170, 5640] in Canada found that virtually all the hydride-forming elements could be preconcentrated at relatively low temperature if the graphite tube had been treated with palladium beforehand. Next to palladium, which was employed successfully by a number of working groups [170, 956, 1220, 1561, 2435, 5771, 6543], coating the tubes with zirconium was also proposed [4344, 6439, 6442]. The real breakthrough came when it was discovered that coating with iridium was not only highly effective in sequestering the hydrides but also gave high durability. If a temperature of 2400 °C is not exceeded, more than 300 preconcentration and atomization cycles can be performed without a noticeable loss in the sensitivity [2437, 5155, 5331, 5910].

For *atomization of the gaseous hydrides*, the argon (or nitrogen)-hydrogen diffusion flame was initially used almost exclusively. The auxiliary inert gas and the hydrogen liberated by the reaction can be conducted into such flames without significantly altering the combustion characteristics. If the hydride had been previously trapped by freezing out, even the hydrogen is removed and only the hydride with carrier gas is subsequently conducted into the flame. Despite the low temperature of the diffusion flame the hydride is completely atomized, most probably via the large number of hydrogen radicals formed in this flame [410, 918, 1453, 6216, 6234]. Further, diffusion flames have sufficient transparency in the far UV range to permit the determination of arsenic and selenium with a favorable S/N ratio.

Soon after the development of the hydride-generation technique the use of electrically heated [1279] or flame-heated [5830] quartz tubes was proposed for the atomization of the hydride. Compared to a flame, quartz tube atomizers (QTAs) offer the advantage of higher sensitivity and negligible background attenuation. For this reason QTAs have completely replaced flames in the hydride-generation technique.

Instead of an externally heated QTA, SIEMER and HAGEMAN [5339] used an unheated quartz tube with a small flame burning in the side entry tube (flame-in-tube, FIT, atomizer). In this system hydrogen was used as the carrier gas and a small amount of oxygen was introduced via a capillary. This quartz tube atomizer in the modified form used by DĚDINA and RUBEŠKA [1453] for their investigations on the atomization mechanisms in the hydride-generation technique is depicted in Figure 1-46. This atomizer gained a high degree of importance through the systematic work of DĚDINA [1458] on interference and atomization mechanisms (see Sections 8.3.2.1 and 8.3.3.3).and is nowadays used preferentially for speciation analysis (see Chapter 7).

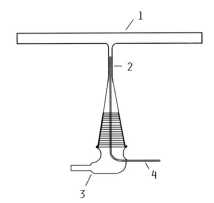

Figure 1-46. FIT atomizer developed by DĚDINA and RUBEŠKA. **1** – quartz tube; **2** – side entry tube; **3** – inlet for the hydride and hydrogen; **4** – inlet capillary for oxygen (from [1458]).

In 1973, KNUDSON and CHRISTIAN [3173] were the first to use a graphite furnace as the atomizer for gaseous hydrides; they collected the hydride in a cold trap and then conducted it directly into the graphite tube preheated to the atomization temperature. They found that the sensitivity was ten times higher compared to flame atomization. Various other working groups used the same principle in subsequent years to determine antimony [198], arsenic [2844, 5969, 6162], germanium [197], and tin [200]. As is shown in Section 8.3.3.3, part 2, the advantage of atomization in a graphite furnace is the considerably lower mutual interferences of the hydride-forming elements compared to atomization in a QTA. The major disadvantage is the lower sensitivity due to the smaller tube dimensions. This disadvantage can, at least in part, be made good by collecting the hydride in a cold trap. This procedure together with atomization in a graphite furnace has the decisive advantage that hydrogen produced during the reaction is removed beforehand. Hydrogen namely reacts at the usual temperatures of 2200–2500 °C used for atomization with the carbon of the graphite tube, leading to premature burn out and higher background attenuation.

Only a few working groups have attempted to conduct the hydride directly into a graphite furnace without prior preconcentration or separation of the hydrogen. INUI *et al.* determined antimony [2794], arsenic [2791], germanium [2793], and selenium [2792] by dispensing 10–25 µL measurement solution with a micropipet onto a sodium tetrahydroborate pellet and conducting the gaseous products with carrier gas directly into a preheated graphite furnace. The quantity of hydrogen is in this case so low that it practically does not cause problems. On the other hand the relative sensitivity is very unsatisfactory owing to the small sample volume used.

MAHER [3798, 3799] used a relatively large sample volume but maintained the atomization temperature at below 1600 °C. As DĚDINA *et al.* [1459] later showed, under given prerequisites atomization even at such low temperatures is possible. DITTRICH *et al.* [1543-1545] used a similar principle but at the same time overcame the problem of poor sensitivity by employing a 92 mm long atomizer with an internal diameter of 9 mm (Figure 1-47). These dimensions were possible through the use of 'graphite paper' which has much lower power requirements than normal graphite tubes.

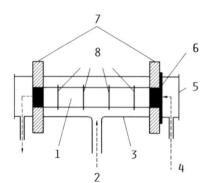

Figure 1-47. Graphite paper atomizer used by DITTRICH. **1** – electrically heated tube made of 'graphite paper'; **2** – argon protective gas; **3** – quartz tube; **4** – inlet for the hydride; **5** – quartz windows; **6** – graphite ring; **7** – water-cooled brass contacts; **8** – graphite stabilizer rings (from [1545]).

Although very interesting interference studies were carried out with a number of furnaces of this type, and it was shown that the mutual interference of the hydride-forming elements was orders of magnitude lower that in QTAs, they never came into general use. The real breakthrough came (as mentioned earlier) after the introduction of the flow injection technique when the collection and preconcentration of the hydrides on graphite tubes pretreated with iridium had been developed to a reproducible and automated technique [1559, 2437, 2441, 5155, 5331, 5910, 5913, 5914]. MATUSIEWICZ and STURGEON [3977] described the development of this technique in a review article. DĚDINA and TSALEV [1464] dedicated a monograph to the hydride-generation technique in which historical aspects are also considered.

1.8.3 Further Techniques of Chemical Vapor Generation

Besides the generation of atomic vapor (cold vapor technique) and of hydrides, investigations were also made on other ways to selectively vaporize the analytes and conduct them free from the mass of the concomitants in gaseous form into the atomizer. One of the oldest procedures of this type was described by SKOGERBOE *et al.* [5397] who vaporized Bi, Cd, Ge, Mo, Pb, Sn, Tl, and Zn as their *chlorides*. The test sample was heated in an HCl atmosphere and the gaseous chlorides produced were conducted into a flame. The procedure was also used later for the separation and preconcentration of arsenic [5804] and aluminium [5805]. Germanium can even be volatilized from solution with hydrochloric acid [2289].

Using a specially designed small furnace, MALLETT *et al.* [3830] were able to volatilize *osmium*, presumably as OsO_4, and conduct it into a nitrous oxide-acetylene flame. The LOD of this technique was 0.2 µg Os.

The technique of volatilizing *nickel* as its *carbonyl* with subsequent determination by AAS, first described by VIJAN [6077] in 1980, was investigated and modified by a number of authors. Nickel is initially reduced to the elemental state with sodium tetrahydroborate and then carbon monoxide is passed through the suspension and nickel is volatilized as the carbonyl, $Ni(CO)_4$. LEE [3481] preconcentrated the nickel carbonyl in a cold trap with liquid nitrogen prior to atomization in a preheated QTA. ALARY *et al.* [105] abstained from collecting the carbonyl and passed it directly into a QTA preheated to 950 °C; the LOD was 10 ng Ni for the sample aliquot taken. STURGEON *et al.* [5639] preconcentrated the carbonyl in a graphite tube at 500 °C, followed by atomization of the collected nickel at 2700 °C. They were able to determine nickel in marine samples in the range < 1 ng/g with an LOD of 4 ng/L, based on 10 mL sample.

BRUEGGEMEYER and CARUSO [895] determined traces of lead by converting it to *tetramethyl-lead* with methyl-lithium and collecting it on a column. After heating, it was conducted directly to a QTA at 980 °C. The LOD was 5 ng Pb. STURGEON *et al.* [5641] simplified and improved the procedure by generating tetraethyl-lead through the reaction with sodium tetraethylborate and preconcentrating the lead on a graphite tube at 400 °C. They attained an LOD of 14 pg Pb. Using the same reagent, EBDON *et al.* [1662] were able to ethylate cadmium and determine it in a QTA.

Since the end of the 1980s the volatilization of metals as their *chelates*, especially as the β-diketonates, has found increasing application. This technique was first proposed by KAWAMURA *et al.* [3057] in 1970 and was further refined by KNAPP *et al.* [3153]. In addition, dithiocarbamates [1114] and, especially, trifluoroacetylacetonates [1107, 1109, 1112–1114] were employed. Originally the chelate complex was collected on a short column which was then heated and the complex conducted to a flame [1113, 1114, 3153]. Later the thermospray technique was used [2920], and atomization in a QTA [1107, 1109, 1112, 2920]. Mostly the determinations of chromium [1107, 1109, 1113, 2920], iron [1112, 1113], and cobalt [1113, 1114, 2920] were described. The increase in sensitivity was typically one to two orders of magnitude. AN *et al.* [169] collected the volatile chromium-β-diketonate in a graphite tube preheated to 900–1000 °C and attained an LOD of 0.1 µg/L Cr.

1.9 Analysis of Solid Samples

At first a distinction must be made between special tools, procedures, or techniques for the introduction of solid samples into conventional atomizers designed for solution analysis, and atomizers designed essentially for the analysis of solid samples. The furnaces first reported in the literature for spectroscopic investigations, such as that by Lockyer (Figure 1-22) or the furnace by KING [3106], were employed in the first instance for solid samples. Even L'VOV [3720] performed his first AAS experiments with a pinch of salt. The reason for this is to be found in the large volume of these furnaces, regardless of whether they were constant temperature or pulse heated, which made them very suitable for such analyses. Even with relatively simple means, astonishingly good results can be achieved, as shown in Figure 1-48 on an earlier application from our laboratory.

In these investigations, powdered rock samples in quantities of around 1 mg were introduced with a spatula into a graphite furnace and heated. A marked dependence of the atomization signals on the temperature was observed. If the temperature was too low (< 1200 °C) the sample sintered and the analyte was slowly and incompletely released. At temperatures around 1400 °C the sample melted smoothly and the analyte was rapidly and quantitatively atomized. At higher temperatures the silicate matrix also

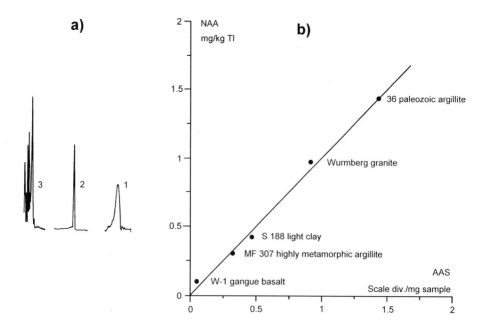

Figure 1-48. Determination of thallium in rock samples by direct solids analysis. **a** – atomization signals in dependence on the tube temperature; **1** – 1100 °C, sample sinters; **2** – 1400 °C, sample melts; **3** – 1700 °C, sample begins to volatilize. **b** – correlation of the AAS signals with neutron activation analysis (NAA).

volatilized and caused considerable interferences, since at that time a spectrometer with a BC was not available. Considering the primitive conditions under which the investigations were performed, the correlation of the AAS signals—even if merely peak height is shown for each weighing—with NAA is very good.

Later, special tools were developed which allowed reliable dispensing of solid samples into conventional graphite furnaces [2227]. A tool of this type, working on the principle of a micropipet and with an interchangeable glass capillary for taking up the sample, is depicted in Figure 1-49. Subsequently, the development moved more in the direction of special tubes, platforms, cups, or probes, conceived for direct solids analysis. This is treated in more detail in Section 4.2.7.2.

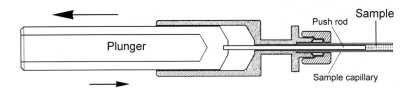

Figure 1-49. Powder sampler for introducing powdered samples into the graphite furnace (Perkin-Elmer).

WOODRIFF and co-workers [4353, 6378] developed a completely different technique for the analysis of solid samples. They sealed the solid sample in a capsule made of porous graphite and then introduced this into a furnace preheated to high temperature (up to 3000 °C). They could use up to 50 mg sample since only atoms are able to diffuse through the graphite, while concomitants, which normally cause high background attenuation, remain in the capsule.

L'VOV and co-workers [3699] made intense studies on direct solids analysis. Next to his original furnace, which he also used for the direct analysis of solids [3684, 3690], L'vov additionally investigated a number of atomizers specially conceived for solids analysis. Initially he used a graphite rod with a cavity into which the sample was introduced. The rod was located in a flame and the sample was volatilized by heating the rod electrically. The flame provided an adequate temperature to prevent recombination of the atoms [3688].

In the following years L'vov investigated the disadvantages of conventional furnaces, such as weighing errors and inhomogeneity, which are imposed by the fact that only a few milligrams of sample can be placed into the furnace. Similar to Woodriff, he proposed that the sample should be volatilized from a hollow space within walls permeable to gases. In an initial version, L'vov used a *graphite capsule* of 45–60 mm³ volume, made by boring out a graphite rod. About 40 mg of sample are mixed with graphite powder and placed into the capsule, which is then sealed with a graphite plug. The capsule is then placed between two graphite contacts and heated in the air-acetylene or nitrous oxide-acetylene flame of a Meker burner (Figure 1-50). The sample is atomized by an electrical impulse of 0.3–2.6 kW, depending on the analyte element. With this arrangement numerous elements could be determined with limits of detection of 1–100 ng/g [3045, 3046, 3699].

Figure 1-50. Graphite capsule proposed by L'vov for the direct analysis of solid samples. The capsule is located beneath the radiation beam of an AA spectrometer.

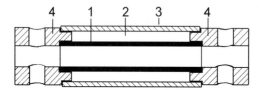

Figure 1-51. Double-walled graphite tube constructed by L'vov for direct solids analysis. **1** – inner tube of porous graphite; **2** – space for the sample; **3** – outer tube of TPG; **4** – EG contacts (from [3699]).

While in the first version the atoms that diffused out of the capsule were measured in the flame above the capsule, in a second version L'vov used a double-walled graphite tube, as shown in Figure 1-51, through which the radiation beam passed in the normal manner. The outer tube was coated with PG so that it was gas-tight, while the inner tube was made of porous graphite. The gap between the tubes had a volume of about 150 mm³ and could accommodate approximately 100 mg sample mixed with graphite powder.

Since the mass of the double-walled tube was substantially greater and the tube was not additionally heated in a flame, higher electrical power of between 1.0 kW (cadmium) and 3.7 kW (nickel) was required for atomization. Difficult-to-atomize elements, such as titanium, vanadium, and molybdenum, could not be determined in this furnace. For the easily atomized elements, on the other hand, LODs of 0.001–1 ng/g could be attained, which was at least two orders of magnitude better than with the capsule in the flame.

A further interesting advantage of both systems was the much lower background attenuation compared to volatilization in open systems, which results from the fact that atoms can diffuse far more easily through porous graphite than molecules. The idea of a double-walled graphite tube for the analysis of solid samples was later taken up in modified form, such as the double cavity tube by HADEISHI and MCLAUGHLIN [2309] or the ring chamber tube by SCHMIDT and FALK [5159].

In 1973 LANGMYHR and THOMASSEN [3427] proposed a different approach and used a *furnace heated by a radio frequency induction generator* for the direct analysis of solids. Although the heating rate of this furnace was much too slow for solutions, it appeared to have had advantages for solid samples. In the following years, LANGMYHR and co-workers [3428–3433, 3438] published a whole series of applications using this furnace.

ROBINSON and co-workers [4900, 4902–4904] also employed the principle of a radio frequency induction generator to heat a 'graphite bed atomizer' for the analysis of solid biological samples such as hair. HEADRIDGE *et al.* [206, 207, 324, 359, 360, 978, 979, 2457, 2459] likewise used an induction heated furnace for the analysis of metallic samples. However, like Woodriff, they introduced the sample into a furnace preheated to constant temperature. For the analysis of metallic samples, LUNDBERG and FRECH [3661] used a similar principle, but worked with a small, preheated graphite cup. Although in subsequent years many attempts were made to develop the direct analysis of solids into a routine procedure, the technique nevertheless did not find general acceptance, as is discussed in detail in Section 4.2.7.

BRADY *et al.* [757, 758] were the first in 1974 to analyze a *suspension* of leaves and sediments. In the following years most authors were concerned with the major problem of this technique, namely how to prepare and stabilize an homogeneous suspension. In aqueous solution, suspensions sediment out very rapidly with a sedimentation velocity that depends on the viscosity of the liquid phase and the particle size of the solid phase. Suspensions can be stabilized with highly viscous liquids such as glycerol [2606, 4100], with non-ionic surfactants [5828], and with organic solvents of higher viscosity [6489]. FULLER and THOMPSON [2020] used a thixotropic reagent to stabilize suspensions of rock samples. MAJIDI and HOLCOMBE [3815] reported that suspensions attained maximum stability when the density of the highly viscous liquid phase is similar to that of the suspended particles.

Since all of these procedures largely required an individual optimization of the liquid phase for the nature of the suspended substance, initially this technique only enjoyed very limited success. The breakthrough came when, instead of attempting to adapt the liquid medium, the suspension was rehomogenized directly prior to dispensing into the atomizer. After a number of early failures, for example with magnetic stirrers [2556], it was found that mixing using ultrasonics was optimum [4100]. MILLER-IHLI [4102] automated the procedure and a commercial system was introduced [1077]. Further details are provided in Section 4.2.7.3.

At an earlier stage, KASHIKI and OSHIMA [3036] and HARRISON and JULIANO [2408] attempted to analyze solids directly by introducing them as *suspensions into a flame*. Water-insoluble tin compounds could be maintained in suspension without stabilizers for as long as required for the measurement [2408], while aluminium oxide catalysts could be held in suspension in methanol with a special stirrer [3036]. The results were dependent on the physical nature of the samples, showed no interferences however, and by calibrating against similar suspensions, good agreement with results determined by digestion methods could be obtained.

WILLIS [6343] sprayed suspensions of geological samples directly into a flame and determined elements such as cobalt, copper, lead, manganese, nickel, and zinc. He found that only particles smaller than 12 µm contributed substantially to the measured absorbance and that the efficiency of atomization increased rapidly with decreasing particle size.

FULLER [2019] determined copper, iron, lead, and manganese in aqueous suspensions of titanium dioxide pigments. Since the particle size of these materials was highly consistent (10.0 ± 0.3 µm), he obtained good repeatability. Nevertheless, he found that when suspensions were aspirated continuously the nebulizer quickly clogged, so he changed to the direct flame injection technique proposed by SEBASTIANI et al. [5219].

Although flames are not particularly suited for direct solids analysis, various attempts were nevertheless made in this direction. COUDERT and VERGNAUD [1338] used a transport screw to introduce pulverized samples into a flame, and GOVINDARAJU et al. [2190] used an iron screw. The sample boat technique [2995] and the Delves technique were also occasionally used for the analysis of solid samples [2498, 3463, 4139].

Using either an arc or a graphite furnace, KÁNTOR et al. [3022-3024] volatilized solid samples and conducted the aerosol to a flame. They found that the sensitivity was markedly higher than with a conventional nebulizer, but they used the technique mainly for the investigation of reaction mechanisms.

As well as these essentially conventional techniques, other procedures of solid sample analysis have been proposed whose application was and probably will remain very restricted. Among these is the technique proposed by VENGHIATTIS [6020] of mixing the sample with a gunpowder-like substance ('solid mix'), pressing it into a tablet and igniting it under the beam of an atomic absorption spectrometer.

In 1967, MOSSOTTI et al. [4204] reported the use of *lasers* for atomization, an idea that was later used by VULFSON et al. [6111]. The main advantage of this procedure was the small sample requirement (10–100 µg) and the excellent absolute LODs (10 pg for copper and silver, and 100 pg for manganese). KÖNIG and NEUMANN [3236] employed a continuous wave ion laser for sample volatilization and investigated the possibility of micro scanning analysis of solid samples by AAS. ISHIZUKA et al. [2807] examined steel, brass, and aluminium alloys using volatilization with a ruby laser in argon and determined Al, Cr, Cu, Fe, Mn, Mo, Ni, and V.

Even by the end of the 1950s, WALSH and co-workers [2085, 5011] had begun to use the hollow cathode lamp not only as a radiation source but also as an atomizer. When a metal is bombarded with fast ions, metal atoms are expelled from the surface. If the metal is located in the cathode of a low pressure discharge, the process is termed *cathodic sputtering*. RUSSEL and WALSH [5011] demonstrated that a high proportion of atoms in the ground state is generated by cathodic sputtering which are available for AAS measurements. As well as for direct solids analysis, this technique offers in particular the possibility of performing measurements in the vacuum UV range with the chance of determining carbon, sulfur, and phosphorus by AAS. WALSH [6136] reported the determination of phosphorus in copper, and silicon and aluminium in steel. GOLEB [2153–2155] used a similar system successfully for elemental and isotope determinations. GOUGH et al. [2182–2184] constructed a sputtering chamber (Figure 1-52) which they used for a whole variety of applications. Further details can be found in a review article by HANNAFORD and WALSH [2360].

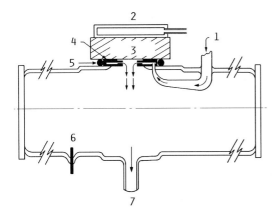

Figure 1-52. Schematic of GOUGH's cathodic sputtering chamber. **1** – argon inlet; **2** – water cooling; **3** – sample (cathode); **4** – quartz ring; **5** – O-ring gasket; **6** – anode; **7** – to pump (from [2183]).

At the end of the 1980s a commercial system was even introduced that worked according to this principle. A number of working groups tested this system and published a series of applications [3101, 4408, 4423, 6355, 6356]. In particular CHAKRABARTI *et al.* [1149, 1151, 1152] and also other groups [1265] examined this system very critically and proposed a number of improvements [1154, 2735, 4585].

2 Physical Principles

2.1 Atomic Structure and Spectra

Atomic spectra are line spectra and are specific to the absorbing or emitting atoms (elements), i.e. the spectra contain information on the atomic structure. The origin of line spectra will be discussed briefly in this section mainly on the example of sodium. For a detailed explanation the interested reader is referred to the scientific literature [1053, 1339, 2463, 2530, 3328, 4137, 5319, 5488, 5493].

2.1.1 Atomic Structure

The sodium atom consists of a nucleus surrounded by 11 electrons. The electrons travel in 'orbits' around the nucleus; in the classical quantum theory these orbits take the form of 'orbitals' while in the quantum mechanical theory they represent the probability-density distributions. Since the potential energy of the electrons in these orbitals increases with increasing distance from the nucleus, they can also be represented as energy levels. Each of these energy levels can be characterized by three quantum numbers: the *principal quantum number n*, the *orbital angular momentum (or azimuthal) quantum number ℓ*, and the *inner quantum number j*. These quantum numbers arise from the fact that the energy levels can only absorb well defined amounts of energy, i.e. they are quantized, since the wave functions must fulfill determined conditions of symmetry.

The *principal quantum number n* defines the shell in which the electron is located and is thus a measure for the relative distance of the electron shell from the nucleus. Electrons can be taken up in a shell in increasing order according to given selection rules; the maximum number of electrons that can exist in a shell is $2n^2$. Within the shells the electrons reside in orbitals (wave functions) of differing symmetry that are described by the *angular momentum quantum number ℓ*, which can take values of 0, 1, 2...$n-1$. Each characterized orbital can take up to two electrons which must have opposite spins. Each of these electrons is characterized by the *inner quantum number j*, which can take values of $j = \ell \pm \frac{1}{2}$. For the sodium atom with 11 electrons this means that the first two shells with $2 + 8 = 10$ electrons are completely filled. The eleventh electron is located on its own in the third shell and is termed the valence electron. Since its transitions are responsible for the occurrence of spectral lines it is also termed the photoelectron. In the *ground state* the valence electron is at the lowest possible energy level, which in the sodium atom is the 3s state.

2.1.2 Atomic Spectra

Next to the ground state, atoms can exist in numerous *excited states* which can be obtained through the addition of energy which causes the valence electron to undergo transitions to orbitals not occupied in the ground state. These excited states are by nature unstable and the absorbed energy is quickly released. According to the law of the conservation of energy, the energy difference associated with the transition of an electron between various energy levels must be exchanged between the atom and its environment. This exchange of energy can take place for example through the exchange of kinetic energy with a collision partner (atom or molecule) or through the exchange of radiant energy. If the energy difference of an electronic transition is exchanged via a radiative process, we can follow this process spectroscopically since, according to equation 1.1, the exchanged energy ΔE corresponds to a determined wavelength λ or frequency ν which is seen as a spectral line in the spectrum of the radiation.

Each spectral line can be regarded as the difference between two atomic states or terms $\tilde{\nu}$:

$$\frac{1}{\lambda} = \tilde{\nu}_k - \tilde{\nu}_j .\tag{2.1}$$

The differences between two such terms are the energy differences between the excited atomic states multiplied by $1/h \cdot c$:

$$\tilde{\nu}_k - \tilde{\nu}_j = \frac{1}{h \cdot c}\left(E_k - E_j\right).\tag{2.2}$$

Three forms of radiative transition are possible between the energy levels E_k and E_j:

i)   *Spontaneous emission* for the transition from a higher excited state to a lower state.
ii) *Induced emission* for the transition from a higher excited state to a lower state stimulated by external radiation of the corresponding frequency.
iii) *Absorption of radiation* (the reversed process to induced emission) with a corresponding transition of a lower state to a higher excited state.

Spectral transitions in absorption or emission are not possible between all the numerous energy levels of an atom, but only according to selection rules, which in the case of alkali metal absorption spectra, for example, require an increase in the angular momentum quantum number ℓ by one unit. The principal quantum number n, on the other hand, can alter by any amount. The permitted transitions of the electrons can be compiled in *term series* in which the principal quantum number n precedes the term symbol as a number. The series in which $\ell = 0, 1, 2, 3$ are designated by the letters s, p, d, f. These designations derive from the historical observation of spectral lines—sharp, principal, diffuse, fundamental. For values of ℓ greater than three, letters are used in alphabetical sequence following f with the omission of j.

When observed through a high resolution spectral apparatus it can be seen that the spectral lines themselves have a multiplet structure consisting of several very closely spaced lines. This *fine structure* is based on the fact that through the *spin* of the electrons and the associated magnetic field, splitting of the energy levels takes place; this is described by the inner quantum number j. For identification, the value of j is written as a subscript following the symbol for the energy level. When the sense of rotation of the electron is parallel to the orbital angular momentum, j takes the value $\ell = +\frac{1}{2}$, and when it is antiparallel, j takes the value $= -\frac{1}{2}$. The spin angular momenta of all the electrons in a fully filled shell give a total spin of zero. For the sodium atom the quantum number S thus assumes the value of the valence electron, i.e. $S = \frac{1}{2}$. The number of possible states of the valence electron, the *multiplicity of terms*, is derived from $(2S + 1)$ and is written as a superscript preceding the term symbol. In this way an atomic state is described as follows: $n^{(2S + 1)} \ell_j$. The valence electron of the sodium atom thus has a multiplicity of two, i.e. each of the sodium lines splits into a doublet. The two components of the sodium D line, for example, are at 589.593 nm and 588.966 nm. As an example the possible terms in an alkali metal spectrum are presented in Table 2-1.

Table 2-1. Possible terms in alkali metal spectra.

	$\ell = 0$ (s)	$\ell = 1$ (p)	$\ell = 2$ (d)	$\ell = 3$ (f)
$n = 1$	$1\,^2S_{1/2}$			
$n = 2$	$2\,^2S_{1/2}$	$2\,^2P_{1/2}$; $2\,^2P_{3/2}$		
$n = 3$	$3\,^2S_{1/2}$	$3\,^2P_{1/2}$; $3\,^2P_{3/2}$	$3\,^2D_{3/2}$; $3\,^2D_{5/2}$	
$n = 4$	$4\,^2S_{1/2}$	$4\,^2P_{1/2}$; $4\,^2P_{3/2}$	$4\,^2D_{3/2}$; $4\,^2D_{5/2}$	$4\,^2F_{5/2}$; $4\,^2F_{7/2}$
n	$n\,^2S_{1/2}$	$n\,^2P_{1/2}$; $n\,^2P_{3/2}$	$n\,^2D_{3/2}$; $n\,^2D_{5/2}$	$n\,^2F_{5/2}$; $n\,^2F_{7/2}$

Using suitable spectroscopic techniques it is possible to follow the radiative absorption and emission of energy by an atom. The differences in the observed spectra can be explained by the various excitation processes.

i) If excitation is via optical radiation, the atoms only absorb defined amounts of energy (i.e. radiation of a given frequency) and an *absorption spectrum* is observed.
ii) If the energy taken up by radiation absorption is emitted by at least a part of the atoms as radiation, a *fluorescence spectrum* is observed.
iii) If excitation is via thermal or electrical energy, i.e. through collisions with other particles, and at least a portion of the atoms emit the absorbed energy as radiation, an *emission spectrum* is observed.

2.1.3 Selection of the Spectral Lines

In order to illustrate the numerous possible spectral transitions in absorption and emission it is usual to use a *term* or *Grotrian diagram*. If we scrutinize the term diagram of sodium in Figure 2-1 it is clear that the thermal energy generated in the atomizers normally used in AAS is not sufficient to excite a larger number of atoms to higher energy levels (terms). Since the valence electron corresponds to the $3\,^2S_{1/2}$ term, *all lines occurring in the absorption spectrum* arise from transitions from this term to P terms as a result of the selection rules:

$$3\,^2S_{1/2} \rightarrow 3\,^2P_{1/2};_{3/2} \qquad\qquad (589.593 \text{ nm}/588.966 \text{ nm})$$

$$3\,^2S_{1/2} \rightarrow 4\,^2P_{1/2};_{3/2} \qquad\qquad (330.294 \text{ nm}/330.234 \text{ nm})$$

$$3\,^2S_{1/2} \rightarrow n\,^2P_{1/2};_{3/2}$$

These lines, which emanate from the ground term of the neutral atom, are referred to as *resonance lines*; they form the main series.

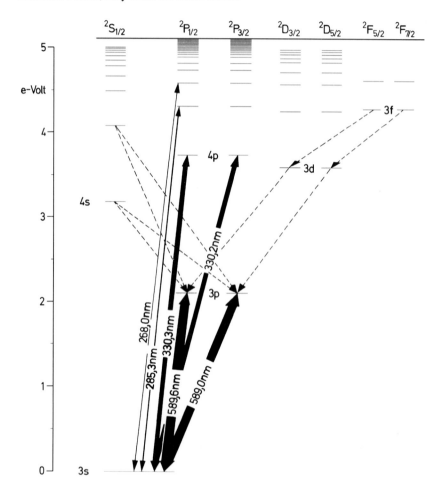

Figure 2-1. Term (Grotrian) diagram for sodium. The continuous lines with double arrows represent transitions for the main series and occur in both absorption and emission. The broken lines are for the transitions of secondary series and at the temperatures of interest occur only in emission. Thick transition lines indicate strong spectral lines.

The absorption spectrum of sodium is depicted in simplified form in Figure 2-2a. Toward the short wavelength end the lines of the main series appear closer and closer together and decrease regularly in intensity until a 'convergence limit' is reached; that is, a value beyond which there are no more lines and only continuous absorption is observed. This convergence limit in the spectrum corresponds to the total stripping of the valence electron from the shell of the sodium atom and thus represents the energy of ionization. Resulting from the changed electronic structure the sodium ion has a different, characteristic spectrum. The relatively simple relationships that can be deduced from the term diagram indicate that absorption spectra have comparatively few lines and are thus clear. If the atoms are excited by collisions with other particles, terms with arbitrary values for n and ℓ can be attained and the *emission spectrum* can contain all lines that are possible starting from these terms in compliance with various selection rules. A complete emission series can be ascribed to each secondary quantum number and the overlapping of these series leads to the multiplicity of lines as depicted in Figure 2-2b. Since the intensity of the lines depends among other things on the population density of the electrons in the energy levels, as given by the Boltzmann distribution (equation 2.16, Section 2.2), the intensities of the lines decrease rapidly toward the shorter wavelength end. As a rule the transitions from the first excited level to the ground state are the strongest. The line associated with the most intense transition is often referred to as the *last line*, since with increasing dilution this line is the last that can be observed.

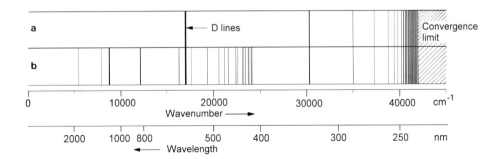

Figure 2-2. Spectra of sodium. **a** – absorption and **b** – emission (refer to the text for an explanation).

While emission depends on the population density in the energy levels and is thus critically dependent on the temperature (exponentially), absorption depends only on the population density in the ground state, and in the first approximation is independent of the temperature. The high proportion of atoms in the ground state and their low dependence on the temperature has frequently been claimed as an advantage of AAS over OES with respect to the power of detection. Such coarse simplifications however, which only take excitation into consideration, completely ignore the complex processes taking place in an atomizer and thus lead to the false conclusions. As will be shown in Section 2.2, the number of free atoms in the absorption (or observation) volume is dependent on the temperature in a number of ways, regardless of whether the absorption or emission of these atoms is being observed.

Nevertheless the clear-cut situation as presented above for the sodium atom is encountered only with the alkali and alkaline-earth metals owing to their very simple atomic structures. For numerous other elements, especially the transition elements, whose ground states consist of a large number of close-lying energy levels, the term function is dependent on the temperature and the number of atoms in the ground state is thus significantly lower than the total number of atoms. DE GALAN [2037] calculated the dependence of the term function on the temperature. PARSONS et al. [4545] pointed out that of the 70 or so elements examined by AAS, only 15 have a single intense line which does not allow any other choice when maximum sensitivity is required. For many elements the energy levels of the 'ground state' comprise a multiplet, not to mention the multiplets of the excited states which offer a large number of transitions even in the flames of relatively low temperature used analytically.

Since the intensity of an absorption line is dependent on the temperature, not only via the term function but also on other factors such as the line profile, which lines are the more intense can often only be determined by experiment. ALLAN [136, 137], DAVID [1429], and MOSSOTTI and FASSEL [4203] examined the absorption spectra of a large number of transition elements and lanthanoids by means of an emission spectrograph in front of which a flame containing high concentrations of the respective elements was mounted.

Frequently, *alternate lines* are used to avoid spectral interferences (seldom in AAS) or to extend the working range to higher concentrations (see Section 5.3.3). In cases where the excited states are less than 1 eV above the ground state, non-resonance lines can also be used; as in emission spectrometry their intensities are strongly dependent on the temperature. An example is the determination of phosphorus at 213.6 nm.

In principle, all elements can be determined by AAS since the atoms of any element can be excited and are therefore capable of absorption. The limitations lie practically only in the field of instrumentation. Measurements below 200 nm in the vacuum UV range are difficult owing to the incipient absorption of atmospheric oxygen and even more so of the hot flame gases. This range contains for example the resonance lines of selenium (189.1 nm and 196.1 nm) and arsenic (189.0 nm and 193.7 nm), which can still be determined with a good AA spectrometer, and also the resonance lines of all gases and typical non-metals. With modified instruments and a shielded flame [3127] or a graphite furnace it is possible to determine such elements as iodine at 183.0 nm [3130, 3685, 3686], sulfur at 180.7 nm [31, 3685], and phosphorus at 177.5 nm, 178.3 nm, and 178.8 nm [3685, 3687].

Apart from the remaining non-metals, all metals and metalloids can be determined by AAS, although with considerably varying efforts and with clearly differing sensitivities. Artificial and strongly radioactive elements are rarely examined for obvious reasons.

2.2 Thermal Equilibrium

Up till now we have regarded an atom that can absorb or emit radiation in isolation. Since for every transition from one energy level to the next it exchanges energy with its environment, it is necessary to include this environment in our observations. For the spectroscopic observation of the exchange of energy of an atom with its environment an 'observation cell' is required in which the atoms can reside for a certain time period. In AAS this observation cell is termed the *absorption volume* or *observation volume*. Since the atoms we wish to observe must be released from their chemical compounds, a hot environment is required that can provide the necessary *dissociation energy* (except for the mercury cold vapor technique). With increasing temperature the kinetic energy of the gas entities in the absorption volume increases. As a result of their thermal movement the number of collisions increases with increasing temperature, and kinetic energy is exchanged at the collisions. If *thermal equilibrium* is reached the velocities of the entities can be calculated by means of statistical mechanics and described by the Maxwell distribution of speeds:

$$F(v)\,\mathrm{d}v = 4\pi\, n \left(\frac{m}{2\pi\, kT} \right)^{3/2} v^2\, \exp\left(-\frac{mv^2}{2kT} \right) \mathrm{d}v\,, \qquad (2.3)$$

where $F(v)\mathrm{d}v$ is the fraction of the entities with velocities between v and $v + \mathrm{d}v$, n the number density of entities, m the mass of the observed entities, k the Boltzmann constant, and T the absolute temperature.

As can be seen from Figure 2-3, in which the Maxwell speed distributions for the entities at various temperatures are shown, the majority of the entities have medium speeds while only very few have very small or very large speeds. With increasing temperature the maximum of the distribution is shifted toward higher speeds while at the same time the distribution becomes flatter. If the energy transferred to a molecule during a collision is sufficient to overcome the *bonding energy*, the molecule can dissociate into atoms. If a thermodynamic equilibrium is attained in which every process is in equilibrium with its reverse process, then the fraction of dissociated molecules can be calculated via the *law of mass action*. A detailed discussion is provided by EWING *et al.* [1786].

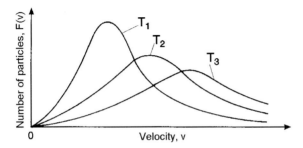

Figure 2-3. Maxwell distribution of speeds at three temperatures $T_1 < T_2 < T_3$.

For a reaction of the type:

$$AX \underset{}{\overset{K_D}{\rightleftharpoons}} A + X \tag{2.4}$$

the dissociation constant K_D for the temperature T is given by:

$$K_D(T) = \frac{[A][X]}{[AX]}. \tag{2.5}$$

The degree of dissociation α_D is defined as:

$$\alpha_D = \frac{[A]}{[A] + [AX]}, \tag{2.6}$$

and is related to the dissociation constant K_D via:

$$K_D(T) = \frac{\alpha_D^2}{1 - \alpha_D} c, \tag{2.7}$$

where: $c = [A] + [AX]$. \tag{2.8}

The dissociation constant is dependent on the temperature as given by the Maxwell distribution of speeds:

$$K_D(T) = \frac{(2\pi \mu \, kT)^{3/2}}{h^3} \frac{U_A \, U_X}{U_{AX}} e^{-E_D/kT}, \tag{2.9}$$

where U is the term sum of the entities involved, E_D is the dissociation energy and μ is the reduced mass and is given by:

$$\mu = \frac{M_A + M_X}{M_{AX}}. \tag{2.10}$$

The term sum of the atom, U_A, is defined as:

$$U_A = \sum_{n=0}^{n=\infty} P_n \, e^{-E_n/kT}, \tag{2.11}$$

where the sum of the P terms is taken over all possible energy levels of the atom including the ground state. The same considerations apply to all other entities.

Without going into details of the various processes, we can discern a number of trends from the law of mass action.

i)   The degree of dissociation increases with temperature; for dissociation energies of around 5 eV an increase in the temperature of 250 K leads to an increase of K_D by roughly a factor of ten.

ii) The degree of dissociation is dependent on the concentration, whereby the dissociation decreases with increasing concentration (Oswald's dilution law).

iii) The degree of dissociation varies for different compounds of an element and decreases with increasing dissociation energy.

iv) The degree of dissociation can be influenced by reaction products from different sources (e.g. from the flame gases) if a common reaction partner is present.

v)   Since the term sum for molecules can often be very large owing to the multitude of vibrational-rotational states, a large fraction of the molecules can remain undissociated even at high temperatures.

The atomizers used in AAS (flames, graphite tubes, quartz tubes) do not have sufficient energy to achieve complete *thermal dissociation* for many elements and compounds. In such cases *chemical influences* according to the law of mass action can be used to increase the degree of dissociation. If the reaction products are also constituents of other reactions in the atomizer, or if they are made as such, for example by modifying the flame chemistry, this can enhance the formation of free atoms. The possibilities of using chemical influences to enhance the formation of free atoms is discussed in Chapter 8.

The degree of dissociation of molecules can be increased by raising the temperature. However, with increasing temperature another reaction can predominate which reduces the fraction of free atoms, namely *thermal ionization*:

$$A \overset{K_I}{\rightleftharpoons} A^+ + e^- \tag{2.12}$$

If the energy transferred to an atom is sufficient to strip the valence electron completely from the shell a charged ion is formed which exhibits a changed spectrum owing to the changed electronic structure. If thermodynamic equilibrium is established the reaction can be described by the *Saha equation*, which is a form of the law of mass action for the dissociation of neutral particles into ionized particles:

$$K_I(T) = \frac{[A^+][e^-]}{[A]} = \frac{\alpha_I^2}{1 - \alpha_I} c_A, \tag{2.13}$$

where K_I is the ionization constant, α_I is the degree of ionization, and c_A is given by:

$$c_A = [A] + [A^+]. \tag{2.14}$$

In analogy to equation 2.9 and by rearrangement we obtain:

$$K_I(T) = \frac{2(2\pi\, m\, kT)^{3/2}}{h^3} \frac{U_{A^+}(T)}{U_A(T)} e^{-E_I/kT}, \tag{2.15}$$

where a factor of two accounts for the two spin states of the electron.

The same considerations apply to ionization as to the dissociation equilibrium discussed above:

i)   The degree of ionization increases with decreasing analyte concentration.
ii)   The degree of ionization can be suppressed by other sources of electrons.
iii) The degree of ionization increases with temperature and with decreasing energy of ionization.

Although the thermal energy of normal atomizers is insufficient in many cases to completely dissociate molecules, ionization can nevertheless play an important role. This is due to the fact that the ionization energies of a number of elements are significantly lower than the dissociation energies of refractory compounds. Since the dissociation equilibrium (atomization) and the ionization equilibrium (ionization) both contain a common term, namely the atom in the ground state, these processes are not independent of each other and must be considered as coupled processes. A shift of the ionization equilibrium thus takes place with a shift of the atomization equilibrium and *vice versa* [6518].

For our further considerations, only the fraction of free atoms is of interest. If we observe a number of free atoms N in a volume of temperature T, atoms in the ground state can be raised to an excited state through transfer of kinetic energy by collisions with other entities (e.g. flame gas molecules). On the other hand, since thermodynamic processes are completely reversible, excited atoms can transfer back their energy of excitation through collisions with other entities. The transfer of the translational, kinetic energy of the collision partner into electronic excitation energy is a very ineffective process, especially with increasing mass of the collision partner compared to the mass of an electron. Thermal excitation by inert gases, for example, which only have translational energy, can be neglected in flames. Excitation through collisions with electrons is a very effective process, but it can be virtually ruled out in the atomizers used in AAS (flames, graphite tubes), since the concentration of electrons is negligible [127, 129, 5623]. In certain radiation sources, however, collisional excitation by electrons plays an important role, as is discussed in Section 3.1.

Much more effective than energy transfer from inert gas atoms is the transfer of energy from molecules with their close-lying rotational and vibrational energy levels [2088, 3144, 5919], since small differences between the electronic excitation energy and the vibrational energy of the molecule can be compensated by additive transfer of translational or rotational energy. The same is naturally true for the de-excitation process, i.e. molecules are much more effective in radiation quenching than inert gases. In addition collision partners in the sample can be considered for the transfer of energy, especially when they are present in large excess. ZARANYIKA *et al.* [6518] were able to explain the increase of the rubidium signal in the flame in the presence of an excess of other alkali elements through charge transfer.

If thermal equilibrium is established between the excitation and quenching processes, i.e. radiationless transfer of energy, the fraction of atoms N_j in an excited state j can be calculated by statistical mechanics. The fraction of excited atoms at a certain energy level is given by the *Boltzmann distribution*:

$$\frac{N_j}{N_0} = \frac{P_j}{U_A} e^{-E_j/kT}, \tag{2.16}$$

where P_J is the statistical weight of the excited state j, U_A is the term sum of the atom, E_j is the energy of excitation, k is the Boltzmann constant, and T is the absolute temperature.

In equation 2.16 the exponent is inversely proportional to the absolute temperature, which means that the increase in the relative number of excited atoms N_j / N_0 is exponential with increasing temperature. WALSH [6135] calculated the ratio N_j/N_0 for a number of elements at various temperatures (Table 2-2). For atoms in which the energy increase to the first excited level is substantially larger than kT (3000 K corresponds to an excitation energy of 0.26 eV), the number of atoms in the ground state N_0 is virtually identical to the total number of atoms N. For such atoms the term sum U_A in equation 2.16 is in the first approximation very close to the term function of the ground state P_0. This situation is valid for about 25 elements.

Table 2-2. Temperature and wavelength dependence of the ratio N_j /N_0—according to WALSH [6135].

Element	Excitation energy	Wavelength	$\dfrac{P_j}{P_0}$	N_j/N_0		
	eV	nm		2000 K	3000 K	4000 K
Zn	5.80	213.9	3	$7.29 \cdot 10^{-15}$	$5.58 \cdot 10^{-10}$	$1.48 \cdot 10^{-7}$
Ca	2.93	422.7	3	$1.21 \cdot 10^{-7}$	$3.69 \cdot 10^{-5}$	$6.03 \cdot 10^{-4}$
Na	2.11	589.0	2	$0.86 \cdot 10^{-4}$	$5.88 \cdot 10^{-4}$	$4.44 \cdot 10^{-3}$
Cs	1.46	852.1	2	$4.44 \cdot 10^{-4}$	$7.24 \cdot 10^{-3}$	$2.98 \cdot 10^{-2}$

Equilibrium thermodynamics provide a relatively simple system for describing the various processes taking place in the observation volume. Nevertheless we must pose the question whether and to what degree can we assume that thermal equilibrium exists in the system to be observed. A system in *thermodynamic* equilibrium is characterized by a single temperature. This temperature is described as a parameter in the Maxwell distribution for the entities involved, in the Boltzmann distribution of the energy states, in the Saha equation for the electron concentration, and in the law of mass action for chemical reactions. A radiation source in thermal equilibrium emits a continuum in accordance with *Planck's radiation law*; this merely gives information on the temperature of the system (black body radiator) but no information whatsoever on its atomic structure.

Strictly speaking the observation of line spectra is a deviation from thermodynamic equilibrium since radiative transitions of emitting atoms are not balanced by the same number of absorbing transitions. However, even though a process deviates from thermodynamic equilibrium, it can still be very useful to regard the process as being in *thermal* equilibrium (e.g. Boltzmann distribution of the population densities) even if other processes (e.g. dissociation) are not in equilibrium. In the thermal flames typically used in atomic spectroscopy the quenching rate caused by collisions, i.e. the radiationless transfer of energy, is very much higher than spontaneous emission. A Boltzmann distribution over the various excitation states can thus be assumed, provided that the number of radiative transitions is negligible compared to the total number of atoms in the various states.

For a chemical reaction the question arises as to whether the residence time of the reaction partners in the observation volume is sufficient for thermodynamic equilibrium to be attained. This question can only be answered in individual cases, taking into account the concentrations of the reaction partners, a knowledge of the activation energies etc., and the residence time. Nevertheless, equilibrium thermodynamics are always useful in explaining trends and phenomena, even when a given process deviates from equilibrium.

2.3 Line Width and Line Profile

When discussing spectral lines we have up to now tacitly assumed that the observed line has an exactly defined frequency. This is in fact not true since every atomic line, even when using a spectral apparatus of infinitely high resolution, and assuming that phenomena yet to be discussed are absent, has a line profile as depicted in Figure 2-4. This profile is characterized by the central frequency, ν_0, the peak amplitude, I_p, and the frequency distribution (line profile) with a width of $\Delta\nu_{eff}$. The latter is generally quoted as the width of the profile at half maximum (FWHM = Full Width at Half Maximum), $I_{p/2}$, or *half-width* for brevity. The spectral range within the half-width is the *line core* and the ranges to either side are the *line wings*.

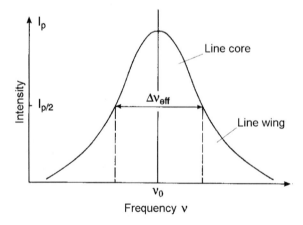

Figure 2-4. Line profile and half-width of a spectral line; refer to the text for an explanation.

Since an emitting or absorbing atom is not an isolated atom, but is in interaction with its environment, the emitted or absorbed line is broadened in its frequency distribution by a number of mechanisms. The most important broadening mechanisms will be described in this section; for a detailed discussion the interested reader is referred to monographs [784, 785] and the literature citations in them.

2.3.1 Natural Line Width

The minimum possible half-width is termed the *natural line width*. An excited atom only remains for a very short period τ in the excited state (typically 10^{-9} to 10^{-8} s) before it releases the energy of excitation, for example as a photon, and according to the Heisenberg uncertainty principle the energy levels of the transition can only be determined with an uncertainty of ΔE over the observation time Δt.

$$\Delta E \Delta t = \frac{h}{2\pi} .$$

(2.17)

From this uncertainty of the energy levels of a radiative transition we obtain via:

$$\Delta E = h\nu ,$$

(2.18)

an uncertainty of the frequency ν of the respective spectral line:

$$\delta\nu = \frac{\Delta E}{h} = \frac{1}{2\pi\tau} .$$

(2.19)

If we observe the transition between two energy levels E_k and E_j, the times of relaxation of each level (which represent the inverse values of all Einstein transition probabilities for the levels) enter the uncertainty relationship, as depicted schematically in Figure 2-5, and we obtain an expression for the natural half-width $\Delta\nu_N$:

$$\Delta\nu_N = \frac{\nu_0^2}{2\pi}\left(\frac{1}{\tau_k}+\frac{1}{\tau_j}\right),$$

(2.20)

where ν_0 is the frequency of the line, τ_k the time of relaxation of the upper (excited) state, and τ_j the time of relaxation of the lower state.

If we observe a transition to a stable ground state (resonance line), only the time of relaxation of the excited state enters the expression, since the ground state is stable ($\tau_j = \infty$). For the alkaline-earth metals, for example, the half-widths are in the order of 0.01 pm (strontium) to 0.14 pm (beryllium), which is negligible compared to other mechanisms of broadening in AAS [892, 1808, 3453, 4542, 4648, 6118]. The same applies to the other elements determinable by AAS.

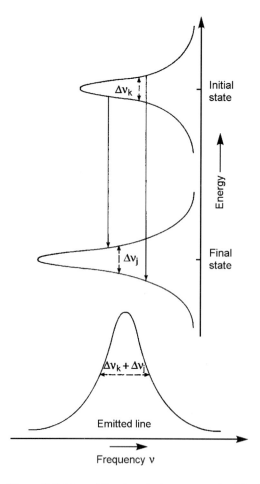

Figure 2-5. Natural line broadening as a result of the broadened energy levels of the transition.

2.3.2 Doppler Broadening

Spectral lines also undergo broadening due to the random thermal movement of the atoms. This movement can be described by the Maxwell distribution (see Section 2.2). Although the velocity of the atoms is in the order of 1000 m/s and thus far below the velocity of light (300 000 000 m/s), it nevertheless has an influence that cannot be neglected. As a result of the positive and negative velocity vectors along the line of observation relative to the observer, the emitted or absorbed photons receive an additive velocity component which brings about a corresponding positive or negative shift in the frequency. A *Doppler profile* thus results from the Maxwell distribution and this can be described by a Gaussian function. The half-width, $\Delta\lambda_D$, of a spectral line influenced by the Doppler effect is given by:

$$\Delta\lambda_D = \frac{v}{c}\sqrt{\frac{2RT}{M}} = 7.16\cdot10^{-7}\,\lambda_0\sqrt{\frac{T}{M}}\,, \tag{2.21}$$

i.e. the Doppler width $\Delta\lambda_D$ is directly proportional to the wavelength λ_0 and the root of the absolute temperature T, and inversely proportional to the root of the relative atomic mass M of the emitting species. For the first resonance doublet of sodium at a temperature of 2500 K and at atmospheric pressure, for example, this leads to a Doppler broadening of 4.5 pm, which is two orders of magnitude greater than the natural line width.

2.3.3 Collisional Broadening

At atmospheric pressure and temperatures of 2500–3000 K, which are typical for the atomizers used in AAS, an atom undergoes in the order of ten collisions per nanosecond with other particles. The lifetime of a collision is generally less than a few picoseconds. Thus in normal flames an excited atom undergoes numerous collisions with other particles during its natural lifetime of a few nanoseconds, and this leads to a change in its velocity [2660]. As a result of interactions of the excited atom with its collisional partners, a further broadening of the spectral line takes place, termed collisional broadening.

For collisional broadening we make a distinction between elastic and inelastic collisions of the excited atoms with other atoms or molecules. In an inelastic collision the excitation energy of the atom is transferred partially or completely to the collision partner, resulting in a *radiationless transition* (quenching), while in an elastic collision no energy of inner degrees of freedom is transferred to the collision partner. Broadening of the line profile by elastic collisions is generally stronger than by inelastic collisions.

Through *inelastic*, quenching collisions the lifetime of the excited state is reduced, and the line width is thus broadened further according to the Heisenberg uncertainty principle (see Section 2.3.1). Since this collision-induced additional line broadening is dependent on the pressure, it is often referred to as *pressure broadening*. Since the lifetime of a collision is generally short compared to the time period between collisions, the greater part of the emitted radiation corresponds to the unpeturbed frequency distribution, and the degree of broadening is inversely proportional to the time between collisions. The frequency distribution taken over the total number of peturbed emissions gives rise to the *Lorentz profile*; the wings of the profile provide the most information on the nature of the collisions while the line core provides little information. Pressure broadening in normal atomizers is of the order of 1 pm.

The probability that the excitation energy will be transferred to the collision partner is greatest when this energy corresponds exactly to a permitted transition of the partner. This situation is most probable when the partner is a molecule since the large number of vibrational and rotational levels permits a large number of transitions. The probability for a transfer of the excitation energy is also particularly large if the collision partner is an identical atom in the ground state. In this case the excited atom, before it can emit a photon, transfers its excitation energy to the atom in the ground state. The shortened lifetime of the excited state in the respective transition leads to a broadening of the line termed *resonance broadening*. This effect, which was interpreted by HOLTZMARK [2653] in 1925, does not however play a role in analytical atomic spectrometry, despite the relatively large collisional cross-section, because the atoms being observed do not attain a notable partial pressure in the atomizers typically used.

If, however, the collision partners are electrically charged particles (electrons, ions), the influence of the local electrical fields on the energy levels of the excited atoms must be taken into consideration. This form of broadening, termed *Stark broadening*, has only an insignificant effect in flames due to the low degree of ionization, but it must be taken into consideration in the high pressure gas discharges used occasionally as radiation sources.

Through *elastic* collisions of the excited atoms with other particles, not only are the emitted lines broadened but also shifted. The degree and direction of this so-called *Lorentz shift* are dependent on the type of interaction and on the particles involved. If the interactions of the collision partners can be described by van der Waals parameters the classical collision theory predicts a value of 2.78 for the ratio of the broadening to the shift. For the resonance lines of the alkaline-earth metals experimental values of between 2.2 and 4 have been obtained for this ratio in typical flames [2652, 2885, 3693, 3700], and values of between 2 and 10 have been obtained for most other lines examined [6116, 6118, 6119]. Heavy gases such as argon or nitrogen cause a red shift, while light gases such as neon, hydrogen or helium cause a blue shift [469, 470, 2224]. The experimental values mostly agree better with the theoretical values when the interactions are described via Lennard-Jones potentials. The observed shift can be explained by the fact that the interaction between the collision partners leads to a shift in the energy level of the emitting species. As depicted schematically in Figure 2-6, the energy shift is dependent on the electron shell of the collision partner and on its distance, and mostly varies for different energy levels.

For a radiative transition between the energy levels E_k and E_j the frequency of the absorbed or emitted radiation during the collision is dependent on the distance $R_{A,B}$ between the partners. For a mixture of atoms or molecules this distance is distributed statistically about a mean value and is dependent on the partial pressures of the collision partners and on the temperature, and can be described in the first approximation by the kinetic gas theory. The broadening and shift of the line resulting from the interaction of the collision partners often leads to an asymmetrical line profile which can no longer be described exactly by a Lorentz profile [2652]. If there is a position on the profile at which $\delta E/\delta R = 0$, a *satellite line* appears in the peturbed line profile [3096]. For sodium, with hydrogen and helium as the collision partners, it has been possible to calculate the line broadening and shift in all its complexity [3504]. Line shifts of this nature have a particularly large influence on the sensitivity and linearity of the calibration functions in AAS [3693, 3700, 5189, 5620, 6118–6120], since due to the low gas pressure and low temperature in the line radiation sources used the observed line shift is often much lower than in the atomizers. Resulting from the varying collision partners (fill gases in line sources, flame gases, purge gas in graphite furnaces, or inert gas and hydrogen in quartz tube atomizers), the shifts can even be in opposite directions.

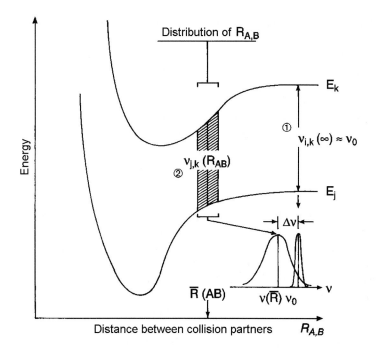

Figure 2-6. Shift of the energy level on the approach of a collision partner. (**1**) – emitted line of the non-peturbed atom; (**2**) – broadened and shifted line in the presence of a collision partner.

As early as in the 1960s L'VOV [3690] investigated the *influence of increased pressure* on the atomization process, since it was easy to vary the conditions in his graphite cuvette (refer to Section 1.7.1). At increased pressure the residence time of the atoms in the absorption volume increases, leading to a broadening of the absorption peak. At the same time the linearity of the calibration curve is improved and the peak height (absorbance) is reduced, two effects that can be traced to the broadening of the absorption line (Lorentz effect). In theory the integrated absorbance is independent of the pressure in an isothermal atomizer.

Using a Massmann furnace with atomization from the tube wall, STURGEON *et al.* [5620] found that with increasing pressure the absorbance and the integrated absorbance both decreased and only the linearity was improved. The authors attributed this effect to increasing condensation of the atoms at cold spots in the non-isothermal atomizer. FAZAKAS *et al.* [1863, 1868, 1871] confirmed this finding for a number of elements, but for other elements they observed an increase in the integrated absorbance and even in the peak height [1865, 1866, 1869]. These authors found that the nature and pressure of the inert gas determined the degree and direction of the Lorentz shift [1867, 1870, 1871]. A red shift was observed in argon while a blue shift was observed in hydrogen. By using a suitable combination of gases and pressures it should be possible to achieve better overlapping of the absorption and emission profiles.

2.3.4 Self-absorption and Self-reversal

A photon emitted by an atom in a radiation source can be absorbed by an atom of the same species in the ground state when the emitted radiation emanates from a resonance line. This is also true for atoms with low-lying excited states provided they are sufficiently populated. This phenomenon, termed *self-absorption*, leads to the attenuation of the emitted radiation. It is also the reason why in emission spectrometry the most intense resonance lines frequently cannot be used and less intense non-resonance lines must be employed.

Particularly in radiation sources with high atom concentrations or with a temperature gradient along the axis of observation a further phenomenon occurs that is termed *self-reversal*. High atom concentrations and high temperatures lead to a broadening of the emitted line profile. If the absorbing atoms are located in areas of lower temperature or atom concentration, the absorption profile is narrower than the emission profile, leading to preferential absorption from the line core, as depicted in Figure 2-7.

If the probability of reabsorption is very high it is usual to talk of an optically dense medium. Since in analytical AAS we are normally dealing with trace quantities of the analyte in the absorption volume the region of an optically dense medium will hardly be reached. Nevertheless, in line sources the effects of self-absorption and self-reversal cannot always be avoided. These stem from the finite expansion of the atom cloud in the source through which the emitted photons must pass, and the concentration gradient

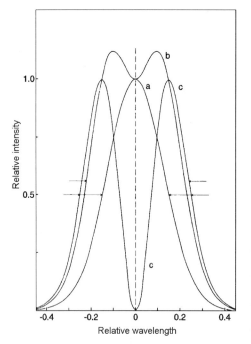

Figure 2-7. Calculated line profiles for self-absorption and self-reversal for the cadmium resonance line at 228.8 nm. For easier comparison the three profiles were expanded to approximately the same height (a = 1, b = 3, c = 4). The ratio of the half-widths (arrows) a : b : c = 1 : 1.53 : 1.67 (from [3794].

of atoms in the ground state within the atom cloud caused by temperature gradients. The problems caused by self-reversal and the means of eliminating them are discussed in Section 3.1.1.

2.3.5 Convolution of the Various Broadening Mechanisms

The total observed line broadening of a spectral line is a convolution product of all broadening mechanisms. If the various broadening processes can be regarded as statistically independent of each other, the resulting half-widths can be calculated as simple addition products. From the convolution of two Doppler profiles with half-widths of $\Delta\delta_{D1}$ and $\Delta\delta_{D2}$, respectively, we obtain via:

$$\Delta\lambda_D^2 = \Delta\lambda_{D1}^2 + \Delta\lambda_{D2}^2 \tag{2.22}$$

a Doppler profile with the half-width $\Delta\delta_D$.

The resulting line width $\Delta\delta_L$ from the convolution of several Lorentz profiles (e.g. for collisional, quenching, Stark, and resonance broadening) is, on the other hand, obtained by simple addition:

$$\Delta\lambda_L = \Delta\lambda_{Collision} + \Delta\lambda_{Quenching} + \Delta\lambda_{Stark} + \Delta\lambda_{Resonance} \,. \tag{2.23}$$

When Doppler and Lorentz broadening operate simultaneously, as is the case in a flame, the line profile can no longer be described by a simple function. The line profile resulting from the convolution of both profiles is described by an integral expression known as the *Voigt profile*. The half-width of the Voigt profile can be calculated in the first approximation from:

$$\Delta\lambda_V = \frac{1}{2}\Delta\lambda_L + \sqrt{\frac{1}{4}\Delta\lambda_L^2 + \Delta\lambda_D^2} \,. \tag{2.24}$$

Since the Doppler profile decreases more rapidly toward the line wings than the Lorentz profile, as depicted schematically in Figure 2-8, the Voigt profile approaches the Lorentz profile with increasing distance from the line center. The ratio α of the half-widths emanating from Lorentz and Doppler broadening is hardly dependent on the temperature and is determined largely by the nature and pressure of the foreign gas. This α-parameter is frequently used to describe the absorption and self-absorption and is given by:

$$\alpha = \sqrt{\ln 2}\,\frac{\Delta\lambda_L}{\Delta\lambda_D} \,. \tag{2.25}$$

Various possibilities of determining α-values and the relevant literature through 1980 have been discussed by ALKEMADE [132]. The α-values for the resonance lines of the alkali and alkaline-earth metals are in the magnitude of one, while most other values are

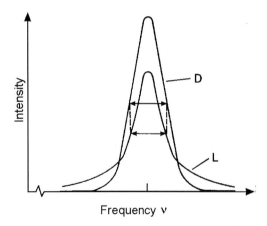

Figure 2-8. Comparison of a Lorentz profile (L) with a Doppler profile (D) with identical half-widths (from [4608]).

between 0.3 and 2.0. Nevertheless, strictly speaking, Doppler and Lorentz broadenings cannot be regarded as statistically independent of each other but are correlated [6167]. Thus the collision frequency and the distribution of the changes in velocity after a collision are dependent on the velocity of the collision partner *relative to the radiation source*. Theoretical considerations taking these dependencies into account [6167] lead to asymmetric profiles with a shift in the direction of the Lorentz shift by a quantity that is smaller than predicted by the Lorentz theory. Otherwise such *correlation effects* lead only to significant deviations in parameters that are insignificant for the characterization of the line profiles in AAS and are thus mostly ignored.

The above mentioned processes lead to a broadening of both the emission and the absorption profiles. Lines emitted by *sources* such as HCLs and EDLs operated at low currents, and thus low temperatures, and at low pressures exhibit mostly only Doppler broadening, corresponding to the working temperatures of 350–600 K [892, 2722, 2884, 6116, 6119]. In the *atomizers* typically used in AAS, which are operated at atmospheric pressure and temperatures of 1000–3000 K, the effects of pressure broadening and Doppler broadening are of the same magnitude. Because of these pressure and temperature differences between the source and the atomizer the absorption profiles are mostly broader than the emission profiles, but they are also shifted with respect to each other. The first effect is definitely desirable while the second is not.

DĚDINA [1460] showed on the example of the selenium resonance line that the line profile and the Lorentz shift depend strongly on the type of atomizer used, its temperature, and the gas atmosphere. This is shown on several examples in Figure 2-9. Selenium is especially suitable for studies of this type since it is one of the very few elements whose resonance lines do not exhibit a fine structure. The significance of the half-widths of emission and absorption lines and the linearity of the calibration function are discussed in more detail in Section 2.5.

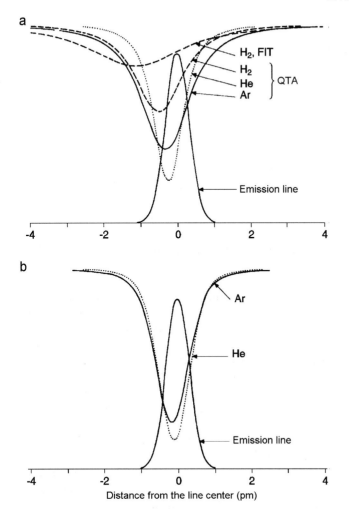

Figure 2-9. Absorption and emission profiles of the selenium resonance line at 196.0 nm. **a** – Emission line from an EDL (6 W) and the calculated absorption profiles for an unheated flame-in-tube (FIT) atomizer (41 °C) and an externally heated (700 °C) quartz tube atomizer (QTA) with different gas atmospheres. **b** – Calculated absorption profiles for graphite tube atomizers (2400 °C) with argon (solid line) and helium (dotted line) as the inert gas (from [1460]).

2.4 Hyperfine Structure

The discussions made up to now on the profile of spectral lines are only valid for 'isolated' lines. Relatively frequently, however, lines exhibit a hyperfine structure consisting of several components with differing intensities that are so close-lying that they can only be resolved by using instruments of high resolving power under special experimental conditions. The experimental conditions must be so chosen that the Lorentz and Doppler broadenings do not exceed the distances between the components. This has become possible in recent years through the use of Doppler-free laser spectroscopy.

For single-isotope elements the hyperfine structure stems from the magnetic interaction of electrons with nuclear moments; these moments comprise the magnetic moment of the atomic nucleus and the less important electric quadrupole moment. The latter results from the deviation of the nucleus from a spherical form. The magnitude of the nuclear moment depends on the structure of the nucleus; nuclei with an even number of protons and neutrons do not have a nuclear moment. Splitting patterns and intensity distributions can be calculated in the first approximation using equations analog to those used to calculate fine structure (see Section 2.1.2) by replacing the resulting electron spin angular momentum S through the nuclear spin angular momentum I, and the orbital moment L of the electron shell through the total angular moment J. Since nuclear moments are in the order of 1000 less than orbital moments, the total hyperfine splitting is less than the fine structure by this factor. Thus, for example, the $3^2S_{1/2}$ ground state of sodium splits into two very close-lying sublevels whereby each of the two sodium D lines splits into two components without, however, shifting the center point of the components. Since this splitting is only about 2 pm while line broadening in a flame is approximately 5 pm, it cannot be observed under normal analytical conditions [5892]. In lamps of relatively low working temperature and low gas pressure, however, in which line broadening is correspondingly narrow, hyperfine structure is nevertheless measurable [503]. Other lines, such as those from copper at 324.7 nm, indium at 410.2 nm, and bismuth at 472.2 nm, exhibit such a high degree of splitting that hyperfine structure can even be observed in flames [6116–6118].

A further cause of hyperfine structure is to be found in *mass effects* caused by different isotopes of an element. In contrast to the effects of hyperfine structure described above, *isotope effects* do not affect a single, isolated atom, but arise from the differing spectra of the various isotopes. If an element has several isotopes, each of these isotopes has its own spectrum. The isotope splitting of the mercury line at 253.65 nm is depicted schematically in Figure 2-10 as an example.

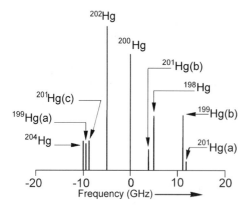

Figure 2-10. The individual isotope lines of the mercury line at 253.65 nm.

Isotope effects in optical spectroscopy can be subdivided into normal and specific mass effects, nuclear volume effects (field isotope effects), and hyperfine splitting. For a detailed discussion on these effects the interested reader is referred to textbooks on spectroscopy [784, 3110].

The *normal mass effect* stems from the movement of the nucleus and the electron shell about the mass center. The resulting wavelength shift, $\Delta\delta$, between two isotopes of an element is derived from:

$$\Delta\lambda = \lambda \frac{m_e}{m_p}\left(\frac{A_1 - A_2}{A_1 \cdot A_2}\right), \tag{2.26}$$

where m_e is the mass of the electron, m_p is the mass of the proton, and A_1, A_2 are the relative atomic masses of the observed isotopes. This effect of isotope shift due to the movement of the nuclei can be measured spectroscopically for atoms up to an atomic number of $Z = 40$. The lines for the heavier isotope are shifted toward shorter wavelengths. Examples of elements for which the isotope ratios can be determined by AAS are lithium [1193, 1196, 2155, 3838, 4773, 6295, 6512] and boron [1196, 2157, 2359, 4042].

The *nuclear volume effect* (or field isotope effect) stems from the deviation of the electric field produced by the nucleus from a point charge due to the finite volume of the nucleus. Only the electrons in close proximity to the nucleus (*s* electrons and to a lesser extent $p_{1/2}$ electrons) are influenced by these small deviations and the term shifts are proportional to the probability of their location in the nucleus. The shift between neighboring isotopes increases with mass and the lines of the heavier isotope are mostly at shorter wavelengths. The nuclear volume effect can thus be observed for elements of high relative atomic masses. The utilization of this effect to determine isotopes by AAS has been described for lead [790, 1486, 3120, 3450], mercury [2057, 4491], and uranium [2153, 2156].

WAGENAAR and DE GALAN [6116] have shown that for the majority of the elements the profile of the resonance lines is determined by the hyperfine structure. For this reason it is not possible to describe an atomic transition by a single Voigt function, as a number of authors attempted [4720, 4982]. The profile of an individual hyperfine component can nevertheless be described in the first approximation by a symmetrical function. A Gaussian function is suitable for the lines emitted from an HCL at low current while a Voigt function is suitable for lines in flames, provided that no self-absorption takes place. Within a transition all components are broadened and shifted to the same magnitude. Typical half-widths for the emission lines from an HCL are in the range 0.5–1.5 pm, while for the absorption lines in a flame they are 1.5–10 pm. A number of profiles for resonance lines emitted by HCLs and absorption lines in flames are depicted in Figure 2-11.

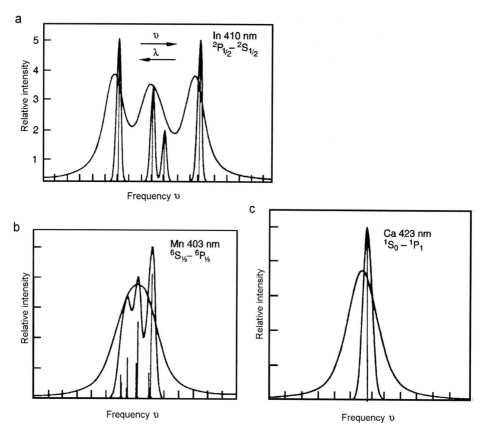

Figure 2-11. Line profiles of spectral lines from HCLs and a nitrous oxide-acetylene flame. The hyperfine structure with the relative intensities is also shown. **a** – indium, 410.2 nm; **b** – manganese, 403.1 nm; **c** – calcium, 422.7 nm. A scale division corresponds to 0.1 cm^{-1} (from [6116]).

2.5 Measuring the Absorption

2.5.1 The Absorption Coefficient

Free atoms in the ground state are able to absorb radiant energy of exactly defined frequency (light quantum $h\nu$) with concomitant transformation into an excited state. The amount of energy absorbed, E_{abs}, per unit time and volume is proportional to the number N of free atoms per unit volume, the radiant energy $h\nu_{jk}$, and the spectral radiant intensity S_ν at the resonance frequency:

$$E_{abs} = B_{jk} \cdot N S_\nu \cdot h\nu_{jk} \, . \tag{2.27}$$

The proportionality factor B_{jk} is the Einstein probability coefficient of absorption for the transition $j \rightarrow k$. The product $B_{jk} \cdot S_v$ is an expression for the fraction of all the atoms present in the ground state that can absorb a photon of energy $h\nu_{jk}$ per unit time.

In unit time a radiation unit of cS_v (c = speed of light), or $cS_v / h\nu$ photons respectively, passes through a unit volume. The fraction of the photons that is absorbed by atoms in the ground state is proportional to the total number N of free atoms and to the 'effective cross-section' of an atom, the so-called *absorption coefficient* κ_{jk}. The total amount of energy absorbed per unit volume can then be expressed as the product of the number of absorbed photons and their energy:

$$E_{abs} = \kappa_{jk} \cdot N c S_v .\tag{2.28}$$

By equating the energies in equations 2.27 and 2.28, we can express the absorption coefficient as:

$$\kappa_{jk} = \frac{h\nu}{c} B_{jk} .\tag{2.29}$$

An atom can also be regarded as an oscillating electrical dipole where the electrons circling round the nucleus represent the oscillators in equilibrium with the radiation. In the electromagnetic field of an optical radiation beam these can be excited to a motion of higher frequency.

According to the laws of electrodynamics, the total amount of energy absorbed by such an harmonic oscillator per unit time can be expressed as:

$$E_{abs} = f \frac{\pi e^2}{m} S_v ,\tag{2.30}$$

where e is the charge and m the mass of an electron, and f is a dimensionless factor, the so-called *oscillator strength*, which merely represents the effective number of classical electron oscillators corresponding to the transition $j \rightarrow k$ for the absorption effect of an atom. Expressed simply, f is the probability that an atom can be excited by the incident radiation ν_{jk}. An oscillator strength of $f = 0.19$ means that statistically 19 from 100 atoms can absorb radiation.

By equating the energies in equations 2.27 and 2.30, and considering that equation 2.30 is valid for one atom, we obtain an expression for the absorption coefficient:

$$\kappa_{jk} = \frac{\pi e^2}{mc} f_{jk} .\tag{2.31}$$

The absorption coefficient κ is a measure of the quantity of radiation of frequency ν that can be absorbed by an atom.

In practice it is more convenient to use the absorption coefficient K (dimension: length^{-1}) referring to the unit of volume rather than the absorption coefficient κ referring to an atom:

Table 2-3. Absolute oscillator strengths f for a number of spectral transitions (according to DOIDGE [1565]).

Element	Spectral line, nm	Transition	f
Ag	328.1	$5\ ^2S_{1/2} - 5\ ^2S_{3/2}$	0.476 ± 0.002
As	193.696	$5\ ^4S_{3/2} - 6\ ^4P_{1/2}$	0.139 ± 0.025
Au	242.8	$5\ ^2S_{1/2} - 5\ ^2S_{3/2}$	0.351 ± 0.015
Ba	553.548	$2\ ^1S_0 - 2\ ^1P_1$	1.64 ± 0.016
Be	234.861	$2\ ^1S_0 - 2\ ^1P_1$	1.34 ± 0.05
Bi	306.772	$6\ ^4S_{3/2} - 7\ ^4P_{1/2}$	0.118 ± 0.012
Ca	422.673	$5\ ^1S_0 - 5\ ^1P_1$	1.77 ± 0.035
Cd	228.802	$5\ ^1S_0 - 5\ ^1P_1$	1.42 ± 0.04
Co	240.7249	$4\ ^4F_{9/2} - 4\ ^4G_{11/2}$	0.38 ± 0.08
Cu	324.7	$5\ ^2S_{1/2} - 5\ ^2S_{3/2}$	0.434 ± 0.004
Fe	248.3271	$2\ ^1S_0 - 2\ ^1P_1$	0.543 ± 0.054
Ga	287.424	$3\ ^3P_{1/2} - 4\ ^2D_{3/2}$	0.285 ± 0.028
In	303.936	$5\ ^2P_{1/2} - 5\ ^2D_{3/2}$	0.31 ± 0.025
Li	670.8	$5\ ^4S_{3/2} - 6\ ^4P_{1/2}$	0.7415 ± 0.0009
Mg	285.213	$2\ ^1S_0 - 2\ ^1P_1$	1.83 ± 0.04
Mo	313.259	$4\ ^7S_3 - 5\ ^7P_4$	0.338 ± 0.019
Na	589	$5\ ^4S_{3/2} - 6\ ^4P_{1/2}$	0.6343 ± 0.0012
Ni	232.0026	$4\ ^3F_4 - 6\ ^5S_2$	0.69 ± 0.076
Pb	283.306	$6\ ^3P_0 - 7\ ^3P_1$	0.19 ± 0.02
Sb	231.147	$5\ ^4S_{3/2} - 6\ ^4P_{1/2}$	0.068 ± 0.013
Se	196.026	$5\ ^3P_0 - 6\ ^3P_1$	0.126 ± 0.018
Sn	286.332	$5\ ^3P_0 - 6\ ^3P_1$	0.230 ± 0.005
Sr	460.733	$5\ ^1S_0 - 5\ ^1P_1$	1.94 ± 0.06
Te	225.904	$5\ ^3P_2 - 6\ ^5S_2$	0.0098 ± 0.0003
Ti	364.2675	$5\ ^3P_2 - 6\ ^5S_2$	0.219 ± 0.005
Tl	276.787	$6\ ^2P_{1/2} - 6\ ^4P_{1/2}$	0.29 ± 0.02
Zn	307.590	$4\ ^1S_0 - 4\ ^3P_1$	0.00019

$$\text{K}_{jk} = N\kappa_{jk} .$$ (2.32)

Equation 2.31 then takes the form:

$$\text{K}_{jk} = N\frac{\pi e^2}{mc} f_{jk} .$$ (2.33)

The probability of radiation absorption per unit volume, K_{jk}, is now a measurable quantity, however attempts to measure the absorption coefficient have rarely been undertaken because of the experimental difficulties [1465]. The other side of the equation is made up of an unequivocally calculable constant and two unknowns, namely the total number of atoms per unit volume N and the oscillator strength f. If N can be measured, f can be calculated. On the other hand, if f is known, the absolute number N of free atoms can be calculated. This procedure, originally proposed by WALSH [6135], has been successfully used by various authors for the calculation of f [882, 3680, 3682, 3684, 3692, 5010, 6070, 6071]. A number of typical values for the oscillator strength f are presented in Table 2-3. Further details are provided in papers by KIRKBRIGHT [3131], HANNAFORD [2361], and DOIDGE [1565]. Nevertheless, even nowadays the oscillator strengths for many lines are still unknown with a precision suitable for performing an absolute calibration in AAS.

Finally we should mention that in the above discussions on the absorption coefficient it was tacitly assumed that the spectral radiant intensity S_ν is constant in a finite frequency interval $\Delta\nu$ and that the frequency ν_{jk} lies within this interval. However, as discussed in Section 2.4, both the emission lines and the absorption lines have a line profile. The expression for the absorption coefficient in equation 2.31 represents, for a spectral line of finite width, an integral over the line width (or the frequency interval):

$$\kappa_{jk} = \int \kappa_\nu \cdot \mathrm{d}\nu .$$ (2.34)

In principle, however, this does not change the above deliberations.

2.5.2 The Beer-Lambert Law

On its passage through a cloud of free atoms in the gaseous state a beam of radiation with a wavelength that can be absorbed by these atoms is increasingly attenuated. The absorption by individual atoms can be described by a particle model [545] in which an absorption cross-section, the *absorption coefficient* κ, can be ascribed to the absorbing species. If we visualize this absorption cross-section as a circle, a photon $h\nu$ can only pass unhindered through the cloud of atoms over a path of length ℓ when the absorption cross-section of an atom is not located in this path, as depicted schematically in Figure 2-12.

The probability that a photon will pass unhindered through a cylinder of length ℓ and cross-section κ thus corresponds to the probability that there is no atom in this cylinder. Assuming that N atoms are evenly distributed over volume v, the probability of finding atoms in cylinder k can be described by a binomial expression:

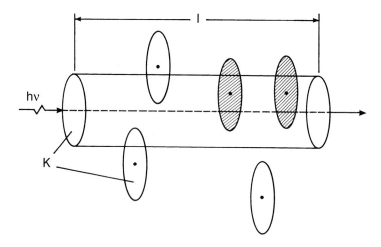

Figure 2-12. Schematic representation of the absorption of a photon when passing through an absorbing medium. The atoms are depicted as circles, corresponding to their absorption cross-sections (according to [545]).

$$P(k) = \frac{N!}{(N-k)!k!}\left(\frac{v}{V}\right)^k \left(1 - \frac{v}{V}\right)^{N-k}, \tag{2.35}$$

where V is the absorption volume and $v = \ell k$. $\tag{2.36}$

The probability of not finding an atom in the cylinder of volume v is thus given by:

$$P(0) = \left(1 - \frac{v}{V}\right)^N, \tag{2.37}$$

and under the prerequirement that v/V is very small:

$$P(0) \cong e^{-\frac{N_V}{V}}. \tag{2.38}$$

Up to now we have considered only the absorption of radiation per unit volume, a quantity that is difficult to measure in practice. If we use the quantity normally employed in absorption measurements, the radiant power Φ, instead of the probability $P(0)$—which represents nothing other than the fraction of transmitted photons—we obtain via:

$$P(0) = \frac{\Phi_{tr}(\lambda)}{\Phi_0(\lambda)}, \tag{2.39}$$

and rearrangement of equations 2.36 and 2.38 the *Beer-Lambert law* in the form:

$$\frac{\Phi_{tr}(\lambda)}{\Phi_0(\lambda)} = e^{-N\ell\kappa(\lambda)}, \tag{2.40}$$

where $\Phi_{tr}(\lambda)$ is the radiant power leaving the absorption volume, $\Phi_0(\lambda)$ is the incident radiant power entering the absorption volume, and ℓ is the length of the absorbing layer. $\kappa(\lambda)$ is the spectral atomic absorption coefficient and N the total number of free atoms.

From this it is clear why the transmittance $\Phi_{tr}(\lambda)/\Phi_0(\lambda)$ does not decrease linearly with the concentration, as would be the case if the total absorption cross-section was the sum of all individual absorption cross-sections. Since a photon is absorbed as soon as an absorbing species is in the path, it is irrelevant whether other absorbing species are in the 'shadow' of the first one, as depicted in Figure 2-12. The effective absorption cross-section of the individual absorbing species thus decreases with increasing number of these species. By logarithmizing equation 2.40 we obtain the expression:

$$A \equiv \log\frac{\Phi_0(\lambda)}{\Phi_{tr}(\lambda)} = 0.43N\ell\kappa(\lambda), \tag{2.41}$$

which states that the *absorbance* A is proportional to the total number of free atoms N and to the length ℓ of the absorbing layer. The quantity A is defined as the negative decadic logarithm of the *transmittance* (spectral transmission factor) $\tau_i(\lambda)$:

$$A = -\log\tau_i(\lambda), \tag{2.42}$$

which in turn is associated with the *absorptance* (spectral absorption factor) $\alpha_i(\lambda)$ via the relationship:

$$\alpha_i(\lambda) = 1 - \tau_i(\lambda). \tag{2.43}$$

If the atomic absorption coefficient is known, it is possible using equation 2.41 to perform absolute measurements on N. This has in fact been done for diagnostic purposes in a limited number of cases [377, 1798, 5373, 6007]. However, in the majority of cases the absorption coefficient is not known with sufficient accuracy to be able to perform absolute measurements. This is not important for routine measurements since the analyst is not interested in the *absolute concentration of atoms in the absorption volume*, but in the *quantity of the analyte in the sample*.

Nevertheless the number of absorbing atoms does not stand in simple relation to the concentration of the analyte in the sample. A number of phase changes must take place on the way to the formation of gaseous atoms. In a flame, for example, these comprise nebulization of the sample, desolvation, volatilization of the aerosol particles, and finally dissociation of the molecules. Equations describing all these functions can in principle be established, but the majority of the required parameters include a high level of uncertainty. Although absolute calibration would certainly be a desirable aim (refer for example to [1798, 1972, 2038, 2040, 2717, 3724, 3730, 3732, 3735, 3795, 4802, 4803, 5427, 5429, 5884, 6007, 6565]), analytical AAS has found another solution to this problem. Under the assumption that these parameters (which are not known with sufficient accuracy) remain constant under constant experimental conditions, they can be described by an 'effective factor' which can be determined via calibration samples. Like most other spectrometric techniques, AAS is thus a relative technique in which the relationship between quantity or mass of the analyte and the measured value is determined by *calibration*, for example by using calibration samples (calibration solutions). Calibration is discussed in Section 5.2.

2.5.3 Deviations from the Linearity of the Calibration Function

As can clearly be seen from equation 2.41, a prerequirement of the Beer-Lambert law is that the absorbance A is linearly proportional to the number N of free atoms. If this law is to be used without limitations for measurements in AAS, a number of prerequirements must be met [3624, 6522]:

i)   All absorbing species must have the same absorption cross-section for the observed radiation, i.e. the radiation must be absolutely monochromatic, the absorption volume must be isothermal, and the orientation of an absorbing species to a plane of polarization of the radiation should not have an influence on its absorption cross-section.
ii) The concentration of the absorbing species should be sufficiently low so that its absorption cross-section is not reduced due to mutual interactions (resonance broadening, see Section 2.3.3).
iii) Other mechanisms besides atomic absorption that can attenuate the radiant power, e.g. scattering, must be negligible.
iv) The absorbing species must be distributed homogeneously throughout the absorption volume.
v)   Every photon must have the same probability of striking an absorbing species, i.e. the radiation beam must be absolutely parallel and the radiant intensity must be homogeneous over the cross-section of the radiation beam, especially when the absorbing species is not distributed homogeneously throughout the absorption volume.

In practice these prerequirements are mostly not met, or only partially, so that as a general rule the relationship between absorbance and the quantity or mass of the analyte is non-linear, especially at high absorbance values. The most important reasons for the deviations, which are due to the spectral characteristics of analytical lines or to the inhomogeneous distribution of the temperature or the atom density in the absorption volume, will be discussed briefly.

Basically two *spectral causes* have been determined for the deviation of the calibration (or analytical) function from linearity. DE GALAN and SAMAEY [2039] have shown, both theoretically and experimentally, that the calibration function cannot be linear if lines from more than one transition of the element pass the spectral bandpass of the monochromator. In principle this is also the case if the observed spectral line exhibits hyperfine structure [2092, 6118, 6357]. GILMUTDINOV et al. [2120] made a detailed examination on the effect of hyperfine structure on the linearity and, as to be expected, found that the deviation was stronger the further the hyperfine components were separated from each other. The influence of broadening of the individual components on the gradient and the curvature of the calibration function is less pronounced and only plays a role when the components are very close-lying. In general the influence of hyperfine structure on the curvature of the calibration function is mostly small and is frequently blanketed by other, much stronger influences.

A much more frequent cause for non-linearity of the calibration function is *non-absorbable radiation* or *stray radiation* from the radiation source [2092, 4057, 5408]. WAGENAAR and DE GALAN [6119] for example checked the influence of the lamp current, and thus the line width, on the linearity of the calibration function. LARKINS [3452] performed similar investigations on high current pulsing, which leads to strong line broadening and hence to considerable non-linearity of the calibration function. GILMUTDINOV *et al.* [2120] showed that the Lorentz component of the broadening of emission lines from HCLs and EDLs leads to α-values of up to 0.05 (refer to Section 2.3.5). High analyte concentrations in the absorbing layer at which absorption from the wings of the emission line also becomes significant (e.g. $A = 1.5$) can lead to deviations from linearity of up to 10%. A further prerequirement for the application of the Beer-Lambert law in AAS is that the half-width of the emission line from the radiation source is significantly narrower than that of the absorption line and that the shifts of the lines with respect to each other are small (refer to Section 2.3.3). This topic and the resulting consequences are discussed in Chapter 3.

GILMUTDINOV *et al.* [2120] place the major factor for non-linearity on *non-spectral causes*. These authors showed that for GF AAS a spatially inhomogeneous distribution of the atoms as well as their temperature throughout the absorption volume can have a much stronger influence on the linearity of the calibration function than hyperfine structure, line broadening, and line shifts. In principle these effects are based on non-absorbed radiation. The influence of temperature on the absorption coefficient is the stronger, for example, the greater pressure broadening (α-parameter) contributes to the line broadening of the absorption profile. The absorbance of a homogeneous absorption layer decreases with increasing *temperature gradient* within the layer. For lines where $\alpha > 1$ a change in temperature of 10% can cause a corresponding change in the absorption coefficient [2120]. Such temperature gradients within the absorption volume can be observed both in flames [6342] and in graphite furnaces [6247]. A quantitative observation in furnaces is impeded since the volume of the atom cloud changes during the measurement.

Furthermore, the atomic absorption is influenced by the concentration distribution of the free atoms. *Concentration gradients* have been observed in flames [1135, 1340, 1854, 2011, 2342, 2692, 4801, 4954, 5188, 6141, 6283] and, more especially, in graphite furnaces [1156, 1614, 2118, 2119, 2122, 2639, 2712, 2713, 4828, 5539, 6469]. Using theoretical models GILMUTDINOV *et al.* [2120] found that concentration gradients over the vertical cross-section of the absorption volume in particular make a significant contribution to deviations from the Beer-Lambert law and thus to considerable curvature of the calibration function.

On the other hand, deviations from a homogeneous distribution along the optical axis cause a loss in sensitivity but they do not have a marked influence on the non-linearity. This is in part due to a compensation of conflicting effects, for example the dependence of the absorption coefficient on temperature gradients in a longitudinally-heated graphite tube. The conditions should be much clearer in transversely-heated atomizers, but investigations in this direction have still to be performed.

In electrothermal atomizers the radial and longitudinal distribution patterns change while the atom cloud is expanding in the absorption volume and thus during *recording of the transient signals*. Especially in the early days of GF AAS, recording the signals led to a distortion of the signal shape due to the poor time-resolved resolution [4033, 4034] and thus to a deviation from linearity [2117, 2214, 3662, 3784, 3836, 4273, 4647, 4971, 5343]. In modern instruments the problems associated with time-resolved signal recording have largely been overcome, but the influences of temperature and concentration gradients on the absorption coefficient can nevertheless lead to deviations from linearity of the calibration function due to the transient nature of these influences.

Photomultipliers are mostly used for the *detection of atomic absorption*. These do not allow any zonal resolution of the radiant intensity, but at the best permit integration of the total radiation incident on the photocathode. The absorbance A is calculated from the sum of the intensity signals for the absorption and reference radiation according to:

$$A = \log \frac{\sum\limits_{0}^{n} \Phi_{0,n}}{\sum\limits_{0}^{n} \Phi_{tr,n}}, \tag{2.44}$$

where n is the number of imaginary segments on the surface of the photocathode. Simple integration (summation) of the intensity distribution over the slit area cannot however fulfill the requirements of the Beer-Lambert law, since this does not correspond to integration of the absorbance corresponding to the signal proportional to the number N of absorbing species:

$$N \propto A = \sum\limits_{0}^{n} \log \frac{\Phi_{0,n}}{\Phi_{tr,n}} \neq \log \sum\limits_{0}^{n} \frac{\Phi_{0,n}}{\Phi_{tr,n}}. \tag{2.45}$$

In addition, photomultipliers frequently exhibit a pronounced sensitivity distribution over the photocathode, as shown in Figure 2-13, so that simple integration of the intensities over the photocathode surface does not take place but rather weighted integration. Signal recording of this type, which is used in principle in all conventional AA spectrometers, can only lead to non-linearity of the calibration function.

As well as spatial concentration gradients within the atom cloud, the *spatial distribution of the radiant flux* has an influence on the atomic absorption. The Beer-Lambert law assumes that a parallel radiation beam with a homogeneous intensity distribution is employed, but for technical reasons the image of the radiation source in the spectrometer is formed in the atomizer (graphite tube, flame). Deviations from a parallel beam mean that the radiant intensity referred to a given cross-section is dependent on the location. Thus, for example, an atom in the middle of a graphite tube is irradiated with a different intensity than one at the tube end [2145]. Furthermore it has been shown that the intensity distribution over the cross-section of an HCL or EDL

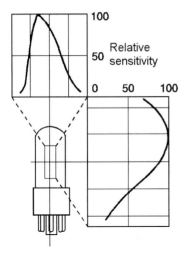

Figure 2-13. Typical sensitivity distribution over the photocathode surface of a side window photomultiplier (by kind permission of Hamamatsu).

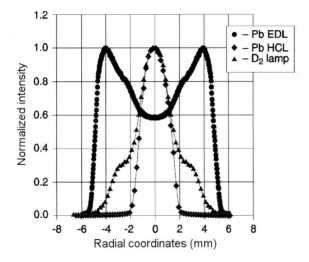

Figure 2-14. Distribution of the radiant intensity of three radiation sources over the cross-section of the absorption volume on the example of lead at 405.7 nm. ◆ HCL, 10 mA; ● EDL, 445 mA; ▲ deuterium lamp (from [2127]).

exhibits pronounced gradients [2125, 2128]. This effect can differ for different lamps, as shown on three examples in Figure 2-14. For the majority of the HCLs investigated the distribution pattern could be described relatively well by a Gaussian function, but the EDLs investigated emitted more strongly near the wall than in the middle (see Section 3.1).

By taking all these concentration gradients and inhomogeneities of the primary radiant intensity into account, but assuming that the radiation beam is parallel, the Beer-Lambert law can be presented in the form [2120]:

$$A \equiv \log \frac{\Phi_0(\lambda)}{\Phi_{\mathrm{tr}}(\lambda)} =$$

$$= \log \frac{\displaystyle\int_{-\lambda*}^{+\lambda*} \int_{-a/2}^{a/2} \int_{0}^{h} J(x,y)J(\lambda)\,\mathrm{d}x\,\mathrm{d}y\,\mathrm{d}\lambda}{\displaystyle\int_{-\lambda*}^{+\lambda*} \int_{-a/2}^{+a/2} \int_{0}^{h} J(x,y)J(\lambda)\exp\left\{-\int_{0}^{l} n(x,\,y,\,z)k(\lambda,T)\,\mathrm{d}z\right\}\mathrm{d}x\,\mathrm{d}y\,\mathrm{d}\lambda} \qquad (2.46)$$

where l is the length of the absorption volume, a the image of the slit width in the atomizer, h the image of the slit height in the atomizer, $J(x,\ y)$ the spatial intensity distribution of the primary radiation, $J(\lambda)$ the spectral line profile (intensity in dependence on the wavelength), $n(x,\ y,\ z)$ the spatial distribution of the analyte atoms, $k(\lambda,T)$ the temperature dependence of the absorption profile, and $\lambda*$ the spectral bandpass of the monochromator.

How far these deviations from a homogeneous distribution, both of the atom cloud and the radiant flux, really have an influence on the applicability of the Beer-Lambert law in its simple form is a topic of current research. Nevertheless it is correct to say that the use of the Beer-Lambert law in the situation that actually exists in AAS represents a coarse simplification. Deviations from linearity that can be observed for the calibration function cannot in many cases be correctly explained. Initial model calculations by HOLCOMBE and TISTEN [2645] nevertheless confirm that the inhomogeneous distribution of the atoms in the atomizer and also the intensity distribution of the primary radiation source are significant sources of non-linearity, which can be markedly reduced through spatially resolved absorbance measurements.

2.6 The Zeeman Effect

In 1897 the Dutch physicist PIETER ZEEMAN discovered the phenomenon that emission lines of atoms split under the influence of a magnetic field [6523]. This effect, named the Zeeman effect after its discoverer, arises from the interaction of the external magnetic field with the magnetic moment of the emitting (direct Zeeman effect) or absorbing (inverse Zeeman effect) atoms. A detailed description of this effect and the history of its discovery are contained in a treatise by BOSCH [723]. In this section, only the basic principles required for an understanding of the processes are discussed.

The magnetic moments of atoms stem from the movement of the electrons in the orbitals and the spin of the electrons. As a result of the interaction with the external magnetic field the terms of the atom depend on its orientation in the field. The relative orientations of the atoms in a magnetic field obey selection rules and are quantized. The resulting shift ΔE of the energy level M_j can be described by:

$$\Delta E = M_j \cdot g\beta \boldsymbol{H} , \qquad (2.47)$$

where M_j is the magnetic quantum number, g the Landé factor, β the Bohr magneton, and $\boldsymbol{H}$ the magnetic field strength. The splitting of the levels increases proportionally to the magnetic field strength $\boldsymbol{H}$, and the spaces are equidistant since M can only take values between $-J$ and $+J$ symmetrically distributed about the degenerate level without magnetic field. Electronic transitions between these energy levels are only permitted according to the selection rule $\Delta M_j = 0, \pm 1$. As depicted in Figure 2-15, the resulting multiplet pattern comprises an inner group of lines with $\Delta M_j = 0$, termed the π components, and two symmetrical outer groups with $\Delta M_j = +1$ and $\Delta M_j = -1$, termed the σ^+ and σ^- components, respectively.

A special case arises when the energy splitting of both levels is the same, i.e. $\Delta E_1 = \Delta E_2$, or if one energy level does not exhibit splitting, i.e. $\Delta E_1 = 0$ or $\Delta E_2 = 0$. In this case there is only *one* π component and only *one* σ^+ and σ^- component, respectively. This situation is termed the *normal Zeeman effect* since it can easily be explained by classical physics. The general situation of multiple splitting is termed the *anomalous Zeeman effect*. The normal Zeeman effect occurs only for singlet lines (terms with $S = 0$), for example the main resonance lines of the elements in the second main and subgroups of the periodic system.

Further distinction must be made between splitting into an *uneven* number of π components in which the original wavelength is retained in at least *one* component, and splitting into an *even* number of components in which the original wavelength *disappears* completely from the spectrum. The Zeeman splitting patterns for the analytical lines of a number of common elements are depicted in Figure 2-16. It should

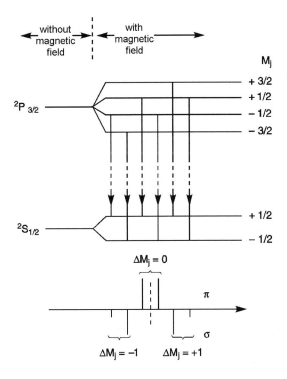

Figure 2-15. Splitting of the energy levels in a magnetic field with the resulting splitting pattern for the transition $^2S_{1/2} \leftrightarrow {}^2P_{3/2}$.

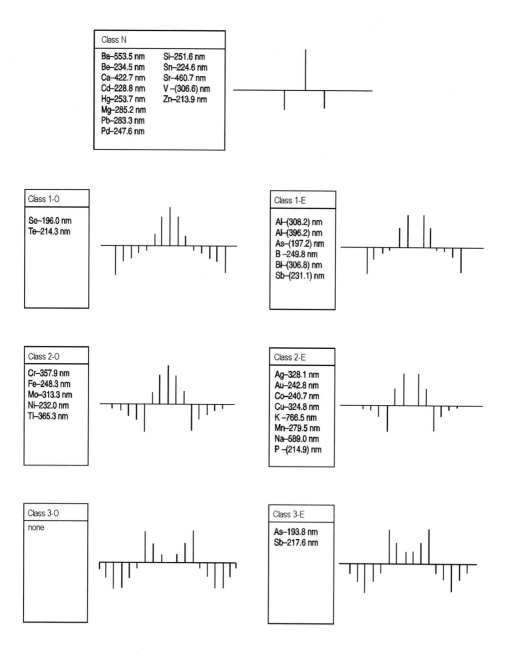

Figure 2-16. Normal and anomalous Zeeman splitting patterns of the resonance lines of 35 elements commonly determined by AAS. Class N: Normal Zeeman splitting. Classes 1-E, 2-E, 3-E: Anomalous Zeeman splitting into an even number of components. Classes 1-O, 2-O, 3-O: Anomalous Zeeman splitting into an odd number of components. Secondary lines are given in brackets.

additionally be mentioned that for elements with a fine structure, caused for example by isotope shifts, each of the isotope lines exhibits its own Zeeman splitting pattern. The experimental determination of splitting patterns under the conditions normally prevailing in AAS is only possible with extremely highly resolving spectrometers [2483].

As well as the splitting of the spectral lines into π and σ components, the radiation is also *polarized*. The π components are linearly polarized in a direction parallel to the direction of the magnetic field, while the σ components are circularly polarized in a direction perpendicular to the magnetic field. The visible intensities of the individual components depend on the orientation of the direction of view with respect to the magnetic field and obey quantum selection rules which describe the transitional probabilities. The two most important rules are:

i)   The intensities of the components are distributed symmetrically about the position of the non-influenced line.

ii)   Since the radiation of the non-influenced line is not polarized, the law of conservation of energy requires that the radiation in every direction in the magnetic field is also not polarized.

In other words, the sum of the intensities of all π components must equal the sum of the intensities of all σ components. Disregarding the inhomogeneous distribution of the radiant intensities mentioned in Section 2.5.3, the following applies for the radiant power:

$$\tfrac{1}{2}\sum \Phi_{\pi} = \sum \Phi_{\sigma^-} = \sum \Phi_{\sigma^+} , \qquad (2.48)$$

and:

$$\Phi_0 = \sum \Phi_{\pi} + \sum \Phi_{\sigma^-} + \sum \Phi_{\sigma^+} . \qquad (2.49)$$

If the direction of observation is perpendicular to the magnetic field (transverse), as depicted in Figure 2-17a, we can only 'see' the circularly polarized σ components at right angles to the magnetic field with half the intensity. The other half of the intensity can be observed in a direction parallel to the magnetic field (longitudinal). In this direction of observation we cannot 'see' the π components since they are polarized parallel to the magnetic field and thus in the axis of observation (electromagnetic waves cannot propagate in the direction of the electric vector). This situation is shown in Figure 2-17b.

The Zeeman effect is applied in AAS for the measurement and correction of background absorption. As already mentioned in Section 1.5, there are a number of principal differences, depending on the configuration of the magnet (at the source or at the atomizer, parallel or perpendicular to the radiation beam) and the type of magnetic field (constant or alternating). This topic is discussed in detail in Section 3.4.2.

a)

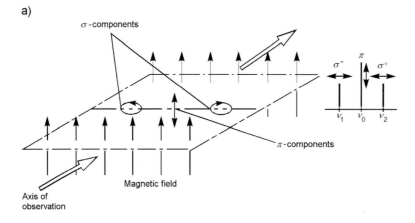

b)

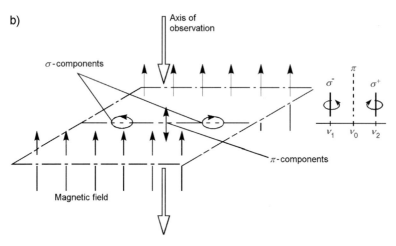

Figure 2-17. The Zeeman effect observed in varying configurations of the magnet to the axis of observation. **a** – In the transverse configuration the circularly polarized σ components appear to be horizontally polarized; **b** – in the longitudinal configuration the π components polarized linearly in the direction of observation cannot be seen. (By kind permission of Perkin-Elmer.)

In all systems utilizing the Zeeman effect for BC, measurements are performed in the same way: The radiant powers of two beams are measured alternately. The radiant power of one beam is measured either at the non-shifted analytical line (with an alternating magnetic field) when the magnetic field is off ($H = 0$) or at the minimally shifted π component (with a constant magnetic field). This is given by:

$$\Phi_{tr,1} = \Phi_{0,1} \exp\left(-K_1^a\right) \exp\left(-K_1^b\right), \tag{2.50}$$

where $\Phi_{0,1}$ is the radiant power incident upon the absorption volume, K_1^a is the absorption coefficient for atomic absorption and K_1^b the absorption coefficient for background absorption. The radiant power of the other beam derives from the $\sigma^{\pm}$ components which are shifted under the influence of the magnetic field. This is given by:

$$\Phi_{tr,2} = \Phi_{0,2} \exp\left(- K_2^a\right) \exp\left(- K_2^b\right), \tag{2.51}$$

where $\Phi_{0,2}$ is the radiant power incident upon the absorption volume, K_2^b the absorption coefficient for background absorption, and K_2^a is the *residual* absorption coefficient for atomic absorption at the shifted $\sigma^{\pm}$ lines. The contribution of the latter component depends, upon other things, on the magnetic field strength and should be as small as possible.

Simple subtraction of both absorbance values (and thus subtraction of the background absorption), and taking the logarithmic relationship into account, leads to:

$$\log \frac{\Phi_{tr,2}}{\Phi_{tr,1}} = \left(K_1^a - K_2^a\right) + \left(K_1^b - K_2^b\right) + \log \frac{\Phi_{0,2}}{\Phi_{0,1}}. \tag{2.52}$$

This expression is the basis for all Φ systems utilizing Zeeman-effect BC.

If the incident radiant powers $\Phi_{0,1}$ and $\Phi_{0,2}$ are identical and if the background absorption in each of them is the same, the net absorbance of the analyte after BC is given by:

$$A = \log \frac{\Phi_{tr,2}}{\Phi_{tr,1}} = 0.43 \left(K_1^a - K_2^a\right). \tag{2.53}$$

In other words the absorbance is proportional to the number of analyte atoms as given by the Beer-Lambert law.

Under the assumption that the (non-absorbable) stray radiation from the source is as low as in conventional AAS and that the above prerequirements are met (i.e. the incident radiant powers $\Phi_{0,1}$ and $\Phi_{0,2}$ are identical and the background absorption in each of them is the same), we obtain the following expressions for the absorbance with Zeeman-effect BC:

i) In an alternating magnetic field:

$$A = \log \frac{\Phi_\sigma}{\Phi_{H=0}} = 0.43 \left(K^a - K_\sigma^a\right). \tag{2.54}$$

ii) In a constant magnetic field:

$$A = \log \frac{\Phi_\sigma}{\Phi_\pi} = 0.43 \left(K_\pi^a - K_\sigma^a\right). \tag{2.55}$$

It is clear from equations 2.54 and 2.55 that spectrometers utilizing the Zeeman effect for background correction operate on the double-beam principle since the ratio of the two radiant powers is formed in rapid sequence ($\Phi_\sigma/\Phi_{H=0}$ or Φ_σ/Φ_π), even if the spectrometers are optical single-beam instruments.

At this point we should nevertheless mention that the assumptions made for deriving equations 2.54 and 2.55 do not necessarily have to apply. If the magnet is located at the radiation source, regardless of whether a constant or an alternating magnetic field is employed, measurement of the background attenuation is at a wavelength shifted by about ±10 pm. Background measurement and correction can only be accurate if the background does not change within this range. If the magnet is located at the atomizer, the background is always measured at the same wavelength, i.e. 'under' the analytical line. However, if an alternating magnetic field is applied at the atomizer the requirement must be met that the background attenuation is the same during the field-on and field-off phases. This requirement is discussed more fully in Sections 5.4.1.3 and 8.2.4.2.

3 Spectrometers

The combination of all optical and mechanical assemblies required for the generation, conductance, dispersion, isolation, and detection of radiant energy is termed a spectrometer. The quality of a spectrometer is determined largely by the signal-to-noise (S/N) ratio, which in itself derives mainly from the radiation (optical) conductance. This depends on the optimum adaptation of the radiation train to the dimensions of the radiation source, the atomizer, and the dispersive spectral apparatus, as well as on the number and quality of the optical assemblies.

The usable wavelength range of an atomic absorption spectrometer depends on the radiation source, the optical components used in the radiation train, and the detector. In practice this is usually from 852.1 nm, the most sensitive wavelength of cesium, to 193.7 nm, the most widely used wavelength of arsenic at the beginning of the vacuum UV.

3.1 Radiation Sources

In AAS the radiation source is a device to generate electromagnetic radiation with given properties; the radiation can be generated for example in a low pressure electrical discharge or in a low pressure plasma. Sources in sealed containers with stationary fill gases are termed lamps or spectral lamps [986]. In AAS, line sources are preferred which emit the spectral lines of one or more elements. Hollow cathode lamps (HCLs) and electrodeless discharge lamps (EDLs) are the main types of lamp employed. Next to line sources, continuum sources also find application as primary radiation sources in AAS under given prerequisites. However, they are used much more frequently for background correction. General recommendations for the characterization of radiation sources are to be found in a IUPAC document published in 1985 [986, 987].

3.1.1 Line Sources

Line sources are spectral radiation sources in which the analyte element is volatilized and excited so that it emits its spectrum. Excitation can be caused by a low pressure electrical (glow) discharge, by micro or radio waves, or by thermal energy.

By using line sources in AAS it is possible to do without high-resolution monochromators, since concomitant elements cannot in principle absorb radiation from the element-specific radiation source. A prerequisite is nevertheless that the analyte must be present in high spectral purity in the source. If there are several elements in the radiation source, spectral interferences in the area of the analytical line, caused by more than one line passing the exit slit of the monochromator and falling on the detector, must be avoided.

Since the 'resolution' in AAS is largely determined by the width of the emission lines from the radiation source, the line width has a major influence on the presence of spectral interferences caused by line overlapping (refer to Section 5.4.1.1). The width of the emission lines also has an influence on the linearity of the calibration function (see Section 2.5.3). One of the prerequirements for the applicability of the Beer-Lambert law is absolutely monochromatic radiation [3624], i.e. an 'infinitely narrow' line. This is of course not possible, as explained in Section 2.3, but nevertheless the full line width at half maximum intensity (half-width) should be as narrow as possible.

The emission intensity from the radiation source does not have a direct influence on the sensitivity in AAS, since the absorbance depends on the *ratio* of the radiant power leaving the absorption volume to that entering it (see equation 2.41). Nevertheless the emission intensity has a very marked influence on the S/N ratio and thus on the precision of a measurement, as well as on the detection and quantification limits. Generally for all sources, within given limits the emission intensity increases with increasing energy input (current, power, etc.). At the same time, however, higher energy input leads to higher temperatures within the source and thus to stronger collisional broadening of the lines. In addition, a higher energy input leads frequently to self-absorption or self-reversal (refer to Section 2.3.4), i.e. to absorption of the emitted radiation within the source. Self-absorption leads to a further broadening of the physical line profile. Self-reversal is caused by the absorption of the emitted radiation by atomic vapor of lower temperature at the perimeter of the radiation source. Self-reversal leads to a reduction of the emitted radiation at the center of the line compared to the wings (see Figure 2-7).

The correlation between self-absorption, line broadening, and the deviation of the calibration function from linearity has been investigated by a number of authors [6099, 6116, 6119]. The tendency to self-absorption and self-reversal is normally greater for the volatile elements than for the non-volatile elements [3451]. L'vov *et al.* [3739] found that differences in the characteristic masses obtained under identical atomization conditions were connected with differing self-absorption of the emission lines from the HCL.

A further consideration for the choice of sources for AAS is that the radiation of the primary source after spectral dispersion falls directly onto the detector. The prerequisites of the Beer-Lambert law are only approached in the first approximation if the radiant power $\Phi_{tr}(\lambda)$ leaving the absorption volume, and thus the transmittance $\tau_i(\lambda)$, approaches zero for a sufficiently high analyte concentration. This presupposes a very high spectral purity of the primary radiation source within the observed spectral bandwidth.

Various authors mentioned the connection between radiation background from spectral lamps and the non-linearity of the calibration function [1882, 2039, 5408]. The influence of the emission lines of the fill gas used in the source is particularly important. For example, a serious interference by neon II lines at 193.89 nm and 193.01 nm in an arsenic HCL filled with neon was reported [1981]. SOMER *et al.* [5492] were able to improve the S/N ratio by a factor of eight by introducing a filter to absorb the stray radiation caused by strong fill gas lines. FLEMING [1927] described two test procedures for determining the relative background emission from an HCL.

The *hollow cathode lamp* is the source type used most frequently and for the longest period in AAS. An HCL is a spectral lamp with a hollow, usually cylindrical, cathode containing one or more analyte elements. The anode is mostly made of tungsten or nickel. The glass cylinder of the lamp is filled with an inert gas (usually neon or argon) at a pressure of about 1 kPa.

If a voltage of 100–200 V is applied across the electrodes, a glow discharge takes place in the reduced gas atmosphere. The complex processes that take place in the cathode are depicted schematically in Figure 3-1. Two plasma regions can clearly be distinguished: the quasi-neutral negative glow (NG) region and the very narrow cathode fall (CS) region. While the NG region is virtually field free, almost the entire anode-cathode potential drop takes place in the CS region. For a hollow cathode with a diameter of 2 mm, this region is only about 0.1 mm thick. Electrons emitted from the cathode are accelerated by the strong electrical field and penetrate the NG region with an energy corresponding to the CS voltage. On their passage through the CS region a number of electrons undergo inelastic collisions with the fill gas or metal atoms. The higher the CS voltage and the lower the pressure of the fill gas, so the greater is the fraction of electrons that reach the NG region without collision, i.e. with full energy [1070]. Electrons that collide with heavy atoms ionize their collision partner (inert gas or metal atoms) and reduce their energy to thermal values while they oscillate between the potential walls of the CS region. Once the electrons have reduced their energy in this manner they diffuse to the anode.

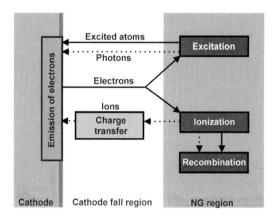

Figure 3-1. Schematic representation of the processes in a hollow cathode lamp.

The fill gas and metal atoms move by means of ambipolar diffusion through the NG region prior to being accelerated through the CS region by the strong field to the cathode. When they impinge upon the cathode they eject metal atoms from the surface. The sputtering rate of the metal ions is more effective by a factor of about five compared to the inert gas ions [3185]. In addition, via the collisions of ions, excited metastable atoms and high energy UV photons, further primary electrons are released from the cathode. If the cathode is in the form of a hollow cylinder, as in an HCL, the discharge takes place almost completely within the cathode. The analyte atoms released from the cathode thus pass into the region of the intense discharge where they collide with electrons or ions and are excited to radiation. Due to the differing energies of the electrons in the various zones of the discharge, there is a pronounced spatial inhomogeneity of the radiation, as

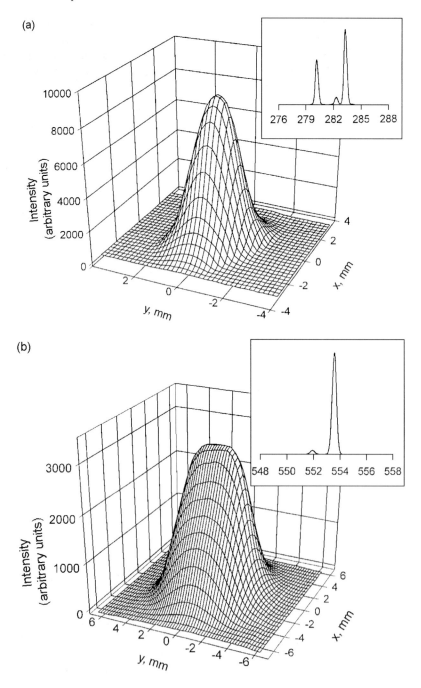

Figure 3-2. Spatial distribution of the radiant intensity of an HCL for two different cathode diameters. **a** – Lead HCL in the region of the analytical line at 238 nm, operating current 12 mA; **b** – barium HCL in the region of the analytical line at 554 nm, operating current 25 mA (from [2124]).

shown by GILMUTDINOV *et al.* [2124]. As depicted in two examples in Figure 3-2, the nature of this distribution depends mainly on the dimensions of the hollow cathode (see Figure 3-3) and to a lesser extent on the operating conditions, such as fill gas, gas pressure, cathode material, and operating voltage. The effects of an inhomogeneous distribution of the radiation on the calibration function is discussed in Section 2.5. PILLOW [4654] has written a critical overview on the spectral and other physical properties of hollow cathode discharges, while the theoretical basis is discussed by CHAPMAN [1192].

Hollow cathode lamps can be manufactured for virtually all elements determinable by AAS. Their design was optimized at an early stage in the development of AAS so that nowadays reliable lamps are available to the analyst. The lifetime of these lamps depends largely on the purity of the cathode material and the quantity of the fill gas. Depending on the operating conditions, cathode material is deposited as a mirror on the glass cylinder and is thus removed from the region of discharge. The construction of a modern HCL is depicted in Figure 3-3. The cathode is carefully isolated from the anode and the glow discharge is very localized and stabilized, so that an intense, narrow beam of high spectral purity is emitted. The low operating temperatures of around 350–450 K [892, 5570, 6116] and the low gas pressure ensure that collisional broadening of the spectral lines is negligible. If the selected lamp current is not too high, self-absorption remains within tolerable limits and the lines emitted have relatively narrow half-widths. BRUCE and HANNAFORD [892] determined for example that the half-widths for the calcium resonance line at 422.7 nm emitted from an HCL were 0.9 pm and 1.5 pm at lamp currents of 5 mA and 15 mA, respectively.

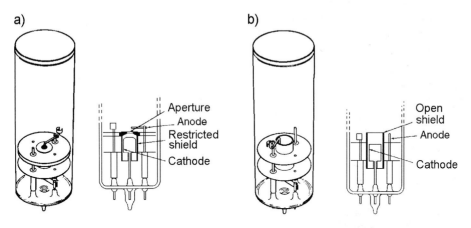

Figure 3-3. Design principle of a modern HCL with isolated cathode and anode to stabilize and localize the glow discharge. **a** – HCL with restricted cathode opening for volatile elements; **b** – HCL with open cathode for non-volatile elements (from [3849]).

With *multielement lamps* care must be taken to ensure that no spectral interferences caused by overlapping of lines occur in the regions of the analytical lines [2899]. It is also important that the emission intensities of the various elements do not vary too much from each other and are not significantly lower than for single-element lamps. While lamps containing combinations of two or three elements can typically be used without problems, lamps with more elements cannot be recommended for all applications. With these multielement lamps the radiant intensity of the individual analytical lines may be considerably lower than with single-element lamps. This leads to an unfavorable S/N ratio which can influence the precision and the limits of detection. Further, the calibration curves are often more strongly curved due to the higher background radiation resulting from the multitude of spectral lines, so that the linear working range is restricted compared to single-element lamps.

Electrodeless discharge lamps (EDLs) are among the radiation sources exhibiting the highest radiant intensity and the narrowest line widths. The radiation from an EDL is based on an inductively coupled discharge that is generated in an electromagnetic radio frequency field (e.g. 27.12 MHz). The physical processes leading to a discharge are depicted schematically in Figure 3-4. The radio frequency current, $i_\phi(t)$, through the coil generates an alternating magnetic field of intensity $H_z(t)$. This magnetic field then induces an azimuthal field, $E_\phi(t)$, which is the driving force behind the inductively coupled discharge. In contrast to the well-known inductively coupled plasma (ICP), which operates at atmospheric pressure, a low-pressure plasma is generated in the sealed quartz bulb in the EDL.

Although the first low-pressure plasma was observed and described more than a hundred years ago [2592], the processes in an EDL and the structure of the plasma have been far less investigated than in an ICP or HCL. Detailed investigations by GILMUTDINOV *et al.* [2125] have shown that the excited species which emit the analytical radiation are located in a very narrow area close to the walls of the lamp bulb. These authors concluded from their investigations that this 'optical skin effect' is caused by the radial cataphoresis effect. As depicted in Figure 3-5 for a lead EDL, this effect causes an unusual distribution of the radiant intensity which exhibits an annular maximum close to the inner wall of the lamp and a pronounced minimum in the optical axis.

An EDL is depicted in Figure 3-6; the actual lamp, comprising a quartz bulb containing the analyte element, is prealigned and permanently mounted in an RF coil. In newer versions the lamp and the RF coil are separate. This permits the use of a single driver for a number of different lamps; next to the RF coil, this driver can also contain a starter and a ventilator.

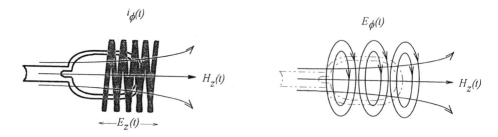

Figure 3-4. Schematic of the function of an EDL. **a** –The radio frequency current $i_\phi(t)$ generates an alternating magnetic field of intensity $H_z(t)$; **b** – this magnetic field induces an azimuthal electric field $E_\phi(t)$ (from [2125]).

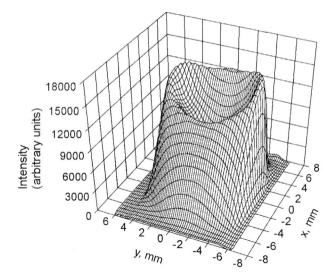

Figure 3-5. Spatial distribution of the radiant intensity of a lead EDL in the region of the analytical line at 283 nm; operating current 440 mA (from [2125]).

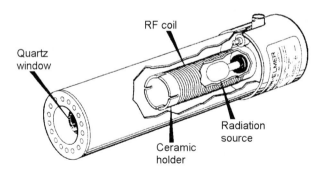

Figure 3-6. Electrodeless discharge lamp (by kind permission of Perkin-Elmer).

For volatile elements and in the far UV range, an EDL with RF excitation brings a significant improvement in the S/N ratio and thus also in the limits of detection and quantitation. The greater spectral purity also gives rise to calibration curves of notably better linearity. Nowadays the EDL is well established in AAS and has more or less totally replaced other lamp types for a number of elements. This is typically true for arsenic, where the EDL provides a twofold improvement in the sensitivity and gives a better limit of detection by an order of magnitude compared to the equivalent HCL (see Section 9.3). For cesium and rubidium the EDLs have replaced vapor discharge lamps and at the same time brought limits of detection that were previously only attainable with flame emission spectrometry [390]. The determination of phosphorus only became feasible after the introduction of the EDL [390]. EDLs are nowadays available for all the volatile elements and complement the range of HCLs. They typically provide limits of detection that are a factor of two to three better. Also, HCLs for the volatile elements are occasionally unstable and have short lifetimes, while for these elements the EDLs are especially stable and have long lifetimes.

3.1.2 Continuum Sources

In a continuum source the radiation is distributed continuously over a greater wavelength range. In AAS, continuum sources are used mainly for sequential or quasi-simultaneous background measurement and correction (refer to Section 3.4.1). Deuterium lamps and halogen lamps are mostly used for this purpose.

The *deuterium lamp* is a spectral lamp with deuterium as the discharge gas in a quartz bulb. For some applications hydrogen has also been used as the discharge gas. The deuterium lamp emits a sufficiently high radiant power in the short wavelength range from about 190 nm to 330 nm.

A *halogen lamp* consists of an electrically heated metal coil or band, usually of tungsten, in a quartz or glass bulb. The blackening of the bulb by metallic deposits is prevented by gaseous halogenated additives. The halogen lamp emits a sufficiently high radiant power in the spectral range above 300 nm. Deuterium lamps and halogen lamps thus complement their respective working ranges.

Continuum sources can also be used in principle to measure atomic absorption. Nevertheless they require a spectral apparatus with high resolution, such as an Echelle polychromator, to retain the specificity of AAS that is normally provided by line sources. The use of continuum sources is only practicable for simultaneous multielement determinations and is discussed in detail in Section 3.7.

3.2 The Radiation Train

In AAS it is normal to distinguish between single-beam, double-beam, single-channel, and multichannel spectrometers. The principle of construction of a single-beam and a double-beam spectrometer is depicted in Figure 3-7. In a single-beam spectrometer the primary radiation is conducted through the absorption volume without geometric beam splitting, while in a double-beam spectrometer the radiation beam is split. A portion of the radiation, the *sampling radiation* (or sample beam), is passed through the absorption volume, while the other portion, the *reference radiation* (or reference beam), bypasses the absorption volume.

Single-beam spectrometers have the advantage that they contain fewer optical components and thus radiation losses are lower. The major advantage claimed for *double-beam spectrometers* is better long-term stability since they compensate for changes in the intensity of the source and the sensitivity of the detector. Nevertheless, this advantage is frequently overvalued, since during the warmup phase not only the radiant intensity of the source but also the line profile change, and thereby the sensitivity [1357]. The double-beam system is neither capable of recognizing nor of compensating for changes in the atomizer, such as flame drift during the warmup phase of the burner. Hence the claim that a double-beam spectrometer does not require a warmup phase for the lamp and burner is not true.

Next to the classical single- and double-beam spectrometers, *single-beam spectrometers with double-beam characteristics* (pseudo double-beam spectrometers) have come increasingly onto the market in recent years. Instead of a reference beam that is conducted around the atomizer compartment, either the sample beam is conducted around the atomizer for a short period by a swing-in mirror assembly or the atomizer is driven out of the sample beam. Such a system is depicted in Figure 3-8. Since after a short warmup phase the radiant energy of the sources used in AAS only changes very slowly, it is sufficient to determine the baseline directly prior to the measurement (refer to Section 3.8.3).

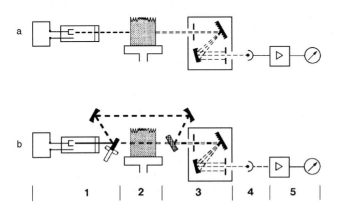

Figure 3-7. Principle of construction of atomic absorption spectrometers. **a** – Single-beam spectrometer with electrically modulated lamp radiation; **b** – double-beam spectrometer with modulation and splitting of the primary radiation by a rotating, partially mirrored quartz disk (chopper). **1** – Radiation source; **2** – atomizer; **3** – monochromator; **4** – detector; **5** – electrical measuring system with readout.

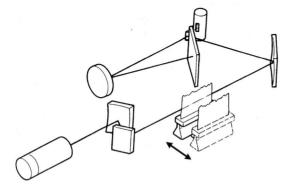

Figure 3-8. Single-beam spectrometer with double-beam characteristics. Directly prior to the measurement the atomizer (flame) is driven out of the sample beam, which then behaves as a reference beam (by kind permission of Perkin-Elmer).

By utilizing the Zeeman effect (see Section 2.6) it is also possible to obtain an (optical) single-beam instrument with double-beam characteristics. When the magnetic field at the atomizer is switched on the analyte atoms cannot absorb the radiation from the source, while they can absorb normally when the magnetic field is switched off. Since the radiation with and without magnetic field has the same intensity, drift in the source or detector are eliminated, as with a double-beam instrument.

Single-channel spectrometers are mostly used in AAS. At any given time, measurements can only be made at a single wavelength. Single-channel spectrometers in which the radiation sources and the wavelength are changed automatically in sequence are termed *sequential spectrometers*. A *multichannel spectrometer* permits measurements at several wavelengths to be performed simultaneously. A multichannel spectrometer either consists of several single-channel spectrometers (monochromators) or is constructed as a simultaneous spectrometer (polychromator). Multichannel spectrometers are discussed in more detail in Section 3.7.

Mirrors and/or lenses are used in the *radiation train*. To keep radiation losses and optical aberration to a minimum, a combination of plane, spherical, and torroidal mirrors is mostly employed. To increase the reflectivity in the UV range and as a protection against environmental influences, mirrors are frequently coated with a protective coating, such as silica. Lenses, if used, should be made of quartz. Because of their wavelength-dependent transmission characteristics, they should either be achromatized or have a mechanical means of focusing. Exceptions are only possible when a lens is mounted very close to the detector.

The geometric splitting and recombining of the beam in double-beam spectrometers, as well as the combination of the radiation from a continuum source for BC, can be effected by means of semitransparent mirrors made from partially mirrored quartz plates, or a rotating chopper made from a partially mirrored rotating quartz disk, or by a combination of these. The reflecting surfaces are usually in the form of stripes or chess-board patterns. Splitting ratios are usually in the range 1:1 to 1:3.

Choppers or rotating sector disks are used for *modulation of the radiation* from the primary radiation source and as beam splitters in double-beam instruments. The frequency of mechanical modulation is restricted to about 100 Hz. Modulators in the form of oscillating quartz plates or prisms are frequently used for wavelength modulation in polychromators in simultaneous spectrometers (refer to Section 3.7).

Aperture stops are frequently employed to restrict the radiation emitted from the atomizer, and especially from electrothermal atomizers. To match differences in energy between the various primary sources, mechanical, swing-in radiation attenuators can be used as an alternative to energy balancing via the lamp current. When an atomizer is changed in the spectrometer, swing-in apertures can be used to accommodate the radiation beam to the absorption volume of the atomizer.

With atomic absorption spectrometers employing Zeeman-effect BC, fixed or rotating polarizers must be used to remove a given polarized radiation component if the magnetic field is at right-angles to the direction of the radiation beam. Additional optical components are not required if the magnetic field is parallel to the direction of the radiation beam (refer to Section 3.4.2).

Depending on the type of spectrometer, the atomizer can be permanently mounted in the atomizer compartment or it can be interchangeable. The radiation beam is conducted by means of mirrors (or lenses) through the atomizer compartment such that the absorption volume of the atomizer is optimally illuminated. To obtain a radiation beam that is as parallel as possible, apertures or slits are located in the plane of the absorption volume. The apertures from the atomizer compartment into the interior of the spectrometer are mostly fitted with quartz windows to protect the instrument from environmental influences. The open front of the atomizer compartment is generally fitted with a door to protect the operator from UV radiation from the atomizer, especially from the flame.

3.3 Dispersion and Separation of the Radiation

When line sources are used the analytical lines are separated from other emission lines from the source and from broad band emission from the atomizer by monochromators or, occasionally, polychromators [988]. Technical parameters such as optical conductance, resolution, and reciprocal linear dispersion have a decisive influence on the analytical parameters such as sensitivity, S/N ratio, and linearity of the calibration curve.

For dispersion of the radiation, diffraction gratings are mostly used in AAS. Prisms are less suitable since the optical conductance of prism monochromators deteriorates by orders of magnitude toward longer wavelengths. The spectral slitwidth (Section 3.3.1) and the reciprocal linear dispersion (Section 3.3.2) of a grating monochromator depend on the grating constants and improve the closer together the rulings on the grating are, i.e. the more rulings per millimeter that a grating has. Both quantities only exhibit limited wavelength dependence, which means that at a fixed slit width a grating monochromator has essentially the same optical conductance over the entire wavelength range.

Holographic or mechanically ruled gratings or grating replicates are mostly used in AAS. Holographic gratings generally exhibit fewer irregularities and therefore produce less stray radiation; although this criterion is less important in AAS. When radiation strikes a grating it is reflected and 'fanned out' to either side of the incident radiation to a degree that is wavelength dependent. Gratings with an asymmetric, saw-tooth groove profile reflect the radiation largely in one direction, which leads to a better yield. However, optimal concentration of the radiation in a given direction depends on the blaze angle and the blaze wavelength. The efficiency of a grating is poorer the further removed a wavelength is from the blaze wavelength; the efficiency falls off much faster toward short wavelengths than toward longer.

Because of the importance of the UV range in AAS, it is advantageous to use gratings that have blaze wavelengths as far as possible in the UV range. Since a relatively wide wavelength range has to be covered by an AA spectrometer, configurations employing two gratings or gratings with several blaze wavelengths frequently are employed to improve the efficiency. The gratings used in AAS usually have a ruling density of 1200–2880 lines per millimeter, so that a good grating can have up to 100 000 lines. At a typical focal length of 200–500 mm, reciprocal linear dispersions of 0.5–5 nm/mm can be attained (the significance of reciprocal linear dispersion is discussed in Section 3.3.2).

The most frequently used monochromator configurations in AAS are the Littrow and the Czerny-Turner; these are depicted schematically in Figure 3-9. For a point radiation source the symmetrical reflection on two identical concave spherical mirrors in the

Czerny-Turner mounting produces a strongly astigmatic image for complete compensation of the coma error. If the exit slit is located in the tangential focus, the spectrum can be moved over the slit by slewing the grating and the angular spread with the best image is always at the location of the exit slit. The astigmatic lengthening of the spectral image is around 4.2 mm for a configuration of 500 mm focal length with 70 mm diameter concave mirrors and 20° angle of reflection at the mirrors. To achieve the highest possible optical conductance the slit heights of the entrance and exit slits of a Czerny-Turner monochromator can be of the same order as the photocathode diameter of the photomultiplier, without a deterioration of the spectral resolution. In comparison to most other monochromator mountings, in a Czerny-Turner monochromator the agreement between the optical system and the detector is almost perfect.

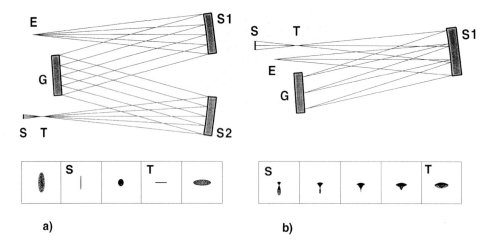

Figure 3-9. Typical basic configurations for monochromators in AAS and their spot diagrams (from [458]). **a** – Symmetrical Czerny-Turner mounting; **b** – asymmetrical Littrow mounting. **E** – Entrance slit; **G** – plane grating; **S1, S2** – spherical mirrors with identical radius; **T** – tangential focus; **S** – sagittal focus.

When continuum sources are used as the primary radiation source, monochromators or polychromators of very high resolution must be used. The resolution should be in the order of the half-width of the atomic line of the analyte, i.e. 1–3 pm. Such requirements can be most conveniently met with an Echelle polychromator. This is a plane grating spectrograph with an Echelle grating as the dispersive element. This type of grating has a low ruling density and a large blaze angle (approx. 60°) and is used to produce spectra of higher order. The dispersed spectrum of the Echelle grating and that of a conventional grating or prism are crossed so that a two-dimensional spread is obtained, as depicted in Figure 3-10. KELIHER and WOHLERS [3071] have published a good overview on the use of Echelle spectrometers in analytical spectroscopy. Further details on multichannel spectrometers and Echelle polychromators are discussed in Section 3.7 with respect to simultaneous spectrometers.

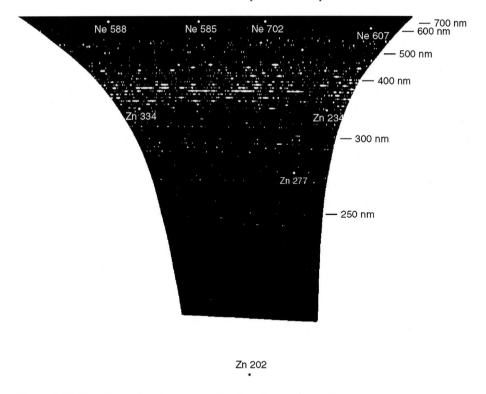

Figure 3-10. Two-dimensional spectrum of an Echelle spectrograph.

3.3.1 Spectral Slitwidth

In AAS the *resolution of the monochromator* is of less significance than with other spectroscopic techniques. Through the use of element-specific line sources and modulation of the radiation, the selectivity and specificity of AAS depend solely on the half-widths of the emission line (approx. 1–3 pm) and of the absorption line (approx. 2–5 pm), in other words on values beyond the resolving power of normal monochromators. The monochromator in an atomic absorption spectrometer has the sole task of separating the analytical line from other emission lines of the source. Spectral slitwidths of 0.2–2 nm are generally adequate for this purpose.

The spectral slitwidth $\Delta\lambda$ is the product of the geometric width s of the slit and the reciprocal linear dispersion $d\lambda/dx$ in the plane of the slit:

$$\Delta\lambda = s\left(d\,\lambda/d\,x\right). \tag{3.1}$$

If optical aberration and diffraction at the slit are negligible, the profile of the radiation passed by the monochromator is in the form of a triangle whose half-width is equal to the spectral slitwidth (the term spectral bandpass is also used). If the widths of the entrance and exit slits are the same, the spectral slitwidth is termed the spectral bandwidth.

There is no advantage in using a spectral slitwidth smaller than that required to separate the analyte line from other emission lines emitted by the radiation source since this will reduce the radiant power and thereby worsen the S/N ratio. If a larger slitwidth is used, AAS does not lose any of its specificity and selectivity, provided that the resonance lines of two elements do not fall simultaneously on the detector when multielement lamps are being used. The disadvantages brought by a slitwidth that is too large are a reduction in the sensitivity and an increasing non-linearity of the calibration curve. If a further non-absorbable emission line beside the analytical line passes the exit slit of the monochromator and falls on the detector, this always 'sees' the radiation of both lines. With increasing absorption of the radiation from the analytical line, the intensity of the second line remains unaltered so that the absorption curve does not approach 100% absorption asymptotically, but approaches a value corresponding to the percentage proportion of the second line. DE GALAN and SAMAEY [2039] showed that the causes for non-linearity of calibration curves are mostly trivial. *Non-resolved multiplets* (several analytical lines passing through the exit slit) and *non-absorbable lines* (apart from the analytical line a non-absorbable line passes the exit slit) were recognized as the two main causes and this was verified by calculation.

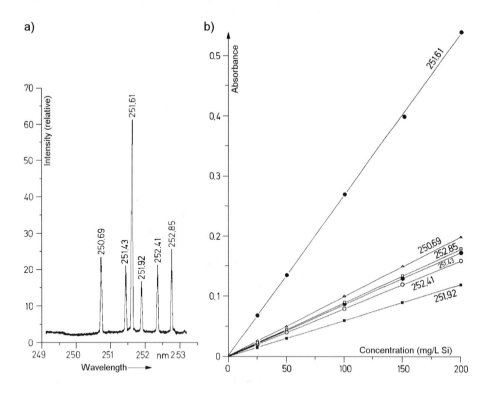

Figure 3-11. Analytical lines for silicon in the range 250–253 nm. **a** – Emission spectrum of a silicon hollow cathode lamp recorded at a spectral slitwidth of 0.07 nm. **b** – Analytical curves of the six analytical lines measured at a spectral slitwidth of 0.07 nm.

The following figures illustrate this behavior and show the influence of the spectral slit-width on the sensitivity, the S/N ratio, and the curvature of the calibration curve. Figure 3-11 shows the six analytical lines for silicon in the range 250 nm to 253 nm and their respective sensitivities in AAS. With a sufficiently small spectral slitwidth every analytical line gives a linear calibration curve.

The influence of the spectral slitwidth of the monochromator on the calibration curve and on the S/N ratio for the 251.6 nm analytical line of silicon is shown in Figure 3-12. With decreasing spectral slitwidth, the sensitivity clearly increases and the linearity of the calibration curve improves. When the optimum slit width is reached (i.e. the spectral slitwidth at which only one analytical line reaches the detector)—in this case 0.2 nm—a further decrease of the slit width brings no noticeable advantages. As shown in Figure 3-12a, the linearity of the calibration curve and the sensitivity can still be improved somewhat since in practice there is always a small amount of residual stray radiation present which can be further eliminated by reducing the slit width. Nevertheless, this slight improvement in the linearity is only obtained at the expense of an increase in the noise and a noticeably poorer S/N ratio, as shown in Figure 3-12b.

In the case of silicon the effect is caused by the influence of a non-resolved multiplet when spectral slitwidths of greater than 0.2 nm are used. The non-linearity of the calibration curve is due to the different absorption coefficients, κ_{jk}, of the individual analytical lines (see equation 2.31). This means that one of the prerequisites for the applica

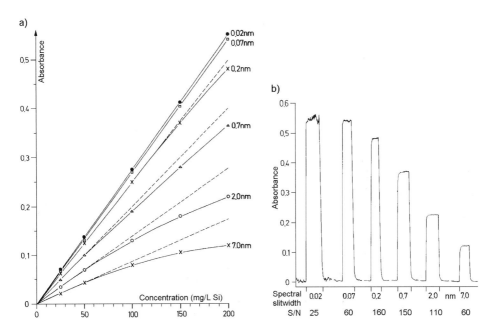

Figure 3-12. Influence of the spectral slitwidth on the determination of silicon at the 251.6 nm analytical line. **a** – The sensitivity decreases noticeably at slitwidths greater than 0.2 nm and the curvature of the calibration curve increases; **b** – the S/N ratio is optimum at a spectral slitwidth of 0.2 nm and clearly decreases to either side. (Calibration solution 200 mg/L Si; nitrous oxide-acetylene flame.)

tion of the Beer-Lambert law, namely a constant absorption coefficient in equation 2.41, is not met when several analytical lines fall on the detector.

To illustrate this further, the effect of the spectral slitwidth on the calibration curve of antimony is shown in Figure 3-13. Beside the analytical line of interest at 217.6 nm, an antimony HCL also emits two further lines at 217.9 nm and 217.0 nm. Neither of these two lines is measurably absorbed even by high antimony concentrations, so that the radiation from them can be regarded as 'non-absorbable'.

As can be seen from Figure 3-13b, with slitwidths up to 0.2 nm antimony behaves in a similar manner to silicon, i.e. decreasing the slit width brings no further advantage

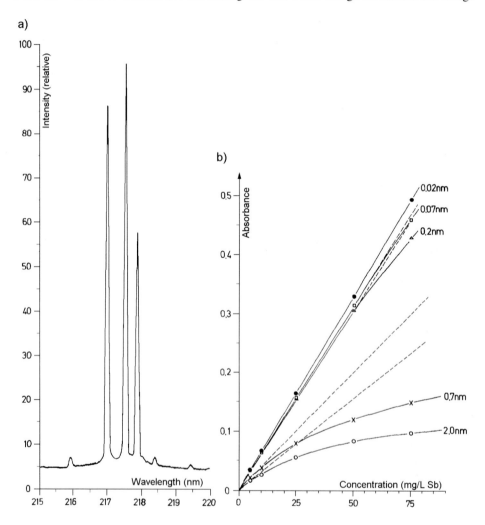

Figure 3-13. Influence of the spectral slitwidth on the determination of antimony. **a** – Emission spectrum of an antimony HCL recorded at a spectral slitwidth of 0.07 nm; **b** – at increasing slit widths the sensitivity decreases dramatically as soon as the non-absorbable lines pass the exit slit.

once the analytical line has been separated. With an increase in the spectral slitwidth there is a sudden decrease in the sensitivity and a strong increase in the curvature as soon as non-absorbable radiation falls on the detector. This is not surprising since the degree of the curvature is determined by the difference between the individual absorption coefficients. The strongest curvature is to be expected when a non-absorbable line falls on the detector. In this case, the calibration curve asymptotically approaches a value corresponding to the percentage transmission of the non-absorbable line. DE GALAN and SAMAEY [2039] calculated this value to be 55% transmission (absorbance 0.254) for antimony and confirmed this 'limiting absorbance' experimentally for a 1000 mg/L Sb solution.

While it can be seen that for a number of elements a spectral slitwidth of 0.2 nm is necessary to obtain good sensitivity and linearity of the calibration curve, for other elements the analytical line is more or less isolated so that larger slit widths can be used without disadvantage. It is obvious that in such cases the widest slit width at which the analytical line is just separated from other lines should be used since the S/N ratio is favorably influenced, as shown in Figure 3-12b.

3.3.2 Reciprocal Linear Dispersion

As indicated in equation 3.1, the spectral slitwidth $\Delta\lambda$ and the geometric slitwidth s are associated via the reciprocal linear dispersion $d\lambda/dx$ of the monochromator. Since the maximum usable spectral slitwidth is determined by the spectrum emitted by the primary source, the reciprocal linear dispersion determines the geometric slit width. A reciprocal linear dispersion of 2 nm/mm means that at a geometric slit width of 1 mm the spectral slitwidth is 2 nm, or a geometric slit width of 0.1 mm is necessary to obtain the desired spectral slitwidth of 0.2 nm.

In an atomic absorption spectrometer the image of the radiation source is formed on the entrance slit, i.e. a radiation beam of several millimeters diameter falls on the slit. It can clearly be seen from Figure 3-14 that the geometric width of the entrance slit determines the amount of radiation falling on the dispersing element and subsequently on the detector. With a wide entrance slit, a relatively large amount of radiant energy therefore falls on the detector; this means that the noise always present in the signal (the shot

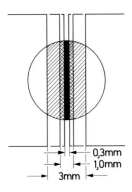

0,3mm
1,0mm
3mm

Figure 3-14. The image of the radiation beam is formed on the entrance slit. Its geometric width determines the amount of radiation that falls on the dispersing element.

noise of the photon current for example) is relatively small compared to the signal. Thus, lower gain can be employed so that possible contributions of the electrical measuring system to the noise are reduced. For the analyst, low noise means a stable signal and hence good precision and low limits of detection and quantitation.

From the above considerations it is thus desirable to have as small a numerical reciprocal linear dispersion as possible, i.e. to use a strongly dispersive element, since a relatively large amount of radiation then passes the monochromator to the detector, even when the spectral slitwidth is small. The significance of the geometric slit width, on the other hand, means that the largest spectral slitwidth just meeting the requirement for the isolation of the resonance line should always be chosen. When a continuum source is used as the primary radiation source, the spectral slitwidth is of far greater significance for the analytical performance of the monochromator. Investigations into this complex subject have been undertaken by HARNLY and SMITH [2391, 5461] and by BECKER-ROSS et al. [460, 461].

3.4 The Measurement and Correction of Background Attenuation

As mentioned earlier, the high selectivity and specificity of AAS derive from the use of element-specific line radiation sources, modulation of this radiation, and selective amplifiers. As a result, far fewer spectral interferences occur than with optical emission spectroscopy (OES). Nevertheless, the radiation passing through the atomizer can be attenuated not only by atomic absorption but also by a number of other effects that are frequently termed collectively as 'non-specific' or 'background absorption' or 'non-specific radiation losses'. Typically these can be absorption by gaseous molecules or radiation scattering on particles. Further details are provided in Section 5.4.1.1.

Since to date there is no technique that allows exclusively the atomic absorption of the analyte to be measured, it is necessary to measure the total absorption (specific and non-specific), then the background attenuation, and then to subtract this from the total absorption. Essentially there are four techniques for the quasi-simultaneous (in rapid sequence) measurement and correction of background attenuation in AAS:

–   with continuum sources,
–   utilizing the Zeeman effect,
–   using high current pulsing,
–   measurement next to the analytical line.

The latter technique is used virtually only together with simultaneous multielement determinations and is discussed with this technique (Section 3.7).

On the basis of theoretical considerations it is possible to place a number of principal requirements on an ideal system for the measurement and correction of background attenuation:

i)   The background attenuation should be measured exactly at (under) the analytical wavelength and with the same line profile as the total absorption.

ii)   The background attenuation should be measured at exactly the same location in the atomizer as the total absorption.

iii)   If the absorption changes with time, the total absorption and the background attenuation should be measured at the same time, or at least in such rapid sequence that the changes in absorption from one measurement phase to the next are very low.

iv)   The total absorption and the background attenuation should be measured with a beam of the same intensity and geometry.

v)   Background correction should not cause a worsening of the S/N ratio.

vi)   The technique employed for background measurement and correction should be applicable for all elements (for the entire spectral range).

None of the techniques mentioned above for the measurement and correction of background attenuation meets all of these requirements. In fact, some of these requirements even mutually exclude each other from the standpoint of their technical realization. Nevertheless this 'catalog of wishes' is a useful means of evaluating the individual techniques. Depending on the principle selected for background measurement and correction, additional optical components are frequently required, such as lenses or mirrors, radiation sources, or polarizers, which often lead to a deterioration of the S/N ratio.

3.4.1 Background Correction with Continuum Sources

Background correction (BC) with continuum sources is based on the assumption that background attenuation, in contrast to atomic absorption, is a broad band phenomenon which does not change within the range of the selected spectral slitwidth. The mode of function is depicted schematically in Figure 3-15. The exit slit of the monochromator isolates the analytical line from the spectrum emitted by the element-specific radiation source. A band of radiation corresponding to the selected spectral slitwidth (e.g. 0.2 nm) is isolated from the spectrum emitted by the continuum source. Analyte atoms absorb a portion of the radiation, proportional to their content, from the element-specific radiation source at the analytical line. Radiation from the continuum source, on the other hand, is only attenuated in the very narrow wavelength range of a few picometers in which the analyte atoms absorb. Depending on the selected spectral slitwidth, this amounts to 1–2% maximum of the continuous radiation passing the exit slit and is negligible. Continuous background attenuation, such as that caused by the dissociation continua of molecules (refer to Section 5.4.1.1), attenuates the radiation from the line and continuum sources to equal degrees, so that the ratio of the radiant power of each beam does not change.

Continuum source BC deviates in virtually all aspects from the requirements stated above for an ideal system:

i)   In continuum source BC the background is measured as a broad band to either side of the analytical line. If the correction is to be correct at (under) the analytical line, the mean attenuation over the observed spectral range must be the same as the background attenuation at the center of the analytical line. This requirement is generally fulfilled if the background is continuous within a range of the selected spectral slitwidth. If this is not the case, over- or undercorrection will result (see Sections 5.4 and 8.2.4.2).

ii)   Even if it is possible to align the line source and the continuum source exactly along the same optical axis, nevertheless differences in the geometry and distribution of the radiant power of each source mean that different absorption volumes are irradiated [2127, 5347] (refer to Figure 2-14).

iii) The measurement of the total absorption and the background attenuation is sequential. Errors in correction cannot be excluded if one or the other is changing faster than the measurement frequency (see Section 8.2.4.2).

iv) By definition, with continuum source BC two radiation sources with differing geometries and radiant intensities are employed. The deuterium lamp emits sufficient radiant power in the short wavelength range from 190 nm to 330 nm. Above this range the halogen lamp provides the better radiant power. If a continuum source is used outside its optimum range, the radiant intensity of the primary source must be reduced, leading to a deterioration of the S/N ratio and limited BC.

v) Through the use of two radiation sources the S/N ratio is worsened since each source contributes its own noise spectrum. The use of additional optical components for beam splitting and recombination also leads to a worsening of the S/N ratio.

Despite these weaknesses, continuum source BC can still be successfully utilized, provided that the requirements are not too high. This is generally the case for flame AAS and the hydride-generation and cold vapor techniques. For electrothermal AAS, on the other hand, continuum source BC can only be employed after thorough investigation (refer to Section 8.2.4.2).

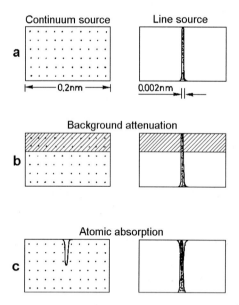

Figure 3-15. Mode of function of continuum source background correction. **a** – The radiant intensity, represented schematically by dots, for the continuum source is distributed over the entire width of the spectral band isolated by the slit (e.g. 0.2–2 nm), while for the line source it is limited to a few picometers. **b** – Broad band background attenuates the radiation emitted by both sources to equal degrees. **c** – Atomic absorption, which again is limited to a few picometers, in the first approximation attenuates only the radiation from the line source.

3.4.2 Background Correction Utilizing the Zeeman Effect

As mentioned earlier (Section 2.6), a number of technical solutions can be conceived for the utilization of the Zeeman effect for BC. They nevertheless differ, in part considerably, in their realization, application, and performance. An historical review on the attempts to utilize various technical solutions is given in Section 1.5. The magnet can be mounted at either the radiation source or the atomizer; the magnetic field can be orientated either parallel or perpendicular to the radiation beam, and a constant or alternating magnetic field can be applied. These possibilities give eight different configurations as shown in Table 3-1.

Table 3-1. Various configurations for the application of the Zeeman effect for BC in AAS.

Location of magnet	Orientation of magnet to radiation beam	Type of magnetic field	Optical components additionally required	Remarks
Source (direct)	parallel (longitudinal)	constant	rotating polarizer	only applicable in exceptional cases
		alternating	none	
	perpendicular (transverse)	constant	rotating polarizer	
		alternating	fixed polarizer	
Atomizer (inverse)	parallel (longitudinal)	constant	—	not applicable in AAS
		alternating	none	
	perpendicular (transverse)	constant	rotating polarizer	
		alternating	fixed polarizer	

Two of these configurations, namely the use of a longitudinal constant field at either the atomizer or the source, cannot in effect be applied in AAS. In each case the π components disappear completely from the spectrum, so that in the inverse configuration the analyte atoms cannot absorb radiation at the analytical line, while in the direct configuration no radiation is emitted at the analytical line. The only exceptions described in the literature [2306, 2307] are mentioned in Section 1.5 and will therefore not be discussed here further.

In order to compare the remaining six configurations a number of aspects must be taken into consideration. The first is the instrumental aspect, i.e. which additional components are required and what is the influence on existing components. Further, the spectroscopic aspects must be considered, as well as the influence of the various configurations on such analytical parameters as the sensitivity, the limits of detection, and the curvature of the calibration curve. Last not least the efficiency of BC must be compared with respect to the principal requirements mentioned above for an ideal system for the measurement and correction of background attenuation.

With regard to the *instrumental aspects*, it was already mentioned in Section 1.5 that the operation of line sources in a magnetic field causes considerable difficulties and probably always will [2311]. Even though new lamp designs were introduced, the applications were always restricted to a few volatile elements. The necessity of using special radiation sources remains the greatest disadvantage of direct Zeeman-effect BC.

When the magnet is located at the atomizer, conventional radiation sources can naturally be used. How far the atomizer and the magnet are compatible with each other depends on the atomizer assembly. The flames typically used in AAS of about 10 cm length and also the quartz tube atomizers used in the hydride-generation and cold vapor techniques are not especially suitable, since a magnet of the appropriate dimensions would have an energy requirement that could hardly be justified. To apply Zeeman-effect BC to these techniques it would be necessary to reduce significantly the size of the atomizers, which in turn would lead to a loss in sensitivity. This would make little sense since continuum source BC can mostly cope with the background attenuation occurring in these techniques.

It is much easier to bring electrothermal atomizers between the pole shoes of a magnet. Further, owing to the pulse operation of this technique, the thermal load on the magnet is very much lower than with a flame. Longitudinally-heated atomizers are restricted to the transverse configuration of the magnet because of their dimensions. Longitudinal fields can only be applied to atomizers that have been optimized for this purpose.

A constant magnetic field can be generated either by a permanent magnet or by a direct current electromagnet. The latter has the advantage that the magnetic field strength can be varied, if necessary, to meet any special requirements of a determination. An alternating magnetic field is generated by using an alternating current electromagnet. Since in this mode of operation measurements are taken in rapid sequence with the field on and the field off, it is important to use a system of low magnetic inertia; the magnetism must dissipate very rapidly when the current is switched off. The electronic measuring system must be tuned to the electromagnet such that the first measurement is taken when the magnetic field strength is at its highest and the second measurement is taken when the magnetism has completely dissipated.

The supplementary optical components required for operation of the various configurations for Zeeman-effect BC are listed in Table 3-1. A longitudinal alternating field is especially favorable from this point of view since no supplementary optical components are required. With a transverse alternating field a polarizer is required, which however is fixed. A transverse constant field is the most complicated since a rotating polarizer is required to discriminate the various polarized Zeeman components. An alternative to this is a phase shifter working on the principle of voltage double refraction [2308] and thus having no moving parts.

With regard to the *spectroscopic aspects* a longitudinal alternating field is also the most practical configuration since supplementary optical components are not required which cause radiation losses. When the magnetic field is switched off, the total absorption is measured as in normal AAS. When the magnetic field is switched on the π components disappear from the spectrum and the background attenuation is measured under the analytical line. With the transverse configuration of the magnet a Zeeman component of the radiation must always be filtered out. This means that there is a radiation loss of more than 50%, often significantly more in the far UV range, regardless of the optical component used (polarizer, filter). When rotating polarizers are used there is a further problem caused by unequal reflection of the π and σ radiation components at the grating of the monochromator. This preference of one direction of polarization in the mono-chromator can also be strongly wavelength dependent. Instruments operating with a constant field thus require an adjustable polarizer, and both components must be newly equalized at each wavelength [2042].

In addition, every polarizer produces a slight offset between both planes of polarization. For BC this means that the background attenuation and the total absorption are measured in different absorption volumes. This can lead to a substantial, periodic base-line offset which can influence both the precision and the trueness of BC [392]. These differences in the transmission effectivity cannot be observed with a phase shifter, since the π and σ components reach the monochromator in the same plane of polarization [6361].

The *sensitivity* that can be attained using Zeeman-effect BC compared to conventional AAS is derived from equations 2.54 and 2.55 (refer to Section 2.6). It depends on how well the analytical line overlaps the π component during measurement of the total absorption and how little it overlaps the σ components during measurement of the background attenuation. If an alternating field is used, the total absorption is measured during the field-off phase. In other words, the same measure is obtained as with conventional AAS. It is thus merely necessary during the second phase to switch on a magnetic field of sufficient strength to shift the σ components completely from the emission profile of the source (inverse) or the absorption profile of the analyte atoms (direct). As shown in Figure 3-16 on the example of chromium, this can be achieved with lower magnetic field strengths when the magnet is at the atomizer since the emission profile of the source is narrower than the absorption profile and the σ components do not therefore have to be shifted so far.

If the separation is not complete, a portion of the atomic absorption is measured with the σ components and—in addition to the background absorption—subtracted from the total absorption, leading to a reduction in sensitivity. The loss in sensitivity observed for a number of elements in an alternating magnetic field of 0.9 Tesla at the atomizer compared to conventional AAS, the so-called Zeeman factor, is shown in Table 3-2.

The same can be said for a constant magnetic field if the line exhibits normal Zeeman splitting, i.e. there is only one π component. As shown in Figure 3-17 on the example of cadmium, the profile of the π component is hardly influenced by the magnetic field strength, so that larger losses in sensitivity are not to be expected.

The situation is much more complex for elements that exhibit anomalous Zeeman splitting, i.e. the majority of the elements, since in a permanent magnetic field the π components used to measure the total absorption are shifted further apart with increasing field strength and thus out of the absorption zone. This can be seen in Figure 3-16 for chro mium. At the optimum field strength of about 0.4 Tesla the π component is hardly broadened so that there is good overlapping of the absorption and emission lines. Nevertheless, the σ components are insufficiently separated so that there is substantial over-lapping and hence absorption. The relative absorbances of both Zeeman components for three elements at two different field strengths are shown in Table 3-3. The situation is particularly unfavorable for elements that have an even number of π components, e.g. silver, since none of the π components coincides with the original analytical line. At high field strengths a type of 'self-reversal' of the π component can take place, leading to a substantial loss in sensitivity.

An electromagnet with adjustable constant field allows the magnetic field strength to be optimized so that maximum sensitivity can be attained for most elements. Nevertheless the maximum sensitivity is still well below that for normal AAS and also below the sensitivities that can be attained with an alternating magnetic field. A permanent magnet or an electromagnet with a fixed field strength does not allow this optimization and hence gives sensitivities that—in dependence on the field strength—are often less than half the sensitivities attainable with normal AAS.

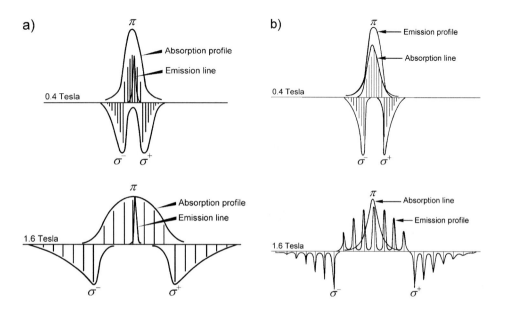

Figure 3-16. Line profiles for chromium at 357.9 nm with magnetic fields of 0.4 Tesla and 1.6 Tesla. **a** – at the atomizer; **b** – at the source.

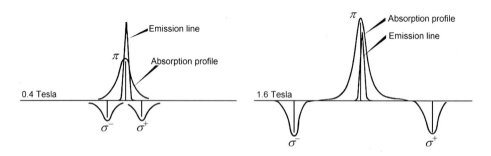

Figure 3-17. Line profiles for cadmium (normal Zeeman splitting) at 228.8 nm with magnetic fields of 0.4 Tesla and 1.6 Tesla, respectively, at the atomizer.

Table 3-2. Zeeman factors (sensitivity ratio Zeeman-effect BC/without BC) measured in integrated absorbance in an alternating magnetic field of 0.9 Tesla at the atomizer (by kind permission of Perkin-Elmer).

Element	Wavelength (nm)	Zeeman factor
Ag	328.1	0.91
Al	309.3	0.90
As	193.7	0.89
Au	242.8	0.84
Ba	553.6	0.96
Be	234.9	0.51
Bi	223.1	0.63
Ca	422.7	0.95
Cd	228.8	0.98
Co	240.7	0.85
Cr	357.9	0.88
Cu	324.8	0.53
Fe	248.3	0.92
Hg	253.7	0.61
In	325.6	0.91
Ir	264.0	0.96
Li	670.7	0.85
Mg	285.2	0.91
Mn	279.5	0.91
Mo	313.3	0.95
Ni	232.0	0.91
P	213.6	0.70
Pb	283.3	0.83
Pd	247.6	0.91
Pt	265.9	0.80
Ru	349.9	0.90
Sb	217.6	0.95
Se	196.0	0.88
Si	251.6	0.98
Sn	286.3	0.94
Sr	460.7	0.98
Te	214.3	0.93
Ti	365.3	0.96
Tl	276.8	0.66
V	318.4	0.77
Zn	213.9	0.88

Table 3-3. Relative absorbance of the π and σ components at different field strengths, and relative sensitivities (Zeeman factors) in a constant field and an alternating field, respectively (from [3593]).

Element	Zeeman splitting	Field strength (Tesla)	Relative absorbance*		Zeeman factor*	
			π	σ	Constant field	Alternating field
Ca	normal	0.4	1.00	0.32	0.68	0.68
		1.0	1.00	0.05	0.95	0.95
Ag	anomalous	0.4	0.78	0.13	0.65	0.87
		1.0	0.24	0.01	0.23	0.99
Fe	anomalous	0.4	0.97	0.45	0.52	0.55
		1.0	0.84	0.08	0.76	0.92

* normal AAS = 1.00

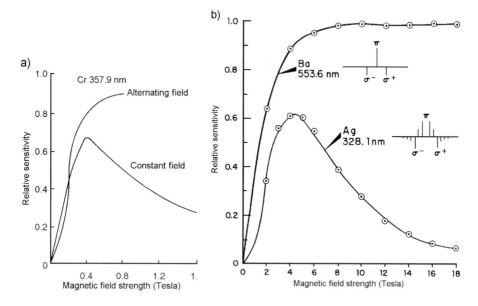

Figure 3-18. Sensitivity for Zeeman-effect BC at various field strengths in comparison to AAS without BC. **a** – Chromium with an alternating and a constant magnetic field at the atomizer. **b** – Barium (normal Zeeman splitting) and silver (anomalous Zeeman splitting) in a constant magnetic field at the atomizer (from [1892]).

The influence of the magnetic field strength on the sensitivity of chromium at 357.9 nm is depicted in Figure 3-18a; an alternating field is compared to a constant magnetic field. The difference between normal and anomalous Zeeman splitting in a constant magnetic field is shown in Figure 3-18b. Significant losses in sensitivity must be expected for anomalous Zeeman splitting in a constant magnetic field, especially if the field strength is not optimum, while for normal Zeeman splitting there is no difference in the sensitivity between an alternating field and a constant field. With normal Zeeman splitting the 'normal' sensitivity of conventional AAS is attained with all types of magnet, provided that the magnetic field strength is high enough to separate the π and σ components. The absorbances of the π and σ components of lead and silver in dependence on the field strength are shown in Figure 3-19. For both elements the absorbance of the σ component approaches zero at a field strength of about 1 Tesla. The absorbance of the π component only remains constant for lead (normal Zeeman splitting), while for silver (anomalous Zeeman splitting) it decreases with increasing field strength.

The S/N ratio is the decisive factor for the *limit of detection*. It can therefore be expected that systems containing supplementary optical components which reduce the radiant power or introduce sources of noise exhibit poorer limits of detection. Such sources of noise, for example, are rotating polarizers [392] or instabilities in sources caused by magnetic fields. In this respect a longitudinal alternating field at the atomizer is the optimum configuration.

Any exact analysis of the S/N ratio of a system must also include the electrical measuring system and the response time (see Section 3.6). Since additional noise components are always present due to the Zeeman effect, there will be a slight loss in detecting power compared to the same system without background corrector [5570]. However, since with Zeeman-effect BC no additional source is required, the noise will always be lower than with continuum source (deuterium or halogen lamp) BC.

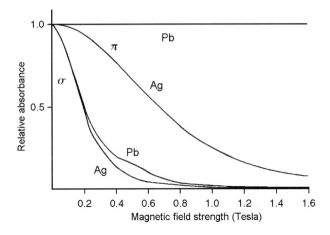

Figure 3-19. Relative absorbance of the π and σ components of lead (283.3 nm) and silver (328.1 nm) in dependence on the magnetic field strength at the atomizer (from [3593]).

Thus for the same sensitivity (normal Zeeman splitting or anomalous Zeeman splitting in a sufficiently strong alternating magnetic field) better limits of detection will be attained compared to continuum source BC. If in addition the improved efficiency of Zeeman-effect BC is taken into account, then considerably better limits of quantitation can be expected for 'real samples'. If, on the other hand, a constant field or, more especially, a permanent magnet with fixed magnetic field strength is employed, this advantage cannot be fully exploited due to the considerably poorer sensitivity [1890].

The *linearity of calibration curves* when utilizing the Zeeman effect has been discussed fully by DE LOOS-VOLLEBREGT and DE GALAN in a series of publications [3586, 3588, 3589, 3592, 3597] and a review article [3593]. Since in principle the calibration curve is the sensitivity represented as a function of the concentration or mass of the analyte, the considerations made for the sensitivity can also be applied. The basis is again equations 2.54 and 2.55. Similar to the sensitivity of Zeeman-effect BC, which is derived as the difference between two absorbance readings, the calibration curve obtained under the application of the Zeeman effect results from the difference of the absorption curve measured at the analytical line or the π component and the absorption curve measured with the σ components (residual overlapping). And like the sensitivity, the absorption curves depend on the magnetic field strength. The simplest situation is again the normal Zeeman effect. The same linearity as for conventional AAS can be achieved for Zeeman-effect BC if the magnetic field is strong enough to shift the σ components out of the absorption zone of the analytical line. This is valid for all forms of Zeeman-effect BC. For elements and wavelengths with anomalous Zeeman splitting the same linearity as for normal AAS can be achieved if an alternating magnetic field of sufficient strength is applied to shift the σ components out of the absorption zone of the analytical line.

The situation becomes more difficult in cases where the σ components cannot be completely separated (because of insufficient field strength) or not separated at all (anomalous Zeeman effect and permanent magnetic field). In such cases the final calibration curve is derived by subtraction of the absorbance for the σ components from the absorbance of either the π component or the analytical line for each analyte concentration, respectively. Since the sensitivity for the π component or the analytical line is always higher than for the σ components, the calibration curve for the former inevitably starts to become non-linear at lower analyte concentrations than the latter. This means that with increasing analyte concentration or mass the absorbance for the π component increases relatively more slowly than for the σ component. This means that not only the slope of the resulting calibration curve is lower than for normal AAS but that it is also more curved.

When the point is reached at which the slope of both curves is the same, after subtraction there is no further increase and the curve runs parallel to the concentration axis. If the slope of the absorption curve for the σ components is steeper at higher analyte masses or concentrations than that for the π components, a phenomenon can be observed that is termed 'rollover' of the resulting calibration curve. This means that if the analyte mass or concentration increases further, the calibration curve decreases. It then becomes ambiguous since two concentration or mass values can be assigned to each absorbance reading, as shown in Figure 3-20.

Just as with the sensitivity, the curvature of the calibration curve and the phenomenon of rollover are most pronounced with a constant magnetic field and for elements that exhibit anomalous Zeeman splitting. The problems are much less severe for normal Zeeman splitting and for an alternating magnetic field. If the magnetic field strength is high enough, the effects even disappear completely.

In electrothermal AAS rollover of the calibration curve can be recognized from the signal shape (Figure 3-21). The interference-free signal for, say, 10 ng Cu reflects the change in the atom concentration over time. At higher analyte masses under application of Zeeman-effect BC with insufficient separation of the σ components, the signal at first rises proportional to the increasing atom concentration until the rollover concentration is reached, after which it begins to fall. As the atom concentration begins to decrease, the signal rises to a maximum once more and then falls toward zero with the decreasing atom concentration.

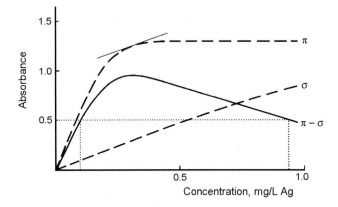

Figure 3-20. Calculated calibration curve for silver (328.1 nm) in the presence of 5% stray radiation in a constant magnetic field of 0.4 Tesla at the atomizer. The dashed lines are the analytical curves for the π and σ components. The dotted lines indicate the two concentrations corresponding to an absorbance of 0.5 (from [3593]).

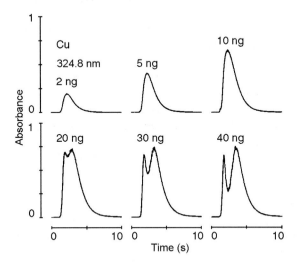

Figure 3-21. Effect of rollover of the calibration curve on the signal shape for electrothermal AAS. Determination of increasing masses of copper in an alternating magnetic field of 0.8 Tesla at the atomizer (from [1892]).

Although rollover of the calibration curve is an undesirable adjunct of the Zeeman effect, the phenomenon is nevertheless frequently overvalued. While the effect of rollover for continuous signals—such as in flame AAS—really leads to a decrease in the absorbance at high concentrations, it is less pronounced for electrothermal AAS. As can be seen in Figure 3-21, only the top of the peak corresponding to the maximum atom concentration exhibits rollover. This negative 'point' of the peak will not be registered by the signal detection system. In electrothermal AAS it is the peak height (absorbance) or, more usually, the peak area (integrated absorbance) that is recorded. Despite rollover the maximum absorbance remains constant and the integrated absorbance even increases further. Thus from the constant absorbance value and the signal shape, the analyst can recognize that he is in the rollover range, however he can even perform measurements with sufficient accuracy using integrated absorbance.

With regard to the *trueness of Zeeman-effect BC* the information given in Section 2.6 is repeated only briefly since this subject is treated in detail together with spectral interferences (Sections 5.4.3.1 and 8.2.4.2). If the magnet is *located at the radiation source* the background attenuation is measured at a wavelength that is shifted by approximately ± 10 pm. Zeeman-effect BC can only be correct if the background does not change within this range. If the magnet is *located at the atomizer*, the background is always measured at the same wavelength as the total absorption, i.e. 'under' the analytical line. Nevertheless, for absolutely correct BC with an alternating field at the atomizer the condition must be met that the background is the same with the field on as with the field off.

STEPHENS and MURPHY [5571] and also MASSMANN [3941] pointed out that molecules can exhibit the Zeeman effect, giving rise to a further potential source of error. However, molecular absorption bands are hardly influenced in magnetic fields of less than 1.5 Tesla [6465], so that the inverse Zeeman configuration will always give the better results.

If the various configurations for Zeeman-effect BC are compared with the requirements placed in Section 3.4 for an ideal system, one system at least comes very close to this ideal. Background correction with a longitudinal alternating field at the atomizer fulfills five of the six requirements: The background attenuation is measured exactly under the analytical line with the same line profile and it is measured at exactly the same location in the atomizer as the total absorption with a beam of the same intensity and geometry. Also, BC does not cause a significant worsening of the S/N ratio since no additional optical components are required and the technique is applicable for all elements (for the entire spectral range). The only requirement that it does not meet—like all other configurations—is the simultaneous measurement of background and total absorption.

At this point it is worth mentioning that all other configurations for Zeeman-effect BC deviate from the maximum requirements in more than one point. Supplementary components, such as polarizers, cause a deterioration of the S/N ratio; this is particularly true for constant magnetic fields since the sensitivity is also markedly influenced. Configurations with the magnet at the radiation source do not meet the requirement for measurement exactly at the analytical wavelength and on occasion also not at the same location. The problems with the lamps also means that this configuration cannot be applied for all elements.

3.4.3 Background Correction with High Current Pulsing

In 1983 SMITH and HIEFTJE [5474] took up the proposals of BARRINGER [395] and LING [3533] for BC using high current pulsing. The technique is based on the strong line broadening and self-reversal of the resonance lines observed in HCLs at high operating currents. The total absorption is measured at normal operating currents and thus with normal line profiles, while the background is measured next to the analytical line with the strongly broadened profile caused by high current pulsing.

In many respects BC with high current pulsing has great similarity to Zeeman-effect BC with the *magnet at the source*. Both systems operate with *one* source, in both systems the analytical line is 'split' so that the background attenuation is measured *next* to the analytical line, and the performance of both systems is strongly dependent on the behavior of the radiation source under the special conditions prevailing.

In a theoretical study DE GALAN and DE LOOS-VOLLEBREGT [2045] came to the conclusion that for BC with high current pulsing, rollover of the calibration curve should be lower than with Zeeman-effect BC, but at the cost of the analytical sensitivity. The loss in sensitivity and rollover both depend on what proportion of the atomic absorption is measured in the high current phase. And this proportion is strongly dependent on how effective the self-reversal is, i.e. on what proportion of the radiation in this phase overlaps the absorption profile of the analyte atoms. It can be assumed that there is a direct interrelation between the volatility of an element and the degree of self-reversal, and this has been confirmed experimentally [5434].

The characteristic concentrations obtained with continuum source BC and high current pulsing for a number of elements as well as the loss in sensitivity for the latter technique are listed in Table 3-4. This loss in sensitivity is directly related to the volatility of the analyte and lies between about 10% for cadmium and 85% for vanadium. It is therefore hardly surprising that the limits of detection measured by SMITH and HIEFTJE [5474] using high current pulsing were mostly poorer than those measured with continuum source BC, even though high current pulsing operates with only one radiation source and lower noise should thus be expected.

The calibration curves determined experimentally with normal lamp currents and with high current pulsing for copper and aluminium are depicted in Figure 3-22 together with the curve calculated from the difference for high lamp current operation. It can be seen that there is a substantial loss in sensitivity, and rollover of the calibration curve also occurs, due to the high residual absorption caused by insufficient self-reversal. At very high analyte concentrations the loss in sensitivity is moderated by the high proportion of stray radiation caused by high current operation [2045, 3595]. In practice however this is of no significance since in this range it is not possible to work analytically.

SIEMER [5345] drew attention to another disadvantage of the 'Smith-Hieftje' system. This is the slow dissipation of the line broadening after a high current pulse. Following a high current pulse with a duration of only 0.3 ms a delay of 40 ms is necessary during which the radiation source must be operated at the lowest current and during which no measurements can be taken. For this reason this system can only be operated at a frequency of 20 Hz, which is certainly insufficient for the rapid signals from electrothermal AAS. SIEMER [5347] further predicted that a baseline offset is to be expected due to the geometric differences in the hollow cathode discharges during both phases of this technique.

To conclude, the performance of BC using high current pulsing should be compared to the requirements for an ideal system presented in Section 3.4. The result of this comparison is that high current pulsing, like continuum source BC, does not meet the requirements in virtually all aspects, even though only one radiation source is used.

i)   Background attenuation is measured next to the analytical line and with a significantly changed profile.

ii)   Although only one radiation source is used there are differences in the beam geometry [5347].

iii) Background attenuation is not measured at the same moment as the total absorption, and with a particularly unfavorable (low) frequency.

iv)   There are significant differences in the radiant intensities in the two phases of measurement.

v)   The S/N ratio is in part markedly worsened due to the loss in sensitivity for BC.

vi)   The technique can only be applied to those elements (sources) that exhibit sufficient self-reversal.

Table 3-4. Characteristic concentrations, β_0, (µg/L) for BC with a continuum source (D_2) and with high current pulsing (S-H), and the relative sensitivity of high current pulsing compared to continuum source BC (from [5434]).

| Element | β_0 (µg/L) | | Relative sensitivity |
	D_2	S-H	D_2/S-H
Al	425	1750	0.25
As	430	625	0.69
Cd	17	19	0.89
Cr	45	85	0.53
Cu	35	55	0.64
Fe	40	80	0.50
Pb	275	400	0.69
Mg	3	6	0.50
Ni	55	80	0.69
V	650	4500	0.14
Zn	8	11	0.73

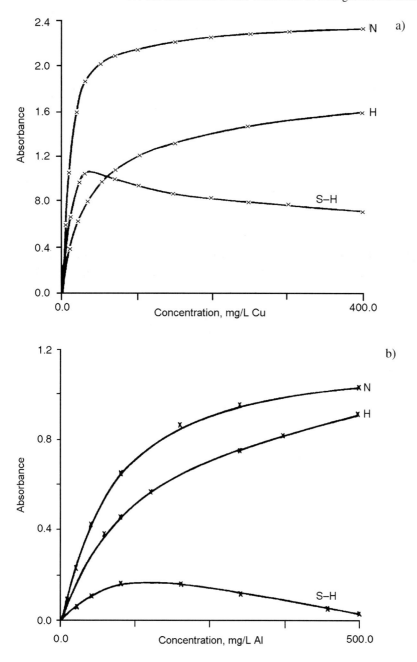

Figure 3-22. Analytical curves for **a** – copper and **b** – aluminium at low (L) and high (H) lamp currents, and for BC using high current pulsing (S-H). The L and S-H curves were determined experimentally, the H curve was calculated by difference (from [5434]).

3.5 The Detection of Radiation

Detectors operating on the photoelectric principle are used exclusively for the detection of radiation in AAS [3447, 3448]. The requirements of AAS are best met by broad band photomultipliers (multialkali types). The photomultiplier tube (PMT) is a radiation detector in which the incident radiation falling on a photocathode causes the emission of primary electrons (outer photoelectric effect) which are released into the surrounding vacuum. Resulting from the applied dynode voltage each primary electron is accelerated so rapidly that when it strikes a dynode two to ten secondary electrons are emitted, leading to a cascade effect as shown schematically in Figure 3-23. Nevertheless there are limits to the voltage that can be applied, since this leads to a higher dark current and thus to increased noise. In extreme cases the dynodes can become saturated.

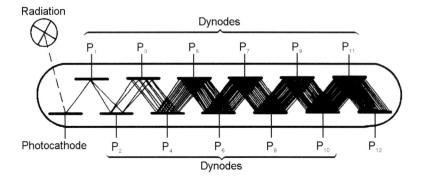

Figure 3-23. Schematic of a photomultiplier with 11 dynodes.

The gain g depends on the voltage U across the electrodes and on the number of stages n according to:

$$g = kU^{\alpha n},\tag{3.2}$$

where α is a coefficient characteristic for the dynode material; it typically lies between 0.7 and 0.8.

PMTs with 9 to 12 dynodes are mostly used in AAS so that the anode output varies as the sixth to tenth power of changes in the applied voltage. Since the output signal is so extremely sensitive to changes in the applied voltage, the stability and freedom from noise of this source are especially important. In modern AAS instruments microprocessors are mostly used to automatically regulate and control the applied voltage.

In principle semiconductor barrier layer detectors can also be used in AAS. Here the incident radiation produces electron-cavity pairs (inner photoelectric effect) which are separated by the electric field at the barrier layer. Recombination takes place via the electric current (photocurrent) through the external circuit. Photoelements, photodiodes, and phototransistors belong to this category. These solid state detectors generally are used preferentially in instruments with fixed wavelength channels. Radiation detectors for the spatially resolved measurement of radiant quantities, such as Vidicon detectors, diode arrays, or charge coupled devices (CCDs), are mainly used in simultaneous multi-element spectrometers (see Section 3.7). These are detectors based on a one- or two-dimensional array of a multitude of individual silicon elements (barrier layer detectors) with enhanced UV sensitivity. Frequently the individual elements are arranged spatially on the surface of the detector at the image position of the exit slit or of the analytical lines. In principle linear detectors can be so arranged that they detect a limited wavelength range or resolve the radiation over the height of the exit slit. In the first case it is then possible to measure simultaneously at the analytical line, on the flank of the analytical line (reduced sensitivity), and next to the line (background). In the second case it is possible to eliminate the influence of an inhomogeneous distribution of the analyte atoms over the observation height in the atomizer [2126, 2127].

The most important characteristics of all photoelectric detectors are the spectral sensitivity, the quantum efficiency, the usable wavelength range, the linear range, the noise or the S/N ratio, the response time, and the dark current.

The *spectral sensitivity* $R(\lambda)$ is given by the ratio of the output signal to the input signal. A most important prerequisite of AAS detectors is a high spectral sensitivity over the entire wavelength range with a sensitivity maximum in the far UV (approx. 200–300 nm).

The *quantum efficiency* η is the ratio of the number of charged entities emitted to the number of incident photons. Spectral sensitivity and quantum efficiency are related by:

$$\eta = \frac{R(\lambda)\,hc}{e\,\lambda},$$
(3.3)

where h is Planck's constant, c is the speed of light, and e is the elementary charge.

The *usable wavelength range* is the range in which the spectral sensitivity does not fall beneath a stipulated value. The lowest value that can be stipulated for the relative sensitivity is limited by the noise. In AAS the spectral sensitivity in the range 190 nm to 860 nm should be at least two to three orders of magnitude above the noise.

The *linear range* or dynamic range is the range where the detector output signal is proportional to the incident radiant power within a stipulated limiting value. The non-linearity (NL) is given by:

$$NL = \frac{R(z) - R(z_0)}{R(z_0)},$$
(3.4)

where $R(z)$ is the sensitivity at value z and $R(z_0)$ is the sensitivity at the reference value z_0. In AAS, as in OES, a range of at least 10^6 is necessary. This requirement is considered in more detail in Section 5.4.3, especially with respect to the S/N ratio.

Noise is regarded as all arbitrary variations that cause an electrical output signal that is superimposed on the measurement signal. The distribution of noise against frequency is described by the noise spectrum. Noise that is independent of the frequency is termed 'white noise'. The most important noise components are the low frequency 'flicker' noise and the 'shot' noise. These components result from a series of independent chance processes, such as the emission of electrons from a photocathode. Shot noise comprises the photon noise and the dark current noise. The dark current is the sum of all currents flowing over the electrodes when no radiation falls on the photocathode.

Noise is mostly quoted as the effective value of an electrical quantity, usually the current. This is equivalent to the standard deviation. The ratio of the measure for the signal to the measure for the noise is termed the *signal-to-noise* (*S/N*) *ratio*. This ratio is especially suitable to characterize a radiation detector for the measurement of low radiant power. In AAS it must be possible to reliably measure changes in the absorbance signal of at least $\Delta A = 0.001$ This requirement is also discussed further in Section 5.2.3.4.

Of especial importance with respect to time-dependent signals, e.g. in GF AAS, is the *response time* of the detector. This consists of the rise time and the fall time, i.e. the time in which the output signal reaches a stipulated fraction of the final value (e.g. 90%) and the time in which the signal falls again to a stipulated fraction of the final value (e.g. 10%). For AAS the response time must be in the millisecond range.

3.6 The Modulation of Radiation

Using electronic measuring techniques, modulation of the radiation allows the absorption at the analytical line to be discriminated from other radiation. If, within the selected spectral slitwidth, only the analytical line is emitted by the radiation source, then only this is modulated and thus only radiation absorption within the emission profile of this line is evaluated. Other radiation, especially that from the atomizer, such as atomic or molecular emission from the flame or radiation emission from the graphite atomizer, is not evaluated. The modulation of the radiation thus leads to the high selectivity of AAS; it is also the reason why relatively large spectral slitwidths can be used in AAS.

The periodic change in the radiant power is usually achieved by either modulation of the discharge current of the radiation source or by rotating choppers. Modulation can also be achieved by alternating magnetic fields (Zeeman effect). Choppers are predominantly used for *mechanical modulation* of the primary radiation; these are generally in the form of rotating sector disks or partially mirrored quartz disks. In double-beam spectrometers the modulating device can be used to both modulate the radiation and split the radiation into the sample beam and the reference beam.

The *modulation frequency* is of significant importance for techniques generating time-dependent signals, especially if the absorbance may change rapidly as in GF AAS. A modulation frequency that is too low can lead to a systematic deviation of the measure. Modulation frequencies of up to several hundred Herz can be achieved electronically. However, the maximum frequency is often limited by properties of the lamp, such as oscillation, unstable ignition, or lifetime. The frequency of opto-mechanical modulation is limited to about 100 Hz for technical reasons. On the other hand, high modulation frequencies demand short response times and these lead to increased noise and thus to a deterioration of the S/N ratio. The optimum response times at which the distortion of time-dependent signals is low and the noise is not too high are around 10–20 ms [2510, 5622], corresponding to a modulation frequency of 50–100 Hz. It has also proven of advantage to select modulation frequencies that are different from the line power frequency or multiples thereof to avoid interferences from the line power supply (interference noise) [2545]. The subject of response time for the measurement of transient signals is discussed in more detail in Section 3.8.

The application of the *Zeeman effect* with an alternating magnetic field also leads to modulation of the radiation beam, either via the periodic line splitting and polarization (transverse magnet configuration) or through the periodic removal of the analytical line (longitudinal magnet configuration). This permits continuous operation of the line source with constant radiant power.

By the use of a *selective amplifier* that is tuned to the modulation frequency, only the radiation modulated at that frequency is processed. Other radiation, especially the non-modulated emission from the atomizer, is received continuously, even when no radiation from the source reaches the detector, and is thus subtracted by the electrical measuring system in all measurement phases. Nevertheless, even with this arrangement a very high proportion of non-modulated radiation can lead to interferences such as emission noise, which is discussed further in Section 5.4.1.1.

3.7 Simultaneous Spectrometers

The strength of AAS lies in its selectivity and specificity. However, the classical configuration: line radiation source – atomizer – monochromator – detector, only allows different elements to be determined one after the other. For this reason attempts were made even at an early stage in the development of AAS to overcome this restriction and to determine several elements simultaneously or quasi simultaneously. An overview of these attempts is provided in Section 1.4. Recent review articles dealing with this subject and the various commercial offerings and their application have been published by HARNLY [2393] and FARAH and SNEDDON [1842].

Even though every multielement technique requires compromises compared to single element determinations and thus certain reductions in the performance, nevertheless a catalog of requirements can be established which should be fulfilled if disadvantages are to be avoided.

i)   The costs for a simultaneous spectrometer should be such that a much more favorable price-to-performance ratio is obtained per element. The costs should also be favorable compared to other multielement techniques (e.g. OES).

ii) The operating convenience of a simultaneous spectrometer should be comparable to that of a single-channel spectrometer.

iii) It should be possible to determine nearly all elements (i.e. virtually any combination of elements) without having to make notable constructive changes.

iv) The sensitivity, S/N ratio, and the limits of detection for a simultaneous spectrometer should not be markedly poorer than for a single-channel spectrometer.

v)   Interferences caused by concomitants and, more especially, by line overlapping should not be significantly more pronounced with a simultaneous spectrometer than for a single-channel spectrometer.

vi) It should be possible to determine analytes with considerably varying concentrations simultaneously.

Compared to OES, atomic absorption spectrometry has the principal disadvantage that the dynamic range is significantly smaller (refer to Section 5.2.3.4). This disadvantage can be of considerable importance for multielement determinations if the individual elements are present in considerably varying concentrations. For this reason it will always be difficult to determine a large number of elements in a single sample by AAS. A further problem is the often substantially differing conditions of atomization that are optimum for each individual element. For these reasons most working groups restricted their investigations on simultaneous spectrometers to two to ten elements. However, it is questionable whether a double-channel spectrometer is useful from the cost/benefit point of view since only minimal gains can be expected. The earlier claims for the advantage of simultaneous BC at a second line are no longer relevant after the introduction of high performance systems for BC (refer to Section 3.4). Further, measurements against a reference element (internal standard) are rarely performed nowadays because of the difficulties encountered (see Section 5.2.2.4). In general, the costs per channel decrease with the increasing number of channels, and a double-channel spectrometer represents the most unfavorable case.

On the basis of a cost/benefit analysis it is also easy to see that a multichannel spectrometer with a *flame as the atomizer* is not especially useful. ICP OES is an alternative technique offering a whole range of advantages, and the cost of a multichannel atomic absorption spectrometer is not markedly different from that of an ICP emission spectrometer. More elements can be determined simultaneously with ICP OES and, in addition, the dynamic range is considerably wider and the limits of detection are noticeably better for most elements.

Since the 1980s, most work on multielement AAS has therefore been restricted to *electrothermal atomization*. It is easy to recognize the reasons for this since GF AAS is one of the most powerful techniques with the lowest limits of detection for trace element determinations. On the other hand, the sample throughput is low so that the simultaneous determination of several elements represents a genuine gain. A prerequirement is naturally that the conditions stipulated above are met. A further advantage of multielement GF AAS is that the only comparable technique, ICP-MS, is still much more expensive and places considerable demands on the expertise of the operating personnel.

As mentioned in Section 1.4, there are two principal ways of constructing a multielement atomic absorption spectrometer. The 'conventional' design employs line sources, mono- or polychromators, and PMT detectors. The alternative is the use of a continuum source, a high resolution Echelle polychromator, and solid state detectors for the spatial measurement of the radiation (Vidicons, photodiode arrays, CIDs, or CCDs). Since the development of high performance continuum sources and solid state detectors, which cover the whole wavelength range of AAS and especially the far UV range, progressed only very slowly, their application was considered to be unpracticable for a long period of time. Conventional, tried and tested components were thus employed in most simultaneous spectrometers. However, in practice this 'simple solution' turned out to be anything other than simple since frequently only given element combinations were possible and considerable losses in sensitivity had to be accepted [414, 4275, 4854]. All in all, very few applications using spectrometers of this type have been published.

The use of line radiation sources initially appears very attractive since they emit high radiant energy at the analytical line and could thus bring the advantages of AAS to a simultaneous spectrometer. The number of lamps that could be used is limited by space to about four to six. Additional beam recombiners are required as well as careful alignment at a change of element. A polychromator with a number of exit slits can be used for spectral dispersion, and a detector must be located behind each slit. It is also possible to use several monochromators or at least several gratings [414, 4854]; a detector is required for each channel. Neither of these solutions is entirely simple, since the first is relatively complex and inflexible, while the second demands careful realignment at each change of element.

When several line sources are used, care must always be taken that each source is modulated at its own frequency, otherwise one of the major advantages of AAS, namely its selectivity, is lost. If the use of any arbitrary element combination should be permitted, then the possibility of line overlapping cannot be excluded and the risk that the lines of several elements pass the exit slit and reach the detector is large. This means that if line radiation sources are to be used a high resolution spectrometer, e.g. an Echelle spectrometer, is a prerequirement if spectral interferences due to line overlapping are to be avoided. On the other hand, the use of an Echelle spectrometer makes the use of solid state detectors virtually mandatory, so that in end effect the demands on a simultaneous spectrometer are largely independent of whether a continuum source or several line sources are employed. A typical design for a simultaneous spectrometer with Echelle polychromator and solid state detector is depicted in Figure 3-24.

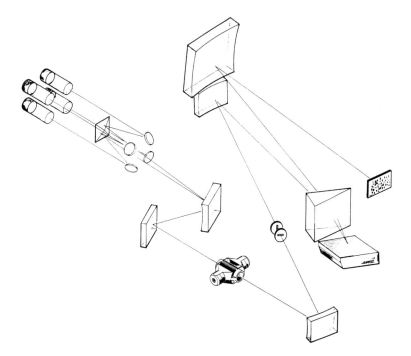

Figure 3-24. Schematic of the optical system of the SIMAA 6000, a simultaneous spectrometer with Echelle polychromator and a two-dimensional solid-state detector (by kind permission of Perkin-Elmer).

Two review articles on the development and the principle of multielement AAS using continuum sources and high resolution Echelle monochromators were published by HARNLY [2389] and O'HAVER and MESSMANN [2743] in 1986. Initially Harnly and O'Haver used a row of PMT detectors and an oscillating quartz plate for wavelength modulation. With this apparatus they were able to not only measure the background next to the analytical line [3902, 4411, 6515, 6516], but also the absorbance on the flanks of the absorption profile to reduce the sensitivity at high analyte concentrations [2386, 2389].

In the meantime a number of working groups have carried out investigations on the subject of continuum source AAS. WINEFORDNER and co-workers used a xenon arc lamp and a photodiode matrix as the detector; they used a graphite furnace [2935], a flame [2936], and an ICP [4078] as the atomizers. PARDUE and co-workers [2688, 3942, 3943] investigated the use of various solid state detectors and an Echelle spectrometer for continuum source AAS.

The performance, and thus the future, of continuum source AAS depends to a very large degree on the developments taking place in the areas of radiation sources and solid state detectors, especially their radiant power and sensitivity in the far UV, respectively. HIEFTJE [2546] published a critical comparison on sources and detectors for atomic spectrometry and BUSCH and BUSCH [977] published an overview on multielement detectors. Certainly the developments in the last few years give good reason to hope that considerable progress has been made for both sources and detectors. The pulsing of continuum sources has proven to be particularly effective [459, 4209, 4210, 5160]. For example, characteristic masses that were only a factor of 1.5 to 2 times worse than with line sources were measured in the far UV for antimony (217.6 nm), arsenic (193.7 nm), cadmium (228.8 nm), and zinc (213.8 nm), using an apparatus comprising a flash lamp, a graphite atomizer, an Echelle spectrometer and a CCD detector.

As well as allowing in principle the determination of all elements measurable by AAS, the configuration comprising continuum source, Echelle polychromator, and solid state detector has two further advantages. Firstly it permits the background to be measured simultaneously, even without an oscillating quartz plate, at a number of points directly neighboring the analytical line, resulting in a very flexible system for BC [3902]. Moreover it is possible to measure the absorption on the flanks of the absorption profile and not only at the line maximum [2386]. This enables one of the major problems of AAS, namely the restricted dynamic range, to be avoided. Depending on the analyte content, it is possible to preselect the sensitivity by measuring at the maximum of the analytical line or at smaller or greater distances from it [2394].

3.8 Data Acquisition and Output

In this section those parameters are discussed over which the operator has no or only very little influence. The quantities and functions that the operator can select are treated in Chapter 5. A clear distinction between these functions is however not always possible, so that occasional repetition and cross-references cannot be avoided.

3.8.1 Measured Quantities

The incident radiant power entering the absorption volume, Φ_0, and the radiant power leaving the absorption volume, Φ_{tr}, are converted, after spectral dispersion and separation, by the detector into electrical signals and amplified. Ratioing both signals leads to the calculation of the transmittance (spectral transmission factor), $\tau_i(\lambda)$, as a measure of the transmission:

$$\tau_i(\lambda) = \frac{\Phi_{tr}(\lambda)}{\Phi_0(\lambda)}, \tag{3.5}$$

and of the absorptance (spectral absorption factor), $\alpha_i(\lambda)$, as a measure for the absorption of the radiant energy in the absorption volume:

$$\alpha_i(\lambda) = \frac{\Delta\Phi(\lambda)}{\Phi_0(\lambda)} = \frac{\Phi_0(\lambda) - \Phi_{tr}(\lambda)}{\Phi_0(\lambda)} = 1 - \tau_i(\lambda). \tag{3.6}$$

The correlation between these quantities and the concentration or mass of the analyte is given by the Beer-Lambert law (Section 2.5.2), which states that for an ideal dilution the absorbance A (or the integrated absorbance A_{int}), given as the negative decadic logarithm of the transmittance, is proportional to the product of the length of the absorbing layer ℓ and the concentration or mass of the analyte. A distinction is made here between the amount-of-substance concentration c (e.g. mol/L), the mass concentration β (e.g. mg/L), and the mass m (e.g. µg) of the dissolved substance i:

$$A = \varepsilon(\lambda) \cdot c(i) \cdot \ell \tag{3.7}$$

$$A' = \varepsilon'(\lambda) \cdot \beta(i) \cdot \ell \tag{3.8}$$

$$A''_{int} = \varepsilon''(\lambda) \cdot m(i) \cdot \ell \tag{3.9}$$

The molar absorbance coefficient ε and the specific absorbance coefficients ε' and ε'', as the proportionality factors, are a function of the wavelength λ and in practice are influenced by a number of marginal conditions such as dissociation in the atomizer and the physical properties of the measurement solution. The pathlength ℓ is the distance the absorption beam passes through the absorption volume. It is determined by the geometry of the atomizers, e.g. the flame or the graphite furnace (see Chapter 4).

As already mentioned in Chapter 2, the requirements of the Beer-Lambert law can only be met in the first approximation. In practice there are deviations, for example due to the low resolution of the monochromator or to an inhomogeneous distribution of the atomic vapor over the cross-section of the absorption volume. In the optimum working range, however, there is a linear relationship between the absorbance A or the integrated absorbance A_{int} and the concentration c or β, or the mass m of the analyte, respectively. This is of advantage for the evaluation of the measurement (see Section 5.2). For following *steady state* signals as absorbance values, as is usual in flame AAS, a distinction is made between the *momentary value* (the output of which is dependent on the instrument) and the *mean value* which is formed from a series of momentary values. Further, it is possible to output measured quantities as an *integrated value*, that is the momentary values integrated over a given time period. For atomization techniques with discrete sample dispensing, such as GF AAS, all the analyte species or a constant proportion thereof contained in the sample are atomized over a defined time period. The output of the signal as the *(time-)integrated absorbance* is of advantage in this case.

3.8.2 Signal Handling

In an atomic absorption spectrometer the source radiation attenuated by atomic absorption or by other radiation losses falls onto the detector. The first quantity that is measured is the transmittance. However, in practice it is the absorbance A that is required since this is linearly proportional to the concentration or mass of the analyte in the sample solution. Atomic absorption spectrometers thus contain a logarithmic amplifier to convert the experimentally observed transmittance into absorbance.

As long as a steady state signal is generated by the atomizer (e.g. in a flame) the conversion is relatively easy. It is only necessary to follow the steady state transmittance for as long as necessary for an adequate approximation. The logarithmic conversion can then be performed in any way desired and the resulting absorbance can be further processed, for example by scale expansion, time integration, and/or curvature correction.

All atomic absorption spectrometers manufactured in earlier years were designed to measure steady state signals. This was one of the problems that hindered the successful application of the graphite furnace technique with its rapid, time-dependent signals. It was not until the middle of the 1970s that increasing numbers of publications dealing with the special requirements of this technique started to appear, especially with respect to signal handling [32, 1444, 2689, 4647, 5988] and to the errors that can occur through the use of too high response times [3662, 5343].

For a transient atomization peak the conversion to absorbance is complicated by two contradictory requirements:

i)   The measurement of the true transmission requires an instrument with a very short response time.

ii)   For every measurement the shot noise should be as low as possible, and this requires long response times, leading to a falsification of the transmission values.

A way out of this dilemma can be found in the radiant power. The more energy that falls on the detector, the better will be the S/N ratio, under otherwise unchanged conditions, and the noise will be low in comparison to the signal. This means that the entire atomic absorption spectrometer must be optimized for measuring fast, transient signals. This requires high intensity radiation sources, high radiance optics, and as few optical components as possible that can cause radiation losses (e.g. polarizers). At the same time all unnecessary sources of noise should be avoided, such as a second source for continuum source BC, rotating polarizers, or high current pulsing.

A further means of optimization is in the analog circuit of the electrical measuring system. As mentioned in Section 3.6, response times greater than 20 ms increasingly cause distortion of the transmission signal, while the noise increases markedly with response times of less than 10 ms [2510]. A modulation frequency of 50–100 Hz therefore appears to be ideal. Nevertheless the line power frequency (e.g. 50 Hz) and multiples thereof (e.g. 100 Hz) should be avoided because of the increased interference noise. Next to the radiant power falling on the detector and the measurement frequency, the *effective measurement time* during a measurement cycle has an important influence on the S/N ratio. The longer the measurement time and the shorter the dark period, the better will be the S/N ratio. The timing diagram for an atomic absorption spectrometer optimized for GF AAS with Zeeman-effect BC is shown in Figure 3-25. The effective measurement time per cycle is 12 ms, i.e. 2/3 of the total cycle time of 18 ms. Two periods of 3 ms each are used for the dark period during which the magnetic field builds up and dissipates. The dark periods are used to ensure that all measurements are made relative to a known zero point and that all non-modulated radiation, such as the emission from the atomizer, is eliminated. A further exceptional feature of the analog circuit of this instrument is the microprocessor-controlled preamplifier. If background attenuation reduces the transmittance markedly, so that the radiant power falling on the detector is significantly lower, the gain is automatically increased.

In the analog electronics the signal is converted into a series of digital values (A/D conversion) which are more suitable for processing by computer. At the same time as A/D conversion, the signal is logarithmized. After A/D conversion noisy signals can be smoothed relatively easily, e.g. using a Savitzky-Golay filter [392], without influencing the peak profile. This means that after digitalization the S/N ratio can be improved without distorting the information contained in the signal sequence.

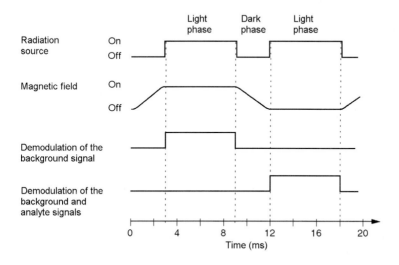

Figure 3-25. Schematic of the individual phases of a measurement cycle of 18 ms for the Model 4110ZL atomic absorption spectrometer with Zeeman-effect BC (by kind permission of Perkin-Elmer).

3.8.3 Baseline Offset Correction

During a measurement cycle the spectrometer generates several measures which derive for example from the radiant power of the sample beam and the reference beam or from the magnet on and magnet off phases. During signal processing the measures for the various phases are used for calculation so that the resulting signal corresponds to the concentration or mass of the analyte atoms in the atomizer. However, due to drift in the electronics, radiation losses at optical components, etc. the absorbance resulting from both signals will never be $A = 0$, even when there are no analyte atoms in the atomizer. It is thus necessary to set the readout of the instrument to zero prior to the start of a series of measurements and at regular intervals during the measurements.

One of the major reasons for constructing double-beam spectrometers was to compensate for drift phenomena (flicker noise) and thus to avoid the necessity of having to reset the baseline frequently (refer to Sections 1.4 and 3.2). However, it has been shown that during the warmup of a lamp, not only the emitted radiant power but also the line profile or the intensity ratio of two lines change [4647]. Correspondingly the absorption profile of the analyte lines in the atomizer, e.g. in a flame, can also change. All these changes lead to drift phenomena which even a double-beam spectrometer cannot compensate. For this reason electronic circuits for drift correction were proposed at an early stage in the development of AAS [1342, 3659, 3985, 5716]. During these investigations it was also shown that single-beam spectrometers were practically independent of variations in the operating conditions and from drift phenomena, provided a sufficiently large number of reference measurements were made [3418, 5716].

The technique that has proven to be the most reliable is automatic baseline offset correction (BOC) immediately prior to atomization of the analyte. In this way, not only constant offsets, but also minor drift phenomena as described above, or the slow contamination of windows, will be automatically corrected. RÖHL [4933] established that the BOC technique is extremely useful for evaluating transient atomization pulses, especially when it is possible to forecast when the signal appears and thus be able to establish the baseline immediately prior to atomization. The *measurement period* for the baseline should be at least 50% of the measurement period for the atomization signal. BARNETT *et al.* [392] found that the BOC measurement period had a not inconsiderable influence on the attainable limits of detection. An increase in the BOC measurement period from 1 s to 2 s brought a significant improvement in the values; measurement periods of greater than 4 s however had no further influence on the precision.

BARNETT *et al.* [392] also drew attention to the significance of the *accuracy* of the value used for BOC. If concomitants or even analyte atoms are volatilized during measurement of the baseline, an incorrect value for the baseline will be subtracted for the subsequent correction of the analyte measure. To avoid this source of error, which in principle is an operating mistake, most modern spectrometers issue an error message if the baseline changes systematically during the BOC measurement. Baseline correction is especially important for the integration of transient signals, such as occur in GF AAS. Small baseline offsets in this case can lead to large errors in integration. On occasions some authors have preferred peak height measurements to peak area measurements (integrated absorbance) when the spectrometer used did not have a BOC function [632,

1564, 2604, 2607]. PRUSZKOWSKA *et al.* [4723] have shown that BOC is a fundamental component of the STPF concept (see Section 8.2.4.1).

3.8.4 Integrated Absorbance

The *analytical* advantages of time integration compared to peak height measurement are mentioned in Section 1.7 and are treated in detail in Section 8.2.4. In the current section, time integration will be considered with respect to *signal handling*. The question of whether transient signals generated by GF AAS could be evaluated better via the maximum absorbance value or via the integrated signal was investigated by a number of authors as early as the 1970s. The inadequate spectrometers of that period naturally played a not inconsiderable role. Nevertheless, all authors were in agreement that integrated absorbance brought calibration curves that were linear over a wider concentration or mass range than those evaluated via the peak height absorbance [4647, 5181, 5617].

PIEPMEIER and DE GALAN [4647] pointed out that smoothing of the signal prior to A/D conversion changed the signal shape and thus the signal height, but not the time integral. In addition, time integration often improves the S/N ratio. RÖHL [4933] found that although the absorbance maximum can be used in individual cases as a measure for the signal strength (and thus the mass or concentration), the determination of the net peak area, i.e. the time integral of the absorbance pulse over the baseline, is the most generally applicable method. VOIGTMAN [6089] emphasized that evaluation via the maximum absorbance only brings satisfactory results when the S/N ratio is high. Beneath a certain S/N ratio, i.e. when approaching the limit of detection, time integration is always superior. This clearly refutes the often made, but never proven, claim that for elements with very narrow absorption pulses (< 1 s) peak height evaluation close to the limit of detection is better because it generates the 'numerically higher' values.

Nevertheless two points appear to be of importance for time integration. PIEPMEIER and DE GALAN [4647] pointed out that integration should not be performed until ***after*** conversion of the transmittance into absorbance since otherwise measurement errors can occur. If the transmittance includes an error resulting from the shot noise, this error naturally is transferred directly to the absorbance. The S/N ratio depends on the absolute value of the absorbance—the best values are found in the range $A = 0.3–1.0$ (refer to Sections 5.2.3.2 and 5.2.3.4)—while for $A = 0$ the S/N ratio also reaches a value of zero. This means that integration in time periods when the absorbance is close to zero, i.e. at the start and end of a peak, contributes very little to the time integral, but quite a lot to the noise, i.e. to the measurement uncertainty.

On the basis of theoretical considerations, RÖHL [4933] stated the requirement that the atomization pulse should be as narrow as experimentally possible so that the time window for integration could be correspondingly short. A very successful step in this direction is the use of a transversely-heated, spatially isothermal atomizer (refer to Section 4.2.2.5). VOIGTMAN [6089] went a step further and proposed that the atomization pulse should be only partially integrated. Optimum S/N ratios were obtained for various (theoretical) peak shapes if time integration was started and ended at about one third of the maximum value.

Nevertheless, such techniques can hardly be applied in practice since the appearance time and the pulse length of the atomization pulse almost certainly can change in the presence of concomitants [5424]. Setting the exact time point for an integration window is thus problematical. The use of automatic peak recognition circuits is not simple, particularly for low absorbance values, due to the relatively high noise level and can lead to additional errors [391]. For practical applications it cannot be expected that algorithms for peak recognition will lead to better values than the use of an integration window based on measured, time-resolved absorption pulses [392]. The advice to select as short a time window as possible for the integration should nevertheless be taken into account in all cases.

4 Atomizers and Atomizer Units

The *atomizer* is the 'place' in which the analyte is atomized, i.e. the flame, the graphite tube, or the quartz tube. The *atomizer unit* encompasses, in addition to the atomizer, all assemblies required for operation, for example a burner with nebulizer and gas supply, or a graphite furnace with power supply. The portion of the atomizer through which the measurement radiation beam passes is termed the *absorption volume* or *observation volume*.

The task of the atomizer unit is to generate as many free atoms in the ground state as possible and to maintain them in the absorption volume for as long as possible. To meet the requirements of the Beer-Lambert law the distribution of the atoms should be as homogeneous as possible in the absorption volume, i.e. over the length and cross-section of the atomizer. The atomization step, i.e. the transfer of the sample, and especially the analyte, into free atoms in the gaseous state, is without doubt the most important process in an analysis by AAS. The success or failure of a determination is dependent on the atomization step. The sensitivity of a determination is directly proportional to the degree of atomization of the analyte and to the residence time of the analyte atoms in the absorption volume. Ultimately, all known non-spectral interferences in AAS are nothing more than influences on the number of analyte atoms generated, either absolute or per time unit, or on their spatial distribution in the atomizer.

The most important criteria for the selection of a suitable atomizer for a given analytical task are the concentration of the analyte in the analytical sample, the amount of sample available, and the state of the sample (solid, liquid). Furnace techniques exhibit higher sensitivity than flames. A further important criterion is the analyte itself, since atomizers vary considerably in their suitability for atomizing individual analytes as a result of the temperature and chemical environment of various atomizer types.

4.1 Flame Atomization

The flame technique is the oldest of the AAS techniques. For many years it was the 'workhorse' for the determination of secondary and trace elements, and also for main constituents, and even nowadays it is difficult to imagine a routine analytical laboratory without this technique. Due to its simplicity and economy, from 1960 onward the flame technique rapidly displaced OES with arcs and sparks. The development of the flame technique as well as flames and atomizer units no longer employed are discussed in Section 1.6.

In flame atomization, either an indeterminate volume or a fixed aliquot of the measurement solution is converted into an aerosol in a nebulizer and transported into the flame. The flame must possess enough energy not only to vaporize but also to atomize the sample. The chemical composition of the flame can have a major influence on these processes. Since nowadays premix burners and air-acetylene or nitrous oxide-acetylene flames are used almost exclusively, the following discussions are limited to these flames.

4.1.1 Spectroscopic Flames

The task of the flame is to vaporize and convert the entire sample as far as possible into gaseous atoms, i.e. not only the analyte but also the concomitants (the processes involved are discussed in more detail in Section 8.1.1). At the same time the flame is also the absorption volume, a fact that places a number of prerequirements on an ideal flame.

i)   The flame must supply sufficient thermal energy to rapidly atomize the sample independent of the nature and quantity of the concomitants, but without causing noticeable thermal ionization of the analyte.

ii)   The flame must provide a suitable chemical environment that is advantageous for atomization, and should be variable to allow both reducing and oxidizing conditions.

iii)   The flame should be transparent to the absorption radiation and should not emit too strongly to prevent saturation of the detector.

iv)   The flame should allow low gas flowrates (burning velocity) so that the atoms remain in the absorption volume for as long as possible [3396].

v)   The flame should be non-turbulent, i.e. laminar combustion, so that the time-resolved distribution of the atoms in the absorption volume is constant.

vi)   The flame should be as long as possible to provide high sensitivity.

vii)   Flame operation should be safe, and both the flame gases and the combustion products should not pose health and safety risks.

viii)   Flame operation should be inexpensive.

It goes without saying that there are no real flames that meet all of these requirements. A number of flames used during the development phase of AAS, and particularly those operated with a direct injection burner, hardly fulfilled even one of the above requirements (refer to Section 1.6).

Nowadays two flame types are used almost exclusively: the air-acetylene and the nitrous oxide-acetylene flames. These two flame types complement each other in an ideal manner and come amazingly close to the requirements of an ideal flame. The properties of these two flames, and also two further flame types that are occasionally used, are presented in Table 4-1.

The flame most well known and used in AAS is the *air-acetylene flame*. For many elements it offers an environment and a temperature suitable for atomization; the alkali metals are noticeably ionized. The flame is completely transparent over a wide spectral range and only starts to absorb radiation below 230 nm, increasing to about 65% absorption at the wavelength of arsenic (193.7 nm), as shown in Figure 4-1. The emission of the air-acetylene flame is very low so that ideal conditions are given for many elements. Normally this flame is operated stoichiometrically or slightly oxidizing; however, the ratio of the flame gases is variable over a wide range, thus further increasing its applicability. A number of noble metals such as gold, iridium, palladium, platinum, and rhodium may be determined with the highest sensitivity and largely free of interferences in a strongly oxidizing air-acetylene flame. The alkaline-earth metals, on the other hand, are determined more favorably in a slightly reducing flame (i.e. a slight excess of fuel gas).

Table 4-1. Spectroscopic flames for AAS with their properties (from [1501]).

Oxidant	Fuel Gas	Maximum Flame Temperature (°C)	Maximum Burning Velocity (cm/s)	Remarks
Air	Acetylene	2250	158	Most commonly used flame
Nitrous oxide	Acetylene	2700	160	For difficultly volatilized and atomized substances
Air	Hydrogen	2050	310	Flame of high transparency; for easily ionized elements
Air	Propane/butane	1920	82	For easily ionized elements

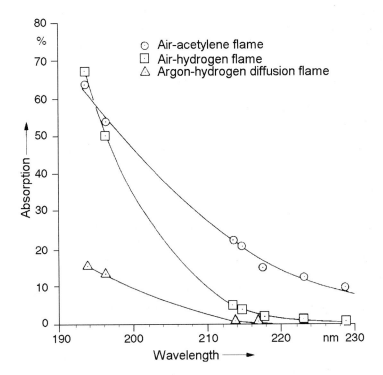

Figure 4-1. Radiation absorption of a number of flames in dependence on the wavelength.

Due to its low partial pressure of oxygen, the fuel-rich, reducing air-acetylene flame is used for the determination of elements that exhibit a high affinity for oxygen. To these elements belong the alkaline-earth metals and also chromium and tin, and even molybdenum [1430, 4801]. Compared to the stoichiometric or oxidizing flame the reducing flame exhibits not only a lower partial pressure of oxygen but also a lower temperature

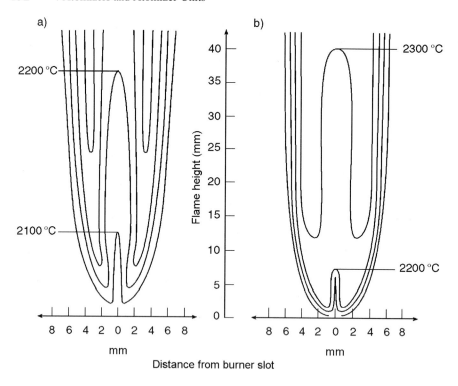

Figure 4-2. Temperature gradients in **a** – a fuel-rich (C:O = 0.9) and **b** – a stoichiometric (C:O = 0.4) air-acetylene flame (from [4693]).

and a steeper temperature gradient (see Figure 4-2), factors that can favor interferences (refer to Section 8.1.2). Further disadvantages of this flame are an increasing emission and a decreasing transparency. In other words it does not meet the requirements of an ideal flame so well as the stoichiometric or oxidizing air-acetylene flame.

There are about thirty elements that in effect cannot be determined in the air-acetylene flame since they form refractory oxides. For these elements the *nitrous oxide-acetylene flame*, introduced by WILLIS [6340] in 1965, offers a highly suitable thermal and chemical environment for high degrees of atomization. The usual nitrous oxide-acetylene flame is operated with a slight excess of fuel gas and exhibits a 2–4 mm high blue-white primary reaction zone above which is a characteristic 5–50 mm high red reducing zone. Above this is then a pale blue-violet secondary reaction zone in which the oxidation of the fuel gas takes place. The red reducing zone is the one of analytical interest. The dissociation of the sample into atoms takes place here and no noticeable oxidation of the metal atoms occurs.

Next to the observation height, the ratio of oxidant to fuel gas has the greatest influence on the degree of atomization and on interferences in the nitrous oxide-acetylene flame. By choosing these parameters correctly most interferences can be avoided. To be able to describe the flame conditions as exactly as possible, MARKS and WELCHER [4115] introduced a parameter ρ which represents the molar oxidant/fuel gas ratio as a fraction of the stoichiometric ratio 3:1 (3 N_2O + C_2H_2 → 2 CO + 3 N_2 + H_2O). Various authors also established that a larger number of free atoms are generated in the nitrous oxide-acetylene flame than in all other flame types [1854, 2900, 3199], but that the atom concentration exhibits a distinct distribution profile.

Nevertheless, the nitrous oxide-acetylene flame has two disadvantages that must be taken into consideration. Firstly, many elements are more or less strongly ionized in the hot flame and thereby show reduced sensitivity. Secondly, the flame exhibits a relatively strong emission [3844], as shown in Figure 4-3. While the ionization interferences can often easily be suppressed by adding an excess of another easily ionized element (refer to Section 8.1.2), the emission can on occasions cause genuine difficulties [637]. The CN, CH, and NH bands, which occur over a wide spectral range and are often very intense, can cause 'emission noise' if they coincide with the resonance line of the analyte element, thus influencing the precision of a determination significantly.

The *air-hydrogen flame* provides a markedly higher sensitivity for tin than either the air-acetylene or nitrous oxide-acetylene flames [1054, 1056]. This flame also offers a number of advantages for the alkali metals, such as lower ionization compared to the air-acetylene flame [5131]. It also exhibits higher transparency down to about 200 nm, as shown in Figure 4-1. On the other hand, considerably more interferences must be taken into account with this flame due to its lower temperature and its higher burning velocity (see Table 4-1). The use of organic solvents is also not without problems [5786].

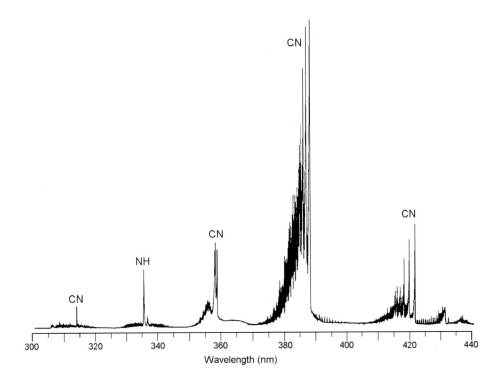

Figure 4-3. Emission spectrum of a fuel-rich nitrous oxide-acetylene flame (from [3844]).

4.1.2 Nebulizer-Burner Systems

Nowadays, premix burners are used almost exclusively in AAS because the laminar flame offers excellent conditions for performing determinations with minimum interference. In systems of this type the measurement solution is aspirated by a pneumatic nebulizer and sprayed into a spray chamber. The sample aerosol is mixed thoroughly with the fuel gas and the auxiliary oxidant in the chamber before leaving the burner slot, above which the flame is burning. Depending on the burner head, the flame is generally 5 cm to 10 cm long and a few millimeters wide. Normally the radiation beam passes through the entire length of the flame. The burner head is usually interchangeable to permit the use of various flame types. A typical premix burner is depicted in Figure 4-4.

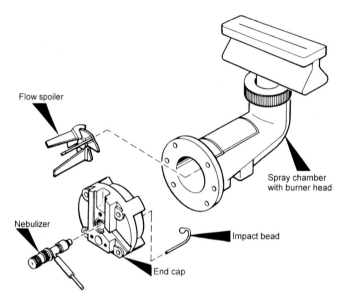

Figure 4-4. Premix burner optionally permitting the use of a flow spoiler or an impact bead (by kind permission of Perkin-Elmer).

4.1.2.1 Burner Heads

The task of the burner head is to conduct the premixed gases to the combustion reaction in such a way that a stable flame is produced. The major requirements placed on a burner head can be summarized as follows:

i)   Safe operation of the flame over a wide range of oxidant-to-fuel gas ratios.
ii)   High thermal stability of the burner head with only a short warmup time.
iii) Good corrosion resistance to the sample solutions.
iv)   Stable flame geometry accommodated to the geometry of the spectrometer's radiation beam.

v)   The longest feasible burner slot to provide a large 'cell length'.

vi) A low tendency to form encrustations from the fuel gas or sample constituents, even with a high solids content in the solutions.

Nevertheless, the geometry of the burner head must be suitable for the oxidants and fuel gases being used, otherwise compromises in the performance must be taken into account. The dimensions of the burner slot must be such that the exit flow velocity of the flame gases is always greater than the burning velocity of the flame. If this condition is not met the flame can flash back into the spray chamber and cause explosive combustion of the gas mixture.

Numerous workers have been occupied with the safe operation of chemical flames and the avoidance of flashbacks [114, 2544, 5674]. In principle, a flame is a stationary explosion. A dynamic equilibrium must be attained in order to produce a stable flame. On the one hand the flame gases streaming from the burner slot tend to blow the flame away from the burner head, while on the other hand the flame spreads back at its characteristic burning velocity into the unburned gases. In an unlimited system a stable flame front can only be obtained by exactly matching the flowrate of the flame gases to the burning velocity. Fortunately the conditions for this sensitive equilibrium are less critical when the flame burns over a small opening or port. The flame is stabilized at the port (burner slot) by the entry of atmospheric air into the burning gases and by the quenching effect caused by the port walls. This latter effect is in particular very important for the prevention of a flashback.

The walls of the port act as a heat exchanger and effectively reduce the burning velocity of the gases in their direct vicinity to zero. Away from the walls the burning velocity quickly reaches its maximum value, as depicted schematically in Figure 4-5. The flow of the flame gases, on the other hand, assumes a parabolic velocity profile that is described by the Poiseuille equation. A flashback of the flame can be avoided when the flow velocity of the flame gases is everywhere higher than the burning velocity.

The limiting velocity at which a flashback of the flame is just avoided, i.e. the minimum flow velocity of the flame gases, depends on a number of factors; the dimensions of the burner port and the stoichiometry of the flame are the most important. The dependence of this limiting velocity on the oxidant-to-fuel gas ratio for an air-acetylene flame in a cylindrical burner port is depicted in Figure 4-6 [2544]. Similar profiles have also been published by other authors [114, 5674].

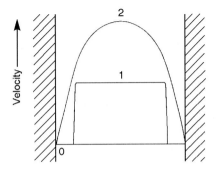

Figure 4-5. Schematic of **1** – burning velocity and **2** – flow velocity of the flame gases in a cylindrical burner port for a stable flame; the walls of the burner port are hatched (from [2544]).

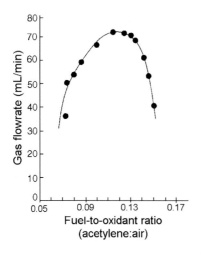

Figure 4-6. Limiting velocity for the flame gases at the burner port required to prevent a flashback of an air-acetylene flame of varying stoichiometry; diameter of the burner port 1.2 mm (from [2544]).

Nevertheless, when designing a burner and deciding on the dimensions of the burner port (slot), other aspects have to be taken into consideration as well as the flame gases and the avoidance of a flashback. It must be remembered that the aerosol from the measurement solution is transported through the burner port into the flame, and the smaller the opening, the greater is the risk that it will become clogged by salt deposits. Thus in practice a compromise must be made between a burner slot that is narrow enough to prevent a flashback of the flame but on the other hand is wide enough to avoid encrustation of sample constituents.

A further important factor is the temperature a burner attains during operation. Numerous authors have demonstrated that the risk of a flashback of the flame increases with the temperature at the burner slot and have thus proposed that the burner head should be cooled [114, 1908, 2149, 2544, 5674]. Further advantages of cooling or thermostatting the burner head include a substantially lower drift of the sensitivity, an increase of the zone of highest atom concentration and thus of the sensitivity, better linearity of the calibration curves, and fewer interferences [2149, 2761]. The risk of increased salt encrustations at the burner slot has however been mentioned as a disadvantage of cooling [1908]. As well as water cooling or the use of cooling fins on the burner head, the desired effects can also be achieved by using materials of lower thermal conductivity [2544]. By a suitable combination of materials for the construction of a burner head it is possible to minimize both the risk of a flashback and the formation of deposits.

4.1.2.2 Nebulizers and Spray Chambers

Normally the measurement solution is transported into the flame as a fine aerosol mixed with the flame gases. Aerosols are a particularly suitable form of presentation for the sample since the drying and volatilization processes can be favorably influenced due to the very large surface-to-mass ratio [891]. The formation and transport of an aerosol in a premix burner can be divided into three phases [889]:

i)   A *primary aerosol* with a polydispersive droplet distribution and a substantial excess of energy is generated by a nebulizer.

ii)   A *secondary aerosol* with a reduced range of droplet sizes is produced by post nebulization when the primary aerosol strikes an impact surface. At the same time the larger droplets are separated.

iii) A *tertiary aerosol*, which is transported into the flame, is formed by the processes of sedimentation, vaporization, impact, and turbulent and centrifugal separation.

A number of supplementary conditions must be placed on the tertiary aerosol in order that in the flame it is vaporized and atomized as quickly, as completely, and as uniformly as possible, so that interference-free and sensitive detection can be attained. Among these are a high aerosol concentration for high sensitivity, a small range of droplet sizes for a high atom concentration in the absorption zone, and small droplets for high atomization efficiency. We must bear in mind that an improvement of the transport efficiency (aerosol concentration) cannot be obtained by using larger droplets. The reason is obvious since in the flame the vaporization of the solvent from the larger droplets leads to the formation of larger particles:

$$d^3_{\text{solid}} = d^3_{\text{liquid}} \cdot \left(\frac{c}{\rho}\right) \cdot 10^6 \,, \tag{4.1}$$

where c is the solids content of the sprayed solution in mg/L, ρ is the density of the solid in g/cm^3, and d is the diameter of the particles in µm.

A rough calculation shows that a droplet of a sodium chloride solution with a concentration of 100 mg/L and a diameter of 1 µm dries to a particle of 0.036 µm diameter, while a droplet of 10 µm diameter dries to a particle with a diameter that is ten times as large. If the concomitants include compounds of low volatility, the volatilization of the analyte from particles of larger size can be influenced to such an extent that volatilization interferences occur (more information in Section 8.1.2.4). Which droplet sizes contribute to the generation of the signal depends on the efficiency of atomization and this is different for different elements. For easily atomized elements such as copper, droplets with a diameter up to 10 µm still contribute to the signal, but difficultly atomized elements such as chromium cannot be atomized from droplets larger than 5 µm. With a well adjusted nebulizer the number of droplets having a diameter greater than 10 µm is very low, but these droplets contain the greatest proportion of the nebulized mass. An estimate indicates that a single droplet of 20 µm diameter contains the same mass as 10 000 droplets of 1 µm diameter. If they reach the flame, larger droplets lead to lower rates of atomization and to solute-volatilization interferences (see Section 8.1.2), and cause earlier encrustation of the burner slot [6344].

To provide optimum analytical performance with respect to precision and freedom from interferences, a nebulizer/spray chamber system must meet the following requirements:

i)   The ratio of the mass of aerosol with sufficiently small droplet sizes to the quantity of solution sprayed must be as large as possible (efficiency of nebulization).

ii)   The mass of aerosol with sufficiently small droplet sizes that reaches the flame per unit time should be as constant as possible, both short term (noise) and long term (drift).

iii) The range of droplet sizes (droplet spectrum) and the mass of aerosol that reaches the flame per unit time should as far as possible be independent of the properties and composition of the sprayed solution (matrix effects).

iv)   The time required for the aerosol to stabilize after the solution is sprayed should be as short as possible (low sample consumption; high sample throughput).

v)   The chemical resistance of the nebulizer/spray chamber system against acids and aggressive digestion solutions should be as high as possible.

vi)   The time required to flush out high analyte or matrix concentrations until a stable signal is attained should be as short as possible (memory effects).

vii) The whole system should exhibit a low tendency to clogging and the formation of encrustations, and should be of simple design with high mechanical stability and should require only a minimum of maintenance.

In practice it is difficult to realize all of these requirements in parallel so that compromises must be made. The sample introduction system has thus justifiably been termed the 'Achilles heel' of atomic spectrometry [890].

The task of generating the primary aerosol, in which energy is required to divide a quantity of liquid into fine droplets, is performed by the *nebulizer*. The amount of work required, W (in Nm), depends largely on the surface area, A_2, of the liquid droplets generated:

$$W = \sigma \left(A_2 - A_1 \right) \approx \sigma A_2 , \tag{4.2}$$

where σ is the surface tension of the liquid in N/m, and A_1, A_2 are the surface areas of the solution prior to and after nebulization, respectively.

The surface tensions of liquids lie in the range from $\sigma > 0.02$ N/m for organic solvents to $\sigma = 0.07$ N/m for water. A droplet with a volume of 1 μL, for example, has a diameter of 1.2 mm and a surface-to-volume ratio of 48 cm^{-1}. If this droplet is nebulized into an aerosol of finer droplets with a diameter of 1.2 μm, the surface-to-volume ratio increases by a factor of 1000 (proportional to $1/d$) and the number of droplets increases to 1 000 000 000 (proportional to $1/d^3$). The increase in the energy required is inversely proportional to the root of the droplet diameter and finally reaches a limiting value after which further energy cannot be introduced by external processes, even though the theoretical energy requirement is relatively low. The energy required to nebulize 1 mL of water at 20 °C into droplets of 1 μm diameter is accordingly 0.427 Joule, which is a relatively low amount of work compared to the 2250 Joule required to vaporize the same amount.

The work required to break down the liquid into droplets can be provided by a number of different means. The nebulizer itself can transfer the work to the liquid [5161, 5162], for example by accelerating the liquid in a jet (monojet) [592, 2833, 4694, 4695], by accelerating a liquid film off a surface (rotating disk nebulizer) [39, 5210, 6143], by accelerating a liquid film by a vibrating membrane (ultrasonic nebulizer) [1487, 2813, 2814, 4907, 4908], or by accelerating droplets in a strong electric field (electrostatic nebulizer) [2181]. Alternatively a liquid column can be accelerated by producing gas bubbles (thermospray) [1487, 3253, 3456, 4186, 4909]. The work can also be provided by the nebulizer through transfer of shearing forces from a second medium, usually the aerosol transport gas in a binary jet system, by ultrasonic waves (Galton pipe, impact bead) or by a turbulent gas stream (pneumatic nebulizer).

All the methods of nebulization mentioned above have been tested at various times for their suitability [1487, 3983], but only the pneumatic nebulizers have come into general use in AAS. In a pneumatic nebulizer the energy of nebulization is provided by the nebulizing gas, which in F AAS is mostly the oxidant or a part of it. According to the arrangement of the gas and liquid jets one can distinguish between *concentric* or *split-ring nebulizers* with concentric gas and liquid jets, *cross-flow nebulizers* in which the jets are at right-angles to each other, *Babington nebulizers* in which the sample is introduced via a liquid film, and *fritted nebulizers* with a bundle of gas jets. Only the first two types can be so designed that they aspirate the sample solution themselves. This saves the need for expensive pump systems, which is an added reason for their wide acceptance. Nevertheless, a self-aspirating nebulizer must then meet the requirements placed on a pump system for the measurement solution:

i)   The transport rate should be independent of the properties of the measurement solution, such as viscosity, surface tension, density, and temperature.

ii)   The transport rate should be independent of the manner in which the solution is presented, such as depth of immersion of the sample capillary in the sample container or the height of the liquid in the container.

iii) The transport rate should be independent of other parameters of the atomizer unit, such as gas flowrates.

iv)   The transport rate should be reproducible and constant without short or long term changes (pulsing, drift).

A number of the problems of F AAS derive from this double function of pneumatic nebulizers, since they do not represent ideal pump systems [3603].

In F AAS, *concentric nebulizers* (earlier termed *split-ring nebulizers*) have been used preferentially from the earliest times. Nowadays they are made of corrosion-resistant metals (Pt, Pt/Ir, Ti, etc.) or polymers. The design details are mostly the result of empirical investigations. In principle they do not differ from earlier designs, such as that used by GOUY [2188] in the last century. Their popularity stems from their relatively simple design, which allows the manufacture of nebulizers with reproducible properties, their mechanical stability, and the fact that they can aspirate the measurement solution unaided. Compared to cross-flow nebulizers, concentric nebulizers have the advantage that it is easier to adjust the gas nozzle and liquid capillary with respect to each other and that they are generally more efficient than the former [1051]. Their efficiency improves with decreasing diameter of the gas nozzle [1051]. A typical concentric nebulizer as used in F AAS is depicted in Figure 4-7.

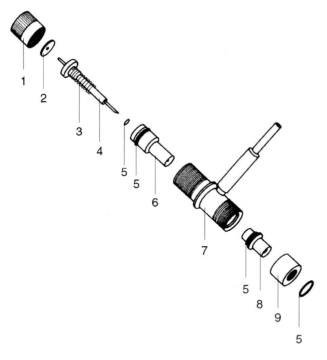

Figure 4-7. Design of a typical concentric nebulizer. **1** – regulator, **2** – PTFE disk, **3** – spring, **4** – nebulizer capillary, **5** – O-ring, **6** – insert, **7** – body, **8** – venturi, **9** – end cap (by kind permission of Perkin-Elmer).

The aspiration rate of concentric nebulizers is determined by the pressure drop at the liquid capillary. The aspiration rate is described by the Hagen-Poiseuille equation:

$$Q = \frac{P \pi r^4}{800 \eta \ell},$$

(4.3)

where Q is the aspiration rate in mL/s, P is the pressure in Pa, r is the radius of the capillary in mm, η is the viscosity of the aspirated liquid in poise, and ℓ is the length of the capillary in mm.

The only parameter that depends on the sample is the viscosity, which must be maintained constant (see Section 8.1.2.4). It must nevertheless be emphasized that the Hagen-Poiseuille equation requires a laminar flow of liquid in the capillary; in addition it must be taken into account that the reduced pressure at the capillary depends on the aspiration rate [4402]. As the liquid stream increases, losses in pressure must be expected, the levels of which depend on the design such as the variations in the diameter between the gas nozzle and the liquid capillary [4402]. The influence of the transport rate on the pressure loss leads to the conclusion that the nebulization itself has an influence on the aspiration rate. This influence can be explained by the variations in the flow and pressure profiles of the nebulized liquid at the nebulizer jet. Such pressure losses with increasing aspiration rate can even have a positive, stabilizing effect, since they reduce the influence of varying viscosity of the aspirated solution on the aspiration rate [5647].

The nebulizers normally used in F AAS generate a pressure difference of 2.5–25 kPa, which is dependent on the preset pressure of the nebulizing gas, the sample capillary being used, and the position of the jet in the venturi [4402]. Nebulizers in which the capillary can be adjusted within the venturi have the advantage that they can be optimized with respect to the experimental conditions and the properties of the measurement solution. A change in height of the measurement solution of 10 cm can lead to a hydrostatic pressure change of 1 kPa, so that the hydrostatic pressure of the liquid column can make a substantial contribution to the pressure difference generated in the nebulizer [5647]. All solutions should thus have approximately the same liquid height during measurements. Narrow-necked flasks, measuring cylinders, and test tubes in which the liquid height changes significantly during repetitive measurements should not be used as sample containers.

The actual mechanism of nebulization has been investigated by a number of authors [647, 3415, 4395], but a generally valid theory still does not exist. As already mentioned, the gas stream at the cross-section of the liquid capillary in a pneumatic nebulizer is the driving force for the aspiration of the liquid. The gas streaming out of the nozzle with high velocity expands and thus generates a drop in pressure at the liquid capillary, resulting in a liquid stream. In agreement with Bernoulli's law, the negative pressure at the jet is equal to the sum of the static and dynamic pressures of the liquid in the sample capillary:

$$\Delta P = 8\, j_f \, \frac{\ell \rho u^2}{20 d_i}, \tag{4.4}$$

where ΔP is the pressure drop at the liquid jet in Pa, $j_f = 8/N_{RE}$ is a dimensionless friction factor for laminar streams where N_{RE} is the Reynold number, ℓ is the length of the sample capillary in mm, ρ the density in g/mL, u the gas flowrate in mL/s and d_i the internal diameter of the capillary in mm.

Stroboscopic observations [3415] have shown that the liquid stream emerging from the capillary is brought to oscillation by the surrounding gas stream of higher flow velocity and the resulting shearing forces cause it to divide into individual droplets, as depicted schematically in Figure 4-8.

Resulting from the surface tension of the nebulized liquid, which is a force acting into the mass of the liquid, an internal pressure P_σ is generated in the droplet which is given by:

$$P_\sigma = 4\sigma/d, \tag{4.5}$$

where σ is the surface tension in N/m.

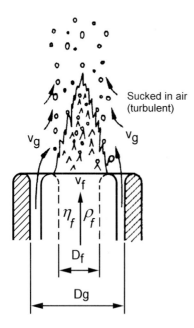

Figure 4-8. Nebulization of the liquid stream at the outlet of the sample capillary (from [3281]). D_g and D_f are the inner diameters of the gas jet and liquid capillary, v_g and v_f are the velocities of the gas and liquid streams, and η_f and ρ_f are the viscosity and density of the solution, respectively.

For a droplet of 1 μm diameter this leads to a pressure of 2.84 bar. Resulting from the streaming effect of the nebulizing gas a dynamic pressure forms on the surface of a droplet. If the dynamic pressure is greater than the internal pressure of the droplet, the gas can penetrate the droplet, causing it to inflate and finally disrupt. The greater the difference is between the flow velocities of the liquid and gas streams, so the smaller are the droplets.

The mean droplet diameter of the resulting primary aerosol is frequently given by the empirical equation developed by NUKIYAMA and TANASAWA [4395]:

$$d_s = \frac{585}{V}\left(\frac{\sigma}{\rho}\right)^{0.5} + 597\left(\frac{\eta}{(\sigma\rho)^{0.5}}\right)^{0.45}\left(\frac{10^3 Q_L}{Q_G}\right)^{1.5}, \tag{4.6}$$

where d_s is the mean droplet diameter according to Sauter (volume-to-surface ratio) in μm, V the velocity difference between the liquid and gas streams in m/s, σ the surface tension of the nebulized solution in dyne/cm, ρ the density of the nebulized solution in g/cm^3, η the viscosity of the nebulized solution in dyne s/cm^2, Q_L and Q_G the volume flowrates of the solution (L) and the nebulizing gas (G) in cm^3/s.

GUSTAVSSON [2293] has pointed out that equation 4.6 is valid only for droplet diameters of 15–90 μm, while for the concentric nebulizers used in F AAS the equation actually predicts diameters of 10–30 μm. Since equation 4.6 does not contain any parameters for the design of the nebulizer under investigation, it is only possible to make a few qualitative predictions. Among other things the Nukiyama-Tanasawa equation forecasts that the size of the droplet will increase with increasing liquid stream, viscosity and surface tension, but will decrease with increasing flow velocity of the nebulizing gas.

Despite the limited applicability of equation 4.6 to nebulizers as the means of sample introduction, nevertheless a number of the predictions agree with the results of empirical observations. For example it has been observed that the sensitivity of F AAS is not proportional to the rate of aspiration, but rather that better nebulizer efficiency is attained at low aspiration rates [1353, 1355, 2939, 4258, 5463, 5726]. Improvements in sensitivity compared to aqueous solutions can be achieved when using organic solvents whose viscosity or surface tension is lower than those of water [310, 1356, 4185]. Improvements can also be obtained by mixing organic solvents with aqueous solutions [3214], even in some cases where they form emulsions [351]. Results indicating that the addition of surfactants (tensides etc.) improves the range of droplet sizes [3190, 5005, 6445] are still disputed [4184, 5093]. It is very questionable whether the time available for the formation of the droplets is in fact sufficient to allow the micelles to orientate with respect to the surface and thus lead to a reduction of the surface tension. Small improvements in the aerosol production through the addition of surfactants can be explained by a reduction in the rate of separation in the spray chamber resulting from the smoother surface of the spray chamber and flow spoiler due to a reduced droplet formation (see also Sections 8.1.1.5 and 8.1.2.2).

Aerosols produced by the nebulizers used in AAS always have a finer droplet spectrum than predicted by the equation 4.6 [1051]. In some cases the equation even predicts false trends; for example, for the use of organic solvents the role of the viscosity is exaggerated while the role of the surface tension is underestimated in the determination of the mean droplet diameter [1051, 4185]. It must be emphasized that for all observations with respect to the influence of the aerosol on the signal it is the tertiary aerosol that is decisive, whereas equation 4.6 is modelled on the properties of the primary aerosol.

The requirements placed on an ideal aerosol are not met by pneumatic nebulizers in a number of points. The primary aerosol generated by the nebulizer generally has a wide range of droplet sizes and in the time available the larger droplets cannot be completely vaporized and atomized in the flame. Furthermore, in pneumatic nebulizers the aerosol is generated with a large excess of energy that must be reduced by calming the turbulent gas streams if negative influences on the flame stability, and thus the S/N ratio, are to be avoided. This 'conditioning' of the aerosol is performed in the spray chamber with its various inserts or baffles; post nebulization or separation of large droplets takes place, the aerosol is thoroughly mixed with the flame gases, and partial vaporization of the solvent occurs.

Few systematic investigations have been made to date on these processes and on the influence of the design of the spray chamber and its inserts [889, 1354, 1358, 2291, 2340, 4404, 4405, 5615]. Usually the design of the spray chamber and the inserts is the result of empirical trials. For the separation of the droplets, sedimentation, impact, or turbulent separation via flow disruption are applied to varying degrees. Moreover the range of droplet sizes of the primary aerosol changes constantly due to mutual interactions (collision, agglomeration) and interactions with the gas phase (vaporization, condensation).

A whole series of possibilities is theoretically conceivable for the transport of the aerosol through the spray chamber. From investigations carried out by SMITH and BROWNER [5463] on solutions of varying salt content, it could be concluded that the primary aerosol at the moment of production vaporizes very rapidly and that spontaneous saturation of the gas stream with water vapor takes place. As long as no external source of heat is present (such as a heated spray chamber), no further solvent from the aerosol droplets on their passage through the spray chamber or from the walls of the chamber can vaporize due to the spontaneous saturation by water vapor. This means that the filter function of the spray chamber cannot be influenced by properties of the measurement solution, such as vapor pressure.

The tertiary aerosol that finally reaches the flame should not contain any droplets greater than 10 μm. However, the discrimination varies according to the type of spray chamber and values in the range 2–15 μm are obtained, depending on the type of baffle used [2340, 5464]. Very frequently *impact beads* are used in spray chambers. These serve to fragment larger droplets through the high velocity impact with the surface of the bead and also to separate droplets from the aerosol stream that were not sufficiently reduced through post nebulization.

The extent of these processes depends among other things strongly on the distance of the impact bead from the nebulizer jet. Above a certain impact velocity, the collision of a liquid droplet with a fixed surface causes the droplet to disrupt. This process depends on the ratio of the inertial forces to the surface forces and is characterized by the 'Weber number (We)'; it is determined by the surface tension, the density, and the impact velocity of the droplets. A distinction must also be made whether the liquid wets the impact surface of not. A dry impact surface requires high We numbers for the disruption of the droplets, while wetted surfaces require smaller We numbers or lower impact velocities since the formation of droplets takes place in part from the liquid film. The disruption limit for the impact of a droplet with diameter d on an impact surface which is covered by a liquid film of thickness δ is, for a relative film thickness of δ/d:

$$\text{We} = 2.5 \cdot 10^3 \left(d\rho\sigma/\eta^2 \right)^{-0.2}, \tag{4.7}$$

where $\text{We} = d\rho v^2/\sigma$, the dimensionless Weber number, v is the impact velocity of the droplet with the surface in m/s, ρ the density of the liquid, and d the diameter of the droplet in μm.

The range of droplet sizes produced upon impact is still the subject of investigation. Only when the impact bead is very close to the nebulizer jet (1–2 mm) are the droplets produced by secondary fragmentation sufficiently small. Albeit at this distance the separation rate is the largest, so that a suitable range of droplet sizes can only be obtained at the cost of aerosol consumption. The most efficient use of the aerosol is obtained with the impact bead at a distance of about 1 mm from the nebulizer jet [6345]. Since the rate of aspiration of a concentric nebulizer is also dependent on the rate at which the aerosol is transported away, the position of the impact bead will influence the aspiration rate when it is closer than about 0.5 mm from the nebulizer [6345].

The extent of the separation of liquid at an impact body is also dependent on its shape; spherical bodies are the most suitable for the dual task of separation and post nebulization. The discrimination of the separation is less steep for impact beads than for plane surfaces. *Plane impact surfaces* reduce the utilization of the aerosol strongly and also lead to a degradation of the signal stability due to increased turbulence in the spray chamber; their use brings no advantages compared to impact beads.

If the portion of the aerosol to be utilized should be reduced, for example to introduce solutions with high analyte concentrations or organic solvents with high vapor pressures, an *impact cup* can be used with advantage [6345]. It allows the portion of aerosol utilized to be varied in the ratio 1:30, while impact beads only allow the ratio to be varied by about 1:5 [4406, 6345]. The range of droplet sizes is improved due to the removal of the larger droplets and this reduces interferences in the flame [1853, 2340, 4405, 4695, 5464, 6341], which is an advantage compared to other measures of reducing the sensitivity, such as turning the burner head, dilution, or the use of secondary lines (see Section 5.3.3). Simply reducing the aspiration rate has the advantage that it reduces the consumption of sample, but at the cost of sample throughput since a reduction of the aspiration rate leads to an increase in the equilibration time [6345].

The larger droplets can also be removed from the aerosol stream by deflecting the gas stream in the spray chamber by means of *flow spoilers* since as a result of their inertia the larger droplets cannot follow rapid changes of direction of the gas stream. In addition, flow spoilers enhance mixing of the aerosol with the flame gases, leading to an improvement of the flame stability. Even though the use of flow spoilers reduces the utilization of the aerosol and thus lowers the sensitivity, the improvement in the stability and range of droplet sizes leads to an increase in the S/N ratio and to a further reduction of interferences.

The range of droplet sizes of the tertiary aerosol that finally enters the flame is very difficult to determine experimentally. Since the sample introduction system, comprising nebulizer, spray chamber and burner, forms a complete entity, small changes lead to a change of the total characteristics [889, 2291, 4404]. Determining the range of droplet sizes when the flame is not ignited, or even without a burner head, does not necessarily represent the situation in analytical operation.

The total effectiveness of the system, referring to the fraction of the tertiary aerosol compared to the total sample quantity, is relatively low as a result of the unfavorable range of droplet sizes of the primary aerosol, and only exceeds values of 5% for low sample aspiration rates of under 2 mL/minute. This low efficiency of the system was the driving force for numerous investigations in the area of sample introduction; they led to gradual improvements, especially with respect to interferences, but did not markedly change the total efficiency. The greater part of the measurement solution (95%) is conducted from the spray chamber via a drain to a waste receiver. Since the spray chamber must be closed to prevent the escape of the flame gases, the drain must be fitted with an interlock (siphon or similar). It is important that the outflow of liquid from the spray chamber is uniform to prevent oscillations in the liquid level in the chamber which could lead to pressure fluctuations and thus to a noise component.

For safety reasons the volume of the spray chamber should be small so that in the event of a flashback of the flame a dangerous explosion cannot take place in the chamber. To prevent destruction of the chamber or other damage in the event of a flashback, some manufacturers include bursting disks or the like. Nowadays, however, modern atomic absorption spectrometers have electronically controlled gas supply systems that reliably prevent a flashback of the flame under virtually all operating conditions.

4.1.3 Special Introduction Techniques for the Measurement Solution

During the last two decades numerous and very varying attempts have been made to circumvent or eliminate the well-known problems of pneumatic nebulizers. Typical starting points were, for example, the relatively high consumption of measurement solution before a stable signal is attained and the double function of the nebulizer as both transport pump for the measurement solution and aerosol generator (this double function is very welcome but not very efficient). It would be outside the scope of this monograph to mention all of these investigations in detail, especially as many of them were made for special applications. Various investigations on different types of nebulizer have already been mentioned briefly in the preceding section. In this section we shall describe a number of selected techniques that have, for one reason or another, gained a degree of importance.

4.1.3.1 The Injection Technique

SEBASTIANI et al. [5219] introduced the flame injection technique in 1973 with the aim of analyzing measurement solutions with high salt concentrations over long periods of time without negatively influencing the function of the nebulizer and burner. In this technique a fixed sample aliquot of typically 25 µL to 500 µL is injected into a small plastic funnel connected to the sample capillary of the nebulizer. The principle of this technique is depicted in Figure 4-9. The authors found that despite the loss in sensitivity due to the small sample volume the absolute limit of detection increased up to threefold.

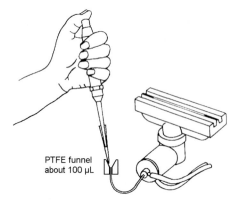

PTFE funnel
about 100 µL

Figure 4-9. Principle of the Injection Technique (from [579]).

In subsequent years the technique was investigated especially by BERNDT et al. [577–579] and a number of other authors [1995, 2773, 3855, 5902] and in particular applied to microanalysis. The practicable lower limit for the volume of measurement solution was found to be 50–200 µL at which relatively good sensitivity and a usable S/N ratio were attained. Like Sebastiani et al. other authors also found noticeably improved absolute sensitivities and limits of detection [577, 1995], and also higher efficiency of nebulization when small volumes are sprayed [623]. FUTEKOV *et al.* [2028] even reported an overall improved S/N ratio and lower limits of detection.

BEINROHR [476] made theoretical and experimental comparisons between continuous aspiration of a measurement solution and the injection technique. He found that the advantages of the injection technique with respect to the improvement in the limits of detection were not as unambiguous as generally assumed. Rather more, Beinrohr found that for the analysis of a given solution with a determined absolute quantity of analyte there is an optimum dilution and an optimum sample volume that give the best precision. In many cases dilution leads to better results than the injection of a small volume of a concentrated solution.

4.1.3.2 The Use of Pumps

The seemingly obvious solution to the problems arising from the double function of the nebulizer as transport agent for the measurement solution and aerosol generator is the use of pumps. If the measurement solution is introduced into the nebulizer the sample matrix can influence only the nebulization itself, i.e. the range of droplet sizes. The influence of a change in viscosity on the range of droplet sizes, for example, is far less than on the rate of aspiration.

MUNY [4241] controlled the flow of measurement solution to the nebulizer by applying a slight excess pressure to the sample container from a compressed air line. JONES [2934] used a syringe mounted on an infusion pump to attain a constant flowrate. FRY and DENTON [1994, 1997] applied a peristaltic pump and a Babington nebulizer to nebulize highly viscous solutions such as blood, urine, and tomato sauce. ROHLEDER *et al.* [4934] found when using peristaltic pumps that it was possible to have measurement solution flowrates that were significantly lower or higher than the aspiration rates attainable with a nebulizer. In this way they were able to vary the sensitivity by more than an order of magnitude. Nevertheless, pumps of this type exhibit pulsation which can have an influence on the measurement signal [3616].

4.1.3.3 Flow Injection

Pumps, mostly peristaltic pumps, are used to propel liquids in flow injection (FI) techniques. Nevertheless, there is a significant difference between the continuous transport of the *measurement solution*, as described in Section 4.1.3.2, and the continuous transport of a *carrier solution* into which a small volume of the measurement solution is injected. If the measurement solution is pumped continuously to the nebulizer a reduction in the sample throughput compared to manual operation will invariably result. The reason is to be found in the substantially longer tubes which must first be filled with solution and then be flushed free again after the measurement. Furthermore, air which enters the tube at a change of measurement solution causes an alteration in the flowrate and influences the stability of the flame. An additional delay time is thus necessary until stable conditions for the measurement have again been attained [4846]. With FI systems, which are discussed in detail in Chapter 6, the measurement frequency is significantly higher than with manual operation since markedly smaller volumes of measurement solution can be employed. Since the measurement solution is injected into a continuously flowing carrier stream, the problems of instability of the flowrate and the flame do not arise.

Using an optimized FI system, FANG et al. [1825] were able to obtain the same sensitivity with only 65 µL of measurement solution as for continuous aspiration of the same solution. Using such a system it would theoretically be possible to perform more than 1000 determinations per hour. At the same time the absolute limit of detection for lead could be improved by a factor of more than 25. Another advantage in using such small volumes of measurement solution in a carrier stream is the possibility of measuring solutions with high salt contents over long periods of time without the usual encrustations on the nebulizer or burner [1822]. FANG et al. [1825] were able to analyze a saturated lithium borate solution, for example, at a frequency of 360/h for more than an hour with a relative standard deviation of better than 1%.

FANG and WELZ [1821] showed that the flowrate of the carrier solution could be markedly lower than the aspiration rate of the nebulizer. As illustrated in Figure 4-10, better sensitivity could be attained than with conventional aspiration of the solution. Despite the small volumes of the measurement solution the precision and limits of detection had practically identical values to those attained with continuous aspiration. The volume and design of the sample coil (knotted reactor, simple loop), however, have a marked influence on the attainable sensitivity and precision [1825, 3759].

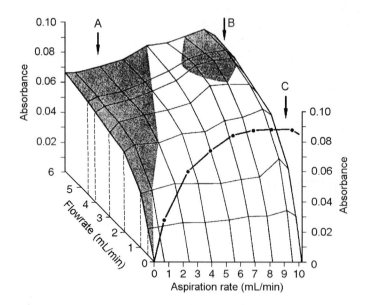

Figure 4-10. Influence of aspiration rate of the nebulizer and flowrate of the carrier solution on the absorbance signal for 10 mg/L Pb; injected volume 430 µL. The shaded areas **A** and **B** indicate conditions under which the FI signal is larger than for continuous aspiration, which is shown in curve **C** (from [1821]).

4.1.3.4 Hydraulic High Pressure Nebulization

Particular mention must be made at this point of a nebulizer proposed by BERNDT [592] that uses a single nozzle. This special nebulizer nozzle has an aperture in the micrometer range (10–30 µm) and carrier liquid (water or an organic solvent) is pumped continuously from a high pressure pump. The measurement solution, in a similar manner to the FI technique, is injected into the high pressure liquid stream via an injection valve. At typical flowrates of 1–2 mL/min, a working pressure of 10–40 MPa (100–400 bar) is generated at the nozzle, depending on the viscosity of the measurement solution. A filter with a pore size of about 3 µm is interposed in front of the nozzle to prevent blockage due to possible particles in the measurement solution. As a result of the high pressure drop at the outlet of the nozzle the turbulent liquid jet spontaneously forms an aerosol of very fine droplets. To prevent droplets recombining again and to attain high utilization of the aerosol, the resulting low pressure aerosol jet must be expanded to an aerosol cloud by impact with a surface (see Figure 4-11). This aerosol can then be mixed with the flame gases in a slightly modified spray chamber and conducted to the burner head. Since the flame gases are merely used to transport the aerosol but not for its generation, the flame stoichiometry can be optimized virtually independent of aerosol production.

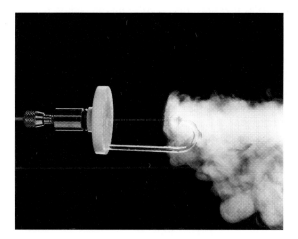

Figure 4-11. Expansion of the aerosol jet generated by a hydraulic high pressure nebulizer at an impact bead (by kind permission of Knauer).

BERNDT *et al.* were able to show that the utilization of the aerosol was largely independent of the viscosity of the measurement solution and that highly viscous liquids [594, 600] and solutions with high salt contents [2833, 4694] could be successfully nebulized. Despite the low flowrate compared to pneumatic nebulizers an aerosol mass transport rate of up to 70% higher can be obtained due to the improved efficiency of nebulization and thereby a concomitant improvement in the sensitivity [592]. Depending on the analyte and the operating conditions, the enhanced exploitation of the aerosol of up to 50% [592] leads to an improvement of the detection and quantitation limits by a factor of up to five compared to conventional concentric nebulizers. Since the greater exploitation of the aerosol does not depend on a larger fraction of coarser droplets, higher tolerance to potential interferents has frequently been observed [593, 4695, 4696]. A further development is the high-temperature, high-pressure nebulizer in which the liquid is heated in the sample capillary so that it is largely spontaneously vaporized on exiting the jet [601, 602]. An impact surface is thus no longer required for the generation of the aerosol. Since sample introduction is by means of a high pressure pump, this type of nebulizer is particularly suitable for the analysis of highly viscous liquids, such as oils [600], and also for the direct coupling of HPLC with F AAS and hence for speciation analysis [596, 597, 599, 602, 4697, 6179–6181].

4.2 Electrothermal Atomization

Although the graphite furnace technique developed by L'vov was made known only three years after the rediscovery of AAS, it took nearly ten years longer than with F AAS before the first commercial accessory for GF AAS was introduced onto the market. One reason for this was certainly the fact that the need for improved limits of detection by two to three orders of magnitude was not discernible until the end of the 1960s. Thereafter a similarly rapid development set in as had been observed in the preceding decade for F AAS (see Section 1.7).

For electrothermal atomization a measured volume of the sample solution, usually 10–50 µL, is dispensed into the atomizer and the temperature is increased stepwise to remove the solvent and concomitants as completely as possible before atomization. Since the entire aliquot introduced into the graphite tube is atomized within a short time (typically 1 s), a peak-shaped, time-dependent signal is generated whose area (integrated absorbance) is proportional to the mass of the analyte in the measurement solution.

Despite the rather harsh criticism that L'vov made about the 'method of simplification' that Massmann introduced (refer to Section 1.7.1), virtually all commercial electrothermal atomizers are based on Massmann's principle. However it must also clearly be said that the majority of efforts put into GF AAS after its introduction were aimed at making the performance characteristics approach more and more those of the L'vov graphite cuvette. The following requirements can thus be placed on an ideal electrothermal atomizer:

i)   At the moment of atomization of the sample the absorption volume should be iso-thermal with respect to both space and time, and its temperature should not change during atomization.

ii)   The substrate of the atomizer and the gas phase should support the atomization (e.g. by reducing oxides) but should not react with the analyte atoms.

iii) The atomizer should provide high absolute sensitivity. This requires rapid atomization and a long residence time in the absorption volume. The tube dimensions (length, diameter) also have a decisive influence.

iv)   The relative sensitivity of the atomizer should also be high. The atomizer must be able to accept a sufficiently large volume of measurement solution to meet this re-quirement.

v)   Of even greater importance than the sensitivity is the S/N ratio. This alone deter-
     mines the precision, the detection and quantitation limits, and the extension of the
     lower linear measurement range. An improvement of the S/N ratio is to be preferred
     to an improvement in the sensitivity (absolute or relative).

vi) The atomizer should emit minimal radiation and should not attenuate the primary
     radiation (S/N ratio).

vii) Within the atomizer, concomitants should exercise the lowest possible influence on
     the analyte atoms.

viii) No residues from the analyte or the concomitants should remain in the atomizer
     after a determination.

ix) The total time per determination should be as short as possible to allow a high sam-
     ple throughput.

x)   The power requirements of the atomizer should be within reasonable limits.

xi) The atomizer (graphite tube) should have the longest possible lifetime and should
     change as little as possible during this time.

   The above requirements are naturally maximum requirements that are not fulfilled by
any atomizer. These requirements are in part contradictory, such as high sensitivity and
low power consumption. Also, a number of the requirements have 'natural' limits since
the apparatus must be operable. Nevertheless, in the following sections we shall examine
the various proposals described in the literature with respect to the above criteria, and
especially how close they come to the ideal time-resolved and spatially-resolved isothermal atomizer. STURGEON [5644] has published an overview on the application of
various graphite furnaces in atomic spectroscopy.

4.2.1 Graphite Structure and Reactivity

Discounting platforms and probes made of total pyrolytic graphite (TPG), nowadays
atomizers are made largely of polycrystalline electrographite (EG) with a typically
50 μm thick coating of pyrolytic graphite (PG). The other graphite and carbon materials
mentioned in Section 1.7.4 are only used very sporadically for special applications or
mechanistic investigations.

   The graphite crystal is an allotropic form of carbon formed by sp^2 hybridization of
the orbitals. The σ electrons form strong covalent bonds in the basal plane, while the π
electrons form weak van der Waals bonds between the planes. The lattices are arranged
largely in the hexagonal ABAB structure system. The hexagonal graphite structure is
depicted schematically in Figure 4-12 in the form of an ideally ordered monocrystal and
also in the form of a 'real' disordered crystal. In the latter the lattice is folded and the
distance between the layers is irregular; within the layers there are flaws. All these flaws
are possible points of attack for oxidizing reagents.

   In addition to the flaws within the lattice all atoms located at the ends of the basal
planes are also in a disordered and incompletely coordinated state. Distinction is made
between 'arm chair', 'zig-zag' formations, and 'fragments' as depicted schematically in
Figure 4-13. All these formations differ in their reactivity. Various authors have shown
that the attack by molecular oxygen takes place exclusively at boundary carbon atoms

[2502, 5822]. TREMBLAY *et al.* [5887] found that there were two significantly different reaction velocities for the desorption of CO from oxidized graphite; they ascribed these reactions to different carbon formations. HUETTNER and BUSCHE [2706] have published a survey on reactivity, structural aspects, and oxidation behavior.

a)   b)

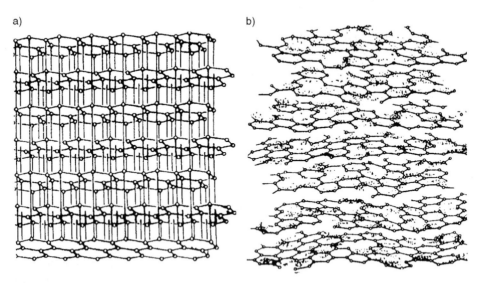

Figure 4-12. Graphite structure. a – perfect graphite lattice, b – schematic model of disordered graphite (from [696]).

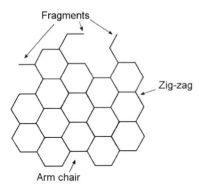

Figure 4-13. Schematic diagram of the graphitic basal plane, illustrating various possible active sites on the surface.

Polygranular graphites, such as coarse- and fine-grained polycrystalline EG, consist of graphite grains held together by a binder of amorphous carbon, as depicted in Figure 4-14. The grains (also termed primary carbon) consist of graphite crystallites that extend between 1 nm and 60 nm in the L_a crystallographic direction. Micropores with dimensions of less than 2 nm are to be found within the grains, while macropores with dimensions up to the millimeter range can be found in the carbon binder bridges. Normally, however, these pores are much smaller, especially in the high density graphite used in GF AAS.

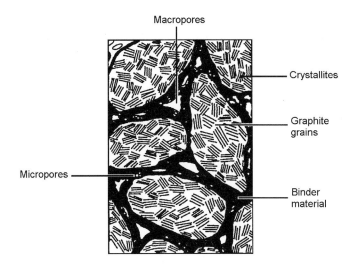

Macropores

Crystallites

Graphite grains

Micropores

Binder material

Figure 4-14. Schematic model of a polycrystalline EG consisting of graphite grains and binder material (from [4055]).

a)

b)

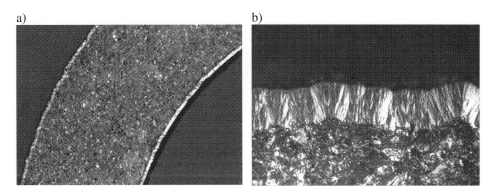

Figure 4-15. Polished section through an EG tube with a 50 μm coating of PG viewed in polarized light. **a** – 50x magnification, **b** – 500x magnification (by kind permission of Ringsdorff-Werke GmbH).

The inner surface of an unused graphite tube made of uncoated EG has already been shown in Figure 1-34 (see Section 1.7.4). The 'open' structure of this type of graphite is obvious and it is therefore hardly surprising that EG, particularly at higher temperatures, is rather reactive, making it unsuitable for GF AAS. Numerous investigations were thus made in earlier years to protect and seal the surface of EG using various coatings, as described in Section 1.7.4. Of these various coatings, nowadays only PG is used since it can be reproducibly made with high reliability, as the sections in Figure 4-15 indicate.

Due to the extremely low permeability of PG, a coating of this material markedly reduces the reactivity of the surface of EG. All open pores of the EG substrate are sealed, and the susceptibility of PG to oxidation vertically to the planes is low due to the high crystalline order. Nevertheless the resistance of PG to oxidation depends on its lattice orientation and on the direction of the oxidative attack. The lattice orientation is a function of the deposition parameters, such as temperature, pressure, and nature of the hydrocarbon gas.

The structure and appearance of a PG coating depend to a large degree on how the sample is prepared and viewed. If a tube is cut and polished, the fine grain structure of EG and the regular crystal growth of PG are visible, particularly when viewed in polarized light (Figure 4-15). If, on the other hand, a tube is broken mechanically, the layer structure can be seen, as shown in Figure 4-16. However, the recognizable layers in this illustration, with separations of several tens of nanometers, are not in fact individual graphite layers but larger 'packets' of layers.

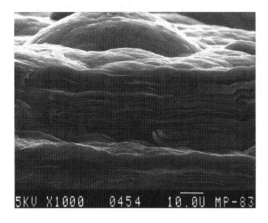

Figure 4-16. Scanning electron micrograph of a PG coating on a deliberately broken graphite tube, magnification 1000x (from [4482]).

The processes that lead to breakage of a graphite tube are also of interest. Tubes that have the electrical contacts at the ends and a sample dispensing hole in the middle break regularly at this position around the entire tube circumference. A thinning of the substrate material is immediately obvious, and this is frequently accompanied by nodular deposits in the direct vicinity. The mechanism of this attack becomes clear in Figure 4-17 by scrutinizing the lower half of a graphite tube with inserted platform that has not quite fully disintegrated. The substrate material at the center of the tube has become porous over its entire width, without however losing much of its form or strength, while nodular deposits have formed on the adjacent platform. This observation suggests the following mechanism: It is known that EG is stable up to about 3000 °C and that sublimation increases markedly above 3200 °C. EG consists of polycrystalline graphite grains that are held together by a binder. If a high current is applied to the ends of the tube, a somewhat higher resistance can arise at the boundary layer between the crystalline grains and the virtually amorphous binder, leading to local overheating. Under the considerable thermal stress caused by repeatedly heating the tube rapidly to temperatures of, say, 2650 °C, this local overheating leads to volatilization little by little of the graphite binder and the smaller graphite grains. After several hundred firings only the coarser graphite grains remain. As to be expected, this overheating effect is strongest at the middle of the tube since the cross-section is about 10% less due to the sample dispensing hole and the resistance is thus higher. Normally the thinning of the tube is restricted to a few millimeters, corresponding roughly to the diameter of the dispensing hole. The nodular carbon deposits from the gas phase onto the adjacent platform are a further indication for this mechanism.

a)   b)

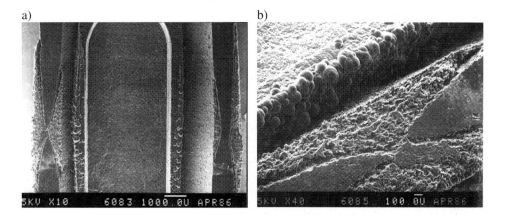

Figure 4-17. Lower half of an EG tube with PG coating and inserted TPG platform after 350 atomization cycles with 1 mg phosphorus. **a** – center part of the tube with inserted platform, magnification 10x; **b** – detail of the tube wall with thinning and coarse grain graphite, and the outer platform wall with nodular deposits of PG caused by deposition from the gas phase (from [6229]).

4.2.2 Graphite Atomizers

4.2.2.1 Dimensions of Graphite Tubes

The dimensions and shape of graphite tubes used in GF AAS, especially in the early years, exhibited significant differences, as shown in the overview in Figure 4-18 (which naturally is incomplete).

At an early stage a number of working groups investigated the ideal dimensions for atomizers in GF AAS. Like L'VOV [3690] before them, STURGEON and CHAKRABARTI [5622] showed that the height of an absorption signal measured in an atomizer depends on the density of the atom cloud. The density is influenced by the velocity of atomization (i.e. the rate of generation of the atoms from the sample) and the residence time of the atoms in the graphite tube (i.e. the rate of loss of atoms from the absorption volume).

If the loss of atoms due to a forced gas stream is temporarily neglected, then the most important factor is the loss of atoms from the tube due to diffusion. According to equation 4.8 the relative mass loss of atoms is inversely proportional to the square of the tube length ℓ and independent of its diameter (D = diffusion coefficient):

$$\frac{\mathrm{d}\,M}{M} = \frac{8D}{\ell^2}\,\mathrm{d}\,\tau\,. \tag{4.8}$$

Figure 4-18. Overview of a number of the graphite tubes used in GF AAS (by kind permission of Ringsdorff-Werke GmbH).

This means that the length of the tube plays a decisive role in the residence time of the atoms in the absorption volume. To what extent other transport mechanisms, such as convection currents or displacement due to expansion [93, 159, 1143, 2113, 2259, 2640], play a role next to diffusion depends on the design of the atomizer [1977, 2313, 2962] and on the experimental conditions, such as the mass of sample vaporized and the time point of atomization. From equation 4.8 one can derive that the mean residence time τ of the atoms in the absorption volume is directly proportional to the square of the length ℓ of the graphite tube:

$$\tau = \frac{\ell^2}{8D} .$$

(4.9)

Thus, to achieve the longest possible residence time and thereby optimum sensitivity, the graphite tube should be as long as possible. However, the tube length cannot be increased at random, because with increasing length the energy requirement for heating also increases.

As well as the tube length the tube diameter also plays a certain role. Admittedly it does not influence the diffusion, and thereby the residence time of the atoms in the tube, but since the volume increases as the square of the diameter, the cloud of atoms will be correspondingly diluted. In other words, the absolute sensitivity in a graphite tube furnace is inversely proportional to the tube diameter. Therefore, a long, narrow tube is required for good *absolute sensitivity*.

However, since a larger graphite tube can normally contain a larger sample volume, the *relative sensitivity* (i.e. the sensitivity in concentration units) is independent of the tube diameter in the first approximation. The same proportionality is valid for the tube length as for the absolute sensitivity, i.e. the requirement for a long tube.

GÜELL and HOLCOMBE [2259, 2260, 2264] made a series of model calculations on the theme of tube dimensions. They found that the geometry of the atomizer, i.e. the interaction of length, cross-section, and size of the sample introduction hole, played a key part in the optimization of GF AAS. The tube dimensions determine the highest atom cloud concentration during atomization just as much as the surface that is available for desorption and adsorption processes between the graphite and the gas phase. The tube dimensions can also have an influence on the radiation flux that passes through the atomizer and reaches the detector. It is not possible to make simple predictions, since changing one parameter almost invariably causes a change in another. If the length is changed the heating characteristics and the radiation flux to the spectrometer are also usually changed. The most important conclusions from these model calculations are summarized below:

i)   The integrated absorbance (peak area) increases linearly and almost proportionally with increasing tube length.

ii) The absorbance (peak height) also increases with increasing tube length, but not so markedly as with integrated absorbance.

iii) An increase in the tube diameter causes a larger fraction of the atoms to stay in the gas phase (fewer are adsorbed) but at the same time lowers both the absorbance and the integrated absorbance.

iv) A substantial proportion of the gaseous sample escapes through the introduction hole. This proportion increases with increasing length and decreasing diameter of the graphite tube.

v)   The proportion that is lost through the introduction hole is higher than predicted from a simple comparison between the areas of the introduction hole and the openings at the end of the tube. The reason for this appears to be that the position in the tube onto which the sample is deposited is closer to the introduction hole and the atoms thus reach this hole first.

vi) A longer atomizer with a narrower diameter would substantially increase the signal, but would also increase the (negative) influence of the sample introduction hole. Geometric changes of this nature must be accompanied by a decrease in the size of the sample introduction hole.

vii) The calculated improvement in the sensitivity due to a longer atomizer with a narrower diameter can only be attained if other factors such as heating rate and non-isothermal behavior do not also change.

These authors came to the conclusion that the optimum diameter for an atomizer for GF AAS is approximately 5 mm, while the optimum length is maximum 40 mm. The majority of commercial atomizers have diameters that are very close to the optimum, but are shorter. The main reason for this is the higher power consumption required for the greater graphite mass of a longer tube, and the longer heating times that would decrease tube lifetime. It is important to point out that these calculations were made for *atomization from the tube wall* and for graphite tubes that had *electrical contact at the tube ends*.

The question of the influence of the *openings in the atomizer* (which are naturally indispensable) on the loss of analyte atoms and how this can be minimized was investigated by other working groups [1811, 5750]. In the absence of a forced gas stream the majority of the analyte atoms escape through the openings in the graphite tube [855, 3690]. In L'VOV's model [3690] the sample introduction hole was not taken into account since the sample is introduced on an electrode that seals this opening during atomization.

L'vov also made the proposal to further narrow the tube ends with graphite rings and thus increase the residence time of the atoms. The practical advantage of these end caps was nevertheless low since due to the design of the furnace the end caps were much cooler than the rest of the atomizer. Thus there was the risk that atoms could condense, leading to a loss in sensitivity.

SIEMER and FRECH [5346] later took up this idea and investigated small, transversely-heated graphite tubes of 7 mm length and 4 mm inner diameter, with and without sample introduction hole, masks, and platforms. End caps with apertures of 2 mm to allow the radiation beam to pass increased the effective gas temperature in the atomizer by 500 °C and doubled the integrated absorbance with lead as the analyte. An additionally introduced platform further increased both the temperature of the atomic vapor and the signal. The idea of the end caps was also again taken up by FRECH and L'VOV [1977] (see Section 4.2.2.5).

4.2.2.2 Profiled Tubes

Numerous working groups have attempted to reduce the problem of *spatial non-isothermal conditions* in atomizers with electrical contacts at the ends by changing the *profile of the tube*. This does not eliminate the cold tube ends, but it can extend the zone of higher temperature. The first tube of this type was introduced in 1972 [1885]. Internally it had a shallow cavity to prevent the uncontrolled spread of the sample along the length of the tube. In addition the wall was thicker in the middle to achieve a more even temperature distribution.

All subsequent design proposals followed the same principle, i.e. either the tube wall was thickened at the center so that it was not heated so strongly, or the tube wall was thinned at the ends so that it was heated more strongly. A contoured tube of this type is depicted in Figure 4-19 in comparison to a standard tube. The change in the temperature profile over the tube length obtained by contouring the tube is also shown.

Various authors followed with similar proposals and were able to report certain improvements in individual cases [76, 1990, 3864, 5498]. The temperature effect clearly had a stronger influence on atomic emission than on absorption [3539–3541, 4502]. HUMAN *et al.* [2724, 2725] treated the problem mathematically and developed a computer program with which they could calculate the temperature profile attainable for a given tube form. GÜELL *et al.* [2265] developed these calculations further and came to the conclusion that only very small improvements in the performance could be obtained by changing the profile of the tube. There is no doubt whatsoever that analyte atoms condense in tubes whose ends are colder, causing memory effects. Furthermore, only minimal improvements in the interference caused by the formation of gaseous molecules could be observed for contoured tubes that were *more isothermal* than standard tubes.

All of these investigations were performed with atomization from the tube wall. In the view of these authors the use of a platform has a much stronger influence on gas-phase interferences than the influence that can be obtained by changing the geometry of the graphite tube. The reduction of interferences is always dependent on the *retardation of heating* between the platform and the tube wall and on the type of chemical reaction, but hardly ever on the type of tube used.

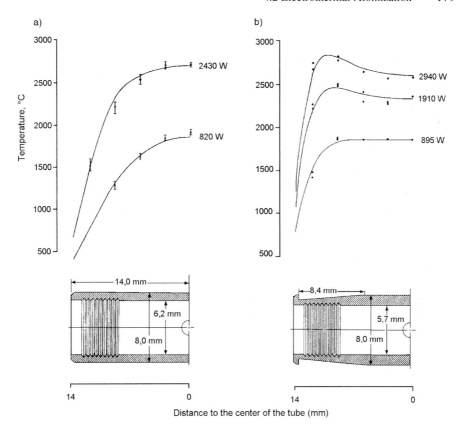

Figure 4-19. Graphite tube profiles and the associated temperature distributions. **a** – Standard tube and **b** – contoured tube for GF AAS (from [5417]).

4.2.2.3 Platforms

While the use of specially contoured tubes brought no noticeable improvements in the performance of GF AAS since they merely extended the hot zone of the tube, the proposal by L'vov [3704] to atomize from a platform placed loosely in the graphite tube was expressly aimed at improving the *temporal isothermal conditions* in the tube. Due to its minimal contact and the anisotropy of the TPG material, the L'vov platform is heated virtually only by radiation from the hot tube wall. As a result of the *retarded heating* of the platform, the analyte is not atomized until the tube and the gas phase have attained a largely stable final temperature. In this way quasi isothermal conditions can be obtained, even when the tube ends are still cooler (refer to Sections 1.7.1 and 1.7.2). A further advantage of platform atomization is that reproducible positioning of the sample solution is made easier and that the spreading of the solution can be more easily prevented than with wall atomization.

L'vov's original proposal of placing a platform loosely in the tube did not prove to be suitable in practice because it is difficult to position the platform reproducibly. Loose platforms can be moved by the adhesion forces to the pipet tip during sample dispensing or by the vibrations caused by the electromagnet of a Zeeman system. Because of these difficulties, numerous attempts were made to secure the platform reproducibly in the tube. It is particularly important to prevent an electric current from passing through the platform and heating it actively since this would negatively influence the 'platform effect', i.e. the retarded heating of the platform compared to the tube.

FRECH *et al.* [1973] made comparative investigations on a number of platform designs, as depicted in Figure 4-20. The platforms differ in the way in which they are secured in the tube. The standard platform is fixed into grooves in the graphite tube; the T-platform rests in slots at the end of the tube; the fork platform rests loosely on two tapered rings and is clamped by the prongs at the end of the tube; and the pin platform is fixed in a small hole drilled in the bottom of the tube wall but does not otherwise contact the tube.

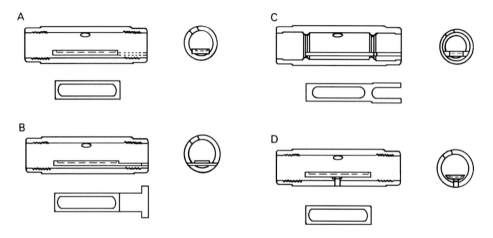

Figure 4-20. Schematic diagram showing four designs of platform. **A** – standard platform, **B** – T-platform, **C** – fork platform, and **D** – pin platform (from [1973]).

From these investigations FRECH *et al.* [1973] did not come to any conclusive results since the various platforms behaved very differently according to the given tasks. The pin platform led to very high gas-phase temperatures but did not provide the required freedom from interferences in the presence of higher concentrations of concomitants. This was not due to its design but rather to its material since it could only be made of glassy carbon and not of TPG. The forked platform provided the highest sensitivity and the best precision, and it also exhibited the greatest retardation in atomization. In the presence of high matrix contents however the T-platform brought even better results. LI *et al.* [6541] also carried out a comparative test using the standard, the forked, and also a loose platform. The forked platform brought the best results. In a platform test on the determination of lead in blood, SHUTTLER *et al.* [5325–5327] established that the contact between the platform and the tube was of decisive importance for the reproducibility of the results.

The greatest technical problem in the platform technique would appear to be the requirements for a *stable, reproducible fixing* of the platform in the tube and *minimal contact* with the tube. The cause of this problem lies in the action of the platform, which is based on the retardation of heating compared to the tube wall and the gas phase. The less the platform is in contact with the tube, the greater is the retardation; contact in the direction of current flow must be avoided at all costs since this would lead to active ohmic heating of the platform. However, the smaller the contact, the less stable is the fixing of the platform. Each of the platforms depicted in Figure 4-20 is thus a compromise between stable fixing and optimum retardation.

An essentially optimum solution to this problem was achieved by making a tube with integrated platform from a single piece of EG and coating it with PG; similar to the pin platform, the platform is attached to the tube at only one point as shown in Figure 4-21. The requirement for minimal contact and the prevention of the direct flow of current through the platform is thus met in an almost ideal manner. In addition, very high mechanical stability is attained through the shape of the platform and, especially, the fact that the tube and platform are manufactured from a single piece of EG. The improvement in the signal shape obtained with an integrated platform compared to a standard L'vov platform can clearly be seen in Figure 4-22.

Figure 4-21. Graphite tube with integrated platform (by kind permission of Perkin-Elmer).

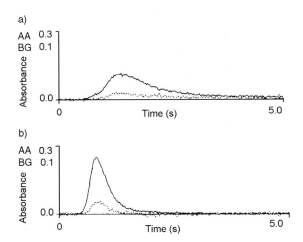

Figure 4-22. Absorbance signals for 400 pg Co for atomization from **a** – a standard L'vov platform and **b** – an integrated platform (by kind permission of Perkin-Elmer).

4.2.2.4 Probes

As an alternative to platforms with their delayed heating, a high level of thermal equilibrium during atomization can also be achieved by the use of probes which are inserted into the preheated atomizer. The initial investigations in this direction by SLAVIN *et al.* [3860, 3861] using a tungsten spiral are described in Section 1.7.1. Other authors also used similar tungsten probes [1680], and BERNDT and MESSERSCHMIDT [588], by additionally heating the probe, were able to determine even carbide-forming elements such as titanium, vanadium, and uranium with good sensitivity.

Later SLAVIN and MANNING [5423] used a graphite probe made of TPG, which they inserted into the preheated graphite tube, and reported improved limits of detection and less carryover for vanadium, titanium, and molybdenum. In subsequent years, OTTAWAY and co-workers [1087, 1088, 2135, 3542, 3544] in particular investigated probe atomization very thoroughly. They developed a large number of models, including tube-shaped probes [1088], but always used TPG as the material. These investigations are summarized in a review article [4503] in which the important characteristics of probe atomization are discussed. These include the direction in which the probe is inserted into the tube, and the dimensions, shape and material of the probe. The procedure of probe atomization is shown in Figure 4-23. The measurement solution is injected onto the probe in the graphite tube, dried and pyrolyzed. The probe is then withdrawn from the tube, the tube is heated to the atomization temperature, and then the probe is reinserted into the tube for atomization.

Unfortunately there are no comparative measurements between platform and probe atomization in the same furnace system, so it is difficult to judge the relative advantages. OTTAWAY and co-workers [3544, 4503] only compared probe atomization with wall atomization. Nevertheless, there are many indications that platform and probe atomization produce very similar results; the use of a platform is naturally easier since it remains

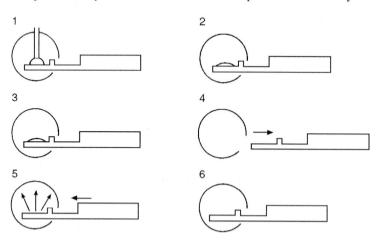

Figure 4-23. Procedure for probe atomization. **1** – inject measurement solution; **2** – dry; **3** – pyrolyze; **4** – withdraw probe, preheat furnace to atomization temperature; **5** – reinsert probe into the tube and measure the integrated absorbance; **6** – tube and probe cleanout (from [877]).

in the tube for the latter's lifetime and must not be moved. An ideal local thermal equilibrium is not obtained in either system, since the introduction of the cold probe into the preheated furnace also disturbs the prevailing local thermal equilibrium [1158]. Both systems also exhibit the same spatial non-isothermal characteristics when the electrical contacts are at the ends of the tube. The major disadvantages of probe atomization compared to platform atomization are the more complicated apparatus and the reduced sensitivity by up to 50%. The latter results from the increased diffusional losses of the atom cloud through the additional aperture required to introduce the probe into the tube.

4.2.2.5 Temporally and Spatially Isothermal Atomizers

The development of electrothermal atomizers was mostly characterized by the desire to provide an environment for the analyte atoms that was as isothermal as possible. Recent endeavors in this direction are compiled in a review article published by FRECH [1979]. As emphasized in Section 1.7, very few atomizers approach ideal temporal and spatial isothermal characteristics. Even L'VOV's [3690] graphite cuvette exhibited a temperature gradient along its length since the electrical contacts were at the tube ends. Without doubt one of the nearest approaches was the two-step furnace developed by FRECH et al. [1964], especially in the version used by LUNDGREN et al. [3667] in which the graphite tube and the graphite cup were provided with integrated contacts (refer to Section 1.7.3). The only problem with this atomizer was its complexity, which made it unsuitable for routine operation.

In a number of comparative investigations, FRECH and co-workers [553, 1970, 3667] found that a transversely-heated atomizer with integrated contacts, in which the sample is atomized from a platform, produced very similar results to a two-step furnace, in which the sample is volatilized from a cup having its own power supply. They found that the spatial thermal equilibrium, offered to the same extent by both atomizers, made the greatest contribution to the performance of either system. The major characteristics compared to atomizers having electrical contacts at the tube ends were low interferences due to concomitants, high peak symmetry (low tailing), and strongly reduced memory effects. The use of a platform clearly retards atomization to the necessary extent, so that comparable isothermal conditions prevail as when the sample is volatilized from an independently heated graphite cup.

The first commercial transversely-heated atomizer with integrated contacts, which is depicted in Figure 4-24, was introduced in 1990. This atomizer is made from a single piece of high-density EG and is coated with PG, and includes an integrated platform which has only one point of contact with the graphite tube. This thus meets the requirements stated in Section 4.2.2.3 for stable fixing of the platform with minimal contact with the graphite tube. This atomizer permits the use of lower atomization temperatures for all elements, and even difficultly volatilized elements like vanadium and molybdenum can be atomized from the platform without difficulties. BERGLUND et al. [553] were even able to determine elements with widely differing volatilities such as cadmium and molybdenum under the same atomization conditions using a suitable modifier. The special characteristics of these atomizers will be described in more detail in the following

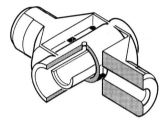

Figure 4-24. Transversely-heated atomizer with integrated contacts and integrated platform (by kind permission of Perkin-Elmer).

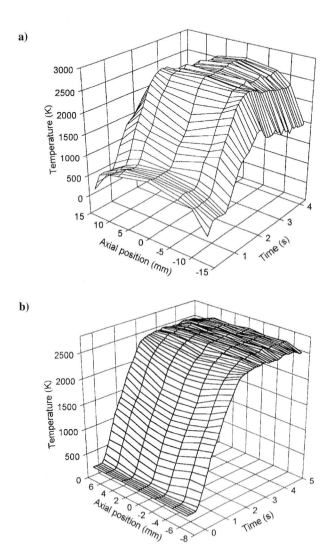

Figure 4-25. Temperature distribution in the gas phase over the length of the graphite tube during atomization. **a** – longitudinally-heated graphite tube; **b** – transversely-heated graphite tube with end caps (from [5520]).

sections. FRECH and L'VOV [1977] further proposed constricting the ends of the tube with end caps having apertures of 3 mm to further reduce diffusion losses and to extend the isothermal zone in the gas phase. On average, the sensitivity was increased by a factor of 1.7 for volatile elements, thus further increasing the detection capabilities [2617], and gas-phase interferences were further reduced.

Figure 4-25 shows how even the distribution of the gas-phase temperature is over the length of a transversely-heated graphite tube with end caps in comparison to a longitudinally-heated tube. During the rapid heating phase the gas-phase temperature in the middle of the tube is lower than at the tube ends due to the platform effect. This gradient disappears when the maximum temperature is attained and the temperature difference over the length of the tube is less than 50 K. In contrast, the temperature gradient over the length of a longitudinally-heated graphite tube can be up to 1500 K.

4.2.3 Heating Rate and Temperature Program

Heating rate is a subject that has already been mentioned repeatedly. L'VOV [3690] placed great emphasis on this parameter and made the stipulation that the rate of supply of atoms into the absorption volume must be much higher than their rate of loss through diffusion. As mentioned in Section 1.7.1, L'vov later amended this view and placed less emphasis on the heating rate, provided that integrated absorbance was used for signal evaluation [3683]. Investigations on solving the problems of non-isothermal conditions in the absorption zone of Massmann-type furnaces by utilizing the very high heating rates of capacitive discharge did not prove to be very successful [1139–1142, 1188]. It was found that even with this technique there were enormous temperature gradients [4008, 6249], but the major obstacle was without doubt the technically complex nature of the apparatus.

The heating rate continued to play an important role in subsequent years. During heating to the atomization temperature the gas in the atomizer expands a considerable amount and escapes through the various openings. If the sample is atomized from the *tube wall* it is volatilized during the heating phase and thus transported from the absorption volume by the expanding gas. A number of authors attempted to compensate this unavoidable disadvantage by applying very high heating rates [1889, 6576]. In this way they obtained higher (in absorbance) but very narrow signals. This effect will be discussed in connection with metal atomizers (Section 4.2.5). Nevertheless, the precision was influenced by the high heating rate and the problems of susceptibility to interferences were not solved.

A possible exception in this respect is the system proposed by LUNDGREN *et al.* [3669] in which a very volatile element can be determined virtually free of interferences in the presence of non-volatile concomitants. The sample is heated very rapidly to a temperature just above the atomization temperature of the analyte. Heating control is performed by an infrared detector that switches off the heating as soon as the atomization temperature is reached. In this way it was possible to determine cadmium, for example, in the presence of higher sodium chloride concentrations virtually without interferences. Other authors also reported the successful determination of cadmium and lead in matrices such as blood using this principle [1467, 4857].

The heating rate assumes a totally different function for *platform atomization*. As discussed in Section 1.7.2, the purpose of the platform is to retard atomization of the analyte until the graphite tube has attained its final temperature and the thermal expansion of the gas phase is practically finished. In this way the analyte is volatilized into a largely isothermal environment and losses due to premature removal are avoided. These conditions can be achieved more easily the higher the heating rate of the tube and the more strongly the heating of the platform is retarded.

A transversely-heated tube with integrated platform offers virtually optimum conditions for isothermal atomization. The temperature profile in the gas phase of an atomizer of this type has been recorded both spatially and temporally resolved using CARS thermometry (coherent anti-Stokes Raman spectroscopy). As can be seen from Figure 4-26, the temperature in the gas phase (which follows the surface temperature very closely) exhibits a maximum temperature difference between platform and tube wall of nearly 1000 K. By choosing a suitable starting temperature the maximum of the temperature difference is at the point where the graphite tube has not only attained its maximum temperature but also exhibits no marked temperature gradient. In the present example this would be an appearance temperature of 2000–2200 K.

It has been found as a further particularity of a transversely-heated tube with integrated platform that during the rapid heating phase a positive longitudinal temperature gradient forms along the platform (see Figure 4-25b). This means that during the heating phase the temperature of the gas phase at the ends of the platform is higher than directly over the platform. This enhances atomization of volatilized sample constituents and prevents their condensation on cold spots, since there are no cold spots present.

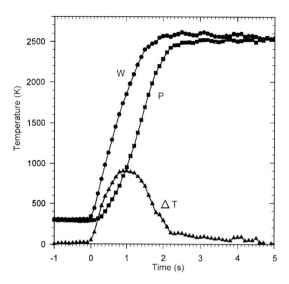

Figure 4-26. Temperature increase for the gas phase close to the platform (P) and the wall (W) of a transversely-heated graphite tube and the temperature difference ΔT between the platform and the wall, measured by CARS thermometry. The temperature difference is up to 1000 K, but disappears just after the tube wall has attained the maximum temperature (from [5520]).

For atomizers of the Massmann type, i.e. tubes with electrical contacts at the ends, the conditions are not quite so simple. In principle temperature profiles similar to those shown in Figure 4-26 form in the middle of the tube, but the temperature difference is mostly somewhat lower. Normally these atomizers exhibit an isothermal temperature distribution, at least over the length of the platform (see Figure 4-25a), so that in this case it is also correct to speak of time-resolved isothermal atomization. The freedom from interferences and the analytical performance confirm this. Nevertheless it must be taken into account that with all these atomizers there is a marked temperature profile over their length and that the analyte and concomitants can condense at the cooler ends. This leads to the often-described tailing of the atomization signal and the well-known memory effects for refractory and carbide-forming elements. A typical examples is shown in Figure 4-27 on the example of molybdenum.

a)

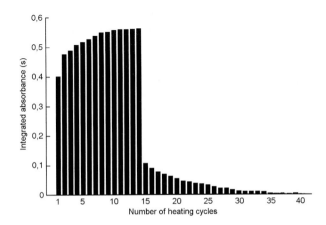

b)

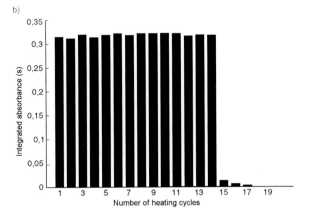

Figure 4-27. Memory effects for 0.8 ng Mo (signals 1–14), followed by a series of measurements on blank solutions in various atomizers. **a** – longitudinally-heated atomizer, atomization from the tube wall at 2650 °C; **b** – transversely-heated atomizer, atomization at 2500 °C from the platform (by kind permission of Perkin-Elmer).

FALK *et al.* [1809] attempted to smooth the marked temperature profile of tubes with electrical contacts at the ends, at least during the rapid heating phase. They found that for heating rates of > 1000 K/s the initial profile was retained for about one second, after which the cooling effect of the contacts became apparent. They therefore proposed that the tube should be allowed to cool to ambient temperature prior to atomization, since a flat temperature profile then prevails, and thereafter to heat rapidly. A number of analysts applied this cooling step routinely in their work [2522, 2611, 3864, 6125], but up to now there have been no reports on the convincing success of this technique. The reason may be that the main mass of the sample is volatilized at a time point at which the tube ends are already cooling so that condensation cannot be effectively prevented.

For *probe atomization* the heating rate of the atomizer is clearly of no significance since the probe is not introduced into the tube until it has attained its final temperature. This has the advantage that a power supply of lower performance and a less precise temperature control system are adequate. There is certainly the disadvantage, however, that at equilibrium the temperature profile over the tube length is very marked and thus tailing and memory effects are more pronounced. In addition, every probe has a cold end, so that even along the probe there is a temperature gradient that can also contribute to tailing and memory effects. For this reason a number of authors proposed actively heating the probe to accelerate atomization and to avoid condensation [588].

Prior to atomization the sample must normally be subjected to a *temperature program* in the graphite tube that serves to largely separate the concomitants. A high heating rate during this phase of the temperature program brings no advantages. The significance of the temperature program in preventing interferences due to concomitants is discussed in detail in Section 8.2.4. Typical program steps include drying, pyrolysis, atomization, and heating out.

4.2.4 Protective Gas and Purge Gas

Carbon reacts with oxygen at increased temperature with the formation of carbon monoxide or carbon dioxide, i.e. it burns. To prevent this reaction, graphite atomizers must be operated in a *protective gas atmosphere*. The majority of analyte atoms also have a high affinity for oxygen; they form oxides in the presence of oxygen, leading to a reduction of free atoms and thus to a loss in sensitivity. For these reasons graphite atomizers must be operated in an inert gas atmosphere and the ingress of atmospheric oxygen must be prevented as completely as possible.

Furthermore one should not forget that for every GF AAS analysis, large quantities (relative to the mass of analyte) of concomitants are volatilized in the atomizer. This should take place during the drying and pyrolysis steps so that volatile concomitants do not cause interferences during atomization. However, it is not sufficient just to volatilize the concomitants; they must be reliably removed from the absorption volume prior to the atomization step. For this purpose a *purge gas* stream that is independent from the protective gas stream has proved ideal; these gas streams for an atomizer unit are shown schematically in Figure 4-28. Atomizers with electrical contacts at the ends exhibit a temperature profile over their length, so it is especially important that the purge gas stream prevents the condensation of volatilized concomitants at cooler spots in the atomizer unit. The gas stream must therefore flow from the cooler spots to the hottest point. If condensation is not prevented, sample constituents can volatilize again during atomization, leading to substantial interferences.

The significance of the purge gas stream is made clear in Figure 4-29. Urine was used as the test substance and the background signal and the signal obtained with continuum source background correction were recorded at the lead line at 283.3 nm. The difference between the two sets of measurements is solely that for one set the purge gas stream from one end of the tube was reduced by half, so that the stream was asymmetric. The results show that in this case volatilized sample constituents were not successfully removed from the absorption volume, leading to interferences during atomization (in this case overcompensation).

The control of the purge gas stream might not be of such significance in an isothermal, transversely-heated atomizer, but nevertheless the rapid removal of volatilized concomitants should be aimed for. The more rapidly this can be achieved, the shorter can be the individual program steps and thus the higher the sample frequency. One should also bear in mind that even though an isothermal atomizer itself has no cold spots, the atomizer unit does. For obvious reasons the condensation of acids, salts, etc. on other places on the atomizer unit should clearly be avoided. How effectively the prevention of condensation of concomitants can be achieved in an isothermal atomizer, and thus the occurrence of interferences, is demonstrated in Figure 4-30 on the example of the ammonium phosphate modifier. The background signal generated in a spatially non-isothermal atomizer under otherwise identical conditions is of a totally different magnitude and demonstrates that even with considerable care condensation on cold spots cannot be avoided.

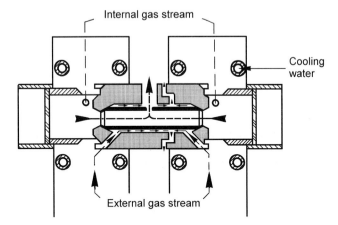

Figure 4-28. Internal purge gas and external protective gas streams in the HGA-800 graphite furnace (Perkin-Elmer). The internal purge gas is introduced from the ends of the graphite tube and leaves it via the sample introduction hole. Volatilized sample constituents are thereby transported from the tube by the shortest possible route. The external protective gas stream passes continuously around the graphite tube, efficiently preventing the ingress of atmospheric oxygen, even when the purge gas stream is shut down.

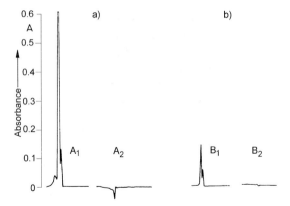

Figure 4-29. Influence of the symmetry of the purge gas stream on the background signal for 20 μL of urine in the graphite furnace. **a** – With an asymmetric gas stream a high background signal (A$_1$) is generated that leads to overcompensation (A$_2$) when a continuum source background corrector is used. **b** – With a symmetric gas stream the background signal (B$_1$) is much smaller and can easily be corrected (B$_2$). Wavelength 283.3 nm.

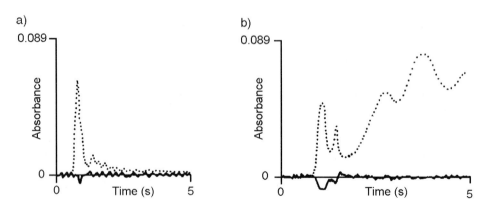

Figure 4-30. Background signals (dotted) and the baseline recorded after Zeeman-effect background correction (solid) for 50 μg NH$_4$H$_2$PO$_4$ at the silver resonance line at 328.1 nm; atomization temperature 1800 °C. **a** – transversely-heated atomizer; **b** – longitudinally-heated atomizer (by kind permission of Perkin-Elmer).

During atomization the purge gas stream is shut down briefly since it would otherwise interfere with the formation of a local thermal equilibrium in the atomizer. When the protective gas and the purge gas are clearly separated in the atomizer unit, the purge gas can be shut off even for longer periods without the risk of the ingress of atmospheric oxygen.

On occasions it is recommended that a reduced gas stream ('mini-flow') of typically 50 mL/min. be passed through the atomizer during atomization. The purpose of this is to reduce the sensitivity, which can occasionally be of advantage for direct solids analysis (see Section 4.2.7). A further reason is when considerable quantities of concomitants are volatilized during atomization and there is a risk that these will condense on cold spots of the atomizer. The use of a mini-flow during atomization requires careful consideration of the advantages and disadvantages. The isothermal area will be further restricted and the risk of gas-phase interferences in the cooler environment increases.

At the end of atomization and during the heat-out phase the purge gas stream must be turned on again to completely remove the analyte and any possible concomitants volatilized during atomization.

Argon of high purity (typically 99.996%) is used almost exclusively as the purge and protective gas. Nitrogen is used occasionally for reasons of cost, but it must be borne in mind that at the high temperatures used in GF AAS it is no longer an inert gas [4619]. MANNING and FERNANDEZ [3107] reported that the determination of aluminium was three times more sensitive in argon than in nitrogen, which indicates the formation of a cyanide. CERNIK [1122] found that the precision for the determination of lead was better in argon than in nitrogen. L'VOV and PELIEVA [3708] reported the presence of CN bands and the spectra of the monocyanides of the alkali and alkaline-earth metals when these were atomized in a nitrogen atmosphere. These spectra could not be observed in argon, which clearly indicates that nitrogen reacts at high temperature with the carbon of the graphite tube and a number of analytes.

A number of authors systematically investigated the use of *alternate gases* during determined program steps to obtain desired reactions or processes. An example is the introduction of air or oxygen during pyrolysis to ash organic materials. The use of alternate gases is discussed further in connection with interferences and their elimination (Section 8.2.4.1).

For reasons of easier handling, electrothermal atomizers are operated exclusively at atmospheric pressure. Nevertheless, even at an early stage L'VOV [3690] investigated the influence of *increased pressure* on the atomization process, since such conditions could be easily realized in his closed system. L'vov found that increased pressure lengthened the residence time of the atom cloud in the absorption volume, leading to broadened absorption peaks. At the same time the peak height (absorbance) was lowered and the linearity of the calibration curve was improved. Both effects can be ascribed to broadening of the line profile of the absorption line (Lorentz effect; Section 2.3.3). In an isothermal atomizer the integrated absorbance should theoretically be independent of the pressure. STURGEON et al. [5620] found in a Massmann-type furnace with atomization from the tube wall that with increasing pressure both the absorbance and the integrated absorbance decreased, which they attributed to increasing condensation of atoms on cold spots on the atomizer.

FALK and TILCH [1810] investigated the effect of *reduced pressure*. They found that the atomization efficiency in their low-pressure FANES system was less than 5%, while for GF AAS at normal pressure it is over 50%. HASSELL et al. [2422] confirmed this finding and found that at a pressure of 0.15 Torr the sensitivity was worse by two orders of magnitude.

4.2.5 Metal Atomizers

Some of the aspects of using refractory metals for the construction of atomizers have already been mentioned in Section 1.7.4. In this section we shall discuss the major differences between atomizers made of graphite and those made of high melting point metals.

The publications on metal atomizers are in general restricted to only a few working groups; OHTA, with more than 35 publications since 1976, takes the leading position. OHTA and co-workers [4428, 4438] preferentially used molybdenum tubes with internal diameters of 1.5–2 mm and a length of 20–25 mm. Due to the relatively low melting point of molybdenum (approx. 2620 °C) they restricted their investigations to elements of high to medium volatility. Since 1979 they also used a tungsten atomizer for the determination of elements such as germanium [4430], vanadium [4434], and titanium [4444].

At the beginning of the 1980s SYCHRA and co-workers [4736, 5721] introduced a tungsten furnace, which they further improved ten years later [5723, 6112]. This atomizer consisted of two half shells which, when fitted together, gave a transversely-heated tube with integrated contacts. In this case it is possible to speak of a spatially isothermal atomizer. CHAKRABARTI and co-workers [1145, 1150, 2121], ČERNOHORSKÝ and KOTRLÝ [1125, 1126], DOČEKAL and KRIVAN [1555], KRAKOVSKÁ et al.[3272–3277], KOMÁREK and GANOCZY [3233], KUŽUŠNIKOVÁ [3376], and SHAN et al. [5275, 5276, 5280] have also performed investigations with this furnace.

The major advantages quoted for metal atomizers were the very fast heating rate (> 10 K/ms) and the fact that the risk of carbide formation no longer existed since carbon was not used in these systems. The principal advantages and disadvantages of *very high heating rates* have already been discussed in Section 4.2.3. Theoretically, greater *freedom from interferences* should be achieved with very high heating rates. However, up to the present this has never been convincingly presented in any publication on metal atomizers. On the contrary, OHTA and co-workers are mostly concerned with interferences, which could often only be eliminated by extracting the analyte [4430–4432, 4436].

The greatest advantage mentioned for high heating rates was usually the significant *increase in sensitivity* achieved in particular for difficultly volatilized elements [3272, 4736, 5723, 6112]. However, it must be taken into account that this increase in sensitivity is exclusively based on the absorbance, but not on the integrated absorbance. Using a tungsten atomizer, SHAN et al. [5276] found for a number of volatile elements that the sensitivity, based on integrated absorbance, was a factor of 3–10 times *worse* than in a graphite atomizer. An additional problem associated with the very tall, narrow signals generated by fast heating rates is the poor precision, quite apart from the fact that most spectrometers are not designed to measure such fast signals.

The fact that elements that form stable carbides cannot do this in a system free from carbon means that, in combination with the fast heating rate, a number of elements (e.g. the lanthanoids and several others) can be better determined in a tungsten atomizer. However, this is usually only possible when hydrogen is mixed with the protective gas to provide a reducing atmosphere [3233, 4435, 4439, 4440, 4442, 4444, 4445, 6112]. The addition of hydrogen also has a positive effect on the lifetime of the tungsten components.

The advantage of the absence of interferences due to the formation of carbides can naturally only be achieved when the sample itself does not contain carbon. This is certainly not the case for all biological materials and other organic samples, since large quantities of carbon are formed during pyrolysis. Such samples must be *completely digested*. This is in fact generally true for all metal atomizers when biological samples are to be analyzed, even when the analyte does not form a carbide. The metals used for the atomizers, namely molybdenum and tungsten, themselves form carbides and at the temperatures used for atomization would immediately react with carbon residues. Tungsten carbide, for example, has a much lower melting point (2860 °C) than tungsten (3410 °C), so that the atomizer would quickly be destroyed in the presence of carbon.

A further problem with metal atomizers is that alloys and intermetallic phases can form when other metals are present in higher concentrations in the atomizer. This can be particularly critical with metallic modifiers such as palladium since the intermetallic phases exhibit totally different heating characteristics to the pure metals. It is thus easy to understand that OHTA and co-workers made detailed investigations on non-metallic modifiers such as thiourea [4429, 4433, 5710, 5712, 5714], thiocyanate [4446], and even elementary sulfur in carbon disulfide [4441, 4443].

Summarizing it can be said that metal atomizers only offer advantages for special applications. They should certainly not be seen as an alternative but, at the best, as a complement to graphite atomizers. Since they are not compatible with graphite atomizers both in respect to their power requirements and signal measurement, they can only be operated in spectrometers especially designed for this technique. This fact has also hindered their broader application.

4.2.6 Atomizer Units for GF AAS

The most important components of atomizer units for GF AAS have been discussed in the preceding sections. Among these are the atomizer itself, the apparatus for heating the atomizer, and the purge and protective gas streams.

At the high temperatures of the atomizers used, a further important component is the *cooling* of the atomizer unit. The cooling performs three major functions: Firstly, it maintains the *housing of the atomizer unit* at a suitably low temperature so that there is no risk to the operator, and components such as gaskets, tubes, etc., which are not temperature resistant, will not be damaged. Secondly, the graphite-to-metal interface through which the current passes should not reach too high a temperature since carbide formation can take place, leading to the destruction of important components. Thirdly, after heating, the atomizer should be cooled rapidly to a temperature at which the next sample can be dispensed. Among other things the *sampling frequency* depends on the efficiency of the cooling.

Nevertheless, too much emphasis should not be placed on this latter point to the extent that the atomizer unit is over-cooled. It has been shown that the temperature and flowrate of the cooling water can have a substantial influence on the drying and pyrolysis temperatures. This explains, for example, why drying temperatures of 300–400 °C are quoted by some authors, while for the same type of atomizer unit other authors require only 130–150 °C. Cooling that is too strong leads not only to poorly reproducible drying and pyrolysis temperatures, but also to an enhancement of the temperature profile in atomizers that have electrical contacts at the tube ends, resulting in condensation and memory effects. The best results with respect to day-to-day reproducibility can be obtained by using a *thermostatted circulatory cooling system*; for long-term operation such a system is also more economical.

A further important aspect is that the atomizer unit is closed at the ends with *quartz windows* that allow the radiation to pass through. Firstly, this measure is necessary to guarantee perfect control of the gas streams through the atomizer, which is essential to remove sample constituents from the absorption volume. Secondly, the ends of the atomizer must be sealed to prevent air from diffusing into the atomizer during normal operation and also when the purge gas stream is shut down. For difficultly atomized elements, STURGEON and CHAKRABARTI [5622] found that there was a substantial loss in sensitivity when the windows were removed from the graphite furnace, since atoms could then diffuse more easily to the colder tube ends and condense.

A *controlled fume extraction* in the vicinity of the sample introduction hole can also have a stabilizing influence on the purge gas stream in the atomizer. The main purpose of such an extraction is naturally the controlled removal of vapors generated during the drying and pyrolysis steps, which are often corrosive or toxic. A controlled extraction that is adapted to the gas flow conditions in the atomizer unit is thus an important safety factor for the instrument and the operator.

Atomizer units for GF AAS are frequently equipped with a *temperature control and measurement system*. Because of the temperatures prevailing in the atomizer and the speed with which they are reached, only contact-free, optical pyrometers are suitable for this task. A characteristic of such systems is that they only work reliably at higher temperatures. However, this is not a problem in practice since normal power control is adequate for the drying and pyrolysis temperatures when the resistance of the graphite atomizer can be monitored with sufficient accuracy.

High performance spectrometers for GF AAS frequently employ the Zeeman effect for background correction. In such cases the *pole shoes for the magnet* are mounted at the atomizer and are thus part of the atomizer unit. The various possible locations of the magnet at the atomizer with respect to the radiation beam are discussed in Section 3.4.2, and thus will not be mentioned here further. We must merely mention that certain configurations of the magnet and the atomizer heating mutually preclude each other. For example, a Massmann configuration with the electrical contacts at the tube ends virtually only allows a transverse configuration of the magnet. An atomizer unit utilizing the alternative configuration, namely a transversely-heated atomizer and a longitudinal magnetic field, is depicted in Figure 4-31. The pole shoes of the magnet have apertures to allow the radiation beam to pass through and are closed with windows.

Although *Sample Introduction and Mechanization* are subjects that are treated in a later chapter (see Section 6.2), at this stage it is worth mentioning certain aspects specific to GF AAS. In the early years of GF AAS dissolved and liquid samples always had to be dispensed into the graphite tube using micropipets. Since the sample introduction hole has to be as small as possible (for reasons discussed in Section 4.2.2.1), this op-

Figure 4-31. Atomizer unit with a transversely-heated atomizer and magnet pole shoes in the longitudinal configuration with windows to allow the passage of the radiation beam (by kind permission of Perkin-Elmer).

eration was rather difficult and demanded a high degree of care on the part of the operator. In addition there is a relatively long delay time between each sample, making manual operation unproductive.

Of even greater importance, however, are the considerations of precision and accuracy with respect to sample dispensing. The place in the graphite tube onto which the sample is dispensed and the way the sample droplet is brought into contact with the graphite surface are of considerable significance. It is virtually impossible to keep these factors fully under control with manual pipetting. Further, there is the risk that particles around the edge of the sample introduction hole in the graphite tube can be introduced into the tube with the pipet tip, leading to contamination. The reproducibility for a series that can be achieved with manual dispensing is at best 3–5%.

The introduction of the first autosampler for GF AAS in 1975 was a major contribution to the general acceptance of this technique [5579]. A comparison between manual dispensing and automatic dispensing, taken from the first publication on this subject [6206], is shown in Figure 4-32 to demonstrate the improvement that can be achieved by mechanization of sample introduction.

A further problem mentioned in many publications is the contamination of pipet tips, which cannot always be eliminated entirely, even after thorough leaching in acid and long rinsing [539, 4844, 5060, 5494]. In contrast, STOEPPLER *et al.* [5579] found with the aid of radiotracers that when using the autosampler depicted in Figure 4-33 the carryover was only 10^{-7}. The advance into the ultratrace range below 1 µg/L, which can be achieved by GF AAS, only became feasible in practice after the introduction of an autosampler with the necessary reliability. For this reason an autosampler should be considered as an essential part of an atomizer unit.

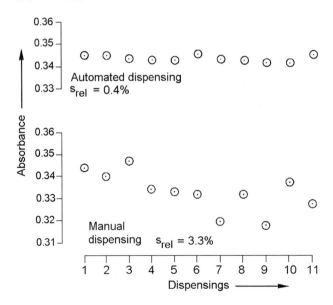

Figure 4-32. Series reproducibility for the determination of Pb (20 μL, 0.2 mg/L Pb) in the graphite furnace with automated and manual dispensing, respectively (from [6206]).

Figure 4-33. AS-72 autosampler for the graphite furnace technique (by kind permission of Perkin-Elmer).

4.2.7 Analysis of Solid Samples

Of all the various AAS techniques discussed in this monograph, the graphite furnace technique is the only one that is really suitable for the analysis of solid samples. In fact, electrothermal atomizers have been used for this purpose since their introduction (see Section 1.9). In the following section we shall make a distinction between the *direct* analysis of solid samples and the analysis of *suspensions* (slurry technique). Both types of analysis clearly have similarities, but they differ markedly in a number of other aspects [3369].

4.2.7.1 Direct Analysis of Solid Samples

The main reasons for performing the direct analysis of solid samples without prior digestion or dissolution can be summarized as follows:

i)   Higher sensitivity for the determination of trace elements, since every solubilization procedure causes substantial dilution.
ii) Less time is required for sample preparation when dealing with samples that cannot be dissolved or digested easily, and thus the total analysis time is reduced.
iii) The risk of contamination from reagents, apparatus, and the laboratory atmosphere is much lower; this is a very important factor for ultratrace analyses.
iv) The risk of loss of the analyte during sample preparation due to volatilization, sorption on the apparatus, or incomplete solubilization is lower.
v)   The use of corrosive or toxic reagents is avoided.
vi) A micro-distribution analysis or homogeneity check can be performed.

Nevertheless, the direct analysis of solid samples is not as advantageous as it would at first appear since there are a number of marked disadvantages and the technique places considerable requirements on the skill of the operator. The majority of disadvantages and problems of direct solids analysis mentioned in this section are not in fact specific to the graphite furnace technique. Many of these problems are typical for the analysis of solid samples in general, and have been known for years, for example, in optical emission spectrochemical analysis using arcs and sparks. An example is the relatively poor repeatability precision which is largely due to the inhomogeneity of natural samples. The disadvantages of direct solids analysis are summarized below; the significance of each point depends on the given situation:

i)   It is usually more difficult to introduce a solid sample into the atomizer than a liquid sample. Directly introducing the sample requires practice and the greatest of care. Losses or contamination of the sample can take place both during weighing and transfer to the atomizer.
ii) As opposed to liquid samples, a new sample weighing must be performed for every replicate measurement. If the inhomogeneity of the sample or the exclusion of random errors makes numerous replicate measurements necessary, the effort required can be unacceptable for routine analyses.

iii) Frequently there are calibration problems since the atomization signals are much more strongly influenced by concomitants than with solutions. Individual species of the analyte are often not or only incompletely volatilized from the sample. Suitable calibration samples are frequently not available and aqueous calibration solutions are often not suitable for calibration.

iv) Chemical modification is often less effective for solid samples since the modifier does not come sufficiently into contact with analyte species that are enclosed within sample particles.

v) It is not usually possible to 'dilute' solid samples when the analyte concentration is too high. Mixing with graphite powder is time-consuming and not very accurate.

vi) Virtually without exception the precision of solid sample analysis is markedly poorer than for solution analysis. The standard deviation is typically around 10%. The repeatability deteriorates with decreasing sample mass, which is normally 0.1–10 mg. Certified reference materials are generally better homogenized than real samples, so the repeatability for the latter is mostly much worse.

vii) Sample residues, e.g. carbon, can remain in the atomizer after the atomization step and influence the following determinations.

viii) Since far greater masses of concomitants are introduced into the atomizer for solids analysis than for solution analysis, much higher background attenuation can often be observed, which then places higher requirements on the correction system.

ix) Much higher atomization temperatures may be required for refractory matrices in order to completely volatilize the analyte from the sample. This can substantially reduce the lifetime of the atomizer.

x) It is very difficult to produce blank test samples, especially for rock and environmental samples.

xi) The use of the analyte addition technique is made more complex by the variable weighings and requires multivariate evaluation algorithms.

It has already been emphasized that a number of these disadvantages are associated with solids analysis and can thus hardly be influenced. A number of others, however, can certainly be influenced and can be brought under control by the careful selection of the methodology. Again here the STPF concept brought the breakthrough, and its correct application can decide between success and failure. Mention must only be made of the differing volatilization with respect to time (temperature) and kinetics of various analyte species that is found only for solids analysis. This can lead to considerable errors of measurement when atomization takes place under non-isothermal conditions. These errors can be minimized when atomization is under quasi isothermal conditions, e.g. from a platform, and the integrated absorbance signals are used for measurement (see Section 8.2.4.1).

4.2.7.2 Aids for Direct Solids Analysis

The incredible number of devices described in the literature for direct solids analysis is an indication of the complexity of the problem. Many of these devices were used merely by the 'inventor' and never found application in other laboratories. For this reason it is not possible to discuss all these solutions in this monograph. A number of devices of historical interest are described in Section 1.9, and overview articles with detailed information on this subject are available for the interested reader [523, 3442].

Although it can be assumed that atomization units designed specifically for direct solids analysis would provide particularly high performance, nevertheless they never came into general use. This is true for the tube-in-flame proposed by L'VOV [3699] as well as the induction furnace proposed by LANGMYHR and co-workers [3437]. The simplest explanation for this phenomenon is that there were no instrument manufacturers prepared to take the risk of constructing an atomizer unit solely for solid samples. And certainly the demand for such instruments was never very high. Currently available atomizer units offer compromise solutions that allow the analysis of both solid and dissolved samples.

A number of atomizer types designed specifically for solids analysis, but which can be operated in the same atomizer units as conventional graphite tubes for solution analysis, have gained a degree of significance. A number of these special atomizers are depicted in Figure 4-34. In a number of these atomizers for solids analysis the tube is replaced by a cup, such as in the carbon rod atomizer from Varian. These graphite cups were also used for the direct analysis of airborne particles; air is drawn through the cups, which are made of porous graphite, and the particles are thus collected [5337, 5338]. According to calculations performed by FALK and TILCH [1810] the efficiency of atomization from such cups is mostly less than 5%, while in graphite tubes it is over 50%. In their two-step furnace, FRECH and co-workers [438, 1964] used a combination of a tube and a cup with separate heating circuits in their two-step atomizer; this increased the performance of the system significantly but also increased its complexity (see Section 1.7.3).

A simpler system was used by VÖLLKOPF et al. [6096] in the cup-in-tube technique (depicted in Figure 4-34f). The cup functions like a platform, but due to its greater mass

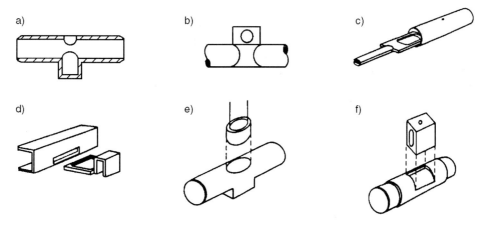

Figure 4-34. Special atomizers designed for solids analysis. **a** – cup cuvette atomizer; **b** – carbon rod atomizer with graphite cup; **c** – boat atomizer; **d** – microboat atomizer; **e** – miniature cup atomizer; **f** – cup-in-tube atomizer (from [523]).

the heating is retarded even more markedly and the rate of heating is slowed, both effects of advantage for solids analysis. An alternative to this technique is the cup cuvette atomizer depicted in Figure 4-34a [5746–5749]. ATSUYA *et al.* [302, 303] placed a separate small container with the sample in the cup under the tube. A disadvantage of this design, however, is that the container with the sample is heated more slowly and does not attain the same temperature as the tube.

A further variation is to introduce the solid sample into the tube on a platform or probe, either from the side or from the end. PRICE *et al.* [4715] introduced the sample into the end of the tube on a cup-shaped platform only 0.2 mm thick. A similar system but with a greater platform mass and larger graphite tubes was later employed by KURFÜRST [3360, 3361]. KURFÜRST *et al.* [3362] even mechanized this system by means of a laboratory robot from weighing of the sample to transport into the graphite tube. NOWKA *et al.* [4393] modified a transversely-heated graphite furnace for this type of sample introduction.

BROWN *et al.* [876] introduced the sample on a probe through a corresponding port into the graphite tube. Even though for probe atomization [4503] the sample is introduced into an atomizer preheated to the atomization temperature, in practice no significant difference could be found between platform and probe atomization [1147]. The reason is probably that the introduction of the cold probe into the hot atomizer causes a considerable perturbation of the local thermal equilibrium and the additional port leads to a loss in sensitivity.

4.2.7.3 Analysis of Suspensions

The analysis of suspensions (slurry technique) combines the advantages of *solid sample dispensing* with *solution analysis* [5565]. In comparison to the disadvantages listed in Section 4.2.7.1, the situation for the slurry technique can be summarized as follows:

i)   Dispensing suspensions into the graphite tube is in most cases as simple as dispensing solutions.

ii) It is easy to take several aliquots in sequence from a homogeneous suspension, so that only one sample weighing is required for replicate determinations.

iii) Suspensions can frequently be measured against aqueous calibration solutions using standard calibration techniques. In cases of doubt the analyte addition technique can be used without problems.

iv) Chemical modifiers can be used in the same way as for solutions; their effectiveness can however be restricted for larger particles.

v)   Suspensions can be diluted just like solutions. Nevertheless care must be taken to ensure that a statistically relevant number of particles remain in each dispensed sample aliquot.

vi) In principle the precision for the analysis of suspensions cannot be higher than for the analysis of solids when the same number of particles is dispensed into the atomizer. The precision can however be considerably better when a substantial portion of the analyte is taken into solution, which is frequently the case.

vii) Problems associated with possible residues in the atomizer, with high background attenuation, with refractory materials, and with the preparation of blank test samples are comparable to those for direct solids analysis.

viii) The only problem that is additionally new for slurry analysis is the production of a homogeneous and stable suspension at the time of dispensing.

A further decisive advantage of slurry analysis is that no changes to the atomizer are required; this fact has played a major role in the general acceptance of this technique. Just like solutions, suspensions can be analyzed in conventional graphite tubes with platforms. But just as mentioned for direct solids analysis, STPF conditions must be strictly maintained if errors are to be avoided.

For more than a decade the major problem was, as mentioned in point viii) above, maintaining a homogeneous suspension up to the time of dispensing. As described in Section 1.9, for a long period attempts were made to increase the viscosity of the liquid phase and thus reduce the rate of sedimentation. However, this technique did not come into general use since optimization of the liquid phase was required for each type of suspended substance. The breakthrough came when attempts to optimize the liquid phase were abandoned and the suspension was homogenized once again, using ultrasonics for example [4100], immediately prior to dispensing. MILLER-IHLI [4102] automated the technique and CARNRICK et al. [1077] described the first commercial system in which the suspension is thoroughly mixed and homogenized in the autosampler directly prior to transfer to the atomizer.

Even though other mixing techniques were investigated, such as passing an argon stream through the suspension [521], magnetic stirrers [1552, 3304], or a vortex homogenizer [1749, 2561], in end effect the ultrasonics technique has established itself as the most successful [756, 1749, 2951]. It can be said without doubt that the analysis of suspensions exhibits far higher repeatability than the direct analysis of solid samples. The concentration of the analyte in a suspension can be easily changed by dilution, and calibration is much more simple than for solid samples. Of greatest importance for routine applications is the fact that the technique can be easily automated, and the greatest problem, namely the homogenization of the suspension, is solved at the same time. Nevertheless, with very dense materials errors can occur due to sedimentation between the end of ultrasonic homogenization and withdrawal of the sample; this can be minimized by optimizing the depth of immersion of the sampling capillary [4112]. As shown by international interlaboratory trials, nowadays suspensions can be analyzed virtually free of problems, just like solutions, when the special conditions are taken into account and provided that STPF conditions are maintained and that certain precautions are taken [4109, 4111]. STOEPPLER and KURFÜRST [5588] have published a review article on the application of this technique.

Nevertheless, the direct analysis of solids can still be utilized for those types of sample that cannot be taken into suspension. For example, these can be plastic materials that cannot be suitably ground down or satisfactorily suspended, or large particles that can be placed in one piece into the atomizer. A further area reserved for direct solids analysis is distribution and homogeneity analysis; KURFÜRST et al. [3363–3365] in particular have made detailed investigations in this direction.

4.2.8 Simultaneous Multielement Determinations

Although instruments for multielement determinations using F AAS were developed at an early stage (refer to Section 1.4), for a long period it was considered that such determinations were not possible by GF AAS. The main reason for this was that when using this technique the conditions with respect to pyrolysis and atomization temperatures, modifier, and even whether wall or platform atomization was required frequently had to be optimized for each analyte. Multielement determinations require 'compromise conditions' which can lead to a significant deterioration in the sensitivity and other data for a number of elements [414].

LUNDBERG et al. [3666] have shown that compromise conditions have the lowest negative influence when atomization takes place under largely isothermal conditions. For conventional graphite furnaces this signifies atomization from the platform, and for longitudinally-heated atomizers additionally a restriction to analytes of high to medium volatility. A further complication is the relatively limited working range of GF AAS that only permits the simultaneous determination of similar analyte concentrations. It is therefore hardly surprising that initial investigations to develop instruments and methods for simultaneous multielement determinations by GF AAS only met with moderate success. Typical applications were limited to the determination of volatile elements such as cadmium and lead [2764, 3982], or antimony, arsenic, lead, and selenium [4451]. As soon as it was attempted to simultaneously determine less volatile elements such as chromium, atomization temperatures of up to 3000 °C were required [2585], which had a negative influence on the tube lifetime and also on the determination of the volatile elements.

The introduction of the transversely-heated graphite atomizer changed the situation decisively. Since the temperature distribution is homogeneous over the length of the tube, significantly lower atomization temperatures are possible and all analytes are atomized from the platform under STPF conditions. The development of universal modifiers also means that very similar conditions for pyrolysis and atomization can be selected for large groups of elements without noticeably influencing the characteristic mass or the S/N ratio [5150, 6265]. These aspects are discussed in detail in Section 8.2.4.1.

A large number of methods for the simultaneous determination of up to six elements were published after the introduction of the first truly simultaneous spectrometer for GF AAS with transversely-heated atomizer, integrated platform, a high-performance Echelle polychromator, and a specially developed photodiode-matrix detector [4783, 4784]. As well as relatively easy to handle combinations of elements of high to medium volatility, such as antimony, cadmium, lead and silver [3459], arsenic, lead, selenium and thallium [5332], copper, iron and nickel [904], or cobalt and manganese [2835], elements of high and low volatility can also be determined simultaneously [2392]. As with all simultaneous multielement determinations by GF AAS, the most volatile element determines the pyrolysis temperature while the least volatile element determines the atomization temperature. The losses in detection capability feared due to the reduced sensitivity resulting from too high atomization temperatures are hardly observed and are in part compensated by the improved S/N ratio [2392]. It could be shown that simultaneous multielement determinations by GF AAS lead to a significant increase in the productivity without having to accept greater compromises in the detection capability for individual elements [2392].

4.3 Chemical Vapor Generation

In this section we shall discuss atomizers and atomizer units used to determine those analytes that can be vaporized in the atomic state or as molecules (e.g. as hydrides or ethyl compounds) by chemical reaction at ambient temperature. These analytes are essentially mercury (by CV AAS), and antimony, arsenic, bismuth, germanium, lead, selenium, tellurium, tin, and also more recently cadmium and thallium (by HG AAS). Since the same, or very similar, apparatus is used for both cold vapor AAS (CV AAS) and hydride-generation AAS (HG AAS), these techniques are treated together in this section. This similarity is not, however, valid for collection and preconcentration techniques and most certainly not for atomization. Since mercury in the course of chemical vapor generation is released as the atomic vapor, an atomizer is not required for this element, but merely an absorption cell. However, the same quartz tube that serves as the atomizer in HG AAS is mostly used for this purpose.

Since the historical development of CV AAS and HG AAS is treated in detail in Section 1.8, the large number of systems used in the investigation of these techniques is not discussed in this section. We shall restrict ourselves to a discussion on the state of the art and only mention historical matters when absolutely necessary.

4.3.1 Systems for Chemical Vapor Generation

4.3.1.1 Batch Systems

The earlier systems used for CV AAS and HG AAS consisted of normal laboratory apparatus, such as flasks, dropping funnels, gas inlet tubes, etc., assembled to meet the requirements. A system of this type is depicted schematically in Figure 4-35. The measurement solution is placed in the flask (volume approx. 50–250 mL), the air is driven out by an inert gas (not necessary for CV AAS), the reductant (e.g. $NaBH_4$, $SnCl_2$) is added, and the gaseous analyte species is transferred in the inert gas stream to the atomizer or absorption cell. A time-dependent signal is generated in this technique and the profile is largely determined by the kinetics of the release of the gaseous analyte from solution.

All commercially available batch systems operate on this principle. Improvements to these systems consist mainly in the ease of operation and the efficiency with which the analyte is driven out of solution. Conical reactor vessels in which the reductant and the inert gas are introduced at the bottom of the cone have proved to be particularly good [2861]. A system of this type in which the reductant and the inert gas are introduced via two concentric immersion tubes is depicted in Figure 4-36; this arrangement guarantees that the gaseous analyte species is driven rapidly from solution. Especially for CV AAS

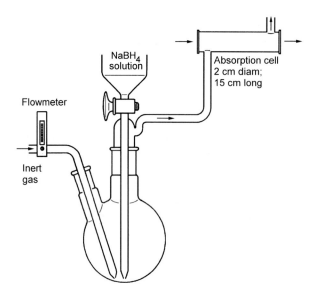

Figure 4-35. Accessory for CV AAS or HG AAS using sodium tetrahydroborate or tin(II) chloride as reductant (from [1793]).

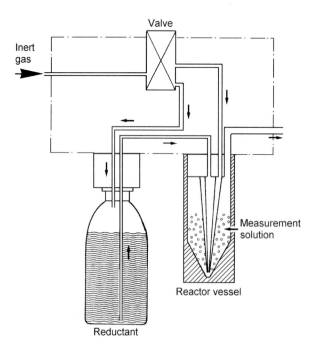

Figure 4-36. Batch system for CV AAS and HG AAS using conical reactor vessels with introduction of the reductant and inert gas at the lowest position (from [6213]).

using tin(II) chloride, where the release of atomic mercury takes place relatively slowly, the use of intense stirring [1580] or heating the reactor [1067, 1068] brings an improvement in the sensitivity.

With batch systems the measured signal is proportional to the mass of the analyte in the measurement solution and not to its concentration. Thus, as in GF AAS, it should be possible to compensate for influences on the release velocity of the analyte by measuring the integrated absorbance. This is also possible in principle; however, in HG AAS, and also in CV AAS when $NaBH_4$ is used as reductant, the pH value changes during the reaction and thus also the volume of hydrogen generated per time unit. The resulting change in the gas stream during the measurement time makes integration difficult and imprecise when the signal appears at different times and for varying lengths of time. Peak height (absorbance) measurements are thus mostly used for batch systems.

Another peculiarity of batch systems is the relationship between the total measurement volume and the test sample volume. As mentioned above, the measured signal is proportional to the mass of analyte and not to its concentration, but the measurement volume nevertheless plays a certain role as depicted in Figure 4-37 on the example of selenium. This effect naturally has the greatest influence on peak height and results from the slower release of the gaseous analyte species from larger volumes. The extent of the effect also depends on the apparatus; small variations in the measurement volume usually have a negligible influence.

The majority of reactor vessels for batch systems are designed to take relatively large measurement volumes, e.g. 50–100 mL, but also require a certain minimum volume, e.g. 5–10 mL, to ensure that the reaction runs properly. As a result of the high sensitivity of CV AAS and HG AAS, however, it is rarely necessary to use sample volumes of 10–50 mL for the determination; volumes of 1–5 mL are usually sufficient to come into the optimum measurement range.

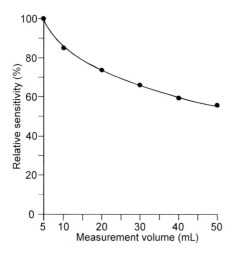

Figure 4-37. Dependence of the selenium signal on the measurement volume for the hydride-generation technique using a batch system.

It is thus usual in practice to add a (not accurately measured) volume of acid, say 10 mL, to the reactor vessel and then to pipet the (accurately measured) sample volume, e.g. 1 mL. The measurement volume can thus be substantially greater than the sample volume. This leads to a considerable number of possibilities to modify the chemical environment for the formation of hydrides so that interfering reactions can be controlled and suppressed over large ranges. In only a few cases, such as the analysis of natural waters, is it necessary to use sample volumes of up to 50 mL. In such cases the test sample volume is the same as the measurement volume.

The greatest advantage of batch systems is the large volumes that can be handled and which lead to a very high *relative sensitivity* (in concentration). Particularly in CV AAS measurement volumes of up to 250 mL [6224] have on occasions been used. The biggest disadvantage of batch systems is the large dead volume that leads to a relatively poor *absolute sensitivity*. These two effects are very closely coupled with each other and in part lead to mutual compensation.

A further disadvantage of batch systems is that in principle they are manual systems with a relatively high requirement in time and effort. Attempts have been made to automate such systems [3311, 3993, 4850], but without a great deal of success. In commercially available systems mechanization is largely limited to the sequence of analysis and to the convenience of operation [6213]; measurement solutions usually have to be changed manually.

4.3.1.2 Flow Systems

The continuous flow systems (CF systems) and flow injection systems (FI systems) discussed in this section have a number of features in common, but on the other hand they also have a number of fundamental differences. A common feature of both systems is that the reaction takes place within tubes and that a separator is thus required to separate the gaseous analyte from the aqueous phase. A further common feature is that both systems can be easily automated and that sample preparation can be entirely or partially incorporated into the automatic analytical sequence. A major advantage of FI systems is that they can easily be used as CF systems if this technique is preferred for a particular application.

4.3.1.2.1 *Continuous Flow Systems*

In CF systems the measurement solution, the reductant, and further reagents as required, are transported continuously in tubes in which they are mixed and, after passing through a reaction zone, the phases are separated in a gas-liquid separator (GLS). This process is very similar to the aspiration and spraying of a measurement solution in F AAS, and likewise, CF systems generate time-independent signals in which the absorbance is proportional to the *concentration* of the analyte in the measurement solution.

As early as 1974 GOULDEN and BROOKSBANK [2187] introduced the first CF system for the automatic determination of antimony, arsenic, and selenium in natural waters.

The system was based largely on a Technicon AutoAnalyzer® that had been appropriately modified. This example was followed by a number of other working groups, most noticeably in north America, no doubt because the AutoAnalyzers were in any case available in the laboratories. Since part of the sample preparation procedure was normally mechanized in these systems, they are discussed in more detail in part 4 of this section.

A number of very simple CF systems were introduced in the 1980s [158, 191, 1453, 1645] in which merely sample and reductant were combined via two tubes and the gas phase introduced into a burner or quartz tube atomizer. A commercial system of this type was introduced by STURMAN [5646] in 1985 and is depicted schematically in Figure 4-38. It is characteristic for systems of this type that the sensitivity increases with increasing flowrate of the sample, but that the efficiency of separation and transfer of the gaseous products into the absorption volume decreases. STURMAN [5646] found that a flowrate of 7 mL/min for the sample solution offered the best compromise. Under the given conditions the signal required about 45 s to attain a stable value and the sample consumption was in the order of 10 mL per determination.

DĚDINA [1458] brought attention to the fact that the concentration and the flowrate of the reductant must also be matched to the sample flowrate. The maximum flowrate that was possible without the reaction from becoming too violent was 15 mg/s NaBH$_4$ (approx. 0.4 mL/s of a 40 g/L NaBH$_4$ solution). The more violent the reaction, the larger must be the dead volume of the GLS, which again has an unfavorable influence on the sensitivity (see part 3 of this section). MUNAF et al. [4234] went to the other extreme and miniaturized a CF system for CV AAS. They used sample and reagent flowrates of 0.05–0.2 mL/min and attained comparable limits of detection of around 0.15 µg/L Hg.

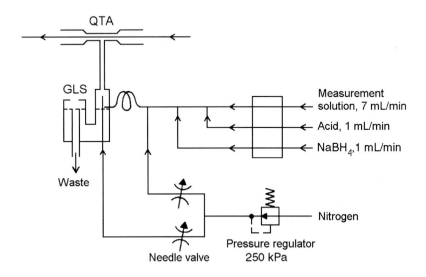

Figure 4-38. CF system for CV AAS and HG AAS (according to [5646]); GLS = gas-liquid separator, QTA = quartz tube atomizer.

4.3.1.2.2 *Flow Injection Systems*

In appearance FI systems differ only from CF systems in that instead of the sample solution a carrier solution (e.g. water or dilute acid) is pumped continuously and a small volume of sample solution (e.g. 0.5 mL) is injected at regular intervals (e.g. every 30 s). A *time-dependent signal* is generated whose form depends on the dispersion of the measurement solution in the carrier solution and not, as with batch systems, on the release of the gaseous analyte species from the solution. Nevertheless, the release of the gaseous analyte species influences the height (absorbance) of the signal, as with CF systems.

ÅSTRÖM [292] employed the FI technique for HG AAS for the first time in 1982. Among other things he found the low sample consumption (0.7 mL), the high sampling frequency (180/h), and the excellent limits of detection (0.1 ng Bi) as particularly remarkable. He especially brought to attention the possibility of modifying the experimental conditions such that interferences could be practically eliminated; this topic is discussed in more detail in Section 8.3.3. A year later DE ANDRADE *et al.* [193] also presented an FI system for CV AAS. Nevertheless, despite the obvious advantages of the FI technique for CV AAS and HG AAS, publications in this area remained relatively scarce [642, 1179, 4514, 6432, 6547] until the first commercial system was introduced at the end of the 1980s [6253, 6256]. This FI system is depicted schematically in Figure 4-39.

The major difference between FI and CF systems is that it is not necessary with FI systems to wait until equilibrium has been attained before taking the measurement. This means that operation is practically always under conditions of thermodynamic non-equilibrium. Nevertheless, because of the excellent temporal repeatability of the FI sig-

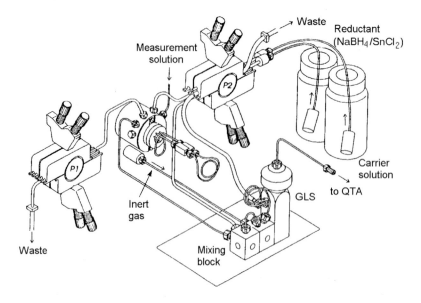

Figure 4-39. FIAS-400 system for CV AAS and HG AAS (by kind permission of Perkin-Elmer).

nals, the technique functions with very high precision. Since it is not necessary to wait for equilibrium to be attained, the major advantages in practice, as already mentioned by ÅSTRÖM [292], are the low sample consumption and the high sample throughput. In addition, FI systems can be easily miniaturized, so that excellent sensitivities can be achieved since the dead volume is very small. A comparison of the absolute and relative sensitivities (m_0 and β_0) for a batch system and an FI system is presented in Table 4-2. For each system the appropriate volumes were used that give the best values. It is quite clear that the FI system is superior to the batch system by 1–2 orders of magnitude in terms of the absolute sensitivity (characteristic mass, m_0). In terms of the relative sensitivity (characteristic concentration, β_0) the two systems are comparable due to the larger volume of the batch system. Nevertheless, owing to the high stability and repeatability of the FI signals the limits of detection with this technique are clearly better.

Table 4-2. Comparison of characteristic mass, m_0 (ng), and characteristic concentration, β_0 (ng/L), for batch and FI systems. The measurement volume is quoted in brackets.

Element	Batch System		FI System	
	m_0 (10 mL)	β_0 (50 mL)	m_0 (0.2 mL)	β_0 (0.5 mL)
Hg	5.5	150	0.10	400
As	0.45	10	0.01	40
Bi	0.65	20	0.04	100
Sb	1.7	50	0.02	90
Se	1.1	30	0.02	100
Sn	1.0	30	0.05	200
Te	0.8	20	0.03	120

Another decisive advantage of not having to wait for equilibrium to be attained with FI systems is that interferences can be eliminated by means of 'kinetic discrimination' (see Section 8.3.3). This technique utilizes the fact that the reduction of the analyte to a gaseous species is almost always the fastest reaction. If the analyte species can be separated before interfering reactions become dominant, interferences can thus be eliminated. This possibility is offered by neither batch nor CF systems [1033, 4327, 6256].

4.3.1.2.3 Gas-liquid Separators (GLS)

In batch systems the gaseous species is separated from the liquid phase in the reactor, but for CF and FI systems a special phase separator is required. These can be either vessels made of glass, silica, or plastic, or membranes that are permeable only for gases.

An unbelievable number of gas-liquid separators are described in the literature; a small selection is depicted in Figure 4-40. They range from simple overflow vessels to 'classical' U-tube separators, which come in a large number of variations, to very complicated assemblies comprising several U-tubes, condensers, scrubbers, etc. It is extremely important that the GLS is large enough to prevent liquid from foaming over and solution droplets from reaching the atomizer. On the other hand, the GLS should be small so that solution is removed rapidly and above all things the dead volume should be as small as possible to guarantee good sensitivity. The simple separator depicted in Figure 4-40c, which can be filled with glass beads to further reduce the dead volume, has proved to be particularly suitable [796]. WELZ and SCHUBERT-JACOBS [6256] proposed fitting this separator with a gas-permeable membrane. Similar double separators were also used by HANNA et al.[2355] and TYSON et al. [5943] to minimize the dead volume but at the same time to guarantee the reliable separation of the liquid phase.

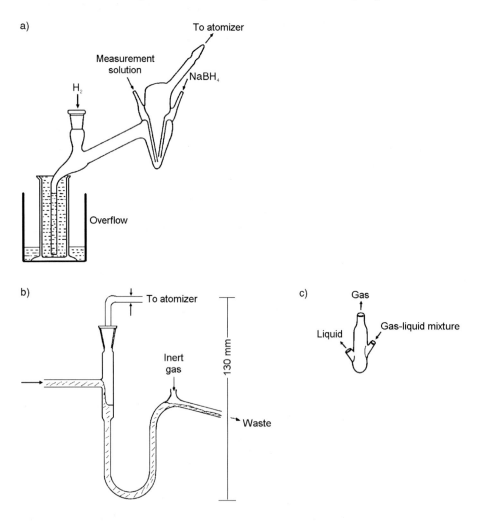

Figure 4-40. Various gas-liquid separators for HG AAS. **a** – overflow system (according to DĚDINA [1455]); **b** – U-tube separator (from [3965]); **c** – phase separator for FI systems (by kind permission of Perkin-Elmer).

As well as classical separators, PTFE membranes or membrane tubes have been investigated as gas-liquid separators for both CV AAS and HG AAS [193, 1183, 4514, 6433]. The greatest advantage of membrane separators is that they cannot foam over and thus prevent sample aerosol or reagent from reaching the atomizer. WELZ and SCHUBERT-JACOBS [6256], however, found that the sensitivity obtained when using such membranes was only about half the value obtained for conventional systems. Also the sensitivity changed with time since the permeability of the membrane changed with the length of use. The best solution appears to be the combination of a phase separator with an additional membrane which does not come into direct contact with the liquid phase, but serves to hold back the aerosol. Semipermeable membranes can also be of advantage to remove residual moisture since they are superior to chemical drying agents.

4.3.1.2.4 *On-line Sample Preparation*

Although automation and mechanization are treated in detail in a later chapter (Chapter 6), a number of special aspects associated with chemical vapor generation are discussed in this section. In CV AAS and HG AAS the actual determination is preceded by a chemical reaction that requires the analyte to be present in a particular chemical form. A certain level of pretreatment of the sample, such as digestion or prereduction, is thus practically always an integral part of the analytical method.

It is therefore hardly surprising that only a few years after the introduction of HG AAS, GOULDEN and BROOKSBANK [2187] presented a system that was coupled to a modified Technicon AutoAnalyzer® for the automated prereduction of As(V) and Sb(V) with potassium iodide. Similar systems were also introduced later by CHAN and VIJAN [1177] and also DRIEHAUS and JEKEL [1611]. FISHMAN and SPENCER [1917] made a step further by mineralizing organic arsenic compounds on-line, either by digestion in sulfuric acid or by UV radiation; in this way they opened the determination of these compounds to HG AAS. Very similar digestion systems with peroxodisulfate were employed by various working groups for the determination of mercury in waste water and other environmental samples [2178, 2179, 3965, 4234]. AGEMIAN et al. [48, 50] presented a system for the determination of mercury in water by CV AAS using flow-through UV digestions.

At the beginning, FI systems followed in the footsteps of CF systems; thus on-line digestion of mercury compounds with persulfate [642, 2355] or on-line photooxidation of organoarsenic compounds [299] were described in the literature. More recently, on-line microwave digestion techniques more suited to the FI technique have been described for the determination of mercury in water [5906, 5907, 6263], urine [2281, 6263], and blood [2282].

LE et al. [3474, 3475] used an on-line microwave-assisted digestion system for organoarsenic compounds in urine. TYSON et al. [5943] combined on-line matrix separation with prereduction under stopped-flow conditions for the determination of arsenic in nickel alloys. WELZ et al. [6270] exploited a number of the possibilities of the FI technique in an on-line digestion system for the determination of arsenic in body fluids. As depicted in Figure 4-41, a small volume of digestion acid is mixed with the sample volume by the merging-zones technique, and the digestion is performed in the microwave cavity under pressure; increased pressure is obtained by placing a flow restrictor at the end of the digestion and reduction zone. The prereduction is conducted in the waste heat of the microwave system and, after passing through the flow restrictor, the sample 'plug' is accelerated by a second carrier stream and transported to the HG AAS section.

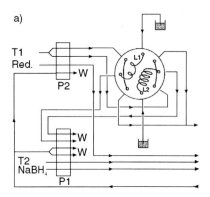

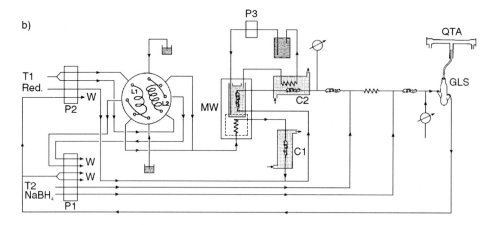

Figure 4-41. Flow injection manifold for on-line digestion of organoarsenic compounds, for prereduction of As(V), and for determination of arsenic by HG AAS. **a** – fill configuration; **b** – inject configuration; T1, T2: carrier solutions; Red: reductant; P1, P2, P3: pumps; L1, L2: loops for test sample solution and digestion reagents; MW: microwave digester; C1, C2: coolers; W: waste (from [6270]).

4.3.1.3 Materials for Containers and Tubing

According to the well-known principle of trace analysis: '*Every contact leaves traces*', in chemical vapor generation all materials that come into contact with the reduced analyte species on its passage from the reactor or reaction zone to the atomizer or absorption volume must be subjected to critical scrutiny. As well as the *materials* and their *surface properties*, the *contact time*, and, more especially, the *ratio of the analyte mass to contact area* play a significant role. The latter aspects, in particular, speak against *batch systems* with their large-volume reactors in which the analyte remains for relatively long periods and thus has intense contact. The circulatory systems used in CV AAS (refer to Section 1.8.1) are particularly bad in this respect, because not only is the time sufficient for equilibrium to be reached between mercury vapor and solution, but also between mercury vapor and the surfaces of containers and tubes. *Flow systems,* because of the substantially smaller contact areas in tubes, are significantly better. The most critical component is probably the GLS with its relatively large surface area (another reason for keeping it as small as possible). *FI systems* have an additional advantage against CF systems in that the contact time is around an order of magnitude shorter, since it is not necessary to wait for equilibrium to be attained. Furthermore a carrier solution can be selected for the FI technique that removes deposited analyte species between the individual measurements. WELZ and co-workers [6259, 6267] have shown for FI-HG AAS that this 'self cleaning' even functions in the vapor phase through the hydrogen radicals that are formed.

Because of its extremely high mobility, *mercury* is particularly prone to exchange reactions with its surroundings. Many authors have dealt with the question of the most suitable materials for the storage of solutions containing mercury. The acquired knowledge can be applied to the reactor for batch systems and also within limits to the GLS. Normal laboratory glassware clearly has the poorest characteristics. PTFE is better, but by no means satisfactory unless it has been previously fumed out with concentrated nitric acid [3001]. In general, quartz and glassy carbon are the most suitable materials. Fuming out with nitric acid reduces the readiness of these materials to adsorb mercury [3002]. WIGFIELD and DANIELS [6319] reported occasional extremely high blank values for mercury in glass apparatus. Mercury which had accumulated on the glass is desorbed by cadmium chloride and thus reaches the solution. LITMAN *et al.* [3537] reported high adsorption rates of mercury on glass, PTFE and, especially, polyethylene (PE) in the concentration range below 1 µg/L. They presumed that the losses could be attributed to a reduction to the metal. KULDVERE [3334] found that tin(II) chloride could be adsorbed onto the walls of PE containers, leading to mercury losses. MACPHERSON and BERMAN [3767] found that noble metals, selenium and tellurium, contained in the samples, sorbed onto polypropylene (PP) surfaces and could not be removed by rinsing with water. This can lead to considerable interferences in the determination of mercury. KOIRTYOHANN and KHALIL [3204] found using PP containers that low results were frequently obtained for calibration solutions, but not for test sample solutions containing an excess of an oxidizing agent remaining from the digestion procedure. The cause was found to be di-t-butyl-methylphenol, which is added to PP as an antioxidant and acted as a reducing agent. The problem could be eliminated by adding an oxidizing agent, such as potassium permanganate, to the calibration solutions. According to current knowledge the most suitable material for the storage of solutions containing mercury and also for reactors is fluorinated engineering polymer (FEP) due to its low wettability and, compared to PTFE, its low permeability.

STUART [5607] drew attention to the fact that every component coming into contact with mercury vapor after its reduction is a potential source of difficulty. To prevent losses, the surfaces coming into contact with mercury should be as small as possible.

Even with the greatest of care, however, some mercury can remain in the system and can be released during the next determination. Frequent blank measurements are thus essential, especially when the mercury concentrations of the samples vary markedly from each other.

TÖLG [5864] drew attention to the possible loss of mercury on the walls of tubes. He particularly mentioned red rubber tubes, which are vulcanized with antimony sulfide and can bind mercury ions to the surface as HgS, and PVC tubes which bind mercury to non-saturated chlorine sites. DANIELS and WIGFIELD [1415] investigated a large number of tubing materials for their adsorption properties for mercury. The losses range from 100% for a silicone tube, to 20–40% for materials such as polyurethane (PU), Tygon®, latex, PVC and PE, to less than 5% for PTFE, Pyrex® and quartz.

As well as for the determination of mercury the materials used for containers and tubes in HG AAS play a role. Using ^{75}Se as a radio tracer, REAMER et al. [4831] found the strongest adsorption on PP and glass. PFA was better than PTFE, the smallest amount of selenium however was retained on silanized glass. PARISIS and HEYNDRICKX [4535] used the sensitivity for the determination of five hydride-forming elements as the criterion for the selection of suitable tube material. Silicone rubber and nylon were the worst, while silanized glass and FEP gave the best values.

4.3.1.4 Special Apparatus for CV AAS

Mention was made in Section 1.8.1 of the particular properties of mercury that led to the construction of special apparatus for the determination of this element, even before the rediscovery of AAS by Walsh. The high toxicity of mercury makes it necessary to determine very low concentrations in widely varying types of sample. This challenge was met by design engineers who developed special systems for the determination of mercury that are discussed briefly in this section.

Since mercury lamps only emit a very limited number of spectral lines and since under the conditions of CV AAS virtually only mercury enters the gas phase, it is possible to dispense with dispersing optical elements and monochromators in systems especially designed for the determination of mercury. HARAGUCHI et al. [2372] described a compact system of this type that could be operated on board ship. The instrument operated in the vacuum UV range and the CsI detector used measured virtually only the radiation of the resonance line at 185.0 nm, while it was not sensitive to longer wavelength radiation. An EDL was used as the radiation source. Chemical vapor generation was performed in a batch system with tin(II) chloride under a stream of nitrogen. The absorption cell was 20 cm long with an internal diameter of 6 mm. The authors reported limits of detection of 0.014 ng and 14 ng/L, respectively, for a 1 mL sample solution.

MCINTOSH et al. [4021, 4023], SCHNEIDER et al. [5172, 5173] and BAASNER and co-workers [2283, 5200] described a similar non-dispersive instrument for the determination of mercury, but operating on the FI principle. This instrument is depicted in Figure 4-42. It operates with a low-pressure mercury lamp, an absorption cell of 24 cm length and 4 mm internal diameter which is heated to 50 °C to avoid condensation, and a UV-sensitive detector. This computer-controlled instrument is fully automatic and has a sampling frequency of greater than 120/hour. The limits of detection are < 0.005 ng and

< 10 ng/L, respectively, for a 500 µL sample volume. An amalgam accessory is offered to further enhance the limit of detection (see also Section 4.3.2.1).

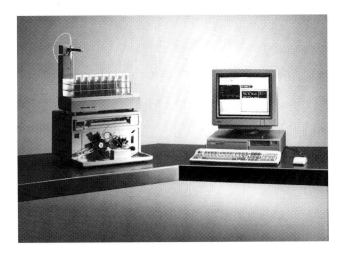

Figure 4-42. FIMS-100—A fully automatic, non-dispersive instrument for the determination of mercury using FI-CV AAS (by kind permission of Perkin-Elmer).

4.3.2 Collection and Preconcentration

Despite the high sensitivity and freedom from interferences of the various techniques of chemical vapor generation, various methods for the collection and preconcentration of gaseous species have been described. These serve to enhance the limits of detection, or to eliminate kinetic effects, or to separate potentially interfering concomitants. Collection can also serve to enable the use of an alternate atomizer, such as a graphite tube atomizer instead of a quartz tube atomizer in HG AAS. The collection of gaseous hydrides under pressure and in cold traps is described in Section 1.8.2, so this topic will not be treated further, particularly since there are no new developments. The only exception is speciation analysis which is treated in Chapter 7.

4.3.2.1 Preconcentration of Mercury

The principle of preconcentrating mercury on noble metals and the historical development of this technique is described in Section 1.8.1. When mercury is collected by amalgamation the resulting absorbance is practically linearly proportional to the test sample volume. In a batch system the influence of the volume of measurement on the sensitivity (Section 4.3.1.1) also disappears, and with CF systems the time-independent signal becomes time dependent. FI systems are generally operated in the same way as CF systems to avoid excessively long sample coils.

The most important parameters for the ultratrace determination of mercury have been described by WELZ et al. [6223]. A gold-platinum gauze has proven to be the most robust and effective collector for mercury. Tin(II) chloride is to be preferred over sodium tetrahydroborate as reductant because of the fewer interferences; a glass fiber filter can be used instead of a drying agent. Helium drives mercury more rapidly from solution than argon or nitrogen, and FEP tubes are superior to all other materials. Under these conditions these authors attained limits of detection of less than 50 pg and less than 1 ng/L, respectively, for a 50 mL test sample volume and a collection time of 2 minutes. The system is shown schematically in Figure 4-43. By applying a 250 mL test sample volume and two independent gas streams to drive out the mercury and to transport it from the collector to the absorption cell, WELZ and MELCHER [6224] attained limits of detection of less than 0.1 ng/L. The purge time was five minutes.

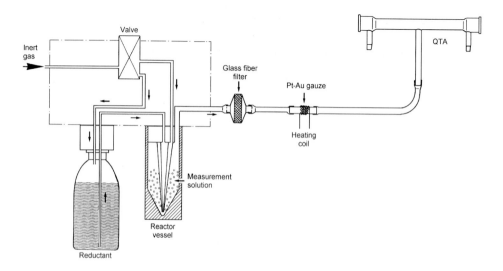

Figure 4-43. Schematic of an apparatus for preconcentrating mercury by amalgamation (from [6223]).

In later work WELZ and SCHUBERT-JACOBS [6249] also successfully employed sodium tetrahydroborate as the reductant for the amalgam technique. They found that it was important that the collector had cooled to a temperature $< 100\ °C$ before collection to prevent it from being poisoned by covaporized hydrides. They also found it of advantage to scrub the gas containing the mercury in sodium hydroxide solution and to dry it over magnesium perchlorate.

MCINTOSH [4022] described an amalgam accessory for an FI system. At a 1 minute collection time, corresponding to an 8.5 mL test sample volume, the limit of detection was 2 ng/L. Considering the optimized volume conditions, this represents a significant improvement compared to batch systems. SINEMUS et al. [5381] placed the gold gauze in a graphite tube to collect the mercury from an FI system. Via the more rapid heating of the graphite tube it was possible to further increase the sensitivity. Additionally, the gold gauze could be heated out after every determination, so that a poisoning could be perma-nently avoided. BAXTER and FRECH [435] placed a platinum gauze or foil in the graphite tube for the same purpose, and YAN *et al.* [6440] collected mercury in a graphite tube coated with palladium chloride.

As well as from solutions, mercury can be preconcentrated on a gold-platinum gauze directly from the air or from gases. Mercury can also be driven out of solid samples by heating and then collected on a noble metal. This type of application is treated in more detail in Chapter 10.

4.3.2.2 Preconcentration of Hydrides

The most successful procedure for the preconcentration of gaseous hydrides on a graph-ite tube coated with palladium was developed independently by STURGEON *et al.* [5640] and ZHANG *et al.* [6540]. Next to the increase in sensitivity, the major advantage of this procedure is that the low stability of certain hydrides no longer plays a role and that a number of interferences cannot be observed in the electrothermal atomizer due to the higher temperature (refer also to Section 8.3.3.3). Using this technique it is also possible to determine elements that only form hydrides with difficulty, such as lead [6439], in-dium [3514], and tin [6543]. Germanium [1561, 5771] can be determined with particular advantage by this technique, since it cannot be atomized in a heated quartz tube and had therefore to be determined in the flame.

For coating the graphite tube a mixture of palladium and iridium [5331] or iridium only [2440, 5155, 5382, 5384, 5385, 5913, 5914] has proven even more effective than palladium. It is not necessary to renew this coating even after more than 300 sequestra-tion and atomization cycles, provided that the atomization temperature does not exceed 2400 °C. The effectiveness of hydride generation and sequestration was investigated by DOČEKAL *et al.* [5331] for palladium-coated graphite tubes and by HAUG *et al.* [2440] on graphite tubes coated with iridium and other stable coatings using radio tracers. It was confirmed that the sequestration of the hydrides of arsenic, antimony, and selenium was virtually quantitative over a wide range of experimental parameters, such as sequestra-tion temperature, mass of modifier, carrier gas flow, or distance of the capillary from the graphite surface (see section 1.8.2). HAUG [2441] found the effectiveness for the seques-tration of lead hydride to be 71%. Other metals such as tungsten and zirconium have also been found as effective for modifying the graphite surface and in a number of cases are superior to an iridium coating in terms of lifetime [2437, 2438, 2441].

4.3.3 Atomization Units

This section deals almost exclusively with the quartz tube atomizers (QTA) used in HG AAS; when not heated they are also frequently used as absorption cells for CV AAS. Flames are no longer used as atomizers in HG AAS since they do not bring the required sensitivity. Historical developments, such as the diffusion flame, are discussed in Sec-tions 1.6 and 1.8.2. Nowadays graphite tube atomizers (refer to Section 4.2.2) are used almost exclusively in combination with the collection of the hydrides (see Section

4.3.2.2). Historically relevant investigations using graphite tube atomizers are described in Section 1.8.2.

4.3.3.1 Absorption Cells for CV AAS

Since mercury is reduced to the element in solution and 'atomized' by chemical vapor generation, merely an absorption cell but not an atomizer unit is required for CV AAS. Frequently the same quartz tube atomizers as used in HG AAS (see Section 4.3.3.2) are employed since in any case the same accessory system is applied. The sole difference to HG AAS is that the cell is either not heated or heated only to 50–100 °C to prevent the condensation of water vapor [1067]. On occasions specially formed tubes [1580, 5646] or tubes geometrically adapted to the profile of the radiation beam [1068] have been used.

Only very seldom has use been made of the fact that in CV AAS the absorption cell can in principle be as long as desired since it does not have to be heated and mercury atoms have an almost unlimited lifetime. The cells are rarely longer than 20–25 cm [2283, 2372, 4023, 5172, 5173, 5200]. Quite frequently, relatively short cells of 3–4 cm length are used [642, 2179, 4234], even though the sensitivity increases practically linearly with the cell length.

Without doubt the simplest 'atomizer' for the determination of mercury in a conventional AAS instrument was described by TONG [5875] and further improved by LAU et al. [3464]. These authors used a standard quartz cell as employed in UV spectroscopy, which they additionally silanized. They placed the test sample solution and the tin(II) chloride solution in the cell, shook the cell manually for 70 s, and then placed it in the sample compartment of the spectrometer such that the radiation beam passed through the vapor phase above the solution. They found that delay times of up to 50 s had no influence on the measure.

4.3.3.2 Atomization Units for HG AAS

Externally heated quartz tubes are discussed exclusively in this section. The flame-in-tube (FIT) atomizer, largely investigated by Dědina and co-workers, in which a fuel-rich oxygen-hydrogen flame burns in the side inlet tube of an unheated quartz tube, was introduced in Section 1.8.2 and is discussed in more detail in Section 8.3.2.1.

For externally heated QTAs a distinction is made between electrical heating [1279] and flame heating [5830]. There is no principal difference in the function of either atomizer. Flame-heated QTAs are cheaper to acquire than electrically-heated QTAs but more expensive to run. The silica tube is mostly mounted in a holder that allows it to be moved into the flame. It is important when operating a flame-heated QTA to ensure that the hydrogen does not ignite at the tube ends, since this leads to poorly reproducible nonspecific absorption that can hardly be corrected [1612]. An ignition can be prevented by mounting graphite cooling rings at the ends of the tube [6212]. A further problem is the rapid devitrification of the silica tube, particularly if the flame is too hot or too oxidizing [6039].

Electrically-heated QTAs are much easier to operate and easier to control; they can be fitted with a temperature measurement and control system [6213]. They can also be closed at the ends with quartz windows [6220, 6234] and thus come much closer to a stable 'absorption cell' than flame-heated QTAs.

The significance of silica (quartz) as the tube material is among other things illustrated by the work of WANG *et al.* [6162]; they found that a temperature of 1200 °C was required to atomize arsine in an aluminium oxide tube but 800–900 °C was sufficient in a quartz tube. CHAMSAZ *et al.* [1176] found for the atomization of tin that for the same temperature the sensitivity was 20% higher with a silica tube compared to aluminium oxide, and graphite had a substantially lower sensitivity. For optimum sensitivity QTAs must be 'conditioned' occasionally. HARTFIELD [2430] proposed sandblasting for this purpose. In the current author's experience bathing in hydrofluoric acid leads to the best results [6216].

Argon is normally used as the carrier gas. Even at an early stage it was nevertheless recognized that the addition of oxygen to the argon could improve the sensitivity, especially at lower atomization temperatures [2187, 6216]. Detailed investigations by DĚDINA and co-workers [1461–1463], however, have shown that the oxygen requirement is essentially a function of the inlet tube of the atomizer. The smaller the internal diameter of the inlet tube, the lower is the oxygen requirement (for details see Section 8.3.2.2). Especially for FI systems, which operate with relatively low gas flowrates, particularly effective inlet tubes can be used, as depicted in Figure 4-44.

Figure 4-44. QTA for the FI technique with an inlet tube of 1 mm internal diameter (from [1461]).

As shown in Figure 4-44, the analyte hydrides are normally introduced at the center of the heated quartz tube. QTAs have also been described where the analyte hydrides enter at one end of the tube and exit at the other [2758, 6430, 6446]. The tube length and diameter have a complex influence on the analyte signal and on interferences due to concomitants [1458]. The reason for this is that the reactions in a QTA are far from thermodynamic equilibrium and wall reactions play a major role. Reaction mechanisms are treated in detail in Section 8.3.

5 The Individual Steps of an Analytical Method

The principal aim of a quantitative analysis is to determine the analyte content in an unknown sample with good *trueness* and *precision*, i.e. with high *accuracy*. In practice, however, this aim is unfortunately made difficult by a variety of problems that are more or less associated with every analytical method.

Under the term *analytical method* we understand a set of written instructions completely defining the procedure to be adopted by the analyst in order to obtain the required analytical result [6349]. Nevertheless the results obtained are influenced by both *random errors* and *systematic errors*. Such errors can occur in every step of an analysis and negatively influence its accuracy to a degree corresponding to their share of the entire method. It is thus of major importance that all steps of an analytical method are documented and that special attention is drawn to steps that are particularly critical. It is also of equal importance that the analyst has all the steps of a method under control since only then can possible problems be recognized, evaluated, and if necessary remedied. To achieve the desired goal it is necessary to adhere to a number rules for the conductance and documentation of an analysis. These rules are generally summarized under the term *good laboratory practice* (GLP). Since errors can occur at every time point in an analysis, controls must also take into account every analytical step. The most important sources of error are summarized in Table 5-1; they are discussed in detail in the sections that follow.

Table 5-1. The individual steps of an analysis and the associated problems.

Analytical step	Associated problems
Sampling	Representativeness of the sample with respect to the objective of sampling; contamination during sampling.
Sample conservation	Stability of the constituents; contamination; losses.
Sample digestion or decomposition	Analyte losses due to volatilization or insufficient digestion; contamination from reagents and containers.
Trace–matrix separation	Efficiency of separation; completeness of preconcentration; contamination from reagents and containers.
Instrumental determination	Suitability and performance of the selected procedure for the given analytical task; condition of the instruments used.
Calibration	Selection of a suitable calibration technique; quality of the calibration samples and calibration solutions employed; adequate regard of the blank values.
Analytical report	Adequate assessment of the results and quality control.

The role of effective methods development is to detect possible sources of error and to eliminate them, or at least to reduce them to an acceptable level. It is the task of *quality control* and *quality assessment* procedures (these terms are defined in Section 5.5) to ensure that an analytical method brings the expected performance in routine operation. If quality assurance is to be effective, it must take all steps of an analytical method into account and should not be limited to just a check of the AAS measurement.

5.1 Sampling and Sample Preparation

In this section we shall discuss the general problems of element determinations in the trace and ultratrace ranges [1379, 2217, 5196, 5865, 5866, 5916, 5917, 6093] and the general measures that can be taken to avoid errors [1378, 1901, 3007, 5019, 5213, 5918, 6350]. Individual techniques and problems specific to the elements and matrices are treated in Chapters 8 through 10.

Nowadays the mg/L concentration range in solutions and the μg/g concentration range in solids are relatively well under control so that with careful work larger systematic errors are hardly to be feared. On the other hand the μg/L concentration range in solutions and the ng/g concentration range in solids are critical; the risk of systematic errors increases exponentially with decreasing analyte concentration. The errors can lead to diminished results if losses of the analyte take place during sample preparation, or to increased results if contamination with the analyte occurs. Since the sources of error always increase with the number of analytical steps, the number of tools and containers used, and the quantity of reagents added, these should as far as possible be kept to a minimum. Thus for ultratrace analysis direct procedures are to be preferred to those that require a larger number of pretreatment steps.

5.1.1 Sampling

Initially a portion is removed from the material to be examined; this is the *laboratory sample* that subsequently will be examined [5961, 5965]. During this procedure, which is termed *sampling*, care must be taken to ensure that the laboratory sample is *representative* of the material from which the sample was taken or of the situation which is to be examined. Even the most accurate analysis can be performed *ad absurdum* if sampling is incorrect [2299]. Unfortunately sampling is frequently performed without the participation of the analyst, so that the control over a very important step is thus lost. The significance here should not be underestimated since the largest systematic errors take place during sampling and cannot subsequently be corrected, even by the most high performance instrumental analysis. THIERS [5816] stated with justification already in 1957 that an analyst should not spend time analyzing a sample whose history is not fully known to him. The responsibilities of the analyst clearly extend beyond the confines of the analytical laboratory [4874].

For sampling to be meaningful the variability of the material or of the situation for which a sample is required must be taken into account. Especially for inhomogeneous materials or markedly changing situations, careful planning and control are prerequirements. For inhomogeneous materials, for example, the *laboratory sample* must have a minimum size for it to be statistically representative of the parent material. If the laboratory sample is too large to be used for analysis, it must be subdivided after suitable homogenization. If a situation for which sampling is required changes with time (i.e. is dynamic), such as river water or waste water streams, the dynamic conditions must be taken into account by taking portions of the parent material, either continuously or at determined intervals. GY [2298, 2300–2302] has developed a theory of sampling that discusses these problems with the aid of statistics.

Sources of error for sampling strategies (place, time, duration and method of sampling), for the selection of apparatus (contamination from the materials used, sorption losses), for execution (sampling, subdividing, filtering, conserving) and delivery (transport, identification, protocol) are manifold and a detailed discussion is not possible within the limits of this monograph. Further information is provided as appropriate under the individual elements (Chapter 9) and under applications (Chapter 10). The problems of sampling are merely indicated superficially in this section.

Problems with the sampling strategy are particularly prevalent for environmental analysis (see Section 10.3) where inhomogeneous systems are examined, such as the soil of a derelict industrial site, or where dynamic systems must be sampled, such as the stack gas plume of an industrial estate or the discharge of waste waters into a river. However good the analytical method might be, it cannot bring the correct results if the sample is not representative for the situation; for example when an air sample is taken while the wind is blowing the stack gas plume away from the sampling location, or when a river water sample is taken while no waste water is being discharged.

Problems with the selection of apparatus for sampling occur most noticeably in the trace and ultratrace ranges since the choice of materials or the design of apparatus has a marked influence on contamination or element losses. Thus for example the determination of chromium or nickel in serum cannot be indicative if the blood sample was taken using a steel needle because of contamination. Sampling sea-water is problematical if a surface sample is taken in the wake of the ship, and sampling deep ocean waters poses problems due to the necessary pressure-resistant metal containers and steel cables.

Even if the few examples presented here appear to be exaggerated, they nevertheless represent daily practice, particularly when sampling is not performed under the control of the analyst. It cannot be repeated often enough that striving to attain relatively high precision in the instrumental determination (which as a rule is better than 10%) is futile if during sampling the given situation has been falsified by orders of magnitude. To prevent this, strategies have been developed that allow the quality of sampling to be judged and assured [5840, 6294]; a detailed discussion is however beyond the scope of this chapter.

5.1.2 Sample Conservation and Storage

In general analyses are performed in a laboratory that has the necessary infrastructure for performing the required manipulations, such as sample preparation rooms, laminar flow cabinets, clean rooms and instrument rooms. If an analysis is to have any meaning, care must be taken to ensure that the laboratory sample is unadulterated during transportation to and storage at the laboratory while awaiting further sample pretreatment. It must also be ensured that sample solutions made during sample preparation, e.g. by dissolving a solid sample, remain stable using suitable *sample conservation* procedures until the measurement is performed [73, 4758]. The problems of sample conservation and storage have been described in a multitude of papers and review articles for natural waters [5532, 5638], ground waters [5648], rain water [1184, 4861], sea-water [2112], environmental samples [1359], biological materials [71, 73, 4472, 5166, 6060, 6124], body fluids [3397, 6132, 4061, 4291], and foodstuffs [2943].

The desire to conserve a sample in all of its complexity is almost futile; rather more, sample conservation must be matched to the subsequent analysis and thus be under the control of the analyst. Depending on the given task, a sample conservation procedure should 'freeze' selected properties of the sample, such as pH and pE values, as well as the contents of main and trace constituents or species. Microbiological activity that can lead to the decomposition of individual constituents must be suppressed just as much as losses due to wall adsorption or contamination through containers, reagents, or the laboratory atmosphere. If a number of constituents must be determined in a laboratory sample, it may be necessary to subdivide this into a number of aliquot portions that are each subjected to different conservation procedures. If a number of constituents are in a dynamic equilibrium (e.g. redox equilibrium) that is part of the examination, such as in speciation analysis (Chapter 7), conservation is exceedingly difficult [422, 1929, 2073, 2076, 4075, 4567].

Analyte losses due to adsorption on materials that come into contact with the sample solution already can occur at relatively high analyte concentrations and can become of substantial magnitude in the ultratrace range. Materials that have been investigated in detail include glass [1684, 4890, 5297, 5606, 5660], polyethylene [4890, 5297, 5606, 5660], and polypropylene [1684, 5606]. Perfluoroalkoxy polymers (PFA) are evidently particularly suitable for the storage of trace contents [1745, 2106, 4489]. The neutral solutions of many metals are not stable and tend to hydrolysis [5455, 5456]; in contrast silicic acid precipitates rapidly from acidic silicon solutions [878]. We should also bear in mind that with dilute solutions the precipitate often cannot be recognized and adheres to the walls of the container. Very dilute solutions, even when acidified, are stable for only short periods; a fact that must be taken into account for calibration solutions. This has been demonstrated very effectively for mercury [3001], but even with less mobile elements such as lead [2811] losses of up to 50% can occur within an hour when neutral aqueous solutions are stored in glass vessels. Similar behavior has also been reported for calcium and magnesium [3812], and it can be more or less expected for all elements. Since various ions compete for the active sites on the walls of the container, a general dependence of the losses on the total concentration of the ions and on the pH value can be observed; the losses increase with decreasing concentration and increasing pH value [533].

Frequently it is sufficient to simply acidify the dilute solution with hydrochloric or nitric acid [2267] to keep losses under control, at least for several hours; for a number of elements, however, it is necessary to add a complexing agent to avoid adsorption effects. Careful selection of the container material is essential, since this frequently has a decisive influence. Since adsorption processes are also proportional to the adsorption area, the latter must be kept as small as possible. As well as adsorption on the walls of the container the reverse process, namely the desorption and contamination by container materials, plays a major role particularly in trace analysis [4891, 5916]. This has been studied in detail for glass [34], silica [2533], plastic containers in general [4178], and also polyethylene [2539, 3116] and PTFE [3116]. Further investigations that mostly refer to selected elements are discussed in Chapter 9.

Normal laboratory glass is generally unsuitable for ultratrace analysis. Materials with relatively inert surfaces, such as FEP, PFA, PTFE, high impact PE, silica, and glassy carbon, are better. But even these materials are not *a priori* uncontaminated and do not remain permanently free of trace elements. Since adsorption processes are reversible, sample contamination can take place when test samples are stored in containers that have had contact with other samples. It is thus essential that containers used for trace determinations are thoroughly cleaned prior to use [3033]. Fuming out with nitric acid has proven to be particularly effective; a suitable apparatus is depicted in Figure 5-1. Experience has shown that rinsing or leaching is not adequate in the ultratrace range [539, 5494]. As an alternative to thoroughly cleaning PFA containers in nitric acid, treatment with dilute hydrogen peroxide solution, ammonia and triethanolamine has been proposed [5205].

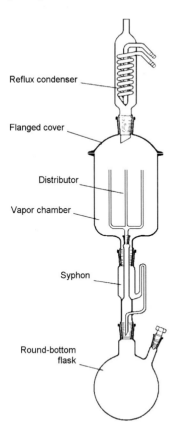

Reflux condenser

Flanged cover

Distributor

Vapor chamber

Syphon

Round-bottom flask

Figure 5-1. Apparatus for fuming out laboratory ware for use in trace analysis (Kürner Analysentechnik, Federal Republic of Germany).

Nevertheless, even the most thorough cleaning of a container that has been used previously to store sample or calibration solutions of higher concentration cannot make it suitable for test samples that contain the same element in the ultratrace range. To prevent contamination of this type, the only remedy is to use disposable laboratory ware or to protocol the history of the containers.

5.1.3 Sample Pretreatment

Since the risk of systematic errors increases with the extent of sample pretreatment, it should be the aim of the analyst when developing a method to keep this to a minimum. Depending on the nature of the sample pretreatment required, one can distinguish between aqueous solutions or suspensions, organic solutions, emulsions or suspensions, and inorganic or organic solids.

Aqueous solutions require the least effort for sample preparation. Samples of this type include natural waters, drinking water, process waters and waste waters, beverages such as wine and beer, and biological fluids such as blood and urine. If the laboratory samples contain solids (suspended matter), either this matter must be separated and, where necessary, analyzed separately, or the sample must be digested, i.e. the solids content must be taken into solution. If solids are separated by filtration, care must be taken to ensure that there are no losses of analyte through sorption on the filter and that there is no contamination. For biological materials such as body fluids, cellular materials are preferentially separated by centrifugation. If the sample contains organic constituents to which the analyte is bound, or if it contains constituents that could cause interferences, a digestion may be necessary. This is very often the case for sample solutions that are to be analyzed by HG AAS or CV AAS since the analyte must be present in a form that is amenable to the chemical reaction. A digestion can often be avoided for F AAS or GF AAS since in principle it takes place *in situ* in the atomizer.

Organic liquids, mostly petrochemical products such as fuels or lubricating oils, can be analyzed by F AAS and GF AAS, either directly or after adjustment of the viscosity by dilution with an organic solvent. The analyte can be in the organic liquid in the form of a complex with organic ligands, as the organometallic compound (e.g. tetraethyl-lead), or in the form of colloidal metal particles (e.g. in used lubricating oils); this can have a major influence on the sensitivity and may make it necessary to calibrate with similar calibration solutions. Even though many metals are available commercially as organic standards, nevertheless it is a prerequirement for calibration that the form or compound in which the analyte is present is known and that this is available as a standard. Also the analyte should not be present in the sample in a number of different compounds that exhibit differing sensitivity. In such cases it is on the safe side to perform a digestion or to extract the constituents into the aqueous phase. For F AAS, techniques of on-line sample preparation using FI can be utilized to eliminate differences in sensitivity by introducing organic solutions with an aqueous carrier as emulsions. With GF AAS the majority of such samples can be analyzed directly since sample preparation is possible in the graphite tube, e.g. *in situ* ashing.

Solid samples can in general only be directly analyzed relatively free of problems by GF AAS, so that a digestion is mostly required at the start of an analysis. Particularly for volatile elements and those that form volatile compounds, there is a risk of losses. Fusions and dry ashing are the most seriously affected due to the higher temperatures employed, but acid digestions are not always without problems. The best protection against vaporization losses is to use composite (hybrid) techniques in closed systems; in recent times a number of such techniques utilizing FI have been proposed (see Section 6.6).

A particular risk of systematic error exists when a technique that functions without problems at higher analyte concentrations is extended to the ultratrace range without prior investigation. This is true for extraction procedures, for example, since in this range it is likely that they are no longer quantitative.

Systematic errors caused by *accidental introduction of the analyte* from reagents, instruments and laboratory ware [34, 2533, 4179, 4891] or the environment [4891] are prevalent in ultratrace analysis. In extreme cases the true content can be falsified by orders of magnitude, so that the analytical result bears no relationship to the original concentration of the analyte in the sample. Reagents in particular must be chosen with especial care for use in the ultratrace range. It is usually easier to obtain pure mineral acids than salts, so that acid digestions are to be preferred to dry fusions. Also the quantity of reagents used should be kept as small as possible, so that microwave-assisted acid digestions, for example, are especially suitable. Such digestions are normally performed in closed containers made of largely inert materials (e.g. quartz, FEP) and with a minimum of reagents.

Due to the high sensitivity of the graphite furnace, hydride-generation, and cold vapor techniques, even chemicals of the highest analytical reagent grade can generate blank values. In Figure 5-2, examples are presented of mercury blank values in proprietary acids of varying purity determined by the amalgam technique. In such cases the purification of reagents for use in ultratrace analysis is essential. Acids are best purified by distillation at a temperature below the boiling point (subboiling distillation); a suitable apparatus is depicted in Figure 5-3. It is essential that purified acids come into contact only with scrupulously clean containers made of suitable materials [3327] and that they are not stored for any length of time since they very quickly take up trace elements from the environment. *Ultrapure water* [5661] is especially critical and should only be permitted to contact materials that have extremely low trace metal contents (PTFE, FEP, high impact PE, silica); since it takes up trace elements very readily from the environment, it should always be prepared freshly when required.

Without doubt *mechanization and automation of analytical procedures* play a major role in the trueness of the results of a trace determination; this aspect is discussed in detail in Chapter 6. A classical example is the dispensing of measurement solutions in GF AAS; with manual dispensing a repeatability of better than 5% could rarely be attained (refer to Figure 4-33, Section 4.2.6). Even more critical, however, were the general problems of contamination which often led to substantial systematic errors. For example, pipet tips were frequently contaminated with trace elements and the contamination was often difficult or even impossible to remove [539, 5494]. The contamination of yellow pipet tips with cadmium is well known [5060]. The problems of poor repeatability and trueness were only brought under control after the introduction of autosamplers [6206]. Carryover from one sample to the next could also be minimized [5579].

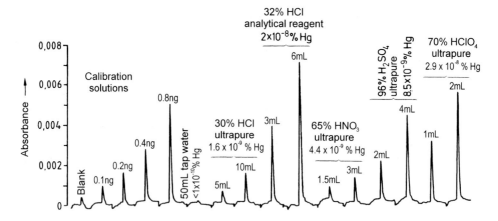

Figure 5-2. Blank values for mercury in various acids of ultrapure and analytical reagent grades, measured by the amalgam technique.

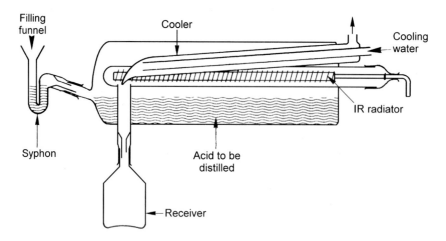

Figure 5-3. Apparatus for subboiling distillation (Kürner Analysentechnik, Federal Republic of Germany).

A recent example of the contribution that well-conceived mechanization and automation can make to the trueness of ultratrace analyses is the application of FI techniques for on-line analyte preconcentration and matrix separation in GF AAS (refer to Section 6.5). By means of fully automatic preconcentration and separation using reagents which are purified on-line, followed by automatic elution of the analyte into the graphite tube, it has been possible in such a composite system to improve the limit of quantitation for real samples such as sea-water by around three orders of magnitude and at the same time to lower the blank values by about the same amount [5510, 5513, 6258].

5.2 Measurement, Calibration, Evaluation

The terminology used in the original German edition is based on the following German Standards: DIN 51 401 'Atomic absorption spectrometry; Terms' [1501], and, where appropriate, DIN 51 009 'Optical atomic spectral analysis - Principles and definitions' and DIN 32 645 'Chemical analysis; Decision limit, limit of detection and determination limit; Estimation in case of repeatability; Terms, methods, evaluation'. Due to certain particularities of the German language, the direct translation of some of the definitions would result in nomenclature contrary to the ISO and IUPAC recommendations. In such cases the appropriate ISO [2786–2789] and IUPAC [6350] nomenclature was used throughout.

5.2.1 Samples and Measurement Solutions

Initially a portion is removed from the material to be examined and is delivered to the laboratory; this is the *laboratory sample* (see Section 5.1.1). The material obtained by suitable treatment or preparation of the laboratory sample (reduction in particle size, homogenization, filtering, stabilization, etc.) is the *test sample* (or, if only analytical chemistry is involved, the *analytical sample*). A *test* (or *analytical*) *portion* is removed from this for analysis. Under given prerequisites it is possible to analyze solid samples directly by GF AAS (refer to Sections 1.9 and 4.2.7), but in most cases the analytical portion must be taken into solution prior to further examination. The *test sample solution* is produced by treating the test portion with solvents, acids, etc. or by digestion. This solution can be used directly or after further pretreatment steps, such as dilution, addition of buffers, etc., for the measurement. A solution used to perform a measurement is termed the *measurement solution.*

The measurement solution must be stable for the duration of the analysis. This is especially important if an autosampler is used, since the measurement solutions can stand for up to several hours before they are analyzed. Solvent extracts with volatile solvents or complexes with unstable metal chelates are particularly critical [5163, 6103]. But even for aqueous solutions it must be borne in mind that the solvent can evaporate to a degree depending on the ambient conditions. Especially at low relative humidity it can be observed that the measurement solutions become more concentrated with time; this effect becomes greater the smaller the volume in the sample container. In an autosampler for GF AAS, SLAVIN [5436] observed a rate of loss of approximately 35 µL/h from a 2 mL container. This could be reduced to 10 µL/h when the autosampler was covered and a container filled with water for humidification was placed with the measurement solutions. Such measures should always be taken to avoid errors due to drift for replicate measurements and calibration. HOFFMANN [2619] modified an autosampler such that the sampling capillary could pierce the PTFE septa used to seal the sample containers.

One or more *calibration samples* are required for the quantitative determination of an element by AAS, usually in the form of *calibration solutions*. Traditionally the term *standard solution* is also used, but this term is deprecated by IUPAC. For the preparation of solutions, only laboratory ware that meets the requirements of titrimetric analysis should be used [3398, 6280]. Prior to use, all laboratory ware should be cleaned in warm dilute nitric acid, e.g. 0.1 mol/L HNO_3, thoroughly rinsed with water, and checked for contamination. Only doubly distilled or water of similar quality should be used. All chemicals should be of analytical reagent or a higher grade of purity. The concentration of the analyte in the chemicals used should be negligible in comparison to the lowest concentration expected in the measurement solutions.

Calibration solutions are normally prepared from a *stock solution* that contains the analyte in an appropriately high, known concentration, frequently 1000 mg/L. Such a stock solution can be prepared by dissolving 1 g of the ultrapure metal or a corresponding weight of an ultrapure salt in a suitable acid etc. and making up to 1 L. For the preparation of stock solutions, only metals and salts are suitable in which the content of the analyte is known accurately to within at least 0.1% [4180]. The content of other elements should also be known and should as far as possible be negligible. Depending on the AAS technique to be used, certain acids or salts may be unsuitable for the preparation of stock solutions, for example sulfates for F AAS or chlorides for GF AAS. With proper storage, stock solutions normally have a shelf-life of about one year.

Calibration solutions are prepared from the stock solution by serial dilution to give a *set of calibration solutions*. When planning a set of calibration solutions, a procedure should be chosen with the lowest possible variance and highest trueness. The dilution error increases with both the number of dilutions and the size of the dilution step [3398, 4593, 6525]. Dilution factors of 1:100 per dilution step should not be exceeded. The rough calculations shown in Table 5-2 give an indication of the errors that can be made even with careful operation when stock solutions are diluted by a factor of 1000 by dispensing 100 μL into a 100 mL volumetric flask and then diluting 1:20 in a second step as practiced by many laboratories.

As can be seen, the true concentration of the 50 μg/L calibration solution prepared in this way can be anywhere between 48.7 μg/L and 51.3 μg/L, thus exhibiting an error of ± 2.6%. This assumes that the given tolerances are maintained (as stipulated in German standard DIN 12664 for Class B volumetric glassware) and that all operations are performed at a temperature of 20 °C, since the quoted volumes of volumetric flasks are valid at this temperature. It is thus clear that considerable errors can be made at this stage. Such errors can easily be increased if the glassware, as is often the case, is washed regularly in a laboratory glass washing machine at increased temperature, and if piston pipets are not regularly checked and maintained or if they are not used properly. A further source of error can be the stock solution itself. The errors can be greater if a weighing of 1.00 g of the analyte element, or if a commercially available stock solution (e.g. 1.000 g ± 0.002 g), is made up to volume in a 1.0 L volumetric flask of class A (1000 ± 0.4 mL). Under these circumstances the only recommendation is either to use volumetric glassware of at least class A and to treat it with the same care as pipets and to make regular gravimetric checks, or, better, to perform the dilutions gravimetrically. We must also mention at this point that many AA spectrometers incorporate software for calibration which uses a least mean squares algorithm for curve fitting. In principle such algorithms can be used only under the premise that the error in the calibration solution is significantly smaller than the error in the AAS measurement. Considering the high precision with which modern instruments operate, this premise is probably in need of revision. Since errors in calibration are transferred in full to the subsequent analyses, it is

Table 5-2. Typical errors in the preparation of a calibration solution of 50 µg/L from a stock solution of 1.000 ± 0.002 g/L in two steps: 100 µL→100 mL and 1000 µL→20 mL.

Parameter	Quantity	Current Value as Influenced by the Tolerance			
Tolerance of pipet		−	−	+	+
Tolerance of volumetric flask		+	−	+	−
Analyte concentration in stock solution: 1.000 ± 0.002 g/L; pipet volume 100 ± 1 µL	Dispensed mass of analyte, µg (nominal 100 µg)	98.8–99.2	98.8–99.2	100.8–101.2	100.8–101.2
First dilution 100 µL → 100 mL; flask volume 100 ± 0.15 mL	Analyte concentration, µg/L (nominal 1000 µg/L)	986.5–990.5	989.5–993.5	1006.5–1010.5	1009.5–1013.5
Pipet volume 1000 ± 10 µL	Dispensed mass of analyte, ng (nominal 1000 ng)	976.7–980.6	979.6–983.5	1016.5–1020.6	1019.6–1023.7
Second dilution 1000 µL → 20 mL; flask volume 20 ± 0.06 mL	Analyte concentration, µg/L (nominal 50 µg/L)	48.69–48.88	49.13–49.32	50.17–50.37	51.13–51.34

futile in the above example to expect an accuracy of ± 1 µg/L when the calibration already contains a higher level of uncertainty.

When performing dilutions, care should always be taken to ensure that the acid concentration is sufficient to effectively prevent adsorption losses. The addition of concomitants in higher concentrations can also increase the stability. Nevertheless, the greatest of care is required with this technique due to the risk of contamination with the analyte. Very dilute solutions in the µg/L range must be prepared daily because of their low stability and the risk of contamination. Calibration solutions should contain the same chemicals as used in the preparation of the test sample solution and also other constituents that can influence the measurement in the same or very similar concentrations to the test sample. For example, if a digestion was used in the preparation of the test sample solution or if a preconcentration step was performed, the calibration solutions should generally be taken through the same procedure. It is essential that the calibration solutions contain the same spectrochemical buffers or modifiers used in the preparation of the test sample solution in the same concentrations as in the test sample solution (see Section 5.4).

A *zeroing solution* is used to set the null signal (baseline) of the spectrometer; usually this is the pure solvent, e.g. deionized water. The zeroing solution is not permitted to contain the analyte in a measurable concentration. If it is not possible to prepare a zeroing solution that meets this requirement (e.g. for GF AAS, HG AAS, or CV AAS), the baseline of the spectrometer can be set without a solution, for example by passing an inert gas through the atomizer.

Under no circumstances may a solution that contains the analyte in a measurable concentration, such as a reagent blank solution (see below), be used for setting the baseline of the spectrometer. Doing so always leads to a deterioration of the precision for calibration (see Section 5.2.3.2) and in the most unfavorable case can lead to systematic errors (see Section 5.2.2). The incorrect use of a reagent blank solution instead of a zeroing solution for setting the baseline can in part be blamed on the instrument manufacturers since in the instrument software simply the term 'blank' is often used. Nevertheless, a general criticism of the manufacturers cannot be justified (e.g. [1756]) since correct treatment of the 'blank problem' is not automatically possible, as will be shown in the following section. With increasing complexity of a method the complexity of the determination of the blank value increases and this can only be duly accounted for by appropriate external action by the analyst.

As stressed repeatedly, for every type of sample pretreatment procedure the risk exists of contaminating the sample with the analyte through reagents, laboratory ware, and the ambient air. In order to recognize contamination of this nature and to keep the resulting errors as low as possible, suitable reagent blank samples or blank samples must be prepared, treated, and measured in the same way as the test sample. The analyte content determined in such samples must be taken into consideration during calibration and when evaluating the results (see also Sections 5.2.2 and 5.2.3).

The *reagent blank solution* contains all of the chemicals used for the preparation of the test sample solution in the same concentrations as in the measurement solution, such as solvents, acids, and especially spectrochemical buffers, modifiers, etc., but it does not contain the concomitants of the test sample (matrix). The analyte is never added to the reagent blank solution, but it can be present as a result of contamination. The analyte content determined in the reagent blank solution is frequently termed the *reagent blank value*. If the test sample is digested or subjected to some other sample preparation procedure, such as separation or preconcentration, the reagent blank solution should be taken through the same procedure. This solution is termed the *blank test solution* and the analyte content determined in the resulting solution is termed the *procedural blank value*. If the *field blank value*, which may occur during sampling, is additionally taken into account, we finally obtain the *method blank value* for the entire analytical method. It can thus be seen that for complex, multistep analytical methods, there is little point in determining the reagent blank value alone since considerably more of the analyte can be introduced during preparation steps, such as reducing the size of the sample, than from the reagents.

It is very important that the *analyte content of the reagent blank solution* or *blank test solution* be determined separately using the same calibration procedure as for the test sample solutions and subtracted from the analyte content of the measurement solutions. In other words, the reagent blank or blank test solution must be calibrated and evaluated in exactly the same way as a test sample solution (see Section 5.2.2). As already emphasized, the uncertainty of measurement is among other things increased if the reagent blank solution is used falsely to set the baseline of the instrument or if, in end effect the same, the absorbance (integrated absorbance) of the reagent blank solution is subtracted from the absorbance of the measurement solutions. Depending on the technique of calibration employed, systematic errors can also be introduced as a result (see Section 5.2.2).

In contrast to the reagent blank solution, the *matrix blank solution* contains all chemicals used in the preparation of the sample solution, such as acids, spectrochemical buffers, modifiers, etc., *together with all constituents of the sample that influence the determination* in the same or very similar concentrations as in the test sample solutions. The analyte is never added to this solution, but like the reagent blank solution it can be contained as a result of contamination. In the same way as the blank test solution the matrix blank solution must be taken through all the sample preparation steps if it is to be representative for the analytical method. The matrix blank solution must also be treated in the same way as a test sample solution for calibration. The matrix blank solution serves primarily for the determination of the limit of detection (LOD) and the limit of quantitation (LOQ) (refer to Section 5.2.3.3). In addition to the reagent blank solution, it also plays an important role for the determination of the analyte content using the standard calibration technique with matrix matching (see Section 5.2.2.1) or the bracketing calibration technique (see Section 5.2.2.2). This is clear if we consider that analyte can be introduced into the matrix blank solution while attempting to match the sample matrix as closely as possible, while additional analyte is not introduced into the test sample solution. The analyte content in the matrix blank solution can then be higher than in the reagent blank solution; the former should not be confused with the latter and subtracted from the analyte content in the sample solutions.

Reference materials are of considerable importance for the accuracy of an analysis. These are materials of which one or more characteristics are known so accurately that they can be used for calibration or for analytical control within a quality assurance program. Reference materials are either synthesized from high purity materials or selected from preanalyzed samples. If the characteristics of a reference material have been determined, documented, and guaranteed by a number of independent and recognized procedures, it is called a *certified reference material* (CRM). Reference materials used for calibration, for example with the standard calibration technique or the bracketing technique, are referred to as *calibration samples*. If a reference material is taken along with the test samples through the entire analytical procedure to check the accuracy of the analytical results, we refer to this as an *analytical control sample* (refer to Section 5.5).

5.2.2 Calibration

The relationship between the amount concentration c, the mass concentration β, or the mass m of the analyte in the measurement solution as the desired quantity and the absorbance or the integrated absorbance as the measured quantity is established and described mathematically in AAS by the use of calibration samples, usually in the form of calibration solutions. This procedure, known as *calibration*, includes the preparation and measurement of the calibration solutions. The same principle applies to the analysis of solid and gaseous samples, but is not discussed in detail. The relationship between the absorbance A or the integrated absorbance A_{int} and the concentration or mass of the analyte is given by the *calibration function* e.g.:

$$A = f(c), \tag{5.1}$$

$$A = f'(b), \tag{5.2}$$

$$A_{\text{int}} = f''(m). \tag{5.3}$$

The graphical representation of the calibration function is the *calibration curve* (or *analytical curve*). In the optimum working range there is a linear relationship between the absorbance A or the integrated absorbance A_{int} and the amount concentration c, the mass concentration β, or the mass m of the dissolved substance (see Section 2.5.2). This is an advantage for the evaluation of the measurement since the linear equation represents the simplest of all possibilities. The calibration function can also be established by using curve fitting of a suitable mathematical model to the determined calibration data. Such models are described by the general function:

$$A = f(c, a, b_1...b_m), \tag{5.4}$$

where a, $b_1...b_m$ are the parameters for the function. These unknown parameters must be so chosen that the model describes the measured data as closely as possible.

Equations 5.1 through 5.4 are valid for the selected procedure and for the atomic absorption spectrometer used for the measurement. They are subject to fluctuations from day-to-day and also within a long series of samples and should thus be established newly every day and checked at regular intervals.

A number of different calibration techniques are available in AAS; these can lead to varying results with respect to precision and accuracy. The most important calibration techniques are the standard calibration technique (with or without matrix matching), the bracketing technique, and the analyte addition technique. Occasionally the reference element technique is also employed. Under given circumstances, interferences can be eliminated or avoided through the choice of a suitable calibration technique. On the other hand, the choice of an unsuitable calibration technique can be the cause of systematic errors (refer to Section 5.4). In addition to these classical calibration techniques, the recording of transient signals opens the possibility of extracting additional information from the signal profile and using this in the evaluation. This information can, for example, be simple peak parameters such as the width at half height, the position of the peak in the time window, or other parameters of the signal profile. The flow injection technique in particular offers a number of completely new calibration strategies which are discussed briefly in Section 6.8. Procedures of this nature are mostly still in the experimental stage and need to be improved for routine use, especially with respect to their reliability, before they can be included in the operating software of commercial instruments.

Although AAS, like all other spectrometric procedures, is in principle a relative technique that can only provide quantitative results through the use of calibration samples, the possibility of 'absolute analyses' has been discussed again and again, especially for GF AAS. L'VOV [3681] in particular took up this subject at an early stage, checked the theoretical principles [3724], and, after the introduction of the STPF concept, improved the basis for realizing this principle [3730, 3732, 3735]. Using longitudinal-heated commercial instruments under STPF conditions, SLAVIN and CARNRICK [5427] showed that a reproducibility of about ±15% could be attained for the characteristic mass, m_0. This finding was largely corroborated through field tests by a number of laboratories [5328]. Temperature gradients over the length of the graphite tube [1972] and poorly reproducible contact of platforms in the graphite tubes [5325, 5326], which were recognized as the limiting factors for the application of 'absolute' AAS, have been eliminated in transverse-heated graphite tubes with integrated platforms [5520]. Variations in the line profiles of primary sources, caused by non-uniform operating conditions, were recognized as additional factors [3739]; further improvements can be achieved by taking this into account [3750]. The application of 'standardless' evaluation [3751] and the attainable results have already been discussed for a number of elements including beryllium [6456], cadmium [1798, 3763, 3765, 6566], chromium [3764, 6565], gold [6457], indium [6568, 6569], selenium [6557], silver [6566, 6569], and thallium [6567, 6569]. The theme of absolute analyses is discussed in somewhat more detail in Section 8.2.5. Nevertheless, this technique is unlikely to find wide application in routine laboratory practice as a result of the prescribed quality assurance programs.

5.2.2.1 Standard Calibration Technique

The standard calibration technique (sometimes referred to as the 'analytical curve technique') is the simplest, fastest, and most commonly used calibration technique. In this technique the absorbance A_x or the integrated absorbance A_{int} of the measurement solution prepared from the test sample is compared directly with the absorbance or integrated absorbance of calibration solutions. The concentration or mass of the analyte in the test sample solution, e.g. the mass concentration β_x, is determined by interpolation; the calibration solutions must cover the entire expected concentration range of the test sample solutions. The evaluation can be performed graphically, in particular for linear calibration functions, as depicted in Figure 5-4. In general though the calibration function is determined by *least mean squares* calculation. The necessary algorithms are included in the software of all modern instruments.

A minimum of five calibration solutions with analyte contents as far as possible in equidistant steps should be used to determine the calibration function. This ensures that deviations from linearity are recognized and that a calibration function with an acceptable curve fit can be calculated. A calibration function determined with four calibration solutions can certainly provide satisfactory precision, nevertheless the confidence band is relatively wide in this case. Having the analyte contents at equidistant steps ensures that the middle of the working range corresponds to the mean content of the calibration solutions [2553]. The *concentration of the analyte in the reagent blank solution* must be determined separately and taken into account in the evaluation.

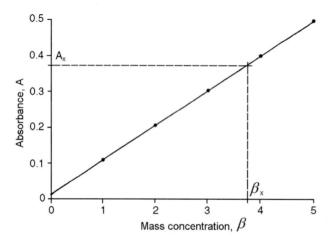

Figure 5-4. Graphical evaluation according to the standard calibration technique.

The use of least mean squares calculations is nevertheless coupled to a number of prerequirements which are not *a priori* met [2933]:

i)   The independent variable (the analyte content) is without error; more realistically, the error in preparing the calibration solutions is significantly smaller than the error in measuring the dependent variables (absorbance, etc.).

ii)   The experimental uncertainty for the determination of the dependent variables follows a normal distribution.

Resulting from the possible sources of error in the preparation of calibration solutions (see Table 5-2 in Section 5.2.1) the first condition must be put in question as soon as the repeatability standard deviation of the measurements is very good (e.g. $s_{rel} = 0,5\%$ for F AAS). The second condition is not met when the test samples are not homogeneous, as can be the case for the direct analysis of solids (refer to Section 4.2.7).

A number of authors have examined the magnitude of errors (i.e. the trueness and precision) of various methods of calculation [57, 58, 913, 5836]. It could be shown for AAS, where the standard deviations for all levels of calibration exhibit very similar values, that the calibration function can often be described by a simple linear regression. More complicated procedures of calculation such as weighted regression can improve the accuracy, but they require more information so that the effort needed for calibration is markedly higher.

Evaluation in the non-linear range can also be performed with suitable methods of calculation. Nevertheless the precision and trueness deteriorate with increasing non-linearity (see Section 5.4), but in fact this deterioration is far less dramatic than is often assumed [4098]. BYSOUTH and TYSON [1009] compared five algorithms used in commercial instruments for fitting non-linear calibration functions. These authors found that there is in general a far greater dependence of the accuracy of the fit on the number of measurement points rather than on the algorithm. In other words, the success of a curve fit is higher the larger the number of calibration solutions used for calibration in the non-linear range, largely independent of the algorithm used. DE GALAN et al. [2046] also found that rather good curve fits could be attained for F AAS with relatively simple polynomials, provided that at least 5–8 calibration solutions are used.

In general, depending on the analyte, the analytical line and the AAS technique, calibration curves in AAS are only linear for absorbance values up to about 0.5–0.8 (refer to Section 2.5.3). For techniques that generate transient signals, the calibration curves are linear over a larger range when the integrated signals are used for evaluation, and the decrease of the gradient with increasing absorbance is less than with evaluation via the peak height. Modern, computer-driven instruments usually allow visual inspection of the calibration curve and interactive checking of various curve fitting functions, such as polynomials of differing order, or the inclusion of selected reference values. Nevertheless, these possibilities should be treated with care and the necessary interpretation applied, since there is little point in forcing a perfect correlation with reduced data using a fitting function of lower order; rather, the aim of the selected calibration strategy must be to describe the analytical situation adequately. Thus it is possible to obtain a perfectly linear curve by excluding four reference values from a five-point calibration, but whether this curve describes the calibration function must be placed in serious doubt.

Since large numbers of unknown samples must normally be measured after calibration by the standard calibration technique, the *stability of calibration* is of the utmost importance. If the sensitivity is strongly influenced by variations in the operating parameters, leading to undesired drift, calibration must be repeated frequently. However, since frequent recalibration has a markedly negative influence on the sample throughput, unexpected drift should be investigated and the cause eliminated (see Section 5.3.1). Although the stability of modern instruments has been substantially improved, reliable statements on the stability of a calibration can only be made by experiment since this factor depends not only on the instrument but also on the analyte and the analytical method. Generally the calibration should be checked after 10–20 measurements using a zeroing solution and a calibration solution of middle concentration, and repeated if necessary. The quality assurance routines frequently included in modern instruments permit automatic recalibration. This is performed when an analytical control sample deviates by more than a preselected tolerance from the expected value.

A prerequirement for the applicability of the standard calibration technique is that the calibration and test sample solutions exhibit absolutely identical behavior in the atomizer unit employed, i.e. the absence of interferences. If concomitants in the sample cause interferences using this calibration technique, this can be compensated by mixing the interferents with the calibration solutions. This is referred to as *matrix matching*. It is naturally of decisive importance that only materials of sufficiently high purity or with the lowest possible, known content of the analyte are used. Any contamination by the analyte caused by applying matrix matching must be determined from a matrix blank solution. The (determined) *absorbance of the matrix blank solution* must be subtracted from the (determined) absorbances of the calibration solutions. The analyte content in the reagent blank solution must be determined independently and taken into account in the evaluation.

If the reagent blank solution is used incorrectly as the zeroing solution for the standard calibration technique, this does not cause a systematic error but most certainly a random error, i.e. the precision deteriorates (see Section 5.2.3.2). If, on the other hand, measurable masses of the analyte are introduced during matrix matching and these are not determined with a matrix blank solution, systematic errors additionally occur since with this technique the analyte is introduced into the calibration solutions but not into the test sample solutions.

5.2.2.2 Bracketing Technique

The bracketing technique is a variant of the standard calibration technique. The calibration function is usually determined using two calibration solutions so selected that their mass concentrations or masses of the analyte closely bracket the expected value for the analyte in the test sample solution. Thus, only a small section of the calibration curve is taken into consideration in this technique. The mass concentration β_x of the analyte in the test sample solution is determined from its absorbance A_x according to:

$$\beta_x = \frac{(A_x - A_1)(\beta_{B1} - \beta_{B2})}{A_2 - A_1} + \beta_{B1}, \tag{5.5}$$

where β_{B1} and β_{B2} are the concentrations of the analyte in the calibration solutions, and A_1 and A_2 are the corresponding absorbance values. Calculation of the mass m_x is performed correspondingly. The integrated absorbance A_{int} can be used instead of the absorbance A for the calculation.

The advantage of the bracketing technique is the improved accuracy that can be obtained. Further, the technique can also be applied to non-linear sections of the calibration curve when the concentrations or masses of the analyte in the calibration solutions neighbor the test sample solution closely enough. In this case the calibration function within the small range observed can be considered as quasi linear, so that complex algorithms are not required for the curve fit. The precision naturally decreases with increasing curvature of the calibration function (refer to Section 5.2.3.5). The analyte concentration in the reagent blank solution must always be determined separately and taken into account in the evaluation.

Since matrix matching is almost always employed with the bracketing technique, the points mentioned in Section 5.2.2.1, above, on the matrix blank value also apply. If measurable masses of the analyte are introduced during matrix matching, these must be determined in a separate matrix blank solution. The mean absorbance in the matrix blank solution must be subtracted from the absorbances of the calibration solutions since analyte contamination is contained in these solutions but not in the test sample solutions.

5.2.2.3 Analyte Addition Technique

The analyte addition technique is often used when the composition of the test sample is unknown and interferences from the concomitants can be expected. In this technique the test sample itself is used to match the matrix. The test sample solution is divided into a number of aliquots; one aliquot remains unspiked while the other aliquots are spiked with increasing masses of the analyte at equidistant intervals. All aliquots are then diluted to the same final volume. With GF AAS the additions can be performed directly in the graphite tube, so that prior mixing and dilution are not required [1323]. Evaluation is performed by extrapolating the calibration function to absorbance $A = 0$ or integrated absorbance $A_{int} = 0s$, as shown schematically in Figure 5-5a. The intercept β_x corresponds to the analyte content in the measurement solution.

The gradient of the calibration function determined by the analyte addition technique is specific for the analyte in the unknown test sample solution. The content of the analyte in the blank solution β_{xB} must always be determined by a separate addition and taken into account in the evaluation, since the gradient of the calibration function mostly differs in this case from that of the test sample solutions, as indicated in Figure 5-5b. If the compositions of the test samples differ significantly, it may be necessary to apply the analyte addition technique separately to every test sample. If a number of samples of similar composition are to be examined, the calibration function determined by the analyte addition technique for a single representative test sample can be used for the other test samples. We refer to this procedure as the *addition calibration technique*.

The analyte addition technique was originally used in F AAS to eliminate transport interferences, caused for example by differing viscosities of the measurement solutions. The analyte addition technique is ideally suited for this application. Problems with this calibration technique first arose with the introduction of GF AAS. When the numerous interferences caused by the inadequacies of the Massmann-type furnace became apparent, the majority of the early users attempted to eliminate them globally by routinely applying the analyte addition technique. This calibration technique was completely overtaxed for this purpose, a fact that was only later fully realized [5432, 6233]. For this reason we shall discuss the advantages and disadvantages of the analyte addition technique and indicate its limitations.

The analyte addition technique has the distinct advantage that every test sample is *calibrated individually*, so that the influence of concomitants can be eliminated. The analyte addition technique compensates *non-specific, multiplicative interferences,* such

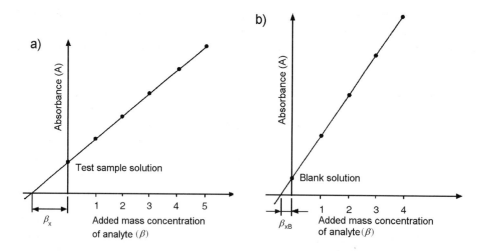

Figure 5-5. Graphical evaluation using the analyte addition technique, **a** – for the test sample solution, and **b** – for the blank solution. Since interferences in the test sample and blank solutions are mostly different, the analyte content in the latter must be determined by a separate addition.

as transport interferences in F AAS, in an ideal manner. On the other hand, it *cannot* eliminate *additive interferences.* These include, in addition to problems associated with contamination and losses of the analyte, in particular all *spectral interferences.* Further, all *concentration-dependent interferences,* such as ionization interferences in F AAS, *cannot* be eliminated by the analyte addition technique. A number of interferences have also been reported that are caused by the formation of compounds in flames, which are concentration dependent, such as the interference of phosphate or aluminium on the determination of calcium, that cannot be corrected properly by the analyte addition technique [2676, 5936]. Further details are provided in Section 5.4.

A prerequirement for the applicability of the analyte addition technique is *identical behavior* of the analyte contained in the test sample solution and the added analyte. For certain techniques, such as GF AAS or HG AAS, it is essential for the accuracy of the determination that the analyte be present in the identical form, e.g. oxidation state or species. This aspect is treated during the discussion of these techniques in Sections 8.2.4 and 8.3.3, respectively. To obtain identical behavior, it may be necessary, for example, to add the analyte prior to the digestion of the test sample, so that it is taken through the entire sample preparation procedure. Proper application of the STPF concept (Section 8.2.4.1) can also make a significant contribution to identical behavior. These are measures that in themselves eliminate interferences in most cases, so that in point of fact the analyte addition technique becomes superfluous. When interferences are really the exception [4026, 5432, 6233], there is in fact no sensible reason to apply the analyte addition technique routinely.

As well as the limitations mentioned so far for the application of the analyte addition technique, which are mostly method specific, there are a number of principal limitations. One of these is that the analyte addition technique should always be used in the *linear range of the calibration function.* A number of attempts have in fact been made using hyperbolic functions to extrapolate the addition curve to absorbance $A = 0$ [1860, 2034, 3255], but for routine operation the restriction to the linear range is realistic. For a linear regression of the measurement values [3457, 4834], too high values for the analyte content in the test sample are always determined when the linear range of the calibration function is exceeded [5936].

The restriction that all measurement values, including the highest addition, must be in the linear range of the calibration curve also means that the *working range is limited* to about a third to a quarter of the range that can be attained with direct calibration [2080]. For this reason, and also from the fact that the test sample solution is always measured in the lower section of the available range and that the zero point is not fixed but is extrapolated, the *precision is generally poorer* [2080, 3457, 5936]. And finally, there is no doubt that the analyte addition technique requires *much more effort for calibration* and should thus be avoided for reasons of economy when it does not bring any advantages. Statements to the effect that the choice between the analyte addition technique and the standard calibration technique depends on 'the confidence the analyst has on the one or the other technique' [2080] most certainly have no place in the area of exact analysis.

5.2.2.4 Reference Element Technique

In this technique a reference element is added to all calibration and test sample solutions in known concentration (this technique is also sometimes referred to as the *internal standard technique*, a term deprecated by IUPAC). The measurement values for the analyte and the reference element are then placed in relation. This technique is based on the prerequirement that in the event of an interference the reference element behaves in an identical manner to the analyte, i.e. it undergoes the same signal enhancement or depression. This prerequirement is only met in principle for non-specific interferences, such as transport interferences in F AAS (refer to Section 8.1.2.2). BARTHO and PHILLIPS [417], for example, used the reference element technique to detect dilution errors. DULUDE and SOTERA [1621] have published an overview on the limited number of papers published on this subject, most of them dating from the 1970s.

There are a large number of reasons why this technique never found favor in AAS, although it is used frequently in OES. One very practical reason is that the reference element technique can only be performed meaningfully with a two- or multichannel spectrometer. Further, its application is limited to F AAS since transport interferences do not occur with the other AAS techniques, and the identical behavior of two elements is even more unlikely than with F AAS. The principal problem, which is a further complication, is that the reference element should not be contained in the sample solutions, or if it is its content must be known exactly. Further, the relatively small measurement range of AAS, compared to OES for example, is a limitation, since a main constituent of known concentration in the sample, such as iron in steel, is not particularly suitable as a reference element for trace analyses. Moreover, all tasks placed on the reference element technique can be met by the analyte addition technique.

5.2.2.5 Selection of the Calibration Function and Its Quality of Fit

The precision of the analytical results depends on the precision of the calibration function. Thus, the starting point of every laboratory internal quality assurance program is the ascertainment, evaluation and, as required, the optimization of the precision of the calibration function [2553].

As well as the calibration function, the number of calibration solutions, their distribution over the working range, and the magnitude of the working range play a major role in the precision and accuracy that can be attained through calibration. Since the time required for calibration is frequently looked upon as unproductive, many laboratories tend to use the lowest possible number of calibration solutions. This way of working must be countered by the objection that the attainable accuracy for the subsequent test samples cannot be higher than that attained for calibration. The effort that must be put into calibration is thus directly proportional to the accuracy required. It is of fundamental importance for the precision and trueness of the analytical results that the selected calibration function adequately describes the relationship between the analyte content of the calibration solutions and the measured absorbance.

The *correlation coefficient* is frequently used to decide whether a linear or a non-linear calibration function should be used, or whether the selected function is adequate. The correlation coefficient r for a pair of measurements, which are written here as x_i and y_i for simplicity, is defined as:

$$r = \frac{\sum (x_i - \bar{x})(y_i - \bar{y})}{\sqrt{\sum (x_i - \bar{x})^2 \sum (y_i - \bar{y})^2}} .$$

(5.6)

The significance of the correlation coefficient is generally overvalued [2730]. Looked at closely, the relationship between the independent variable (content) and the dependent variable (absorbance) cannot be described as a correlation at all, since in statistical theory a correlation can only exist between random variables, while a calibration is based on a functional relationship between two quantities, ***neither of which*** is random.

As can be seen from the examples in Figure 5-6, a high correlation coefficient is suitable neither to judge whether a plot is linear nor whether a non-linear function better describes the measured reference values. The difficulties involved in interpreting the correlation coefficient and its statistical significance can be read in numerous publications [175, 253, 4092, 5837]; the problems discussed in these papers should be taken into consideration before a too high significance is placed on the correlation coefficient. If the linearity of a calibration function is to be proven statistically within a method validation procedure, much more complicated tests are required than just the comparison of correlation coefficients. Such tests are not included to date in the software of commercial spectrometers, so they must be performed using external software.

Of the numerous tests that have been proposed to check linearity [177, 961, 1739, 2702, 3011, 3305, 3911, 3930, 4092, 4921, 5837, 5991] we shall only mention Mandel's test [3835]. It has the advantage that the characteristics being checked are not influenced

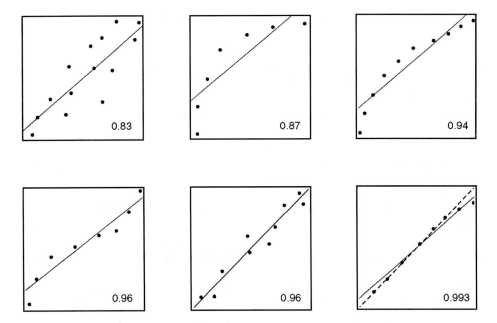

Figure 5-6. Examples of scatterplots with the associated correlation coefficients. In the last example the dashed line y = x describes the curve much better than the linear regression (from [175]).

by varying degrees of freedom of the curve fitting functions [2023]. This test checks the reduction of the residual variance for the transition from the linear fitting function to a function of higher order for significance. For this, the linear calibration function, a calibration function of higher order (for simplicity the quadratic calibration function), and the corresponding residual standard deviations are determined from the concentration and absorbance values of a sufficiently large number of calibration solutions (at least 10). The residual standard deviation s_y is a measure for the scatter of the absorbance values y about the calculated fitting function:

$$s_y = \sqrt{\frac{\sum (y_i - \hat{y}_i)^2}{N - 2}},$$ (5.7)

where y_i represents the mean values of the measured absorbance of the calibration solutions, while $\hat{y}_i$ represents the values predicted from the fitting function. The difference of the variances is calculated from the respective residual standard deviations:

$$Ds^2 = (N - 2)\, s_{y1}^2 - (N - 3)\, s_{y2}^2.$$ (5.8)

A test statistic Q is calculated from this result and used for the statistical F-test:

$$Q = \frac{Ds^2}{s_{y2}^2}.$$ (5.9)

A comparison of the test statistic Q with tabulated F-values ($P = 99\%$, $f_1 = 1, f_2 = N - 3$) gives:

i)   If $Q \le F$, no improvement to the quality of fit can be obtained with a second order fitting function and the calibration function is considered linear.

ii)   If $Q > F$, a second order fitting function will bring a substantially better quality of fit compared to a first order fitting function; the actual calibration function is significantly non-linear.

Determining and checking the quality of calibration is an increasingly important part of quality control procedures [4600]. This is especially true for modern analytical instruments which often run unattended for long periods and even perform sequential analytical tasks with differing methods overnight; automatic calibration is then essential. In order to distinguish between a successful and an unsuccessful calibration, various authors have proposed using a quality criterion for the calibration [2046, 2693, 3931, 5992]. KNECHT and STORK [3157] proposed a quality coefficient, QC, which is defined as:

$$QC = 100 \sqrt{\frac{\sum_{i}^{N} \left(\frac{y_i - \hat{y}_i}{y_i} \right)^2}{N - 1}},$$ (5.10)

where y_i is the measure for each calibration solution, $\hat{y}_i$ is the value predicted from the fitting function, and N is the number of calibration solutions, not including the zeroing solution. If the QC lies within a predefined criterion value, e.g. 5%, the corresponding linear calibration is used for the subsequent evaluation. If the QC is unacceptable, a check should be made to see whether the quality of calibration can be improved by rejecting outliers, limiting the working range, or using a non-linear calibration function [3931]. The use of such a quality criterion in the instrument software has not been described to date, but in view of the increasing level of automation its introduction is only a matter of time. Although the automatic rejection of an unacceptable calibration by assigning a criterion value for the QC certainly can help to improve the quality of a calibration, we should nevertheless not forget that as part of a quality assurance program a check should always be made using independent methods and materials. After all, the use of incorrectly prepared calibration solutions can still lead to a calibration of acceptable quality.

5.2.3 Evaluation

The performance characteristics of an analytical method are given by a set of quantitative, experimentally determinable *statistical quantities* (or *descriptors*). Such characteristic data can be used to compare various analytical methods or techniques, to choose a suitable method for a given task, for quality control, or to judge analytical results. If these statistical quantities are to be truly comparable between different analytical methods, they must be obtained using precisely defined statistical procedures. Such procedures are laid down in national standards (e.g. DIN 32645, DIN 38402, DIN 55350, BS-2846, AS-2850, etc.) and international recommendations (e.g. ISO, IUPAC). Unfortunately there is still a good deal of confusion on the definition of various terms, such as limit of detection and limit of quantitation, not only at the international level but even frequently within national standards.

It is essential that the analyst is aware of the statistical quantities for an analytical method if a given analytical task is to be performed successfully. They are the prerequisites for the correct choice of method and for effective quality assurance (see Section 5.5).

5.2.3.1 Evaluation Function and Sensitivity

The evaluation of the analytical results is performed via the *evaluation function*, which is nothing other than the inverse of the calibration function, i.e.

$$\beta = g(A), \tag{5.11}$$

$$c = g'(A), \tag{5.12}$$

$$m = g''(A_{int}). \tag{5.13}$$

The graphical representation of the evaluation function is the *evaluation curve*.
The slope S of the calibration function is termed the *sensitivity*, i.e.:

$$S = \frac{d A}{d \beta}, \tag{5.14}$$

$$S' = \frac{d A}{d c}, \tag{5.15}$$

$$S'' = \frac{d A_{int}}{d m}. \tag{5.16}$$

Within the linear range of the calibration function the sensitivity is independent of the concentration or mass of the analyte. In the non-linear range it is a function of the concentration or mass. In this case the concentration or mass of the analyte at which the sensitivity was determined must be quoted together with the sensitivity.

To provide a measure for the sensitivity of the analyte under given conditions we use the terms *characteristic concentration*, β_0 or c_0, and *characteristic mass*, m_0 in AAS. This is the concentration or mass of the analyte corresponding to an absorbance $A =$

0.0044 (1% absorption) or an integrated absorbance $A_{int} = 0{,}0044s$. The sensitivity in AAS is thus given by physical quantities such as the absorption coefficient of the analytical line (refer to Section 2.5.1) and also by characteristics of the atomizer unit, e.g. effectivity of the nebulizer or atomizer.

The characteristic concentration and especially the characteristic mass are important quantities in quality assurance. Virtually all manufacturers of atomic absorption spectrometers provide tables of reference values or the like so that the current performance of instruments can be checked. As part of instrument performance verification the sensitivity should be measured regularly and documented, preferably in the form of a control chart (see Section 5.5). The significance of the characteristic mass m_0 for an 'absolute analysis' by GF AAS is discussed in Section 5.2.2.

5.2.3.2 Precision

Under precision we understand in general the closeness of agreement between independent test results obtained under prescribed conditions [PE44620]. Since the prescribed conditions can be very different from case to case, agreement has been reached that two extreme situations should be observed, termed the repeatability conditions and the reproducibility conditions, respectively.

Repeatability conditions are those conditions under which independent test results are obtained with the same method on identical test objects in the **same laboratory** by the **same operator** using the **same equipment** within short intervals of time. The precision obtained under these conditions is termed the *repeatability*. The standard deviation obtained under these conditions is termed the *repeatability standard deviation, s_r*. The *repeatability limit, r*, is the value obtained under repeatability conditions in which the absolute difference between two single test results is expected to be with a probability of 95%.

In contrast to the above, the *reproducibility conditions* are those conditions under which independent test results are obtained with the same method on identical test objects in **different laboratories** with **different operators** using **different equipment**. The statistical values obtained under these conditions are correspondingly termed the *reproducibility*, the *reproducibility standard deviation*, and the *reproducibility limit, R*. Next to the above two extreme situations there are a large number of intermediate situations that have not been defined in any standards. It is thus essential that when the precision or standard deviation of an analytical procedure is quoted, the conditions used to obtain it are also clearly stated.

Generally, the reliability of a measurement result increases with the number of measurements used to obtain it. For this reason measurement results are rarely based on *single values* but rather on *mean values* obtained from a number of measurements. If all the single values only deviate from each other randomly, i.e. they are distributed normally about a single central value in a Gaussian distribution, we can calculate the mean $\bar{x}$ from n single values x_i according to:

$$\bar{x} = \frac{\sum\limits_{i=1}^{N} x_i}{n} .$$
(5.17)

It is an indication for *systematic errors* (*bias*) if the single values do not correspond to a *normal distribution* (Gaussian distribution); the source of such errors should always be determined [5086]. The search can be limited to those steps of the analytical procedure that were repeated.

If merely the instrumental measurement is repeated and an aliquot of the same measurement solution is used, we speak of *replicate measurements*. The scatter of the single values x_i about the mean $\bar{x}$ is a measure for the *repeatability* of the repeated analytical step. The precision of the repeated analytical step can be determined by variance analysis in the form of the *repeatability standard deviation, s_r*:

$$s_r = \sqrt{\frac{\sum\limits_{i=1}^{N} (\bar{x} - x_i)^2}{n-1}} ,$$
(5.18)

where x_i represents the single values and $\bar{x}$ the arithmetic mean of n independent replicate measurements. Frequently the *relative standard deviation* is used, which is given by:

$$s_{rel} = \frac{s}{\bar{x}} .$$
(5.19)

If an estimate is to be made on the precision of the entire method, all steps of the method must be included in the variance analysis, i.e. sampling, sample pretreatment, calibration, and measurement must be repeated as often as required. If the sources of error in the individual steps are independent of each other, their proportion of the *total standard deviation*, s_{total}, is given by the law of propagation of errors according to:

$$s_{total} = \sqrt{(k_1 s_1)^2 + (k_2 s_2)^2 + ... (k_n s_n)^2} ,$$
(5.20)

where $k_1, k_2, ... k_n$ are constants. From this equation it is clear that poor precision in an individual analytical step can be decisive for the total precision of a method. In general the instrumental measurement is the individual step with the best precision, so that it rarely determines the precision of the entire method [13]. Correspondingly the precision of a method cannot be significantly improved by performing a large number of replicate measurements [178]. Frequently, sampling and sample preparation are of decisive significance for the precision (and accuracy). If we compare the attainable precision for the instrumental measurement of an analyte determination by GF AAS in a solution with the determination of the same analyte by direct solids analysis, we can observe that the latter in all probability exhibits the poorer precision. If we take into account that for direct solids analysis we have analyzed independent samples, while for solution analysis we have merely analyzed aliquots of the same sample solution, the results for the total precision of a series of digestion solutions can turn out to be positive for direct solids analysis [3368].

If the contribution that an individual step makes to the total precision needs to be determined, either this step can be examined alone (if this is possible) or an analysis of the variance (ANOVA) can be performed using a test plan (refer to textbooks on statistics). Attempts to theoretically derive the precision of the AAS measurement itself have been undertaken by PRUDNIKOV [4719] and VOIGTMAN [6090] among others. For an analysis, the various sources of noise (radiation source, atomizer, detector, sample introduction procedure) must be identified and both their frequency behavior and their contribution to the total noise must be investigated [334]. It must also be borne in mind that the AAS measurement comprises a number of single measurements, such as element-specific absorption, background attenuation and emission with possibly varying noise characteristics (see Chapter 3).

Additionally, the attainable precision for a measurement value bears a complex relationship to the signal height (absorbance) [4719]. Since various sources of noise contribute to the S/N ratio and these are dependent to varying degrees on the signal height, different sources of noise dominate in different absorbance ranges. The most important sources of noise are listed in Table 5-3.

The *absolute magnitude of the standard deviation* is constant for the smallest absorbance values and increases proportionally for higher absorbance values. One cause for the deterioration of the precision with increasing absorbance is to be found in fluctuations of the atom distribution which becomes more important with increasing concentration. In flames these fluctuations depend on the flame type and gas flowrate [3694–3696, 3698, 5556, 5557]. In graphite furnaces they depend on interactions of the analyte with the graphite surface and thus on the element. The *magnitude of the relative standard deviation* is constant in the middle absorbance range and increases very rapidly for small absorbances. Toward higher absorbance values the relative standard deviation for the analyte concentration or mass also increases due to the increasing curvature of the calibration curve since in this range the determination of concentration or mass becomes more unreliable. If we plot the values for the relative standard deviation against the logarithm of the analyte concentration or mass, we obtain a typical precision pattern such as the example in Figure 5-7.

As emphasized in Section 5.2.1, a solution containing a measurable concentration of the analyte (such as a reagent blank solution) should under no circumstances be used to set the baseline of the spectrometer since this always leads to a deterioration of the precision. The extent of the deterioration in the precision is indicated on several examples in Table 5-4. Since the measured zero value (here the blank value) is automatically subtracted from all measurement values before statistical analysis is performed, the uncertainty of the blank value is not taken into account in the evaluation. The real uncertainty is then given by the law of the propagation of errors (eq. 5.20). It is thus clear that, especially in such cases where the blank value is of a similar magnitude to the analyte concentration in the test sample solution, the total uncertainty is far higher than that calculated from the replicate measurements alone. For calibration solutions the errors are even more severe since via the uncertainty of the calibration they affect the subsequent

Table 5-3. Sources of noise in an atomic absorption spectrometer.

Source of Noise	Cause	Dominance	References
Detector	thermal noise	negligible for photomultipliers	[4719]
	dark current noise	negligible for photomultipliers; for SSD at higher absorbance values	[5554]
	signal reading noise	for SSD at higher absorbance values	[5554]
	amplifier noise	negligible in modern instruments	[4096, 5554]
Radiation source (HCL, EDL, D$_2$ lamp)	shot noise	particularly for GF AAS near the LOD	[4719, 5410, 5554]
	source flicker	low absorbance (A < 0.1)	[744, 2772]
	drift	during warmup, particularly with single-beam instruments	[2991, 6010]
Atomizer	emission noise	particularly in the visible range at higher absorbance or low lamp intensity and wide slit widths; higher for the nitrous oxide-acetylene flame compared to the air-acetylene flame; poor alignment of the graphite furnace in the radiation beam	[740, 743, 2772, 4956]
	transmission noise	low absorbance (A < 0.1), especially in the range < 200 nm, and with flames of low transparency	[740, 743, 744, 2772, 4956]
	radiation scattering, background attenuation	high matrix concentration, poor nebulization, insufficient matrix separation	[3885]
	efficiency of atomization	high matrix concentration, wrong flame or flame stoichiometry	[741, 5556]
	atom distribution	high absorbance; stronger in the nitrous oxide-acetylene flame compared to the air-acetylene flame; in particular in graphite furnaces with elements that react with graphite	[2118–2120, 5556–5558]
	drift	during warmup of the burner head; aging of the graphite tube	[2149] [2605]

Table 5-3 continued

Source of Noise	Cause	Dominance	References
Sample introduction, flame	variations in the sample transport to the nebulizer	partial blockage of the sample capillary, particles in the sample stream, or pulsating pump operation	[1358, 2470, 6371]
	nebulizer efficiency, distribution of aerosol droplets, spray chamber drainage	in the middle working range	[741, 2348, 5557]
Sample introduction, graphite furnace	repeatability of the sample volume	particularly with small volumes or viscous liquids	[1641, 2434, 4642, 5580, 6194, 6206]
	repeatability of the dispensing location	over the entire working range, particularly noticeable with manual dispensing, depends on the wetting properties of the sample solution	[1641, 3783, 4642, 5580, 6206]

measurement values. To be statistically correct, the reagent blank solution must be in-cluded in the calibration and treated by regression like every other calibration solution. Nevertheless we should mention at this point that not every commercial instrument al-lows a calibration curve with an intercept on the ordinate axis, so that high blank values are particularly critical.

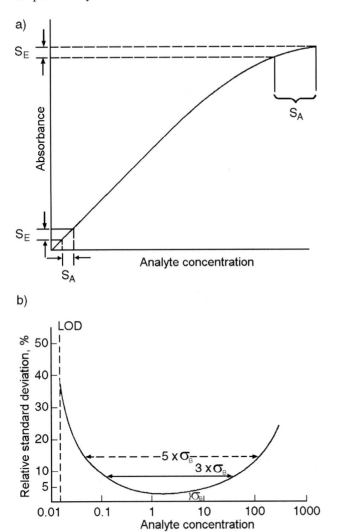

Figure 5-7. Dependence of the precision on the analyte concentration and the gradient of the calibration curve. **a** – Typical calibration curve; **b** – typical profile for the repeatability precision; s_E – standard deviation of the absorbance signal; s_A – resulting standard deviation of the analyte concentration; σ_B – 'best precision'; LOD – limit of detection (from [985]).

Table 5-4. Influence of measurement values on the precision when a reagent blank solution is used as a zeroing solution (from [5333]).

Example	Measured	Mean	Calculated		Real	
		A	s	s_{rel}, %	s	s_{rel}, %
1	Blank	0.010	0.0010	10		
	Sample	0.020	0.0016	8		
	Net signal	0.010			0.0019	18.9
2	Blank	0.100	0.0020	2		
	Sample	0.120	0.0024	2		
	Net signal	0.020			0.0031	15.5
3	Blank	0.005	0.0006	12		
	Sample	0.050	0.0010	2		
	Net signal	0.045			0.0012	2.6
4	Blank	0.002	0.0003	15		
	Sample	0.079	0.0003	0.4		
	Net signal	0.077			0.0004	0.6

5.2.3.3 Limits of Detection, Identification, and Quantitation

An important characteristic of an analytical technique or method is the *limit of detection* (LOD); the alternative designation recommended by IUPAC is *Minimum Detectable (true) Value* (L_D). The limit of detection is a measure for that concentration or mass of the analyte which, when exceeded, allows recognition with a given statistical certainty that the analyte content in the test sample (test solution) is larger than that in the blank test sample (blank test solution). Since the measured signal must be distinguished with a given certainty from the signal for the blank test sample (the blank value), it stands in close relationship to the precision of the determination of the blank test sample, as depicted schematically in Figure 5-8.

Under ideal conditions, the *blank test sample* is a sample that contains all constituents of the test sample with the exception of the analyte (see Section 5.2.1). The difficulties lie in the availability of the blank test sample, since it is impossible to reproduce the matrix of the test sample exactly and it cannot be guaranteed that it is completely free of the analyte. Under real conditions the blank test sample is thus a sample that contains a very low content of the analyte and comes very close to the remaining com position of the test sample. As is often the case with statistical comparisons, we can choose between two alternatives: If we require a high level of certainty for the detection of the analyte, the detection capability deteriorates; if, on the other hand, we require the best possible limit of detection, the probability increases proportionally that a blank value is taken in error for an analytical signal (error of the first kind). If a limit is set at a value $2s$ above the blank value, this risk is 2.3%; if we set a limit at $3s$, the residual risk is acceptably small at 0.1%. Signals with values of more than $3s$ above the blank value can thus be ascribed with high certainty to the analyte. Since signals are not particularly

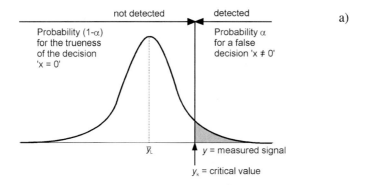

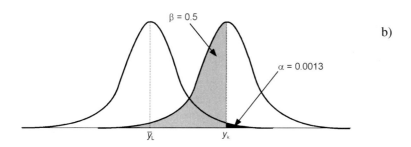

Figure 5-8. The relationship between the distribution of the measurement values for the blank test sample (here: normal distribution, Gaussian curve), the limit of detection, and possible false interpretations. **a** – Error of the first kind (significance α for the interpretation of the measurement value for the blank test sample); **b** – errors of the first and second kinds (β) for the interpretation of the measurement value for a sample whose concentration is $3s$ above the concentration of the blank test sample.

suitable for interlaboratory comparisons, the value of the signal is converted via the calibration function into a concentration or mass of the analyte; this is designated the limit of detection and is defined as:

$$
x_{\text{LOD}} = \frac{s_{\text{blank}}}{S} \, t_{(f,\alpha)} \sqrt{\frac{1}{m} + \frac{1}{n}} \, , \tag{5.21}
$$

where n is the number of measurements on the blank test sample, m the number of measurements on the test sample, and t the tabular value of the quantile of the t-distribution for $f = n-1$ degrees of freedom and the desired level of significance for α.

Depending on the given application, the gradient S of the calibration function can be referred to the mass or concentration of the analyte according to equations 5.14–5.16.

Discounting the technique and the apparatus used, the magnitude of the limit of detection is dependent on the acceptable risk for an error of the first kind, on the number of parallel measurements used to determine it, on the sample matrix, and on external conditions prevailing during its determination. As an example, the limit of detection for lead in waste water is higher than for the basic analytical method in matrix-free solution. A distinction is also made between the limit of detection determined under repeatability conditions, i.e. in the same laboratory with the same instrument, and that determined under reproducibility conditions, i.e. in different laboratories with different instruments etc. (see Section 5.2.3.2).

In order to make a comparison between instruments, the limit of detection is determined under repeatability conditions, and in agreement with customary international rules the following conditions are selected:

i)   Number of measurements on the blank test sample $n = 10$,
ii)   Number of measurements on the test sample $m = 1$,
iii) Probability for an error of the first kind $\alpha = 0.01$.

The parameters used to determine the limit of detection must be clearly stated in the documentation for a method. As a multiple of the repeatability standard deviation, the limit of detection so determined has a relatively wide confidence interval and thus only moderate comparability. For the determination of the limit of detection under reproducibility conditions, differences by a factor of two or more can easily occur, so that there is little point in determining or quoting the limit of detection with higher precision [173].

The smallest limiting factor for the limit of detection of a method is the *instrument limit of detection* (ILOD). This is determined most conveniently by performing replicate measurements on either the zeroing solution (F AAS) or without measurement solution (GF AAS). Investigations have also been made to calculate the ILOD via theoretical models for both F AAS [6357] and GF AAS [3745, 5554]. Under the prerequisite that errors in sample introduction and atomization can be neglected, all that remain for small concentrations are the random errors for the determination of the baseline and the photometric errors for the determination of the atomic absorption, the background attenuation, and the emission. The error components depend among other things on the frequency of measurement and the integration time used for the determination of the baseline and the signals.

The limit of detection **cannot** be applied for the quantitation of the analyte concentration or mass in test samples, which unfortunately is often done. The reason is patently clear since in AAS a calibration is necessary for the determination of the analyte content and the statistical certainty of the calibration is not included in the determination of the limit of detection. Analytical signals close to the limit of detection therefore have nothing to do with a quantitative determination since according to definition the standard deviation of these measurement values is of the same magnitude as their mean.

As a comparative criterion for a qualitative technique of detection, the limit of detection is often interpreted incorrectly as a guarantee that a concentration or mass of the analyte above the limit of detection can be determined reliably, or, on the other hand, that the absence of a detectable signal is the proof for a concentration or mass of the analyte **below** this limit [4216]. Analyte contents in the order of the limit of detection can lead to signals that can be erroneously interpreted as blank signals due to random scatter (error of the second kind). Statements about the analyte content such as 'lower than the limit of detection' cannot be statistically justified since an analyte content that just corresponds to the limit of detection can only be detected with 50% certainty (see Figure 5-8b). If it is necessary to state a limiting content above which an analyte can be detected with high statistical certainty, this value lies approximately as far above the limit of detection as the limit of detection above the background (under the justified assumption that the standard deviation for the measurement signals for the blank solution is of the same magnitude as that for a sample solution with a concentration in the limiting range). A test sample with a concentration or mass of $6s$ above the blank sample can thus be detected with a certainty of 99.9%. This limit is termed the *Erfassungsgrenze (EG)* in German. It was designated *limit of guarantee for purity* by KAISER [3003–3005] and *limit of identification* by BOUMANS [734]. The *EG* and the limit of detection are often confused, not least because the term 'limit of detection' is frequently used for both quantities. Since proof of detection requires the limit of detection to be exceeded, the *EG* cannot exist without the LOD. If the same probability for errors of the first and second kinds ($\alpha = \beta$) is selected for the calculation of *EG* (as assumed above), the *EG* is obtained by doubling the LOD. If a different statistical certainty is required, the *EG* is calculated according to:

$$x_{EG} = x_{LOD} \frac{s_{\text{blank}}}{S} t_{(f,\beta)} \sqrt{\frac{1}{m} + \frac{1}{n}} \tag{5.22}$$

for $f = n-1$ degrees of freedom.

Under *limit of quantitation* (LOQ) we understand the smallest mass or concentration of an analyte that can be determined quantitatively from a single analysis with a risk of error $< 5\%$ with the required statistical certainty (e.g. P $= 95\%$). Although detection is positive in the range between the LOD and the LOQ, quantitative statements are inadmissible. As an alternate the term *determination limit* is frequently found in the literature, but this term is not recommended by IUPAC to avoid ambiguity.)

Since the determination of the LOQ should include a variance analysis of the calibration data and thus requires a good deal of effort, it is frequently calculated solely from the scatter of the values from blank or reagent blank solutions, whereby the statistics of calibration are taken into account by model multiplication factors. A rapid estimation of the limit of quantitation using the recommended factors (k = 3) gives:

$$x_{LOQ} = 9 \frac{s_{\text{blank}}}{S} . \tag{5.23}$$

The limit of quantitation can be calculated by iteration from the squares of the deviation of the calibration curve according to:

$$x_{LOQ} = k \frac{s_{y,x}}{S} t_{(f,\alpha)} \sqrt{\frac{1}{m} + \frac{1}{n} + \frac{(x_{QL} - \bar{x})^2}{Q_x}}, \tag{5.24}$$

with

$$f = n - 2,$$

$$s_{y,x} = \sqrt{\frac{\sum_{i=1}^{n}(\hat{y}_i - y_i)^2}{n-2}},$$

$$\bar{x} = \frac{1}{n}\sum_{i=1}^{n}x_i, \text{ and}$$

$$Q_x = \sum_{i=1}^{n}(x_i - \bar{x})^2,$$

where x_i represents the concentrations of the calibration samples and y_i the corresponding measurement values; $\hat{y}_i$ gives the estimated function values for the calibration. The limit of quantitation can be calculated in good approximation by setting $x_{LOQ} \approx kx_{LOD}$ in the square root expression in equation 5.24. A prerequisite for this calculation is that the variances over the limited range of the calibration curve can be taken as homogeneous. This premise is mostly sufficiently met only over a relatively limited range, so that the calibration curve must be established directly neighboring the LOD. The ratio of the calculated LOD and the highest calibration value should not exceed a factor of ten. The relationship between the LOD, the *EG*, and the LOQ is depicted in Figure 5-9.

5.2.3.4 Measurement Range; Working Range

To obtain a true representation of the precision of an analytical method, we must convert the measurement values into concentrations or masses via the calibration function. As soon as a technique or method is calibrated, we can determine changes in the precision with the analyte content. For this it is necessary to perform measurements over the widest possible concentration or mass range and to calculate s_{rel} for every concentration or mass [742]. If these values are plotted against the logarithm of the concentration or mass, we obtain a precision pattern like the example depicted in Figure 5-7 (Section 5.2.3.2).

The precision curve passes through an optimum for a medium concentration or mass range, which for many elements is in an absorbance range of $A = 0.2–0.8$ [4952]. The increase of s_{rel} for small concentrations or masses is a result of the deteriorating S/N ratio since the measurement signal decreases while the noise remains more or less constant. BOWER and INGLE [740] found that fluctuations in the transmission behavior of the flame were the major noise components for low concentrations. For a concentration or mass of zero, $s_{rel} = \infty$. The increase of s_{rel} at higher concentrations or masses is a result of the decreasing sensitivity due to increasing curvature of the calibration curve.

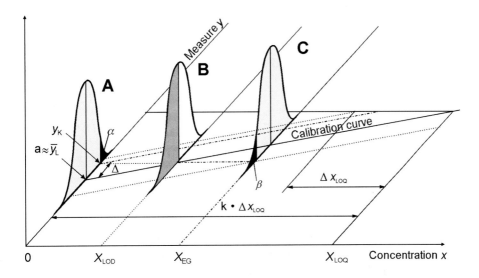

Figure 5-9. The relationship between the calibration curve and the LOD, the *EG*, and the LOQ. **A** – Distribution of the blank test sample measurement values; **B** – distribution of the measurement values for a sample with a concentration corresponding to the LOD; **C** – distribution of the measurement values for a sample with a concentration corresponding to the *EG*. y_k – critical value for the measure (absorbance); a – ordinate intercept of the calibration function; y_B – blank value; Δ – one-sided confidence interval; Δ_{LOQ} – two-sided confidence interval of the LOQ; l/k – relative uncertainty of the result to characterize the determination limit; x_{LOD} – limit of detection; x_{EG} – *Erfassungsgrenze*; x_{LOQ} – limit of quantitation.

As can be seen from Figure 5-7, with increasing curvature of the calibration curve, constant variations in the measurement values (in absorbance) lead to greater uncertainty in the derivation of the concentration or mass [4796, 4952].

The concentration or mass range in which the method or procedure is calibrated and in which we obtain quantitative measurement values of sufficient statistical certainty for this method is termed the *working range* [985]. In essence it is determined by the initial gradient of the calibration function (sensitivity) and by the later decrease of the gradient (curvature of the calibration function). The lower limit of the working range is given by the limit of quantitation, while the upper limit is given by the concentration or mass of the analyte that can be determined with the same statistical certainty as the limit of quantitation. If the measurement values are outside of the range of precision required for a method, various actions can be taken as described in Section 5.3.

As well as the working range we can also define the *linear working range*; the lower limit is again given by the limit of quantitation, while the upper limit is determined by the change in sensitivity (gradient of the calibration function) in comparison to that close to the limit of quantitation. For this we establish a calibration curve with n calibration solutions ($n \geq 5$) and determine the *linearity* according to:

$$L = \left(A_n - A_{n-1}\right)/\left(A_1 - A_B\right), \tag{5.25}$$

where A_n is the absorbance of the highest calibration solution, A_{n-1} the second highest, A_1 the lowest and A_B the reagent blank solution. The calibration function is regarded as linear when $L \geq 0.7$ [2786]. In principle, other, preferably higher, limits for the linearity are possible, depending on the given requirements.

In contrast to the working range, the *measurement range* is not clearly defined. In practice it would be sensible to designate that range in which measurements are still possible (not quantitative determinations) as the measurement range. Correspondingly the lower limit of the measurement range is the limit of detection of the procedure and the upper limit is given by the limiting absorbance A_{lim} at which a defined concentration or mass of the analyte can no longer be assigned (refer to Section 2.5.2).

In AAS, similar to OES, the radiant intensity that can be measured by the detector extends over a range of around six orders of magnitude, at least in the absence of background attenuation. While in OES the emission intensity increases linearly with the analyte concentration, in AAS it is proportional to the logarithm of the ratio of two radiant intensities. The attenuation of the radiant intensity of the primary radiation source by an order of magnitude through absorption corresponds to one unit of absorbance. Resulting from the logarithmic relationship between the concentration or mass and the signal ratio, the working range of AAS techniques is less than that of other optical techniques.

Let us assume that the limit of detection corresponds to an absorbance of $A = 0.001$ under the following conditions: $\Phi_0 / \Phi_{tr} \rightarrow 1$ and $A \rightarrow 0$ (see Section 5.2.3.3). Due to the presence of stray radiation, for example from the radiation source, the limiting absorbance for the majority of analytical lines lies around $A_{lim} = 2–3$. As shown schematically in Figure 5-10, this absorbance range corresponds in the first approximation to a measurement range of three orders of magnitude in concentration or mass of the analyte. By using optics of the highest quality to avoid stray radiation, it might be possible to attain a maximum absorbance of $A = 6$. However, this only increases the corresponding concentration or mass range from 3 to 3.5 orders of magnitude. For this reason there is little value in attempting to extend the measurement range in AAS to higher absorbance values.

Since the extension of the measurement range to higher values is always limited by stray radiation, the only recourse to extend it is to go toward *lower* absorbance values. In other words, the S/N ratio must be improved by reducing the noise. Every action taken to reduce the noise while maintaining the signal (sensitivity) and thus improving the precision also leads directly to an improvement in the LOD and the LOQ, and thus to an extension of the measurement and working ranges.

RADZIUK *et al.* [5554] determined the absorbance noise for a modern graphite furnace AA spectrometer with Zeeman-effect BC and compared the results with calculated values. The results are shown in Table 5-5. The good agreement between the theoretical and the measured values for the photometric fluctuations without graphite furnace operation is an indication that the measurement was performed close to the shot noise limit. As to be expected, the graphite furnace is an additional source of noise during the rapid heating phase with an effective current of several hundred Amperes and an alternating magnetic field of 0–0.8 Tesla at a frequency of 54 Hz. It is still a current topic of inves-

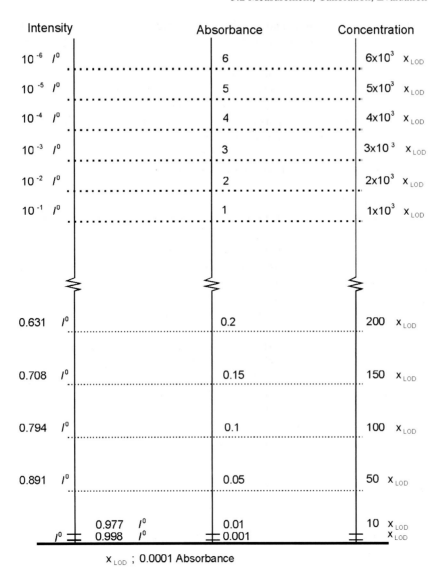

Figure 5-10. Schematic of the relationship between the dynamic measurement range of the detector system and the resulting measurement range in absorbance and concentration. I^0 – original radiant intensity; x_{LOD} – limit of detection.

tigation in how far this is due to electromagnetic radiation or to transmission noise through the hot gas atmosphere. Nevertheless, the contribution of the atomizer to the noise is surprisingly low. It is clear that an improvement in the photometric precision can only be obtained by increasing the radiant flux density or the optical conductance, or increasing the quantum efficiency of the detector, since the noise components of the detector itself or of the electronics are insignificant. Under the condition that the shot noise is the limiting noise component, with increasing radiant flux density the photometric precision is improved by [3745]:

$$\Delta A_{min} \approx \frac{1}{\sqrt{\Phi_0}} \tag{5.26}$$

The logical conclusion from these observations is that every attempt to increase the sensitivity (gradient of the calibration curve) without at the same time improving the S/N ratio will in fact limit the already small measurement range of AAS even further and thus lead to an effective *deterioration in the performance*.

Table 5-5. Comparison between theoretical precision and effective precision for photometric measurements with the Perkin-Elmer Model 4100 ZL AA spectrometer (from [5554]).

Element	Wavelength	Spectral slitwidth	Photons per measurement interval	Calculated shot noise	Measured absorbance noise A	
	nm	nm	(6 ms)	A	without atomization	with atomization
As	193.7	0.7	1.9×10^6	7.6×10^{-4}	5.7×10^{-4}	8.1×10^{-4}
Se	196.0	2.0	5.4×10^6	4.4×10^{-4}	6.2×10^{-4}	9.7×10^{-4}
Cd	228.8	0.7	5.6×10^6	3.1×10^{-4}	2.8×10^{-4}	3.6×10^{-4}
Cr	357.9	0.7	2.6×10^7	1.2×10^{-4}	1.2×10^{-4}	1.7×10^{-4}
Pb	283.3	0.7	1.5×10^7	1.6×10^{-4}	1.8×10^{-4}	3.2×10^{-4}

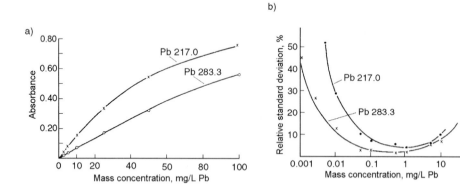

Figure 5-11. Comparison of the determination of lead in a silicate rock in the air-acetylene flame at the analytical lines at 217.0 nm and 283.3 nm. **a** – Calibration curves; **b** – precision of the measurements (from [985]).

The term *sensitivity* has often been misused in AAS. In the form of the characteristic mass or concentration it is a useful criterion for the correct functioning of a system, but these parameters are little suited for making comparisons between different systems. The decisive criterion for the performance capabilities of an AA spectrometer or an atomizer is, just as in OES and other spectroscopic procedures, the S/N ratio or the limit of detection, respectively. This is indicated in Figure 5-11 on the example of the determination of lead by F AAS at both analytical lines at 217.0 nm and 283.3 nm. It can be seen from the calibration curves in Figure 5-11a that the line at 217.0 nm exhibits a much higher sensitivity, whereas in Figure 5-11b it is clear that the precision of this line is poorer over the entire concentration range compared to the line at 283.3 nm. In the lower concentration range this is caused by the much higher flame noise and in the upper range by the more marked non-linearity. Due to the poorer precision the working range at the 217.0 nm line is noticeably smaller than at the 283.3 nm line.

5.2.3.5 Trueness

The most important goal of every quantitative analysis is to obtain the correct analytical result; this goal can only be achieved when no systematic errors are made [1081, 2412, 3876, 4239, 5916, 5917, 5019, 5213, 6051, 6055]. The *trueness* is an indication of the closeness between the 'true' or expected value (concentration or mass) of an element in a test sample and the mean obtained from a series of results of replicate determinations using a given analytical method. The trueness can be calculated from the difference between the true value and the averaged measurement values [1031, 5396]. In contrast to the statistical quantities that we have discussed so far, such as precision which can be determined precisely by performing a sufficiently large number of replicate measurements, trueness is very difficult to obtain by calculation [4931]. The problem is quite simply that in an unknown test sample we do not know the true concentration or mass of the analyte.

Before we treat this problem and its solution further, it would be appropriate to discuss the difference between precision and trueness [289, 1379], and also between *random* (statistical) *errors* and *systematic errors* (bias). These relationships are depicted schematically in Figure 5-12 in the form of hits on a target; the true (expected) analyte content is the center of the target and the 'shots' correspond to the measurement values. It is important to realize that good precision (Figure 5-11 c and d) does not necessarily signify good trueness [3176, 3186]. There are plenty of examples in the literature where values were measured with high precision but exhibited large systematic errors and were thus a long way from the true content (Figure 5-12c). On the other hand, poor precision does not necessarily have to be an indication for poor trueness; the statistical mean of a sufficiently large number of highly scattered single values can in fact be very close to the true content (Figure 5-12a). Nevertheless, the greatest aim of every analyst should be to obtain high trueness **and** good precision (Figure 5-12d).

As we have already discussed, calibration must be performed to correlate the measurement value to the concentration or mass of the analyte in a test sample. If the calibration sample has an identical composition to the unknown test sample (except for the

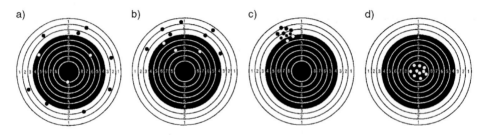

Figure 5-12. Schematic representation of various types of analytical error in the form of hits on a target. **a** and **b** – poor precision; **c** and **d** – good precision; **b** and **c** –systematic errors; **a** and **d** – good trueness.

analyte content) and both are treated in an identical manner, we can assume that the measurement result is correct and that the determined content corresponds to the true content [5022, 5534]. In practice, however, there will almost always be differences in the composition of the calibration sample and the test sample, leading to potential interferences and measurement errors. In order to recognize systematic errors of this type a number of procedures are available that differ widely both in the effort required and in the effectiveness.

Many of the techniques for recognizing measurement errors cannot in fact be applied to the entire analytical method but only to parts of it. These include, for example, the choice of a different technique of calibration (e.g. the analyte addition technique instead of the standard calibration technique), a different digestion technique, or a different AAS technique (e.g. HG AAS instead of GF AAS). Also, the use of a different instrument and operator, or a different laboratory (see reproducibility, Section 5.2.3.2), can bring useful information. If one of these techniques leads to a different result, this is a clear indication for a systematic error; this must then be sought so that it can be eliminated.

A technique used frequently in routine analysis is the *regular analysis of analytical control samples*, i.e. reference samples with a composition as close as possible to that of the test samples [6328]. However, this technique can hardly be used when the sample composition changes frequently and markedly since the effort required is very large and the number of available reference samples is normally limited. In a number of areas, however, such as steel analysis where a large number of similar samples must be examined, analytical control samples are used almost exclusively.

Another frequently-used technique is *spiking with the analyte* with subsequent determination of the *recovery* [3075]. In this technique a known mass of the analyte is added to an aliquot of the test sample. If the added quantity of analyte can be recovered during the analysis of the spiked test sample, this is an indication that the analyte content determined in the non-spiked sample is correct. Since the spiking technique is based on the calibration function determined during calibration, the recovery **cannot** be used as evidence for the trueness within a quality assurance program, and especially not when the same calibration solution as used for calibration is used to spike the test sample. If an error is made during calibration, it cannot be identified by determining the recovery. If the spiking technique is to be of any use at all, the same points as mentioned under the analyte addition technique (Section 5.2.2.3) must be carefully observed. The spiking technique is used preferentially to recognize non-spectral interferences (see Section 5.4.2).

Methods development always requires especial effort [1901, 2750]. Particularly in the trace and ultratrace ranges, if at all, only *interlaboratory calibration* [761, 2675] can provide information on the true concentration or mass, i.e. as many different techniques as possible should be used in a number of different laboratories [174, 676, 2708, 2956, 3105, 4421, 5191, 5706]. The outcome of an interlaboratory calibration is not infrequently a certified reference material [2755, 4761, 5592, 6528], so that the effort put in can be made available to other laboratories. The statistical criteria for trueness and its control are discussed in more detail under quality assurance (see Section 5.5).

5.3 Optimizing Instruments and Methods

In contrast to techniques that utilize optical emission, AAS has a parameter that, under given instrumental and analytical conditions, depends practically only on physical quantities; this is the *sensitivity* (gradient of the calibration function, see Section 5.2.3.1), or its derived quantities *characteristic concentration* or *mass*. The sensitivity thus offers a very useful control criterion for the correct functioning of an instrument and for optimizing a method. On the other hand, by a suitable choice of instrument parameters or analytical conditions the sensitivity can be changed correspondingly so that the *working range* (refer to Section 5.2.3.4) can be shifted to lower or higher analyte concentrations or masses. The second important parameter in an analytical method is the *noise* (refer to Section 5.2.3.2), which can also serve as a control criterion for the correct functioning of the instrument and the proper conductance of the analysis. Via the *signal-to-noise ratio* it also directly enters such quantities as the precision and the working range. The optimization of these important parameters for a given task is the topic of this section.

5.3.1 Error Recognition and Instrument Optimization

Atomic absorption spectrometry is a relative technique in which quantitative determinations are performed by comparison to calibration samples (calibration solutions). Since every change in the instrument parameters, the condition of the instrument, and the conditions of measurement are reflected in the calibration and are thus eliminated (except drift), it is still possible to obtain quantitative results even when the sensitivity is far from optimum. Nevertheless, despite this possibility, we would strongly advise against working in this manner since all control on the condition of the instrument is lost. Apart from the fact that the performance characteristics of an analysis are worse under these conditions, the judgment of the analytical data is questionable since significant changes in the sensitivity or noise are indications of malfunctions of the instrument or errors in the method, which may lead to measurement errors. For this reason a protocol of the performance characteristics of every analysis should be carefully maintained, and any larger deviations from specified values or values in earlier protocols should be a reason to look for the cause.

Frequent causes for a *too low sensitivity* in AAS are in general:

i)   Dilution errors when making up the calibration solutions.
ii)   Analyte losses in the stock or calibration solutions due to insufficient stabilization or unsuitable containers.
iii)   Wrong or faulty lamp; a too high lamp current.
iv)   Incorrect wavelength or a too large slit width.
v)   Matrix effects due to either the use of the wrong buffer or modifier, or the absence of a buffer or modifier.

The following causes additionally can be present in F AAS:

vi)   Wrong flame type or non-optimized flame stoichiometry.
vii)   The horizontal or vertical burner position is not optimized.
viii)   The sampling capillary, the nebulizer, or the burner slot are clogged; incorrect aspiration rate.
ix)   Unsuitable inserts in the spray chamber; misaligned contact bead.

Frequent causes for a sensitivity that is too low in GF AAS (too high characteristic mass) are:

x)   Erroneous sample dispensing.
xi)   Preatomization analyte losses; these can be due to a pyrolysis temperature that is too high or to either an unsuitable modifier or the absence of a modifier.
xii)   Unsuitable (too high) atomization temperature.
xiii)   Erroneous blank value or baseline correction.
xiv)   A too high gas flowrate during atomization or the use of an unsuitable purge gas.
xv)   A too low heating rate; this can be caused by an unsuitable power supply or poor electrical contact of the graphite tube.
xvi)   A too low magnetic field strength for Zeeman-effect BC (e.g. insufficient voltage at the electromagnet).

Among the most frequent causes for a too low sensitivity in HG AAS are:

xvii)   pH value of the measurement solution not optimized.
xviii) Wrong concentration of the reductant solution.
xix)   False gas flowrate of the carrier gas.
xx)   The quartz tube atomizer is contaminated or not conditioned.

Much more seldom than a sensitivity that is too low is a *sensitivity that is too high* (too low characteristic mass). This can be traced almost without exception, especially in trace analysis, to accidental introduction of the analyte. If this occurs a systematic search should be made to determine the cause. All analytical results generated under such conditions, and also all corresponding publications, should be regarded with considerable skepticism.

As possible general causes for *insufficient precision* or too high noise we can consider the following:

i)   Low lamp energy; poor quality lamp.
ii)   Wrong lamp current or poor lamp alignment.

iii)   Wrong wavelength or a too narrow slit width.

iv)   Contaminated optical components (windows, lenses, mirrors, etc.).

v)   Emission noise.

vi)   Poor alignment of the line source and the continuum source for background correction.

vii)   Inhomogeneous sample solution or slurry, the presence of suspended matter in the measurement solution, nugget effect for solids analysis.

For F AAS we can also consider the following specific causes in addition to those above:

viii)   Partially clogged nebulizer or burner; defective nebulizer.

ix)   Contaminated spray chamber or inserts (large droplets on the surface are an indication of wetting problems).

x)   Irregular drain flow from the spray chamber.

xi)   Irregular gas flow (e.g. too low gas pressure).

xii)   Unsteady flame caused by drafts, strong ventilation system, change of solvent, etc.

xiii)   A too short integration time or insufficient damping.

Specific causes for insufficient precision or too high noise in GF AAS include the following:

xiv)   Unreproducible dispensing of the test sample solution.

xv)   Spattering of the solution caused by either too rapid or incomplete drying.

xvi)   Irregular creeping of the measurement solution for wall atomization (wetting problem).

xvii)   Aged graphite tube or contacts.

xviii)   A too long integration time or incorrect integration window.

xix)   Baseline offset correction (BOC) too short or erroneous.

Finally we must also mention a number of frequent causes for *drift*:

i)   Changes in the radiant intensity of the lamp (or lamps) or of the line profile during the warming-up phase or in the event of a defect; changes in the ratio of the radiant intensities of the line source and the continuum source for background correction.

ii)   Evaporation of the solvent from the measurement solution.

iii)   For F AAS; warming or cooling of the burner, or the formation of encrustations on the burner slot.

iv)   For F AAS; changes in the rate of aspiration or in the efficiency of nebulization due to physical changes in the sample capillary or the nebulizer.

v)   For GF AAS; aging of the graphite tube or the accumulation of residues (e.g. carbon from biological materials).

vi)   For HG AAS; increasing contamination of the quartz tube atomizer.

This list of frequent causes for marked deviations in the most important performance characteristics can only serve as a starting point for troubleshooting. Nevertheless, we wish to emphasize that deviations of this sort are nearly always an indication of a malfunction or an operational error and must therefore be taken seriously. Further details on the possible interferences to be found in the individual techniques and their elimination are discussed in Chapters 4 and 8. The important performance characteristics such as sensitivity (or characteristic mass or concentration) and the S/N ratio should always be documented with the analytical results so that malfunctions or operational errors can be traced or eliminated. Increasingly, functions such as monitoring the condition of the instrument and maintaining a protocol are performed automatically in modern instruments in which the corresponding algorithms are implemented in the instrument software (refer to Section 6.8).

SLAVIN [5436] described a procedure for monitoring further instrumental functions in GF AAS in addition to the sensitivity and S/N ratio. Using a certified reference material (Trace Elements in Water, NIST 1643 b), the characteristic masses and the Zeeman factors for chromium, copper and silver are determined without a chemical modifier and under easily reproducible conditions; additionally the lamp energy is recorded. Copper exhibits the greatest loss in sensitivity at the 324.8 nm line when Zeeman-effect BC is applied and is very sensitive to changes in the magnetic field strength. This element can thus serve as an indicator for an insufficiently strong magnetic field. Chromium is a difficultly volatilized element and requires a relatively high heating rate; at the 257.9 nm line its determination is sensitive to emission interferences, which can indicate a poorly aligned graphite tube atomizer. Silver, on the other hand, is not noticeably influenced by the magnetic field strength, the heating rate, or the alignment of the graphite tube. If the sensitivity of all three elements is influenced to the same degree, this is an indication of dispensing problems or that the actual temperature deviates markedly from the rated value. If this test is performed at regular intervals it serves as a powerful tool to document the state of the instrument.

It may be necessary to accept compromises when optimizing an instrument or a method since different performance characteristics can exhibit varying optima. It is quite often the case that the conditions for optimum sensitivity are not the same as the conditions for the best possible tolerance to constituents of the matrix. The required task must therefore be defined before optimization can be undertaken. Frequently the task is to optimize analytical performance characteristics such as detection capability, precision, working range, or tolerance to matrix constituents, but occasionally considerations of economics, such as high sample throughput, minimal manual effort, low sample or reagent consumption, or low costs, are more important.

In GF AAS the stabilizing effect of a modifier (e.g. palladium) can lead to a lower rate of atomization, so that broader signals are obtained than if a modifier were not used. During signal processing a broadening of the signal can lead to a deterioration of the S/N ratio and to a poorer LOD. On the other hand, the linear working range can be extended upward since the limiting absorbance is only reached at higher analyte masses, and the tolerance against concomitants is significantly improved due to the more effective separation during the pyrolysis step and the higher effective atomization temperature (see Section 8.2.4.1).

In F AAS we can observe that a burner position which is optimized for maximum sensitivity can exhibit very low tolerance to matrix constituents. Procedures for optimizing the instrument parameters have been described by numerous authors; for F AAS: [267, 1120, 1410, 1642, 1666, 1821, 2427, 2498, 4053, 4120, 4464, 4544, 5558, 6010]; for GF AAS: [246, 874, 2427, 3490, 4863, 5540]; for HG AAS: [861, 1455, 6038]. Various strategies have been applied for optimization, such as methods of experimental design [246, 3490, 4464, 6508], simplex optimization [1666, 1841, 4606, 5084, 6508], or complete response-surface mapping [3438]. The recommendations in the ISO standards for F AAS [2786, 2787] and GF AAS [2788, 2789] can also serve as a basis for methods development.

Particular attention must be paid to optimization for multielement determinations by F AAS [1841, 3019] or GF AAS [2387]. The highly varying conditions required for pyrolysis and atomization in graphite furnaces would appear to stand in the way of attaining acceptable compromise conditions. However, looked at closely the choice of conditions is relatively simple; the most volatile analyte determines the maximum pyrolysis temperature and the least volatile analyte determines the atomization temperature. Nevertheless the use of a transversely-heated atomizer with integrated platform is decisive since it offers optimum conditions for both volatile and non-volatile elements and allows the same atomizer to be used under STPF conditions (refer to Section 8.2.4.1). It has further been shown that the losses in sensitivity and selectivity under multielement conditions are far less than generally assumed when a modern instrument concept is applied. For example, the loss in sensitivity for an increase in the atomization temperature above the optimum is far less than the sole observation of the increasing diffusional losses would predict [2392]. Losses in sensitivity due to a reduction of the residence time in the graphite tube are partially compensated by an increase in the absorption coefficient with increasing temperature; this results from a decrease in the share that the pressure broadening contributes to the broadening of the absorption profile. A reduction of the integration time has a positive influence on the limit of detection since shorter integration times lead to an improvement in the S/N ratio [2392]. All in all, losses in the detection capability by a factor of 2–3 must be expected for multielement determinations, but this can be tolerated in most cases.

5.3.2 Determining Smaller Concentrations or Masses

The endeavor to be able to determine ever lower analyte concentrations or masses has always accompanied the development of AAS and has had a decisive influence on that development. The various efforts made to increase the sensitivity of F AAS via design measures to either increase the portion of the analyte reaching the atomizer or to increase the residence time of the analyte atoms in the observation volume are described in detail in Sections 1.6 and 4.1. These developments have of course been superseded by the graphite furnace, hydride-generation, and cold vapor techniques. Nevertheless, techniques that can improve the limit of detection or limit of quantitation of a procedure or analytical method are still of considerable interest. These can be based on an increase of the sensitivity or the measurement signal, or in a decrease of the noise; ideally we should consider both aspects.

One possibility to obtain a larger measurement signal and thus to increase the detection capability is to *increase the analyte concentration in the measurement solution*. In principle, this aim can be attained via two different strategies:

i)   By avoiding unnecessary dilutions, for example with digestion acids, solvents, etc.
ii)   By preconcentration of the analyte by extraction, precipitation, etc.

For techniques such as F AAS, which only function satisfactorily with solutions where the viscosity is not too high, relatively narrow limits are placed on the first strategy. For GF AAS, which allows the use of solutions of higher viscosity, slurries, and solid samples, improvements in the detection capability of 1–2 orders of magnitude can be achieved by these means (see Section 4.2.7).

The second strategy is based on composite (hybrid) techniques, i.e. the combination of physical-chemical *trace–matrix separation* with simultaneous preconcentration of the analyte and instrumental detection. In this respect sorbent-extraction techniques have proven highly successful for analyte preconcentration since they can very easily be combined with FI for on-line automation (refer to Section 6.5). Using these techniques, an increase in the sensitivity by 1–2 orders of magnitude can be achieved, even with relatively short enrichment times which do not noticeably impede the analysis sequence. For samples with high matrix contents, the matrix separation that takes place at the same time as the extraction allows the detection capability to be increased by more than three powers of ten.

For techniques with discontinuous sample introduction, such as GF AAS or HG AAS with batch systems, the measurement signal is proportional to the analyte mass and not to the concentration. With these techniques, the detection capability can be improved by *dispensing larger volumes* of the measurement solution or larger quantities of the test sample. In GF AAS the upper limit for a single dispensing is restricted by the atomizer design to maximal 50–100 µL solution or 1 mg solid sample. For test samples with a low matrix content the detection capability can be improved by multiple injections, each followed by a drying step and, if required, a pyrolysis step [558, 1281, 2230, 2690, 3357]; this is a technique that is supported by modern autosamplers. The additional time required for these steps can be reduced by injecting into a preheated graphite tube so that drying already begins during dispensing [236, 558, 3171, 3355, 3356, 4996]. Even more effective is *in situ preconcentration from the gas phase* such as the amalgam technique for mercury (see Section 4.3.2.1) or the collection of hydrides in the graphite tube (see Section 4.3.2.2).

For techniques with continuous sample introduction, mainly F AAS, there is the possibility of increasing the detection capability by improving the notoriously low efficiency of pneumatic nebulizers (approx. 5%; see Section 4.1.2.2). There have been numerous attempts to improve this step and investigations are still continuing. A number of different nebulizer types are available that differ slightly in their effectiveness and which must be optimized for the required task with respect to aspiration rate and droplet size. A recent development in this area is hydraulic high pressure nebulization, which offers distinct advantages especially for viscous solutions (refer to Section 4.1.3.4). Further means of increasing the effectiveness of this step include the use of organic solvents [310, 1356, 3052, 3053, 3828, 6345], solvent mixtures [231, 1692, 1953, 4494], or emulsions [349–351, 5562, 6068], which have a positive influence on the effectiveness of nebulization. [1846, 4185] The influence of organic solvents is not restricted to nebulization, however, but also continues during atomization in the flame [225, 227–229] (see Section 8.1.1.5).

A further means of extending the measurement range to lower analyte concentrations is to *reduce the noise* and thus improve the S/N ratio. An important quantity that can have a substantial influence on the noise is the radiant flux density and thus the *intensity of the primary radiation source*. Intensity of course has no influence on the sensitivity, since in AAS it is the *ratio of intensities* that is measured; nevertheless via the noise it has a direct influence on the precision and the limit of detection of a procedure. As a rough rule of thumb the detection capability increases with the square root of the radiant intensity. An increase of the radiant intensity by a factor of four leads to a doubling of the S/N ratio. Unfortunately the qualitative radiation parameters, namely the intensity and the radiation distribution with respect to wavelength and beam cross-section, are governed by the lamp current and cannot be varied independently from each other. In general an increase in the intensity by increasing the lamp current also leads to a broadening of the line profile. Since both parameters influence the detection capability—higher radiant intensity leads to lower noise while line broadening leads to poorer sensitivity—the maximum detection capability cannot be attained by simply maximizing the intensity but must be achieved by optimization. While simply increasing the lamp current only leads to a conditional improvement in the detection capability, the use of radiation sources that emit a significantly higher radiant flux density (e.g. EDLs) is a reliable way of achieving this aim.

For the sake of completeness we must also mention that for continuum source BC the radiant flux density of the continuum source is just as important as that of the line source. We should also mention that for Zeeman-effect BC the noise is lower than for continuum source BC since only one radiation source is required. The longitudinal configuration of the magnetic field is particularly advantageous since an additional polarizer is not needed and there is thus one source of noise less.

As well as the radiant flux density of the radiation source the *radiation conductance* of the spectrometer (refer to Section 3.3) plays a major role in reducing the noise and thus leading to an improvement in the detection capability. Although the analyst cannot influence this parameter, it is an important criterion for the choice of an instrument when good precision and high detection capability are required. Spectrometers utilizing Echelle polychromators offer a completely new level of performance with respect to radiant power due to the wide entry slits permitted by the excellent resolution. It is also clearly important to maintain the instrument in good condition, i.e. mirrors, lenses, beam splitters, windows, etc. must be free of dust, and surface coatings should not be oxidized.

A further component over which the analyst has no influence but which should be decisive for the choice of an instrument is the detector. Photomultipliers are used as detectors in the majority of modern AA spectrometers (refer to Section 3.5). Recent investigations have shown that photodiode arrays such as those used in multielement spectrometers exhibit a significantly better S/N ratio than photomultipliers, especially in the far UV range [633, 2937, 2938, 3247, 4784, 4783, 5717].

As well as these instrumental measures to reduce the noise and thus improve the S/N ratio, the *signal processing* can also make a substantial contribution. It is possible to utilize the differing frequency behaviors of the signal and the noise to separate these components (filter) or to selectively amplify the signal. The situation is relatively simple for a steady-state signal X which is overlaid by fluctuations $x(t)$:

$$X(t) = X + x(t). \tag{5.27}$$

The repeatability precision of such signals can be increased by *integration* or *averaging*. In these cases the improvement is obtained by averaging the individual measurements against time. The time required for averaging is limited by economic aspects (sample consumption, allowable time) and by the drift behavior of the spectrometer with respect to sensitivity and baseline. Measurement times of 4–5 s have proven suitable for F AAS with continuous sample introduction. Even in those cases where sample consumption does not limit the measurement time, times of greater than 5–10 s do not bring any improvement in the S/N ratio since additional low-frequency noise components (drift) occur. If the available quantity of sample is the limiting factor, it is possible to utilize the injection technique or FI with volumes of 50–100 μL and synchronized signal integration.

For techniques that generate transient signals the situation is somewhat more complex. The transient signal only briefly reaches a maximum so that for integration individual signals $X(t)$ with varying S/N ratios are integrated. If the integration time includes periods in which the signal is low or in which there is no signal at all, the S/N ratio of the integrated signal is deteriorated. The integration times for transient signals should thus not exceed the signal time. For long signals that only return slowly to the baseline, such as can be observed for the atomization of refractory elements like molybdenum or vanadium in longitudinally-heated atomizers, it can be of advantage to terminate integration before the signal has returned to the baseline. Nevertheless, this requires that the appearance time be reproducible and is not influenced by varying matrix constituents. The stability of the baseline is more critical for signal integration in comparison to peak height measurements since small deviations when integrated over a longer time period can lead to substantial errors. For instruments that do not incorporate suitable baseline offset correction (BOC) facilities which can be programmed by the user with respect to time point and period of integration (refer to Section 3.8.3), the use of peak height rather than peak area measurements is to be preferred [632, 1564, 2604, 2607].

If the conditions of measurement that lead to transient signals (e.g. for the injection technique, FI, or GF AAS) can be exactly reproduced, the S/N ratio can be improved by *ensemble summation*. For this, every measurement time point is assigned its own channel in which the replicate measurements are summed (e.g. a signal of 5 s duration at a measurement frequency of 54 Hz gives 270 channels). While the signals increase proportionally with every measurement, in the ideal case the distribution of the fluctuations does not change so that the noise only increases by $\sqrt{N}$. This leads to a net gain in the S/N ratio since the standard deviation of the measurement decreases with the number of measurements:

$$s_{\text{sum}} = \frac{s_0}{\sqrt{N}} .$$
(5.28)

While the time required for the measurements increases proportionally to N, the noise is only improved by $\sqrt{N}$, so that a reasonable gain (factor 4) can only be expected for up to 16 measurements. In reality the gain is mostly lower since other noise components like drift additionally occur. Nevertheless, in contrast to the technique of integrating a continuous signal, with summation it is possible to correct a baseline offset between individual measurements; however, a correction of the drift in sensitivity would be much more complicated. Although the relatively high time requirement for measurements of several seconds such as in the injection technique [484] or FI [5512, 5516] would appear to be acceptable, for GF AAS [408, 591, 1044, 2503, 4962] with measurement cycles lasting longer than a minute it would only be suitable in exceptional cases.

5.3.3 Determining Higher Analyte Contents

Although AAS is primarily a technique for trace determinations it is not exclusively so and there are numerous examples for its successful application to the determination of main constituents (see Chapter 10). Frequently in practice the situation arises where a number of test samples with highly varying analyte contents must be examined, but since the content is *a priori* unknown, the samples are treated and diluted in the same manner. Since the working range of AAS is limited to 2–3 powers of ten (refer to Section 5.2.3.4), the question arises of what to do with test sample solutions whose analyte concentrations exceed the working range. The simplest apparent remedy, namely diluting the test sample solution, frequently cannot be considered since it requires additional effort and is a further source of error [3283, 3398, 4593, 5761, 6525].

In principle there are three possible ways of influencing the sensitivity and thus the working range of a procedure as a result of the Beer-Lambert law (see equation 2.41, Section 2.5.2). Since the absorbance A or the integrated absorbance A_{int} is directly proportional to the number of atoms N in the absorption volume, the length ℓ of the absorbing layer, and the absorption coefficient $\kappa(\lambda)$, each of these parameters is a suitable starting point for influencing the sensitivity.

The way in which the *number N of atoms in the absorption volume* can be reduced, absolute or per time unit, depends largely on the atomizer unit. For techniques that use a measured volume of the test sample solution, such as GF AAS, injecting smaller volumes leads directly to the corresponding reduction of the sensitivity and thus to an extension of the working range toward higher concentrations. Nevertheless this procedure is limited by the decreasing precision and accuracy for the injection of smaller volumes. The HG technique and CV AAS with batch systems are exceptions since the sensitivity can be varied over a range of about four powers of ten via the sample volume (see Section 4.3.1.1).

A further possibility of influencing the sensitivity in GF AAS and HG AAS is by varying the inert gas stream. This does not influence the total number N of analyte atoms of course, but it does influence their residence time in the absorption volume, which has the same effect on the sensitivity and working range [1864]. Nevertheless, increasing the gas flowrate through the atomizer is not without problems for either technique and should therefore be applied with caution (refer to Sections 4.2.4 and 9.3.2.2, respectively). Yet another possibility is to control the pressure of the atmosphere in graphite tube atomizers [1863, 1865, 1866, 1868, 2422, 2603, 2644, 4648, 5620, 5630], which allows a working range of five orders of magnitude [6157]; however this technique is not supported by commercial atomizers.

When FI systems are used to dispense the measurement solution in HG AAS and CV AAS, the same possibilities exist in principle for the control of the sensitivity, but the volume of the measurement solution can only be varied by around a power of ten. Nevertheless the FI technique offers a number of other means, such as automatic dilution of the sample solution, which are discussed in Section 6.4. For dispensing the measurement solution with CF systems in which a time-independent signal is generated, the sensitivity can only be influenced by adding the measurement solution more slowly. This technique is also limited, however, since the pulsations caused at low revolutions of the peristaltic pumps normally used can markedly influence the precision.

There are a number of ways of influencing the sensitivity of F AAS by reducing the number N of analyte atoms per time unit in the absorption volume, although none of them is particularly effective [1322, 1357, 2999, 4120]. One of these is to influence the mass flow to the atomizer; there are a number of ways of achieving this aim: Reducing the aspiration rate of the nebulizer is subject to narrow limits since the range of droplet sizes is correspondingly affected (see Section 4.1.2.2). The application of FI systems also has little influence on the number of analyte atoms—except when automatic dilution is utilized. One of the reasons for this is that the efficiency of nebulization increases when less solution per time unit is conducted to the nebulizer [1355, 2292, 5726]. For FI-F AAS the number of analyte atoms per time unit in the absorption volume can solely be significantly influenced by increasing the dispersion, which is achieved by lengthening the mixing zone [1833, 1837, 5979]; nevertheless in principle this is the same as dilution. It is better to influence the exploitation of the tertiary aerosol, for example by placing a contact beaker in front of the nebulizer jet, so that the aerosol can be reduced by a factor of up to five [4406, 6345]. An advantage of this technique is that the range of droplet sizes is shifted toward smaller droplets, thus reducing interferences and improving the precision. Albeit this technique is not yet offered by any AA instrument manufacturer.

The absorption volume in the flame can be altered without any special measures by simply changing the height of the burner. This procedure is supported in particular by modern burner systems which can be driven both horizontally and vertically by motors under program control, thus allowing a variation in the sensitivity over a wide range.

A change in the *length ℓ of the absorbing layer* brings about a proportional change in the absorbance and thus in the working range. This possibility is hardly practicable for tube atomizers, such as used in GF AAS and HG AAS, since shortening the atomizer has a major influence on the atomizer unit and on its heating characteristics. But for F AAS this approach is used relatively frequently since it is easy to carry out. The burner heads for premix burners are usually removable and can mostly be turned through 90° so that the radiation beam passes across the flame rather than along it. In this case ℓ is reduced by around a power of ten and the working range is correspondingly increased upward without having to change any parameters and without having to incur any disadvantages [1351, 1685, 1747, 1855, 2992, 4246, 5781].

The third possibility of extending the working range toward higher analyte concentrations is to effect a change in the absorption coefficient $\kappa(\lambda)$. This is most easily achieved in practice by *selecting a different analytical line*. The greater majority of all elements exhibit a relatively large number of analytical lines with considerably varying sensitivities, so that it is frequently possible to find a line suitable for the given task or analyte concentration. This technique has been frequently reported for GF AAS for example [1075, 2375, 2612, 2613, 2951, 3364]. The most important alternate analytical lines and their relative sensitivities are quoted in the discussions on the individual elements (Chapter 9).

Simultaneous spectrometers are of especial advantage when the analyte content of test sample solutions or solid test samples varies substantially. With such spectrometers it is possible to measure the analyte simultaneously at a number of analytical lines of varying sensitivity. It is then possible to select the line with the appropriate sensitivity for the analyte concentration for evaluation. HARNLY and O'HAVER [2386, 2389, 4099, 4410] used a different approach with a simultaneous spectrometer with an array detector by measuring the absorption at different positions on the line profile (refer to Section 2.5.1). Since the absorption coefficient changes along the line profile, it is possible for example to measure low concentrations with high sensitivity at the center of the line, while high concentrations can be measured at the line wings.

A further procedure that can be applied together with Zeeman-effect BC and which is in principle based on an influence on the absorption coefficient is the *three-field technique* proposed by DE LOOS-VOLLEBREGT and co-workers [3585, 3589, 3592, 3594, 3597, 3598]. This procedure is based on the fact that for Zeeman-effect BC the absorbance derives from the difference of two absorption coefficients (refer to equations 2.54 and 2.55, Section 2.6). In addition to the usual 'magnet on' and 'magnet off' phases in an alternating field, in the three-field technique there is an additional phase with a weaker magnetic field. Since the σ components are separated less effectively from the line profile of the primary source during this phase, a higher absorption coefficient κ_σ is subtracted, resulting in a lower sensitivity for the measurement. This is shown on the example of copper in Figure 5-13. The three-field technique allows determinations with

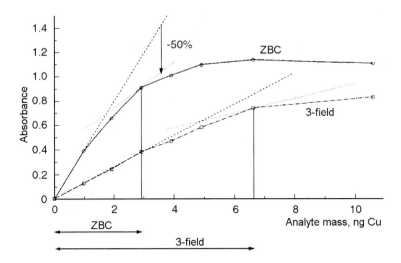

Figure 5-13. Analytical curves for copper at 324.8 nm for measurements with Zeeman-effect BC (longitudinal configuration of the magnet at the atomizer) and the three-field technique. The dotted line shows the gradient corresponding to 50% of the original gradient of the calibration curve (from [3598]).

two different sensitivities to be performed during one measurement phase. This is of especial advantage for Zeeman-effect BC since the working range for this technique is mostly even more strongly restricted than for normal AAS (see Section 3.4.2). Nevertheless, as we can see from Figure 5-13, the three-field effect can only increase the working range by factors of 2–5 and thus in effect can merely correct the additional restrictions brought about by Zeeman-effect BC.

5.4 Interferences in AAS

The presence of concomitants in the test sample can cause interferences in the determination of the analyte. The interference can influence both the attainable precision (see Section 5.2.3.2) and the trueness [3531]. However, an interference only leads to a measurement error if it is not eliminated, or if it is not taken into consideration in the evaluation process, by suitable measures. Such measures can be instrumental, such as background correction, or the choice of a suitable calibration technique, such as the analyte addition technique, or the use of chemical additives. An influence on the analyte signal by the atomizer, for example by the flame gases or the graphite material, or by the solvent, is not considered as an interference since the effect is usually the same for test sample and calibration solutions.

5.4.1 Classification of Interferences

Interferences in spectrochemical analysis can be classified according to a number of aspects. The various categories are not mutually exclusive but rather complement each other in characterizing the interference. Frequently the distinction is made between specific and non-specific or between additive and multiplicative interferences. In the following we shall largely adhere to the IUPAC recommendations [2827]; these recommendations have been accepted into various national and international standards. Accordingly, interferences are divided generally into spectral and non-spectral interferences. Spectral interferences are due to the incomplete isolation of the radiation absorbed by the analyte from other radiation or radiation absorption detected and processed by the electrical measuring system. With non-spectral interferences the number of free analyte atoms, and thus the measurement signal, is affected directly. Non-spectral interferences are most conveniently classified according to the place or time of their occurrence.

5.4.1.1 Spectral Interferences

In AAS, spectral interferences are encountered far less frequently than in OES. We make a distinction between the following causes of spectral interferences in AAS:

i)   Direct overlapping of the analytical line with the absorption line of another element.

ii)   Absorption of the radiation of the analytical line by gaseous molecules.
iii) Absorption by concomitants of other radiation from the radiation source (or sources) that is not separated by the monochromator.
iv)   Radiation scattering caused by particles in the absorption volume.

The most frequently observed spectral interference in OES is *interference by thermal emission of concomitants*. This emission is passed by the exit slit of the monochromator or reaches the detector as stray radiation. In AAS it is corrected instrumentally by *modulating the radiation* from the radiation source and using a selective amplifier tuned to the modulation frequency so that all radiation generated in the atomizer not having the same frequency is eliminated (refer to Section 3.6). This is a decisive advantage of AAS compared to all spectrochemical techniques that are based on emission.

If the portion of non-modulated radiation reaching the detector is relatively large compared to the modulated portion, saturation of the photomultiplier can occur in extreme cases. This leads to increased noise, termed *emission noise*, which although it does not cause systematic errors, it does influence the precision of the measurement. Under particularly unfavorable conditions a high portion of non-modulated radiation can lead to an offset of the baseline which might be interpreted as a measurement signal. This effect is nevertheless always accompanied by excessive noise and is strongly dependent on the selected slit width so that it can easily be recognized.

At the temperatures prevailing in the atomizers used in AAS the number of absorption lines is much lower than the number of emission lines in a flame or in a plasma. For this reason only a limited number of cases of *direct overlapping of atomic lines* has been observed in AAS. These are summarized in Table 5-6. The extent of the interference is dependent on the degree of overlapping of the emission line of the analyte with the absorbing line of the interfering element, on the absorption coefficient of the interfering line, and on the concentration of the interfering atoms in the absorption volume. If the interfering element is atomized to only a limited degree under the prevailing analytical conditions, then even strong overlapping of the lines will merely lead to a minor interference. As we can see from Table 5-6, in most cases the interfering element must be present in a large excess to cause false measurements. The majority of these spectral interferences are thus of little analytical significance. FASSEL [1852] has nevertheless drawn to attention the fact that absorption lines can become relatively broad in the presence of high element concentrations. For calcium, for example, at 30 pm from the line center he was able to measure 10% of the absorption measured at the center. The total width of an absorption line including the wings can occasionally increase up to 100 pm, which is more than an order of magnitude above the normal half-width of 1.5–10 pm. PANDAY and GANGULAY [4524] reported overlapping of this sort that could be ascribed to broadening of the absorption line.

Spectral interferences can also occur if the radiation source emits radiation other than the analytical line and this radiation is not separated by the exit slit of the monochromator. Depending on whether it is absorbed by the analyte or not, this additional radiation causes curvature of the calibration curve to a greater or lesser degree and a reduction in the sensitivity (see also Section 3.3.1). Furthermore a genuine spectral interference can occur if this radiation is absorbed by concomitant elements. Next to the line at 217.589 nm, antimony for example exhibits a non-absorbable line at 217.023 nm that only occurs in emission (see Figure 3-11 in Section 3.3.1). This line partially overlaps the lead line at 216.994 nm [5403]. This interference can become significant at slit widths greater than 0.5 nm.

A spectral interference can also occur if the line of a second element falls on the detector; this can be the case when a multielement lamp is used and the selected spectral

Table 5-6. Spectral interferences in AAS caused by direct overlapping of analytical lines. Primary analytical lines are indicated by an asterisk (*).

Analyte	Wavelength nm	Interfering element	Wavelength nm	Sensitivity ratio	References
Al	308.215	V	308.211	200	[1852, 4991]
As	189.042	Cr	189.055		[2600]
As	228.812	Cd	228.802		[2021]
Ca	422.673*	Ge	422.657		[3691, 4991]
Cd	228.802*	As	228.812		[2021, 4991]
Co	252.136	In	252.137		[3627]
Cu	324.754*	Eu	324.753	500	[1852, 4991, 5524]
Fe	271.903	Pt	271.904	500	[1852, 4991, 5573]
Fe	279.470	Mn	279.482		[4385]
Fe	285.213	Mg	285.213		[4385]
Fe	287.417	Ga	287.424		[4385]
Fe	302.064	Cr	302.067		[4097]
Fe	324.728	Cu	324.754		[4385]
Fe	327.445	Cu	327.396		[4385]
Fe	338.241	Ag	338.289		[4385]
Fe	352.424	Ni	352.454		[4385]
Fe	396.114	Al	396.153		[4385]
Fe	460.765	Sr	460.733		[4385]
Ga	403.298	Mn	403.307	3	[141, 4991]
Ge	422.657	Ca	422.673		[4385]
Hg	253.652*	Co	253.649	8	[3851, 4991]
Hg	285.242	Mg	285.213		[4385]
Hg	359.348	Cr	359.349		[4385]
I	206.163	Bi	206.170		[4385]
Mg	285.213*	Fe	285.179		[4097]
Mg	285.213*	Tb	285.214		[4524]
Mn	279.482*	Fe	279.501		[4097]
Mn	403.307	Ga	403.298		[141]
Os	290.906*	Cr	290.905		[4524]
Pb	241.173	Co	241.162		[4385]
Pb	247.638	Pd	247.643		[4385]
Pr	492.495	Nd	492.453		[4991]
Sb	217.919	Cu	217.894		[4385]
Sb	231.147	Ni	231.097		[3213, 3691, 4385]
Sb	323.252	Li	323.261		[4385]
Si	250.690	V	250.690	8	[1852, 4991]
Zn	213.856*	Fe	213.859		[3072, 4097, 4991]
Zn	213.856*	Cu	213.85		[6401]

slitwidth is too large. The absorption of both elements is then measured; the degree of the interference again depends on the absorption coefficient of the second line and the concentration of free atoms of this element in the absorption volume. As well as with multielement lamps, an interference of this sort can also occur if for reasons of stability the cathode is made from an alloy or an intermetallic compound, or if a volatile metal is contained in a support cathode made of another metal. In most cases this second element is not stated by the lamp manufacturer, but it emits its spectrum nevertheless. Especial care is thus recommended when secondary analytical lines are used.

If a continuum source is used as the primary radiation source, every element that exhibits an absorption line within the spectral slitwidth of the monochromator can generate a measurement signal in dependence on the absorption coefficient of the line. In this case, as with OES, the freedom from interferences depends on the resolution of the spectrometer (see Section 3.3). The same problem also occurs, occasionally with inverted sign, when a continuum source is used for background correction. If a concomitant element exhibits an analytical line within the spectral slitwidth of the monochromator, it can absorb radiation from the continuum source to a degree corresponding to its concentration. This supposed background absorption is then subtracted from the atomic absorption, leading to *overcompensation* and thus to a measurement error. Potential interferences of this type are summarized in Table 5-7; the radiation of the continuum source is only significantly attenuated when the interfering element is present in very high concentrations. This type of interference occurs commonly in trace determinations in metallurgical samples such as alloys.

If a portion of the test sample or measurement solution introduced into the atomizer is not completely vaporized before it reaches the absorption volume, *scattering of the radiation* from the radiation source on solid or liquid particles can occur. In the first approximation, this effect obeys Rayleigh's law of scattering; the coefficient of scattering τ is given by:

$$\tau = \frac{\Phi_{tr}}{\Phi_0} = 24\pi^3 \frac{N \cdot v^2}{\lambda^4} . \tag{5.29}$$

Radiation scattering is thus directly proportional to the number N of scattering particles per unit volume and to the square of the particle volume v, but inversely proportional to the fourth power of the wavelength λ. Radiation scattering therefore occurs more strongly with increasing particle size (τ increases by a factor of 64 for a doubling of the particle radius) and decreasing wavelength (from 800 nm in the visible range to 200 nm in the UV range, τ increases by a factor of 256).

In HG AAS and CV AAS the analyte is separated from concomitants by vaporization, so that with these techniques virtually no particles can reach the absorption volume. In a well-designed premix burner (refer to Section 4.1.2.2) and an air- or nitrous oxide-acetylene flame sample particles rarely reach the absorption volume unvaporized, at least as long as the total salt concentration does not exceed 5 g/L. In flames of lower temperature, e.g. the air-propane flame, and at wavelengths below 300 nm substantial radiation scattering has nevertheless been observed [636]. In GF AAS the occurrence of radiation scattering depends to a very large degree on whether concomitants can be successfully separated or destroyed prior to the atomization step. Radiation scattering is far more pronounced in atomizers that exhibit a temperature profile over the length of the tube compared to isothermal atomizers.

Table 5-7. Potential spectral interferences at the primary analytical line due to overcompensation when using continuum source BC.

Analyte	Wavelength nm	Matrix	Wavelength nm	Line separation, nm	References
Ag	328.1	Cu	327.4	0.7	
		Rh	328.1	0.1	
		Zn	328.2	0.1	
As	193.7	Ag	193.2	0.5	
	193.7	Al	193.47	0.2	[439, 3925, 4726, 4870, 4871]
Au	242.8	Co	242.5	0.3	[437, 1893]
		Sn	242.9	0.1	
Be	234.9	As	235.0	0.1	
Bi	223.1	Cu	223.0	0.1	[1893, 5973]
Ca	422.7	Dy	422.5	0.2	
Cd	228.8	As	228.9	0.1	
		Fe	228.72	0.1	[424, 3483]
Co	240.7	Ru	240.3	0.4	
Cr	357.9	Fe	358.1	0.2	
		Nb	358.0	0.1	
Cu	324.7	Ni	324.3	0.4	
		Pd	324.3	0.4	
Dy	421.2	Rb	421.5	0.3	
Er	400.8	W	400.9	0.1	
Eu	459.4	Cs	459.3	0.1	
Fe	248.3	Pt	248.7	0.4	
Ga	294.4	Fe	294.8	0.4	
		V	294.1	0.3	
Gd	368.4	V	368.8	0.4	
		Pb	368.3	0.1	
Hg	253.6	Co	253.6	0.1	[5571]
		Fe	253.6/254.1	0.1	
Ho	410.4	In	410.2	0.2	
		Tm	410.6	0.2	
		Y	410.2	0.2	
In	303.9	Fe	303.7	0.2	
		Ni	303.8	0.1	
		Sn	303.4	0.5	
Ir	208.9	B	209.0	0.1	
Mn	279.5	Mg	279.5	0.1	
		Pb	280.2	0.7	
Mo	313.3	Ni	313.4	0.1	
		Ir	314.0	0.7	
Ni	232.0	Sn	231.7	0.3	

Table 5-7 continued

Analyte	Wavelength nm	Matrix	Wavelength nm	Line separation, nm	References
Pb	217.0	Cu	216.5	0.5	[5973]
		Fe	216.7	0.3	[3483]
		Ni	216.6	0.4	
		Pt	216.5	0.5	
		Sb	217.6	0.6	
Pd	244.8	Cu	244.2	0.6	
		Ru	245.5	0.7	
Pt	265.9	Ir	266.5	0.6	
		Sn	266.1	0.2	
Re	346	Co	345.5/346.6	0.6	
Rh	343.5	Ni	343.3	0.2	
		Ru	343.7	0.2	
Ru	349.9	Co	349.6/350.2	0.3	
		Ba	501.1	0.2	
		Ni	349.3	0.6	
Sb	217.6	Pb	217.0	0.6	[3213, 6426]
		Cu	217.9	0.3	[5973, 6426]
		Co	217.5	0.1	[1893]
		Fe	217.8	0.2	[1893, 3497, 6426]
		Al	217.4	0.2	[3497]
Sc	391.2	Eu	390.7	0.5	
Se	196.0	Bi	196.0	0.1	
		Fe	196.0	0.0	[1895, 2550, 2963, 3347, 3857, 4592, 4781]
Si	251.6	Fe	251.8/252.3	0.7	
		Co	252.1	0.5	
		V	252.0	0.4	
Sn	224.6	Cu	224.4	0.2	
		Pb	224.7	0.1	
Te	214.3	Sn	214.9	0.6	
		Zn	213.9	0.4	[5973]
Tl	276.8	Pd	276.3	0.5	
Tm	371.8	Fe	372.0	0.2	
Y	410.2	Ho	410.4	0.2	
		In	410.2	0.1	
		Tm	410.6	0.4	
Yb	398.8	Ti	399.0	0.2	
Zn	213.9	Fe	213.9	0.1	[3072, 6305]
		Te	214.3	0.4	[5973]

If the concomitants in a test sample are vaporized but not atomized, the gaseous molecules can absorb radiation in a similar manner to atoms. In contrast to atomic absorption, this *molecular absorption* takes place over a larger spectral range; we make a distinction between *dissociation continua* and *electron excitation spectra*. If molecules absorb radiation having a quantum energy higher than the dissociation energy, the final state is not a discrete energy state. Above the boundary wavelength we thus observe a spectral continuum. MASSMANN and co-workers [2256, 3938] undertook detailed studies on these types of molecular spectra in graphite furnaces and scanned the dissociation continua of numerous molecules; a number of examples are depicted in Figure 5-14. Electron excitation spectra, on the other hand, derive from electron transitions in the molecules and exhibit a characteristic fine structure, as depicted on an example in Figure 5-15. The structure of the bands is determined by transitions from the rotational and vibrational levels of one electronic state to those of a different electronic state, i.e. transitions between discrete energy states of the molecule [3939].

It is clear that molecular absorption will only cause a spectral interference when it coincides with the analytical line. This is fairly easy to ascertain for broad-band dissociation continua, but for electron excitation spectra the degree of interference depends on the effective overlap with the rotational lines and can only be measured with high resolution spectrographs.

Molecular absorption and radiation scattering always cause a too high measurement signal, which then comprises the analyte specific atomic absorption and the analyte non-specific radiation attenuation. Separation of both signals cannot be achieved by simple means. In practice it is difficult to distinguish between molecular absorption and radiation scattering, and their effect on the measurement signal is very similar, so that both effects are combined under the general term *background attenuation*, even though their causes and physical mechanisms are different.

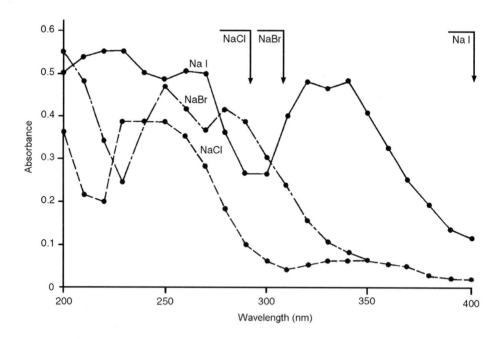

Figure 5-14. Dissociation continua of NaCl, NaBr and NaI, respectively, in a graphite furnace. The boundary wavelengths for photodissociation are marked by arrows (from [2256]).

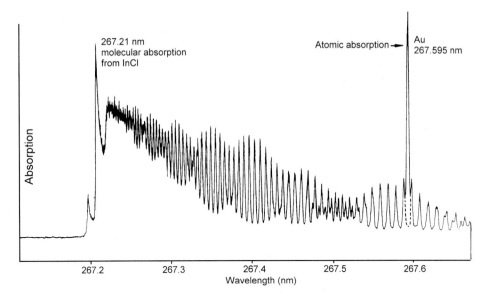

Figure 5-15. Highly resolved absorption spectrum of InCl in the neighborhood of the gold analytical line at 267.6 nm; the analytical line is exactly in the middle between two rotational lines of the InCl molecule (from [3939]).

5.4.1.2 Non-spectral Interferences

Non-spectral interferences are those which cause an influence on the number of analyte atoms (absolute or per time unit) in the absorption volume (and thus on the measurement signal). Non-spectral interferences are most conveniently classified according to the place or time of their occurrence. The earlier classification into 'physical' and 'chemical' interferences is discouraged because it is not appropriate. If an interference cannot be classified, perhaps because its cause is unknown or is of a complex nature, we speak of an effect. Under *matrix effect*, for example, we understand a composite interference due to concomitants in the test sample. If a solvent other than water is used, the influence is not termed an interference since the calibration solutions must always be made up using the same solvent. Nevertheless we can speak of a solvent effect. The same is true for all additives to the test sample solution used to eliminate interferences or to achieve a certain result. Their influence is not termed an interference since they must also be added in the same quantity to the calibration solutions.

If an interference occurs in different ways with a number of different analytes we speak of a *specific interference*, otherwise we refer to a *non-specific interference*. As mentioned above, the most convenient classification is according to the place or time of their occurrence. Both the type and the extent of non-spectral interferences depend to a very high degree on the technique applied and the atomizer unit employed. For this reason a detailed discussion of these interferences is given in Chapter 8 under the various AAS techniques.

If the actual determination of the analyte is preceded by a chemical reaction, such as in the HG AAS and CV AAS techniques, an *interference in the liquid phase* can occur. In most cases this is a reaction of the reduced analyte (the hydride or elemental mercury) with a reduced form of the interfering element (metal, metal boride). In this way a portion of the analyte is bound and remains in the liquid phase, thus influencing the efficiency of chemical vapor generation. Interferences of this type are specific interferences.

All techniques in which the test sample, the measurement solution, or the analyte are transported to the atomizer ***during*** the measurement can exhibit *transport interferences*. In the flame technique these are the interferences that influence the mass transport per time unit of the desolvated test sample through the horizontal flame cross-section at the observation height, e.g. viscosity, surface tension, aspiration rate of the nebulizer, effectiveness of nebulization, or the portion desolvated. Transport interferences are normally non-specific interferences, i.e. independent of the analyte. Techniques in which the sample is introduced into the atomizer ***prior*** to atomization, as is mostly the case for GF AAS, do not exhibit transport interferences. To date transport interferences have not been observed for the HG AAS and CV AAS techniques, either.

Solute-volatilization interferences can occur both in F AAS and GF AAS. In the flame technique these interferences are based on a changed rate of volatilization of the aerosol particles if the volatilization of the analyte in the presence, or absence, of a concomitant is incomplete. This interference is specific if the analyte and the interfering element form a new phase (compound) with a different thermal stability. The interference can also be non-specific if the analyte is enclosed in a large excess of a difficultly melted or volatilized substance and therefore volatilizes more slowly.

In the graphite furnace technique solute-volatilization interferences are based on a change of the volatilization temperature or time of the analyte in the presence of a concomitant. The most commonly occurring form is prior loss of the analyte during the pyrolysis step when chemical modifiers are not used. Another possible interference is the formation of difficult-to-volatilize compounds between the analyte and the matrix, e.g. the formation of carbides with residues of organic materials. Also belonging to this group of interferences is the too rapid or too early volatilization of the analyte with concomitants that decompose during the atomization step. Except for this latter case, volatilization interferences in GF AAS are mostly specific.

Vapor-phase interferences can occur in all AAS techniques. They are caused by a change in the fraction of the analyte that is dissociated, ionized, or excited in the vapor phase. Vapor-phase interferences are always specific interferences. *Dissociation interferences* are a form of vapor-phase interference found fairly commonly both in F AAS and GF AAS caused by a change in the dissociation equilibrium in the absorption volume by concomitants. In flames, these vapor-phase interferences are frequently caused by other cations that form a difficult-to-dissociate mixed oxide with the analyte. In the graphite furnace technique, on the other hand, the interference is caused by anions, especially the formation of difficult-to-dissociate chlorides.

Due to the relatively low temperatures prevailing in graphite tube atomizers and quartz tube atomizers, no noticeable ionization can be observed, so that *ionization interferences* cannot occur. In principle this should also be true for the flames normally used in AAS, provided that we only consider thermal ionization. In the primary reaction zones of air-acetylene and especially nitrous oxide-acetylene flames the analyte atoms are mainly atomized by charge transfer from molecular ions of the flame gases (see Section 8.1.2.5). Considerable deviations from the Saha equilibrium are possible (refer to Section 2.2). The presence of ionization, and thus the risk of ionization interferences, can be recognized by curvature of the calibration curve away from the concentration axis, especially at lower absorbance values (further details see Section 8.1.2.5).

Interferences found specifically in the hydride-generation technique with atomization in a QTA are those that influence the *lifetime of the hydrogen radicals* or the *lifetime of the analyte atoms* in the absorption volume. Both reactions are based on the fact that under normal analytical conditions the conditions prevailing in the QTA are far removed from thermodynamic equilibrium. Substances that reach the atomizer in addition to the analyte can accelerate the attainment of equilibrium and thus influence either the lifetime of the radicals responsible for atomization or the analyte atoms themselves.

Excitation interferences, which are theoretically possible due to a shift in the equilibrium between atoms in the ground state and those in the excited state, do not play any role in the atomizers used in AAS. Due to the exceedingly short residence time of around 10^{-8} s in the excited state, statistically more than 99% of all analyte atoms are in the ground state. Any changes in this area are thus without significance in practice.

As well as the interferences mentioned so far, *spatial-distribution interferences* can occur in all atomizers. The effects nevertheless vary considerably for each atomizer type. In flames a change in the mass flowrate or mass flow pattern of the analyte in the absorption volume in particular leads to false measurements when these effects cause an uneven distribution of the analyte in the atomizer. Normally an irregular radial distribution of the analyte atoms in the observation volume leads merely to an increased curvature of the calibration curve [3694, 3695, 3698].

Occasionally *memory interferences* are observed, especially for chemical vapor generation, i.e. with HG AAS and CV AAS, and for GF AAS in spatially non-isothermal atomizers. These interferences are not caused by concomitants in the current sample but by residues from an earlier sample. Interferences of this type are easy to recognize since they also influence pure, matrix-free calibration solutions. Such interferences are often very difficult to eliminate and require the entire atomizer unit to be thoroughly cleaned.

5.4.2 Recognizing Interferences

With a certain level of experience and the necessary attention to detail, it is usually possible to recognize coarse interferences rapidly. Smaller systematic errors can mostly only be detected when the analysis is under statistical control (see Section 5.5). For this purpose a criterion can be introduced that lies significantly above the expected precision. Since the actual AAS measurement can be performed in the optimum working range with a precision $s_{rel} < 5\%$, an interference can be detected with high certainty if the deviation is > 10%.

Spectral and non-spectral interferences can mostly be recognized by *diluting the test sample solution*, since the majority of the interferences decrease with increasing dilution. Spectral interferences can usually be recognized by an overproportional decrease of the measurement value, while non-spectral interferences cause an underproportional decrease.

Spectral interferences can almost always be recognized by repeating the measurement *at another analytical line*, since the magnitude of the interference is generally more or less wavelength dependent. This technique is more reliable and more informative the further the analytical lines are separated from each other. A *change of the slit width* can also be utilized to recognize spectral interferences. In particular for molecular absorption caused by electron excitation spectra, a change of the slit width causes a significant influence on the background attenuation. In GF AAS spectral interferences can often be recognized by an influence on the *time-resolved absorption signals* (see Section 8.2.4.2).

A further very reliable and generally applicable procedure for recognizing spectral interferences is to *measure a matrix blank solution*. This solution is identical to the test sample solution except that it does not contain the analyte. If this solution generates a signal it can only be due to concomitants and is therefore non-specific. The biggest handicap to this procedure is that matrix blank solutions are often not available, especially when the composition of the test sample or the test sample solution is not known. Frequently it suffices, however, to investigate the main constituents or the most important additives of a digestion in their approximate concentrations in order to obtain indications of possible interferences.

Non-spectral interferences can often be recognized using the *spiking technique*. Here, the analyte is added in a known mass concentration β or mass m to the test sample solution. By comparing the concentration of the analyte determined in the spiked test sample solution with that in the (correspondingly diluted) solution used for spiking it is possible to calculate the *recovery*:

$$Recovery = \frac{\beta_{\text{spiked solution}} - \beta_{\text{original solution}}}{\beta_{\text{spiking solution}}} \times 100\% . \qquad (5.30)$$

A further possibility is to compare the gradient of the calibration curve for the spiked test sample solution with that of the calibration solutions. If the recovery clearly deviates from 100% or if there is a significant difference between the gradients of the calibration curves, in all probability a non-spectral interference is present. Since the test of recovery is based on a comparison of calibration functions, it is not suitable as an independent test for trueness (see Section 5.2.3.5); it is merely an indicator for interferences and their magnitude. To have any significance, the spiked test sample must be typical for the test samples being analyzed and the spiked mass or mass concentration should be of the same magnitude as the analyte content in the test sample solution. If necessary, experiments must be performed in various sectors of the working range; all measurement values must be in the linear working range.

Tests of this type can be included in the working sequence and data processing of the atomic absorption spectrometer if a corresponding quality assurance routine is included in the instrument software; this is the case for a number of newer instruments from various manufacturers. For a number of techniques such as GF AAS and especially HG AAS and CV AAS, these procedures depend very strongly on the oxidation state or the bonding form of the analyte. A quantitative recovery of the added analyte does not necessarily mean that interferences are absent. In such cases it is advantageous to perform the spiking prior to digestion or sample preparation.

A further important aid in recognizing interferences is to record the transient signals. This is especially the case for the diagnosis of spectral interferences in GF AAS by comparing the specific and the background signals [590, 4072, 4598, 6185–6187]; further information is provided in Sections 8.2.2.1 and 8.2.4.2. The same is true for double peaks [2766, 2837, 4027, 5064] which often indicate varying compounds [2765] or phases [2589, 5746], or poor mixing of the sample and the modifier [5792], or derive from atomization from different surfaces, e.g. wall and platform [2763].

The flow injection technique offers a number of possibilities for recognizing interferences. For example, by adding potential interferents to the test sample and carrier solutions it is possible to draw conclusions about the magnitude and even the mechanism of interferences by scrutinizing the transient signals [1284, 5939]. Examples of such diagnostic techniques can be found for F AAS [5509] and for HG AAS [6259] and CV AAS [1416].

5.4.3 Avoiding and Eliminating Interferences

The principal ways of avoiding and eliminating interferences are merely summarized in this section. Since the magnitude of the interferences and the approaches that can be applied to eliminate them depend very strongly on the respective AAS technique, details are discussed in Chapter 8.

5.4.3.1 Eliminating Spectral Interferences

Spectral interferences caused by direct line overlap are best avoided by *changing to an alternate analytical line*. A sufficiently large number of alternate lines is available, especially for the transition elements (refer to the individual elements in Chapter 9). In the case of background attenuation it is also possible to use an alternate analytical line to avoid or eliminate the interference. The interference frequently decreases markedly with increasing wavelength.

Like non-spectral interferences, spectral interferences can be reduced, or on occasions even eliminated, by *diluting* the test sample solution. Nevertheless, there are limits to this technique since the analyte is also always diluted.

If it is really possible to prepare a true matrix blank solution for any given test sample solution, the chance is given not only to recognize the interference but also to eliminate it. It is merely necessary to subtract the measurement value for the matrix blank solution, which can only derive from background attenuation, from that of the test sample solution.

Since background attenuation is caused by non-volatilized or non-dissociated sample constituents, this interference can be reduced or even eliminated by raising the effective temperature in the atomizer. This can be achieved in F AAS for example by changing from the air-acetylene to the nitrous oxide-acetylene flame. A similar effect can be achieved in GF AAS by using the platform technique or the STPF concept (see Section 8.2.4.1).

Nowadays instrumental techniques are used almost exclusively to eliminate interferences caused by background attenuation. The two most important techniques are continuum source BC and Zeeman-effect BC. These, and a number of other rarely used techniques, are described in Section 3.4. Continuum source BC usually suffices for the interferences that occur in F AAS; for GF AAS, on the other hand, it is often insufficient. Zeeman-effect BC is thus recommended for GF AAS to increase the reliability; in addition it mostly leads to a better S/N ratio.

We must emphasize that the use of continuum source BC alone cannot guarantee correct results; on the contrary, it may even introduce errors. With unknown samples it is important to work with and without background correction or to record the background signal. In this way it is possible to obtain important information about the appearance and magnitude of the background attenuation that would be lost if work was performed with background correction alone. An important prerequisite for the correct functioning of continuum source BC is that the background is continuous over a range of twice the spectral band passed by the exit slit of the monochromator. This prerequisite is met for radiation scattering on particles and for dissociation continua (refer to Section 5.4.1.1).

Continuum source BC, on the other hand, is not able to correct electron excitation spectra since these consist of many narrow lines. The actual interference to the measurement, i.e. the effectiveness of background correction, depends on the degree of overlap between the analytical line and the individual molecular rotational line or lines [2494, 3938]; the resolution of the monochromators normally used in AAS is completely inadequate in this situation. In the InCl spectrum depicted in Figure 5-14 (Section 5.4.1.1) the analytical line for gold at 267.6 nm is exactly in the middle between two rotational lines so that the actual background attenuation is relatively low. If we now apply a continuum source background corrector, this determines the background attenuation measured over the observed spectral range, which is naturally larger than the effective background attenuation at the gold line. Subtraction of the absorbance values leads to a measurement error due to overcompensation [2628].

Although the limitations of continuum source BC have been known for years, many analysts nevertheless have a blind faith in the infallibility of background correction. A check on the reliability of the technique should always be performed when unknown samples are being analyzed. A qualitative check can be performed by measuring the background at a close-lying non-absorbable line (refer to Section 1.5). It is an indication of an error in background correction if a measurement signal is obtained.

Background correction with high current pulsing offers better correction than with continuum sources, provided that the background does not exhibit rotational fine structure. The main reason for this is that the background attenuation is measured in the direct neighborhood of the analytical line so that the demands on the continuity of the background are lower. For a *background with fine structure* we must make a clear distinction between techniques in which the background is measured ***near*** to the analytical line (e.g. high current pulsing and the Zeeman effect with the magnet at the radiation source) and those in which the background is measured ***at*** the analytical line (Zeeman effect with the magnet at the atomizer).

MASSMANN and EL GOHARI [3940] showed for numerous examples that the distance between the individual lines of a rotational-vibrational spectrum is frequently in the order of 10 pm and thus of the same magnitude as line splitting for high current pulsing or the application of the Zeeman effect. It is thus possible that the analytical line lies exactly between two rotational lines of the background spectrum, as depicted in Figure 5-14, while the background measurement is performed at the maxima of two rotational lines. In this situation the error in background correction near to the line is larger than the error with a continuum source. The accuracy of background correction with measurement near to the line is never guaranteed when the background is structured.

With Zeeman-effect BC with the magnet at the atomizer the analyte absorption and the background attenuation are measured with the same line profile and at exactly the same wavelength. It should thus be possible to attain highest accuracy with this configuration. A prerequirement is nevertheless that the background attenuation does not change under the influence of the magnetic field. MASSMANN and EL GOHARI [3940] have drawn attention to the fact that molecules can also exhibit the Zeeman effect. For example they measured the background absorption of the OH radical using a bismuth lamp at the 306.77 nm line. With a magnetic field of 1.3 Tesla they found a value that was 20% too high for the background 'under' the line.

The OH spectrum does not occur in graphite tube atomizers so that this interference can only be observed in flames. For investigations on the background absorption of SO_2, caused by the vaporization of sulfates, in the wavelength ranges 200–220 nm and 260–310 nm in which the strongest bands are to be found, MASSMANN and EL GOHARI [3940] were unable to detect the Zeeman effect. At a magnetic flux density of 1.3 Tesla the differences in the absorption were less than 1%, indicating that Zeeman-effect BC provides accurate values. The same can be assumed with high probability for other polyatomic molecules such as NO_2. These findings also largely agree with practical experience obtained for the application of the Zeeman effect with the magnet at the atomizer. Further details are provided in connection with the graphite furnace technique (Section 8.2.4.2).

5.4.3.2 Elimination of Non-spectral Interferences

Non-spectral interferences, i.e. direct influences on the sensitivity, can generally be eliminated by matching the test sample solutions and the calibration solutions as closely as possible. In the ideal case a *calibration solution* contains not only the same solvent and the same additives as the test sample solution, but also the same concomitants. A calibration solution should thus according to definition correspond to the test sample solution except for the concentration of the analyte. If this were really the case we would not observe any interferences since the matrix effects influence the analyte in the test sample and calibration solutions to the same extent.

Albeit this ideal case can rarely be realized in practice. In the first place the exact composition of a test sample frequently is not known and in many cases cannot be reproduced. Even if these requirements are met, reagents of the highest purity and often considerable effort are required to prepare a calibration solution that corresponds exactly to the test sample solution. Since in many laboratories large numbers of samples with varying compositions must be analyzed, this procedure is in any case unrealistic.

Fortunately the interferences in AAS are rarely so pronounced that it is necessary to match the calibration solution to the test sample solution. It often suffices to use the same solvent and to match a main constituent of the test sample. In the flame technique in particular routine determinations can frequently be performed against simple calibration solutions.

The application of the *analyte addition technique* is recommended when the composition of the test samples changes frequently or complex matrix effects are present. Since an individual calibration curve is established for each test sample by adding calibration solutions, the ideal case described above is virtually attained. All non-spectral interferences should in effect be eliminated using this technique, except those that depend on the concentration of the analyte.

Ionization is most definitely dependent on the concentration, so that the analyte addition technique cannot be applied to eliminate this interference. In HG AAS the oxidation state of the analyte or the species frequently plays a decisive role for the sensitivity; this fact must be taken into account when the analyte addition technique is applied. In GF AAS the bonding form and the ligands also have a considerable influence on the thermal behavior and thus on the volatility of the element. If the analyte in the addition has a different bonding form its behavior can be totally different so that the interference is not eliminated. The application of the analyte addition technique in GF AAS must be regarded very critically. The technique can only be successful when all conditions of the STPF concept are fulfilled. Under these conditions, however, the interferences that can be observed under other conditions are usually eliminated, so that the application of the analyte addition technique is superfluous [6233].

Ideally the analyte addition technique can be applied to eliminate all non-specific interferences, such as transport interferences in F AAS. If this technique is used to eliminate specific interferences, then great care must be taken to ensure that the influences on the analyte are the same in both the test sample and the added solutions.

The *reference element technique*, in which another element is added to the sample in known concentration, is, by definition, only suitable for the elimination of non-specific interferences, such as transport interferences. Since logically all specific interferences are specific to the given analyte, it cannot be expected that they can be eliminated by measuring another element. As already explained (see Section 5.2.2.4) this technique makes very high demands on the instrument and the analytical technique, and is thus little suited for AAS.

An important prerequisite of this technique is that the signal depression caused by concomitants is not too severe. If the interferences are so large that the calibration curve or the addition curve is too shallow, suitable measurements often cannot be made. The loss in sensitivity also leads to a deterioration in the S/N ratio.

A technique used widely for the elimination of interferences is the addition of *spectrochemical buffers* to the test sample and calibration solutions. Depending on the type of interference, different buffers with varying modes of action can be employed. *Nebulization aids*, for example, can be used to bring the physical properties of the solutions, such as viscosity or surface tension, into line and thereby eliminate transport interferences.

Volatilizers are a further group of spectrochemical buffers that improve the volatilization or atomization of the analyte by converting it into a more suitable form.

Releasers are frequently used to prevent the analyte from entering a thermally stable compound. This technique for the elimination of solute-volatilization and vapor-phase interferences is based on the knowledge that these interferences occur during desolvation, during initial precipitation from the saturated solution, or through rearrangement reactions in the solid particles.

An *ionization buffer* is added to suppress ionization; this is an easily ionized element, such as cesium or potassium, that increases the concentration of free electrons in the atomizer, thereby suppressing and stabilizing ionization of the analyte.

In GF AAS *chemical modification* is a procedure for reducing or eliminating volatilization and gas-phase interferences. A reagent (*modifier*) is added in high concentration to bring the physical and chemical properties of the measurement solutions into line. The reagent frequently serves to convert the analyte into a less volatile form, so that higher pyrolysis temperatures can be applied, and/or the concomitants into a more volatile form. Either measure serves to bring about more effective separation of the analyte from concomitants during the pyrolysis step.

When using flames as atomizers it should always be borne in mind that the dissociation of molecules into atoms is a temperature-dependent equilibrium and that this reaction can be influenced in magnitude and kinetic by parallel and secondary reactions. Since the various reactions take place successively and thus in various zones of the flame, the choice of the observation height is of decisive importance [130]. As discussed in Section 8.1.1.4, all processes from desolvation to the release of free atoms from the molecular bonds must occur within a few milliseconds, since this is the rise time from the burner slot to the absorption volume. Any influence on the reaction velocity can lead to interferences when the absorption volume is located in a flame zone in which equilibrium has not yet been attained.

Vapor-phase interferences can sometimes be avoided or reduced by measuring in higher flame zones. Interferences caused by reaction of the analyte atoms with flame gas components, such as oxygen, can occasionally be avoided by measuring nearer to the burner slot [4954]. The higher the affinity of the analyte for oxygen, the earlier and more rapid is the decrease in the atom concentration with the observation height.

A flame temperature that is too low or an unsuitable chemical environment leads to a lower degree of atomization and to increased solute-volatilization and vapor-phase interferences. This problem can be significantly reduced by using a nitrous oxide-acetylene flame since, next to the relatively high temperature (approx. 2750 °C), this flame offers an ideal reducing environment for the dissociation of numerous compounds. The even hotter oxygen-acetylene flame (approx. 3050 °C), on the other hand, causes many interferences due to its strongly oxidizing characteristics.

Many interferences can thus be eliminated by raising the flame temperature or changing the chemical environment. Similar principles also apply to the graphite furnace technique where solute-volatilization interferences can be eliminated by chemical modification and gas-phase interferences can be avoided by atomizing into a thermally stable environment. As early as 1970 L'VOV [3690] had drawn attention to the significance of isothermal conditions for atomization since under these conditions the remaining nonspectral interferences can be eliminated by integrating the peak area. The majority of gas-phase interferences observed in GF AAS can be attributed to the fact that in Mass-mann-type graphite furnaces the test sample is volatilized into a non-isothermal environment. Atomization takes place during the heating phase when the temperature is changing markedly and the gas in the graphite tube is expanding. Since at the moment of atomization or volatilization of the analyte from the tube wall the inert gas is colder than the tube surface, considerable recombination of atoms or the formation of compounds takes place in the gas phase. By using an atomizer that is in temporal thermal equilibrium, at least at the moment of atomization, or even better is in both temporal and spatial thermal equilibrium, many gas-phase interferences disappear.

5.5 Methods Development, Quality Control and Quality Assessment

The accuracy of a determination depends not only on the selection of a suitable procedure, but also on the stability of the performance characteristics of that procedure. If analytical results are to be informative, they must always be accompanied by a quantitative assessment of their level of uncertainty [3141]. In addition, assurance is always required that the analytical data provided by a laboratory are true. We have already emphasized that the information content of analyses depends to a very large extent on the sampling and that this requires the greatest of care (refer to Section 5.1.1). In this section we shall describe a number of measures that serve to control and assess the quality of the results generated in an analytical laboratory.

The first requirement for implementing a quality assurance program is the organization of events in the laboratory. This organization should follow certain rules that are generally summarized as *Good Laboratory Practice* (GLP) and *Good Analytical Practice* (GAP). GLP involves the organization of events in the laboratory and the conditions under which laboratory investigations are planned, conducted, and assessed, as well as maintaining a protocol and issuing a report of the investigation [4064, 4732]. The aim of GLP is to make procedures in the laboratory so transparent that they can be verified by third parties (e.g. regulatory authorities, clients). The scope of these rules is laid down in the chemical legislation in force in most of the industrial nations. A detailed discussion on this topic is beyond our intended scope and the interested reader is referred to introductory monographs, e.g. [2082]. Maintaining all the rules of GLP is both expensive and requires a considerable increase in personnel; on the other hand, GLP alone cannot guarantee the trueness and quality of analytical data. The final responsibility for quality assurance still remains with the analyst [4064]. Quality control and assessment programs must be implemented and the rules of GAP must be followed in a laboratory to obtain high quality results with sufficient accuracy and precision. The rules of GAP include the following:

i)   The use of validated methods.
ii)   Analytical instruments that are calibrated and maintained in good condition.
iii) The use of certified reference materials for calibration.
iv)   An effective internal quality assurance program.
v)   Regular participation in interlaboratory ('round-robin') trials.
vi)   Quality assessment by an independent quality assurance unit.

vii) Accreditation or other certification of the laboratory by regulatory authorities.

viii) Last not least, an adequate number of adequately trained staff.

In the following, quality control and assessment are discussed briefly within the framework required for performing determinations by AAS. These two terms are frequently used synonymously in the literature; in this monograph we have used the definition proposed by TAYLOR [5787]. Accordingly *quality control* is the mechanism established to control errors, while *quality assessment* is the mechanism to verify that the system is operating within acceptable limits. In principle, two different strategies are applied to meet these goals. On the one hand, various forms of replicate measurements provide information on the expected precision, while on the other hand the analysis of certified reference samples provides information on the expected trueness. Accordingly, external auditing or the participation in interlaboratory trials are measures included in a quality assessment program, while quality control comprises the internal control procedures. For a more detailed discussion we refer our readers to the comprehensive literature: Monographs [2024, 2278, 3043, 5100], leading articles [176, 178, 851, 2541, 3141, 5333], and discursive articles [1332, 4843].

5.5.1 Methods Development and Validation

A suitable analytical method must be selected and adequately characterized before procedures for quality control can be defined, i.e. measures must be taken to ensure that major problems such as interferences or instrument malfunctions are recognized and eliminated, and that the method generates reasonably stable results. For this, the method must be characterized by the parameters discussed in Sections 5.1 through 5.4. During methods development and validation these include:

i)   Check for contamination.

ii)   Determining the limits of detection and quantitation.

iii)   Check for possible interferences.

iv)   Comparison of the analytical results with those from alternate methods.

v)   Comparative analyses with other experienced laboratories.

vi)   Determination of the precision within a series and between series.

vii)   Analysis of certified reference materials.

Further parameters that should be applied during application of a method include:

viii) Regular checks for contamination.

ix)   Internal quality controls.

x)   Continuous statistical control.

xi)   Participation in external quality control programs.

Since a contract laboratory can only select methods that have been adequately characterized, in principle it cannot accept chance analytical tasks.

As far as possible, effective quality control should be incorporated directly into the analytical program, since analytical results cannot first be generated and then deliberations subsequently made about their trueness. The further the test sample, the expected concentration range, or the analyte deviates from the routine tasks performed by a laboratory, the higher the effort the laboratory should make with respect to quality control. Quality control always requires a high level of effort and must thus be accommodated to the analytical task at hand, i.e. the higher the requirements on the accuracy of a method, the greater is the effort required for quality control. Nevertheless, higher costs for quality control can in end effect be justified when costs in other areas are thereby saved, such as those caused by repeated measurements or by false results [6293]. Depending on the given requirements, when a quality control program is planned either the probability of error recognition must be maximized or the probability of a false warning must be minimized. By excluding contamination, drift, and other errors, a quality control program can lead to an improvement in the performance of a method.

Every internal procedure used by a laboratory for quality assessment should begin with the definition, the assessment, and as required the optimization of the *precision of the calibration function* [2553], since the precision of the analytical results depends directly on it (see Section 5.2.2.5). From collaborative tests and certification actions it is known that 25–30% of all false results derive from incorrect calibration [4756]. We cannot therefore stress often enough that only chemicals of exactly known composition and purity should be used for calibration [4180].

Thereafter the *instrumental repeatability* should be determined (refer to Section 5.2.3.2). If it is already known it can easily be integrated as a criterion in the quality control routine of the instrument software to control the sequence of replicate measurements [5329, 5333]. Although the precision increases with the number of replicate measurements, the sample throughput decreases, so that limiting the number of replicates to the necessary minimum is a requirement for optimization of the costs. Stipulating the required precision can help to identify measurements in which the precision is unacceptable and to initiate further replicates until the required precision is attained. A typical application for a routine of this type is to perform a fifth replicate measurement when an outlier has caused poor precision with four replicates. Routines of this type avoid having to repeat a test sample at the end of a measurement series. Nevertheless special cases can pose problems for such routines, for example when the concentration is below the limit of quantitation or when particles are present in the sample. It is thus important to include a termination criterion (e.g. a maximum number of replicate measurements) or subsidiary parameters (e.g. minimum signal height) in the routine. Otherwise the incorporation of a quality control routine during the measurements can become a liability if the analytical sequence is repeated *ad infinitum* due to poor performance of the quality control software. Furthermore it should be borne in mind that the imprecision of the AAS determination is mostly the smallest part of the total uncertainty (refer to Section 5.2.3.2).

The method must be under statistical control before investigations on systematic errors (interferences) can be undertaken, otherwise it is hardly possible to distinguish whether the result is 'good' or 'bad' [2726]. This check can be performed with reference samples or certified reference materials, or the test sample can be spiked [4599]. Since interferences mostly depend on the content of the interferent or the interferent/analyte ratio, tests should always be performed at the limits of the working range [2726]. Under the assumption that the interference increases with the concentration of the interferent, work can be facilitated by using a concentration that is somewhat higher than the maximum concentration expected in the sample. Since interferences can occur in all steps of the analysis, the analytical control sample should be taken through all steps. A statistical test can be applied in order to decide whether the mean value for the analytical control sample deviates significantly from the 'known' concentration. During methods development for techniques that generate transient signals, not only the analytical values but also the complete signal profile should be documented. In this way a diagnostic means is provided by evaluating the time-resolved signal profiles for the specific and the non-specific signals [590].

During methods development it is of advantage to check the results by using alternate methods. Method comparisons can be performed either in the laboratory, if alternate methods are available, or with the aid of an external laboratory. Many laboratories are prepared to share preanalyzed samples with other laboratories, or to analyze their samples in the course of internal controls. For the final comparison, however, a regression calculation using the method of least mean squares cannot be applied since the results of a method comparison contain errors [334, 711, 2377, 2411, 6330].

A major component of any methods development is the *analysis of certified reference materials*. CRMs are available for metallurgical analyses, environmental analyses, geochemistry, and increasingly for foodstuffs and biological materials. Further information on CRMs can be obtained from the catalogs of the various vendors (e.g. BCR [1778], NIST [5886]) or from databanks (e.g. COMAR [3146]). Current trends and problems are the subject of regular symposia (e.g. BERM [4928–4930]) and are published regularly in the analytical journals (e.g. Geostandards Newsletter, publications from BCR, IAEA, NIST); information on special materials is provided in Chapter 10. The analysis of CRMs is particularly suitable in methods development since they are independent of the calibration. The greater the likeness of the CRM to the test sample, the higher is the information content of this type of validation. Considering the multitude of materials that have to be analyzed, it is not surprising that CRMs are not available for all of them [2756]. Nevertheless, this should not prevent the analyst from maintaining a broad spectrum of reference samples, since during method validation it is more advantageous to be able to analyze one of them correctly, even if it is not identical in composition with the test sample, than not to analyze a reference sample at all. Since the time and effort required to certify reference materials is substantial, they are not cheap [457, 1376, 1750, 2783, 4816]. For this reason they should only be used in exceptional cases for routine applications (e.g. calibration) [2903]. Extensive use of reference samples means that they are used up more rapidly and this can cause problems with long-term comparative studies. Nevertheless, the regular use of CRMs during methods development and method validation, and also during cyclical quality control, is an essential component of GAP. Validation should be performed under all circumstances before a method is published and made available to a wider audience. Unfortunately in many cases validation is performed inadequately or not at all and is not documented (see [3601]). It would be a great aid to the scientific community if the publishers of scientific journals refused to publish methods that are not supported by minimum evidence of validation.

5.5.2 Internal Quality Control

All actions taken during methods development and validation to recognize and eliminate sources of error, or at least to reduce them to an acceptable level, certainly aid in lending confidence to a method, but they cannot guarantee that the method brings the expected quality characteristics over a longer period of time. It is thus essential to check a method regularly during use to ensure that the quality characteristics are being maintained. This is most conveniently done by regularly analyzing *analytical control samples*. The results are documented in the form of *control charts* [2079, 2684, 4230, 5300]; they can be plotted by hand, or by using spreadsheets [3496, 4505], or with special software [1274, 1291, 4843]. Control charts serve multiple functions [2079, 4230]:

i)   Recognition of drift, caused for example by aged standards.
ii)   Recognition of problems such as damage to or misalignment of the instrument.
iii) To uncover errors in sampling or sample preparation.
iv)   To support the analyst with respect to his clients by assuring his results.

Control charting is an excellent way of documenting measurements over longer periods of time; a control chart should also contain notes about any special occurrences and the name of the analyst. Nevertheless, control charts are solely a means of detecting deteriorations and deviations from the normal. A control chart cannot change anything, and will not issue a warning, when a measurement system generally exhibits poor precision, which is merely expressed as wider control limits.

A control chart, such as the example depicted in Figure 5-16, is established by taking an analytical control sample through the method under routine conditions just like any other sample. If samples with highly varying analyte contents are to be analyzed by the method being controlled, it can be advantageous to establish several control charts using the appropriate analytical control samples. During method validation the method is characterized by observing the repeatability over a longer period of time. The number of replicate measurements is selected according to the strictness of the control; the more strict, the greater the number of replicates [2684]. The mean value $\bar{x}$ of the replicate measurements and the standard deviation for the control chart are calculated after an adequate number (e.g. 20–30) of control determinations. The random distribution of the values taken as the basis can be checked by the Wald-Wolfowitz test [2684].

This phase of characterization is then followed by the phase of statistical control in which analytical control samples are measured at regular intervals and the results entered into the control chart. How often 'regular' means is a question of cost/benefit analysis and must be decided in each individual case. It is a sensible rule to place at least one analytical control sample in each group of contract analyses (e.g. sample tray of the auto-sampler). A larger number of analytical control samples may be required if it is desired to perform a very strict control or if the stability of the method being controlled is known to be especially low. The analytical control sample must also be a random member of the sample contingent to prevent the periodic occurrence of possible errors. As well as the control data, particular situations, problems, instrumental changes, etc. should be entered in the control chart. In GF AAS, for example, such situations can be the opening of a new batch of graphite tubes (charge number), the installation of new graphite

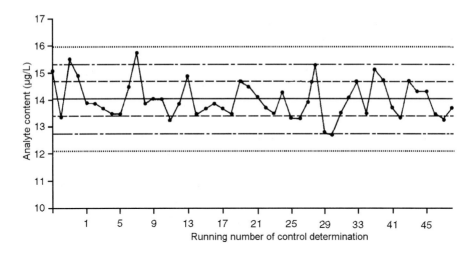

Figure 5-16. Example of a control chart. — — — — ± 1*s* limit; — · — · — ± 2*s* limit;
·············· ± 3*s* limit (from [5333]).

contacts, or a new sampling capillary, the use of new hollow cathode lamps, etc. Notes of this type can be very useful in tracing the cause of problems if the method gets out of control. Various rules can be established with which to ascertain problems [2079]:

i)   One control value outside the 3*s* range.
ii) Two successive values on one side of the middle line outside the 2*s* range.
iii) Seven rising or falling values in sequence.

If any of these rules is broken, the cause is in all probability not a random error but has a specifiable cause. If required, further rules can also be set [4230], for example:

iv)   The difference between two successive values exceeds 4*s*.
v)   8–10 successive values on one side of the middle line.
vi)   14 successive values alternately above/below the middle line.
vii) 2 (of 3) successive values in the range > ±2*s*.
viii) 4 (of 5) successive values in the range > 1*s* on one side of the middle line.
ix)   8 (15) successive values in the range > ±1*s*.

However, such extensive rules considerably increase the complexity of the application and also conceal the risk of a false alarm. If the control chart indicates an out-of-control situation, planned activities must be performed with the aim of finding and eliminating the cause. In AAS these activities for example can be:

i)   Checking the characteristic mass for the respective element.
ii) Checking the energy of the primary radiation source.
iii) Visual inspection of the recorded signal profiles.
iv) Checking the alignment of the primary source, continuum source, and atomizer.

The actions performed must naturally be entered on the control chart.

The use of a limited number of rules and the direct maintenance of the control chart by the analyst has the advantage of high transparency; this becomes obvious when a problem can be directly recognized and eliminated. The use of computers at the workplace also supports the maintenance of control charts, allowing the application of further rules (e.g. rules iv–viii) or a floating mean value. Continually updating the mean value and the control limits permits the learning effect to be documented, i.e. the improvement of the precision and accuracy with the length of application, and at the same time gives earlier warning of potential problems. Suitable routines are included in commercial statistical software packages or can be utilized as macro routines in the usual spreadsheet programs (e.g. Microsoft Excel™ or Lotus 1-2-3™). While personnel are being trained it is of advantage to maintain control charts graphically by hand before this task is given over to software support.

The analytical control sample can be a CRM or it can be a sample prepared in the laboratory. A major prerequirement for a material that is to be used for control purposes is that it is adequately stable over the time period of application [176]. Further, this material should be as similar as possible to the test samples and it should be treated just like an unknown sample. If the analytical control sample is accorded special treatment, the value of the control chart is severely limited.

6 Mechanization and Automation

The desire to mechanize and automate processes is as old as mankind himself, because mechanization makes work easier and man has always attempted to make his life easier and more comfortable. In the discipline of analysis, well-conceived automation not only facilitates work, it can also lead to a reduction in costs, improve the trueness and precision of the results, and even open up new possibilities. In the strictest sense [5975], only those processes can be considered as automatic that are under complete control of a computer incorporating a feedback control system and with processing of the data, either from the spectrometer itself or from special sensors. In the wider sense, however, processes can also be considered as automatic if control is via a fixed or a user-defined program. In contrast to mechanization, automation always requires a degree of automatic monitoring and control.

In AAS, mechanization and automation have undergone a number of different stages over the years. *Mechanical sample changers* (*autosamplers*) have been used for several decades; these are devices that transport the measurement solution to the atomizer unit. The benefits brought by this type of mechanization depend very strongly on the application and on the technique to which they are applied. This aspect is treated in Section 6.2.

Automation is naturally not limited to just the transport of the measurement solution—rather more it should encompass all procedures from sample preparation to data management. In this connection, we must mention three developments that have contributed to automation in AAS, even if to widely varying degrees. *Computers* must be regarded as the most important since they have had a revolutionary effect not only on data processing and management, but also on the monitoring and control of spectrometers. Surprisingly, *laboratory robots* have gained little significance; OWENS and ECKSTEIN [4506] reported on their application in an AAS laboratory. One reason for this lack of interest could be that the requirements placed on both the apparatus and the program are high, even to automate relatively simple processes, so that the net gain is comparatively small. The situation is just the reverse for *flow injection* (*FI*), which is quite simple to operate, utilizes new techniques, and opens up completely new possibilities. Since we mention FI repeatedly in this monograph, it would be expedient to discuss the basic principles in the following section before going on to automation of the individual steps of an analysis.

6.1 Flow Injection

In the first edition of their book 'Flow Injection Analysis', RUZICKA and HANSEN [5015] defined flow injection as a technique 'based on the injection of a liquid sample into a moving, nonsegmented continuous carrier stream of a suitable liquid. The injected sample forms a zone, which is then transported toward a detector that continuously records the absorbance or another physical parameter as it continuously changes due to the passage of the sample material through the flowcell.' In the meantime virtually all points of this definition have been superseded as a result of the rapid developments that have taken place in this technique. For example, segmenting of the carrier stream is used increasingly to prevent dispersion under given circumstances. The carrier stream can also be interrupted (*stop flow*) or even reversed (*reverse flow*) to allow a reaction to proceed or to perform better mixing of two solutions.

For this reason FANG [1831] proposed a new definition of FI as 'a non-chromatographic flow technique for quantitative analysis through reproducible manipulation of sample and reagent zones in a flowing system under thermodynamically non-equilibrated conditions.' This definition contains at least three decisive aspects of FI which are of considerable importance for our further discussions.

The first is the *non-chromatographic character of FI*. All separations performed with this technique, even if chromatographic materials are used, serve exclusively to *separate individual components from each other* and not to *separate a mixture into its constituents*. The second aspect is the *high repeatability and reproducibility of all processes* in FI systems with respect to time. Only under these conditions of reproducibility is it possible to perform a quantitative analysis under *thermodynamically non-equilibrated conditions*. If a reaction product is always measured at exactly the same time point after injection, i.e. from the beginning of the reaction, it is irrelevant if the reaction is complete or to which degree it has proceeded provided a relative technique is used.

In practice, not having to wait for equilibrium to be attained leads to a considerable *saving in time* since it is not necessary to wait until the reaction is complete. Further, *unstable intermediates* can be measured, or reagents can be applied under conditions under which they are not thermodynamically stable. As will be shown repeatedly, time plays a decisive role in the FI technique. By using suitably short reaction zones, it is possible to maintain reaction times of less than one second and hence to promote the faster reactions through *kinetic discrimination* and to suppress the slower ones.

Although the original definition of FI by Ruzicka and Hansen has now been superseded in all points, nevertheless there are still large numbers of systems that function on the classical principle of injection and continuous transport to the detector, and thus on the *dispersion* associated with this transport. Under dispersion we understand the progressive mixing of the test sample and carrier solutions, as depicted in Figure 6-1, which leads to blurring of the measurement signal. Since dilution of the measurement solution takes place during dispersion, the effect is not always desirable. Nevertheless for many applications it is an essential part of the procedure. However, the dispersion is as highly reproducible as all other reactions in FI. This means that at a given time point after injection the degree of dilution is practically the same and depends only to a limited amount on the characteristics of the sample. This is true not only for the dilution at peak maximum but also for all points on the dispersion curve shown in Figure 6-1; this is reflected in the measurement signal.

If we compare various geometries of FI reactors with respect to dispersion, we can observe that this is strongest in a *straight tube* (Figure 6-2A); the dispersion increases with increasing diameter of the tube. A frequent means of reducing dispersion to an acceptable level and at the same time to enhance mixing of the solution is to roll the tube into a *coil*, as shown in Figure 6-2B; the dispersion decreases with decreasing tube diameter and at the same time the intensity of mixing increases. Even more effective than reaction coils are *knotted reactors*; a typical example is shown in Figure 6-2E. Substantial centrifugal forces occur due to the permanently changing direction of the carrier stream, leading to intense radial mixing and to minimal dispersion [1825, 5016]. The mixing chamber and single-bead string reactors shown in Figure 6-2 C and D, respectively, are seldom used in AAS.

A decisive advantage of FI systems, especially for trace analysis, is that all reactions take place in a *closed composite system*, thus reducing the risk of contamination. A further contribution in this direction is the small surface area with which the sample comes into contact, and that mostly inert and practically contamination-free materials such as PTFE or FEP are used. In addition it is also possible to purify reagents on-line, using precolumns for example, so that extremely low blank values can often be attained with FI systems. TYSON [5942] has published a review article on the numerous applications of FI in AAS and the resulting advantages. FANG devoted befitting space to this topic in a monograph [1836] and discussed newer developments in a review article [1837]. The monograph by RUZICKA and HANSEN [5016] provides detailed information on the basics and applications of FI.

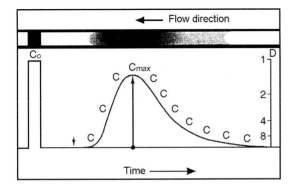

Figure 6-1. An originally homogeneous sample zone (upper left) disperses during its transport through a tubular reactor (upper center), thus changing from an original square profile (lower left) to a continuous concentration gradient with maximum concentration C_{max} at the apex of the peak (lower center). D is the dispersion and is given by $D = C_0/C$ (from [5016]).

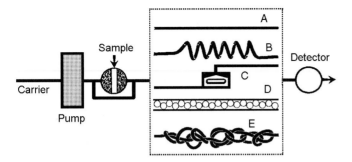

Figure 6-2. Geometries of reactors used in FI. **A** – Straight, open tube; **B** – coiled tube; **C** – mixing chamber reactor; **D** – single-bead string reactor; **E** – knotted reactor (from [5016]).

6.2 Automatic Transport and Change of Measurement Solutions

The mechanization of measurement solution transport and change using *autosamplers* for AAS was described already in the 1960s [2087] and commercial systems were offered [5402]. For F AAS, however, autosamplers have only proven to be of advantage in limited cases since manual operation is in any case quick and simple, and sample preparation is invariably the most time-consuming step in an analysis. For GF AAS, on the other hand, where microliter volumes must be dispensed into the graphite tube, and a furnace program takes 2–3 minutes, autosamplers have made a major contribution to the acceptance of this technique. They have relieved the operator of a difficult operation, improved the trueness and precision of the determination, and significantly reduced the risk of contamination [5579, 6206]. This is even more true for the dispensing of slurries, since manual operation requires considerably more effort and the precision is limited (see Section 4.2.7.3). The analysis of solid samples could only be applied routinely after the development of autosamplers in which a slurry is homogenized by ultrasonics directly prior to dispensing [1077, 1749, 4102].

Autosamplers generally have either dedicated locations or freely selectable locations for the blank solutions, calibration solutions and modifiers. The sequence of the measurement solutions and the number of replicate measurements can usually be programmed. Additional pretreatment functions of autosamplers, such as automatic dilution, addition of reagents, and automatic addition of the analyte for the addition technique, are discussed in Sections 6.3 and 6.4.

Using an FI system, transport of the measurement solution for F AAS can in the simplest case be via a single line, as depicted schematically in Figure 6-3. The major differences of this technique compared to manual operation or to the use of autosamplers are that the measurement solution is transported in a *continuous stream* of a carrier solution and that *propulsion is by means of a pump* and thus independent of the aspiration capability of the nebulizer. These principal advantages of this technique are discussed in detail in Section 4.1.3.3. A further advantage is the *improved stability of the flame* since air is not drawn in between measurement solutions; this leads to a better S/N ratio. The usual sensitivity can be attained with volumes of measurement solution < 100 µL and, in theory, over 1000 measurements/h can be achieved [1825].

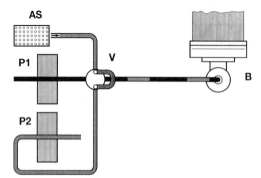

Figure 6-3. Simplest principle using FI for the transport of measurement solution for F AAS. **P1** – pump for carrier solution; **P2** – pump for measurement solution; **AS** – autosampler; **V** – injection valve; **B** – AAS burner (by kind permission of Perkin-Elmer).

A further important advantage of the transport of measurement solutions by FI is the considerably lower *influence of the viscosity and the salt content of the solutions* on the sensitivity. Firstly, propulsion by means of a pump eliminates or strongly reduces the influence of viscosity on the aspiration rate and thus possible transport interferences. Secondly, through the continuous transport of a carrier solution, the nebulizer and burner are rinsed automatically and thoroughly after every determination. It is thus possible to analyze test sample solutions with high total salt contents over several hours without problems, whereas with conventional aspiration the burner would be clogged in the shortest time [333, 3323, 5938]. This has been demonstrated for solutions containing 300 g/L NaCl, 1000 g/L K_2HPO_4, or saturated lithium borate solution [1822]. ZHOU *et al.* [6575] found that markedly lower limits of detection could be obtained for the analysis of steel since solutions of considerably higher total salt concentration could be applied.

FI systems have also been used for the *direct transfer of slurries* for F AAS [194, 944, 3606]. In some cases surfactants were added to stabilize the slurries [3914] or a mixing chamber reactor (see Figure 6-2C) was used to homogenize the slurry. ARRUDA *et al.* [277] even used an FI system to inject slurries directly into a graphite furnace. Normally however, because of its continuous mode of operation, the conventional FI technique is not particularly suitable for injecting measurement solutions for the discontinuous GF AAS technique. An exception is the coupling of on-line preconcentration, which is also discontinuous, with GF AAS; this topic is discussed in more detail in Section 6.5.

6.3 Automatic Addition of Reagents or Calibration Solutions

With most autosamplers, reagents, such as spectroscopic buffers or modifiers, have special containers at predefined locations in the autosampler. Calibration solutions either have predefined locations or the locations can be programmed. Reagents or calibration solutions for the analyte addition technique can be added in the sample container or into a dilution container, or for GF AAS into the atomizer by sequential dispensing.

For the FI technique there are a large number of possibilities for adding reagents to the measurement solution. In most cases the reagent is transported in a *separate tube line* and is mixed with the carrier and measurement solutions in a *manifold*, followed by a reaction zone of varying length. A system of this type for HG AAS is depicted in Figure 4-40 (Section 4.3.1.2). In a system of this type the reagent is propelled in a separate channel by the same pump as the carrier solution and the two solutions are continuously mixed. This is of advantage for HG AAS, for example, since if the acid concentration of the carrier is sufficiently high the generation of hydrogen is even and constant. The reagent can also be propelled by a second pump that is only switched on when the addition of reagent is required. A further possibility is *to add the reagent to the carrier* prior to the injection of the measurement solution. High dispersion is a prerequirement for this technique, however, since only then is thorough mixing of the measurement solution and the reagent guaranteed. At the same time this causes considerable dilution of the measurement solution and broadening of the signal, which in turn reduces the measurement frequency.

The reagent can also be added via a *second loop* (*merging zones*), which can be mounted either on a second injection valve or on the same injection valve, as shown in Figure 4-42 (Section 4.3.1.2). The advantages of this technique include optimum mixing of the measurement solution and reagent, minimum dispersion, and low reagent consumption. The technique was first described by ZAGATTO *et al.* [2676] who added lanthanum for the determination of calcium, potassium and magnesium in plant materials by F AAS.

6.4 Automatic Dilution

Since the measurement range of AAS is relatively limited, it is often necessary to dilute the measurement solutions. With mechanical autosamplers dilution is usually performed automatically batchwise, either directly in the sample container or in a separate dilution container, using a piston pump for example. The first automatic diluters for F AAS were described already in the 1960s [5402].

FI techniques have also been used for automatic dilution, even from an early stage; dilution is performed on-line or in-line. TYSON [5942] has published an overview article on the numerous investigations involving on-line dilution procedures using FI. Procedures for on-line dilution using mixing zones or mixing chambers include merging streams [2675], flow splitting [4126], partial overlapping of dispersing zones (zone penetration) [2674], cascade dilution [2286, 6303], time-based binary sampling [2129], and controlled dispersion [1833, 5299]. Dilutions of up to around 130 000 can be achieved by combining these techniques [1701]. Despite the large number of proposed techniques, most of which were developed for specific applications, no one technique has come into general use. For this reason, we shall only discuss those techniques that have a real chance of coming into routine use.

A very promising technique is *zone penetration*, which has been investigated by a number of authors [1832, 2674, 5940]. A configuration in which two sample loops of differing volume and an intermediate loop for carrier solution (e.g. water) are connected to the same valve is depicted in Figure 6-4. Both volumes of measurement solution, separated by the intermediate volume, are transported by the carrier to the atomizer unit. As a result of dispersion the volumes overlap slightly, leading to two maxima and a minimum, as shown in Figure 6-5. In this example using volumes of measurement solution of 100 μL and 20 μL, respectively, and an intermediate volume of 150 μL, the dilution ratio obtained was 1 : 20 : 4 (maximum : minimum : maximum); the relative standard deviation (n = 10) was approximately 0.6%, 2.2% and 0.6% [1832]. Since three measures are obtained for every injection when using this technique, the most suitable dilution can be selected for evaluation. In principle it would also be possible to use three volumes of measurement solution, leading to five different dilutions [2674]. The greatest obstacle to the application of this technique using commercial instruments is that generally they are not able to read more than one measure per measurement cycle. Also, data handling programs can recognize a maximum but not a minimum, so that manual evaluation of the tracings is necessary.

Even higher dilution factors can be obtained by *injecting very small volumes* of test sample solution in the lower µL range with *controlled dispersion*. This technique was investigated in detail by FANG *et al.* [1833]. The sample loop on the injection valve is only filled to a small part with measurement solution. Computer-controlled, stepper-motor-driven peristaltic pumps permit the smallest volumes to be dispensed exactly and reproducibly [1833, 3613]. For a pump operating time of only one second at a speed of 10 revolutions/min, a volume of 0.7 µL sample solution was dispensed with a repeatability of about 2% relative (n = 8). By increasing the pump operating time to three seconds, corresponding to a dispensed volume of measurement solution of 2.1 µL, the repeatability improved to values around 0.5% relative. These values could even be bettered by using a stepper-motor-driven piston pump [1840].

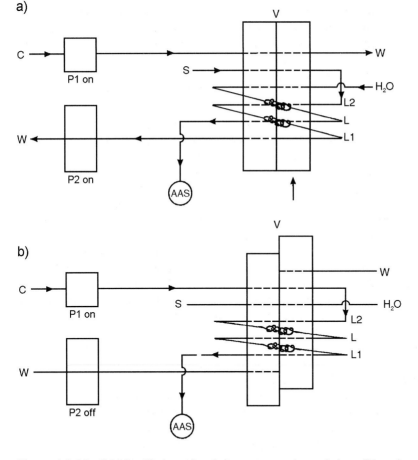

Figure 6-4. Manifold for dilution using the zone penetration technique. **P1** – piston pump; **P2** – peristaltic pump; **V** – injection valve; **S** – measurement solution; **L1**, **L2** – sample loops; **L** – dilution loop; **C** – carrier; **W** – waste. **a** – loading; **b** – dilution (from [1832]).

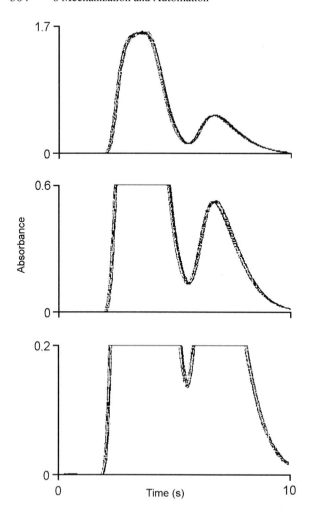

Figure 6-5. Dilution of a 40 mg/L Ca calibration solution by zone penetration using the manifold shown in Figure 6-4. **L1** = 100 μL; **L2** = 20 μL; **L** = 150 μL. Ten determinations superimposed; three different absorbance scales (from [1832]).

Next to the volume of measurement solution, the degree of dilution obtained also depends on the length of the mixing coil, i.e. on the dispersion, as shown in Figure 6-6. By selecting a suitable volume of measurement solution the dilution factor can be varied by more than two orders of magnitude, and by a further order of magnitude via the length of the dispersion zone. Dilution factors of more than a 1000 and with high repeatability can thus be obtained. In addition, as shown in Figure 6-6, a relatively high sampling frequency can be attained with this technique, which is not possible with most other techniques. FANG et al. [1833] also draw attention to the fact that by suitably programming the computer, dilutions that bring the measurement signal into the optimum range can be calculated automatically.

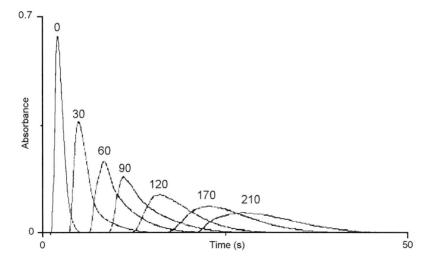

Figure 6-6. Influence of mixing coil length (1.3 mm i.d., 10 cm coil diameter) on signals for a 100 mg/L Mg calibration solution using 1.4 μL injections (2 s, 10 rpm) and a carrier flowrate of 5.7 mL/min. Tube lengths (cm): 0, 30, 60, 90, 120, 170, and 210 (from [1833]).

6.5 Automatic Separation and Preconcentration

With conventional autosamplers, preconcentration of the analyte, and thus an extension of the measurement range down to lower concentrations, is only possible for GF AAS by repeatedly dispensing, drying, and pyrolyzing aliquots of the measurement solution prior to atomization. Nevertheless larger increases in the sensitivity can practically only be attained for matrix-free measurement solutions.

The technique of solvent extraction, used frequently in earlier times to extend the measurement range of F AAS, will not be discussed in this section since it is mostly an off-line procedure and is only employed nowadays for sporadic applications. Techniques for preconcentrating hydrides on iridium-coated graphite tubes by sequestering and mercury on gold by amalgamation are discussed in detail in Section 4.3.2. Automated instruments or accessories are offered commercially for both techniques.

Flow injection techniques are employed in a large variety of ways for on-line preconcentration and separation; these include liquid-liquid or sorbent (solid-phase) extraction, gas-liquid separation, dialysis, precipitation, or coprecipitation [5978]. A detailed overview of these techniques is presented in the monograph *Flow Injection Separation and Preconcentration* by FANG [1834]. This author has also published good review articles on FI on-line column preconcentration for atomic spectroscopy in general [1828] and for GF AAS in particular [1839]. Of the numerous possibilities, three that have been relatively well examined by AAS—namely sorbent (solid-phase) extraction, coprecipitation, and solvent extraction—are discussed in the following sections.

6.5.1 FI On-line Sorbent Extraction

FANG [1828] summarized the advantages of FI on-line sorbent extraction compared to the corresponding off-line techniques as follows:

i)   Higher efficiency with respect to sample throughput by 1–2 orders of magnitude.
ii)   Lower consumption of sample and reagents by 1–2 orders of magnitude.
iii)   Better precision; s_{rel} typically 1–2%.
iv)   Lower risk of contamination since a closed system made of inert materials is used.
v)   Easy to automate.

Through the application of FI techniques, preconcentration procedures that are normally the time-determining step of an analysis can be made compatible with F AAS and GF AAS, so that on-line operation is practicable.

In order to better compare the various techniques and procedures, we can use a number of criteria with which we can make comparisons. The most frequently used criterion is the *enrichment factor* (EF); however, this is defined in different ways by different authors. The easiest way is to compare the gradient of the linear section of the calibration curve prior to and after preconcentration. Alternatively we can calculate the enrichment factor from the analyte concentration of an untreated test sample solution which gives the same signal as a test sample solution after preconcentration. In F AAS a clear distinction must be made between the enrichment factor and the *enhancement factor*. The latter additionally contains the solvent effect (refer to Section 8.1.1.5) for elution with organic solvents and is normally larger than the former.

A further, very important criterion, especially with respect to the sample throughput, is the *concentration efficiency* (CE). This is defined as the product of EF and the number of samples that can be determined per minute. The unit of this quantity is thus 1/min. The CE values for typical manual batch preconcentrations, even when several extractions are performed in parallel, are mostly less than 4 EF/min, while for FI systems CE values of close to 100 EF/min can be attained.

A further criterion that takes the volume of the analytical sample into account and is required to attain a given EF value is the *consumptive index* (CI). This is defined as:

$$CI = \frac{V}{EF},$$ (6.1)

where V is the volume of the test sample solution in mL. Manual batch procedures often have CI values of greater than 5 mL, while for on-line column preconcentration procedures with FI, values down to 0.03 mL can be achieved.

The final important criterion is the *retention efficiency* (%E), which gives the percentage of the analyte that is retained during preconcentration. This factor is particularly important for FI techniques which are very frequently employed under conditions of non-equilibrium. Low %E values can be caused not only by too short reaction times, which prevent equilibrium being attained, but also by concomitants that cause a 'breakthrough' of the analyte.

Although similar column materials to liquid chromatography (LC) are used for FI sorbent extraction, as already pointed out in Section 6.1 it must nevertheless be emphasized that this is a *non-chromatographic technique* with significant differences to LC. In LC a test sample is separated into its constituents which are measured sequentially in a flow-through detector. In contrast, the separation in FI sorbent extraction can be compared to the collection and subsequent dissolution of a precipitate on a filter. When a test sample is introduced onto an FI column, only the quantitative or reproducible sorption of the analyte is of interest, while the concomitants pass through the column and are rejected. At subsequent elution it is consequently of importance that the analyte is rapidly and quantitatively released from the column and transported to the atomizer.

Frequently, additional measures are taken to counteract the possible chromatographic behavior of the stationary phase. These include washing of the column after sample loading and prior to elution to remove concomitants that might also have been sorbed. Frequently the *column washing step* is performed in a *direction counter to the direction of sample loading*. This counteracts the dispersion of the analyte along the column. Similarly, *reversed-flow elution* is employed so that the analyte is concentrated in the smallest possible volume, thus giving the highest possible signal at the detector. For the same reason the *strongest eluent* is used to elute the analyte as quickly as possible from the column.

In their first paper on FI sorbent extraction, OLSEN *et al.* [4469] used an ion exchanger (Chelex-100) as the column packing material and eluted the sorbed elements cadmium, copper, lead, and zinc with dilute nitric acid directly to the burner of an AA spectrometer. In subsequent years F AAS remained the preferred detector, and ion exchangers were mostly used as column packing materials. A preconcentration system using 8-hydroxyquinoline on controlled pore glass (CPG-8Q) as the exchanger is depicted in Figure 6-7a. As can be seen from Figure 6-7b a sampling frequency of 120/h was attained, and using only 1.6 mL of test sample solution per extraction EF values of between 21 and 35 were obtained for cadmium, copper, and lead [1823]. Compared to ion exchange resins, such exchangers immobilized on glass or silica gel have the advantage that they must not be regenerated and that they do not swell or shrink under changing conditions [2213].

In 1989 RUZICKA and ARNDAL [5017] introduced C_{18} immobilized on silica gel as the sorbent for diethyldithiocarbamate (DDTC) complexes of numerous heavy metals. A major advantage of this material is that the alkali and alkaline-earth elements are retained to a much lower extent than on the usual ion exchangers [1826]. A further advantage in conjunction with F AAS is that organic solvents such as ethanol or methanol can be used as the eluent and these additionally enhance the sensitivity as a result of the solvent effect. FANG [1828, 1834] has published good overviews on the column packing materials mentioned in the literature up to 1991.

Occasionally FI on-line sorbent extraction with F AAS detection has been looked upon as a rival to GF AAS. For this reason attempts were often made to attain the highest possible enrichment factors; this naturally markedly deteriorates the sampling frequency and increases the risk of false measurements due to a breakthrough of the column. Later it was recognized that reducing the enrichment time to around 30 seconds and thus maintaining an acceptable sampling frequency of 100–120/h for flame operation was more sensible [1823].

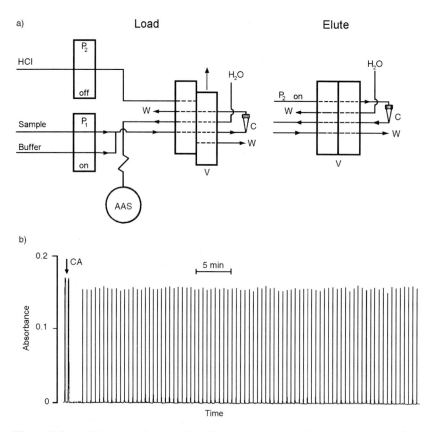

Figure 6-7. a – Schematic diagram of an FI system for ion-exchange preconcentration and analyte determination by F AAS. **P1, P2** – peristaltic pumps; **C** – conical CPG-8Q ion-exchange column; **V** – multi-functional valve; **W** – waste. Sample loading, 20 s; elution, 10 s. **b** – Recorder traces of 0.1 mg/L Cu after preconcentration using the above system; **CA** – signal from 3 mg/L Cu by conventional aspiration. EF = 25 (from [1823])

While the combination of FI with F AAS is generally without problems and merely requires the flowrates to be matched, the combination with GF AAS, which is a discontinuous technique, is without doubt much more complex. Nevertheless, for the processes of separation and preconcentration on packed columns, FI is also a cyclical procedure, so that the sample loading and elution phases can be synchronized to the temperature program of the graphite furnace without too much difficulty. The first coupling of FI on-line sorbent extraction with GF AAS detection was described by FANG *et al.* [1824].

The relatively small volume of eluent that can be accommodated in the graphite tube and the sensitivity to concomitants when a modifier is not used are a source of some difficulty for the combination of sorbent extraction with GF AAS. The volume of eluent must be limited to about 40–50 μL and a wash step after loading the column is indispensable. The program is thus more complicated and can encompass 7 to 9 steps; however, since GF AAS programs are more time-consuming than flame atomization, this is mostly irrelevant.

A typical manifold for sorbent extraction in combination with GF AAS is depicted in Figure 6-8. The columns generally have a volume of 10–15 µL [1824, 5510, 5513, 5517, 6266]. Mostly C_{18} immobilized on silica gel is used as the solid sorbent packing material. The pH of the test sample solution is adjusted to a value suitable for the formation of a complex by the addition of an acid or a buffer, either in advance or on-line. The sodium diethyldithiocarbamate (NaDDTC) complexing agent is run on-line through a precolumn to remove possible contamination and then added to the measurement solution. Since DDTC is unstable at low pH values compared to the metal complexes, the metal complex is retained preferentially on the column even after only a short distance, while excess reagent is dissipated. The distance between the point of mixing of the complexing agent with the measurement solution and the column must be short since the formation of the complex is very rapid and the metal complexes tend to accumulate on the walls of the tubing. Accumulations of this type can be avoided by loading the column with complexing agent first and then conducting the sample through the modified column in a second step [5204]. The interaction of complexing agents with plastic materials can also be utilized for preconcentration since the complex formed can be sorbed on the walls of the tubing (see Section 6.5.2).

After sorption of the analyte a *wash step* is required to remove residual matrix constituents remaining on the column and complexing agent sorbed onto the column. Through a suitable choice of acid and pH value for the wash fluid other sorbed concomitants can be removed from the column [5510, 5514]. The wash step is particularly effective when it is performed in a direction counter to sample loading [5514]. It has proven of advantage, particularly to control the dispersion, to displace the wash fluid from the column and tubing by air following the wash step [5522, 6266].

Elution is also performed in a direction counter to sample loading. In this way dispersion of the analyte in the column can be counteracted. The disparity between the volume of eluate, 200–300 µL, and the volume to be dispensed into the graphite tube of maximum 50 µL can be obviated by dispensing the zone of eluate containing the highest portion of analyte. This technique, which is based on the high repeatability of all processes occurring in FI systems, is depicted schematically in Figure 6-9. By selecting a strongly polar eluent, such as methanol or ethanol, and a slow flowrate for elution, it is possible to concentrate about half the sorbed analyte in 40 µL of eluate. The eluate volumes prior to and after this concentrated zone contain relatively little analyte. By means of exact time control [5510, 6262], possibly combined with the collection of a measured volume [1824, 6262], very high repeatability can be attained; albeit the long-term stability is not so good when peristaltic pumps are used. Occasionally an air bubble is introduced prior to and after elution to totally suppress further dispersion of the organic phase [6266]. Elution by means of a measured volume of eluent, which is confined by air as it passes through the column, eliminates the stability problems. Moreover 26 µL ethanol are adequate using this technique to quantitatively elute the analyte from a 10 µL column and transport it into the graphite tube [5522].

As we can see from the example of the determination of lead in sea-water shown in Figure 6-10, determinations can be performed virtually free of blank values, despite the enormous sensitivity of the technique. The reason for this can be found not only in the on-line prepurification of reagents that are subject to contamination, but also in the fully

a)

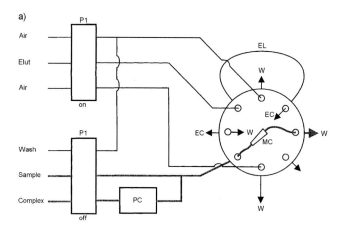

b)

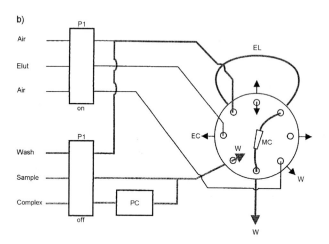

c)

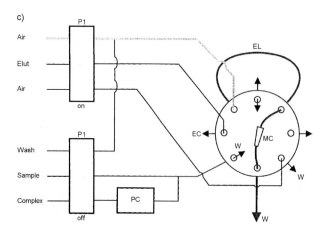

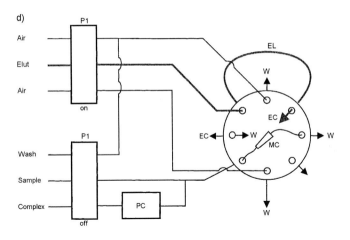

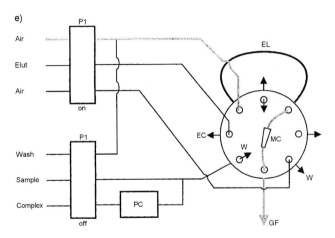

Figure 6-8. Simplified schematic of a manifold for on-line preconcentration by sorbent extraction for GF AAS. **a** – sample loading; **b** – wash step; **c** – displacing the acid with air; **d** – filling the elution loop; **e** – elution into the graphite tube. **P1**, **P2** – peristaltic pumps; **PC** – precolumn; **MC** – microcolumn; **EL** – elution loop; **EC** – eluate container; **W** – waste; **Elut** – eluent; **Wash** – wash fluid; **GF** – graphite furnace.

automatic analytical sequence in a closed system. The fact that the background signal for the sea-water sample is not higher than for the aqueous calibration solutions demonstrates how well the matrix has been separated. This also explains why the limits of detection for real samples are three orders of magnitude better compared to direct determinations without separation of the matrix, although the EF values are only around 25 [5510, 5514]. If required, by using longer preconcentration times of up to 10 minutes it is possible to attain EF values of 100 or more [5513].

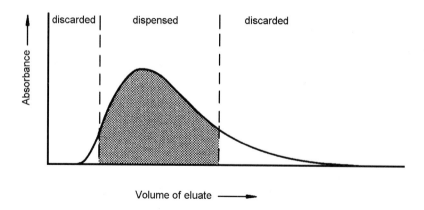

Figure 6-9. Principle of eluate zone sampling. Only the middle zone of eluate containing the greatest portion of the analyte is dispensed into the graphite tube. The eluate segments either side of the collected zone are discarded (from [1824]).

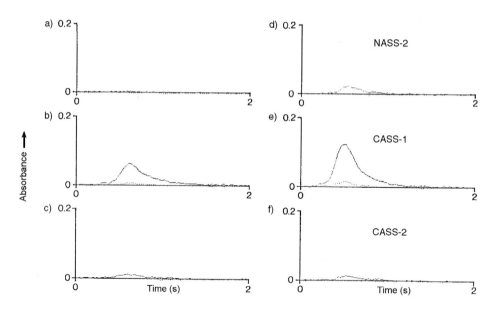

Figure 6-10. Lead signals by GF AAS after a preconcentration time of 1 min. **a** – blank; **b** – 100 ng/L Pb calibration solution; **c** – deionized water; **d, e, f** – sea-water reference materials: NASS-2 = 39 ng/L; CASS-1 = 251 ng/L; CASS-2=19 ng/L (from [5510]).

For elements that exist in several oxidation states, the tendency of these states to form complexes and to sorption is mostly very different and depends among other things on the pH value. This offers the possibility for the *selective determination of oxidation states.* Since the species with the highest toxicity, such as As(III), Se(IV), Cr(VI), etc., form the most stable complexes with DDTC and can thus be determined directly via sorbent extraction with C_{18}, this opens interesting possibilities for speciation analysis (see also Section 7.1.2). With the development of new, highly selective sorbents, such as *immobilized crown ethers* [897, 2850], a whole range of application areas is opened to this on-line preconcentration technique [5522] that at the current time cannot be fully estimated.

6.5.2 FI On-line Precipitation and Coprecipitation

The preconcentration of elements by coprecipitation is characterized by very high EF values, but is very time-consuming, requires considerable effort, and is highly subject to the risk of contamination. The use of filters is a problem when coprecipitation techniques are transferred to FI. The size of the filter must be selected such that the flow is not blocked even for the maximum quantity of precipitate. Large volume filters, on the other hand, impair the attainable enrichment factor through dispersion and retarded dissolution of the precipitate. DITTFURTH *et al.* [1536] thus attained an enrichment factor of 55 for manganese after a collection time of four minutes, while DEBRAH *et al.* [1450] reached an enrichment factor of 12 for copper after a collection time of 2.5 minutes. Based on a batch procedure originally proposed by EIDECKER and JACKWERTH [1686, 1687] using coprecipitation with iron(II)-hexamethylenedithiocarbamate (Fe(II)-HMDTC), FANG *et al.* [1827] successfully automated the procedure for FI without the use of a filter. While retaining all the advantages of the manual procedure, with automation the sampling frequency could be increased by nearly two orders of magnitude and the sample consumption reduced by a factor of 40. The efficiency of the procedure was demonstrated on the determination of cadmium, cobalt, nickel, and lead in blood and liver tissue using both F AAS [1827, 6257] and GF AAS [1830]. This procedure was also taken up, modified and improved by other working groups [4122].

The manifold used for on-line coprecipitation and its mode of function are depicted schematically in Figure 6-11. The heart of the system, and also its most important component, is the *knotted reactor* of 100–150 cm length and 0.5 mm inner diameter in which the precipitate is collected without the use of a filter. Resulting from the centrifugal forces that develop due to the secondary flows in the three-dimensionally disordered system, the precipitate is transported to the inner wall of the reactor and deposited. This assumed mechanism is supported by the observation that in straight or coiled reactors made of the same material the precipitate is collected far less effectively. A further contribution to the retention of the precipitate is the hydrophobic properties of the tube material and the precipitate. The advantages of a knotted reactor for collecting the precipitate include:

i)   The relatively high capacity resulting from the large internal surface area. This is particularly important for coprecipitation since a relatively large quantity of precipitate is produced.
ii) Since the system is open and does not contain any filters, the back pressure is low even for high flowrates of 5–6 mL/min [1835, 5521, 6443, 6444].
iii) Because of the three-dimensionally disordered configuration the loss of sensitivity due to dispersion can be neglected.

iv)   Freedom from contamination due to the inert tube material (e.g. PTFE).
v)   The reactor is easy and cheap to make.
vi)   The lifetime is virtually unlimited and the reactor requires no maintenance.

Since adhesion forces are responsible for the retention of precipitates on the tube walls, the properties of the precipitate play a major role for the effectiveness of preconcentration. Certainly not all precipitates can be collected in this manner, but as well as HMDTC it has also been possible to collect dithizone and APDC complexes [1592] and even inorganic precipitates such as lanthanum hydroxide and hafnium hydroxide [4260] on the walls of knotted reactors.

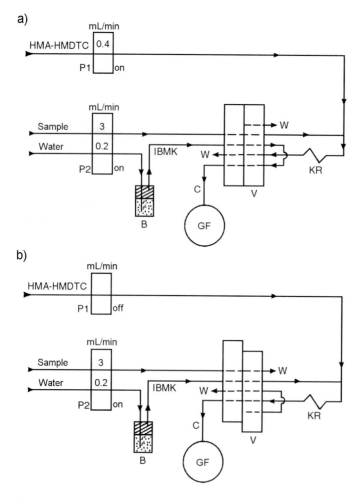

Figure 6-11. Schematic function of an FI manifold for on-line coprecipitation of trace elements with Fe(II)-HMDTC and their determination by GF AAS after dissolution in IBMK. **a** – coprecipitation with HMA-HMDTC; **b** – dissolution and introduction of the concentrate into the atomizer. **P1, P2** – peristaltic pumps; **KR** – knotted reactor; **B** – solvent container; **V** – injection valve; **C** – collecting tube for the concentrate; **GF** – graphite furnace; **W** – waste (from [1834]).

Next to the effective retention of the precipitate, effective dissolution is of major significance for a high enrichment factor. The precipitates can be very effectively dissolved in a small volume of isobutyl methyl ketone (IBMK), so that for F AAS a relatively sharp signal of high sensitivity can be recorded, while for GF AAS a high proportion of the precipitate can be introduced into the atomizer. As shown in Figure 6-11, the IBMK is not pumped directly since it attacks the pump tubing, but is displaced from the reservoir bottle by water. A further important design feature is that the IBMK is injected at the same point at which the sample solution and the complexing agent meet, so that no deposits can occur in the system. The performance of on-line coprecipitation of this type is amply demonstrated by the enhancement factor of 135 that FANG et al. [1829] were able to attain for lead by F AAS with a coprecipitation time of 30 s. For coupling on-line coprecipitation with GF AAS it has proven of advantage to dissolve the sorbed precipitate with a measured volume of solvent confined by air in a similar manner to preconcentration on microcolumns (refer to Figure 6-8) [5521].

6.5.3 FI On-line Solvent Extraction

Solvent (or liquid-liquid) extraction is one of the most widely used separation techniques in analytical chemistry and also occupies an important place in the FI technique. Nevertheless, the number of published applications for AAS is surprisingly low and, with a few exceptions, limited to F AAS [3315]. Next to the preconcentration and separation of metal ions [663, 1314, 3322, 3523, 3524, 4047, 4048, 4380, 4418, 5718, 6485, 6558], on-line extraction [4379] is applied in particular for the indirect determination of anions [2050, 5365], organic compounds [664, 665, 3922] or pharmaceuticals [1694, 1695, 4175, 5089]. The reason that this technique is used relatively seldom is to be found in the low flowrate of extraction compared to the aspiration rate of the nebulizer. For solvent extraction the enrichment factor depends mainly on the ratio of the flowrates of the aqueous and organic phases. Assuming complete extraction and phase separation, this means that for a flowrate of 10 mL/min for the aqueous phase, the flowrate for the organic phase will be 0.5 mL/min if an EF of 20 should be attained. However, this is an order of magnitude lower than the normal aspiration rate of nebulizers.

For this reason the organic phase is almost always introduced discontinuously into the nebulizer-burner system. The organic phase is mostly collected in an injection valve, while the nebulizer aspirates air or water [1820]. As a result, changes in the flame conditions are unavoidable, negatively influencing the stability. Also the sampling frequency suffers under discontinuous operation, so that EF and CI values are not particularly good in comparison to column preconcentration techniques [4380].

On-line solvent extraction for GF AAS was described by BÄCKSTRÖM and DANIELSON [341–343, 1418] and TAO and FANG [5774]. Initially the dithiocarbamate complexes of various trace elements are extracted with Freon 113. After phase separation an acidic solution of Hg(II) is added which forms a stronger dithiocarbamate complex and thus displaces the other trace elements. After a second phase separation the organic phase is discarded and the aqueous phase with the trace elements is collected in a tube and then dispensed into the graphite furnace. The system was used for the determination of Cd, Co, Cu, Ni, and Pb; EF values of 50–100 and a sampling frequency of 30/h were attained. This system has also been used to determine aluminium species by complexing and extracting reactive aluminium [1417].

6.6 On-line Sample Pretreatment

The pretreatment of a test sample, especially when a digestion is involved, is frequently the time-determining step of an analysis. Although the introduction of microwave-assisted acid digestion procedures significantly shortened the digestion time, the total effort from weighing of the test sample portion until the diluted test sample solution is ready is not noticeably less. Sample preparation still contains a large number of manual steps with the concomitant risk of contamination. One solution to this problem is direct solids analysis, but this is only practicable with GF AAS (refer to Sections 1.9 and 4.2.7). The other alternative is to automate procedures as far as possible and to use on-line digestions.

The number of publications on this subject is an indication of the requirement for on-line sample pretreatment for a given AAS technique and also an indication of the problems encountered. For GF AAS the requirements for sample pretreatment are mostly low, except that the risk of contamination must be taken into account. Numerous liquid test samples, such as water or body fluids, do not require digestion since they can be thermally decomposed or ashed *in situ*. It is therefore hardly surprising that there are only a few publications on the microwave-assisted digestion of biological materials for determinations by GF AAS [947].

In contrast, a digestion for CV AAS and HG AAS is indispensable, even for the analysis of water samples, and is therefore an integral part of the analytical method. With HG AAS a prereduction must also frequently be performed after the digestion. For this reason, and because water samples can be very easily treated on-line, automated on-line pretreatment techniques were described at an early stage (see Section 4.3.1.2). The most widely used were digestions with persulfate and also under UV radiation; these procedures are still regularly used nowadays [299, 1659, 1812, 2682]. The first microwave-assisted on-line digestions for CV AAS and HG AAS were only described at the beginning of the 1990s [2281, 2282, 3474, 3475, 5906, 5907, 6263, 6270]; all these digestion procedures are based on the FI principle. A manifold in which the on-line sequential digestion and prereduction of arsenic is performed is depicted in Figure 4-42 (Section 4.3.1.2).

With only very few exceptions [4650], all publications on on-line sample digestion for F AAS originated in the period after 1985 and employed FI techniques. The majority of authors used microwave-assisted digestions in which the test sample was introduced into the microwave oven in the form of a slurry in a suitable acid. In the simplest case the residence time is determined by the length of the tube, i.e. through the flow-through time [948, 951, 954, 1060, 2247, 2252]. However, systems were also described in which the test sample was circulated in a loop until digestion was complete [1062]. Alternatively, the flow can be interrupted for a given period to increase the digestion time [3032]. Some authors additionally performed digestions under pressure; the usual PTFE tubes can still be used at pressures of up to 5–6 bar [2428], but for pressures of 25–30 bar glass reactors for example are required in which the test sample is subjected to the microwave field for around five minutes [2146].

The high pressure digestion of cocoa powder at 210 °C is one of the few investigations described without the use of microwave support [2145]; the test sample was dissolved, while the matrix was not fully digested. LÁZARO et al. [3472] merely treated slurried analytical samples on-line by means of ultrasonics; the slurries were not digested but the elements of interest were leached out. BERGAMIN et al. [547] utilized an electrolytic procedure based on FI for the on-line determination of soluble aluminium in steel, while YUAN et al. [6495] later used a similar technique for the determination of copper in aluminium alloys.

There is no doubt that automated on-line digestion will gain considerably more significance in future since it brings not only a substantial reduction of effort and saving of time, but can also improve the trueness of a determination. FI and microwave techniques will play an important role in this development. Nonetheless, there is still much work to be done until such systems can be used routinely without problems for large numbers of varying samples. Initial investigations have already been made on the feasibility of offering commercially microwave-assisted digestion apparatus with flow-through reactors [671, 1905], but the applications are still very restricted.

6.7 Automatic Setup and Optimization of Instrument Functions

In principle, all parameters and functions of an atomic absorption spectrometer can be automatically set up, changed, optimized, and monitored. *Automatic lamp change* is a feature of sequential spectrometers. When the radiation source is changed, the changed operating conditions are set up automatically. Spectral lamps are frequently fitted with coded plugs so that the optimum lamp current is set. With computer-controlled spectrometers the *analytical line and the slit* are set automatically via the stored analytical program, and in many cases are also monitored. Filters and stops are moved in and out of the radiation beam as required, and a grating change can also be performed.

Automatic operation with flames includes lighting and extinguishing the flame, changeover of gases, and optimization and monitoring of the gas flowrates. All these procedures must be performed in a sequence that guarantees the highest level of operational safety. In addition, risk situations are automatically monitored, such as loss of power, incorrectly installed or no burner head, whether the flame is burning, the liquid level in the waste container, and the correct fuel gas to oxidant ratio. Even the *burner position* can be automatically optimized, both with respect to the maximum sensitivity and to the stored data determined for a given element. In this way, for example, it is possible to program both the burner position and the flame gases for minimal interferences rather than maximum sensitivity if these two settings are not identical. For single-beam instruments with double-beam characteristics, the burner can be driven automatically out of the radiation beam after every measurement; the beam then serves as the reference beam and automatic baseline offset correction can be performed.

With *electrothermal atomizers* the preselected temperature program is carried out automatically. Furthermore, switch-over and monitoring of the protective, purge and alternate gases is normally performed automatically. Risk situations are also monitored, such as insufficient flowrate of the protective gas, insufficient cooling, overheating of the furnace or the Zeeman magnet, breakage of the tube, or when no tube is installed. Even automatic monitoring and control of the graphite tube temperature over the entire range from 100–2600 °C on the basis of the measurement data of the graphite tube emission and the cooling water temperature has been realized.

With *systems for chemical vapor generation*, dispensing of sample solutions, addition of reagents and gases, and, where applicable, collection and vaporization of the analyte are performed automatically. Flow systems generally allow a much higher level of automation than batch systems.

6.8 Automatic Data Handling

The minimum requirements placed on automatic data handling systems are direct output of the concentration or mass of the analyte, and the mean value and standard deviation of a series of measurements. Further possibilities offered by data handling systems include automatic selection of the optimum algorithm for the evaluation of non-linear calibration functions, outlier tests, statistical quality and plausibility controls, generation of an analytical report, and communication with laboratory information management systems [393, 2729, 4413, 5329].

In addition to the calibration techniques mentioned in Chapter 5, FI offers a number of additional possibilities for calibration, especially for F AAS [5941, 5942]; several of these are mentioned below. They are based, at least in part, on the special characteristics of FI which are described in Section 6.1.

TYSON and APPLETON [5937] proposed the *technique of continuous dilution*, in which a concentrated calibration solution is continuously diluted in a mixing chamber reactor; this results in an exponentially increasing concentration gradient. The concentration-time profile follows the relation:

$$\beta(t) = \beta_1 \left[1 - \exp\left(-\frac{Qt}{V} \right) \right], \tag{6.2}$$

where $\beta(t)$ is the mass concentration at time t, β_1 is the original concentration prior to dilution, Q is the flowrate, and V is the volume of the mixing chamber reactor. During the calibration step the effluent from the mixing chamber reactor is conducted directly to the nebulizer of the atomic absorption spectrometer and the absorbance-time profile is recorded with a microcomputer. The absorbance of the unknown test sample solution is compared with the stored data. From this a value for time t can be obtained at which the calibration solution has the same concentration. By substituting into equation 6.2 the concentration of the test sample solution can be calculated. The accuracy of the technique is around ±3%.

A very similar technique, proposed by OLSEN et al. [4468], is *electronic dilution*. In this technique the absorbance is measured on the falling flank of the peak at a predetermined time point after the maximum value has been reached. As a result of the dispersion an analyte concentration β_1 can be assigned. In this way the signals from relatively concentrated measurement solutions can be evaluated without prior dilution.

A further procedure for measuring highly concentrated test sample solutions without prior dilution, *evaluation of the absorbance peak at half peak height*, was proposed by BYSOUTH and TYSON [1010]. Nevertheless the errors obtained with this procedure were greater than can normally be tolerated for AAS calibration techniques, so that it is only suitable as a screening method to determine the dilution factor for an exact measurement.

TYSON [5935] also proposed a *reversed flow analyte addition technique* (reversed flow injection) in which the test sample solution is used as the carrier and a constant signal is thus obtained. Calibration solutions of varying concentration are then injected into this 'carrier solution'. Depending on whether the content of analyte in the calibration solution is higher or lower than in the test sample solution, a positive or a negative signal is obtained, as shown in Figure 6-12. The author implied that the uncertainties in using this procedure to determine the analyte concentration are lower than with the conventional analyte addition technique, since this is an interpolative and not an extrapolative procedure. Closer scrutiny of this procedure nevertheless indicates that this is not true and that the risk of systematic errors exists. For reversed flow injection the test sample solution is diluted by a calibration solution at every addition, but when interpolating, the measure for the undiluted test sample solution is used. The prerequirement for proper correction of matrix effects is that these do not depend on the concentration of the matrix constituents. In this special case, however, it would be possible in principle to dispense with the addition technique and to use a less complicated calibration technique.

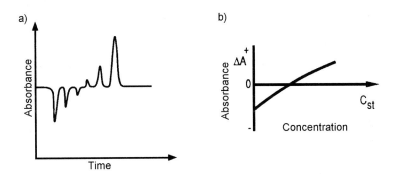

Figure 6-12. Reversed flow analyte addition technique. **a** – Calibration solutions of increasing concentration are injected into the test sample solution, which functions as the carrier; **b** – plot of ΔA against the concentrations of the injected calibration solutions (from [5942]).

A powerful and versatile calibration technique has been proposed by SPERLING *et al.* [3256, 5509, 5512], termed *gradient ratio calibration*. Here, not just the peak height or area is evaluated, but the entire transient signal is stored in the computer and used for the calibration. The ratio of the absorbance of the calibration solution and the test sample solution is formed at the reading frequency of the spectrometer, i.e. every 18 ms or so. If no interferences are present and the measurement solutions are all in the linear range, the ratio does not change during the entire measurement. If the maximum absorbance of a test sample solution is outside the linear range, however, the absorbance ratio changes as soon as the linear range is exceeded. This is recognized and corrected by the computer. The same situation occurs if an interference is present which decreases with increasing dilution (dispersion), as shown in Figure 6-13a for the interference of phosphate on the determination of calcium. If the computer program is used to extrapolate the signal ratio over the entire absorbance profile against absorbance $A = 0$, the interference can be eliminated by calculation (Figure 6-13b).

Nevertheless, there are interferences that initially increase with increasing dilution until they finally disappear at infinite dilution. Behavior of this type has been observed, for example, for the interference of aluminium on the determination of the alkaline-earth elements [6271]. In this case the algorithm fails, but it is sufficient to add a small quantity of a buffer, such as lanthanum chloride or strontium chloride, to be able to eliminate

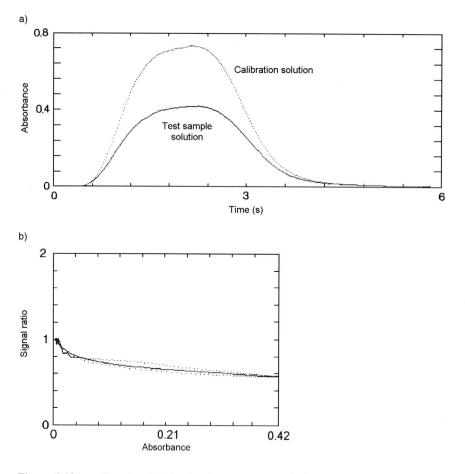

Figure 6-13. a – Transient FI signals of measurement solutions containing 16 mg/L Ca without (calibration solution) and with a concentration of 0.01 mol/L PO_4^{3-} (test sample solution); **b** – signal ratio of the calibration solution to the test sample solution as a function of the absorbance of the test sample solution (from [5509]).

the interference by calculation. As shown schematically in Figure 6-14, the addition of buffer influences the interference only marginally—a concentration 2–3 orders of magnitude higher would be required to eliminate it—but it brings about 'normal behavior' of the interference. A small addition is sufficient, nevertheless, to provide the requirement for chemometric elimination of the interference.

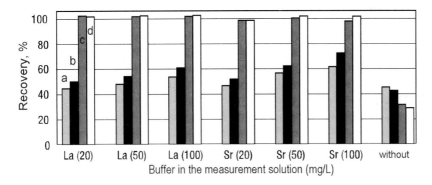

Figure 6-14. The determination of 10 mg/L Ca in the presence of 80 mg/L Al with lanthanum or strontium as buffer and the chemometric elimination of the interference. Concentration of the buffer in mg/L. Evaluation: **a** – in absorbance; **b** – in integrated absorbance; gradient ratio calibration in **c** – absorbance and **d** – integrated absorbance.

A major significance of this procedure is certainly that not only can non-spectral interferences in many cases largely be eliminated by calculation, but more importantly that it can *recognize interferences*. The possibility is thus given of issuing a warning signal that permits the analyst to thoroughly examine the trueness of the result determined by the computer. This opens completely new perspectives for quality assurance with respect to the trueness of analytical results.

It can be seen that this procedure already belongs in the area of *expert systems* (although their definition and potential extent have not yet been clearly defined) [773, 869, 3389, 3511, 4666, 6174, 6175, 6312]. A minimum requirement of expert systems should be that they provide a guide for working correctly and that they monitor the analytical procedure. Initial investigations in this direction have been performed by VANKEERBERGHEN *et al.* [5993] for GF AAS, by LAHIRI *et al.* [3389] for F AAS, and for FI by CHEN *et al.* [1226], PERIS *et al.* [4607], WU *et al.* [6394], and BRANDT and HITZMANN [781]. Expert systems serve to avoid errors and should issue feedback and warnings. First moves in this direction are already to be found in commercial instrument software [4414]. There is clearly an enormous requirement in this area and it will take quite a number of years until systems exist that can fully exploit the capabilities of current computer systems.

7 Speciation Analysis

The form in which a trace element occurs largely determines its absorbability by organisms and in end effect also determines its toxicity. Thus the identification and quantitation of, say, fat-soluble metal compounds and easily reducible inorganic ions that are readily assimilated by organisms is of far greater interest than the determination of the total content of the element. The quantitation of individual forms of an analyte is termed speciation analysis (or species determination); it should be noted that the term *species* is not used uniformly in the literature. Species can be distinguished, for example, by their oxidation state (redox species), by their molecular form (organometallic compounds), by coordination (complexes), by the phase in which the analyte exists (dissolved, colloidal, suspended, sorbed, etc.), or also the solubility (leachability) in selected solvents (e.g., water, salt solutions, acids). The instrumental aspects involved in the determination of the oxidation state or the molecular form are the main topics in this chapter.

Since only very few atomic spectrometric techniques are selective for the determination of species, speciation analysis mostly involves two steps: separation and determination. The most important combinations of separation techniques and spectrometric detection that are described in the literature are presented schematically in Figure 7-1. In this chapter we shall only discuss couplings in which AAS is used for detection. For the separation techniques we must make a distinction between *chromatographic* and *non-chromatographic techniques*. In the former the various species are separated, for example by GC or HPLC, and determined one after the other in the sequence of their retention times by AAS. In the latter either a single species, for example a given oxidation state of the analyte, or a group of species, for example all inorganic or all water-soluble forms, is separated from the test sample and determined. In this case only those techniques are discussed in which the separation and determination are performed on-line or quasi on-line.

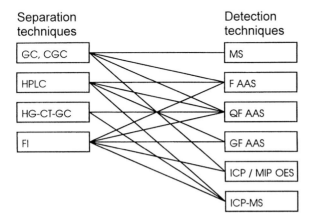

Figure 7-1. The most important separation and detection techniques described in the literature for the spectrometric determination of element species.

Since the concentrations of the various compounds or oxidation states in which trace elements can occur are always lower than the total content of the analyte, speciation analysis is normally an ultratrace determination in the ng/L range for solutions and the ng/g range for solids. The requirements placed on the detector are thus correspondingly high if extensive preconcentration should be avoided. As atomizers, *flames* have the advantage that they operate continuously and can thus be used on-line with all chromatographic and non-chromatographic techniques. Their disadvantage is that the sensitivity is often insufficient and extensive preconcentration is then necessary. *Graphite furnaces* exhibit high sensitivity, but in practice cannot be operated continuously over long periods of time, making a coupling with chromatographic techniques rather difficult. Their application is thus largely limited to combination with non-chromatographic techniques [1424]. *Quartz tube atomizers* exhibit high sensitivity and are operated continuously. Their sole disadvantage is that they are only suitable for gaseous species. The choice of an atomizer unit thus requires careful assessment of various factors that can in part mutually influence each other. The instrumental aspects of speciation analysis are discussed in this chapter, while important compounds or forms in which an element can occur are discussed under each individual element (see Chapter 9). Details of methodology, such as sample pretreatment for speciation analysis and the best determination technique, are treated in more detail under applications (see Chapter 10).

7.1 Non-chromatographic Separation Techniques

7.1.1 The Hydride-generation Technique

When using sodium tetrahydroborate as the reductant in the hydride-generation technique, the individual oxidation states of the analytes exhibit varying sensitivities since they are not reduced to the same degree or with the same velocity to the hydride (see Section 8.3.1.4). In the case of selenium and tellurium, only the +4 oxidation state forms a hydride under normal analytical conditions, while the +6 oxidation state is hardly reduced [5375]. With antimony and arsenic the +3 oxidation state forms the hydride much faster than the +5 oxidation state and the level of hydride formation for these two elements depends very strongly on the pH value of the measurement solution [54, 191]. This varying behavior of the individual oxidation states can be utilized for their selective vaporization and determination by HG AAS. After appropriate pretreatment, such as reduction, digestion, etc., other oxidation states or organic compounds of the analyte can be determined. Quartz tube atomizers and graphite furnaces, especially after *in situ* preconcentration (see Section 4.3.2.2), are suitable atomizers.

7.1.2 Preconcentration on Packed Columns

The preconcentration of trace elements on packed columns and their elution directly into a flame or a graphite furnace for on-line determination is discussed in detail in Section 6.5.1. If an element occurs in more than one oxidation state or molecular form, processes such as complex formation, sorption, etc. are mostly very different for the various species. This leads to the possibility of preconcentrating a single species and thus to a differentiated determination.

Although similar materials (sorbents) and components (columns, pumps) to chromatography are frequently used for solid-phase preconcentration, the procedure is significantly different. In chromatography a sample is injected onto a column and the individual constituents are separated as a result of their differing affinities for the column material and the eluent; the constituents can be identified via their retention times. In non-chromatographic techniques we attempt to suppress the chromatographic characteristics of the column packing material. This can be achieved, for example, by eluting the species of interest from the column in the counter direction to sample loading. This counteracts chromatographic separation and concentrates the analyte in a minimum volume of eluent. A further action to increase the selectivity is to wash the column between loading and elution to remove unwanted species that may be retained on the column. A strongly polar eluent is mostly used to elute the collected species rapidly and quantitatively from the column. The procedure is thus much more similar to separation via extraction or precipitation than to a chromatographic separation.

A characteristic of the FI technique is that it can function with accuracy under conditions of thermodynamic non-equilibrium (see Section 6.1). The reason for this is that the time-dependent processes can be exceedingly well controlled and are highly repeatable. This property is of particular advantage for speciation analysis. In the first place it allows a high sample frequency. But more than that time plays a major role for the trueness of a determination, especially for redox species, since the various oxidation states transform very rapidly if the pH of the solution changes or if a species is removed from equilibrium. The slower a procedure functions, the greater is the danger that during preparation for measurement the original conditions existing in the test sample change and systematic errors can thus occur. With FI procedures, pretreatment steps such as acidification, buffering, complexing, or sorption take place on-line and within fractions of a second. An influence on the natural redox equilibrium and thus on the species distribution during measurement is significantly less likely with this technique than with time-consuming chromatographic techniques. A prerequirement for the trueness of a species determination is naturally that no changes have taken place to the species distribution during collection and storage of the sample; this is a problem to which great attention must be paid for the determination of species.

Preconcentration on packed columns is used preferentially for the determination of redox species such as As(III)/As(V) [5511, 5514], Se(IV)/Se(VI) [5523], Sb(III)/Sb(V) [2065b], Fe(II)/Fe(III) [5784] or Cr(III)/Cr(VI) [5511, 5515, 5518]. A further advantage of this technique is that the most toxic species usually forms the more stable complex and can be sorbed more easily. In this way the species of greatest interest for the environment can be determined rapidly, directly, and on-line with high trueness. Since this technique can be operated on-line not only with F AAS but also with GF AAS, a highly sensitive detector is available should the need arise. In environmental analysis it is often sufficient, next to the determination of the most toxic species, to determine the total content of the analyte to be able to estimate the level of pollution, for example of water samples. This procedure is naturally only practicable when a limited number of easily separable species is to be determined. The separation of a large number of constituents must normally be carried out by chromatographic procedures.

7.2 Chromatographic Separation Techniques

Techniques involving the coupling of chromatographic separation and spectrometric detection are termed *hyphenated techniques*, since according to IUPAC recommendations the short forms are written with a hyphen, for example GC-F AAS or HPLC-GF AAS. A number of requirements must be placed on the performance and operating convenience of an ideal hyphenated technique:

i)   As far as possible we should be able to detect all species of an element, including the inorganic (redox species) and organic, and from the volatile, the methylated and ethylated to the long-chain alkylated and to the complex, high molecular mass species.
ii)   The concentration and distribution of the individual species should not change during the analysis (trueness).
iii) The sample requirement should be as small as possible.
iv)   The sensitivity of the procedure should be as high as possible.
v)   A minimum of effort should be required for sample preparation.
vi)   As far as possible the procedure should operate on-line, be easy to automate, and permit high sample throughput.

It goes without saying that none of the procedures described in the literature meets all of these requirements. This means that in practice it is necessary to make compromises, i.e., the analyst must decide which criteria are the most important for a particular analysis. In the following section we shall present criteria that should aid in the decision as to which is the most suitable procedure for a given situation.

7.2.1 Gas Chromatographic (GC) Separation

The first coupling of a gas chromatograph with the nebulizer of a flame AA spectrometer was described in 1966 by KOLB *et al.* [3224] for the determination of lead alkyls in gasoline. These authors found that 'the disadvantage of the lower sensitivity of F AAS compared to ionization detectors is largely compensated, especially for trace analysis, by the absolute selectivity of this detector since the chromatographic column can be loaded with a sample quantity far above its normal limit'. Initially this technique of speciation analysis found only sporadic interest [3584], but in more recent times interest has increased markedly [2405] and up to 1996 more than 350 papers have appeared on this subject.

One reason for the relatively late acceptance of GC-AAS was almost certainly to be found in the low sensitivity of the *flame* which was used exclusively as the detector in earlier years. In the mid 1970s SEGAR [5222] and ROBINSON *et al.* [4895, 4897, 4898] reported the first use of *graphite furnaces* as detectors. These were heated continuously to 1700–2000 °C. At the beginning of the 1970s HEY [2538] and LONGBOTTOM [3576] used an AA spectrometer with a *quartz tube* as the detector for mercury species which had been separated by GC and converted into atomic mercury vapor in an FID. VAN LOON and RADZIUK [3580, 4779] developed this into a *heated quartz tube atomizer* (QTA) that was heated to around 1000 °C and was suitable for a large number of elements. CHAN and co-workers [1210, 1212, 1213, 1215–1217] employed this system intensively in subsequent years.

Compared to graphite furnaces, the QTA is characterized by its high sensitivity, which results from the longer absorption path, its much simpler design, and its lower power consumption for heating. Heated QTAs in which a hydrogen-oxygen flame burns [1644], similar to an FIT atomizer (refer to Section 1.8.2), are particularly effective. This type of atomizer, which is designated QF in the following, is used nowadays almost exclusively for GC-AAS. BAXTER and FRECH [441] made theoretical studies on the characteristics of tubular atomizers as detectors for GC-AAS. They came to the conclusion that as a result of their larger volume quartz tube atomizers are superior to graphite furnaces. These authors found that the absorption cell used by JIANG et al. [2910] (180 mm long by 7 mm internal diameter), which was however used for mercury determinations and was thus not heated, came closest to the theoretical optimum. Together with JOHANSSON [2929] they checked their model on the basis of experimental data to optimize an atomizer for the determination of organic lead species under various operating conditions.

The connection between the GC column and the AAS detector is critical. It should be as short as possible and also heated to prevent condensation and tailing of the peaks. Transfer lines made of steel or quartz are described in the literature. These are heated to between 150 °C [1313] and 900 °C. The optimum temperature at which condensation is avoided but at which decomposition does not take place depends on the analyte. For organotin compounds, for example, a temperature of 225–250 °C is suitable [1942, 2893], while for mercury temperatures of 700–900 °C are used [910, 2910] since the compounds must be decomposed (atomized) before they enter the absorption cell. A typical design for the GC-to-QF AAS interface is depicted in Figure 7-2.

Although packed columns are mostly used for separation, open tubular capillary columns would appear to offer a number of advantages for speciation analysis. The better separation of those constituents that might be eluted with the organometallic species avoids potential interferences during atomization and detection. During their theoretical investigations on the performance characteristics of GC-AAS systems, BAXTER and FRECH [441] found that capillary columns provided sharper peaks and could be operated with lower carrier gas flowrates than packed columns. Both factors lead to an improvement in the sensitivity.

The major problem associated with the coupling of GC with AAS is the limited volume of measurement solution that can be injected onto the column; the maximum is around 100 μL [1585]. We must emphasize that this limitation lies with GC and is independent of the detector, i.e. the same limitation applies to other hyphenated techniques such as GC-MS or GC-MIP. All couplings with GC therefore require a relatively high level of preconcentration in dependence on the detector sensitivity. *Extraction of the analyte species* from the test sample and its subsequent preconcentration are always

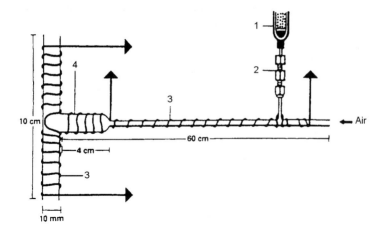

Figure 7-2. Principle of a GC-QF AAS coupling. **1** – GC column; **2** – transfer fitting; **3** – heating wire; **4** – QF (from [963]).

performed off-line and mostly manually. In particular, ionic inorganic and organometallic species are mostly not extracted quantitatively, even in the presence of complexing agents.

Following extraction a *derivatization step* is required that in general comprises a number of substeps and is very time-consuming. Derivatization is performed most rapidly with sodium tetraethylborate and hydride-generation with sodium tetrahydroborate, both in aqueous medium. BERGMANN and NEIDHART [554] described the derivatization of organolead compounds with tetrabutylammonium-tetrabutylborate, also directly in aqueous medium. Organic lead and tin compounds, however, are mostly converted with a Grignard reagent into a derivative suitable for GC. After completion of the reaction the excess of reagent must be destroyed, and the organic phase separated and purified over silica gel. Finally the analyte is eluted with an organic solvent and the volume correspondingly reduced before the extract is injected onto the column. With such a large number of steps, the risk of analyte loss is naturally very large, especially for volatile compounds. For this reason the *trueness* attained with GC-AAS techniques is often not particularly good.

This GC procedure allows neither the detection of redox species nor, depending on the method, inorganic species. Volatile species are often lost during sample pretreatment. Taking into account the above restrictions, higher alkylated and phenylated compounds can be measured relatively well. A typical chromatogram for organotin species generated via GC-AAS is depicted in Figure 7-3. Organometallic compounds with very high molecular masses which do not form volatile derivatives also cannot be detected by GC-AAS.

In recent times a number of papers have been published on speciation analysis utilizing coupling with capillary gas chromatography (CGC); in terms of the time required for sample preparation compared to the techniques described above, this technique can only be described as revolutionary. SZPUNAR *et al.* [5729, 5730] described *microwave-assisted leaching* of organometallic species from sediments and biological materials that only required 1–5 minutes, followed by simultaneous derivatization with sodium tetra-ethylborate and extraction in hexane in only 5 minutes. These authors reported a sample throughput of 5/h using this sample preparation technique and the coupling of CGC with QF AAS. PEREIRO *et al.* [4924] simplified and accelerated this technique using *simultaneous leaching, derivatization and extraction* in a focused microwave field in only 3 minutes. The organic phase can be injected directly onto the capillary column.

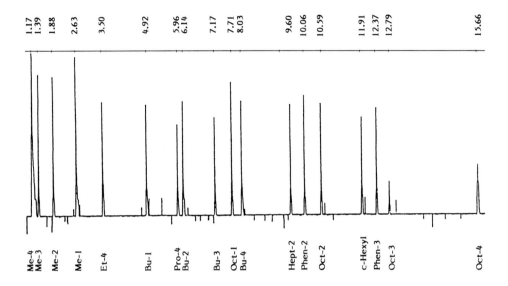

Figure 7-3. GC-QF AAS chromatogram of 18 organotin species; 2.5 ng Sn absolute per component (from [2894]).

7.2.2 Liquid Chromatographic (HPLC) Separation

Liquid chromatography, and especially HPLC, is also frequently used for the separation of organometallic species. This should not be seen as a rival to GC techniques, but rather as a complement, increasing the type and number of species that can be made available for determination. Additionally, LC techniques offer a number of advantages compared to GC techniques. One of the most important is that HPLC allows the separation of species without prior derivatization. In addition, the numerous chromatographic techniques, such as adsorption, ion exchange, gel permeation, normal and reversed phase chromatography, permit the separation of ionic, volatile, complexed, and high molecular mass organometallic and biological species.

The polarity of the inorganic and also of the majority of the organic arsenic species makes them amenable to speciation analysis by ion exchange and reversed phase HPLC. Hyphenated techniques with HPLC are used preferentially for the speciation analysis of organometallic molecules in soils, oil, and other biological materials. Although derivatization is not necessary, all HPLC techniques require a purification step after extraction and substantial preconcentration, so that the procedure is slow and requires considerable effort.

The possibilities of coupling HPLC with AAS have been discussed in detail by EBDON *et al.* [1653, 1657]. The combination of LC with F AAS usually results in relatively insensitive systems [3227]; the interface often makes a not inconsiderable contribution to the poor sensitivity. The main problem is the low efficiency of nebulization and desolvation of the mobile phase, and thus a low transport rate of the analyte species into the atomizer. A further problem is the considerable difference in the flowrate of the mobile phase of the HPLC system and the aspiration rate of the nebulizer of F AAS. The organic solvents used for HPLC are also not always compatible with F AAS and can lead to interferences.

There has been no shortage of efforts to improve the poor sensitivity of HPLC-F AAS, for example sample introduction by thermospray [2548, 3456]. SLAVIN and SCHMIDT [5414] employed the injection technique (see Section 4.1.3.1) for the coupling of HPLC with F AAS. The mobile phase dripped into a funnel that was attached to the end of the nebulizer sample capillary. Every droplet caused a peak-shaped signal, the sum of which generated the chromatogram. EBDON *et al.* [1652, 1654] proposed a different technique; as depicted schematically in Figure 7-4, the droplets are collected on a platinum spiral, dried, and then introduced into the flame.

The direct coupling of HPLC with GF AAS is highly problematical due to the short lifetimes of the graphite tubes under the influence of the continuous inflow of solvent and water, and is also not particularly sensitive [4397, 5616]. By the use of a fraction collector it was possible to couple continuous HPLC separation with the discontinuous atomization of GF AAS. This technique found considerable interest due to the excellent

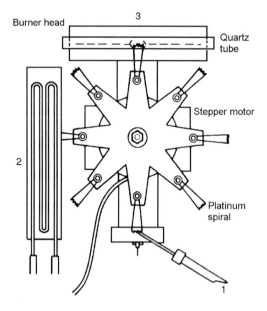

Figure 7-4. Interface with rotating platinum spirals for HPLC-F AAS. **1** – HPLC column; **2** – drier; **3** – burner head with quartz tube (from [1652]).

detection limits of the latter [293, 5578, 6382]. The eluate from the HPLC column is usually collected in fractions in containers in an autosampler, or collected in a capillary [6065], and then injected into the graphite tube. A chromatogram generated in this manner, which again comprises a multitude of individual peaks, is depicted in Figure 7-5. A further possibility is the use of a flowcell from which the autosampler withdraws aliquots of the mobile phase at fixed time intervals [296–298, 792, 793, 1160, 1162, 1911, 2035, 2739, 2907, 3228, 3229, 3385, 3906, 4537, 4539, 4860, 6382, 6383]. Since the time-resolved resolution of the GF AAS detector is low at about one determination per minute maximum, the elution times must be carefully matched to the measurement frequency of the detector [294]. The coupling of LC with GF AAS has been automated by a number of authors [2425, 4657, 5578], and KÖLBL et al. [3226] developed a program for chromatographic data evaluation.

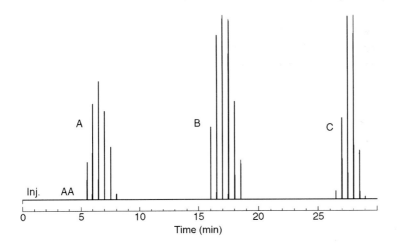

Figure 7-5. Chromatogram of various arsenic species generated via HPLC-GF AAS using a fraction collector. **A** – inorganic arsenic; **B** – arsenobetaine; **C** – arsenocholine; 1 µg As absolute per component (from [5578]).

The direct coupling of HPLC and QF AAS has been described [656, 963, 1185, 5933], but the application of the QF, which is particularly efficient for GC-AAS, to HPLC-AAS is problematical. BLAIS et al. [656] thus proposed thermochemical hydride generation for this purpose; this technique was successfully applied by MOMPLAISIR et al. [4162] for the determination of arsenic species in marine organisms. In this technique the eluate is introduced by means of a thermospray into a pyrolysis chamber in which the sample and solvent are burnt in an oxygen-hydrogen flame, which results in significant dilution in the gas phase. Apart from the correspondingly low sensitivity, safety aspects speak against this flame. The coupling of LC to QF AAS is performed more easily via conventional hydride generation with sodium tetrahydroborate since the analyte is separated from the solvent [639, 3946, 4162]. Since not all organic compounds of the hydride-forming elements can be hydrogenated, however, and the hydrides exhibit varying sensitivities, it is necessary to destroy organic molecules prior to hydride formation. For this purpose conventional [3600] or microwave-heated [3402, 4520, 5906, 5907] reactors, or UV photolysis [1812, 1814–1816, 2682, 4999], have been proposed. Applications for such LC-HG AAS couplings have been described for arsenic [252, 1185, 1240, 1726, 1847, 2322, 2424, 2492, 2682, 3285, 3286, 3402, 3477, 3599, 3600, 3810, 3837, 3910, 4520, 4859, 4998, 4999, 5609, 5610, 5933, 6016], antimony [242, 5454], selenium [1311, 1720, 2173, 3879, 4068, 4238], and tin [963, 5199]. Corresponding methods have also been developed for the coupling of LC with CV AAS for the determination of mercury [1812–1816, 2635, 4232, 4299, 5106, 5107, 6391].

7.2.3 The Coupling of Hydride-generation, Preconcentration, and GC Separation

The coupling of hydride-generation, preconcentration in a cold trap, and GC separation was proposed by RAPSOMANIKIS *et al.* [4810] in 1986 and was developed in the following years, especially by DONARD and co-workers [1582, 1584, 1587, 4752], for the speciation analysis of numerous elements. This procedure, abbreviated to HG-CT-GC-QF AAS, consists in principle of four steps:

i)   On-line derivatization in aqueous solution; depending on the analyte and the given task, hydride-generation (HG) with sodium tetrahydroborate or ethylation (Et) with sodium tetraethylborate can be performed.
ii) Collection and preconcentration of the volatile analyte species in a cold trap (cryo-trapping, CT), usually on the chromatographic column.
iii) Chromatographic separation (GC) by slowly warming the column.
iv) Detection in an electrically heated quartz tube atomizer under the addition of hydrogen and oxygen as the reaction gases (QF AAS).

The apparatus for this technique is depicted schematically in Figure 7-6. Compared to GC and HPLC hyphenated techniques, this procedure offers a number of principal advantages:

i)   Relatively large sample volumes can be applied, e.g. 50 mL; if necessary, even larger volumes are possible.
ii) Liquid test samples can be used directly without prior extraction, preconcentration, or derivatization.
iii) All steps in the procedure, such as derivatization, preconcentration, separation and detection, are performed on-line in a closed system and can be easily automated.
iv) The high sensitivity of the procedure meets all requirements of speciation analysis.

Hydride-generation with sodium tetrahydroborate for elements such as As, Bi, Ge, Sb, Se, Sn, and Te has been thoroughly investigated and is well documented in the literature (see Section 8.3.1.1). Moreover it can be used not only for the inorganic but also for the partially alkylated species; this is shown in equation 7.1 on the example of an organotin species:

$$R_xSn^{(4-x)+} \xrightarrow{\text{NaBH}_4, \text{H}^+} R_x SnH_{(4-x)} + H_2 \, , \tag{7.1}$$

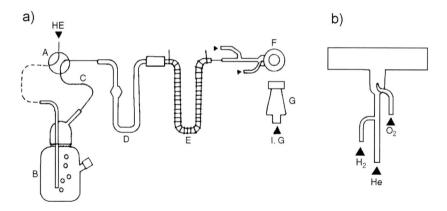

Figure 7-6. Apparatus for speciation analysis by HG-CT-GC-QF AAS. **a** – Schematic of the complete apparatus: **A** – valve; **B** – reactor; **C** – transfer line; **D** – water trap; **E** – cold trap with heating coil; **F** – quartz tube atomizer; **G** – burner. **b** – Detail of the QF (from [1582]).

where x = 1, 2, or 3 and R = methyl, ethyl, or butyl. By selecting a suitable pH range, it is also possible to use this technique to selectively determine redox species of antimony, arsenic, and selenium.

Important species of lead and mercury, which cannot be determined by hydride-generation in analogy to equation 7.1, can be detected via alkylation. Both inorganic lead and mercury and also their alkyl derivatives are quantitatively ethylated by sodium tetra-ethylborate in aqueous solution at pH 5 according to:

$$Pb^{2+} \xrightarrow{\quad NaBEt_4 \quad} Et_2\, Pb(II) \qquad\qquad (7.2)$$

$$2\, Et_2\, Pb(II) \rightarrow Et_4\, Pb(IV) + Pb(0) \qquad\qquad (7.3)$$

$$Me_2\, Pb^{2+} \xrightarrow{\quad NaBEt_4 \quad} Me_2\, Et_2\, Pb \qquad\qquad (7.4)$$

$$Hg^{2+} \xrightarrow{\quad NaBEt_4 \quad} Et_2\, Hg\, . \qquad\qquad (7.5)$$

Using cryofocusing, enrichment factors of 50–100 can be attained. This is performed directly in a small packed column coated with a non-polar methylsilicone phase which is immersed in liquid nitrogen. Separation is performed by warming the column, and the species are eluted according to their boiling points and their chromatographic properties. The column offers a limited separating potential of about 3000 theoretical plates. Nevertheless, in combination with highly specific AAS detection, it offers an extremely selective and sensitive technique for speciation analysis which meets many of the requirements listed in Section 7.2.

A typical chromatogram showing the separation of several organotin species is depicted in Figure 7-7. A comparison with the chromatogram shown in Figure 7-3 (Section 7.2.1) generated via a GC-QF AAS coupling is interesting. This illustration also gives an impression of the speed of the technique with a time requirement of less than 15 minutes for the entire analytical sequence. The procedure has the advantage that all inorganic species, including redox species, and all volatile and medium-volatile organometallic species are detected. Losses of volatile species are effectively prevented and high accuracy is guaranteed. The procedure is especially suitable for the determination of highly toxic species that are environmentally relevant in a multitude of sample materials. The technique cannot be used, however, for the determination of species of low volatility such as phenylated tin compounds or compounds of high molecular mass such as arsenobetain or selenomethionine. Special applications of this technique for speciation analysis are discussed in more detail in Chapter 10.

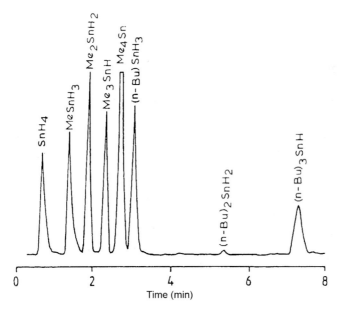

Figure 7-7. Determination of inorganic tin and methyl and various n-butyl tin compounds by HG-CT-GC-QF AAS. The tin content in the individual compounds is between 0.2 ng for $(n\text{-Bu})_2Sn^{2+}$ and 7.5 ng for $(n\text{-Bu})^3Sn^+$ (from [1582]).

8 The Techniques of Atomic Absorption Spectrometry

The flame, graphite furnace, hydride-generation, and cold vapor techniques are discussed in some detail in this chapter, especially with respect to their particular characteristics. The typical processes taking place in each type of atomizer are treated, in as far as they are not already discussed in Chapter 4, and especial attention is paid to the mechanisms of atomization. Emphasis is placed on practical analytical aspects, such as, for example, the interferences characteristic for each technique and their avoidance or elimination. In keeping with this practical emphasis we shall restrict our discourse to commonly available atomizers; 'exotic' flames and non-commercial atomizers are merely mentioned as exceptions.

8.1 The Flame Technique

Keeping the above restrictions in mind, in this section we shall treat the analysis of *solutions* which are sprayed into a *premix burner* by means of a *pneumatic nebulizer* and atomized in an *air-acetylene* or *nitrous oxide-acetylene flame*. Despite these restrictions we must point out that the *experimental conditions* can vary considerably from laboratory to laboratory. Since in the majority of publications the experimental details, such as stoichiometry of the flame, aspiration rate and efficiency of the nebulizer, the size of the aerosol droplets, complete information on the ions in solution etc., are not usually given in sufficient depth to be repeated, it is not always possible to judge the occasionally significant differences or even contrary observations. For this reason we shall not offer a complete survey of the relevant literature, but shall restrict ourselves to an important selection. The review article published by RUBEŠKA and MUSIL [4991] served as a useful guide.

8.1.1 Atomization in Flames

The processes that lead from aspiration of the measurement solution to the formation of free atoms in the vapor phase are depicted in Figure 8-1; nebulization and the formation of the aerosol are described in detail in Section 4.1.2.2. The tertiary aerosol enters the flame where it is dried, i.e. the remaining solvent is vaporized. The particles are volatilized and the gaseous molecules dissociate into free atoms.

We shall not consider the notoriously poor efficiency of the pneumatic nebulizer (about 5%, refer to Section 4.1.2.2) in our discussions on atomization. As our starting point we shall consider the aerosol that effectively reaches the flame so that we can distinguish between improvements in the degree of nebulization and improvements in the degree of atomization. Moreover, particularly for matrix-free solutions, we can assume that with a premix burner and a flame of sufficient temperature the analyte is transferred completely to the vapor phase, i.e. all particles are volatilized.

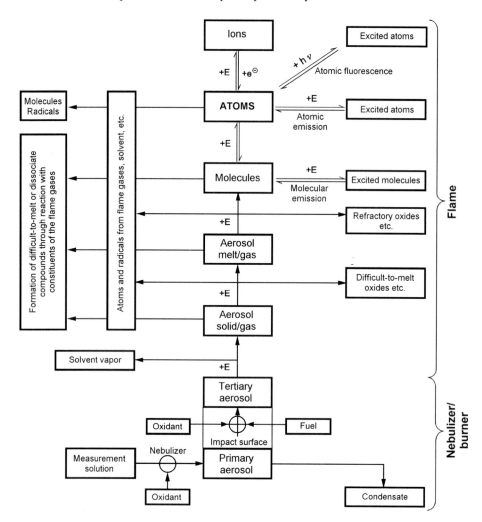

Figure 8-1. Schematic representation of the most important processes possible in a flame. As well as the introduction of thermal energy E the chemical processes taking place in the flame have a major influence.

8.1.1.1 The Process of Volatilization

In the vapor phase the analyte M can exist as atoms, M_a, or in molecular form, MY_i; both forms can be ionized or excited to varying degrees. The partial pressures p of the components follow the relationships:

$$p \sum M = \sum_i p_{MY_i} + p_{M_a} \qquad (8.1)$$

and

$$p_{M_a} = p_{M^0} + p_{M^*} + \sum_n p_{M^{n+}} ,$$ (8.2)

where M^0, M^*, and M^{n+} are free atoms in the ground state, in the excited state, and in the nth ionized state, respectively. From the approximate relationship (equation 8.3) for premixed flames [4991]:

$$p\sum M \approx 10 c_M ,$$ (8.3)

where c_M is the molar analyte concentration in the measurement solution and $p\sum M$ is expressed in Pa, we can estimate the partial pressure of the analyte in all its forms to be 10^{-1} through 10 Pa. This is a negligibly small value in comparison to the partial pressures of the flame gas components. Assuming that the analyte is completely volatilized, then:

$$\beta_a = \frac{p_{Ma}}{p\sum M} ,$$ (8.4)

where β_a represents the atomized portion of the analyte.

The composition of the flame gases depends largely on the gas mixture and the design of the burner [2088]. If a matrix-free analyte solution is nebulized, the composition of the components in the flame can change as a result of the solvent, but not as a result of the analyte since its concentration is too small in comparison to the other natural components in the flame.

8.1.1.2 The Process of Dissociation

Diatomic molecules predominate in the vapor phase; triatomic species are limited to the monohydroxides of a few alkali and alkaline-earth elements, to monocyanides [3705], and to oxides such as Cu_2O. The majority of other polyatomic compounds dissociate very rapidly at temperatures well below those prevailing in analytical flames. The process of thermal dissociation is treated in detail in Section 2.2.

The dissociation energies, E_D, of diatomic molecules are generally in the order of 3–7 eV. In the usual analytical flames, molecules are mostly atomized completely at $E_D < 3.5$ eV; compounds with $E_D > 6$ eV are classed as difficult-to-dissociate. The degree of dissociation can change significantly with the flame temperature; the solvent can also have an influence on the flame temperature (refer to Section 8.1.1.5). The influence of the sample matrix on the flame temperature is minimal and the effect on the dissociation can be neglected.

The description of equilibria by means of dissociation constants is valid, independent of the mechanism; in other words, provided that the system is in equilibrium, it is immaterial whether it is a true thermal dissociation or whether chemical reactions are involved. Merely the *velocity* with which equilibrium is attained depends on the reaction mechanism and the reaction speed.

Dissociation equilibria involving *components of the flame gases* must be accorded special attention. Typical examples are the dissociation of oxides, hydroxides, cyanides, and hydrides. The concentrations of O, OH, CN, and H in a flame are determined by the reactions between the natural components of the flame. The influence of any sample constituents on these components is negligible, since every reaction that leads to a reduction in the concentration of a species is immediately counteracted by infinitesimal shifts in the equilibria of the main components. This *buffer effect of the flame gases* with respect to the concentrations of O, OH, CN, and H still takes place even when the components of the flame gases are not in equilibrium. Among the known deviations from a state of equilibrium are the *increased* concentrations of these radicals in the primary reaction zone, so that the buffer capacity in this zone is in fact even higher.

In the primary reaction zone of a nitrous oxide-acetylene flame the concentration of atomic oxygen is around three orders of magnitude higher than the thermodynamically predicted value [4991]. For this reason mechanisms of interference cannot be explained by 'competition in the flame for available oxygen' [4817]. The same is also valid for other reaction equilibria that involve components of the flame gases, such as the dissociation of cyanides [3705].

We must emphasize that under conditions of equilibrium the actual mechanism has no influence on the equilibrium itself; a typical mechanism is the reduction of oxides by components of the flame gases (C, C_2, CH, CO, H, CN) to free analyte atoms. This is easily explained by equation 2.4 (refer to Section 2.2) since it contains no statement on how equilibrium is attained [2332]. Under these conditions the dissociation equilibria of oxides, hydroxides, and similar compounds with components of the flame are not influenced by concomitants of the sample. If vapor-phase interferences can be observed under conditions of equilibrium, there is little point in considering the mechanism of atomization.

The situation with respect to the *dissociation of halides* is totally different since these are not components of the flame gases; their concentration in the flame depends on their concentration in the measurement solution. The quantity of halide bound to the analyte can be neglected when the concentration of the halogenic acid is ≥ 1 mol/L. By inserting the corresponding contents n_x, p_x, or c_x for the degree of dissociation α_D into equation 2.6 (see Section 2.2) we obtain:

$$\alpha_D = \frac{1}{1+n_x/K_D} = \frac{1}{1+p_x/K_D} = \frac{1}{1+10c_x/K_D}, \tag{8.5}$$

where n_x/K_D is given in cm^{-3}, p_x/K_D in Pa, and c_x/K_D in mol/L.

Analyte halides are normally completely vaporized, so that equation 8.5 represents the gradient of the calibration curve as a function of the halide concentration c_x in the measurement solution. Halide concentrations of > 1 mol/L thus tend to depress the analyte signal for those elements that have a dissociation constant $K_D < 10^3$ Pa. This possible interference is discussed in more detail in Section 8.1.2.5.

8.1.1.3 Deviations from Equilibrium

For a number of elements the experimentally determined values for the degree of atomization do not agree with the calculated values for the thermal dissociation of monoxides. According to thermodynamic calculations, for example, it should be possible to determine aluminium in a fuel-rich air-acetylene flame; CHESTER et al. [1254] calculated the degree of atomization as $\beta_a = 0.9$, while the experimentally determined value is only around $\beta_a = 0.05$ [2041]. Similarly for lithium the calculated value for a fuel-rich flame of $\beta_a > 0.9$ is relatively far removed from the experimental value of $\beta_a < 0.3$. For magnesium, on the other hand, the theoretical value is lower than the experimentally determined value [2332]. These apparent discrepancies can be explained, at least in part, by the *participation of free radicals* in the reduction of metal oxides [3697].

The absorbance of many elements increases with an increasing fuel gas-to-oxidant ratio to a given level, even though the flame temperature changes only slightly. The gradient of the function $A(\rho)$, the dependence of the absorbance A on the fuel-to-oxidant ratio ρ, increases in the sequence Ni < Co < Fe < Cr, just like the dissociation of the respective oxides. This indicates that the atomization of these elements is determined not only by thermal dissociation [3697].

In order to describe the atomization process quantitatively it is necessary to know the concentrations of the radicals. The concentrations of species such as H, O, and OH can reach values of 10^{10} cm^{-3} in the primary reaction zone of hydrocarbon flames. These values decrease rapidly with increasing flame height, however; the equilibrium concentrations are only attained 40–50 mm above the burner slot [2332], a height at which analytical measurements are not normally performed. This relatively slow decrease in the radical concentration can be explained by the fact that in the absence of wall reactions the radicals can only recombine via triple collisions, which are relatively improbable.

It has long been known that oxides with dissociation energies of > 6.3 eV (523 kJ/mol) exhibit refractory behavior. This value is close to the bonding energies of CO and C_2, so that the following reactions are conceivable:

$$MO + CO \rightarrow M + CO_2 \tag{8.6}$$

and

$$MO + C_2 \rightarrow M + CO + C. \tag{8.7}$$

Nevertheless, the change in entropy has not been taken into account when comparing the bonding energies. A species can only bring about reduction when the resulting change in free energy ΔG is negative, i.e. when ΔG of the reaction:

$$2CO + O_2 \rightarrow 2CO_2 \qquad \left(\Delta G = -120\,\text{kJ at 2400 K}\right) \tag{8.8}$$

is smaller than the free energy ΔG_f for the formation of metal oxides MO. Numerical calculations have shown that for difficult-to-dissociate monoxides $\Delta G_f < -600$ kJ, while oxides with $\Delta G_f > -480$ kJ are easy to dissociate. Possible reduction reactions should thus exhibit ΔG values that lie between these two (at 2400 K). Conceivable reactions include:

$$4\,H+O_2 \rightarrow 2\,H_2O \qquad\qquad \left(\Delta G = -560\,kJ\right) \qquad\qquad (8.9)$$

or

$$\tfrac{1}{2}C_2 +O_2 \rightarrow CO_2 \qquad\qquad \left(\Delta G = -580\,kJ\right). \qquad\qquad (8.10)$$

Reduction by CO is thus improbable, but reduction by C_2 and H radicals is very likely [2332].

As indicated above, the reduction process is hardly influenced by concomitants of the sample, independent of whether the radical concentration is in equilibrium or not. Solely constituents that change the number of C_2 or H radicals can influence reduction in the vapor phase. This aspect is discussed under the treatment of the solvent effect (see Section 8.1.1.5).

8.1.1.4 The Kinetic Aspect

In the essentially thermodynamic treatment thus far we have assumed that, with the exception of radicals, equilibrium conditions exist. The mechanism that takes place after equilibrium is attained does not play any role. This view is only justified if equilibrium is attained within a time that is shorter than the time it takes for the analyte to rise from the burner slot to the absorption volume. Since the burning velocity of normal flames is in the order of 10^3 cm/s (see Table 4-1 in Section 4.1.1), the time available to reach equilibrium is only a few *milliseconds*. We must therefore pose the question whether the available time is sufficient for equilibrium to be reached. In cases of doubt we must therefore take the kinetics of a process into consideration. This means that the *reaction mechanisms* can also play a role.

Deviations from equilibrium can occur when the reaction speed of the process being investigated is too slow. The ionization equilibrium, for example, which is reached very rapidly in the primary reaction zone as a result of the very fast charge transfer reactions, deviates more and more from equilibrium in higher flame zones due to the slower collisional ionization with its relatively long time of relaxation.

If a condition of equilibrium is reached only slowly, concomitants can have a catalytic action. This can lead to signal enhancement, for example when the dissociation equilibrium is attained more rapidly in the presence of a catalyst, but it can also lead to signal suppression if competitive reactions, such as radical recombination, are accelerated.

If H radicals are involved in the atomization of tin, for example, the degree of atomization of tin should be proportional to the concentration of H radicals above equilibrium. This would explain the high effectiveness of the hydrogen diffusion flame for the atomization of tin, since this flame exhibits a far higher concentration of H radicals than either the air-hydrogen or air-acetylene flames. All elements and compounds that catalyze the recombination of H radicals, such as SO_2, NO_x, should depress the signal for tin, which can be observed in practice [4160, 4987, 4988].

8.1.1.5 The Solvent Effect

According to IUPAC the influence of the atomizer itself (e.g. the flame gases) or the solvent on the degree of atomization or the sensitivity of an element is not considered to be an interference [2824, 2827]. The reason for this is that in principle the same atomizer and the same solvent are used for all test sample and calibration solutions. Nevertheless, the influence of organic solvents in F AAS goes far beyond the effect normally expected, so that further discussion is required.

Organic solvents were used even in the early years of flame emission spectrometry to increase the sensitivity [682, 683] and were described for the first time for F AAS by ALLAN [139]. Most organic solvents have a lower viscosity and surface tension than water and are thus more easily aspirated and more finely nebulized. FARINO and BROWNER [1846] observed that organic solvents generally shift the range of droplet sizes toward smaller droplets. This results in higher nebulization efficiency, so that a higher proportion of the measurement solution reaches the flame. As can be seen from Table 8-1, this transport effect is not the same for all elements due to the additional influence of the solvent on the flame parameters, such as temperature.

While the dissociation of water (which always takes place in the flame when aqueous solutions are nebulized) is a strong endothermic reaction that noticeably reduces the flame temperature, the combustion of an organic solvent (except for highly halogenated) is generally an exothermic reaction which increases the flame temperature. Both a

Table 8-1. The effect of various solvents on the absorbance signals of copper and zinc in an air-acetylene flame relative to an aqueous solution (according to [310]).

Solvent	Relative Absorbance [a]	
	Cu	Zn
Methanol	6.8	5.0
Ethanol	5.9	3.8
n-Propanol	5.1	2.8
n-Butanol	3.6	2.3
n-Pentanol	2.4	1.7
2-Propanol	5.2	3.6
2-Butanol	4.2	2.7
3-Pentanol	3.5	1.8
Acetone	7.4	6.3
Butanone	8.6	5.8
3-Pentanone	7.1	4.8
Isobutyl-methyl-ketone	7.4	4.4
Formaldehyde	1.6	0.92
Propanaldehyde	2.1	4.0

[a] The absorbance of an aqueous solution is set to 1

higher flame temperature and the lower thermal stability of organic molecules can lead to an enhancement of the degree of atomization.

Temperature effects of this type explain why the signal enhancements for the individual elements are different, since an increase in temperature has a different effect on each element. Nevertheless, the observation that butanone enhances the signal more strongly for copper than acetone, while for zinc the effect is reversed cannot be explained by the temperature alone (see Table 8-1). These effects indicate that individual organic solvents influence the formation of free radicals in the flame and thus additionally intervene in the reaction mechanism and the effectiveness of atomization.

Finally we must mention that the solvent effect can be strongly dependent on the experimental conditions and the type of burner used. The published data can differ markedly [1846] and the values in Table 8-1 should merely be used for orientation. A further important factor is the water content of the organic solvent. Solutions saturated with water, such as occur with solvent extractions, frequently enhance the signal far less than water-free solutions.

8.1.2 Interferences

8.1.2.1 Spectral Interferences

Spectral interferences caused by direct overlapping of the analytical line emitted by the radiation source and the absorption line of a concomitant element are limited to a few individual cases, which are listed in Table 5-6 (Section 5.4.1.1). Moreover, no real interferences are known on the main resonance lines, so that care is only required for the occasional use of less sensitive, alternate lines. In the presence of very high concentrations of concomitant elements (several g/L) interferences can occur due to considerable broadening of the absorption profile [3891], as discussed in Section 5.4.1.1.

With proper use of a premix burner and an air-acetylene or nitrous oxide-acetylene flame, we can assume that after nebulization of the measurement solution the particles are largely volatilized. Interferences caused by radiation scattering on particles in flames are thus seldom.

The background attenuation observed in F AAS is more or less exclusively caused by molecular absorption; nevertheless it rarely reaches such a magnitude that it causes interferences. This type of interference occurs most frequently when an easily atomized element is determined in the presence of a concomitant element that for example forms a difficult-to-dissociate oxide or hydroxide in the air-acetylene flame. A prerequirement for an interference is naturally that the molecular absorption coincides with the analytical line. For example, strong absorption of the CaOH radical at the barium line at 553.6 nm in an air-acetylene flame has been reported [1055]. Nevertheless, since barium is determined almost exclusively in the nitrous oxide-acetylene flame, this interference is without analytical significance.

The alkali halides are further examples of molecular spectra in flames (discussed in Section 5.4.1.1). FRY and DENTON [1996] made thorough investigations on background attenuation for a number of matrices in the spectral range 190–300 nm. For urine they found that the background signal was hardly higher than the noise. These authors were able to scan the molecular spectra of 50 g/L solutions of sodium chloride and calcium chloride in an air-acetylene flame with absorbance maxima of $A = 0.02$ and 0.03, respectively, between 230 nm and 240 nm. Undiluted sea-water exhibited a maximum absorbance of $A = 0.015$. In a nitrous oxide-acetylene flame the molecular absorption could be reduced by around a power of ten, thus becoming more or less without significance. Albeit the sensitivity of a number of volatile elements is markedly poorer in the hotter flame, so that it cannot always be used.

In the relatively few cases in which spectral interferences caused by the absorption of molecules or radicals in a premixed, laminar flame occur, they are best eliminated by applying continuum source BC. Since background attenuation in flames is never very high and in addition a static signal is generated, the background can mostly be corrected without problems by a continuum source corrector.

HÖHN and JACKWERTH [2628] reported an interference for the determination of gold in indium which could be attributed to molecular absorption by InCl, which is known to exhibit fine structure (see Figure 5-15, Section 5.4.1.1). In this case the use of the hotter nitrous oxide-acetylene flame is recommended, or the use of a more sensitive technique, such as GF AAS, since it is possible to work with substantially greater dilutions and hence with reduced interferences.

The application of the Zeeman effect for background correction in F AAS has been reported in a number of special cases [6465]. However, interferences occurring in the flame can be eliminated by much simpler means. There is thus no real cause for using Zeeman-effect BC in F AAS since the disadvantages are greater than the advantages (refer to Section 3.4.2).

8.1.2.2 Transport Interferences

The transport of the measurement solution to the atomizer is the process that has the greatest influence on the sensitivity of F AAS. As discussed in detail in Section 4.1.2.2, it is hardly possible to introduce more than 5% of the measurement solution into the flame when using a pneumatic nebulizer if interferences caused by larger droplets cannot be tolerated. Transport interferences occur when the *mass flow of the analyte* through the horizontal cross-section of the flame at the height of the absorption volume is changed by the presence of concomitants in the measurement solution. An excellent mathematical treatment of the individual processes and effects that lead to transport interferences has been published by RUBEŠKA and MUSIL [4991]. We shall limit ourselves to a more qualitative discussion of the individual processes in this section.

Transport interferences depend on the aspiration rate of the nebulizer and the range of droplet sizes of the aerosol; the latter is decisive for the portion that actually reaches the flame. Under constant experimental conditions these factors depend on physical characteristics of the measurement solution, such as density, viscosity, and surface tension.

A change in the *density* of a solution brought about by inorganic salts has no influence on the rate of aspiration and only little influence on the droplet size of the aerosol. On the other hand, changes to the *viscosity* brought about by inorganic salts or free acids can be 10% or higher. If the viscosity of a solution is increased by adding increasingly greater quantities of a salt, we can observe that, at a fixed nebulizer setting, less and less solution is aspirated and that the aerosol droplets become larger. This means that a larger portion of the aerosol is condensed in the spray chamber, so that only a smaller portion of the measurement solution reaches the flame. This interference becomes noticeable for total salt contents of above around 10 g/L; organic macromolecules, such as proteins and sugars, have a stronger influence than pure inorganic salts [3200].

Surface tension has virtually no influence on the rate of aspiration, but it does influence nebulization and has a decisive effect on the size of the droplets. The surface tension of aqueous solutions is little affected by inorganic salts, but organic materials have a marked influence. The influence of surfactants has been investigated by a number of working groups [3189, 3190, 3251], but the results vary considerably. One reason for this seems to lie in the mode of action of these substances. Surfactants do not appear to significantly change the range of droplet sizes [1846]. The effect appears to depend on the *charge*; a charge opposite to that of the analyte enhances the signal. KORNAHRENS *et al.* [3251] have proposed a mechanism based on a distribution of the analyte ions in which the analyte is preferentially concentrated in droplets of smaller diameter. A further important factor would appear to be the chain length of the surfactant. The shorter the chain length, the greater is the effectiveness [5005]. Long chain reagents lower the surface tension more strongly, but they need longer to migrate to a new surface, so that they do not have sufficient time during nebulization to develop their full effectiveness [4184].

All in all, the influence of surfactants is slight and the signal enhancement is rarely more than 10–20% [1846, 4634]. The *addition* of such substances to increase the sensitivity can therefore hardly be justified. An *interference* caused by the presence of surfactants in the test sample is conceivable, but is only likely to require correction in exceptional cases.

In real samples several physical properties, such as viscosity and surface tension, often change simultaneously so that the effect of a change of one property cancels the effect of another. A change in the *temperature of the measurement solution*, which is supposed to have a substantial influence on the sensitivity of the measurement, is an example of this. CRESSER and BROWNER [1353] found that when the temperature of a measurement solution is reduced from +25 °C to 0 °C, less solution is aspirated due to the higher viscosity, but at the same time the droplets are smaller. The changes resulting from these two effects on the measured absorbance were so small that in routine operation they play no major role. O'GRADY *et al.* [4403] also found for a temperature change of 40 °C a much lower effect on the measured signal than would be expected for a change in the aspiration rate alone.

Taken altogether, the influences to be expected on the measured absorbance in F AAS resulting from changes in the physical properties of the measurement solution are not very significant. Without doubt organic solvents have the greatest effect, but as explained in Section 8.1.1.5 this is not counted as an interference since all solutions in a series of measurements must be prepared using the same solvent. In all cases transport interferences can be eliminated relatively easily; this is done most conveniently by matching the physical properties of the test sample and calibration solutions, for example by diluting the test sample solutions or by adding the interferent to the calibration solutions. If this is not possible the analyte addition technique can be used since it eliminates transport interferences reliably and completely.

Transport interferences caused by a change in the aspiration rate of the nebulizer can easily be avoided by pumping the measurement solution to the nebulizer [2934, 4934]. This technique is discussed in connection with flow injection in Section 6.2. It is interesting to note that pumping rates well below the normal aspiration rate lead to an improved range of droplet sizes, so that the efficiency of nebulization increases and interferences are further reduced [1821].

8.1.2.3 Spatial-distribution Interferences

Spatial-distribution interferences are caused by a change in the *spatial distribution of the mass transport* of the analyte in the atomizer under the influence of concomitants. RUBEŠKA and MUSIL [4991] established that this type of interference is caused by the movement of solid particles in flowing gases and the corresponding change in the spatial distribution of the analyte.

Changes in the spatial distribution of the analyte can be caused, for example, by alterations in the quantity of combustion products and thereby by the changes in volume of the flame, or by changes in the flowrate or flow direction of the analyte within the flame. BOSS and HIEFTJE [728] and also L'VOV *et al.* [3700] examined flow patterns of this type in detail. The results showed that the patterns are influenced mainly by the size and rate of volatilization of the particles, while the diffusion of molecules and atoms is of less significance.

Alterations in the spatial distribution of the atoms can lead to an increase or a decrease of the sensitivity, depending on the selected absorption volume [4714, 4801]. Since such interferences can be observed more easily in smaller absorption volumes, the magnitude of the interference depends on the instrument used, its optical system, and the burner system.

The gas mixture containing the sample aerosol streams vertically from the burner slot at a given velocity. The gas volume changes substantially in the primary reaction zone, mainly due to its thermal expansion. Consequently its velocity and flow pattern also change; as well as the vertical flow, a horizontal component becomes noticeable. This sudden change in the flow pattern exerts a horizontal force on the particles that causes a deviation of the linear movement. Because of their mass, however, the lateral spread of the particles is less than that of the flame gases and becomes smaller with increasing size of the particles. A spatial-distribution interference can occur when an interferent in the sample causes the particles to be bigger after desolvation, so that with a sufficiently long residence they are concentrated more in the middle of the flame than if no matrix was present.

L'VOV *et al.* [3694, 3695] examined this effect quantitatively and found that it could be neglected for normal, lower analytical concentrations. In extreme cases, however, with high total salt contents of refractory compounds it can attain a considerable extent. Interferences of the spatial distribution of atoms in the hot nitrous oxide-acetylene flame are much more frequent since this flame has a lateral expansion that is five times as large as that of an air-acetylene flame. Spatial-distribution interferences were reported even for the very first work on flames of higher temperature [166]. KOIRTYOHANN and PICKETT [3198] explained the signal enhancement caused by perchloric acid on Al, Ba, Ca, Li, and Sr as an interference of this type. These authors found that the interference disappeared when the slot burner was turned through 90° and verified their supposition by measuring the lateral absorption profile for strontium.

Subsequently L'VOV et al. [3695] measured the lateral flame profiles for numerous elements using an absorption volume of only 0.3 mm diameter. On closer scrutiny of the problem there is a strong dependence of the observed effects on the experimental conditions and the technical details [3890, 4989, 6281, 6283], as well as a close relationship to solute-volatilization interferences [440]. To classify an interference as a spatial-distribution interference, two criteria must be met: the interference must disappear when the burner is turned through 90° to the optical axis, and the relative increase should become smaller with increasing observation height. Using this procedure, WEST et al. [440] and also RUBEŠKA [4990] were able to show that the interference caused by molybdenum and tungsten on the determination of aluminium was a spatial-distribution interference, but that the interference caused by aluminium or vanadium on the determination of molybdenum was not.

8.1.2.4 Solute-volatilization Interferences

8.1.2.4.1 *The Volatilization of Particles*

A solute-volatilization interference is based on a *change in the rate of volatilization* of the aerosol particles, caused by the presence or absence of a concomitant. The volatilization of solid particles in hot gases is one of the least researched processes in flame spectrometry and is thus based largely on conjecture. In the most favorable cases using simplified suppositions and estimated numerical values, it has been possible to obtain results mathematically that are comparable with the experimental results. The degree of reliability that can be expected with calculations of this type is nevertheless rather low. Since, however, the completeness of volatilization of the aerosol particles is of the greatest significance in flame spectrometry, considerable efforts have been made despite the difficulties to investigate these processes [131, 416, 1293, 1294, 2543, 2637]. Fortunately, simplified models are sufficient to estimate interferences in F AAS [4991].

Mathematical models for calculating volatilization rates assume that above the surface of the particle a diffusion layer is formed in which, for the given temperature, the vapor pressure of the volatilizing substance attains saturation. The thickness of the saturated layer corresponds to the mean free pathlength, λ, of the volatilizing species. It is determined by diffusion of the volatilizing species out from the saturated layer into the surrounding atmosphere, as depicted schematically in Figure 8-2.

If the vapor pressure of the substance at the given temperature is small, the heat of volatilization required can be easily supplied to the particle by conduction or radiation. The rate of volatilization is then determined by the diffusion of the species from the

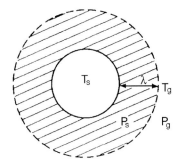

Figure 8-2. Schematic of a volatilizing particle (from [4991]). λ – mean free pathlength of the analyte species; p_s – partial pressure of the analyte in the diffusion layer; p_g – partial pressure of the analyte in the flame gases; T_s – temperature of the particle surface; T_g – temperature of the flame gases.

saturated diffusion layer into the surrounding gases. In this case we speak of *mass-transfer controlled volatilization*. In this mechanism the temperature of the particle surface, T_s, is very close to the gas temperature T_g [1293].

If, on the other hand, the vapor pressure of the volatilizing species is high, the heat of volatilization required will lower the temperature of the particle. T_s is thus lower than T_g and the transfer of heat becomes the time-determining step. In such cases we speak of *heat-transfer controlled volatilization*. The temperature difference is given by:

$$T_g - T_s = \frac{L\,M\,D\,p_s}{h\,R\,T_g},$$ (8.11)

where L is the heat of volatilization, M is the molecular mass of the volatilizing species, p_s is its partial pressure, D is the diffusion coefficient, h is the thermal conductance of the flame gases, and R is the gas constant [1922].

As mentioned above, a solute-volatilization interference occurs when the volatilization of the aerosol particles is incomplete in either the presence or the absence of a concomitant substance. The vapor pressure of the substance containing the analyte must therefore be low at the prevailing flame temperature. In this case we can expect a mass-transfer controlled volatilization. Under these conditions the rate of volatilization depends on the relative particle diameter and the mean free pathlength, λ, of the analyte, i.e. the thickness of the diffusion layer. For large particles $(2r > \lambda)$ the rate of volatilization can be described by:

$$-\frac{d\,m}{d\,t} = 4\,\pi\,r\,\frac{M\,D\,p_s}{R\,T},$$ (8.12)

while for small particles $(2r < \lambda)$ it can be described by:

$$-\frac{d\,m}{d\,t} = 4\pi\,r^2\,p_s\,\sqrt{\frac{M}{2\pi\,R\,T}},$$ (8.13)

where t is the time, m is the mass of the volatilizing particle, r is its radius, p_s is the vapor pressure of the volatilizing species at T_s, M is the molecular mass, R is the gas constant, and T $(T_s = T_g)$ is the temperature in Kelvin.

The size (radius) of the solid aerosol particle, r_s, naturally depends on the size of the liquid droplet, r_ℓ, produced at nebulization and on the salt content, β, of the sprayed solution. Assuming that the initial mean radius of the particle reaching the burner is 5 µm (see Section 4.1.2.2) and that the solution contains 1 g/L salt with a density of $\rho = 3$ g/cm^3, the dry aerosol particle would have a radius of:

$$r_s = r_\ell \sqrt[3]{\frac{\beta}{\rho}} = 0.35 \text{ µm.} \qquad (8.14)$$

The mean free pathlength can then be calculated according to [3702]:

$$\lambda = 3.07 \cdot 10^{23} \frac{T}{d^2}, \qquad (8.15)$$

where d is the gas kinetic diameter of the volatilizing species. For a temperature of 2500 K we obtain a value of $\lambda = 10^{-4}$ cm. This means that we can use equation 8.13 to calculate the rate of volatilization.

8.1.2.4.2 Reactions during Volatilization

As well as the physical aspects discussed above, chemical reactions can have an equally large influence on volatilization. We must naturally take all reactions into account that can play a role prior to the actual volatilization. These include:

i)   The formation of compounds of the analyte with concomitants during desolvation; the compounds with the lowest solubility product precipitate first.
ii)   Dehydration of hydrated salts, which is often accompanied by hydrolysis.
iii) Thermal decomposition of salts.
iv)   Reactions between the analyte and concomitants that can lead to new phases with differing thermal stability.
v)   Heterogeneous reactions at the surface of the particle with flame gases, in particular reduction to the element or to carbides.

Volatilization can take place simultaneously with reactions iii)–v) and compete with them. Different forms of the analyte usually have differing volatilities so that the rate of volatilization can change, depending on the reaction route by which the analyte reaches the vapor phase. All of these competitive reactions can be influenced by concomitants, thus opening the way for interferences. The anions present in the solution have a major influence. The analyte always crystallizes with the ion that gives the least soluble compound.

The most volatile compounds are generally the halides. The oxides that are formed through the decomposition of nitrates, sulfates, etc. are mostly less volatile. Nevertheless we must also take into account that the halides of many cations crystallize out as hydrates, so that oxides can be formed during thermal decomposition. We can determine from thermoanalytical data whether a hydrated chloride loses its water of crystallization during thermal decomposition and volatilizes as the chloride, or whether it splits off HCl to form the oxide. The reaction route can also be determined experimentally from the vertical flame profiles. The rising flank of the flame profile is an indication of the relative rate of volatilization of the various forms of the analyte. If there is a difference between the chloride and the nitrate, the analyte is volatilized as the chloride, while if there is no difference the chloride is converted to an oxide.

Two examples of this are the elements magnesium and calcium, whose flame profiles are depicted in Figure 8-3. In the absence of concomitants, magnesium chloride decomposes to give magnesium oxide according to:

$$MgCl_2 \cdot H_2O \xrightarrow{\ 581\,°C\ } MgO + 2\,HCl\,. \tag{8.16}$$

The same compound is also formed from the nitrate, so that both salts behave the same in the flame. Calcium chloride (and also strontium chloride), on the other hand, initially loses its water of crystallization to form largely the dehydrated chloride. The flame profiles of the chloride and nitrate of calcium thus differ.

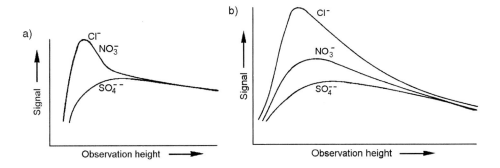

Figure 8-3. Changes in the absorbance with the observation height in an air-acetylene flame for the chlorides, nitrates, and sulfates of **a** – magnesium and **b** – calcium (according to [4991]).

All elements that form stable oxides (i.e. the elements that must be determined in a nitrous oxide-acetylene flame) also form oxides from hydrochloric acid solution. The oxides of the elements in Groups III through V of the periodic table, in particular, are very difficult to volatilize because they form large, three-dimensional, polymer structures [706, 707]. The degree of atomization of these elements can be markedly increased if the formation of a metal-oxygen bond is prevented during crystallization of the final form of the analyte [5108, 5618].

As well as the formation of oxides, the formation of carbides must be considered in both the air-acetylene [3697] and the nitrous oxide-acetylene [4989] flames. The reduction of the analyte species begins simultaneously with the volatilization of the dry aerosol particles in the flame. If the reduced form of the analyte is far less volatile, these two competing processes determine which proportion of the analyte is volatilized and then atomized and which proportion is reduced to the non-volatile form. These competitive reactions can also be influenced by concomitants, so that numerous interferences are possible. High temperature reactions, such as the oxidation of metals or carbides by refractory oxides of other elements, are relatively common. Since the thermodynamic stability of the carbides of Groups IV through VI of the periodic system, and also within a group, decreases from heavier toward lighter elements, certain trends can be predicted.

RUBEŠKA [4990] developed a general scheme for interferences of this sort in the nitrous oxide-acetylene flame based on the volatility of oxide-metal-carbide (O-M-C) of these elements. In this way, three groups of elements can be distinguished. The first group exhibits increasing volatility in the sequence O-M-C (the carbide is the most volatile compound); Be, Al, Sc, Y, La, and several of the rare earth elements belong to this group. The second group exhibits decreasing O-M-C volatility, in other words the opposite behavior; B, Mo, Si, Ti, V, and W belong to this group. The third group includes elements in which all forms are refractory, such as Hf, Nb, Ta, Th, and Zr.

The elements of the first group generally only exhibit limited interferences. The dissociation energies of the gaseous dicarbides are always less than those of the corresponding oxides, so that dissociation is facilitated by reduction. This is confirmed by the influence of concomitant elements described in the literature.

As to be expected, the situation is totally different for the second group of elements: molybdenum and tungsten, where the boiling points of the oxide and the carbide differ by several thousand degrees, exhibit the greatest interferences, which can even lead to total signal suppression. Silicon, on the other hand, where the boiling point of the carbide is only 600 °C above that of the oxide, exhibits relatively few interferences. For most of the elements of this group the interferences can be eliminated by adding an excess of aluminium; this has been reported for titanium [3578, 6201] and vanadium [5032]. Scandium, yttrium, and lanthanum are said to have the same positive effect as aluminium [5032].

The behavior of elements in the third group is not controlled by oxidation or reduction reactions since the oxidic and carbidic species are equally refractory. In this case the reactions that take place during crystallization from aqueous solution determine in which form the analyte is present after thermal conversion. Oxygen forms large, three-dimensional, polyatomic structures with these elements that can comprise the entire aerosol particle [707, 5109]. The vaporization of the analyte from these particles is far more difficult than from monomeric forms. A typical reagent that maintains the element in the monomeric form is hydrofluoric acid, since it forms relatively volatile oxyfluorides and thus eliminates numerous interferences [706, 708, 4638].

8.1.2.4.3 *The Alkaline-earth Elements*

As already mentioned, many elements form an oxide at the end of the drying process, irrespective of the anions originally present in the solution. At the relatively high temperatures in the flame, oxides can react with one another and form phases of even higher thermal stability. The volatility of the analyte is thereby further reduced. In particular, spinel types ($MO \cdot M_2O_3$), or ilmenite and perowskite types ($MO \cdot MO_2$), form very stable lattices that do not melt at the temperature of an air-acetylene flame. Interferences by aluminium, phosphate, or silicon on the determination of the alkaline-earth elements in the air-acetylene flame are caused by such reactions. References to this subject in the

literature, especially for calcium, are overwhelming. For this reason we shall only refer
to selected papers.

We can distinguish two types of interferences, corresponding to the volatilities of the
oxides of the interferents. If the oxide is volatile, such as is the case with P_2O_5, even
fairly large particles volatilize relatively quickly, and the influence of the interferent
decreases the signal with increasing addition just to the point that corresponds to a stoichiometric compound of the analyte and the interferent. Beyond this point there is no
further signal suppression. If, on the other hand, the interfering oxide is refractory, such
as Al_2O_3, an excess delays the volatilization of the analyte even further and thus suppresses the analyte signal [128]. As shown in Figure 8-4 on the example of calcium-
aluminium, the interference nevertheless increases slowly, so that by extrapolating from
a point of inflection it is possible to determine the stoichiometry of the thermostable
phase. In the example shown the extrapolated values are close to the mole ratio
Al:Ca = 2:1, which corresponds to the mixed oxide $CaAl_2O_4$ (= CaO · Al_2O_3). RUBEŠKA
and MOLDAN [4984] identified the analogous magnesium compound $MgAl_2O_4$ by X-ray
diffraction analysis after collecting the non-volatilized particles from the flame.

Releasers are mostly used to eliminate these interferences; lanthanum is by far the
most widely applied reagent. According to the generally accepted view on the action of
releasers they preferentially form a thermally stable phase with the interferent, thus allowing the analyte to volatilize without hindrance. In 1987 KÁNTOR [3025] was the first
to express doubts about this explanation since he observed that only the chloride, but not
the nitrate, of lanthanum or strontium was effective as a releaser in eliminating the

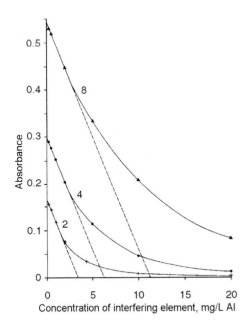

Figure 8-4. Influence of aluminium nitrate on the absorbance signals of 2, 4 and 8 mg/L Ca in an
air-acetylene flame (from [6271]).

interference of aluminium and phosphate on the determination of calcium. Lanthanum-aluminate, which according to the supposed explanation is the compound that prevents the interference by aluminium, is thus excluded as the active component. KÁNTOR [3025] explained the action of the releaser in that excess lanthanum chloride enhances the volatilization of calcium chloride at relatively low temperature, thus preventing the formation of calcium aluminate. Indications in the literature that the interference by aluminium chloride, for example, was less pronounced than that by aluminium nitrate [4984] received little attention. Reports were published on interference curves that first passed through a minimum and then a maximum with increasing aluminium chloride concentration [648, 3690]. This behavior was systematically investigated by LÜCKE [3646, 3647, 3649]. Subsequently WELZ and LÜCKE [6271] were able to explain the observed phenomena by taking both chemical reactions and the particle size into account.

The difference in the interfering behavior of aluminium chloride and aluminium nitrate is shown in Figure 8-5 (logarithmic scale). With aluminium nitrate the signal for calcium approaches zero asymptotically with increasing aluminium concentration, as depicted in Figure 8-4. Aluminium chloride causes a less pronounced interference which reaches a maximum and then decreases again with increasing aluminium concentration.

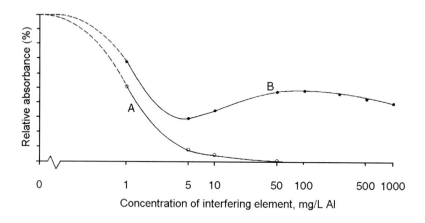

Figure 8-5. The influence of aluminium on the relative absorbance signal of 1 mg/L Ca. **A** – aluminium nitrate; **B** – aluminium chloride (from [6271]).

The interpretation of the interference on the atomization of calcium in the presence of aluminium was brought about by making the following deliberations: the actual cause for the occurrence of the interference is the formation of the refractory mixed oxide $CaAl_2O_4$ which is formed from the oxides of calcium and aluminium. Calcium chloride crystallizes from aqueous solution as the hexahydrate and preferentially reacts to give the anhydrous chloride according to:

$$CaCl_2 \cdot 6H_2O \rightarrow CaCl_2 + 6H_2O \,. \tag{8.17}$$

Additionally, hydrolysis is also possible to a limited extent in analogy to magnesium (equation 8.16):

$$CaCl_2 \cdot 6H_2O \rightarrow CaO + 2HCl + 5H_2O. \tag{8.18}$$

The level to which these reactions take place depends to a very large degree on the 'environment' of the calcium and thus also on the concomitants. In the presence of an excess of aluminium oxide, which is very rapidly formed from the nitrate, reaction 8.18 evidently takes place more readily, so that calcium is quantitatively bound as the mixed oxide.

Aluminium oxide is also formed from aluminium chloride in a two-stage reaction:

$$\left[Al(H_2O)_6\right]Cl_3 \rightarrow \left[Al(OH)_3(H_2O)_3\right] + 3HCl \quad \text{and} \tag{8.19}$$

$$2\left[Al(OH)_3(H_2O)_3\right] \rightarrow Al_2O_3 + 9H_2O. \tag{8.20}$$

HCl is released in the first phase so that reaction 8.18 is suppressed and the formation of anhydrous $CaCl_2$ is favored; this is easily dissociated into atoms. The rise in the curve in Figure 8-5, i.e. the decrease in the interference with increasing aluminium concentration, cannot be explained by a shift in the equilibrium alone. Here it is necessary to take the particle size into consideration.

The influence of the droplet size, and hence the particle size, on the extent of the interferences has been investigated by a large number of working groups [2340, 3309, 4405, 5464]. It could be shown that many interferences, such as the interference of phosphate on the determination of calcium, could be completely eliminated by rigorously limiting the droplet volume [4405, 5464].

For the interference by aluminium the increase in the particle size with increasing aluminium concentration does not play a role up to a concentration of about 5 mg/L Al; this can be deduced from Figures 8-4 and 8-5. However, at higher concentrations the effect becomes increasingly noticeable. In the case of aluminium nitrate a branched Al_2O_3 lattice is formed that increasingly hinders and then finally prevents the volatilization of calcium. With aluminium chloride, HCl gas is formed inside the particle according to reaction 8.19 so that the formation of anhydrous $CaCl_2$ is favored (reaction 8.17). As the particle size increases the influence of oxidizing flame gases on the analyte becomes less, so that the atomization of calcium is enhanced with increasing particle size. Nevertheless, if the particles become too large it is increasingly difficult for $CaCl_2$ to diffuse to the surface and be atomized before it reaches the absorption volume. This explains the maximum in curve B in Figure 8-5.

This effect of the particle size and the reactions taking place within the particle were supported by two further experiments by WELZ and LÜCKE [6271]. First, cesium as the chloride was added in the same concentration of 2.5 g/L to all solutions, and then the flow spoiler was removed from the spray chamber. These two measures were aimed at increasing the particle size, even if by different means. The results of the two experiments are shown in Figure 8-6.

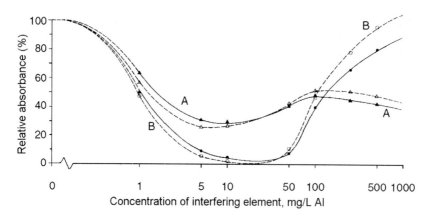

Figure 8-6. Influence of aluminium chloride on the relative absorbance signal of 1 mg/L Ca. **A** – without cesium; **B** – with the addition of 2.5 mg/L Cs as the chloride. Solid lines with flow spoiler, broken lines without flow spoiler (from [6271]).

In this case cesium chloride can be regarded as an 'inert' salt that hardly influences the equilibrium reactions between calcium, aluminium, and HCl. The addition of cesium serves primarily to enlarge the particles. Cesium chloride is an easily volatilized salt whose volatilization is probably controlled by heat transfer (equation 8.11), so that its influence should be to cool the particles and thus further enhance the interferences. Exactly this behavior can be seen in Figure 8-6 where in the presence of cesium the interference curve in its falling flank is practically identical to the interference curve for aluminium nitrate (curve A in Figure 8-5), i.e. the effect of chloride in reducing interferences is more or less eliminated. But the higher the quantity of aluminium chloride in the particle (for a constant mass of cesium), the more dominant is the influence of the HCl released by hydrolysis, so that the interference decreases again.

If the mass of aluminium chloride is increased further an additional effect comes into play. In the presence of aluminium alone (curve A in Figure 8-6) the interference increases again slowly because the particles become too large and $CaCl_2$ can no longer reach the surface and volatilize. Cesium chloride has a boiling point of 1290 °C and if it is additionally present it volatilizes and causes the large particles to disintegrate and thus supports the volatilization and atomization of $CaCl_2$. The fact that all of these effects, both the enhancing and the suppressing, are amplified when a flow spoiler is not installed in the spray chamber, i.e. more larger particles reach the flame, is a further indication that this interpretation is correct.

The mechanisms discussed above are also in complete agreement with the interpretation proposed by KÁNTOR [3025] for the action of lanthanum chloride as a releaser. It is interesting to note that not only the higher alkaline-earth elements strontium and barium behave in a similar manner to calcium but also magnesium [6271]. This is interesting because magnesium chloride normally hydrolyses to the oxide according to reaction 8.16 and does not form an anhydrous chloride. Anhydrous $MgCl_2$ can only be formed in an HCl gas atmosphere. The analogous behavior of magnesium shows that the HCl atmosphere inside the particle plays a major role in the stated reactions.

In addition to releasers, various complexing agents and surfactants have also been proposed to eliminate interferences in the determination of the alkaline-earth elements. The action of complexing agents is easy to explain: they prevent the oxidation of the analyte and thus the formation of mixed oxides. The action of surfactants is certainly more complicated and has not yet been fully explained [2738, 4634]. It has been proposed that surfactants enclose the analyte in a micelle and that interfering ions are prevented from entering due to their opposite charge [6445]. However, this explanation is questionable because the time available during the nebulization process is too short to allow the formation of such structures.

Finally we should like to mention the work of SMETS [5448] who clearly demonstrated that many of the above interferences, such as those caused by sulfate and phosphate, are dependent on *the concentration of the analyte*. In the presence of these interferents there is thus *no linear relationship* between the measured absorbance and the concentration of the analyte. These interferences *cannot be eliminated by the analyte addition technique*, at least not completely.

8.1.2.4.4 *The Flame Effect*

A further 'classical' interference is that of iron on the determination of chromium [4089]. A number of authors have attributed this to the formation of chromite ($FeCr_2O_4$) [6447]. However, this can hardly be taken into consideration since both iron oxide and chromium oxide are readily reduced to the metal or the carbide in an air-acetylene flame. These interferences, which are generally only observed in reducing flames, then probably derive from the formation of mixed carbides. The addition of ammonium chloride eliminates this interference since it increases the volatility of chromium [4955]. This can be explained by the form in which chromium crystallizes out of solution. In the presence of large quantities of ammonium chloride, chromium crystallizes out of hydrochloric acid solution as an ammonium complex, which on warming forms chromium chloride with a volatilization point of 1300 °C. The same complex is also formed when potassium dichromate is heated with ammonium chloride. In the absence of ammonium chloride, chromium forms the refractory chromium oxide, which is responsible for the interference.

A large number of other interferences have also been described for chromium, such as those caused by phosphate and molybdenum. These have been explained by the intercalation of the analyte in the lattice of the hetero poly acids [14]. Chromium is also the only element for which a dependence of the absorbance signal on the oxidation state, +3 or +6, has been reported [55, 4089], in addition to the more frequently reported dependence on the anion.

The strong dependence of the above interferences on the stoichiometry of the flame is mentioned in virtually all publications on chromium. The fundamental problem for the determination of chromium is that the sensitivity in a fuel-lean air-acetylene flame is rather poor. For this reason the determination is performed in a fuel-rich flame that is usually optimized for maximum sensitivity rather than minimum interferences—which in this case appears to be an irreconcilable conflict [55, 4089].

POSTA and SZÜCS [4693] showed that this phenomenon is not specific for chromium, so that for this reason we shall discuss it in somewhat more detail. As can be seen from Figure 8-7, numerous elements exhibit similar effects to chromium in fuel-rich flames, namely that the sensitivity depends on the anion, i.e. on the species that is formed on drying of the aerosol droplets. This figure clearly shows that for the three elements depicted there are virtually no anion effects in an almost stoichiometric flame.

However, in fuel-rich flames it is not only the anions bound to the analyte that cause interferences but also to an even greater extent the concomitants; this is indicated in Figure 8-8 for cobalt, chromium, and copper. In a stoichiometric flame neither cobalt nor copper is noticeably influenced by the various salts [4693].

From these observations it is clear that a fuel-rich air-acetylene flame is not particularly suitable for measurements with F AAS. This fact is further confirmed by a number of other papers dealing for example with the influence of organic complexing agents on the determination of cobalt and iron. These interferences were strongly dependent on the flame and disappeared in a stoichiometric flame [4156, 4641]. Much more effective than trying to optimize the air-acetylene flame to avoid interferences is to change to the nitrous oxide-acetylene flame [3811, 4253]; nevertheless losses in analyte sensitivity must be taken into account in a number of cases. The determination of chromium in the nitrous oxide-acetylene flame, for example, is virtually free of interferences. Likewise for the alkaline-earth elements the numerous interferences reported in the air-acetylene flame do not occur in the nitrous oxide-acetylene flame.

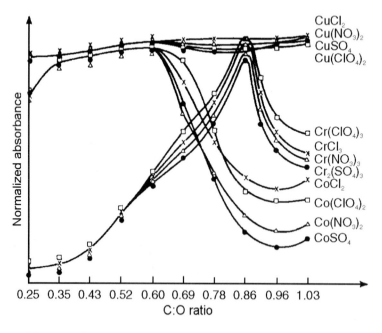

Figure 8-7. Normalized sensitivities for various salts of cobalt, chromium, and copper in an air-acetylene flame of varying C:O ratio. A stoichiometric flame is given when C:O = 0.43. The observation height for all measurements was 5 mm above the inner flame cone (from [4693]).

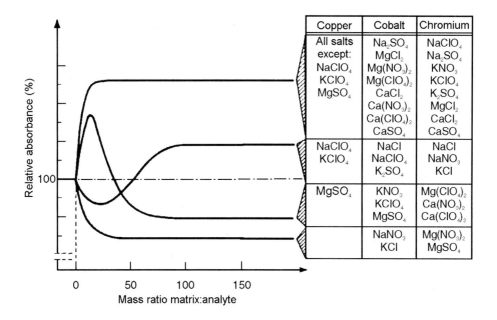

	Copper	Cobalt	Chromium
	All salts except: NaClO$_4$ KClO$_4$ MgSO$_4$	Na$_2$SO$_4$ MgCl$_2$ Mg(NO$_3$)$_2$ Mg(ClO$_4$)$_2$ CaCl$_2$ Ca(NO$_3$)$_2$ Ca(ClO$_4$)$_2$ CaSO$_4$	NaClO$_4$ Na$_2$SO$_4$ KNO$_3$ KClO$_4$ K$_2$SO$_4$ MgCl$_2$ CaCl$_2$ CaSO$_4$
	NaClO$_4$ KClO$_4$	NaCl NaClO$_4$ K$_2$SO$_4$	NaCl NaNO$_3$ KCl
	MgSO$_4$	KNO$_3$ KClO$_4$ MgSO$_4$	Mg(ClO$_4$)$_2$ Ca(NO$_3$)$_2$ Ca(ClO$_4$)$_2$
		NaNO$_3$ KCl	Mg(NO$_3$)$_2$ MgSO$_4$

Figure 8-8. Dependence of the absorbance signal for cobalt, chromium, and copper on the concomitants in a fuel-rich (C:O = 0.86) air-acetylene flame (from [4693]).

8.1.2.5 Vapor-phase Interferences

A flame is a dynamic system, but we can nevertheless assume that the residence time of the sample in the absorption volume is long enough in the first approximation to permit equilibrium between free atoms and compounds (dissociation equilibrium), ions (ionization equilibrium), and excited atoms (excitation equilibrium) to be reached. These equilibria can be described by the law of mass action, the Saha equation, and the Boltzmann distribution law. All the processes are temperature dependent, and since temperature varies within the flame, the concentrations of individual components change with flame height [4842] (see also Section 8.1.2.4, part 2).

As long as AAS remains a relative technique, we cannot consider the incomplete conversion of the analyte into gaseous atoms as a result of the above equilibria to be an interference. Merely the *change in an equilibrium* brought about by concomitants constitutes an interference. A vapor-phase interference presupposes that the analyte and the interferent are *simultaneously* present in the vapor phase and that they have a *common third partner*, such as a common anion or a free electron, which can influence the equilibrium.

Since the excitation equilibrium is shifted completely in the direction of the non-excited state under the normal conditions used for AAS, we do not observe *excitation interferences*. The only possible interferences in the vapor phase are thus dissociation and ionization interferences.

8.1.2.5.1 *Dissociation Interferences*

We can experimentally distinguish whether a concomitant is causing a vapor-phase or a solute-volatilization interference by using a spray chamber with twin nebulizers. The analyte is sprayed through one nebulizer while the interferent is sprayed through the other. If a dissociation interference is present there is no change compared to aspirating the analyte and interferent through the same nebulizer, while a solute-volatilization interference can no longer be observed in this situation.

The process of dissociation is discussed in detail in connection with atomization in Section 8.1.1.2. We should like to mention again that all dissociation equilibria that contain components of the flame gases, e.g. O, OH, CN, or H, cannot be influenced by concomitants. Every change in one of these components is immediately compensated by minimal shifts in the equilibria in the flame gases. This is also true when the component in question is not in equilibrium with the other flame components. Interferences cannot thus be explained by competition for one of the flame gas components, such as oxygen.

On the other hand, dissociation interferences can be expected when equilibria contain components that are not part of the flame gases, such as halides. If the analyte is present as the halide it is usually completely volatilized, but dissociation into atoms does not by any means have to be complete. The existence of non-dissociated halides, especially in cooler flames, has been demonstrated repeatedly through the molecular spectra of mono-halides. Since the dissociation constants of halides are temperature dependent, the strongest dissociation interferences are found in flames of lower temperature. In general we can say that when the acid concentration is > 1 mol/L this determines the anions in the environment around the analyte.

The dissociation energies of the halides increase with decreasing atomic mass of the halogens. The determination of indium and copper are examples in which the increase in the dissociation energies in the sequence HI < HBr < HCl agrees with the level of depression [2010]. Often, though, the interferences caused by halogenic acids are much more complex since they can also influence volatilization (refer to Section 8.1.2.4, part 3).

A further possible influence through concomitants is given when a dissociation equilibrium is attained only relatively slowly. The concomitant can then accelerate the rate at which equilibrium is attained and cause signal enhancement. On the other hand, concomitants can also accelerate competitive reactions, such as the recombination of radicals, and thus cause signal depression. An example for this second possibility is the catalytic effect of a number of elements, such as Ba, Ca, Sr, Cr, and U, and molecules like SO_2 and NO_x on the recombination of H radicals and the thus resulting signal suppression for tin [4160, 4987, 4988].

8.1.2.5.2 *Ionization Interferences*

Thermal ionization is unlikely at the temperatures prevailing in the flames normally used in AAS. On the other hand, ionization occurs in the primary reaction zone due to charge transfer from molecular ions such as $C_2H_3^+$ or H_3O^+ and collisional ionization with relatively long times of relaxation takes place in higher reaction zones. This can lead to a shift in the Saha equilibrium. All ionization reactions involving species produced in the primary reaction zone come to equilibrium slowly, so that the concentration of neutral analyte atoms increases with increasing observation height. Ionization can be increased in the presence of electron acceptors such as cyanides since these capture free electrons [4817]:

$$CN + e^- \rightarrow CN^- . \tag{8.21}$$

In contrast, ions are produced in the primary reaction zone through reactions with the flame gases, such as:

$$CH + O \rightarrow CH^+ + O^- . \tag{8.22}$$

These ions suppress ionization of the analyte element.

The question as to whether noticeable ionization will occur and whether an ionization interference is to be expected depends on the ionization energy of the analyte, on possible interfering concomitants, and on the flame used. In the hot nitrous oxide-acetylene flame ionization normally occurs for elements having an ionization energy of less than 7.5 eV. In flames of lower temperature, ionization is practically limited to the alkali metals. Moreover, the extent of ionization for a given element and temperature is dependent on the concentration of the analyte. As shown in Figure 8-9 on the example of barium, ionization is stronger at lower concentrations than at higher. This results in a concave calibration curve which can be explained by the more rapid recombination of ions and electrons at higher concentrations.

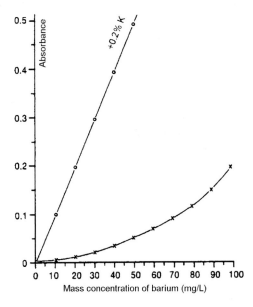

Figure 8-9. Effect caused by ionization in the determination of barium in a nitrous oxide-acetylene flame. The lower curve (x) was obtained for measurement of aqueous solutions, while the upper curve (o) was obtained after addition of 2 g/L K (as KCl) to suppress ionization.

Non-linearity of the calibration curve caused by ionization has two major disadvantages for practical analysis: Firstly, numerous calibration solutions are required to establish the curve. Secondly, the loss of sensitivity in the lower range precludes trace analysis. For these reasons elimination of ionization is very desirable.

In principle there are two ways of suppressing ionization. Firstly, we can determine the analyte in a flame of lower temperature. This is possible for the alkali metals; they are noticeably ionized in the air-acetylene flame, but hardly ionized in the cooler air-hydrogen flame (Table 8-2). This is not practicable for the majority of elements, however, since in cooler flames either they cannot be determined (e.g. rare earth elements) or considerable interferences must be taken into account, e.g. solute-volatilization or dissociation interferences.

The second, and usually only amenable, way to suppress ionization is to add a large excess of a very easily ionized element (usually potassium or cesium) which acts as an ionization buffer by producing a large excess of electrons in the flame:

$$M \rightleftharpoons M^+ + e^-. \tag{8.23}$$

The corresponding increase in the analyte signal can be explained by the fact that the partial pressure of electrons generated by the ionization buffer shifts the ionization equi-

Table 8-2. Ionization of a number of elements in the air-acetylene and nitrous oxide-acetylene flames.

Element	Concentration mg/L	% Ionization air-acetylene	% Ionization nitrous oxide-acetylene
Li	2	0	
Na	2	22	
K	5	30	
Rb	10	47	
Cs	20	85	
Be	2	–	0
Mg	1	0	6
Ca	5	3	43
Sr	5	13	84
Ba	30	–	88
Y	100		25
SEE			35–80
Tm	50		57
Yb	15		20
Lu	1000		48
Ti	50		15
Zr	500		10
Hf	1000		10
V	50		10
U	5000		45
Al	100		10

librium of the analyte in favor of the uncharged atoms [1950]. The ease with which the ionization equilibrium can be shifted by concomitants makes ionization a potentially real interference and is a good reason why it should be suppressed. If a calibration solution is used that contains only the analyte, this is ionized to a degree dependent on its ionization energy and concentration. If the measurement solution contains elements other than the analyte, which is to be expected, and if these are also ionized, the ionization of the analyte will be suppressed, i.e. a relatively higher signal is generated. In this way systematic errors often of considerable magnitude are possible.

Since solute-volatilization interferences and also dissociation interferences decrease with increasing flame temperature but ionization interferences increase, we must often judge which one is the more serious. The choice often falls to the flame of higher temperature since ionization is in general much easier to control. The decision also often depends on the matrix; the choice of the best means of atomization in dependence on the matrix is therefore discussed as appropriate in later chapters.

KORNBLUM and DE GALAN [3252] found that elements with an ionization potential of less than 5.5 eV were more or less completely ionized in the nitrous oxide-acetylene flame. An addition of at least 10 g/L of another easily ionized element is required to permit determinations free of ionization interferences. Elements with an ionization potential of up to about 6.5 eV require the addition of 1–2 g/L of an ionization buffer, while elements with an ionization potential of more than 6.5 eV require only about 0.2 g/L. In determining the necessary quantity of an ionization buffer it is usual to try to reach the plateau of the curve and at the same time to optimize the sensitivity. The quantity determined for matrix-free solutions may in fact be unnecessarily high for real sample solutions since concomitants can also have a buffer action [4326].

8.2 The Graphite Furnace Technique

The discussions in this section are largely restricted to atomizers made of graphite in the form of tubes; we shall refrain from discussing electrothermal atomizers made of other materials (e.g. high melting point metals) and in other forms (e.g. rods). Although graphite tube atomizers are by far the most widely used and have been investigated in more detail, there are still a number of open questions with respect to these systems. However, it is beyond the scope of this monograph to discuss all the theories of atomization and the mechanisms of interferences. Instead we shall attempt to present a simplified representation of the most commonly discussed mechanisms that basically can be applied for a large number of elements.

A major difference between a flame and a heated graphite tube as the atomizer is the composition of their vapor phases. While in flames the hot combustion products of the flame gases determine the chemical environment and the reactions that take place, in a graphite atomizer (this abbreviated name is used in the following) there is an inert gas atmosphere. Together with the reducing action of incandescent graphite, this inert atmosphere should provide ideal conditions for the atomization of the majority of analytes. Strictly speaking, though, this is only valid for the pure analyte. As soon as concomitants are present in excess, after volatilization they also determine the composition of the gas atmosphere and influence reactions. To reduce or eliminate this dominance of the concomitants, which can vary for every test sample, in GF AAS we use the technique of *chemical modification*, which is discussed in more detail in a later section (Section 8.2.4.1, part 1).

A further major distinction between F AAS and GF AAS is the residence time of the atoms in the absorption volume. In F AAS the measurement solution is usually aspirated continuously, vaporized within a few milliseconds, and the analyte atomized. The residence time of the atoms in the absorption volume is of the same order as the time required to reach the absorption volume. In GF AAS, on the other hand, an aliquot of the sample is first freed as far as possible from concomitants before the analyte is atomized. The residence time of the atomic cloud in the absorption volume is 2–3 orders of magnitude longer than in a flame, since the mass transport takes place largely by diffusion. In principle this should lead to greater freedom from interferences [3690], ***provided*** that *atomization takes place under isothermal conditions*. As discussed in detail in Section 4.2, numerous efforts have been made to come as close as possible to the ideal state of isothermal atomization. The results obtained from these efforts were nevertheless very different and the reports on the degree of freedom from interferences were correspondingly varied. In order to present a lucid discussion on interferences, we shall largely restrict ourselves to systems that come relatively close to the concepts of isothermal atomization.

8.2.1 The Atomization Signal

Despite the enormous advances that have been made in GF AAS since its introduction by L'vov at the end of the 1950s, there are still a number of fundamental mechanisms that have not been fully explained. These include the actual process of atomization and the transport of the atoms to and from the absorption volume; in other words, the processes that determine the shape of the atomization signal. Without going into the process of atomization itself, L'vov proposed a simple mathematical model that described the time dependence of the atom population in the absorption volume. The change in the number N of atoms in the absorption volume over the time t is given by:

$$\frac{dN}{dt} = n_1(t) - n_2(t),\tag{8.24}$$

where $n_1(t)$ is the number of atoms entering the absorption volume per time unit and $n_2(t)$ is the number leaving it per time unit.

These two processes have often been treated separately in order to facilitate their description, although in practice they are naturally inseparable. The first attempts to mathematically describe the supply of atoms to the absorption volume were made by L'vov *et al.* [3714] and by PAVERI-FONTANA *et al.* [4583]. The supply of atoms depends on the temperature of the atomizer, the frequency factor, and the activation energy of the reaction that determines the velocity of the transfer of the analyte into free atoms. For a constantly increasing atomization temperature, which can be assumed for the majority of atomizers at least at the beginning of atomization, and $A = dT/dt$, then:

$$n_1(t) = At \qquad (8.25)$$

and

$$\int_0^{\tau_1} n_1(t)\,\mathrm{d}t = N_0, \qquad (8.26)$$

where τ_1 is the time required for the total number of atoms N_0 to be transferred to the absorption volume. Thus:

$$n_1(t) = \frac{2N_0\,t}{\tau_1^2}. \qquad (8.27)$$

If we assume that no atoms are removed from the absorption volume at the start of atomization, the second expression on the right-hand side of equation 8.24 can be ignored. By integration of equation 8.27 we then obtain:

$$N = \frac{N_0}{\tau_1^2}\,t^2 . \qquad (8.28)$$

The start of the absorption signal has been used in various ways to calculate the activation energy; we can assume that the start of the signal is determined by the conversion of the analyte into free atoms and is not affected by their transport from the absorption volume. This aspect is discussed in Section 8.2.2.4, part 1.

The start of the atomization signal could be described very well by simplified mathematical models [90, 91], but the descending flank caused considerable difficulties. In 'open' atomizers, e.g. graphite rods or plates, which are not surrounded by a graphite tube, it was fairly easy to describe the loss of analyte by diffusion [1806, 3711, 5806, 5882], but the agreement for tube atomizers was less good. VAN DEN BROEK and DE GALAN [855] found that in the absence of a forced gas flow the simple diffusion model gave a loss of analyte from the absorption volume that was too rapid compared to experimental observations. They thus concluded that the signal is codetermined by adsorption-desorption processes in the tube.

Much of the theoretical work done on atomization and signal formation was carried out in the early days of GF AAS [1136, 5621, 6562]. Nevertheless, even in more recent work simplifications are assumed that cannot be transferred to normal operation. MATOUŠEK [3947], for example, introduced his test samples on a thin tungsten wire into a preheated tube to obtain quasi isothermal conditions. Also, analyte losses through the sample introduction hole were often not taken sufficiently into consideration [1797, 1811, 1932, 2260, 5621, 5622]. GILMUTDINOV and FISHMAN [2113] developed a model that takes both the spatial and the temporal non-isothermality of tube atomizers into account. Moreover, various diffusion mechanisms, convective gas expansion, the final expansion of the sample, and gas-phase reactions are also considered.

The greatest problem with the publications mentioned so far is that they are all based on atomization from the tube wall and thus not on STPF conditions. FAGIOLI et al. [1797] have pointed out that under STPF conditions the integrated absorbance A_{int} is independent of the rate of formation of the atoms. In addition, apart from diffusion, no other transport mechanisms occur since the analyte is not atomized until the gas expansion is practically complete. Analyte losses through the sample introduction hole are thus substantially reduced. Under these conditions we can assume that atomization takes place under isothermal conditions, at least with respect to time and, within certain limits, also to volume, even if the model is simplified.

According to VAN DEN BROEK and DE GALAN [855] the number of atoms contained in the volume of the atomizer can be described by a convoluted integral:

$$N(t) = \int_0^t S(t') R(t-t') \, dt', \qquad (8.29)$$

where $S(t)$ is the function of atom supply, $R(t)$ is the function of atom removal, and t' is the transition variable. The absorbance measured at any given time t by an atomic absorption spectrometer is practically proportional to $N(t)$ [855]. Of the various equations that can be derived from equation 8.29, two are of particular interest. The first is the equation for the integrated absorbance:

$$A_{int} = \int_0^\infty CN(t) \, dt = CN_0 t_r, \qquad (8.30)$$

where C is a proportionality factor and t_r is the mean residence time of the atoms.

The second equation represents the measurement of $N(t)$ when $S(t)$ is practically zero. In this case:

$$N(t) = N_{t_i} R(t), \qquad (8.31)$$

where N_{ti} is the number of atoms present at time t_i, where $t_i < t$. Assuming that diffusion from the ends of the tube is the only mechanism of analyte loss, equation 8.31 changes to:

$$N(t) = N_{t_i} e^{-k(t-t_i)}, \qquad (8.32)$$

where $k = 8 \, (D\ell^{-2})$, D is the coefficient of diffusion in cm^2/s and ℓ is the length of the graphite tube in cm.

From their data FAGIOLI et al. [1797] concluded that under STPF conditions diffusion is the main mechanism for the removal of atoms. For a number of elements they also found very satisfactory agreement between theory and practice for the dependence of D_T and A_{int} on the temperature.

FRECH and L'VOV [1977] developed an initial model for the loss of analyte through diffusion and convection in a transversely-heated, isothermal atomizer. The convective flow patterns presumed by these authors are depicted schematically in Figure 8-10. By reducing the openings at the tube ends a greatly improved signal form and sensitivity could be obtained, thus confirming the theory; this is depicted in Figure 8-11.

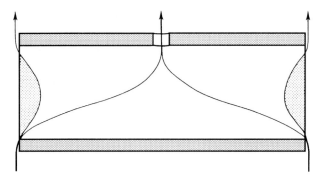

Figure 8-10. Schematic illustration of convective flow patterns in an isothermal atomizer (from [1977]).

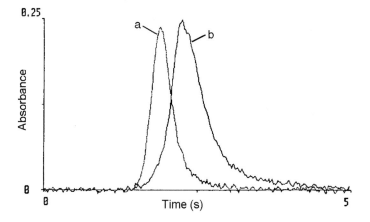

Figure 8-11. Atomization signal from 500 pg cobalt in an isothermal atomizer. **a** – without, **b** – with restriction of the tube ends (by kind permission of Perkin-Elmer).

Most of the models discussed here assume that the analyte atoms are homogeneously distributed over the cross-section of the atomizer at every time point and that merely diffusion along the graphite tube requires a finite time. Nevertheless, GILMUTDINOV and co-workers [2116, 2118, 2119, 2122] have shown repeatedly that this is not the case (refer to Section 8.2.2.1, part 4) and that the uneven spatial distribution of the analyte atoms has a direct influence on the absorption signal. HARNLY and RADZIUK [2392] also found that in a transversely-heated, spatially isothermal atomizer the decrease in sensitivity (integrated absorbance) was much less with increasing atomization temperature than would be expected on the basis of increased diffusion losses. GILMUTDINOV [2120] explained this by a broadening of the absorption profile at higher temperatures and the thus improved overlapping with the emission profile of the radiation source.

8.2.2 Methods to Elucidate Mechanisms

As early as the 1960s, L'VOV [3690] undertook investigations on the mechanisms of atomization in GF AAS, and since the middle of the 1970s various working groups have made intense efforts to explain the mechanisms of atomization and interferences. Nevertheless, up to the present not all analytes and only a few model matrices have been extensively investigated, and the mechanisms are still often the subject of dispute. The reasons for this are the complexity of the systems, the low analyte mass, the high temperatures, and the transient nature of the processes taking place. Pyrolysis and atomization curves have certainly been recorded for all the elements determinable by GF AAS, but in most cases these cannot provide conclusive information on the mechanisms. The main problem of AAS procedures when used to determine reaction mechanisms is that they only allow the measurement of *one* species, namely the free analyte atoms, whereas to explain mechanisms it is absolutely necessary to have a knowledge of the precursors to the atoms and of the intermediate products [5626]. The calculation of thermodynamic equilibria also frequently does not lead to the correct results since a number of important reactions, such as those between oxygen and carbon, are not thermodynamically but kinetically controlled [3716].

In order to elucidate atomization and interference mechanisms it is thus essential to use independent diagnostic techniques. We must also make a distinction between investigations in the condensed phase, for which techniques of surface analysis are mostly used, and investigations in the gas phase. In the former the heating process is normally interrupted at a given time point and, after cooling, the surface of the atomizer is examined. Since measurements of this type also include all the effects that resulted during cooling of the atomizer, such as condensation and reversed reactions on the surface, extrapolation of the events taking place under high temperature conditions is problematical. On the other hand, measurements in the gas phase during heating and atomization only permit suppositions to be made on the participation of condensed species without being able to prove them.

Frequently it is necessary to transport gas phase species from the atomizer to the measuring apparatus in order to measure them, and work must often be performed *in vacuo*. Moreover, we must expect numerous heterogeneous reactions in a graphite furnace for which the procedures employed do not provide any conclusive results. An additional problem is that most of the independent techniques require quantities of analyte that are orders of magnitude above the quantities normally used in GF AAS. Extrapolation to analytical masses is then often not possible or at the least very questionable. Conclusive evidence of the processes taking place in a graphite atomizer can only be provided by a combination of techniques which detect all species simultaneously *in situ* and at high temperature in the condensed phase and the gas phase, using normal analytical masses. Until this is possible, all proposed mechanisms must in principle wait for final confirmation.

8.2.2.1 Atomic and Molecular Spectrometric Techniques

8.2.2.1.1 *Pyrolysis and Atomization Curves*

One of the simplest and longest used techniques was originally proposed by WELZ [6207] and tested for numerous elements; the technique is normally used to optimize the pyrolysis and atomization temperatures, but also permits conclusions to be drawn on the atomization and interference mechanisms. As shown in Figure 8-12, two complementary curves are established in which the integrated absorbance is plotted against the temperature. In the *pyrolysis curve* the integrated absorbance measured for atomization at the optimum temperature is plotted against the pyrolysis temperature as the variable. In the *atomization curve*, on the other hand, at optimum pyrolysis temperature the atomization temperature is varied and the integrated absorbance is measured in each case. The first curve indicates to which temperature we can thermally pretreat an element under the prevailing conditions without losses taking place during pyrolysis. From the first curve we can also determine the temperature at which the analyte vaporizes quantitatively in the selected time interval. From the profile of this curve between these two points we can also occasionally draw conclusions about conversion reactions in the sample. From the second curve it is possible to read off the temperature at which atomization is first evident, the *appearance temperature*, and also the optimum atomization temperature at which the maximum atom cloud density is attained.

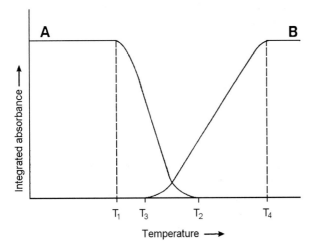

Figure 8-12. Pyrolysis and atomization curves in a graphite furnace. In the pyrolysis curve **A** the integrated absorbance measured for atomization at the optimum atomization temperature T_4 is plotted against the pyrolysis temperature as the variable. T_1 is the maximum temperature to which the analyte can be thermally pretreated in the given matrix without losses. T_2 is the lowest temperature at which the analyte quantitatively vaporizes. The atomization curve **B** shows the integrated absorbance in dependence on the atomization temperature. T_3 is the appearance temperature at which the first atomization signal can be observed and T_4 is the optimum atomization temperature.

If we enter physical data, such as the melting point and decomposition point, of the analyte and its compounds into these experimentally determined curves, it is frequently possible to draw conclusions about the atomization mechanism. This is a purely empirical method, but it has led to some very useful results [6205]. These pyrolysis and atomization curves are further discussed in some detail in Section 8.2.4 in connection with interferences and their elimination.

8.2.2.1.2 *Time-resolved Signals*

The *time-resolved atomization signals* are a further very important source of information in GF AAS and can normally be obtained very easily since most atomic absorption spectrometers display and output them largely free of distortions. If they are regularly evaluated and compared, these signals can be a very useful diagnostic tool, especially for recognizing interferences.

In Figure 8-13 the time-resolved atomization signals for silicon are depicted, without modifier and with a palladium-magnesium nitrate modifier. The atomization signal for the aqueous solution exhibits two maxima, indicating either that two species are present with varying atomization characteristics, or that there are two atomization mechanisms. After chemical modification there is only one maximum, indicating the stabilizing influence of the modifier. Frequently we can observe that when a modifier is added there is a shift of the atomization signal toward higher appearance temperatures.

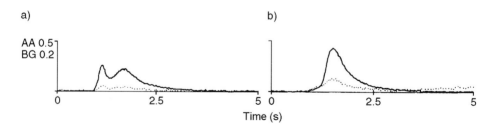

Figure 8-13. Atomization signals for 2 ng Si; atomization temperature 2500 °C. **a** – without modifier; **b** –15 μg Pd and 10 μg Mg(NO$_3$)$_2$ (from [6265]). Continuous signal: atomic absorption (AA); dotted signal: background (BG).

The influence of concomitants and thus the risk of false measurements can often be recognized from the time-resolved atomization signals. The influence of increasing amounts of copper on the atomization signal of selenium is shown in Figure 8-14. The matrix causes a higher appearance temperature, substantial signal broadening, and the formation of two maxima. Under STPF conditions, however, the integrated absorbance remains virtually unchanged despite the markedly changed signal form.

A dramatic change in the atomization signal is depicted in Figure 8-15; this shows the interference of very high masses of potassium sulfate on the determination of selenium. Alkali metal sulfates only become noticeably volatile above 1000 °C, so that even with the application of a modifier they cannot be separated in the pyrolysis step. The matrix is thus vaporized at the start of the atomization step. As a result of the high mass of sulfate the vaporization is quite violent and the analyte is carried more or less quantitatively out of the absorption volume and only incompletely atomized. From the time-resolved atomization signal we can see that evaluation is meaningless.

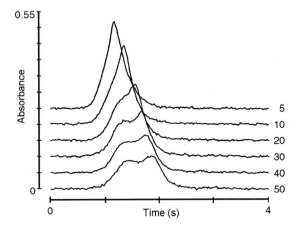

Figure 8-14. Atomization signals for 0.5 ng Se(IV) in the presence of 5–50 μg Cu; pyrolysis temperature: 1000 °C, atomization temperature: 2100 °C (from [6219]).

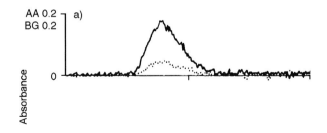

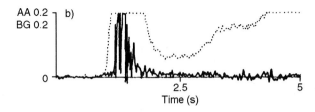

Figure 8-15. Atomization signals for 1.5 ng Se in the presence of 15 μg Pd and 10 μg Mg(NO₃)₂; pyrolysis temperature: 1000 °C, atomization temperature: 2100 °C. **a** –without matrix; **b** – in the presence of 60 μg sulfate as K₂SO₄ (from [6265]). Continuous signal: atomic absorption (AA); dotted signal: background (BG).

The evaluation of time-resolved signals has proven to be especially useful in recognizing spectral interferences. This aspect is discussed in more detail in Section 8.2.4.2. Frequently incorrectable background can be recognized in that the baseline falls to negative absorbance values, as shown in Figure 8-16. Since 'negative absorbance' is impossible, behavior of this nature is a clear indication of an erroneous function, in this case 'over-correction' of the structured background. In this particular case the error only occurs with continuum source BC, but not with Zeeman-effect BC.

Signal diagnostics also play a decisive role in investigations on interferences using a *dual-cavity platform*. This type of platform with two separate cavities was originally proposed by WELZ *et al.* [6227, 6240] to distinguish between interactions in the condensed phase and those in the gas phase (see Section 8.2.4.3, part 1). Based on the time-resolved atomization signals it was also possible to detect a whole series of heterogeneous reactions, such as the 'wandering' of antimony and lead from one cavity to the other. As shown in Figure 8-17, after pretreatment at 400 °C antimony was located almost completely in the cavity which contained nickel and was thus stabilized. Although the dual-cavity platform only allows qualitative interpretations to be made on the processes taking place in the graphite furnace, it has also been successfully applied by other working groups [95–97, 1457, 1589, 1590, 1969, 5791, 6421]. Similar 'wanderings' during pyrolysis have also been observed by CHEN and JACKSON [1229] for lead, selenium, and thallium; they dispensed the analyte and the modifier at different sides of the platform.

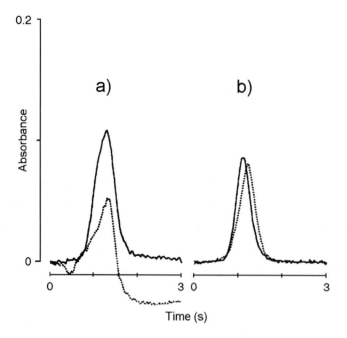

Figure 8-16. Determination of selenium in serum. **a** – continuum source BC; **b** – Zeeman-effect BC. Continuous signal: aqueous calibration solution; dotted signal: serum sample (from [6217]).

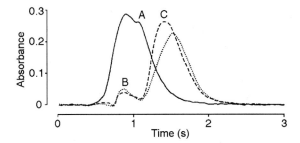

Figure 8-17. Time-resolved absorption signals for 1 ng Sb in aqueous solution; pyrolysis temperature: 400 °C (20 s ramp, 20 s hold). **A** – antimony alone; **B** – antimony mixed with 10 µg Ni as chloride; **C** – antimony and nickel in separate cavities of the dual-cavity platform (from [6227]).

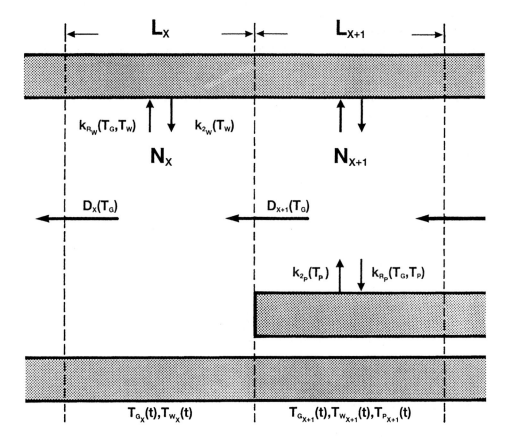

Figure 8-18. Schematic representation of the adsorption, revolatilization, and diffusion of analyte atoms for two segments of a graphite tube at the end of the platform (from [6246]). **x** is the number of the segment, **L** its length, **T** its temperature, **D** the diffusion coefficient, **k** the rate constant, and **N** the number of atoms in the gas phase in each segment. The indices **W**, **P**, and **G** signify wall, platform, and gas.

As well as these purely empirical, but very useful methods of signal diagnostics, WELZ *et al.* [6246] developed a mathematical model for peak interpretation based on a proposal by MUSIL and RUBEŠKA [4257]. In this, attempts are made to draw conclusions from the falling flank of the atomization signal on interactions of gaseous analyte atoms with the tube surface after atomization. These authors divided the tube into a number of segments, each of different temperature, and for every segment calculated processes such as adsorption, revolatilization, and diffusion in dependence on the measured temperature. This principle is depicted schematically in Figure 8-18 for two segments. By combining various tube and platform materials it was shown that the surface from which the analyte was vaporized (the platform) frequently had a far smaller influence on the atomization signal than the tube surface with which the analyte came into contact after atomization.

8.2.2.1.3 *Molecular Absorption*

Important additional information for the determination of reaction mechanisms can often be gathered if molecular absorption is used as a diagnostic tool alongside atomic absorption. Due to the low sensitivity of this procedure, however, it is more suited to investigating interference mechanisms than atomization mechanisms. OHLSSON [4426], for example, scanned the molecular spectrum of AlH and found that with decreasing mass of aluminium the ratio AlH : Al also decreased. He concluded that a model for atomization based on µg masses of aluminium possibly could not be transposed to the ng and pg ranges. This is just as valid for all other diagnostic techniques.

A further problem with molecular absorption is that the high temperature spectra of numerous molecules and radicals are not sufficiently known and described, so that they can be interpreted in different ways. At this point we must merely mention the almost classical discussion as to whether the spectra observed in graphite tubes in the presence of large masses of aluminium and a number of other elements are those of oxides or dicarbides [1371, 3047, 3709, 3712, 3713, 3725, 3727].

Molecular absorption spectra can be most easily scanned using a continuum source such as the deuterium lamp of the background corrector. The procedure is nevertheless time-consuming since the transient signals require numerous single measurements to be made at different wavelengths. This can usually be performed in relatively large steps for the photodissociation spectra of alkali metal and other halides [133]. Electron excitation spectra, such as the spectrum of SO_2 depicted in Figure 8-19 [6261] or of the monohalides of aluminium, gallium, and indium [5926], require a great deal of effort.

Molecular spectra can of course be scanned far more elegantly in simultaneous spectrometers. OHLSSON and FRECH [4425], for example, used a high-resolution Echelle spectrograph to determine possible overlapping of atomic lines with a copper nitrate-ammonium phosphate matrix. MAJIDI and co-workers [3816, 4821] used a laser-induced plasma as the radiation source and a high-resolution spectrograph together with an additional monochromator in order to scan atomic and molecular absorption simultaneously. TITTARELLI and BIFFI [5856, 5858] used a UV spectrometer with a diode-array detector for the same purpose.

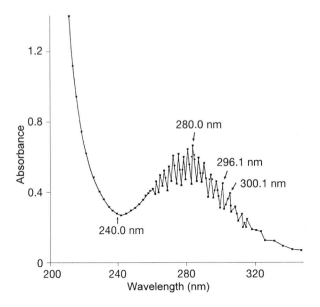

Figure 8-19. Molecular absorption spectrum of SO_2 scanned when 100 µg $MgSO_4$ are volatilized at 2100 °C in a graphite furnace. Every measurement point is the mean of three volatilization and measurement cycles (from [6261]).

8.2.2.1.4 Time-resolved and Spatially-resolved Measurements

SALMON and HOLCOMBE [5063] developed an optical system with which it was possible to resolve the absorbance signal in a graphite atomizer vertically into nine zones and to measure quasi simultaneously. Each zone was 0.3 mm high and 0.2 mm wide. The time- and spatial-resolution of the absorption measurement was achieved with a rotating disk which had 60 evenly spaced radial slits. In this way all nine zones could be measured within 20 ms or 40 ms. An example of the data gathered with this system is shown in Figure 8-20.

HOLCOMBE and co-workers [1614, 2639, 4828, 4829] used this system to investigate a whole series of elements and calculated the rates of atomization from the signal distribution and drew conclusions about interactions of the analyte with graphite. IWAMOTO *et al.* [2838] also performed spatially resolved measurements in different graphite atomizers to elucidate the atomization mechanism for aluminium. For this purpose they measured not only the atomic absorption of aluminium but also the molecular absorption of CN and C_2.

Using a pulsed schlieren system, STAFFORD and HOLCOMBE [5539] were able to observe the analyte distribution over the cross-sectional area of a graphite atomizer during the course of atomization. They used a tunable laser as the radiation source and a miniature camera as the 'detector'. These authors found that at certain time points during atomization the concentration of sodium atoms close to the tube wall was less and con-

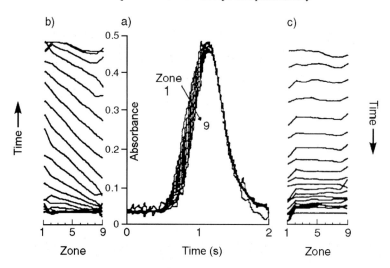

Figure 8-20. Time-resolved and vertically-resolved absorbance signals for 0.2 ng Cr at 40 ms time intervals. **a** – absorbance signals for each of the nine zones along the vertical cross-section; **b** and **c** – absorbance over the vertical cross-section at given time points; **b** – prior to and **c** – after peak maximum, respectively (from [2639]).

cluded that there was a strong interaction between sodium atoms and graphite. HUIE and CURRAN [2712, 2713] also used a laser to investigate the distribution of gaseous species during the vaporization of sodium in a graphite furnace. They used a Vidicon camera as the detector and also found a relatively marked asymmetric distribution of the sodium atoms at the start of atomization and a significantly lower analyte concentration close to the tube wall. However, both groups were only able to 'freeze' momentary events, i.e. they could not make time-resolved measurements and merely investigated atomization from the tube wall. Because a laser was used as the radiation source the measurements were not element-specific.

GILMUTDINOV *et al.* [2116, 2118] introduced the technique of spectral shadow filming (SSF). Initially they used a conventional film camera and were able to directly record the absorption and emission processes with high time resolution. Quantitative estimates could be made about the atom cloud density based on the blackening of the film. The example in Figure 8-21a shows the atomization of gallium from a platform. It is interesting to note in this example that atomization does not start from the platform, but from the opposite tube wall. The authors explained this phenomenon by the fact that gallium is initially volatilized in molecular form—which is confirmed by the molecular absorption shown in Figure 8-21b—and the atomization begins on the hot graphite tube wall and then continues from there. The area in the direct vicinity of the platform remains free of gallium atoms for a longer period; this can be explained by the condensation of gallium on the cooler platform. The fact that no atoms appear under the platform for an extended period signifies that the atom cloud does not expand longitudinally over the platform.

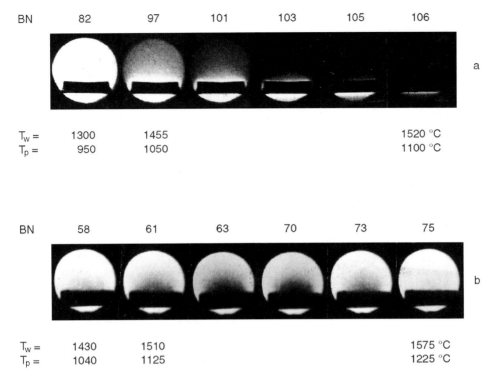

BN	82	97	101	103	105	106	
							a

T_w =   1300   1455   1520 °C
T_p =   950   1050   1100 °C

BN	58	61	63	70	73	75	
							b

T_w =   1430   1510   1575 °C
T_p =   1040   1125   1225 °C

Figure 8-21. Dynamics of the formation of absorbing species during the atomization of gallium from a platform. **a** – atomic absorption at the resonance line at 403.3 nm; 50 ng Ga in HCl (from [2118]). **b** – molecular absorption at 253.7 nm; 1 μg Ga in HCl (from [2119]). In all figures: T_w = wall temperature, T_p = platform temperature, and BN = frame number for a sequence of **a** – 24 and **b** – 12 frames per second.

GILMUTDINOV and co-workers [1156, 2119, 2122] used this technique to investigate the atomization behavior of a large number of elements, e.g. Ag, Al, Ga, In, and Tl. For aluminium in particular the distribution of the atoms was extended by the time-resolved and spatial distribution of molecular species and particles. When using high masses of aluminium, characteristic condensation phenomena were observed; these phenomena had also been described by L'VOV and co-workers [1975, 3743] and were ascribed to an inhomogeneous temperature distribution in the tube.

Subsequently CHAKRABARTI *et al.* [1156] used a CCD camera for signal detection, thus permitting a much more elegant method of working and quantitative evaluation with a computer; the atomization of aluminium from a platform is shown in Figure 8-22 as an example. Atomization begins at positions on the wall of the graphite tube that are the furthest from the sample introduction hole (upper left). Simultaneously we can observe atomization **under** the platform. We can thus conclude that aluminium is first volatilized in molecular form and that these gaseous molecules are then atomized on contact with the hot tube wall. The influence of the sample introduction hole is maintained during the entire atomization process and we may assume that through the entry of atmospheric oxygen aluminium atoms are oxidized. It is also interesting to note that after reaching peak maximum a redistribution of the atom density takes place and that this is highest *under* the platform. This is without doubt the place with the lowest oxygen concentration and thus has the highest reducing effect.

HUGHES *et al.* [2707] used the same technique to investigate the behavior of matrices and modifiers in the gas phase. They observed a temporal and spatial change in the radiation scattering which they ascribed to the formation of condensed particles in the gas phase. The inhomogeneous distribution of these particles was explained by gas flows and temperature gradients in the atomizer.

Using coherent forward scattering, YASUDA and MURAYAMA [6469] were able to measure the *longitudinal distribution of the atom density* in a graphite furnace for a number of elements.

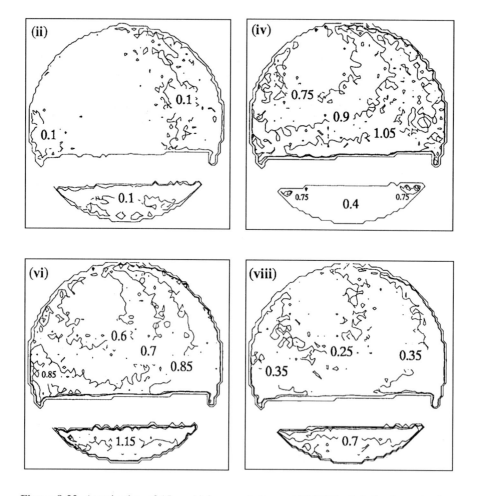

Figure 8-22. Atomization of 15 ng Al from a platform at 2250 °C; digitally determined contour map for the distribution of the absorbance readings over the cross-section of the tube, taken with a CCD camera at 396.1 nm (from [1156]).

8.2.2.2 Techniques for Investigating the Gas Phase

Techniques for performing *in situ* measurements in the graphite tube were discussed in the preceding section. In this section techniques that require *transport of the gaseous species* into an independent detection system are discussed. In this case the atomizer serves merely for the *electrothermal vaporization* (ETV) of the sample. A disadvantage of these techniques is that losses of and changes to the gaseous species can take place during transport. With these techniques it is frequently necessary to use analyte masses that are orders of magnitude higher than the normal analytical quantities, so that the results are not always conclusive.

8.2.2.2.1 *ETV-F AAS and ETV-ICP-MS*

KÁNTOR *et al.* [3023, 3026] transferred substances vaporized in a graphite furnace into a flame and were thereby able to draw conclusions about preatomization losses since species vaporized in molecular form were atomized in the flame. These authors also proposed that the signals obtained by slow heating made a pyrolysis curve superfluous [3026]. The major disadvantage of this procedure is the low sensitivity of F AAS. This problem was largely eliminated by combining ETV with ICP-MS as proposed by BYRNE *et al.* [1006, 1008]. These authors investigated the interference of chloride on the determination of manganese and found that it was in part a vapor-phase interference. GRÉGOIRE *et al.* [2216] draw particular attention to the possibilities offered by the combination of the data from GF AAS and ETV-ICP-MS. A distinction between reactions in the condensed phase and those in the vapor phase which lead to vaporization and atomization can contribute substantially to the elucidation of matrix effects.

8.2.2.2.2 *Mass Spectrometric (MS) Investigations*

Mass spectrometric measurements of gaseous reaction products provide one of the most successful and conclusive methods of investigation of reactions taking place during atomization. Here we must make a distinction between systems in which the atomizer is in a vacuum and systems that operate at atmospheric pressure. In the first case homogeneous and heterogeneous vapor-phase reactions are largely suppressed due to the extremely long mean free pathlengths through which the neutral, gaseous species travel, while in the second case the vapor-phase reactions of GF AAS are retained. In both cases transient signals are recorded which, as usual for GF AAS, are plotted as a function of temperature, as depicted in Figure 8-23.

The first mass spectrometer for performing simultaneous measurements on atomic and molecular species was developed by STYRIS and KAYE [5649]; the species were generated under controlled electrothermal conditions of atomization *in vacuo*. Similar systems were also later used by STURGEON *et al.* [5626], BASS and HOLCOMBE [411], and HAM and MCALLISTER [2341]. STYRIS [2377] also developed the first system in which the graphite furnace was operated at atmospheric pressure and the gaseous species passed via a two-stage pumping system to the mass spectrometer. The mode of function of this system is depicted schematically in Figure 8-24.

We shall not go into further details of the results obtained from MS investigations at this stage since they are discussed in detail in connection with the atomization mechanisms for the individual elements in Section 8.2.3.

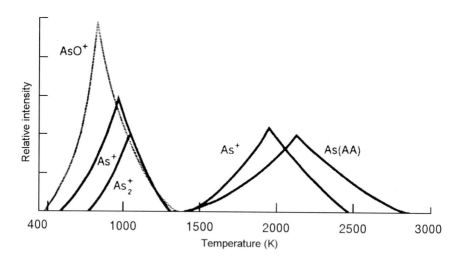

Figure 8-23. Example for data from an MS experiment on the electrothermal vaporization of arsenic nitrate at atmospheric pressure (from [5655]).

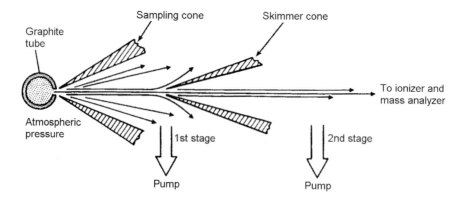

Figure 8-24. Schematic representation of the nested-pair of cones used for sampling from the graphite tube and for the formation of a molecular beam. Only those species having momenta directed along the cone axis enter the skimmer cone and pass into high vacuum (from [5653]).

8.2.2.3 Techniques for Investigating the Condensed Phase

As mentioned at the start of this chapter, investigations on the condensed phase are performed almost without exception *after* the process of interest has been run. The graphite tube is initially heated to the desired temperature, e.g. the drying, pyrolysis, or atomization temperature, and is then examined after cooling. Possible sources of error in this technique include condensation of gaseous products on the surface, reversed reactions in the condensed phase, and changes to the surface, for example due to the action of atmospheric oxygen. Another problem is to be found in the low sensitivity of many surface analysis techniques, thus making the use of large analyte masses necessary. Under conditions of this nature, totally different reactions can take place since the behavior of the analyte with respect to graphite, for example, is completely different to that under analytical conditions.

8.2.2.3.1 *The Use of Radiotracers*

The application of radiotracers for fundamental studies and for determining possible sources of error offers a number of advantages in GF AAS; these are summarized in a review article published by KRIVAN [3300]. The most important characteristics of this technique are the extremely high sensitivity and the very simple detection of the tracers via their radioactive emission, which is independent of their physical state and chemical bonding form. Analyte masses in the femtogram range can be followed very easily through the various steps of a temperature program and can be localized exactly. A further advantage of the tracer technique is that contamination is virtually excluded. With this technique care must merely be taken that the radioactive isotope is added in the same form as the analyte, i.e. it must be representative for the analyte.

Using radiotracers, L'VOV and KHARTSYZOV [3689] established that 20–40% of the analyte atoms can diffuse through the wall of an uncoated EG tube. They found that these losses could be markedly reduced by lining the tube with a tantalum foil. Radiotracers were also used to optimize furnace design [5291] or to investigate the influence of the tube surface on the retention of the analyte [271, 3248].

Analyte losses during drying and pyrolysis [1223, 1525, 3295, 5156], as well as the effect of modifiers [118, 1117, 3299, 3493, 5037, 5156], were investigated most frequently using radiotracers as a complement to pyrolysis curves. The completeness of atomization under different conditions was also investigated [1221]; VEILLON et al. [6011], for example, established that a portion of chromium was irreversibly retained in the atomizer that they were using. For a number of analytes it was also possible to demonstrate differing behavior of the various oxidation states or of organic species. ARPADJAN and KRIVAN [268, 271] found distinct differences between Cr(III) and Cr(VI), for example, which under certain conditions could even be used to separate the two oxidation states. THOMASSEN and co-workers [118, 5037] and CEDERGREN et al. [1117] found that the stabilization of selenium by a number of different modifiers depended very strongly on the chemical bonding form; the greatest differences were found between inorganically and organically bound selenium. Similar behavior was also observed for arsenic [3297].

8.2.2.3.2 *Electron Microscopy*

The most comprehensive investigations on graphite surfaces using *scanning electron microscopy* (SEM) were carried out by ORTNER, WELZ and co-workers [4482, 4483, 6229, 6230, 6251, 6252, 6255]. The results of some of this work are discussed in earlier sections. For example, the difference between an uncoated and a PG-coated surface of an EG tube are shown in Section 1.7.4. The corrosion of TPG tubes by pitting and the delamination of the layers of glassy carbon tubes due to catalytic graphitization, i.e. crystallization of amorphous carbon, could clearly be demonstrated. In Section 4.2.1 it was shown on the basis of SEM photographs that longitudinally-heated graphite tubes break due to local overheating which causes volatilization of the binder material.

A further process that was discovered during these investigations was the recoating of the graphite surface (tubes and platforms) through deposition of carbon from the gas phase. As a result of this permanent recoating (which is shown on two examples in Figure 8-25) the number of active sites on the graphite surface can be drastically influenced. This also explains the reason for the dramatically reduced lifetimes of some tubes caused by strongly oxidizing reagents since these prevent recoating. The effects of intercalation and of heterogeneous reactions could also be demonstrated by means of SEM.

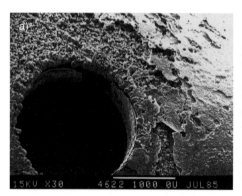 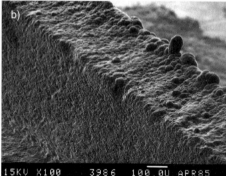

Figure 8-25. a – Recoating of a graphite tube under the destroyed primary pyrolytic coating after 500 atomizations of molybdenum in the presence of 2 g/L Fe(III) as the chloride; **b** – coating of the inner wall of a TPG platform after 1260 atomizations of vanadium (from [6230]).

In addition to purely morphological investigations, SEM has been used increasingly in recent years to determine the distribution of the analyte and modifier in the tube or on the platform in dependence on the experimental conditions. *Energy-dispersive X-ray spectrometry* (EDX) is a complementary technique to SEM that is used to eliminate false interpretations due to artifacts; the combination of both techniques has enabled observed phenomena to be ascribed to a given element. The distribution of the palladium modifier, for example, on the platform in dependence on the concentration [1560], or on other reagents [4743, 6108], or on the matrix [5034], or on the manner of introduction [2685, 4744] has been investigated relatively frequently. The common occurrence of the analyte and modifier, which indicates the formation of a compound or generally the stabilization mechanism, has been the subject of a number of papers [4743, 5789].

Completely new possibilities are offered by *scanning tunneling microscopy* (STM), *transmission electron microscopy* (TEM), and *atomic force microscopy* (AFM) which allow observations to be made at the atomic level [3788]. Using TEM, YASUDA *et al.* [6466–6468] were able to observe the formation of intermetallic compounds between analytes (In, Pb and Sn) and modifiers (Pd, Ni, Mn) at the atomic level during pyrolysis and the release of the analytes during atomization. HABICHT *et al.* [2305] observed topographical changes to the graphite surface after only a few atomization cycles using AFM.

8.2.2.3.3 *X-ray Diffraction Analysis (XRD)*

X-ray diffraction analysis can be used to determine the chemical compound in which an analyte is present in the solid phase. With XRD it is possible to detect conversion products of the analyte during thermal pretreatment, i.e. possible precursors of the atoms, and thus gain important information for the determination of atomization mechanisms. Unfortunately the technique is relatively insensitive and requires masses of the analyte that are several orders of magnitude above those normally used in GF AAS. We cannot automatically assume that compounds formed in the µg range are also formed in the ng or pg ranges since the relationship to other reaction parameters, such as the partial pressure of oxygen or the number of active sites on the surface of the graphite, are changed by orders of magnitude.

Nevertheless, despite these limitations, XRD analyses are useful in the determination of reaction mechanisms. The results of such analyses are discussed in detail in connection with the atomization mechanisms for the individual elements (refer to Section 8.2.3). In particular, MÜLLER-VOGT, WENDL and co-workers [2317, 3223, 4228, 6273, 6274, 6276, 6277] have performed a whole series of investigations in this area and thus made a substantial contribution to an understanding of the reaction mechanisms. The most frequently investigated elements include molybdenum [1146, 4228, 5476, 6273, 6393], tin [2317, 5048, 6272, 6273], germanium [3223, 6273, 6421], and vanadium [3948, 6273, 6274], as well as a number of other, mostly refractory, analytes.

8.2.2.3.4 *Miscellaneous Techniques of Surface Analysis*

Electron spectroscopy for chemical analysis (ESCA; sometimes referred to as X-ray photoelectron spectroscopy, XPS) with a penetration depth of 2–4 nm is a typical technique of surface analysis that has been applied to clarify reactions taking place during the pyrolysis step. For this a test sample is heated on a platform to the given temperature and then, after cooling, analyzed in high vacuum. EKLUND and HOLCOMBE [1699] applied ESCA as early as 1979 and found that $CuSO_4$ is reduced via CuO to $Cu_{(s,\ell)}$; this mechanism was also verified by WANG *et al.* [6155]. In a similar manner SABBATINI and TESSARI [5030] investigated the conversion of lead oxides, and WANG and DENG [6151] described the reduction of iron. SHAN and WANG [5269] noted indications that in the presence of palladium as modifier, compounds of lead or bismuth with palladium are formed. YANG et al. [6454] also investigated the analyte-palladium system and found that arsenic, lead, and zinc form intermetallic solid solutions in the presence of an excess of palladium.

In contrast to ESCA, which is a purely surface analysis technique, with *Auger electron spectroscopy* (AES) it is possible to generate depth profiles and thus make a distinction between surface effects and changes within a crystal. Using AES, WU et al. [6393] investigated the pyrolysis of molybdenum, and WANG and DENG [6151] researched the processes taking place with iron.

A further technique that provides information not only on the nature of the compounds present on the surface but also on their depth of penetration into the substrate (to several μm) is *Rutherford backscattering spectroscopy* (RBS). This technique was first applied by MAJIDI and co-workers [1727–1729, 3817] to elucidate processes taking place in GF AAS. A major advantage of this technique compared to those mentioned above is the improved sensitivity which allows analytical masses to be used. These authors found that the palladium modifier, as well as several analytes, penetrated the PG coating up to 3 μm, even at ambient temperature. The stabilization of selenium appears to take place via the formation of a stoichiometric compound of selenium, oxygen, and palladium. The stabilization of cadmium and lead with the ammonium hydrogen phosphate modifier takes place via the formation of a phosphate glass.

HASSELL et al. [2423] were the first to apply temperature-programmed, static *secondary ion mass spectrometry* (SIMS) for the investigation of surface chemical reactions. Next to the very high sensitivity, this technique also offers the possibility of following surface reactions at high temperature, i.e. of performing measurements **during** pyrolysis and not until after the substrate has cooled. These authors observed the formation of cadmium oxyphosphorous compounds in the presence of the ammonium hydrogen phosphate modifier, while no such reactions could be observed for silver.

ROHR [4937] and ORTNER et al. [4487, 4488] used dynamic SIMS to perform a depth profile analysis in the range 1–50 μm during investigations on the mechanism of action of the Pd-Mg modifier. Compared to RBS, dynamic SIMS has the advantage of much higher sensitivity and a significantly higher depth of penetration. The lateral resolution of dynamic SIMS is also excellent and is in the micrometer to submicrometer range. In the course of *dispersive X-ray fluorescence analysis* these same authors were able to successfully utilize peak shift and change of profile of X-ray lines for locally-resolved analysis with an *electron beam microprobe* to determine the state of bonding of the palladium modifier in graphite. The intercalation of PdO and Pd^0 with covalent bonding of the palladium with a part of the corresponding cluster on the boundary of the graphite lattice could be demonstrated. ORTNER et al. [4486] have published a good overview on the potential of locally-resolved analysis in GF AAS investigations. MAJIDI et al. [3818] applied *laser-desorption time-of-flight mass spectrometry* (LD-TOF-MS) and also methods of *thermogravimetry* and *calorimetry* to elucidate reactions on the graphite surface during pyrolysis.

8.2.2.4 Derived Procedures and Techniques

Reactions involving solid phases can be classified generally as solid-solid, solid-liquid, solid-gaseous, and catalysis. At least two of the following fundamental mechanisms are involved in each of these reactions: adsorption/desorption, homogeneous or heterogeneous solid-solid interactions, nucleation on the surface or in the mass (e.g. in the graphite substrate), and solid phase diffusion from the surface or from the mass. The reaction velocity is determined by the slowest of the mechanisms involved; this mechanism can depend on parameters such as lattice structure and defects of the most varying types.

The procedures described in the preceding sections can be used to observe either the gaseous species or the reaction products in the condensed phase. These techniques only permit suppositions to be made about reactions between the gas and condensed phases, or they provide material for model calculations.

8.2.2.4.1 *Arrhenius Diagrams*

Adsorption mechanisms can be studied by analyzing the desorption spectra. We could thus reasonably expect that the atomization signals in GF AAS could under suitable conditions provide similar information about the adsorption of the analyte on the surface of the atomizer. For this purpose TESSARI and co-workers [4583, 5806, 5882] developed a procedure to obtain Arrhenius diagrams from absorption signals. These were later investigated in detail by STURGEON *et al.* [5619]. The basis of this procedure is the assumption that at the start of atomization the atoms are transported into the gas phase but that no losses due to diffusion take place (refer to Section 8.2.1). It should thus be possible to calculate the activation energy for the atomization process from the first part of the absorption signal and to obtain indications about the atomization mechanism. In subsequent years this procedure was repeatedly modified, for example by SMETS [5449], CHUNG [1283], MCNALLY and HOLCOMBE [4031], ROJAS and OLIVARES [4939, 4941, 4942], YAN *et al.* [6441], and IMAI *et al.* [2769]; the procedure was even partly automated [447]. This procedure was in fact widely applied in GF AAS to obtain information on the kinetics of atom formation [90, 6590], on reaction and atomization mechanisms [92, 1116, 1155, 4939, 4941, 4942, 5619, 5630, 5633], and on the role of the graphite surface [1153, 4031] or modifier [3998, 4942]. However, even at an early stage FRECH *et al.* [1965] drew attention to the problems that can arise when trying to derive gas-phase reactions from such diagrams, and HOLCOMBE [2642] published evidence about a number of false conclusions. All in all, physical interpretations of activation energies based on Arrhenius diagrams in GF AAS are questionable, except in such cases where desorption processes actually predominate. Even in such simple situations high requirements must be placed on the experiments [413] since the poor temporal and spatial resolution of the absorption measurements can lead to substantial errors [1965, 2115, 4829].

8.2.2.4.2 *High-temperature Equilibria Calculations*

FRECH *et al.* [1962, 1967] applied high-temperature equilibria calculations to predict the mechanisms for the vaporization of 21 elements from a graphite surface in the presence of argon. McALLISTER [4001] later further completed these calculations for a number of elements. STYRIS and REDFIELD [5653] performed thermodynamic equilibria calculations taking into account the adsorbed phase; they assumed that the adsorbed phase was represented by a two-dimensional gas on the adsorbing surface. HOLCOMBE *et al.* [2638] performed studies on homogeneous gas phase reactions which influence the population of free atoms. They came to the conclusion that although *thermal* equilibrium was probably attained in the atomizer, the frequency of collision of the analyte with the interferent is so low that *chemical* equilibrium is not guaranteed. Due to the very low analyte concentration, such reactions can certainly be kinetically controlled, so that calculations on the basis of equilibria can lead to false results.

Nevertheless, thermodynamic equilibria calculations have proved to be very useful, even if they are not always suitable for making predictions about reactions. These calculations were very valuable in comparing various systems with differing effective gas temperatures with respect to the expected gas-phase interferences. These aspects are discussed in more detail in connection with gas-phase interferences (Section 8.2.4.4).

8.2.2.4.3 *Monte Carlo Simulation Techniques*

Monte Carlo techniques investigate chemical systems by observing the random characteristics that are inherent in many chemical and physical processes. They thus frequently offer a much better view of nature with a considerably simplified mathematical approach than is offered by conventional theories [2261]. In principle the Monte Carlo technique can be interpreted as a method to evaluate integrals. The model of the process or the simulation is frequently used simply to draw conclusions about the process by averaging the individual results. In a given sense, this technique is a procedure to evaluate certain types of integral that can be regarded as an expectation value.

HOLCOMBE and co-workers used Monte Carlo techniques to solve a large number of questions with respect to atomization in graphite tubes, e.g. the temperature dependence of the diffusion coefficient [650, 2261], the readsorption of analyte atoms on the graphite surface [650, 2263], or the influence of non-isothermality of electrothermal atomizers [2262, 2263, 2265]. GÜELL and HOLCOMBE [2259, 2260] also attempted to determine the optimum geometric dimensions of the atomizer using this technique and to optimize the S/N ratio [2264]. ROHRER and WEGSCHEIDER [4938] proposed a simplified method with which Monte Carlo simulations can be performed using a normal PC and HISTEN *et al.* [2591] later made available PC-based software for general use.

8.2.3 Atomization Mechanisms

Although research into atomization mechanisms was initiated by L'VOV [3690] in the 1960s, by no means have all elements been investigated up to the present, and the mechanisms are often discussed with some controversy [1371]. Even interpretations of the results obtained by MS have been put into question, not to speak of the results from molecular spectroscopy or even Arrhenius diagrams. We shall make our contribution to this controversial discussion by presenting briefly the most important interpretations and giving references to the literature. The results obtained by independent techniques are largely discussed under the individual elements; the interpretations of MS results are given a certain priority since these appear to be the most reliable.

Comprehensive studies by L'VOV and co-workers [3705, 3708–3710, 3712, 3713] on the 'thermochemistry of gaseous media' were summarized by the authors as follows: 'For a long period of time we have been living in an "oxidizing" world and have some-what got used to the idea that monoxides and hydroxides represent the only obstacles on our way to solving the problem of complete and overall atomization. With the advent of high-temperature reducing flames and graphite furnaces a hope began to dawn that free carbon present in atomizers of these kinds will help to solve this problem. However, this did not happen. Having eliminated the previous obstacles we stumbled upon others. The reducing medium did not remain inert with respect to free atoms and "issued" in place of monoxides and hydroxides their carbon analogs, i.e. dicarbides and monocyanides.' L'vov and co-workers thus drew attention to the long-known similarity between the CN radical and halogen atoms. The dissociation energies of monocyanides lie between those of chlorides and fluorides. Due to the high electronegativity of the CN radical compared to the OH radical the dissociation energies of monocyanides are higher than those of hydroxides.

L'VOV and PELIEVA [3705, 4488] investigated 42 elements and found that 30 of them form monocyanides. The spectra of numerous monocyanides could be clearly identified in a nitrogen atmosphere, while they could not be observed in an argon atmosphere [3708]. L'VOV also expressed the opinion that the spectra of oxides and hydroxides scanned by various authors in graphite furnaces were in reality incorrectly interpreted spectra of monocyanides or halides [3709].

The numerous observations made by L'VOV [3715, 3717, 3718, 3722, 3727, 3728, 3733] during the 1980s led to the proposal of the ROC model which describes the reduction of oxides by carbon. As L'vov mentioned, the reduction of the oxides of a number of elements to the metal (e.g., Co, Cu, Fe, Ge, Ni, Pb, etc.) or to the carbide (e.g., molybdenum and vanadium) by carbon had been accepted at an early stage since this could be confirmed by analysis of the products remaining in the tube or on the platform after pyrolysis, and also by kinetic studies performed during atomization. These same kinetic studies had also indicated for a further series of elements that these elements are atomized via dissociation of their oxides. Nevertheless, these studies did not take the thermo-dynamically favored reduction of oxides by carbon into account, even though this should take place at temperatures 500–800 K below the dissociation of oxides:

$$M_xO_{y(s,\ell)} + yC_{(s)} \rightarrow xM_{(g)} + yCO . \tag{8.33}$$

The main problem with all these theories is that only the end product is visible, but not the process itself.

A major supporting factor for L'vov's ROC model was the observation of 'spikes' when he slowly heated (5–50 K/s) larger masses (> 1 µg) of aluminium oxide, or a number of other metals. For this observation, L'vov [3736] proposed an autocatalytic, two-stage process according to:

$$M_{(g)} + zC_{(s)} \rightarrow MC_{z(g)},$$
(8.34)

$$MC_{z(g)} + \left(z/y\right) M_x O_{y(s,\ell)} \rightarrow \left(1 + xz/y\right) M_{(g)} + zCO.$$
(8.35)

Yet a further support for L'vov's ROC model was the observation of 'carbon blisters' made by WELZ et al. [6252] during their SEM investigations on carbon surfaces; an example is depicted in Figure 8-26. L'vov had already come to the conclusion that the ROC process comes to a stop because the carbides decompose on the cooler tube surfaces and a thin carbon coating is formed. When the temperature is increased further and the threshold value is again attained, the carbon coating is oxidized, resulting in a further cycle of autocatalytic reactions. The observed carbon blisters can remain when large analyte quantities are involved.

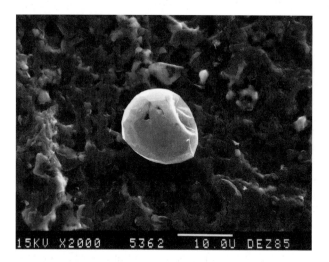

Figure 8-26. Carbon blisters observed in a TPG tube after 550 determinations of phosphorus with lanthanum as modifier.

The theory of autocatalytic reactions was also supported by OHLSSON et al. [4427] who found that in an isothermal atomizer only one 'spike' could be observed. Nevertheless we should take into account that these experiments were performed with extremely high analyte masses and that the heating rates are in no way comparable with those normally used for the atomization of an analyte. Extrapolation to 'analytical conditions' is certainly exceedingly risky. It is thus hardly surprising that numerous working groups have investigated the ROC model and the 'spikes', and in some cases have attacked the theory violently [520, 2118, 2643, 3047]. This discussion, which was not always objective, was taken up again by L'vov in a review article in which he again defended his theory [3749].

On the other hand, L'VOV *et al.* [3738] have expressed serious doubts about the adsorption-desorption model; this aspect is discussed in more detail under the individual elements. Although the authors of this monograph do not wish to become involved in this controversy, the question of adsorption-desorption or condensation-vaporization may in end effect be a question of the analyte concentration. The lower the analyte concentration, the greater is the chance of a monoatomic or submonoatomic layer, so that adsorption-desorption mechanisms will prevail. The higher the analyte mass, on the other hand, the greater is the chance that microcrystallites or microdroplets will form and thus condensation or vaporization.

STYRIS and REDFIELD [5657] also drew to attention the numerous reaction possibilities given by the collision of gaseous species with the surface of the atomizer. The macroscopic interactions between the gas phase and the surface can also influence the diffusion of gaseous species from the surface into the absorption volume. Such interactions can lead to the formation of a *Langmuir film*, i.e. a more dense gas layer above the surface, within which an equilibrium gas pressure can form. This film is formed when equilibrium between the pressure and the gas velocity is attained, and the diffusion of particles out of this film is determined by the concentration gradient. For a given initial composition of the condensed phases and a given number of reactions, the partial pressures and the mass fractions of the condensed phases can be estimated [1967].

8.2.3.1 The Alkali Metals

Currently, of the alkali metals, only sodium and rubidium have been investigated; for sodium merely the analyte distribution in the graphite tube has been measured. For this purpose STAFFORD and HOLCOMBE [5539] used schlieren photographs, while HUIE and CURRAN [2712, 2713] applied a Vidicon camera (refer to Section 8.2.2.1, part 4). Both of these groups found a strong gradient within the graphite tube and a significantly lower atom concentration close to the tube wall. The gradient was much more pronounced in EG tubes than in PG-coated tubes. Both groups ascribed the gradient to a strong interaction of the sodium atoms with graphite. They concluded that sodium is either adsorbed or follows the adsorption/desorption mechanism first proposed by SMETS [5449].

STYRIS and KAYE [5649] investigated the atomization of rubidium by MS, but only in a tantalum atomizer *in vacuo*. Under these conditions free rubidium appeared initially at 670 K, exhibited an initial peak maximum at 1400 K and a second peak maximum at 2100 K. Together with this second peak further signals for $Ta_{(g)}$, $TaO_{(g)}$, and $TaO_{2(g)}$ were measured. These authors concluded that the first signal was generated by the thermal dissociation of $RbO_{2(s)}$ while the second derived from sorbed rubidium. For this element these results also indicate a strong interaction with the atomizer, even though it was made of a totally different material.

8.2.3.2 The Alkaline-earth Elements

Although the elements of this group are determined relatively seldom by GF AAS, their atomization mechanism has been investigated very thoroughly and nowadays can be regarded as largely known. From their absorption measurements, GREGOIRE and CHAKRABARTI [2215] concluded that $Mg(NO_3)_2$ decomposes to $MgO_{(s)}$ and is vaporized as $MgO_{(g)}$. STURGEON et al. [5619] proposed the thermal dissociation of the chlorides or the oxides as the atomization mechanism of the alkaline-earth elements. The relatively poor sensitivity of barium was attributed to the formation of a refractory carbide that was intercalated in the graphite lattice [3700]. From an estimation of the heats of formation it is probable that all of the elements in this group form refractory carbides [3704].

During their MS investigations on the atomization of beryllium nitrate in the gas phase, STYRIS and REDFIELD [5654] observed the monoxide BeO, the dimer and the tetramer, as well as a number of carbides. Both the oxides and the carbides appear in the same temperature interval in which beryllium atoms can be observed. By comparing calculated partial pressures with thermodynamic data, these authors excluded a reduction reaction as the mechanism of atomization. They proposed that the thermal dissociation of the adsorbed monoxide was the most important source of free beryllium atoms. The monoxide is repeatedly replenished by the thermal dissociation of higher oxides.

The formation of gaseous beryllium carbide is explained by the following heterogeneous reaction:

$$xBeO_{(g)} + (x+y)C_{(s)} \rightarrow Be_xC_{y(g)} + xCO_{(g)}. \tag{8.36}$$

As we shall show, the atomization mechanism for beryllium is different to the other alkaline-earth elements.

When magnesium, calcium, strontium, and barium are vaporized, gaseous carbides are formed not only during atomization but also at markedly lower temperatures [4710, 5651]. Gaseous hydroxides and oxides have also been observed below the atomization temperature for all of these elements. PRELL et al. [4708, 4710] made the requirement that adsorption takes place at two different reactive sites on the graphite surface (refer to Section 4.2.1) to explain why the metal oxides form free atoms at one temperature and gaseous carbides at a lower temperature. Adsorption of the oxides with simultaneous dissociation should take place at both sites (type 1 and type 3). The analyte is volatilized as the gaseous carbide at lower temperature from the type 1 site since presumably the bonding of the carbon on the lattice is weak at this site. The formation of free atoms takes place according to the following equation by desorption at type 3 sites which have stronger bonding of the carbon to the lattice:

$$M_{(ad, type 3)} \rightarrow M_{(g)}. \tag{8.37}$$

For strontium the requirement is made for the formation of gaseous carbides, which appear at the same or a higher temperature than the gaseous analyte atoms, according to the following heterogeneous reaction [4710]:

$$Sr_{(g)} + 2C_{(s)} \rightarrow SrC_{2(g)}. \tag{8.38}$$

For the other alkaline-earth elements, on the other hand, Frank-Condon splitting of the surface states has been proposed as the cause for the formation of high-temperature carbides [4710, 5657]. The oxides that have been observed as the precursors of the atomization of magnesium, calcium, strontium, and barium can be explained by associating adsorption on type 2 sites with subsequent desorption according to:

$$MO_{(ad, type 2)} \rightarrow MO_{(g)}. \tag{8.39}$$

The results of these studies on the alkaline-earth elements emphasize the significance of the graphite surface for atomization. They also support the claims made by other groups that the formation of carbides competes with the formation of gaseous atoms of these elements [3708, 3731, 3784, 4437, 5711].

8.2.3.3 Yttrium and the Rare-earth Elements (REE)

PRELL and STYRIS [4709] investigated the atomization of Y_2O_3 in a PG-coated graphite tube using MS *in vacuo* and at atmospheric pressure. For atomization at atmospheric pressure all gaseous species appear at high temperature: free $Y_{(g)}$ and $YC_{3(g)}$ near to 2700 K; $YO_{(g)}$, $Y_2O_{2(g)}$, and $Y(OH)_{3(g)}$ close to 2400 K. In experiments performed *in vacuo* no free $Y_{(g)}$ was observed, which indicates that the atomization of yttrium takes place via a gas-phase reaction. The fact that $YC_{3(g)}$ appears both at atmospheric pressure and *in vacuo* excludes the participation of the carbide in the atomization.

During the volatilization of $Y_2O_{3(s)}$ it has been proposed that the monoxide is formed according to:

$$Y_2O_{3(s)} \rightarrow 2YO_{(g)} + O_{(g)}. \tag{8.40}$$

$Y_2O_{2(g)}$ only appeared in experiments performed at atmospheric pressure. This species must thus be formed from $YO_{(g)}$ in the Langmuir film. It has been proposed that the oxides are dissociatively chemisorbed due to the numerous collisions with the graphite surface. Free $Y_{(g)}$ is then formed when adsorbed yttrium is desorbed at temperatures that are sufficient to overcome the corresponding adsorption energies.

This mechanism is similar to the one proposed by WAHAB and CHAKRABARTI [6128] in which the authors investigated the influence of the pyrolysis temperature on the appearance temperature in tantalum and tungsten atomizers. They presumed that the decomposition of $YO_{(g)}$, which is formed by vaporization from $Y_2O_{3(s)}$, is responsible for the atomization of yttrium.

During XRD investigations on a number of REEs, WENDL et al. [6275] always found the corresponding oxides, but never the metals. In analogy to similar elements (e.g. aluminium), they assumed that atomization is based on a thermal dissociation of the oxide. The very poor sensitivity of a number of the REEs (e.g. dysprosium, yttrium) can be explained by the high thermal stability of the respective oxides.

L'VOV and co-workers [3703, 3729, 3731, 3737] found that the REEs can be determined much better when they are atomized from a tantalum platform. They attributed this to interactions between the analyte and graphite in the condensed phase. They found only a very slight interaction with the graphite tube in the gas phase, so that lining the tube with tantalum brought no further improvements.

8.2.3.4 Vanadium, Chromium, and Molybdenum

Although these three elements are not in the same group of the periodic table, they are treated here together since they exhibit very similar behavior. STYRIS and KAYE [5650] investigated the atomization of V_2O_5 in tubes made of tantalum and glassy carbon using vacuum MS. The gas-phase oxides were observed in both types of atomizer; $VO_{(g)}$ and $VO_{2(g)}$ in glassy carbon, while $VO_{(g)}$ only was observed in tantalum. In both atomizers the molecular species appeared at the same time as free $V_{(g)}$; the appearance temperature was 100 K lower in glassy carbon than in tantalum. Since free $V_{(g)}$ appeared at a markedly lower temperature than the melting point of V_2O_5 (943 K), these authors concluded that V_2O_5 is either reduced to a species of higher melting point or is dissociated. The species observed, $VO_{(g)}$ and $VO_{2(g)}$, must therefore either be reaction products of the first generation, or they must derive from the reduction or dissociation of another oxide in the condensed phase. For a number of reasons the authors ascribed the gaseous oxides to the dissociation of the reduced oxide $V_2O_{3(s)}$.

This hypothetical mechanism for atomization *in vacuo* is distinctly different from the mechanism proposed by STURGEON *et al.* [5619] based on activation energies for atomization in a PG-coated tube at atmospheric pressure. While the MS data make the sublimation of $V_{(s)}$ as the most likely source for $V_{(g)}$, STURGEON *et al.* [5619] proposed the thermal decomposition of $VO_{(g)}$ which is formed by the decomposition of $V_2O_{5(s)}$. Although neither of these mechanisms has been verified, we would draw to attention HOLCOMBE's [2642] doubts about the application of Arrehenius diagrams for the derivation of gas-phase reactions. A clear argument in favor of the sublimation theory is that the appearance temperature of $V_{(g)}$ at atmospheric pressure should be close to the melting point of $V_{(s)}$, which agrees well with the appearance temperature of 2200 K found by STURGEON *et al.* [5619].

WENDL and MÜLLER-VOGT [6273, 6274] observed both V_2O_3 and VC from XRD investigations at 1300 K, but only the carbide VC above 1500 K. They thus proposed the thermal decomposition of the carbide as the atomization mechanism according to:

$$V_2O_{3(s)} \xrightarrow{\ 1300\,K\ } V_2C_{(s)} \xrightarrow{\ 1500\,K\ } VC_{(s)} \xrightarrow{\ 2000\,K\ } V_{(g)} + C_{(s)} \qquad (8.41)$$

These authors presumed that the relatively high heating rate used by STYRIS and KAYE [5650] in their investigations hindered the complete reduction of vanadium to the carbide and thus explained the appearance of the oxide species VO and VO_2.

WENDL and MÜLLER-VOGT [6273, 6274] proposed a very similar mechanism for *chromium*. Using XRD analysis they observed solely Cr_2O_3 at 1300 K. At 1600 K after 30 s pyrolysis they observed a mixture of Cr_2O_3 and Cr_3C_2, but after 3 minutes pyrolysis only Cr_3C_2. They thus proposed the thermal decomposition of the carbide, which is unstable at higher temperatures, as the atomization mechanism according to:

$$Cr_2O_{3(s)} \xrightarrow{\ 1500\,K\ } Cr_3C_{2(s)} \xrightarrow{\ 1700-1800\,K\ } Cr_{(g)} + C_{(s)}. \qquad (8.42)$$

STURGEON *et al.* [5619] and GENÇ *et al.* [2095] proposed the reduction of chromium by carbon and vaporization of the free metal as the mechanism. However, this mechanism cannot explain why a substantial portion of the chromium is retained irreversibly in the graphite tube [6011]. This retention is explained by CASTILLO et al. [1106, 1108] as being in part due to the formation of higher carbides which only decompose very slowly according to:

$$Cr_2O_{3(s)} + C_{(s)} \rightarrow Cr_3C_{2(s)} + CO_{(g)} \rightarrow Cr_{(g)} + C_{(s)} \qquad (8.43)$$

CHAKRABARTI and co-workers [1146, 6393] made a thorough-going investigation of the atomization of *molybdenum* using XRD, AES, and SEM with EDX. Three oxides are formed during pyrolysis at temperatures < 1500 K: $MoO_{2(s)}$, $MoO_{3(s)}$, and $Mo_4O_{11(s)}$. Provided higher pyrolysis temperatures are not applied, the following possible reactions can lead to atomization:

$$MoO_{2(s)} + 2C_{(s)} \rightarrow Mo_{(s)} + 2CO_{(g)} \qquad (8.44)$$

$$MoO_{2(s)} + 2CO_{(g)} \rightarrow Mo_{(s)} + 2CO_{2(g)} \qquad (8.45)$$

$$Mo_{(s)} \rightarrow Mo_{(g)} \qquad (8.46)$$

$$2\,Mo_{(s)} + C_{(s)} \rightarrow Mo_2C_{(s)}. \qquad (8.47)$$

At temperatures > 1500 K they observed virtually only $Mo_{(s)}$, $MoC_{(s)}$, and $Mo_2C_{(s)}$, and at temperatures > 2000 K all the $Mo_2C_{(s)}$ had converted to $MoC_{(s)}$. They proposed the following reactions:

$$Mo_2C_{(s)} \rightarrow Mo_{(s)} + MoC_{(s)} \qquad (8.48)$$

$$Mo_2C_{(s)} \rightarrow 2\,Mo_{(s)} + C_{(s)} \qquad (8.49)$$

$$MoC_{(s)} \rightarrow Mo_{(s)} + C_{(s)} \qquad (8.50)$$

$$Mo_{(s)} \rightarrow Mo_{(g)} \qquad (8.51)$$

but also

$$MoC_{(s)} \rightarrow MoC_{(g)} \rightarrow Mo_{(g)} + C_{(s)}. \qquad (8.52)$$

The latter reaction is attributed as the cause for molybdenum losses during atomization since a part of the $MoC_{(g)}$ can escape undissociated from the absorption volume. The reaction products and mechanisms described above are in good agreement with the findings of MÜLLER-VOGT et al. [4228] and also WENDL and MÜLLER-VOGT [6273], so we shall not go into these investigations further.

As an especially interesting phenomenon, the pyrolysis curves for molybdenum depicted in Figure 8-27 indicate that analyte losses occur between 1200 K and 1800 K when PG-coated graphite surfaces are involved. CHAKRABARTI and co-workers [1146, 6393] found that on PG molybdenum forms relatively large, well-shaped crystals of MoO_2 during pyrolysis, while on uncoated EG the same substance forms needle-like crystals. Due to the smaller surface area of the larger MoO_2 crystallites, heterogeneous reactions are kinetically hampered so that reduction by $C_{(s)}$ or $CO_{(g)}$ according to equations 8.44 and 8.45 can only proceed slowly. Non-reduced $MoO_{2(s)}$ was thus always detected on PG-coated graphite. $MoO_{2(s)}$ is unstable above 1270 K and decomposes according to:

$$3\,MoO_{2(s)} \rightarrow Mo_{(s)} + 2\,MoO_{3(g)}. \tag{8.53}$$

The gaseous $MoO_{3(g)}$ is driven off, thus accounting for the losses.

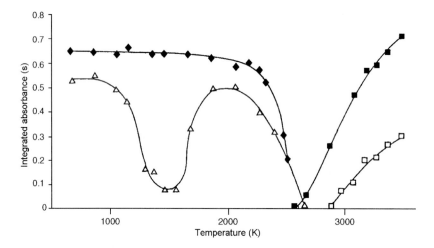

Figure 8-27. Pyrolysis and atomization curves for molybdenum in uncoated and PG-coated EG tubes. ◆ pyrolysis curve for 3 ng Mo in an uncoated EG tube; Δ pyrolysis curve for 1 ng Mo in a PG-coated tube; atomization of 1 ng Mo in a PG-coated tube after pyrolysis at temperatures of ■ 670 K and □ 1400 K.

8.2.3.5 Manganese and the Iron Group

Up to now the atomization mechanism for *manganese* has only been determined from Arrhenius diagrams and equilibria calculations, but not supported by independent measurements. From spatially- and time-resolved absorption measurements, MCNALLY and HOLCOMBE [4031] concluded that manganese is atomized by volatilization from small 'heaps' of a manganese compound on the surface. The mechanism they proposed is in agreement with the results from thermodynamic calculations performed by FRECH et al. [1967] and STURGEON et al. [5619]:

$$Mn_3O_{4(s)} \rightarrow MnO_{(s)} \rightarrow Mn_{(g)}. \tag{8.54}$$

BYRNE et al. [1005] found that the addition of oxygen to the inert gas did not influence the appearance temperature of manganese and they thus excluded a gas-phase equilibrium under participation of $MnO_{(g)}$. These authors also proposed vaporization with dissociation according to equation 8.54 as the atomization mechanism.

AKMAN et al. [92, 93] nevertheless observed that the activation energy calculated from Arrhenius diagrams depended on the experimental conditions such as analyte mass and heating rate. These authors concluded that with high analyte masses and slow heating rates, an equilibrium between $MnO_{(s)}$ and $MnO_{(g)}$ can be established and atomization takes place according to equation 8.54. With low analyte masses and fast heating rates, on the other hand, this equilibrium does not come about, so that $MnO_{(s)}$ is vaporized without being dissociated and atomization is via the gas-phase dissociation of the oxide according to:

$$MnO_{(s)} \rightarrow MnO_{(g)} \rightarrow Mn_{(g)} + O_{(g)}. \tag{8.55}$$

For the atomization of *nickel*, ROJAS [4940] came to analogous results. For higher analyte masses (4–8 ng Ni) she determined a second order reaction and an activation energy corresponding to the reaction:

$$NiO_{(s)} \rightarrow Ni_{(g)}. \tag{8.56}$$

That is to say, vaporization with concurrent dissociation. STURGEON et al. [5619] and BYRNE et al. [1005] had also come to the same result. For lower analyte masses (0.8–3 ng Ni), on the other hand, Rojas determined a first order kinetic and an activation energy corresponding to gas-phase dissociation of the oxide according to:

$$NiO_{(s)} \rightarrow NiO_{(g)} \rightarrow Ni_{(g)} + O_{(g)}. \tag{8.57}$$

This explanation appears very obvious since with decreasing analyte mass the distribution in a submono-layer becomes increasingly likely, while for higher analyte masses small heaps can be more easily formed. These findings naturally bring into line all other results obtained with higher analyte masses.

HAM and MCALLISTER [2341] investigated the atomization mechanism of *cobalt* in a PG-coated tube using MS *in vacuo* and by thermochemical equilibria calculations. Cobalt was introduced into the tube as the nitrate and the chloride and, after drying and pyrolysis, was heated at a rate of 500 K/s. For the nitrate, NO_2 was detected during drying and $Co_{(g)}$ was the only species present during the atomization step. For the chloride there were considerable losses of $CoCl_{2(g)}$ during pyrolysis. During atomization a smaller, but otherwise identical, signal for $Co_{(g)}$ was recorded.

The results of the calculations indicate that $CoO_{(s)}$ is reduced by carbon at 500 K and that $CoCl_{2(s)}$ is reduced by carbon and water at 750 K to give $Co_{(s)}$. If, however, $HCl_{(g)}$ is retained in the tube it can recombine to give $CoCl_{2(g)}$, which dissociates above 1600 K to give $Co_{(g)}$. The $CoO_{(s)}$ is probably formed by the decomposition of the nitrate, and the $Co_{(s)}$, which is produced during this reduction, sublimes at around 1700 K.

Based upon the spatially- and time-resolved atomization signals, MCNALLY and HOLCOMBE [4030] made a continuous adsorption-desorption mechanism a requirement for this element. This is preceded by reduction of the oxide by carbon according to:

$$CoO_{(s)} \xrightarrow{\ \ C\ \ } Co_{(s,\ell)} \rightarrow Co_{(g)}. \qquad (8.58)$$

This atomization mechanism is in agreement with the results obtained by FRECH *et al.* [1967], L'VOV *et al.* [3710], and STURGEON *et al.* [5619]. The existence of a weakly-bound carbide as claimed by L'vov [3704] could be neither confirmed nor refuted by these experiments.

A mechanism analogous to that for cobalt has been proposed as the atomization mechanism for *iron*. WANG and DENG [6151] performed investigations using SEM, XRD, XPS, and AES. They identified $Fe_2O_{3(s)}$, $Fe_3O_{4(s)}$, and $Fe_{(s)}$. The initial appearance of $Fe_{(s)}$ at 1600 K agreed satisfactorily with the appearance temperature of around 1500 K for iron in AAS, taking into account that the XRD measurements were performed with masses of iron that were several orders of magnitude above the analytical masses. Above about 1900 K $Fe_{(s)}$ was the only species that could be detected by XRD. The reduction of the oxide to metal by carbon and atomization via volatilization of the metal as proposed by these authors was in agreement with the findings of CHUNG [1283], L'VOV *et al.* [3710], and STURGEON *et al.* [5619].

8.2.3.6 The Noble Metals

In addition to the platinum metals, copper, silver, and gold are discussed in this section since the atomization mechanisms for all these elements exhibit great similarity. As early as 1979, ROWSTON and OTTAWAY [4974] proposed that the atomization of Ag, Au, Ru, Rh, Pd, Ir, and Pt was via direct vaporization of the metal. L'VOV *et al.* [3710] also found that the experimental results for the heats of atomization were in good agreement with the heats of vaporization of these elements. Other authors confirmed these findings [5400]. Although there is general agreement that atomization takes place from the metal in the condensed phase, there are considerable differences in the views on the actual course of atomization.

The sole element in this group that has been investigated by MS is copper. WANG *et al.* [6156] observed that during volatilization *in vacuo* $CuO_{(g)}$ appears close to 400 K and $Cu_{(g)}$ appears near to 720 K. The simultaneous appearance of NO_2 and O_2 with the CuO indicates the decomposition of $Cu(NO_3)_2$. If the nitrate was heated to 523 K at atmospheric pressure, the signal for the oxide became smaller while the analyte signal increased. They concluded that at normal pressure the gaseous oxide is adsorbed onto the graphite tube. Atomization is then most likely according to:

$$Cu(NO_3)_{2(s)} \rightarrow CuO_{(ad)} \rightarrow Cu_{(ad)} \rightarrow Cu_{(g)}. \qquad (8.59$$

L'VOV *et al.* [3738] expressed doubts as to whether a sufficiently large quantity of oxide could be adsorbed at relatively low temperatures and they thus put the adsorption-desorption mechanism for atomization in question. The activation energies determined from Arrhenius diagrams by a number of different working groups are in part contradictory [1931, 3754, 4030, 6156]; this is generally valid for this whole group of elements.

Based on activation energies, MCNALLY and HOLCOMBE [4030, 4031] suggested that copper and palladium are desorbed from the graphite surface in the form of individual atoms, while gold forms microdroplets which are then volatilized. For silver and copper CHUNG [1283] found that the volatilization of microdroplets was the likely mechanism; albeit for copper only when a sufficiently high pyrolysis temperature was applied. At a lower pyrolysis temperature this author found the dissociation of the dimer $Cu_{2(g)}$ to be the mechanism. AKMAN et al. [93] came to the same results for copper, iridium, and platinum; these authors merely applied a drying step and then atomized directly using a high rate of heating.

These apparent discrepancies were a reason for ROHRER and WEGSCHNEIDER [4938] to investigate the causes of the often varying values for the activation energy of silver. These authors established that virtually all experimental parameters, from the analyte concentration, and thus the covering of the surface, to the heating rate, to the dimensions of the atomizer, had an influence on the values determined. This fact was also ascertained by FONSECA et al. [1933], at least empirically, since they determined that the activation energies for silver and gold are dependent on the mass, and thus the atomization mechanisms also. At higher analyte masses both elements formed microdroplets and atomization was via vaporization. If the analyte mass was reduced, both elements formed 'infinitely small droplets', i.e. individual atoms adsorbed on the graphite surface, and atomization was a pure desorption. From this we can conclude that the varying mechanisms in this entire group can be ascribed to differing analyte masses and experimental conditions.

8.2.3.7 Zinc and Cadmium

L'VOV and YATSENKO [3719] made a carbothermal reduction of the oxide a requirement for the atomization mechanism of zinc and cadmium. STURGEON et al. [5619], on the other hand, were of the opinion that based on the activation energies obtained from Arrhenius diagrams there was more in favor of the dissociation of gaseous oxides. CHUNG [1283] and HULANICKI et al. [2714] also came to the same conclusion. However, no final conclusions can be drawn for these two elements due to the very sparse data available and to the fact that no corroboration through independent techniques is available.

The effect of modifiers has been investigated for both elements. YANG et al. [6454] found that zinc forms a solid solution with excess palladium. HASSELL et al. [2423] and ELOI et al. [1727] observed a reaction between cadmium and a phosphate modifier that was presumably based on the formation of a phosphate glass.

8.2.3.8 Aluminium and Boron

Aluminium is one of those elements for which the atomization mechanism was thoroughly investigated at a very early stage; the older publications are all based on F AAS measurements. Based on the agreement of the experimentally determined activation energy and the dissociation energy of $AlO_{(g)}$ SMETS [5449] proposed the dissociation of this compound as the atomization mechanism. STURGEON et al. [5619] found two activation energies for the atomization and concluded that $AlO_{(g)}$ is formed by the thermal decomposition of $Al_2O_{3(g)}$. From the thermodynamic equilibria calculations performed by FRECH et al. [1967] it was found that in the condensed phase at a partial pressure of oxygen of 10^{-8} Pa the oxide is stable up to 1800 K; at higher oxygen concentrations gaseous aluminium oxide is stable up to 2000 K.

Even at the start of the 1980s, L'VOV and co-workers [3712, 3717, 3718, 3725] reported the explosive atomization of aluminium that manifested itself as 'spikes' when larger masses of aluminium were heated slowly (5–10 K/s) in graphite tubes. They explained this by an autocatalytic reduction of Al_2O_3 by carbon [3733, 3736] according to:

$$2\,Al_{(g)} + 2\,C_{(s)} \rightarrow Al_2C_{2(g)} \tag{8.60}$$

$$2\,Al_2O_{3(s)} + 3\,Al_2C_{2(g)} \rightarrow 10\,Al_{(g)} + 6\,CO_{(g)}. \tag{8.61}$$

This phenomenon of spikes led numerous working groups to perform investigations [520, 2114, 2643] and a very controversial discussion was the outcome [1371, 3047]; STYRIS and REDFIELD [5657] briefly summarized the arguments. Since this phenomenon is only observed with analyte masses that are at least four orders of magnitude greater than for normal analytical work and at heating rates that are more than two orders of magnitude slower, we shall not go into further details. It is very likely that under these extreme conditions other mechanisms are effective than under analytical conditions [4426].

STYRIS and REDFIELD [5653] investigated the atomization of aluminium in a graphite atomizer at atmospheric pressure using MS. They observed various aluminium carbides (Al_2C, Al_2C_2, Al_2C_3, Al_2C_5, Al_4C_5) and suboxides (AlO, AlO_2, Al_2O, Al_2O_2), either at lower temperatures (i.e. prior to atomization) or together with free $Al_{(g)}$. Thermodynamic calculations, which also took the adsorbed species into account, indicated that the appearance temperature of 2100 K for free aluminium atoms was not compatible with a reduction of the oxide in the condensed phase. As a result of these calculations and also due to the strong presence of suboxides, these authors thus made a thermal decomposition of $Al_2O_{3(s)}$ a requirement for the atomization mechanism (this mechanism had also been proposed earlier by STURGEON et al. [5619]). They further concluded that the carbides are formed by reaction of adsorbed aluminium with carbon; this is a reaction that is thermodynamically possible, in contrast to the reduction of suboxides with the formation of carbides.

In their SSF experiments, GILMUTDINOV et al. [2118] observed that aluminium atoms are preferentially formed close to the wall of the graphite atomizer and under the platform (refer to Figure 8-22 in Section 8.2.2.1). This finding is in agreement with the radially resolved AAS measurements performed by MCNALLY and HOLCOMBE [4031]. GILMUTDINOV et al. [2118] were however unable to detect molecular absorption close to the wall of the graphite tube and concluded that gaseous oxides are reduced by graphite.

They thus proposed the following reactions as being involved in the atomization of aluminium:

The thermal dissociation of aluminium oxide in the condensed phase according to:

$$Al_2O_{3(s)} \rightarrow Al_{(g)} , Al_2O_{(g)} , AlO_{(g)} \tag{8.62}$$

follows a homogeneous gas phase oxidation according to:

$$Al_{(g)} + O \rightarrow AlO_{(g)} \tag{8.63}$$

and

$$2\,Al_{(g)} + O \rightarrow Al_2O_{(g)} . \tag{8.64}$$

In the presence of higher analyte contents and sufficient oxygen (in the middle of the tube), aluminium oxide particles can be formed in the gas phase:

$$Al_2O_{(g)} + O_2 \rightarrow Al_2O_{3(s,\ell)}, \tag{8.65}$$

while close to the tube wall the suboxide is reduced to the metal:

$$Al_2O_{(g)} + C_{(s)} \rightarrow 2\,Al_{(g)} + CO_{(g)} . \tag{8.66}$$

The dissociation of aluminium oxide (eq. 8.62) is in agreement with the atomization mechanism proposed on the basis of MS measurements. The proposed reduction of the suboxide (eq. 8.66) has not yet been verified by other experiments. This again demonstrates how important it is to use as many independent techniques as possible to clarify these difficult tasks. LAMOUREUX et al. [3405] combined the diagnostic possibilities of SSF experiments with ETV-ICP-MS and confirmed that the aluminium atoms derive from gaseous precursors of Al_2O.

Magnesium nitrate is mostly used to stabilize aluminium [5421]. In the presence of excess MgO, STYRIS and REDFIELD [5653] observed a shift of the MS profile for $Al_{(g)}$ to higher temperatures; this then appeared in the descending flank of the MS spectra for $Mg_{(g)}$ and $MgO_{(g)}$. They ascribed this to the oxidation of adsorbed aluminium and its suboxides, which are formed by the rapid decomposition of $Al_2O_{(s)}$, by $MgO_{(g)}$ according to:

$$MgO_{(g)} + 2\,Al_{(ad)} \rightarrow Al_2O_{(ad)} + Mg_{(g)} \tag{8.67}$$

$$MgO_{(g)} + Al_2O_{(ad)} \rightarrow Al_2O_{2(ad)} + Mg_{(g)} \tag{8.68}$$

$$MgO_{(g)} + Al_2O_{2(ad)} \rightarrow Al_2O_{3(ad)} + Mg_{(g)}. \tag{8.69}$$

As soon as the concentration of $MgO_{(g)}$ becomes sufficiently low, free $Al_{(g)}$ can be formed at the now higher temperature.

The magnesium nitrate modifier is also able to stabilize aluminium during pyrolysis. In the absence of magnesium an aluminium hydroxide appeared in the gas phase at 1300 K. In the presence of magnesium, $HMgOH_{(g)}$ and $MgOH_{(g)}$ appeared close to 500 K, while the signal for $Al(OH)_{2(g)}$ disappeared. REDFIELD and FRECH [4837] found that in the absence of magnesium, up to 20% of the aluminium can be lost in this form. Magnesium oxide is apparently hydrated at lower temperature by residual H_2O and can thus prevent analyte losses.

The 'established' atomization mechanism for aluminium, which can by no means be seen as proven, was put into question by OHLSSON [4426] since in the course of *in situ* spectroscopic measurements he was able to detect AlH in relatively high concentration but no AlO. He supposed that $AlH_{(g)}$ could be a precursor of $Al_{(g)}$ since both appeared simultaneously at 2160–2200 K and the AlH disappeared before the aluminium. A particularly important observation was that the ratio AlH:Al was clearly dependent on the analyte mass. OHLSSON [4426] pointed out that an atomization mechanism based on µg-masses of aluminium may no longer be valid in the ng or pg range.

Relatively few investigations have been made on the atomization mechanism of *boron*, but the results nevertheless fit to the general picture of the elements in Group III of the periodic table. Using ETV-ICP-MS, BYRNE *et al.* [1008b] were able to show that the major cause for the poor sensitivity was not the formation of refractory carbides, as was frequently assumed, but was due to the preatomization loss of molecular boron compounds which began at around 800 °C. Model thermodynamic calculations by Frech showed that HBO_2 and BO as gaseous species could be responsible for such losses. GOLTZ *et al.* [2160] used the SSF technique with a CCD camera to observe the spatial distribution of atoms and molecular boron species in the gas phase of the graphite furnace. These experiments confirmed the preatomization losses of molecular species; BO was observed first in the upper region of the graphite tube close to the sample introduction hole and not above the location of the sample. The fact that the lowest concentration of BO was always measured at the bottom of the graphite tube was interpreted by these authors as being due to the low concentration of oxygen prevailing in this reducing environment further away from the sample introduction hole. Additionally a gradient with decreasing concentration toward the wall of the graphite tube was observed which indicated a heterogeneous reaction between the gas phase and the hot graphite. Atomic boron was first observed from 2400 °C, while molecular species no longer occurred above 2200 °C. The distribution of boron atoms exhibited far less pronounced gradients than those of molecular species with a somewhat higher concentration toward the wall of the graphite tube.

From these observations we can derive the following mechanism: At temperatures < 1800 °C early volatilization of B_2O_3 starts presumably with the formation of a suboxide:

$$B_2O_{3(s)} \rightarrow B_2O_{3(g)}, BO_{(g)} . \tag{8.70}$$

When the atomizer reaches a temperature of about 1800–2000 °C boron oxide is dissociated by homogeneous gas-phase reactions according to:

$$B_2O_{3(g)} \rightarrow 2\,BO_{(g)} + \tfrac{1}{2}\,O_{2(g)} \tag{8.71}$$

$$2\ HBO_{2(g)} \rightarrow 2\ BO_{(g)} + H_2O_{(g)} + \tfrac{1}{2}\ O_{2(g)}. \tag{8.72}$$

At the atomization temperature of 2400 °C there is only a small portion of boron ($< 5\%$) remaining in the tube, presumably as the oxide, the element, or the carbide. Beside the reduction of the remaining oxide, the desorption of boron from the carbide could also contribute to atomization according to:

$$B_4C_{(s)} \rightarrow 4\ B_{(g)} + C_{(g)}. \tag{8.73}$$

The formation of the carbide and the slow release of boron are further causes for the low analytical sensitivity and the tailing of the signal.

8.2.3.9 Gallium, Indium, Thallium

These three elements exhibit great similarity and have often been investigated together, so it is convenient to treat them together in this section. L'VOV et al. [3725] introduced mg-quantities of gallium or indium as the metal into a graphite atomizer and ascribed the observed molecular absorption to the gaseous carbides $GaC_{2(g)}$ or $InC_{2(g)}$, respectively. Based on semiquantitative estimates, L'VOV and YATSENKO [3723] came to the conclusion that the concentration of $GaC_{2(g)}$ was substantially higher than that of $Ga_{(g)}$, while that of $InC_{2(g)}$ was about the same as $In_{(g)}$. They rated these findings as confirmation of the ROC model for the atomization of these elements [3727].

GILMUTDINOV et al. [2119] introduced µg-quantities of gallium nitrate or indium nitrate into PG-coated graphite tubes with platform and into tungsten-lined graphite tubes and investigated the molecular absorption using their SSF technique. The distribution of the molecules in both tube types was very similar for both elements: the molecular density was highest close to the sample and lowest near to the opposite tube wall. It is thus unlikely that the molecular absorption is caused by a carbide. The MS investigations performed by MCALLISTER [4001] indicate that in all probability it is caused by the oxides $Ga_2O_{(g)}$ or $In_2O_{(g)}$, respectively. In earlier SSF experiments with atomic absorption, GILMUTDINOV et al. [2118] showed that free $Ga_{(g)}$, $In_{(g)}$, or $Tl_{(g)}$ are formed close to the wall in the upper half of the atomizer, precisely where molecular species are not to be found (refer to Figure 8-21 in Section 8.2.2.1). These authors thus concluded that $Ga_2O_{(g)}$ or $In_2O_{(g)}$ are dissociation products of $Ga_2O_{3(s)}$ or $In_2O_{3(s,\ell)}$, respectively, and either dissociate close to the atomizer wall or are reduced there to form free atoms. For investigations with organic modifiers, IMAI et al. [2768] also found that carbon plays an active role in these processes.

The MS investigations performed by MCALLISTER [4001] are in good agreement with the results obtained by GILMUTDINOV et al [2118, 2119]. Reportedly $Ga_2O_{3(s)}$ is reduced in the condensed phase between 800 K and 1000 K to give Ga_2O and $Ga_{(s)}$. The condensed phase vaporizes near to 1200 K, and $Ga_2O_{(s)}$ decomposes above 1500 K to give $Ga_{(g)}$. McAllister drew special attention to possible losses of the analyte in the form of $Ga_2O_{(g)}$. The reduction of $In_2O_{3(s)}$ already takes place at 600–800 K and $In_{(g)}$ should appear at 1100 K. Like gallium, there is the risk of losses of the analyte in the form of $In_2O_{(g)}$.

WENDL et al. [6277] were able to show that losses of gallium in the form of volatile oxides could be substantially reduced by adding oxygen to the protective gas; this observation was also confirmed by IMAI et al. [2767]. HAHN et al. [2318] made thoroughgoing investigations on the influence of oxygen on thallium. They showed that the effects caused by oxygen, such as higher appearance temperature and improved sensitivity, derived from a change in the surface properties of the atomizer. After drying, thallium is present as Tl_2O_3 on the surface. In the absence of oxygen the reduction to $Tl_{(g)}$ is via the volatile suboxide Tl_2O, which is the cause for losses of thallium. If the graphite tube is treated with oxygen, the formation of the volatile suboxide is prevented by a change in the graphite surface. This takes place via the chemisorption of oxygen at active sites, which is then removed at higher temperature. This brings about a shift of the reduction process toward higher temperatures at which thallium can be atomized with lower losses [786].

The stabilizing influence of the palladium modifier on indium was investigated by YASUDA et al. [6468] using SEM during pyrolysis and atomization; they attributed this to the formation of the intermetallic compound $PdIn_3$.

8.2.3.10 Silicon, Germanium, Tin

MÜLLER-VOGT and WENDL [4227] studied the reactions of sodium silicate in the graphite tube. The reduction to $SiO_{(s)}$ and $Si_{(s)}$ starts from 1200 °C, and they observed the formation of $SiC_{(s)}$ above 1700 °C. The authors noticed that at pyrolysis temperatures below 1550 °C the pyrolysis time had a marked influence on the sensitivity of the determination of silicon. They concluded that at lower temperatures a longer period of time was required to reduce the silicate to $Si_{(s)}$.

Building upon this work, RADEMEYER and VERMAAK [4774] attempted to identify the molecular species formed from silicon during atomization. They assigned the observed spectra to the molecules $SiO_{(g)}$ and $SiC_{2(g)}$. The authors suggested that at the start of atomization $SiO_{2(s)}$ decomposes to form $Si_{(g)}$ according to:

$$SiO_{2(s)} \rightarrow Si_{(g)} + O_{2(g)} . \tag{8.74}$$

This mechanism is supported by the observation that the silicon signal is strongly depressed in the presence of oxygen. Depending on the conditions prevailing in the tube, one of the following reactions can occur: If the partial pressure of oxygen in the tube is below a critical value, $Si_{(g)}$ reacts with the carbon of the tube to give $SiC_{2(g)}$, which then reacts with SiO to give $Si_{(g)}$ according to:

$$Si_{(g)} \xrightarrow{\ C_{(s)}\ } SiC_{2(g)} \xrightarrow{\ SiO\ } Si_{(g)} + CO_{(g)} . \tag{8.75}$$

The first step is supported by the observed influence of the tube surface and the second by the fact that CO suppresses the atomization signal for silicon. In the presence of higher oxygen concentrations the following reaction is probable:

$$Si_{(g)} \xrightarrow{O} SiO_{(g)} \xrightarrow{C_{(s)}} Si_{(g)} + CO_{(g)}. \tag{8.76}$$

Using comparative MS investigations on aqueous and solid gold samples, BROWN *et al.* [880] were able to detect different atomization mechanisms for silicon. While the occurrence of Si and SiO in the gas phase could be confirmed, SiC_2 could be found for aqueous samples only.

For the determination of *germanium*, KOLB *et al.* [3223] used XRD among other techniques to investigate the chemical reactions in the graphite tube. The $Na_2GeO_{3(s)}$ formed during the drying step remained unchanged up to 1100 K. At higher temperatures $Ge_{(s)}$ is formed. By means of molecular absorption the authors were able to detect $GeO_{(g)}$ above 1100 K, reaching a maximum at about 1450 K. The addition of NaOH increased the germanium signal by a factor of two. The authors attributed this to a reduction of GeO to $Ge_{(s)}$ by metallic sodium at temperatures above 1500 K.

SOHRIN *et al.* [5489] also found that the main problem for the determination of germanium was the volatility of GeO. As well as the addition of sodium hydroxide, these authors proposed reducing the activity of the graphite surface by an oxidizing acid or by treating with tantalum. XUAN and LI [6421] attempted to stabilize germanium with the nitrates of nickel and iron so that the pyrolysis temperature could be raised and losses of GeO prevented. The authors suspected that the formation of intermetallic compounds was responsible for the stabilizing effect. XUAN [6423] finally employed a mixture of palladium and magnesium nitrates to stabilize germanium. XPS analyses showed that both Ge-Pd and Ge-Mg compounds were formed. Intermetallic compounds and germanates could be detected by XRD. It can be assumed that compounds such as $GePd_2$ dissociate directly to $Ge_{(g)}$, while germanates such as $MgGeO_3$ are reduced to $Ge_{(g)}$ via GeO_2 and GeO.

Similar to germanium, analyte losses for *tin* can be ascribed to the formation of the volatile $SnO_{(g)}$ molecule. RAYSON and HOLCOMBE [4828] found that oxygen, which can find its way into the atomizer, can substantially influence the fraction of atomic tin through the formation of $SnO_{(g)}$. The ability of graphite to reduce oxidates in the gas phase or molecular tin compounds is only given at increased temperatures and close to the wall of the graphite tube. A further problem in the presence of sulfates is the formation of volatile $SnS_{(g)}$.

On the basis of high-temperature equilibria calculations, LUNDBERG *et al.* [3663] showed that tin forms volatile oxides, halides, and sulfides that are stable to relatively high temperatures. Losses of these molecules can easily happen in non-isothermal atomizers, while this interference is less in an isothermal atomizer.

BROWN and STYRIS [879] investigated the atomization of $SnCl_2$ by MS *in vacuo* and at atmospheric pressure. They solely observed $SnO_{(g)}$ and $SnCl_{2(g)}$ *in vacuo* but no free $Sn_{(g)}$. The results were very similar at atmospheric pressure, i.e. only molecular species can be detected by MS but not $Sn_{(g)}$. This leads to the conclusion that atomic tin, which can be detected by AAS, is only formed on the graphite surface. The authors proposed that initially at lower temperature $SnO_{(g)}$ is adsorbed according to:

$$SnO_{(g)} \rightarrow SnO_{(ad)},$$ (8.77)

which is then reduced at increased temperature by $CO_{(g)}$,

$$SnO_{(ad)} + CO_{(g)} \rightarrow Sn_{(ad)} + CO_{2(g)}.$$ (8.78)

A corresponding reaction with the participation of $S_{(s)}$ is also conceivable [4229b]. At even higher temperature the adsorbed species is desorbed with the formation of free tin atoms according to:

$$Sn_{(ad)} \rightarrow Sn_{(g)}.$$ (4.79)

The stabilization of tin through interaction with the graphite surface is higher for uncoated tubes or tubes treated with oxygen than for pyrolytically coated tubes [786, 4229b]. As well as the predicted losses through the formation of SnO, in the presence of sulfate TITTARELLI et al. [5858] also detected SnS by molecular spectroscopy.

8.2.3.11 Lead

During the vaporization of lead nitrate in a graphite furnace at 700 K *in vacuo*, STURGEON et al. [5626] observed a narrow MS signal for PbO and a much broader signal for lead; the appearance temperature for lead was 680 K but the signal maximum was reached later than the PbO signal. The authors presumed that $PbO_{(s)}$ sublimes to give $PbO_{(g)}$, thus accounting for the narrow signal since at higher temperature thermal dissociation of $PbO_{(s)}$ sets in to give $Pb_{(g)}$. The reduction of $PbO_{(s)}$ by carbon to give $Pb_{(g)}$ can also be considered as a further possible mechanism.

BASS and HOLCOMBE [411] investigated the atomization of lead *in vacuo* by heating either a solution of $Pb(NO_3)_2$ or solid PbO on a graphite substrate. For lead nitrate, MS signals for PbO, NO_2, O_2, CO_2, and CO appeared simultaneously at 550 K; free $Pb_{(g)}$ was not observed until a temperature of 200 K higher. For $PbO_{(s)}$, in contrast, no signal for $PbO_{(g)}$ was observed prior to the atomization of $Pb_{(g)}$. This means that $PbO_{(g)}$ is not formed by the sublimation of $PbO_{(s)}$ but by the decomposition of $Pb(NO_3)_2$ with the release of NO_2 and O_2. This reaction does not contribute to the atomization of lead, however, since the signal for $Pb_{(g)}$ appears on its own at a much higher temperature.

The conversion of lead nitrate into the oxide was supported by XPS investigations performed by SABBATINI and TESSARI [5030]. They found that the original three-dimensional aggregate spread over the graphite surface with increasing temperature, leading to a two-dimensional distribution of two different oxide species. Using XRD analysis, WENDL and MÜLLER-VOGT [6276] observed PbO, PbO_2, and Pb_3O_4 up to 500 K, but only PbO above this temperature. LYNCH et al. [3754] also ascribed the observed atomization signals to a two-dimensional distribution of lead on differing graphite materials.

BASS and HOLCOMBE [411] presumed that thermally induced changes in the crystalline nature of the nitrate were responsible for the volatilization of $PbO_{(g)}$. One reason for this supposition was, as STURGEON *et al.* [5626] had noticed, that the ratio $PbO : Pb$ decreased when the concentration of lead was lower in the same volume of measurement solution. This points to an influence of the crystal volume on the volatilization of $PbO_{(g)}$. The absence of the early signal for $PbO_{(g)}$ during the vaporization of $PbO_{(s)}$ is a further indication.

The simultaneous appearance of CO and CO_2 with free $Pb_{(g)}$ indicates a reduction of PbO on the surface as the atomization mechanism according to:

$$PbO_{(ad)} + C_{(s)} \rightarrow Pb_{(g)} + CO_{2(g)} + CO_{(g)}. \tag{8.80}$$

In conventional GF AAS where atomization takes place at atmospheric pressure, $PbO_{(g)}$ either can be adsorbed on the tube wall and reduced to $Pb_{(g)}$, or the oxide is thermally decomposed in the gas phase. This is similar to the mechanism proposed by GENÇ *et al.* [2095] in which the nitrate decomposes and the resulting oxide sublimes and dissociates in the gas phase to free lead.

MS experiments performed at atmospheric pressure to investigate the influence of the pretreatment temperature showed no change in comparison to the spectra measured *in vacuo* provided that the pyrolysis temperature (T_p) remained below the decomposition temperature (T_d) of the nitrate [412]. If $T_d < T_p$, however, the signal for $PbO_{(g)}$ became smaller while that for $Pb_{(g)}$ increased. Under the same conditions *in vacuo* the signal for $PbO_{(g)}$ likewise became smaller, but that for $Pb_{(g)}$ remained constant. This is a clear indication for the readsorption of $PbO_{(g)}$ at atmospheric pressure and atomization according to equation 8.80.

Already in the late 1970s, SALMON and HOLCOMBE [5063] showed that the signal for lead appeared at significantly higher temperatures if 1% O_2 is mixed with the argon purge gas. A similar effect could be observed when the graphite surface was pretreated with oxygen at 1000 K and then atomization took place in pure argon. L'VOV and RYABCHUCK [3716] proposed a heterogeneous reaction between O_2 and the condensed metal as the mechanism for this shift. On the other hand, BYRNE *et al.* [1006] presumed that a homogeneous gas-phase reaction between oxygen and the free analyte was the cause. Interestingly, BASS and HOLCOMBE [412] found no difference in the mass spectra from a graphite platform pretreated with O_2 and an untreated platform when they worked *in vacuo*. From this we can conclude that the temperature shift is based on an interaction with the gas phase. BASS and HOLCOMBE found an increase in the MS signals for CO and CO_2 when the surface was pretreated with O_2. MÜLLER-VOGT and co-workers [786, 2318, 4229] were able to show that through the treatment with oxygen mainly the graphite surface was changed. The most probable mechanism is thus an increased readsorption of $PbO_{(g)}$ on the wall of the graphite tube and stronger binding of the reduced analyte on the active sites on the surface, and thereby a delayed release of $Pb_{(g)}$.

BASS and HOLCOMBE [411] also investigated the stabilization of lead by phosphate. In this case they found no signal for $PbO_{(g)}$, and free $Pb_{(g)}$ appeared at 1150 K together with signals for PO and PO_2. From this coincidence the authors concluded that a thermal decomposition of a lead phosphate took place in the condensed phase.

8.2.3.12 Arsenic

MCALLISTER [4001] recorded the mass spectra measured when heating As_2O_3 in a graphite atomizer *in vacuo* and then took account of the observed species in equilibria calculations. These calculations differ from those made by FRECH *et al.* [1967] in that they really only contain observed species and do not limit the oxygen content. During the atomization step McAllister observed the spectra of As^+ and As_2^+; during the drying step he also observed $As_3O_4^+$ which he assumed was a fragment of $As_4O_{6(g)}$. He used this higher oxide for his calculations. The calculations predicted that this oxide should occur in the gas phase at around 500 K and $As_{2(g)}$ and $As_{4(g)}$ above this temperature. The calculations further predicted that the $As_4O_{6(g)}$ dissociates above 1000 K to give $AsO_{(g)}$. Above 2300 K the $AsO_{(g)}$ should decrease rapidly and $As_{(g)}$ should become the predominant species.

In their MS investigations, STYRIS *et al.* [5655] showed that $As_{2(g)}$ and $AsO_{(g)}$ appeared before the atomization of As_2O_3 in a PG-coated graphite tube, both at atmospheric pressure and *in vacuo*. A carbide $AsC_{(g)}$ was only observed when heating *in vacuo*. This fact and the presence of $As_{2(g)}$ in the vacuum experiments indicates the existence of elementary arsenic in the condensed phase. The strong signal for $AsO_{(g)}$ prior to atomization was used as an argument against the assumption that elementary arsenic was produced by reduction at these low temperatures (around 500 K). Instead, STYRIS *et al.* [5655] suggested that elementary arsenic and the oxide are formed by thermal decomposition according to:

$$4\,As_2O_{3(g)} \rightarrow 2\,As_{(s,\ell)} + (3-2)O_2 + As_4O_{4-6(g)}. \tag{8.81}$$

The $As_{2(g)}$ is then formed by sublimation from the condensed phase according to:

$$As_{(s,\ell)} \xrightarrow{\ 770\ K\ } As_{2(g)}. \tag{8.82}$$

Since free arsenic atoms were not observed at atmospheric pressure until all other species had disappeared (see Figure 8-23 in Section 8.2.2.2), the conclusion was drawn that the energy of release must be greater than the heat of sublimation for reaction 8.82. In other words, in the experiments at atmospheric pressure the arsenic must have been very strongly adsorbed. However, since free arsenic atoms appear at lower temperatures *in vacuo*, arsenic cannot be adsorbed from the condensed phase or from the decomposition of $As_2O_{3(s)}$. The adsorption must therefore be the result of a heterogeneous reaction. The dissociative adsorption of $AsO_{(g)}$ and $As_{2(g)}$ is presumably augmented by the numerous collisions with the graphite surface induced in the Langmuir film. Free arsenic can thus be formed according to:

$$AsO_{(g)} + C^*_{(s)} \rightarrow As_{(ad)} \xrightarrow{\ 1400\ K\ } As_{(g)} \tag{8.83}$$

or

$$As_{2(g)} + C^*_{(s)} \rightarrow 2\,As_{(ad)} \xrightarrow{\;1400\ K\;} 2\,As_{(g)}, \tag{8.84}$$

where $C^*_{(s)}$ signifies an active site on the graphite.

We should also like to mention the parallels to selenium (see Section 8.2.3.13) both in respect to atomization and also to precursor reactions, such as the formation of a gaseous carbide *in vacuo*. These parallels to selenium become even more obvious when we consider stabilization by palladium (although certain differences then become more noticeable).

STYRIS *et al.* [5655] found when heating a solution of palladium nitrate and arsenic nitrate in a PG-coated graphite atomizer that the signal for $AsO_{(g)}$ was an order of magnitude lower than without palladium and that the signal for $As_{2(g)}$ disappeared completely. The precursors of atomic $As_{(g)}$ also disappeared completely when the palladium was prereduced at 1300 K before the arsenic was dispensed into the graphite tube. The authors proposed the formation of a compound or a solid solution between palladium, arsenic, and oxygen according to:

$$x PdO_{(s)} + y As_{(s,\ell)} \rightarrow \left[Pd_x As_y O_z\right] + (x-z)O_{(g)} \tag{8.85}$$

$$x PdO_{(s)} + y AsO_{(g)} \rightarrow \left[Pd_x As_y O_z\right] + (x+y-z)O_{(g)}. \tag{8.86}$$

The same compound could also be formed with reduced palladium. However, since in this case no gaseous oxide could be observed, the authors proposed a direct reaction with the oxide in the condensed phase:

$$2\,x Pd^0_{(s)} + y As_2 O_{3(s)} \rightarrow 2\left[Pd_x As_y O_z\right] + (3y-2z)O_{(g)}. \tag{8.87}$$

These reactions also explain why no $AsC_{(g)}$ is observed *in vacuo* in the presence of palladium. The carbide is formed in the condensed phase and the presence of palladium prevents a reaction of the condensed phase with graphite according to reactions 8.83 or 8.84. Similar to selenium (see section 8.2.3.13), $As_{(g)}$ appears at a higher temperature (1600 K) in the presence of palladium than without. This can be explained by the greater thermal stability of $[Pd_x As_y O_z]$ in comparison to $As_{(ad)}$.

A completely new mechanism of fixation of arsenic was demonstrated by ROHR [4937] and ORTNER *et al.* [4488] utilizing the combined application of dynamic SIMS and electron beam microprobes (EMP) with quantitative particle analysis and X-ray peak offset measurements for the bonding analysis of palladium as well as ion chromatographic elution experiments. This is summarized briefly below and is further mentioned in Section 8.2.4.1, part 4. As a result of strong adsorption forces the analyte and palladium are distributed evenly over the entire platform surface already during the drying step ($\leq 120\ °C$). A good portion penetrates the surface area of the platform to a depth of 10 μm, another portion forms particles on the surface. The PdO intercalated in the graphite during drying forms covalent palladium-graphite bonds at the boundary layer of

PdO clusters to the graphite lattice. During penetration palladium nitrate is largely converted to PdO. These 'activated' palladium atoms are able to bind the analyte covalently. During the pyrolysis step (1300 °C) PdO is converted quantitatively into Pd^0 in the surface particles. During this process further diffusion of the modifier and the analyte into the graphite surface and thus further fixation of the analyte is observed. Thermal decomposition of the covalent fixation of the analyte on palladium takes place during the atomization step (2200 °C), followed by diffusion of both the analyte and palladium to the surface and thence entry into the gas phase.

8.2.3.13 Selenium

Selenium is one of the elements that has been investigated in the greatest detail. Varying modifiers, including palladium, have played a major role in these investigations. This is of considerable significance since without the use of modifiers selenium cannot be determined without losses.

MS studies on the vaporization and atomization of selenium (as H_2SO_3) in a graphite tube show clear differences between vaporization *in vacuo* and at atmospheric pressure [5652, 5656]. The most important compounds appearing in a vacuum are the carbide SeC_2 at lower temperature, the dimer Se_2, and the oxide SeO_2, but no atomic selenium. On the other hand, the carbide disappeared at atmospheric pressure but atomic selenium was observed. STYRIS [5652] also used metal atomizers in his investigations and found that when free selenium atoms were generated the carbide could no longer be observed. L'VOV [3727] found that the appearance temperature of 400–500 K for the carbide agreed very well with the calculated temperature for the start of reduction of SeO_2 by carbon. Studies by DROESSLER and HOLCOMBE [1613, 1614] led to results similar to those of Styris with the exception that no carbide occurred. This can possibly be ascribed to the higher pressure by 2–3 orders of magnitude under which the experiments were performed [5657].

STYRIS [5652] concluded from the MS studies that gaseous selenium is formed at atmospheric pressure through the thermal dissociation of the carbide which had been adsorbed on the graphite substrate. This carbide itself is formed from adsorbed selenium which was produced from the reduction of SeO_2 by carbon. In simplified form the reaction can be expressed as:

$$SeO_{2(s)} + C^*_{(s)} \rightarrow Se_{(ad)} \xrightarrow{1600 \text{ K}} Se_{(g)}, \tag{8.88}$$

where $C^*_{(s)}$ signifies an active site on the graphite.

In the absence of a modifier selenium is mainly vaporized in elementary form as the dimer Se_2 and as the monoxide. Gaseous polymers are formed by sublimation or vaporization of selenium from the condensed phase according to:

$$Se_{(s,\ell)} \xrightarrow{400 \text{ K}} \tfrac{1}{2} Se_{2(g)}. \tag{8.89}$$

Solid or liquid selenium is formed in the same manner as the observed gaseous oxides and hydroxides by the heterogeneous, or even possibly homogeneous, reaction of adsorbed SeO_2 with water:

$$SeO_{2(ad)} + H_2O \xrightarrow{400\ K} SeO_{2(g)} + SeO_{(g)} + Se_{(s,\ell)} + Se(OH)_{2(g)}. \qquad (8.90)$$

Both STYRIS [5652] and DROESSLER and HOLCOMBE [1613] investigated the volatilization of selenium in the presence of nickel as modifier. While Styris was able to observe complete suppression of the molecular species (Se_2, SeO_2, SeO) in the presence of nickel, Droessler and Holcombe merely observed a higher appearance temperature for these species in the presence of nickel. As a result of this discrepancy and also due to the fact that nickel did not prove to be a very good modifier for selenium, these investigations will not be discussed further.

Much more interesting are the experiments performed with selenium in the presence of palladium as modifier [5656]; the palladium was either added to the solution (referred to as PdO in the following) or reduced in advance by heating in the graphite tube (reduced palladium Pd^0). For vaporization *in vacuo* in the presence of Pd^0, both $Se_{2(g)}$ and $SeC_{2(g)}$ occurred, leading to the conclusion that Pd^0 practically does not react with $Se_{(s,\ell)}$. Stabilization is based largely on a reaction of Pd^0 with SeO_2. From the fact that free selenium and free palladium appear together *in vacuo* at 1150 K, the authors concluded that a stoichiometric compound of [Pd, Se, O] is formed, but they did not investigate its exact composition. At atmospheric pressure the appearance temperatures for $Se_{(g)}$ and $Pd_{(g)}$ were 400 K and 800 K higher, respectively, than *in vacuo*. This was attributed to a 'retaining mechanism' at high energy sites on the graphite substrate produced by the presence of palladium.

STYRIS *et al.* [5656] proposed the following mechanisms for the stabilization of selenium by Pd^0:

$$Pd^0_{(s)} + SeO_{2(s,\ell)} \xrightarrow{< 400\ K} [Pd, Se, O] + SeO_{(g)} \qquad (8.91)$$

$$Pd^0_{(s)} + SeO_{2(g)} \xrightarrow{> 400\ K} [Pd, Se, O] + SeO_{(g)} \qquad (8.92)$$

$$[Pd, Se, O] \xrightarrow{1200\ K} Se_{(g)} + Pd_{(g)} \rightarrow (Se-Pd)_{(ad)}. \qquad (8.93)$$

The dissociation products $Se_{(g)}$ and $Pd_{(g)}$ are retained in the Langmuir film at atmospheric pressure and are readsorbed at sites on the graphite surface that are produced in the presence of palladium (we shall go into this in detail later). With a further increase in temperature the atomic species are desorbed and released:

$$(Se-Pd)_{(ad)} \xrightarrow{1550\ K} Se_{(g)} + Pd_{(ad)} \qquad (8.94)$$

$$Pd_{(ad)} \xrightarrow{\text{1900 K}} Pd_{(g)}. \tag{8.95}$$

When palladium was introduced together with selenium in solution into the graphite tube, similarly to Pd^0, SeO and SeO_2 were observed, both *in vacuo* and at atmospheric pressure. In contrast to Pd^0, however, signals for Se_2 and $SeC_{2(g)}$ could not be observed. This means that PdO prevents either the formation of condensed-phase selenium or its vaporization. Additionally, no $Se(OH)_2$ was observed, indicating a competing hydration of PdO. The following reactions were proposed when palladium is present in solution:

$$PdO_{(s)} + H_2O_{(g)} \rightarrow Pd(OH)_{2(s)} \tag{8.96}$$

$$SeO_{2(s)} \rightarrow SeO_{2(ad)} + SeO_{(g)} + Se_{(s,\ell)} \tag{8.97}$$

$$PdO_{(s)} + Se_{(s,\ell)} \rightarrow [Pd, Se, O] \tag{8.98}$$

$$PdO_{(s)} + SeO_{2(s,\ell)} \rightarrow [Pd, Se, O]. \tag{8.99}$$

Since the same compound [Pd,Se,O] is formed when the analyte and the modifier are dispensed together as for Pd^0, the subsequent reactions leading to atomization are identical to 8.93 and 8.94.

The formation of a compound or a solid solution from selenium and palladium was also indicated by the SEM investigations performed by TEAGUE-NISHIMURA *et al.* [5789]. These authors found a largely spatially-resolved coincidence of selenium and palladium on a PG-coated graphite substrate; energy-dispersive measurements indicated a Pd:Se ratio of 1. QUIAO and JACKSON [4743] reported that 100 ng Pd are required to stabilize 2 ng Se up to a temperature of 1270 K when the selenium is present as the nitrate on a PG platform. Higher masses of palladium had little further effect. These authors thus came to the conclusion that the stabilizing effect of palladium was largely of a 'physical nature' and that selenium is dissolved in droplets of molten palladium or palladium oxide and possibly forms a compound.

MAJIDI and ROBERTSON [3817] investigated the stabilizing mechanism of palladium on selenium by means of RBS and confirmed both the formation of a compound and the 'physical' effect of palladium. In the absence of selenium, palladium began to diffuse into the PG substrate above 372 K and the diffusion increased with increasing temperature. If palladium was pretreated at 1000 K (Pd^0) before selenium was added, this diffused instantly into the graphite. This was attributed to diffusion along channels that had previously been formed by palladium. At temperatures above 770 K palladium and selenium diffuse to the surface, and above 1500 K it was observed that palladium diffuses back into the graphite.

If the palladium was not thermally pretreated but was dispensed together with the selenium, selenium could not diffuse into the graphite at lower temperature, but it could at higher temperature. Since oxygen was also found together with the diffusing species, the authors proposed the formation of a compound $Pd_xSe_yO_z$ in the channels that had been formed by the diffusion of palladium. The compound migrated to the surface at around 1100 K and decomposed at higher temperatures with the formation of free selenium atoms. The palladium then migrated back into the graphite mass.

Although the RBS data, similar to those from SEM, were not obtained until the substrate had cooled, it is very unlikely that an artifact was observed, especially since the information gathered by different techniques was in excellent agreement.

8.2.4 Interferences

Virtually no transport interferences occur in GF AAS since the measurement solution is normally dispensed as a predefined volume into the atomizer, usually with an autosampler, and not with a nebulizer. Also, almost no references to ionization interferences are to be found in the literature (refer to Section 5.4.1.2). In contrast, however, there appears to be an almost endless number of publications on solute-volatilization and gas-phase interferences in GF AAS, many of which are contradictory. Many of these interferences are caused by the incorrect use of this technique and are greatly reduced or eliminated when the STPF concept is applied consistently. We do not consider it to be of any great use to make a detailed presentation of these avoidable interferences and we shall therefore only cite the corresponding references in exceptional situations; these papers have been discussed in detail by SLAVIN [5428]. In the following we shall discuss those interferences and their elimination that cannot be avoided by the use of the STPF concept. Further, we shall once again treat the function and the correct use of the STPF concept, and especially the use of chemical modifiers. We should also again like to point out that in GF AAS an exact analysis of the time-resolved signals greatly facilitates the recognition of interferences (see Section 8.2.2.1, part 2) [6186].

8.2.4.1 The STPF Concept

The principle behind the STPF concept is described in Section 1.7.2. We shall thus examine more closely those aspects that are of special importance for the avoidance of interferences. The STPF concept is a 'package' of measures to achieve three aims, or to put it another way, to achieve interference-free determinations in three steps:

i)   To separate concomitants as far as possible prior to atomization.
ii)   To control the reactions in the condensed phase to prevent analyte losses and to better separate concomitants.
iii) To minimize the influence of non-separated concomitants on the analyte in the gas phase by as complete atomization as possible.

The STPF concept includes four major measures to achieve these aims: chemical modification, well-controlled graphite surfaces, atomization under largely isothermal conditions, and effective background correction. This latter aspect is discussed in detail under spectral interferences (see Section 8.2.4.2). In addition, and especially for methods development and for monitoring, fast electronics and an undistorted, time-resolved presentation of the atomization and background signals are essential (refer to Section 8.2.2.1, part 2).

8.2.4.1.1 *Chemical Modification*

In GF AAS we use a temperature program to separate the analyte and concomitants *in situ* prior to the atomization step. The highest possible pyrolysis temperature is required to effectively separate the concomitants. However, since the analyte must not be volatilized during the pyrolysis step, there are limitations to the maximum temperature. To determine the maximum temperature we establish pyrolysis curves (see Section 8.2.2.1, part 1).

However, since every element can occur in a large number of chemical compounds, which often differ substantially in their physical properties and thus in their volatilities, the pyrolysis curve depends on the element species present. The species present in a test sample or which is formed during pyrolysis depends very strongly on the concomitants and is very often not known. If a pyrolysis curve is established using matrix-free, aqueous calibration solutions, we have absolutely no guarantee that the analyte species present in the test sample will behave in the same manner. To eliminate this uncertainty and to gain control over the form in which the analyte is present, we use chemical additives that bring the physical and chemical properties of the sample and calibration solutions as close together as possible. This procedure is termed *chemical modification*. In most cases the aim is to form the most stable compound or phase of the analyte. In the ideal case the thermal stability of the concomitants is at the same time lowered. Under no circumstances, however, should the thermal stability of concomitants be increased since the aim of chemical modification is to achieve *separation of the analyte and the matrix during the pyrolysis step*.

SCHLEMMER and WELZ [5150] have made a list of selection criteria for an ideal modifier:

i)   It should be possible to thermally pretreat the analyte to the highest feasible temperature. In many cases large quantities of salts, such a sodium chloride, or an organic matrix must be removed. A pyrolysis temperature of ≥ 1000 °C is often required to reduce the bulk of the concomitants significantly.

ii)   The modifier should stabilize as many elements as possible. The more similar are the conditions or the pyrolysis and atomization of the individual elements, the easier it is to establish methods that allow several elements in a test sample to be determined simultaneously or in sequence.

iii) The reagent should be available in high purity to prevent the introduction of blank values.

iv)   The modifier, which is normally added in large excess, should not contain any element that might later be determined in the trace range.

v)   The modifier should not shorten the lifetime of the graphite tube and platform.

vi)   The modifier should only make the lowest possible contribution to the background attenuation.

The mixed palladium nitrate-magnesium nitrate modifier (Pd-Mg modifier) proposed by SCHLEMMER and WELZ [5150] does not meet these criteria in all points (we shall discuss this later), but it meets them better than any other modifier that has been described to date. WELZ et al. [6265] investigated 21 elements which could be stabilized with this modifier. The attainable maximum pyrolysis temperatures and optimum atomization temperatures are presented in Table 8-3. We must naturally point out that the maximum pyrolysis temperature determined without modifier was measured for matrix-free calibration solutions and may be considerably lower in the presence of concomitants. On the other hand, the maximum pyrolysis temperature determined with modifier does not change noticeably even in the presence of higher concentrations of concomitants [6265].

The relatively small differences in the atomization temperatures of the various elements in the presence of the Pd-Mg modifier is an indication that the stabilization mechanisms are similar. This aspect is discussed in Section 8.2.3 for a number of elements. From MS data STYRIS et al. [5655, 5656] concluded that for arsenic and selenium a compound of the type [Pd-M-O] is formed, but they did not investigate the stoichiometry. STURGEON et al. [5640] confirmed this conclusion for arsenic via AES measurements, and LIAO and LI [3514] found an indication for the formation of a [Pd-In-O] species using XPS analysis. The formation of a compound or a solid solution

Table 8-3. Maximum usable pyrolysis temperatures without modifier and with the Pd-Mg modifier, and the optimum atomization temperatures with the Pd-Mg modifier (from [6265]).

Element	Maximum pyrolysis temperature, °C		Optimum atomization temperature, °C
	without modifier	Pd-Mg modifier	
Ag	650	1000	1600
Al	1400	1700	2350
As	300	1400	2200
Au	700	1000	1800
Bi	600	1200	1900
Cd	300	900	1700
Cu	1100	1100	2600
Ga	800	1300	2200
Ge	800	1500	2550
Hg	<100	250	1000
In	700	1500	2300
Mn	1100	1400	2300
P	200	1350	2600
Pb	600	1200	2000
Sb	900	1200	1900
Se	200	1000	2100
Si	1100	1200	2500
Sn	800	1200	2400
Te	500	1200	2250
Tl	600	1000	1650
Zn	600	1000	1900

between selenium and palladium was also indicated by the SEM investigations carried out by TEAGUE-NISHIMURA et al. [5789]. QIAO and JACKSON [4743] found that 100 ng Pd were necessary to stabilize 2 ng Se; in the view of the authors this pointed to a more 'physical' stabilization. Using SEM, XRD, and XPS, YANG et al. [6454] investigated the stabilization of arsenic, lead, and zinc by palladium; they found that with excess palladium all three elements formed an intermetallic, solid solution. The diffusion of palladium into the graphite structure, which was detected by MAJIDI and ROBERTSON [3817], without doubt plays an important role in stabilization.

Recent investigations by ROHR [4937] and ORTNER et al. [4488] show that, at least for arsenic and probably for other elements as well, stabilization by palladium or the Pd-Mg modifier derives from the formation of a covalent compound with palladium during the drying step. The palladium is activated by covalent bonding to the graphite lattice (intercalation compound). The 'active range' of this phenomenon is at the surface of the platform and penetrates to a depth of up to 10 µm as shown by SIMS depth profile measurements. With SIMS it is possible to detect analyte contents in the ng/g range, which is not possible with any other topochemical technique. We can assume that this principle is valid for numerous volatile elements which are retained to amazingly high atomization temperatures by palladium. Retention of the analyte by particles on the surface at temperatures exceeding 800 °C is not possible from considerations on the diffusion of analytes in the corresponding matrices (e.g. PdO, MgO). Likewise the formation of intermetallic compounds between arsenic and palladium does not occur due to the extreme concentration ratio (Pd:analyte $\geq$ 1000:1).

This high stabilizing power of the Pd-Mg modifier can nevertheless lead to lower sensitivity for a number of volatile elements. In the presence of this modifier the atomization signals are frequently broader, so that the absorbance is lower. This is of little significance, however, since integrated absorbance should always be used in GF AAS. Nevertheless the integrated absorbance can also decrease for a number of elements when the optimum atomization temperature with modifier is noticeably higher than without. As shown in Figure 8-28 on the example of bismuth, the diffusion losses are greater at the higher temperature made necessary by the use of the modifier.

In this connection, however, we would clearly point out that sensitivity should not be taken as the decisive criterion for GF AAS. Freedom from interferences and trueness of the determinations have a far higher value. Effective stabilization of the analyte and a higher optimum atomization temperature are thus decisive factors that are far more important than better sensitivity. On the other hand, there are also examples where a significantly higher sensitivity can be attained with the use of the Pd-Mg modifier. We can mention silicon and tin in particular since the modifier prevents analyte losses, leading to a higher atomization efficiency [2165, 6265].

As well as the Pd-Mg modifier, reduced palladium has proven to be advantageous in a number of cases [4853, 6105, 6108, 6241, 6260]. This is further discussed in connection with interferences and their elimination (refer to Section 8.2.4.3). As mentioned in Section 1.7.2, an unbelievable variety of modifiers are mentioned in the literature, but we shall not go into details. TSALEV et al. [5904, 5915] have published a comprehensive overview on the most varying modifiers and their effect. We would strongly advise against the use of modifiers containing frequently-determined elements like copper

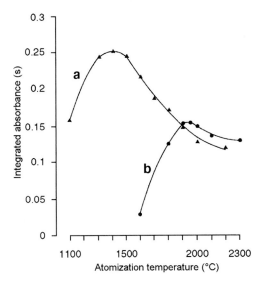

Figure 8-28. Atomization curves for 1 ng Bi. **a** – without modifier; **b** – with Pd-Mg modifier (from [6265]).

[1285, 3956, 3957, 6083] or nickel [233, 1174, 1257, 1677, 2167, 2967, 3512, 3513, 3556, 3557, 3615, 3617, 3956] since the determination of the elements following the use of the modifier is not possible without complete decontamination of the atomization unit, usually involving replacement of the graphite tube and contacts.

At this point we would briefly mention two further modifiers: magnesium nitrate and ammonium phosphate. *Magnesium nitrate* was proposed by SLAVIN *et al.* [5421] for elements of medium volatility, such as aluminium, beryllium, chromium, and manganese. On the basis of MS investigations, STYRIS and REDFIELD [5653] suggested two mechanisms for the stabilization of aluminium. In the one, the MgO produced oxidizes adsorbed (reduced) aluminium and lower aluminium oxides to higher oxides and thereby shifts atomization to higher temperatures (refer to Section 8.2.3.8). In the other, magnesium binds residual water by forming $MgOH_{(g)}$ and $HMgOH_{(g)}$, thus preventing the formation of $Al(OH)_{2(g)}$ and hence losses of aluminium. Very similar behavior was also found for beryllium [5654].

Phosphate in the form of $NH_4H_2PO_4$ or $(NH_4)_2HPO_4$ has been used extensively as a modifier for the determination of lead and cadmium. Nevertheless, its use is not completely free of problems since the $PO_{(g)}$ formed by vaporization during the atomization step exhibits an absorption spectrum with pronounced fine structure (Figure 8-29) that can lead to spectral interferences (refer to Section 8.2.4.2). BASS and HOLCOMBE [411] used MS to investigate the stabilization of lead by phosphate and found that the rate-determining step for the atomization of lead was the reduction of PbO to $Pb_{(g)}$, and that losses of $PbO_{(g)}$ could thus take place. In the presence of phosphate, $Pb_2P_2O_7$ is formed which decomposes at 1150 K, so that losses of $PbO_{(g)}$ are prevented. Using SIMS, HASSELL *et al.* [2423] observed the formation of cadmium-oxyphosphorous compounds

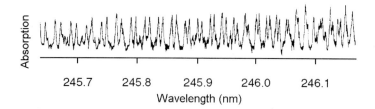

Figure 8-29. Absorption spectrum of PO during the vaporization of phosphate-containing reagents in the graphite furnace (from [3940]).

that shifted the atomization of cadmium to higher temperatures. The formation of comparable silver compounds was however not observed, so that this element is not stabilized. In the course of their RBS investigations, ELOI *et al.* [1727] found that both lead and cadmium form a metal-oxyphosphorous compound, probably a phosphate glass of the type $xMO \cdot P_2O_5$. Both elements also migrate with the modifier into the graphite lattice, which could further enhance stabilization.

Occasionally, gases have also been used as 'modifiers' in GF AAS. This possibility is discussed in a later section (refer to Section 8.2.4.1, part 3). Nevertheless, one of these gaseous modifiers is mentioned at this point since it often plays a significant role in separating the analyte from the matrix, namely oxygen. Since graphite atomizers are operated in an inert gas atmosphere, organic materials can only be charred in the pyrolysis step so that often a thick layer of carbon remains in the tube as depicted in Figure 8-30. In order to prevent this, we can introduce oxygen or air into the graphite tube during the pyrolysis step so that the organic sample constituents are ashed and the carbon can thus be removed [452]. If the temperature is not allowed to exceed 500–600 °C in this step the PG coating is not attacked so that the lifetime of the tube and platform is not influenced.

By removing concomitants as far as possible during the pyrolysis step, spectral and non-spectral interferences are eliminated or at least reduced.

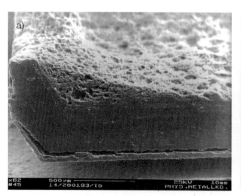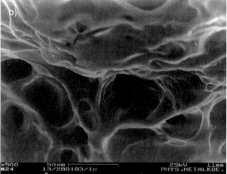

Figure 8-30. SEM photographs of residues of organic materials after pyrolysis in an argon atmosphere. **a** – cross-section of a platform after 100 blood analyses; **b** – residue from a sugar solution magnified 500x (from [4937]).

8.2.4.1.2 *Isothermal Atomization*

The influences of non-separated concomitants on the atomization of the analyte are further reduced by isothermal atomization. The major aspects that contribute to isothermal atomization have already been listed and discussed in detail. They are repeated briefly here and the connections are elucidated.

The most important, visible component of the isothermal atomizer is the platform onto which the sample is dispensed and which is responsible for delayed atomization of the analyte (refer to Section 4.2.2.3). However, delaying atomization to the time point at which the graphite tube and gas atmosphere have reached their final temperature is only achieved when the platform has the smallest possible contact with the graphite tube and the heating rate is sufficiently fast (see Section 4.2.3). This requirement is met particularly well by 'integrated' platforms that have a single point of contact with the graphite tube and heating rates of 1500–2000 °C/s. Similar effects to platform atomization can also be obtained with probe atomization (see Section 4.2.2.4). The prerequirements for isothermal atomization are met especially well by transversely-heated tubes since these offer not only temporal but also spatial isothermality (refer to Section 4.2.2.5).

The gas stream through the graphite tube makes a decisive contribution to isothermality (refer to Section 4.2.4). During the drying and pyrolysis steps a gas stream through the graphite tube is essential to purge the vaporized concomitants from the absorption volume as quickly as possible. However, a forced gas stream during the atomization step would not only contribute to the removal of the analyte atoms but would also disturb the thermal equilibrium. FRECH and L'VOV [1977] even showed that convection currents influence the isothermality and can lead to interferences. The recognition of this fact led to the development of the transversely-heated tube with end caps (see Section 4.2.2.5), which comes very close to the ideal conception of isothermal atomization.

For some period of time there was a certain lack of clarity about the temperature step from pyrolysis to atomization. It is clear that the selected atomization temperature should not be too high. The best temperature is several hundred degrees above the appearance temperature. In this case atomization begins at a time point at which the graphite tube has really attained its preselected final temperature; this is shown in the temperature diagrams in Figure 4-26 (see Section 4.2.3). Originally it was assumed that the temperature step from pyrolysis to atomization should be as small as possible (< 1000 °C) to keep the expansion of the gas during heating as low as possible. FALK *et al.* [1809] nevertheless claimed that the temperature distribution along the tube is more homogeneous, at least at the start of atomization, if the tube is heated from ambient temperature. Numerous analysts therefore included a cooling step between pyrolysis and atomization. The usefulness of this step has however never been conclusively demonstrated. In a transversely-heated atomizer the greater is the temperature difference measured between the platform and the wall (platform effect), the lower is the initial temperature of the heating step [5520].

All of the measures required to attain isothermal atomization must naturally be supported by the application of an effective modifier. Only together with chemical modification can a high pyrolysis temperature and an atomization temperature that is merely a few hundred degrees above the appearance temperature be reliably and reproducibly applied. Only under conditions of isothermal atomization is the evaluation of the atomization signals as integrated absorbance meaningful. Then only under conditions of isothermal atomization can differing behavior during atomization be successfully compensated; a different time point and a different rate of vaporization are without influence on the integral. A typical example for this is the determination of antimony in the presence of a 25 000fold excess of nickel as shown in Figure 8-31.

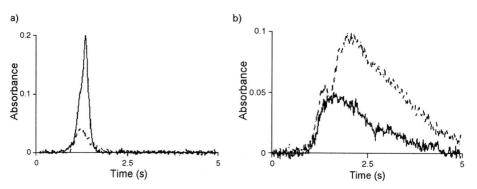

Figure 8-31. Time-resolved signals for 40 µg/L Sb in a transversely-heated atomizer with integrated platform and Zeeman-effect BC. Pyrolysis temperature 1300 °C; atomization temperature 2100 °C; Pd-Mg modifier. **a** – calibration solution; $A_{int} = 0.071$ s; **b** – calibration solution + 1000 mg/L Ni; $A_{int} = 0.073$. The dashed signal shows the background absorption.

At this point we must again mention that the STPF concept can only bring the required results when it is applied in its entirety. Occasionally it may be possible under given prerequisites to obtain correct results when using only parts of the STPF concept, such as the use of a platform, but evaluation via the peak height. Nevertheless such techniques are exceedingly risky and can lead to systematic errors even for the following test sample. We have intentionally desisted from citing negative examples from the literature since there are too many of them and it would not be correct to pick single examples. We must emphasize that a heating rate that is too slow (< 1000 °C/s), an uncontrolled gas stream, or atomization from the tube wall prevent isothermal atomization and thus *significantly increase the susceptibility to interferences.* Also, many of the interferences that occur when isothermal atomization is not applied cannot be corrected by the analyte addition technique. The added analyte species often behaves differently to the species present in the test sample when non-STPF conditions are applied and the basis of the analyte addition technique is thus removed.

8.2.4.1.3 *Control of the Atmosphere in the Atomizer*

GF AAS determinations are performed in an inert gas atmosphere which, in contrast to F AAS, has absolutely no buffer action. This means that the concomitants vaporizing from the sample matrix govern the gas phase and can substantially influence the atomization of the analyte. These influences can be largely kept under control by the use of chemical modification and the best possible separation from the matrix.

A further, very important parameter is the nature and quality of the graphite surface. Sample constituents can remain in uncoated or poorly coated EG tubes often to very high temperatures and thus interfere in the determination. FRECH and CEDERGREN [1959, 1960] found for example that even after 15 minutes in an uncoated graphite tube at 1200 °C *in vacuo* water can be retained in quantities sufficient to influence reactions. The water gas equilibrium is quickly established at higher temperatures so that there is a relatively high partial pressure of hydrogen in an uncoated tube:

$$CO + H_2O \rightleftharpoons CO_2 + H_2 \tag{8.100}$$

Similar behavior has also been observed for a number of acids, in particular sulfuric acid and perchloric acid. MANNING *et al.* [3862] found that many of the chloride interferences described in the literature could be attributed to a poor quality or a nonexistent tube coating. Chlorine cannot be retained in tubes with a good PG coating and these interferences are not observed. For the determination of manganese in sea-water, CARNRICK *et al.* [1072] found that tubes with a good PG coating exhibited long lifetimes and high signal stability. These authors observed intercalation in poorly coated or uncoated tubes, i.e. retention of sodium or NaCl in the graphite structure, that led to considerable losses in sensitivity for the determination of manganese and to early failure of the tubes.

Through the controlled admixture of gases it is possible to change the conditions in the atomizer to meet a given requirement. The addition of oxygen to ash organic materials during the pyrolysis step is an example of this. In a series of studies HOLCOMBE and co-workers [412, 1614, 5064, 5065] showed that it is possible to change the graphite surface with oxygen to such an extent that the appearance time and signal form of the analyte are changed. HAHN *et al.* [2318] showed that treatment with oxygen primarily brings about an increase in the number of active sites on the surface of the graphite. These sites bind the analyte better and thus reduce losses.

WELZ and SCHLEMMER [6242] found that in the presence of 1% methane in argon a number of refractory elements could be determined better and that several interferences were less pronounced. These authors also reported that memory effects for the determination of molybdenum were reduced when Freon was added during the tube cleaning step [6243].

For the determination of lead in steel, FRECH and CEDERGREN [1960] showed that interfering chloride could be removed as $HCl_{(g)}$ by hydrogen. WELZ *et al.* [6241] found that by introducing hydrogen in the pyrolysis step in addition to chemical modification, the last remaining interferences for the determination of thallium in sea-water could be eliminated. A similar effect to hydrogen can also be achieved with a number of organic modifiers, such as ascorbic acid. During pyrolysis ascorbic acid releases H_2 and CO [1007] which can lead to a substantial reduction in the partial pressure of free oxygen in the gas phase of a graphite tube during atomization [6104]. GILCHRIST *et al.* [2110] found that the addition of H_2 or CO to the argon purge gas had the same effect as the addition of ascorbic acid to the aqueous analyte solutions. The absorption signals for As, Cr, Pb, Se, Sn, and Zn were in all cases shifted to lower temperatures and the interference by HCl on the determination of these elements was eliminated. These changes can be explained by an influence on the dissociation equilibrium of the analyte oxides in the gas phase brought about by H_2 or CO. BYRNE *et al.* [1006, 1008] described very similar effects brought about by ascorbic acid for the determination of manganese in a chloride matrix.

8.2.4.2 Spectral Interferences

GF AAS is a technique for extreme trace analysis. In practice this means that a concentration difference of 6–7 orders of magnitude between the analyte and concomitants is no rarity, especially for the analysis of solids. This vast excess of concomitants automatically leads to an increased risk of interferences. As explained in the preceding section, this problem can be reduced by separating the concomitants as much as possible prior to the atomization step.

The completeness of this separation depends in the first place on the volatility of the analyte and the concomitants. It is not always possible to modify these parameters by chemical modification to such an extent that largely complete separation can be achieved in the pyrolysis step. As well as the volatility of the constituents, the design of the atomizer unit plays an important role. With a graphite tube with electrical contacts at the ends there is a temperature profile along the tube and thus the risk that vaporized concomitants will condense on the cooler parts and then be revaporized during atomization. This risk is greater when the purge gas stream is not under good control (refer to Figure 4-29 in Section 4.2.4).

Even if concomitants can be largely separated during the pyrolysis step, the remaining quantity still makes background correction necessary. Continuum source BC, which was developed for F AAS and is perfectly adequate for this technique, is frequently overtaxed in GF AAS, especially when the background corrector cannot follow rapid changes in the background signal particularly at the start of atomization, or when the background attenuation exceeds the correctable range with absorbance values > 1, or when the background exhibits fine structure which cannot be properly corrected by the use of continuum sources.

The problems caused by the correction of rapidly changing background signals were thoroughly investigated by HARNLY and HOLCOMBE [2388, 2390, 2641]. The resulting signal interferences and the artifacts introduced by the type of signal processing employed are depicted in Figure 8-32. In agreement with the results of Harnly and Holcombe, BARNETT *et al.* [392] were able to show that the bracketing technique (i.e. the subtraction of the mean value of the background signal measured prior to and after the analyte signal) at a frequency of 50–100 Hz largely eliminated this interference. However, this is only valid for Zeeman-effect BC. For continuum source BC two radiation sources are employed which exhibit different radiant flux densities, distributions and geometries, and it is not possible to obtain exact coincidence. The level of this imperfection is expressed in the magnitude of the artifacts.

Problems caused by deviations in the radiation geometry between the two radiation sources also occur with higher background absorption. The correction range quoted by the instrument manufacturer can only be attained with optimum matching of both sources. Small, unavoidable changes in the alignment can lead to a substantial deteriora-

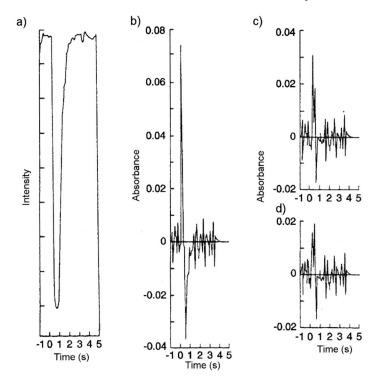

Figure 8-32. Background signal from 20 μL of a solution of 20 g/L NaCl in a graphite atomizer at 2700 °C at the resonance line of lead at 283.3 nm. **a** – background signal; **b** – asymmetric correction; **c** – bracketing technique; **d** – non-linear calculation with three measurement points (from [2641]).

tion in the correction performance. And lastly we must also remember that continuum source BC leads to a significant deterioration in the S/N ratio and that this deteriorates even further at high background absorption values.

The fact that background absorption with a fine structure cannot be corrected properly with continuum source BC is discussed in Section 5.4.3.1. In the early 1980s, SAEED and THOMASSEN [5150] and BAUSLAUGH *et al.* [429] described the spectral interferences caused by iron and phosphate on the determination of antimony, arsenic, selenium, and tellurium. Attempts, even recently, to eliminate these interferences, either by chemical modification to such an extent that continuum source BC could be applied [2610, 3347, 3766, 4592] or by optimizing the program to separate the analyte and background signals [590], mostly resulted in the authors having to resort to peak height evaluation and often to the analyte addition technique, thus abandoning the STPF concept.

It is naturally to be expected that a technique that measures the background next to the analytical line must fail when the background exhibits fine structure (see Section 5.4.3.1). It is therefore hardly surprising that spectral interferences, such as those on the determination of selenium, cannot be eliminated by correction with high-current pulsing [147].

After the introduction of Zeeman-effect BC in the 1980s, numerous papers were published demonstrating the superiority of this technique in comparison to continuum source BC. The best-known examples are the interferences caused by iron and phosphate on the determination of selenium [1073, 1892, 1893]; refer to Figure 8-16 in Section 8.2.2.1. Strong overcompensation can be observed when using continuum source BC, while no interference can be ascertained with Zeeman-effect BC.

Without Zeeman-effect BC, these interferences make the determination of selenium virtually impossible in biological materials [1893, 4334, 6095, 6231], in environmental samples [3497, 6094], and in metallurgical samples [1895]. Apart from iron and phosphate, when using continuum source BC the determination of selenium is also subject to interferences by aluminium [3497, 4726] and cobalt, chromium, and nickel [1895]. Similar interferences to those for selenium have also been reported for arsenic [1893, 3497, 6095, 6231], antimony [1893, 3497, 6426], bismuth, gold, and tellurium [1893], and thallium [1893, 3497]. These examples in particular demonstrate the superiority of Zeeman-effect BC since no interferences can be observed when this mode of correction is applied.

The situation is very similar for the determination of phosphorus in steel [6214], which is shown in Figure 8-33, or the determination of cadmium in fish tissue [6231] or waste water [6094]. A spectral interference by palladium on the determination of copper and thallium is seen when continuum source BC is applied but not with Zeeman-effect BC [6245], so that the use of this modifier is prohibited if continuum source BC is used. All in all, Zeeman-effect BC is without doubt the system of choice for GF AAS.

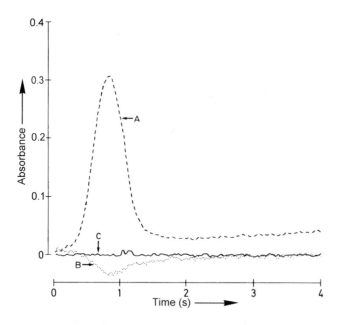

Figure 8-33. Determination of phosphorus in pure iron (20 g/L Fe). **A** – non-specific absorption; **B** – atomization signal with continuum source BC; **C** – atomization signal with Zeeman-effect BC (from [6214]).

However, the impression should not be given that Zeeman-effect BC is completely without interferences; nevertheless, any interferences are powers of magnitude lower than with continuum source BC and can be eliminated or avoided by simple means. In principle there are two possible sources of spectral interferences with Zeeman-effect BC: atomic lines that are less than 10 pm from the analytical line can cause overlapping due to their own Zeeman splitting, and molecular bands with a rotational fine structure which exhibit the Zeeman effect. The first category was investigated thoroughly in particular by WIBETOE and LANGMYHR [6305–6307]. All interferences caused by atomic lines in Zeeman-effect BC which are mentioned in the literature are presented in Table 8-4. Of the 22 cases of line overlap, merely seven are at recommended primary analytical lines. Since the most sensitive analytical lines are used almost exclusively in GF AAS, we shall not go into the other cases. Suitable care is recommended for the analysis of solid samples where occasionally less sensitive lines are employed.

Table 8-4. All spectral interferences in Zeeman-effect BC quoted in the literature and means of avoiding them.

Analyte	Wavelength nm	Nature of wavelength[a]	Interferent	References	Avoidance[b], Notes
Ag	328.1	P	Rh	[6279]	temperature program
Al	308.2	s 1.5	V	[6307]	308.2 P
Au	267.6	s 2	Co	[6306]	242.8 P
B	249.7	P	Co	[6306]	minimal interference
Bi	227.6	s 15	Co	[6307]	222.8 P
Co	243.6	s 10	Pt	[6307]	240.7 P
Cr	360.5	s 2	Co	[6307]	357.9 P
Eu	459.4	P	V, Cs	[6307]	462.7 s 1.3
Fe	271.9	s 3	Pt	[1076]	248.3 P
Ga	287.4	P	Fe	[6305]	294.4 s 1
				[5374]	temperature program
Hg	253.7	P	Co	[6306, 1076]	temperature program
Ni	341.5	s 3	Co	[6307]	232.0 P
	305.1	s 4	V	[6307]	232.0 P
Pb	261.4	s 25	Co	[6307]	283.3 P
Pd	247.6	s 1	Pb	[1988]	244.8 P, temperature program
Pt	265.9	P	Eu	[6307]	very seldom
	273.4	s 3	Fe	[6307]	265.9 P
	306.5	s 1.5	Ni	[6307]	265.9 P
Si	250.5	s 3	Co, V	[6307]	251.6 P
Sn	300.9	s 3	Ca	[6307]	286.3 P
	303.4	s 2	Cr	[6307]	286.3 P
Zn	213.9	P	Fe	[6305]	temperature program

[a] P = primary resonance line; s = secondary resonance line; the figure indicates the factor by which the sensitivity is less than at P.
[b] Recommended alternate wavelength; symbols used as under a).

A suitable temperature program can be used to avoid the interference of rhodium on the determination of silver, of cobalt on the determination of mercury, of lead on the determination of palladium, and of iron on the determination of zinc. Volatile interfering elements such as lead can be separated in the pyrolysis step and volatile analytes can be determined at atomization temperatures below 2000 °C without the interfering element being noticeably volatilized. The interference of cobalt on the determination of boron is minimal [6306]. Europium exhibits a large number of alternate lines with virtually the same sensitivity at which the interference does not occur. Gallium also exhibits an alternate line with the same sensitivity but with slightly lower linearity. For this element separation using a suitable temperature program is also possible [6305]. The interference of europium on the determination of platinum is hardly likely to be of any significance in practice, but numerous alternate lines are available if required.

For the sake of completeness we must mention that the interferences listed in Table 8-4 cannot be successfully avoided by turning to continuum source BC [6305] or to correction using high current pulsing. In the case of continuum source BC the mean background attenuation measured over the spectral bandpass is subtracted (refer to Section 3.4.1); with background correction using high current pulsing the rotational lines are completely detected because of their closeness to the analytical line, thus leading to interferences that are the same or even worse than with Zeeman-effect BC.

MASSMANN [3941] was the first to point out the possibility of interferences in Zeeman-effect BC by molecular spectra; the excitation spectra with a large number of lines can undergo splitting in a magnetic field. This risk is mostly present for light, diatomic molecules. OHLSSON and FRECH [4425] investigated the PO bands produced by the vaporization of $NH_4H_2PO_4$ using a high resolution spectrograph and established that the absorption profiles of a number of PO bands could be influenced by a magnetic field. All the spectral interferences by PO in Zeeman-effect BC mentioned in the literature are compiled in Table 8-5.

Table 8-5. Spectral interferences caused by phosphate (PO) in Zeeman-effect BC and means of avoiding them.

Analyte	Wavelength (nm)	References	Avoidance
Ag	328.1	[4425]	minimal interference
Cd	326.1	[1076, 4425, 6308]	primary analytical line at 228.8 nm, ~ 300 x more sensitive
Cu	244.2 / 247.3	[6308]	primary analytical line at 324.7 nm ~ 300 x more sensitive
Fe	246.3	[4425]	primary analytical line at 248.3 nm ~ 10 x more sensitive
Hg	253.7	[4425]	temperature program
In	325.8	[6308]	alternate wavelength at 304.0 nm with same sensitivity
Pb	217.0	[2464]	wavelength at 283.3 nm
Pd	247.6 / 244.8	[4425, 6308]	wavelength at 276.3 nm

The interference of PO on the determination of silver is very low and is without significance in practice, unless large masses of phosphate are used as a modifier. The interferences on the determination of cadmium, copper, and iron are only of academic interest since the lines at which the interferences are observed are less sensitive by factors of 10–300 than the respective primary analytical lines. The interference on the determination of mercury is also of a more theoretical nature since no phosphate is volatilized at the atomization temperature of 1000 °C. For indium, lead, and palladium interference-free alternate lines are available with the same sensitivity.

In these cases the use of another technique of background correction is also not a solution. The interferences caused by a structured background are much higher with continuum source BC or with high current pulsing than with Zeeman-effect BC. In addition, many more spectral interferences occur with continuum source BC that cannot be observed with Zeeman-effect BC [1076].

Nevertheless, the interference caused by PO on Zeeman-effect BC in the presence of very high phosphate concentrations, such as the analysis of bones, cannot be ignored [2464]. For the analysis of biological materials, RADZIUK and THOMASSEN [4782] attempted to correlate the absorption of phosphorus atoms with the magnitude of the observed interference. They established that the interference was the weakest under conditions at which a maximum of phosphorus atoms was produced and thus a minimum of PO molecules.

A frequent cause for the occurrence of interferences by PO is the use of phosphate for chemical modification. For the determination of low cadmium concentrations in urine in the presence of a phosphate modifier, YIN et al. [6481] even observed a spectral interference at the primary analytical line for cadmium. As we can see from Figure 8-34, this interference can no longer be seen when a palladium modifier is used. Because of the relatively high occurrence of interferences caused by a phosphate modifier, the use of such a buffer can only be recommended in exceptional cases.

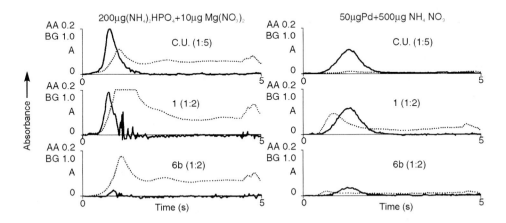

Figure 8-34. Atomization signals for cadmium in urine. **C.U.** – control urine with 6.5 μg/L Cd; **1** – urine from a smoker with 3 μg/L Cd; **6b** – urine from a non-smoker with 0.2 μg/L Cd; continuous line: atomic absorption (AA); dotted line: background signal (BG) (from [6481]).

Apart from the interferences caused by PO compiled in Table 8-5, very few cases are reported in the literature in which the splitting of molecular bands leads to false measurements. The interference of CN on the determination of chromium reported by DOIDGE [1562] could not be confirmed by OHLSSON and FRECH [4425] and can presumably be attributed to the lamp employed which contained iron. WENNRICH et al. [6279] reported on spectral interferences caused by molecules such as AlBr, InBr, AlO, and BaO, but these can all be avoided by using a suitable temperature program.

All in all, Zeeman-effect BC is without doubt the best alternative for avoiding spectral interferences in GF AAS. Naturally, Zeeman-effect BC must be seen as part of the STPF concept, i.e. concomitants must be removed as completely as possible prior to atomization using suitable modifiers and atomization must take place under isothermal conditions. Only in this way is it possible to reduce the background attenuation for many samples to such a level that it can be corrected without problems.

8.2.4.3 Non-spectral Interferences

8.2.4.3.1 *Classification of Non-spectral Interferences*

Customarily we classify non-spectral interferences in GF AAS into *solute-volatilization interferences* and *gas-phase interferences*. The majority of solute-volatilization interferences are caused by the formation of volatile compounds of the analyte in the condensed phase, which are then vaporized during the pyrolysis step. Nevertheless, it is not always easy to assign an interference to the one or the other category. By using a dual-cavity platform, WELZ et al. [6227, 6240] attempted to distinguish between interferences in the condensed phase and those in the gas phase. Simplified, these authors assumed that the absence of an interference when the analyte and the interferent were in different cavities was an indication for a condensed-phase interference; if the interference was still present it was an indication for a gas-phase interference. This simple distinction was only valid in a limited number of cases, however, since a large number of heterogeneous reactions between the gas phase and the condensed phase also played a role. STYRIS and REDFIELD [5657] have emphasized the importance of the Langmuir film on the surface which has a decisive significance for many reactions.

If the analyte forms a stable carbide that is not completely atomized at the selected atomization temperature and time, this is not an interference provided that all measurement solutions are affected to the same degree. If, on the other hand, concomitants cause the formation of the carbide to be more or less pronounced than in the calibration solutions, a solute-volatilization interference is present. The situation is very similar for analytes that form a volatile compound, such as a suboxide or a dimer, during reduction to gaseous atoms since some of this compound can be lost during the pyrolysis step.

A further interference that can take place during the atomization step, but which is not considered as a gas-phase interference, is *covolatilization or expulsion*. This can be observed, for example, when a concomitant spontaneously decomposes at the start of the atomization step, because the pyrolysis temperature was too low, with the formation of a large volume of gas, e.g. HCl, which then partly or fully carries the analyte with it and thus removes it from the determination [2640].

During the 1970s and at the start of the 1980s, the majority of papers on GF AAS were occupied with interferences. Since the STPF concept had not been developed at that time, we consider it to be of little value to go into these publications. A complete over-view of this early work is presented in a monograph by SLAVIN [5428]. The number of papers dealing with interferences in GF AAS has declined drastically since the middle of the 1980s, which can be ascribed to the increasing use of the STPF concept. Even nowa-days, though, it is still possible to find papers in which studies are made as to whether it is better to atomize a volatile element from the wall or platform and whether signal evaluation should be via peak height or integrated absorbance [1683, 2715]. If the result is then that an element such as lead or cadmium should be atomized from the tube wall and the signal evaluated via peak height, we can only say that either unsuitable instru-ments or experimental conditions, or incorrectly selected parameters for the decision, were chosen.

8.2.4.3.2 *Interference Mechanisms*

There are relatively few well-founded studies on interference mechanisms in GF AAS, and for obvious reasons only model compounds were selected for these investigations. Next to these diagnostic studies there are pure applications in which the interference-free determination of individual analytes or analyte groups is reported and in which the find-ings from the studies on interference mechanisms are translated into practice. In this section we shall deal mainly with the fundamental studies, while the application work is reported in Chapter 10.

The most notorious and best investigated interferents in GF AAS are the halogens, in particular the chlorides. Sulfates then follow in second place. The magnitude of the inter-ference as well as its mechanism depend on the cation to which the interferent is bound after the drying step. FRECH et al. [1967] investigated the influence of copper chloride, copper sulfate, sodium chloride, and sodium sulfate on 21 analytes. For these investiga-tions they used a conventional graphite atomizer operating under STPF conditions and a two-step atomizer. They compared the experimental results with the expected gas-phase interferences based on high temperature equilibria calculations. For 13 elements these authors observed slight signal suppression in the presence of copper chloride or sodium chloride which they attributed to preatomization losses, i.e. to a solute-volatilization interference. The difference between the calculated and the experimental data was most noticeable for aluminium and barium. The thermodynamic equilibria calculations pre-dicted complete signal suppression for both aluminium and barium in the presence of 1 g/L $CuCl_2$, while in practice the suppression for aluminium was negligible and for barium was only a few percent. The discrepancy could be attributed to the very effective removal of the chloride during the pyrolysis step. A further cause that these authors quote for poor agreement between calculated and experimental results is the fact that solutions were acidified with nitric acid so that a part of the chloride was removed at low temperature. Further, the analyte and the interferent do not necessarily have to be vaporized together, so that there is no perfect overlapping and thus no maximum gas-phase interference.

The only element whose signal was practically fully suppressed even by 2 µg sulfate was silicon. For about half of the remaining elements the signal suppression in the presence of 100 µg copper sulfate or sodium sulfate was less than 10%. The authors thus concluded that the platform technique and a suitable modifier effectively eliminate interferences due to sulfate. For arsenic and manganese, which exhibited relatively strong signal suppression in the presence of sulfate, the calculations indicated that this suppression was due to a solute-volatilization interference. For silicon even a very small amount of sulfur was sufficient to cause a gas-phase interference.

In individual cases, FRECH et al. [1967] were able to show that independent optimization of the atomization temperature prior to analyte vaporization, such as is possible with a two-step atomizer, reduces the gas-phase interferences. On the other hand, the interferences caused by simultaneous vaporization of the analyte and the interferent cannot always be counteracted by increasing the effective atomization temperature. Careful separation of interfering concomitants prior to the atomization step is thus still the most effective technique for avoiding interferences. A further important point is to avoid the simultaneous vaporization of non-separable concomitants with the analyte so that overlapping in the gas phase does not take place.

In their investigations of the influence of sodium chloride and nickel chloride on lead using a dual-cavity platform, WELZ et al. [6240] found that it was *not* a gas-phase interference. In the absence of a modifier losses of lead occurred above 500 °C in the presence of NaCl, probably in the form of $PbCl_{2(g)}$. The difficult-to-dissolve $PbCl_{2(s)}$ is formed during the drying step. Its melting point of 501 °C agrees well with the onset of the losses. Lead is practically quantitatively volatilized as molecules at 700 °C. This mechanism for the volatilization of lead chloride was verified by WENDL and MÜLLER-VOGT [6276]. In the presence of nickel chloride lead can no longer be detected when the pyrolysis temperature is merely 600 °C. In this case the expulsion of lead is mainly by $HCl_{(g)}$ which is formed by the hydrolysis of nickel chloride according to:

$$NiCl_2 \cdot 6H_2O_{(s)} \rightarrow NiO_{(s)} + 2HCl_{(g)} + 5H_2O_{(g)}.\tag{8.101}$$

Depending on the pyrolysis temperature, the lead is volatilized in molecular form and carried from the tube either during the pyrolysis step or at the start of the atomization step. It was shown that $HCl_{(g)}$ can undergo a heterogeneous reaction with any $PbO_{(s)}$ or $Pb_{(s)}$ that might be produced to form a chloride that can then be carried from the tube. The determination of lead was free of interferences in the presence of these substances when a suitable modifier was used that prevented the volatilization of $PbCl_2$ or the attack of $HCl_{(g)}$ so that the chloride matrix could be separated in the pyrolysis step.

WELZ et al. [6227] also found completely analogous reactions for antimony in the presence of nickel chloride. Depending on the pyrolysis temperature, the expulsion of the analyte with the hydrolytically decomposing matrix or via a heterogeneous reaction of $HCl_{(g)}$ with $Sb_2O_{3(s)}$ was dominant. Also in this case with the use of a suitable modifier and STPF conditions it was possible to eliminate or avoid all interferences.

BYRNE et al. [1006] came to very similar conclusions for the interference of magnesium chloride on the determination of manganese. For their investigations these authors used a combination of electrothermal vaporization (ETV) with ICP-MS. As discussed in the cases above, the interference mechanism depends on the pyrolysis temperature. The manganese is lost at pyrolysis temperatures above 700 °C. At this temperature magnesium chloride decomposes under hydrolysis according to:

$$MgCl_2 \cdot 6\,H_2O_{(s)} \rightarrow MgO_{(s)} + 2\,HCl_{(g)} + 5\,H_2O_{(g)}. \qquad (8.102)$$

The manganese is carried from the tube by the $HCl_{(g)}$. At pyrolysis temperatures below 700 °C the manganese signal is suppressed by the formation of manganese chloride in the gas phase. These authors eliminated both interferences by adding ascorbic acid which prevented the losses of manganese during the pyrolysis step, presumably by delaying the hydrolysis of magnesium chloride.

BYRNE et al. [1008] attributed the interference of NaCl on the determination of manganese to the formation of manganese chloride during atomization and not to losses during pyrolysis. This gas-phase interference could also be eliminated by adding ascorbic acid; the authors attributed this to the fact that the chloride is vaporized at a temperature 250 °C lower than manganese so that no overlapping in the gas phase takes place. We should nevertheless mention that these investigations were not conducted in a graphite tube but on an open platform from which the gases were conducted to the ICP-MS.

SHEKIRO et al. [5292] came to the conclusion that the interferences caused by calcium chloride and magnesium chloride on the determinations of copper and manganese were caused by the formation of a chloride in the gas phase. These experiments were performed in the g/L range, however, so that they are hardly transferable to the analyte concentration which is 4–5 orders of magnitude lower.

HAM and MCALLISTER [2341] used MS and thermodynamic equilibria calculations to investigate the influence of chloride on the determination of cobalt. The authors found that both $CoCl_{(g)}$ and $CoCl_{2(g)}$ are formed during pyrolysis at 500 K. The calculations indicated that above 1000 K $CoCl_2$ can be formed in the presence of HCl. Albeit $CoCl_2$ decomposes to $Co_{(g)}$ close to 1600 K, so that above this temperature an interference due to chloride should not be present. Under the experimental conditions used, Ham and McAllister nevertheless found that there were substantial losses of $CoCl_{2(g)}$.

Due to the low sublimation points of SnS (1020 K) and $SnCl_2$ (925 K), RAYSON and HOLCOMBE [4828] concluded that these molecules are preferentially volatilized during pyrolysis. On the other hand, LUNDBERG et al. [3663] found that the interferences of Cl and S on the determination of tin were practically eliminated in a two-step furnace. Although these results initially appear to contradict each other, they merely differ in the pyrolysis temperatures applied. If a pyrolysis temperature is selected that is above the sublimation point of the chloride or the sulfide, these molecules are volatilized prior to atomization. If a lower pyrolysis temperature is selected, these molecules are not volatilized until the atomization step; the effective temperature prevailing in the atomizer then decides whether these thermally stable molecules are dissociated into atoms or not.

Next to the frequent case that a molecular analyte species leaves the absorption volume without being atomized, there is also the possibility that the analyte is irreversibly bound in the graphite tube, usually as the carbide. In this, as in similar cases, we only speak of an interference if the fraction of analyte retained is different with and without matrix. Using radiotracers, VEILLON et al. [6011] found that the quantity of chromium remaining irreversibly in the graphite tube was dependent on the matrix. CASTILLO et al. [1106] found for the analysis of chromium-chelate complexes that the formation of chromium carbide with the carbon residues of the organic constituents led to a loss of sensitivity.

MÜLLER-VOGT et al. [4228] explained the interferences of elements such as Ca, La, Ti, Zr, V, Nb, Cr, and W on the determination of molybdenum via the atomization mechanism. At around 1300 K Mo_2C is formed by the reduction of MoO_3. Above 1900 K Mo_2C is very slowly converted to MoC. With rapid heating rates both MoC and Mo_2C thus reach the atomization phase. The interference of the above elements is based on the influence of their carbides on the decomposition reaction of the molybdenum carbide phase. A similar mechanism, namely the formation of mixed carbides, was also proposed by MANNING and SLAVIN [3864] for the interference on the determination of vanadium by other carbide-forming elements such as lanthanum, molybdenum, tungsten, and zirconium.

8.2.4.3.3 Avoiding Non-spectral Interferences

While discussing atomization mechanisms we also mentioned that a number of elements form relatively volatile intermediate compounds on their way to gaseous atoms (see Section 8.2.3). Examples include beryllium [5654] and aluminium [5653] which form volatile hydroxides; silicon forms both a volatile oxide and a gaseous carbide [4774]. Chromium [1108], germanium [3223, 6273, 6421], thallium [2318], and tin [879, 3663, 4828] form volatile suboxides, and arsenic [5655] and selenium [5656] form both volatile oxides and dimers.

We have mentioned repeatedly that the STPF concept, and especially chemical modification, are indispensable for reliable analyses with GF AAS. One of the aims of chemical modification is to prevent analyte losses. Magnesium oxide (which is formed from the magnesium nitrate modifier), for example, is preferentially hydrolyzed to the oxide and thus prevents the formation of volatile hydroxides of aluminium [5653] and beryllium [5654]. Lead [411] and cadmium [1727] form very stable compounds with ammonium hydrogen phosphate of the type $(MO)_xP_2O_5$, so-called phosphate glasses. It has been reported that numerous elements form solid solutions or intermetallic compounds with palladium, such as Pd_3Sn_2 [2165], Pd_3Pb_2 or $PdIn_3$ [4452]. These investigations were not, however, performed with analytical masses; the formation of phases or compounds with the analyte are improbable considering the extreme excess of modifier that is normally used. Arsenic and selenium, on the other hand, form compounds of the type $Pd_xM_yO_z$ which can be ascribed to the penetration of palladium into the upper graphite layers. Resulting from the formation of covalent palladium-graphite bonds the palladium atoms are 'activated' and can then bind the analyte atoms covalently [3817, 4488, 4937, 5655].

Since the test samples in GF AAS, especially when they are liquids such as water or body fluids, often can only be incompletely digested or not at all, it is a further important function of chemical modification to stabilize all species to the same level. Thus for example it was not possible to stabilize all oxidation states of selenium with nickel, a previously frequently used modifier [3299]. Selenomethionine could not be stabilized with nickel at all [3299]. Generally, uncontrollable losses of selenium occurred in the presence of excess organic materials when nickel was used as modifier [1457]. The Pd-Mg modifier exhibits the best stabilizing properties for all oxidation states of selenium [1560, 3299]. Similar results were also found for various oxidation states of arsenic [3297].

The greatest significance of the STPF concept is to be found in the elimination or avoidance of gas-phase interferences. In the course of their comprehensive studies on 21 elements, FRECH et al. [1967] established two essential points. Firstly they found that generally there were fewer interferences in integrated absorbance, with the exception of a few carbide-forming elements. They pointed out that in contrast to the signal area, the peak height is a function of the rate of atom formation and that this is very dependent on the matrix. Secondly they found that in a two-step furnace, i.e. in an isothermal atomizer in which the atomization temperature can be optimized independently of the volatility of the analyte, interferences due to chlorides could not always be prevented. The signal suppressions in the presence of copper chloride at 3000 K calculated by these authors are listed in Table 8-6. From this we can see that with *simultaneous volatilization* of the analyte and the matrix, interferences often cannot be avoided even by increasing the effective atomization temperature. This makes clear the significance of *separation of the analyte and matrix* prior to the atomization step. If this separation is not successful, at least simultaneous volatilization of the analyte and interferent should be avoided to reduce overlapping in the gas phase to a minimum. Chemical modification must therefore take over a key role in this task.

Table 8-6. Calculated signal suppression in % at 3000 K in the presence of various concentrations of copper chloride (from [1967]).

Analyte	Signal suppression in % caused by different concentrations of $CuCl_2$			
	10 g/L	3 g/L	1 g/L	0.3 g/L
Ag	47	29	15	5
Al	100	100	98	92
As	0	0	0	0
Ba	100	100	100	100
Be	100	100	98	88
Bi	15	7	3	1
Co	79	47	17	3
Cr	0	0	0	0
Cu	74	56	35	14
Fe	95	80	42	7
Mn	96	83	46	7
Ni	100	99	97	91
Pb	51	31	16	5
Sb	0	0	0	0
Se	16	16	16	16
Si	98	93	78	56
Sn	36	10	2	0
Te	0	0	0	0
V	0	0	0	0

FRECH and CEDERGREN [1960] found that *hydrogen* was suitable to eliminate the interference of chloride on the determination of lead in steel; the chloride is removed as $HCl_{(g)}$. A number of organic modifiers such as ascorbic acid or oxalic acid that have occasionally been used appear to have the same effect since they form hydrogen and carbon monoxide on decomposition [1007, 2110, 6104]. 'Modifiers' of this type should nevertheless only be used with the greatest of caution and for clearly defined matrices since they are mostly unable to stabilize the analyte itself (refer to Section 8.2.4.1, part 3). In many cases the analyte is even atomized at a lower temperature than without modifier [2110], and for a number of analytes increased losses can occur due to the formation of volatile hydrides [1457].

As mentioned in Section 1.7.2, EDIGER *et al.* [1676] proposed the use of *ammonium nitrate* to remove sodium chloride in the form of sodium nitrate and ammonium chloride. CHANDHRY and LITTLEJOHN [1222] found that about 85–90% of the chloride could be removed from a platform at 200 °C in this way, but that the remaining chloride did not volatilize until 1000 °C. The sodium nitrate or the resulting sodium oxide, which is reduced to metallic sodium, has a certain stabilizing effect on a number of elements [1676], but it certainly cannot be considered as an effective modifier. For this reason ammonium nitrate should only be used as a stabilizer in exceptional cases, preferably together with palladium [6481, 6455].

Ammonium hydrogen phosphate or *ammonium dihydrogen phosphate* has been used very successfully to stabilize lead and cadmium. For silver, on the other hand, this modifier exhibits no stabilizing effect [1727], despite claims to the contrary [1389]. The stabilization mechanism is discussed in detail in Section 8.2.4.1, part 1. During their studies with a dual-cavity platform, WELZ *et al.* [6240] found that the interference due to chloride on the determination of lead was eliminated with the phosphate modifier. WENDL and MÜLLER-VOGT [6276] confirmed this with XRD and MAS measurements, but they pointed out that a large excess of modifier was necessary. Using radiotracers, SCHMID and KRIVAN [5156] found that the combination of a phosphate modifier and platform atomization brought optimal results for the determination of lead in a variety of matrices, including blood and urine.

On the other hand, BAXTER and FRECH [438] found that the Pd-Mg modifier was superior to the phosphate modifier for the determination of lead and cadmium in solid samples. PENNINCKX *et al.* [4596] confirmed this finding for the determination of lead in various biological materials and in foodstuffs. We should nevertheless mention that in the cited work on the phosphate modifier the Pd-Mg modifier was not investigated, so that a contradiction must not necessarily be present. In a number of papers on the determination of lead in slurried soil samples, HINDS and co-workers [2558–2560] found that a relatively small amount of palladium was sufficient to stabilize lead. Additionally, the atomization was delayed until practically isothermal conditions existed in the graphite tube. In order to eliminate all interferences and to obtain a similar signal shape for suspensions and aqueous solutions, however, it was necessary to use more palladium and the additional presence of magnesium nitrate.

WELZ *et al.* [6265] investigated the interference-free range for 20 elements in the presence of NaCl and K_2SO_4 as model substances using the Pd-Mg modifier. The results are summarized in Table 8-7. The individual values serve in this case to demonstrate the effectiveness of the Pd-Mg modifier, not only in respect to its stabilizing power, but also to which level it contributes to the separation of the matrix and thus to freedom from interferences. It is also interesting to make a comparison with Table 8-6 and the calculated signal suppression by copper chloride. In Table 8-7 100 µg NaCl correspond to a concentration of 10 g/L.

The Pd-Mg modifier is currently the most powerful and universal modifier, not only in respect to its ability to stabilize the analyte, but also in providing freedom from interferences. Palladium alone can stabilize the analyte in a similar manner, but it has only a limited effect or even no effect for eliminating interferences due to chloride. Additionally, palladium should always be used as the nitrate and not as the chloride, since this would introduce additional chloride and thus increase the susceptibility to interferences.

At this point it would be relevant to discuss 'reduced palladium' (Pd^0) which was mentioned in connection with atomization mechanisms (Section 8.2.3) and chemical modification (Section 8.2.4.1, part 1). Unfortunately there are virtually no direct comparisons between Pd^0 and the Pd-Mg modifier, since most comparisons were between reduced and non-reduced palladium without magnesium nitrate. A further difficulty in making a comparison is that some of the investigations on Pd^0 were not performed under STPF conditions [4853] and are therefore not especially expressive. The manner in which the palladium is reduced also appears to have an influence on its effect

Table 8-7. Interference-free measurement range (< 10% signal suppression) in the presence of NaCl and K_2SO_4 using the Pd-Mg modifier (from [6265]).

Analyte	Pyrolysis temperature	Atomization temperature	Tolerable mass of interferent (µg)*	
	(°C)	(°C)	NaCl	SO_4^{2-}
Ag	900	1600	300	100
Al	1700	2350	300	100
As	1400	2200	200	10
Au	1000	1800	10	5
Bi	1200	1900	10	100
Cd	900	1700	5	5
Cu	1100	2300	300	300
Ga	1300	2200	200	10
Ge	1500	2550	100	3
Hg	250	1000	300	50
In	1500	2300	300	50
Mn	1400	2300	200	200
Pb	1200	2000	200	10
Sb	1250	1900	300	10
Se	1200	2100	50	6
Si	1200	2500	200	1
Sn	1200	2400	50	50
Te	1200	2250	200	30
Tl+	1000	1650	100	10
Zn	1000	1900	100	20

* 10 µL measurement solution
+ reduced palladium

[6105, 6108, 6579]. We can obtain Pd⁰ most easily by heating a palladium salt on a plat-form in a graphite tube to 900–1000 °C. This procedure can be accelerated and per-formed at a lower temperature when hydrogen is mixed with the inert purge gas. Organic substances such as ascorbic acid or citric acid have a similar effect since they release hydrogen when they decompose. A further possibility is the application of metallic pal-ladium in finely dispersed form [6105].

In our experience Pd^0 merely is of advantage for the determination of thallium in the presence of high chloride concentrations [6241]. QIAO et al. [4745] attribute the in-creased stabilization to an adsorption of the analyte through underpotential deposition on the Pd^0. The analyte remains adsorbed on the palladium until the sodium chloride is volatilized at 1100 °C in the pyrolysis step. Upon further heating the thallium is embed-ded in the molten palladium and diffuses out during atomization. These authors also found that the addition of magnesium improved the determination of thallium even fur-ther. This aspect is discussed in further detail under thallium (Section 9.51).

8.2.5 Absolute Analyses

Even in his first paper in 1955, WALSH [6135] pointed out the theoretical possibility of performing absolute analyses by AAS. The use of a flame as the atomizer, with the nu-merous difficulties, retarded the realization of this idea, but the development by L'vov of the graphite cuvette as the atomizer brought this idea within reach. L'VOV [3679] only a few years later compared experimental data with theoretically calculated absorbance values; he attributed the poor agreement to incomplete spectroscopic data. With his paper *Electrothermal atomization—the way towards absolute methods of atomic absorption analysis*, published in 1978, L'VOV [3704] established the basis of modern GF AAS and also of the STPF concept. Some time later, SLAVIN and CARNRICK [5427] were able to show that using the STPF concept, GF AAS offered the possibility of performing deter-minations with an accuracy of 10–20% in varying and complex matrices without cali-bration. As the reference quantity they used the characteristic mass, m_0, which they cor-respondingly corrected for the use of differing tube dimensions. The authors quoted especially varying atomization temperatures, spectral bandpasses, and lamp currents as being the factors responsible for the scatter.

L'VOV et al. [3724] attempted to calculate m_0 on the basis of a simplified model ac-cording to:

$$m_0 = 5.08 \cdot 10^{-13} \frac{MD\Delta\tilde{v}_D}{H(\alpha,\omega)\gamma f} \frac{Z(T)}{g_1 \exp(-E_1/kT)} \frac{r^2}{\ell^2} . \tag{8.103}$$

This model assumes that the gaseous atoms diffuse from the center of the tube to the ends in an atomizer that is isothermal both with respect to time and space. M is the mo-lecular mass of the analyte, D is the diffusion coefficient of the atoms in argon at tem-perature T, $\Delta\tilde{v}_D$ is the double line width, $H(\alpha,\omega)$ is the Voigt integral for a point on the profile of the absorption line that is removed from the line center by $\omega = 0.72\,\alpha$, γ is a coefficient that takes the hyperfine structure of the analytical line into account, and f is the oscillator strength. $Z(T)$ is the sum of the states at temperature T, g_1 and E_1 are the statistical weight and the energy of the lower energy level of the analytical line, and r and ℓ are the internal radius and the length of the graphite tube.

For 30 elements that were measured under STPF conditions, L'VOV et al. [3724] found a mean value of 0.90 with a standard deviation of 0.25 for the ratio of calculated to measured characteristic mass, $m_0(\text{calc}) / m_0(\text{exp})$. For 10 elements that were atomized from the tube wall the agreement was poorer, as was to be expected. The authors were of the view that the results obtained indicated that absolute analyses for AAS based on the STPF concept could be developed.

BAXTER and FRECH [434] calculated the corresponding m_0 values for 21 elements for their transversely-heated, spatially isothermal atomizer on the basis of equation 8.103 and compared these values with the experimental results. They showed that for a given tube length a spatially-dependent isothermal atomizer gives better $m_0(\text{exp})$ values than an atomizer with a temperature gradient since this in effect reduces the tube length of a Massmann-type furnace.

Due to its notable stability, the characteristic mass m_0 is a useful control criterion in GF AAS that in the meantime is universally recognized. Within a laboratory the m_0 can serve to check the correct functioning of an instrument and the correctness of the selected parameters. It can also be used to provide information on trueness during the development of new methods and serves as a criterion in interlaboratory trials. Values for m_0 that are too high mostly indicate that the method or instrument is not optimized, while values that are too low are nearly always an indication of contamination. When comparing different atomizers the tube dimensions must naturally be taken into account as given in equation 8.103. Also, the splitting and broadening of the lines caused by the Zeeman effect has an influence, as mentioned in Section 3.4.2.

Independent of the recognized uses of the characteristic mass m_0, the topic of 'absolute analyses' has been a subject of extremely controversial discussion for many years [1371]. Various authors obviously have differing preconceptions about the practical application of the concept. In how far 'standardless analysis' as proposed by SLAVIN and CARNRICK [5427] should really do without the use of calibration samples and calibration solutions is not always clear.

The stability and reproducibility of m_0 under STPF conditions is $\pm$ 10–20%, depending on the element. A major factor for this relatively high level of uncertainty are the emission profiles of the analytical lines which depend on the quality of the lamp and the lamp current [3739]. Since m_0 is influenced by a number of further factors, attempts to correct such influences by means of empirical models, for example by parameters derived from the Zeeman ratio and the roll-over absorption [3740–3742, 3744, 3747, 3748, 3750, 6506], must remain of limited applicability.

Even if the remaining uncertainty is acceptable within a laboratory, the functionality of the instrument must nevertheless be regularly checked (at least once daily), so that at least for this purpose it is not possible to do without calibration samples. Quite apart from the fact that nowadays the majority of laboratories are required by regulation to use a given number of calibration samples at given intervals to check the accuracy of their analytical results (refer to Section 5.5).

8.3 The Hydride-generation Technique

The bulk of the publications on the hydride-generation technique are application-based and only a relatively limited number of working groups have been involved with the question of mechanisms. An additional problem in comparing the results is that the majority of studies were performed using homemade apparatus which often differed in numerous details. These differences in the apparatus frequently have a considerable influence not only on the atomization, and thus on the sensitivity, but also on the interferences observed. In addition, considerably differing experimental conditions were used, such as the sample volume, the concentration, volume and composition of the acid, and the nature, volume and concentration of the reductant, and many more parameters besides.

In this section we shall simplify the discussions by referring solely to those systems and conditions that have contributed substantially to an understanding of the events taking place. For the same reason we shall restrict ourselves to sodium tetrahydroborate as the reductant since this is used almost exclusively nowadays. Further, we shall only consider quartz tube and graphite tube atomizers since the flame is nowadays hardly used.

8.3.1 Generation and Transport of the Hydride

8.3.1.1 Reduction to the Hydride

The net reaction of sodium tetrahydroborate in acidic solution and the simultaneous reduction of the hydride-forming element can be described quite simply as follows:

$$BH_4^- + H_3O^+ + 2\,H_2O \rightarrow H_3BO_3 + 4\,H_2 \tag{8.104}$$

and

$$3\,BH_4^- + 3\,H^+ + 4\,H_2SeO_3 \rightarrow 4\,SeH_2 + 3\,H_2O + 3\,H_3BO_3, \tag{8.105}$$

where equation 8.105 merely serves as an example for all the hydride-forming elements. Nevertheless, equations 8.104 and 8.105 are coarse simplifications of the true processes, which to date are still unknown in detail. We only have to mention the term 'nascent hydrogen', which is still not explained, and a number of other observations in connection with the decomposition of sodium tetrahydroborate in acidic solution, which we shall discuss briefly below.

The finding that under optimum conditions the generation of the hydride and the transport to the atomizer is in principle virtually quantitative for all the hydride-forming elements is of major significance in practice [1458]. By using radiotracers, for example, various authors have shown that the conversion of selenium into SeH_2 and its transport to the atomizer takes place with a yield of 95% [1455, 3293]. Similar figures have also been determined for bismuth [3482], and even lead, without doubt the most difficult element, is converted into lead hydride to about 80% [2922]. This is important for the following considerations, because practically quantitative transfer of the analyte to the atomizer means that all subreactions must also proceed equally quantitative. 'Optimum conditions' can mean, for example, that an oxidizing reagent must be added to lead [1103] or that a buffer must be added to germanium and tin [197, 5646, 5776]. It can also mean that an element must be converted to a lower oxidation state prior to the determination, such as is essential for selenium and tellurium, and recommended for arsenic and antimony. These matters are discussed in part in this section and in detail under the individual elements.

8.3.1.2 Driving the Hydride out of Solution

Based on kinetic observations of the peak form, VAN WAGENEN et al. [6121] came to the conclusion that As(III) is quantitatively converted to AsH_3 in less than 0.1 s. The rate-determining step is thus *driving out* the arsine from solution; for this purpose the hydrogen generated is far more effective than the purge gas. DĚDINA [1455] came to very similar results in his studies on the formation and release of selenium hydride.

In a series of interesting experiments on this topic, AGTERDENBOS and co-workers [63, 64, 4296] showed that in a flow system 70% of the sodium tetrahydroborate solution had decomposed 0.3 ms after it had come into contact with 1 mol/L sulfuric acid; 80% had decomposed after 1 ms and no further $NaBH_4$ could be detected after 10 ms [62]. Neither the concentration of the tetrahydroborate nor of the acid had any noticeable influence. The authors concluded that the formation of the hydride, at least for Se(IV) and As(III), was complete within this period. The fact that a reaction zone interposed after the reactor increased the sensitivity for both elements led them to the conclusion that the release of the hydride was the rate-determining step. This agreed with the results obtained by VAN WAGENEN et al. [6121]. NARSITO et al. [4296] also found that the hydrogen generated by the reaction with sodium tetrahydroborate played a major role in driving out the hydride.

For As(V), on the other hand, NARSITO et al. [4296] found that the reaction was much slower and that this species continued to be reduced in the reaction loop. The authors explained this by the fact that $NaBH_4$ is not converted directly into H_3BO_3 according to equation 8.105 but into an intermediate product that is still able to produce AsH_3 from As(V). This observation signifies that equation 8.105 is a coarse simplification of the true situation (as we have already mentioned). This fact is further confirmed by an observation by DĚDINA [1455] that 'old', reacted tetrahydroborate solution can retain a large quantity of selenium. He further found that hydrochloric acid which had been 'purified' with a little $NaBH_4$ led to substantial losses in sensitivity for selenium. Since boric acid, which is the end product of the reaction, had no influence, he concluded that an unknown reaction product between sodium tetrahydroborate and hydrochloric acid is formed.

In a series of investigations, SANZ-MEDEL and co-workers [1896, 5094, 5095] showed that in the presence of surface-active organic media, such as micelles or vesicles, the characteristics of chemical vapor generation are noticeably improved. This can be mainly ascribed to efficient expulsion, the stabilization of unstable compounds, and improved transport of the volatile species.

8.3.1.3 The Influence of the Hydride-generation System

The design of the hydride generator and its mode of function can also have a substantial influence on the release of the hydride. In batch systems in which the acidic measurement solution is present and the alkaline sodium tetrahydroborate solution is added over a given time period, the pH value for example can change markedly during the reaction. AGTERDENBOS et al. [64] showed that selenium hydride is reabsorbed by the solution according to equations 8.106–8.108 when the pH value is raised after the generation of the hydride:

$$SeH_{2(g)} \rightleftharpoons SeH_{2(\ell)} \tag{8.106}$$

$$SeH_{2(\ell)} \rightleftharpoons SeH^- + H^+ \tag{8.107}$$

$$SeH^- \rightleftharpoons Se^{2-} + H^+. \tag{8.108}$$

Along with the change in the pH value during the reaction in batch systems the development of hydrogen can also change. This can lead to a retarded release of the hydride, for example, and to tailing of the atomization signal, as was observed by VAN WAGENEN et al. [6121].

In flow systems, on the other hand, the solutions are mostly pumped and mixed continuously so that a certain reaction equilibrium can be established. In contrast to batch systems, in which the reaction and separation of the hydride take place in the same container, in flow systems these processes mostly take place in separated zones. The reaction time, during which phase separation naturally already takes place, can be influenced by the length of the reaction zone and the flowrate. By using short reaction zones and pumping off the reaction solution rapidly following phase separation gives the possibility of largely suppressing kinetically slow reactions. This technique is termed *kinetic discrimination* and is used especially in FI systems to eliminate interferences or for speciation analysis [2365].

A further possibility of influencing the reaction in the FI technique, in which a relatively small volume of the measurement solution (< 1 mL) is injected into the carrier solution, is to select substantially differing pH values for the measurement and carrier solutions. In this way, for example, the development of hydrogen can be increased or terminated toward the end of the measurement signal by either correspondingly raising or lowering the pH value (carrier solution acid or water).

8.3.1.4 The Influence of the Oxidation State of the Analyte

The differences between batch and flow systems mentioned above are also of considerable importance when observing the influence of differing oxidation states of hydride-forming elements on the kinetics of the reaction. For the elements in Group IV of the periodic system a difference has merely been reported for tin; Sn(IV) exhibits the highest sensitivity [1111]. This fact should cause no analytical problems, however, since Sn(II) compounds are unstable and rapidly oxidized to Sn(IV). For germanium there are no references in the literature to influences due to the oxidation state, whereas lead must always be determined in the presence of a strong oxidant [1101, 1103], from which we can conclude that the reaction proceeds from the tetravalent state.

The situation is clear for the elements in Group VI, selenium and tellurium, since the hexavalent oxidation state does not generate a measurable signal in any situation. Reduction to the tetravalent state prior to determination by HG AAS is thus mandatory for these two elements. For this purpose boiling hydrochloric acid in the concentration range 5–6 mol/L is used almost exclusively [1384, 4840, 4841]. While a short period of heating is adequate for the reduction of Te(VI) [5375], at least 15 minutes are required for the reduction of Se(VI) [1000, 5375].

For the elements in Group V, arsenic and antimony (bismuth is practically only present in the trivalent state), the difference between the trivalent and pentavalent oxidation states depends very strongly on the system used and the experimental conditions. In batch systems and at a pH ≤ 1 arsine is formed more slowly from As(V) than from As(III), as shown in Figure 8-35. This results in a peak height that is 25–30% lower, but the peak areas are more or less the same [3020, 5341]; these differences additionally depend on the concentrations of the acid and the sodium tetrahydroborate solution [2569, 5830]. These conditions are also quite similar for antimony except that the difference in the peak heights for the trivalent and pentavalent states is somewhat more pronounced.

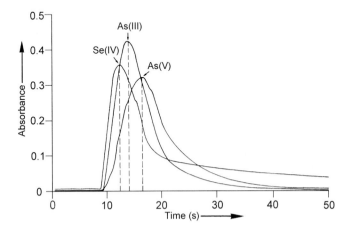

Figure 8-35. Atomization signals for As(III), As(V), and Se(IV) using a batch hydride-generation system.

The situation is significantly different in flow systems where the slower reduction of the pentvalent oxidation state leads to greater differences in sensitivity due to kinetic discrimination. This is shown in Figure 8-36 for an FI system with a short reaction coil. The difference in sensitivity for arsenic is an order of magnitude and for antimony it is even a factor of 25 [6268]. These differences can be reduced but not completely eliminated by using longer reaction zones [6256].

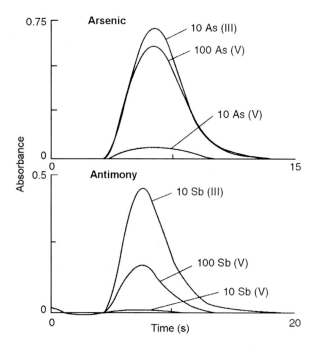

Figure 8-36. Absorbance signals for different concentrations of As(III)/(V) and Sb(III)/(V) using FI-HG AAS; 500 µL sample loop. All values in µg/L (from [6268]).

Using batch systems, a number of authors omitted the prereduction of As(V) and Sb(V) for simplicity when these species were present after a digestion [1780, 6210, 6212]. This practice cannot be recommended with flow systems due to the loss in sensitivity. To perform this reduction it was usual earlier to use potassium iodide, either alone [3986, 5375] or mixed with ascorbic acid [2601, 5342]. A certain disadvantage is that relatively high concentrations of reductant (30–100 g/L KI) and acid (5–10 mol/L HCl) are required and that potassium iodide solutions are unstable. Under these conditions antimony is reduced almost spontaneously while arsenic reacts more slowly so that heating the solution (in a reactor if it is a part of the FI system [4361]) and a longer delay period are required [5375]. BRINDLE and co-workers [795, 1230, 1231] proposed L-cysteine as an alternative reductant to potassium iodide. The advantages of this reductant include the fact that it is already effective at concentrations of 5–10 g/L and that the optimum concentration of acid is 0.1–1 mol/L HCl [6268]. The disposal of residues after the analysis is thus far less of a problem. A further advantage of L-cysteine is that it stabilizes the reduced analytes arsenic and antimony in the trivalent state for several days [6268].

Although the differing behavior of the individual oxidation states to reduction with sodium tetrahydroborate is troublesome for the determination of the total concentration, it can be utilized for speciation analysis. This is particularly simple for the elements in Group VI since without prereduction only the tetravalent oxidation state can be selectively determined, while after prereduction the total concentration of ionic selenium or tellurium is determined. For arsenic and antimony we can make use of the fact that at pH values of 5–7, instead of the usual pH of < 1, the trivalent state is reduced to the hydride but not the pentavalent state. Using a whole series of buffers, ANDERSON *et al.* [191] were able to selectively determine As(III), dimethylarsinic acid, the total inorganic arsenic, and the total arsenic.

8.3.1.5 Transport of the Hydride

In this section we shall discuss the transport of the gaseous hydride to the atomizer *after phase separation*, not the transport of solutions in flow systems. In addition to the hydrogen released during the reaction a further purge gas is used in most cases. This can be auxiliary hydrogen or an inert gas such as argon or nitrogen. Helium must be used as the purge gas if it is intended to trap the hydride in a cold trap since both argon and nitrogen can condense out with the hydride in the cold trap.

DĚDINA [1455] found substantial differences between the calculated and the observed atomization signals for the hydride-generation technique. The peaks are broader and appear later than predicted by theory. As a possible cause he suggested sorption-desorption interactions during transport of the hydrides to the atomizer. At the same time, however, he established that the transport was virtually quantitative when the hydride was not collected. As we mentioned earlier, selenium hydride [1455] and especially bismuth hydride [2008] are relatively unstable and decompose fairly quickly in particular during collection under pressure, but also in a cold trap [1455]. For this reason direct transfer to the atomizer or collection in the atomizer (graphite tube) is to be preferred.

The magnitude of the interactions of the hydride with the surfaces it comes into contact with should be inversely proportional to the gas flowrate and directly proportional to the cross-section of the hydride generator and the tubes [1458]. Losses of hydrides should thus be lower in systems with a small dead volume and high gas flowrates. Nevertheless, since the gas flowrate has a considerable influence on the atomization of the hydride, it cannot be increased at random. Taken generally, these considerations speak very strongly for miniaturization, such as to be found in FI systems, and against batch systems with their large-volume reactors. The above points also explain the difficulties reported by a number of authors since they were using systems that were far from optimum [4535, 4831].

8.3.2 The Atomization of Hydrides

In the following section we shall discuss atomization in quartz tube atomizers and graphite tube atomizers. In addition to the externally heated QTA, which is the most widely used type, we shall also mention the flame-in-tube atomizer in which a small oxygen-hydrogen flame burns in the inlet. Discoveries of decisive importance were made at an early stage using this type of atomizer so that for this reason alone it should be mentioned. Additionally it is used quite widely for speciation analysis since it has a number of advantages over the QTA. We shall not discuss atomization in nitrogen- or argon-hydrogen diffusion flames since this type of flame is no longer used nowadays. In addition it is inferior in virtually all respects to QTAs and graphite tube atomizers.

8.3.2.1 Non-heated Flame-in-tube Atomizers (FIT Atomizers)

The FIT atomizer is mentioned briefly in Section 1.8.2 and depicted in Figure 1-46. Fundamental conclusions about atomization in the hydride-generation technique were elucidated by DĚDINA and RUBEŠKA [1453] using this apparatus in 1980. As was later established, the observations they made on the atomization of selenium hydride in the FIT atomizer could be largely transferred directly to the QTA.

Free radicals are generated in the reaction zone of the fuel-rich oxygen-hydrogen flame burning in the inlet tube of the atomizer according to:

$$H + O_2 \rightleftharpoons OH + O \tag{8.109}$$

$$O + H_2 \rightleftharpoons OH + H \tag{8.110}$$

$$OH + H_2 \rightleftharpoons H_2O + H. \tag{8.111}$$

In the presence of excess hydrogen only OH and H radicals are formed and in quantities corresponding to the total quantity of oxygen, i.e. two H radicals per oxygen molecule. Since the recombination of radicals proceeds much more slowly than their formation, a quantity far above the equilibrium concentration can be formed in the secondary reaction zone. As a result of the high reaction rate of reaction 8.110 a state of equilibrium between H and OH radicals is very quickly established. Since the hydrogen concentration in the atomizer is additionally much higher than the water concentration, the concentration of H radicals should be several orders of magnitude higher than of the OH radicals.

The actual atomization proceeds via a two-step mechanism with the prevailing radicals:

$$SeH_2 + H \rightarrow SeH + H_2 \quad \left(\Delta H = -189 \text{ kJ/ mol} \right) \tag{8.112}$$

$$SeH + H \rightarrow Se + H_2 \quad \left(\Delta H = -131 \text{ kJ/ mol} \right). \tag{8.113}$$

It is also possible to conceive the corresponding reactions with the OH radicals, but the role played by these reactions is probably negligible considering the low concentration of OH radicals. A further reaction that must be considered is recombination:

$$Se + H \rightarrow SeH \qquad\qquad \left(\Delta H = -131 \ kJ/mol \right). \qquad\qquad\qquad (8.114)$$

If we merely consider reactions 8.112–8.114, we can easily show that an equilibrium is attained after a sufficiently large number of collisions with H radicals. There is no SeH_2 present in this equilibrium and the ratio between SeH and selenium corresponds to the ratio of the constants of formation of the reactions 8.113 and 8.114. Since recombination (equation 8.114) is strongly exothermic and requires a third partner to take up the energy released, we can assume that this reaction is much slower than the formation of selenium atoms. The probability for the formation of free selenium atoms from SeH_2 is thus proportional to the number of collisions with free radicals, and the efficiency of atomization should increase with the increasing number of free radicals. At an optimum flowrate of oxygen all the selenium is present as free atoms.

The H radicals form a spatially limited cloud which does not reach the absorption volume when the oxygen flowrate is optimum. The cloud of H radicals nevertheless stretches far beyond the hot zone of the flame, which is limited to the direct neighborhood around the end of the capillary for the introduction of oxygen. It thus follows that the radical density in this cloud is much higher than under conditions of thermodynamic equilibrium. This means that the fate of the H radicals is controlled by kinetic rules rather than by thermodynamic rules.

From the flow velocity of the gases in the FIT atomizer we can deduce that the lifetime of the H radicals is less than 1 ms. The total number of collisions of the analyte hydride with H radicals, and thus the local atomization efficiency in the radical cloud, does not depend on the *number* of H radicals but on their *density*. This conclusion is supported by the observation that the oxygen requirement for maximum sensitivity is higher in a QTA atomizer with a wider side inlet tube.

The most probable route for the attrition of the H radicals is assumed in reaction 8.115:

$$H + O_2 + X \rightarrow HO_2 + X, \qquad\qquad\qquad\qquad\qquad (8.115)$$

where X is a third body that can take up the released energy. This can be either a water molecule or the quartz surface of the atomizer. The HO_2 radical then combines with a further H radical to form O_2 and H_2.

The distribution pattern of the free atoms in the atomizer is derived from the supply of the analyte to and the removal of the free atoms from the absorption volume. For the latter, in addition to the forced gas stream there are further reactions the nature of which is in part still unknown. The fact that free selenium atoms can be observed in the absorption volume but that their existence in a hydrogen atmosphere under 1200 °C is thermodynamically forbidden [1117] is an indication that this system is far removed from a state of equilibrium, i.e. the attrition of the free selenium atoms is too slow for equilibrium to be attained.

Various observations lead to the conclusion that the attrition of free analyte atoms takes place mainly at the wall of the quartz tube. We can determine a limiting value for the gas flowrate under which all analyte atoms react onward within the absorption volume. Likewise we can determine an upper limiting value for the gas flowrate above which practically no onward reactions in the absorption volume take place. The reaction mechanism for free atoms at the quartz surface is still unknown. It could conceivably comprise of the following steps:

i)   Free analyte atoms lose part of their energy through interactions with the quartz surface.

ii)   Analyte atoms bound to the surface form compounds such as dimers, polymers, or hydrides.

iii) Since no noticeable quantities of analytes can be detected in the quartz tube even after prolonged use of FIT atomizers, we can assume that the compounds formed according to ii) leave the quartz surface after a given residence time.

8.3.2.2 Atomization in Heated Quartz Tubes (QTAs)

Similar to CHU et al. [1279], who were the first to use an electrically heated quartz tube as the atomizer, the majority of other authors paid little attention to the mechanisms taking place in such systems that led to atomization. Based on comments published in a number of papers [1780, 4067, 5827], the prevailing opinion was that atomization took place via a simple thermal dissociation according to:

$$2\,AsH_3 \rightarrow 2\,As + 3\,H_2 \,. \tag{8.116}$$

Nevertheless for a number of reasons this mechanism is highly improbable. In the first place atomization temperatures of around 2200–2600 °C are required in graphite furnaces to obtain optimum sensitivity [197, 198, 200, 2791–2793, 5259], while in QTAs temperatures of 800–900 °C are adequate. In the second place the addition of a small amount of air or oxygen to the purge gas has a substantial influence on the sensitivity and makes complete atomization possible even at temperatures of around 700 °C [195, 3864, 4651, 6076]. And in the third place the considerable influence that the quartz surface has on the atomization signal is conspicuous [1780, 4067, 6039, 6559]. Based on thermodynamic equilibria calculations, DĚDINA and co-workers [1459, 1463] then showed that neither arsenic nor antimony can exist as gaseous atoms at temperatures below 1000 °C.

EVANS et al. [1780] observed that the atomizer attained its final sensitivity for an element after a prolonged period of use. Without going into the actual atomization mechanism, they concluded that a 'catalytic film' of the analyte on the quartz surface was necessary for reliable atomization. Similar theories were proposed by ZHAO and LI [6559] who attempted to produce such an 'analyte film' on the quartz surface. At this point we should remember that the *establishment of an equilibrium* can merely be accelerated by catalysis. Since thermodynamically analyte atoms are not capable of existing at the prevailing temperatures, atomization can only be influenced negatively by a catalytic process but never positively.

WELZ and MELCHER [6216] performed a whole series of experiments, including some with pure arsine in argon, in which they demonstrated that at 900 °C in the absence of hydrogen practically no arsenic atoms were formed. For atomization in a QTA, next to hydrogen a small amount of oxygen is necessary; the oxygen requirement is higher at lower atomization temperatures. Under normal analytical conditions the oxygen requirement can be met by the air dissolved in the measurement solution. From this observation WELZ and MELCHER [6216] concluded that in a QTA the hydride is atomized via the same mechanisms as in an FIT atomizer, i.e. with H radicals that are produced by the reaction of hydrogen with oxygen. In analogy to selenium (equations 8.112 and 8.113) they proposed a three-step mechanism for the atomization of arsenic according to:

$$AsH_3 + H \rightarrow AsH_2 + H_2 \qquad \left(\Delta H = -196 \text{ kJ/ mol} \right) \qquad (8.117)$$

$$AsH_2 + H \rightarrow AsH + H_2 \qquad \left(\Delta H = -163 \text{ kJ/ mol} \right) \qquad (8.118)$$

$$AsH + H \rightarrow As + H_2 \qquad \left(\Delta H = -163 \text{ kJ/ mol} \right). \qquad (8.119)$$

This mechanism was confirmed in the following years by a large number of further experiments performed by WELZ and co-workers [6234, 6254], by DĚDINA [1462], and by DĚDINA and WELZ [1461, 1463]. Among other things an MS detector was applied to investigate the products appearing under varying conditions in the QTA [6254]. From all these investigations we can obtain the following impression of the processes occurring in the QTA:

A cloud of H radicals is formed in the hot zone of the atomizer through the reaction of hydrogen and oxygen. The exact location of this cloud of radicals is determined by the temperature profile inside the atomizer, by the flowrate and composition of the purge gas, and by the design of the atomizer. The sole function of the QTA heating is to start and maintain the reaction between hydrogen and oxygen. There is nevertheless a synergetic effect between the supply of oxygen and the temperature on the sensitivity over the temperature range from about 600 °C to over 1000 °C. The oxygen requirement for optimum sensitivity is lower at higher temperatures.

The number of H radicals is largely determined by the oxygen supply to the atomizer. In a QTA, like in an FIT atomizer, it is not the *number* of H radicals that is decisive for effective atomization but rather the *cross-sectional density* of the radical cloud. Consequently the oxygen requirement is determined mainly by the internal *diameter* of that section of the atomizer in which the radical cloud is located. The gas flowrate is selected such that the radical cloud is located in the *side inlet tube* of the atomizer, i.e. in the same place where it is located in an FIT atomizer due to the oxygen supply capillary. This side inlet tube should have an inner diameter that is as small as possible.

Analog to the FIT atomizer the atoms are removed from the absorption volume by forced convection and by attrition. Thermodynamically, free arsenic or selenium atoms, for example, are forbidden outside the radical cloud [1459, 1463]. This is the driving force for their attrition, probably a first-order reaction, which takes place at the quartz surface. This assumption is supported by the observation made by a number of authors that the surface in HG AAS has a major influence on the sensitivity. Summarizing we can say that in the QTA analyte atoms can be observed merely because the attrition of the H radicals and also the analyte atoms is kinetically hampered.

The velocity of the attrition of the analyte atoms depends on the contact of free atoms with the quartz surface, i.e. on the flow profile of the gases in the atomizer and on the reactivity of the surface. While laminar flows prevail in the larger part of the atomizer, considerable turbulence occurs where the side inlet tube joins the absorption tube. A considerable loss of analyte atoms takes place in the turbulent zones, while in the laminar zones such losses only become significant at low gas flowrates. Losses through wall reactions are substantially accelerated at impurities on the surface, especially at elevated temperatures, which explains the differences in sensitivity frequently observed for atomizers of identical dimensions.

A suitable procedure for reducing contamination at the surface, first described by WELZ and MELCHER [6216], is cleaning with hydrofluoric acid which substantially reduces the reactivity of the surface. Further passivation appears to be difficult since sample residues continuously change the surface. This reflects the 'fragile nature of the catalytic film' described by EVANS et al. [1780]. Analyte atoms or H radicals are however apparently able under given prerequirements to passivate the surface further via wall reactions. This fact was incorrectly interpreted by EVANS et al. [1780] and also by ZHAO and LI [6559] as improved atomization due to a 'catalytic film'.

The species that is formed by analyte atom attrition is not known. For all the hydride-forming elements that have been investigated this species appears to be more or less volatile, at least in the presence of hydrogen, and can be atomized again through interactions with H radicals. AGTERDENBOS et al. [59] proposed the formation of a dimer as the possible mechanism. It is not certain whether this species, which was observed by the authors at analyte concentrations of 2–3 orders of magnitude above those usually applied, is formed at analytical concentrations. From thermodynamic considerations DĚDINA [1462] considered the formation of SeH_2 as more likely since this is the only stable compound in the presence of hydrogen [1459]. During their MS investigations WELZ et al. [6254] found for higher concentrations of arsenic above 900 °C both the dimer As_2 and the formation of arsine.

Results that are obtained for extremely high analyte concentrations or under conditions that are far from optimum cannot with certainty be extrapolated simply to normal conditions, but nevertheless they can provide a certain level of information on mechanisms. For example, DĚDINA and WELZ [1461] performed determinations of arsenic under an extreme deficiency of oxygen. They degassed all solutions and prevented the diffusion of atmospheric oxygen through the tubing. The calibration curves obtained for arsenic under these conditions were very strongly curved and exhibited a pronounced dependence on the purge gas. As depicted in Figure 8-37, when hydrogen was used as the purge gas a limiting absorbance was very rapidly attained whose value depended on the supply rate of oxygen. From this we can conclude that reactions 8.118 and 8.119 are much faster than reaction 8.117, so that AsH and AsH_2 molecules can compete successfully with AsH_3 for the insufficient number of H radicals. When argon was used as the purge gas the calibration curves rolled over as depicted in Figure 8-38. This indicates that under these conditions reaction 8.117 or 8.118 is much faster than reaction 8.119. An increasing portion of the analyte is transported from the radical cloud in a non-atomic, intermediate form, i.e. as AsH_2 or AsH.

These results are in excellent agreement with the work published by WELZ and GUO [6259] who performed measurements on very high analyte concentrations in an FI system under non-optimum conditions. They observed double peaks as shown in Figure

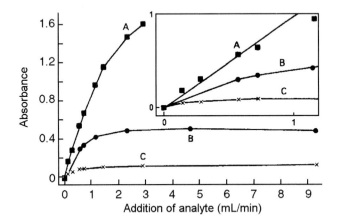

Figure 8-37. Calibration curves for arsenic with hydrogen as the purge gas. Continuous hydride generation; atomizer temperature 950 °C; hydrogen flowrate 93 mL/min. **A** – oxygen flowrate 2.4 mL/min; **B** – oxygen flowrate 0.03 mL/min; **C** – no oxygen flow (from [1461]).

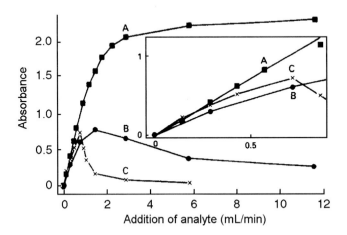

Figure 8-38. Calibration curves for arsenic with argon as the purge gas. Continuous hydride generation; hydrogen flowrate 8 mL/min; argon flowrate 50 mL/min. **A** – oxygen flowrate 0.4 mL/min; atomizer temperature 950 °C; **B** – oxygen flowrate 0.008 mL/min; atomizer temperature 950 °C; **C** – oxygen flowrate 0.008 mL/min; atomizer temperature 700 °C (from [1461]).

8-39; there was frequently a shift in time of the second peak compared to the peak measured under optimum conditions. Such time shifts of signals contradict one of the fundamental principles of FI—its highly time-resolved repeatability. Nevertheless they can easily be explained on the basis of the hydride-generation technique and atomization via H radicals. Under the selected conditions—atomizer temperature 750 °C, oxygen supply solely from the solutions used—there is certainly a deficit in H radicals prevailing in the system and the density of the radicals is not adequate to atomize large quantities of hydride. Clearly for antimony, similar to arsenic, the analogous reactions to 8.117 or 8.118 are much faster than reaction 8.119 which leads to atomization.

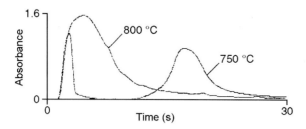

Figure 8-39. Influence of temperature on the atomization signal for 100 µg/L Sb in an FI system (from [6259]).

Provided the analyte concentration is not too high, antimony is completely atomized in the initial phase (the rising peak flank is identical at 750 °C and 800 °C). As soon as all the H radicals have been consumed by this reaction, however, a further increase in the analyte concentration causes only reactions 8.117 or 8.117 and 8.118 to proceed and no further atoms are formed. The signal returns to the baseline. In this phase the analyte is increasingly transported from the radical cloud in a non-atomic form, e.g. SbH_2 or SbH. This intermediate form is unstable and thermally decomposes in the atomizer and is deposited on the surface, presumably in elemental form. No atomic absorption is measured during this phase. As soon as the analyte concentration falls to below the critical limit again (toward the end of the peak at 800 °C) there is once again an excess of H radicals. These are now capable of volatilizing the analyte deposited in the atomizer and at least partially to atomize it; this explains the time-shifted second peak.

Similar phenomena were also observed by AGTERDENBOS et al. [59, 60] and NARSITO et al. [4296] for antimony and selenium, but were contradictorily interpreted. The thermal decomposition of intermediate, molecular species of the analyte under conditions of radical deficiency that leads to deposition of the analyte in the atomizer has been demonstrated by WELZ and co-workers [6216, 6234, 6254] under a large number of differing conditions. Based on interference studies, WELZ and STAUSS [6267] showed that in principle all hydride-forming elements exhibit the same behavior. Merely the speed of removal is markedly different; tin appears to react very slowly with H radicals.

Although the hydrogen radical mechanism is able to explain all experimental observations consistently, it has not received undivided recognition. In a series of papers Agterdenbos and co-workers have cast doubt on this mechanism of atomization and have proposed alternates. Since the atomization of arsenic, antimony, and selenium begins at about the same temperature, they do admit that 'a common factor initiates the formation of atoms', and that this is possibly radicals [4296]. Nevertheless they still proposed two very different mechanisms for the atomization of arsenic and selenium. For selenium hydride they proposed thermal decomposition according to:

$$SeH_{2(g)} \rightarrow Se_{(g)} + H_{2(g)}. \tag{8.120}$$

They based this reaction on thermodynamic calculations [59], but it could later be shown that they used incorrect enthalpy values for both selenium and SeH_2 [1459].

Similar thermodynamic calculations were also performed by BAX *et al.* [432] for the thermal decomposition of AsH_3. The results showed that atomization takes place at much lower temperatures than would be expected from the calculations. NARSITO and AGTERDENBOS [4295] also found that, in contrast to selenium, the signal for arsenic did not decrease with increasing hydrogen concentration. They therefore proposed reaction 8.121 as the atomization mechanism for arsenic:

$$4\, AsH_3 + 3\, O_2 \rightarrow 4\, As + 6\, H_2O. \tag{8.121}$$

The main argument of these authors against the hydrogen radical mechanism was that the partial pressure of H radicals must be at least twice as high as that of the SeH_2 molecules or three times as high as that of the AsH_3 molecules, respectively, i.e. $> 10^{-7}$ atm [430]. The *equilibrium concentration* of H radicals from the reaction:

$$H_2 \rightleftharpoons 2\, H \tag{8.122}$$

at 700 °C is however about three orders of magnitude below this value [62] and 'there is no indication that in situations like the one observed here that the actual concentration of H radicals exceeds the equilibrium concentration by the necessary factor' [4295].

STAUSS [5555] then performed a series of investigations to provide the 'missing link' for the hydrogen radical mechanism. On the first hand he attempted to measure the H radicals directly and to determine their concentration under analytical conditions. On the other hand he attempted to detect the intermediate products from reactions 8.117–8.119, which could also serve as evidence for the hydrogen radical mechanism.

Measurements of the H radicals at the Lyman α-line at 121.57 nm in the far vacuum UV, a line that is clearly separated from H_2 molecular bands, were subject to interferences from As_2 and Sb_2 molecules [5555] but were nevertheless very informative. If a QTA heated to 800–1000 °C was operated alternately with pure helium and a mixture of 90% He + 10% H_2, no difference could be observed in the transmission. However, during *switchover* of the gases a high, transient absorption was observed. This can be explained by atmospheric oxygen that diffuses through the tubing and which in the presence of hydrogen initiates reactions 8.109–8.111. It was also interesting to determine at which temperature in the presence of H_2 and O_2 reactions 8.109–8.111 started. A marked decrease of the absorption was observed at 750–775 °C with a gas mixture of 89.9% He + 10% H_2 + 0.1% O_2 at a flowrate of 1 L/min, while with continuous hydrogen generation in a flow system and a purge gas stream of 200 mL/min the absorption became very intense even at temperatures < 600 °C. In the first case the radical cloud was located with certainty in the absorption volume due to the high gas flowrate so that the oxygen requirement was high as a result of the large cross-section of the tube, and the reaction thus did not start until a relatively high temperature. In the second case there was more oxygen available from the solutions and the radical cloud was located in the narrow inlet tube due to the low flowrate, so that the reaction started at markedly lower temperatures.

The attempts to measure the concentration and distribution of the radicals in the QTA using a platinum probe were highly successful [5555]. This technique enabled the radical concentration to be determined via the heat of recombination. Although this technique is not specific to H radicals, in the presence of H_2 and O_2 these are the only radicals that can be expected in higher concentrations [1458]. Figure 8-40 depicts a typical radical distribution in a QTA with an inlet tube of 1 mm internal diameter. At a purge gas flowrate of 100 mL/min, which is typical for analytical work with this type of QTA, no radicals could be measured in the observation volume since, as to be expected, the radical cloud was located completely in the inlet tube of the atomizer. At a flowrate of 300 mL/min the radical cloud is located in a spatially limited area directly in front of and also probably partly within the inlet tube. With a further increase in the gas flowrate the radical distribution becomes increasingly flatter and more diffuse, and the maximum wanders slowly from the inlet tube to the ends of the cell. Similar observations have also been made with atomizers of different design, but we shall not go into details here.

It is interesting to note in Figure 8-40 the slightly increasing radical concentration toward the ends of the cell. This can possibly be attributed to the secondary formation of radicals with atmospheric oxygen that entered by diffusion. At a correspondingly high

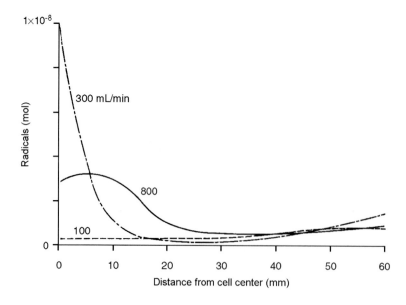

Figure 8-40. Spatial distribution of the radicals along the optical axis in a QTA heated to 900 °C. Inlet tube i.d.: 1 mm; position 0 = center of the cell in front of the inlet tube. H_2 flowrate: 43 mL/min; O_2 flowrate: 2 mL/min; Ar flowrate: 100, 300, and 800 mL/min (from [5555]).

gas flowrate, STAUSS [5555] further observed a significant decrease in the radical concentration close to the quartz surface; he attributed this to wall reactions in the region of strong turbulence. This radial asymmetry disappeared in the laminar flow region at greater distances from the inlet tube. As can be seen in Figure 8-41, the profile of the gas temperature in the inlet tube is exactly opposite to the radical concentration. While the gas is warmed close to the quartz surface, the radical concentration decreases due to wall reactions.

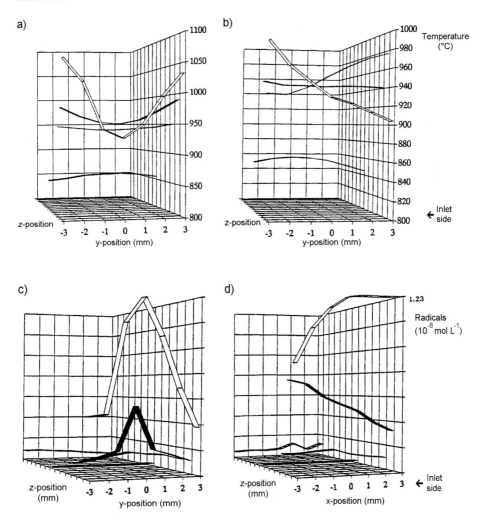

Figure 8-41. Changes in the gas temperature (**a** and **b**) and the radical concentration (**c** and **d**) over the cross-section and the length of a QTA; gas flowrates: Ar: 500 mL/min; H$_2$: 43 mL/min; O$_2$: 2 mL/min. **a** and **c** − vertical, and **b** and **d** − horizontal section through the quartz tube (from [5555]).

Based on these measurements, STAUSS [5555] calculated that the quantity of radicals formed in the effective zone of the probe was $> 10^{-7}$ mol. Although the effective zone of the probe is unknown, it is certainly very restricted, so that the concentration of radicals in the atomizer is in the order of 10^{-2} mol/L and thus at least four orders of magnitude above the concentration required to atomize analytical masses of hydride-forming elements. This means that the prevailing radical concentration exceeds the equilibrium concentration by at least six and probably by 7–8 orders of magnitude. We feel that these findings should dispel the remaining doubts about the hydrogen radical mechanism of atomization of the hydrides in the QTA.

To conclude we shall mention a further experiment by STAUSS [5555] that provided interesting information about the location of the radical cloud. DĚDINA and WELZ [1461] were the first to report luminescence that appeared in the gas stream under the controlled addition of oxygen and which could be stabilized as a small, blue flame in the inlet tube about 50 mm upstream of the T-junction. The interesting thing about this observation was that the temperature at this location was maximum 70 °C. STAUSS [5555] was able to reproduce this phenomenon and to scan the spectra under varying conditions.

Without analyte the spectrum of this small flame consisted more or less exclusively of OH radical transitions, which is hardly surprising considering the large excess of oxygen. In the presence of AsH_3 these authors were able to identify with reasonable certainty the chemiluminescence reactions 8.123 and 8.124:

$$As + 2\,OH \rightarrow AsO^{*B} + H_2O \qquad \left(\Delta H_f^o = -550\,kJ/\,mol\right)$$
$$AsO^{*B} \rightarrow AsO + h\nu_B \tag{8.123}$$

$$As + O \rightarrow AsO^{*A} \qquad \left(\Delta H_f^o = -482\,kJ/\,mol\right)$$
$$AsO^{*A} \rightarrow AsO + h\nu_A. \tag{8.124}$$

Since arsenic atoms are involved in both reactions there can be no doubt that the small flame indicates the location of the radical cloud. Although the flame is not usually visible with the much lower amounts of oxygen normally used, the location of the radical cloud is not likely to change to any extent. In addition to the transitions described above, the authors also observed a relatively broad band system in the visible spectral range which could be ascribed to the AsH_2 radical. This is a further indication for the radical mechanism.

Cadmium is an exception. It was originally determined by CACHO et al. [1023] in a QTA, but subsequently SANZ-MEDEL et al. [5094, 5095] established that this was not necessary. In the reaction with $NaBH_4$, cadmium in all probability initially forms a hydride, which is very unstable and decomposes very rapidly according to:

$$(CdH_2)_{unstable} \rightleftharpoons Cd^0 + H_2. \tag{8.125}$$

Atomic cadmium vapor is formed which, like mercury vapor, must merely be conducted into a non-heated quartz cell for measurement.

8.3.2.3 Atomization in Graphite Furnaces

If the hydride is passed, either directly or after collection in a cold trap, into a graphite furnace for atomization, atomization temperatures of 2200–2600 °C are normally applied [197, 198, 200, 2791–2794]. A problem, which we shall not go into here, is the transfer of the hydride into the graphite furnace. If the hydride is conducted into the graphite tube with the inert gas supply it must pass metal and graphite parts on which it can be decomposed and sorbed. A graphite tube mounted at the sample introduction hole [4920, 5259] has similar disadvantages since the hydride can also decompose there. A cooled quartz capillary inserted directly into the sample introduction hole has proven to be relatively successful [1456].

Only a limited number of the papers published give an indication of the presumed atomization mechanism of the hydride in the graphite tube. SHAIKH and TALLMAN [5259] and WANG et al. [6162] assumed that arsine decomposes in the tube, the arsenic is deposited on the surface, and is then volatilized and atomized. DITTRICH et al. [1545] found in their graphite paper atomizer (temperature about 1800 °C) that the sensitivity for arsenic sank to 70% *in the absence of hydrogen*. From this they concluded that a thermal atomization mechanism played the major role. For purely thermal atomization according to equation 8.116, however, just the opposite should happen; the *presence of hydrogen* should reduce the effectiveness of atomization due to the formation of arsine. From these observations we can conclude that at 1800 °C next to thermal atomization other mechanisms also play a role.

DĚDINA et al. [1459] proposed two independent mechanisms for the atomization of selenium hydride in graphite furnaces: one mechanism is effective from about 1200 °C, but only in the presence of traces of oxygen. This mechanism is similar to the one in a QTA and is based on an interaction with H radicals which are formed by the reaction of hydrogen with oxygen. The other mechanism, which is effective at temperatures above 1600 °C, is independent of the supply of oxygen to the atomizer. The hydride is presumably thermally atomized either in the gas phase or on the graphite surface. The removal of free analyte atoms from the absorption volume has not yet been explained, but the graphite surface most certainly plays a decisive role.

The major disadvantage of direct atomization of hydrides in the graphite tube is the low sensitivity compared to the QTA. The major advantage is the low mutual influence of the hydride-forming elements, which is several orders of magnitude lower than in the QTA, as is discussed in Section 8.3.3.4. The disadvantage of low sensitivity can be eliminated while the freedom from interferences is retained when the hydride is collected by sequestration in the graphite tube prior to atomization. Graphite tubes coated with iridium have proven to be very suitable for this purpose, as described in Section 4.3.2.2.

Virtually nothing is known about the mechanism of sequestration in the graphite tube and the initial investigations threw up more questions than they provided answers [2487]. Nevertheless we can assume that the hydride is initially sorbed without decomposition at the relatively low temperature of 250 °C [5331] and then forms a compound with the noble metal. Atomization is probably similar to that from solution under the application of a noble metal modifier. For very large analyte masses (1 μg AsH_3), STURGEON et al. [5640] found an irregular surface coating on palladium which indicates preferred sites for deposition and nucleation. Using AES, these authors also found a clear indication for a [Pd-As-O] species, which again points to the similarity to the analysis of solutions [5655]. ROHR [4937] and ORTNER *et al.* [4488] found that even during the drying step PdO is intercalated in the graphite and the thus activated palladium is able to bind the analyte covalently (refer to Sections 8.2.3.12 and 8.2.4.2). Using XPS analysis, LIAO and LI [3514] also found indications for the formation of a [Pd-In-O] species when they collected indium hydride on palladium. The major advantage of this technique is the temporal separation of hydride generation and atomization of the analyte. MATUSIEWICZ and STURGEON [3977] have published an overview on the development of this technique covering papers up to 1996. The conditions for preconcentrating hydride-forming elements on modified graphite surfaces are compiled in Table 8-8.

Table 8-8. Publications on the *in situ* preconcentration of hydride-forming elements on modified graphite surfaces. T_{dep} = deposition temperature; m_0 = characteristic mass in peak height (H) or peak area (A). When several surface modifiers were investigated, the one preferred by the authors is underlined.

Analyte	Surface modifier	T_{dep}, °C	m_0, pg	Lifetime*	References
As	Pd	100–1000	10 (H)	1	[6539]
	Pd	600	11 (H)	1	[170, 1561]
	Pd	200	23 (H)	1	[5640]
	Pd	400	31 (A)	1	[6131]
	Pd	300	17 (A)	1	[6543]
	Pd	200–900	–	–	[1220]
	Ir/Pd	250	45 (A)	300	[5331]
	Ir/Pd	350	60 (A)	–	[1527]
	Ir/Zr	400–500	35 (A)	600–700	[5913]
	Ir	400	17 (H), 30 (A)	–	[5384]
Bi	Pd	1000	16 (H)	1	[1561]
	Pd	200	28 (H)	1	[6540]
	Pd	300	16 (A)	1	[6543]
	Pd, Pt, Rh, Ru	200	34 (H)	–	[5640]
	Ir/Pd	–	76 (A)	300	[5331]
	Ir/Pd	200–1000	–	200	[3976]
	Ir/Zr	400–500	35 (A)	600–700	[5913]
Cd	Pd	150	–	1	[2148]
Ge	Pd	1000	12 (H)	1	[1561]
	Pd	600	26 (H)	1	[6540]
	Pd	700–800	10 (H)	1	[2432,
		300–1000	40 (A)		2435, 2433]
	Pd	600	32 (A)	1	[6543]
	Pd	800	46 (A)	1	[4349]
	Pd	400–800	–	1	[5771, 1220,
					2487]

Table 8-8 continued

Analyte	Surface modifier	T_{dep}, °C	m_0, pg	Lifetime*	References
Ge	Zr, Nb, Ta, W	400/500	54 (H)	> 400	[2437,
		600/700	170 (A)		2439]
	Zr, Pd, Pd/Zr	800	–	–	[4350]
In	Pd	800	630 (H)	1	[3514]
Pb	PG	450–600	368 (H)	> 4000	[1758]
	Zr	300	53 (H)	150–200	[6439]
	Zr, Pd	300	–	–	[6442]
	Ir, Ir/Mg, Ir/Pd,	200–300	21 (H)	> 400	[2441]
	W, Zr	50–600	75 (A)		
	EG	50–750	46 (A)	–	[5380]
	Pd, EG	600	57 (A)	1	[6131]
	Pd	350	71 (A)	1	[1528]
Pb	Pd	100-1100	13 (H)	1	[6539]
	Pd	600	13 (H)	1	[1561]
	Pd, Pt, Rh, Ru	200	14 (H)	1	[5640]
	Pd	300	28 (A)	1	[6543]
	Ir/Zr	400–500	83 (A)	600–700	[5913]
Pb	Ir	400	31 (H), 52 (A)	-	[5384]
	Pd	100–1000	15 (H)	1	[6539, 6542]
	Pd	800	17 (H)	1	[1561]
	Pd, Pt, Rh, Ru	200	23 (H)	1	[5640]
Se	Ir	250	85 (A)	> 300	[2357]
	Pd	300	30 (A)	1	[6543]
	Pd	150–700		1	[1559]
	Ag, Pd	200–400	17 (H)	1	[4347]
	Ir/Zr	400–500	43 (A)	600–700	[5913]
	Ir	300	48 (H), 80 (A)	-	[5384]
Sn	Pd	200	16 (H)	1	[5640]
	Pd	100–800	27 (A)	1	[6543]
	Pd	300	127 (A)	1	[5773]
	Pd	600	-	1	[1561]
	Zr	500	14 (H)	150–200	[4344]
	Zr, Nb, Ta, W	500–600	17–20 (H)	> 400	[2438]
	Ir/Zr	400–500	104 (A)	600–700	[5913]
Te	Pd	200	22 (H)	1	[6540]
	Pd	800	18 (H)	1	[1561]
	Pd	300	20 (A)	1	[6543]
	Ag	200–800	18 (H)	1	[4347]
	Ir/Zr	400–500	48 (A)	600–700	[5913]

* Number of determinations before retreatment of the surface with modifier

8.3.3 Interferences

The interferences in the hydride-generation technique can be of widely varying nature and often depend very strongly on the apparatus used. In this section we shall use the classification proposed by DĚDINA [1454], which is depicted schematically in Figure 8-42.

Interferences in the hydride-generation technique can take place during *hydride generation*, during *transport*, and in the *atomizer*. Under hydride generation we understand all processes taking place up to phase separation, while the interferences occurring during transport refer solely to the transport of the gaseous products after phase separation. Interferences can be either *direct*, i.e. caused by concomitants in the actual sample, or they can be caused by *residues* from a previous determination (memory). For both hydride generation and transport we make a distinction between an influence on the *effectiveness* of generation or transport and an influence on the *velocity* (kinetics). Depending on the type of hydride system (batch system or flow system) and signal processing (absorbance or integrated absorbance), these interferences can assume very different magnitudes. The interferences during *atomization* depend very strongly on the type of atomizer used. Here we make a distinction between interferences caused by a *deficiency of radicals* and those caused by a reduction of the *lifetime of the analyte atoms*.

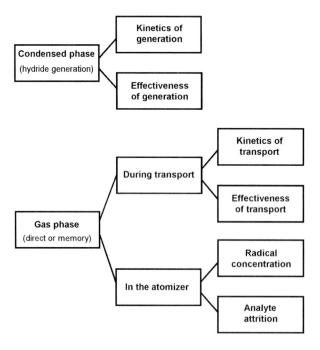

Figure 8-42. Classification of interferences in the hydride-generation technique according to DĚDINA [1454].

8.3.3.1 Spectral Interferences

Spectral interferences caused by background absorption are seldom in HG AAS since the concomitants are normally retained in the condensed phase and only small amounts of gaseous products are formed. When flames were used—we shall not consider this further—variations in the transparency frequently occurred when there were changes in the gas composition, such as with batch systems. When quartz tubes are used as atomizers there are no interferences due to background absorption when proper procedures are applied, except for quartz tubes that are open at the ends where the hydrogen can ignite [5830]. Spectral interferences can occasionally be observed in the presence of very high concentrations of other hydride-forming elements [1454, 6267]. As will be shown, these interferences are negligible in comparison to the atomization interferences that can be expected.

For *atomization in graphite furnaces* strong background absorption can be observed when hydrogen generated in the reaction reaches the graphite tube together with the hydride. This background absorption increases markedly with increasing atomization temperature [3298]. As well as background absorption, the introduction of hydrogen reduces the lifetime of the graphite tube; this can be attributed to a reaction between hydrogen and graphite in which acetylene is formed. This interference naturally does not occur when the hydride is collected prior to atomization and the hydrogen is separated.

8.3.3.2 Interferences during Hydride Generation

8.3.3.2.1 *Influence of the Analyte Species*

In contrast to F AAS and GF AAS, in HG AAS there is a chemical reaction prior to the actual determination step: namely generation of the hydride. If the analyte in the sample is not in exactly the same form as the analyte in the calibration solution, the generation of the hydride may not be with either the same speed or the same yield. Once again we should like to emphasize that this interference **cannot** be eliminated by the analyte addition technique since the *added* analyte species behaves 'normally' but the species *present* in the test sample does not (refer to Section 5.4.3.2).

The influence of the oxidation state on the determination by HG AAS is discussed in Section 8.3.1.4. We merely recapitulate that for selenium and tellurium reduction of the hexavalent oxidation state to the tetravalent state is essential and that for arsenic and selenium reduction of the pentavalent state to the trivalent state is strongly recommended. With the exception of purely inorganic samples, e.g. metallurgical samples, the hydride-forming elements can be present not only in various oxidation states but also in a multitude of organic compounds. A number of these compounds form volatile hydrides, but their behavior is significantly different during atomization and they can exhibit differing sensitivity. Numerous organic compounds of the hydride-forming elements do not react at all with sodium tetrahydroborate and cannot thus be determined directly by HG AAS. The fact that such compounds can be present even in 'simple' water samples is demonstrated by the publications on 'hidden' arsenic [607, 2680], a species that could not be determined for a relatively long period. In the hydride-generation technique digestion of the samples is thus mandatory and should be part of the method, more espe-cially for the analysis of biological materials and environmental samples, but also for water analyses. The digestion procedure must guarantee that the *total content* of the respective analyte is determined.

For the determination of *arsenic in water* photooxidation by UV radiation is mostly sufficient [848, 1373, 1917, 2680]. The digestion with sulfuric acid and hydrogen per-oxide has also proven to be very convenient and reliable [1502]. Strong oxidants such as perchloric acid are essential for the determination of *arsenic in biological materials* and *environmental samples* to ensure that difficult-to-digest substances such as oxy-bis-dimethylarsine [2745], arsenobetaine and arsenocholine [3801, 6228] are completely broken down. A digestion in nitric acid-sulfuric acid-perchloric acid at increased tem-perature [2745] or microwave-assisted [2694, 5184] has proven to be suitable. Dry ash-ing with magnesium nitrate [901, 5342] is an alternative, but the risk of contamination and the effort required are higher.

The situation for selenium is very similar to arsenic. In this case selenomethionine, selenocysteine and the trimethylselenonium ion are especially resistant to digestion. For the determination of *selenium in water,* digestion in sulfuric acid with an additive of vanadium pentoxide has proven to be very reliable [1508]. For the determination of *selenium in biological samples,* such as blood, urine, or foodstuffs, the use of perchloric acid appears to be indispensable [2319, 2799]. Stepwise heating with nitric acid-sulfuric acid-perchloric acid up to high temperatures is the most successful [6225, 6239], even if papers are now and again published in which it is claimed that the correct results can be obtained with the use of 'milder' digestions. In most of these cases, however, the sam-ples were not too difficult [928, 6176], so that the results are not necessarily representa-tive. Solely dry ashing with magnesium nitrate appears to be an alternative to perchloric acid [901, 2369].

There is very little published in the literature on the digestion of the other hydride-forming elements. Since we can discern no clear trend, we shall report further in Chapter 10. We should like to mention once again that the differing reactivities of the individual species of an element have been used successfully in many cases for speciation analysis (refer to Chapter 7).

8.3.3.2.2 *Interferences Caused by Concomitants*

Interferences caused by concomitants influence almost exclusively the *effectiveness of hydride generation* or the release of the hydride from solution. In contrast to the influ-ence of the analyte species discussed above, this interference can frequently be corrected by the analyte addition technique. Since the interferents are usually present in large ex-cess compared to the analyte, the interference is a first order process. This means that the interference does not depend on the analyte concentration or the ratio of the analyte to the interferent, but on the *concentration of the interferent*. This fact has been verified by a number of working groups [433, 2531, 4067, 6212]. The most well-known interfer-ences of this type and which have been investigated in the greatest detail are caused by the transition metals, especially the iron and copper groups, and also by the platinum metals.

The fact that the signal suppression does not depend on either the ratio of the hydride-forming element to the interfering element or on the contact time of the reaction partners prior to reduction excludes the formation of difficult-to-reduce compounds of the analyte and the interferent prior to reduction. MEYER et al. [4067] thus came to the conclusion that after formation the hydride reacts with the free interfering ions in acidic solution at the gas/solution phase boundary to give insoluble arsenides, selenides, tellurides, etc. A reaction of this type depends only on the diffusion velocity of the hydride from inside the bubble to the phase boundary.

PIERCE and BROWN [4652] found that the interferences were more pronounced when sodium tetrahydroborate was first added to the samples followed by the acid than the other way round. At the same time they also observed that a precipitate formed when sodium tetrahydroborate was added to the neutralized test sample solution. They associated the formation of the precipitate directly with the interferences as already postulated in earlier work [5457].

KIRKBRIGHT and TADDIA [3132] drew to attention the fact that nickel and the elements of the platinum group are hydrogenation catalysts and can adsorb large quantities of hydrogen. These metals can also trap and decompose the hydride, especially when they are in finely dispersed form such as is the case after this type of reduction. The addition of 500 mg nickel powder, for example, to the test sample solution prior to hydride generation brings about complete signal suppression. A direct connection between the precipitation of finely dispersed metal and the observed interference thus appears obvious. WELZ and MELCHER [6220] proposed that selenium hydride after formation is sorbed by the finely dispersed precipitate and then decomposed.

The observation that the interference by copper, iron, and nickel on the determination of As(V) was much stronger than on the determination of As(III) was a further confirmation of this mechanism. WELZ and MELCHER [6221] attributed this to the slower reduction of As(V) and the thus much more intimate contact of the arsine with the precipitate. BYE [997] confirmed this reaction mechanism, but pointed out that in all probability the precipitate was not the metal but the metal boride.

AGTERDENBOS and BAX [62, 63] initially contradicted this mechanism and were of the opinion that the interferences due to heavy metals were caused by the catalytic decomposition of sodium tetrahydroborate. Nevertheless, in later work BAX et al. [433] and also NARSITO et al. [4297] largely revised their view. They found that even if catalytic decomposition of the sodium tetrahydroborate took place, there was still sufficient time for hydride generation since this takes place much faster (see Section 8.3.1). These authors also finally confirmed that the reaction of the hydride with the reduced form of the interferent played the decisive role.

Based on these mechanisms of interference, we can develop systematic measures to reduce or even to eliminate these interferences. Some of these measures were determined empirically before there was clarity about the mechanism of interference. Thus an *increase in the acid concentration* proved to be an effective means to reduce interferences caused by the transition metals [583, 1926, 2527, 6212, 6226]. The better solubility of the precipitate under these conditions and the thus retarded formation of the interfering species, as well as the increased formation of hydrogen and the thus more rapid expulsion of the hydride, have been used to explain the increased interference-free range of often up to several orders of magnitude [6232]. A further effective measure is a *reduc-*

tion of the sodium tetrahydroborate concentration [6232, 6430]; this also reduces the reduction and precipitation of the interfering species. The effects of both of these measures on the interference of nickel on the determination of arsenic are shown in Figure 8-43; the interference-free range has been increased by a factor of more than 1000.

An interesting observation was described by FLEMING and IDE [1926] for the analysis of metallurgical samples. While nickel interfered in the determination of all of the hydride-forming elements at even relatively low concentrations, this influence was much lower in the presence of higher concentrations of iron. These authors were thus able to perform interference-free determinations of As, Bi, Sb, Se, Sn, and Te in steel even in the presence of higher nickel contents. For the analysis of high-alloyed steels they even added Fe(III) to eliminate the influence of nickel and other interfering elements. WELZ and MELCHER [6222] investigated this phenomenon and came to the conclusion that the *buffer action of trivalent iron* was based on its relatively high electrochemical potential. The potentials of possible reduction reactions that can take place when sodium tetrahydroborate is added to a solution containing Fe(III) and Ni(II) are listed in Table 8-9. As a result of the electrochemical potentials Fe(III) is initially reduced to Fe(II) before Ni(II) is reduced to the metal and then precipitated either as the metal or as the boride. BYE [998] later described Fe(III) as the most effective reagent for the suppression of the interference of copper on the determination of selenium. WICKSTROM *et al.* [6310] also used this buffer to reduce the interferences of copper and nickel on the determination of tellurium.

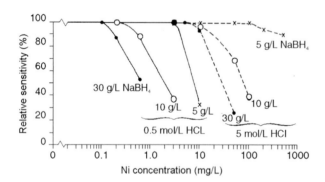

Figure 8-43. Influence of the concentration of hydrochloric acid (0.5 and 5 mol/L) and sodium tetrahydroborate (5, 10, and 30 g/L) on the interference of nickel on the determination of 5 µg/L As(V) (from [6232]).

Table 8-9. Electrochemical potentials for the reduction of Fe(III), Fe(II), and Ni(II) by sodium tetrahydroborate (from [6222]).

Reaction			Potential
$Fe^{3+} + e^-$	$\rightleftharpoons$	Fe^{2+}	+0.77 V
$Fe^{2+} + 2e^-$	$\rightleftharpoons$	Fe^0	−0.41 V
$Ni^{2+} + 2e^-$	$\rightleftharpoons$	Ni^0	−0.23 V

In this case, as with increasing the acid concentration or lowering the sodium tetrahydroborate concentration, the buffer mainly has a *delaying action* so that all of the hydride can leave the solution before interfering species are precipitated. With flow systems, especially for FI compared to batch procedures, this mode of action allows the interference-free measurement range to be markedly increased through *kinetic discrimination*. By selecting a suitably short reaction zone between the addition of the sodium tetrahydroborate solution and the gas/liquid separator the hydride can be separated before the interfering element is reduced. A further advantage of FI is that due to miniaturization and the extremely small dead volume the concentration of the sodium tetrahydroborate solution can often be reduced to ≤ 1 g/L, thus bringing about a further extension of the interference-free range. The techniques of coprecipitation [462, 463, 3532, 5835, 6074] and ion exchange [2528, 2792] used earlier in extreme cases to separate the analyte from the interferent have thus become more or less superfluous.

A further possibility of markedly reducing interferences is based on the differing stabilities of the compounds that the hydride-forming elements form with elements of Group VIII and Subgroup I. MEYER *et al.* [4067] established that the influence of silver on the determination of selenium is different in the presence of tellurium. While silver in concentrations above 25 µg/L (in 0.3 mol/L HCl) interferes already in the determination of selenium when tellurium is absent, in the presence of 200 µg Te this interference cannot be observed below 5 mg/L Ag. This corresponds to an increase of the interference-free range by a factor of 200. Based on this behavior, KIRKBRIGHT and TADDIA [3133] developed a procedure to reduce the interferences of various elements on the determination of selenium. The tellurium hydride formed by reduction with sodium tetrahydroborate can form very stable tellurides with numerous interfering ions. The stability constants of the tellurides of copper, nickel, palladium, or platinum are noticeably lower than those of the corresponding selenides. The interference-free range for the determination of selenium in the presence of any of these elements can be substantially increased by adding an excess of Te(IV).

A large number of working groups investigated the use of organic *complexing agents* to eliminate interferences by preventing the reduction and precipitation of the interfering element. In the presence of *ethylendiamine-tetraactic acid* (EDTA) it was possible to determine bismuth in nickel alloys [1612] and in cobalt [3532], arsenic and selenium in the presence of cobalt and nickel [3981], and arsenic in the presence of copper, iron, and nickel [5969]. *Thiourea* was used as a suitable complexing agent for copper in the determination of antimony [3532] and selenium [992]. *Citric acid* was proposed to mask high concentrations of nickel in the determination of selenium [994]. The influence of copper and nickel on the determination of tin was reduced by *sodium oxalate* [6074]. To eliminate varying heavy metal interferences on the determination of arsenic, *pyridine-2-aldoxime* [1600], *thiosemicarbazide* [3132], and *1,10-phenanthroline* [2791, 3132] were applied; the latter is also suitable for the determination of antimony [2794] and germanium [5164].

In this respect *L-cysteine* assumes a special role since it appears to act as a quasi universal reagent, at least for arsenic and antimony. For these two elements L-cysteine serves as a prereductant, as a stabilizer, and as a buffer to more effectively eliminate interferences due to transition elements than the majority of other reagents investigated [6268, 6269]. *Potassium iodide* is the reductant that has been used most frequently for arsenic and antimony up to the present and it is reported that it reduces the interferences caused by transition elements such as copper, iron, nickel, and silver [1177, 4947, 6435]. Nevertheless potassium iodide has a number of disadvantages: it is only effective in strongly acidic solution (> 5 mol/L HCl), its solutions are light-sensitive and relatively unstable, and insoluble copper(I) iodide precipitates in the presence of higher concentrations of copper. The latter can cause considerable contamination in flow systems.

BRINDLE and co-workers investigated L-cysteine for the determination of antimony [1231, 3473], arsenic [795, 1230, 3473], germanium [794], and tin [3473] using the hydride technique. For his investigations Brindle used a DC plasma and determined these elements by OES, so that the detection limits he attained were 1–2 orders of magnitude poorer than is usual in AAS. A thorough-going investigation of this buffer and reductant showed that it is superior to potassium iodide in virtually all situations [6268, 6269]. A further advantage of this reagent is that it can be much better tolerated by the environment since it is virtually non-toxic and only relatively low hydrochloric acid concentrations (0.1–1 mol/L) are required.

The interferences due to transition metals, which are caused by only a limited number of elements, have been thoroughly investigated. The majority of the other elements in the periodic system do not interfere, even in high concentrations. Apart from these interferences, only a very limited number of other interferences are known. The majority of acids have no noticeable influence on the hydride-forming elements, except for the positive effect of increasing the freedom from interferences described above. Solely tin is relatively sensitive to changes in the acid concentration, so that it is usual to determine it in a strongly buffered solution. This aspect is discussed under the treatise on this element.

The only acid which appears to cause certain problems is hydrofluoric acid. PETRICK and KRIVAN [4626] used radiotracers to investigate the influence of this acid on the determination of arsenic and antimony. In the presence of HF concentrations > 1%, $[AsF_5OH]^-$ is formed and this does not react with sodium tetrahydroborate. After hydrolyzing this ion it was possible to eliminate the interference by complexing the HF with boric acid. In the presence of HF Sb(V) does not form a hydride and the reduction to Sb(III) with potassium iodide is strongly hampered and requires careful optimization of the acid composition.

A number of authors have reported interferences to the determination of selenium by nitrose gases in the hydride-generation technique [61, 240, 883, 6133]. Investigations conducted by SINEMUS et al. [5376, 5377] showed that the interference could to attributed to nitrite, which is an interference that had already been described by CUTTER [1386]. In the presence of high concentrations of HCl nitrite reacts to form nitrosyl chloride, NOCl, which causes the interference. Cutter recommended sulfanilamide and this appears to eliminate the interference completely.

We should finally like to mention some further possible interferences in HG AAS that are caused by a number of different organic constituents. The presence of a large portion of non-digested organic constituents, especially biological samples such as blood and urine, can lead to strong foaming. Up to a certain level this interference can be controlled by the addition of an anti-foaming agent. Nevertheless, strong foaming should be avoided since there is the risk that sample droplets will be carried over into the atomizer and cause contamination, which is subsequently very difficult to remove.

Volatile organic solvents, such as hydrocarbons, alcohols, and ketones, have a strongly depressive influence on the sensitivity of all of the hydride-forming elements; they must be removed prior to analysis. Presumably these organic molecules reduce the lifetime of the radicals in the atomizer and thus directly influence the atomization of the hydride.

Carbon particles, which can remain in the test sample solution after an incomplete digestion, interfere strongly in the determination of selenium [1195]. Humic acid [6384] and other water-soluble organic compounds [4920] also interfere in the generation of the hydride. All of these interferences disappear after a complete digestion. We have already stressed the importance of a complete digestion at the start of this section.

8.3.3.2.3 *Memory Interferences*

Under memory interferences we understand those influences on the atomization signal that are not caused by concomitants of the actual sample being analyzed but by residues from earlier samples. In principle these are the same types of interferences as discussed in the preceding section. Interferences that cause a precipitate are particularly predestined to influence subsequent measurements when the precipitate is not completely removed. Some of these precipitates adhere tenaciously to the walls of tubes and containers, for example copper(I) iodide which is produced from potassium iodide in the presence of high copper concentrations.

The larger the surface area and the longer the contact times, the greater naturally is the risk that these residues remain in the system. In this respect batch systems are more subject to this problem than flow systems. The materials used for containers also play a certain role. Smooth, non-wetting plastic surfaces sorb precipitates far less readily than rough surfaces or glass. In the event of contamination, on the other hand, batch systems are easier to clean than flow systems. It is thus particularly important to avoid precipitates from forming in flow systems. Short reaction zones are preferable since the formation of a precipitate is kinetically slower than the generation of the hydride [6256].

Compared to CF systems, FI systems have the advantage that much smaller test sample volumes are required. The risk that precipitates can accumulate in the system is thus much lower, quite apart from the fact that FI systems generally exhibit much lower interferences compared to those with continuous sample introduction [4327]. Through the choice of a suitable carrier solution it is also possible in FI systems to thoroughly rinse the entire system after every measurement solution since the volume of the carrier solution is mostly much greater.

A typical example for a memory interference is depicted in Figure 8-44. Without the addition of a buffer nickel interferes relatively strongly in the determination of antimony. The signals for a calibration solution not containing nickel that was measured prior to and after a test sample solution are virtually identical, i.e. there is a direct interference but no memory interference (Figure 8-44a). The direct interference of copper on the determination of antimony is noticeably lower than by nickel (Figure 8-44b). However, a memory interference can be observed in the presence of copper since the signal for the subsequent calibration solution without copper is much more strongly influenced

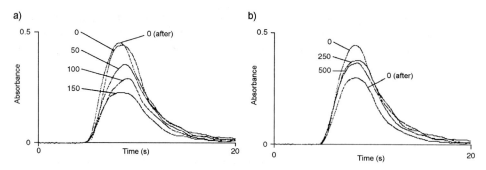

Figure 8-44. Influence of nickel and copper on the signal for 10 µg/L Sb(III) in an FI system. Sample solution in 0.1 mol/L HCl without buffer; carrier solution 1 mol/L HCl; '0 (after)' is the first calibration solution with 10 µg/L Sb(III) following the highest concentration of the interfering element. **a** – 0, 50, 100, 150, 0 mg/L Ni(II) as interferent; **b** – 0, 250, 500, 0 mg/L Cu(II) as interferent (from [6269]).

than the preceding signal was influenced by the direct interference. This can be explained by the fact that sometime during the course of the preceding reaction the precipitate formed and thus only interfered partially in the determination. The following signal was then totally affected by the system contamination. The fact that after 15 further measurements without copper the signal for the calibration solution still had not reached the value obtained in the presence of 500 mg/L copper is an indication of how tenacious precipitates of this sort can be [6269].

The interference caused by copper was much more pronounced when potassium iodide was used as the reductant and 'buffer', since at concentrations > 100 mg/L Cu insoluble copper(I) iodide is precipitated and prevents further analyses. In the presence of L-cysteine as the reductant and buffer, on the other hand, no memory interferences could be observed, even in the presence of concentrations > 1000 mg/L Cu [6269]. It is also interesting to note that an antimony calibration solution buffered with L-cysteine could be measured in a copper-contaminated system without interference. A subsequent calibration solution without buffer once again exhibited the original interference. This means that the L-cysteine buffer permits an interference-free determination in a contaminated system, not because it eliminates the cause of the interference, namely the precipitate, but because it prevents the analyte from reacting with the precipitate, for example by forming a complex [6269].

The influence of decomposition products of the sodium tetrahydroborate solution on the determination of selenium, described by DĚDINA [1455], belong most definitely in the category of memory interferences. If hydrochloric acid is treated with sodium tetrahydroborate prior to the determination of selenium, the quantity of selenium hydride subsequently generated falls to below 20%. An interference of this type can occur if the reactor is not thoroughly cleaned prior to the next determination, or if the HCl is 'purified' with sodium tetrahydroborate to reduce blank values. It is not known which compound causes the interference, but it is certain that it is not boric acid, which is the end product of the decomposition of sodium tetrahydroborate.

Memory interferences can be most easily *recognized* by frequently measuring a calibration solution, or occasionally from a distorted signal form. They can most easily be *avoided* by preventing the formation of a precipitate by one of the measures described above. It is often very difficult to *eliminate* system contamination. The entire system must be very thoroughly cleaned and for flow systems in most cases the tubes must be exchanged.

8.3.3.3 Interferences in the Atomizer

It is logical that the nature and extent of interferences occurring in the atomizer depend to a very large part on the type of atomizer unit used. Most of the interferences are caused by other hydride-forming elements which are also vaporized during the reaction. To date there have been no reports that interferences have been caused by mercury which would also be vaporized during the reaction.

8.3.3.3.1 *Quartz Tube Atomizers*

Because of the great similarity of atomization in FIT atomizers and QTAs (see Section 8.3.2) we shall treat both of them together. If we consider the atomization mechanism and the transport of free atoms in the absorption volume, then there are two possibilities for interferences in the atomizer. Either the interferent influences (reduces) the *concentration of H radicals* to such an extent that the analyte can no longer be completely atomized, or it accelerates the *attrition of free atoms*. In the first case the interference should be reduced when more radicals are generated, e.g. by increasing the rate of supply of oxygen. In the second case the interference should decrease when the probability of contact of the analyte atoms with the quartz surface (if the attrition takes place there) is reduced or if the probability of contact with the interferent in the gas phase (if the interference is based on a reaction analyte–interferent in the gas phase) is reduced.

In the FIT atomizer only the interferences of the other hydride-forming elements on the determination of selenium have been investigated in detail. DĚDINA [1454] found that when the supply of oxygen was adequate, merely tin caused an interference due to the inadequate concentration of radicals while antimony, arsenic, bismuth, and tellurium reduced the lifetime of the analyte atoms. With the minimum supply of oxygen antimony also caused an interference due to the inadequate concentration of radicals. Higher concentrations of arsenic, antimony, bismuth, and tin also caused a memory interference, possibly because of the low temperature in the non-heated quartz tube.

DĚDINA [1454] concluded from the results of these investigations that the interference caused by the radical concentration was primarily due to the accelerated recombination of radicals at the inner surface of the inlet tube. The recombination of radicals was catalyzed by *deposition of the interfering element in the inlet tube and the resulting change in the surface*. If the interference due to the radical concentration could be attributed to the direct consumption of the H radicals by the interfering hydride, then it should not be so very different for each individual hydride. And furthermore we would not observe any memory interferences. Similarly the occurrence of memory interferences through accelerated attrition of the analyte atoms indicates that these are primarily surface reactions and not gas-phase reactions.

Selenium is also the element that has been investigated in most detail in the QTA. The interferences caused by the other hydride-forming elements on the determination of selenium are compiled in Table 8-10; these results, however, are very inconsistent. As WELZ and STAUSS [6267] were able to show, these often substantial differences are based on the variety of systems used for hydride generation and the varying experimental conditions. On the other hand, these differences gave useful indications about possible mechanisms of interference.

Table 8-10. Literature citations on interferences caused by other hydride-forming elements in the determination of selenium in a QTA. The tolerance limit is defined as the reduction of the absorbance (peak height) by 10% in the presence of the interferent.

Acid	Concentration (mol/L)	Tolerance limit for interfering elements (mg/L)						References
		As(III)	As(V)	Bi(III)	Sb(III)	Sn(IV)	Te(IV)	
HCl	0.5		0.12	0.1	0.03	0.02	0.1	[402]
HCl	1		10	180	0.15	0.1	15	[1540]*
HCl	0.3	0.04		0.2	0.04	0.1	15	[4067]*
HCl	6			800	0.1–0.4§	0.04		[4287]
HCl	6	0.15		0.1	> 0.1			[5663]+
H$_2$SO$_4$	0.35			0.03	0.2	0.03	1	[6036]
HCl	6		1	2	1			[6076]+
HCl	0.6	0.1	0.3					[6213]*
HCl	0.5	0.02	0.1	0.1	0.03	0.02	0.2	[6267]†
HCl	1	0.06	2	0.1	0.1	0.5	10	[6267]‡

* Taken from curves
+ Values determined by interpolation or extrapolation
§ Depends on the selenium concentration
† Batch system
‡ FI system

In batch systems reactors with a volume of about 100 mL are typically employed, so that there is a large dead volume. Systems of this type require a relatively high sodium tetrahydroborate concentration of 5–50 g/L and a purge gas flowrate of several hundred mL/min to attain optimum sensitivity. These conditions demand a relatively wide inlet tube (approx. 3–6 mm i.d.) for the QTA in which the cross-sectional density of the H radicals is relatively low when oxygen is not mixed with the purge gas. Furthermore, batch systems must be purged with an inert gas prior to every determination to remove air from the dead volume. As a result of this purging quite a lot of air is removed from the measurement solution [6216]. A shortage of radicals is thus the most common problem in batch systems and the tolerance limit to other hydride-forming elements is correspondingly low. A further reason for reduced tolerance to other hydride-forming elements in batch systems is that the other hydrides, which are generated kinetically much more slowly such as As(V), Sb(V), and Sn(IV), can reach the atomizer quantitatively and lead to interferences.

In FI systems, where test sample volumes of less than 1 mL and very small dead volumes are normal, sodium tetrahydroborate concentrations of 0.5–2 g/L and gas flowrates of around 100 mL/min are optimum. It is thus possible to use atomizers with inlet tubes having an inner diameter of only 1 mm in which a much higher cross-sectional density of H radicals is generated. In addition the solutions are normally saturated with atmospheric oxygen since the system does not have to be prepurged with an inert gas. In systems of this type a shortage of radicals is not a major problem, so that interferences are mostly caused by a reduced lifetime of the analyte atoms. These interferences are primarily based on the fact that the hydride of the interferent, which is present in relatively high concentration, is preferentially decomposed in the turbulent zone at the inlet to the QTA, is deposited at this location, and then enhances the recombination or attrition of the analyte atoms. The magnitude of this interference depends among other things on the stability of the hydride of the interferent. The interference can be reduced by increasing the flowrate of the purge gas. In FI systems interferents that generate a hydride more slowly can be eliminated via kinetic discrimination, so that much less of the interfering hydride reaches the atomizer.

Important information with respect to mechanisms of interference could be obtained from the time-resolved signals measured by WELZ and STAUSS [6267] in an FI system. In Figures 8-45a and b two sets of selenium signals generated in the presence of an interferent and a 10fold or 100fold excess of Sb(III), respectively, are compared. It is clear that at the start of hydride generation there is no interference on the selenium signal but that the interference starts during the course of the reaction. Indications about the mechanism of interference can be obtained from Figure 8-45c when we look at the concentration of the antimony interferent corresponding to the interfered atomization signal for selenium. As long as the antimony signal increases normally the signal for selenium also develops without interference. As soon as the antimony signal reaches its maximum and starts to fall, the interference on the selenium signal begins. From earlier investigations [6259] it is known that at this time point the reservoir of radicals is exhausted so that the antimony hydride cannot be properly atomized (refer to Section 8.3.2.2). The resulting intermediate products are thermally decomposed and deposited in the atomizer. At the end of hydride generation when the antimony hydride concentration has decreased and a sufficient quantity of H radicals is again available, the deposited antimony species is volatilized and partially atomized, which explains the second peak.

From this we can conclude that in the presence of concentrations of antimony that under the prevailing conditions (cross-sectional density of H radicals) can still be completely atomized, there is no significant interference to the determination of selenium. From Figure 8-45a it is clear that the course of the interference is 'asymmetrical', i.e. it takes about 2 s from the start of hydride generation until the interference starts, but that it then prevails until the end of the signal. The only mechanism that can satisfactorily explain this behavior is that already at a small excess of antimony the interferent is deposited on the surface of the QTA, e.g. near to the inlet, and causes enhanced (earlier) attrition of the analyte atoms. The interfering species is rapidly removed at the end of hydride generation, so that the next atomization signal, at least at the beginning, exhibits no interferences. A major interference is observed at the moment when the quantity of radicals is no longer sufficient for complete atomization, so that increased thermal de-

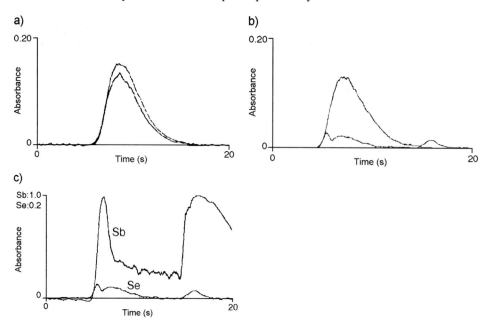

Figure 8-45. Time-resolved, superimposed atomization signals for 10 μg/L Se(IV) in an FI system with a 250 μL sample loop. **a** – without interferent and with 0.1 mg/L Sb(III); **b** – without interferent and with 1 mg/L Sb(III); **c** – selenium signal in the presence of 1 mg/L Sb(III) and the atomization signal for 1 mg/L Sb(III) (from [6267]).

composition and deposition of the interferent in the atomizer takes place. Analogous behavior has also been observed for the other hydride-forming elements [6267].

The mechanism proposed by DITTRICH and co-workers [1543, 1545], namely the formation of diatomic molecules between the analyte and the interferent of the type AsSb or SbSe, is improbable, at least for the analyte selenium, based on the findings of WELZ and STAUSS [6267]. Molecular formation of this type should lead to a 'symmetrical' change in the atomization signal and should also decrease when the atomization of the interferent decreases, and not increase as shown in Figure 8-45c. Initially the 'late' small selenium peak recorded in Figures 8-45b and c would appear to be an indication that an SbSe compound is formed and deposited, and then partially atomized by H radicals toward the end of the reaction. A detailed investigation, however, showed that this signal is caused by background absorption of volatilized antimony species (probably Sb_2 or SbH_3) and disappears when background correction is applied.

A further indication about the prevailing interference mechanism, especially for the change of mechanism with increasing concentration of the interferent, was given by investigations with changed gas flowrates and composition of the gas [6267]. This is shown for the influence of the increasing antimony concentration on selenium in Figure 8-46. Initially the interference increases slowly with increasing antimony concentration; the interference is lower at a higher gas flowrate, but the addition of oxygen to the purge gas does not influence the interference. In this range the interference is caused by accelerated atom attrition. Beyond a given concentration of the interferent the signal suppression becomes very intense and is hardly influenced by an increase in the purge gas flowrate. Based on other observations (compare Figure 8-45c), a deficiency of radicals was determined as the cause for the interference in this range. The fact that the interfering influence could be significantly reduced by adding oxygen to the purge gas confirmed the previous assumption. The course of the interference in the presence of higher oxygen concentrations corresponds to an accelerated attrition through increased surface contamination in the presence of higher concentrations of the interferent. As is known, this interference is not influenced by oxygen, i.e. by the H radical concentration.

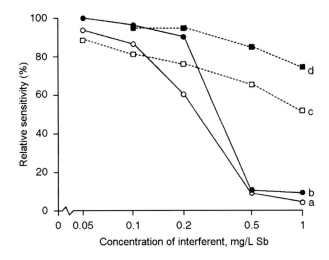

Figure 8-46. Influence of an increasing Sb(III) concentration on the signal for selenium in an FI system. Purge gas: **a** – 55 mL/min Ar; **b** – 100 mL/min Ar; **c** – 55 mL/min Ar + O_2 (99.5 + 0.5%); **d** – 100 mL/min Ar + O_2 (99.5 + 0.5%) (from [6267]).

The memory interferences observed in an FIT atomizer by DĚDINA [1454] for nearly all of the hydride-forming elements at higher concentrations could not be confirmed by WELZ and STAUSS [6267] for the QTA. Merely bismuth and tin caused a minor memory interference. In a continuous flow system, however, WALCERZ et al. [6130] found that bismuth and tellurium also caused an interference in the condensed phase. This finding was confirmed by PETRICK and KRIVAN [4627] who used radiotracers in a batch system.

As well as the possibilities of reducing the gas-phase interferences mentioned above—increasing the gas flowrate reduces the interference caused by an accelerated atom attrition and adding oxygen reduces the interference caused by the radical concentration—and the fact that fewer interferences are observed in FI systems compared to batch systems, a number of working groups attempted to reduce the interferences by chemical means. The basic idea is that an interference in the gas phase can only take place when the interfering element is vaporized. If the formation of the hydride of the interfering element can be selectively prevented, then the interfering influence can be eliminated.

WELZ and MELCHER [6213] utilized the highly differing influence of copper on the generation of the hydrides of selenium and arsenic, for example, to eliminate the interference of selenium on the determination of arsenic. The addition of 50 mg/L Cu suppressed the generation of selenium hydride, but not that of arsine, so that the interference-free range for the determination of arsenic could be extended by nearly three orders of magnitude. A similar effect was achieved by BARTH et al. [402] by adding potassium iodide which reduces arsenic and antimony to the trivalent state, but selenium and tellurium to the element, so that hydride generation is also prevented. For the determination of arsenic, DE LA CALLE-GUNTIÑAS et al. [1034] found that a mixture of lactic acid and potassium iodide in 1 mol/L HCl completely eliminated the interference caused by a tenfold excess of antimony and a hundredfold excess of selenium.

8.3.3.3.2 Graphite Tube Atomizers

In their graphite paper atomizer, DITTRICH and MANDRY [1543] observed substantially fewer interferences on the determination of antimony by arsenic, bismuth, and selenium. In the range 1600–2000 °C the interferences decreased systematically with increasing temperature. This can be taken as the direct proof for the positive influence of higher temperatures on the reduction of interferences in atomizers. The clearly lower mutual interferences of the hydride-forming elements in the graphite furnace [1543–1545, 2974] in comparison to the QTA may in part be due to the differing materials and the composition of the gas phase [1458]. An example of this is the influence of bismuth, antimony, and selenium on the determination of arsenic, which is very pronounced in the QTA, but which virtually disappears in the graphite furnace at 1600 °C [1543].

If the hydride is collected prior to atomization a similar reduction of the interfering influences as for direct atomization is to be expected [5632, 6332, 6543]. AN et al. [170] nevertheless ascertained that there was a sort of competition for the active sites in the tube if there was not enough palladium present for sequestration. This then led to an increase in the mutual interferences of the hydride-forming elements. If the hydrides are collected in the graphite tube prior to atomization, a number of interferences in the condensed phase should also be indirectly reduced. If the rate of hydride generation is not included in the measurement, the concentrations of acid or sodium tetrahydroborate can be optimized independently, for example. Up to the present, however, no systematic work on this subject has been reported. Solely AN et al. [170] reported that under optimized conditions As(III) and As(V) generate identical signals. These authors also found lower interferences with diluted sodium tetrahydroborate solutions, which agreed with the earlier results of other working groups [3020, 5341, 6232, 6430].

8.4 The Cold Vapor Technique

The cold vapor technique (CV AAS) can only be applied for the determination of mercury and is based on a number of typical characteristics of this element—the sole exception is the element cadmium which can also be determined by this technique, as discussed in Section 8.3 [5094, 5095]. Since mercury can be easily reduced from its compounds to the metal and exhibits a marked vapor pressure of 0.0016 mbar at 20 °C, it can be determined directly without a special atomizer unit. After reduction it must merely be transported into the vapor phase by a stream of gas. For the CV AAS technique a discussion on the atomization mechanism is thus superfluous as is a discussion of possible interferences to the atomization process.

Virtually no spectral interferences occur with the cold vapor technique. There were occasional reports in earlier papers that water vapor caused background attenuation. However, the H_2O molecule does not exhibit any absorption bands at the mercury line. The observed interferences were due to droplets of solution carried along with the gas stream or to condensation of water vapor in the absorption cell [5607]. This can be prevented by a suitable drying agent or more simply by warming the absorption cell.

The causes for the systematic errors that frequently can be observed for the determination of mercury all depend to a greater or lesser extent on the mobility of this element and its compounds. These errors include blank values and contamination due to reagents, the laboratory ware, and the atmosphere, and losses due to vaporization, adsorption, or chemical conversions. In the extreme trace range, in which mercury must often be determined, these phenomena can lead to results that are wrong by several orders of magnitude. Even in the 'normal' range, substantial errors can occur if extreme care is not taken.

A further problem is that mercury cannot be reduced from numerous compounds, in particular organic compounds, to the element with the usual reductants, tin(II) chloride or sodium tetrahydroborate. The determination must thus often be preceded by a digestion that can have a decisive influence on the trueness of the determination. For such digestions, next to the completeness of digestion, parameters such as analyte losses and contamination play a major role. These problems are therefore discussed in somewhat more detail in the following.

8.4.1 Mobility, Contamination, and Losses

Even the storage of samples prior to the determination of mercury can be problematical. For example, soil samples that had been stored in air to dry at 20 °C for 30 days took up mercury from the atmosphere in quantities many times over the original content; the values increased by up to a hundredfold, but without any particular regularity [3002]. If test samples are stored at locations at which there is a mercury concentration gradient between the environment and the sample, an exchange of mercury can even take place through plastic foils; this diffusion process depends very strongly on the material of the foil, its thickness, and the temperature.

A further source of systematic errors in the determination of mercury are losses due to vaporization, adsorption, or chemical conversion. These losses can take place during sampling, sample preparation, digestion, storage, and also during the actual measurement. Due to the high mobility of mercury most of these losses are caused by exchange reactions. A test sample that contains a relatively high concentration of mercury can deposit some on the walls of the container. If the same container is reused without prior cleaning for a sample with a relatively low mercury concentration, a reversed exchange takes place and mercury from the walls of the container passes into solution. The loss of mercury from one sample can be the cause of contamination in the next. The question of suitable materials for containers used for the storage of mercury and as the reactor for mercury determinations is mentioned in Section 4.3.1.3 and discussed in detail in Section 9.27.

Mercury(I) ions disproportion very readily to Hg(II) and Hg^0; elementary mercury can then be lost under a variety of conditions, either by adsorption or by vaporization [6317]. Suspended matter in liquid test samples should not be removed by filtration since the filter can adsorb a considerable quantity of mercury due to its large surface area. Likewise with centrifugation a portion of the mercury can be adsorbed on the precipitate and this can be substantial for determinations in the lower ng/g range [3002].

With solid samples systematic errors mostly occur during particle reduction, sieving, and homogenizing. Great care must be taken to ensure that the materials used for cutting tools, for example, do not cause losses through amalgamation. Containers made of inert plastic and cooled in liquid nitrogen have proven to be suitable for homogenization [3002]. The reports on losses during freeze drying are conflicting [3537].

Severe systematic errors can be caused by contamination of containers and laboratory ware from the laboratory atmosphere. In non-contaminated areas the concentration of mercury in the air is normally very low and rarely exceeds a few ng/m^3. Mercury can nevertheless be strongly enriched in laboratory atmospheres and values of several hundred ng/m^3 are not seldom [2078]. A further problem is that mercury, unlike the majority of the other contaminating elements, is not bound to dust particles but is present in gaseous form and cannot thus be held back even by the high-performance filters of a clean room. This can easily lead to uncontrollable blank values and to contamination of the test samples, the surfaces of containers, and reagents [5864].

Thorough cleaning of containers, for example by fuming out with nitric acid and water, should naturally be a part of every laboratory procedure, but even then not all mercury traces can be removed. The attempt should thus always be made to use as little laboratory apparatus as possible; this apparatus should be made of inert materials and should have the smallest possible surface area. In this respect flow systems, especially the FI technique, have considerable advantages compared to batch systems. Furthermore, on-line techniques are superior to classical digestion techniques (refer to Chapter 6).

The reagents used are the best-known source of elemental blank values. In the first place there are very few reagents that can be prepared in a very pure form without considerable effort. Among these are water and some acids; all other reagents hardly meet the demands of extreme trace analysis [5864].

Acids can be very effectively purified by distillation at a temperature below their boiling point (subboiling distillation; see Section 5.1.3), but we must always bear in mind that such solutions do not remain pure over prolonged periods. Mercury obeys an effective distribution function, and an equilibrium between a gas and an aqueous phase is rapidly established [3204]. This means that in an open container a high concentration of mercury is quickly obtained when mercury is present in the laboratory atmosphere. It has even been found that opening a bottle of ultrapure acid briefly can markedly increase the blank value.

8.4.2 Digestion Procedures

The experimental conditions and the reductant used play a major role in establishing to which level organic mercury compounds can be detected for direct determinations without prior digestion. MARGEL and HIRSH [3883] found that ionic mercury and methylmercury, for example, could be determined directly using sodium tetrahydroborate in the presence of 10 μmol/L copper sulfate at a pH of 9.3–9.5, a temperature of > 25 °C and a reaction time of > 1 minute. Tin(II) chloride, on the other hand, is not able to release elemental mercury from organic mercury compounds [5139]. It has even been reported that tin(II) chloride is not able to reduce mercury hydroxide complexes in water to the element [224]. For this reason a digestion prior to the determination is mostly indispensable, even for water samples, and most certainly for the examination of biological materials. It is important during the digestion to maintain a high oxidation potential to prevent mercury from being reduced to the element with the consequent losses.

Acidic digestions with nitric acid and sulfuric acid under the addition of potassium permanganate have frequently been used [1633, 3560, 4496, 5321]. Nevertheless, even this digestion procedure is apparently not able to break down all organic mercury compounds [4471]. A digestion in alkaline permanganate appears to be more effective [1197]. The biggest problem with this digestion, however, is the potassium permanganate since it represents a substantial source of contamination [5381]. As an alternative vanadium pentoxide has frequently been added as a catalyst to the digestion [1068, 3825, 3926], and the digestion performed under pressure [3254].

Particularly in the 1970s, digestions were performed by combustion in oxygen in a Schöninger flask to keep contamination by reagents as low as possible [685, 2379, 4286, 4454, 4533]. A major disadvantage of this technique was the large surface area of the flask, which was often the cause for carryover and contamination. The same problem was also repeatedly observed over a long period of time in water analysis using the popular technique of digestion by UV radiation [48, 50, 800, 1917].

TÖLG [4507] further developed the technique of ashing the sample in a stream of oxygen using a closed system. The vaporized mercury is collected on a cold finger, after which it is dissolved by refluxing with a small quantity of nitric acid. The technique was used successfully for determining mercury in a number of biological materials [3243, 6228], but it was found to be too complex and too slow for routine use.

KNAPP et al. [3151, 3152] reported a mechanized system for digesting organic materials in a chloric acid-nitric acid mixture. The very high oxidation potential of this mixture appears to be the reason why mercury remains in solution even at digestion temperatures of over 100 °C. Nevertheless the shape of the digestion vessel plays a signifi-

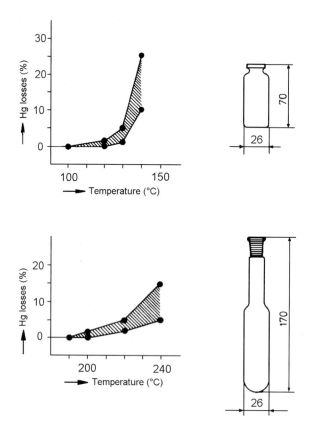

Figure 8-47. Losses of mercury during digestion in chloric acid and nitric acid in dependence on the temperature and the shape of the digestion flask (from [3002]).

cant role [3002]. Figure 8-47 shows the dependence of mercury losses on the heating block temperature and the shape of the digestion flask. While mercury losses are significant at 120 °C in a short flask, the temperature can be increased to 200 °C when a long-necked flask is used. A portion of the digestion mixture vaporizes at these temperatures, but there is no loss of mercury, presumably due to the oxidation of mercury in the vapor phase by the ClO_2 released from the chloric acid [3002].

A technique used very successfully in recent years for the digestion of practically all organic mercury compounds in water and biological materials is digestion in potassium permanganate and potassium peroxodisulfate in acidic solution [436, 1625, 1793]. It has been found especially advantageous to perform the digestion in an ultrasonic bath at 50 °C since the reagent requirement can be reduced substantially and thus the risk of contamination also [436, 1793, 5379]. HANNA et al. [2355] described a similar digestion procedure with persulfate and sulfuric acid which they were able to perform on-line with an FI system.

Microwave-assisted digestion is a further technique that has been applied success-fully in recent years for the digestion of varying materials for the determination of mercury [4309, 5176, 6043]. A major advantage, next to the saving of time, is that substantially lower acid concentrations are required. SUN et al. [5692] described a digestion in potassium bromate-potassium iodide in which BrCl is released and this leads to a rapid digestion of organic mercury compounds in water and urine. TSALEV and co-workers [5906, 5907, 6263] modified this procedure to a microwave-assisted, on-line digestion in an FI system. The technique has been used successfully for the on-line determination of mercury in water [5907, 6263], urine [2281, 6263], and blood [2282].

8.4.3 Interferences

8.4.3.1 Interferences on the Release of Mercury

In this section we shall summarize all of the interferences that take place in the condensed phase and which are responsible for the non-quantitative release of mercury from solution. The observed interferences differ for the two reductants. The tolerable mass concentrations of concomitant elements under varying conditions are tabulated in Table 8-11

When using tin(II) chloride as the reductant, merely iodide and Se(IV) interfere more strongly. YAMADA et al. [6425] eliminated the interference due to iodide by mixing some sodium tetrahydroborate to the tin(II) chloride. Iodide does not interfere when sodium tetrahydroborate is used as reductant, but a number of heavy metals do since they can be reduced by the sodium tetrahydroborate and then react in the reduced form with mercury. In principle, these interferences are analog to those that can be observed in the hydride-generation technique. Reduced noble metals can react with mercury via the formation of amalgams, metals that are less noble can bind mercury by cementation. Since the reactions that lead to the interferences are similar, the same measures for their elimination as applied in HG AAS usually help (refer to Section 8.3.3.2, part 2). Next to an increased acid concentration and the use of Fe(III) ions as a buffer, masking rea-

Table 8-11. Tolerable mass concentrations of concomitant elements (in mg/L) in the measurement solution for interference-free determination of mercury by CV AAS (from [1793]).

Reductant	$SnCl_2$	$NaBH_4$	$NaBH_4$	$NaBH_4$
Technique	direct/amalgam	direct	direct	amalgam
Medium	0.5 mol/L HCl	0.5 mol/L HCl	5 mol/L HCl	5 mol/L HCl
			+ 0.2 g/L Fe(III)	+ 0.2 g/L Fe(III)
Concomitant element				
Cu(II)	500	10	10	2
Ni(II)	500	1	500	500
Ag(I)	1	0.1	10	1
I(I)	0.1	100	10	10
As(V)	0.5	0.5	0.5	0.5
Bi(III)	0.5	0.05	0.5	0.02
Pb(II)	0.5	0.5	0.5	0.02
Se(IV)	0.05	0.005	0.05	0.05

gents such as cyanide can be applied to eliminate the interferences caused by heavy metals [1707]. Also in analogy to the hydride-generation technique, interferences can be reduced in FI systems by kinetic discrimination. The interferences presented in Table 8-11 were determined in batch systems and can be noticeably lower in flow systems.

STUART [5608] drew attention to several sources of error that can arise from the use of common reagents. Hydroxylamine hydrochloride, which is added to solutions to reduce excess permanganate, can have a substantial influence on the determination of mercury at higher concentrations; 25 mg cause a signal depression of 15%, while 100 mg cause a depression of 65%, both in peak area and peak height. A further interference, which is only observed in peak height however, is caused by SH-groups. An example is cysteine, which is often added to complex mercury to obtain better reproducibility. Cysteine is also present in numerous biological matrices, but it only interferes in this case if the sample is not completely ashed. Reagents containing SH-groups retard the release of mercury from the measurement solution; this is an interference that can be eliminated by peak area integration or by amalgamation. An interference caused by nitrogen oxides after a digestion in nitric acid can be eliminated by passing argon through the solution [4943].

8.4.3.2 Interferences to Amalgamation

If mercury is preconcentrated by amalgamation on a noble metal prior to determination, the same interferences due to heavy metals will occur, but not the kinetic interferences, provided the collection time is long enough. A further interference in this technique is partial or complete coating or poisoning of the gold surface by other gaseous reaction products which prevents complete amalgamation. The risk is much higher for sodium tetrahydroborate than for tin(II) chloride, since more substances are released using the stronger reductant. Due to the more violent reaction of sodium tetrahydroborate the probability that solution droplets will be carried to the amalgamation surface is also greater [4043]. Frequent conditioning of the gold adsorber in nitric acid is recommended to obtain optimum amalgamation. Many of the materials proposed as adsorbers, such as silver or gold wool, or quartz wool coated with gold, do not withstand frequent conditioning. A gold-platinum gauze has proven to be very effective due to its mechanical stability [3002, 6223].

Among the substances vaporized by sodium tetrahydroborate are the hydrides of As, Bi, Sb, Se, Sn, and Te. The influence of arsenic on the determination of mercury as observed in work by WELZ et al. [6223] is shown in Figure 8-48. Nevertheless, it could be shown that this interference only occurred so strongly when the gold-platinum gauze had not cooled sufficiently prior to the next collection [6249]. When the gold-platinum gauze is at a temperature < 100 °C it is not contaminated by relatively large amounts of arsine so that the amalgamation of mercury is without interferences.

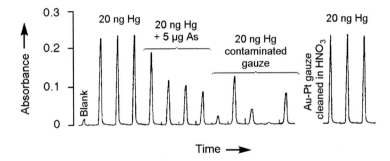

Figure 8-48. Deactivation of the gold-platinum gauze by arsine. This interference is only observed when the temperature of the gauze during collection is above 100 °C (from [6249]).

9 The Individual Elements

For each of the elements described in this chapter we shall begin by mentioning its occurrence, use, biological significance, environmental relevance, and toxicity. Most of this information was culled from *Römpps Chemie-Lexikon* [1803]. Next to the element itself, we shall discuss individual species if these are of biological or toxicological significance. We shall further mention any particular features of an element that can influence the accuracy of an analysis, such as the stability of the solutions, the volatility of individual compounds, or the increased risk of contamination.

Following this general introduction we then present AAS-specific details, such as wavelength, slit width, flame type and stoichiometry, and the attainable sensitivity. A comparative overview of the limits of detection (LOD) attainable for each element with different atomization techniques is presented in Table 9-1. For numerous elements alternate analytical lines are quoted which normally exhibit reduced sensitivity and are thus suitable for the determination of higher analyte concentrations. Occasionally some of these alternate lines can be employed to avoid spectral interferences (see Sections 5.4.1.1 and 5.4.3.1). It should be noted that the *relationship of the line sensitivities* depends very strongly on the atomizer (this can clearly be seen in Tables 9-40 and 9-41, Section 9.52). Next to the ground state , the majority of elements exhibit a number of lower excited states whose populations depend on the temperature of the atomizer. These relationships have been investigated in detail for F AAS, but there are few reports on the use of secondary lines in GF AAS, except for the analysis of solid samples and slurries.

The information on *spectral interferences*, which is mostly presented in tabular form, is cited independent of whether it was calculated by means of a computer program [5973] or even if it was checked in practice. Many of these interferences only become noticeable in the presence of extremely high matrix contents, such as trace analyses in metallurgical samples under the application of continuum source BC.

For determinations by GF AAS details are provided on the maximum pyrolysis temperature, the optimum atomization temperature, the recommended modifier, and the characteristic mass. The use of STPF conditions is assumed unless other details are given, i.e. atomization from a platform, maximum heating rate, the use of a modifier and integration of the peak area.

For the elements determinable by HG AAS and CV AAS we have provided the corresponding information on optimum conditions and the attainable sensitivity. For all techniques we have provided brief details about possible interferences of a general nature, and the recommended chemical additives. Sample-specific interferences and special applications are discussed in Chapter 10. The information presented in the current chapter is taken in part from a number of applications manuals published by the PERKIN-ELMER CORPORATION [684, 4609–4611]. We have carefully checked and extended this basic information, where necessary.

Table 9-1. Comparison of relative instrumental limits of detection[a] ($3s$) in µg/L for F AAS, GF AAS, HG AAS, and CV AAS (by kind permission of Perkin-Elmer).

Element	F AAS	GF AAS[b]	HG/CV AAS[c]
Ag	1.5	0.02	
Al	45	0.1	
As	150	0.2*	0.01
Au	9	0.15	
B	1000	20	
Ba	15	0.35	
Be	1.5	0.01	
Bi	30	0.25	0.03
Ca	1.5	0.01	
Cd	0.8	0.003*	
Co	9	0.15	
Cr	3	0.01*	
Cs	15		
Cu	1.5	0.1*	
Dy	50		
Er	60		
Eu	30		
Fe	5	0.1	
Ga	75		
Gd	1800		
Ge	300		
Hf	300		
Hg	300	1.0	0.009
Ho	60		
In	30		
Ir	900	3	
K	3	0.008	
La	3000		
Li	0.8	0.06	
Lu	1000		
Mg	0.15	0.004	
Mn	1.5	0.01*	
Mo	45	0.03*	
Na	0.3	0.02	
Nb	1500		
Nd	1500		
Ni	6	0.3	
Os	120		
P	75000	130	
Pb	15	0.1	
Pd	30	0.8	
Pr	7500		
Pt	60	2	
Rb	3	0.04	
Re	750		

Table 9-1 continued

Element	F AAS	GF AAS[b)]	HH/CV AAS[c)]
Rh	6		
Ru	100	1.5	
Sb	45	0.15	0.05
Sc	30		
Se	100	0.25*	0.03
Si	90	1	
Sm	3000		
Sn	150	0.2	0.5
Sr	3	0.025	
Ta	1500		
Tb	900		
Te	30	0.4	0.03
Ti	75	0.35	
Tl	15	0.25*	
Tm	15		
U	15000		
V	60	0.06*	
W	1500		
Y	75		
Yb	8		
Zn	1.5	0.1	
Zr	450		

a)   All limits of detection were determined using matrix-free calibration solutions.
b)   Measured under STPF conditions with 50 µL measurement solution; * measured with a SIMAA 6000 spectrometer, all others measured with a Zeeman 4100 or Zeeman 5100 spectrometer.
c)   Measured with FI and 500 µL measurement solution.

9.1 Aluminium (Al)

More than 8 % of the Earth's crust comprises aluminium so that after oxygen and silicon it is the third most abundant element. Due to its strong affinity for oxygen it is never found native but always in the form of its compounds, mainly feldspars, micas, and their detritus the clays. Bauxite is the most important mineral for the extraction of Al; it is mainly a mixture of aluminium hydroxide minerals.

Aluminium is the most important light metal and it is used in a multitude of ways in industry. Aluminium alloys are especially important due to their low density and find particular use in the aerospace and automotive industries. Aluminium compounds find a variety of industrial applications, e.g. as flocculants, adsorbents, water repellents, fillers, and thickeners.

It is difficult to envisage that the third most abundant element in the Earth's crust is toxic at even relatively low concentrations. Nevertheless the absorption of Al is associated with arteriosclerosis and Alzheimer's disease. The toxicity of Al is closely related to the resorbability of the respective compound or species. The greater part of the Al ingested with the nutrition (10–40 mg) is of a mineral nature and is excreted without being resorbed. Aluminium hydroxides, which are frequently taken to alleviate slight stomach complaints, can be very much more easily resorbed. Beside the solid phase, which is largely inert, numerous complexes and phases with greatly differing reactivities have been found in water; in particular 'fast reacting Al' has been investigated thoroughly in recent years [546, 1298].

Aqueous Al solutions must be acidified to stabilize them and to prevent adsorption losses, which become noticeable at pH values > 1.5 [5456, 5660, 6323]. The losses are lower with urine and serum samples than with waters [6323], but even for these samples the addition of acid is recommended. Solutions with Al concentrations < 50 µg/L are stable in high-density polyethylene (HD-PE) bottles when the solutions contain 0.1 mol/L nitric acid or citric acid [1799].

Aluminium forms very stable bonds with oxygen than can hardly be severed in the air-acetylene flame. Prior to 1966, a number of authors attempted to determine Al in oxygen-acetylene flames of varying stoichiometry; organic solvents were frequently used to increase the sensitivity. A completely satisfactory determination of Al was first possible after the introduction of the nitrous oxide-acetylene flame by WILLIS [6340]. Practically no interferences are present in this flame; merely the presence of silicon is said to have a slightly suppressive influence on the absorption of Al [1306, 1900]. The flame stoichiometry also has an influence on the sensitivity; the best results are obtained in a fuel-rich flame which should exhibit a red plume. The characteristic concentration of Al in such a nitrous oxide-acetylene flame is approximately 1 mg/L at the 309.3 nm line; comparative values for other analytical lines are compiled in Table 9-2.

Since Al is ionized by about 10 % in the nitrous oxide-acetylene flame, about 1–5 g/L K as the chloride or another easily ionized metal should be added to the test sample and calibration solutions. Iron, silicon, and hydrochloric acid in concentrations above 2 g/L and also sulfuric acid depress the sensitivity; the addition of titanium enhances the sensitivity of Al.

Table 9-2. Al analytical lines.

Wavelength (nm)	Energy level (K)	Slit width (nm)	Characteristic concentration* (mg/L)	Characteristic mass [3865] (pg)	Zeeman factor	Spectral interferences
309.27/ 309.28	112–32435	0.7	1.1	11	0.85	
396.15	112–25348	0.7	1.1	14	0.9	
308.22	0–32435	0.7	1.5	22	0.72	V [1852, 4991]
394.40	0–25348	0.7	2.5	29	0.95	
237.31	112–42234	0.2	3.3			
236.71	0–42234	0.7	4.8			
257.51	112–38394	0.2	6.7			
256.79	0–38929	0.2	7.8			

* Nitrous oxide-acetylene flame, reducing, addition of 1–2 g/L K.

The greatest problem for the determination of Al by GF AAS is without doubt its omnipresence and the thus resulting risk of contamination, e.g. from dust particles and reagents [1118, 1349, 1764, 2270]. Depending on the source of the contamination this is reflected as poor repeatability or high blank values. A lower ('better') characteristic mass is a reliable indication for contamination and not of an improved atomization efficiency. Glass is not recommended as a container material due to the risk of contamination [2244, 6323], but even bottles made of PP or PS [1799, 2106] can cause contamination due to the elution of catalyst residues [123, 1300, 2106, 2269, 2898, 3397, 6078]. Containers made of PFA are easy to clean due to their smooth, inert surfaces and are therefore particularly suitable for dilute solutions [2106]. Accurate determinations in the trace and ultratrace ranges require a clean room or at least a laminar-flow cabinet [123, 124, 4381, 5394]. For the determination of Al species the analysis should be performed immediately since Al hydrolysis reactions are very sensitive to changes in the temperature and the pH value. If it is necessary to store a sample, solids and colloids should be removed by filtration and the sample should be stored at the original temperature [3572].

For the determination of Al by GF AAS argon should always be used a the purge gas [3853], since Al forms a stable monocyanide in the presence of nitrogen [3704]; this leads to a substantial loss in sensitivity. In the early literature there were frequent reports about strong interferences in the determination of Al by GF AAS. Based on thermodynamic equilibria calculations, PERSSON et al. [4616, 4617] found that even small quantities of oxygen, hydrogen, chlorine, nitrogen, and sulfur could cause interferences during atomization; hydrogen and chlorine can also cause interferences during pyrolysis. They found that the interferences were smallest when the highest possible pyrolysis temperatures were used. The presence of hydrogen reduces the usable pyrolysis temperature markedly. MANNING et al. [3862] investigated the determination of Al under STPF conditions. They used 50 µg magnesium nitrate as the modifier and were thereby able to apply a pyrolysis temperature of 1700 °C [5421]. MANNING et al. [3862] also found that the quality of the PG layer on the graphite tubes had a major influence on the interferences observed, e.g. by perchloric acid. Hardly any interferences were observed in a tube with a perfect PG coating under STPF conditions.

WELZ et al. [6265] investigated the Pd-Mg modifier for the determination of Al and were also able to apply a pyrolysis temperature of 1700 °C. One of the advantages of this mixed modifier is that much less reagent is required and the problems of contamination are thus smaller. Using the Pd-Mg modifier even 10 g/L chloride (as NaCl) or sulfate caused no interferences. TANG et al. [5767] compared various modifiers and came to the conclusion that calcium nitrate was the best. They were unable to observe interferences from hydrochloric acid or sulfuric acid when using integrated signals. The optimum atomization temperature is 2500 °C in a longitudinally-heated atomizer and 2300 °C in a transversely-heated atomizer. The characteristic mass at the 309.3 nm line with and without Zeeman-effect BC is 10 pg Al in a longitudinally-heated atomizer and 30 pg Al in a transversely-heated atomizer. The 396.2 nm line can be of advantage, particularly for the application of Zeeman-effect BC, when a wide linear range is required [3865]. Due to the high portion of radiation emission from the hot graphite tube, critical alignment of the graphite furnace in the radiation beam is required when the 396.2 nm line is used.

9.2 Antimony (Sb)

Approximately $10^{-4}\%$ of the Earth's crust is composed of Sb; it is thus more rare than some of the REEs but nevertheless more abundant than the noble metals. Antimony is very brittle and is practically only used as a component of alloys with lead or tin, for example in type metal, except in the semiconductor industry. Physiologically Sb salts are less toxic than the corresponding As salts since they are less easily transported though the walls of the stomach and intestines and are excreted more rapidly.

Aqueous Sb solutions should be stabilized by the addition of acid to prevent adsorption losses, which can become significant after only a few hours at pH values > 1.5 [5455]. The problems of sample conservation are more complex when Sb species are to be determined. SUN et al. [5689] observed that in natural waters Sb(III) is oxidized to Sb(V) after a few hours. This oxidation could be prevented for at least five days by adding tartaric acid ($w = 1\%$). ANDREAE et al. [198] added potassium antimonyltartrate to the calibration solutions; they reported that solutions with a concentration of 1 mg/L were very stable. This fact was verified by CALLE-GUNTIÑAS et al. [1035] who found that calibration solutions containing both Sb(III) and Sb(V) could be stored in PE bottles for up to 12 months at 0–4 °C without the ratio Sb(III)/Sb(V) from changing. They advised against an addition of ascorbic acid since with increasing decomposition it promotes the oxidation of Sb(III) [1035, 1038].

Antimony can be determined in the air-acetylene flame practically free of interferences. However, little has been published about the determination of this element by F AAS since it is mostly present in very low concentrations and must therefore be determined using more sensitive techniques. For the determination of Sb, three analytical lines of similar sensitivity are available as listed in Table 9-3. The 217.6 nm line is employed for the majority of determinations; high concentrations of iron, copper, and lead can cause slight spectral interferences at this line. A slit width of 0.2 nm must always be used for the determination of Sb at the 217.6 nm line since otherwise these spectral interferences can reach a substantial magnitude. In the presence of these interfering elements, and also for higher Sb concentrations, the 231.1 nm line is recommended. Nevertheless, a spectral interference by nickel must be taken into account at this line.

Table 9-3. Sb analytical lines.

Wavelength (nm)	Energy level (K)	Slit width (nm)	Characteristic concentration* (mg/L)	Spectral interferences[§]
217.58	0–45945	0.2	0.55	Cu [4208, 4385, 5973] Pb [3213, 5403, 6426] Fe [6426]
206.83	0–48332	0.2	0.85	Bi [5973]
231.15	0–43249	0.7	1.3	Ni [3213, 3691, 4385]
212.74	0–46991	0.7	12	

* Air-acetylene flame, oxidizing
§ Depend on the mode of BC

Antimony can be determined by GF AAS under STPF conditions and using the Pd-Mg modifier practically free of interferences [6265]. Concentrations of NaCl of up to 30 g/L have no influence on the Sb signal under these conditions. Solely sulfate contents of > 20 mg/L cause a slight gas-phase interference, which can nevertheless be largely eliminated by optimizing the thermal pretreatment conditions [6265]. The Pd-Mg modifier has proven to be far more reliable than the earlier-used nickel modifier. The double peaks occurring during the atomization of Sb without a modifier are also no longer observed when the Pd-Mg modifier is applied [6265]. Under the above conditions the maximum pyrolysis temperature is 1200 °C and the optimum atomization temperature is about 2000 °C. The characteristic mass in a longitudinally-heated atomizer with Zeeman-effect BC is 19–20 pg and slightly lower without Zeeman-effect BC. The characteristic mass in a transversely-heated atomizer is 55 pg.

Antimony can be determined with advantage by HG AAS. The atomization signal is nevertheless dependent on the oxidation state and the hydride system used, as discussed in Section 8.3.1.4. In batch systems Sb(III) generates a signal that is nearly twice as high as that for the same mass of Sb(V) [6210]. In flow systems this difference can be more than one order of magnitude, depending on the length of the reaction zone (see Figure 8-36). These differences are also pH dependent; this offers the possibility, for example, of selectively determining Sb(III) at pH 8 in the presence of a large excess of Sb(V) [6428].

For the determination of the total Sb content it is nevertheless necessary to perform a prereduction to ensure that all of the antimony is present as Sb(III). A solution of potassium iodide in hydrochloric acid is frequently used for this purpose [5375]. Various interferences caused by transition elements such as iron are reduced by orders of magnitude under these conditions [1177, 3923, 6212]. However, we must consider that the high acid concentration (semiconcentrated hydrochloric acid) and also the potassium iodide pose a substantial disposal problem. In addition, potassium iodide solutions only have limited stability. For these reasons L-cysteine is highly recommended as the reductant; L-cysteine was first proposed by BRINDLE and co-workers [1231, 3473] and thoroughly investigated by WELZ and ŠUCMANOVÁ [6268]. The major advantages of this reagent are that it is virtually non-toxic, it is fully effective in acid concentrations of only 1 mol/L, and the solutions are stable for several weeks. Moreover, interferences caused by transition metals, and especially copper and nickel, can be far more effectively avoided with L-cysteine than with potassium iodide [6269]. The attainable LOD for Sb in the presence of L-cysteine is 25 pg absolute or 0.05 µg/L (500 µL measurement solution), and the calibration curve is linear up to about 10 µg/L.

9.3 Arsenic (As)

Approximately $5 \times 10^{-4}\%$ of the Earth's crust is composed of As, so that it is about as abundant as beryllium and germanium. The element occurs as native arsenic and also forms intermetallic compounds with antimony and copper. It is to be found most frequently in sulfidic and mixed ores; metals that are won from sulfidic ores almost always contain traces of As, which is very difficult to remove. Arsenic is distributed in trace amounts throughout the entire crust of the Earth; it is found in solidified rocks, coal, sea and mineral waters, and also in fauna and flora. Arsenic is used as a constituent of alloys to increase the hardness, e.g. in lead alloys. High purity As is used in the production of GaAs and InAs semiconductors. In the glass industry As compounds are used as cleaning and bleaching agents.

Arsenic is present ubiquitously in all organic tissues of human and animal organisms. Its biological significance as a trace element has not been completely clarified, but it is generally recognized as an essential element. The toxic effect of As and its compounds decreases in the sequence As(III) > As(V) > MMA > DMA > AsB, where MMA = monomethylarsonate, DME = dimethylarsinate, and AsB = arsenobetaine. The toxic, inorganic forms are also preferentially accumulated in organisms.

The major uses of organic As compounds are in medicine and in agriculture as pesticides and biocides; in recent years there has been a clear decrease in the production of organic As compounds. The extraction of As from soils influences its presence in water and in the atmosphere, so that it is distributed in the most varying areas of the ecosystem. Arsenic compounds in biological systems are complex molecules that are difficult to extract and to determine, such as arsenobetaine, arsenocholine, and also As-containing sugars [2680]. The inorganic and methylated As species are equally to be found in the environment and take part in the biochemical cycles.

Oxidation, reduction, and methylation are among the chemical and biological processes that this element can undergo in the natural environment; these processes also change the behavior of the species being investigated. The methylation of As has been observed in fresh waters and in marine systems. For terrestrial organisms the biomethylation of As leads to its elimination by vaporization as methylarsine, CH_3AsH_2, dimethylarsine, $(CH_3)_2AsH$, or trimethylarsine, $(CH_3)_3As$, or by excretion as monomethylarsonic acid, $CH_3AsO(OH)_2$, or dimethylarsinic acid, $CH_2AsO(OH)$. Marine organisms excrete As in the form of water-soluble species, or the compounds are incorporated directly in their sugars and lipids [5811]. In algae and also via anaerobic decomposition, sedimentation, methylation, and oxidation, inorganic arsenate is converted to arsenobetaine, $(CH_3)_3As^+CH_2COO^-$, [1374, 1679].

Since As is present in natural waters in anionic form, losses due to complexation with humic materials or sorption losses on the walls of containers are relatively uncritical [1737, 6115]. Water samples for the determination of total As can thus be stored for several months in glass or plastic bottles without losses [99, 1737]. Sample conservation is much more of a problem when As species are to be determined. TALLMANN and SHAIK [5760] found that the oxidation of As(III) to As(V) in various aqueous samples was so slow that storage without a conserving agent was possible for up to three weeks. AGGETT and KRIEGMAN [56] investigated the stability of sediment interstitial water samples in PE and glass bottles for a period of 42 days. They preferred conservation by acidification with hydrochloric acid to a pH value of 2 and storage in glass bottles close to 0 °C under the exclusion of oxygen. The thermodynamically stable forms of inorganic As in oxygen-rich water are $HAsO_4^{2-}$ (sea-water) and $H_2AsO_4^-$ (fresh water) [1374], while in anoxic waters the equilibrium is moved to the trivalent form [201]. Although arsenate is the thermodynamically more stable form, VAN ELTEREN et al. [1737] observed that in deionized water it is reduced to the trivalent state within a week. They attributed the cause to organic traces from preparation of the water (ion exchange resin) and to the release of reductants from the PE containers. To prevent such redox reactions from taking place on the walls, the ratio of the sample volume and the surface area of the container should be as large as possible and materials with very smooth surfaces should be chosen, such as PFA which releases virtually no organic constituents [1745]. For natural fresh water and sea-water samples, on the other hand, slow oxidation of As(III) was observed for storage in PE bottles without the addition of a conservative which proceeded faster at 20 °C than under cooling [1737]. While acidification to pH 2 almost totally stopped the oxidation of As(III) in fresh water, in sea-water As(III) was completely oxidized to As(V) within three days independent of the storage temperature [1737]. The storage of sea-water for the determination of As(III)/As(V) species cannot therefore be recommended; the determination should be performed on the non-conserved sample within a day. Other natural waters (river water, ground water) could be stored for up to 125 days without a change in the As(III)/As(V) ratio by acidification to pH 2. Under the exclusion of oxygen, filtered water samples cooled to 4 °C could be stored for up to 14 days without further additions [2344]. The stability of organoarsenic compounds has not been as thoroughly investigated. ANDREAE [195, 196] found that methylated species were relatively stable in filtered, acidified water samples. To avoid the problems of sample storage for speciation analysis, the analysis should be performed within a few hours of sampling. Immediate shock freezing and storage at –20 °C has frequently been proposed [196, 1302, 1388, 6431]. Dilute solutions of organic species, on the other hand, can be stored for over six months in the dark at 4 °C [168]. PALACIOS [4520] published information on the stability of urine samples with respect to their contents of various arsenic species.

The main analytical line for As is at 193.7 nm, which makes it the element furthest in the vacuum UV range that can be determined with normal atomic absorption spectrometers. Since the flames normally used in AAS exhibit substantial radiation absorption in this range (refer to Figure 4-1 in Section 4.1.1), the sensitivity of the detectors deteriorates markedly, and radiation losses on optical components increase, the attainable S/N ratio for this element depends very strongly on the quality of the spectrometer. The radiation source also has a major influence; EDLs definitely provide better results [389]. Neon is unsuitable as a fill gas for As lamps [1981].

In the air-acetylene flame a characteristic concentration of 0.5 mg/L can be attained; in an argon-hydrogen diffusion flame it is about 0.2 mg/L and the LOD is 0.02 mg/L. However, since considerable spectral and non-spectral interferences can be expected in this flame, the improved LOD is of little practical significance. Spectral interferences caused by aluminium [4870], lead [1751], and silver [5973] have also been reported in the air-acetylene flame. Because of these difficulties and the fact that As is mostly present in very low concentrations, it is usually determined by GF AAS or HG AAS.

The determination of As by GF AAS under STPF conditions is relatively free of problems [2967]. The original nickel modifier proposed by EDIGER [1677] and used by numerous working groups does not, however, always bring optimum results. The Pd-Mg modifier allows the use of pyrolysis temperatures up to 1300–1400 °C and an optimum atomization temperature of 2200–2500 °C. Concentrations of NaCl of up to 20 g/L and sulfate of up to 1 g/L have no influence on the As signal under these conditions; after careful optimization of the pyrolysis step even 10 g/L of sulfate can be tolerated [6265]. While palladium nitrate stabilizes both inorganic and organic arsenic, palladium chloride fails in the presence of organic As species [5401]. Rhodium has been proposed as an alternative to palladium since it is said to permit a pyrolysis temperature of 1600 °C at which temperature even calcium phosphate can be volatilized [4351]. The characteristic mass in a longitudinally-heated atomizer is 15 pg, while in a transversely-heated atomizer it is about 40 pg at an atomization temperature of 2000 °C.

Arsenic can be determined with advantage by HG AAS. The atomization signal is nevertheless dependent on the oxidation state and the hydride system used, as discussed in Section 8.3.1.4 [3042]. In batch systems the difference between As(III) and As(V) is about 25–30% (refer to Figure 8-35) and disappears almost completely with peak-area integration [3020, 5341]. In flow systems this difference can be up to one order of magnitude, depending on the length of the reaction zone (see Figure 8-36 in Section 8.3.1.4). These differences are also pH dependent [4295]; this offers the possibility, for example, of selectively determining As(III) at pH 4–5 [54, 2679, 3173, 3599, 4279, 5259, 5903].

For the determination of the total As content it is therefore necessary to ensure that all of the As is present in a common oxidation state. If maximum sensitivity is not required a sample preparation procedure can be chosen after which all of the As is present as As(V) [1001]. Most commonly, however, a prereduction is performed to ensure that As is present in the trivalent state. This prereduction can be part of the sample preparation procedure or it can also be performed on-line [795, 1611, 3901, 5519, 5943, 6270, 6416]. A solution of potassium iodide in semiconcentrated hydrochloric acid is frequently used for this purpose [703, 1001, 1325, 1938, 4224, 4737, 5342, 5375, 5496]. Various interferences caused by transition elements are also reduced by orders of magnitude under these conditions. However, we must consider that the high acid concentration (semiconcentrated hydrochloric acid) and also the potassium iodide pose a substantial disposal problem. In addition, potassium iodide solutions only have limited stability. For these reasons L-cysteine is highly recommended as the reductant; L-cysteine was first proposed by BRINDLE and co-workers [795, 1230, 3473] and thoroughly investigated by WELZ and ŠUCMANOVÁ [6268, 6269]. The major advantages of this reagent are described under antimony (Section 9.2). The attainable LOD for As in the presence of L-cysteine is 5 pg absolute or 0.01 µg/L (500 µL measurement solution), and the calibration curve is linear up to about 5 µg/L. Occasionally other reductants have been used for the prereduction, such as thiourea [5956, 6392], thiosulfate [4737, 5662], sodium sulfite [2699], or titanium(II) chloride [1282, 1575, 3250, 4936], particularly when several hydride-forming elements are to be determined simultaneously.

9.4 Barium (Ba)

Barium makes up about 0.04–0.05% of the Earth's crust; granites also contain 0.05% Ba on average. As a result of its very non-noble character and its high reactivity, Ba only occurs in the form of its compounds. The most important Ba mineral is barytes ($BaSO_4$). Pure Ba, or alloyed with aluminium or magnesium, serves as a getter in electronic tubes and for activating electrodes. Metallic Ba is used as an additive in bearing metals since, like calcium, it causes strong hardening of the lead. Barium salts are used in fireworks, for the manufacture of cathode ray tubes, and in the glass industry. Water-soluble Ba compounds are poisonous and cause muscular cramp and heart interferences.

The stability of Ba solutions was investigated by SHENDRIKAR *et al.* [5297] for deionized water in the pH range 1–7 in glass and PE bottles over 15 days and by EICHHOLTZ *et al.* [1684] in the pH range 3.3–10.3 in glass, PE, and PP containers over 24 hours. SMITH [5455] followed the losses of Ba from solutions containing NaCl in glass and PE bottles over 28 days. To prevent losses, acidification with hydrochloric acid or nitric acid to pH values < 1.5 is necessary. In water Ba can be sorbed to a significant level on natural suspended matter and be remobilized by the addition of acid or NaCl [1039].

Barium forms thermally very stable oxides which are split to only a limited extent in the air-acetylene flame. Correspondingly Ba exhibits poor sensitivity and also numerous solute-volatilization and vapor-phase interferences in the air-acetylene flame [1290, 3063, 3080] which are markedly reduced or even completely absent in the nitrous oxide-acetylene flame [3786]. The strong emission of the nitrous oxide-acetylene flame near to the Ba line can on occasions interfere (see Figure 4-3 in Section 4.1.1) since it is a considerable noise component (Figure 9-1). Increasing the lamp current and reducing the slit width is a partial remedy.

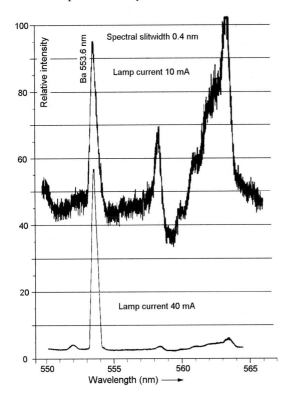

Figure 9-1. Influence of the lamp current on the determination of Ba. The upper spectrum shows the high proportion of the emission radiation that is contributed by the nitrous oxide-acetylene flame when a low-intensity HCL is used. The lower spectrum was scanned under identical conditions with a high-intensity HCL. Zero balance was carried out in each case before the flame was ignited.

Barium is ionized to 80–90 % in the nitrous oxide-acetylene flame (see Figure 8-9 in Section 8.1.2.5) so that it is necessary to add 2–5 g/L K as the chloride or another easily ionized metal to all test sample and calibration solutions. The characteristic concentration at the 553.6 nm analytical line in the nitrous oxide-acetylene flame is 0.5 mg/L. The line at 350.1 nm can be used for higher Ba concentrations since it exhibits a sensitivity that is about a power of ten lower.

The radiation emission of the graphite tube can interfere in the determination of Ba by GF AAS. For this reason a too high atomization temperature should be avoided; transversely-heated atomizers are especially advantageous because of their intrinsically lower atomization temperatures. A further problem is that a deuterium lamp only emits very low energy at the Ba wavelength and thus cannot be used for BC. A logical consequence is that only spectrometers with a halogen lamp as the continuum source or that operate with Zeeman-effect BC can be used.

In longitudinally-heated atomizers Ba can only be determined satisfactorily by atomization from the tube wall; the required atomization temperature is 2550 °C. A characteristic mass of 6.5 pg can be attained [643]. Since Ba forms a relatively stable carbide, the quality of the PG coating is of primary importance [4169]. Without modifier the maximum pyrolysis temperature is 1200 °C. Modifiers have been applied to better separate the matrix, such as ammonium nitrate for the analysis of sea-water [4927], or to suppress carbide formation [643, 3069]. In a transversely-heated atomizer Ba can be determined under STPF conditions by atomizing from the platform. The required atomization temperature is 2300 °C. Although the characteristic mass of 15 pg is considerably higher, a significantly better S/N ratio can be achieved.

9.5 Beryllium (Be)

Beryllium is a rare element and only makes up about 0.006 % of the Earth's crust. Due to its non-noble character it only occurs in the form of its compounds. The most important mineral is beryl ($3BeO \cdot Al_2O_3 \cdot 6SiO_2$). Pure Be metal is used for windows that allow X-rays to pass through about 17 times more strongly than Al. The hardness, strength, and the temperature and corrosion durability of aluminium, cobalt, copper, iron, and nickel can be substantially increased by alloying with Be. Intermetallic compounds of Be are suitable for watch springs, surgical instruments, valve springs, etc. Beryllium bronzes are used in electrical engineering due to their good electrical and thermal conductivity.

Beryllium and its compounds are highly toxic and are also listed as carcinogenic materials. In the form of dust and vapors they can cause severe, irreversible lung damage (berylliosis), frequently resulting in death. The skin and mucous membranes are severely attacked, and chronic exposure causes liver damage and spleenic enlargement. Beryllium is accumulated in organisms and can thus lead to long term damage.

The stability of Ba solutions was investigated by BENEŠ et al. [535, 536] and by SHENDRIKAR et al. [5297]. Significant losses due to sorption were observed on glass and PE surfaces and must be suppressed by acidification.

Beryllium can only be determined very poorly in the air-acetylene flame, but exhibits excellent sensitivity in the nitrous oxide-acetylene flame with a characteristic concentration of about 0.03 mg/L. The most important analytical line is at 234.8 nm; the use of other lines has not been reported up to the present. The determination of Be appears to be largely free of interferences in the nitrous oxide-acetylene flame [1922]. BOKOWSKI [695] reported a slight interference in the presence of high concentrations of silicon and aluminium, but found no influences from phosphate or sulfate. RAMAKRISHNA [4794] eliminated the depressive effect of aluminium by adding about 1.5 g/L fluoride. He found a further signal increase of about 20% by adding acetic acid. As well as fluoride, chloride, and EDTA, LUECKE [3646] added 8-hydroxyquinoline in particular to suppress the interference from aluminium. A spectral interference due to arsenic has also been reported [5973]. Beryllium is not noticeably ionized in the nitrous oxide-acetylene flame so that the addition of alkali is not required.

The determination of Be by GF AAS also appears to be free of problems. Using magnesium nitrate as the modifier [238, 2231, 2374, 4548, 5421, 5654, 5990], a pyrolysis temperature of 1500 °C can be applied. As well as magnesium nitrate, calcium nitrate [4316, 5831, 6424, 6464], aluminium nitrate [3674], and EDTA [3506, 3953] have been used as modifiers. Using ammonium molybdophosphate-ascorbic acid as the modifier, SHAN et al. [5274] observed a 50% increase in sensitivity compared to magnesium nitrate. The optimum atomization temperature under STPF conditions is 2300–2500 °C. The characteristic mass using Zeeman-effect BC is 1.0 pg; in a non-Zeeman instrument it is 0.5 pg [6456], and in a transversely-heated atomizer using Zeeman-effect BC it is 2.5 pg.

9.6 Bismuth (Bi)

Bismuth is one of the rarest of elements; it only makes up an estimated $2 \times 10^{-5}\%$ of the Earth's crust. Bismuth is used for the production of alloys; Bi telluride is used in Peltier cooling elements. The major use of Bi is in pharmaceutical applications; Bi compounds have astringent, antiseptic, and diuretic properties, a fact that has been known since ancient times. The metal is classed as non-toxic.

The stability of Bi solutions in various container materials was investigated by SMITH [5455] and GREENLAND and CAMPBELL [2212]. Significant losses were observed even after only 24 hours when the pH value was > 1.5 [5455], so that all solutions should have a pH value ≤ 1 [5455].

Bismuth can be determined without interferences in the air-acetylene flame. The characteristic concentration at the 222.8 nm resonance line is 0.2 mg/L; various other analytical lines are compiled in Table 9-4. An improved S/N ratio can be obtained in the air-hydrogen flame with an LOD of about 0.015 mg/L. Next to its applications in metallurgy [3136, 6298], Bi is found in biological and environmental samples where its concentration is too low to be determined by F AAS without prior preconcentration. In such cases it is determined by GF AAS or HG AAS.

Table 9-4. Bi analytical lines.

Wavelength	Energy level	Slit width	Characteristic concentration*	Spectral interferences
(nm)	(K)	(nm)	(mg/L)	
222.83	0–44865	0.2	0.2	Fe [5894]**
306.77	0–32588	0.7	0.6	OH[3941]§
206.17	0–48489	0.7	1.6	
227.66	0–43912	0.7	2.7	

*   Air-acetylene flame, oxidizing (lean, blue)
** With Zeeman-effect BC
§   With Zeeman-effect BC in flames

For the determination of Bi by GF AAS under STPF conditions using the Pd-Mg modifier a pyrolysis temperature of 1200 °C can be applied [2924]. The optimum atomization temperature under these conditions is 1900 °C; the characteristic mass with Zeeman-effect BC is 28 pg, while in a non-Zeeman instrument it is about 20 pg. As indicated in Figure 8-28 (Section 8.2.4.1), without a modifier Bi can be determined at an optimum atomization temperature of only 1400 °C with the sensitivity improved by about 60 %. Nevertheless, this is hardly to be recommended in practice since the maximum pyrolysis temperature is only 600 °C [2140] and the risk of interferences is drastically increased [3954]. In a transversely-heated atomizer the optimum atomization temperature is 1700 °C and the characteristic mass with Zeeman-effect BC is 60 pg.

Under STPF conditions using the Pd-Mg modifier, 30 g/L sulfate and 1 g/L NaCl do not interfere in the determination of Bi. Higher concentrations of NaCl cause a loss in sensitivity by about 20 %, which however does not increase further up to a concentration of 30 g/L NaCl [6265]. A spectral interference has been observed at the 223.1 nm line caused by high iron concentrations, such as are found in metallurgical samples [5894].

Bismuth can also be determined with excellent sensitivity by HG AAS; the LOD of 0.03 µg/L is more than an order of magnitude better than with GF AAS. Various interferences by transition metals have been described in the literature which can be significant, especially for the analysis of metallurgical samples. Measures recommended to eliminate these interferences include the addition of potassium iodide [4947], 8-hydroxyquinoline [6420], thiosemicarbazide [5752], or EDTA [1612], performing the determination at a higher acid concentration [1926, 6212], and the addition of iron as a buffer [1926, 6398].

9.7 Boron (B)

Boron only occurs in nature in the form of compounds with oxygen. It is estimated that B makes up about 0.001 % of the Earth's crust. Amorphous boron is used as an additive in pyrotechnical mixtures and in solid fuels for rockets. It is also used in alloys to produce steels of especial hardness which are used as neutron absorbers in nuclear reactors. High-purity boron is used in thermistors, for example, and for doping silicon and germanium semiconductors. Boron fibers have been developed for aerospace applications to strengthen light metals and resins. Borax is used in glazes for earthenware and porcelain, in the production of enamels, for the manufacture of glass, and in cosmetics.

The stability of acidified boron solutions is low. Losses from solutions in hydrofluoric acid and hydrochloric acid in particular can be observed at ambient temperature; these losses can be ascribed to the formation of volatile halogen compounds whose boiling points are below ambient temperature: -101 °C for BF and $+12.5$ °C for BCl_3. Losses from solutions in nitric acid, sulfuric acid, or perchloric acid are lower but still significant. These losses can effectively be suppressed by adding an excess of mannitol [1873, 2805]. Contamination must be taken into account when borosilicate glass apparatus is used [2100, 4730, 6015].

Boron can be determined by F AAS practically only in the nitrous oxide-acetylene flame at 249.7/249.8 nm; with a characteristic concentration of about 15 mg/L it is one of the least sensitive elements in AAS. The low sensitivity derives from the fact that boron forms very stable oxides, nitrides, and carbides, and is thus only atomized to a very low degree, even in a fuel-rich nitrous oxide-acetylene flame. The sensitivity for the determination of boron can be markedly increased by converting it to the volatile boric acid methyl ester and conducting this into the nitrous oxide-acetylene flame [1102, 1738], or by vaporization as the fluoride produced by the reaction with copper hydroxyfluoride [1194]. Spectral interferences due to copper and iridium have been reported for the application of continuum source BC at both the primary analytical line at 249.7 nm and the secondary line at 208.9 nm [5973].

The determination of boron by GF AAS is only possible in tubes with a perfect PG coating. A rapid rate of heating for atomization increases the sensitivity and reduces tailing. Nevertheless, the characteristic mass of 1 ng [6353] attained in a longitudinally-heated atomizer with atomization from the wall at 2650–2700 °C is unsatisfactory. A pyrolysis temperature of 1000 °C can be applied in the presence of 0.5 mg/L Ca as the modifier and the characteristic mass improved to 0.25 ng [2917]. In a transversely-heated atomizer with atomization from the platform under STPF conditions the atomization temperature can be lowered to 2500 °C and the tailing is significantly reduced. Under these conditions the characteristic mass is 0.6 ng. The sensitivity can be improved by coating the graphite surface with carbides such as those of zirconium [3651], tantalum, niobium [3375], titanium or tungsten [3959]. Memory effects can be reduced by using special cleanout steps under the use of fluoride [387].

Isotope analysis by AAS is only possible for very light or very heavy atoms. Attempts to separate the isotope lines of ^{10}B and ^{11}B, for example by cooling hollow cathode lamps in liquid air [4213] or by using isotope lamps [2157], were unsuccessful. The isotope shift is about 1 pm, while the line width in the nitrous oxide-acetylene flame is about 6 pm, so that separation is impossible. HANNAFORD and LOWE [2359] finally managed to determine the isotope ratio in a water-cooled cathodic sputtering chamber as the absorption cell at the doublet at 208.89/208.96 nm, since this exhibits a higher isotope shift.

9.8 Cadmium (Cd)

Cadmium makes up about $5 \times 10^{-5}\%$ of the Earth's crust and is thus one of the rarer metals. In very low concentrations Cd is nowadays ubiquitous. The source of some occurrences, e.g. in sea-water, is unknown. The distribution of Cd on land is a consequence of emissions from industrial plants, especially zinc smelters and iron and steel works, and also from waste incineration plants and brown-coal-fired power stations. About a third of the Cd production is used in the manufacture of batteries. The use of Cd as an anticorrosive for iron and other metals is in second place. The use of Cd pigments and Cd soaps as stabilizers for PVC is also a major application.

Cadmium and numerous of its compounds are toxic. Cadmium first came into disrepute in Japan with the occurrence of Itai-Itai disease, which causes severe skeletal changes and often death. The body of a human adult contains about 30 mg Cd, but it has not yet been proven that it is an essential element. In the organism cadmium is transported as metallothionine. The accumulation of Cd in the liver and kidneys is critical. Significantly higher concentrations of Cd are found in smokers compared to non-smokers. Since a carcinogenic potential is suspected, Cd and a number of its compounds are included in various national and international lists stipulating the maximum concentration at the workplace.

The stability of Cd solutions has been the subject of numerous investigations [1184, 2070, 2106, 2687, 3109, 3932, 4512, 4692, 5056, 5138, 5297, 5455, 5606, 5660, 5666]. Sorption losses take place at pH values > 4. These losses can be effectively suppressed by acidifying the solutions to pH 2 with nitric acid; acidification should be performed at sampling since adsorption losses are in part irreversible [5455]. Polyethylene containers have proven to be suitable for storage; PP containers occasionally exhibit contamination [5606]. PFA is a suitable material for the long-term storage of dilute solutions [2106]. Investigations on the stability of Cd in biological samples have been performed on urine [5666] and on whole blood conserved with heparin [4061, 5665]; no losses were observed for urine in PE containers at ambient temperature over 28 days [5666].

Cadmium can be easily atomized and thus determined without any noteworthy interferences in the air-acetylene flame. The characteristic concentration is 0.02 mg/L at the 228.8 nm resonance line. With continuum source BC a spectral interference can take place at this line in the presence of high iron concentrations due to overcorrection [3483]. The analytical line at 326.1 nm is suitable for determining higher Cd concentrations; the characteristic concentration is about 6 mg/L, so that excessive dilution can be avoided [2968].

For a long period the greatest difficulty for determining Cd by GF AAS was its high volatility. By using a fast heating rate, LUNDGREN et al. [3669] utilized this volatility to atomize Cd at 800 °C before sodium chloride became noticeably volatile. EDIGER [1677] investigated several ammonium salts as modifiers to thermally stabilize Cd up to about 900 °C. SLAVIN et al. [5416, 5419] and HINDERBERGER et al. [2555] found that it was possible to determine Cd free of interferences under STPF conditions with ammonium phosphate as the modifier. For the determination of very low concentrations of Cd in the presence of high chloride concentrations, such as Cd in urine, the phosphate modifier causes a spectral interference [4425] that not even Zeeman-effect BC can eliminate

[6481]. Spectral interferences have also been reported for iron [424, 3483] and arsenic [2021, 4991, 5973]. The Pd-Mg modifier, possibly with the addition of ammonium nitrate, has proven to be especially good for determinations of this type [6481]. Under these conditions up to 3 g/L NaCl or sulfate did not interfere in the determination of Cd. The maximum pyrolysis temperature is about 800 °C. In a longitudinally-heated atomizer at an atomization temperature of 1500–1700°C the characteristic mass is about 0.4 pg, while in a transversely-heated atomizer it is 1.3 pg. If the attainable sensitivity is too high at the primary resonance line, as is often the case for direct solids analysis, the only usable alternate line is at 326.1 nm with a lower sensitivity by a factor of about 300 [1075, 4368]. A spectral interference due to phosphate must be taken into account at this line [1076].

CACHO et al. [1023] were the first to report the determination of Cd by HG AAS; using sodium tetrahydroborate in DMF they generated a 'volatile Cd species, presumably a hydride' and atomized in a QTA. EBDON et al. [1662] later investigated ethylation with sodium tetraethylborate in aqueous solution to volatilize Cd and likewise determined it in a QTA. SANZ-MEDEL et al. [5094, 5095] established that Cd can be determined by CV AAS. They reduced Cd^{2+} with sodium tetrahydroborate in vesicles of didodecyl-dimethylammonium-bromide (DDAB), whereby presumably the unstable cadmium hydride is formed, which decomposes to give atomic Cd vapor, which can be measured by AAS in an unheated quartz cell. These authors reported an LOD of 80 ng/L, which can be further improved by working at low temperature. The sensitivity can be increased by collecting the atomic Cd vapor in a graphite tube pretreated with palladium at 150 °C and then reatomizing at 1600 °C [2148].

9.9 Calcium (Ca)

Calcium constitutes about 3.63 % of the Earth's crust. After iron and aluminium it is the third most abundant element in our environment. Due to its reactivity it only occurs in the form of its compounds; in certain regions of the Earth limestone ($CaCO_3$) exists in huge deposits. Calcium is of major significance for the animal and plant worlds. Calcium compounds are for example the basis of bones and shells, and are required for cell division, muscle contraction, etc.

The stability of aqueous Ca solutions was investigated by SMITH [5456] for deionized water and NaCl solution (w = 0.5 %), and by MAJER and KHALIL [3812] for EDTA solutions (w = 0.03 %); the sorption losses for Ca were lower and occurred more slowly than for magnesium. The losses were higher in glass containers than in PE bottles and at a pH value of 1.5 were significant after 30 days (3–15 %); at a pH value of 11 they were up to 85 %. At pH values > 5 SMITH [5456] observed losses after 24 hours. It is therefore advisable to acidify solutions to a pH value of 1.5 and to store them in suitable plastic bottles. The selection of the container material depends on the Ca concentration of the solutions, since blank values of 4 µg/L have been measured even for high purity PFA [2106].

Calcium was one of the first elements to be determined by F AAS and is still one of the elements most frequently determined by this technique. The number of publications is correspondingly large. The early work published by WILLIS [6335, 6337] and DAVID [1427, 1428] between 1959 and 1961 was concerned with the determination of Ca in serum and urine, and in plants and soils, respectively.

Calcium can be determined in the nitrous oxide-acetylene flame [3787] practically free of interferences—except for an ionization interference [3787, 4244]—at the 422.7 nm resonance line with a characteristic concentration of 0.09 mg/L. In this flame interferences can only be observed for very high concentrations of silicon and aluminium, and the slight ionization can easily be eliminated by adding a sufficient quantity of alkali [3845]. Despite this fact, Ca was very often determined in the air-acetylene flame and then served as the prototype for elements subject to interferences by numerous concomitants (see Section 8.1.2.4, part 3). The interferences due to phosphate [1704, 2738, 3309, 5103, 5195, 5448, 5527, 5589, 6198, 6283], sulfate [5195, 5448], aluminium [4984], and silicon [1704, 5590] can be largely controlled by adding 10 g/L La as the chloride [3025, 6198] or 10 g/L EDTA. A spectral interference due to germanium has been reported [3691, 4991]. Although a larger spectral slitwidth (approx. 2 nm) is used for the determination in the air-acetylene flame, this cannot be recommended for

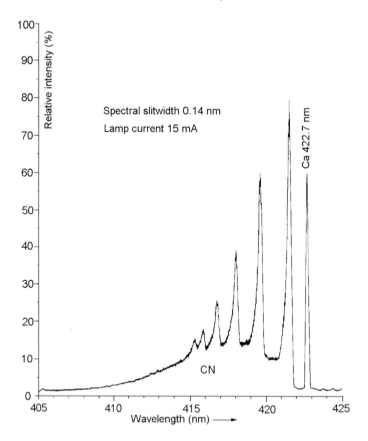

Figure 9-2. Emission of the nitrous oxide-acetylene flame in the neighborhood of the calcium resonance line at 422.7 nm. Lamp current 15 mA; spectral slitwidth 0.14 nm.

the nitrous oxide-acetylene flame. As shown in Figure 9-2, there is an intense CN emission band between 422 nm and 410 nm. If this is not separated by the monochromator, the S/N ratio deteriorates markedly.

Higher Ca concentrations can be determined at the analytical line at 239.9 nm with a characteristic concentration of about 13 mg/L. The S/N ratio is not so good at this line however, so that it cannot be recommended for precision determinations. BRAMALL and THOMPSON [770] investigated a number of non-resonance lines for their suitability in AAS and recommended the line at 430.3 nm which exhibits a characteristic concentration of 10 mg/L in the nitrous oxide-acetylene flame.

Since Ca is mostly present in higher concentrations, the sensitivity of GF AAS is rarely required for its determination. The majority of such applications involve the microanalysis of very small sample amounts [691, 1787, 2098, 4377]. Modifiers have not been described for Ca since a pyrolysis temperature of 1100 °C can be applied without modifier. For atomization at 2600 °C in a longitudinally-heated atomizer a characteristic mass of 0.8 pg can be attained. A similar characteristic mass of about 1 pg can be attained in a transversely-heated atomizer at a lower atomization temperature of 2500 °C; in this case the S/N ratio is significantly better.

9.10 Cesium (Cs)

Cesium is one of the rarer elements; its proportion of the Earth's crust is about $7 \times 10^{-4}\%$. Sea-water contains about 1 µg/L Cs. With a half-life of 30 years, ^{137}Cs is one of the most dangerous radionucleides. It is formed in nuclear fission processes and, along with other fission products, it was released into the atmosphere as a result of the Chernobyl reactor catastrophe. It is accumulated in meat, milk, and dairy products, and is completely resorbed from the stomach and intestines.

Little is known about the stability of aqueous Cs solutions. EICHHOLZ et al. [1684] found that the sorption losses for Cs were higher in glass than in PP. ROBERTSON [4890] investigated sorption losses from sea-water samples in glass and PE containers and recommended acidification to pH 1.5 immediately after sampling and storage in PE bottles.

The main resonance line for Cs is at 852.1 nm at the start of the infrared and is thus outside of the spectral working range of a number of AA spectrometers. An EDL is the only suitable radiation source [390]. Cesium can be determined without any noteworthy interferences in the air-acetylene flame with a characteristic concentration of 0.1 mg/L. Nevertheless Cs is substantially ionized in this flame so that a large quantity of another alkali salt must be added [2880]. Much lower ionization is observed in the air-hydrogen flame which is easier to work with for this element. For the determination of higher Cs concentrations several further analytical lines are available which are compiled in Table 9-5.

The determination of Cs by GF AAS was first described by BARNETT et al. [390] in connection with their investigations on EDLs. Interferences due to chloride must always be reckoned with [1255]. Non-linear calibration curves with increasing gradient have also occasionally been reported [189, 750]. Whether this effect can be ascribed to ionization is a cause of contention since it cannot be observed under STPF conditions [189].

Potassium nitrate [1198], nitric acid [557], and sulfuric acid [2228] have been recommended as modifiers for the determination of Cs. A pyrolysis temperature of 900 °C can be applied under STPF conditions; the characteristic mass in a longitudinally-heated atomizer at an atomization temperature of 1900 °C is about 5 pg, while in a transversely-heated atomizer it is 12 pg. Solely instruments with Zeeman-effect BC or using a halogen lamp for BC are suitable for eliminating interferences due to background attenuation.

Table 9-5. Cs resonance lines.

Wavelength (nm)	Energy level (K)	Slit width (nm)	Characteristic concentration* (mg/L)
852.11	0–11732	0.4	0.1
894.35	0–11178	0.4	0.2
455.54	0–21947	0.4	14
459.32	0–21766	0.4	54

* Air-acetylene flame, oxidizing (lean, blue)

9.11 Chromium (Cr)

Chromium constitutes an estimated 0.02 % of the Earth's crust and is therefore among the more common elements. In nature Cr is virtually only found in the form of its compounds. The most important Cr ore is chromite, $(Fe,Mn)Cr_2O_4$. Chromium is used mostly for the production of stainless steel and chromium alloys, for chromizing and for chromium plating. Chromium(III) is an essential trace element and plays a significant role in glucose metabolism. Compounds of toxic significance are Cr(VI) compounds, and especially Cr(VI) oxide and alkali metal chromates.

The stability of aqueous Cr solutions has been the subject of numerous papers; a number are cited in Table 9-6. Although acidification with nitric acid to a pH value < 1.5 can be recommended for the storage of samples for the determination of the total Cr content [5456, 5660], this form of sample conservation is not suitable for the determination of Cr species. Particularly in the presence of organic traces, such as are present in natural waters, Cr(VI) is reduced to Cr(III); the reaction rate increases with decreasing pH value [5591]. At pH values < 2 Cr(VI) is reduced even in deionized water, albeit slowly. Thus for natural waters immediate analysis of the non-acidified samples is recommended; the samples should be filtered and cooled [5591]. Calibration solutions of both redox species can be stabilized with a carbonate buffer for a longer period (at least 160 days) [1331, 1638, 4582, 4771].

Chromium can be determined in both a fuel-rich air-acetylene flame and a nitrous oxide-acetylene flame. The better sensitivity is attained in the former, but the determination is much more subject to interferences. Moreover, Cr(III) and Cr(VI) exhibit different sensitivities in a reducing air-acetylene flame; this aspect is discussed in Section 8.1.2.4, part 4. The addition of 20 g/L ammonium chloride [386, 2104, 2210, 4955],

Table 9-6. Stability of aqueous sample and calibration solutions with respect to their Cr concentration.

Species	Matrix	pH range	Materials*	Time period	Recommended storage	References
Cr	river water	1.5–8	BG, PE	30 days	PE; acidification with HNO_3 to pH < 1.5	[5660]
	sea-water (spiked)	natural	PE, PP, Pyrex	14 days	$t_{1/2} = 1.8 \pm 0.3$ days, little difference between materials	[2109]
	deionized water, NaCl solution (w = 0.5%)	1.5–12.0	BG	24 hours	Cr losses at pH values > 3.5; adjust pH value < 1.5 for safety	[5456]
	sea-water	natural	BG	6 months	no losses	[5206]
Cr(III/VI)	deionized water	6.95	PE, FG, Pyrex	15 days	after 15 days at pH 6.95 up to 25% Cr(III) had been adsorbed on PE, but only 1% Cr(VI)	[5295]
Cr(III/VI)	carbonate buffer solution	6.4	S	160 days	50 mmol/L carbonate buffer, CO_2 protective gas at 5 °C, or stored freeze-dried at −20 °C	[1331, 1640, 1638]
Cr(VI)	deionized water, waste water	2.1–12	BG, PE	21 days	BG; carbonate buffer at pH 9, EDTA	[4582]
Cr(VI)	0.01–0.02 mol/L NaCl solution, various acids	1.2	–	60 days	storage becomes more critical the lower are the Cr(VI) concentration and the pH value	[250]
Cr(VI)	natural water	< 1.5 and natural	PE	43 days	water samples should be filtered and cooled and then determined immediately without acidification	[5591]
Cr(VI)	carbonate buffer solution	9.6	S	160 days	50 mmol/L carbonate buffer, CO_2 protective gas at 5 °C, or stored freeze-dried at −20 °C	[1640]

* BG: borosilicate glass; FG: flint glass; S: silica; PC: polycarbonate; PE: polyethylene; PP: polypropylene

4 g/L potassium thiocyanate [3142], 20 g/L potassium persulfate [6088], or also 10 g/L ammonium hydrogen fluoride, either alone or together with 2 g/L sodium sulfate [2732, 4735], to all measurement solutions has been proposed to eliminate this effect. More recently organic additives such as amines [2976, 4964] have also been used successfully.

These difficult-to-control effects in the air-acetylene flame can most easily be eliminated by using the nitrous oxide-acetylene flame [77, 1308, 2733, 2746]. Virtually no interferences are known in this flame and both oxidation states exhibit the same sensitivity [1352, 3269]. Merely the addition of potassium is recommended for determinations in complex matrices.

As shown in Table 9-7, chromium exhibits a multitude of analytical lines of similar sensitivity; the line at 357.9 nm is used for most determinations. At this line the characteristic concentration in a fuel-rich air-acetylene flame is 0.04 mg/L. In the nitrous oxide-acetylene flame a characteristic concentration of about 0.3 mg/L can be attained at the 357.9 nm line. The line at 425.4 nm with a characteristic concentration of 0.1 mg/L is more suitable for the determination of higher concentrations of Cr.

For a long period the determination of Cr by GF AAS was considered to be difficult. One reason for this is that at 357.9 nm the deuterium lamp used for BC does not exhibit adequate radiant intensity. We must advise very strongly against the use of this background correction technique for the determination of Cr. The reports on spectral interferences caused by overcorrection using continuum source BC are therefore only of theoretical significance [1549, 4108, 4109, 5557, 5973, 6085]. This problem was eliminated with the introduction of Zeeman-effect BC. In addition there were reports about the high volatility of a number of Cr compounds [471], which could lead to considerable losses during pyrolysis; other authors were unable to confirm this, however [5221].

HINDERBERGER et al. [2555] were able to perform interference-free Cr determinations in body fluids and tissues by adding primary ammonium phosphate as modifier and working under STPF conditions. SLAVIN et al. [5419, 5421] proposed magnesium nitrate as the modifier, which allows the pyrolysis temperature to be increased to 1650 °C so that the majority of concomitants can largely be removed. Virtually no interferences are observed for the determination of Cr under STPF conditions. The optimum atomization temperature in a longitudinally-heated atomizer is 2500 °C and in a transversely-heated atomizer it is 2300 °C. The characteristic mass in a longitudinally-heated atomizer is 3 pg and in a transversely-heated atomizer is 7 pg.

Table 9-7. Cr resonance lines.

Wavelength (nm)	Energy level (K)	Slit width (nm)	Characteristic concentration* (mg/L)
357.87	0–27935	0.7	0.04
359.35	0–27820	0.7	0.05
360.53	0–27729	0.7	0.07
425.44	0–23499	0.7	0.11
427.48	0–23386	0.7	0.14
428.97	0–23305	0.7	0.20

* Air-acetylene flame, reducing (yellow)

9.12 Cobalt (Co)

Cobalt nearly always occurs together with nickel; the average ratio Co:Ni is around 1:4. Cobalt comprises $2 \times 10^{-3}\%$ of the Earth's crust and is thus the 32nd most abundant element. Cobalt is present as a trace element in most soils and occurs in many minerals. Manganese nodules contain up to 1% Co, making up an estimated Co resource of six billion tons. Cobalt is used in the manufacture of high heat-resisting alloys for machine parts and for cutting tools, for the production of magnetic alloys, to harden tungsten carbide for cutting tools, for the manufacture of pigments in the glass, enamel, and ceramic industries (cobalt blue), and in mixed-bed catalysts (e.g. Fischer-Tropsch process). Cobalt is an essential trace element; it is the central atom in vitamin B_{12}, which is largely responsible for producing red blood corpuscles. Ruminants require Co so that vitamin B_{12} can be synthesized by bacteria in the first stomach. Deficiency diseases can occur in regions with a lack of Co in the soil.

The stability of Co solutions has been investigated in deionized water [537, 5227, 5456], in NaCl solutions [5456], and in natural waters [4890, 5206, 5660]. Losses of Co occur at pH values > 5 [5456] and reach up to 95% on glass surfaces at pH 9 [5227]. The adsorption can be reduced in the presence of high ion concentrations, so that sea-water samples stored at natural pH values even for six months [5206] and river water samples stored in PE bottles at pH 8 for 30 days showed no losses, while in pyrex glass bottles the losses were > 10% after only five days [5660]. The simplest means of sample conservation is acidification to pH values < 1.5, preferably with nitric acid, and storage in PE bottles [4890, 5660]. Even solutions with Co concentrations of 5 μg/L conserved with nitric acid and stored in PFA bottles are stable for over 300 days [2106].

Cobalt can be determined without difficulty in the air-acetylene flame; optimum results are obtained in an oxidizing, fuel-lean flame. Although it exhibits a large number of analytical lines, the most important of which are listed in Table 9-8, only a limited number are suitable for analytical purposes. In multielement lamps that also contain nickel and at spectral slitwidths of > 0.2 nm, significant spectral interferences can occur at a number of the lines listed in Table 9-8. The line at 240.7 nm is the best-suited for lower Co concentrations; the characteristic concentration is 0.1 mg/L. At higher absorbance values there is marked curvature of the calibration curve due to slight self-absorption and a number of weak ion lines that cannot be separated from this resonance line. For higher Co concentrations the line at 352.7 nm is more suitable; it exhibits good linearity and a favorable S/N ratio, so that good precision is possible.

The following elements do not interfere in the determination of Co in the air-acetylene flame [4032]: 2 g/L Cr, Ni, and W; 1 g/L Cu and Mo; 500 mg/L Si; 200 mg/L Mn and V; 100 mg/L Ti; 50 mg/L P and S. In the presence of high matrix concentrations, matching of the main constituents can nevertheless be necessary. With a refractory matrix it can be advantageous to determine Co in the nitrous oxide-acetylene flame; the characteristic concentration is around 0.7 mg/L in this flame at the 240.7 nm line. The sensitivity of F AAS is inadequate for the majority of environmental samples and biological samples so that either the sample must be preconcentrated or the GF AAS technique must be used.

Table 9-8. Co analytical lines.

Wavelength (nm)	Energy level (K)	Slit width (nm)	Characteristic concentration* (mg/L)	Spectral interferences
240.73	0–41529	0.2	0.08	Ru$^+$ [5973]
242.49	0–41226	0.2	0.10	Au, Sn$^+$ [5973]
241.16	816–42269	0.2	0.14	
252.14	0–39649	0.2	0.15	In [3627]
243.58	0–41041	0.2	0.6	Fe, V$^+$ [5973]
304.40	0–32842	0.2	1.0	
352.69	0–28346	0.2	1.8	
345.35	3483–32431	0.2	2.0	
346.58	0–28845	0.2	2.6	
347.40	0–28777	0.2	4.2	

* Air-acetylene flame, oxidizing (lean, blue)
$^+$ Interferences with deuterium BC possible

SLAVIN et al. [5419, 5421] proposed magnesium nitrate as the modifier for the determination of Co by GF AAS; the maximum attainable pyrolysis temperature is 1400 °C. The application of the Pd-Mg modifier has also been recommended [921, 2868, 4604, 5073, 5074, 5153]. The determination of Co is practically interference-free under STPF conditions. If very low concentrations of Co are to be determined in the presence of a complex matrix, prior separation and preconcentration can be advantageous, e.g. by solid phase extraction [5513]. The optimum atomization temperature under STPF conditions is 2400–2500 °C. The characteristic mass in a longitudinally-heated atomizer is 6–7 pg and in a transversely-heated atomizer it is 17 pg.

9.13 Copper (Cu)

Copper makes up about 0.007 % of the Earth's crust, so that it is the 25th most abundant element. As a semi-noble metal Cu occasionally occurs native, but mostly it is in the form of copper minerals, e.g. as copper pyrites ($CuFeS_2$) or copper glance (Cu_2S). About 40 % of the annual production of Cu is used for the manufacture of alloys. Copper is one of the few elements that plays an important technical role in the pure, non-alloyed state, mainly in the electrical industry. Because of its excellent heat conductivity it is used for brewing vats, vacuum pans, soldering irons, stills, heating and cooling coils, etc. Because of their fungicidal properties, Cu salts have been used since ancient times for crop protection and as wood preservatives.

Copper is an essential trace element for humans and higher mammals, and also for numerous plants; it is a constituent of copper proteins with enzyme activity. The blood of marine mollusks and crabs contains the Cu-containing hemocyanin rather than the Fe-containing hemoglobin which is taken up from the sea-water and acts as a respiratory catalyst.

The stability of aqueous Cu solutions has been investigated in deionized water [2106, 2539, 5056, 5456], NaCl solutions (w = 0.5 %) [5456], and for natural waters [1184, 4594, 5660]. Strongly curved calibration curves are frequently obtained if the Cu calibration solutions are insufficiently acidified [5056]. Adsorption losses already occur from dilute solutions at pH values 1.5–3.5 [5456]. On the other hand, Cu traces are released from the surface of various materials such as glass [5472], PE [2539, 3116], PTFE [3116], and even PFA [2106]. For sample conservation, solutions must be acidified to a pH value < 1.5, preferably with nitric acid, and stored in carefully precleaned PE or better FEP or PFA bottles. MÉRANGER et al. [4061] investigated the stability of blood treated with heparin. In this case contamination from the containers used could be observed and made up to 30–50 % of the concentration present in the sample. The worst contamination was caused by PP, followed by glass and pyrex. Frequently the instruments and containers used for sample collection are a source of contamination [1300, 4291, 4570].

Copper is among those elements determined frequently by AAS. It can be atomized relatively easily [3138] and exhibits practically no interferences in the air-acetylene flame [3118, 4667], is virtually independent of the stoichiometry of the flame and the lamp current, and is therefore frequently used as a standard to test an instrument or procedure. It exhibits a number of resonance lines, all of which are analytically suitable. The most important analytical lines with their characteristic concentrations are compiled in Table 9-9. By selecting a suitable analytical line, even higher Cu concentrations can be determined without excessive dilution [4959]. For lower Cu concentrations the 324.8 nm line with a characteristic concentration of 0.03 mg/L is the most suitable. FUJIWARA et al. [2010] made detailed investigations on the distribution of Cu atoms in the air-acetylene flame in the presence of various acids.

Table 9-9. Cu resonance lines.

Wavelength (nm)	Energy level (K)	Slit width (nm)	β_0* (mg/L)	m_0§ (pg)	Spectral interferences
324.75	0–30784	0.7	0.03	4(8)	Pd [6245] Eu [1852, 4991, 5524] Ni, Pd+ [5973]
327.40	0–30535	0.7	0.07	12(15)	Ag, Rh+ [5973]
216.51	0–46173	0.2	0.2	(60)	Fe, Pb, Pt+ [5973]
222.57	0–44916	0.2	0.5	(120)	
249.22	0–40114	0.7	2.4	(420)	
202.44	0–49383	0.2	5.8		
244.16	0–40944	0.7	10	(2200)	

* Characteristic concentration, air-acetylene flame, oxidizing
§ Characteristic mass, longitudinally-heated atomizer, platform atomization at 2300 °C, without (with) Zeeman-effect BC [1075]
+ Interference with deuterium BC possible

Copper is one of the few elements that was determined successfully by GF AAS at a very early stage since it allows a pyrolysis temperature of 1100 °C without the use of a modifier. Even without the application of the STPF concept, a number of authors only reported slight interferences [1090, 1782, 4628, 5780]. Nevertheless, in the following years there were increased indications about interferences, especially by chlorides [747, 2823, 2879, 2966, 3111, 5369], so that the STPF concept was increasingly applied for the determination of Cu [569, 1065, 1494, 2002, 2053, 2734, 2892]. With the Pd-Mg modifier the pyrolysis temperature can be increased to 1300 °C. In a detailed investigation on interferences it was found that a slightly lower pyrolysis temperature of 1100–1200 °C was more suitable. Under these conditions and applying the STPF concept, up to 30 g/L NaCl or sulfate did not cause any interferences. The optimum atomization temperature for Cu in the presence of the Pd-Mg modifier is 2600 °C in a longitudinally-heated atomizer and 2000 °C in a transversely-heated atomizer.

Copper is one of the few elements that exhibits a loss in sensitivity of about 50% when Zeeman-effect BC is applied at the main analytical line at 324.8 nm. In a longitudinally-heated atomizer under these conditions the characteristic mass is 8 pg, while in non-Zeeman instruments it is about 4 pg. When a continuum source background corrector is used a spectral interference from the palladium modifier is observed at the 324.8 nm line [6245], so that this line cannot be used under these conditions. The interference is not observed at the secondary line at 327.4 nm, but the characteristic mass is only 12 pg, so that at the primary analytical line even with Zeeman-effect BC the sensitivity is better by a factor of two. The loss in sensitivity due to Zeeman-effect BC is significantly lower and thus the linearity better at the 327.4 nm line, so that this line is better suited for the determination of higher Cu concentrations. In a transversely-heated atomizer the characteristic mass is higher by a factor of about two. The relatively low interferences observed for the determination of Cu by GF AAS together with the multitude of analytical lines allow a fairly simple direct analysis of solids [2, 234, 306, 2093, 4276, 5751] and slurries [347, 1648, 4148, 4198, 4199, 6081].

STURGEON et al. [5645] reported that an unidentified, volatile Cu species is formed (which they determined by ICP OES) when aqueous Cu solutions are treated with sodium tetrahydroborate.

9.14 Gallium (Ga)

Gallium comprises about 0.0015% of the Earth's crust, so that it is about as abundant as lead. Since Ga has a similar ionic radius to aluminium it can easily penetrate the crystal lattice of aluminium compounds such as bauxite and the clays. The major part of the production of Ga is used in the manufacture of compounds of the type GaAs which are used as semiconductors. The intermetallic compound V_3Ga is a superconductor. Temperatures between −15 °C and +1200 °C can be measured with Ga thermometers.

The low melting point of Ga of only 30 °C originally caused considerable difficulties in the manufacture of HCLs. MULFORD [4218] described a hollow cathode in which the metal was present in liquid form.

Relatively little has been reported to date on the determination of Ga. GUPTA *et al.* [2290] investigated the influence of 43 different ions on the absorption of Ga in an air-acetylene flame and found slight influences from 12 of them. On the whole it appears that Ga can be determined largely free of interferences. ALLAN [141] reported a spectral interference caused by manganese at the 403.3 nm resonance line. Gallium can be determined in the air-acetylene flame at the 287.4 nm line with a characteristic concentration of 1.3 mg/L. With a nitrous oxide-acetylene flame the best sensitivity is observed at the 294.4 nm line with a characteristic concentration of 1.1 mg/L. Further analytical lines are compiled in Table 9-10.

Table 9-10. Ga analytical lines and characteristic concentrations.

Wavelength (nm)	Energy level (K)	Slit width (nm)	Characteristic concentration* (mg/L)	Spectral interferences
287.42	0–34782	0.2	1.3	Fe [6305], Pb$^+$ [5973]
294.36	826–34782	0.2	1.1	Fe, V$^+$ [5973]
417.21	826–24788	0.2	1.5	
403.30	0–24788	0.7	2.8	Mn [141, 4991]
250.01	826–40811	0.7	10	
245.01	0–40803	2.0	12	

* Nitrous oxide-acetylene flame, reducing (red)
$^+$ Interference with deuterium BC possible

There is also little information on the determination of Ga by GF AAS in the literature. Severe interferences due to halides were observed when a modifier was not used [1539]. A 45-fold increase in the sensitivity was achieved by the use of a mixed modifier [3955] containing EDTA, nickel nitrate and aluminium nitrate. A pyrolysis temperature of 1300 °C can be applied in the presence of the Pd-Mg modifier under STPF conditions, and up to 20 g/L NaCl had practically no influence on the integrated absorbance for Ga [6265]. Sulfates also did not interfere up to a concentration of 1 g/L; higher concentrations caused signal depression, presumably due to the formation of volatile GaO [6265]. The optimum atomization temperature for Ga under STPF conditions is 2200–2300 °C. The characteristic mass for instruments with Zeeman-effect BC and a longitudinally-heated atomizer is around 20 pg; it improves to 12 pg in non-Zeeman instruments. In a transversely-heated atomizer with Zeeman-effect BC the characteristic mass is 50 pg.

9.15 Germanium (Ge)

Germanium is a fairly rare element that only makes up about 5 x 10^{-4}% of the Earth's crust. In some minerals it is found in concentrations of up to 8%; it is present in trace quantities in some types of coal and is concentrated in the fly ash and dust during zinc smelting. The major use of Ge is in transistors, although it is being increasingly replaced by Si, GaP, or GaAs. Germanium is transparent to infrared radiation and is used in IR spectrometers and other optical instruments in the form of Ge windows and lenses.

The discovery of mono- and dimethylgermanium in natural waters in the absence of anthropogenic sources is taken as an indication for the biomethylation of Ge [5810]. The methylated species comprise about 70% of the known total concentration of Ge in sea-water and are evenly distributed throughout the oceans. The trimethylated species, which are produced in sediments, have not been observed in either fresh waters or sea-water. The biomethylation of Ge takes place preferentially in polluted areas [3502].

Germanium can be determined in the nitrous oxide-acetylene flame at the 265.1 nm line with a characteristic concentration of 2.2 mg/L [3846]. A number of further analytical lines with their characteristic concentrations are compiled in Table 9-11. Germanium can be determined with markedly higher sensitivity by HG AAS using $NaBH_4$ as the reductant [1887]. Albeit a flame must be used as the atomizer since GeH_4 is insufficiently atomized in a QTA. Much higher sensitivity can be obtained, and the use of a flame avoided, by collecting the hydride in a graphite tube prior to atomization [2435, 5771, 6540].

There is only little information in the literature on the determination of Ge by GF AAS. Without a modifier a pyrolysis temperature of only 800 °C can be applied, whereas with the Pd-Mg modifier it can be increased to 1400 °C. Up to 10 g/L NaCl and 2 g/L sulfate do not interfere in the determination of Ge under STPF conditions [6265]. Preatomization losses due to volatile oxides [1563] can be reduced by adding an oxidant or an alkali [6564]. The optimum atomization temperature in a longitudinally-heated atomizer is 2500 °C; the characteristic mass without Zeeman-effect BC is around 26 pg while with Zeeman-effect BC it is about 30 pg. In a transversely-heated atomizer the optimum atomization temperature is 2300 °C and the characteristic mass with Zeeman-effect BC is 25 pg.

Table 9-11. Ge analytical lines and characteristic concentrations.

Wavelength (nm)	Energy level (K)	Slit width (nm)	Characteristic concentration* (mg/L)	Spectral interferences
265.16	0–37702	0.2	2.2	Fe [6305]
270.96	557–37452	0.2	5.0	
259.25	557–39118	0.2	5.2	
275.46	1410–37702	0.7	6.1	Mn [141, 4991]
269.13	557–37702	0.7	8.6	
303.91	7125–40803	2.0	170	

* Nitrous oxide-acetylene flame, reducing (red)

9.16 Gold (Au)

Gold is one of the rarest elements in our environment; its amount in the solid crust of the Earth is about 4 mg/t. In sea-water Au can be present in concentrations up to 10 $\mu g/m^3$.

The majority of Au is hoarded in the form of gold bars and coins. Gold is also used to a large extent in electronics and electrical engineering.

The stability of gold solutions has been investigated by a number of authors [528, 530, 1190, 1745, 4718, 5455, 5852]. Gold losses of 5 % occurred after 24 hours in glass bottles for NaCl solutions (w = 0.5 %) acidified to pH 3 with hydrochloric acid [5455]. BENEŠ [528, 530] observed quantitative adsorption in PE containers at a pH value of 3.9. Acidification of solutions with hydrochloric acid to a pH value ≤ 1 is thus recommended. The Au concentration in multielement calibration solutions acidified with hydrochloric acid was not stable even for storage in PFA bottles [1745], whereas single-element calibration solutions in hydrochloric acid were stable. The Au concentration in serum samples was stable for at least three days when 0.1 % Triton X-100 and 0.1 mol/L hydrochloric acid were added [2547].

Gold is frequently determined by AAS in ore samples and plating baths. The most sensitive analytical line is at 242.8 nm with a characteristic concentration of 0.3 mg/L; spectral interferences caused by cobalt and tin can occur when continuum source BC is applied [5973]. The slightly less sensitive line at 267.6 nm with a characteristic concentration of 0.6 mg/L often enables better and more precise determinations due to its higher intensity. The sensitivity of F AAS is however inadequate for environmental and the majority of geological samples, so that either sample preconcentration or the use of a more sensitive technique is required.

Gold is normally determined in a sharp, fuel-lean air-acetylene flame in which no noteworthy interferences are observed. Slight differences can merely be observed when cyanidic Au solutions are compared with those in hydrochloric acid; for this reason test sample and calibration solutions should be matched. The determination of Au appears to be somewhat difficult in the presence of other noble metals [1667, 5595]. ADRIAENSENS and VERBEEK [41, 42] found that in a 20 g/L KCN solution the majority of interelemental interferences observed in acidic solution did not occur. SEN GUPTA [5232, 5233] eliminated the same interferences by adding a Cu-Cd buffer solution, while VAN LOON [3583] used lanthanum. A number of these interferences are reduced in the nitrous oxide-acetylene flame but the sensitivity of the determination is reduced by a factor of six [3829].

Without the use of a modifier considerable interferences occur in the determination of Au by GF AAS due to the alkali and alkaline-earth elements [5428]. With the Pd-Mg modifier a maximum pyrolysis temperature of 1000 °C can be applied. Both NaCl and K_2SO_4 caused a slight loss in sensitivity in integrated absorbance which in all probability could be attributed to a gas-phase interference [6265]. The signal depression reached a value of about 10 % in the presence of 1 g/L NaCl or K_2SO_4.

Under STPF conditions the optimum atomization temperature is 1800 °C; the characteristic mass in a longitudinally-heated atomizer is around 10 pg and in a transversely-heated atomizer about 18 pg. But even this sensitivity seldom meets the requirements of geological prospecting or environmental analysis so that frequently composite techniques with preconcentration are used, which in recent times are increasingly performed on-line by FI [1521].

9.17 Hafnium (Hf)

Hafnium comprises about 4 x $10^{-4}\%$ of the Earth's crust; this element is thus more abundant than for example silver, mercury, antimony, or uranium. Hafnium is present in zirconium minerals by about 1–2.5 %. Because of its large capture cross-section for thermal neutrons Hf is used in control rods for nuclear reactors and as a neutron absorber in the recycling of nuclear fuels. Hafnium is also used as a getter in high-frequency engineering and as a hardener in alloys.

Hafnium forms very stable oxides and can thus only be determined with moderate sensitivity even in a reducing nitrous oxide-acetylene flame at the 286.6 nm line. The effectiveness of atomization can be increased when the analyte is complexed with oxygen-free ligands such as hydrofluoric acid [5108]. The best sensitivity is attained when test sample and calibration solutions contain at least 0.1 % hydrofluoric acid. BOND [708] preferred adding 0.1 mol/L ammonium fluoride to enhance the signal. WALLACE et al. [6134] proposed adding aluminium to improve the linearity, precision, and LOD. A number of further analytical lines with their characteristic concentrations are listed in Table 9-12.

Since Hf forms not only very stable oxides but also stable carbides there are few references in the literature to the determination by GF AAS [509]; atomization must always be performed from metallic surfaces.

Table 9-12. Hf analytical lines.

Wavelength (nm)	Energy level (K)	Slit width (nm)	Characteristic concentration* (mg/L)	Spectral interferences
286.64	0–34877	0.2	15	
307.29	0–32533	0.2	25	Zn^+ [5973]
289.83	2357–36850	0.2	70	
296.49	2357–36075	0.2	50	
295.07	2357–36237	0.2	100	
294.08	0–33995	0.2	150	
290.48	2357–36773	0.2	150	
377.76	0–26464	0.2	150	

* Nitrous oxide-acetylene flame, reducing; addition of 0.1 % HF
+ Interference with deuterium BC possible

9.18 Indium (In)

With a content of around 0.1 g/t in the upper crust of the Earth, In is one of the rarer metals; its abundance is between that of bismuth and the majority of the noble metals. Similar to gallium it is present in traces in all zinc blendes and in Mansfeld copper ores. Indium is used as an additive in bearing metals, in dental fillings, in low-melting solders, etc. It also finds application in electrical contacts, fuses, diodes, semiconductors, transistors, in metallic fillers and binders, and color glass.

When stored in glass containers losses from aqueous In solutions due to wall adsorption take place at pH values > 3. At pH values > 4 the losses are already 67% after 24 hours [5455]. Without sample conservation by acidification the losses from sea-water stored in PE bottles are virtually quantitative [4890].

Indium can be determined at the 303.9 nm line in an oxidizing air-acetylene flame with a characteristic concentration of around 0.4 mg/L. Further analytical lines are summarized in Table 9-13. As with most of the low melting point elements, the S/N ratio, and thus the attainable LOD, depends on the design and quality of the HCL [4218]. EDLs provide markedly better results.

Table 9-13. In analytical lines.

Wavelength (nm)	Energy level (K)	Slit width (nm)	Characteristic concentration* (mg/L)	Spectral interferences
303.94	0–32892	0.7	0.42	Fe, Ni+ [5973]
325.61	221–32916	0.2	0.46	
410.48	0–24373	0.7	1.3	
451.13	2213–24373	1.4	1.5	
256.02	0–39048	2.0	5.1	
275.39	0–36302	2.0	11	

* Air-acetylene flame, oxidizing
+ Interference with deuterium BC possible

There is little information on the determination of In by F AAS in the literature, but there appear to be no serious interferences [5116]. MULFORD [4218] nevertheless found that various elements, especially in higher concentrations, depressed the absorption for In. FUJIWARA et al. [2010] made a detailed investigation on the influence of various acids on the atomization of In in the air-acetylene flame. The addition of lanthanum was recommended to suppress interferences, particularly from aluminium [576].

For the determination of In by GF AAS, DITTRICH et al. [1539] found severe gas-phase interferences in the presence of high chloride concentrations. These authors did not use a modifier so that a maximum pyrolysis temperature of only 700 °C could be applied. In the presence of the Pd-Mg modifier the pyrolysis temperature can be raised to about 1500 °C. Under STPF conditions up to 30 g/L NaCl and 5 g/L sulfate do not interfere in the determination of In [6265]. The optimum atomization temperature under STPF conditions in the presence of the Pd-Mg modifier is 2300 °C in a longitudinally-heated atomizer and 2100 °C in a transversely-heated atomizer. The characteristic mass in longitudinally-heated atomizers with and without Zeeman-effect BC is around 20 pg and in transversely-heated atomizers about 80 pg.

9.19 Iodine (I)

With a content of about $3 \times 10^{-5}\%$ in the solid crust of the Earth, iodine is one of the rarer elements. Sea-water contains about 0.05 g/t ; a number of marine organisms accumulate iodine and can reach concentrations of close to 20 g/kg dry weight. Iodine functions as an essential trace element for vertebrates. The human body contains 10–30 mg iodine, of which 99% is present in the thyroid glands. Iodine deficiency frequently causes hypothyreosis, often with goiter, and in children it can lead to cretinism. To avoid such problems iodized table salt is used in many countries. High concentrations of iodine, on the other hand, can have a toxic effect, so that an overdose must be avoided.

The resonance line for iodine is at 183.0 nm and hence so far in the vacuum UV that a determination with commercial instruments is hardly possible. KIRKBRIGHT et al. [3127, 3129] nevertheless performed the determination of iodine by flushing the monochromator and parts of the sample area with nitrogen. They employed a nitrous oxide-acetylene flame shielded with nitrogen and an EDL and were thereby able to attain a characteristic concentration of 12 mg/L and an LOD of about 6 mg/L. The determination was free of interferences and exhibited no differences between the different bonding forms of iodine.

L'VOV and KHARTSYZOV [3686] determined iodine in a graphite furnace at the 206.2 nm line; this is an excited line with a lower level of 0.94 eV above the ground state. The LOD was about 2 ng at an atomization temperature of 2400 °C. KIRKBRIGHT and WILSON [3130] determined iodine in a commercial double-beam instrument by GF AAS at 183.0 nm an obtained an LOD of 1 ng. Numerous ions caused signal enhancement by up to 100%; the authors presumed that this was caused by background attenuation which they could not correct at this wavelength. NORRIS and WEST [4385] reported a possible spectral interference due to bismuth.

NOMURA and KARASAWA [4378] observed that by heating solutions that contained mercury(II) nitrate and iodine in a graphite furnace, two peaks were recorded for mercury. The second peak could be assigned to the stable HgI_2. In this way these authors were able to perform an indirect determination of iodine; nevertheless cyanide, sulfide, and thiosulfate interfered in the determination.

9.20 Iridium (Ir)

Iridium is one of the very rare elements; it makes up about $10^{-7}\%$ of the solid crust of the Earth, while higher concentrations are present in the Earth's core. Since it is a noble metal it only occurs native, often alloyed with platinum and osmium. Because of its extreme brittleness Ir is only used in alloys, for example for the manufacture of fountain pen nibs, corrosion-resistant injection needles, contacts, and sparking plugs for aero engines.

GLADNEY and APT [2138] used containers made of glass, silica, PE, PVC, PP, PC, and PTFE over a period of four months and were unable to observe any adsorption losses for Ir, even in non-acidified deionized water. ENGLERT et al. [1745] used PFA containers and could not find any losses after 56 days.

For a long period Ir was considered to be one of those elements that could not be determined by AAS. The first successful determination was performed by MULFORD [4219] who, on Willis' suggestion, investigated this element more exactly. Later MANNING and FERNANDEZ [3848] made a thorough study and found that in a somewhat fuel-rich air-acetylene flame the 208.9 nm line exhibited the best sensitivity but that the 264.0 nm line provided the better S/N ratio and LOD. A number of further analytical lines are summarized in Table 9-14.

Table 9-14. Ir analytical lines.

Wavelength (nm)	Energy level (K)	Slit width (nm)	Characteristic concentration* (mg/L)	Spectral interferences
208.88	0–47858	0.07	1.5	B$^+$ [5973]
263.97	0–37872	0.2	4	
266.48	0–37515	0.2	5	
237.28	0–42132	0.2	6	
284.97	0–35081	0.2	6.5	Mg$^+$ [5973]
250.30	0–39940	0.2	7	
254.40	2835–42132	0.2	9	
351.36	0–28452	0.2	35.	

* Air-acetylene flame, reducing (yellow)
$^+$ Interference with deuterium BC possible

FUHRMAN [2004] found that the sensitivity of the Ir determination could be doubled by adding 100 mg/L K plus 1000 mg/L La; the addition of sodium or potassium alone, on the other hand, reduced the signal considerably. Fuhrman further reported that in the presence of potassium and lanthanum, no interferences due to platinum or titanium could be observed on the absorption of Ir. JANSSEN and UMLAND [2886] favored the addition of 10 g/L each of sodium and copper for the determination of Ir in the presence of other platinum metals. This is said to have increased the sensitivity by a factor of ten and to have eliminated the interferences due to other platinum metals. GRIMALDI and SCHNEPFE [2223] used the same additions for the determination of Ir in rocks. To eliminate the same interferences, TOFFOLI and PANNETIER [5862] preferred the addition of 5 g/L Li, under certain circumstances with the addition of some copper, and SEN GUPTA [5232] used 5 g/L Cu and 5 g/L Cd.

LÜCKE and ZIELKE [3644] found that in the nitrous oxide-acetylene flame the numerous interferences present during the determination of Ir disappeared completely. The reduction of the sensitivity by about 50% could be increased to the original value by chlorination. As the hexachloro complex, Ir exhibits the same sensitivity as in the air-acetylene flame. In later work [6582], however, these authors found that various Ir complexes could exhibit markedly differing sensitivities in this flame.

For the determination of Ir by GF AAS, MAZZUCOTELLI et al. [3996] observed substantial interferences from the ubiquitous elements present in environmentally relevant samples: calcium, iron, potassium, and sodium. Moreover the concentration of Ir to be expected in such samples is frequently too low to permit a direct determination; separation and preconcentration by extraction, ion exchange, or sorption is thus required [124, 214, 769, 2602, 3041, 3292, 3996, 5202, 6062]. A maximum pyrolysis temperature of 1300 °C can be applied without a modifier. Iridium can only be atomized from the tube wall in a longitudinally-heated atomizer; the atomization temperature is 2650 °C and the characteristic mass is 250 pg. In a transversely-heated atomizer under STPF conditions Ir can be atomized at 2400 °C and the characteristic mass of 230 pg is slightly better.

9.21 Iron (Fe)

Iron is probably the most abundant element on our planet and the isotope ^{56}Fe is the most common type of atom. While the top 16 km of the Earth's crust contain about 5% Fe, the Fe content of the entire planet is estimated at 37%. Based on the composition of meteorites, about half of which comprise around 90% Fe, we can assume that Fe is a major constituent of all the other celestial bodies. Iron is without doubt the most important industrial metal. Iron and steel are so universal in all fields of technology and manufacturing that we shall not make a listing. Physiologically Fe is an essential trace element for animal and vegetable organisms. An adult human with a body weight of 70 kg contains 4.2 g Fe, of which the majority is bound to the hemoglobin. Iron is also an important micronutrient for plants which influences photosynthesis and the formation of chlorophyll and carbohydrates.

The stability of Fe solutions in glass and PE bottles has been investigated for aqueous calibration solutions [99, 531, 532, 5456] and also for natural water samples [1184, 4594, 4890, 5286, 5660]. The losses from dilute solutions in glass bottles start at pH values > 1.5 and attain a value of 34% at a pH value of 3.5 after only 24 hours [5456]. Even for river water samples that were stored in PE bottles the adsorption losses at a pH value of 2.5 were over 10% after ten days. Water samples for the determination of the total Fe content should be stabilized by acidification to a pH value < 1.5, preferentially directly after sampling [4890]. The losses of Fe from acidified samples (0.06 mol/L HCl) were under 2% after two months in glass and PE containers [99]. This form of sample conservation cannot be used if Fe(II)/Fe(III) species are to be determined since the pH value of the sample has a very marked influence on the equilibrium distribution of Fe species [5959].

Iron is among those elements most frequently determined by AAS. The determination of Fe in an oxidizing, fuel-lean air-acetylene flame appears to be largely free of interferences. ALLAN [136] merely found an interference due to silicon, which could be eliminated by adding 2 g/L Ca [4667]. TERASHIMA [5796] found that in addition to silicon, strontium, aluminium, and manganese, as well as citric acid and tartaric acid also depressed the signal for Fe; the effect decreased with increasing distance from the burner slot. ROOS and PRICE [4953] also found a severe influence on the Fe signal by citric acid, which they eliminated by adding phosphoric acid or sodium chloride. OTTAWAY et al. [4501] and also CUNNINGHAM [1375] observed severe depression of the Fe signal by cobalt, copper, and nickel. With the exception of metallurgical samples, however, these elements are rarely present in concentrations that make this interference problematical. The extent of interference depended however on the flame conditions, such as the fuel-to-oxidant ratio, the observation height in the flame, and the anion used. The addition of 8-hydroxyquinoline [4501] or ammonium chloride [1375] was recommended to eliminate this interference. In a thorough-going study, THOMPSON and WAGSTAFF [5833] found that although the best sensitivity could be obtained in a fuel-rich air-acetylene flame, there were distinct differences between the two oxidation states of Fe and that interferences due to silicon and calcium occurred. They recommended a fuel-lean flame in which no interferences could be observed.

Numerous analytical lines of differing sensitivity are available for the determination of Fe; the most important are summarized in Table 9-15. The most frequently used analytical line is at 248.3 nm. Due to the large number of lines in the spectrum of Fe, a spectral slitwidth of 0.2 nm is required for this and for most of the other lines to avoid losses in sensitivity and excess curvature of the calibration curve. The setting of the wavelength must also be performed exactly to ensure that the correct line is used.

Iron is frequently present in higher concentrations in airborne dust. The determination of Fe is thus strongly subject to the risk of contamination. This is already a problem for F AAS and much more so for GF AAS. All containers must be thoroughly cleaned, for example by fuming out (refer to Section 5.1.2), and the use of an autosampler is strongly recommended (see Sections 4.2.6 and 6.2).

The use of magnesium nitrate as the modifier is recommended for the determination of Fe by GF AAS since it allows a pyrolysis temperature of 1400 °C to be applied. Inter-

Table 9-15. Fe analytical lines.

Wavelength (nm)	Energy level (K)	Slit width (nm)	Characteristic concentration* (mg/L)	Spectral interferences
248.33	0–40257	0.2	0.04	Cu, Pt+ [5973]
248.82	416–40594	0.2	0.07	Co§
252.29	0–39626	0.2	0.07	
271.90	0–36767	0.2	0.12	Pt [1852, 4991, 5573]
				Ga, Ta+ [5973]
302.06	0–33096	0.2	0.14	Cr [4097]
372.00	0–26875	0.2	0.27	
296.69	0–33695	0.2	0.32	
250.11	0–39970	0.2	0.41	
246.26	0–40594	0.2	0.48	
385.99	0–27167	0.2	0.49	
344.06	0–29056	0.2	0.65	
293.69	0–34040	0.2	0.8	
382.44	0–26140	0.2	5.0	

*Air-acetylene flame, oxidizing (lean, blue)
+ Interference with deuterium BC possible
§ With multielement lamps that contain Co

ferences are hardly to be expected when these conditions are applied under the application of the STPF concept. In particular, the interferences due to compounds of chlorine such as perchloric acid described in earlier papers [2966, 3206] are no longer observed. The optimum atomization temperature under STPF conditions in a longitudinally-heated atomizer is 2400 °C and the characteristic mass at the 248.3 nm line is about 5 pg. In a transversely-heated atomizer the optimum atomization temperature is 2100 °C and the characteristic mass is 12 pg.

9.22 Lanthanum (La) and the Rare-earth Elements (REE)

The elements in Group III of the periodic system with atomic numbers from 57–71 (i.e. La–Lu) are referred to collectively as the lanthanoids. It is also common to speak of lanthanum and the lanthanides, where the lanthanides are elements 58–71 (i.e. Ce–Lu). Together with scandium and yttrium, the lanthanoids comprise the rare-earth elements. In this chapter we have treated scandium and yttrium separately from the other rare-earth elements.

Lanthanum comprises about 0.0018 % of the Earth's crust and is thus the 35th most abundant element. The REEs make up an estimated 0.01–0.02 % of the solid crust of the Earth and are thus not among the rarest of elements. In keeping with Harkin's rule the lanthanides with even atomic numbers are more abundant than those with odd atomic numbers. The REEs have a substantially larger ionic radius than the elements that normally form rocks and thus became concentrated in the residual melt during solidification of the magma. The most important ores are monazite, bastnesite, xenotime, and gadolinite.

A number of the REEs are used to improve the properties of alloyed steels, in magnetic materials, and in high-temperature superconductors. The oxides of La and the REEs are also used for dyeing and bleaching, and to improve the properties of glass. A large part of the production of REEs is used in catalyzers.

The stability of aqueous sample and calibration solutions has been investigated for La [537, 1684], Ce [1684], and Eu [2742]. Adsorption losses were more pronounced on PP surfaces than on glass [1684] and had to be suppressed by acidification.

In their behavior in AAS, the REEs are very similar, even if they can only be determined with markedly varying sensitivities. A reducing nitrous oxide-acetylene flame is required for the determination of all of the REEs; these elements are more or less strongly ionized in this flame so that an addition of 1–3 g/L K is always required to suppress this effect. The most important analytical lines of La and the REEs are compiled in Table 9-16 along with the attainable characteristic concentrations.

Since the individual REEs do not influence each other mutually in F AAS and there are no noteworthy interferences in the hot flame, we have an ideal means for a specific determination of these otherwise very difficult-to-separate elements. Atomic emission techniques are generally more sensitive for the REEs, but because of the colossal number of spectral lines spectral interferences are frequent. AMOS and WILLIS [166] have published a comprehensive study on the determination of the lanthanoids in the nitrous oxide-acetylene flame. Detailed information, especially on the various analytical lines, is

Table 9-16. Analytical lines of the REEs.

Element	Wavelength (nm)	Energy level (K)	Slit width (nm)	Characteristic concentration* (mg/L)
La	550.06	3010–22804	0.4	48
	418.73	0–23875	0.4	63
	494.98	0–20197	0.4	72
	403.72	0–24763	0.4	170
Dy	421.17	0–23736	0.2	0.7
	404.60	0–24708	0.2	1
	418.68	0–23877	0.2	1
	419.49	0–23832	0.2	1.3
	416.80	0–23985	0.2	7
Er	400.80	–	0.2	0.7
	415.11	–	0.2	1.2
	389.27	–	0.2	2.3
	408.77	–	0.2	3.4
	393.70	–	0.2	3.6
	460.66	–	0.4	12
	390.55	–	0.2	13
	402.05	–	0.2	22
Eu	459.40	0–21761	0.2	0.6
	462.72	0–21605	0.2	0.8
	466.19	0–21445	0.2	0.9
	321.06	0–31138	0.2	7
	321.28	0–31116	0.2	9
	311.14	0–32130	0.2	9
	333.43	0–29982	0.2	12
Gd	368.41	0–27136	0.2	19
	407.87	533–25044	0.2	19
	405.82	215–24850	0.2	22
	405.36	999–25661	0.2	25
	371.36	215–27136	0.2	28
	404.50	0–24715	0.2	47
	367.41	215–27425	0.2	47
Ho	410.38	–	0.2	0.9
	405.39	–	0.2	1.1
	416.30	–	0.2	1.4
	417.32	–	0.2	5
	404.08	–	0.2	7
	410.86	–	0.2	14
	412.72	–	0.2	15
	422.70	–	0.2	29
	425.44	–	0.2	48
	395.57	–	0.2	57
Lu	335.96	1994–31751	0.2	6
	331.21	0–30184	0.2	11
	337.65	0–29608	0.2	12

Table 9-16 continued

Element	Wavelength (nm)	Energy level (K)	Slit width (nm)	Characteristic concentration* (mg/L)
Lu	356.78	0–28020	0.2	13
	298.93	0–33443	0.2	55
	451.86	0–22125	0.2	66
Nd	492.45	0–20301	0.4	7
	463.42	0–21572	0.4	11
	471.90	–	0.4	19
	489.69	1128–21543	0.4	35
Pr	495.14	–	0.4	39
	513.34	–	0.4	61
	492.46	–	0.4	79
	505.34	–	0.4	100
	504.55	–	0.4	110
	502.70	–	0.4	110
	503.34	–	1.4	150
	491.40	–	1.4	190
Sm	429.67	4021–27288	0.2	7
	476.03	812–21813	0.4	12
	511.72	–	0.4	14
	472.84	1490–22632	0.4	16
	520.06	1490–20713	0.4	17
	528.29	1490–20413	0.4	28
	478.31	293–21194	0.4	29
	458.16	812–22632	0.4	34
Tb	432.65	–	0.2	8
	431.89	–	0.2	10
	390.14	–	0.2	13
	406.16	–	0.2	14
	433.85	–	0.2	16
	410.54	–	0.2	29
Tm	371.79	0–26889	0.2	0.5
	410.58	0–24349	0.2	0.7
	374.41	0–26701	0.2	0.7
	409.42	0–24418	0.2	0.8
	420.37	0–23782	0.2	1.5
	375.18	0–26646	0.2	2.8
	436.00	0–22930	0.2	4.4
	530.71	0–18837	0.4	10
Yb	398.80	0–25068	0.2	0.1
	346.44	0–28857	0.2	0.5
	246.45	0–40564	0.2	1
	267.20	0–37415	0.2	5

* Nitrous oxide-acetylene flame, reducing

given by MANNING [3843, 3844], FERNANDEZ [1880], and MOSSOTTI [4203]. PICKETT *et al.* [4641] found an interference due to high-boiling oxygen-containing mineral acids and organic compounds which had also been observed earlier [5724]. THOMERSON and PRICE [5826] discussed interelemental interferences.

For a long period it was not possible to determine *cerium* by F AAS [2899] despite various attempts [166, 4203]. There were indications that cerium exhibited minimal absorption at 522.4 nm and 569.7 nm with an LOD of about 100 mg/L. JOHNSON *et al.* [2932] proposed an indirect determination via the heteropoly-molybdo-cerium-phosphoric acid which contains six atoms of molybdenum for each atom of cerium. After extraction in isobutyl acetate the molybdenum is determined and as a result of the enrichment effect a characteristic concentration of 0.1 mg/L Ce is obtained. Recently this type of indirect determination has been performed by on-line coupling with flow injection [3918].

Many of the REEs can be determined by GF AAS with good sensitivity. GROBENSKI [2226] investigated the influence of a PG coating on the graphite tube and found a direct parallel between the dissociation energy of the gaseous monoxide and the sensitivity for GF AAS. L'VOV and PELIEVA [3703] and SEN GUPTA [5235] determined the characteristic concentrations for numerous lanthanoids; Sen Gupta used a PG-coated graphite tube, while L'vov and Pelieva used a graphite tube lined with tantalum. L'VOV *et al.* [3707] also determined cerium by GF AAS; they were able to attribute the difficulties associated with the determination of this element to the formation of a gaseous mono-cyanide. SEN GUPTA [5237] determined cerium in 19 international geological reference materials by GF AAS. Cerium is atomized to 5 % in an argon atmosphere at 2700 °C but only to 0.4 % in a nitrogen atmosphere. Cerium is completely atomized from a tungsten probe in a furnace lined with tantalum. Under these conditions the LOD at the 569.7 nm line is 4 ng. Using surface analysis techniques, GAO and DENG [2060] determined that poor sensitivity and carryover for the determination of lanthanum and samarium was caused by the formation of refractory carbides. JIE and GUO [2919] were able to determine ytterbium with a characteristic mass of 1.3 pg when they used europium as the modifier.

A number of the difficulties and the observed interferences for the determination of the REEs by GF AAS can definitely be ascribed to the use of longitudinally-heated furnaces with atomization from the tube wall. The lanthanoids can be determined in a transversely-heated atomizer with atomization from the platform under STPF conditions. This leads to an improvement of the results, but more detailed investigations in this direction still have to be performed. Table 9-17 presents the optimum atomization temperatures and characteristic masses for the lanthanoids in longitudinally-heated and transversely-heated atomizers, respectively. From this table we can see the superiority of the latter atomizer for this group of elements. While for the majority of other elements the characteristic mass obtained in a transversely-heated atomizer is higher on average by a factor of three due to the tube dimensions, for the lanthanoids the same or even better (lower) values by a factor of three are found. This can only be explained by a far higher atomization efficiency, which is usually accompanied by a reduction of the interferences.

Table 9-17. Optimum atomization temperatures T_{at} and attainable characteristic masses m_0 for some REEs in a longitudinally-heated atomizer (HGA) with wall atomization and a transversely-heated atomizer (THGA) under STPF conditions (all values by kind permission of Perkin-Elmer).

Element	T_{at}, °C		m_0, pg	
	HGA	THGA	HGA	THGA
La	2650	2500	26000	7500
Dy	2650	2500	40	40
Er	2650	2450	70	70
Eu	2600	2350	20	18
Gd	2650	2450	11000	3500
Nd	2650	2350	1800	900
Pr	—	2500	—	4400
Sm	2600	2450	240	230
Tm	2650	2450	13	13
Yb	2500	2300	2.5	3

9.23 Lead (Pb)

Lead makes up only 0.0018 % of the Earth's crust and is therefore rarer than for example cerium, tungsten, vanadium, or yttrium. Nevertheless it was known to the ancient Egyptians 5000 years ago because it was concentrated in a few large deposits in the form of easily reduced galena. About half of the Pb production is used for accumulators, a quarter for chemicals such a tetraethyl-lead and other organolead compounds, and the rest for mould castings, cables, alloys, and for pigments.

Metallic Pb and its compounds are toxic; they can enter the body by ingestion, inhalation, or resorption through the skin. Acute lead poisoning is nevertheless seldom due to the low resorption. The continuous assimilation of small quantities of lead is far more dangerous. The lead is initially bound loosely to the erythrocytes and only a small part is excreted in the urine, the majority being accumulated in the bones.

Of all organolead compounds the trialkylated forms have the strongest neurotoxic effect for mammals. The tetraalkylated forms are rapidly metabolized to trialkylated after assimilation. The main source of organolead compounds in the environment is their use as antiknocking agents for gasoline. As a result of legislation in many countries, however, this source of pollution is decreasing. As soon as they enter the atmosphere, tetraalkyl-lead compounds are subject to photocatalytic decomposition, which under given conditions can be very rapid. The decomposition products are tri-, di-, and mono-alkylated Pb compounds. Lead is a ubiquitous pollution element that is present in all areas of the biosphere.

The stability of Pb solutions has been investigated under various conditions and in a number of container materials [2811, 3103, 4183, 4511, 5055, 5056, 5297, 5456, 5606, 5660]. In non-acidified, dilute solutions, Pb losses of 15–40 % were observed within the first four days in both glass and PP [4511]. Losses due to sorption on glass, PE, or PTFE surfaces are reduced by the presence of other cationic main constituents (e.g. Al^{3+}

[5138]) or by complexation [2687], but are best reduced to a minimum by 0.1 mol/L nitric acid [2811, 4511, 5297, 5456, 5606]. For the storage of water samples under these conditions both PE [5660] and brown glass [4511] are suitable; losses from solutions with Pb concentrations > 0.1 mg/L were lower for brown glass, while for Pb concentrations < 0.1 mg/L storage in PE bottles was advantageous [4511]. The stability of the Pb content in blood samples was investigated by WANG and PETER [6158], SUBRAMANIAN et al. [5665], MÉRANGER et al. [4061], and MOORE and MEREDITH [4183]. No losses could be observed in PP containers or Vacutainers [6158] after the addition of EDTA or heparin. Likewise, no losses were observed after 60 days for blood samples treated with heparin in PC, PE, PP, PS, borosilicate, or pyrex containers if the samples were stored at –10 °C [5665]. At temperatures around freezing the losses in PS and pyrex were significant after only a day. At ambient temperature only PE and borosilicate glass are suitable for storage for a maximum of six days [5665].

The acidification of samples has an influence on the distribution of Pb species and must therefore be avoided when a distinction is to be made between dissolved and other portions (particulate, colloidal). The stability of solutions for the determination of organolead species was investigated by QUEVAUVILLER et al. [4757], CLEUVENBERGEN et al. [1304], NYGREN [4399], and BAXTER and FRECH [442]. To prevent photochemical reactions that can lead to the decomposition of alkyl-lead compounds [1304] it is recommended to store the samples out of light and under cooling, or better to extract these species at or directly following sampling. While tetraalkyl-lead compounds (TTAL) tend to adsorption losses on glass surfaces [608, 2402, 4777] and decompose at active sites [2896], such losses are insignificant for ionic alkyl-lead compounds [4777]. The decomposition of all organolead compounds is accelerated by light; TTAL compounds exhibit the lowest sensitivity and decompose within a few days. Methylated compounds are more stable than ethylated [442], and dialkylated are more stable than trialkylated [3569]. It would appear that acidification does not disturb the distribution of organolead species [3569]. The extraction of organolead compounds from aqueous samples should be performed as soon after sampling as possible and ideally should be done in the sampling bottle so that the portion adsorbed to the walls is also extracted.

Lead is the element determined most frequently by AAS. The most important analytical lines of Pb are summarized in Table 9-18. Although the 217.0 nm line is noticeably more sensitive than the 283.3 nm line, it does not give a better LOD owing to a poorer S/N ratio (see Figure 5-11 in Section 5.2.3.4). In addition, more background absorption effects occur at the 217.0 nm line so that the 283.3 nm line is to be recommended.

Lead can be determined in the air-acetylene flame largely free of interferences. Interferences caused by aluminium or iron can be eliminated by adding ascorbic acid, citric acid, and EDTA [1712]. Since Pb must often be determined in very low concentrations, the LOD of about 0.01 mg/L attainable by F AAS is often inadequate. Even at an early stage techniques such as the sample boat [2995] or the Delves system [1474] were applied to increase the sensitivity. The real breakthrough came with the introduction of GF AAS, which is nowadays used principally for the determination of Pb.

Early papers on the determination of Pb reported numerous spectral and non-spectral interferences that partly depended on the high volatility of Pb and on the influence of

Table 9-18. Pb analytical lines.

Wavelength (nm)	Energy level (K)	Characteristic concentration* (mg/L)	Characteristic mass§ (pg)	Spectral interferences
217.00	0–46069	0.08		AlO [6279] Fe [3483, 5973] Cu, Ni, Pt, Sb [5973] Phosphate [2464]
283.31	0–35287	0.2	10	BaO [6279] S [3367] Os, Pt, Sn [5973]
205.33	0–48687	3.5		
202.20	0–49440	4.5	230	
261.42	7819–46061	5	1000+	
368.35	7819–34960	17	2400+	
363.96	7819–35287	40	6500+	

* Air-acetylene flame, oxidizing (lean, blue); slit width 0.7 nm
§ GF AAS with Zeeman-effect BC, no modifier [1075]
+ Matrix influences for direct solids analysis caused by strong dependence on the atomization temperature [438, 5340]

concomitants. Most of these interferences were no longer observed with proper use of the STPF concept [5416], so that Pb can be determined free of interferences in biological samples including urine [4473].

The Pd-Mg modifier brings the best results for the determination of Pb. It allows a pyrolysis temperature of 1200–1400 °C, which enables the separation of most interfering concomitants. The stabilizing effect of this modifier also causes a relatively high atomization temperature of 2000 °C at which the characteristic mass in a longitudinally-heated atomizer is about 16 pg (see Section 8.2.4.1). In a transversely-heated atomizer the characteristic mass is about 30 pg. The ammonium phosphate modifier that is also frequently used permits a lower atomization temperature of about 1600 °C and gives a better characteristic mass of 12 pg; nevertheless, the lower maximum pyrolysis temperature of only 600–700 °C hardly allows interfering concomitants to be separated. A Pd-Sr modifier has also been applied in the presence of high sulfate concentrations.

Lead can also be determined by HG AAS [1047], but the sensitivity is not particularly good. THOMPSON and THOMERSON [5830] obtained an LOD of 0.1 mg/L in a QTA. FLEMING and IDE [1926] were able to increase the sensitivity significantly by adding tartaric acid and potassium dichromate, but they reported severe interferences from copper and nickel. BRINDLE et al. [1232] investigated the influence of several chelating agents which also have a positive influence on the sensitivity. By use of the nitroso-R salt FANG et al. [3551] were able to improve the LOD for FI-HG AAS to 1 µg/L.

9.24 Lithium (Li)

Lithium is present in the upper crust of the Earth on average to 0.006%, making it the 27th most abundant element. It is found preferentially in acidic volcanic rocks and is often concentrated in granitic pegmatites and in tin formations. Lithium is used in nuclear technology for the manufacture of tritium, as a screening agent, for the detection of thermal neutrons, and as a reactor coolant. As an additive to alloys Li can improve the mechanical properties of aluminium and lead and the resistance to corrosion of magnesium. Lithium soaps, along with those of sodium and calcium, are used as a basis for the production of greases. Lithium is also required in organic syntheses that proceed via organolithium compounds. For humans Li does not appear to play any role as a trace element. A number of Li salts are administered for the therapy of depressive conditions. Lithium is toxic in large doses.

The stability of aqueous Li solutions was investigated by SMITH [5455], and GIEBENHAIN and RATH [2106], among others. Lithium was the sole element that did not exhibit any losses over the pH range 1–11.

Lithium can be determined easily and practically free of interferences in the air-acetylene flame. The most important analytical lines for Li with their characteristic concentrations are compiled in Table 9-19.

Table 9-19. Li analytical lines.

Wavelength (nm)	Energy level (K)	Slit width (nm)	Characteristic concentration* (mg/L)
670.78	0–14904	1.4	0.03
323.26	0–30925	0.7	7.5
610.36	14904–31283	1.4	130

* Air-acetylene flame, oxidizing (lean, blue)

Although Li is in Group I of the periodic system, in its chemical behavior it belongs more to the alkaline-earth metals than to the alkali metals [3199]. In contrast to the alkali metals, in the air-acetylene flame Li is hardly ionized [4563] and can only be determined in cooler flames with markedly reduced sensitivity. Included in the few interferences reported in the literature is a slight signal enhancement due to strontium [1916, 5413]. For the determination of Li in rocks and minerals using the air-acetylene flame, LUECKE [3645] observed slight signal depressions of up to 10% in the presence of higher concentrations of aluminium, phosphate, fluoride, and perchlorate; these all disappeared in the nitrous oxide-acetylene flame however. For this reason Luecke recommended the use of the hotter flame for the determination of Li in geological samples.

Not a great deal has been reported on the determination of Li by GF AAS [686, 688, 737, 1877, 3159, 4094, 4212, 4875, 5075, 5261, 5294, 5888, 6540]. The maximum pyrolysis temperature without a modifier is 900 °C. Ammonium nitrate has been used as a modifier to eliminate the interference due to chloride [4094, 5888, 6563] and various phosphates [516, 1877, 5294, 6540], also in combination with ammonium nitrate [5888,

6563]. A number of authors used tubes coated with tantalum carbide [5075, 5294] or lined with tantalum foil [686, 688, 5261, 6459] to suppress the formation of thermally stable lithium carbide and thus to enhance the sensitivity. The optimum atomization temperature in a longitudinally-heated atomizer under STPF conditions is 2600 °C; a noticeably lower atomization temperature of 2200 °C can be used in a transversely-heated atomizer. The characteristic mass for Li in a longitudinally-heated atomizer is 1.4 pg and in a transversely-heated atomizer it is 5.5 pg.

It is possible to determine the isotope ratio ^6Li to ^7Li by AAS. This was initially proposed by WALSH [6135] and later carried out by ZAIDEL and KORENNOI [6512] and a number of other working groups [1193, 1196, 2155, 3838, 4773]. The basis of the determination of isotopes by AAS is the isotope shift in the absorption spectrum, which in the case of Li amounts to 15 pm, i.e. is much greater than the natural line width of an absorption line. A disadvantage however is that the resonance line at 670.8 nm is a doublet with a separation of 15 pm, so that the weaker doublet line of the commoner isotope ^7Li coincides with the more intense doublet line of the rarer isotope ^6Li. As depicted in Figure 9-3, this situation does not really allow the independent determination of both isotopes. Nevertheless, as work by WHEAT [6295] and also CHAPMAN and DALE [1193, 1196] showed, by using a dual-channel AA spectrometer and Li isotope HCLs it is easily possible to determine the isotope ratio with a precision of better than 2%. This makes AAS to an economically very attractive alternative to MS.

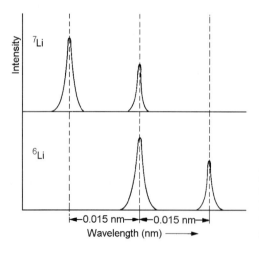

Figure 9-3. Doublet separation of both Li isotopes ^6Li and ^7Li. The weaker doublet line of the commoner isotope ^7Li coincides with the more intense doublet line of the rarer isotope ^6Li.

9.25 Magnesium (Mg)

Approximately 1.95% of the Earth's upper crust is made of Mg, so that it is the eighth most abundant element. Magnesium is most widely distributed in the silicates. The most important sources for the production of Mg are magnesite ($MgCO_3$), dolomite ($CaCO_3 \cdot MgCO_3$), natural and artificial salt sols, and sea-water. On average, sea-water contains 3.8 g/kg $MgCl_2$, 1.66 g/kg $MgSO_4$, and 0.076 g/kg $MgBr_2$.

The greatest quantity of Mg is used for the production of alloys, especially with aluminium. The application of Mg for the removal of sulfur and as a deoxidant in the iron and steel industry has also gained considerable significance. Magnesium is also used in pyrotechnics for flares and magnesium torches. In organic chemistry Mg is used to dry and prepare absolute alcohol, and as a reactant in numerous syntheses.

The human organism contains around 20–25 g Mg, about half of which is present in the bones and teeth. The remainder is largely dissolved in cells, and in part also bound to proteins and then functions as an activator for the disassimilation of sugars and as a calcium antagonist. Magnesium deficiency is manifested as cramp and appears to promote arteriosclerosis and heart infarction. On the other hand an increased Mg level causes a reduction of the erethism of nerves and muscles, even leading to complete paralysis. In plants Mg is a particularly important constituent of chlorophyll.

The stability of aqueous Mg solutions has been investigated for a number of storage conditions [2106, 3812, 5456]. Adsorption losses are higher on glass surfaces than on plastic since Mg ions can be irreversibly bound to the silicate lattice via exchange processes [3812]. For solutions containing 0.5 mg/L Mg the adsorption losses can be substantial, for example 43% after two weeks in glass, and are significant after 24 hours at pH values > 6.

Magnesium is one of the elements determined most frequently by F AAS, which in part can be ascribed to the very high sensitivity that can be attained for this element. The characteristic concentration in the air-acetylene flame at the 285.2 nm resonance line is 0.003 mg/L. Higher Mg concentrations can better be determined at the 202.5 nm line, which exhibits a characteristic concentration of around 0.1 mg/L. In the air-acetylene flame the determination of Mg appears to be largely free of interferences.

The numerous investigations on matrix influences, which are discussed in detail in Section 8.1.2.4, part 3, showed that for the determination of Mg only higher concentrations of silicon and aluminium interfere. RUBEŠKA and MOLDAN [4984] showed that the interference due to aluminium is based on the formation of Mg-spinel ($MgO \cdot Al_2O_3$). HARRISON and WADLIN [2407] found that this interference depends very strongly on the observation height above the burner slot. WELZ and LÜCKE [6271] established that this interference was substantially lower in the presence of chloride and could be eliminated by adding lanthanum chloride.

AMOS and WILLIS [166] found that the sensitivity was reduced by up to 50% in the nitrous oxide-acetylene flame. Furthermore Mg is ionized by about 6% in this flame so that the addition of 1 g/L K is recommended. Although this flame is not normally required for the determination of Mg, it may be useful for eliminating the interferences mentioned above [745]. This possibility was confirmed by NESBITT [4326] for the interference due to aluminium on the determination of Mg in silicate rocks. FLEMING [1925] found a dependence on the formation of spinel on the stoichiometry of the nitrous oxide-acetylene flame, which again was confirmed by HARRISON and WADLIN [2407].

In GF AAS Mg is among the most sensitive of elements. Since it is present in the majority of samples in concentrations that can be determined by F AAS, the sensitivity of GF AAS is rarely needed. The biggest problem for the application of this technique is contamination. For this reason, and because Mg does not form any particularly volatile compounds, the use of a modifier is usually avoided. Under these conditions a maximum pyrolysis temperature of 900 °C can be applied. The optimum atomization temperature under STPF conditions is 1700–1900 °C. A similar characteristic mass of 0.3–0.4 pg is obtained in both longitudinally-heated and transversely-heated atomizers. The high sensitivity is mostly a disadvantage if Mg is to be determined by direct solids analysis. In this case a non-resonance line of lower sensitivity can be used, such as the line at 383.8 nm [493].

9.26 Manganese (Mn)

Manganese is the second most abundant heavy metal; it is present to about 0.1% in the Earth's crust and is in twelfth position in the abundance list of the elements. Far higher concentrations are present in the Earth's core. As a non-noble metal it is only present in nature in the form of its compounds in the +1 through +7 oxidation states. It is to be found as a trace element in virtually all soils, and frequently together with iron in ground waters. Large quantities of Mn are to be found in nodules on the seabed.

Manganese is an essential trace element that is present in all living cells. It is resorbed by plants in the form of Mn(II) salts and plays a major role in photosynthesis. It is also an essential trace element in animals and is present in numerous oxidoreductases and other enzymes. The human body contains about 20 mg Mn, mainly in the mitochondria, cell nuclei, and bones. A daily ingestion of at least 3 mg is considered as necessary.

The stability of aqueous Mn solutions has been investigated in various container materials and under varying conditions of storage [529, 537, 2106, 2539, 2904, 5297, 5456, 5660]. Losses of Mn start at pH values > 2.5 [5660], increase with increasing pH value, and are virtually quantitative at a pH value of 8. The losses on glass surfaces are significantly higher than on PE [5660]. In PFA bottles dilute, acidified solutions could be stored for over a year [2106].

Manganese can be determined in the air-acetylene flame without major interferences. The best characteristic concentration of 0.03 mg/L can be attained at the 279.5 nm analytical line. In practice, however, the triplet at 279.5/279.8/280.1 nm is used since the wider slit width brings a marked improvement in the S/N ratio. The characteristic concentration is then 0.1 mg/L. Manganese is an exception in that several spectral lines can be used simultaneously without the usual disadvantages becoming dominant (refer to Section 3.3.1). The three triplet lines exhibit almost the same sensitivity, as can be seen from Table 9-20, so that greater losses in sensitivity and strong curvature of the calibration curve are avoided. Each individual line naturally exhibits much better linearity.

ALLAN [136] found that sodium, potassium, calcium, magnesium, and phosphate do not cause interferences in the air-acetylene flame. GROSSMANN and MÜLLER [2236] investigated interferences on the determination of Mn by various constituents of alloys and found no influences due to iron, cobalt, and molybdenum, slight signal depression in the presence of nickel and aluminium, and slight enhancement in the presence of copper. PLATTE and MARCY [4667] reported that even small amounts of silicon depress the Mn signal; they eliminated this interference by adding 200 mg/L Ca (as the chloride). EL-DEFRAWY et al. [1703, 1706] observed signal suppression in the presence of phosphorus, borate, tungstate, dichromate, silicate, and cyanide. Complexing ligands such as

Table 9-20. Mn resonance lines.

Wavelength (nm)	Energy level (K)	Slit width (nm)	Characteristic concentration* (mg/L)	Spectral interferences
279.48	0–35770	0.2	0.03	Fe^{++} [4097, 4385] Mg, Pb^+ [5431]
279.83	0–35726	0.2	0.04	Pb^+ [5431]
280.11	0–35690	0.2	0.06	Pb^+ [5431]
403.08	0–24802	0.2	0.3	Ga^{++} [141, 4385]

* Air-acetylene flame, oxidizing (lean, blue)
+ Interference with deuterium BC possible
++ Direct line overlap

EDTA, DCTA, and NTA, on the other hand, caused signal enhancement, which was explained by the formation of volatile complexes. ABDALLAH *et al.* [4206] observed a signal depression in the presence of molybdenum and rhenium which they attributed to the formation of heteropoly anions which trap Mn in the lattice. Various complexing agents, such as sulfosalicylic acid, help to prevent such interferences [15, 1706, 5285]. Such interference can also be avoided by using the nitrous oxide-acetylene flame [388, 2733]; the characteristic concentration in this flame is about 0.3 mg/L.

For the determination of Mn by GF AAS, SLAVIN *et al.* [5419, 5421] proposed magnesium nitrate as the modifier and found that a pyrolysis temperature of 1400 °C could then be applied. At an optimum atomization temperature of 2200 °C the characteristic mass in a longitudinally-heated atomizer is 2 pg. In a transversely-heated atomizer the optimum atomization temperature is lowered to 1900 °C, and the characteristic mass is 6 pg when the Pd-Mg modifier is used. An interference caused by chloride [1006, 3861], which leads to losses during atomization [1008, 5791], can be controlled by the addition of ascorbic acid, either alone or together with ammonium nitrate [3344].

Under STPF conditions Mn can be determined free of interferences in virtually all matrices; the application of Zeeman-effect BC is essential in certain circumstances, particularly for metallurgical samples and for trace analyses in biological materials. By using the Pd-Mg modifier, WELZ *et al.* [6245] were able to establish a set of common parameters for the determination of As, Cu, Mn, Pb, Sb, and Se.

9.27 Mercury (Hg)

Mercury is one of the rarer elements; its content in the Earth's crust is estimated to be $5 \times 10^{-5}\%$, so that it is at position 62 in the abundance list of the elements. This low natural occurrence bears no relation to the significance that this element has acquired in recent years. The outstanding property of Hg is its volatility. All of its compounds are volatile at temperatures below 500 °C and decompose easily—particularly in the presence of reductants—to the free metal. This has the remarkably high vapor pressure of 0.0017 mbar at 20 °C, corresponding to a saturation concentration in the atmosphere of about 14 mg/m^3. The major sources of Hg are weathering and volcanic activity. Resulting from these natural processes, between 55 000 t and 180 000 t Hg are released annually and taken up by water and the atmosphere; a further 8000–38 000 t/year are released from anthropogenic sources. The particular characteristics of Hg, especially its mobility, are discussed in Sections 1.8.1 and 8.4.

In contrast to liquid Hg, mercury vapors are extremely toxic. Among Hg compounds the divalent forms are generally more toxic than the monovalent; the toxicity is directly dependent on the solubility of the respective compound. Of even greater significance than the inorganic compounds are the organomercury compounds. These are without doubt the organometallic compounds in our environment that have been investigated the most thoroughly. They have been responsible for a number of significant ecological catastrophes. The toxicity of the short-chain monoalkyl Hg compounds, particularly methylmercury CH_3Hg^+, is especially pronounced due to their strong tendency to bioaccumulation and low disassimilation. The methyl-Hg bond is very stable in the majority of organisms and the high lipid solubility allows it to cross the cell membrane and the blood-brain barrier. In addition, methylmercury forms compounds with biological ligands that have much longer half-lives in humans (60–70 days) than inorganic compounds alone (3–4 days) [1347]. In particular methylmercury attacks the motor function of the central nervous system. In addition to the damage caused by organomercury compounds, their decomposition to inorganic Hg(II) leads to secondary effects that are typical for this species.

Among the numerous catastrophes that are associated with Hg, the one at Minamata (Japan) in the 1950s is the most notorious. Methylmercury from a chemical plant was discharged into Minamata bay from whence it entered the nutritional chain. It was directly absorbed by microorganisms, further concentrated in fish, and then ingested by the local population whose major source of food was fish. Part of the Hg was also discharged in inorganic form, which in the natural environment can also be methylated. As a result of this catastrophe more than one hundred people died [3326].

Due to the high toxicity of methylmercury compounds, numerous studies on the biomethylation of Hg have been undertaken [1347]; biomethylation can take place in nature in sediments, interstitial waters and soils, by fulvic and humic acids, and in organisms. The methylation of Hg can take place in aerobic [3928] and anaerobic environments; methylmercury is formed preferentially in neutral and acidic media while dimethylmercury is formed under basic conditions [3326]. The latter is volatile and can thus enter the atmosphere.

Methyl-, ethyl-, and phenylmercury compounds are used to treat seed-corn, as fungicides, bactericides, and as catalysts in the chemical industry. Other, more complex, organomercury compounds are contained in pharmaceuticals since they exhibit very interesting properties [1347].

The stability of Hg solutions is a major problem since Hg must usually be determined in the trace and ultratrace ranges where losses and contamination are especially frequent. The problem becomes even more complex when a speciation analysis of Hg is required. The large number of single and comparative investigations [4764] is an indication of this problem; several of these investigations are cited in Tables 9-21a and b. This topic is also discussed in a number of review articles [346, 2220, 2346, 2905, 3633, 5313, 5322].

Table 9-21a. Stability of aqueous test sample and calibration solutions with respect to their total Hg concentration.

Matrix	pH range	Material*	Period	Recommended storage	References
Sea-water	1.5	BG		BG; HCl (stable for several months, glass recommended since Hg vapor can penetrate PE)	[730]
Deionized water	0–2	BG, PE	7 days	5% HNO_3 + 0.01% $K_2Cr_2O_7$ or 1% HNO_3 + 16 µg/L Au(III) (< 3% loss after 1 week)	[1278]
Distilled water, glacial water	natural; 2	PE, LPE, FEP	50 days	BG; Hg vapor can penetrate PE bottles	[1345]
Deionized water	0–2	BG, PE	10 days	solutions stabilized with 5% HNO_3 + 0.01% $K_2Cr_2O_7$ were stable in glass containers for at least 5 months	[1874]
Distilled water	< 1	BG, PE	3 months	HNO_3 + Cl^-	[3160]
Deionized water, natural water	3–9	PE, PP	30 days	complexing with humic acid exhibited lower losses than with 5 % HNO_3 + 0.05% $K_2Cr_2O_7$	[2466]
Deionized water, river water	5–8.1	PE	20 days	PE bottles must be conditioned with the water sample. Analysis must be performed within 10 h. Sample must be saturated with O_2	[3797]
Deionized water, river and sea water	< 1	PE	10 months	0.1 mol/L HNO_3 + 0.001 mol/L NH_4SCN or NaCl (no losses over 10 months)	[5083]
Drinking water		BG, PET	16 days	PET; acidification with HCl or stabilization with $K_2Cr_2O_7$	[1324]
Mains water, piped water	< 3	PE	45 days	PE; 0.5% HNO_3 + 0.05% $KMnO_4$ or $K_2Cr_2O_7$	[4637]
Deionized water	1–5	BG, PE	18 months	BG; acidification to pH values < 1, addition of 2% NaCl reduces losses	[162]
Deionized water, $CaCl_2$ solution, HNO_3	3.5–8.2	BG, PE, PP	15 days	BG; Hg losses from $CaCl_2$ solution were higher than from deionized water; no losses from HNO_3	[4337]
Sea-water	< 3	PE	4 months	0.1 mol/L HNO_3 (no losses after 4 months)	[5082]
Deionized water	0–14	BG			[534]

Table 9-21a continued

Matrix	pH range	Material*	Period	Recommended storage	References
Deionized water	1–5	BG, PE, PP, PTFE	18 months	H_2SO_4 (pH value 1) + 3% NaCl (no losses after 18 months)	[2465]
Deionized water	0–7	PE	21 days	HNO_3 (pH value 0.5) + 0.05% $K_2Cr_2O_7$ or 0.2 mg/L Au(III) (< 2% loss after 21 days)	[3566]
Deionized water, mains water	2.6–10	BG	12 days	$SnCl_2$ as reductant is not adequate to release Hg from its hydroxy complexes which are formed when water is not acidified with HCl	[224]
Natural water, deionized water	2, 7	BG, PE, PVC	16 days	BG; acidification with nitric acid to pH value 0.5	[4958]
Deionized water	< 2	PE, PP, BG, S	57 days	PE; $HCl + H_2O_2$ or $HNO_3 + H_2O_2$ (no losses after 57 days)	[3296]
Stream water, deionized water	natural; < 1	PE	19 days	HNO_3 (pH 1), acid should be present in container prior to sample to prevent losses	[1343]
Lake and river water		PE	30 days	1–100 mg/L-cysteine, 3% NaCl (no losses after 30 days)	[6196]

* BG: borosilicate glass; PE: polyethylene; PP: polypropylene; PTFE: polytetrafluoroethylene; PVC: polyvinyl chloride; S: silica; LPE: linear poly ethylene; FEP: fluorinated engineering polymer; PET: polyethyleneterephthalate.

Table 9-21b. Stability of aqueous test sample and calibration solutions with respect to their content of Hg species.

Matrix	pH range	Material*	Period	Recommended storage	References
Sea-water	4–10	BG	3 months	2% HCl (no losses of MeHg$^+$ but losses of Hg^{2+}) or 2% HNO$_3$ (no losses of Hg^{2+} but losses of MeHg$^+$)	[68]
Deionized water, sea-water	< 3	BG, FG, PE, PP, PTFE	4 months	BG; 1% H$_2$SO$_4$ + 0.05% K$_2$Cr$_2$O$_7$ (no losses over 4 months)	[1089]
Sea-water		BG, PE	2 months	BG; 0.2 mol/L H$_2$SO$_4$ (chloride suppresses adsorption losses, no losses within 2 months, PE bottles unsuitable because of contamination)	[3951]
Sea-water	natural; 2.5	BG, PE	3 months	brown BG; HCl (pH value 2.5) (total Hg was stable for 2.5 months, MeHg$^+$ 30 days, rapid conversion of MeHg$^+$ to Hg^{2+} when HNO$_3$ used instead of HCl)	[5581]
Deionized water		BG, PE, PTFE	2 months	BG; 5% NaCl (10% losses of MeHg$^+$ in 35 days when light excluded; 12.5% losses when 1% HNO$_3$ used)	[66]
Deionized water, sea-water	1, 6	BG, PE, PTFE	5 days	MeHg$^+$ in deionized water was stable in PTFE for 20 days; acidification to pH value 1 necessary in glass containers; 60% losses in acidified sea-water after 2 weeks; solutions of Hg^{2+} acidified to pH value 1 in glass or PTFE were stable for 30–40 days	[3487]
Deionized water		BG, PTFE	6 months	no losses after 6 months in PTFE; losses observed in glass containers, which could be reduced by cooling (5° C) and pretreatment with HNO$_3$	[3444]
Sea-water, natural water	natural	BG, PE	1 month	storage in brown glass bottles, cooled; Hg losses due to adsorption on the container walls and diffusion through the walls	[3987]

* BG: borosilicate glass; FG: flint glass PE: polyethylene; PP: polypropylene; PTFE: polytetrafluoroethylene

FELDMAN [1874] quoted the following mechanisms for Hg losses: (i) volatilization of Hg compounds, (ii) reduction of Hg compounds with subsequent vaporization of elemental Hg, (iii) adsorption of Hg on the walls of the container, (iv) adsorption of Hg on suspended particles and colloids, (v) inclusion of Hg in stable complexes that elude the determination, and (vi) inclusion of Hg in stable amalgams. Sample conservation for the determination of total Hg usually requires the addition of an acid and an oxidizing agent; the acid suppresses adsorption losses and biological activity such as the growth of algae, while the oxidizing agent should prevent the reduction of Hg(II) to volatile Hg^0. Reductants can be constituents of the sample, such as bacteria, algae, or humic acid [106, 314, 906, 5179, 5398], or also active sites on the surface of the container, trace impurities in the container material [5231], or contamination from earlier use [3334]. Occasionally an oxidizing acid is added alone for sample conservation, such as nitric acid [1343, 3160, 4337, 4958] or sulfuric acid; nitric acid is preferred to sulfuric acid [1278]. The addition of a strong oxidizing agent is better, such as dichromate [1278, 2346, 3566] or bromide/bromate [1843], since it has been shown that nitric acid or sulfuric acid alone is not adequate [1089]. The use of permanganate [1089, 1874, 4637, 5880] is problematical due to the precipitation of manganese dioxide [1089, 1874] and to the frequently high blank values [1089, 1278]. The best combination is nitric acid and dichromate. Stabilized samples should be stored in glass containers [2346] since many plastics, such as PE and PTFE, are permeable to Hg.

A further problem is the absorption of Hg from the laboratory atmosphere [730, 1345, 2539, 3002, 4196, 5081, 5864]; this increases with increasing oxidation potential of the solution [4196]. Recently polyethyleneterephthalate (PET) has been recommended as an alternative to glass [1324]. It is essential that the containers are thoroughly cleaned to prevent contamination [6319]. Since Hg can diffuse into and through the walls of numerous plastics, cleaning plastic containers is more of a problem than glass. The use of apparatus that finds application in electroanalytical laboratories must be avoided due to the uncontrolled contamination from the mercury electrodes.

The stabilization of samples for Hg speciation analysis is far more difficult than of those for the determination of total Hg since strong acids and oxidizing agents cannot be used. As well as the container material, the acid concentration, and the oxidation potential, the history and pretreatment of the containers [3657], the temperature and the influence of light are further factors that must be taken into account. While AHMED and STOEPPLER [66] found that methylmercury was rapidly decomposed by photolysis, LEERMAKERS et al. [3487] were unable to find any influence by light on the rate of decomposition of methylmercury. Even strong heating (80 °C for 2.5 hours) had only a minor influence, and yet under drastic conditions (oxidizing acids, 200 °C, pressure digestion) only about a third of the methylmercury was digested [66]. As a means of stabilization the addition of 50 g/L NaCl and acidification with hydrochloric acid can be considered, with storage in glass containers [3487]. When stored in the dark in glass containers, dilute aqueous calibration solutions of methylmercury are stable for about five days [3487]. After the addition of 50 g/L NaCl they are further stabilized, so that after 10 days more than 95 % of the methylmercury could be recovered and after 30 days more than 90 % [66], while without the addition of NaCl the recovery was only 10 % after 19 days storage [3487]. Containers made of PTFE are even better than glass for methylmercury since loss-free storage is possible over several months [3487]. Storage of this type is not suitable for inorganic Hg, however, since losses due to adsorption take place. For sea-water samples, on the other hand, losses take place when they are stored under these conditions; these losses can be ascribed to the formation of compounds that elude determination by CV AAS [68]. When sea-water samples are acidified the conversion of methylmercury to inorganic Hg(II) is enhanced.

Since Hg must mainly be determined in the trace range, F AAS is little suited because the characteristic concentration at the 253.7 nm analytical line is only 5 mg/L. This low sensitivity is due to the fact that the resonance line for the transition from the ground state to the first excited state is at 184.9 nm in the vacuum UV and is thus not amenable for the determination in normal instruments.

Nowadays Hg is determined almost exclusively by CV AAS. The development of this technique and a number of earlier methods are discussed in Section 1.8.1, while the instrumental aspects are treated in detail in Section 4.3; mechanisms and interferences are described in Section 8.4. At this point we should again like to emphasize that a digestion is indispensable when the total Hg concentration is to be determined in a test sample. Since only inorganic Hg(II) is quantitatively detected by CV AAS, all organomercury compounds must be converted to Hg^{2+} prior to the determination. The most usual technique is an oxidative digestion with potassium permanganate or potassium dichromate in sulfuric acid solution (see Section 8.4.2). The speciation analysis of Hg compounds in environmental samples is treated in Section 10.3.1.6.

Using Pd^0 as the modifier, Hg can be determined by GF AAS. The characteristic mass in a longitudinally-heated atomizer is 70–100 pg [6260], while in a transversely-heated atomizer it is 220 pg. The maximum pyrolysis temperature is between 250 °C and 400 °C [6260], depending on the conditions used, and the optimum atomization temperature is 1000–1300 °C. If a temperature of 1500 °C is not exceeded during the heat-out step the modifier is not volatilized, so that only one addition of modifier is required for about 50 atomization cycles [924, 6260].

9.28 Molybdenum (Mo)

Molybdenum is at position 38 in the abundance list of the elements and is thus one of the rarer elements. With a melting point of 2620 °C and a boiling point of about 5560 °C, Mo is one of the refractory elements. It also forms very stable oxides and carbides, although MoO_3 sublimes above 800 °C. Molybdenum occurs in its compounds in the divalent through hexavalent oxidation states; Mo(VI) compounds are the more stable. More than 80% of the Mo production is used in the steel industry for the production of refractory metals and high-temperature superalloys. It also finds application in pigments and catalyzers.

Molybdenum is an essential trace element; it is a constituent in the enzymes nitrogenase and nitratereductase, which participate in the fixation of nitrogen by certain algae and nodule bacteria, and in the assimilation of nitrate in green plants and bacteria. A deficiency of Mo in various higher plants thus results in deficiency disorders. In animal organisms organic Mo compounds are respiration catalysts. Nevertheless, the ingestion of large quantities of Mo by animals leads to diarrhea and growth inhibition.

The stability of aqueous Mo solutions was investigated by SMITH [5456]; losses of Mo start already at pH values between 1.5 and 3.5, so that sample conservation by acidification to pH values < 1.5 is required.

In earlier times Mo was frequently determined in a fuel-rich air-acetylene flame [1430]; however, this cannot be recommended since the sensitivity is a power of ten lower than in the nitrous oxide-acetylene flame and numerous interferences must be expected [1433] (see Section 4.1.1). In the nitrous oxide-acetylene flame at the 313.3 nm resonance line a characteristic concentration of 0.1 mg/L can be attained and the determination is largely free of interferences. VAN LOON [3579] found that the few interferences observed in this flame could easily be eliminated by the addition of 1–3 g/L Al to the measurement solutions; this can be explained by a reduction of the lateral diffusion, thus leading to an increase in the analyte concentration in the absorption volume [6282]. A number of further analytical lines of lower sensitivity are listed in Table 9-22; they are mainly of interest for the determination of major constituents in alloys [6404].

Table 9-22. Mo analytical lines.

Wavelength (nm)	Energy level (K)	Slit width (nm)	Characteristic concentration* (mg/L)	Spectral interferences
313.26	0–31913	0.7	0.07	Ir, Ni+ [5431, 5973]
317.04	0–31533	0.7	1.1	
319.40	0–31300	0.7	1.4	
390.30	0–25614	0.7	2.9	
315.82	0–31655	0.7	3.5	
320.88	0–31155	0.2	7.4	
311.21	0–32123	0.2	18	

* Nitrous oxide-acetylene flame, reducing (fat, red)
+ Interference with deuterium BC possible

The determination of Mo by GF AAS is discussed in detail in Section 8.2.3.4. Owing to the volatility of MoO_3, analyte losses can occur during the pyrolysis step at temperatures above 800 °C (refer to Figure 8-27). Nevertheless, if the temperature range 800–1500 °C is covered rapidly, pyrolysis temperatures of 1700–1800 °C can be applied. The use of magnesium nitrate as the modifier additionally aids in avoiding analyte losses. Despite its refractory character and a tendency to forming carbides, Mo can be determined with relatively good sensitivity by GF AAS. In a longitudinally-heated atomizer the characteristic mass is 9 pg. Nevertheless tailing and memory effects occur due to carbide formation and condensation. By the introduction of Freon to the inert gas during the heating-out step Mo can be removed more easily from the atomizer and thus a prolonged heating-out step can be avoided [6243]. Molybdenum can be determined particularly favorably in a transversely-heated atomizer, as shown in Figure 4-28 (Section 4.2.3); the characteristic mass is 12 pg. Since the risk of condensation of volatilized analyte on cooler parts of the atomizer is eliminated in a transversely-heated atomizer, Mo can be atomized from the platform at 2400 °C, while in a longitudinally-heated atomizer it can only be atomized from the wall at 2650 °C at the lowest. As can be seen in

Figure 9-4, the interferences on the determination of Mo by copper chloride in a longitu-dinally-heated atomizer can largely be eliminated by platform atomization, but the sig-nals are difficult to evaluate. Solely the application of STPF conditions and a trans-versely-heated atomizer bring optimum results. The peak shift in Figure 9-4d results from a type of modifier effect due to copper.

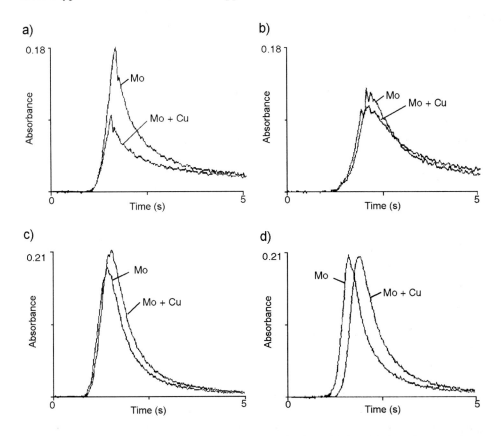

Figure 9-4. Influence of 200 mg/L CuCl$_2$ on the determination of 0.4 ng Mo in various atomizers. **a** and **b** – longitudinally-heated atomizer, 2650 °C; **c** and **d** – transversely-heated atomizer, 2400 °C; **a** and **c** – wall atomization; **b** and **d** – platform atomization.

9.29 Nickel (Ni)

The proportion of Ni in the Earth's crust is estimated to be 0.015 %; Ni is thus at position 22 in the abundance list of the elements. Since iron meteorites contain on average 8–9 % Ni, it is surmised that larger quantities of Ni are contained in the Earth's core. On the surface of the Earth Ni is almost always bound to sulfur, silicic acid, arsenic, or anti-mony. The largest deposits of Ni are to be found in the oceanic manganese nodules, which contain about 1 % Ni. The greatest part of the production of Ni is used in the steel industry and for the production of alloys. Organonickel compounds play an important role in a number of polymerization processes.

Physiologically Ni is one of the trace elements; the human body contains about 10 mg. Little is known about the biological role that Ni plays, but it appears to participate in carbohydrate metabolism; the concentration in serum and urine is around 0.5 µg/L. Inhalable dusts or aerosols of Ni metal or compounds, such as are produced during production and processing, are classified as hazardous substances and are carcinogenic. Nickel is thus an important element in the areas of toxicology and industrial hygiene.

The stability of aqueous Ni solutions has been investigated under varying storage conditions and in a variety of container materials [2106, 5456, 5606, 5660]. Sorption losses were significant after 24 hours in borosilicate glass if the pH value of calibration solutions acidified with hydrochloric acid was > 5 [5456]. Slightly acidified solutions containing 5 and 50 µg/L Ni could be stored in PFA bottles for up to 50 days without losses [2106]. KIILUNEN et al. [3097] and NIEBOER and JUSYS [4358] investigated the stability of urine samples with respect to their Ni content.

Nickel is among the elements determined frequently by AAS. The best characteristic concentration of 0.04 mg/L is obtained in the air-acetylene flame at the 232.0 nm resonance line. In order to use this line a spectral slitwidth of 0.2 nm is necessary since otherwise the strong emission lines at 231.716 nm and 232.138 nm cause considerable curvature of the calibration curve and a reduction of the sensitivity. Since the 232.0 nm line exhibits a noticeably non-linear calibration curve, even with a slitwidth of 0.2 nm, it is not particularly suitable for determining higher Ni concentrations. For this purpose the line at 341.5 nm with a characteristic concentration of 0.2 mg/L is more suitable. This line even allows the use of wider spectral slitwidths of up to 0.7 nm and thus provides a good S/N ratio. Further analytical lines are compiled in Table 9-23; slight spectral interferences can occur at a number of these lines, particularly with multielement lamps that contain cobalt, so that critical scrutiny of the lines and the signals is recommended.

In an oxidizing (fuel-lean) air-acetylene flame the determination of Ni appears to be largely free of interferences. KINSON and BELCHER [3117] found that high concentrations of Al, Co, Cr, Cu, Mn, Mo, V, and W did not cause interferences. Likewise, HCl, HNO_3, H_2SO_4 and H_3PO_4, had no influence. These findings were confirmed by

Table 9-23. Ni analytical lines.

Wavelength (nm)	Energy level (K)	Slit width (nm)	Characteristic concentration* (mg/L)	Spectral interferences
232.00	0–43090	0.2	0.04	Sn+ [5431]
231.10	0–43259	0.2	0.07	
341.48	205–29481	0.2	0.2	
305.08	205–32973	0.2	0.25	
346.17	205–29084	0.2	0.35	

* Air-acetylene flame, oxidizing (lean, blue)
+ Interference with deuterium BC possible

PLATTE and MARCY [4667]; nevertheless, careful optimization of the burner height and the flame stoichiometry are required [388, 5695]. In a less oxidizing flame, interferences due to iron and chromium become noticeable. SUNDBERG [5695] related the various effects (enhancing and suppressing) of other concomitant elements found with non-optimum gas settings and observation height to the dissociation energies of the corresponding monoxides. The potential influences of iron and chromium can be eliminated by the use of the nitrous oxide-acetylene flame; the characteristic concentration in this flame at the 232.0 nm resonance line is only around 2 mg/L, however.

The trace determination of Ni in body fluids and tissues by GF AAS has especial significance. HINDERBERGER et al. [2555] used primary ammonium phosphate as the modifier and atomized from the L'vov platform. SLAVIN et al. [5419, 5421] investigated magnesium nitrate as the modifier and found that Ni could be thermally stabilized up to 1400 °C. The determination of Ni is virtually free of interferences under STPF conditions. The optimum atomization temperature for Ni is then 2500 °C in a longitudinally-heated atomizer and 2300 °C in a transversely-heated atomizer. The characteristic mass with and without Zeeman-effect BC is 13 pg and 20 pg, respectively. Since Ni must often be determined in very low concentrations and in complex matrices, various on-line separation and preconcentration procedures have been described, e.g. liquid-liquid extraction [340, 343, 5774], sorbent extraction [20, 323, 4689, 5510, 5517, 6262], or co-precipitation [1830]. If it is necessary to determine Ni traces fairly regularly, the use of this element as a modifier should generally be avoided since an effective control of contamination is otherwise not possible.

VIJAN [6077] attempted to volatilize Ni as the carbonyl by reducing it to the metal from its compounds with sodium tetrahydroborate and then applying carbon monoxide under pressure. He conducted the resulting gas into a heated quartz tube and obtained a characteristic mass of 20 pg Ni. However, he found that the formation of nickel carbonyl was not quantitative. This method was later revived by LEE [3481] and ALARY et al. [105], and particularly in combination with in situ preconcentration in the graphite tube could gain new significance.

9.30 Niobium (Nb)

With a melting point of about 2468 °C and a boiling point of 4927 °C, Nb is one of the refractory elements. Niobium occurs in its compounds in the +2, +3, +4, and, most commonly, +5 oxidation states. Niobium is used in the form of ferro-niobium and ferro-niobium-tantalum in the production of structural and high-temperature steels. It is also a constituent of numerous superalloys which are used in gas turbines and jet engines due to their high thermal stability.

The stability of aqueous Nb solutions was investigated by EICHHOLZ et al. [1684] and DAVYDOV [1441]. Sorption losses on glass surfaces are more pronounced than on PE [1684] and can amount to 14 % even on silica at a pH value of 2 [1441]. Acidification to a pH value of 1 is thus recommended.

Niobium can be determined in the nitrous oxide-acetylene flame with a characteristic concentration of 40 mg/L at the 334.3 nm analytical line and a spectral bandpass of

0.2 nm [3846]. Since Nb is noticeably ionized in the hot flame, the addition of 1 g/L K as the chloride to all measurement solutions is recommended. WALLACE *et al.* [6134] found that the addition of 1 % HF and 2 g/L Al markedly improved the linearity of the calibration curve as well as the sensitivity and precision. Aluminium could reduce the tendency to form refractory oxides and facilitate atomization. It could also reduce the lateral diffusion and thus enhance the sensitivity. Little has been reported on interferences in the determination of Nb.

Due to its refractory character, Nb cannot be determined appropriately by GF AAS so that little information is available [509].

9.31 Non-metals

Iodine, phosphorus, and sulfur are treated in separated sections since direct AAS methods are available for these non-metals. The remaining non-metals cannot be determined directly by AAS since their analytical lines lie in the vacuum UV range.

For the indirect determination of non-metals by AAS, two different routes have been followed: the utilization of specific interferences and the use of precipitation reactions or complex formation of metals. CHRISTIAN and FELDMAN [1277] used the first technique to determine *orthophosphate, sulfate, sulfide, iodate, iodide, glucose, protein, 8-hydroxyquinoline*, and other complexing agents by examining their influence on the determination of calcium, chromium, and iron and placing the extent of this influence in a relationship to the concentrations of the interfering ions. KUNISHI and OHNO [3354] made a thorough investigation on the indirect determination of sulfate and utilized the suppressive effect on the determination of iron and its elimination by lanthanum. When the ion ratio $La^{3+}:SO_4^{2-} = 2:3$ the original sensitivity for iron is restored. These authors spoke of a 'sharp change' that makes very precise determinations possible. BOND and O'DONNEL [706] determined *fluoride* in a similar manner by quantitatively evaluating their influence on the determination of magnesium in an air-coal gas flame and on zirconium in a nitrous oxide-acetylene flame. SAND *et al.* [5079] determined *silicate, phosphate*, and *sulfate* via their influence on the atomization of calcium.

The second procedure, the formation of an insoluble precipitate with a metal cation with subsequent determination of this metal, was employed by EZELL [1788] and WESTERLUND-HELMERSON [6291] for the determination of *chloride* in various materials. The chloride ion is precipitated with silver nitrate and either the precipitate is dissolved in ammonia solution and analyzed for its metal content, or the excess metal is determined in the solution after an exactly measured addition of silver nitrate [398]. GAMBRELL [2052] determined chloride in water by end-point determination with silver nitrate by AAS; two different quantities of silver nitrate are added to the test sample solution, the absorbance of silver is measured, and then extrapolated against zero.

MANAHAN and KUNKEL [3833] determined *cyanide* by utilizing the solubility of Cu(II) from basic copper carbonate in alkaline medium. The complexed copper, as $[Cu(CN)_3]^-$, was determined by F AAS; the characteristic concentration was quoted as 2×10^{-5} mol/L CN^-. JUNGREIS and AIN [2971] also determined cyanide by filtering the solution through silver wool and then determining the dissolved complexed silver by

GF AAS. A further method for determining cyanide is based on the extraction of an ion-pair complex between copper cyanide or silver cyanide and benzyldimethylhexadecyl-ammonium [2015]. KOVATSIS [3265] determined *carbon disulfide* via reaction with zinc acetate and N,N-dibenzylamine; the zinc dibenzyldithiocarbamate which is formed is extracted in toluene and the zinc then determined. MANAHAN and JONES [3832] described a specific LC detection system for *chelating agents*. A solution containing a chelating agent is conducted through a short column containing a chelate ion exchanger in the copper form and then the copper concentration is determined by F AAS. The copper concentration is directly proportional to the quantity of the chelating agent; the LOD was quoted as 5×10^{-7} mmol chelating agent.

KUMAMARU *et al.* determined *nitrate* [3340] and organic molecules via coextraction with metal complexes. *Nitrate* and *phthalic acid* [3341, 3342] were extracted as ion pairs from bis-(neocuproin)-Cu(I) with IBMK. *Pentachlorophenol* [6434] was also extracted as an ion pair with tris-(1,10-phenanthroline)-Fe(II) in nitrobenzene; the concentrations of copper and iron, respectively, in the extracts are proportional to the concentrations of the ions and molecules under study and can be determined by AAS.

WOODIS *et al.* [6375] determined *biuret* in fertilizer by treating an alcoholic solution of biuret and copper with a strong base. A copper-biuret complex is formed, the excess copper is precipitated, and the copper in the complex determined. OLES and SIGGIA [4461] determined *aldehydes* by oxidizing with a silver-ammonia complex (Tollen's reagent), separating the reduced silver, dissolving in nitric acid, and determining by AAS. They also determined *1,2-diols* [4462] by oxidizing with periodic acid; the iodate formed is precipitated as silver iodate, which is then dissolved in ammonium hydroxide, and the silver determined by AAS.

A number of these indirect procedures for the determination of anions have in the meantime been modified for FI and automated [2364, 3324]. This has led to a noticeable revival of interest in these indirect methods for the determination of molecules and ions. Flow injection is used for the liquid-liquid extraction, precipitation, and filtration of ion pair complexes [5974, 5976], or for the mobilization of metals from solid bed reactors for the determination of the following substances: fungicides and pesticides [663–665]; alkaloids [1694–1696, 1698]; other pharmacologically active substances [1597, 2068, 3391, 3392, 3449, 3916, 4171–4175, 4879, 5089, 5977]; surfactants [2049, 3922]; saccharin [6472]; anions such as ammonium [1768]; carbonate [1767, 1772]; chloride, bromide, and iodide [1767, 1769, 1770, 3919–3921, 5197]; cyanide [1768, 1771, 2321]; EDTA [4121]; nitrate [2050, 5365]; nitrite [2050, 5365]; oxalate [1772, 3921]; perchlorate [2048]; sulfate [2051]; sulfide [4625]; thiosulfate [1768]; and thiocyanate [1771].

9.32 Osmium (Os)

With a melting point of approximately 3045 °C and a boiling point of 5500 °C, Os is one of the refractory metals. When heated in air it forms the volatile, toxic osmium tetraoxide OsO_4. Osmium occurs in the oxidation states 0 and +2 through +8; the tetravalent and hexavalent states are the most stable.

The stability of aqueous Os solutions was investigated by GLADNEY and APT [2138], VAN LOON [3581], and ENGLERT *et al.* [1745]. Sorption losses from distilled water were observed for concentrations of 1–1000 mg/L Os on various materials such as borosilicate glass, silica, PE, PTFE, PVC, PP, PC [2138] and PFA [1745], and were between 50% after several months [2138] and 20% after only two days [1745]. Even when acidified with 0.1 mol/L hydrochloric acid losses could not be prevented on the plastic containers mentioned above. When acidified with nitric acid Os is vaporized as the tetroxide. Osmium solutions are stable for up to two months in 1 mol/L hydrochloric acid when they are stored in glass or silica containers [2138].

In 1968 SLAVIN [5413] reported that Willis had detected strong absorption for Os at a number of wavelengths. FERNANDEZ [1881] investigated the sensitivity of ten resonance lines and obtained a characteristic concentration of 1 mg/L at the 290.9 nm line in the nitrous oxide-acetylene flame. The sensitivity is poorer in a fuel-rich air-acetylene flame by a factor of five. A number of further analytical lines with the attainable sensitivities are compiled in Table 9-24. The high volatility of osmium tetroxide can be utilized to improve the transfer efficiency through sample introduction in gaseous form via an OsO_4 generator [3830].

Despite its refractory character, Os can be determined by GF AAS [1562, 3232, 4974]. Due to the volatility of the tetroxide a pyrolysis temperature in excess of 200 °C cannot be applied, making a determination in the presence of higher concentrations of concomitants more difficult. To date, a modifier that could stabilize Os has not been described. At an atomization temperature of 2650 °C and evaluation via the peak height a characteristic mass of only 1.4 ng is attained, so that Os is among the elements exhibiting the poorest sensitivity.

Table 9-24. Os resonance lines (according to [1881]).

Wavelength (nm)	Energy level (K)	Slit width (nm)	Characteristic concentration (mg/L)		Spectral interferences
			N_2O-acetylene	air-acetylene	
290.9	0–34365	0.2	1.0	5.0	Cr [4524]
305.9	0–32685	0.2	1.6	6.4	
263.7	0–37909	0.2	1.8		
301.8	0–33124	0.2	3.2		
330.2	0–30280	0.2	3.6		
271.5	0–36826	0.2	4.2		
280.7	0–35616	0.2	4.6		
264.4	0–37809	0.2	4.8		
442.0	0–22616	0.2	20		
426.1	0–23463	0.2	30		

9.33 Palladium (Pd)

Palladium occurs frequently in association with other noble metals such as platinum and gold, but also with arsenic and antimony, in the 0 and +2 oxidation states, and less frequently in the +3 and +4 oxidation states. Palladium is frequently a constituent in jewelry metals and it is also often included in a number of other high-grade alloys. It is also frequently applied as a catalyst.

Even dilute Pd solutions are relatively stable. Acidified calibration solutions containing 10 µg/L Pd were stable in PFA bottles for over 56 days [1745].

Palladium can be determined in a very sharp, fuel-lean air-acetylene flame with a characteristic concentration of 0.15 mg/L; the analytical lines at 247.6 nm and 244.8 nm exhibit approximately the same sensitivity. The latter line gives a highly non-linear calibration curve, even at a spectral slitwidth of 0.2 nm, so that the line at 247.6 nm is to be preferred. Some further analytical lines with the attainable characteristic concentrations are compiled in Table 9-25.

Table 9-25. Pd analytical lines.

Wavelength (nm)	Energy level (K)	Slit width (nm)	Characteristic concentration* (mg/L)	Spectral interferences
244.79	0–40839	0.2	0.22	Pb^{++} [1988] PO^{++} [4425, 6308] Cu^+ [5431] Ru^+ [5431]
247.64	0–40369	0.2	0.25	Pb^{++} [1988, 6307] PO^{++} [4425, 6308] Fe^+ [5431]
276.31	0–36181	0.2	0.74	
340.46	6564–35928	0.2	0.72	

* Air-acetylene flame, strongly oxidizing (lean, blue)
+ Interference with deuterium BC possible
++ Interference with Zeeman-effect BC possible

JANSSEN and UMLAND [2886] eliminated the interferences that occur during the determination of noble metals by adding 10 g/L each of Na and Cu; PANNETIER and TOFFOLI [4529] preferred 5 g/L Li, with the possible addition of copper. To eliminate interelemental interferences, SCHNEPFE and GRIMALDI [5175] and SEN GUPTA [5232] added 5 g/L each of Cu and Cd. ADRIAENSSENS and VERBEEK [41] found that most of the interelemental interferences occurring in acidic solution did not occur in 20 g/L KCN solution. In a recent study, EL-DEFRAWY [1705] was able to confirm the positive effect of KCN. Merely gold and platinum influenced Pd somewhat; this effect disappeared by adding silver. HEINEMANN [2468] found that the determination of Pd is little influenced by rhodium and not by platinum. The best conditions are obtained when lanthanum is added as a buffer [3583]. Beside inorganic buffers, organic buffers such as butylamine [2975, 3522] have been applied more recently since they help to suppress the formation of refractory components in the flame. The sensitivity of F AAS is insufficient in most cases to determine Pd traces, so that composite procedures which include preconcentration or GF AAS must be applied.

Palladium can be determined by GF AAS in a longitudinally-heated atomizer with a characteristic mass of 22–24 pg at an atomization temperature of 2650 °C. In a transversely-heated atomizer Pd can be atomized at 2200 °C and the characteristic mass is then 50 pg. The maximum pyrolysis temperature is 900 °C at which numerous matrix constituents have already volatilized and the remaining interferences can be controlled under STPF conditions [5148]. If very high requirements must be met, GF AAS must be preceded by preconcentration, which can also be performed on-line by FI techniques [5204].

9.34 Phosphorus (P)

Phosphorus is present to about 0.1 % in the structure of the upper, 16 km thick crust of the Earth and is at position 12 in the abundance list of the elements between chlorine and carbon. Due to its reactivity phosphorus never occurs in elemental form but only in the form of stable phosphates.

White phosphorus is used for the manufacture of phosphoric acid and phosphates, and also to a limited extent in alloys. Red phosphorus serves for the manufacture of phosphides and other P compounds which find numerous applications in inorganic and organic syntheses.

Phosphates are indispensable for the normal metabolic processes in plants and animals. Phosphorus is present in organisms as calcium phosphate in bones, in the form of phosphoric acid esters as building blocks of nucleic acids, and as a constituent of phospholipids in living cells.

HERCZYNSKA and CAMPBELL [2517] observed significant sorption losses of up to 60 % at a pH value of 4. Acidification of solutions to a pH value < 1.5 is thus recommended.

Phosphorus exhibits three resonance lines at 177.5 nm, 178.3 nm, and 178.8 nm in the vacuum UV which are not accessible to normal atomic absorption spectrometers. KIRKBRIGHT [3128] successfully determined P using a modified atomic absorption spectrometer with a flushed monochromator and a nitrous oxide-acetylene flame shielded with nitrogen. The line at 178.3 nm gave the best characteristic concentration of 5.4 mg/L. WALSH [6136] determined P by cathodic sputtering in a vacuum spectrometer.

L'VOV and KHARTSYZOV [3687] were the first to report the successful determination of P at the lines at 213.55/213.62 nm and 214.91 nm; these lines emanate from metastable states. Using an HCL and the nitrous oxide-acetylene flame, MANNING and SLAVIN [3852] found a characteristic concentration of 290 mg/L at the 213.6 nm line and 540 mg/L at the 214.9 nm line. After the introduction of a high-performance EDL for P

[390], EDIGER [1678] established standard conditions for the determination of P by GF AAS using lanthanum as the modifier.

PERSSON and FRECH [4619] investigated both theoretically and experimentally the factors that influence the determination of P by GF AAS. They found that the risk of losses of P during pyrolysis was very large. At temperatures above 1300 °C the formation of gaseous PO is highly probable and depends on the partial pressure of oxygen in the graphite tube. Above about 1350 °C losses of the dimer P_2 were observed, so that an especially high rate of heating for atomization is particularly important. The authors found that reproducible results could only be obtained when atomization took place under isothermal conditions.

Next to the formation of volatile compounds and the concomitant risk of analyte losses, one of the major problems for the determination of P by GF AAS is that the analytical line does not emanate from the ground state but from a high-energy, excited state. Since the sensitivity is directly dependent on the population in this excited state, which again is dependent on the temperature, the modifier and STPF conditions play a key role in the determination. Without a modifier the sensitivity for P is very low, even when atomization is from a platform [1381]. The lanthanum modifier proposed by EDIGER [1678] does improve the sensitivity significantly, but it has a very negative influence on the lifetimes of graphite tubes and platforms [6229, 6251]. CURTIUS et al. [1382] found that a mixture of palladium nitrate and calcium nitrate brought the best results. With this modifier a pyrolysis temperature of 1350–1400 °C can be applied and a characteristic mass of about 5 ng attained; this permits the sensitive determination of P in vegetable materials, for example [1383]. Iron causes an interference with continuum source BC at the 213.6 nm line, so that Zeeman-effect BC must be used in the analysis of iron or steel.

9.35 Platinum (Pt)

Platinum is one of the rarest of the noble metals; it is at position 76 in the abundance list of the elements. It occurs mostly together with other platinum-group metals, either native or as a mineral, frequently with arsenic, antimony, or sulfur. Platinum has a multitude of uses for electrical contacts, heat conductors, thermoelements, resistance thermometers, spinning nozzles, medical instruments, crucibles, dishes, etc., as well as jewelry alloys and as a catalyst. A compound of pharmacological significance, albeit very toxic, is cis-diammine-dichloroplatinum(II), known as Cisplatin, which is used as a cytostaticum against tumors. Due to its use in catalysts and its pharmacological activity, the determination of platinum in the trace range has become increasingly important in recent times. BAREFOOT [383] has published an overview on various methods in this area.

The stability of aqueous Pt solutions is relatively critical [1745, 3115, 4718, 5072, 5455]. Sorption losses could be observed after only 24 hours from solutions containing 1–10 mg/L when the pH value was > 1.5 [5455]. When stored in PFA bottles, slightly acidified solutions containing 10 µg/L Pt were stable for over 56 days [1745].

Platinum is best determined in a sharp, oxidizing air-acetylene flame. It exhibits a multitude of analytical lines, the most important of which are compiled in Table 9-26. The best characteristic concentration of around 1 mg/L is obtained at the 265.9 nm line.

JANSSEN and UMLAND [2886] found numerous interelemental effects for the determination of Pt in the presence of other noble metals; they eliminated these by adding 10 g/L each of Na and Cu. At the same time they found that this additive enhanced the sensitivity for Pt by 50%. PANNETIER and TOFFOLI [4529] preferred the addition of 5 g/L Li with the possible addition of copper, while SCHNEPFE and GRIMALDI [5175], and SEN GUPTA [5232] used 5 g/L each of Cu and Cd as a buffer. ADRIAENSSENS and VERBEEK [41] found that most of the interelemental interferences observed in acidic solution during the determination of noble metals did not occur in 20 g/L KCN solution. PITTS et al. [4664, 4667] proposed the use of the nitrous oxide-acetylene flame as an alternative; the sensitivity is reduced by a factor of five but no interferences occur.

Platinum can be determined relatively free of problems by GF AAS, although the sensitivity is rather low. Without a modifier a pyrolysis temperature of 1300 °C can be applied. In a longitudinally-heated atomizer the characteristic mass is 100 pg at an atomization temperature of 2650 °C. In a transversely-heated atomizer a temperature of 2200 °C is sufficient for atomization from the platform, and the characteristic mass is 220 pg.

Table 9-26. Pt analytical lines.

Wavelength (nm)	Energy level (K)	Slit width (nm)	Characteristic concentration* (mg/L)	Spectral interferences
265.95	0–37591	0.7	1.0	Ir, Sn+ [5431, 5973] Eu++ [6307]
306.47	0–32620	0.7	2.0	
262.80	776–38816	0.7	2.7	
283.03	0–35322	0.2	3.3	
292.98	0–34122	0.7	3.8	
273.40	776–37342	0.2	4.0	
299.80	776–34122	0.2	5.3	Fe, Ni, Pd+ [5431]
270.24	776–37769	0.2	5.6	
244.01	0–40970	0.2	7.3	
271.90	824–37591	0.2	12	Fe+++ [4385]
304.26	824–33681	0.7	18	

* Air-acetylene flame, oxidizing (lean, blue)
+ Interference with deuterium BC possible
++ Interference with Zeeman-effect BC possible
+++ Direct line overlap

9.36 Potassium (K)

Potassium constitutes about 2.4% of the Earth's crust and is thus one of the ten most abundant elements. Potash feldspar and potash mica are major constituents of numerous igneous rocks. During weathering the K bound to the feldspars is largely converted into soluble compounds, but unlike sodium compounds they do not end up in the sea but rather are bound by calcium zeolites in the soil and exchanged for calcium. For this reason there is about 2.3% K in sedimentary rocks but only 0.04% in sea-water.

Potassium plays a major role in animal and vegetable organisms. It is taken up to a much higher degree than other cations by the roots of plants. Part of the K is present as free K^+ in the sap and influences osmotic processes. In the human body about 98% of the K is found inside the cells in contrast to sodium which is found outside. The K/Na concentration gradient in the cell is established and maintained by the 'potassium-sodium pump'. This gradient is responsible for the osmotic pressure in the cells, for the excitability of the muscles and nerves, and for the secretion of liquids. Potassium is an essential element due to its manifold actions and it must be ingested by the body, if necessary therapeutically.

For routine applications K can be determined with adequate sensitivity and accuracy in simple flame photometers. Nevertheless AAS, with the choice of various analytical lines and flames, offers greater variability and hence optimization for each analytical task. As well as the most sensitive line at 766.5 nm, with a characteristic concentration of 0.03 mg/L, the second doublet line at 769.9 nm is also advantageous; it exhibits about half the sensitivity ($\beta_0 = 0.05$ mg/L) and a calibration curve with better linearity. For high concentrations of K the doublet at 404.4 nm and 404.7 nm is available with a characteristic concentration of about 5 mg/L.

Potassium is frequently determined in the air-acetylene flame; there is noticeable ionization which should be eliminated by adding 1 g/L Cs. The determination of K is sometimes simpler in a cooler flame in which no ionization can be observed. The air-hydrogen flame has proven very advantageous; especially at the secondary doublet at 404.4/404.7 nm it gives a greatly improved S/N ratio and thereby better precision.

The determination of K by GF AAS is strongly subject to the risk of contamination since K is present in dust at relatively high concentrations. Signal depression for K in the presence of excess sodium and iron [4564] and also of several acids has been reported. The application of GF AAS should thus only be considered under clean-room conditions, or in special cases for microanalysis [687, 4298, 4517, 4564, 5253, 5705, 6358] and ultratrace analysis [418], such as a purity check on ultrapure chemicals [3526, 6163], water [2620], semiconductor materials [478], raw materials for the production of semiconductors [3303, 5001, 5756], ceramic materials or ultrapure metals [1553, 1556, 1558, 2417, 3257, 5174, 6322]. The maximum pyrolysis temperature without the addition of a modifier (which is advantageous due to the contamination risk) is 950 °C. The optimum atomization temperature under STPF conditions is 1500 °C. In a longitudinally-heated atomizer the characteristic mass is around 1 pg and in a transversely-heated atomizer it is about 2 pg.

9.37 Rhenium (Re)

Rhenium is one of the very rare elements in the Earth's crust. Its most important use is in bimetallic reformer catalysts for the production non-leaded petrol (gasoline). It is also used as a constituent in superalloys and in the manufacture of thermoelements, heating coils, and electrical contacts.

ENGLERT et al. [1745] were able to show that dilute solutions with a concentration of 10 µg/L Re were stable for over 56 days in PFA bottles, and also no losses could be observed for 5 mg/L Re solutions in glass and PE containers [99].

Rhenium can be determined in the nitrous oxide-acetylene flame with a characteristic concentration of 14 mg/L at the 346.0 nm resonance line. Further analytical lines with their characteristic concentrations are compiled in Table 9-27. There are no reports to date on interferences in the nitrous oxide-acetylene flame.

Due to its extremely high melting point of 3186 °C and boiling point of around 5900 °C, Re cannot be practically determined by GF AAS since an adequate vapor pressure cannot be attained at the prevailing temperatures. KOIDE et al. [3193] observed strongly broadened peaks and a sensitivity that was highly dependent on the atomization temperature. For atomization from the wall of a PG-coated graphite tube at 2800 °C they attained an instrumental LOD of 0.5 ng. To avoid memory effects several heat-out cycles at 2850 °C were necessary; this is very unfavorable for the lifetime of the graphite tube (see Section 4.2.1).

Table 9-27. Re analytical lines.

Wavelength (nm)	Energy level (K)	Slit width (nm)	Characteristic concentration* (mg/L)	Spectral interferences
346.05	0–28890	0.2	14	
346.47	0–28854	0.2	24	Co^+ [5431]
345.19	0–28962	0.2	36	Co^+ [5431]

* Nitrous oxide-acetylene flame, reducing (fuel-rich, red)
+ Interference with deuterium BC possible

9.38 Rhodium (Rh)

Rhodium is one of the rarest elements and occurs mainly in elemental form together with other noble metals. It is mainly used in catalysts and for high-grade electroplating.

The stability of Rh solutions is relatively critical. SMITH [5455] observed significant losses after only 24 hours when the pH value was > 1.5. EICHHOLZ et al. [1684] found that losses on glass surfaces were higher than on plastic surfaces. In PFA containers even dilute calibration solutions with concentrations of 10 µg/L Re were nevertheless stable for 56 days [1745].

Rhodium can be determined in the nitrous oxide-acetylene flame with a characteristic concentration of 0.8 mg/L at the 343.5 nm resonance line. A sharp, fuel-lean air-acetylene flame exhibits markedly better sensitivity, but substantial matrix effects must be expected. Further analytical lines and the attainable characteristic concentrations in both flame types are compiled in Table 9-28.

ATWELL and HEBERT [311] investigated the influence of 14 cations in concentrations of up to 3 g/L and of four acids on Rh in the air-acetylene and nitrous oxide-acetylene flames. All elements interfered to greater or lesser degrees in the air-acetylene flame, while in the nitrous oxide-acetylene flame only iridium and ruthenium interfered; all other elements and the four acids had no influence. The addition of 5 g/L Zn to the measurement solutions eliminated the interferences completely, and the stoichiometry of the flame, the burner height, and similar parameters were not critical.

GARSKA [2083] and HEINEMANN [2469] found that the majority of interferences in the air-acetylene flame could be eliminated by using lanthanum in hydrochloric acid as a buffer. Nevertheless, the correct setting of the acetylene flowrate and the burner height were critical parameters with respect to the sensitivity and the linearity of the calibration curve. In a sharp, fuel-lean air-acetylene flame KALLMANN and HOBART [3013] found no interferences due to phosphorus, but they found strong signal enhancement due to the sulfates of alkali metals, aluminium and zinc which they attributed to the formation of an Rh alum. JANSSEN and UMLAND [2886] used an addition of 10 g/L each of Na and Cu to enhance the sensitivity for Rh and to simultaneously eliminate the interelemental interferences caused by the other noble metals. PANNETIER and TOFFOLI [4529] preferred 5 g/L Li and a little copper to eliminate the same interferences, and SEN GUPTA [5232] used a buffer of 5 g/L Cu and 5 g/L Cd.

Since the melting point of Rh of 1970 °C is not especially high it can be determined relatively well by GF AAS. The characteristic mass in a longitudinally-heated atomizer is 10 pg and in a transversely-heated atomizer it is 24 pg. A pyrolysis temperature of 1300 °C can be applied without modifier and the optimum atomization temperature is 2400 °C. Interelemental interferences are observed with the other noble metals and are attributed to the formation of alloys. ARPADJAN et al. [273] thus observed strong signal depression in the presence of platinum. Nickel caused a signal enhancement of

Table 9-28. Rh analytical lines.

Wavelength	Energy level	Slit width	Characteristic concentration (mg/L)		Spectral interferences
(nm)	(K)	(nm)	air-acetylene	N₂O-acetylene [311]	
343.49	0–29105	0.2	0.1	0.8	Ni, Ru* [5431]
369.24	0–27075	0.2	0.1	1.4	
339.69	0–29431	0.2	0.2	1.7	
350.25	0–28543	0.2	0.3	2.5	
365.80	1530–28860	0.2	0.5	2.5	
370.09	1530–28543	0.2	1	5.5	
350.73	2598–31102	0.2	2.5	2.5	

* Interference with deuterium BC possible

42% [692]. Due to the complex nature of the interferences and the low concentration of Rh that must be determined, it is recommended to separate and preconcentrate the analyte [214, 550, 1722, 3041, 4370, 5242, 5442, 6062, 6063]; this can also be performed on-line by FI [3484, 5203].

9.39 Rubidium (Rb)

Although Rb is at position 17 in the abundance list of the elements between nitrogen and fluorine, and is ten times more abundant than lead, it has acquired relatively little significance. This fact is reflected in the relatively low number of publications on this element.

ROBERTSON [4890] investigated the stability of sea-water samples with respect to their concentration of Rb and observed significant losses from non-acidified samples.

In the air-acetylene flame a characteristic concentration of around 0.1 mg/L can be attained at the 780.0 nm resonance line. The sensitivity can be enhanced by a factor of about two in flames of lower temperature. A number of further analytical lines with the attainable characteristic concentrations are compiled in Table 9-29.

Table 9-29. Rb analytical lines.

Wavelength (nm)	Energy level (K)	Slit width (nm)	Characteristic concentration* (mg/L)
780.02	0–12817	1.4	0.1
794.76	0–12579	1.4	0.2
420.19	0–23793	0.7	8.7
421.56	0–23715	0.7	19

* Air-acetylene flame, oxidizing (lean, blue)

Rubidium, like the other alkali metals, is markedly ionized in the air-acetylene flame so that an easily ionized element—preferably cesium—must be added to suppress this effect. In the air-hydrogen flame Rb is practically no longer ionized so that this flame may be preferable.

LUECKE [3645] investigated the determination of Rb in rocks and minerals and found in part substantial signal suppression due to large excesses of aluminium, phosphate, fluoride, and perchlorate. The addition of lanthanum could only partially correct these interferences. The interferences could be eliminated by changing to the nitrous oxide-acetylene flame and using EDTA as a buffer. JOSEPH et al. [2961] described the determination of [87]Rb in various natural samples by AAS.

Rubidium can be determined by GF AAS; under STPF conditions in a longitudinally-heated atomizer the characteristic mass is 2.3 pg while in a transversely-heated atomizer it is 10 pg. Contamination problems may occasionally occur due to the good sensitivity [2228]. Little has been reported on a modifier for this element; ascorbic acid has been proposed as a modifier for the determination of Rb in water [1408]. Without modifier a pyrolysis temperature of 800 °C can be applied and the optimum atomization temperature is 1800–1900 °C.

9.40 Ruthenium (Ru)

Ruthenium is the rarest and lightest element in the platinum-metal group. It occurs in nature mostly in very small quantities in company with platinum. It is used mainly in catalysts for a variety of applications.

The stability of Ru solutions is very critical [507, 1684, 1745, 2138, 3079, 5455]. GLADNEY and APT [2138] reported that neutral Ru solutions were not stable for longer than a day. Even solutions acidified to 1 mol/L with nitric acid exhibited losses of up to 25 % after four months. SMITH [5455] observed significant losses from solutions in hydrochloric acid after 24 hours when the pH value was > 1.5. The solutions were stable for up to four months, however, if the hydrochloric acid concentration was > 0.1 mol/L and storage was in glass, silica, or PE containers [2138].

Ruthenium can be determined in the air-acetylene flame with a characteristic concentration of 0.7 mg/L and in the nitrous oxide-acetylene flame with a characteristic concentration of 2.5 mg/L at the 349.9 nm resonance line. Further analytical lines and their characteristic concentrations are compiled in Table 9-30. SCHWAB and HEMBREE [5207] found that the addition of 20 g/L La and 1 mol/L HCl doubled the sensitivity for Ru and also eliminated all interferences. ROWSTON and OTTAWAY [4973] used a mixture of 5 g/L Cu and 5 g/L Cd as the sulfates to eliminate the numerous interferences. EL-DEFRAWY et al. [1702] found that KCN in the presence of hydrochloric and sulfuric acids eliminated practically all interferences; they attributed this to its strongly complexing nature.

The sensitivity of elements in the platinum group in GF AAS depends largely on their melting points. Owing to its relatively high melting point of 2310 °C, Ru can only be determined with a characteristic concentration of 31 pg in a longitudinally-heated atomizer and 45 pg in a transversely-heated atomizer. Without a modifier a pyrolysis temperature of 1400 °C can be applied and the optimum atomization temperature is 2400–2500 °C. HAMID [2345] investigated interferences from 22 elements; he was able to eliminate them by coating the tube with tantalum carbide.

Table 9-30. Ru analytical lines.

Wavelength (nm)	Energy level (K)	Slit width (nm)	Characteristic concentration* (mg/L)	Spectral interferences
349.90	0–28572	0.2	0.65	Co, Ba, Ni+ [5431]
372.80	0–26816	0.2	0.85	
379.94	0–26313	0.2	1.6	
392.59	0–25465	0.2	7.5	

* Air-acetylene flame, oxidizing (lean, blue) under addition of 1 g/L La and 1 mol/L HCl
+ Interference with deuterium BC possible

9.41 Scandium (Sc)

Scandium is a light metal that occurs in its compounds exclusively in the trivalent oxidation state. Although it is more abundant than silver, gold, or mercury, scandium-rich minerals are very rare. Scandium has virtually no technical significance so that reports in the literature about this element are correspondingly low.

The stability of Sc solutions has been investigated for both fresh water samples [537] and sea-water samples [4890]. ROBERTSON [4890] recommended acidification to a pH value < 1.5 to prevent losses for storage in PE bottles.

Scandium can be determined in the nitrous oxide-acetylene flame with a characteristic concentration of 0.3 mg/L at the 391.2 nm analytical line. AMOS and WILLIS [166] and also MANNING [3844] investigated the sensitivity at numerous other analytical lines; a number of which are compiled in Table 9-31. Scandium is noticeably ionized in the nitrous oxide-acetylene flame so that 1–2 g/L K as the chloride should be added to all measurement solutions to eliminate this effect.

SEN GUPTA [5236, 5238, 5241, 5243] determined Sc and a number of other rare-earth elements by GF AAS in geological materials. At an atomization temperature of 2600 °C in a PG-coated graphite tube he attained a characteristic mass of 130 pg [5238]. Using a graphite tube lined with tantalum foil as proposed by L'VOV and PELIEVA [3703], it was possible to improve the sensitivity by a factor of 10 [5238]. As well as the use of tantalum [6400, 6463] and tungsten [6399] foils, the use of tantalum platforms [3729] and lanthanum as a modifier [2454] have been proposed to improve the sensitivity. BET-TINELLI et al. [612] showed that pyrolysis temperatures of 1700 °C are possible without analyte losses and that no significant interferences are to be expected. The characteristic mass for wall atomization in a PG-coated tube was around 30 pg.

Table 9-31. Sc analytical lines.

Wavelength (nm)	Energy level (K)	Slit width (nm)	Characteristic concentration* (mg/L)	Spectral interferences
391.18	168–25725	0.2	0.3	Eu+ [5431]
390.75	0–25585	0.2	0.4	Eu+ [5431]
402.37	168–25014	0.2	0.4	
402.04	0–24866	0.2	0.6	
327.00	0–30573	0.2	1.0	
327.36	168–30707	0.2	1.5	Ag, Cu, Rh+ [5431]
408.24	168–24657	0.2	2.1	

* Nitrous oxide-acetylene flame, reducing (fuel-rich, red)
+ Interference with deuterium BC possible

9.42 Selenium (Se)

Selenium is one of the less common elements; it makes up about $10^{-5}\%$ of the Earth's crust and is thus close to iodine and silver. In the periodic system Se is immediately below sulfur and like sulfur it occurs in the -2, 0, $+2$, $+4$, and $+6$ oxidation states. About half of the production of Se is used in electrical technology where it finds many applications due to its semiconducting properties. Further important areas of use are in the manufacture of pigments and in the glass and ceramic industries. FISHBEIN [1915] has published a short overview.

Although the toxicity of Se and its compounds has been known for a long time, its physiological significance as an essential element has only been known for a few decades. Like arsenic, the inorganic forms of Se have the highest toxicity; Se(IV) is more toxic than Se(VI). Among the organic species the trimethylselenonium ion, $(CH_3)_3Se^+$, has the highest toxicity. On the other hand, Se is an essential element for humans and numerous other animal species. Selenium is a constituent of glutathione-peroxidase, which is necessary for the metabolism and excretion of H_2O_2, and the lipid-peroxidases in the cells. The biochemistry of Se is closely related to that of sulfur; selenomethionine, CH_3-Se-CH_2-CH_2-$CH(NH_2)$-COOH, can replace methionine for the promotion of cell growth. Nevertheless, in large doses selenomethionine is cytotoxic. With Se the boundary between deficiency and toxic effects is very narrow, so that the addition of Se to feeds or to fertilizers, or also as a dietary supplement, must be done with great caution.

Since the ingestion of Se is largely from plants, which take it up from the soil, Se deficiency is often regionally limited. An example is the Keshan disease, a debility of the coronary muscles, which occurs in certain regions of China and can be attributed to an Se deficiency. The most common species present in humans and mammals is selenocysteine. This and also selenomethionine and other organoselenium species are formed in plants and can react further in animal organisms after the intake of nutrition.

The biomethylation of Se has acquired increasing attention in recent years. Although the chemistry of Se is similar to that of arsenic, the redox conditions and the pH value influence the biomethylation of Se far less than that of arsenic. The main products of biomethylation are the volatile dimethylselenium, $(CH_3)_2Se$, dimethyldiselenium, $(CH_3)_2Se_2$, and the water-soluble trimethylselenonium ion. Biomethylation can be regarded as a form of detoxification since the inorganic species are far more toxic than the organic species. Synthetic organoselenium compounds are used for their therapeutic, anti-inflammatory, and antibiotic properties, and also in chemotherapy [1218].

The stability of Se solutions has been investigated in a number of studies, some of which are cited in Table 9-32. Losses due to wall adsorption and also to volatilization must be taken into account. We must also make a distinction whether total Se or species are to be determined and which AAS technique is to be used. The American EPA recommends acidification with nitric acid to a pH value < 2 for the conservation of Se in natural waters, although under these conditions when using PE containers losses of Se occur and nitric acid interferes in the determination by HG AAS [1386, 5376, 5377]. Acidification with sulfuric acid to a pH value of 1.5 and storage in PE bottles is suitable for the stabilization of total Se in water samples [1225]. Nevertheless, sulfuric acid inter-

Table 9-32. Stability of aqueous test sample and calibration solutions with respect to their Se concentrations.

Analyte	Matrix	pH range	Materials*	Time period	Recommended storage	References
Se	distilled water, simulated sea-water	1–8.5	BG, PE, PTFE	28 days	no losses due to sorption	[3932]
Se	distilled water	0–7	BG, PE	15 days	BG; 4% loss of Se(IV) after 15 days in glass at a pH value of 7	[5296]
Se	sea-water	natural		40 hours	significant losses after 40 hours at a pH value of 8.1	[2143]
Se(IV)	acidified solutions	< 1.5	BG, PE, FEP	50 days	BG, PE; losses of Se(IV) could be reduced by acidification with 15% HCl–5% H_2SO_4	[3992]
Se(IV)/(VI)	distilled water, NaCl solution	2–6	PE, PTFE	12 months	PE; stable at –20 °C without acidification; at ambient temperature the losses at a pH value of 6 are less than at a pH value of 2	[1310]
Se(IV)/(VI)	distilled water, harbor water	1–7.2	BG, PE	4 months	PE; acidification to a pH value of 1.5 is optimal for stabilizing both species for 125 days; alternatively storage is possible at 4 °C without acidification	[1225]
Se(IV)/(VI)	sea-water	2	BG, PE	4.5 months	no losses after acidification to a pH value of 2 with HCl	[4036]
Se(IV)/(VI), SeMet	NaCl solution		PTFE	50 days	samples should not be acidified; ion strength should be increased with NaCl; cooling to 4 °C is helpful	[2497]

* BG: borosilicate glass; PE: polyethylene; PTFE: polytetrafluoroethylene; FEP: fluorinated engineering polymer

feres in the determination by GF AAS. Conservation with hydrochloric acid is far more compatible with a determination by AAS. REAMER *et al.* [4831] investigated sorption losses on PP, PTFE, glass, and silanated glass; PP and untreated glass were both unsuitable. As well as the pH value and the container material, the surface/volume ratio [5296] and the salt content play a part in the stability of solutions; no losses were observed in sea-water samples, even without conservation [5206].

For a subsequent speciation analysis, strong acidification with oxidizing acids should be avoided. The use of hydrochloric acid also cannot be recommended without reservations since it can reduce Se(VI) to Se(IV), at least when oxygen is present [2497]. CUTTER [1384] observed that in 4 mol/L HCl 60 % of the Se(VI) had converted to Se(IV) after seven days. In general the stability of the various species decreases in the sequence Se(VI) > Se(IV) >> Se(0) > Se(–II) > methyl selenides [4886]. CÁMARA *et al.* [1043] found that with the addition of 5 g/L chloride Se(IV) and Se(VI) were stable for four months in PE bottles at both a pH value of 6 and a pH value of 2, while in PTFE bottles at a pH value of 6 a change in the Se(IV)/Se(VI) distribution in favor of the Se(VI) species occurred that was not observed at a pH value of 2. By cooling to –20 °C calibration solutions could be stored for 12 months without changes. The freezing of samples, especially for the analysis of organoselenium compounds, is the method of choice [4886].

The most sensitive analytical line of Se is at 196.0 nm and thus at the start of the vacuum UV range where the flame gases absorb a substantial part of the radiation (see Figure 4-1 in Section 4.1.1). A characteristic concentration of 0.4 mg/L can be attained in the air-acetylene flame at this line. This value can be improved to 0.25 mg/L in the argon-hydrogen diffusion flame, but at the same time the risk of spectral and non-spectral interferences increases markedly. Several other lines are available for the determination of higher Se concentrations; these are compiled with the attainable characteristic concentrations in Table 9-33.

Table 9-33. Se analytical lines.

Wavelength (nm)	Energy level (K)	Slit width (nm)	Characteristic concentration* (mg/L)	Spectral interferences
196.03	0–50997	2.0	0.4	PO^{++} [4782] NO^{++} [631] Co^+ [1895] Fe^+ [1895, 3857] Bi^+ [5973]
203.99	1989–50997	0.7	1.6	PO^{++} [4782] NO^{++} [631] Cr, Ni^+ [1895]
206.28	2534–50997	0.7	6.7	
207.48	0–48182	0.7	22	

* Air-acetylene flame, oxidizing (lean, blue)
$^+$ Interference with deuterium BC possible
$^{++}$ Interference with Zeeman-effect BC possible

Since Se must normally be determined in the trace range, F AAS is not suitable for the majority of applications. Suitable alternatives are GF AAS and HG AAS. A comparison of these two techniques with respect to sample consumption, and relative and absolute sensitivity is given in Table 9-34 so that the most suitable technique can be selected for the given application. GF AAS has by far the lowest sample requirement; nevertheless, GF AAS and HG AAS exhibit the same absolute sensitivity (m_0), which means that equally small volumes of measurement solution can be used in the latter technique when necessary. On the other hand, FI-HG AAS has the advantage that significantly larger volumes of measurement solution can be used so that the relative sensitivity (β_0) can be improved. The characteristic concentration attainable with FI-HG AAS is around a power of ten better than with GF AAS. With batch HG AAS procedures it is possible to achieve similarly good characteristic concentrations, but the volume of measurement solution required is a hundredfold greater, so that such techniques are not particularly attractive.

Table 9-34. Comparison between the characteristic mass (m_0) and the characteristic concentration (β_0) of Se for HG AAS with a batch system and an FI system, and for GF AAS (from [6256]).

System	Volume (mL)	m_0 (ng)	β_0 (µg/L)
HG AAS, batch	10	1.1	0.11
HG AAS, FI	0.2	0.025	0.12
GF AAS	0.02	0.025	1.2

The determination of Se by GF AAS is discussed in detail in Section 8.2.3.13. We should just like to mention that the nickel modifier originally proposed by EDIGER [1677], and which is widely used, is less suitable since not all oxidation states and compounds of Se are stabilized to the same degree. This modifier can thus be the cause of systematic errors, which have been repeatedly observed for the determination of Se.

On the other hand, palladium, either as the Pd-Mg modifier or, better, as reduced Pd^0 [6105], has proven to be very successful and reliably prevents losses of Se from all compounds up to a pyrolysis temperature of at least 1000 °C [1560, 2054] and interferences caused by sulfate [149, 749, 4348, 6261]. Likewise in a direct comparison of various platinum-group metals, palladium was found to be the most effective modifier [6106]. The optimum atomization temperature in a longitudinally-heated atomizer is about 2100 °C and the characteristic mass is around 25 pg. In a transversely-heated atomizer the atomization temperature is 1900 °C and the characteristic mass is about 45 pg. The proper application of STPF conditions is of decisive significance for the determination of Se.

The determination of Se by HG AAS is discussed in detail in Section 8.3. A very important aspect for the application of this technique is that numerous organoselenium compounds, such as selenocysteine, selenomethionine, and the trimethylselenonium ion, are very difficult to digest [2881, 4330, 5853, 6040, 6237]. Correspondingly strong oxidizing conditions must thus be selected, such as digestion in perchloric acid, sulfuric acid and nitric acid [4330, 4331, 5853, 6040, 6237], or sulfuric acid under the addition of vanadium pentoxide [1508], or hydrobromic acid with an excess of bromine [1619], or peroxodisulfate under strongly acidic conditions [1311, 4477]. With this type of digestion the greatest care must be taken to ensure that no losses of Se occur [464, 1619, 2370, 4832, 4885]. An important point also is that only Se(IV) can be determined by HG AAS, while Se(VI), which is frequently formed during oxidative digestions, does not generate a measurable signal. A reduction step must thus be interposed prior to every determination of Se by HG AAS. This is normally done by heating with 5–6 mol/L HCl for 15–30 minutes under reflux [791, 993, 1000]; the acid concentration, temperature, and length of reaction are critical [4632]. Under certain conditions free chlorine can be formed which, if it is not removed by purging, will quickly oxidize the Se(IV) back to Se(VI) [3293].

Since even the sensitivity of HG AAS is on occasions inadequate, a number of on-line preconcentration techniques have been described for Se. We can mention preconcentration on a packed column prior to the determination by GF AAS [5523] or HG AAS [4478, 6547], and especially the preconcentration of hydrides in the graphite tube by sequestration [2357, 3977, 4347, 5331, 5632, 5640, 5910, 6006, 6332, 6539, 6542]. The selective determination of individual Se species is discussed in Sections 10.1.2, 10.2.6, 10.3.1.6, and 10.3.3.5.

9.43 Silicon (Si)

After oxygen, silicon is the most widely distributed element on Earth. It occurs almost exclusively in inorganic minerals such as clays, sand, and rocks, and only in traces in vegetable and animal organisms. Silicon is used as a constituent of alloys, mainly in the steel industry in the form of ferrosilicon, in the semiconductor industry for the manufacture of integrated circuits, and in solar cells. Silicon is also the base element for a large group of synthetic polymers, the silicones, and numerous other polymers.

The omnipresence of Si represents a severe risk of contamination during its determination which is only alleviated somewhat by the fact that many Si compounds are very difficult to dissolve. Nevertheless extreme care is required, especially for the determination of Si traces. A further problem is that acidic Si solutions are unstable and tend to precipitate. DE VINE and SUHR [1445] established that in aqueous solution in the range pH 1–8 and in concentrations up to about 120 mg/L the monomeric silicate is the most stable form of Si, while at higher concentrations polymerization takes place. These colloidal forms can be determined by F AAS, but there is a risk that with time they are sorbed onto the walls of the container, leading to low results. This problem does not occur at higher pH values. The use of freshly prepared solutions and frequent checks of the sensitivity are indispensable if systematic errors are to be avoided; such errors have nothing to do with AAS, however. PREWETT and PROMPHUTHA [4712] proposed the use of water-soluble $(NH_4)_2SiF_6$ as the starting compound for the preparation of stable Si calibration solutions. PARALUSZ [4534] investigated various organosilicon compounds for their suitability as calibration solutions in organic solvents. MEDLIN et al. [4039]

proposed a lithium metaborate fusion for the analysis of silicates which maintains Si and other cations in solution.

Silicon can be determined relatively free of problems and with good sensitivity in the nitrous oxide-acetylene flame. The characteristic concentration at the 251.6 nm analytical line is around 2 mg/L; further analytical lines of lower sensitivity are compiled in Table 9-35.

Since Si exhibits a large number of close-lying analytical lines (refer to Figures 3-9 and 3-10 in Section 3.3.1), a spectral slitwidth of 0.2 nm is necessary to attain the quoted characteristic concentrations and good linearity of the calibration curves.

The determination of Si in the nitrous oxide-acetylene flame is practically free of interferences. MUSIL and NEHASILOVÁ [4256] investigated the influence of numerous substances on the determination of Si and found that sulfuric acid causes signal depression while the alkali elements, aluminium, and a number of further elements in large excess cause signal enhancement. CHAPMAN and DALE [1194] improved the sensitivity for Si by heating with copper hydroxyfluoride to obtain the volatile SiF_4 and then conducting this directly to the flame.

The determination of Si by GF AAS is discussed in detail in Section 8.2.3.10. The greatest of care must be taken to avoid contamination when ultratraces are to be determined; for example, silicone hoses should not be used for the gas supply lines. Under STPF conditions and using the Pd-Mg modifier, pyrolysis temperatures of 1200–1400 °C can be applied and losses due to prior volatilization of SiO and SiC effectively suppressed [6580]. The characteristic mass in a longitudinally-heated atomizer is about 30 pg at an atomization temperature of 2500–2600 °C, while in a transversely-heated atomizer is around 120 pg at an atomization temperature of 2350 °C.

Table 9-35. Si analytical lines.

Wavelength	Energy level	Slit width	Characteristic concentration*	Spectral interferences
(nm)	(K)	(nm)	(mg/L)	
251.61	223–39955	0.2	2	Fe, Co, V+ [5431, 5973]
251.92	77–39760	0.2	3	
250.69	77–39955	0.7	6	Fe+ [5973] V++ [1852]
252.85	223–39760	0.2	6	
251.43	0–39760	0.2	6	Fe+ [5431, 5973]
252.41	77–39683	0.2	7	Fe, Co+ [5431, 5973]
221.67	223–45322	0.2	8	
221.09	77–45294	0.2	14	
220.80	0–45276	0.2	25	

* Nitrous oxide-acetylene flame, reducing (fuel-rich, red)
+ Interference with deuterium BC possible
++ Direct line overlap

9.44 Silver (Ag)

Silver is one of the rarer elements; it is estimated that it makes up about $10^{-6}\%$ of the Earth's crust. In nature it is mainly found as silver sulfide collectively with other sulfides. Silver is the noble metal that is most heavily produced and used. It is mostly alloyed with copper to increase its hardness. Its use as a coinage metal and also in jewelry has played a major role since ancient times. Other areas of application include photography, silver plating, and for hard solders.

The stability of Ag solutions is very critical and losses have frequently been described through sorption, the formation of colloids, and precipitation [1189, 1632, 1637, 2106, 2506, 2944, 3932, 4091, 4718, 4730, 4890, 5072, 5606, 5660, 5852, 6285–6287]. The losses due to sorption from a 1 µg/L Ag solution in distilled water were 80% after 11 days [2944]; they were quantitative in PE bottles [3932] and must be prevented by acidification with nitric acid. Solutions that are acidified to 0.3 mol/L nitric acid can be stored for at least 11 days in glass [2944] and 36 days in borosilicate glass [5606]. Massive losses take place at a pH value of 4.5 and reach 50% after 24 days [5606]. Losses cannot be prevented even by acidification if the solutions are stored in PP containers [5606]. Next to the pH value and the container material, light and storage temperature play a decisive role for the stability of samples [5606]. Great care must be taken when preparing Ag calibration solutions that no chloride ions are present, since otherwise precipitation can easily take place. The use of multielement calibration solutions that contain Ag is thus mostly problematical [4730]. On the other hand, in highly concentrated chloride solutions Ag is stabilized through the formation of the $AgCl_2^-$complex [3932, 6286, 6287]. In semiconcentrated hydrochloric acid about 25 mg/L Ag thus remain in solution [4825], and sea-water samples exhibit higher stability than freshwater samples or aqueous calibration solutions [4890, 5206]. Since the losses of Ag from chloride-containing solutions through sorption on PE and PTFE are lower at lower pH values, the pH value should be < 2 [3932]. Calibration solutions can also be stabilized by complexing the Ag with diethylenetriamine [2208] or thiourea [4091]. In PFA bottles acidified calibration solutions containing 5 µg/L and 50 µg/L Ag could be stored for over a year without significant losses [2106].

Silver can be determined easily by F AAS; the characteristic concentration at the 328.1 nm line in the air-acetylene flame is 0.02 mg/L. According to existing experience the determination of Ag is free of interferences. WILSON [6352] and BELCHER et al. [501] investigated a large number of cations and anions for their influence on the determination of Ag and found no noticeable interferences even in flames of lower temperature. Information on possible spectral interferences and a second resonance line are compiled in Table 9-36.

Since Ag must frequently be determined in the lowest traces, the sensitivity of F AAS is often inadequate. Silver can also be easily determined by GF AAS with good sensitivity. The characteristic mass under STPF conditions in a longitudinally-heated atomizer is around 1.5 pg at an atomization temperature of 1800 °C, while in a transversely-heated atomizer it is 4.5 pg at an atomization temperature of 1500 °C. The Pd-Mg modifier allows a maximum pyrolysis temperature of 1000 °C to be applied. Since the determination of Ag by GF AAS is uncomplicated and free of interferences, it is also used to check the condition of the instrument [3866].

Table 9-36. Ag resonance lines.

Wavelength	Energy level	Slit width	Characteristic concentration*	Spectral interferences
(nm)	(K)	(nm)	(mg/L)	
328.07	0–30473	0.7	0.02	Rh⁺ [6279] PO⁺⁺ [3866, 4425]
338.29	0–29552	0.7	0.04	Fe⁺ [4385]

* Air-acetylene flame, oxidizing (lean, blue)
⁺ Interference with deuterium BC possible
⁺⁺ Interference with Zeeman-effect BC possible

9.45 Sodium (Na)

Sodium is the sixth most abundant element. It occurs mainly in rocks in the form of silicates and in sea-water as sodium chloride. Sea-water contains on average 27 g/L sodium chloride, i.e. 77 % of all the salts present. While Na is a micronutrient for the majority of plants, for animal organisms it is a macroelement that is also involved in numerous reactions and functions in the human organism. Sodium occurs in nature almost exclusively in the monovalent form.

Sodium is normally determined at the 589.0/589.6 nm doublet; in an oxidizing air-acetylene flame the characteristic concentration is 0.01 mg/L. Due to the slightly varying sensitivities of the two lines the calibration curve is not quite linear, but the S/N ratio is very favorable. The more sensitive doublet line at 589.0 nm ($\beta_0 = 0.006$ mg/L) can be separated by using a slit width of 0.2–0.4 nm when better linearity of the calibration curve is required. Higher Na concentrations can be determined with advantage at the 330.2/330.3 nm doublet where the characteristic concentration is around 2 mg/L.

MANNING [3847] found that it is possible to determine Na using a zinc lamp and thus reduce the sensitivity even further. Zinc exhibits a doublet at 330.3 nm in which the second line at 330.294 nm is only separated by 5 pm from the second line of the Na doublet at 330.299 nm. Owing to this narrow separation there is a slight overlap of the line wings so that Na atoms can partly absorb this radiation. The characteristic concentration at this line is about 140 mg/L. Since zinc does not absorb at this line there is no risk of a spectral interference.

Sodium is totally atomized [2333] and markedly ionized in the air-acetylene flame so that another easily ionized element must be added to the test sample and calibration solutions to suppress this effect; this is done most effectively by adding about 0.5 g/L Cs. Various authors nevertheless preferred a cooler flame, such as the air-propane or air-hydrogen flame [1490, 3841], to avoid ionization. The latter flame also exhibits a much better S/N ratio, especially at the 330.2/330.3 nm doublet.

For the determination of Na by GF AAS a pyrolysis temperature of only about 900 °C can be applied without the addition of a modifier. The optimum atomization temperature in longitudinally-heated and transversely-heated atomizers is 1500 °C. The characteristic mass with and without Zeeman-effect BC is around 1.0 pg. Markedly better values should be taken as an indication of contamination. Because of the high risk of contamination, chemical modification is also usually avoided. When the analyte concentration is sufficiently large, performing the determination at the 330.2/330.3 nm doublet is to be preferred to diluting the test sample solution. For the analysis of solid samples the use of the less sensitive doublet is often more practicable.

In the laboratory Na is one of the ubiquitous elements and great care is therefore required, even for F AAS. Sodium can be easily dissolved out of normal laboratory glassware, so that 'distilled' water, test sample solutions, and calibration solutions should not be permitted to contact glass. Water should only be distilled in quartz apparatus or obtained from mixed-bed ion exchangers. Especial care is required with normal laboratory cleaning agents since they frequently contain large amounts of Na. Only plastic containers are suitable for storing measurement solutions. Care is also required with numerous chemicals which either contain Na or, especially with acids, take up Na from glass containers (blank values).

9.46 Strontium (Sr)

Being a non-noble metal, strontium only occurs in nature in the form of its compounds in the divalent oxidation state; it is at position 18 in the abundance list of the elements. In contrast to barium and its compounds, Sr is not toxic; the human body contains about 100–200 mg Sr which is taken up with the nutrition. Technically Sr has found little application except in the electronic tube industry as a getter, to harden the lead plates in accumulators, for the production of specially hardened steels, to reduce the grain of Al-Si castings, and in pyrotechnics.

The stability of Sr solutions is not particularly critical [5456, 5660]. Adsorption losses become increasingly evident at pH values > 5 [5456] and for neutral solutions are higher on PE surfaces than on glass [1684]. These losses can easily be prevented by acidification with nitric acid. In PFA bottles acidified calibration solutions even in the concentration range 5–50 µg/L Sr are stable for about a year [2106].

In its behavior in F AAS Sr lies between calcium and barium; it can be determined in both the air-acetylene and the nitrous oxide-acetylene flames. As to be expected, numerous interferences can be expected in the cooler flame. These interferences have been investigated by a number of authors. Strontium is markedly ionized in the hotter flame. The most sensitive analytical line is at 460.7 nm; the characteristic concentration in the air-acetylene flame is 0.15 mg/L while in the nitrous oxide-acetylene flame a value of 0.1 mg/L can be attained when the ionization is suppressed.

LOKEN et al. [3574] and PARKER [4536] found that in the air-acetylene flame Sr is ionized to about 10 %. INTONTI and STACCHINI [2790] established that this effect could only be reliably suppressed by rubidium; an addition of 50–100 mg/L Rb is adequate. All the other alkali metals exhibited irregular effects with increasing concentration. AMOS and WILLIS [166] found that Sr is ionized to 80 % in the nitrous oxide-acetylene flame; this can be eliminated by an addition of 1–2 g/L K as the chloride. Detailed studies on the interference to the determination of Sr by aluminium have been carried out by

LUECKE [3649] and WELZ and LUECKE [6271]. Strontium behaves in a very similar manner to calcium so that we refer our readers to the information in Section 9.9.

Strontium can be determined with very high sensitivity by GF AAS provided that PG-coated tubes are used; the condition of the surfaces plays an important role in the atomization [3729, 3731, 4710, 5418]. Interferences have been observed from calcium [2490, 3997, 4708] and perchlorate [2490, 3049]. A characteristic mass of 1.4 pg can be attained in a longitudinally-heated atomizer at an atomization temperature of 2600 °C, while in a transversely-heated atomizer it is around 4 pg at an atomization temperature of 2400 °C. Without modifier a pyrolysis temperature of 1300 °C can be applied.

9.47 Sulfur (S)

The estimated proportion of sulfur in the Earth's crust is 0.048 %, placing it at position 15 in the abundance list of the elements. It occurs in elemental form and as sulfides and sulfates in many places on the Earth. Sulfur occurs in the –2 through +6 oxidation states; compounds of the divalent and hexavalent forms are the most common and stable. The greater part of the world production of sulfur is won from the desulfurization of natural gas and crude oil. Around 85–90 % of the production is used for the manufacture of sulfuric acid and its byproducts, mainly fertilizers. Sulfur is also used to vulcanize rubber, for the production of plastics, matches, gunpowder and fireworks, and in fungicides, insecticides, dyes, and disinfectants.

Sulfur cannot be determined directly by AAS in normal commercial instruments since its main resonance line is at 180.7 nm in the vacuum UV. KIRKBRIGHT [3128] determined S at this line using a modified atomic absorption spectrometer with a purged monochromator, a nitrous oxide-acetylene flame shielded by nitrogen, and an EDL. He reported a characteristic concentration of 9 mg/L and a limit of detection of around 5 mg/L; he did not observe any influences from the bonding form and used this technique to determine S directly in crude oil.

Using a similar apparatus, ADAMS and KIRKBRIGHT [31] also determined S directly by GF AAS. At the 180.7 nm line they found a characteristic mass of 420 pg; the other two lines at 182.0 nm and 182.6 nm, with characteristic masses of 680 pg and 1500 pg, respectively, were somewhat less sensitive. The authors obtained identical calibration curves for sulfate, thiocyanate, and thiourea.

A number of indirect methods have also been described which can be widely applied. ROE et al. [4926] converted organically-bound sulfur into sulfate using Schöninger or Benedikt digestion techniques, precipitated the sulfate as barium sulfate, and then determined the barium concentration in the precipitate after dissolution in EDTA. Other authors [1046, 1629, 5999] determined sulfate by adding a known quantity of barium chloride and then determining the excess barium remaining in solution by F AAS. ROSE and BOLTZ [4961] determined sulfur dioxide after oxidation to sulfate by precipitating as lead sulfate, and after centrifugation, determining the excess lead in the supernatant liquid by AAS. CHRISTIAN and FELDMAN [1277] exploited the occurrence of specific interferences for the determination of sulfate. They found that there was a direct relationship between the absorption of calcium and the concentration of sulfate in the solution being investigated. Additionally, thiosulfate [410, 1169, 2102], sulfite [2970, 3899], and sulfide [2206, 3054, 3219, 3264, 4827] have been determined by indirect methods. Recently a number of indirect methods for the determination of sulfate [1187, 1227, 2051, 2277, 2364, 6588], sulfide [1227, 4625], and thiosulfate [1768] have been modified for FI and automated.

9.48 Tantalum (Ta)

Tantalum is one of the rarer elements; it is found in association with niobium in various deposits scattered around the whole world. Tantalum is used in the manufacture of chemical apparatus and in electronics for the production of capacitors. It finds further application for the manufacture of spinning nozzles, in laboratory equipment, and in surgery as bone pins and joint implants.

The stability of Ta solutions is relatively critical. Significant losses were observed after 10 days from calibration solutions acidified to a pH value of 1 with nitric acid and stored in glass containers. The losses were higher the lower the concentration of Ta, and reached over 60 % after 30 days from a 10 µg/L Ta solution [4730].

Tantalum can be determined in the nitrous oxide-acetylene flame at the 271.4 nm resonance line with a characteristic concentration of 16 mg/L. The absorbance of Ta is markedly enhanced in the presence of hydrofluoric acid and iron, so that test sample and calibration solutions should contain at least 0.1 % HF to eliminate this interference and to attain the best sensitivity. BOND [708] preferred the addition of 0.1 mol/L ammonium fluoride to enhance the sensitivity.

VAN LUIPEN [3652] found that sulfuric acid and phosphoric acid as well as titanium and vanadium interfered in the determination of Ta in the nitrous oxide-acetylene flame. The most probable explanation for these interferences is the varying volatilities of compounds such as $Ta_2O_5 \cdot Ti_2O_3$, $Ta_2O_5 \cdot V_2O_3$, or $Ta_2O_2(SO_4)_3 \cdot H_2O$, which are presumably formed in the nitrous oxide flame. WALLACE et al. [6134] found improved linearity, sensitivity, precision, and limit of detection for Ta when they added aluminium in excess; this presumably reduces the tendency of forming refractory oxides and promotes atomization [5108]. A reduction of the lateral diffusion could also be an explanation for this phenomenon, which has also been observed by other authors [3827].

Owing to its melting point of 3020 °C and the tendency of forming stable carbides (melting point TaC = 3985 °C), Ta cannot be determined by GF AAS.

9.49 Technetium (Tc)

Technetium is a radioactive, artificial heavy metal. The isotope ^{99}Tc is produced in kilogram quantities during the fission of uranium. Next to its use as a β-emitter in radiopharmaceuticals and for the calibration of radiation measurement instruments, its particular corrosion-inhibiting and super-conducting properties are of interest.

HARELAND et al. [2376] were the first to report the determination of Tc and they investigated a multitude of analytical lines. In a fuel-rich air-acetylene flame at the

261.4/261.6 nm doublet they attained a characteristic concentration of 3 mg/L. Interferences from alkaline-earth metals could be eliminated by adding aluminium (about 100 mg/L Al per 50 mg/L interfering ion). IHSANULLAH [2757] systematically investigated Tc losses during sample preparation and storage.

BAUDIN *et al.* [425] determined Tc by GF AAS and found the best characteristic mass of 10 ng at the 429.7 nm line; the atomization temperature was 2200 °C. KAYE and BALLOU [3060] obtained markedly better sensitivity and an LOD of 0.09 ng at the non-resolved doublet at 261.4/261.6 nm. They used a demountable HCL as the radiation source and an atomization temperature of 3200 °C, which they reported as the 'optimum compromise between sensitivity and tube lifetime'. This information must nevertheless be seriously doubted since the lifetime of graphite tubes is significantly influenced at temperatures > 2700 °C (see Section 4.2.1), and at 3200 °C is likely to be only a few atomization cycles. A practical problem for the determination of Tc is that currently no radiation source is offered for this element. HAUG [2436] reported the use of lamps for other elements for the determination of Tc. Using a palladium HCL at the 371.9 nm line he attained a characteristic mass of 0.8 ng. A modifier was not required since a pyrolysis temperature of 1800°C could be applied without Tc losses.

9.50 Tellurium (Te)

Tellurium is one of the rarest elements; it makes up about 10^{-7} % of the Earth's crust and is thus about as abundant as gold, palladium and platinum. Like sulfur and selenium it occurs in its compounds in the −2, +2, +4 and +6 oxidation states; Te(IV) compounds are the most common and stable. The bulk of Te is used in the iron and non-ferrous metal industries to improve the workability and to increase the temperature resistance and hardness of steels and copper and lead alloys. Tellurium and its compounds are noticeably less toxic than those of selenium since as a largely insoluble element it cannot pass the walls of the stomach.

Tellurium can be determined in the air-acetylene flame practically free of interferences and with good sensitivity. The characteristic concentration at the 214.3 nm resonance line is 0.3 mg/L. The sensitivity is only slightly better in the air-hydrogen flame. NAKAHARA and MUSHA [4271] determined Te in an argon-hydrogen diffusion flame with a characteristic concentration of 0.13 mg/L. There are also very few interferences in this flame. A number of analytical lines of lower sensitivity are available for the determination of higher Te concentrations; these are compiled in Table 9-37.

In analogy to selenium, EDIGER [1677] proposed nickel as the modifier for the determination of Te by GF AAS. He reported a maximum pyrolysis temperature of 1200 °C. WEIBUST *et al.* [6191] found that Cd, Cu, Pd, Pt, and Zn were equally suitable as modifiers for inorganic Te. On the other hand, organotellurium compounds were only stabilized by silver, palladium, and platinum, so that Te behaves in a very similar manner to selenium. Using the Pd-Mg modifier Te can be stabilized to a pyrolysis temperature of 1200 °C. Under STPF conditions a characteristic mass of around 17 pg can be attained in a longitudinally-heated atomizer at an atomization temperature of 2000–2200 °C. In a transversely-heated atomizer it is around 50 pg at an atomization temperature of 1800 °C.

Table 9-37. Te analytical lines.

Wavelength	Energy level	Slit width	Characteristic concentration*	Spectral interferences
(nm)	(K)	(nm)	(mg/L)	
214.28	0–46653	0.2	0.3	Sn, Zn$^+$ [5431, 5973] PO$^+$ [1893, 5038]
225.90	0–44253	0.2	3.7	PO$^+$ [5038]
238.58	4751–46653	0.2	13	

* Air-acetylene flame, oxidizing (lean, blue)
$^+$ Interference with deuterium BC possible

Approximately the same characteristic mass can be attained for Te(IV) using the FI-HG AAS technique. Due to the much higher volumes of measurement solution that can be used in this technique, the characteristic concentration of 0.07 µg/L attainable with FI-HG AAS is a power of ten better than with GF AAS. Like selenium, only the tetravalent oxidation state forms a hydride, so that all test samples must be digested and reduced prior to the determination. Heating briefly in 5–6 mol/L HCl is sufficient for the prereduction.

9.51 Thallium (Tl)

Making up about 10^{-5} % of the Earth's crust, thallium is one of the rarer elements. It can be found in very small amounts in many places on the Earth and occurs mostly in company with lead, iron, copper, or zinc. Thallium is a trace element that is regularly present in animals and plants. Larger quantities are contained in airborne dust and stack gases. The Tl emission from cement works has been particularly well investigated since Tl poisoning of plants and animals was observed at Lengerich in Germany in 1979. Thallium has only a limited technical importance, such as the production of photocells and IR detectors, and for the production of special glasses with high refractive index.

The stability of Tl solutions is relatively uncritical. Adsorption losses from hydrochloric acid solutions commence at pH values > 4 on glass surfaces and are higher for 10 mg/L solutions than for 1 mg/L solutions, where losses first start at pH values > 6.5 [5455]. For even more dilute solutions (10–100 µg/L), KIM and HILL [3103] found stability to glass and PE up to pH values of 7.

Thallium can be determined easily and without interferences in the air-acetylene flame. The characteristic concentration at the 276.8 nm resonance line is 0.2 mg/L. Further analytical lines are available for the determination of higher Tl concentrations; these are compiled in Table 9-38.

The determination of Tl by GF AAS suffers from a persistent chloride interference [3492] which even the Pd-Mg modifier cannot completely eliminate [6241, 6455]. As shown in Figure 9-5, this interference exhibits two anomalies. In the first place, largely

Table 9-38. Tl analytical lines.

Wavelength (nm)	Energy level (K)	Slit width (nm)	Characteristic concentration* (mg/L)	Spectral interferences
276.79	0–36118	0.7	0.2	CS++ [3359] Pd+ [5431, 5973]
377.57	0–26478	0.7	0.6	
237.97	0–42011	0.7	1.3	
258.01	0–38746	0.7	4.8	

* Air-acetylene flame, oxidizing (lean, blue)
+ Interference with deuterium BC possible
++ Interference with Zeeman-effect BC possible

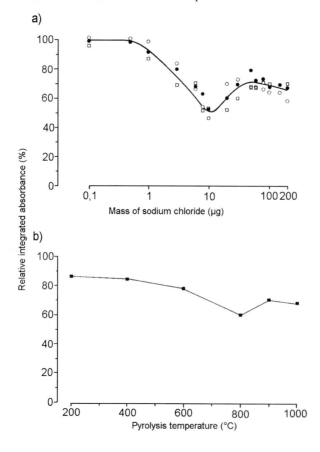

Figure 9.5. Influence of sodium chloride on the determination of thallium in the presence of palladium modifiers. **a** – Influence of increasing masses of sodium chloride in the presence of ● Pd(NO₃)₂; ○ Pd(NO₃)₂–Mg(NO₃)₂; □ Pd(NO₃)₂–ascorbic acid; pyrolysis temperature 800 °C; atomization temperature 1650 °C. **b** – Influence of the pyrolysis temperature in the presence of 10 µg NaCl and Pd-Mg modifier (from [6241]).

independent of the modifier used, it commences at very low chloride concentrations but then decreases in the presence of higher chloride concentrations (Figure 9-5a). This indicates a solute-volatilization interference, i.e. the volatilization of TlCl during pyrolysis, but which is largely prevented due to embedding in the presence of high salt concentrations. That this is not a gas-phase interference can be deduced from the results of ZHANG et al. [6544] who worked without a modifier or a pyrolysis step and determined Tl free of interferences in the presence of covaporized chloride. The second anomaly can be elucidated from the pyrolysis curve (Figure 9-5b) which exhibits Tl losses in the presence of chloride even at relatively low pyrolysis temperatures; this has been verified by other working groups [3805]. At high pyrolysis temperatures these losses decrease: the stabilization in this case is most probably due to stronger bonding to palladium intercalated in the graphite.

It was finally possible to bring this chloride interference under control by the use of reduced Pd^0 as the modifier [6241]. This is achieved by dispensing the modifier onto the platform and heating to 1000 °C prior to dispensing the measurement solution, or using hydrogen as the purge gas during pyrolysis. Under these conditions it is possible to apply a pyrolysis temperature of 1000 °C, even in the presence of higher chloride contents. The optimum atomization temperature is 1600 °C, independent of the type of atomizer used. A characteristic mass of 13 pg can be attained in a longitudinally-heated atomizer, but which deteriorates to 19 pg when Zeeman-effect BC, which is frequently required for this element, is applied. In a transversely-heated atomizer with Zeeman-effect BC the characteristic mass is 50 pg.

EBDON et al. [1663] reported the determination of Tl by HG AAS. Using a CF system with a manifold cooled to 0 °C, they mixed the test sample solution with sodium tetrahydroborate and atomized the TlH in a QTA. These authors attained a characteristic concentration of 4 µg/L.

9.52 Tin (Sn)

Tin makes up about 0.0035 % of the Earth's crust and is close to cobalt, cerium, and yttrium in the abundance list of the elements. Tin occurs in its compounds in the divalent and tetravalent oxidation states. By far the most important tin ore is cassiterite, SnO_2, but it also occurs in sulfidic ores. The major use of Sn is in the production of tin plate for the canning industry and in soft solders. Large quantities are also used in bronzes, bearing metals, phosphor-bronze, and special alloys for fuses, printing plates, and bells. Organotin compounds find use in a wide range of applications. As a result of a number of environmental catastrophes organotin compounds have become as notorious as mercury; this problem was particularly severe in France. The first case occurred in 1954, caused by the medicine Stalinon, and led to the deaths of over 100 people and to severe poisoning of over 200 more. During the 1980s the use of tributyltin compounds as antifouling paints for boats drew to attention the ecological risks of organotin compounds. These biocidal substances directly entered the aquatic system and led to the total destruction of numerous oyster beds. As a result of this problem the speciation analysis of organotin compounds was discussed very intensely and further developed.

It was shown that tributyltin can have grave consequences when it enters the aquatic system. Even concentrations of 1 ng/L can cause characteristic diseases in female mussels through bioaccumulation [909]. After the appalling destruction of the oyster beds, France was the first country in 1982 to ban the use of antifouling paints containing tributyltin for coastal vessels. In general the level of toxicity of organotin compounds increases with the degree of alkylation [678] according to $R_4Sn = R_3SnX > R_2SnX_2 > RSnX_3 >> SnX_4$. The toxicity also depends on the alkyl group, R, and decreases in the sequence ethyl > methyl > butyl > octyl > phenyl.

For industrial applications the di- and trisubstituted Sn compounds are of the greatest interest; the applications are very diverse. They include catalyzers, stabilizers—especially for polymers such as PVC—, biocides and fungicides in agriculture, and in paints and wood preservatives [679]. Due to the manifold applications of organotin compounds there are a correspondingly large number of routes through which they can enter the environment. They reach the atmosphere via the application of biocides and the combustion of materials that were treated or stabilized with organotin compounds. They enter the soil via agricultural applications, through the treatment of wood, and through the storage of waste. They reach the aquatic system through antifouling paints for boats, the leaching of soils, and from PVC materials. The biocide tributyltin enters the aquatic system very rapidly, is quickly adsorbed on particles, accumulates in sediments, and can reach high concentrations in the microlayer at the surface [2331]. Values for the half-life of tributyltin range from four months to nearly two years, depending on the nature of the sediment [3792, 4187, 5229]. For humans, tin and the majority of its compounds are largely non-toxic with the exception of tin hydride. In fact Sn has been recognized as an essential element for a number of years; a deficiency can lead to a loss of appetite, loss of hair, and acne.

A number of publications dealing with the stability of Sn solutions under varying conditions and in different materials are cited in Table 9-39.

Tin can be determined in the air-hydrogen flame, in a fuel-rich air-acetylene flame, and in the nitrous oxide-acetylene flame with comparable sensitivity. The corresponding values for the three most important analytical lines are compiled in Table 9-40. The best characteristic concentration of 1.1 mg/L can be attained in the air-hydrogen flame at the 224.6 nm resonance line.

On the basis of Sn, CAPACHO-DELGADO and MANNING [1056] were able to clearly demonstrate the influence of the flame temperature on the populations of the individual energy levels and thus on the sensitivity that can be attained at the individual analytical lines. As we can see from Table 9-41, the absorbance for resonance lines that emanate from the ground state is higher by a factor of 2.6 in the air-hydrogen flame, which is only about 200 °C cooler, than in the air-acetylene flame. The factors for the analytical lines that emanate from the levels of higher energy, 1692 K and 3428 K, are only 1.7 and 1.0, respectively, which reflects the denser populations at these levels in the hotter air-acetylene flame.

The best sensitivity for Sn can be attained in the argon-hydrogen diffusion flame; nevertheless numerous acids and virtually all metallic elements cause strong interferences in this flame [4987]. Numerous interferences have also been reported in the air-hydrogen and air-acetylene flames [1056, 3500]. HARRISON and JULIANO [2406] inves-

Table 9-39. Stability of solutions with respect to their Sn concentration.

Species	Matrix	pH range	Materials*	Time period	Recommended storage	References
Sn	NaCl (w = 0.5%) in HCl solution	1.5–11	BG	24 hours	losses start at pH values > 1.5	[5455]
Sn	distilled water, 0.05 + 0.5 mol/L H_2SO_4, 0.1 + 1 mol/L HNO_3, 0.1 + 1 mol/L HCl	0–7	BG, PE, PTFE, PVC, PP, PC	60 days	no losses found for acidified samples when stored at ambient temperature in the dark, independent of the acid used; for non-acidified samples the losses were highest in glass	[2139]
Organo-Sn	mussel extracts		PG	1 year	–20 °C in the dark	[1066]
TBT	sea-water		PC	678 days	15% loss after 60 days	[5984]
MeT, BT	river water		PC	77 days	+10 °C or –20 °C for 5 days	[5135]
MBT, DBT, TBT	sea-water		BG	4 months	acidification to pH 2, storage in glass bottles in the dark. No losses of MBT, DBT and TBT at 4 °C; losses of MBT and DBT at 20–25 °C	[4749]

*BG: borosilicate glass; PC: polycarbonate; PE: polyethylene; PG: Pyrex glass; PP: polypropylene; PTFE: polytetrafluoroethylene; PVC: polyvinyl chloride.

Table 9-40. Sn analytical lines and the characteristic concentrations attainable in various flames.

Wavelength (nm)	Energy level (K)	Slit width (nm)	Characteristic concentration* (mg/L)			Spectral interferences
			Air-H$_2$	Air-acetylene	N$_2$O-acetylene	
224.61	0–44509	0.2	1.1	2.0	3.0	Pb$^+$ [5973]
286.33	0–34914	0.7	1.8	3.5	5.4	InBr^{++} [6279]
235.48	1692–44145	0.7	2.2	2.5	3.8	As$^+$ [5973]

$^+$ Interference with deuterium BC possible
$^{++}$ Interference with Zeeman-effect BC possible

Table 9-41. Absorbance values at various Sn analytical lines in the air-hydrogen and air-acetylene flames (from [1056]).

Wavelength (nm)	Energy level (K)	Absorbance		Ratio H$_2$: C$_2$H$_2$
		Air-H$_2$	Air-C$_2$H$_2$	
224.61	0–44509	0.821	0.337	2.6
286.33	0–34914	0.492	0.193	2.7
254.66	0–39257	0.178	0.074	2.5
207.31	0–48222	0.013	0.005	2.6
			Mean:	2.6
235.48	1692–44145	0.398	0.239	1.7
270.65	1692–38629	0.201	0.194	1.7
303.41	1692–34641	0.148	0.102	1.5
300.91	1692–34914	0.087	0.052	1.8
219.93	1692–47146	0.042	0.028	1.5
233.48	1692–44509	0.041	0.023	1.8
266.12	1692–39257	0.028	0.013	2.2
			Mean:	1.7
284.00	3428–38629	0.108	0.119	0.9
242.95	3428–44576	0.105	0.109	1.0
226.89	3428–47488	0.065	0.075	0.9
317.51	3428–34914	0.053	0.055	1.0
220.97	3428–48670	0.023	0.021	1.0
228.67	3428–47146	0.006	0.005	1.2
			Mean:	1.0

tigated the influence of organic solvents and found that even small quantities of alcohols, ketones, and organic acids markedly reduced the absorption of Sn in the air-hydrogen flame. This effect becomes stronger with increasing chain length and branching of the respective class of compound. The authors attributed this effect to a reduction of the concentration of hydrogen atoms in the flame, which play an important role in the atomization mechanism for Sn, by organic solvents.

This effect excludes the determination of Sn in organic solvents in the air-hydrogen flame. Since the fuel-rich air-acetylene flame is difficult to operate with organic solvents, only the nitrous oxide-acetylene flame remains. Nevertheless, ionization is observed in this flame which must be eliminated by the addition of alkali. Despite the somewhat lower sensitivity, the nitrous oxide-acetylene flame can generally be recommended when Sn is to be determined in more complex matrices since interferences are clearly the lowest in this flame.

Tin is often looked upon as a difficult element in GF AAS; there are numerous papers reporting severe interferences in the presence of chloride and sulfate [5428]. PRUSZKOWSKA et al. [4724] proposed a mixture of ammonium phosphate, magnesium nitrate, and nitric acid as a modifier; under STPF conditions it should eliminate the majority of the interferences reported in the literature. Nevertheless, large quantities of chloride should be avoided, even when this mixed modifier is applied. Using the Pd-Mg modifier and a pyrolysis temperature of 1200–1400 °C, 1–5 g/L chloride or sulfate do not cause any interferences [6265].

Tin is among those elements whose sensitivity is strongly influenced by Zeeman-effect BC. While the characteristic mass in a non-Zeeman instrument with a longitudinally-heated atomizer is 10 pg, this value deteriorates to around 25 pg when Zeeman-effect BC is applied. The optimum atomization temperature for Sn is 2300–2400 °C. The atomization temperature can be reduced to 2200 °C in a transversely-heated atomizer; the characteristic mass with Zeeman-effect BC is 90 pg. BELARRA et al. [496] reported the use of secondary lines to reduce the sensitivity by factors of up to 140.

Tin can also be determined by the hydride-generation technique; in contrast to the other hydride-forming elements the sensitivity is relatively strongly dependent on the pH value. The determination is best performed in a saturated boric acid solution; under these conditions an LOQ of around 0.5 ng absolute or 0.5 µg/L can be attained.

9.53 Titanium (Ti)

It is estimated that Ti makes up about 0.56 % of the Earth's crust and it is thus at position 10 of the abundance list of the elements, more abundant than chlorine, phosphorus, carbon, sulfur, or nitrogen. Along with aluminium, iron, and magnesium, it is one of the most abundant metals occurring in nature. Titanium occurs in its compounds in the +2, +3, and +4 oxidation states. Due to its properties it is a very valuable material in aerospace and deep sea technology. It is used in the production of special steels since in iron it forms carbides and nitrides. Titanium alloys and non-alloyed Ti are used in the production of chemical apparatus. Titanium plays an important role in medicine for the production of bone pins and artificial limbs.

The stability of Ti solutions is relatively critical and the precipitation of hydroxides takes place even at low pH values. Virtually quantitative losses of Ti have been observed from hydrochloric acid solution at a pH value of 3.5 after 24 hours in glass containers [5456]. The acidification of solutions to pH values < 1.5 is thus indispensable.

Titanium can be determined in the nitrous oxide-acetylene flame at the 364.3 nm line with a characteristic concentration of 2 mg/L. A multitude of further analytical lines was investigated by CARTWRIGHT *et al.* [1092], several of which are compiled in Table 9-42.

AMOS and WILLIS [166] found that the signal for Ti was markedly enhanced in the presence of iron and hydrofluoric acid. An increase in the hydrofluoric acid concentration from 0.2 % to 4 % brought about a 50 % increase in the signal; 2 g/L Fe doubled the absorbance. The concentrations of these concomitants should thus be maintained as constant as possible; the effect seems to be largely under control when the concentrations of iron and hydrofluoric acid are high. BOND [708] found that an 0.1 mol/L solution of ammonium fluoride had an even greater influence. The strongest signal enhancement was observed in the presence of aluminium [4270]; 0.5 g/L Al caused a fivefold increase in the absorbance [4527] and is thus very suitable as a buffer [1307, 1791].

RAO [4806] determined Ti in ores free of interferences after a lithium metaborate fusion by maintaining a concentration of 20 g/L lithium metaborate, 8 g/L ammonium fluoride, and 0.5–2 g/L SiO_2 (corresponding to 12.5–50 % SiO_2 in the original test sample). The maximum sensitivity for Ti was attained under these conditions.

The determination of Ti by GF AAS depends very strongly on the quality of the pyrolytic coating of the graphite tube [3856, 5418] and on the rate of heating for atomization. This can be easily explained by the fact that although the melting point of Ti is relatively low at 1610 °C, the melting point of the carbide is very high at around 3150 °C. This means that the formation of TiC must be prevented as much as possible. For the same reason a similar characteristic mass of 70 pg can be attained in a transversely-heated atomizer due to the higher effective gas temperature under STPF conditions at an atomization temperature of 2500 °C as in a longitudinally-heated atomizer. In the latter atomizer a characteristic mass of around 45 pg can be attained with atomization from the wall at 2650 °C. The maximum pyrolysis temperature in both types of atomizer is 1400 °C.

Table 9-42. Ti analytical lines.

Wavelength (nm)	Energy level (K)	Slit width (nm)	Characteristic concentration* (mg/L)	Spectral interferences
364.27	170–27615	0.2	2	Pb+ [5431, 5973]
335.46	170–29971	0.2	3	
399.86	387–25388	0.2	3	
398.98	170–25227	0.2	4	
395.63	170–25439	0.2	4	
394.87	0–25318	0.2	8	

* Nitrous oxide-acetylene flame, reducing (fuel-rich, red)
+ Interference with deuterium BC possible

9.54 Tungsten (W)

About $10^{-3}\%$ of the Earth's crust is composed of tungsten so that it is close to copper in the abundance list of the elements. In its compounds it occurs in the +2 through +6 oxidation states. In nature it is mostly found in the form of tungstates. Tungsten has become indispensable in the illumination industry due to its high melting point of 3400 °C; virtually all electric light bulbs use filaments made of W. It is used in numerous special alloys, in rocket jets and as heat shields in space travel.

Tungsten can be determined in the nitrous oxide-acetylene flame at the 255.1 nm line with a characteristic concentration of around 10 mg/L. Although AMOS and WILLIS [166] and MANNING [3844] mention other lines of higher sensitivity, the 400.9 nm line is nevertheless mostly used, even though the sensitivity is halved, due to the much more favorable S/N ratio. Numerous metallic ions interfere in the determination of W [1672]. The addition of sodium sulfate [1473, 1672] or potassium peroxodisulfate [4890] not only reduces these effects but also improves the sensitivity by 40–75%.

Tungsten cannot be determined by GF AAS due to its high melting point and its tendency to form carbides. CHONG and BOLTZ [1266] described an indirect determination via precipitation as lead tungstate and the determination of lead in the supernatant solution.

9.55 Uranium (U)

The proportion of uranium in the Earth's crust is estimated to be about 4 mg/kg, so that it is more abundant than mercury, silver, or gold. Sea-water contains 3 µg/L U. It occurs in its compounds in the +3 through +6 oxidation states. Currently U is the most important nuclear fuel. Enriched U is used as a material of high density in the aeronautics industry, as radiation shielding, as an additive for catalysts or steels, and in the glass and ceramics industries. Uranium compounds are extremely poisonous and cause kidney and liver damage. Some plants, such as olives and certain fungi, are able to store U contents of several g/kg.

Uranium can be determined in the nitrous oxide-acetylene flame at the 351.5 nm line with a characteristic concentration of 50 mg/L. The analytical lines at 358.5 nm and 356.7 nm exhibit better sensitivity, but an unfavorable S/N ratio. MANNING [3844] investigated further lines and determined the characteristic concentrations. Since U is noticeably ionized in the nitrous oxide-acetylene flame [166], about 1 g/L K as the chloride should be added to all measurement solutions to suppress this effect. KEIL [3065] found that numerous metals more or less strongly enhance the signal for U; gallium exhibits the most pronounced effect. The addition of 10 g/L Ga produced a characteristic concentration of 16 mg/L at the 358.5 nm analytical line and a characteristic concentration of 26 mg/L at the 351.5 nm analytical line. SEYMOUR et al. [5252] reported an enhancement of the U signal by a factor of four in the presence of 800 mg/L Al.

In GF AAS uranium is one of the elements with the lowest sensitivity. A pyrolysis temperature of only 1000 °C can be applied due to the low melting point of U of 1132 °C. The fact that the determination of U has such low sensitivity despite its low melting point is in part due to the high stability of its oxides, which are in part volatile at temperatures above 1100 °C [2159]. Prerequisites for the determination of U are a high quality PG coating of the graphite tube and a fast heating rate. A characteristic mass of 12 ng can be attained in a longitudinally-heated atomizer at an atomization temperature of 2650 °C. In a transversely-heated isothermal atomizer under STPF conditions the atomization temperature can be lowered to 2550 °C and the characteristic mass is 40 ng.

GOLEB [2153, 2156] determined the $^{235}U/^{238}U$ isotope ratio using a cooled HCL and GF AAS. ALDER and DAS [109, 110, 112] described an indirect procedure for the determination of U based on the oxidation of U(IV) by Cu(II), which is reduced to Cu(I) and then determined after extraction as the neocuproin complex.

9.56 Vanadium (V)

It is estimated that V makes up 0.014 % of the upper crust of the Earth so that it is close to nickel and zinc in the abundance list of the elements. In its compounds it occurs in the +2 through +5 oxidation states, and in organovanadium complexes in the +1, 0, and –1 oxidation states. Vanadium is used almost exclusively as a constituent of alloys in the steel industry; it also plays a certain role in non-ferrous metallurgy. Vanadium is an essential trace element for plants and animals which stimulates the synthesis of chlorophyll and promotes the growth of young animals. Vanadium compounds count as toxic in high concentrations or after long periods of action. There are also V compounds that exhibit chemotherapeutic significance in the treatment of leukemia.

The stability of V solutions is relatively critical due to the low solubility of oxygen-containing species; the minimum is at a pH value of around 7 [5456]. Substantial losses take place from hydrochloric acid solutions in glass containers at a pH value of 3.5, and are practically quantitative at a pH value of 5 after 24 hours [5456].

Vanadium can be determined in the nitrous oxide-acetylene flame at the 318.3/318.4/318.5 nm triplet with a characteristic concentration of around 2 mg/L. CARTWRIGHT [1092] investigated numerous other analytical lines, a number of which are compiled in Table 9-43.

CAPACHO-DELGADO and MANNING [1057] observed an enhancement of the V signal in the presence of phosphoric acid. GOECKE [2147] established that concentrations of more than 1 g/L Fe interfered and he removed this from strong hydrochloric acid solution by extraction in isopropyl ether. He further found that 0.2 g/L Al enhanced the V signal by a third; by adding aluminium to all measurement solutions and by optimizing the nitrous oxide-acetylene flame it was possible to improve the characteristic concentration to 0.2 mg/L. WEST et al. [6282] explained this increase in the absorbance by a reduction of the lateral diffusion in the flame. KRAGTEN [3271] observed that the sensitivity for V depended on the oxidation state; the pentavalent state exhibits the best sensitivity. He attributed poorly reproducible results to the degree of polymerization in solution. Stable results could be obtained after the addition of 10 g/L ammonium chloride and 0.1 % hydrogen peroxide.

The atomization mechanism for V in GF AAS is described in detail in Section 8.2.3.4. Despite its relatively high melting point of 1929 °C, V can be determined very well by GF AAS. The probable reason for this is that the carbides and oxides are not particularly stable. The prerequisites for a good determination are nevertheless a perfect PG coating and the highest possible heating rate. Magnesium should be added as the modifier to prevent losses due to volatile oxides during the pyrolysis step. A pyrolysis temperature of 1100 °C can be applied under these conditions. An atomization temperature of 2650 °C is necessary in a longitudinally-heated atomizer; the characteristic mass is around 30 pg for atomization from the tube wall. This deteriorates to 40 pg when Zeeman-effect BC is applied. Practically the same sensitivity can be attained in a transversely-heated atomizer but at a much lower atomization temperature of 2400 °C under STPF conditions. Beside magnesium nitrate [558, 3864], palladium [553, 2195, 2197, 5743], platinum [5820], rhodium [5820], chromium [1172], ascorbic acid [5820, 5869, 5870], and citric acid [2198] have been applied as modifiers for the determination of V.

Table 9-43. V analytical lines.

Wavelength (nm)	Energy level (K)	Slit width (nm)	Characteristic concentration* (mg/L)	Spectral interferences
318.34	137–31541 ⎫			
318.40	323–31722 ⎬	0.7	2	
318.54	553–31937 ⎭			
306.64	553–33155	0.2	5	Bi, Pt$^+$ [5431]
306.05	323–32989	0.2	5	
437.92	2425–25254	0.2	8	
438.47	2311–25112	0.2	9	
370.47	2311–29296	0.2	11	
390.23	553–26172	0.2	13	
320.24	323–31541	0.2	13	
257.40	553–39391	0.2	20	
251.71	323–40039	0.2	35	Si$^+$ [5973]
253.02	553–40064	0.2	35	
250.78	137–40001	0.2	40	

* Nitrous oxide-acetylene flame, reducing (fuel-rich, red)
$^+$ Interference with deuterium BC possible

9.57 Yttrium (Y)

Yttrium is the major constituent of the yttrium group earths and is at position 32 in the abundance list of the elements; it is thus more abundant than lead, for example. Yttrium is used in alloys to reduce the grain size of chromium, molybdenum, titanium, and zirconium, and to increase the hardness of aluminium and magnesium alloys. Its compounds find a multitude of applications in electronics and laser technology. The major application of Y in the form of the oxide, oxydisulfide, or vanadate is as a red phosphor for color television and for fluorescent lights.

Losses of Y due to the formation of colloids are possible [1684] and occur more readily from hard waters than from dilute solutions since seeds are required to start colloid formation.

Yttrium can be determined in the nitrous oxide-acetylene flame at the 410.2 nm line with a characteristic concentration of 2 mg/L. The sensitivity is only slightly lower at the 407.7 nm, 412.8 nm, and 414.3 nm analytical lines. About 1–2 g/L K as the chloride should be added to all measurement solutions to suppress ionization.

With a characteristic mass of only 13 ng in a longitudinally-heated atomizer, Y is one of the least sensitive elements determinable by GF AAS; this can in part be ascribed to the stability of gaseous oxides [4709]. The graphite surface [6127, 6128] and the heating rate play a major role. In a transversely-heated atomizer under STPF conditions a significantly better characteristic mass of 4 ng can be attained at a lower atomization temperature due to the higher effective gas temperature. Nevertheless, this is still inadequate for trace analyses.

9.58 Zinc (Zn)

Making up about 0.012 % of the Earth's crust, zinc is close to copper, strontium, and vanadium in the abundance list of the elements. As a very non-noble metal it only occurs in nature in the form of its compounds, mostly in the divalent oxidation state, in company with lead and cadmium. The major part of the Zn produced is used to galvanize steel. Large quantities are also used for the production of brass, bronze, solders, and other zinc alloys. Zinc oxide is used in glass, ceramics, and dyes. Further uses of zinc compounds are found in the soap, cosmetic, pharmaceutical, rubber, and plastics industries.

Zinc is an essential trace element for humans, animals, plants, and microorganisms; the Zn content in humans is 2–4 g. Larger quantities of zinc salts can nevertheless cause external caustic burns and very painful internal inflammation of the digestive organs. The toxic limits are nevertheless much higher than for the other essential elements such as copper.

The stability of Zn solutions has been investigated for calibration solutions [3932, 5297, 5456, 5606], sea-water [1567, 4629, 4890, 5206], rain water [1184], pond water [1566], and other surface waters [5660]. Sorption losses have been observed from calibration solutions stored in glass at pH values > 5 [5456, 5606], while no such losses were observed in PE containers under the same conditions. These losses can easily be controlled by acidification. MÉRANGER et al. [4061] investigated the stability of whole blood stabilized with heparin in respect to its Zn content. As well as sorption losses, contamination by Zn must be taken into account since it is used in the manufacture of a number of plastics. Polypropylene containers are unsuitable for the storage of Zn solutions due to the risk of contamination [5606] and borosilicate glass releases noteworthy quantities [5606]. Materials used for the manufacture of vacutainers and other sampling containers for body fluids [1300, 2303, 2304, 2488, 2711, 3480, 6015, 6192, 6329], stoppers

[1784], pipet tips [5494], parafilm [4844], paper towels and filter materials [4844] must be critically checked for Zn contamination when Zn is to be determined in the trace range.

Zinc is one of those elements determined most frequently by AAS. A characteristic concentration of 0.01 mg/L can be attained in the air-acetylene flame at the 213.9 nm resonance line; Zn is thus the most sensitive element in F AAS. High Zn concentrations can be determined with advantage at the 307.6 nm resonance line with a characteristic concentration of around 50 mg/L [2052].

In the early days of AAS DAVID [1426] and ALLAN [138] made a thorough investigation on the determination of Zn and found that F AAS was superior to all other techniques. The determination of Zn in the air-acetylene flame appears to be free of interferences, as SPRAGUE and SLAVIN [5530] found for the analysis of biological materials. PLATTE and MARCY [4667] found no interferences due to 1 g/L sulfate, phosphate, nitrite, nitrate, bicarbonate, silicate, EDTA, and nine various cations.

Zinc is also one of the most sensitive elements in GF AAS. The characteristic mass in a longitudinally-heated atomizer is 0.4 pg and in a transversely-heated atomizer it is around 1 pg; the optimum atomization temperature is 1600–1800 °C, independent of the type of atomizer. Zinc can be stabilized with the Pd-Mg modifier to a pyrolysis temperature of 1000 °C. Due to the omnipresence of Zn in numerous reagents, attempts are often made to use a minimum of modifier, such as 5 µg $Mg(NO_3)_2$. In this case it is only possible to apply a pyrolysis temperature of 600–700 °C. In the presence of high chloride contents, such as in sea-water samples or soil extracts, a number of interferences have been observed [95–97, 689, 1014, 1015, 1590] which could be ascribed to losses during pyrolysis or at the start of atomization [1590] and which can be controlled by the addition of nitric acid [1589].

Since the dissociation constant of zinc chloride is much less than that of other chlorides, no gas-phase interferences from chlorides are to be expected so that a separation is not absolutely necessary. This offers the possibility for Zn of avoiding the use of a modifier, provided that concomitant elements are not present in too high concentrations, and atomizing directly after the drying step. Nevertheless the use of Zeeman-effect BC is indispensable since a correspondingly high background attenuation must be taken into account. Moreover a number of potential spectral interferences have been reported at the 213.9 nm analytical line, as listed in Table 9-44.

The problem of contamination naturally does not start to occur during the determination, but already begins during sampling and sample preparation. The use of containers and reagents that are as free of blank values as possible is essential for the trueness of the analyses. The number of preparatory steps should be as small as possible and should preferably be carried out in closed laboratory ware made of inert materials. Frequent control of the blank value is indispensable. An unexpectedly good (low) characteristic mass is a reliable indication for contamination.

Table 9-44. Zn resonance lines.

Wavelength (nm)	Energy level (K)	Slit width (nm)	Characteristic concentration* (mg/L)	Spectral interferences
213.86	0–46745	0.7	0.01	Cu^{+++} [6401] Te^+ [5431] Fe^{+++} [3072, 6305] PO^{++} [4425] NO^+ [631]
307.59	0–32502	0.7	50	

* Air-acetylene flame, oxidizing (lean, blue)
+ Interference with deuterium BC possible
++ Interference with Zeeman-effect BC possible
+++ Direct line overlap

9.59 Zirconium (Zr)

It is estimated that zirconium makes up 0.02 % of the Earth's crust, putting it at place 20 in the abundance list of the elements between barium and chromium; it is therefore much more abundant than lead, copper, nickel, or tin. Nevertheless Zr was only discovered at a late stage since it occurs solely in low concentrations and larger accumulations in ores are rare.

Due to its high corrosion resistance, Zr is a suitable material in the chemical industry for the construction of spinning jets, valves, and pumps. It is used in the electronics industry for components of vacuum tubes, and in pyrotechnics for the manufacture of flares and tracer ammunition. In metallurgy it has proven of value as a binder for oxygen, nitrogen, and sulfur. It is further used as a jacket for nuclear fuel rods, for impregnating textiles, and for the manufacture of pigments and refractory ceramics.

There is little reported in the literature on the stability of Zr solutions. Sorption losses are more pronounced on glass surfaces than on PE, which is an indication for the formation of colloids [1684].

Zirconium can be determined in the nitrous oxide-acetylene flame at the 360.1 nm analytical line with a characteristic concentration of 10 mg/L; further analytical lines are compiled in Table 9-45. Zirconium is ionized by about 10 % in the nitrous oxide-acetylene flame [3844]; 1 g/L K as the chloride should be added to all measurement solutions to suppress this effect [5411].

AMOS and WILLIS [166] found that the absorption of Zr is noticeably increased in the presence of hydrofluoric acid and high iron concentrations; they thus added 2 % HF to all measurement solutions. SLAVIN et al. [5411] found that hydrochloric acid had a similar influence; 10 % HCl increased the Zr signal fourfold compared to an acid-free solution. BOND [708] obtained the strongest effect by the addition of 0.1 mol/L ammonium fluoride solution, which caused an eightfold signal enhancement. WALLACE et al. [6134]

found improved linearity, sensitivity, precision, and limit of detection for Zr through the addition of an excess of aluminium. Aluminium could reduce the tendency to forming refractory oxides and promote atomization. It is also conceivable that aluminium reduces the lateral diffusion so that the atoms are more concentrated in the middle of the flame.

Due to its very strong tendency to forming carbides, the determination of Zr by GF AAS [509, 2443, 3703] has gained hardly any significance and can practically only be performed in tubes lined with tantalum foil [3703].

Table 9-45. Zr analytical lines.

Wavelength	Energy level	Slit width	Characteristic concentration*	Spectral interferences
(nm)	(K)	(nm)	(mg/L)	
360.12	1241–29002	0.2	10	Cr, Ni[+] [5973]
354.77	570–28750	0.2	15	
303.09	1241–34240	0.2	15	
301.18	570–33764	0.2	17	Ni[++] [3627]
298.54	0–33487	0.2	17	
362.39	570–28157	0.2	19	

* Nitrous oxide-acetylene flame, reducing (fuel-rich, red)
[+] Interference with deuterium BC possible
[++] Direct line overlap

10 Applications of AAS

Because AAS is specific and exhibits great freedom from interferences, its range of applications is correspondingly large. Although the actual analytical methods are often largely similar, the problems posed by different branches of analysis justify a separate treatment of each area. In this chapter we shall place emphasis on the applications of AAS, including sample preparation, such as dissolution or digestion of solid samples. Within the scope of this monograph, however, we can merely touch upon suitable procedures for the selected AAS technique for given materials; the interested reader is referred to the original publications for more detailed information. We have also provided information on specific sources of error, taking into account the whole analytical method from sampling through to the determination by AAS. For our discussions on interferences we have assumed that the information provided in previous chapters can be taken as known, so that we can restrict ourselves to sample- or matrix-specific interferences and their elimination.

We have divided this chapter into the following main sections: Body fluids and tissues (human domain); biological materials (vegetable and animal); environmental analysis (water, soil, air); geochemistry and prospecting (rocks, minerals, ores); crude oils and petrochemical products; metallurgical products; and miscellaneous industrial products. Classification in some areas is nevertheless difficult. Thus, for example, the *analysis of soils* is of interest in environmental analysis, in prospecting, and in the analysis of plants and animal feeding stuffs. The *analysis of coal* can be assigned to geochemistry, to petrochemistry, and to the environment. To prevent unnecessary repetitions we have thus mostly assigned classifications to the major area of application, but have also taken analytical aspects into account. *Bioindicators*, for example, clearly belong in the environmental domain, but we have treated them under biological materials because the techniques of digestion and determination are the same as for foodstuffs analysis. Individual terms are listed in the index.

Regular reports on advances in techniques of analytical atomic spectrometry appear yearly in *The Journal of Analytical Atomic Spectrometry* and every two years in *Analytical Chemistry*.

10.1 Body Fluids and Tissues

Metal ions influence the well-being of humans in a multitude of ways. A large number of these elements are essential and nature regulates their intake, metabolism, and excretion. Their concentration in various parts of the human body is exactly defined. Other elements and their compounds are inert or their function is not yet known. Others again are toxic at even very low concentrations. These various effects are summarized in Figure 10-1 in very simplified form. It can also be seen that even essential elements can have adverse effects if they are present in too high concentrations. This knowledge has led to the establishment of recommended limits for the intake of individual elements in the nutrition. These limits are summarized for a number of elements in Table 10-1. This table also contains information on the average contents of these elements in the human body and their concentration in blood plasma.

Due to the considerably differing analytical and instrumental requirements with respect to sensitivity, precision, sample throughput, etc., we shall distinguish in the following between the electrolytes, the trace elements, and the therapeutically administered elements. In the case of the trace elements there is a major difference between the clinical-biochemical domain, where essential concentrations and deficiency symptoms are of interest, and toxicology and industrial hygiene, where markedly increased contents are of importance. Nowadays we no longer speak of toxic elements, but of toxic quantities

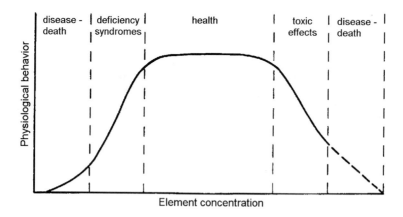

Figure 10-1. Ambivalent physiological behavior of trace elements (from [5868]).

Table 10-1. Recommended daily ingestion of electrolytes and essential trace elements (WHO) and their average contents in the human body and blood plasma.

Element	Recommended daily intake	Total content in the body	Concentration in blood plasma	References
Na	1–2 g	100 g	3.25 g/L	[3077, 5863]
K	2–5 g	140 g	160 mg/L	[641]
Mg	0.7 g	19 g	22 mg/L	[188, 3077, 5356]
Ca	0.8 g	1000 g	80 mg/L	[1526, 3077, 5354]
Cr	0.1 mg	6 mg	0.15 µg/L	[2524, 3077]
Mo	0.3 mg	9 mg	0.6 µg/L	[220, 3077]
Mn	4 mg	12 mg	0.5 µg/L	[1262, 3077]
Fe	10–20 mg	4.2 g	1.1 mg/L	[119, 3077]
Co	3 µg	1 mg	0.1–0.2 µg/L	[3077, 5847]
Ni	–	1 mg	0.3–0.6 µg/L	[3077, 5130, 5355]
Cu	3 mg	72 mg	1.1 mg/L	[3077, 5102, 5351, 5352]
Zn	15 mg	2.3 g	1 mg/L	[5353, 5848, 5986]
Se	0.1 g	15 mg	100 µg/L	[2736, 3785, 5538]

and concentrations. It has been shown that many elements which are toxic in higher concentrations are essential in low concentrations and that deficiency syndromes occur if they are not present in the body in a sufficiently high concentration. Table 10-2 presents an overview of trace elements listed in the sequence of the discovery of their essentiality with their biological functions and toxic effects. For a number of elements their essentiality has only been demonstrated in animal experiments. The resorbability often plays a major role in the toxicity, and this again depends on the route of ingestion (oral, inhalation, or skin contact) and the chemical and physical form (species) of the element.

In the majority of cases, serum, plasma, or whole blood are used for examinations in clinical chemistry, industrial hygiene, and toxicology. The excretion in urine is also frequently of interest for toxicological and therapeutical investigations. Other body fluids are generally only of minor interest. Sampling is of major significance for the accuracy and the level of information of the analytical results; this aspect is treated in detail in Section 5.1. In addition, we should like to draw our readers' attention to the numerous publications that deal with the problems of sampling biological samples in general [472, 1330, 2842, 3050, 4222, 4540, 5120, 5226, 5584, 6012, 6051, 6060], as well as body fluids [71, 72, 1334, 1763, 2000, 3397, 4130, 4291, 4656, 6025, 6053] and human tissues [428, 2663, 2703, 2704, 6527]. Contamination during the sampling of body fluids and tissues has in particular been described for aluminium [479, 1300, 1764, 3260, 3397, 4963], chromium [853, 4504], copper [1300], lead [1364], manganese [2703, 3397], nickel [853, 4358], vanadium [3397], and zinc [1300, 2711, 5466].

In the following sections we shall deal with the analysis of body fluids, while the analysis of tissue is treated in Section 10.1.5.

10.1.1 The Electrolytes

The electrolytes sodium, potassium, calcium, and magnesium are present as cations in body fluids, in part bound to proteins and other organic constituents, and are in equilibrium with the anions. The major task of these ions is to control the osmotic pressure of water equilibrium. In the extracellular region the most important element is sodium with the anion chloride, and in the intracellular region potassium with the anion bicarbonate is the most important. Calcium and magnesium have a variety of functions in biochemical processes with respect to the activity of various enzymes and in neuromuscular activity. The concentrations of calcium and magnesium in serum or plasma give an indication of numerous diseases.

The determination of these four electrolytes in body fluids was one of the earliest applications of F AAS [6346]. For the determination of sodium and potassium, flame photometers (emission instruments with a flame as atomizer and excitation source, and with optical filters to separate the analytical lines) are still preferentially used; with lithium as reference element ('internal standard') these instruments provide correct results and meet the requirements for routine analyses. Ion-selective electrodes are also used, although these measure the activity and not the concentration. The determination of sodium and potassium by F AAS has only been reported in a limited number of cases. It is best performed on a 1:50 dilution of serum in an air-hydrogen flame; sodium should

Table 10-2. Essentiality of trace elements, their biological function, and their toxic effects (from [2848]).

Element and discovery of essentiality	Biological activity	Deficiency syndromes (toxic effects)	Particularly endangered groups
Iron (17th c.)	oxygen and electron transport	anemia, hemochromatosis	
Iodine (1850)	constituent of the thyroid hormone	goiter and thyroid hypofunction (thyrotoxicosis)	in regions of iodine deficiency
Copper (1928)	oxidative enzymes, interaction with iron, cross-linking of elastin	anemia, changes in osteogenesis	underfed, TPN patients, pregnant women
Manganese (1931)	participation in mucopolysaccharide metabolism, associated with superoxide-dismutase	none known (neurological disorders)	
Zinc (1934)	constituent of more than 100 enzymes	disturbances of growth, sexual immaturity	vegetarians, underfed, TPN patients
Cobalt (1935)	constituent of vitamin B12		vegetarians
Molybdenum (1953)	associated with xanthine-, aldehyde- and sulfide-oxidases	no deficiency disorders known (diseases similar to gout)	(has occurred in Russia)
Selenium (1957)	constituent of gluthathione-peroxidase, interaction with heavy metals	cardiomyopathy (toxic effects detected in animals)	in regions of selenium deficiency such as China or Finland
Chromium (1959)	participation in insulin activity surmised	reduced glucose tolerance factor, (Cr(VI) is carcinogenic)	TPN patients, elderly people (metal workers, welders)
Tin (1970)	participation in promotion of growth surmised	unknown	
Vanadium (1971)	participation in promotion of growth surmised	unknown (interference of iron resorption)	(metal workers)
Fluorine (1971)	participation in growth of teeth and bones	caries, osteoporosis (?)	

Table 10-2 continued

Element and discovery of essentiality	Biological activity	Deficiency syndromes (toxic effects)	Particularly endangered groups
Silicon (1972)	participation in growth of bones and sustentacular tissue	unknown and improbable	
Nickel (1976)	largely unknown	(interference of iron resorption, nickel allergy, cancer)	(metal workers)
Arsenic (1977)	largely unknown	largely unknown (As(III) more toxic than As(V))	

be determined at the secondary line at 330.2 nm to avoid excessive dilutions. The micromethod described by PASCHEN [4556] for the determination of four electrolytes in a single serum dilution is particularly elegant; 100 µL serum are diluted with 2.5 g/L strontium chloride solution and analyzed directly for the four electrolytes. The addition of strontium eliminates both the influence of phosphate on calcium and the ionization of sodium and potassium in the air-acetylene flame. The repeatability precision of the determination is between 0.3% and 0.5%.

Flame photometers are less suitable for the determination of calcium since the results are dependent on the matrix. Flame AAS provides correct results and serves as a reference method [1526, 2177]. Flame AAS is also the method of choice in clinical laboratories for the determination of magnesium [188]. The simplest and most frequently used procedure for the determination of calcium and magnesium in serum and urine is the direct analysis of the 1:20 to 1:50 diluted test samples in the air-acetylene flame [2234]. To suppress the interference of phosphate, 10 g/L EDTA, 5 g/L La in hydrochloric acid solution, or 2.5 g/L Sr must be added to all measurement solutions. Lanthanum should only be added to the already diluted serum or urine since otherwise the protein coagulates. For this reason EDTA is occasionally preferred. Although proteins do not interfere in the determination of calcium and magnesium in the presence of lanthanum at these high dilutions, a number of authors found that deproteinized test samples gave better repeatability [5119, 5336]. We must nevertheless note that trichloroacetic acid and hydrochloric acid influence the determination of both of these elements [361, 4163]. The calibration solutions must therefore contain the same amount of acid.

The use of the nitrous oxide-acetylene flame has also been proposed for the determination of calcium and magnesium in urine since it is then only necessary to suppress the ionization by diluting the test samples with a potassium buffer [624]. A micromethod has also been proposed for the determination of sodium, potassium, calcium, and mag-

Table 10-3. Selected literature on the determination of electrolytes in body fluids by F AAS.

Body fluids	Sodium	Potassium	Calcium	Magnesium
Cerebrospinal fluid	[938]	[938]	[754, 938, 3143]	[754, 938, 1452]
Erythrocytes	[1956]	[1956, 5214]		[1956, 5214]
Lymphocytes	[6197]	[5214, 6197]	[6197]	[5214, 6197]
Milk	[2235, 3999, 5031]	[5031]	[2235, 3999, 5031]	[2235, 3999, 5031]
Plasma			[5299]	[5299]
Serum	[248, 446, 582, 937, 943, 2234, 2525, 4556, 6336]	[446, 582, 937, 943, 2234, 2525, 4556, 6336]	[248, 446, 582, 943, 1030, 2234, 3339, 3781, 4556, 4644, 4645, 4919, 5195, 5299, 5897]	[247, 446, 582, 943, 2234, 3339, 3781, 4556, 4849, 5299]
Urine	[247]	[247]	[247, 5119, 5897, 6198]	[5119]
Whole blood			[4950, 5282]	[4950, 5282]

nesium in 0.1 mL urine [2242]; the urine is diluted 1:100 under the addition of lanthanum and cesium and analyzed directly in the air-acetylene flame. Table 10-3 provides an overview of selected publications on the determination of electrolytes in body fluids.

Further information on the determination of electrolytes in body fluids by F AAS is given by TSALEV [5911], SLAVIN [1875, 5433], and DELVES [1477]. In the last couple of years FI-F AAS has been applied increasingly since dilution can be performed on-line so that the effort required for manual sample preparation can be noticeably reduced [246, 936–938, 943, 949, 3781, 4919, 4950, 5299, 6114].

Occasionally GF AAS has also been used for the determination of electrolytes in body fluids, especially when the sample volumes were very limited. Microanalytical procedures have been described for calcium [2098, 4377], magnesium [1280, 2719, 5248], potassium [687, 4298, 4447, 5260, 6358], and sodium [687, 4298, 5260, 6358].

10.1.2 Trace Elements

In this section we shall essentially deal with the determination of trace elements in the clinical-chemical domain, i.e. in the concentration range observed in non-exposed subjects. Three of the elements belonging to this category—iron, copper, and zinc—are present in body fluids in concentrations that can be determined with adequate accuracy by F AAS. All other trace elements are present in concentrations in the human body that are orders of magnitude lower. The requirements placed on contamination- and loss-free procedures, beginning with sampling via transport and storage to the actual measurement, are correspondingly high [6053, 6060]. Regrettably in a large number of papers the importance of quality control was reprehensibly neglected by over-confident researchers. In this way false data for a large number of elements were accumulated, which other workers then used as a reference for their own (false) results. The agreement of results with published data is in this case not proof for the trueness [6058]. An example is the published contents of chromium in the serum and plasma of non-exposed subjects, shown in Figure 10-2. From values around 200 µg/L in the early 1950s the content has fallen to 0.1–0.2 µg/L, which is nowadays taken as the correct value [4357]. In the following we shall only report on work that, at the current state of knowledge, has given rise to more or less correct results.

Nowadays GF AAS is the preferred technique for the determination of trace elements in body fluids; it allows *in situ* pretreatment so that a digestion is mostly not required. This does not mean of course that the necessary precautions during sampling, transport, storage, and during the determination can be ignored. In a review article DELVES [1477] expressed the view that although GF AAS is easy to use, it is also one of the simplest ways of producing false results with good repeatability. Nevertheless this is based either on errors during sampling or on a disregard of the basic rules for the proper operation of this technique (refer to Sections 1.7.2, 5.4.3.2, and 8.2.4). It has been shown that GF AAS under the consequent application of the STPF concept is most certainly a technique highly capable of determining trace elements in body fluids [5433, 5668, 5672]. The fact that body fluids require little or no pretreatment contributes to the trueness of the results obtained by GF AAS. A further important advantage, especially for clinical applications, is the low sample requirement of only a few microliters.

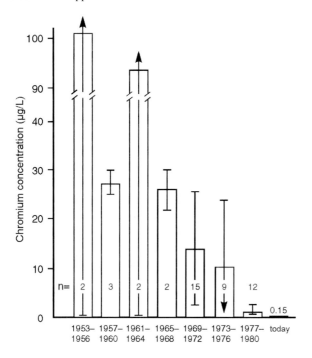

Figure 10-2. Chronological decrease in the published concentration of chromium in the serum and plasma of non-exposed subjects (from [4357]).

Aluminium: It is not known whether aluminium is an essential element and its toxicity is still discussed with controversy [5129]. The greatest problem for the determination of this ubiquitous element is without doubt contamination, as demonstrated impressively by CEDERGREN and FRECH [1118]. For the determination of the aluminium concentration in the blood of 43 subjects in a clinic and observing all of the usual precautions for the prevention of contamination, they found a mean value of 7.5 ± 6.4 µg/L and a blank value of 1.0 ± 0.6 µg/L. When the experiment was repeated on 11 subjects under identical conditions in a clean room the mean value was 1.6 ± 1.3 µg/L and thus very close to the blank value and only about a power of ten above the limit of detection. Aluminium is determined far more frequently in dialysis solutions for extrarenal dialysis than in body fluids since there is a risk of accumulation in the body. A selection of publications on the determination of aluminium by GF AAS in various clinical samples is presented in Table 10-4. Further information is provided in review articles by TAYLOR and WALKER [5785], SCHALLER et al. [5129], BEINROHR et al. [479], OSTER et al. [4497], FROMENT and ALFREY [1992], CEDERGREN and FRECH [1118], SAVORY et al. [5122], and SLAVIN [5430]. Special information on the avoidance of contamination and other problems of sample preparation is given by ERICSON [1764], SKELLY and DISTEFANO [5394], and RÖSICK et al. [4963]; WILHELM and OHNESORGE [6323] provide suggestions on sample storage. In recent times the speciation analysis of aluminium in body fluids by GF AAS after prior separation by HPLC [1519, 1857, 2170, 3067, 3411, 3498, 6389] or ultrafiltration [2170, 6388] has found increasing interest.

Table 10-4. Selected publications on the determination of aluminium in body fluids by GF AAS.

Sample	Procedure	Atomization	Modifier	References
Milk	aqueous calibration solutions; sample diluted 1+3 with 0.2% HNO₃; integrated absorbance	platform, 2600 °C	Pd-Mg	[278]
Plasma	matrix-matched calibration solutions	THGA	Mg(NO₃)₂–Triton X-100	[755]
Plasma, erythrocytes	aqueous calibration solutions	tube wall	Triton X-100	[366, 2197]
Saliva	matrix-matched calibration solutions	tube wall	HNO₃	[5358]
Serum	aqueous calibration solutions	platform	Mg(NO₃)₂ or K₂Cr₂O₇	[618]
Serum	comparison of direct determination with various digestions	tube wall	Triton X-100	[6110]
Serum	396.2 nm line used preferentially because of better linearity	tube wall	Triton X-100–Mg(NO₃)₂	[2074]
Serum	aqueous calibration solutions	tube wall	Triton X-100–K₂Cr₂O₇	[5273]
Serum	comparison of various methods	platform, probe, or tube wall	Triton X-100	[3877]
Serum	calibration solutions matched with 5% albumin, serum diluted 1+4	tube wall	Mg(NO₃)₂	[922]
Serum	matrix-matched calibration solutions, Zeeman-effect BC	platform	Triton X-100	[6159]
Serum	influence of the pyrolysis temperature on the sensitivity	tube wall	—	[3099]
Urine	matrix-matched calibration solutions	THGA	Mg(NO₃)₂–Triton X-100	[755]
Urine	matrix-matched calibration solutions	platform	Triton X-100	[6159]

Arsenic: It has not yet been proven whether arsenic is an essential element for humans, but it is probable [2744]. It is present in all organs and body fluids; its concentration depends to a large degree on the nutrition. Fish, for example, contains relatively large quantities of non-toxic arsenobetaine. Since arsenic is of mainly toxicological interest its determination is discussed in Section 10.1.3.

Chromium: Like aluminium, the greatest source of error for chromium is contamination during sampling, transport, and storage [1763, 3097, 4504, 6012, 6052]. These problems are mentioned at the start of this chapter (see Figure 10-2). In routine analysis GF AAS is used almost exclusively for the determination of chromium [2524, 3353, 4357]. The generally accepted normal value of 0.1–0.2 µg/L in serum and plasma is so close to the detection limit of this technique that the quality of the system used can have a decisive influence on the analysis and the results. The requirements for a direct analysis without a digestion or preconcentration can only be met by a transversely-heated atomizer with an integrated platform. *In situ* ashing in the graphite tube with air or oxygen reduces the background signal substantially and improves the long-term stability. Contamination must definitely be taken into consideration for the use of modifiers [1200]. Continuum source BC is also a problem [590, 6014] since the deuterium lamp exhibits low radiant intensity at a wavelength of 357.9 nm [2336]. The application of Zeeman-effect BC is thus indispensable for this element, unless continuum source BC with a halogen lamp is available [3501, 5819]. A selection of recent papers on the direct determination of chromium in body fluids under the application of Zeeman-effect BC is summarized in Table 10-5; further information can be taken from review articles [2524, 3353, 4995, 6013].

Table 10-5. Selected publications on the determination of chromium in body fluids under the application of Zeeman-effect BC and the standard calibration technique.

Sample	Atomization	Modifier	References
Serum	tube wall	Triton X-100–HNO$_3$	[1200]
Serum, plasma, whole blood, blood constituents	THGA	none	[2200]
Serum, whole blood	platform	none	[5137]
Serum, whole blood	tube wall	Triton X-100–HNO$_3$	[1271]
Urine	THGA	Mg(NO$_3$)$_2$	[752]
Urine	tube wall	none	[4996]
Urine	THGA, without pyrolysis	none	[6544]
Urine	platform	Pd-Mg + H$_2$	[5117]
Urine	tube wall	Triton X-100–HNO$_3$	[1923, 4554]
Whole blood	platform	Pd–Triton X-100–Antifoam	[2585]
Whole blood, plasma, erythrocytes	tube wall	Triton X-100	[3501]

Cobalt: Cobalt is an essential element for humans since it is a constituent of vitamin B_{12}. The normal value for cobalt in serum is probably around 0.1 µg/L [2476, 3352, 5847, 6058] and is thus below the quantification limit of GF AAS. Nevertheless this technique is the most frequently used for the determination of cobalt in body fluids and tissues since NAA, the only technique that is sensitive enough for a direct determination, has a dissipation period of one month [6058], which is mostly too long, and is in any case not generally available. The determination of cobalt in body fluids by GF AAS usually requires a digestion and preconcentration. Dry ashing and dissolution of the residue in a little acid is particularly suitable for this purpose since there is a minimum risk of contamination [37]. Higher cobalt concentrations, such as found in exposed subjects, can be determined directly; the use of Zeeman-effect BC is recommended. Table 10-6 presents a selection of recent applications.

Copper: Copper is to be found throughout the entire body; the concentration is highest in the liver, brain, the heart, and the kidneys. The concentration of copper in serum and plasma is between 0.815 and 1.37 mg/L with a mean of 1.1 mg/L [5102, 6058]. In urine the normal copper values are between 15 µg/24 h and 36 µg/24 h. AAS is by far the most-used technique for the determination of copper in body fluids [1478]. A number of authors attempted to avoid the problems of viscosity in the analysis of serum by F AAS by deproteinizing with trichloroacetic acid; however, systematic errors occurred, mostly due to changes in volume.

In 1971 MERET and HENKIN [4063] described a technique for the direct determination of copper in body fluids which was characterized by its accuracy and simplicity. 0.5 mL serum is diluted 1 + 4 with a 6% aqueous solution of butan-1-ol to eliminate transport interferences. Using the air-acetylene flame the authors obtained a repeatability of 0.4–1.0% in the series and 1.0–1.8% from day-to-day, as well as excellent agreement with colorimetric techniques. DELVES [1478] modified the technique slightly and used it for over 15 000 determinations of copper even in pathological samples with less than 0.3 mg/L.

MANNING [3855] reduced the sample requirement for the determination of copper in serum to 100 µL by applying the injection technique. Using this technique, WEINSTOCK and UHLEMANN [6193] routinely analyzed undiluted and untreated serum, and even after 500 determinations they found neither carryover nor clogging of the burner or nebulizer. They used a pool serum for calibration. The repeatability was 1.8% in the series and 2.2% from day-to-day. SHERWOOD *et al.* [5299] improved this procedure further by applying FI techniques and obtained a repeatability in the series of < 1% for an injection of 120 µL. A selection from recent papers on the determination of copper by F AAS is compiled in Table 10-7.

The most important applications of GF AAS for the determination of copper are the analysis of samples with low copper concentration (e.g. urine), samples of high viscosity (e.g. semen), or for when the sample volume is limited (e.g. serum from newborn babies).

Table 10-8 presents a number of recent publications on this subject. Further information can be obtained from review articles [1478–1480, 5121, 5433, 5435, 5672].

Table 10-6. Selected publications on the determination of cobalt in body fluids by GF AAS.

Sample	Procedure	Atomization	Modifier	References
Milk	acid digestion, Zeeman-effect BC	tube wall	none	[2272]
Plasma	extraction with APDC / IBMK	tube wall	none	[180]
Plasma from patients with metallic implants	direct determination, calibration with reference samples	tube wall	none	[5167]
Plasma, urine	direct determination	tube wall	Pd–Triton X-100–Antifoam	[5073, 5074]
Serum	acid digestion	tube wall	Pd or HNO$_3$	[921]
Serum, urine	acid digestion, extraction with APDC / IBMK	tube wall	none	[407]
Serum, whole blood	1+7 dilution with modifier	tube wall	0.01% Triton X-100	[217]
Synovial fluid	comparison of direct determination and acid digestion	tube wall	Triton X-100 for direct determination	[3650]
Urine	1+1 dilution with modifier, simultaneous determination with Mn	THGA	HNO$_3$–Triton X-100	[2835]
Urine	extraction with DPTH	platform	none	[1317]
Urine	simultaneous determination with Ag, Ni and Cr	THGA	Mg(NO$_3$)$_2$	[752]
Urine	direct determination	platform	Mg(NO$_3$)$_2$	[4132]
Urine	direct determination, comparison with HMA-HMDC extraction	platform	HNO$_3$–Triton X-100	[733]
Urine from exposed workers	extraction with HMDC / xylene + diisopropylketone, aqueous calibration solutions	tube wall	none	[217]
Urine from exposed workers	1+1 dilution with modifier, direct determination, Zeeman-effect BC	tube wall	HNO$_3$–Mg(NO$_3$)$_2$	[3104]
Urine, whole blood	extraction with HMA-HMDTC / diisopropylketone	tube wall	none	[215, 2475]
Whole blood	comparison of direct determination and deproteinization	tube wall	none	[2476]

Table 10-7. Selected publications on the determination of copper in body fluids by F AAS.

Sample	Procedure	Sample introduction	References
Cerebrospinal fluid	matrix-matched calibration solutions	FI	[938]
Erythrocytes	hemolysis or HNO_3 extraction	conventional	[45]
Intraocular fluid	direct determination	FI	[940, 942]
Milk	pressure digestion	FI	[953]
Milk	dilution with detergents	conventional	[265]
Perspiration	direct determination	conventional	[2629]
Plasma	direct determination	injection technique	[3821]
Plasma	various	conventional or injection technique	[356]
Plasma, serum	various	various	[5466]
Plasma, serum	on-line dilution	FI	[5299]
Saliva	direct determination	injection technique	[952]
Serum	direct determination	conventional	[4492]
Serum	direct determination after 1 + 9 dilution with 6% butanol	conventional	[4665]
Serum	direct determination	injection technique	[581, 585, 5945, 6193]
Serum	direct determination	FI	[4917]
Serum	direct determination	conventional	[5062]
Serum, urine	on-line dilution	FI	[247]
Urine	direct determination	conventional	[4925]
Urine	'atom trapping'	conventional	[873]
Urine	metal-protein species	HPLC-F AAS	[6178]
Urine	Australian Standard Method AS 4202.2	conventional	[5553]
Urine	on-line sorbent extraction	FI	[6414]
Urine	'atom trapping'	FI	[6413]
Urine	direct determination	conventional	[2847]
Various	various	various	[1478]
Whole blood	3 replicate measurements from 800 µL	FI	[6417]
Whole blood	ultrasonic treatment with Triton X-100	FI	[4950]
Whole blood	direct determination	FI	[939, 3781]
Whole blood	direct determination	injection technique	[5282]
Whole blood	on-line digestion	FI	[951]

Iron: About 67% of the iron is bound to hemoglobin in the red blood cells, about 27% is bound to ferritin in tissues (liver, bone marrow), and only about 0.8% is to be found bound to transferrin in serum. Of especial diagnostic significance is the determination of iron in serum (650–1750 µg/L) and in liver tissue, and also the iron binding capacity (2500–4500 µg/L) in serum [119]. Although F AAS exhibits adequate sensitivity to determine iron in about 1:5 diluted serum with sufficient sensitivity, this technique finds relatively little application. The reason is that with this direct technique higher values are systematically found compared to colorimetric techniques. Various authors have proposed deproteinization [542, 5783, 6533] or extraction [1475, 5407, 6533], but this requires too much effort for routine analyses. On the other hand, the iron binding capacity [4470, 6513, 6533] and iron in hemoglobin [2526, 6534, 6535] can be determined with advantage by F AAS.

Since conventional F AAS and also colorimetric techniques require about 2 mL sample, alternate sample introduction techniques are preferred when this quantity of serum is not available, such as the injection technique or, even better, flow injection. With FI,

Table 10-8. Selected publications on the determination of copper in body fluids by GF AAS.

Sample	Procedure	Atomization	Modifier	References
Cerebrospinal fluid	direct determination	tube wall	none	[44, 506]
Corpuscular constituents of blood	Zeeman-effect BC	tube wall	none	[5168, 5170]
Milk	direct determination	tube wall	Triton X-100	[262]
Milk	direct determination	platform	$Mg(NO_3)_2$–Triton X-100	[394]
Milk	direct determination	platform	Mg–Ni	[5071]
Plasma	direct determination, Zeeman-effect BC	tube wall	none	[5167]
Saliva	matrix-matched calibration solutions	tube wall	none	[5358]
Seminal fluid	acid digestion	platform	none	[155]
Serum	direct determination	tube wall	Triton X-100–NH_4NO_3–Mg	[2927]
Serum proteins	HPLC separation	platform	none	[2142]
Serum proteins	HPLC separation	tube wall	none	[1329]
Serum, plasma	direct determination	platform	$Mg(NO_3)_2$–$NH_4H_2PO_4$	[3446]
Serum, urine	direct determination, rapid program	tube wall	Triton X-100–HNO_3	[6160]
Urine	direct determination, 'hot injection'	tube wall	Pd	[2609]
Urine	matrix-matched calibration solutions	tube wall	HNO_3	[1615]
Urine	direct determination	THGA	none	[6544]
Urine	direct determination, rapid program	tube wall	none	[2335]
Various	direct determination	tube wall	Pd	[3356]
Various	direct determination	platform	none	[4013]
Whole blood	matrix-matched calibration solutions	tube wall	Triton X-100–EDTA–propanol	[1615]

sample preparation steps such as digestion, dilution, and the addition of buffers can be integrated in the system. Table 10-9 summarizes a number of recent publications on the determination of iron in body fluids by F AAS.

Graphite furnace AAS can be applied when only microliter quantities of body fluids are available. For the determination of iron in serum, OLSEN et al. [4467] proposed a technique for deproteinization at a micro scale which provides correct results even for only 20 μL serum [6480]. Micro techniques have also been established for the determination of the iron binding capacity by GF AAS [555, 6480] which exhibit very good agreement with the macro method of OLSON and HAMLIN [4470]. The sample requirement is also only 20–40 μL serum. Using a two-step furnace and time-separated atomization, SMITH and HARNLY [5462] were able to distinguish between iron bound to heme and iron not bound to heme. D'HAESE et al. [1519] reported the speciation analysis of iron following HPLC separation. Table 10-10 summarizes a number of papers on the determination of iron in body fluids.

Table 10-9. Selected publications on the determination of iron in body fluids by F AAS.

Sample	Procedure	Sample introduction	References
Cerebrospinal fluid	direct determination, matrix-matched calibration solutions	FI	[938]
Intraocular fluid	direct determination, on-line dilution	FI	[102, 940, 942]
Milk	pressure digestion	FI	[953]
Milk	direct determination after homogenization with detergents	conventional	[265]
Plasma	direct determination	injection technique	[3822]
Saliva	direct determination	FI	[952]
Serum	direct determination	conventional	[3824]
Serum	direct determination and iron binding capacity, comparison with other methods	conventional	[6024]
Serum	direct determination	injection technique	[585, 5945]
Serum	on-line dilution	FI	[941]
Serum	protein precipitation, total iron and iron binding capacity	FI	[4918]
Serum, saliva	on-line dilution	FI	[101]
Serum, urine	direct determination	conventional	[1442]
Urine	direct determination	conventional	[4925]
Urine	speciation analysis metal complexes	HPLC-F AAS coupling	[6178]
Various	on-line dilution	FI	[949]
Whole blood	microwave-assisted on-line digestion	FI	[951]
Whole blood	direct determination	injection technique	[5282]

Table 10-10. Selected publications on the determination of iron in body fluids by GF AAS.

Sample	Procedure	Atomization	Modifier	References
Cerebrospinal fluid	matrix-matched calibration solutions	tube wall	Mg	[2240]
Granulocytes	oxygen ashing	tube wall	Na_2HPO_4–Na_2EDTA	[3966]
Intraocular fluid	direct determination	tube wall	HNO_3	[2059]
Intraocular fluid	total iron and iron binding capacity	tube wall	Triton X-100	[4014]
Milk	direct determination	platform	none	[5071]
Milk	direct determination	tube wall	n-octanol	[4127]
Milk (colostrum)	direct determination	tube wall	Triton X-100	[262]
Saliva	matrix-matched calibration solutions	tube wall	HNO_3	[5358]
Serum	Fe-protein speciation after HPLC separation	tube wall	none	[3411]
Serum	direct determination or deproteinization	tube wall	none	[542]
Urine	'hot injection'	tube wall	Pd–H_2	[3356]

Table 10-11. Selected publications on the determination of manganese in body fluids by GF AAS.

Sample	Procedure	Atomization	Modifier	References
Corpuscular blood constituents	acid digestion	tube wall	none	[4118]
Milk	direct determination	tube wall	Triton X-100	[262]
Milk	direct determination, ZBC	platform	$Mg(NO_3)_2$	[5071]
Perspiration	direct determination	tube wall	none	[1449]
Saliva	matrix-matched in reference solutions	tube wall	none	[5358]
Serum	direct determination, ZBC	platform	none	[4550]
Serum	direct determination, ZBC	tube wall	Triton X-100–EDTA	[4335]
Serum	direct determination	tube wall	Triton X-100	[4551]
Serum	deproteinization with trichloroacetic acid	tube wall	none	[4711]
Serum	direct determination	tube wall	HNO_3	[2716]
Urine	direct determination	platform	HNO_3	[1968]
Urine, whole blood	direct determination	platform	Mg	[237]
Whole blood	direct determination, oxygen ashing	platform	Triton X-100–HNO_3	[2349]
Whole blood	direct determination	tube wall	n-octanol	[134]

Manganese: The greater part of the manganese is bound to erythrocytes, so that even slight hemolysis leads to increased manganese values in serum. The blood sample should be taken with a plastic needle since contamination has been observed with steel needles [6049, 6058]. The normal value for manganese in serum or plasma is 0.5 µg/L [6058] and in whole blood about 8 µg/L [6049, 6058]. The normal value in urine at around 1 µg/L is likewise very low. The sensitivity of F AAS is thus insufficient, even though this technique was used frequently earlier for the determination of this element. Manganese can be determined with good accuracy in 1+1 diluted serum by GF AAS [2334]; the use of an efficient background corrector is essential. The use of a transversely-heated atomizer with integrated platform allows noticeably lower limits of detection and thus increases the measurement reliability. BARUTHIO et al. [406] have written a good review article on the determination of manganese in a multitude of biological samples by GF AAS. Contamination also plays a major role in the determination of manganese [1485, 4291, 6053, 6057], and many of the values published in the literature, for example values > 0.65 µg/L in serum, are probably not correct [6057]. Manganese deficiency symptoms appear to be very seldom and intoxication is limited mainly to the inhalation of manganese vapors or dust [1262]. Table 10-11 presents a number of recent publications on the determination of manganese in body fluids.

Molybdenum: The average ingestion of molybdenum in the nutrition is markedly higher than the calculated requirement of 25 µg/day, so that molybdenum deficiency symptoms are highly improbable. Nevertheless, in certain regions of the Earth a correlation has been established between an increased molybdenum concentration in the soil (and thus in the nutrition) and the occurrence of certain diseases [220].

The normal concentration of molybdenum in human plasma, serum, or whole blood is with certainty < 1 µg/L and the probable value is around 0.6 µg/L [6058]. Apart from RNAA, only GF AAS has provided reliable values in this range [1763]. Significantly higher values that have been reported by numerous authors are due either to contamination, inadequate BC, or other errors in the analysis. Without doubt a direct determination by GF AAS in a transversely-heated atomizer with integrated platform under STPF conditions and oxygen ashing offers the best prospects of a correct determination with adequate sensitivity. Under these conditions the doubts expressed a few years ago by the IUPAC commission about the suitability of GF AAS [3880] should be allayed and the stipulated conditions be met. A selection from recent publications on the determination of molybdenum in body fluids by GF AAS is compiled in Table 10-12.

Table 10-12. Selected publications on the determination of molybdenum in body fluids by GF AAS.

Sample	Atomization	Modifier	References
Milk	tube wall	BaF_2–octan-1-ol	[559]
Plasma	tube wall	none	[5169]
Plasma	tube wall	none	[5167]
Serum	tube wall	CaF_2	[2062]
Serum	tube wall	Triton X-100–$Mg(NO_3)_2$	[4662]
Serum	tube wall	$Mg(NO_3)_2$ or BaF_2 or HNO_3 or Pd–$OHNH_2HCl$	[561]
Synovial fluid	tube wall	Triton X-100	[3650]
Urine		Pd–Mg	[4663]
Urine	tube wall	none	[4658]

Nickel: In 1984, SUNDERMAN [5696] reported in a review article that the nickel concentration in the serum and plasma of healthy, non-exposed subjects was 2 µg/L, and that there was excellent agreement of the values obtained by various groups in Germany, Japan, Spain, and the U.S.A. A short time later SUNDERMAN *et al.* [5697] published a much lower normal value of 0.46 ± 0.26 µg/L for nickel in serum, which they then corrected downward to 0.3 ± 0.3 µg/L in the following year [3478]. This development resulted from new knowledge about the avoidance of contamination [5699], from improvements in the GF AAS system, and especially from the application of Zeeman-effect BC. In the meantime the normal value proposed by Sunderman has been verified by other working groups [183, 6348] and is nowadays considered as reliable [5130, 5794]. Of decisive importance for the accuracy of the GF AAS determination of nickel in serum or plasma are the exclusion of contamination during sampling, a minimum of sample preparation, the application of *in situ* oxygen ashing, the STPF concept, and Zeeman-effect BC [6058]; the best results are obtained in a transversely-heated atomizer with integrated platform. A selection from recent publications on the determination of nickel in body fluids is compiled in Table 10-13.

Table 10-13. Selected publications on the determination of nickel in body fluids by GF AAS.

Sample	Procedure	Atomization	Modifier	References
Plasma	direct determination	tube wall	Triton X-100–HNO$_3$	[183]
Serum	direct determination, ZBC	tube wall	Triton X-100–HNO$_3$	[4372]
Serum, whole blood	FI on-line extraction after digestion	tube wall	none	[5774]
Urine	direct determination	THGA	none	[752]
Urine	direct determination	tube wall	Triton X-100	[4474]
Urine	direct determination	tube wall	Triton X-100–HNO$_3$	[4552]
Urine	direct determination, hot injection	tube wall	Pd–H$_2$	[3356]
Urine	direct determination	tube wall	HNO$_3$	[5698, 6299]
Whole blood	FI on-line coprecipitation after digestion	platform	none	[1830]
Whole blood	multistep pyrolysis for the injection of 50 µL	tube wall	Triton X-100–octanol	[218]

Selenium: Originally selenium was mainly of interest because of its toxicity, which had been known for a relatively long period and which is also naturally still of interest nowadays [6321]. A large number of working groups has been involved with this element since the discovery that it is essential [5209] and that it plays a role in glutathione peroxidase activity [4970]. A number of observations led to this discovery: It was shown in animal experiments that selenium had an anti-carcinogenic activity on tumors [2209, 2221]; a number of authors pointed to the connection between the ingestion of selenium in the nutrition and the frequency of cancer in various population groups [5186, 5187, 6327]. Further, selenium plays an important role in the prevention of diseases of the coronary vessels and of cardiac infarction. Epidemically an inverse relationship between the selenium concentration in the environment or in blood and the occurrence of heart diseases has been observed [5067, 5262]. The Keshan disease, a type of cardiomyopathy which is observed in parts of China and especially among children, could be practically eliminated by a large-scale action to supplement selenium in the diet [1244, 3087]. Moreover selenium has an antagonistic action against a number of toxic metals such as arsenic, cadmium, lead, mercury, and methylmercury compounds [649, 2366, 3191, 4401]. A number of symptoms in patients with parenteral nutrition could be attributed to selenium deficiency [5021].

The concentration of selenium in serum or plasma has been thoroughly investigated by numerous working groups in many interlaboratory trials and can nowadays be considered as reliable. The 'normal' range is 70–130 µg/L. The differences in the values [6058] found by different working groups could be ascribed to geographical and nutritional causes. The selenium concentration in the urine of non-exposed subjects is around 30 µg/L (range 20–200 µg/day), but can increase to values > 1000 µg/L in regions with a high selenium level [4887, 4888, 5091].

The risk of contaminating a serum sample with selenium appears to be relatively low. Nevertheless for both GF AAS and HG AAS method-specific problems have been observed that can lead to systematic deviations of up to 50% [2749]. In the meantime the causes for these systematic errors have been recognized and means for their avoidance have been described [6250].

In the case of GF AAS, selenium losses during pyrolysis due to the use of unsuitable modifiers as well as errors in BC were responsible for values that were too low. As explained in Section 8.2.3.13, the nickel modifier used earlier in the determination of selenium was not able to stabilize varying oxidation states of selenium (selenite, selenate) nor organic selenium compounds (selenomethionine, selenocysteine, etc.). Copper was originally proposed as an alternative [6219], but nowadays the Pd-Mg modifier is used almost exclusively to stabilize selenium and prevents such losses. The second problem— the spectral interferences caused by iron and phosphate that lead to overcorrection and thus to reduced values when continuum source BC is applied [2963, 5039]—are discussed fully in Section 8.2.4.2 (see also Figure 8-16). In the meantime it has been demonstrated that by applying the STPF concept with an optimized modifier and Zeeman-effect BC, correct values for selenium in body fluids can be obtained [4333, 6250] and that GF AAS is suitable for routine use in the clinical laboratory [3499, 4195]. Selected recent publications on the determination of selenium in body fluids by GF AAS are compiled in Table 10-14.

The main problem with HG AAS is that only ionic selenium(IV) can be determined (see Section 8.3.3.2, part 1) and that without proper precautions the digestion of organic selenium compounds is often not complete [6040, 6225]. Compounds such as selenocysteine, selenomethionine (most important compound in blood and serum), or the trimethylselenonium ion (in urine) are exceedingly stable [4330, 6040] and require strong oxidizing conditions for digestion, for example with nitric acid, sulfuric acid, and perchloric acid at high temperature [6225, 6239]. Under such conditions reproducible and true values for selenium in various body fluids can be obtained by HG AAS [6239]. The major advantage of HG AAS compared to GF AAS is the higher sensitivity by

Table 10-14. Selected publications on the determination of selenium in body fluids by GF AAS using Zeeman-effect BC.

Sample	Procedure	Atomization	Modifier	References
Corpuscular blood constituents	direct determination	platform	Pd–Mg	[5007]
Intraocular fluid	direct determination	platform	Pd	[4015]
Seminal fluid, sperm	direct determination, oxygen ashing	platform	Cu–Mg–Triton X-100	[4334]
Serum	analyte addition technique	THGA	Pd–Mg	[1399]
Serum	calibration with reference samples	THGA	Rh-Mg	[2323]
Serum	direct determination	tube wall	Pd–ascorbic acid–Triton X-100	[3172]
Serum	oxygen ashing	platform	Cu–Mg	[4332]
Serum, urine	direct determination	platform	comparison of various modifiers	[2928]
Urine	rapid program without pyrolysis	THGA	none	[6544]
Whole blood	direct determination	platform	Pd–Triton X-100	[1286]
Whole blood	oxygen ashing	platform	Ir–Mg, injected prior to the sample	[2614]
Whole blood	direct determination	tube wall	Pd–ascorbic acid	[3170, 4673, 4674]

around an order of magnitude. The low values associated with selenium deficiency of around and below 10 µg/L in serum can thus be reliably determined. When the FI technique is applied the sample requirement is hardly higher than for GF AAS [759]. Table 10-15 presents a number of recent publications on the determination of selenium in body fluids by HG AAS.

In recent years the determination of selenium species in biological samples has gained in importance [1329, 4888]. For this purpose LC separations with off-line analysis of the individual fractions by GF AAS [85, 1949, 3384, 6041] or by HG AAS [6037] have been used, or also directly by LC-GF AAS [3385] and LC-HG AAS [657]. A simple distinction can be made between inorganic and organic selenium compounds by HG AAS since without prior digestion only the inorganic species can be detected.

Silicon: Up to the present there are hardly any references in the literature on silicon in serum. Lo and CHRISTIAN [3564] reported a mean value of 770 µg/L (n = 5) using GF AAS. Nevertheless, due to the exceedingly high risk of contamination in the determination of silicon, this unexpectedly high value should be regarded with extreme caution [6058]. Because of the ubiquitous presence of this element, the risk of contamination exists particularly in all containers, in the ambient air, and also from tubing made of silicones that are frequently used in graphite furnaces [2136]. Selected publications on the determination of silicon in body fluids by GF AAS are compiled in Table 10-16.

Table 10-15. Selected publications on the determination of selenium in body fluids by HG AAS.

Sample	Digestion	System	References
Corpuscular blood constituents	HNO_3–HCl–$Mg(NO_3)_2$	FI	[2371]
Milk	high pressure ashing	batch	[5183]
Milk	HNO_3–$HClO_4$	batch	[1493]
Plasma, serum	HNO_3–H_2SO_4–$HClO_4$	FI	[4028]
Plasma, urine, whole blood	HNO_3–$HClO_4$	batch	[3771]
Serum	HNO_3–$HClO_4$	FI-HG-GF AAS	[1399]
Serum	HNO_3–H_2SO_4–$HClO_4$	FI	[759]
Serum	microwave-assisted on-line digestion	FI	[4069]
Serum, urine, whole blood	HNO_3–H_2SO_4–$HClO_4$	batch	[6225, 6235, 6239]
Urine	pressure digestion with HNO_3–$HClO_4$	FI-HG-GF AAS	[6539]
Urine	multistep digestion with HNO_3, H_2SO_4, $HClO_4$, H_2O_2, HCl	FI-HG-GF AAS	[4346]
Urine	HNO_3–H_2SO_4–$HClO_4$	batch	[5125]
Urine, whole blood	HNO_3–H_2SO_4–$HClO_4$ or HNO_3–HCl + $Mg(NO_3)_2$	batch	[2369]
Whole blood	HNO_3–H_2SO_4–$HClO_4$	FI	[6256]
Whole blood	HNO_3–H_2SO_4–$HClO_4$	batch	[6250]

Table 10-16. Selected publications on the determination of silicon in body fluids by GF AAS.

Sample	Procedure	Atomization	Modifier	References
Plasma	direct determination	platform	$Ca(NO_3)_2$	[2646]
Plasma, urine	direct determination	W-coated platform	K_2EDTA–KH_2PO_4	[2136]
Saliva	matrix-matched calibration solutions	tube wall	none	[5358]
Serum, urine	direct determination	tube wall	$CaCl_2$–$La(NO_3)_2$–$NH_4H_2PO_4$	[2701]
Serum, urine	direct determination	W-coated tube wall	none	[4605]
Urine	1+349 dilution	tube wall	$NiCl_2$	[3182]

D'HAESE *et al.* [1519] reported the determination of silicon species following HPLC separation.

Tin: Although it was established in 1970 that tin is an essential element for animals [5171], up to the present its biological action is still largely unknown. The normal value for tin in serum is probably 0.5 µg/L [6059]; nevertheless there are few reliable publications. Both GF AAS and HG AAS can be used for the determination of tin [739]. Since the limits of detection are close to the normal value for tin in serum, preconcentration is required for both techniques. Because of the associated sources of error, the determination by HG AAS with *in situ* preconcentration in the graphite tube [5635, 5773, 6543] would appear to offer the best prerequirements for good accuracy. The direct determinations of tin in urine by HG AAS after $1 + 5$ dilution [6483] or in whole blood by GF AAS after acid digestion [1257] are only suitable for the determination of markedly increased concentrations.

Vanadium: The content of vanadium in body fluids such as serum, blood, or urine is still the subject of controversial discussions [677]. In a comprehensive review article, HEYDORN [2540] established that due to the attainable limits of detection only GF AAS and RNAA were suitable for this determination. The mean value taken from the concentrations determined by 12 working groups is 0.2 µg/L V. In contrast to this, VERSIERCK and co-workers [1004, 6054, 1327] using RNAA found values of < 0.1 µg/L in pool serum and also in a group of healthy subjects. The limit of detection and the quality of BC as well as the freedom from contamination play a decisive role in the GF AAS determination of vanadium. The application of Zeeman-effect BC and the use of a transversely-heated atomizer with integrated platform certainly make a major contribution to the accuracy. Selected publications on the determination of vanadium in body fluids by GF AAS are compiled in Table 10-17.

Table 10-17. Selected publications on the determination of vanadium in body fluids by GF AAS.

Sample	Procedure	Atomization	Modifier	References
Serum, urine	direct determination	tube wall	Pd	[2195]
Urine	direct determination	THGA	HNO_3	[5330]
Urine	direct determination	tube wall	Pd–Mg–NaF	[1924]
Urine	direct determination	tube wall	Triton X-100–HNO_3	[4553]
Urine	rapid determination without pyrolysis	THGA	none	[6544]
Urine	extraction with cupferron/IBMK	tube wall	none	[236, 249]
Urine, whole blood	direct determination, air ashing	platform	Pd–citric acid–HNO_3–Triton X-100	[5743]

Zinc: The human body contains on average 2.5 g Zn, of which the main part is in the muscles (60%) and the skeleton (30%). Nowadays the zinc level is used for the early diagnosis of various diseases, so that an accurate determination of this element is required. Serum and plasma are the two body fluids analyzed most frequently. The usual normal values nowadays lie between 0.8 and 1.2 µg/L; they depend on the age, sex, and state of health of the subject [5848, 6058]. Contamination is a frequently observed problem in the determination of zinc and can be responsible for a number of the higher 'normal' values. VERSIECK *et al.* [6056] introduced a contamination-free reference serum in 1988 with a zinc concentration of 0.873 ± 0.018 mg/L, which probably comes very close to the true normal value.

Because the zinc concentration in the majority of body fluids is relatively high, the risk of contamination during sampling from stainless steel needles or from the syringe is not very great provided sampling is performed carefully. Potential sources of contamination are especially reagents, the laboratory air, and rubber bungs [4844, 6053]. For the determination of zinc the smallest amounts of reagents and only ultrapure reagents should generally be used. The greatest danger is from anticoagulants which are frequently contaminated with zinc. A further risk is the high zinc concentration in red blood cells; even partially hemolyzed serum or plasma must be discarded [5848].

Due to the high zinc concentration in body fluids and tissues F AAS is normally used for the determination of this element. Body fluids can usually be analyzed after simple dilution and without prior mineralization [4010], as shown in the summary in Table 10-18. In a review article, DELVES [1478] discussed various techniques of sample pretreat-

Table 10-18. Selected publications on the determination of zinc in body fluids by F AAS.

Sample	Procedure	Sample introduction	References
Cerebrospinal fluid	matrix-matched calibration solutions	FI	[938]
Cerebrospinal fluid	analyte addition technique	injection technique	[44, 4521]
Lacrimal fluid	direct determination	conventional	[5024]
Milk	direct determination after dilution with Triton X-100	conventional	[262]
Milk	pressure digestion	conventional	[5350]
Milk	homogenization with detergents	conventional	[265]
Ocular fluids	direct determination after on-line dilution	FI	[940]
Plasma	direct determination after 1+4 dilution	conventional	[4922]
Plasma	digestion with HNO_3	conventional	[4559]
Plasma, serum	various	conventional	[5497]
Plasma, whole blood	direct determination	FI	[945]
Saliva	direct determination	FI	[952]
Serum	direct determination after 1+9 dilution	conventional	[4665]
Serum	direct determination after 1+5 dilution with Brij-35	conventional	[4614]
Serum	direct determination	injection technique	[5945, 6313]
Serum	direct determination	conventional	[5062]
Serum	deproteinization with trichloroacetic acid	FI	[309]
Serum, urine	direct determination after on-line dilution	FI	[247]
Serum, whole blood	digestion with HNO_3 + $HClO_4$	FI	[666]
Urine	digestion with HNO_3 + $HClO_4$	conventional	[4925]
Urine, whole blood	digestion with HNO_3 + H_2O_2	conventional	[1105]
Whole blood	direct determination	injection technique	[5282]
Whole blood	ultrasonic treatment with Triton X-100	FI	[4950]
Whole blood	microwave-assisted on-line digestion	FI	[951]

ment and analysis in great detail. The application of GF AAS is only sensible for the determination of extremely low zinc concentrations or for very small sample amounts, such as is the case for speciation analysis in protein fractions after prior separation [261, 355, 1856, 1934–1936, 1937, 2071, 2072, 2142, 6302]. The contamination problems in GF AAS are much more severe than in F AAS. Selected publications on the determination of zinc by GF AAS are compiled in Table 10-19.

Table 10-19. Selected publications on the determination of zinc in body fluids by GF AAS.

Sample	Procedure	Atomization	Modifier	References
Corpuscular blood constituents	direct determination, Zeeman-effect BC	tube wall	none	[5170]
Milk	direct determination after 1+99 dilution with Triton X-100	tube wall	Triton X-100	[259, 260]
Saliva	matrix-matched calibration solutions	tube wall	HNO$_3$	[5358]
Seminal fluid	microwave-assisted digestion with HNO$_3$, analyte addition technique	platform	none	[155]
Seminal fluid	pressure digestion	platform	Pd–Mg	[467]
Serum	direct determination after 1+19 dilution	tube wall	Triton X-100 + (NH$_4$)$_3$PO$_4$	[27]
Serum, plasma, whole blood	direct determination, Zeeman-effect BC	platform	Mg–NH$_4$NO$_3$– Triton X-100	[1372, 2927]

10.1.3 Industrial Hygiene and Toxicology

As we mentioned earlier, there are no toxic elements but only toxic concentrations. Even essential trace elements can cause damage to health or even death at increased concentrations (refer to Figure 10-1). The form (species) in which an element is ingested also plays a major role in its resorbability or toxicity.

A number of the elements discussed in Section 10.1.2 also play a major role in toxicology and industrial hygiene, such as chromium, nickel, and selenium. We shall not discuss these elements again, however, since chemical analysis is not noticeably different from that discussed above, except that the concentrations are higher in toxicology and industrial hygiene. This means that for body fluids sample preparation is not required and we can perform a direct determination by GF AAS. Nevertheless, all precautions against contamination during sampling must still be rigorously observed.

A further difference is frequently in the types of samples investigated. While mostly serum or plasma are investigated in the clinical domain, for toxicology it is the element concentration in whole blood and tissue samples, and the excretion in urine, that is of interest, particularly for screening and during detoxification. Urine can mostly be analyzed directly without sample pretreatment by GF AAS, while tissue samples usually require a digestion (this is discussed in detail in Section 10.1.5).

Arsenic: We mentioned in Section 10.1.2 that arsenic is possibly an essential element. Nevertheless, arsenic is notorious for the toxicity of a number of its compounds; from the late middle ages until well into the 19th century white arsenic (As_2O_3) was widely used to solve private and political problems. All soluble compounds of trivalent and pentavalent arsenic have toxic effects; these include the oxides and halides, arsenites and arsenates, the gaseous arsine (AsH_3), and mono- and dimethylated compounds. A number of organic arsenic compounds, such as methyl- and ethyldichloroarsine or diphenylchloroarsine, have been used as poison gases. On the other hand, metallic arsenic, metal arsenides, compounds with sulfur, and arsenobetaine, which is mainly ingested with fish and shellfish, are not toxic [2744].

As we mentioned in Section 10.1.2, the 'normal' value for arsenic depends strongly on the nutrition, and particularly on the consumption of fish. Owing to the marked differences in the toxicity of the individual arsenic species [2806, 4301, 5533], the total concentration of arsenic is not an indication of possible undue ingestion. For this reason the publications compiled in Table 10-20 on the determination of arsenic by GF AAS are only conditionally suitable for toxicological tasks and can solely be regarded as screening procedures. This is also true for the determinations of total arsenic by HG AAS after complete digestion in the citations listed in Table 10-21. The question of

Table 10-20. Selected publications on the determination of total arsenic in urine by GF AAS.

Procedure	Atomization	Modifier	References
Direct determination without pyrolysis	THGA	none	[6544]
Direct determination	platform	Pd–persulfate	[4373]
Direct determination for total arsenic, extraction protocol for species	platform	Pd or Pd–ascorbic acid	[5671]
Direct determination, Zeeman-effect BC	platform	Ni–Mg–HNO_3	[4549]

Table 10-21. Selected publications on the determination of total arsenic and toxic or 'hydride-forming' arsenic in body fluids by HG AAS.

Sample	Procedure	Species	System	References
Serum	microwave-assisted digestion	total arsenic	batch	[3995]
Urine	microwave-assisted on-line digestion	total arsenic	FI	[5519]
Urine	HNO_3–$HClO_4$ digestion	total arsenic	*in-situ* preconcentration in the graphite tube	[6539]
Urine	direct determination	hydride-forming arsenic	batch	[6209]
Urine	direct determination	hydride-forming arsenic	FI	[4245]
Urine	microwave-assisted on-line digestion	toxic arsenic	FI	[3475]
Urine, whole blood	various	total arsenic	batch	[74]
Whole blood	microwave-assisted on-line digestion	total arsenic	FI	[6270]

an intoxication can best be answered by HG AAS without a prior digestion, or with only a 'mild' digestion, since arsenobetaine is not detected. Several selected publications on this simplified speciation analysis are also cited in Table 10-21. This distinction between 'toxic' and non-toxic arsenic is adequate for the majority of applications, but techniques have also been described in which individual species or groups of species have been selectively detected by HG AAS [191]. Separation analysis of toxic and non-toxic arsenic can also be performed by selective extraction [567] or on microcolumns in an FI system, either off-line [4374] or on-line [2356]. A true separation analysis of individual species is only possible by HPLC [4197] and detection by GF AAS [526, 792, 793, 1162, 1911, 2739, 5578, 6382, 6383] or HG AAS [3477, 299, 1185, 4162, 6017, 6551].

Cadmium: Acute poisoning at the workplace is caused by inhaling cadmium vapors; during the 1940s there were also reports about acute poisoning from foodstuffs [1982]. Such cases are seldom nowadays due to the improvement in hygienic conditions. Chronically increased cadmium values are nevertheless quite common and depend on the working environment and the way of life. The major sources of cadmium are food (rice, wheat, mussels, kidneys) and especially smoking. In numerous publications a clear relationship has been established between the number of cigarettes smoked and the cadmium concentration in blood and urine [1982, 2514, 6048].

The cadmium concentration in blood is an indication of the intake of cadmium during the preceding few months and is thus very useful for control purposes. The kidneys are the most critical organ with respect to cadmium accumulation, since the major quantity is deposited there. The cadmium concentration in urine reflects the total impact on the body very well, but it can also indicate short-term acute accumulation, for example from the workplace [1982, 2514, 3467]. The cadmium concentration in the blood and urine of healthy, non-exposed adults is 0.1–4 µg/L and 0.05–2 µg/L, respectively [1785]. The values for non-smokers are usually < 1 µg/L.

By far the most-used technique for the determination of cadmium in blood and urine is GF AAS [693, 1239, 2514]. Limits of detection of 0.005 µg/L can be attained in a transversely-heated atomizer with integrated platform under STPF conditions, so that the prerequirements are met for the reliable detection of even the lowest cadmium concentrations. The risk of contamination on the one hand and the volatility of cadmium and its compounds on the other mean that procedures must be performed with painstaking care and that all parameters must be carefully optimized.

The most frequently used modifier for the determination of cadmium in blood or urine is ammonium phosphate [1476, 2555, 4725], occasionally with the addition of nitric acid [4565] or magnesium nitrate [4565]. Cadmium is thermally stabilized up to 500–600 °C with this modifier. For the analysis of blood, *in situ* oxygen ashing is recommended since otherwise extremely high background signals occur and residues build up very rapidly in the graphite tube (refer to Figure 8-30), which can influence the trueness and precision. For the determination of cadmium in urine a palladium nitrate-ammonium nitrate mixed modifier has proven to be better [5451, 5485, 6481], since a clearly lower background signal is obtained (see Figure 8-34) and a higher pyrolysis temperature of about 900 °C can be applied. For the determination of cadmium in biological materials, Zeeman-effect BC and strict maintenance of STPF conditions are essential prerequisites for the accuracy of the values obtained since it is not possible to separate concomitants completely due to the volatility of this element. A selection from publications on the determination of cadmium in body fluids by GF AAS is listed in Table 10-22.

Lead: Investigations on bones have shown that chronic lead poisoning due to the use of lead-containing pewter vessels was common in ancient Rome. In our century the major sources of lead in our environment come from leaded fuels, from lead paints, and from lead water pipes. As a result of sharpened regulations, the level of pollution has decreased steadily over the last years [1272]. Lead is nevertheless a widely distributed metal that finds a multitude of applications [5669], so that the actual impact depends very much on the living and working environment and also on the way of life. Much lower levels of lead are found in rural areas compared to towns, for example. Smoking and the consumption of alcohol (especially wine) contribute significantly to an increase of the lead concentration in blood [1272]. Despite the steadily decreasing impact, lead is still one of the most-investigated elements, especially in the U.S.A. The reason for this is that in children a relationship has been established between lead and the memory and the ability to learn [505, 854, 3443, 6505].

Since lead is mostly bound to erythrocytes its determination is performed preferentially on whole blood. Occasionally the excretion in urine is also used as an indicator for lead pollution. Lead is accumulated to the greatest extent in bones and teeth with a biological half-life of more than 10 years and serves in this case as an indicator for long-term impact. The analysis of teeth plays a major role [1878] since they are relatively easily available, especially from children.

It is assumed that the natural lead concentration prior to the onset of pollution was 2 µg/L [4566]. Nowadays in industrial areas the mean values are around 40–200 µg/L, in other words values that are far removed from a possible 'normal' value.

The most widely used technique for the determination of lead in blood is GF AAS since it permits a direct determination from only a few microliters of blood. The first direct technique, in which blood was merely diluted with Triton X-100, was described by FERNANDEZ [1888]. This author also later modified the technique for STPF conditions and applied ammonium phosphate as the modifier [1894]. In principle this technique is still generally used nowadays; occasionally, apart from Triton X-100 and ammonium phosphate, nitric acid [4090] or magnesium nitrate [2927] is used as the modifier and calibration is performed against spiked blood samples [2927, 4090, 5324]. Zeeman-effect BC is also preferred by many authors compared to continuum source BC [2865, 2927, 4090, 5324], even though good results have been attained with the latter [1518]. Table 10-23 presents a selection of publications taken from the profusion on the determination of lead in body fluids by GF AAS.

Mercury is one of the classical toxic elements; it is also discussed in detail in other sections (see Sections 8.4 and 9.27). An increased ingestion of mercury frequently comes from nutrition [6385, 6386], where particularly fish and fish products can frequently exhibit high concentrations. Locally, air and drinking water can contribute substantially to the ingestion of mercury [6386]. A further source of mercury that has been recognized in recent years is dental amalgam used in tooth fillings [1607, 1608]; a linear relationship has been found between the number of filled teeth and the concentration of mercury in the body. Mercury poisoning can occur at the workplace through the inhalation of mercury vapor; organic compounds such as methylmercury are particularly dangerous [6385].

Table 10-22. Selected publications on the determination of cadmium in body fluids by GF AAS.

Sample	Procedure	Atomization	Modifier	References
Saliva	direct determination after dilution with modifier	platform	Triton X-100–HNO$_3$	[6300]
Seminal fluid	direct determination	platform	Pd–Mg	[467]
Serum	direct determination after dilution with modifier	platform	NH$_4$NO$_3$–Triton X-100–HNO$_3$	[2927]
Serum	direct determination	platform	Pd–Mg	[919]
Urine	direct determination without pyrolysis	THGA	none	[6544]
Urine	direct determination after dilution with modifier	platform	Triton X-100–HNO$_3$	[1952]
Urine	direct determination	platform	Pd–NH$_4$NO$_3$	[5451]
Urine	HNO$_3$ digestion	THGA	Pd–NH$_4$NO$_3$	[5485]
Urine	direct determination after 1+1 dilution with modifier, without pyrolysis	platform	HNO$_3$	[2338]
Urine	direct determination	platform	(NH$_4$)$_2$HPO$_4$–HNO$_3$	[5664]
Urine	direct determination	platform	Pd–NH$_4$NO$_3$	[6481]
Urine	microwave-assisted digestion, preconcentration on 8-hydroxyquinoline on silica, rapid heating program	platform	none	[1239]
Urine	direct determination	platform	Pd–Mg + H$_2$	[5117]
Whole blood	various (comparative study)	various	various	[2511]
Whole blood	calibration solutions stabilized with (NH$_4$)$_2$HPO$_4$ and Mg(NO$_3$)$_2$, whole blood diluted with modifier	platform	Triton X-100	[544]
Whole blood	direct determination after 1+6 dilution with Triton X-100, oxygen ashing	platform	H$_3$PO$_4$–Mg–HNO$_3$	[4565]
Whole blood	direct determination after 1+3 dilution with modifier (Australian Standard Method AS-3503)	–	(NH$_4$)$_2$HPO$_4$–Triton X-100	[5545]

Table 10-23. Selected publications on the determination of lead in body fluids by GF AAS.

Sample	Procedure	Atomization	Modifier	References
Milk	direct determination after dilution with modifier, Zeeman-effect BC	platform	Triton X-100	[4293]
Milk, urine, whole blood	microwave-assisted digestion, Zeeman-effect BC	platform	Pd–Mg	[4596]
Seminal fluid	deproteinization with HNO_3, Zeeman-effect BC, matrix-matched calibration solutions	platform	HNO_3	[2972]
Seminal fluid	microwave-assisted digestion	platform	none	[155]
Seminal fluid, sperm	direct determination after dilution with modifier	platform	Pd–Mg	[467]
Urine	direct determination after 1+3 dilution	probe	none	[1243, 3878]
Urine	direct determination without pyrolysis	THGA	none	[6544]
Urine	direct determination after dilution with modifier	platform	$NH_4H_2PO_4$–Triton X-100–HNO_3	[3819]
Urine	direct determination after 1+1 dilution with modifier	platform	$NH_4H_2PO_4$	[4547]
Urine	direct determination, Zeeman-effect BC	platform	$NH_4H_2PO_4$	[5250]
Whole blood	comparison of background correction: continuum source to Zeeman-effect	platform	$NH_4H_2PO_4$–Triton X-100–HNO_3	[4746]
Whole blood	matrix-matched calibration solutions, Zeeman-effect BC	platform	$NH_4H_2PO_4$–Triton X-100–HNO_3	[371]
Whole blood	direct determination, air ashing	platform	Pd–citric acid	[5743]
Whole blood	direct determination (Australian Standard Method AS-4090)	—	$(NH_4)_2HPO_4$–Triton X-100	[5551]
Whole blood	direct determination	THGA	$NH_4H_2PO_4$–EDTA–NH_4OH	[222]
Whole blood	direct determination after dilution with 1% Triton X-100 + 0.2% HNO_3	THGA	$(NH_4)_3PO_4$–$Mg(NO_3)_2$	[6482]

Table 10-23 continued

Sample	Procedure	Atomization	Modifier	References
Whole blood	direct determination	platform	$(NH_4)_3PO_4$–Triton X-100	[725]
Whole blood	direct determination	platform	Pd–citric acid–Triton X-100	[2196]
Whole blood	direct determination after dilution with modifier	platform	NH_4NO_3–Mg–Triton X-100	[2927]
Whole blood	direct determination after 1+9 dilution with modifier	platform	$(NH_4)_2HPO_4$–Triton X-100–HNO_3	[4090]
Whole blood	direct determination after 1+9 dilution with modifier	platform	$NH_4H_2PO_4$–Mg or $NH_4H_2PO_4$–Mg–HNO_3	[1337, 4307]
Whole blood	direct determination after dilution with modifier	platform	EDTA–$NH_4H_2PO_4$–NH_4OH	[5324]

Whole blood and urine are used for the determination of the mercury burden; the latter is subject to stronger variations through increased excretion, for example after a fish meal. For a correct evaluation the time point of sampling is important [504, 1299]. Nowadays the 'normal' values that can be regarded as harmless are < 3 µg/L mercury in blood and < 5 µg/L mercury in urine. Values > 10 µg/L in urine are definitely regarded as increased and can endanger health.

The method of choice for the determination of mercury in biological materials is CV AAS [694, 1608]. As mentioned in Section 8.4.1, all precautions to prevent contamination and losses must be strictly observed. For this reason a complete digestion is often omitted for the analysis of blood or urine, even if the organomercury compounds are not quantitatively detected [5861]. Generally 1 mL of urine or blood is mixed with 10 mL of nitric acid, potassium permanganate, and octanol as an antifoaming agent, the mercury is reduced with sodium tetrahydroborate and then collected by amalgamation on gold [216, 5140]. A disadvantage of this procedure is that it was developed for batch systems and is thus difficult to automate. DRASCH and SCHUPP [1609] found that practically all mercury is extracted from blood and tissue samples when they are allowed to stand under nitric acid for 15 hours at ambient temperature. This extract is suitable both for the determination by FI-CV AAS and for speciation analysis. As an alternative WELZ et al. [6263] proposed microwave-assisted on-line digestion as part of an automated procedure for the determination of mercury in urine by FI-CV AAS. A similar automated procedure was also later applied to the determination of mercury in blood [2282, 2284]. An overview of selected publications on the determination of total mercury in body fluids by CV AAS is presented in Table 10-24.

Next to the determination of total mercury, procedures have also been described for the selective determination of inorganic mercury using $SnCl_2$ as reductant [172, 548, 3790]. Moreover, off-line [3989] and on-line [75, 920, 1519, 2670] techniques have been described for the chromatographic separation of individual mercury species and their determination by AAS.

Table 10-24. Selected publications on the determination of mercury in body fluids by CV AAS.

Sample	Technique	System	References
Milk	high pressure ashing	batch	[5182, 5183]
Milk, urine	treatment with cysteine / NaOH	batch	[6318]
Saliva	bromination and on-line digestion	FI	[2283]
Saliva, urine, whole blood	pressure digestion	FI	[5577]
Saliva, urine, whole blood	direct determination	batch	[4342]
Urine	bromination and on-line digestion	FI	[2281, 5907, 6263]
Urine	off-line / on-line digestion	FI	[2355]
Urine, whole blood	digestion overnight	FI	[955]
Urine, whole blood	pressure digestion	batch	[5742]
Whole blood	microwave-assisted on-line digestion	FI	[2282]

Thallium: These days acute thallium poisoning has become seldom since the use of this element as an insecticide and a rat poison, and also in medicine for the treatment of venereal diseases, has become severely restricted. The main sources of pollution nowadays come from anthropogenic emissions from refineries, coal-fired power stations, metal smelters, and cement works [5459]. The 'normal' values for thallium in non-exposed subjects are 0.3 µg/L in urine, 3 µg/L in blood, and 5–15 µg/kg in hair [3073]. A urine analysis is the best way of recognizing thallium poisoning since the thallium concentration is markedly increased, even after several months.

Since the normal value for thallium in urine is at or below the limit of detection for GF AAS, a direct determination is only possible for increased values (> 2 µg/L). Most procedures thus employ an extraction after complexing with sodium diethyldithiocarbamate or APDC [1186, 3316, 5582]. An oxidative digestion prior to extraction does not appear to be necessary [5127]. For the direct determination of thallium in urine, 'reduced palladium' has proven to be a reliable modifier [6241]. The Pd-Mg modifier can also be successfully used [4133]. The consequent application of the STPF concept, preferably in a transversely-heated atomizer with integrated platform, and Zeeman-effect BC are prerequirements for a satisfactory determination.

Miscellaneous Elements: In 1986 EMMERLING *et al.* [1742] compiled the then prevailing clinical-toxicological knowledge on Bi, Ga, Ge, In, Sb, and Te, and the possible methods for their quantitative determination. For antimony, bismuth, and tellurium HG AAS clearly provides the better limits of detection, while for gallium, germanium, and indium GF AAS is the method of choice. The best results are obtained for gallium and indium using the Pd-Mg modifier [5153, 6265], and strict observance of STPF conditions is a prerequirement for reliable results. A selection of publications on the determination of bismuth, germanium, and tellurium in body fluids is compiled in Tables 10-25 through 10-27.

Table 10-25. Selected publications on the determination of bismuth in body fluids.

Sample	Procedure	Technique	Modifier	References
Serum	direct determination after 1+1 dilution with modifier	GF AAS	Pd	[4675]
Serum, urine	HNO$_3$ digestion	HG AAS	–	[1993]
Serum, urine	direct determination	GF AAS	Pd–Mg	[5153]
Serum, urine	precipitation of protein	GF AAS	Pd	[1448]
Serum, urine, whole blood	extraction with iodide/IBMK	GF AAS	Pd	[4345]
Serum, whole blood	direct determination	GF AAS	Rh	[4498]
Serum, whole blood	extraction with APDC/IBMK	GF AAS	Pt	[5440]
Urine	direct determination	GF AAS	Pd–W	[5912]
Urine	microwave-assisted on-line digestion	FI-HG AAS	–	[5907]
Urine, whole blood	HNO$_3$–HClO$_4$ digestion	HG AAS	–	[4951]
Various	*in-situ* preconcentration	FI-HG-GF AAS	–	[3976]
Various	HNO$_3$ digestion	HG AAS	–	[6144]

Table 10-26. Selected publications on the determination of germanium in body fluids by GF AAS.

Sample	Procedure	Modifier	References
Plasma	direct determination after 1+3 dilution with modifier	Triton X-100–Ca	[2279]
Serum	direct determination	La	[6407]
Serum	direct determination after 1+4 dilution with Triton X-100–HNO₃	Pd	[5302]
Serum, urine, whole blood	platform atomization, Zeeman-effect BC	Pd–Mg	[5153]

Table 10-27. Selected publications on the determination of tellurium in body fluids.

Sample	Procedure	Technique	Modifier	References
Serum, urine	direct determination	GF AAS	Pt	[5335]
Serum, urine, whole blood	direct determination, compromised conditions for the simultaneous determination of several elements	GF AAS	Pd–Mg	[5153]
Urine	HNO₃–HClO₄ digestion	HG AAS	–	[3179–3181]
Urine	HNO₃–H₂O₂ digestion	GF AAS	Pd–Mg	[53]
Urine	extraction with IBMK	GF AAS	none	[3178, 3181]

Table 10-28. Selected publications on the determination of beryllium in body fluids by GF AAS.

Sample	Procedure	Modifier	References
Serum, whole blood	direct determination, STPF, Zeeman-effect BC	Pd–Mg	[5153]
Urine	1+3 dilution with modifier, STPF, Zeeman-effect BC	HNO₃–Mg–Triton X-100	[4548]
Urine	direct determination	Mo–ascorbic acid	[5274]
Urine	1+1 dilution, STPF, Zeeman-effect BC	Mg	[238]
Urine	extraction after acid digestion	none	[5841]

The determination of *beryllium* has also gained interest in the fields of industrial hygiene and toxicology since it is classed in the group of carcinogenic and mutagenic poisons. Recent publications are compiled in Table 10-28.

Strontium is probably not essential, but it is also less poisonous than barium [5092]. It can be determined by F AAS, but GF AAS can also be applied, especially when the sample quantity is limited [401]. Table 10-29 provides an overview of publications on the determination of strontium.

Table 10-29. Selected publications on the determination of strontium in body fluids.

Sample	Procedure	Technique	Modifier	References
Cerebrospinal fluid		GF AAS		[1483]
Plasma	dilution	GF AAS, ZBC	none	[401]
Plasma, urine	dilution (1:20–1:40) with HNO_3	GF AAS	HNO_3	[3488]
Plasma, urine	dilution (1:20) with HNO_3	GF AAS, ZBC	HNO_3	[511]
Saliva	matrix-matched calibration solutions	GF AAS, ZBC	HNO_3	[5358]
Serum	calibration solutions matched with glycerol	F AAS, injection technique		[3967]
Serum	1+3 dilution with Triton X-100–HNO_3	GF AAS, ZBC	none	[1520]
Urine	1+19 dilution with Triton X-100–HNO_3	GF AAS, ZBC	none	[1520]

10.1.4 Therapeutically Administered Elements

A number of elements that are only present in the lowest concentrations in the bodies of healthy persons, and which are not classified as essential, play an important part in the chemotherapy of certain diseases. During treatment the concentration of such an element is measured, usually in blood, or occasionally in tissue, and the excretion is determined in urine or feces, to monitor the resorption on the one hand and to keep toxic side-effects as low as possible on the other hand.

Gold: The most noted use of gold in the clinical domain is for the treatment of rheumatoid arthritis. It can be administered by injection as gold sodium thiomalate (Myocrisin) or orally as Auranofin [2803]. The gold concentration in the blood of a patient undergoing chemotherapy is normally in the range 0.1–10 mg/L. Graphite furnace AAS is the technique used most frequently in clinical laboratories for monitoring the gold concentration in the blood of arthritis patients since the sample requirement is very low (< 20 µL) and operation is simple.

A number of authors have reported interferences in the determination of gold by the electrolytes [5426] which can in part be ascribed to a spectral interference (the spectrum of NaCl has a maximum at 240 nm, close to the gold line at 242.8 nm). High performance BC is thus of decisive importance [1436, 3017, 5287]. The most suitable method for the determination of gold in blood and urine was described by EGILA *et al.* [1683] and SHAN *et al.* [5272]. They used a mixed modifier comprising 4 µg Pd and 10 µg Mo, which allowed a pyrolysis temperature of 1300 °C to be applied. No interferences were observed under STPF conditions using Zeeman-effect BC, so that matrix-free calibration solutions could be used. An *in situ* ashing with oxygen can bring a further improvement in the technique with respect to the long-term stability, especially for blood samples that are minimally diluted. Selected publications on the determination of gold in body fluids by GF AAS are compiled in Table 10-30.

Table 10-30. Selected publications on the determination of gold in body fluids by GF AAS.

Sample	Procedure	Atomization	Modifier	References
Cerebrospinal fluid	from the supernatant liquid after centrifugation	tube wall	HNO$_3$	[1713]
Serum	comparison of various methods	tube wall	Triton X-100	[2547]
Serum	direct determination after dilution	tube wall	Mg(NO$_3$)$_2$– NH$_4$H$_2$PO$_4$	[4194]
Serum, urine	comparison of various modifiers	platform	various	[5817]
Serum, whole blood	direct determination after dilution	tube wall	Ni or Pd–Mo	[5272]

Lithium: The most important clinical application of lithium therapy is to avoid large emotional variations in manic excitement or depression [640]. Lithium is administered orally in the form of lithium carbonate tablets at a dose of up to 2 g per day. The treatment is monitored via regular controls of the lithium concentration in the blood or serum 12 hours after the last dose. The lithium concentration should then be in the range 25–50 mg/L [2176]; higher lithium concentrations can lead to toxic side-effects. The lithium concentration in healthy subjects is substantially lower at around 1 µg/L.

The lithium concentration in serum can be determined with good accuracy by F AAS. Most of the described methods give satisfactory accuracy, except those in which deproteinization is used. The best results are obtained after a simple 1:5 dilution of the serum and calibration against aqueous solutions [4769]. The procedure can be easily automated by applying the FI technique with on-line dilution [247, 4916, 5299]. As well as F AAS, GF AAS has also been used for the determination of lithium in serum [6296]; the substantially higher sensitivity permits the use of significantly smaller volumes and higher dilution. Table 10-31 provides an overview on the determination of lithium in body fluids by GF AAS.

Table 10-31. Selected publications on the determination of lithium in body fluids by GF AAS.

Sample	Procedure	Atomization	Modifier	References
Erythrocytes	direct determination	tube wall	KH$_2$PO$_4$– Triton X-100	[6450]
Serum	direct determination after dilution	tube wall	Triton X-100	[737]
Serum, urine	halogen lamp BC	tube wall	NH$_4$NO$_3$	[4094]
Serum, urine	direct determination after dilution	tube wall	KH$_2$PO$_4$	[516]

Platinum: Cisplatin and, more recently, Carboplatin and Iproplatin have been used very effectively in the chemotherapy of tumors; the action of the Pt(II) complexes is based on the binding of platinum to two nitrogen atoms in the same DNA chain, preferentially to guanine [4002]. Because of the severe toxic side-effects, regular monitoring of the platinum concentration in blood or serum, urine, and tissue is required.

Due to its sensitivity and selectivity, GF AAS is the method-of-choice for the determination of platinum. Numerous working groups have described methods based either on a direct determination [6168] or on wet ashing of the biological materials, followed by solvent extraction after prior complexing with dithizone [152], APDC [2759], or N-acylthiourea [5202]. TÖLG *et al.* [5867] provide a good overview on preconcentration techniques. IKEUCHI *et al.* [2759] described a simplified procedure in which they added a mixture of hydrochloric, nitric and perchloric acids to the plasma or urine and then determined platinum directly by GF AAS. Similar direct procedures, with or without prior acid digestion, have been described by a number of other authors [2661, 5288, 5293, 5450, 6113]. By using a transversely-heated atomizer with integrated platform, *in situ* oxygen ashing, STPF conditions and Zeeman-effect BC, it should be possible to perform interference-free determinations of platinum in body fluids without a great deal of sample preparation. More recently the determination of platinum species has also been described [6113]; free platinum or platinum complexes are separated from protein-bound platinum by ultrafiltration [1691], extraction [3372], or LC [3929, 4867, 4868].

On occasions elements that have no pharmacological activity in themselves but which are used in diagnostic or therapeutic procedures are determined in body fluids, such as contrast media used in X-ray or NMR diagnostics. These include barium sulfate for X-ray diagnostics [5128] or complexes of gadolinium or dysprosium for NMR tomography [3174].

10.1.5 The Analysis of Tissues

In contrast to Sections 10.1.1 through 10.1.4, in this section we shall not discuss the determination of the individual elements since this does not greatly differ from the procedures mentioned above. The analysis of digested tissue samples in which the organic constituents have been oxidatively decomposed is similar in principle to the analysis of inorganic samples. Marked matrix influences occur solely for the determination of trace elements in tissue samples with a high mineral content, such as bones and teeth.

In 1957 THIERS [5816] stated with justification that an analyst should not waste his time analyzing a sample whose history is unknown to him. This warning is particularly relevant for tissue samples since the risk of contamination is much more severe during sampling and sample preparation than with body fluids.

The first problem for the analysis of tissue samples is that practically all organs and many other tissues are highly inhomogeneous. Since only very small tissue samples can normally be removed from the living body, chance often plays a major role during sampling [2844, 2846, 4790]. The second problem is, as mentioned above, contamination from the instruments used during sampling. VERSIECK [6052] investigated the contamination of liver tissue by biopsy needles for a number of elements. For iron, which is present in liver tissue in very high concentration, the error was only 5–10%; for copper, manganese, and zinc it was already 25–40%; for cobalt and especially for chromium and nickel the contamination was many times over the natural content of these elements. The error is greater the smaller the tissue sample that is taken. Instruments made of plastic, titanium, or silica can largely prevent contamination during sampling [5696, 6052]. Further errors can also occur during the washing [2846] and drying [2840] of tissue samples.

Generally, prior to analysis tissue samples must be digested or taken into solution. During dry ashing there is a risk of analyte loss due to volatilization [2840] and also of contamination, in particular from ashing agents such as magnesium nitrate. The risk of losses is much lower for wet digestions, especially if the digestion is performed in a closed container. Likewise the risk of contamination is also lower since acids can be prepared with a much higher level of purity than salts. The acids used primarily for digestion are nitric acid, sulfuric acid, and perchloric acid. Nevertheless we must point out that even ultrapure acids can contain certain elements in concentrations that are higher than in the human body [2840, 6058, 6060]. Subboiling distillation can further reduce the element concentration in acids (refer to Section 5.1.3). Additionally, the smallest possible volumes of acids should only be used. More recently the use of microwave-assisted digestions has been increasingly applied [2108]; this technique offers advantages with respect to both the analysis time and to the control of contamination since plastic or quartz containers are used (see also Section 10.2.2). Selected publications on the analysis of tissue samples after digestion are compiled in Tables 10-32 and 10-33.

Various working groups have investigated the possibility of examining tissue samples directly by GF AAS without prior dissolution [2513]. This technique is not completely free of problems either since the test sample must be homogenized by cutting up and grinding [3363, 3582, 5583] with the inherent risk of contamination. Further problems also arise in transferring the sample quantitatively to the atomizer and in calibration (see Section 4.2.7). These latter problems occur far less frequently for the analysis of slurries than for direct solids analysis, since in principle slurries can be treated in the same way as solutions. Table 10-34 provides an overview of selected publications on the direct solids analysis of human tissue.

The analysis of hair [517–519] and nails assumes something of a special status since these samples are readily available. By analyzing individual segments of hair it is also possible to obtain information on the course of an intoxication. Long-term exposure is manifested by an increased value that is distributed fairly evenly along the length of the hair, while an acute intoxication produces a 'peak'. The biggest problem for the analysis of hair and nails is cleaning them [122, 288, 370, 515, 751, 1263, 5061, 5215], since externally such samples can be severely contaminated due to contact with the environment. The element concentration in hair can also be changed due to bleaching or coloring [518], or to the use of shampoo [122].

Table 10-32. Selected publications on the analysis of human soft tissue by GF AAS after digestion.

Sample	Element	Procedure	Atomization	Modifier	References
Bone marrow	Fe	H_2SO_4–H_2O_2 digestion	platform	none	[1523]
Brain	Al	HNO_3 digestion in closed system	tube wall	$K_2Cr_2O_7$	[6410, 6411]
Brain, bones, liver, kidney	Bi	HNO_3–H_2SO_4 digestion overnight	tube wall	none	[5441]
Brain, liver, kidney	Ag	analyte addition technique, Zeeman-effect BC	platform	none	[1610]
Brain, liver, kidney	Sn	HNO_3–$HClO_4$ digestion	tube wall	ascorbic acid	[2818]
Breast tissue	Si	extraction with heptane	tube wall	none	[6309]
Eye lens	Cr	HNO_3–H_2SO_4 digestion, Zeeman-effect BC	tube wall	none	[4655]
Eye lens	Cu	HNO_3 digestion	tube wall	$NH_4H_2PO_4$	[2130]
Eye tissue	Se	Zeeman-effect BC	platform	Pd	[4015]
Liver	Co	HNO_3 pressure digestion	tube wall	none	[1029]
Liver	Cr, Mn, Ni	HNO_3–H_2O_2 digestion	tube wall	none	[705]
Lung	Cr, Ni	acid digestion, Zeeman-effect BC	tube wall	none	[4790]
Renal cortex	Cd	HNO_3 digestion	platform	none	[3806]
Various	Ag	HNO_3 digestion	tube wall	NH_4NO_3	[6145]
Various	Al	microwave-assisted digestion in HNO_3, clean room	platform	$Mg(NO_3)_2$	[5394]
Various	Al	comparison of various methods	—	—	[1118]
Various	Al	HNO_3 digestion	platform or tube wall	Ca or none	[3509]
Various	Al, Fe	short program without pyrolysis	platform	none	[3508]
Various	Al, Cu, Cr, Mn	HNO_3 digestion under reflux	tube wall	none	[3756]
Various	Mn	HNO_3–H_2O_2 digestion	tube wall	none	[1110]
Various	Ni	comparison of methods, contamination control	—	—	[5699]
Various	Sr	dissolution in tetraethylammonium hydroxide	tube wall	none	[1520]

Table 10-33. Selected publications on the analysis of human tissue by GF AAS after digestion.

Sample	Element	Procedure	Atomization	Modifier	References
Bones	Al	HNO$_3$ digestion	tube wall	none	[2656]
Bones	Pb	microwave-assisted digestion	platform	NH$_4$H$_2$PO$_4$– Ca(NO$_3$)$_2$ or Mg(NO$_3$)$_2$	[6587]
Bones	Sr	HNO$_3$ digestion	tube wall	none	[1520]
Bones	Al	microwave-assisted digestion	platform	Ca(NO$_3$)$_2$	[5768]
Bones (biopsy)	Al, Ca, Fe, Pb, V	high pressure ashing	tube wall	none	[4308]
Bones (biopsy)	Ag, Al, Au, Ba, Be, Bi, Cd, Co, Cr, Cu, Mn, Ni, Pb, Se, V, Zn	microwave-assisted digestion	platform	various	[4134]
Hair	Al	acid digestion	platform	none	[1199]
Hair	Mo	on-line preconcentration with FI	tube wall	none	[1234]
Hair	Ni	on-line extraction with FI	tube wall	none	[5774]
Hair	Tl	acid digestion, extraction with APDC/IBMK	tube wall	none	[514]
Hair, nails	Se	microwave-assisted digestion	platform	Pd	[2399]
Nails	Al, Co, Cr, Mn, Mo, Ni	dissolution in tetraalkylammonium hydroxide, matrix-matched calibration solutions	tube wall	none	[5909]
Teeth	Pb	HNO$_3$ digestion	tube wall	none	[2836, 3061, 4734]
Teeth	Pb, Fe, Cu	HNO$_3$ digestion	tube wall	none	[661]
Teeth	Mo	HNO$_3$–HClO$_4$ pressure digestion	tube wall	hydrazine sulfate	[2897]
Teeth	Cu, Mn, Sr	analyte addition technique	tube wall	none	[4847]
Vesical calculus	Cd, Cr, Cu, Ni, Pb, Zn	pressure digestion	platform	Pd–Mg or Mg(NO$_3$)$_2$	[5604]

Table 10-34. Selected publications on the direct solids or slurry analysis of human tissue by GF AAS.

Sample	Element	Remarks	Modifier	References
Dental calculus	Cd, Pb, Zn	aqueous calibration solutions	none	[5603]
Hair	As	longitudinal distribution	Pd–Mg	[3242]
Hair	Cu	solid standard	none	[306]
Hair	Cr, Co, Mn, Pb	slurry, oxygen ashing	none	[1658]
Hair	Pb	slurry	Pd–Mg	[573]
Liver	Cd, Pb	aqueous calibration solutions	none	[4620]
Liver	Cu	solid standard	Pd–Mg–Triton X-100	[2]
Liver (biopsy)	Se	solid standard	Ni or Ag	[1]
Placenta	Cd	aqueous calibration solutions	none	[2509]
Various	Se	aqueous calibration solutions	Pd	[3530]
Vesical calculus	Cd, Cr, Hg, Ni, Pb	comparison with digestion	none	[5605]

10.2 Biological Materials

In this section we shall discuss the analysis of vegetable and animal tissues and also vegetable and animal products of the most widely varying types. The element concentrations in animal and vegetable organisms are influenced by numerous factors, the most important of which depend on the environment and habitat (see also Figure 10-3 in Section 10-3). For plants and vegetable products these are for example the nature of the soil, the fertilizers, crop protection agents, insecticides and pesticides used, and on the proximity of roads or industrial installations. Animals and animal products are influenced by the nutrition, animal feeds, and the environment, while aquatic species are influenced by their habitat itself. All foodstuffs are further influenced by preparation, storage, packaging, etc. We should further remember that handling procedures can cause not only an increase but also a decrease in the element concentration.

As well as the analysis of *foodstuffs and drinks*, which is clearly of the greatest importance, we shall also discuss the analysis of *animal feeds* for their content of minerals and trace elements. Biological materials are also investigated during *environmental analysis*; plants and animals serve as *bioindicators* for environmental pollution in given regions (traffic arteries, industrial locations, rivers, coastal marshlands, etc.) [2175, 3538, 3889, 4356, 4400, 4636, 5930]. Plants are also frequently analyzed during prospecting [1630, 1759] since they can accumulate certain elements and thus give indications about possible deposits. Since the actual analysis, starting from digestion of the sample through to the actual determination, is determined by the sample material and not by the purpose of the analysis, in the following we shall not distinguish between the various areas of application but between the various types of sample and the techniques applied.

In principle the same elements are mainly determined in animal and vegetable samples, in foodstuffs and animal feeds, etc. as in body fluids and tissues. The principal aspect is *healthy nutrition and a sufficient supply of minerals and trace elements*. The main elements Na, K, Ca, Mg, and P belong in this area and also the trace elements B, Co, Cr, Cu, Fe, Mn, Mo, Si, Sn, V, and Zn, which are present in much lower concentrations. A major difference to the analysis of body fluids, however, is that within a sample type, such as fish and fish products, vegetables or dairy products, the element concentration can differ by 3–4 powers of ten [4788]. Taken over the entire palette of biological materials the contents of a mineral or a trace element can differ by more than five orders of magnitude, which naturally has an influence on the analysis and the technique used.

A further, very important aspect for the analysis of biological materials is the *intake of heavy metals in concentrations injurious to health* from the nutrition, animal feeds, or from the soil. We only need to mention lead, which reaches the fields and meadows alongside major roads, cadmium, mercury, and other heavy metals from contaminated soils or fertilizers (sewage sludge), or arsenic and copper, which find their way into wine from herbicides. The accumulation of metals such as cadmium, lead, and mercury in fungi or mussels and in the nutritional chain (predatory fish) are further known examples. Foodstuffs can also be contaminated during production (chromium and nickel from steel tanks; silicon in oil), during storage (aluminium or tin from tin cans), and during processing or preparation (Al, Cr, Ni, or Fe from saucepans). In addition to toxicity, such aspects as taste or color are also of significance. Copper in concentrations as low as 0.1 mg/L changes the taste of milk [4192] and produces a rancid taste in butter. A number of trace elements cause changes in color in alcoholic beverages. In addition to the determination of the total concentration of mineral and trace elements, nowadays speciation analysis is being increasingly applied (see Section 10.2.6) to provide information on questions of bioavailability, toxicity, and action.

For foodstuffs and animal feeds there is also the aspect of *quality control*, especially when these contain particular additives (e.g. iron, selenium) or for when particular substances are removed ('low iron', 'reduced sodium'). Tablets and similar products that are freely available to supplement the mineral content of the diet play a particular role in this context [1591, 5359]. The determination of trace elements can also serve to identify the origin [445, 1861, 3301, 3460, 5208] or the authenticity [4884] of certain foodstuffs since particularities of the soil, the animal feeds, the environment, or the form of preparation are always reflected in the product.

The analysis of biological materials is one of the oldest applications of AAS, as shown by the papers by DAVID [1426, 1427, 1431] published in the late 1950s and early 1960s, and the review articles by DAVID [1432] and SLAVIN [5404] published in 1962. More recent review articles have been published by IHNAT [2748] and RAINS [4788] on the subject of foodstuffs analysis, by PRITCHARD and LEE [4717] on agricultural samples, and by HOENIG and GUNS [2616] on environmental and biological samples. RUBIO *et al.* [4995] made a detailed study on the determination of chromium in the environment and in biological materials; CERVERA and MONTORO [1129] published an overview on the determination of arsenic in foodstuffs and the environment; MASSEY and TAYLOR [3933] investigated the determination of aluminium in foodstuffs and the environment; MARCZENKO and LOBINSKI [3880] published an overview on the determination of molybdenum in biological materials. Numerous other minerals and trace elements are treated in the book *Quantitative trace analysis of biological materials* edited by MCKENZIE and SMYTHE [4024] and in the monograph *Trace Minerals in Foods* edited by SMITH [4359].

10.2.1 Sampling and Sample Pretreatment

Sampling can be of decisive significance for the accuracy of the analyses of biological materials. This can begin, for example, in foodstuffs analysis with the selection of the sample (which lettuce from a whole field should be examined), the size of the sample (numerous foodstuffs are very inhomogeneous), the mode of sampling (contamination), the timepoint of sampling, and pretreatment (removal of external leaves, surface washing, conservation) before it reaches the laboratory. Once again we must repeat the statement that an analyst should not waste his time in investigating a laboratory sample whose history is not known in all details and is not documented. In the laboratory the sample must be *homogenized* [3515], *conserved*, and frequently *stored*; naturally all precautions described in other chapters to prevent contamination from tools, containers, the laboratory atmosphere, and losses due to volatilization or adsorption must be scrupulously observed (see Sections 5.1 and 5.5). Comprehensive information on this subject has been published by KRATOCHVIL and TAYLOR [3282] and by JONES *et al.* [2941–2943].

MAURER *et al.* [3979] reported on the problems of the representative sampling of tree needles; KRIVAN and SCHALDACH [3294] made a thorough investigation of the influence of cleaning the needles and other preparative steps on the analytical results. MARKERT [3887, 3888] posed the general question whether cleaning plants prior to analysis made any sense at all. WAGNER [6123] reported on the variability of the element concentration in tree foliage in dependence on the mode of sampling. PERRING [4613] drew attention to errors in determining the mineral content of apples when the cores were included in the analysis. MATTER [3962] published an overview on the problems of the sampling and sample preparation of foodstuffs of animal and vegetable origin. SCHLADOT and BACKHAUS [5146] discuss the problems related to samples of marine origin. RAO and KOEHLER [4807] investigated the influence of the number of samples on the representativeness of the results for the quality control of foodstuffs. IYENGAR [2845] discussed sampling strategies for investigations on the uptake of elements from foodstuffs. IHNAT [2750] presented a general discussion on the influence of sampling, sample preparation, and calibration on the accuracy and reliability of elemental analysis in biological materials.

10.2.2 Digestion Procedures for F AAS and GF AAS

Further sample pretreatment steps depend on the nature of the sample, on the analytes and their concentrations, and to a very large degree on the technique of analysis to be used for the determinations. In the 1960s and 1970s *dry ashing* was widely used for the digestion of biological materials [104, 185, 914, 1059, 1909, 4049, 4059, 4392, 5889,

6172, 6378]. The organic constituents are practically destroyed without residue and the analytes are mineralized, which is very advantageous if a preconcentration step is to follow the digestion [2094, 4046]. Additionally, relatively large sample aliquots can be ashed and taken up in a minimum volume of acid; since F AAS was the only technique available at that time, this was accommodating to trace determinations [115, 2363, 2461, 2875, 4707, 6593]. Dry ashing can be performed directly or under the addition of ashing agents such as magnesium nitrate [1983, 3565, 4832, 5342] or nitric acid and sulfuric acid [2462, 4707] usually at 450–550 °C. The ash is then taken up in dilute hydrochloric acid, either directly or after prior fuming off with sulfuric acid [2875]; under certain circumstances the addition of hydrofluoric acid can be useful [104]. Nowadays dry ashing is only applied sporadically [117, 1393, 1395, 1396, 1802, 2036, 2956, 3553, 4105, 4678], since other digestion procedures are faster [5790]. Further, values that are systematically too low are often found after dry ashing; this can be traced to analyte losses [4010, 4949, 4972, 6470, 6583], which are nevertheless in part avoidable [2957, 3774]. High blank values, which are difficult to control, are a further problem for trace analysis following dry ashing. Furthermore, from the end of the 1970s GF AAS has been used increasingly for trace analyses of biological materials and the additives used for dry ashing are a source of interference in this technique. This is especially true for *fusions*, for example with lithium metaborate [4070], sodium tetraborate [738], lithium tetraborate and sodium carbonate [5052], which were used for the digestion of vegetable materials for subsequent determinations exclusively by F AAS.

Oxidative wet digestions are used far more frequently than dry ashing or fusions for the subsequent analysis by F AAS or GF AAS. These can be performed in open vessels, with reflux condensation if required, or in closed containers under pressure (autoclaves). A number of authors compared dry and wet digestions of biological materials and mostly found no significant differences in the results [958, 1098, 1719, 2801], except for the time advantage of wet ashing particularly under pressure [2969].

Perchloric acid was widely used in the past as the digestion acid because of its high oxidation potential [171, 1783], usually together with nitric acid [143, 2058, 2796, 3088, 3188, 4729, 5854, 6519], nitric acid and sulfuric acid [466, 1095, 1435, 2636, 2659, 3064, 5051], or also nitric acid and hydrofluoric acid [6460]. Since these acid mixtures usually guarantee complete mineralization of the analytes, they are particularly suitable if subsequent separation and preconcentration are required, for example by coprecipitation [1096, 2659] or solvent extraction [1095, 1391, 3064, 4541, 5051, 5368, 6517]. Digestions with perchloric acid were also used for subsequent direct determinations by F AAS [143, 1435, 2058, 2796, 6519] and GF AAS [466, 3088, 3188, 4729, 5849, 5854, 6460]; the reduced time requirement mostly played a role in these procedures.

Nevertheless, due to the *in situ* pretreatment inherent in a *direct determination* by F AAS and GF AAS, complete mineralization and thus the use of perchloric acid is not necessary. For GF AAS in particular, the use of mixtures of strongly oxidizing, high-boiling acids as digestants is not recommended since they reduce the lifetime of the graphite tube [6351]. These acids are also more difficult to clean up than lower boiling acids and are thus often more strongly contaminated. For the determination of lead, low results have been reported due to the formation of mixed lead sulfates after digestion in nitric acid, sulfuric acid, and perchloric acid [5793, 5854], and the formation of insoluble silicates which occlude macro and trace elements has been observed for the digestion of vegetable materials in perchloric acid [2747, 2751]. Added to this, perchloric acid, sulfuric acid, and phosphoric acid penetrate the graphite lattice [263, 1083, 2501, 5002] and often remain in the tube after pyrolysis [3924]. Next to the resulting tube corrosion [2507], this can lead to interferences due to the formation of volatile chlorides [1683, 1740, 2490, 2925, 2966, 3206, 4653, 4873, 5270], especially for the main Group III elements (aluminium, gallium, thallium) [5316, 5422, 6498], or to background interferences by molecular absorption [6261].

For these reasons, and also because of the explosion risk associated with the use of perchloric acid, alternative digestion techniques have been sought for biological materials for the direct determination of numerous elements by F AAS and GF AAS. In particular the combination of nitric acid and hydrogen peroxide, possibly with the addition of sulfuric acid, has proven to be very reliable. A number of selected papers on the use of this digestion mixture are cited in Table 10-35. Additionally, digestion mixtures not containing hydrogen peroxide have been applied, for example a mixture of nitric acid and sulfuric acid for the determination of Cd, Cr, Cu, Ni, Pb, and Zn in fish tissue [51] or a mixture of nitric acid, hydrochloric acid, and hydrofluoric acid for the digestion of corks [5486]. Nevertheless, these are not digestions in the accepted sense in which the organic matrix is destroyed, but largely dissolution of the test sample and leaching of the analytes.

Table 10-35. Selected publications on the digestion of biological materials with a nitric acid-hydrogen peroxide mixture for direct determinations by F AAS and GF AAS.

Digestion mixture	Test sample	Analyte	References
$HNO_3–H_2O_2$	vegetable material	Al	[5114]
	animal tissue	As, Se	[3314]
	fish liver	Cd, Cu, Fe, Pb, Zn	[712]
	fish and shellfish	Cu, Fe, Mn, Ni, Pb, Zn	[3008]
$HNO_3–H_2SO_4–H_2O_2$	vegetable material	Al, As, Cr, Cu, Fe, Zn	[244]
	cereal products, fruit, vegetables	Cu, Fe, Mn, Zn	[543]
	animal tissue	As	[2618]
$HNO_3–HF–H_2O_2$	vegetable material	14 elements	[4391]

WIETESKA et al. [6314] showed that numerous elements could be extracted from vegetable materials even with dilute acids. This form of extraction or leaching has been described for a whole series of analyte-matrix combinations, especially under the use of nitric acid, hydrochloric acid, or a mixture of both, mostly under relatively mild conditions. A number of selected publications on this subject are compiled in Table 10-36. It could be shown for the determination of cadmium and lead for instance in the muscles and offal of fat stock that leaching in nitric acid for five minutes at 80–100 °C [1534] brought the same results as after 5–6 hours at 250 °C [5563].

The extraction of rice grains with 0.1 mol/L HCl, on the other hand, does not fulfill the purpose of detecting the entire analyte content but merely the soluble portion from the surface [1288]. The same is true for the treatment of meat and vegetables with a simulated digestive juice [1362] or the extraction of leaves with ethanol-water [5928]. In

Table 10-36. Selected publications on the leaching of analytes with acids from biological materials for direct determinations by F AAS and GF AAS.

Acid / conditions	Test sample	Analyte	References
HNO$_3$ / 40 °C / 5 min, ultrasonics	biological materials	Cd, Cu, Pb, Mn	[4123]
HNO$_3$ / 50°C overnight, 2 h boiling, volume reduced to 20%	fruit, berries, vegetables	Ca, Fe, K, Mg, Mn, Zn	[5745, 5744]
HNO$_3$ / 60 °C	bovine liver	Cu	[4493]
HNO$_3$ / 80 °C / 4 h	bovine liver	Cr	[5561]
HNO$_3$ / 80–100 °C / 5 min	animal tissue	Cd, Pb	[1534]
HNO$_3$ / 90 °C / 1 h	mussels	Cd, Cu, Fe, Pb, Zn	[5490]
HNO$_3$ /100 °C / several hours	tree rings	12 elements	[1898]
HNO$_3$ / 105 °C / 20 min	bovine liver, porcine kidneys	Cu, Fe, Mn, Zn	[4352]
400 μL HNO$_3$	insect larvae	Cd, Cu, Pb, Zn	[4363]
HNO$_3$–H$_2$O$_2$ / 20 °C	rice plants	Cd, Ni	[4191]
HNO$_3$–HCl / 90 °C / 30 min	foodstuffs	various	[4731]
HNO$_3$–HCl / 100 °C / 30 min	biological materials	Ca, Cu, Fe, K, Mg, Mn, Na, Zn	[3978]
HNO$_3$–HCl / ultrasonics	vegetable materials	Cu, Mn, Ni, Zn	[3349]
HNO$_3$–HClO$_4$–H$_2$O / 60–80 °C	animal tissue	Cu, Zn	[3882]
HNO$_3$–HClO$_4$ / 20 °C / 24 h + 70 °C / 3 h	fish tissue, fish eggs	Cd	[541]
H$_2$SO$_4$–H$_2$O$_2$ / 100 °C	wood	As, Cr, Cu	[836]
HF–HCl	vegetable material	Si	[4390]
1 mol/L HCl / 20 °C / 30 min	rat liver	Cu, Mn, Zn	[3678]
1 mol/L HCl / 80 °C / 15 min	coffee, beans, maize, sunflowers, grass	Ca, Cu, K, Mg, Mn, Zn	[4145]
50% H$_2$O$_2$ / 55 °C / 4 h	liver tissue	As, Cu, Mn, Se, Zn	[107]
trichloroacetic acid / ultrasonics	dairy products	Ca, K, Na	[6591]

such cases a similar goal is followed as for the determination of the permeability of crockery to heavy metals by leaching with acetic acid (see Section 10.7.2) and is in end effect a speciation analysis (see Section 10.2.6).

Nevertheless such mild dissolution or leaching techniques can only be applied to given matrices. It was found for example that after a 15-minute extraction of vegetable materials in 1 mol/L HCl at 80 °C for the determination of Ca, Cu, K, Mg, Mn, and Zn there was good agreement with a digestion in nitric acid-perchloric acid, but that the recovery of iron and phosphorus was only about 50–70% [4145]. Generally, vegetable materials appear to give more problems than animal materials; numerous authors reported low results after digestions in hydrochloric and nitric acids [1489, 2211]. Various elements are apparently bound to particles to a considerable extent even though the digestion appears to be complete.

To avoid the problems of incomplete digestion or dissolution of the analytes without having to use perchloric acid, *pressure digestions* were widely applied in the 1970s and 1980s. In this way clear digestion solutions and good trueness of the results were obtained for a large number of biological materials of vegetable and animal origin at a digestion temperature of around 180 °C even with nitric acid alone [255, 344, 908, 1011, 1318, 2199, 2337, 3125, 5194, 5575, 5957, 6583]. WÜRFELS *et al.* [6402, 6403] made a thorough investigation of the residues and reaction products of biological materials after pressure digestions and were able to verify the efficiency of nitric acid under these conditions. The major advantages quoted for pressure digestions included safety (compared to digestions with perchloric acid), the relatively short time requirement and the high recovery, combined with a generally lower risk of contamination [2969].

Even better results, especially for samples with a high fat content such as chocolate or oil, can be obtained using high pressure ashing [1292, 5731]. As an alternative to classical pressure digestions, pressure digestions with acidic vapors have been described [3970, 3972, 3973]. The major advantage quoted for this technique is the extremely low blank values since trace impurities in the acids used remain in the liquid phase and do not reach the test sample.

Since the end of the 1980s these various digestion techniques have been increasingly replaced by *microwave-assisted acid digestions* which are frequently performed in closed containers and under pressure. An overview of microwave-assisted digestions for the analysis of biological materials is presented in Table 10-37. KUSS [3373], MATUSIEWICZ and STURGEON [3969] and also CHAKRABORTI *et al.* [1175] have published review articles on the application of microwave-assisted digestions for elemental analysis. The major advantages given compared to conventional digestion techniques included the reduced time and reagent requirements, the lower blank values, the reduced risk of analyte losses, and the possibility of automation and of being able to digest a number of samples simultaneously [1627, 3154, 3155, 5790]. The methods nevertheless require careful optimization with respect to the matrix; chemometric techniques have been applied for this purpose [3409, 4157, 6573]. Open microwave-assisted digestions are less critical with respect to the digestion conditions since any gases released can be removed and oxidative reagents such as hydrogen peroxide can be added during the digestion. Their particular advantage is that sample quantities of up to 10 g are possible [2601, 3267, 3547, 4463, 5176, 5790], while for most closed systems quantities of only up to 250 µg are allowed.

In recent times successful investigations have been made to incorporate microwave-assisted digestions into flow systems, for instance coupled with FI-F AAS. This can be advantageous due to the short digestion times of only a few minutes. An overview of some of the investigations in this area is presented in Table 10-38.

Next to these widely used digestion procedures a number of techniques have been proposed that have not come into general use but can occasionally be of interest as comparative techniques. These include, for example, combustion in a stream of oxygen in a closed system [344, 1627, 3154, 5281] or by intense IR irradiation [5759]. Digestions have also occasionally been performed in an ultrasonic bath to obtain clear solutions in a comparatively short time at relatively low temperatures [3349, 5076, 6591]. AHLGRÉN *et al.* [65] described the ashing of vegetable materials by means of a CO_2 laser. SAH and

Table 10-37. Selected publications on microwave-assisted off-line digestions of biological materials for analysis by F AAS and GF AAS.

Test sample	Analyte	Digestion system	Acid	References
Biological materials	Cd, Cr, Cu, Fe, Pb, Zn	closed, pressure	H_2SO_4–HNO_3	[321]
	Se	closed, pressure	HNO_3	[2601, 4568]
	25 elements	closed, pressure	HCl–HNO_3–HF	[614]
	As, Ca, Cd, Cu, Fe, K, Mg, Mn, Na, Pb, Zn	closed, pressure	HNO_3	[3387]
	Cd, Cu, Pb	high pressure	HNO_3– HCl	[5673]
	various	open, focused	HNO_3–H_2SO_4–H_2O_2	[3313]
	B, Cd, Cu, Fe, Mn, P, Pb, Zn	closed, pressure	HNO_3–H_2O_2 or HNO_3–$HClO_4$	[4128]
	Cd, Cu, Fe, Mn, Pb	open	HNO_3–HCl	[1168]
Vegetable materials	Ca, Cu, Fe, Mg, Mn, Zn	closed, pressure	HNO_3–HCl	[1469]
	various	closed, pressure	HNO_3–HCl–H_2O_2	[1626]
	various	closed, pressure	HNO_3–HCl or HNO_3–HF–H_2O_2	[3395]
	Ca, Co, Cr, Cu, Fe, K, Mg, Mn, Ni, Pb, Zn	closed, cooled	HNO_3–H_2O_2	[2491]
	Cd, Cr, Cu, Pb	closed, pressure	HNO_3–HF	[659]
	Cd	closed, pressure	HNO_3–HF	[870]
	Ca, Na	closed, pressure	HNO_3	[6474]
	15 elements	high pressure	HNO_3–HF–H_2O_2	[6592]
	Ba, Ca, Cu, Mg, Mn, Zn	temperature controlled	HNO_3–HF	[4125]
Peat	various	closed, pressure	HNO_3–$HClO_4$–HF	[4532]
Rice, sesame	Mo, Se	closed, pressure	HNO_3–H_2O_2	[2909]
Tobacco	Cd, Co, Fe, Ni, Pb	closed, pressure	HNO_3–HF	[156]
Grapes, wine, foliage	Pb	open	HNO_3–H_2O_2	[5790]
Animal tissue	Cu, Fe, Zn	open	HNO_3–H_2SO_4	[322]
	Se	open	HNO_3–H_2O_2	[1617]
	Cd, Cu, Fe, Pb, Zn	closed, pressure	HNO_3	[5602]
	Ni	closed, pressure	HNO_3–HCl–H_2O_2	[540]

Table 10-37 continued

Test sample	Analyte	Digestion system	Acid	References
Liver tissue	Fe, Mn	closed, pressure	$HNO_3–H_2O_2$	[6408]
	Fe	closed, pressure	$HNO_3–H_2O_2$	[6405]
	Cu, Fe, Mg	closed, pressure	$HNO_3–HCl$	[3561]
	Al, Co, Cu, Fe, Pb, Se, Zn	closed, pressure (miniaturized)	HNO_3	[4354]
Liver, muscle tissue	Cu, Fe, Zn	open	HNO_3	[946]
Milk	Ca, Mg	closed, pressure	HNO_3	[2001]
Fish, marine mammals	Cd, Cu, Zn	closed, miniaturized	HNO_3	[362]
	As	closed, pressure	$HNO_3–H_2O_2$	[1328]
	As, Cd, Co, Cr, Cu, Fe, Mn, Ni, Pb, Se, Zn	closed, pressure	$HNO_3–HClO_4$	[4282]
	Cd, Cu, Fe	closed, pressure	HNO_3	[680]
	Cr, Cu, Fe	closed, pressure	$HNO_3–H_2O_2$	[3971]
	Cd, Cr, Cu, Pb, Zn	closed, pressure	HNO_3	[4010]
	Cu, Fe, Zn	high pressure	HNO_3	[5500]
	Cu, Fe, Zn	high pressure	HNO_3	[5501]
	Pb	closed, pressure	$HNO_3–V_2O_5$	[1019]
	various	closed, vapor phase	$HNO_3–HF$	[3973]

Table 10-38. Selected publications on microwave-assisted on-line digestions of biological materials for analysis by FI-F AAS and GF AAS.

Test sample	Analyte	System	Acid	References
Biological materials	—	comparison of various systems	HNO$_3$	[3975]
Biological reference material	Ca, Fe, Mg, Zn	flow system	HNO$_3$	[2428]
Biological reference material	Pb	flow system	HCl–HNO$_3$	[947]
Fish	Al	flow system	HNO$_3$	[279]
Fish	Se	flow system	HNO$_3$	[280]
Cocoa powder	Cu, Fe	stopped-flow, pressure	HNO$_3$	[2145]
Cocoa powder, kidney tissue	Ca, Cd, Fe, Mg, Zn	stopped-flow, pressure	HNO$_3$	[2146]
Foodstuffs	Cu, Mn	circulatory system	HNO$_3$–H$_2$O$_2$	[1062]
Foodstuffs	Cu, Mn	circulatory system	HNO$_3$	[2247]
Liver, kidney tissue	Cd, Zn	flow system	HNO$_3$	[954]

MILLER [5045] investigated a spontaneous, exothermic reaction that is induced by microwave energy and takes place during the digestion of biological materials with nitric acid-hydrogen peroxide in a closed container. Once the reaction has been initiated it continues to complete dissolution of the sample without the need for additional energy input. ZHOU et al. [6572] were able to obtain good results from microwave-assisted dissolution of biological materials with tetraethylammonium hydroxide and EDTA.

10.2.3 Determination Procedures for F AAS

According to the basic principle, which we have mentioned in a number of places in this monograph, that a technique should only be applied in its optimum working range, F AAS is suitable without reservations only for the direct determination of potassium, sodium, calcium, and magnesium, and also copper, iron, and zinc, and under certain circumstances also manganese, in biological materials. Table 10-39 presents a number of applications. Under given conditions it is also possible to determine aluminium [1435, 1898, 2754], chromium [77, 2746, 4135, 5446], nickel [115, 2363], silicon [738, 4070], and tin [1718] directly by F AAS. A dry ashing step, performed with great care, that leaves a relatively problem-free matrix for F AAS, and restricted dilution is often an advantage. The use of the nitrous oxide-acetylene flame is a prerequirement for most of these elements to obtain correct results.

Table 10-39. Selected publications on the direct determination of Ca, Mg, K, Na, Cu, Fe, Mn, and Zn in biological materials by F AAS.

Analyte	Test sample	Remarks	References
Ca	milk, milk powder	—	[2036, 3486]
	foodstuffs	slurry	[5069]
Ca, Mg	milk	digestion	[2001]
		FI, Sr(NO$_3$)$_2$ carrier solution	[5058]
	biological materials	interlaboratory trial—dry	[5971]
	bamboo shoots	ashing	[1876]
	foliage, needles		[5298]
Ca, Mg, K	vegetable materials	FI	[6509]
	foliage	fusion	[35]
Ca, Mg, K, Na	milk	dry ashing	[859]
	animal tissue		[4250]
	fruit	FI, slurry	[944]
	tree needles	slurry	[1085]
Ca, Mg, K, Na, Cu, Fe, Zn	biological reference materials	75 µL direct nebulization	[5944]
	animal feeds	interlaboratory trial	[5004]
Ca, Mg, K, Na, Cu, Fe, Mn, Zn	milk	influence of processing	[6594]
	plants	FI	[6546]
	cocoa beans	—	[4459]
Ca, Mg, K, Na, Cu, Fe, Mn, Zn (Al, Cd, Pb, Rb)	tree rings	F AAS (and GF AAS)	[1898]
Ca, Mg, K, Na, Cu, Fe, Mn, Zn, Al, Sr	foodstuffs	reference materials	[2754]
Ca, Mg, K, Cu, Mn	foliage	slurry	[1795]
Ca, Mg, K, Cu, Mn, Zn	vegetable materials		[4145]
Ca, Mg, K, Si	wild rice	fusion	[4070]
Ca, Mg, Fe, Mn, Zn	vegetables	FI, slurry	[6082]
	Chinese medicinal herbs	injection technique	[6560]
Ca, Mg, Fe, Zn	flour, grain	FI, slurry	[6080]
Ca, Mg, Fe	flour, grain	pressure digestion, without/with microwave assistance	[5077]
Ca, Na	fruit, vegetables	La + K as buffer	[6474]
K	vegetable materials	FI	[4845]
K, Na	biological materials	interlaboratory trial	[5970]
		Cs as buffer	[117]
Mg	plants, dairy products	laurylpyridine chloride as buffer	[5753]
Cu	residues from grapes		[2795]
	bovine serum		[3469]

Table 10-39 continued

Analyte	Test sample	Remarks	References
Cu, Fe	liver tissue		[6069]
	vegetable materials	water-alcohol mixture	[697]
Cu, Fe, Zn	biological materials	FI	[946]
	milk		[265]
	mussels		[5500, 5501]
Cu, Mn, Zn	flour, liver tissue	slurry	[1999]
Cu, Zn	mussels	interlaboratory trial	[2495]
Fe	animal tissue	injection technique	[1273]
Mn	green coffee	determination of origin	[3301]
Rb	biological materials, beverages	overview on Rb in food-stuffs	[221]
Zn	milk	emulsion stabilized with surfactants	[568]
Zn	marine organisms	reference materials	[290]

The direct determination of trace elements such as cadmium or lead by F AAS is only possible in exceptional cases and corresponding publications should be treated with due skepticism. In interlaboratory trials it has been repeatedly shown that the results for these elements are not satisfactory [2495]. The only exceptions are samples that are highly contaminated or in which the analytes have become strongly accumulated [1471, 3009, 4500], or wood samples that have been treated with chemicals [836, 3400].

A number of *indirect procedures* have also been described, for example the determination of *sugar* in vegetable materials by adding copper and determining the non-reduced portion by F AAS [4698]. *Chloride* can be determined in plants or saps by adding silver and measuring the non-precipitated portion in the solution [1788, 4017]. WIFLADT et al. [6315] determined *iodine* in seaweed after a fusion by reacting with Hg(II). CHAKRABORTI and DAS [1171] determined *tungsten* after extracting the ion association complex formed between iron dipyridyl and tungstate in chloroform. GONZALEZ et al. [3917] determined *cationic surfactants* after extraction of the Cu(II)-thiocyanate-surfactant ion pair in benzene.

During the 1960s and 1970s trace elements in biological materials were mostly determined by F AAS after *preconcentration by solvent extraction*. By far the most widely used extraction system was a combination of ammonium-pyrrolidine-dithiocarbamate/isobutyl-methyl-ketone (APDC/IBMK), which can be universally used over a wide pH range. Although nowadays this system has lost substantial significance, a number of publications selected from the profusion in the literature are cited in Table 10-40.

Table 10-40. Selected publications on the extraction of trace elements from biological materials with APDC/IBMK or chloroform for determination by F AAS.

Analyte	Test sample	References
Cd	equine kidneys	[1717]
Cd, Pb	fungi	[3374]
Cd, Co, Pb, Zn	mussel tissue	[1059]
Co	vegetable materials	[2094]
	foodstuffs	[379]
	fish tissue	[2964]
Co, Cr, Ni	vegetables	[5821]
Co, Cu, Ni	vegetable materials	[3216]
Cr	vegetable materials	[1095]
Cu, Fe, Mn, Zn	vegetable materials	[5052]
Cu, Fe, Pb, Zn	tea leaves	[4077]
Cu, Mo	rat teeth	[2489]
Fe, Sn	canned foodstuffs	[2954]
In, Tl	foodstuffs	[1781]
Pb	animal and vegetable tissue	[6517]
	grapes	[1858]

During the 1980s more selective extraction systems were progressively introduced which included β-diketones, hydroxamic acids, and various liquid exchangers. A number of these systems with the corresponding applications are compiled in Table 10-41. Nevertheless we must pose the question whether it makes any sense nowadays to use solvent extractions. For a number of elements, such as zinc, the risk of contamination is so large from reagents, containers, and the laboratory atmosphere during the extensive sample pretreatment steps of digestion and extraction that it is difficult to obtain correct results. For cadmium, on the other hand, very poor recoveries have been reported for the extraction of higher concentrations [1717].

As well as solvent extraction, in recent years preconcentration techniques using sorbent extraction have been described. A number of important publications on this subject are compiled in Table 10-42. The determination of boron in plants and fruit juices is also worth mentioning since it is volatilized as the methylborate and conducted directly into the nitrous oxide-acetylene flame [1102].

These classical preconcentration techniques take on a fully new dimension when they are performed with FI automatically and on-line in a closed system for F AAS determinations. The application of this technique for the analysis of biological materials is presented in Table 10-43. Sorbent extraction on packed microcolumns and precipitation/coprecipitation and sorption in knotted reactors have found wide application. An important aspect of these on-line techniques is not so much high enrichment factors as a rate of sample throughput that is suitable for F AAS. Enrichment factors of 16–40 with a sample throughput of 120–180/h [1233, 4048] appear to be much more sensible than enrichment factors of 90–180 at a sample throughput of only 13/h [2587]. In addition, if the enrichment time is too long the susceptibility to interferences and the risk of errors increase, such as the breakthrough of the column.

Table 10-41. Selected publications on the preconcentration of trace elements in biological materials by solvent extraction for determination by F AAS.

Analyte	Test sample	Extraction system	References
B	vegetable materials	2-ethyl-1,3-hexandiol / IBMK	[4046]
Ba	marine organisms	β-diketone / pentyl alcohol	[646]
Cd	animal tissue	β-diketone / chloroform	[644]
Cd, Cu, Fe, Mn, Pb, Zn	fish tissue	DDTC / IBMK	[2965]
Cd, Cu, Pb	foodstuffs	KI, H$_2$SO$_4$ / IBMK	[5051]
Cd, Zn	animal tissue	Aliquat-336 / IBMK	[367]
Co	vegetable materials	2-nitroso-1-naphthol / chloroform	[5368]
Cr	foodstuffs	KMnO$_4$, HCl / IBMK	[1849]
Cu	marine organisms	Aliquat-336 / benzene	[1422]
	foodstuffs	naphthoquinone-thiosemicarbazone / IBMK	[5364]
	tea	KI, HCl / IBMK	[325]
Fe	vegetable materials	hydroxamic acid / IBMK	[8]
Ge	vegetable materials	hydroxamic acid / phenylfluoron	[4]
In	vegetable materials	hydroxamic acid / IBMK	[7]
Mn	marine plants	β-diketone / chloroform	[645]
Mo	rat organs	benzohydroxamic acid / IBMK	[3875]
Nb	biological materials	hydroxamic acid / IBMK	[5]
U	vegetable materials	hydroxamic acid / chloroform	[11]
W	vegetable materials	hydroxamic acid / IBMK	[9]

Table 10-42. Selected publications on the preconcentration of trace elements in biological materials using sorbent extraction for determination by F AAS.

Analyte	Test sample	Extraction system	References
Cd, Pb	vegetables	chelate complexes on activated carbon	[6437]
Co	vegetable materials	bromo-pyridylazo-diethylamino-phenol and ammoniumtetraphenyl-borate on naphthaline / DMF	[5113]
Co	vegetable materials	8-hydroxyquinoline / activated carbon	[6438]
Cu	biological materials, beverages	tripyridyl-triazine-tetraphenylborate on naphthaline	[5966]
Mn	vegetable materials	Amberlit LA-2	[1204]
Ni	tea leaves	phenanthrenquinone-dioxime on naphthaline	[5112]
Pb	eggs	APDC-polystyrene on Pt wire	[3631]
V	vegetables	oxine complex on activated carbon	[2257]

Table 10-43. Automatic preconcentration procedures for trace elements in biological materials for on-line determination by F AAS.

Analyte	Test sample	Preconcentration procedure	References
Bi, Co, Ni, V	vegetable materials	sorbent extraction: CPG-8-hydroxyquinoline, Chelex-100	[1819]
Cd	vegetable materials	sorbent extraction: Chelex-100	[2586]
	animal tissue	sorbent extraction: dithizone / activated carbon	[4595]
	biological materials	sorption of chelate complexes in a knotted reactor	[1835]
Cd, Co, Ni	biological materials	coprecipitation with HMDTC-Fe complex, collection in a knotted reactor	[6257]
	animal tissue	coprecipitation with APDC-Fe complex, collection in a knotted reactor	[1593]
Cd, Cr, Cu, Fe, Mn, Pb, Zn		sorbent extraction: Muromac A1	[2587]
Cd, Cu	biological reference materials	sorbent extraction: DDTC complexes on Si-C18	[6414]
Co	vegetable materials	sorbent extraction: Al_2O_3 with nitrosonaphthol-disulfonate mod.	[5893]
Co	biological materials	sorbent extraction: ion pair with tetra-butylammonium bromide on Si-C18	[3552]
Cu	vegetable materials	sorbent extraction: activated carbon impregnated with 8-hydroxyquinoline	[4242]
	biological materials	solvent extraction: 8-hydroxyquinoline / toluene-xylene	[4048]
	vegetable materials	sorption of DDTC-chelates in a knotted reactor	[1233]
NO_2^-, NO_3^-	meat	solvent extraction: Cu(I)-neocuproin complex / IBMK	[5365]
Pb	biological materials	coprecipitation with HMDTC-Fe complex, collection in a knotted reactor	[1827]
		sorption of DDTC-chelates on an Si-C18 column	[3521]
		sorption of lead-tetrabutylammonium ion pair on an Si-C18 column	[5775]
Zn	biological materials	solvent extraction: DDTC / IBMK	[4418]

10.2.4 Determination Procedures for GF AAS

During the 1970s, the application of GF AAS to the analysis of biological materials was marked by reports on substantial interferences—as in other areas as well—which could only be partially eliminated. Extraction techniques were thus frequently employed to separate the analytes from the interfering matrix [1599, 2378, 4455, 5505]. As well as these matrix effects, which occurred for atomization from the tube wall with peak height evaluation, unrecognized contamination in the trace range was a frequent cause of error

[5506]. A further source of error was inadequate background correction with continuum sources in instruments with slow-reacting electronics [1392].

The breakthrough came at the beginning of the 1980s with the introduction of the STPF concept. The idea gained acceptance that interference-free measurements were possible by GF AAS, at least for the more volatile elements, if atomization was from a L'vov platform under the addition of a chemical modifier and evaluation was via integrated absorbance [1891, 2555, 3207, 3991, 4787]. Table 10-44 cites a number of publications on the direct determination of trace elements in biological materials in which the requirements of the STPF concept were largely met. A further major contribution to the trueness of trace analyses in biological materials was the introduction of Zeeman-effect BC; as shown in Table 10-44, this found acceptance in numerous methods. Moreover the recognition that accurate values can only be obtained for trace analyses with the consequent exclusion of all forms of contamination steadily gained ground as well as the realization that these values were much lower than had previously been believed [2955, 6049]. Direct determinations by GF AAS using a minimum of reagents are a good starting point for the avoidance of contamination. SLAVIN [5435] published a review article on the development of the analysis of biological materials by GF AAS during the 1980s.

Owing to the clear advantages of the direct determination of trace elements by GF AAS with respect to simplicity, speed, and accuracy, the number of publications on extraction techniques has decreased markedly since the beginning of the 1980s. Table 10-45 presents an overview on more recent publications on the separation and preconcentration of trace elements in biological materials for determinations by GF AAS. In certain situations these preconcentration techniques are indispensable since a number of elements, such as the noble metals gold [2678, 3064], palladium or platinum [152, 3261], are present in concentrations that are not amenable to a direct determination. The same is true for a number of trace elements such as bismuth [3371], cobalt, molybdenum, and occasionally selenium [886, 4541]. However, the following determination by GF AAS can be influenced favorably by the choice of suitable reagents. An example is the preconcentration of selenium by coprecipitation with palladium [1396], which then acts as a modifier in the graphite tube. Nevertheless, not all preconcentration techniques are sensible. Especially when larger quantities of reagents or numerous pretreatment steps are required [1096], we must pose the question as to whether adequate accuracy can be attained. As with F AAS, the application of on-line preconcentration by FI techniques could be a new way since it is far less susceptible to contamination than manual techniques.

Although biological materials are relatively easily dissolved or digested, *direct solids analysis* and especially the *analysis of slurries* have frequently been described. The first papers on the direct solids analysis of biological materials were published in the second half of the 1970s [3622, 4643]; LANGMYHR and co-workers [3434, 3438–3441] in particular determined numerous elements such as Cd, Co, Cr, Cu, Mn, Ni, P, and Pb in a multitude of materials such as plants, liver tissue, fish and fish products, bones, and ivory. The analysis of biological materials in suspensions was then described at the beginning of the 1980s [1859]; initially the suspensions were often stabilized with thixotropic reagents [5565, 5566]. Magnetic stirrers or ultrasonic probes were later used almost exclusively for homogenization; the manner in which the slurry is prepared can nevertheless have an influence on the trueness and precision of the analysis [4108].

Table 10-44. Selected publications on the direct determination of trace elements in biological materials by GF AAS after digestion.

Analyte	Test sample	Conditions, remarks	References
Ag	animal tissue	platform, ZBC, modifier: $NH_4H_2PO_4$	[184]
Ag, Au	rat plasma	platform, ZBC	[5846]
Ag, Co, Si, Zn	port	simple dilution, platform, diverse modifiers	[5487]
Al	milk	platform, diverse modifiers	[278]
	tea, baby food	platform	[444]
	foodstuffs	STPF, modifier: $Mg(NO_3)_2$	[4226, 6164]
	animal tissue	platform, ZBC, modifier: $(NH_4)_2HPO_4$	[4778]
	beer		[5289]
Al, Cd, Cu, Fe, Pb	corks	STPF, modifier: $Mg(NO_3)_2$ for Al, Fe; Pd–Mg for Pb; $NH_4H_2PO_4$ for Cd	[5486]
Al, Co, Cr, Cu, Mn, Ni, Se	meat, liver, kidneys	avoidance of contamination	[2955]
As	mussel tissue	platform, ZBC	[6471]
As, Cd, Co, Cr, Cu, Fe, Mn, Ni, Pb, Se	marine organisms	STPF, ZBC	[1491]
As, Cd, Cr, Cu, Ni, Pb	biological reference materials	STPF, ZBC, rapid analysis	[5437]
As, Cd, Cr, Cu, Pb, Se, V	vegetable materials	platform, ZBC; rapid analysis without pyrolysis and modifier	[6544]
As, Cd, Cu, Pb, Se	marine organisms	STPF, ZBC, diverse modifiers	[5149]
As, Cd, Pb, Se	marine organisms	STPF, ZBC, diverse modifiers	[6236]
As, Cd, Sb	biological materials	STPF, ZBC, modifier: Pd–Mg	[617]
As, Cd, Se	marine organisms	STPF, ZBC	[6231]
As, Se	marine organisms	STPF, ZBC, modifier: Cu–Fe, Ni	[3314]
As, Sb	wine	ZBC, modifier: Pd	[3098]
Au, Pd, Pt	vegetable materials	comparison with ICP-MS	[2328]
B	vegetable materials	modifier: Ca–Mg	[2917]
	vegetable materials	modifier: diverse; $CaCl_2$, ZBC	[729]

Table 10-44 continued

Analyte	Test sample	Conditions, remarks	References
Cd	vegetable materials	STPF	[1020]
	biological materials	STPF, ZBC, modifier: Pd–NH$_4$NO$_3$	[6481]
	foodstuffs	STPF, ZBC, modifier: Pd–NH$_4$NO$_3$	[5451]
Cd, Co, Fe, Ni, Pb	cigarettes, reference materials	STPF	[4044]
Cd, Cr, Cu, Fe, Mn, Mo, Ni, Pb, Zn	milk	chemometric investigation to make distinctions (feeds)	[1862]
Cd, Cr, Cu, Fe, Mn, Ni, Pb, Zn	fish	platform, modifier: Mg(NO$_3$)$_2$, NH$_4$H$_2$PO$_4$	[1377]
Cd, Cr, Cu, Pb	vegetable materials	STPF, ZBC	[659]
Cd, Cu, Fe	marine organisms	STPF, modifier: (NH$_4$)$_3$PO$_4$–Mg(NO$_3$)$_2$	[680]
Cd, Cu, Fe, Mn, Pb	animal tissue	ZBC	[466]
	biological reference materials	STPF, ZBC	[1168]
Cd, Cu, Pb	grain, flour	platform, ZBC	[5141]
Cd, Hg, Pb	game	comparative measurements	[690]
Cd, Pb	milk	platform, ZBC, dilution with Triton X-100	[4293]
		STPF, ZBC, modifier: NH$_4$H$_2$PO$_4$	[3244]
		modifier: (NH$_4$)$_2$HPO$_4$, ashing in O$_2$	[3782]
	foodstuffs	platform, ZBC, modifier: Mg(NO$_3$)$_2$, (NH$_4$)$_2$HPO$_4$	[5535]
	non-vegetable foodstuffs	platform, ZBC	[736]
	Ca-rich matrices	ZBC, modifier: Mg(NO$_3$)$_2$, (NH$_4$)$_2$HPO	[1719]
	biological reference materials	ZBC	[2890]
	feeds		
Co	foodstuffs	better precision than with F AAS and extraction	[380]
	bovine liver	platform, modifier	[1050]
Co, Cr, Fe, Mn, Mo, Ni	milk, baby food	ZBC	[2272]

Table 10-44 continued

Analyte	Test sample	Conditions, remarks	References
Cr	milk, baby food	modifier: $Mg(NO_3)_2$, Triton X-100	[1312]
	meat, liver, kidneys	ZBC, modifier: $Mg(NO_3)_2$	[5142]
	biological reference materials	interlaboratory trial	[5136]
	biological materials	STPF, ZBC, modifier: $Mg(NO_3)_2$	[4105]
		investigation of interferences	[3764]
		review article	[4995]
Cr, Cu, Pb	vegetable materials	STPF, modifier: $(NH_4)_2HPO$	[2337]
Cu	milk	separation of lipids and proteins; ashing in O_2	[714]
Ge	animal tissue	ZBC, modifier: $Co(NO_3)_2$	[5314]
	ginseng	ammonium molybdate impregnated tubes	[372]
	medicinal herbs	modifier: $Ni(NO_3)_2$	[3525]
Hg	vegetable materials	STPF, modifier: Pd^0	[6260]
Li	animal tissue	analyte addition technique	[4876]
Mn	animal tissue	STPF, ZBC, modifier: $Mg(NO_3)_2$	[1602]
	mussel shells	comparison with NAA	[2005]
Mn, Zn	milk	STPF, ZBC, microwave-assisted digestion in HNO_3	[2003]
Mo	milk	modifier: BaF_2	[560]
	foodstuffs	comparison of various methods	[6005]
	vegetable materials	integrated absorbance	[2605]
	feeds	modifier: La, integrated absorbance	[885]
	rice	La-coated tubes, modifier: $CaCl_2$	[6453]
Mo, Se	rice, sesame	platform, modifier: $Ni(NO_3)_2$, ZBC	[2909]
P	biological reference materials	STPF, ZBC, modifier: Pd–Ca	[1383]
	tea, mussel tissue	STPF, modifier Pd–Ca	[6460]
	starch-flour	modifier: $Ni(NO_3)_2$	[2626]
	milk	STPF, ZBC, modifier: La	[1260]

Table 10-44 continued

Analyte	Test sample	Conditions, remarks	References
Pb	milk, baby food	STPF, ZBC, modifier: Mg(NO$_3$)$_2$, (NH$_4$)$_2$HPO$_4$	[181]
	marine organisms	investigation of modifiers	[4038]
	green vegetables	comparison with IDA-ICP-MS	[443]
	biological materials	modifier: Pd–Mg	[4596]
	algae	STPF, ZBC; alternate gas: H$_2$	[4389]
	beer	STPF, ZBC; modifier: NH$_4$H$_2$PO$_4$	[6126]
	sugar	STPF, ZBC; oxygen ashing; modifier: Mg	[4110]
Pt	animal tissue	alternate gas: NH$_3$	[3949]
Ru	vegetable materials	ZBC	[669]
Se	milk	STPF, ZBC, modifier: Pd–Mg	[3245]
	chicken eggs	ZBC	[6147]
	animal blood	STPF, ZBC, modifier: Pd–Ni, oxygen ashing	[1668]
	animal tissue	STPF, ZBC, modifier: Ni	[1681]
	bovine liver	ZBC, modifier: Pd–Cu, alternate gas: H$_2$	[3908]
	animal tissue	Pd–Mg-modifier	[1249]
	meat, liver	wavelength 204 nm; modifier: Pd	[4037]
	grain, grain products	STPF, ZBC, modifier: Pd–Mg	[1752, 1753]
	feeds	ZBC, modifier: Rh(NO$_3$)$_3$	[2891]
		STPF, ZBC, modifier: Cu	[1285]
Si	bones, tissue	ZBC, modifier: Ca–La–Phosphat	[2700]
Sn	biological materials	modifier: Pd–hydroxylammonium chloride	[912]
Yb	feces, digesta	STPF, ZBC, THGA	[3520]

The question as to whether a test sample should be digested prior to analysis or whether the solid can be analyzed directly or after slurrying depends on a number of varying factors which are largely discussed in Section 4.2.7. For the analysis of biological samples the nature of the sample plays a major role in such considerations. Liver tissue, for example, can be digested very easily so that a microwave-assisted digestion is optimum with respect to both the accuracy and the speed of analysis when several elements are to be determined [4101]. For vegetable materials in particular, digestions often require prolonged effort so that a direct analysis can bring substantial advantages [1086, 3331]. A number of analytes, such as boron [4531] and in particular mercury, can be lost relatively easily during digestion so that it is not surprising that the latter element is frequently determined by direct solids analysis (refer to Table 10-46). MILLER-IHLI [4103] and also STURGEON [5643] have published review articles in which the topic of digestions or direct analyses is discussed.

If we scrutinize the publications on the direct analysis of solids or slurries by GF AAS in Tables 10-46 and 10-47, we can immediately observe that very few elements are determined by direct solids analysis, except cadmium, copper, lead, and mercury. Element-dependent difficulties have been reported in interlaboratory trials, such as in the determination of copper [2516], which is indicative that this low number of elements investigated is not arbitrary. Although cadmium, copper, and lead receive more attention than other elements for the analysis of slurries, nevertheless around twice as many elements have been investigated and the distribution is broader.

Independent of whether the test sample is investigated directly or as a slurry, it was very soon established that, similar to the analysis of solutions, the STPF concept led to significantly better results [1137, 1138, 1147]. Next to platform atomization and the addition of modifiers (which is much easier for slurries than for solid samples), numerous working groups also described oxygen ashing directly in the graphite tube [563, 3454, 4158, 5565, 5566], which significantly reduces matrix effects [1147, 1658]. As an alternative the addition of hydrogen peroxide and nitric acid has been proposed to pre-ash the test sample [6086, 6087] and to prevent the buildup of carbon residues in the tube [6085]. Instead of pre-ashing in the tube, a number of authors used external pre-ashing [2522, 3608, 6545]; compared to *in situ* pre-ashing this merely offers an advantage when larger sample quantities are to be analyzed and the sample mass can thus be reduced [307, 308].

Although for the majority of applications the analysis of slurries offers a number of advantages in practice compared to direct solids analysis, particularly due to the ease of handling the samples, there are a number of applications for which only direct solids analysis is suitable. These applications include, for example, the *distribution analysis* of elements such as cadmium and lead in organs [3634, 3639, 3640] or cadmium, copper, and zinc on wheat grains [4646]. The endogenous contamination of muscle tissue by lead due to calcified cysts was also investigated using this technique [3641]. A further area of application is *testing the homogeneity of reference materials* [354, 2752, 3365, 3366, 4159, 4574, 4576, 4578]; KURFÜRST [3363] drew particular attention to the influence of the nugget effect on the statistics. All in all, direct solids analysis has proven to be very useful for the certification [2621], classification [2512], and quality assurance [3636, 3637] of reference materials.

Table 10-45. Selected publications on the determination of trace elements in biological materials by GF AAS after preconcentration.

Analyte	Test sample	Preconcentration procedure	References
As	animal tissue	extraction of the halides in toluene	[2618]
	foodstuffs	coprecipitation with Cu–Fe–APDC	[1397]
Au	animal tissue	extraction in IBMK	[3064]
	vegetable materials	preconcentration on PU foam	[2678]
Be	milk	extraction with liquid exchanger / cyclohexane	[628]
Bi	vegetable materials	sorption on ion exchanger	[3371]
Cd	biological materials	extraction with DPTH	[1773]
Cd, Co, Ni, Pb	foodstuffs	coprecipitation with Cu–Fe–APDC	[1394, 2754]
Cd, Cr, Fe	canned foodstuffs	extraction with APDC / IBMK	[2954]
Cd, Cu, Pb	marine organisms	FI off-line sorbent extraction DDTP / C-18	[3758]
Co	feeds	extraction with 2-nitroso-1-naphthol / xylene	[713]
	fish tissue	extraction with APDC / IBMK	[4029]
Co, Mo, Se	liver tissue	extraction with APDC / chloroform	[4541]
Co, Se	feeds	extraction with APDC / IBMK	[886]
Cr	vegetable materials	coprecipitation with Fe; remove Fe with IBMK	[1096]
Ge	vegetable materials	extraction of $GeCl_4$ with CCl_4	[5147]
		extraction with IBMK / DMF	[373]
Ir, Pt	marine organisms	sorption on Dowex AG 1-X2	[2602]
Mo	vegetable materials	extraction with NH_3SCN / DIBK	[1380]
		extraction with cupferron / $CHCl_3$	[5283]
Ni	biological materials	extraction with BPTH / IBMK	[6026, 6027]
Pb	fish tissue	extraction with dithizone / toluene, back extraction	[2382, 2383]
	sheep blood	extraction with APDC / IBMK after dilution with Triton X-100	[2134]

Table 10-45 continued

Analyte	Test sample	Preconcentration procedure	References
Pd, Pt	vegetable materials	extraction with KI / IBMK	[3261]
Pt	vegetable materials	extraction with dithizone / IBMK	[152]
Se	vegetable materials foodstuffs	extraction as 5-chloropiazselenol in toluene coprecipitation with Pd	[4330] [1396]
Te	foodstuffs	extraction with HCl / IBMK	[3306, 3307]

Table 10-46. Selected publications on the determination of trace elements in biological materials by direct solids analysis by GF AAS.

Analyte	Test sample	Conditions / remarks	References
Al	biological materials	ZBC	[1971]
As	biological materials	ZBC, modifier: Ni–H_2SO_4–HNO_3	[304]
As, Cd, Cr, Cu, Mn, Pb	hay	ZBC, standard calibration technique	[6096]
Cd	wheat	ZBC, platform	[2666]
Cd, Cr, Pb	dairy products	ZBC, platform, alternate gas: O_2	[3454]
Cd, Cu, Hg, Pb	biological materials	interlaboratory trial	[2516]
Cd, Cu, Pb, Zn	biological materials	interlaboratory trial	[5132]
Cd, Hg, Pb	biological materials	ZBC, platform	[4965]
	cigarette tobacco	ZBC, platform	[4621]
	marine organisms	ZBC, platform	[2225]

Table 10-46 continued

Analyte	Test sample	Conditions / remarks	References
Cd, Pb	algae	platform, modifier: $NH_4H_2PO_4$–HNO_3	[876]
	foodstuffs	ZBC, alternate gas: O_2	[4158]
	animal tissue	ZBC, platform	[4620]
	liver tissue	ZBC, platform	[1765, 3638, 4966]
	kidney tissue	ZBC, platform	[3635]
	feathers	ZBC, platform	[2315, 2316]
Cd, Pb, Zn	liver tissue	ZBC, platform, rapid determination	[3149]
Cr	mussel tissue	ZBC	[2820]
Cu	vegetable materials	modifier: Ni–NH_4NO_3–HNO_3, ethanol	[3760]
	rice	ZBC, modifier: EDTA	[4041]
	soil fauna in a vineyard	ZBC, platform	[6364]
	dairy products	platform, modifier: Mg	[394]
Cu, Mn	maize roots	comparison with wet ashing	[1794]
Fe, Zn	rice	alternate wavelength, sensitivity reduced by gas flowrate	[495]
Hg	fungi, worms	ZBC	[1921]
	biological materials	various modifiers	[3031]
	fish	comparison with digestion methods	[721]
Mn	pancreas tissue	alternate wavelength	[4957]
Pb	biological materials	modifier: $NH_4H_2PO_4$	[3558]
	muscle flesh	ZBC, endogenous contamination	[3641]
	liver tissue	platform, alternate gas: O_2	[1147]
Se	biological materials	ZBC, modifier: Cu–Pd–Triton X-100	[4450]
	animal tissue	ZBC, modifier: Pd	[3530]

Table 10-47. Selected publications on the determination of trace elements in biological materials by GF AAS using slurry sampling.

Analyte	Test sample	Conditions / remarks	References
Ag, Cu, Fe, Mn, Pb, Zn	biological reference materials	platform	[1003]
Al	chewing gum	pretreatment with ethanol, HNO_3, $Mg(NO_3)_2$, H_2O_2	[6087]
Al, As, Cd, Fe, Pb, Se	foodstuffs	ZBC, platform	[1097]
Al, Cr	vegetables	modifier: HNO_3, H_2O_2	[6085]
As	vegetable materials	THGA	[3331]
B	vegetable materials	modifier: Ca–Mg–Ni, clean-up: methanol, H_2SO_4, NaF	[387]
	cell suspensions	—	[4531]
Cd	biological materials	platform, modifier: $NH_4H_2PO_4$	[1647]
	foodstuffs	ZBC, platform, modifier: Pd	[3755]
	foodstuffs	platform, modifier: Pd	[5315]
Cd, Cr, Cs, Cu, Mo, Pb, Se, Zn	milk, milk powder	ZBC, modifier: various	[6122]
Cd, Cr, Pb	biological materials	ZBC, modifier: Pd	[5508]
Cd, Cu, Fe, Mn, Pb	cereal products	platform	[1766]
Cd, Cu, Fe, Pb, Se	fruit		[1022]
Cd, Cu, Mn, Pb, Zn	foodstuffs	modifier: $NH_4H_2PO_4$; alternate Gas: O_2	[5566]
Cd, Pb	biological reference materials	precharred material	[1796]
	biological materials	platform, suspended in methanol, glycerol, HNO_3	[2606]
	vegetable materials	modifier: $(NH_4)_2 HPO_4$	[1568]
	cereal products	platform, modifier: Pd for Cd, $NH_4H_2PO_4$ for Pb	[6083]
	vegetables	platform, rapid program without pyrolysis	[6084]
Co, Cr, Mn, Pb	biological materials	alternate gas: O_2	[1658]
Co, Ni	foodstuffs	Triton X-100, HNO_3, H_2O_2	[6086]
Cr, Cu, Fe, Pb	pine needles	comparison with wet ashing	[1086]

Table 10-47 continued

Analyte	Test sample	Conditions / remarks	References
Cr, Cu, Pb	vegetable materials	ZBC, platform	[4108]
Cu	biological microsamples	platform	[1648]
	milk powder	comparison with wet and dry ashing	[3089]
	vegetable materials	—	[347]
	cereal products	rapid program	[6081]
Cu, Fe, Mn, Zn	bovine liver	comparison with digestion	[4101]
Mn	biological materials	rapid program	[2951]
Mn, Pb, Tl	biological materials	comparison with LEAFS	[981]
P	vegetable materials	Zr-coated tube	[3318]
Pb	biological materials	stabilized suspension	[1859]
	biological materials	modifier: ascorbic acid	[1646]
	mussel tissue	platform, modifier: Pd–Mg, alternate gas: O_2	[563]
	foodstuffs	platform, modifier: Pd	[3753]
	vegetable materials	platform, modifier: $NH_4H_2PO_4$	[6494]
	paprika, pre-ashed	platform, modifier: $(NH_4)_3PO_4$	[2522]
	spinach	alternate gas: O_2	[5565]
	flour	modifier: $NH_4H_2PO_4$	[6574]
Se	flour	platform, modifier: Pd	[4742]
		THGA, ZBC, modifier: Pd–Mg	[524]
		rapid program without pyrolysis	[3617]
	biological materials	platform, enzymatic pretreatment, modifier: Pd–citric acid	[5763]
	biological materials		
Ti	vegetable materials	pre-ashing, modifier: hexametaphosphate	[3608]

Finally we should like to mention an application in which biological materials are investigated but which does not really belong to this area. This is the analysis of biomass, bacteria, or algae which act as sorbents for the accumulation of trace elements, in water for example. Such samples can be introduced directly into the graphite tube in the form of slurries. HOLCOMBE and co-workers described the determination of cadmium [3813, 3814], copper [4148], cobalt and nickel [4149], and lead [3796] in river and sea water after sorption on algae; ELMAHADI and GREENWAY [1724, 1725] accumulated cadmium, cobalt, copper, and lead using algae. ROBLES and co-workers used immobilized bacteria to accumulate beryllium [4913], gold [4912], and cadmium [4914] which they then analyzed after slurrying by GF AAS. In further papers they described not only the use of algae [4557, 4558] and bacteria [148, 3873], but also of lichens [4795], yeasts [3780, 3872], chitosan [5570], and turf [5932] as biosorbents.

10.2.5 Digestion Procedures for Analytical Techniques Using Chemical Vapor Generation

In contrast to F AAS and GF AAS, where the test sample is in effect subjected to an *in situ* decomposition so that a complete digestion is mostly not required, for techniques using chemical vapor generation the analyte must be present in ionic form and in a given oxidation state. This means that the test sample must frequently be completely digested prior to the determination and especially that it must be released from organic compounds. The digestion technique is thus determined in the first instance by the compound in which the analyte is present and only in the second instance by the composition of the sample. We shall thus discuss this topic according to element since a number of analytes are present in biological materials as compounds which are exceedingly difficult to digest so that special procedures are required. Moreover a number of the elements are very volatile, making digestion procedures even more difficult.

Arsenic: Among the most difficult-to-digest compounds of this element are arsenobetaine, arsenocholine, and oxy-bis-dimethylarsine; they require strong oxidants, high temperatures, and long digestion times. During the 1970s dry ashing procedures were widely used for the subsequent determination of arsenic by HG AAS; Table 10-48 shows an overview of selected papers on this subject. Owing to the volatility of arsenic and a number of its compounds, dry ashing without the addition of an ashing agent is not possible since up to 80% of the analyte can be lost [6366]. These losses are matrix-dependent and are higher for animal tissues than for vegetable tissues. The most widely used ashing agent is magnesium nitrate hexahydrate, frequently under the addition of some magnesium oxide; an excess of 4- to 8-fold of the sample mass and exceedingly careful heating, especially at the start, are required [6366]. Samples were also frequently pre-ashed with acid or a combined wet/dry digestion procedure was applied, as shown in Table 10-48. Nevertheless there was no shortage of indications that losses of arsenic of 10–20% took place [3801]. Further disadvantages of dry ashing procedures are the risk of contamination due to the large amount of reagents and the fact that such procedures cannot be easily automated.

Table 10-48. Selected publications on dry ashing and combined wet/dry ashing techniques for the determination of arsenic in biological materials by HG AAS.

Digestion procedure	Test sample	References
$Mg(NO_3)_2$: 450 °C	fish	[3386]
500 °C / H: 2–5 h	foodstuffs	[6366]
$Mg(NO_3)_2$–HNO_3: 450 °C / R: 3.5 h; H: 12–14 h	tomato products	[1128]
$Mg(NO_3)_2$: 450 °C / R: 4 h; H: 16 h	beer	[1127]
+ HNO_3: R: 2 h; H: 14 h	orange juice	[1130]
HNO_3–H_2O_2 + $Mg(NO_3)_2$	plant materials	[5612]
HNO_3–$Mg(NO_3)_2$: 70–80 °C / 16 h	fish	[901]
– 500 °C / R: 4 h; H: 2–4 h		

R = ramp time; H = holding time

For these reasons investigations were made at an early stage on wet digestions for the determination of arsenic in biological materials by HG AAS; a selection is presented in Table 10-49. Although a number of authors attempted to manage without the use of perchloric acid, for the majority of animal and vegetable materials this acid appears to be indispensable [2745, 3801, 5097]. Mixtures of nitric acid, sulfuric acid, and perchloric acid in various compositions are the most effective and relatively the most safe. Even for microwave-assisted digestions under pressure (up to 85 bar) the use of this acid mixture appears to be necessary. SCHRAMEL and HASSE [5184] found that with nitric acid alone arsenobetaine and arsenocholine were digested to less than 50%, and with nitric acid and perchloric acid only to 80–90%. Correspondingly these authors were only able to find the true values for arsenic in certified reference materials of marine organisms after digestion in a mixture of nitric, sulfuric, and perchloric acids. A mixture of nitric acid and hydrogen peroxide, even at high pressure or under microwave assistance, is not able to digest these very stable arsenic compounds. The microwave-assisted digestion with nitric acid and vanadium pentoxide as catalyst [4310, 4311] appears to be a possible alternative to perchloric acid, but a thorough investigation has still to be performed. Photooxidation under UV radiation has also been proposed as a digestion technique and has been successfully applied for the determination of arsenic in fish and other marine organisms by HG AAS [1373].

After digestion and corresponding dilution the expected concentrations of transition metals and other hydride-forming elements in biological materials are so low that interferences do not take place. Nevertheless, a number of authors described the use of masking reagents such as 1,10-phenanthroline [1181] or thiourea and ascorbic acid [2655]. On the whole, however, it suffices to keep the concentration of sodium tetrahydroborate low in order to avoid interferences [6430]. BERGHOFF and VON WILLERT [551] drew to attention the risk of oxidation of As(III) to As(V) by Fe(III).

Table 10-49. Selected publications on wet digestions for the determination of arsenic in biological materials by HG AAS.

Digestion procedure	Test sample	References
HNO_3–ClO_4: 3–16 h / 250 °C	vegetable materials	[6075]
60–70 °C	fish tissue	[1377]
HNO_3–H_2SO_4–$HClO_4$: 16 h / 200 °C	vegetable materials	[1181]
(30:1:5) 2 h boiling	foodstuffs	[158]
(10:2:3) 4 h boiling	biological reference	
	material	[3804]
(5:2:3) 4 h boiling	fish tissue	[3801]
(23:1:23) 500 °C almost to dryness	seaweed	[4586]
HNO_3: 1 h / 140 °C pressure	marine organisms	[6228]
+ H_2SO_4–$HClO_4$ 20 min / 310 °C		
HNO_3–H_2SO_4: 4 h reflux + $HClO_4$: 5 h boiling	spinach, tomato leaves	[5097]
HNO_3–$HClO_4$ (5:1): reflux	foodstuffs	[2745]
+ HNO_3–H_2SO_4 (1:1) boiling		
HNO_3: 10 h / 230 °C, pressure	brewing materials	[1594]
HNO_3–H_2SO_4 (2:3) + H_2O_2	seaweed	[3333]
H_2SO_4–H_2O_2–permanganate–persulfate	fish tissue	[49]
HNO_3–H_2SO_4–$HClO_4$ (81:1:6), microwave, pressure	milk powder, marine organisms	[5184]
HNO_3–V_2O_5, microwave	vegetables	[4311]
	fish tissue	[4310]

Selenium: Among the organic selenium compounds present in biological materials, selenomethionine, selenocysteine, and the trimethyl selenonium ion in particular cause difficulties in digestion. Dry ashing has been described relatively seldom for this element [901, 902, 2369, 4649]. In all official methods the test sample is initially heated under reflux, mostly overnight, in a mixture of nitric acid and magnesium nitrate, and then carefully fumed off prior to the actual ashing at 500 °C. Despite all precautions, however, slight losses cannot usually be avoided [901, 3803].

Selected publications on the oxidative wet digestion of biological materials for the determination of selenium by HG AAS are compiled in Table 10-50. As with arsenic, the majority of working groups found that the use of perchloric acid was unavoidable. The use of nitric acid alone or in mixtures with sulfuric acid and hydrogen peroxide usually led to results that were substantially too low [3803, 5097]. The use of vanadium pentoxide as a catalyst might however be an alternative to the use of perchloric acid [1682].

The conditions for pressure digestions are definitely critical and the sample material also plays a certain role. For fish and other marine organisms a pressure digestion with nitric acid at 140 °C is *not* adequate to release all of the selenium [6228], so that the authors introduced post treatment with perchloric acid. For the determination of sele-

Table 10-50. Selected publications on wet digestions for the determination of selenium in biological materials by HG AAS.

Digestion procedure	Test sample	References
HNO_3–$HClO_4$: 4 h / 140 °C + 1 h / 200 °C	biological materials	[3510]
(7:3) 30 min / 225 °C	biological materials	[2047]
50–60 °C	marine organisms	[1377]
(5:2) 200 °C	equine blood	[4690]
(4:1) 16 h / 125 °C	vegetable materials	[6076]
HNO_3–H_2SO_4–$HClO_4$ (6:3:1): 16 h / 200 °C	vegetable materials	[1181]
(30:1:5): 2 h boiling	foodstuffs	[158]
HNO_3–$HClO_4$ (5:1): 2 h boiling + HNO_3–H_2SO_4 (1:1): fumed off	foodstuffs	[2745]
HNO_3–H_2SO_4: 4 h reflux + $HClO_4$: 5 h reflux	spinach, tomato leaves	[5097]
HNO_3: 1 h / 140 °C, pressure + H_2SO_4–$HClO_4$: 20 min / 310 °C	marine organisms	[6228]
HNO_3–$HClO_3$ + $HClO_4$ (2:3): 240 °C	biological materials	[4814]
HNO_3–$HClO_3$–$HClO_4$ (7:2:1): 160 °C	biological materials	[2799]
HNO_3–H_2SO_4 (1:1), V_2O_5: 20 min boiling	fish products	[1682]
H_3PO_4: 3 h / 120 °C + HNO_3–H_2O_2: 20 min / 160 °C + 6 h / 120 °C + 30 min / 160 °C	feeds	[5611]
HNO_3: 10 h / 230 °C, pressure	brewing materials	[1594]
HNO_3: microwave, pressure	fish tissue	[3399]
HNO_3–H_2O_2: microwave, pressure	porcine liver	[6409]
	high-fat foodstuffs	[1627]
HNO_3–H_2SO_4: microwave, pressure	mussel tissue	[3409, 3410]
HNO_3–H_2SO_4–H_2O_2: microwave, pressure	fish tissue	[3407, 3408]

nium in bovine liver, on the other hand, a pressure digestion in nitric acid (90 min/175 °C) appears to be sufficient since the same results are obtained as with dry ashing and wet digestion with perchloric acid [4631]. For the determination of selenium in vegetable materials a pressure digestion in nitric acid for 10 hours at 230 °C is necessary [1594]. Using chemometric methods, LAN et al. [3409, 3410] optimized a microwave-assisted digestion for the determination of selenium in mussel tissues. They found that neither perchloric acid nor hydrogen peroxide was necessary and they used a mixture of nitric acid and sulfuric acid (approx. 4:1). For fish tissue these authors found that reliable results could only be obtained under the addition of hydrogen peroxide [3408]. DUNEMANN and MEINERLING [1627] reported that for the determination of selenium in high-fat foodstuffs only a microwave-assisted digestion under pressure led to the required results, while a digestion at atmospheric pressure was inadequate.

During the 1980s in particular, various procedures were described for the combustion of the test sample in oxygen, for example in a Schöninger flask [2623, 2624], in a Wickbold apparatus [1754, 5225], or in a closed composite system [2350]. The advantage of these techniques is the very low blank values that can be attained, provided that the procedure is performed correctly, while as disadvantages we can mention the low sample throughput and the fact that such techniques cannot be automated at all easily.

There is little reported in the literature on the determination of the other hydride-forming elements in biological materials. SMITH [5458] published an overview on the determination of *antimony* in biological materials. Apparently a digestion with nitric acid, sulfuric acid and perchloric acid is necessary for this element [3802, 6367], but systematic investigations have hardly been performed. Owing to the markedly lower concentration of this element in comparison to arsenic and selenium, interferences are more probable. WOIDICH and PFANNHAUSER [6367] added phosphate to the solution to eliminate an interference from iron, while HON et al. [2655] used thiourea and ascorbic acid as masking reagents.

ROMBACH and KOCK [4947] described the determination of *bismuth* in milk powder by HG AAS after a pressure digestion in nitric acid. They added thiourea to eliminate an interference caused by nitrose gases.

Although *lead* can be determined with good sensitivity and largely free of interferences by GF AAS, relatively thorough investigations have been made on HG AAS as an alternative [3779]. A mixture of nitric acid, sulfuric acid, and perchloric acid [264] was also applied for digestion, but even more frequently mixtures of nitric acid with hydrogen peroxide [710, 1021], ammonium persulfate [3775, 3776], or potassium dichromate [3777, 3778] were used since these provide ideal conditions for the subsequent hydride generation. Of the other hydride-forming elements, only Se(IV) and Te(IV) interfere in the determination of lead [710, 3775]; this can easily be eliminated by oxidizing these elements to the hexavalent state. The addition of oxalate eliminates a potential interference due to copper [3777]. For the analysis of wine, alcohol in particular interferes so that it must be removed [3777]. The direct analysis of slurries of foodstuffs in nitric acid and ammonium persulfate has also been reported [3776]. Lead can be determined free of interferences against aqueous calibration solutions in drinks and in slurries of fruits in a closed composite FI system with microwave-assisted on-line digestion [1021].

In the case of *tin* it is the determination of the organic compounds that is of particular interest, as discussed in detail in Section 10.2.6. Only little has been published on the determination of total tin in foodstuffs and marine organisms. For digestions, mixtures of the following acids have been described: nitric acid, sulfuric acid, and perchloric acid [3800], nitric acid, hydrofluoric acid, and perchloric acid [5635], nitric acid and sulfuric acid with hydrogen peroxide [157], and nitric acid and hydrochloric acid under pressure [2398].

Since the hydride-forming elements are still frequently of interest even at very low concentrations in biological materials, various preconcentration techniques are described in the literature. Preconcentration is also occasionally applied to eliminate interferences. These techniques include classical procedures such as coprecipitation with lanthanum hydroxide [3800], sorbent extraction on packed microcolumns [5637], or typically for HG AAS trapping in a cold trap [251, 3510]. By far the most successful technique is *in situ* preconcentration in the graphite tube atomizer, especially since it can be easily automated [6333]. An overview of publications on this subject is compiled in Table 10-51. This technique is particularly reliable and highly efficient if the graphite tube has been preconditioned with a noble metal prior to preconcentration (refer to Sections 4.3.2.2 and 8.3.2.3). Not only has *in situ* preconcentration in the graphite tube been coupled with the classical hydride-generation technique, but also with the electrochemical hydride generation of antimony [1528], the cold vapor technique for mercury [435, 5383, 6440], with ethylation for the determination of lead [5641, 6333], with carbonyl generation for the determination of nickel [1757, 5639], and with the vaporization of volatile chelates [169].

Table 10-51. Selected publications on *in situ* preconcentration of trace elements in the graphite tube atomizer after chemical vapor generation.

Analyte	Test sample	Conditions	References
As	marine organisms	600 °C, uncoated tubes	[5631]
As, Bi, Ge, Sb, Se, Sn, Te	biological materials	300 °C (Ge: 600 °C), Pd-coated platform	[6543]
As, Sb	vegetable materials	As: 400 °C, Sb: 600 °C, Pd modifier	[6131]
As, Se	marine organisms	600 °C, uncoated tubes	[5632]
Bi, Sb	vegetable materials	250 °C, uncoated tubes	[865]
Bi	biological materials	Pd–Ir-coated tubes	[3976]
Ni	marine organisms	(carbonyl) > 500 °C	[5639]
Pb	biological materials marine organisms	500 °C, used tubes (ethylation) 400 °C (ethylation) 300 °C	[264] [5641] [6333]
Pb	vegetable materials		[1758]
Se	marine organisms	600 °C, uncoated tubes	[6332]
Se	mineral tablets		[2357]
Sn	marine organisms	600 °C, uncoated tubes	[5635]
Sn	biological materials	300°C, Pd-coated tubes	[5773]

Mercury is without doubt one of the elements determined most frequently in biological materials and the method of choice is CV AAS. The most critical step of the analysis is digestion, firstly because of the high volatility of mercury and its compounds, and secondly because of the risk of contamination due to inadvertent introduction and carryover of this element. For these reasons dry ashing of the sample for the subsequent determination of mercury can hardly be considered. In the majority of cases acid digestions are applied for the subsequent determination of mercury in biological materials, as shown in the overview in Table 10-52. It is most important, especially if high temperatures are used for the digestion, that strongly oxidizing conditions are maintained at all times, in particular in the gas phase, to prevent the loss of mercury. The shape of the digestion vessel also plays a role; especially at the start of the reaction; highly efficient cooling is necessary to prevent losses [3413, 4237, 4786].

Table 10-52. Selected publications on acid digestions for the determination of mercury in biological materials by CV AAS.

Digestion procedure	Test sample	References
HNO$_3$–HClO$_4$ (1:1): 3 h / 150 °C, closed vessel	biological materials	[3988]
HNO$_3$–HClO$_4$ (17:3)	medicinal herbs	[5849]
HNO$_3$–H$_2$SO$_4$ (4:5): reflux + HNO$_3$–HClO$_4$ (1:1): 10 min fumed off	fish (interlaboratory trial) biological materials	[4237] [4786]
H$_2$SO$_4$–KMnO$_4$: 2 min.:.1.5 h / 50→80 °C	fish	[3393, 5842, 5968, 6018]
HNO$_3$–H$_2$SO$_4$–KMnO$_4$	biological materials	[2901]
HNO$_3$–H$_2$SO$_4$: 1 h / 180 °C + KMnO$_4$: 20 min / 120 °C	vegetables, fruit	[5766]
HNO$_3$: 10 min / 60 °C + H$_2$SO$_4$–KMnO$_4$: 10 min / 60 °C	fish tissue	[5321]
HNO$_3$–H$_2$SO$_4$–(NH$_4$)$_2$S$_2$O$_8$–KMnO$_4$: 15 min / 120 °C reflux	vegetable materials	[1625]
KMnO$_4$–HNO$_3$–H$_2$SO$_4$–K$_2$S$_2$O$_8$: 30 min / < 90 °C	milk	[3076]
H$_2$O$_2$–H$_2$SO$_4$: 30 min / 60 °C + KMnO$_4$: 1 h / 20 °C + K$_2$S$_2$O$_8$: 16 h / 20 °C	fish tissue	[49]
HNO$_3$–H$_2$SO$_4$: 1 h / 80–90 °C + KMnO$_4$: 1 h / 80–90 °C + K$_2$S$_2$O$_8$: 1 h / 80–90 °C	fish tissue	[2099]
K$_2$Cr$_2$O$_7$–H$_2$SO$_4$: 15 min / 0 °C + 30 min 185 °C	vegetables	[3412]
HNO$_3$–H$_2$SO$_4$–V$_2$O$_5$	fish cereal products	[1634, 1682, 5675] [3825]
HNO$_3$–V$_2$O$_5$	biological materials	[1068]
HNO$_3$–H$_2$SO$_4$: 60–90 min / 90 °C	biological materials	[38]
HNO$_3$: 16 h / 20 °C + H$_2$SO$_4$: 6 h / 250 °C	vegetable materials	[4819]
HNO$_3$–H$_2$SO$_4$ (4:1): 300 °C	animal tissue	[6492]
HCl–HNO$_3$–H$_2$SO$_4$: 30 min / 85–100 °C	fish	[3625]
H$_2$SO$_4$–H$_2$O$_2$	biological materials	[2201, 2202]
HNO$_3$–H$_2$SO$_4$–HClO$_4$–KMnO$_4$: 1 week / –10 °C	marine organisms	[5997, 5998]
HNO$_3$: 30–60 min / 140–150 °C, pressure 16 h / 120 °C, pressure 10 h / 230 °C, pressure	marine organisms biological materials brewing materials	[2631, 6228] [6042] [1594]
HNO$_3$–H$_2$SO$_4$: 100 °C, pressure	fish, seaweed	[4573]
HNO$_3$–V$_2$O$_5$: pressure	biological materials	[3254]
HNO$_3$: microwave, pressure	biological materials vegetables fish	[6043] [4312] [3399, 4309, 4923, 5742]
HNO$_3$–H$_2$O$_2$; +H$_2$SO$_4$, microwave	fish	[4249]
HNO$_3$–H$_2$SO$_4$: microwave, open	fish	[5176]

ADELOJU *et al.* [38] compared four frequently-used digestion procedures for the determination of mercury in fish and kidney tissue. Digestions in nitric acid plus sulfuric acid or in nitric acid plus hydrogen peroxide gave good recoveries for inorganic and organically-bound mercury, while digestions in nitric acid alone or in nitric acid plus perchloric acid only released about half of the organically-bound mercury. For the determination of mercury in vegetables, LANDI *et al.* [3412] obtained even better accuracy using potassium dichromate in dilute sulfuric acid than with the four digestions above. DUMAREY *et al.* [1625] investigated 14 different digestion procedures for the determination of mercury in aquatic plants and obtained the true result using nitric acid plus sulfuric acid with an addition of ammonium peroxodisulfate and potassium permanganate. Albeit the risk of contamination is relatively high due to the large number of reagents used.

Due to the high volatility of mercury and its compounds, a number of authors preferred long digestion times at relatively low temperature over a digestion at higher temperature [49, 2099]. The technique applied by COLINA DE VARGAS and ROMERO [5997, 5998] is particularly worth mentioning; they digested animal and vegetable tissue in a mixture of nitric acid, sulfuric acid, and perchloric acid under the addition of potassium permanganate and hydrogen peroxide in an ice bath at $-10\ °C$ for a week. For a large number of test samples, these authors found the same mercury concentration as for pressure digestions in nitric acid at 130 °C.

Digestions in closed systems, with or without microwave assistance, are an alternative to the above techniques. With appropriate precautions mercury losses can largely be avoided. The advantage of such digestions is the greatly shorter time requirements and that digestions can be performed at only slightly increased pressure [3988]. Microwave-assisted digestions also give good results for mercury when they are performed at atmospheric pressure [5176, 5177]. This reduces the risk of exchange reactions with the walls of the container and allows the digestion to be performed on-line in an FI system [6263].

As well as acid digestions, alkaline digestions have also been described for the determination of mercury in biological materials by CV AAS. CHAPMAN and DALE [1197] warmed numerous vegetable and animal tissue samples in potassium hydroxide solution until they had been dissolved or homogenized; they then allowed potassium permanganate to react, followed by acidification and several hours standing time prior to the determination of mercury. KONISHI and TAKAHASHI [3239] described a similar digestion with alkaline potassium cyanide solution but which only required a few minutes. WIGFIELD and EATOCK [6318] heated biological materials to 80 °C with a 180 g/L sodium hydroxide solution under the addition of cysteine for 10 minutes and diluted the mixture with a 10 g/L sodium chloride solution. The recovery for mercury was virtually 100% in fish, about 90% in milk, and around 85% in liver tissue. GUTIÉREZ *et al.* [2296] described a very similar digestion procedure with sodium hydroxide, cysteine, and sodium chloride. As an alternative to heating they allowed the mixture to stand at 25–30 °C overnight.

Combustion in oxygen in a Schöninger flask for the determination of mercury in fish and eggs by CV AAS was proposed as early as 1966 [4533]. Such combustions in oxygen, for example in a Wickbold apparatus [1755] or in an oxygen bomb [1091], are still occasionally used nowadays, particularly for reference procedures. An alternative is combustion of the sample in a stream of air or oxygen and collection of the vaporized mercury in a cold trap [5035] or, better, by amalgamation on gold [1624].

There are hardly any reports on interferences in the determination of mercury in biological materials, except for low results caused by inadequate digestion. An exception is merely the interference caused by iodine on the determination of mercury in seaweed. This interference only occurred however in the calibration solutions in which iodide was oxidized to iodate; the interference did not occur for the test samples [3332]. Precisely this interference was utilized by a number of authors for the *indirect determination of iodide* in seaweed [1170, 3335, 5688]. The samples are best taken into solution by a fusion with potassium hydroxide [5688]. The interference is strongly dependent on the pH [3335] and is most pronounced at higher acid concentration, so the determination of iodide is best performed under these conditions.

10.2.6 Speciation Analysis

For speciation analysis in biological materials the separate determinations of *inorganic mercury* and *organic mercury*, mainly *methylmercury* (MeHg$^+$), take first place. Methylmercury can enter the environment directly from industrial emissions or it can be produced by the biomethylation of inorganic mercury. Methylmercury accumulates in the nutritional chain and can reach substantial concentrations in the muscle tissue of predatory fish, for example [1347].

For analysis the first important question is whether a distinction must be made between inorganic and organically-bound mercury, or whether the various organic mercury compounds that might be present should be determined separately. Chromatographic separation is indispensable in the second case, while in the first case the varying reactivities or solubilities of both bonding forms can be exploited. Systematic errors can occur in the differential determinations of mercury if the carbon-mercury bond of organic mercury is attacked during the reduction of the inorganic mercury so that the signal assigned to the inorganic mercury is too high. The magnitude of this error depends on the nature and properties of the test sample, on the conditions of the reaction, and on the strength of the reductants used. In order to obtain a correct analytical separation between inorganic and organic mercury by sequential reduction, we must additionally stabilize the carbon-mercury bond, preferably with thiol (mercapto) compounds [2380, 2381]. The determination of the organically-bound mercury fraction is performed after releasing it from the thiomercury complex by lowering the pH value or by an excess of heavy metal ions.

In 1971 MAGOS [3789] was the first to describe the selective determination of organically-bound and inorganic mercury in non-digested biological samples. He found that in alkaline solution under the addition of cysteine only inorganic mercury was reduced by tin(II) chloride, while the organically-bound mercury was reduced by a mixture of tin(II) chloride and cadmium(II) chloride. Magos' procedure was used widely, for example for the analysis of foodstuffs and other biological materials [3869], and a recent interlaboratory trial showed that the results compare very well with newer techniques [3528]. A number of authors modified this technique and used a mixture of tin(II) chloride, sodium chloride, and copper(II) chloride, for instance, for the determination of organically-bound mercury in marine organisms [3950]. A further variation is the determination of inorganic mercury in fish directly in the sulfuric acid digestion solution using tin(II) chloride and cadmium(II) chloride, and then the subsequent determination of organically-bound mercury in the same solution after adding an excess of sodium hydroxide [3393, 6019]. ODA and INGLE [4416] reduced inorganic mercury selectively with tin(II) chloride and released the organically-bound mercury with sodium tetrahydroborate. HARMS and LUCKAS [2381] have published a good overview on the various procedures for the differential determination of inorganic mercury and organically-bound mercury in biological materials.

Another technique for the differential determination of organic mercury compounds is to separate them prior to analysis [2672]. The most frequently described technique is extraction of methylmercury chloride or methylmercury bromide from acidic solution with toluene, for example for fish tissue. The extract can then be analyzed directly by GF AAS after addition of dithizone [5323] or thiosulfate [5145]. For the determination by CV AAS the extraction is followed by a back-extraction with a cysteine acetate solution in which the methylmercury is then oxidized with potassium permanganate in sulfuric acid solution [1440]. REZENDE et al. [4856] also described a determination directly in the organic extract after the addition of nitric acid with sodium tetrahydroborate, both in DMF. A disadvantage of this extraction procedure is the dependence on the matrix and the frequently poor reproducibility of the extraction yield [346, 2380].

MAY et al. [3989, 3990] dissolved inorganic mercury and methylmercury with 6 mol/L hydrochloric acid from biological samples and separated the inorganic mercury with an ion exchanger. Alkylmercury compounds were decomposed by UV oxidation and determined by CV AAS after reduction and amalgamation. HORVAT et al. [2670] proposed steam distillation which allows 95% of the methylmercury to be separated with good reproducibility from a multitude of biological samples. The methylmercury is then mineralized with either nitric acid and perchloric acid or by UV radiation, and after preconcentration on gold is determined by CV AAS. COLLETT et al. [1319] collected the distillate in a potassium persulfate solution in which the methylmercury is immediately decomposed and prepared for determination by CV AAS. PADBERG et al. [4516] compared this distillation technique with the technique of acid extraction and subsequent separation on an ion exchanger. On the whole they found good agreement but with certain advantages for the steam distillation technique.

For the coupling of GC with CV AAS the same separation techniques are applied as described above for the direct selective determination of methylmercury. The organically-bound mercury is mostly released from it protein bonding by the addition of copper(II) at low pH, extracted with benzene or toluene, and after further purification, drying and preconcentration is passed through the column [990, 2171, 2384, 2385, 2911]. The methylmercury can then be decomposed to elemental mercury in a pyrolyzer at 700–900 °C connected to the outlet of the column; the mercury is determined by CV AAS either directly [2911] or after collection on gold [2385]. The steam distillation technique described above has also been successfully applied for the separation of organic mercury compounds prior to their chromatographic separation [346, 2672]. Separation techniques using HPLC have also been combined with CV AAS [1664, 4523]; in particular this simplifies sample pretreatment since no derivatization is required.

Viewed generally, the coupling of GC or HPLC with CV AAS is only required when the various organic mercury compounds must be separated from each other, but it does not make practical sense when a distinction must merely be made between inorganic and organically-bound mercury compounds [6325]. We should also mention that the monomethylated form is virtually the only organomercury species that has been observed to date in biological materials. It has been shown that the results for the determination of organically-bound mercury depend solely on the separation technique (extraction, distillation, etc.), but not on whether the determination is performed directly or after chromatographic separation [2380, 2672, 3528, 5812]. Compared to direct procedures, the chromatographic techniques have the disadvantage that they require more complex sample preparation, can only tolerate relatively small sample volumes and thus exhibit lower sensitivity, and that the column packing must be frequently regenerated.

RAPSOMANKIS et al. [1910, 4812] described a very simple chromatographic procedure for the determination of methylmercury in fish which does not have the above disadvantages. The homogenized fish sample is dissolved in methanolic potassium hydroxide solution in an ultrasonic bath, and after the addition of an acetate buffer is reacted with sodium tetraethylborate. The ethylation products are driven out of solution by a stream of helium and collected on a GC column cooled in liquid nitrogen. The column is then heated from -198 °C to $+120$ °C within two minutes. In just *one* chromatogram it is then possible to detect inorganic mercury (as Et_2Hg), methylmercury (as MeHgEt), and dimethylmercury (as Me_2Hg). Next to simple sample pretreatment, the great advantage of this technique is the possibility of using larger sample volumes and to preconcentrate the analytes *in situ*, and also that it can easily be automated. PUK and WEBER [4733] used the same apparatus, but instead of the ethylation reagent they used sodium tetrahydroborate to generate the hydrides rather than the ethyl compounds. The great advantage is the rapid reaction, since hydride-formation is spontaneous, while ethylation requires 10–15 minutes. Based on a number of interlaboratory trials, QUEVAUVILLER et al. [4755] have drawn attention to the improvements achieved in the meantime for the determination of methylmercury. The stability of methylmercury, even in biological samples, and its transformation during storage can play a major role and must be taken into consideration [2673].

Next to the usual separated or sequential determinations of inorganic mercury and methylmercury, as well as dimethylmercury and diethylmercury, a number of authors have made a distinction between other species. Among these is the determination of ionizable (toxic) and stable (non-toxic) inorganic mercury in animal tissue [5676]. The ionizable mercury was released by tin(II) chloride in the presence of sodium chloride and semiconcentrated sulfuric acid. DANIELS and WIGFIELD [1414] investigated the effect of cysteine on the determination of mercury in acidic and alkaline solution and distinguished between sulfhydryl-complexed mercury and non-complexed mercury.

The determination of the various *arsenic species* also fills a substantial volume in the literature. In a report by the committee for speciation analysis in the Commission for Microchemical Techniques and Trace Analyses of IUPAC, next to inorganic arsenite and arsenate, eleven organic arsenic species are also named which can occur in biological samples [4197]. Arsenite and other easily-dissolved compounds of arsenic(III) are the most toxic, followed by the corresponding compounds of arsenic(V), while a number of organic arsenic compounds, mainly arsenobetaine and arsenocholine, are practically not toxic. A large number of papers are therefore concerned with the *selective determination of As(III)* or of *inorganic or toxic arsenic*. As early as 1976 AGGETT and ASPELL [54] described a procedure for the selective determination of As(III) in vegetable materials based on the fact that at pH 4–5 only the trivalent state reacts with sodium tetrahydroborate to release arsine, which can be determined by HG AAS. These authors determined the total inorganic arsenic after acidifying the sample in 5 mol/L hydrochloric acid and using the same reductant. A problem with this procedure to determine 'inorganic' arsenic is that several organic arsenic compounds, such as monomethylarsenious and dimethylarsinic acids, form volatile hydrides in strongly acidic solution; these cannot be determined with the same sensitivity as inorganic arsenic, however. LE *et al.* [3476] thus proposed the addition of 2% cysteine in the presence of which the above compounds are reduced to As(III) and can then be determined with the same sensitivity [2683]. These authors thus detected the sum of all *'toxic' arsenic compounds.* WILLIE [6334] described a similar procedure for the determination of toxic arsenic in biological materials via *in situ* preconcentration of arsine in a graphite furnace.

ANSTENFELD and BERGHOFF [312] selectively determined As(III) in vegetable materials by separating it from As(V) on an ion exchanger prior to determination by HG AAS. HOLAK and SPECCHIO [2636] extracted As(III) from strong hydrochloric acid solution as the neutral trichloride with chloroform from various foodstuffs samples and determined it following back-extraction in dilute acid by HG AAS. In 1982 FLANJAK [1920] proposed a procedure for the separation of *inorganic arsenic* by distillation of the chloride from 6.6 mol/L hydrochloric acid solution and in this way determined inorganic arsenic in crustaceans by HG AAS. A number of authors later used this technique for the analysis of fish [858, 4384], shellfish and fish products [858], and algae [4254].

MAHER [4513] determined As(III), As(V), monomethylarsenious and dimethylarsinic acids in marine organisms by HG AAS after extraction and separation by ion exchange chromatography. He found that the greater part of the arsenic was **not** present in the form of these species (but as arsenobetaine and arsenocholine, as was found later). BALLIN *et al.* [363] described an extraction scheme for the determination of various arsenic species in fish. They distinguished between inorganic, water-soluble, and fat-soluble arsenic. The arsenobetaine present in the aqueous extract is subjected to an alkaline hydrolysis and after acidification is converted almost entirely to trimethylarsine by sodium tetrahydroborate [3312, 4382]. The trimethylarsine along with the hydrides formed from monomethylarsenious and dimethylarsinic acids and the arsine from inorganic arsenic can be determined by QF AAS after gas chromatographic separation [3312, 4382]. OCHSENKÜHN-PETROPULU *et al.* [4415] utilized selective extraction with methanol and chloroform, followed by an acid digestion, for the determination of arsenobetaine in marine organisms by HG AAS. NORIN *et al.* [4382, 4383] additionally determined arsenobetaine and arsenocholine in crabs after extraction and separation by ion exchange chromatography by GF AAS or QF AAS.

Complete separation of all the arsenic species to be expected in biological materials, particularly in marine organisms, can only be achieved with HPLC; careful optimization of the mobile phase is required. Detection can be by GF AAS, coupled to an overflow container from which aliquots are removed at regular intervals [6382], or on-line by F AAS [3455]. A markedly better HPLC-AAS coupling, which operates on-line with adequate sensitivity, is based on the thermochemical generation of the hydride proposed by BLAIS *et al.* [656]. MOMPLAISIR *et al.* [4162] applied this system successfully for the determination of arsenic species in marine organisms. More recently the coupling of HPLC and HG AAS has been used; the various species are converted on-line into hydride-forming species in a simple heated reactor [3600], or a microwave reactor [3477, 4520], or a UV photolysis reactor [2682, 4999], which is mounted between the separating column and the hydride generator.

For the separation of various *lead species* in biological materials, fractionated wash procedures using organic solvents and acids of varying strengths have been applied [1370, 5212]. The investigation of roots, foliage, and other plant components, as well as the dissection of leaves into subcellular fractions with the corresponding compartmentalization of the lead content, have been described [5212]. Normally GF AAS is used as the determination technique. The sum determination of *volatile lead*, i.e. tetraalkyl-lead compounds, in marine organisms by GF AAS [1370] and QF AAS [1214] has been described.

The majority of work published on the speciation analysis of lead, however, describes the use of a GC-QF AAS coupling since the individual alkyl-lead compounds vary widely in their toxicity. Ethyl compounds are more toxic than methyl compounds, and the toxicity increases with the increasing degree of alkylation. The procedures mostly start with an enzymatic hydrolysis [1943, 1945, 1946] or the dissolution or extraction of the tissue material with tetramethylammonium hydroxide [1217, 1303, 4081], followed by the addition of a complexing agent such as EDTA, and then extraction of the organo-lead compounds in hexane, for instance. Thereafter the di- and trialkylated compounds are derivatized, mostly butylized, followed by corresponding purification and gas chromatographic separation with on-line determination by QF AAS. Such procedures were mainly used for fish, fish products and other marine organisms [1217, 1370, 1943, 1945, 1946, 4080, 4081], but also for grass and tree foliage [1303]. The quantity of coextracted concomitants is reduced by dissolution in tetramethylammonium hydroxide or by enzymatic hydrolysis, but thorough purification of the extract is nevertheless necessary to avoid the rapid contamination of the column. For the analysis of peanut oil and lard, FORSYTH [1944] proposed a purge-and-trap system as an alternative for the purification of the raw organic extracts.

A simplified procedure with which only the *tetraalkyl-lead compounds* are detected was proposed by CHAU *et al.* [1213, 1214] for the analysis of fish and other marine organisms. The homogenized sample is extracted with EDTA and hexane and the extract is injected directly onto the GC column without derivatization. CRUZ *et al.* [1370] vaporized the tetraalkyl-lead compounds by heating, collected them in a cold trap, and conducted them directly by warming onto the GC column.

There is little to be found on the determination of *chromium species* in the literature. PLESSI and MONZANI [4671] determined *bioavailable chromium* in cereal products, vegetables, and oil seeds by extracting with ethanol and determining by GF AAS after a digestion of the extract. MILAČIČ and ŠTUPAR [4084] applied ion exchange HPLC with off-line determination of chromium in the individual fractions by GF AAS for the determination of *chromium(III) complexes* and *chromium(VI)* in vegetable plants. Positively charged chromium(III) species and kinetically labile complexes are not retained on the column, while the relatively stable chromium(III)-EDTA and -oxalate complexes and chromium(V) are retained and separated.

An area that is reserved for HPLC separation with various AAS detectors is the analysis of *metallothioneines*. The major interest is in *cadmium* [2548, 3140, 3456, 5691, 5709, 6561], but other elements such as lead [3456], copper [465, 3456], or zinc [465] are also occasionally determined. Likewise for this application the difficulty of coupling HPLC with AAS detectors is apparent. On-line coupling with F AAS is possible when the nebulizer aspiration rate is reduced to about 2 mL/min, but the sensitivity is relatively poor [465, 3140, 5709]. Graphite furnace AAS offers the required sensitivity but it can only be applied off-line [5691, 6561]. LARSEN and BLAIS [3456] used a thermospray interface for the coupling of HPLC to F AAS and were thus able to improve the absolute limit of detection for cadmium to about 2 ng. HIGH *et al.* [2548] replaced the flame by a pyrolyzer and a QTA and were thus able to attain a limit of detection of 70 pg, which is adequate for the requirements of this analysis.

For *selenium* the determination of the organic compounds as well as the distinction between inorganic selenite and selenate is of interest. CUTTER [1387] determined selenium(IV) in plankton selectively by HG AAS after dissolution in sodium hydroxide and acidification to pH 1.6–1.8. KÖLBL *et al.* [3225] applied ion chromatography to separate selenite and selenate and used F AAS or GF AAS for detection, depending on the required sensitivity, for the analysis of feedstuffs and selenium additives in fodder. CHAU *et al.* [1210] and also RADZIUK and VAN LOON [4779] determined volatile *dimethylselenide* and *dimethyldiselenide*, which are the most important selenium metabolites formed by a number of plants, using a GC-QF AAS coupling.

For *tin* it is only the organic compounds that are of interest because of their toxicity; this applies in particular to *tributyltin oxide* which is used as an antifouling paint. The selective extraction of this compound from mussels and other marine organisms is described in a number of papers; the tin concentration in the extract is determined by GF AAS [1064, 2977, 4025, 4528, 5574]. The homogenized tissue is mixed with hydrochloric acid and extracted with an organic solvent such as n-hexane. Mono- and dibutyltin compounds are removed with dilute sodium hydroxide solution, and the tributyltin oxide is back-extracted in nitric acid and determined by GF AAS. Another tin compound of interest is the pesticide *tricyclohexyltin hydroxide* (Cyhexatin); residues in apples can be determined directly by F AAS [3626] or better by GF AAS [2131] after extraction with chloroform. TSUDA *et al.* [5920] described a procedure for the selective determination of *inorganic tin* in fish by HG AAS after extraction and separation on a silica gel column.

Next to these selective extraction procedures for a single tin species, separation techniques with AAS detection play a major role for the identification and determination of numerous organotin compounds. The coupling of HPLC to F AAS has been described for the determination of *mono-*, *di-*, and *tributyltin* in oysters [1656], but probably does not exhibit sufficient sensitivity for the majority of applications. A much more promising technique that is applied more frequently is the combination of GC with QF AAS; it nevertheless requires more complicated sample preparation. Extraction into an organic solvent is followed by derivatization and thorough purification, and often also further preconcentration, before the extract can be injected onto the GC column. Using a procedure of this type, SHORT [5320] determined *tributyltin* in fish tissue and FORSYTH *et al.* determined *butyl-*, *cyclohexyl-*, *octyl-*, and *phenyltin compounds* in fruit juices [1948] and in wine [1947].

A procedure that requires far less sample preparation is hydrogenization of the organotin compounds with sodium tetrahydroborate, collection of the volatile hydrides on a cooled column, chromatographic separation by warming the column, and detection by QF AAS. For sample preparation it is often sufficient to leach the homogenized sample in 2–7 mol/L hydrochloric acid [2352, 2404, 4811] or better concentrated acetic acid [1492, 4754]; this is very suitable for routine applications and also markedly reduces the risk of false measurements. Additionally this procedure allows the simultaneous determination of soluble inorganic tin as the hydride. The only disadvantage is that non-volatile compounds such as tetrabutyltin or phenyltin compounds cannot be determined. A large number of working groups have been involved in the determination of butyltin compounds in oysters and other marine organisms using this technique [1492, 2352, 2404, 4530, 4811, 4878]. Next to hydrogenization with sodium tetrahydroborate, ethylation with sodium tetraethylborate has also been described for the analysis of fish samples [5290, 5729] and other marine organisms [4813]. In addition the determination of *inorganic tin* and *methyltin compounds* has been reported [2352]. For the determination of butyltin compounds in algae with a high content of chlorophyll, QUEVAUVILLER *et al.* [4754] found an interference which they attributed to the impediment of hydride generation.

Next to the above elements and species that are determined relatively frequently in biological materials, there are a large number of applications that have only been investigated by a limited number of working groups. The element *cadmium* is to be found fairly frequently in this class, no doubt due to its high toxicity. Graphite furnace AAS is mostly used for detection due to its high sensitivity and low sample requirement. It is of particular interest to determine to which protein the cadmium is bound; gel filtration and gel permeation chromatography are especially suitable to separate various metal-binding proteins. The determination of the metal concentrations in the individual fractions is carried out off-line by GF AAS. Metals determined in this way include cadmium in fractions of vegetable foodstuffs [2908], cadmium and copper in lettuce [2508], cadmium and thallium in rape plants [2274], and cadmium and zinc in vegetables [2275]. Gel permeation chromatography in combination with GF AAS has also been applied for the separation of copper-flavanoid complexes in plant extracts [6182] and for the determination of free and bound cadmium, copper, silver, and zinc in the digestion glands of lobsters [1269]. Ultrafiltration and diafiltration also with GF AAS detection have been used for the investigation of the properties and bonding forms of cadmium and nickel in protein extracts from bean kernels [3419, 3420].

WANG and MARSHALL [6150] investigated the differences in the mobility of As, Cd, Cu, Mn, Pb, Se, and Zn species in animal tissues by extraction in supercritical liquid carbon dioxide and on-line determination in an FIT atomizer. MO *et al.* [4147] determined copper and magnesium in pigments of duckweed after HPLC separation and on-line F AAS detection with a thermospray interface. JONES and MANAHAN [2940] separated various copper chelates used in food processing by HPLC and applied F AAS on-line as the detector. AJLEC and ŠTUPAR [78] determined various iron species in wine by F AAS after separation by ion exchange chromatography. VAN STADEN and VAN

RENSBURG [5536, 5537] determined free calcium in milk by F AAS after on-line dialysis in an FI system.

10.3 Environmental Analysis

The trend in environmental analysis, as in other areas, is toward the determination of ever lower concentrations and to the analysis of a greater number of samples for an increasing number of analytes. A particular characteristic of environmental analysis is that all aspects of the analysis, from the frequency of the analyses, to the strategy and methods of sampling, to the pretreatment, and to the instrumental determination, are governed to a large part in the finest detail by laws, regulations, and guidelines. A good overview of prescribed analytical methods, as well as international and German standards (as of 1993), is given in the book by HEIN and KUNZE [2467]. We shall not go into details in this section since in each case the analyst must procure and follow exactly the original method. It is our goal in this section to point to the most important work and to establish the relationships.

In very simplified form Figure 10-3 depicts the areas coming under the global heading 'environment' and the mutual interactions that can be expected between those areas. Traditionally environmental analysis is divided into the analysis of water, soils, and the atmosphere; this 'classical' division is no longer relevant to the multitude of different environmental samples that must be analyzed nowadays. We shall thus only follow this classification very loosely to assign the most widely varying materials into suitable categories. For example, *coal* is discussed together with airborne particles since it is the analysis of the elements emitted from coal-fired power stations and coking works that is of major interest. Because of its heterogeneity, *waste* is hardly ever analyzed directly but via seepage waters or the analysis of residues from waste incineration [2480]. Next to the determination of the total concentration of an analyte, nowadays speciation analysis plays an increasing role due to the varying toxicities of individual species, such as inorganic As(III) and As(V), Cr(III)/Cr(VI), or tributyltin. The requirements for speciation analysis are discussed in Chapter 7, so that in this section we shall only discuss special requirements with respect to sampling, and sample conservation and preparation. DAS [1423] has published an overview on the methods used for solid samples and QUE-VAUVILLER [4762] on the existing legal requirements in Europe (as of 1996).

Sampling strategies are discussed in detail in Section 5.1.1 and also in Sections 10.1 and 10.2. We draw our readers' attention once again to the fact that improper sampling leads to meaningless analyses.

Introductory information on environmental analysis is available in the form of monographs [3895, 4838] and review articles [783, 1361, 2616, 3068, 3345, 5185]. Publications on the determination of individual elements in environmental samples are also available, for instance aluminium [1976], arsenic [2798], cadmium [4882, 4883, 5585, 5587], chromium [4106, 4995, 5420], cobalt [5586], lead [2551], mercury [1605], nickel [219, 5586], platinum [151], selenium [2753, 4815], thallium [5043], and tin [6177].

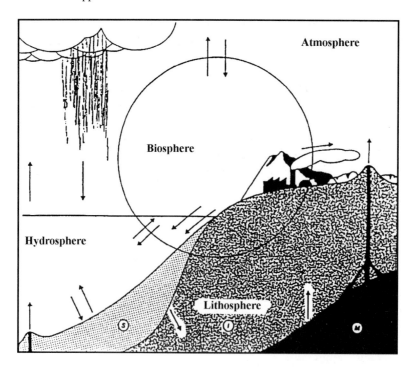

Figure 10-3. Schematic representation of our environment and the interactions that can be expected. **S** – sediments; **I** – soils and rocks; **M** – magma (from [4140]).

10.3.1 Water Analysis

Although the compound H_2O is clearly defined for the chemist, the material 'water' offers an unbelievable multitude of forms for the analyst. This fact naturally has a marked influence on the analytical methods. Depending on the source and the intended use of various waters, they are normally classified by legislators and analysts as follows:

i)   Drinking water and industrial water [3308];
ii)   Mineral and table waters;
iii) Bathing water;
iv)   Waste water [3834];
v)   Seepage and ground waters;
vi)   Precipitation [3279];
vii) Glacial ice;
viii) River and lake waters [2486];
ix)   Sea-water and estuarine waters.

Additionally the term *surface waters* is frequently used to encompass all natural waters such as rivers, lakes, etc. Correspondingly the palette ranges from highly pure waters to salty waters to extremely polluted waters, both in respect to the analyte concentration and the matrix content. The parameters *analyte concentration, total salt concentration,* and *organic content* are decisive for the procedure and the selection of the analytical technique.

Water analysis, at least for non-polluted waters, has undergone a similar development as the analysis of body fluids (see Section 10.1). In 1986 TOPPING [5879] wrote that the majority of published data on the determination of the base concentrations of metals in sea-water or for an understanding of the processes that influence the concentration and distribution of these elements were practically worthless. At around the same time, however, BERMAN and YEATS [556] saw higher reliability for the results, which they attributed firstly to better control of contamination during sampling, storage and analysis, and secondly to the improved separation techniques and advances in the development of the instrumentation. These authors reported that the published concentrations of trace elements in sea-water had decreased by around an order of magnitude within a decade due to the above improvements. The same is true for other non-polluted waters and rain-water.

A number of monographs [1366, 2727, 5008, 6289] and review articles [1845, 4993] generally on water analysis and specifically on sea-water analysis [1365, 2112, 2204, 2373, 5036] have been published. The journals *Water Research* and *Vom Wasser* (in German) deal exclusively with this subject.

10.3.1.1 Sampling and Sample Conservation

The technique of sampling is determined by the type of water that is to be sampled, such as piped water (drinking and industrial waters), springs and wells (drinking and mineral waters), bore holes (seepage waters), canals (surface waters, waste waters), rivers, lakes, seas, and precipitation. The majority of still and flowing waters exhibit substantial variations in their element concentrations over time and in their spatial distribution. It is thus relatively difficult to take a 'representative' sample. The required task must therefore clearly be established before sampling is undertaken. The simplest case is the taking of *random samples*, for example at an outflow location at given time points. Sampling of this type is only meaningful, however, when the place and timepoint of sampling are clearly defined, based on prior detailed information, and refer to a special situation. We can obtain much better information for one or several analytes by determining the *change in concentration with time*, either continuously or at regular time intervals, for example the inflows and outflows of a sewage treatment plant. We can obtain the most detailed information from an analysis when we establish a *network of sampling points*, for example along the length of a river or over the depth profile of still waters, and measure the changes at each sampling point over the short term (days) or over the long term (months, years). For this purpose we can also use multivariate statistical procedures [2091]. Sampling strategies are given in numerous national and international standards [2467] and also in a number of review articles [556, 660, 720, 5638]. STURGEON and BERMAN [5638] have published a detailed overview on containers and container materials and their suitability for trace element analysis.

The collection of wet precipitation (rain, snow, cloud aerosols, etc.) is an especial challenge, particularly when the wet precipitate must be kept separate from the dry precipitate (dust, see Section 10.3.3). For this reason *rainwater collectors* are frequently fitted with a device that covers the collector during dry periods and only opens at the onset of rain [845, 3279, 5217, 6044, 6359]. Combined systems that collect the wet and dry precipitates in two separate containers are of particular interest [4932]. WINKLER *et al.* [6360] were particularly concerned with the problem of suitable materials for precipitation collectors, since substantial errors can occur due to adsorption on the walls of the container.

Moreover we must take into account that a water sample can change from the timepoint of sampling if it is not conserved. Further procedures depend on whether the *total concentration* or the *concentration of dissolved analyte* should be determined in a sample. In the first case the water sample may have to be digested, depending on its properties and the selected analytical technique. In the second case the water sample is filtered after sampling, usually through a membrane filter of pore size 0.45 μm. Immediately following filtration, or sampling if the sample is not filtered, the sample must be stabilized, usually through the addition of 10 mL of nitric acid per liter of water sample. The pH value must be < 2, otherwise there is the risk that the analyte concentration can change due to precipitation or wall reactions. We must also mention briefly that a different procedure may be required if individual element species are to be determined (see Section 10.3.1.6). If the sample has been filtered, it may be of interest to examine the residue separately. In this case the method used is almost the same as for the analysis of sludges or sediments (see Section 10.3.2).

Another strategy for sample conservation, namely preconcentration at the site of sampling by coprecipitation and flotation, ion exchange and adsorption, or by electrode deposition, was proposed by MÉRANGER *et al.* [4060]. Next to the obvious advantage that larger volumes of sample do not need to be transported, these authors mentioned that in a controlled chemical form concentrated samples are more stable and that the risk of contamination during transport and storage is lower. The disadvantage, on the other hand, is that during preconcentration mostly only ionic species and only *one* oxidation state of the analyte are included, unless a comprehensive sample preparation procedure is performed at the sampling site. Furthermore it is hardly possible to include all analytes quantitatively with one preconcentration procedure. If interferences occur during preconcentration, it is very difficult to recognize them subsequently, and in particular the original sample is no longer available, which is a reason why this strategy has not received wider acceptance.

10.3.1.2 Methods for F AAS

The selection of the AAS technique and the analytical method is normally governed by the task at hand or the legal regulations. Generally direct procedures are to be preferred to procedures that require complicated sample preparation, not only for reasons of economy but also because of the better accuracy that can be expected. As a rule of thumb we can say that a direct procedure is suitable when the lowest analyte concentration of interest is a power of ten above the limit of detection. This means that in non-polluted surface waters, rainwater and sea-water, apart from the main elements sodium [1512, 2784], potassium [1513, 2785], calcium and magnesium [417, 1344, 5549, 6166], and under given conditions strontium [1488], none of the trace elements can be determined directly by F AAS [3, 1488, 2203, 2253, 3807, 5143, 5378, 6577].

If the task is to monitor given limiting values, such as in the drinking water regulations or other corresponding international guidelines (see Table 10-53), then it is also possible to determine barium [727, 1488], copper [727, 1515], iron, manganese and zinc [727], and under given conditions cobalt [1500, 1504] and chromium [823, 2782], by F AAS [6521]. For the analysis of waste waters, seepage waters, or more strongly polluted surface waters in general, it is also often possible to determine elements such as cadmium [3458], lithium [1488, 5834], nickel [1516, 3458], or lead [1517, 3458] satisfactorily by F AAS [3458, 6595].

Normally, filtered and stabilized water samples require no further pretreatment for analysis by F AAS, except possibly for the addition of a buffer. Unfiltered water samples can also be measured directly, provided that they do not contain suspended matter that could block the nebulizer or burner. According to most official methods a digestion can be omitted if the analyte can be completely detected without the use of this pretreatment step. This is usually the case for relatively pure water with a low content of suspended matter, but not for waste water, etc. or for when the analyte is bound to suspended matter. In such cases a digestion is indispensable.

By far the most used digestion procedure for the determination of heavy metals in polluted waters by F AAS is digestion in nitric acid-hydrogen peroxide. Nitric acid and

Table 10-53. A number of national and international reference values and limits for trace elements in drinking water (in µg/L).

Element	USA limit	EU reference value	EU limit	WHO reference value	Germany limit
Al	–	50	200	–	200
Ag	50	–	10	–	10
As	50	–	50	50	40
Ba	1000	100	–	–	–
Cd	10	–	5	5	5
Cr	50	–	50	50	50
Cu	1000	100[a] 3000[b]	–	–	–
Fe	300	50	200	–	200
Hg	2	–	1	1	1
Mn	50	20	50	–	50
Ni	–	–	50	–	50
Pb	50	–	50	50	40
Sb	–	–	10	–	–
Se	10	–	10	10	–
Zn	5000	100[a] 5000[b]	–	–	–

[a] at the water treatment plant; [b] after 12 h in the pipe

hydrogen peroxide are added to the water sample in a beaker and the volume is reduced to a wet residue, if required under repeated addition of hydrogen peroxide. The residue is taken up in nitric acid and water and the digestate is analyzed by F AAS [1500, 1503, 1515–1517, 2782].

In addition to the recommendations given for the individual elements in Chapter 9, the following points should be particularly observed for the analysis of strongly polluted waters:

i)   Background correction should be used for all determinations at wavelengths < 350 nm.
ii) Chromium should only be determined using the nitrous oxide-acetylene flame.
iii) The use of the nitrous oxide-acetylene flame is also recommended for the determination of iron and manganese in water samples with complex matrices (e.g. waste water), even if the sensitivity with this flame is a factor of 4–8 poorer.
iv) The addition of at least 2 g/L La as the chloride is required if calcium and magnesium are determined in the air-acetylene flame. If the nitrous oxide-acetylene flame is used, 1 g/L Cs as the chloride must be added.
v)   The addition of 1 g/L Cs as an ionization buffer is recommended for all determinations in the nitrous oxide-acetylene flame and also for the determination of sodium and potassium in the air-acetylene flame.
vi) The flame should be optimized for minimum interferences and not for maximum sensitivity, particularly for strongly polluted waters. If the sensitivity is then inadequate, a more sensitive AAS technique should be chosen, e.g. GF AAS.

During the 1960s and 70s trace elements were also determined in fresh waters and sea-water by F AAS after prior solvent extraction. In this case complete digestion of the sample is essential. The analyte is extracted as the chelate complex with an organic solvent and aspirated directly into the flame. The most commonly used combination was ammonium pyrrolidine dithiocarbamate (APDC) as the complexing agent and isobutyl methyl ketone (IBMK) as the solvent. APDC is able to form complexes with a large number of heavy metals and IBMK has excellent combustion properties, so that the solvent effect (see Section 8.1.1.5) further enhances the sensitivity [789, 1675, 1916, 4220]. Even in the more recent literature there are references to the use of solvent extractions; a number are compiled in Table 10-54. Nevertheless, solvent extraction has been increasingly replaced by sorbent extraction in recent years; examples are presented in Table 10-55. Next to these procedures, ion exchange [6473, 6476], precipitation [4264, 5317], flotation [1522], and freeze drying [2324] have been used for the separation and preconcentration of trace elements.

Although comparisons of various preconcentration procedures have been repeatedly made and it could be shown that in the hands of a competent analyst satisfactory results can be attained [933, 6047], we must nevertheless pose the question as to how far such techniques make sense when combined with F AAS. We must bear in mind that F AAS is the least sensitive AAS technique and also requires a relatively large volume of measurement solution, so that large initial volumes and high enrichment factors are required. Since practically all preconcentration techniques require a high level of time and effort, rarely function quantitatively and reproducibly [2373], and are susceptible to contamination, they have without doubt contributed to a large quantity of false data that has appeared in the literature. When a paper is published in 1994 on the determination of cadmium and silver in surface waters after a 10 000-fold preconcentration by ion exchange followed by a solvent extraction and determination by F AAS using 20 μL microinjections of the extract [5777]—which in itself leads to a loss in sensitivity—then this is clearly far removed from practice. Direct procedures with a more sensitive technique offer a far higher level of reliability. A possible exception is on-line preconcentration, for example by sorption on packed microcolumns [1817, 1823, 3554, 4047, 4154, 4469], and this aspect is discussed in more detail together with GF AAS. The factor of time is also relatively important in F AAS and a determination of lead in drinking water with an enrichment factor of 250 and a sample throughput of 11/h [6554] is hardly likely to find any interest, particularly since the direct GF AAS procedure is more sensitive and faster.

Table 10-54. Selected publications on the preconcentration of trace elements in water by solvent extraction for the subsequent determination by F AAS.

Analyte	Extraction procedure	References
Be	N-benzoyl-N-phenylhydroxylamine / IBMK	[4911]
Cd	Cryptand 221 (macrobicyclic polyether)	[2056]
Co, Pb	sodium hexamethylenimine-carbodithioate / IBMK	[5386]
Cu	1,10-phenanthroline / tetraphenylborate	[5111]
Fe	N-benzoyl-N-phenylhydroxylamine / methyltrioctylammonium ion / IBMK	[5057]
Fe	tributylphosphate / N-phenyl-2-furylacrylhydroxamic acid	[8]
Ge	N-(p-bromphenyl)-2-furylacrylhydroxamic acid / phenylfluorone	[4]
Hf	N-benzoyl-N-phenylhydroxylamine / IBMK	[6]
In	N-(p-methoxyphenyl)-2-furylacrylhydroxamic acid / IBMK	[7]
Mn	N-phenyl-2-furylacrylhydroxamic acid / methyltrioctylammonium ion / IBMK	[12]
Mn	dibenzo-18-crown-6 / thenoyltrifluoracetone	[634]
Nb	N-(p-methoxyphenyl)-2-furylacrylhydroxamic acid / IBMK	[5]
Pb	Cryptand 222 B / eosine	[2055]
Pd	1-phenyl-3-methyl-4-benzoyl-5-pyrazolone / IBMK	[5874]

Table 10-55. Selected publications on the preconcentration of trace elements in water by sorbent extraction for the subsequent determination by F AAS.

Sorbent	Analyte	References
Activated alumina (FI)	Pb	[6554]
Activated carbon	Be, Cd, Co, Cu, Mn, Ni, Pb, Zn	[6165]
Amidoxime resin	Cd, Co, Cu, Ni, Pb, Zn	[375]
Alizarin red-S immobilized on Amberlite XAD-2	Cd, Ni, Pb, Zn	[5124]

Table 55 continued

Sorbent	Analyte	References
APDC and oxine complexes on Amberlite XAD-4	Cd, Co, Cu, Fe, Mn, Ni, Pb	[1714]
Biosorbents (various)	Ag, Al, As, Cd, Co, Cr, Cu, Fe, Hg, Ni, Pb, Zn	[4795]
Carboxymethyl- and hydroxamat -dextrane on SiO_2	Cd, Co, Cu, Mn, Ni, Pb, Zn	[5018]
Chelex-100	Fe	[4994]
Dithiocarbamate-chitin	Ag, Cd, Co, Cu, Ni	[2415]
2-mercaptobenzothiazol on naphthalene	Cu	[5110]
Polyacrylonitrile	Co, Mn, Ni	[257]
Poly(aminophosphoric acid)	Cd, Co, Cr, Cu, Fe, Mn, Ni, Pb, Zn	[6478]
Poly(dithiocarbamate)	Cu, Fe, Zn	[6477]
	Mn	[6475]
Polystyren-azo-3-arsonophenol	Cd, Cu, Pb, Zn	[409]
Pyrocatechol-violet on Amberlite XAD-2	Cd, Zn	[5387]
Pyrocatechol-violet on Amberlite A-26	Cd, Pb	[5388]
Pyrocatechol-violet on DOWEX-2	Cd, Cu	[5389]
2,4,6-tri-(2-pyridyl)-1,3,5-triazin-tetraphenylborate on naphthalene	Cu	[5966]
Sulfhydryl cotton	Cd, Cu, Pb	[6536]
	Ag	[2163]
Cellulose xanthate	Cu, Pb, Zn	[6556]
Cellulose loaded with Fe(III) or In(III)	Al, Mo, Ti, U, V	[931]

10.3.1.3 Methods for GF AAS

On average, GF AAS is 2–3 powers of ten more sensitive than F AAS and thus offers the best prerequirements for the direct determination of a large number of trace elements in water. Since in an analysis by GF AAS *in situ* pretreatment, or even *in situ* ashing in oxygen, is part of the procedure, a digestion is not required for the majority of samples. This is also valid for water samples with a high content of suspended particles, which must merely be homogenized directly prior to dispensing (see Section 4.2.7.3). The quantities of reagents and external pretreatment steps are thus reduced to a minimum for GF AAS, which is a prerequirement for accurate results in the extreme trace range. The often-cited disadvantage of GF AAS that the measurement time is relatively long and the sample throughput is thus low is no longer relevant following the introduction of high-performance simultaneous spectrometers for this technique, and is in any case more than compensated by the reduction of sample pretreatment steps.

Strongly polluted water samples mostly exhibit a higher analyte content and can thus be substantially diluted for an analysis by GF AAS. At the same time this reduces inter-ferences, so that GF AAS is an interesting alternative to F AAS for samples of this type.

Mineral waters and healing waters can nevertheless cause considerable analytical problems on occasion if they have a high matrix content, for example of sulfur compounds [748, 6261], when the analyte concentration is very low. Samples of this type require very careful methods optimization, a point we shall discuss in detail later.

Surface waters, drinking and industrial water, and rainwater can mostly be analyzed directly by GF AAS without problems. A number of selected publications on the direct determination of trace elements in such waters by GF AAS are compiled in Table 10-56. The literature from the 1970s and early 80s is full of reports about interferences and we shall not go into this further. As early as 1983 MANNING and SLAVIN [3863] showed that numerous trace elements in surface waters could be determined free of interferences against matrix-free calibration solutions by applying the STPF concept. This finding was repeatedly confirmed in subsequent years [5378, 6094]. Many standard methods are nowadays based on a direct determination, for example, for chromium [823, 2782], cobalt [1500, 1504], copper [1515], lead [1517], nickel [1516], selenium [1508], and silver [1514], and STPF conditions are recommended even if they are not directly prescribed. An interlaboratory trial conducted in 1985 by the British Analytical Quality Control Committee on the determination of dissolved cadmium in river water [179] represents an almost 'schoolbook' example on the effectiveness of the STPF concept. Of the 11 participating laboratories, only one laboratory attained the set goals with respect to precision, accuracy, and repeatability, and this was the sole laboratory that had applied STPF conditions, while all the others, even if they used a modifier (mostly lanthanum), atomized from the tube wall and in some cases worked with uncoated tubes.

If the sensitivity of GF AAS is inadequate for the direct determination of trace elements in very pure water, the sensitivity can be markedly increased by using multiple injections (refer to Section 5.3.2.1) or by carefully reducing the sample volume by boiling down [1052, 2324, 5851]. Nevertheless such procedures can only be used for waters with low concentrations of inorganic or organic concomitants, while they often fail for waste water, high-salt mineral water, or sea-water. Separation and preconcentration steps must then be introduced for this type of sample.

The trace element concentrations in non-polluted *sea-water* are in the lower µg/L to ng/L range. An additional difficulty with sea-water samples is the total salt concentration of on average 35 g/L, which can lead to spectral and non-spectral interferences. Even in 1991, in a contribution to a book on the application of AAS for the analysis of sea-water, HARAGUCHI and AKAGI [2373] merely mentioned direct procedures on the side and reported essentially on preconcentration techniques. In Table 10-57 the trace element concentrations of two non-polluted sea-water samples of different origins are compared with the best published limits of detection for GF AAS. From this we can conclude that it certainly is possible to determine a number of trace elements directly whose basic concentrations are above the limit of detection by the stated order of magnitude. Table 10-58, in which selected publications on the direct determination of trace elements in sea-water are compiled, demonstrates that this can actually be achieved in practice.

Table 10-56. Selected publications on the direct determination of trace elements in precipitation and surface waters by GF AAS.

Analyte	Test sample	Remarks	References
Ag, Al, Ba, Cd, Co, Cr, Cu, Fe, Li, Mn, Mo, Ni, Pb, Zn	surface water, drinking water	STPF conditions	[5378]
Ag, Al, Ca, Cd, Cu, Fe, K, Mg, Mn, Na, Pb, Zn	high-altitude Alpine snow	clean room conditions	[418]
Al	mineral water		[4672]
Al, As, Be, Cd, Co, Cr, Mn, Ni, Pb, Se, V	surface water	STPF conditions	[3863]
Al, As, Be, Cd, Cr, Pb, Sb, Se, Tl	drinking water, mineral water	STPF conditions	[727]
Al, Cu, Fe, Mn, Pb, Zn	cloud aerosol, Alpine precipitation		[2203]
As, Cd, Ge, Sb, Se	thermal waters rich in Fe	platform, Mo modifier, ZBC	[1363]
As, Se	mineral waters rich in sulfates	STPF conditions, ZBC	[748]
B	surface water	Ca–Mg modifier	[2101]
		Ca modifier	[5967]
Ba	drinking water	ZBC	[4131]
Bi	waste water	Pd modifier	[2924]
Cd, Cr, Cu, Ni, Zn	rainwater	influence through sea-water aerosols	[1014]
Cd, Cr, Pb	waste water	STPF conditions, ZBC	[6094]
Cd, Cu, Fe, Mn, Pb, Zn	rainwater	comparison with PIXE	[2368]
Cd, Cu, Pb	rainwater	comparison with ASV	[3884]
Cd, Cu, Pb, Zn	precipitation	comparison with ICP OES	[4836]
Cd, Fe, Mn, Pb	surface water	platform, HNO_3 modifier	[4266]
Cd, Pb	surface water	platform, La modifier	[2728]
Cr	surface water	V–Mo modifier	[3870]
Mo	surface water	Ca modifier	[1741]
	surface water	Pd^0 modifier, ZBC	[3629]
Se	surface water	Pd–Ni–Mg modifier	[6451]
	surface water	STPF conditions, Pd^0 modifier	[5851]
Sn	surface water	Pd–Ni–Mg modifier	[6452]
Tl	surface water	Pd modifier	[3505]

Table 10-57. Trace element concentrations in non-polluted sea-water reference materials compared to the published limits of detection (LOD) for GF AAS [5332].

Element	NASS-1[1]	ICES-5[2]	LOD
As	1.65	1.4	0.2
Cd	0.029	0.020	0.003
Co	0.004	0.004	0.15
Cr	0.184	0.08	0.01
Cu	0.099	0.123	0.1
Fe	0.192	0.376	0.1
Mn	0.022	0.240	0.01
Mo	11.5	–	0.03
Ni	0.257	0.202	0.3
Pb	0.039	0.049	0.1
Zn	0.159	0.392	0.1

[1] National Research Council Canada [4305]
[2] International Council for the Exploration of the Sea [556]

Table 10-58. Selected publications on the direct determination of trace elements in sea-water by GF AAS.

Element	Modifier	Background correction	References
Ag	Pd–ascorbic acid	deuterium	[570]
Al	$(NH_4)_4EDTA–MgCl_2$	deuterium	[2698]
	$Mg(NO_3)_2$	deuterium	[3862]
As, Cd, Cr, Mn, Mo, Ni	$Mg(NO_3)_2$ for Cr	ZBC	[2230]
Ba	ammonium nitrate	deuterium	[4927]
	V–Si	ZBC	[643]
Cd	$(NH_4)_3PO_4–Mg(NO_3)_2–HNO_3$	ZBC	[4722]
	EDTA	deuterium	[2268]
	none	ZBC	[3656]
	NaOH	ZBC	[3406]
	$(NH_4)_2HPO_4–HNO_3$	ZBC	[1281]
	$Mg(NO_3)_2$	ZBC	[5425]
Cd, Cu, Ni	NH_4NO_3	ZBC	[1041]
Cd, Mn, Pb	Pd–ammonium oxalate	ZBC	[5034]
	$Pd–NH_4H_2PO_4–Mg(NO_3)_2$	deuterium	[2615]
Cr	W	deuterium	[243]
	none	ZBC	[2690]
	$Ca(NO_3)_2–Mg(NO_3)_2$	deuterium	[2164]
	ascorbic acid–triamincitrate	ZBC	[6458]
Cr, Cu, Mn	HNO_3 or oxalic acid	ZBC	[1016]

Table 10-58 continued

Element	Modifier	Background correction	References
Cr, Mn, Mo	Pd–NH$_2$OH·H$_2$O for Cr, Mg(NO$_3$)$_2$ for Mn, Pd for Mo	ZBC	[2697]
Cu	HNO$_3$	ZBC	[1012,1013]
	NH$_4$NO$_3$	ZBC/deuterium	[2695]
Fe, Mn, Zn	none	deuterium	[5223, 5224]
Mn	none	ZBC	[1072]
	Pt	deuterium	[2608]
Mo	none	deuterium	[4272]
Mo, V	ascorbic acid	deuterium	[5871]
Pb	Pd	deuterium	[5279]
	Pd–Mg or oxalic acid	ZBC, THGA	[1018]
Sn	Pd	deuterium	[2839]
Zn	NH$_4$VO$_3$	ZBC	[2696]
	HNO$_3$ or ascorbic acid	ZBC	[1015]

To achieve the goal of the direct determination of trace elements in sea-water it is necessary to attain a good S/N ratio and to reduce interferences as far as possible; in other words to carefully optimize the instrumentation and the methods. For this purpose we must take the following points into consideration:

i)   The STPF concept must be applied consistently to minimize spectral and non-spectral interferences. A transversely-heated atomizer with integrated platform provides the best results.

ii)   As much of the sea-water matrix (NaCl) as possible must be removed prior to atomization. The addition of ammonium nitrate or nitric acid to the modifier aids in vaporizing the chloride as NH$_4$Cl.

iii)   Since high background attenuation is to be expected, even when the matrix is largely removed, the application of Zeeman-effect BC is recommended.

iv)   A polarizer is not required for the longitudinal configuration of the magnet for Zeeman-effect BC. This configuration thus provides a better S/N ratio than the other configurations.

Nevertheless, as we can see from Table 10-57, with the above measures it is not possible to determine all trace elements directly in non-polluted sea-water. For those elements, as with the analysis of fresh water, a preconcentration step prior to the determination is indispensable. It is quite clear that such a step in the concentration range < 1 µg/L requires extreme care due to the enormous risk of contamination [660, 5638]. Below are listed a number of basic prerequirements for accurate working procedures, which naturally apply to the same degree for all non-polluted waters:

i)   All containers should be used exclusively for trace analysis. They must be thoroughly cleaned prior to first use, preferably by fuming out with nitric acid (refer to Section 5.1.2), and checked regularly for blank values.

ii)   All reagents must be of the highest purity; if necessary they must be further puri-
       fied, for example by subboiling distillation (refer to Section 5.1.3). It is essential to
       use the smallest possible volumes of reagents.

iii)   All procedures should be performed in a clean room, or at least in a dust-free envi-
       ronment or in a closed system, to minimize contamination from the laboratory at-
       mosphere or from dust.

iv)   On-line procedures in closed systems are superior to all manual batch procedures
       due to the significantly lower risk of contamination and to their much better repeat-
       ability and reproducibility.

The preconcentration technique used most widely and for the longest period of time
for trace elements in fresh waters and sea-water is *liquid-liquid extraction*. The extraction of gold from natural water as the chloro [860, 1289] or bromo [4019] complex in
IBMK has been described quite frequently; the water sample is often boiled down beforehand to attain limits of detection in the lower ng/L range [2320, 4019]. SHIMIZU
[5312] determined cobalt in natural water after a microextraction in the field with meth-
yltrioctyl ammonium chloride, and GOHDA *et al.* [2151] extracted molybdenum with
oxine in DIBK. While organic solvents have the advantage in F AAS that they enhance
the sensitivity, in GF AAS they tend to be a disadvantage since they often wet graphite
and can spread uncontrollably. For this reason, and for the fact that a number of organic
complexes are unstable, a back-extraction into the aqueous phase (dilute acid) is occasionally performed in GF AAS [660, 2373, 5033, 5306, 5307]; this also leads to a higher
enrichment factor. BRUNLAND *et al.* [899] developed a highly efficient double extraction
procedure with APDC/DDTC in chloroform, followed by a back-extraction with
7.5 mol/L nitric acid. This technique has found many imitations. An overview on a number of recent publications on liquid-liquid extraction of trace elements from sea-water is
given in Table 10-59.

Another very effective procedure for trace preconcentration is *coprecipitation*. An
advantage of this technique is that very often a large number of trace elements can be
preconcentrated at the same time; one disadvantage is the high expenditure in time and
effort required, and that considerable experience is necessary. SKOGERBOE *et al.* [5399]
and NAKASHIMA *et al.* [4281] added 2–5 mg/L Fe and Pd to the water samples and reduced in alkaline medium with sodium tetrahydroborate. The precipitate was collected
after 15–20 hours on a membrane filter and then redissolved. The recovery for Ag,
As(V), Bi, Cd, Co, Cr(VI), Cu, Mn, Ni, Pb, Sb, Se(IV), Sn Te, Tl, and Zn in sea-water
was > 90%, and the limits of detection were all in the lower ng/L range. An interesting
aspect of preconcentration by coprecipitation in combination with GF AAS is the direct
analysis of the precipitate as the solid or a slurry [81, 305, 4278]. An overview on publications dealing with preconcentration by coprecipitation from fresh waters and sea-water
is given in Tables 10-60 and 10-61. For GF AAS, coprecipitation reagents that can be
easily vaporized or decomposed in the pyrolysis step are particularly suitable; these include elements such as indium, mercury, selenium, or tellurium, and organic precipitation reagents. The use of reagents that also act as modifiers, such as magnesium, nickel,
or palladium, can also be advantageous.

Table 10-59. Off-line preconcentration of trace elements from sea-water by liquid-liquid extraction for determination by GF AAS.

Extraction system	Analyte	References
Dithizone / IBMK	Ag, Au	[1268]
Dithizone / benzene	Ag	[5305]
Dithizone / CHCl₃	Co, Cu, Cr, Mn, Ni	[709]
Dithizone / CHCl₃	Cd, Cu, Ni, Zn	[5473]
APDC / DDTC / CHCl₃	Cd, Co, Cu, Fe, Mn, Ni, Pb, Zn	[3567]
APDC / DDTC / Freon	Cd, Cu, Ni, Pb, Zn	[4818]
DDTC / CHCl₃	Cd, Cu, Pb	[900]
DDTC / chlorotoluene	Ni	[5304]
NaDDTC / CCl₄	Cd, Co, Cu, Fe, Ni, Pb	[1166]
DTC (various) / IBMK	Cd, Co, Cr(VI),Cu, Fe, Mn, Ni, Pb	[2691]
TMDTC / CHCl₃	Mo	[4165]
BPHA / xylene	Mo	[5309]
DTC / Freon	Co, Fe, Mn, Ni, Zn	[4221]
	Cd, Mn	[5597]
8-hydroxyquinoline / CHCl₃	Mn	[3147]
Thiocyanate	V	[5303]
TOPO	Ba	[5678]
TOPO / cyclohexane	Cd, Co, Cu, Mn, Ni, Pb	[5954]
PAR / TMDTC	V	[4166]
Kelex-100 / toluene	Cd, Co, Cu, Mn, Ni, Pb, Zn	[4584]

Table 10-60. Selected publications on the off-line preconcentration of trace elements in fresh waters by coprecipitation and the subsequent determination by GF AAS.

Precipitation reagent	Remarks	Analyte	References
Se	ZBC	Au, Pd, Pt, Rh	[1722]
Te	ZBC	Au, Pd, Pt	[1321]
Fe(OH)₃	ground waters rich in Fe	Cu, Pb, Zn	[6199]
In(OH)₃	dissolution in HBr, modifier: thiourea	Cd, Co, Cr, Cu, Fe, Mn, Ni, Pb	[1248]
Sn(OH)₄	dissolution in HNO₃, metastannic acid precipitates	Cd, Co, Cu, Fe, Ni, Pb	[2580]
Zr(OH)₄	direct solids analysis of the precipitate	Be, Cd, Co, Cr, Cu, Fe, Ni, Pb	[4278]
8-hydroxyquinoline	direct solids analysis of the precipitate	Cd, Pb, Zn	[81]
		Co	[82]
		Cu, Mn	[80]
Mg-oxinate	filter dissolved in HNO₃	Co, Cu, Cd, Ni, Pb	[1250]
Nitroso R / benzyl-dimethyltetradecyl-ammonium chloride	filter dissolved in DMF	Co	[2819]
DDTC		Cu	[3018]

Table 10-61. Selected publications on the off-line preconcentration of trace elements in sea-water by coprecipitation and the subsequent determination by GF AAS.

Precipitation reagent	Analyte	References
Pd	Cd, Cu, Pb	[6581]
$Fe(OH)_2$ / $Fe(OH)_3$	Yb	[2006, 2007]
	Co	[4766]
	Al, Pb, V	[6195]
	Cr	[1287]
$Mg(OH)_2$ or $Fe(OH)_2$	Co, Cu, Cr, Mn, Ni	[709]
HgS	Ag	[4767]
Zinc dithiolate $C_7H_6S_2Zn$	Co	[2584]
Ni-DDTC	Pb	[5677]
Co-APDC	Ag	[675]
	Bi, Cd, Cr, Cu, Fe, Mn, Ni, Pb, Zn	[1716]
Ni-DMG / PAN	Cu	[2821]
APDC	As, Cd, Co, Cu, Mo, Pb, Sn	[2926]
Te	Au	[4018]
$Hf(OH)_3$	Ga, In	[5952]
	Bi	[5953]

Within recent years solvent extraction and coprecipitation have been increasingly replaced by *sorbent (solid phase) extraction and ion exchange*. The markedly lower quantity of reagents required is frequently quoted as an advantage, although this is strongly method dependent. A further advantage is that the solid phase is packed into a column and can thus be used repeatedly. A prerequirement is nevertheless that the analyte is retained as selectively and as quantitatively as possible on the column and can subsequently be eluted with the minimum volume of solvent. Only in this way is it possible to achieve high enrichment factors with good separation from the matrix, especially for sea-water. Procedures starting with 10 L water sample, requiring 500 mL ammonium sulfate solution to elute the analyte from the column, evaporating the eluent to dryness and then taking up the residue in acid [5000] are hardly likely to find their place in daily practice.

The problem of good sorption and poor (slow) desorption of the analyte has been treated in a number of papers and widely varying solutions have been proposed. An interesting suggestion was made by LUO and HOU [3675] who achieved enrichment factors of 200–400 with a microwave-assisted elution of thallium. Other authors avoided the problem by not desorbing the analyte but by dispensing the analyte-loaded ion exchange resin or activated carbon as a slurry directly into the graphite tube [235, 6190]. The often substantial and varying blank values, in dependence on the solid phase, are the greatest problems in this procedure. Tables 10-62 and 10-63 present an overview of selected publications on off-line preconcentration from fresh waters and sea-water by sorbent extraction. Further information can be taken from a review article by NICKSON *et al.* [4355].

Table 10-62. Selected publications on the off-line preconcentration of trace elements from fresh waters by sorbent extraction and the subsequent determination by GF AAS.

Sorbent	Remarks	Analyte	References
Activated carbon	extraction in IBMK	Au	[2326]
	acetylacetone complex	Be	[4456]
Activated carbon / acetylacetonate	direct analysis of the slurry, Pd modifier	In	[6190]
Dithizone	sorption using ultrasonics	Ag	[2576]
Dithizonesulfonate on exchange resin	direct analysis of the slurry resin / water	Pd Co, Cu, Ni, Pd	[4124] [1258]
Polyorgs IX		As, Sb	[3040]
Polyorgs XI-N		Ag, Au	[5334]
PU foam	elution in the microwave oven	Tl	[3675]
TOPO	collection on W wire	Au	[3630]
W wire loop	(oxidative surface film)	Cd, Cu, Pb, Zn	[5708]
Cellulose-Hyphan		Be	[929]

Table 10-63. Selected publications on the off-line preconcentration of trace elements from seawater by ion exchange or sorbent extraction.

Complexing agent / sorbent	Analyte	References
Chelex-100	Cd, Cu, Mn	[4572]
	Cd, Cu, Ni, Pb, Zn	[4818]
	Co, Cu, Cr, Mn, Ni	[709]
Cellex-P	Pb	[768]
Muromac A-1	Ga	[4805]
DTP / activated carbon	Ag	[313]
AA / activated carbon	Be	[4456]
Modified silica gel	Cd, Cu	[94]
Macrocycles / XAD-4, XAD-7	Cd, Cu, Mn, Ni, Pb, Zn	[652, 653]
7-dodecenyl-8-hydroxyquinoline / XAD-4	Ag, Al, Bi, Cd, Cu, Fe, Ga, Mn, Ni, Pb, Tl	[2815]
TAR / XAD-4	Co	[2816]
NaHEDC / XAD-4	Ag, Cd, Fe, Mo, Ni, U, V, Zn	[3108]
Kelex-100 / silica-C18	Cd, Cu, Pb	[3602]
APDC / silica C-18	Cd, Cu	[3559]
8-hydroxyquinoline / silica-C18	Cd, Co, Cu, Fe, Mn, Ni,	[5625]
8-hydroxyquinoline / SDVBC polymer	Cr(III) / Cr(VI)	[2817]
8-hydroxyquinoline / silica gel or polymer	Cd, Co, Cu, Mn, Ni, Pb, Zn Cd, Co, Cu, Fe, Mn, Ni, Pb, Zn	[4280] [5624, 6331]
Alumina	Sb	[5452]

Table 10-63 continued

Complexing agent / sorbent	Analyte	References
Bathocuproindisulfonate ion exchanger	Cu	[4449]
Tetraazacyclotetradecane immobilized on polymer	Cu, Zn	[4601]
TOMA loaded onto silicon-coated support	Cd	[5123]
Algae cells	Cd	[3814]
	Cu	[4148]
DOWEX 50W-X8	Ba	[1466]
Ion exchange on ammonium-hexacyan-cobaltoferrate	Cs	[1991]
Ion exchanger	Au, Ir, Pd, Pt, Re, Ru	[2152]
AG-1X8 or AG-1X2	Au	[3194]
	Re	[3193]
	Ir, Pt	[2602]
Anion exchange	Mo	[3370]
	Tl	[4869]
Aliquat 336 / silica-C18	Zn	[84]

In recent years *on-line procedures for sorbent extraction* with GF AAS detection, mostly based on FI, have been increasingly applied. The principle of this preconcentration technique on microcolumns is discussed in Section 6.5.1. As we can see from Table 10-64, usually the same or similar sorbents as for off-line techniques are used. The on-line procedures nevertheless offer a number of advantages:

i)   The substantially shorter time required, which can be measured in minutes and not in hours.

ii) The much lower risk of contamination since preconcentration and elution are performed in a closed flow system, only inert materials and small surface areas are used, and the reagents can be purified on-line.

iii) Since reactions such as acidification, complexing, and sorption take place on-line and can follow each other in rapid sequence, it is also possible to work quantitatively and reproducibly with unstable complexes or intermediate products.

iv) All processes can be easily automated and can, for example, take place unsupervised in a clean room.

Expressed simply, we can say that these on-line preconcentration procedures have made possible the determination of trace elements in the lower ng/L range in sea-water on a more or less routine basis.

Table 10-64. On-line procedures for the separation of trace elements from sea-water with subsequent determination by GF AAS.

Complexing agent / sorbent	Analyte	References
8-hydroxyquinoline on CPG or Amberlite XAD-2	Al	[6496]
8-hydroxyquinoline on silica	Cd, Cu, Fe, Mn, Ni, Pb, Zn	[323]
APDC / silica-C18	Cd, Co, Cu, Fe, Ni, Pb	[4689]
DDTC / silica-C18	Pb	[1824]
	Cd, Cu, Ni, Pb	[5510]
	Cd, Co, Cu, Ni, Pb	[5517]
	Co	[5513]
	Cu, Ni, Pb	[788, 6262]
Silica-C18, loaded with NaDDTC	Cd, Cu	[6154]
Muromac A-1 exchanger	Cu, Mo	[5701]

10.3.1.4 Methods for HG AAS

The major advantage of HG AAS compared to GF AAS is the higher sensitivity by a power of around ten for elements such as As, Bi, Sb, Se, and Sn. Moreover, with HG AAS the analyte is separated as the volatile hydride from the bulk of the matrix, which can be of major significance especially for the analysis of sea-water. Particularly when flow systems are employed the hydride can be collected in a graphite furnace and thus be further preconcentrated free of contamination (refer to Section 4.3.2.2).

Among the disadvantages of HG AAS is that higher concentrations of other hydride-forming elements and also transition metals such as copper and nickel can cause interferences. For strongly polluted waters care must therefore be exercised and if necessary GF AAS used instead. Organic constituents also interfere in HG AAS so that for polluted waters an oxidative digestion is indispensable. Furthermore if the total concentration of an element is to be determined by HG AAS a complete digestion is required for most water samples since a number of organic compounds do not react with sodium tetrahydroborate; these include arsenobetaine, trimethylated arsenic compounds [606, 607], selenomethionine or selenocysteine. After the digestion, antimony and arsenic must be reduced to the trivalent state and selenium and tellurium to the tetravalent state [5375, 6256] (see Section 8.3.1.4). This varying reaction behavior of individual compounds with sodium tetrahydroborate can be very advantageous for the determination of individual species.

Digestions in nitric acid-perchloric acid have proven to be very effective [660, 4283] as well as dry ashing with magnesium oxide-magnesium nitrate [660]. However, since perchloric acid is used less and less owing to the associated risks, and losses have been observed for some elements (e.g. selenium) during dry ashing [1508], investigations were undertaken to find better digestion procedures. *Arsenic* seems to be relatively free of problems since no losses occur during dry ashing; as an alternative the sample can be fumed off with sulfuric acid-hydrogen peroxide until SO_3 fumes appear [1502]. For 'hidden' arsenic in coastal waters, irradiation with UV appears to be sufficient to make it determinable by HG AAS [607, 848, 2680, 2681]; this procedure is also suitable for the determination of arsenic in river water [5642]. *Organotin compounds* can be decomposed by treatment with sulfuric acid-potassium permanganate [1042].

The pretreatment of *selenium* is always very critical since a number of organic selenium compounds are very difficult to digest but during digestion volatile selenium compounds are formed. A digestion in divanadium pentoxide-sulfuric acid in a round-bottom flask under reflux and with a condensate reservoir has proven to be reliable [1508]. The water sample is fumed off until SO_3 fumes appear and then the solution in the condensate reservoir is allowed to run back to prevent selenium losses. *Antimony* can also be digested free of losses in nitric acid-hydrogen peroxide in a similar apparatus [1509].

To *reduce* Se(VI) to Se(IV) the water sample is brought to boiling with semiconcentrated hydrochloric acid under reflux [791, 993, 1000, 2552, 4632]. A stream of nitrogen must be passed over the solution to remove any chlorine that might be formed, since this would rapidly oxidize Se(IV) back to Se(VI) [1508, 3293]. Nitrosyl chloride, which can be formed by the reduction of nitrate or nitrite, also causes a similar interference [1386, 5376, 5377]. Potassium iodide in 5 mol/L HCl [1611, 5943, 6270] or L-cysteine [795, 1230, 6268, 6270] (see Section 8.3.1.4) is suitable to reduce As(V) and Sb(V) to the trivalent state. Table 10-65 presents an overview on selected publications on the direct

Table 10-65. Selected publications on the direct determination of hydride-forming elements in water.

Analyte	Water sample	Remarks	References
As	drinking and industrial water	on-line pretreatment	[1611]
	waste water	minimization of interferences	[6215]
	thermal water		[16]
	river water	immobilized BH_4^-	[4290]
As, Bi, Sb, Se, Sn, Te	surface water	FI, optimization of conditions	[6256]
As, Bi, Sb, Se, Te	lake water	influence of oxidation state	[5375]
As, Ge, Sb	sea-water		[199]
As, Sb, Se	surface and drinking water	influence of nitrite	[5377]
	lake water	measurements over 10 years	[5378]
	cloud water, aerosols	location: Whiteface Mountain, NY, USA	[960]
Bi, Sb	waste water	FI	[4740]
Sb	sea-water	optimization of conditions	[3802]
	sea-water	FI	[1033]
Se	natural water	interference from nitrite	[1386, 5376]
	waste water	minimization of interferences	[6226]
	sea-water	FI, on-line prereduction of Se(VI)	[1309]
Sn	fresh and sea-water	digestion of organotin compounds	[1042]

determination of hydride-forming elements in various water samples. If the water samples are sufficiently acidified, which is usually the case after the mandatory digestion and subsequent reduction, the transition elements which can usually be expected in strongly polluted waste waters do not interfere [6215, 6226], particularly when FI techniques are applied [6256].

Various procedures of analyte preconcentration have also been described for determinations by HG AAS, such as the 'classical' collection of the hydride in a cold trap for the determination of antimony and arsenic [3632], germanium [197], and selenium [4477] in fresh waters and sea-water. The on-line procedure proposed by XU and FANG [6415] for the preconcentration of antimony by ion exchange is also interesting, as is the procedure described by TAO and HANSEN [5772] for the on-line coprecipitation of selenium with lanthanum hydroxide; both procedures utilized FI techniques. Nevertheless, the *in situ* preconcentration of the hydrides in the graphite tube has gained the widest interest in HG AAS, either at increased temperature or, better, after prior coating of the tube with palladium or iridium. Table 10-66 presents a number of applications on this subject. The procedure proposed by STURGEON et al. [5641] for the preconcentration of lead deserves special mention; the lead is vaporized as tetraethyl-lead by reaction with sodium tetraethylborate so that the well-known difficulties for the generation of the hydride are avoided. When FI systems are used for hydride generation, other preconcentration techniques, such as sorbent extraction, can be interposed in the system [5773]. MATUSIEWICZ and STURGEON [3977] have published a review on the development of *in situ* preconcentration techniques.

Table 10-66. Selected publications on the *in situ* preconcentration of hydride-forming elements in graphite tubes.

Analyte	Water sample	Conditions	References
As	sea-water	Pd coating, 600 °C	[956, 5631]
	river water	used tube, 600 °C	[5642]
As, Bi, Cd, Pb, Sb, Se, Sn, Te, Tl	aqueous samples	various Ir coatings	[5910]
As, Sb, Se	natural water, waste water	Pd coating	[6539]
As, Se	mineral water	Pd coating, 300 °C	[6006]
	natural water	Sb coating, 230 °C	[867]
	sea-water	600 °C	[5632]
	fresh and sea-water	Pd coating	[1220]
	aqueous samples	Zr coating	[2065]
Bi	fresh and sea-water, digestion solutions	25–350 °C	[3482]
Bi, Ge, Tl	natural water	Pd coating	[6540]
Bi, Sb	natural water	280–300 °C	[868]
Cd	aqueous samples	Pd coating	[2148]
Ge	aqueous samples	Pd coating	[1561]
	tap water	Pd coating	[5771]

Table 10-66 continued

Analyte	Water sample	Conditions	References
Ge	aqueous samples	various coatings (Zr, Nb, Ta, W)	[2437]
Pb	natural water	ethylation; 400 °C	[5641]
	tap water	Zr coating	[6439]
Sb	drinking and surface water,	uncoated EG tube	[5380]
	sea-water	250 °C	[5629]
Se	sea-water	600 °C	[6332]
Sn	sea-water	Zr coating, 500 °C	[4344]
	tap water	Pd coating	[5773]
Te	natural water	300 °C (150–500 °C)	[202]

10.3.1.5 Methods for CV AAS

Mercury is determined almost exclusively by CV AAS since this is the only technique offering adequate sensitivity. If tin(II) chloride is used as the reductant, only inorganic Hg(II) can be determined; organomercury compounds are in part detected using sodium tetrahydroborate, but a digestion is absolutely essential in order to determine the total mercury content. We have drawn to attention in detail elsewhere the precautions required for the determination of mercury due to its volatility and mobility (refer to Section 8.4.1).

The special position that mercury assumes in comparison to other trace elements already begins during sampling and sample conservation. Since elemental mercury can pass through PE and PTFE, bottles made of these materials are unsuitable [730, 1345, 3987, 5581]. Glass is more suitable [730, 1874, 2077, 4458], particularly because it is relatively easy to clean. A number of new materials that are not wetted by water have proven to be particularly usable, for example polysulfone (PSF) and tetrafluoroethylene-perfluoropropylene (fluorinated engineering polymers, FEP) [1793]. Compared to other trace elements, it is not sufficient merely to acidify the water samples [660, 3951]; an oxidant must be present to prevent losses of mercury. For this purpose a solution of potassium dichromate in sulfuric or nitric acid is widely used [1793, 1874]. This reagent has also been used for on-line digestions in flow systems [2180, 4233, 4234]. Mixtures of potassium permanganate and potassium peroxodisulfate [5947] or a bromination reagent [1843, 1844, 4318] have also been proposed as stabilizers; these reagents can also be used for the subsequent digestion. The presence of NaCl appears to largely prevent losses of mercury, so that for sea-water samples simple acidification is adequate for stabilization [730, 2077, 3951, 4458, 5082]. Sample solutions stabilized with oxidants can however take up mercury from the ambient atmosphere if the walls of the container are permeable to mercury [1345].

A digestion in potassium permanganate and potassium peroxodisulfate in nitric or sulfuric acid solution is used most frequently for the determination of total mercury in water [1793, 3246, 5947]. During this digestion Se(IV) is oxidized to Se(VI) so that it cannot interfere in the determination of mercury [4706]. The digestion of organomercury compounds in potassium peroxodisulfate can be accelerated by adding iron(II) chloride as a catalyst [4235]. The digestion with potassium permanganate and potassium peroxodisulfate cannot be employed in the presence of high NaCl concentrations, on the other hand, since the chloride is rapidly oxidized to elemental chlorine [1700]. An alternative is a digestion with bromide-bromate [1843, 1844, 4318]; this can also be performed on-line in a flow system [6263], or by UV irradiation [48, 50]. A variation of the potassium permanganate-potassium peroxodisulfate digestion procedure, which utilizes ultrasonics, has been proposed for water samples not containing any undissolved organic material [436, 1793, 5379]. The advantages of this procedure are the greater speed and the lower reagent requirement by around a power of ten, which also leads to lower blank values.

The actual determination of mercury can be performed either directly or after collection on gold by amalgamation [67, 658, 674, 2111, 4910, 5157, 5947, 6249, 6436]; the mercury content in the water sample normally determines which procedure is applied [1793]. One aspect in favor of preconcentration by amalgamation is that the water sample can be more strongly diluted so that interferences can be reduced or avoided. Matrix effects are hardly to be expected for the determination of mercury by CV AAS in natural waters, however [6320]. Moreover, special instruments have been developed for the determination of mercury that have such low limits of quantitation that preconcentration is often superfluous [4023]. An alternative to amalgamation and the subsequent determination of mercury in a quartz cell is collection in a palladium-coated graphite tube and determination by GF AAS [6440]. Because of their simplicity and the reduced risk of contamination, *in situ* preconcentration procedures are to be preferred to solvent extraction [625].

10.3.1.6 Methods of Speciation Analysis

The analysis of filtered water samples is in principle a form of speciation analysis, namely the distinction between dissolved species and suspended or sorbed species [2571, 4143, 5054]. This physical separation can be refined to a fractionation of particulate and colloidal species via multistage filtration and ultrafiltration [1157, 1252, 2622]. In the following section we shall however concentrate on the determination of individual oxidation states (redox species) and compounds (particularly organometallics) of analytes. Considerably more detailed deliberations on the term 'species in aquatic systems' can be found in review articles by ANDREAE [203], CRAIG [1346], and FLORENCE *et al.* [1930], and in the monograph by BATLEY [423].

We have emphasized repeatedly in this monograph that sampling, sample preparation and conservation can cause substantial errors in any trace or ultratrace analysis. For speciation analysis there is the additional risk that the chemical species can convert so that a false distribution is thus determined. This danger is particularly large in natural waters in which biological and chemical species are in dynamic equilibrium. This equilibrium can be perturbed simply by taking a sample, for example through contact with the atmosphere or the walls of the container, or by changes in pressure and temperature [422]. To minimize exchange reactions and especially the activity of bacteria and algae, water samples for speciation analysis should be filtered in the normal manner as soon after sampling as possible.

Even more critical than sampling is the conservation and storage of water samples for speciation analysis. For example, while it is sufficient to simply acidify sea-water samples to stabilize the total mercury content for several weeks (refer to Section 10.3.1.5), methylmercury converts quite rapidly into inorganic mercury under these conditions [67, 346, 3487, 5581]. The pH value of the water sample is critical for the determination of redox species. Under acidic conditions Cr(VI) is easily reduced, especially in the presence of organic materials [405]. The stability of Cr(VI) is dependent on the concentration, pH value, the container material, and the temperature [4582]; Cr(III), on the other hand, is readily oxidized in alkaline medium. The various chromium species are comparatively stable against redox reactions solely in neutral solution, due to the slow kinetics and also the oxidation potential of the Cr(III)/Cr(VI) pair in such solutions [2414]. Unfortunately the sorption losses of Cr(III) on the container walls are greatest at pH 7; PE is the least suitable material [2109]. Water samples for chromium speciation analysis are best stored in PTFE containers at 5 °C under CO_2 with the addition of a carbonate buffer [1638, 1640].

For the determination of As(III)/As(V) and Se(IV)/Se(VI), on the other hand, PE is the most suitable material for the storage of water samples, while glass is unsuitable [1224]. To prevent the oxidation of As(III) in anoxic waters it is essential to purge out all oxygen with nitrogen and to store the water sample at 4 °C [56]; the losses of As(III) are greater in acidified samples than in non-acidified samples [195]. QUEVAUVILLER and DONARD [4748] have drawn attention to the sources of error during sampling and sample pretreatment and conservation for the determination of butyltin compounds. Even at 4 °C, sea-water samples lose more than 60% of tributyltin within a week when they are stored in PE containers, while filtered samples are stable for sufficiently long when they are stored in Pyrex glass containers in the dark [4749]. Sea-water samples can also be stabilized by freezing [5984].

These examples cannot by any means give a complete overview, but merely a glance into the problems. As far as possible, a water sample for speciation analysis should not be stored, or only for the minimum period of time, and the analysis should be immediate. More important than storage is the time factor for the analysis itself, since during analysis conditions can rarely be maintained at which the species of interest are stable. This means that procedures in which the species of interest are rapidly separated and converted into a stable form give the best accuracy. Due to the varying ecological-toxic effects of the individual elements, highly different species are of interest, so that widely varying procedures are applied. For this reason we shall discuss the elements individually in the following. Further information can be found in the review article by DONARD and RITSEMA [1586] and in the monograph by HARRISON and RAPSOMANIKIS [2405].

For *aluminium*, next to fractionation by filtration [4376], a distinction is made between slow-reacting, labile, and fast-reacting aluminium mainly by kinetic discrimination via complex formation [546, 1298]. FI techniques are particularly useful due to their high level of time control and repeatability.

For *antimony*, speciation analysis mainly involves the separate determination of the two oxidation states Sb(III) and Sb(V). The technique of choice is HG AAS [198, 239, 1033, 4153, 6429]; either the pH-dependence of hydride generation is exploited—Sb(III) forms a hydride in a weak acidic environment while Sb(V) only reacts in strongly acidic solution—or a citric acid buffer is added to suppress the hydride formation of Sb(V). ANDREAE *et al.* [198] also determined methylantimonic acid by HG AAS. Next to simple speciation analysis by HG AAS, the separation of Sb(III) and Sb(V) by solvent extraction [10], or on an alumina column [5453], or by an immobilized enzyme [1037], with subsequent determination by GF AAS has been described.

For *arsenic*, next to the oxidation states As(III) and As(V), the mono- and dimethylated species are of interest. The selective determination of As(III) is performed most easily by HG AAS at pH values > 5 [956, 6429]. Procedures have also been described for selective extraction [1159, 1164, 4062, 5014], coprecipitation, either off-line [1247, 1903], or on-line by FI-HG AAS [4360]. On-line sorbent extraction has been described for determination by GF AAS [5511, 5514]. By a careful selection of pH value and buffers it is also possible to selectively determine the mono- and dimethylated arsenic species by HG AAS [191, 6142]; nevertheless chromatographic techniques are mostly used for their separation. With the conventional coupling of LC-GF AAS [2194] or LC-HG AAS [1228, 6484], however, inadequate sensitivity is mostly only achieved so that off-line preconcentration of the hydride in a graphite tube has been proposed [2351]. A much simpler and more sensitive procedure is to generate the hydrides from a large sample volume, to collect the hydrides in a cold trap, and to separate them on the basis of their boiling points by warming the trap [1302, 1320, 2416, 2681, 4076, 6142]. Nevertheless it has been shown that not all arsenic species are detected [607]. Particularly in coastal waters and estuarine waters this 'hidden' arsenic can amount to more than 25% [2680, 5642]. In order to determine these, mainly trimethylated, arsenic compounds [606] it is necessary to subject the water sample to UV photolysis [2680, 5642] or to a strong acidic digestion [5642].

For *cadmium* a distinction is mainly made between the negatively and positively charged species, which is an indication of the different bonding forms on humic acids. Various ion exchangers [348, 3555], a combination of coprecipitation and ion exchange [2575], as well as solvent extraction [348] have been used for their separation analysis. GF AAS is the most suitable detector.

A principal distinction is made for *chromium* between the relatively strong toxic effect of Cr(VI) and the practically non-toxic Cr(III). In practice this is none too easy since, for example, in waste waters from tanning works only a small part of the chromium is present in ionic form while the major amount is in the form of very stable complexes or is bound to colloids [5560]. In natural aquatic systems chromium is also present in a large number of various charged and non-charged complexes [5518]; depending on the separation technique used, this can lead to substantial errors . For example, in the extraction of Cr(VI) with organic solvents, which is widely described [2946, 4231, 5670, 6129], Cr(III) compounds are also coextracted, thus leading to systematically high results [4051]. The coprecipitation of dissolved Cr(VI) with lead sulfate [3912], on the other hand, often leads to results for sea-water that are too low [6107]. ZOU *et al.* [6589] described the on-line coprecipitation of Cr(III). A number of authors applied sorbent extraction for the selective preconcentration of Cr(III) [481, 1131, 2809], Cr(VI) [1484, 2817, 4697, 5511, 5515], or for the sequential preconcentration of both oxidation states in dependence of the pH value [3320, 5518]. For the application of anion exchangers [2921, 5680] for the preconcentration of Cr(VI), interferences due to negatively charged

Cr(III) colloids can nevertheless take place since these are also sorbed [2573]. Very good values have been obtained for the sequential sorption of Cr(III) and Cr(VI) on an alumina column; this is a technique that can be easily automated and performed on-line using FI with F AAS as the detector [5518]. A totally different procedure was proposed by FUNG and SHAM [2022] which exploits the varying volatilities of the trifluoroacetyl-acetonate complexes of Cr(III) and Cr(VI) during thermal pretreatment in the graphite tube. VIDAL *et al.* [6067] utilized the differing precipitation potentials of Cr(III) and Cr(VI) for their selective electrolytic preconcentration on a L'vov platform.

For the speciation analysis of *copper*, ion exchangers are mainly used [3555] to determine free Cu(II) [5719], the lability of Cu(II) complexes [3056], and humic acid and other negatively charged colloids [2574], or to separate cations and easily dissociated complexes from particle-bound copper [4144]. Other authors used a combination of coprecipitation [4113] or solvent extraction [3952] with sorbent extraction to distinguish between 'reactive' dissolved copper and organic copper compounds. Organic copper complexes can be preconcentrated directly by solvent extraction [4093], and the use of membranes for the separation of copper species has been investigated [4546]. GF AAS is used almost without exception for detection.

HAMBRICK *et al.* [2343] determined *germanium* and methylgermanium species in water after hydride generation and collection in a cold trap. The various species were conducted directly into a graphite furnace heated to 2700 °C by rapidly heating the cold trap.

For *iron* solely the distinction between free metal ions and labile complexes has been described using Donnan dialysis [1341] and separation of Fe(II) and Fe(III) by ion chromatography [5392]. GF AAS served for detection in both cases.

In the case of *lead* it is the alkyl-lead compounds deriving from leaded gasoline (petrol) that are of interest. These are normally determined by GC-AAS [3568] following extraction and derivatization (hydride formation [1939], butylation [1161, 1216], or ethylation [4810]). The most varying water samples have been investigated: rainwater [142, 1348, 1531, 2402], runoff from roads [654, 1163], river water [4317, 6372], drinking water [1167, 4775], and sea-water [4901]. If only the concentration of 'organic lead' is of interest, this can be extracted with hexane, back-extracted with nitric acid and determined by GF AAS without further chromatographic separation [209].

For *mercury* a distinction is mainly made between the inorganic and the alkylated forms, and also between the various oxidation states and the complexes of varying stability and lability [346]. A simple distinction between inorganic and total mercury is made by CV AAS by working with and without prior digestion [642, 2178]. Inorganic mercury can also be separated by extraction with dithizone-chloroform [5144] and methylmercury by extraction with benzene [5948, 6427]. The separation of species by HPLC has also been described [1665]; in this case preconcentration is necessary to attain adequate sensitivity [4232]. MADRID *et al.* [3780] investigated the biosorption of mercury species on yeast and found that methylmercury was quantitatively sorbed in all cases. The actual determination of mercury was performed by CV AAS, with amalgamation as required, since the required sensitivity could only be attained with this technique.

Speciation analysis for *selenium* is mostly restricted to the selective determination of the two oxidation states Se(IV) and Se(VI). This can be done very easily by HG AAS since only Se(IV) generates a measurable signal with this technique. After reduction of Se(VI) the dissolved inorganic selenium can be determined by the same technique and thus Se(VI) by difference [240, 1309, 4238, 4633]. An alternative is to preconcentrate the Se(IV) on a packed column and to determine it by GF AAS [5627]. Organic selenium compounds have only been sporadically investigated [1385, 4509]. Next to the inorganic selenium species, CUTTER [1384] measured volatile dimethylselenide and dimethyldiselenide in water by driving them out with helium, collecting them in a cold trap, and after chromatographic separation conducting them to a QTA.

For *tellurium* the selective determination of Te(IV) in water and aerosol particles has been described using hydride generation with subsequent preconcentration and atomization in the graphite tube [6487].

In the case of *tin* the methylated and butylated species exhibit the highest toxicity (see Section 9.52). Three different routes have been taken for their determination. If solely the concentration of organotin compounds is of interest without distinguishing these further, they can be extracted from water with toluene, for example, and determined by GF AAS [241, 965, 1398, 2032, 4538]. Alternatively, preconcentration can be performed on activated carbon and the analytes eluted with an organic solvent [1256, 1899]. NI et al. [4344] increased the sensitivity of the procedure further by generating the hydrides after solvent extraction and preconcentrating them in the graphite tube.

If the individual organotin compounds are to be separated, a chromatographic technique is required. The GC-AAS coupling has been used the longest, but it is time-consuming and laborious and susceptible to errors. Extraction, derivatization and preconcentration must be performed prior to the chromatographic separation; derivatization with sodium tetrahydroborate [1939, 3534] or sodium tetraethylborate [6326] has simplified the process in comparison to the Grignard reagents used earlier [3791]. The risk of losing volatile compounds is nevertheless always present. Next to GC, LC has been applied for the separation of organotin compounds; in this case the derivatization step can be omitted, but not the need for preconcentration, for example by sorbent extraction [5199]. EBDON et al. [1659] coupled HPLC separation with in-line photolysis to convert the organotin compounds to Sn(IV), from which the hydride is then generated.

The simplest, quickest, and most reliable procedure for the determination of inorganic and organic tin species was proposed by DONARD and WEBER [1581]; the hydrides are generated with sodium tetrahydroborate, collected in a cold trap, and separated on-line on the basis of their differing boiling points. This procedure has been further refined and applied to a multitude of varying water samples [161, 364, 1235, 1582, 1587, 3909, 4878, 5134, 5541, 5542].

As well as the above element- or even species-specific procedures, a number of other procedures have been published that are suitable for larger groups of elements. HAYASE et al. [2451] determined the association of Cd, Cu, Mn, Mo, Ni, and Pb with organic materials in estuaries by extraction with chloroform at differing pH values. DONAT et al. [1588] investigated the suitability of sorbent extraction with C-18 to isolate organic complexes of Cd, Cu, Fe, Mn, Ni, and Zn from sea-water. KOBAYASHI et al. [3177] separated high molecular mass complexes of Cu, Fe, Mo, V, and Zn from sea-water and lake water by ultrafiltration and LC. HIRAIDE et al. [2577] fractionated humic complexes of Al, Ba, Co, Cu, Fe, Mn, Ni, Sr, and Zn by adsorption on XAD-2 resin pretreated with indium and on DEAE-Sephadex A-25 ion exchanger prior to determination by GF AAS.

BURBA [934] also used an ion exchanger to investigate labile and inert humic complexes of Al, Co, Cu, Fe, Mn, Ni, Pb, and Zn.

10.3.1.7 Indirect Determination Methods

Next to the direct determination of elements and their species, a number of indirect procedures have been described in the literature which in combination with FI techniques go a long way to meeting the requirements of modern routine analysis. On-line extraction or on-line precipitation are frequently applied. Table 10-67 gives an overview on recent publications on this subject. Nevertheless, these indirect determination procedures are not always as specific and selective as we are accustomed to in AAS. For example, the absorbance signal for calcium is depressed not only by phosphate [2276], but also by sulfate, aluminium and a number of other elements. Phosphate interferes in the determination of sulfate via precipitation as barium sulfate [2051]. It is thus essential to know the composition of the water sample if these indirect procedures are to give correct results.

Table 10-67. Procedures for the indirect determination of anions, molecules, and molecular ions in water.

Substance to be determined	Procedure	Analyte	Technique	References
Cationic surfactants	extraction of the ion pair with $[Co(SCN)_4]^{2-}$ in 4-methylpentan-2-one	Co	FI-F AAS	[3922]
	extraction of the ion pair with $[Co(NO_3)_6]^{4-}$ in 1,2-dichlorethane	Co	GF AAS	[1202]
Anionic surfactants	extraction of the ion pair with 1,10-phenanthroline-Cu(II) in IBMK	Cu	FI-F AAS	[2049]
Non-ionic surfactants	precipitation as the phosphotungstic acid-calcium complex/HPLC-microfiltration	Ca	F AAS	[2205]
Phosphate	interference on the absorbance of calcium	Ca	F AAS	[2276]
	flotation as ion pair phosphomolybdate with bis[2-(5-chloro-2-pyridylazo)-5-diethyl-amino-phenolate]Co(III)	Co	F AAS	[5741]
Phosphate, arsenite, arsenate	flotation as ion pair phosphomolybdate with malachite green	Mo	F AAS	[4299]
Sulfate	precipitation as barium sulfate	Ba	FI-F AAS	[2051]
Chloride	precipitation as silver chloride	Ag	FI-F AAS	[5197]
Cyanide	extraction of the Cu-2-benzoylpyridine-thiosemicarbazide-cyano complex	Cu	F AAS	[1201]
Fluoride	dissolution of lead from a column packed with lead zirconate-titanate	Pb	F AAS	[1261]
Phenol	bromination and complexing between cadmium and iodine	Cd	F AAS	[6406]
Thiosulfate	complexing with Pb(II)	Pb	F AAS	[1169]

10.3.2 Soils, Sediments, Sludges

A number of very widely differing materials belong in this domain, beginning with arable soils, to contaminated soils, soils from derelict sites, sediments from rivers, lakes, and seas, to sewage sludge. The procedures for sampling, sample homogenization and preparation are correspondingly manifold, so that we can only treat them in a general manner. Here again we shall concentrate on the analytical aspects. Independent of the nature of the sample, we must distinguish between three procedures: Determination of the *total concentration* of an element; determination of the *extractable portion*; and the determination of individual species, especially *organometallic compounds*; the boundaries between these procedures are naturally flexible. Both for the analyst and for the selection of the most suitable method the decisive question is: *For which purpose are the results required?*. The selection of the determination technique (F AAS, GF AAS, etc.) is simple in comparison; it is largely determined by the concentration of the analyte and also by the nature of the matrix. Introductory monographs and review articles on the analysis of soils [1360, 2974, 5468, 5962], sediments [2668, 3536], sludges and other waste products [2480, 4490], and on speciation analysis in soils and sediments [3086, 3414, 3466, 4877] are available to the interested reader.

10.3.2.1 Sampling

The problem for sampling soils [605, 732, 2161, 3535], sediments [1267, 3270, 5146, 5896], sludges [720, 2161], and waste [4490] is the removal of a few grams that are representative for the situation being investigated from a material present in tons or cubic meters. Usually a *large number of individual samples* are taken from different places from the material being investigated and then combined into a *composite sample*. This is then dried, reduced and sieved so that finally a realistic quantity is available as the test sample. Next to statutory regulations for sampling [1505–1507, 2532, 4066], a number of statistical procedures have been elaborated to facilitate representative sampling, for example in order to characterize contaminated areas [1689, 1690, 3287, 3494, 4798, 5839]. The marked effect that uncontrolled sampling can have on the analytical results was demonstrated by QUEVAUVILLER and DONARD [4748] on the example of sediment samples. As well as the local inhomogeneities in the horizontal plane of the sample and the dependence of the element distribution on the depth of sampling and the particle size [3230, 6061], there can also be seasonal variations in sediments and other solid samples [6521]. The problems of sampling naturally become more complex when element species are to be determined [422, 4997].

10.3.2.2 Determining the Total Concentration of Analytes

The total analyte concentration in a test sample can only be determined after a fusion with sodium carbonate [4702] or lithium metaborate [400, 610, 6023], or after an acid digestion in the presence of hydrofluoric acid and frequently also perchloric acid [52, 6368, 6369]. Nevertheless these digestion procedures are relatively time-consuming and certainly not without problems. A number of minerals, such as chromite, zircon, rutile, or corundum, are only partially dissolved by such digestion procedures [2325, 2329, 3404]. In addition there is the risk of loss of volatile elements. The high matrix concentrations require regular blank controls to be performed, and frequently the calibration solutions must be matched to the matrix contents. Moreover the high salt concentration can lead to problems with nebulization in F AAS and to strong background attenuation in GF AAS. Digestions in hydrofluoric acid or perchloric acid are dangerous and demand special safety measures. Next to silicon (which should be removed in this manner), a number of elements such as antimony, arsenic, boron, and germanium form volatile fluorides which can be lost.

Owing to these numerous problems, instead of a 'total digestion' it is more usual in environmental analysis to perform a *pseudo total digestion* with *boiling aqua regia* under reflux. This is justified on the assumption that heavy metals that are so strongly bound to the mineral matrix that they do not go into solution with such a digestion will also not be taken up by plants and cannot be dissolved by water or bacteria. Even if this is not true for every element and every test sample, it has nevertheless been shown that correct values can be obtained with this method [604, 1275, 1276, 3768, 3896, 4000]. A further advantage is that *aqua regia* digestions can be analyzed by F AAS relatively free of problems. The element concentrations normally to be expected in soils is so high for numerous analytes that the sensitivity of F AAS is fully adequate. The concentrations of the remaining elements, such as cadmium and lead, is also still so sufficiently high that they can be suitably diluted for determination by GF AAS. The relative complexity of the matrix nevertheless requires STPF conditions and frequently the application of Zeeman-effect BC [673, 911, 1491, 1616, 3665, 4726]. In the meantime reference materials have been produced for which, next to the certified total concentration, information is also provided on the fraction that can be extracted in *aqua regia* [5989].

Microwave-assisted digestions have found rapid acceptance in the environmental analysis of solid samples [3222]. Digestions in hydrofluoric acid [614, 615, 3973], perchloric acid [4282], and also *aqua regia* [4116, 4117, 4364] have been investigated; in general the same results were mostly obtained as for conventional digestion procedures. The major advantages of microwave-assisted digestions are given as the substantially shorter digestion times [2271, 3114, 3284], the lower risks of contamination, the reduced reagent requirements [2535], and the lower risk of losses [2694]. Due to the high effectiveness of microwave-assisted digestions it has also been possible to incorporate them on-line in a flow system with F AAS detection [1060, 1062, 2247]. Table 10-68 provides an overview on important publications dealing with the determination of the total concentration of metals and metalloids in solid environmental samples.

As in so many areas, mercury also takes a special position here due to both its toxicity and its particular physical and analytical characteristics. Mercury can be determined free of interferences in environmental samples by CV AAS even after a relatively mild digestion [3337, 4800, 6320], in contrast to biological materials (see Sections 10.1 and 10.2). Alternatively it can be quantitatively vaporized by heating the test sample to 400–500 °C [700, 782] and determined either directly [700] or after collection on gold or silver [330, 782]. Mercury can also be determined by GF AAS in environmental samples; palladium has proven to be a reliable modifier [565, 6260].

Table 10-68. Selected publications on the determination of metals and metalloids in soils, sediments, sludges, and solid waste by AAS.

Analyte	Test sample	Technique	Remarks	References
Ag	soils	F AAS		[5040]
Ag	soils	GF AAS	sorbent extraction with activated carbon	[313]
Ag, As, Bi, Cd, Co, Cu, Hg, Mn, Mo, Ni, Pb, Sb, Se, Te, Tl, Zn	soils, sediments	F AAS, GF AAS, CV AAS, HG AAS	common digestion in *aqua regia* for all methods	[3768]
As	soils, sediments	FI-HG AAS		[701]
As	sediments (sea)	HG-GF AAS	*in-situ* preconcentration in the graphite tube	[5631]
As	sediments (lake)	HG AAS	extraction with $HCl–H_2O_2$	[3386]
As	sediments	GF AAS	modifier: Pd	[5267]
As, Cd, Cr, Cu, Fe, Mn, Ni, Pb, Se	sediments (sea)	GF AAS	STPF conditions	[1491]
As, Cd, Cr, Cu, Pb, Tl, V	sediments, sludges	GF AAS	THGA, rapid temperature program without pyrolysis	[6544]
As, Cd, Tl	soils	GF AAS	digestion in *aqua regia*	[2834]
As, Sb, Se	soils, sediments	HG AAS	Wickbold digestion	[1754]
As, Sb, Se	sediments	FI-HG AAS		[2280]
As, Se	soils, sediments, waste	HG AAS	digestion: $HNO_3–H_2SO_4–HClO_4$; 1,10-phenanthroline to mask Cu and Ni	[1181]
As, Se, Hg	sediments (estuarine)	FI-HG AAS	digestion: $HNO_3–H_2SO_4–HClO_4$ under reflux or $HNO_3–H_2SO_4–HCl$ microwave-assisted	[5099]
Be	soils, sediments, waste	GF AAS	various modifiers, platform atomization, investigation of 'absolute' analyses	[1723, 3674, 6424, 6456]
Bi	sediments	GF AAS, slurry-GF AAS	preconcentration on anionic exchanger or activated carbon	[3371, 4265]
Ca	soils	F AAS	addition of lanthanum	[3871]
Cd	soils, sediments	GF AAS	microwave-assisted digestion or pressure digestion	[1069, 3289, 4117]
Cd, Cr, Pb	sewage sludge	GF AAS	microwave-assisted digestion; modifier: Pt	[5818]
Cd, Cu, Fe, Mn, Pb, Zn	sewage sludge	F AAS	microwave-assisted extraction	[4189]

Table 10-68 continued

Analyte	Test sample	Technique	Remarks	References
Cd, Cu, Pb	soils, sediments	FI-F AAS	on-line sorbent extraction with DDTC / silica–C18	[3757]
Cd, Cu, Pb, Zn	sediments (lake)	F AAS	extraction overnight with HCl–HNO$_3$	[6064]
Cd, Pb	soils	F AAS	background interference by iron	[3483]
Cd, Pb, Hg, As	soils, waste	FI-HG AAS, FI-F AAS	on-line dilution	[1734]
Co	sewage sludge	F AAS, GF AAS	German standard method DIN 38406-24	[1500]
Cr	soils, sediments	F AAS	interferences, optimization of flame conditions	[77, 2947, 4314, 5049]
Cr	sediments	F AAS	digestion problems	[3549]
Cr	sediments	GF AAS	influence of atomization temperature, various digestions	[1173, 5049, 6565]
Cu	sewage sludge	F AAS, GF AAS	German standard method DIN 38406-7	[1515]
Cu, Cr, Ni, Pb, Zn	sewage sludge	F AAS	dry ashing at 450 °C	[2667]
Cu, Mn, Pb	sewage sludge	FI-F AAS	microwave-assisted on-line digestion	[1062, 2247, 3914]
Cu, Zn	soils	F AAS	microwave-assisted pressure digestion	[3284]
Er	sediments	GF AAS	graphite tube with tantalum foil	[3762]
Ga	sediments	GF AAS	modifier: Ni	[5270]
Ge	sediments	HG-GF AAS	*in-situ* preconcentration in Pd-coated graphite tube	[4349]
Hg	soils	CV AAS, GF AAS	comparison of digestion procedures	[923]
Hg	soils, sediments, sludges	CV AAS	various digestions	[3413, 4515, 6571]
Hg	sediments	FI-CV AAS	microwave-assisted on-line digestion	[2358]
Hg	soils, sediments, sludges	FI-CV AAS	digestion in *aqua regia* according to DIN 38414	[2285]
In	sediments	GF AAS	digestion: HNO$_3$–HClO$_4$–HF; modifier: Pd–EDTA; influence of atomization temperature	[6568]

Table 10-68 continued

Analyte	Test sample	Technique	Remarks	References
In	sediments (river)	GF AAS	extraction with NH₄I–IBMK	[5268]
K	sewage sludge	F AAS	German standard method DIN 38406-13	[1513]
Mn	soils, silicate materials	F AAS	digestion: HF–HClO₄	[1206]
Mo, V	soils	GF AAS	digestion in *aqua regia*	[3479]
Na, K, Ca, Mg	soils	FI-F AAS	on-line dilution	[1897]
Ni	sewage sludge	F AAS, GF AAS	German standard method DIN 38406-11	[1516]
Pt, Pd	soils, silicate materials	GF AAS	extraction of the iodo complex with IBMK	[862]
Re	sediments (sea)	GF AAS	preconcentration on anion exchanger	[3193]
Se	soils	GF AAS, HG AAS	comparison of digestion procedures	[4343]
Se	sediments (sea)	HG-GF AAS	*in-situ* preconcentration in the graphite tube	[6332]
Se	sediments	HG AAS	optimization of digestion and prereduction, interferences and their elimination	[2822, 3510]
Se	sediments	GF AAS	preconcentration of Se(IV) on activated carbon	[3319]
Sn	sediments (sea)	HG AAS	pressure digestion: HNO₃–HCl, influence of the digestion acid	[2398, 3489]
Sn	sediments	HG-GF AAS	*in-situ* preconcentration in the graphite tube	[5635, 6543]
Sn	sediments	GF AAS	digestions and modifiers	[912, 1733]
Tl	soils, sediments	F AAS, GF AAS	off-line preconcentration on an anion exchanger	[4869, 5901]
Tl	soils	F AAS	off-line preconcentration through extraction with HBr	[282]
Tl	sediments	GF AAS	direct determination; modifier: Pd–ascorbic acid	[5271]
Zr	soils	F AAS	extraction with liquid ion exchanger	[627]

The direct analysis of slurries by GF AAS has acquired considerable importance for the analysis of solid environmental samples since the *total concentration of an element without prior digestion* can be determined. Procedures for direct solids analysis have also been described [2606, 3358, 5857, 6096], but the precision and the trueness are better for the analysis of slurries [1568], especially when they are homogenized by ultrasonics [89, 735, 1749, 4107, 5080]. Further improvements such as a reduction of the background attenuation can be obtained by *in situ* ashing with oxygen [1658] or the addition of hydrogen to the purge gas [1661]. Using FI techniques, slurries of environmental samples have been successfully analyzed by F AAS [3611, 3914]. The most important applications of the slurry technique and the direct analysis of solid environmental samples are compiled in Table 10-69.

Table 10-69. Selected publications on the direct analysis of solid environmental samples (SS) and slurries by AAS.

Analyte	Test sample	Technique	Remarks	References
Al, Si, Zn	sediments (sea), suspended matter	SS-GF AAS	alternate lines, internal gas flow to reduce the sensitivity	[2613]
Al, Si, Fe	sediments	slurry-GF AAS		[5504]
As, Sb	sediments, soils	slurry-GF AAS	homogenization by magnetic stirrer, modifier: HF	[3620]
As, Cd, Cr Cu, Ni, Pb	sediments	slurry-GF AAS	ultrasonic homogenization, THGA	[89] [3145]
As, Cd, Hg, Pb, Sn	sediments	slurry-GF AAS	modifier Pd or Pd–Mg, aqueous calibration solutions	[572]
As, Cd, Cr, Cu, Pb, Mn	soils	SS-GF AAS	aqueous calibration solutions	[6096]
As, Cd, Cu, Pb, V	soils, minerals	SS-GF AAS or slurry-GF AAS	aqueous calibration solutions	[5857]
As, Fe, Mn, Pb	sediments (river)	slurry-GF AAS	ultrasonic homogenization	[1749]
Ca, Fe, Mg	silicate materials	FI-F AAS	on-line dilution of suspensions	[3611]
Cd	soils	slurry-GF AAS	comparison with Delves cup F AAS	[5020]
Cd	soils	SS-GF AAS		[71]
Cd, Cu, Pb	soils, sewage sludge, suspended materials	SS-GF AAS	continuum source BC	[4967, 5495]
Cd, Pb	soils, sediments	slurry-GF AAS	aqueous calibration solutions	[2556, 5080]
Cd, Pb	sediments	SS-GF AAS or slurry-GF AAS	aqueous calibration solutions	[1568]
Cd, Pb, Tl	soils, sediments	slurry-GF AAS	aqueous calibration solutions for Cd and Pb, analyte addition technique for Tl	[3618]
Cd, Pb, Zn	soils, sediments	SS-GF AAS	calibration by 'extrapolation to zero matrix'	[4579]
Cr, Mn, Pb	soils, sewage sludge	slurry-GF AAS	background reduced by H_2 in the purge gas	[1661]
Cr, Co, Mn, Pb	sewage sludge	slurry-GF AAS	ashing in air in the graphite tube	[1658]

Table 10-69 continued

Analyte	Test sample	Technique	Remarks	References
Cr, Pb	sediments	slurry-GF AAS	comparison between 18 laboratories	[4109]
Cu	soils, sediments	SS-GF AAS	modifier: HF–HNO$_3$–Ni	[3760]
Cu, Cr, Fe	biological and geological samples	slurry-GF AAS	influence of solids content and particle size	[4107]
Cu, Cr, Mn	sediments, dust, plant materials	slurry-GF AAS	rapid program, ultrasonic homogenization	[5438]
Cu, Cr, Pb	sediments	slurry-GF AAS	preparation of the slurry, alternate lines, optimization	[4104, 4108]
Fe, Ca, Mg	soils, sediments	FI-F AAS	suspension in HCl–HF	[3614]
Hg	soils	SS-GF AAS	calibration with spiked sand	[5859]
Hg	sediments	slurry-GF AAS	modifier: Pd	[565]
K	soils	FI-F AAS	suspension in HNO$_3$	[2471]
Pb	soils	slurry-GF AAS	magnetic stirrer or vortex mixer	[2561]
Pb	soils	slurry-GF AAS	rapid temperature program without pyrolysis and without modifier	[2562]
Pb	soils	slurry-GF AAS	comparison of various Pd modifiers	[2557-2560]
Pb	sediments (sea)	slurry-GF AAS	modifier: Pd–Mg; influence of modifier mass and particle size	[564]
Se	soils, sediments	slurry-GF AAS	modifier: HF–Ni	[3615]
Sn	sediments (sea)	slurry-GF AAS	modifier: Pd–Mg	[574]

10.3.2.3 Selective Extraction of Individual Constituents and Bonding Forms

The leaching of soils with a variety of differing reagents and the determination of essential elements (e.g. copper, manganese, zinc) and potentially toxic trace elements (e.g. molybdenum, nickel) in the leachates has been used successfully for decades for the diagnosis of plant growth disturbances and diseases of animals, and as a means for avoiding them [5962]. The goal of these investigations is to determine the *bioavailable proportion of trace elements* rather than their total concentration. Although the majority of these leaching procedures were determined empirically they nevertheless give good correlation for numerous elements. Review articles on various extraction procedures have been published by PICKERING [4639], MORABITO [4188], and KERSTEN and FÖRSTER [3086]. Table 10-70 presents a number of soil fractions and phases together with the usual reagents for the extraction of the trace elements associated with them.

Table 10-70. Sequential extraction scheme for soil samples according to MILLER and MCFEE [4095]. 1 g soil sample is shaken with volume V of the reagent for time t. After centrifugation, the supernatant liquid is analyzed, the residue is washed with 10 mL deionized water, again centrifuged, and then extracted with the next reagent.

Extracted phase	Extraction solvent	V, mL	t, h
1. water soluble	deionized water	10	0.5
2. exchangeable	1 mol/L KNO_3	10	16
3. organically bound	1 mol/L $Na_2P_2O_7$	15	16
4. carbonate / amorphous iron oxide occluded	0.1 mol/L EDTA	10	16
5. manganese oxide occluded	0.1 mol/L $NH_2OH \cdot HCl$ + 0.01 mol/L HNO_3	10	0.5
6. crystalline iron oxide occluded	0.27 mol/L sodium citrate + 0.1 mol/L $NaHCO_3$ + 0.25 g $Na_2S_2O_3$ / 80 °C	10	0.25
7. sulfides	0.1 mol/L HNO_3	10	16
8. residue	digestion with HNO_3 + H_2O_2	15	12

For the determination of the element by AAS it is of no great consequence whether the sample material is a soil, a sediment, or a sludge, and whether it is contaminated or not. Since with the sequential leaching of soils with a variety of differing reagents only relatively small element concentrations are leached out, especially with weak extracting agents, F AAS is only sufficiently sensitive for a few elements [726, 4603, 4785, 4789, 5006, 5823]. Usually it is necessary to apply GF AAS [1575, 1578, 3491, 4479, 4518, 4685, 4789, 5006] or HG AAS [1576, 1579, 2426, 3907]. Due to the considerably lower total salt concentration, the matrix influences are far less pronounced than with a complete digestion [2426]. Attempts have also been made to automate such extraction procedures using FI [5211]. The repeatability and comparability of sequential extraction procedures is frequently problematical, and the exact assignment of the various extracts to the corresponding phases is not always unambiguous since redistribution for example can take place during the sequential extraction [4792]. In more recent times improved control of parameters such as extraction temperature and time, or the application of microwave ovens [3623], has led to an improvement in the reproducibility [2330, 5964]. In addition, reference materials are now available for which not only the total content but also the extractable portion are certified [4267, 4303, 4304, 4759, 4760, 4765]. Table 10-71 presents an overview on the extraction of trace elements from solid environmental samples.

Table 10-71. Selected publications on the extraction of trace elements from solid environmental samples.

Analyte	Test sample	Remarks	References
Ag, Cd	sediment	multistage extraction, F AAS	[3654]
Al, Ca, Co, Cu, Fe, Hg, K, Mg, Mn, Na	soils	multistage extraction, F AAS or GF AAS	[1618]
As	sediment	multistage extraction, F AAS	[3655]
As	sediment	extraction, LC-HG AAS	[3837]
Ca, Cd, Cr, Cu, Fe, Mn, Pb	sediment (river)	reproducibility of the five-stage extraction according to Tessier	[26]
Ca, Mg, Na, K	soils	determination of the cation exchange capacity	[726]
Ca, Cu, Fe, Mn	sediment	microwave-assisted	[2271]
Cd	soils	five-stage extraction according to Tessier, F AAS	[3102]
Cd	soils	six-stage extraction, GF AAS	[1616]
Cd, Cr, Cu, Fe, Mn, Ni, Pb, Zn	sediment (estuarine)	six-stage extraction	[3085]
Cd, Cr, Cu, Ni, Pb, Zn	soils, sediment	ammonium acetate extraction, GF AAS; modifier: Pd	[5963]
Cd, Cr, Cu, Ni, Pb, Zn	sediment (river)	three-stage extraction, F AAS	[5823]
Cd, Cr, Ni, Pb	sewage sludge and soils fertilized with sewage sludge	five-stage extraction, GF AAS	[3491]
Cd, Cu, Fe, Mn, Zn,	calciferous soils	carbonate extraction, F AAS	[1711]
Cd, Cu, Ni, Pb	sediment	six-stage extraction, GF AAS	[4518]
Cd, Cu, Ni, Pb, Zn	soils	multistage extraction, GF AAS	[5201]
Cr, Cu, Fe, Mn, Mo, Ni, Pb, V, Zn	sediment	three-stage extraction according to BCR, F AAS and GF AAS	[1438]
Cr, Mn, Ni, U, V	sediment	three-stage extraction according to BCR, F AAS and GF AAS	[499]
Cu	calciferous soils	optimizing Tessier's extraction procedure, GF AAS	[4479]
Cu, Mn, Ni, Pb, Zn	sediment	five-stage extraction, F AAS	[1045]
Hg	soils, sediment	distinguishing organic and inorganic mercury	[4763, 5872]
Mn	soils	comparison of extraction procedures	[1546]
Pb	contaminated soil, waste	simulation of the extraction by digestive juices	[2084]
Se	soils	multistage extraction and chromatography	[1907]
Se	soils	extraction of selenium available to plants	[4685]

Table 10-71 continued

Analyte	Test sample	Remarks	References
Se	soils	three-stage extraction, HG AAS	[3907]
Sn	soils, sediment	three-stage extraction, HG AAS	[5041]
Tl	sediment (river)	comparison of procedures according to Tessier and Psenner	[5042]
Various	soils	overview	[1360]
Various	soils, sediment	comparison of extraction procedures	[1360, 2588]
Various	sediment, sewage sludge	comparison of Tessier's five-stage extraction with the three-stage extraction according BCR	[3621, 4603]
Various	sediment (river)	Tessier's extraction procedure	[5807]
Various	contaminated sediment	overview	[1040, 1940, 1941, 5066]
Various	synthetic sediment	problems of selectivity	[3093]
Various	sediment, waste	overview	[3086]

10.3.2.4 Speciation Analysis in Soils, Sediments, and Sludges

In Section 10.3.2.3 we discussed the selective extraction of trace elements using various reagents for the determination of bioavailable species. In this section we shall treat separately the determination of redox species and organometallic compounds. In principle the difference is negligible since the species determination is mostly preceded by an extraction. The determination of the individual species is hardly different from the analysis of water (see Section 10.3.1.6).

The *non-chromatographic procedures* for speciation analysis include the extraction of Cr(VI) in weak alkaline medium [405, 2025, 4150], the fractional extraction of various mercury [3989, 5053] and selenium [23, 1387] compounds in dependence on the pH, and the extraction of tributyltin oxide with n-hexane-dichloromethane [1064, 5574], toluene [2168], or isooctane [4202] and its back extraction with nitric acid [1064, 5574]. The selective extraction of chelates has been applied for the species determination of As(III) [566], Sb(III) [1036], and Cr(VI) [2529, 4051, 4150, 4313]. The pH-dependence of the hydride generation of a number of species has been utilized to distinguish between As(III) and As(V) [5105, 5499, 5903]; the selective hydride generation of Sb(III) and Sb(V) from slurries [1032] or solutions of soil and sediment samples [1036] is especially simple. Speciation analysis based on time-resolved thermal vaporization has been described for mercury [330, 629, 630, 700, 2670] and arsenic [5104]. Moreover, ion exchangers have been utilized in batch or column procedures to distinguish between As(III) and As(V) [1904] or Cr(III) and Cr(VI) [1350].

Gas and liquid chromatography are mainly applied for the *chromatographic separation* of species in soil and sediment extracts. Table 10-72 provides an overview on the

Table 10-72. Selected publications on the determination of organometallic compounds in solid environmental samples after chromatographic separation; off-line techniques are shown by a slash (/), on-line techniques by a hyphen (-).

Analyte/species	Sample	Technique	Remarks	References
Al complexes	soils	HPLC / GF AAS	preseparated fractions	[2075]
As	sediments	HG-CT-GC-QF AAS	hydride generation with NaBH$_4$	[1651, 1736]
As(III), As(V), MMA, DMA	soils	HPLC-GF AAS	coupling via flowcell	[2739, 6382, 6383]
As(III), As(V), MMA, DMA	soils	LC / GF AAS	solvent extraction, anion exchange chromatography	[5754]
Cd, Mn	soils	HPLC / GF AAS	preseparated fractions	[884]
Cr (III/VI)	soils	HPLC-F AAS	modified RP-silica-C18 column	[4697]
Cr (III/VI)	soils	HPLC-GF AAS	reversed-phase ion pair chromatography	[4082]
Hg	sediments	various	overview	[6324]
Hg, MeHg	sediments	HPLC-CV AAS	seasonal variations	[2570]
Hg, Sn (various)	sediments	HG-CT-GC-QF AAS	ethylation	[4812]
Pb	sediments	HG-CT-GC-QF AAS	ethylation	[1583]
Pb-alkyl compounds	soils and sediments	HPLC-QF AAS	thermospray microatomizer	[655]
Pb: TAL, DAL	soils and sediments	GC-AAS	DDTC extraction, butylation with Grignard reaction	[654, 1167, 1217, 2402, 4776, 6372]
Pb	various	various	review article	[3568]
DMSe, DESe, TMSe	soils	GC-GF AAS	*in-situ* preconcentration in the graphite tube	[2912]
Se(IV/VI)	soils	HG-CT-GC-QF AAS	NaOH extraction	[1387]
Seleno-methionine	soils	HPLC / HG AAS	adsorption chromatography on XAD-8	[23]
Sn: DBT, TBT	sediments	GC-QF AAS	GC separation after ethylation	[1028, 1533, 2893, 6326, 6548]

Table 10-72 continued

Analyte/species	Sample	Technique	Remarks	References
Sn: MBT, DBT, TBT	sediments	HG-CT-GC-QF AAS	ethylation with NaBEt4	[283, 1026, 1027, 4753]
Sn: MBT, DBT, TBT	sediments	HPLC-GF AAS	tropolon-toluene extraction	[293, 297]
Sn: MBT, DBT, TBT, MMT	sediments	HG-CT-GC-QF AAS	hydride generation with NaBH4	[283, 293, 295, 1025, 1336, 1492, 3909, 4747-4749, 4751, 4799, 5134, 5135, 5229, 5542, 5983]
Sn: various	sediments	GC - various	review article	[1532]
Sn: various	sediments	GC-QF AAS	butylation or pentylation	[1530, 3793]

most important publications from this area. For GC the analytical sample is usually extracted with an organic solvent such as tropolon, pentane, hexane, toluene, or a mixture of these [283, 1219, 1370, 1530, 2893], followed by derivatization and preconcentration (refer to Section 7.2.1). Alternatively for organotin compounds an *in situ* ethylation with sodium tetraethylborate can be performed, which simplifies the procedure considerably [1026–1028, 2893]. Organolead compounds are mostly extracted in the presence of a complexing agent, followed by butylation [654, 1167, 1213, 1214, 1217, 6372]. HPLC is applied preferentially for the separation of organoarsenic compounds; this requires extraction and preconcentration but no derivatization (refer to Section 7.2.2) [5754, 6382, 6383].

The on-line derivatization, preconcentration, and separation of organometallic compounds by hydride formation or ethylation, on the other hand, is simpler, faster, and more accurate (refer to Section 7.2.3). The samples are leached with a weak acid or also with methanolic NaOH, derivatized with sodium tetrahydroborate [295, 1492, 1583] or sodium tetraethylborate [4812] and volatilized, collected on a cooled GC column and separated by warming (HG-CT-GC-QF AAS). This procedure has been mainly used for organotin [295, 1492, 1583, 4799, 4812], but also for organolead [1583] and organomercury [4812] compounds.

10.3.3 Coal, Ash, Dust, and Air

In this section we shall treat the complete emission - pollution cycle and also the impact on the workplace. Looked at globally the bulk of airborne particles (approx. 90%) derive from a number of natural sources. Salt from the oceans with about 10^9 t/a makes the largest contribution. Due to the relatively high population and industrial density in the northern hemisphere, the anthropogenic sources of particle emissions are by no means evenly distributed. The *highest emissions*, mostly from a number of sources, occur in the typical *conurbations* in which human activity is the most pronounced and impact a large proportion of the population in the form of pollution [1419].

Emission monitoring frequently starts at the raw material stage, for example the *coal* used in coal-fired power stations or coke works, or waste for incineration plants or as used in the cement industry. For the mass balance from coal-fired power stations, next to the dust carried with the scrubbed gases, which is important for the actual emission, the analysis of dusts collected in cyclones and electrostatic separators, and of ashes and other residues from the combustion must also be included. Only that portion that actually leaves the stack contributes to air pollution, and in the form of dust deposits or airborne dust influences our environment.

Air pollutants can occur in gaseous, fluid (dissolved or suspended), or solid form. Air must therefore be considered as an inhomogeneous, multiphase aerosol whose composition is subjected to considerable spatial and temporal variations [3148]. The gas phase is only of limited interest for the elemental determination of metals and metalloids. Exceptions here are mercury (see Section 10.3.3.4) and a number of gaseous species (see Section 10.3.3.5), especially close to hot emission sources [1481]. The liquid phase (rain, snow, etc.), also termed *wet deposition*, belongs analytically to the area of *water* and is treated in Section 10.3.1. The analysis of air, except for the above mentioned exceptions, is thus largely limited to the analysis of *airborne dusts*.

As well as the general composition of dusts, we must also make a distinction between the physical and chemical forms (species) [4684]; we shall only partially go into this in this monograph. As depicted in Figure 10-4, the morphological composition of dust can be very heterogeneous and the various forms can have markedly varying physiological activity. Smooth grains similar to fusion beads have a much smaller surface area, and are thus less reactive, than needle-shaped, fine crystalline sorts. A further criterion is the *solubility*, which decides whether an element can directly enter the nutritional chain, or whether it is initially sedimented. Finally we must also consider the *particle size* since next to the morphology it also has a decisive influence on the physiological activity. *Total dust* has a range of particle sizes of about 0.001–150 µm.

10 µm

Figure 10-4. Electron microscope photograph of a dust sample on Mylar foil (W. Dannecker's group, Hamburg).

10.3.3.1 Sampling Dusts

As in many areas of applied analytical chemistry, collecting the dust sample is the most critical step in the analysis. For the analyst, air is a practically infinitely expanded aerosol on which random investigations must be performed according to defined spatial and temporal regulations. The number and size of the random samples, as well as the place,

time, and sequence of sampling, are decisive for a representative statement on the relationship between quantity of material, space and time. A number of authors, in review
articles and book contributions, have treated the subject of sampling dusts very thoroughly, either in general [4661] or in selected areas such as the free atmosphere [30,
3148, 4661, 5477] or at the workplace [1906, 5249].

For the *investigation of trace substances in the free atmosphere*, the temporal and
spatial distribution of sample collection in an area to be investigated must be matched to
the meteorological conditions, such as wind direction, wind speed, air temperature or
humidity, through the selection of the length of an investigation program, the measurement frequency, and the density of monitoring stations. Frequently the concentration of
constituents determined by a measurement program is influenced in a characteristic
manner by the geographical location of the monitoring station. This is typical in areas for
example where dust is whirled up strongly or close to the coast where the constituents of
the sea spray become clearly noticeable. The extent to which meteorological conditions
can influence sampling can be seen when the wind velocity is high, causing turbulence at
the sampling port or even an underpressure so that the collected sample material can be
transported back to the outside.

When sampling aerosols we must make a distinction between procedures that use
suction and those that do not. For the latter various collectors for dust precipitates are
available. In Germany, for example, the pollution values of the Technical Committee for
Air [927] are based on a standard household jar with an internal diameter of 9.5 cm and a
collecting area of 62.2 cm^2 according to DIN 5071. For procedures without suction larger
particles are preferentially collected since they sediment out more easily than smaller
particles.

For sampling under suction, measurement and maintenance of a constant volume
stream are essential components of sampling. For sampling gaseous constituents, frequently *wash bottles with a suitable absorber solution* are used (we shall discuss the
determination of mercury in Section 10.3.3.4). To separate dust from the air *flat filters*
are mostly used and the air is sucked through the filter by a pump or a fan. Membrane
filters made of cellulose materials are the most suitable due to the low and constant element blank values. The use of fiberglass filters permits a larger volume of air to be taken
through, but because of the high and often markedly varying element blank values they
are hardly suitable for trace analysis in airborne dust [1419, 4306]. Next to the sampling
of the total dust in which no grading takes place, single stage grading can be performed
by *preseparating the coarse dust*, for example in a cyclone. Multistage grading can be
performed in a *cascade impactor* [1570, 2895, 3893, 4661].

Sampling for *emission measurements* is usually performed directly in the stack or
chimney, for example of a power station or a waste incineration plant. The difficulties to
be expected in taking a sample representative for the total mass flow under isokinetic
conditions [5895] can be seen from the example of a stack gas stream of several
100 000 m^3/h at a temperature of up to 300 °C and with a varying dust content of mg/m^3
to g/m^3. The true mass flow of particulate matter can frequently only be guessed, even if
we speak of measurement. For trace analysis the problems of the level and scatter of the
blank value as well as contamination during sampling can additionally be derogative to
the results of the examination [1419]. This is the more so the case for emission meas-

urements since cellulose filters cannot be used due to the high temperatures and quartz wool must preferentially be used [603].

For the *measurement of dust at the workplace* we make a distinction between local or stationary and personal dust measuring instruments. Stationary measuring instruments are similar to those used for measuring in the free air and have the advantage that the air throughput is relatively high at about 20 m^3/h, permitting a sampling period of only a few hours. The relatively small personal measuring instruments, on the other hand, have an air throughput of only about 120 L/h (corresponding approximately to the volume of air breathed), so that sampling is required for the entire length of a working day to collect sufficient material for the subsequent chemical analysis [1906, 2534, 2594, 5249]. Membrane filters made of cellulose materials are used preferentially for dust measurements at the workplace.

As well as the collectors mentioned so far, which are largely standardized, there is a wide range of collectors designed for special applications, such as the investigation of automobile exhaust gases [4661] or soot from Diesel engines [1551], or those that are specific for the subsequent analytical technique. Among these in particular are those developed for GF AAS procedures in which a cup [5337] or probe [1148] made of porous graphite is used as a filter and inserted directly into the atomizer. A further procedure that has been described in detail by SNEDDON [5481, 5483, 5484] is the collection of aerosols on the inner wall of a graphite tube by impaction [5478, 5479] or by electrostatic deposition [5482, 5883]. The major advantage given is that the measurement can be performed directly and virtually in real time since the sensitivity is very high and the collection time is thus very short [5479, 5482]. The disadvantages are that only one element can be determined per sample collection—except in simultaneous spectrometers—and that replicate measurements are not possible, so that random errors are very difficult to recognize. Any form of statistical evaluation requires multiple collection and analysis and thus a good deal of effort. In general, procedures of this nature must be seen as complementary to the standardized collection procedures and not as alternatives.

10.3.3.2 Digestion Procedures

Normally a pressure digestion with hydrofluoric acid-nitric acid [2413, 5044] is used for the digestion of coal and coal ash, mostly under the addition of an oxidant such as perchloric acid or ammonium peroxodisulfate [21, 1603, 1623, 2026, 3900]. EBDON and WILKINSON [1650] also described a digestion with perchloric acid (w = 72%) alone, which nevertheless requires a good deal of attention. Substantially simpler and far less dangerous are microwave-assisted digestions with hydrochloric acid-nitric acid [153] alone, or under the addition of hydrofluoric acid [614]; such digestions are much faster and in combination with FI techniques can be performed on-line [2146]. The combustion of coal in an oxygen bomb [3529] or oxygen combustion [4146] as well as fusion with lithium metaborate [609] have also been described; the latter is not necessarily ideal for a subsequent analysis by AAS. In a review article, MILLS and BELCHER [4114] discuss in detail the problems of sample pretreatment and digestion for the analysis of coal and ash.

The majority of *membrane filters* used nowadays in dust collectors have the advantage for further procedures that they are easily subdivided and can thus be subjected to different digestions [4306]. Membrane filters dissolve in nitric acid and also in organic solvents such as benzene [2594]. The *quartz wool* used in dust collectors for emission measurements, on the other hand, usually requires digestion in nitric acid or hydrofluoric acid under the addition of hydrogen peroxide [6029].

A combined *multistage digestion* using nitric acid, perchloric acid, and hydrofluoric acid [549, 1419, 4306] provides the most reliable results for a multitude of elements and the most varying dust samples; if necessary dry ashing can be performed in advance [146, 1603]. If insoluble, fluoride-containing residues are produced during digestion, they can be taken back into solution by the addition of boric acid [4306]. Nevertheless a number of authors preferred a *fusion* with lithium metaborate [610] or sodium carbonate-sodium tetraborate [3027] because of its greater speed. Both digestion techniques have certain disadvantages; the former because of the use of perchloric acid, the latter because of the resulting high total salt concentration which can lead to contamination in trace analysis and to interferences in GF AAS due to the high background signal. *Microwave-assisted acid digestions* are an interesting alternative which permit the determination of numerous elements without the use of perchloric acid but merely with hydrochloric, nitric, and hydrofluoric acids [613, 614]. Also, the required volume of acid is lower, the risk of contamination or losses is reduced, and the resulting solution is well suited for analysis by GF AAS [613, 2674].

A number of authors have investigated the *leaching of dusts*. Depending on the task at hand, this can be in addition to the determination of the total content to establish the leachable portion [1473, 1635, 2367, 2595], or it can be to avoid a digestion. For example, cadmium and lead can be taken quantitatively into solution in nitric acid-perchloric acid without the use of hydrofluoric acid [4306]. *Leaching in an ultrasonic bath* is particularly effective [232, 2353]; between 80% and 93% of the cadmium, copper, lead, and zinc are taken into solution in dilute nitric acid [4519].

For the actual *determination by AAS* we should always observe the maxim that the best results can be expected when the optimum concentration range is used [1419]. This means that F AAS is only suitable for the determination of higher analyte contents, but not for trace analysis. Due to the frequently complex and refractory matrix, particularly when the digestion is complete, the *hottest possible flame* should in general be used and the measurement should be performed *as high above the burner slot as possible*. In other words this means that the measurement should be optimized for minimum interferences and not for maximum sensitivity. Selected publications on the analysis of coal, fly ash, and dusts by F AAS are compiled in Table 10-73. The alternative to avoiding interferences is higher dilution and determination by GF AAS. In this case, particularly after a complete digestion, we must also reckon with matrix effects, so that the risk should be minimized by the application of the STPF concept and the use of Zeeman-effect BC [396, 611, 613, 617, 3571, 3572, 6244]. Table 10-74 presents an overview on the determination of trace elements in coal, fly ash, and dusts by GF AAS after digestion.

Due to the high sensitivity of the technique, HG AAS has also been applied to the analysis of coal, fly ash, and dusts, mainly for arsenic and selenium [61, 1650, 3348, 3510, 5361, 5814, 5815, 6032]. It has also been applied for the other hydride-forming elements [4261] and even for the determination of lead [4324, 5782]. Nevertheless in the

Table 10-73. Selected publications on the analysis of coal, ash, and dusts by F AAS after digestion.

Analyte	Test sample	Digestion / pretreatment	References
Al, Ca, Fe, K, Mg, Na, Si	coal, fly ash	$Li_2B_4O_7$ fusion	[4260]
Ba, Be, Cr, Cu, Li, Mn, Ni, Pb, Sr, V, Zn	coal, coke, fly ash	Australian Standard Method AS-1038	[5543]
Be, Ca, Cd, Co, Cu, K, Li, Mg, Mn, Ni	coal	pressure digestion HNO_3–HF–H_3BO_3	[2413]
Cd, Co, Li, Pb	coal, ash	dry ashing / pressure digestion HNO_3–HF–$HClO_4$	[1603]
Ge	brown coal ash	extraction of $GeCl_4$ in hexane	[1099]

Table 10-74. Selected publications on the determination of trace elements in coal, ash, and dusts by GF AAS after digestion.

Analyte	Test sample	Remarks	References
Al, Cd, Cr, Cu, Fe, Mn, Ni, Pb, Zn	airborne dust	acid digestion with HNO_3–HF–$HClO_4$	[6146]
As	coal	removal of interferences	[4871]
As, Sb	coal, fly ash	pressure digestion HNO_3–HF	[5044]
Fe, Ni, V	coal	microwave-assisted digestion	[153]
Ga	coal, fly ash	modifier: Ni	[5270]
Pb	coal	ZBC, modifier: ascorbic acid	[5284]
Pb	airborne dust	modifier: $NH_4H_2PO_4$–$Mg(NO_3)_2$	[3010]
Se, Te	coal	coprecipitation with arsenic	[6373]
Tl	coal, fly ash	modifier: Pd–ascorbic acid	[5271]

presence of complex matrices GF AAS appears to be easier to control, provided that the rules mentioned above are correctly maintained [613, 6033, 6244]. The decision as to which technique is the more suitable must be made in each case based on the sample to be examined and the required sensitivity.

10.3.3.3 The Direct Analysis of Solid Samples and Slurries

Although dusts and ashes can under no circumstances be considered as uniform, they are nevertheless a particularly suitable sample material for the direct analysis of solid samples or slurries due to their particle size. These techniques are also suitable for the analysis of coal since it can be relatively easily ground and pulverized. The major reasons for the application of these techniques are the avoidance of a digestion, the much lower sample requirement, and the increased sensitivity. The *direct analysis of solid samples is* restricted to GF AAS. The application of the STPF concept and Zeeman-effect BC are quasi prerequirements for accurate analyses [1631, 3628, 5857], among other things because the matrix often has a substantial influence on the appearance time and the signal profile [5152]. A number of authors attempted to simplify introduction of the solid samples by punching small disks from the filter and analyzing directly [3628, 4211]. The precision of this procedure is nevertheless poorer than when the filter is homogenized in a mill [5178].

In contrast to direct solids analysis, the *analysis of slurries* is more flexible, and in the case of dusts has been extended to all AAS techniques. For GF AAS mixing and homogenization by ultrasonics brought about simplification and automation of the procedure [756, 1071], while for F AAS flow injection brought the breakthrough [1879, 3611]. For the analysis of coal slurries, *in situ* ashing with air or oxygen has proven to be particularly advantageous [1655], so that aqueous calibration solutions can be employed. Even for procedures of chemical vapor generation, slurries of dusts have been successfully applied, such as for the determination of arsenic by HG AAS [3612, 4325] and mercury by CV AAS [3612]. In both cases, powerful stirring or sonification is adequate for a virtually quantitative yield. Table 10-75 provides an overview on recent publications on the analysis of coal, ash, and dusts by GF AAS using direct solids analysis or the slurry technique.

Table 10-75. Selected publications on trace analysis in coal, ash, and dusts by GF AAS using direct solids or slurry sampling.

Analyte	Test sample	Procedure / remarks	References
As	coal	slurry in $Ni(NO_3)_2$–$Mg(NO_3)_2$–HNO_3–ethanol	[1649]
As, Cd, Cu, Pb, V	coal, dusts	ZBC	[5857]
As, Pb, Se, Tl	coal, fly ash	rapid analysis without modifier	[756]
Be	coal	STPF, modifier: $Mg(NO_3)_2$	[2374]
Cd, Cr, Ni, V	coal, fly ash, dusts	direct solids analysis, STPF	[5152]
Co, Cu, Mn, Mo, Ni, V	coal	slurry with Triton X-100	[4703]
Ga	coal, fly ash	modifier: Ni	[5277]
Ge	brown coal	direct solids analysis	[1760]
Se	fly ash	ZBC, modifier: Ni	[1631]
Se	coal	slurry in HNO_3–ethanol; O_2 ashing	[1655]
V	brown coal	direct solids analysis	[1761]

10.3.3.4 Methods for the Determination of Mercury

Since all mercury compounds are volatile at temperatures below 500 °C and also easily decompose with the formation of elemental mercury, practically all of the mercury released during thermal processes (combustion of fossil fuels, waste incineration, volcanic activity, etc.) reaches the atmosphere in gaseous form. In addition mercury and its compounds have a very low tendency to accumulate on dust particles, so that for this element the examination of the particulate fraction (total dust) only gives a very incomplete picture of the total pollution. For *emission measurements* in particular a two-stage separation is required in which particle-bound mercury and mercury that passes the filter are both detected. Suitable for this purpose, for example, is a filter probe packed with quartz wool followed by at least two gas washbottles containing an absorption solution (potassium permanganate in sulfuric acid solution) [2493, 6031]. The mercury content in the dust-loaded quartz wool can be determined after leaching in nitric acid, while the absorption solution (after bleaching with hydroxylammonium sulfate) can be examined directly.

A three-stage separation has been proposed for *measurements at the workplace* in which particle-bound mercury is collected on a membrane filter. This is followed by two tubes packed with different solid phases; organic mercury is collected in the first tube while elemental mercury is collected in the second (by amalgamation) [4302]. The mercury is thermally desorbed from each stage and determined by CV AAS. If only the content of gaseous, elemental mercury in air (e.g. in the laboratory atmosphere) is to be determined, it is sufficient to use a simple gold or silver adsorber from which the mercury is subsequently released by heating [384, 1918, 5190]. Various air pollutants interfere in this procedure [384], but this problem can be largely eliminated by a two-stage amalgamation [1918]. FRIESE *et al.* [1986] have made a detailed investigation on the calibration of this procedure. As well as noble metals, activated carbon [2069] or carbon-loaded paper [2887] have been used for the determination of total mercury at the workplace.

Total digestion procedures have been described for the determination of mercury in coal or particle-bound mercury in dust or fly ash, for example with nitric acid, perchloric acid and ammonium peroxodisulfate at 250–300 °C [1623], or with nitric acid, hydrofluoric acid, sulfuric acid and ammonium vanadate under pressure at 160 °C [2493]. Nevertheless, such digestion procedures are unsuitable for daily routine since they can take up to several days [2493] and the results can often be falsified by exchange reactions with the wall of the digestion vessel [336, 3263, 5916]. These digestion procedures make even less sense when we consider that mercury can be simply leached from the analytical sample with nitric acid [2671, 4872] or nitric acid-hydrofluoric acid [336, 337] and taken quantitatively into solution under far less severe conditions. Higher mercury concentrations can be determined by GF AAS using palladium as the modifier [6260]. An alternative to this is the pyrolysis of the analytical sample at > 500 °C [782, 1624, 2014] or combustion in a stream of oxygen [1596, 4146, 5552] and either collection of the pyrolysis or combustion products in permanganate in nitric acid solution [1596, 5552] or collection and preconcentration by amalgamation [782, 1624, 2014, 2671, 4872]. In this way interferences are eliminated and the sensitivity is increased so that it is rarely necessary to add a buffer such as EDTA [603]. A step even further is the direct determination of mercury in solid samples such as coal or ash and dusts by pyrolysis at 1000 °C and measurement by GF AAS [337, 5859].

10.3.3.5 Speciation Analysis

In this section we shall mention the determination of individual oxidation states and of organometallic compounds; the differentiation based on solubility characteristics is described in Section 10.3.3.2.

Arsenic, selenium, tellurium: As described in Section 7.1.1, the individual oxidation states of these elements can be determined by selective reduction by HG AAS. Gaseous species can be preconcentrated for example by adsorption on gold-coated quartz spheres and determined individually after stepwise leaching [772, 4214, 4215]. A number of authors have reported the determination of organoarsenic compounds by GC-AAS after the extraction of dusts with benzene-toluene [242, 4569], or after hydride generation and collection in a cold trap [4217]. Volatile organoselenium compounds were preconcentrated directly by trapping in a cold trap and then separated and determined by GC-AAS [2914, 2915].

Chromium (VI): The determination of Cr(VI) in welding fumes is one of the most important industrial hygiene examinations due to the toxicity of this species [427, 2529, 5249]. The influence of the collecting procedure, the welding fumes themselves, the storage period, and the analytical procedure have been investigated in detail [4590]. While the Cr(VI) content on fiberglass filters did not appear to change over several weeks storage, ROHLING and NEIDHARD [4935] found that during collection Cr(VI) is reduced by SO_2, which can lead to severe losses of Cr(VI), depending on the length of the collection period. GIRARD and HUBERT [2132] showed that the Cr(VI) can be quantitatively recovered from the filter medium by extraction with acetic acid-sodium acetate. BRESCIANINI *et al.* [787] eliminated the interferences caused by large quantities of calcium, iron, and sodium during the determination of chromium by GF AAS by separating the analyte on an ion exchanger.

Lead: Because alkylated lead compounds are used as antiknocking agents in gasoline (petrol) it is hardly surprising that a relatively large number of publications has appeared on the determination of organolead compounds in the atmosphere [2950, 3568, 4065]. The non-chromatographic procedures are relatively simple in which the air, after separation of the particulate portion on a filter, is passed through a solution of iodine monochloride [639, 2947] or through activated carbon [4896, 4978] to selectively preconcentrate the alkyl-lead compounds. For the speciation analysis of sorbed or dissolved organolead compounds the time-consuming procedure of extraction, derivatization, and preconcentration (refer to Section 7.2.1) prior to separation by GC and determination by AAS has been applied [654, 1165, 1531, 2402, 2537]. A much easier procedure is to collect the alkyl-lead compounds directly from the air on cooled Porapak columns and to thermally desorb them for the GC-AAS determination [142, 2401, 2536]. The disadvantage of this procedure is the limited capacity of the column packing material since some water also always condenses [2948]. Numerous authors thus used a cool trap filled with glass beads which allows collection times of 1 h without problems [1211, 2913, 2915, 2948, 4780, 4899].

Manganese, nickel, and tin: COE *et al.* [1313] described the determination of methylcyclopentadienyl-manganese-tricarbonyl, which is frequently added to unleaded gasoline (petrol), by GC-GF AAS after preconcentration in a cold trap. NYGREN [4398] optimized the determination of organotin compounds using an LC-GF AAS coupling. ELLER [1721] collected nickel carbonyl in a solution of iodine in isopropanol, extracted the nickel in nitric acid in an ultrasonic bath and determined it by GF AAS.

10.4 Rocks, Minerals, Ores

Applied geochemistry encompasses essentially two aspects, namely prospecting for mineral reserves and environmental studies. For each application a number of elements must be detected in an area that can cover hundreds of square kilometers, which means that a large number of samples are required. In a second stage anomalies that are observed can be investigated in greater detail; for this purpose the density of the sampling locations must be substantially higher. As well as rocks, soils, and sediments, plants and water are frequently included in the investigations. Since the analysis of these materials is discussed in Sections 10.2 and 10.3.1, we refer our readers to these sections for further information. We shall also not go into the aspect of environmental analysis since this is treated in detail in Section 10.3.

For geochemical prospecting we can assume that perhaps 95% of the analyses indicate that the area being prospected is not of interest [5838]. For this reason, and also because of the large number of samples, the analytical methods in the first instance should be *cheap, fast, simple, accurate, and as far as possible applicable to various materials* [5838, 6073]. Flame AAS has fulfilled these requirements for a large number of elements in an almost ideal manner and continues to do so nowadays due to its simplicity and freedom from interferences [5243, 6073], even though multielement techniques such as ICP OES are finding increased use. For the determination of lower element concentrations, extraction procedures for F AAS 'compete' with direct determinations by GF AAS or HG AAS; the decision between these alternatives can be influenced by a number of factors. This aspect is discussed in the following sections. We should also like to draw our readers' attention to a number of review articles and book contributions on this subject [345, 382, 1144, 2771, 4862, 5838, 6073].

10.4.1 Digestion Procedures

Information and exact procedures for the digestion of geological materials for subsequent analysis by AAS have been published by THOMPSON and BANERJEE [5838], HEINRICHS [2478, 2479], and CHAO and SANZOLONE [1191]. The selection of the most suitable digestion procedure depends essentially on the purpose of the analysis. A total digestion is often not required since partial digestions often allow anomalies to become even more apparent and leach out elements that are associated with a specific mineral phase.

If a *complete digestion* is required, the decision must be taken as to whether silicon should also be determined or not. If silicon is to be determined a *fusion* is required and the procedure for geochemical applications using *lithium metaborate* as proposed by SUHR and INGAMELLS [5679] is the technique of choice. According to the original method 0.1 g of the finely ground test sample is mixed with 0.5 g LiBO$_2$, placed in a baked-out graphite crucible, and heated in a muffle furnace for 10–15 min at 900–1000 °C. The clear melt is agitated in the hot crucible so that small pellets are formed that do not adhere to the walls and then poured rapidly into 40 mL nitric acid (w = 3%) contained in a 200 mL PTFE or PE beaker. The beaker is covered and the contents are stirred to obtain complete solution. This procedure, either as described or slightly modified [450, 893, 1470], is suitable for the determination of numerous major and minor constituents in a large variety of rock and soil samples [3578, 6023]. Due to the large number of samples that often have to be analyzed, a fully automated system for this digestion technique has been described [717, 718, 3084]. The major disadvantage of this technique is the relatively high total salt content in the solution which can lead to interferences for trace analyses.

If silicon does not have to be determined, *acid digestions* with *hydrofluoric acid* are preferred [1104], usually together with hydrochloric, nitric, or even perchloric acids [1955]. During the 1970s these hydrofluoric acid digestions were increasingly performed in autoclaves made of PTFE under pressure, which reduced the digestion time and also contamination and losses. We should particularly like to mention the detailed investigations by LANGMYHR and PAUS [3421, 3422, 3424–3426]. Nowadays these classical digestion procedures are increasingly being replaced by *microwave-assisted digestion procedures* [154, 614, 3074, 4804, 5465, 5715]. The major advantages are a reduction of the total amount of acid required and a shortening of the digestion time to in some cases just a few minutes.

For rocks and soils with a low content of silicates, or for when the portion of the analyte present in the silicate crystal lattice is not of interest, *partial digestions* or *leaching* without the addition of hydrofluoric acid are adequate [6008]. For numerous elements *concentrated nitric acid* is particularly suitable [1955, 5564]. Digestions with phosphoric acid [2362] or hydrochloric acid-hydrogen peroxide [4412] as well as leaching with lithium hydroxide at 180 °C [2150] have also been described for special applications. A digestion with potassium chlorate and hydrochloric acid has been described for the selective determination of elements associated with sulfides in basic rocks [4457]. Procedures for the stepwise leaching of geochemical samples have also been described in order to determine which fraction (water soluble, ion exchangeable, reducible manganese, reducible iron, lime, etc.) the analyte is associated with [2086]. JORDANOV and IVANOVA [2953] described a vacuum extraction method for the determination of As, Bi, Cd, Sb, Tl, and Zn in soils and rocks.

Gold and the associated platinum metals hold a special position in geochemical analysis, which is reflected in the large number of publications on the subject. Although a *digestion in aqua regia* [513, 550, 775, 5800] is frequently used for this determination, it is not suitable in all cases since up to 59% of the gold can remain in the residue [2327]. A combination of *aqua regia* and hydrofluoric acid appears to be more effective [5240, 5802], where required under pressure [3292] or with post treatment with perchloric acid [2067]. A combined treatment with acid and bromine is particularly suited for the subsequent extraction of gold as the bromoaurate (see Section 10.4.3). Worth mentioning is the sequential digestion with *aqua regia* and bromine-hydrobromic acid [1205] or with phosphoric acid-perchloric acid and bromine-hydrobromic acid [6045]. The extraction of gold with bromine and hydrochloric acid at ambient temperature has also been described [4835]. An alternative is extraction with sodium cyanide in alkaline medium, also at ambient temperature [1928].

10.4.2 Direct Determinations from Digestion Solutions

A large number of major and minor constituents of rock and soil samples, such as Ag, Al, Ba, Ca, Co, Cr, Cu, Fe, K, Li, Mg, Mn, Na Ni, Pb, Si, Sr, and Zn, can often be determined directly from the digestion solution by F AAS. Selected publications on this subject are compiled in Table 10-76. In general we can say that, particularly for complete digestions, a very refractory matrix must be expected, which can be decisive for the selection of a suitable flame. LÜCKE [3646, 3647, 3649] drew attention to the interference of aluminium on the determination of the alkaline-earth elements; this can only be eliminated in the nitrous oxide-acetylene flame under the addition of lanthanum (as the chloride) [6271]. The interference-free determination of the refractory elements tungsten [2849, 3826], molybdenum [3579, 5707], tantalum [3827], niobium [3827], hafnium [6], aluminium [626, 3379, 3577], silicon [745, 2189] and titanium [745], and also manganese [3381], beryllium [6046], chromium [454], and a number of other elements [397, 3642], is only possible in this hot, reducing flame. LÜCKE [3648] also observed an interference on the determination of the alkaline-earth elements in silicate materials in the presence of hydrochloric acid and thus generally recommended the use of nitric acid.

Table 10-76. Selected publications on the direct determination of various analytes in digestion solutions of rock and soil samples by F AAS.

Analyte	Flame (oxidant)	Remarks	References
Al, Ca, Fe, K, Mg, Mn, Na, Si, Ti	air or nitrous oxide	comparison between F AAS and ICP OES	[894]
Al, Ca, Fe, Mg, Si	air or nitrous oxide	microwave-assisted digestion	[3074]
Al, Ca, Fe, K, Mg, Na, Si, Ti	air or nitrous oxide	digestion: H_3PO_4; comparison between F AAS and ICP OES	[2362]
Al, Fe, Ca, Mg	nitrous oxide	buffer: KCl–tartaric acid–$SrCl_2$	[1207]
Al, As, Ca, Cd, Cu, Fe, K, Mg, Na, Pb, Sb, Zn	Al: nitrous oxide, all others: air	comparison of digestion methods, open digestion with HF preferred	[1104]
Al, Si	nitrous oxide	digestion: $HF–H_3BO_3$	[3379]
Ba, Ca, K, Li, Mg, Rb, Sr	nitrous oxide	salt solutions of halite	[3968]
Ba, Sr	Ba: nitrous oxide, Sr: air	digestion: $HF–H_2SO_4$; buffer: EDTA	[5239]
Ca	air	digestion: $HF–H_3BO_3$; buffer: HCl–La	[3871]
Ca	nitrous oxide	digestion: $HNO_3–HF$	[2144]

Table 10-76 continued

Analyte	Flame (oxidant)	Remarks	References
Ca, Cd, Co, Cr, Cu, Fe, K, Li, Mg, Mn, Na, Sr, Zn	air or nitrous oxide	universal method	[5143]
Ca, Fe, Mg	air	slurry, on-line dilution with FI-F AAS	[3611]
Cu, Fe, Pb, Zn	air or nitrous oxide	*aqua regia* digestion, addition of ammonium acetate to prevent precipitation of lead sulfate	[160]
Fe	air	digestion: H_2SO_4–HF–HCl	[5725]
Fe, K, Mg, Mn, Na	air	pressure digestion: HNO_3–HF	[5715]
K, Na	air	digestion: HNO_3 to prevent chloride interferences	[3648]
Mg	air	on-line addition of the lanthanum buffer using FI-F AAS	[4419]
Mn	nitrous oxide	comparison with air-acetylene flame	[3381]
Mn	air	pressure digestion: HNO_3–HCl–HF	[3874]
Mn	air	digestion: HF–$HClO_4$; buffer: KCl–tartaric acid–$SrCl_2$	[1206]
Na, K, Cs	air	buffer: 1 g/L La + 2 g/L Rb	[1208, 1209]
Na, K, Mn	air	buffer: KCl–tartaric acid–$SrCl_2$	[1207]
Si	nitrous oxide	digestion: HF in closed PFA bottle	[3137]
Si	nitrous oxide	fusion: $LiCO_3$–H_3BO_3; stabilized with fluoride	[450]
Sr	nitrous oxide	buffer: ascorbic acid–potassium citrate	[4252]
Sr, Zn	Sr: nitrous oxide, Zn: air	fusion with $LiBO_2$	[5059]
Ti	nitrous oxide	combined acid digestion and fusion	[376]
Ti	nitrous oxide	digestion: HF–H_3BO_3	[3380]
Ti	nitrous oxide	buffer: 20 g/L CsCl	[3390]

Because a large number of test samples must usually be investigated during the analysis of geological materials, the *FI technique* (refer to Chapter 6) offers a number of advantages [1818]. These include automatic dilution of measurement solutions [2677, 4285], automatic addition of buffers such as lanthanum [4419], or the analysis of solutions with high total salt content [6497]. The small volume of measurement solution required, the high measurement frequency, and automatic sample introduction are further advantages that meet the demands of rapid routine analysis.

Graphite furnace AAS is the most sensitive analytical technique that can be applied for the direct determination of trace and ultratrace constituents in matrices as complex as those found in geological samples [1144]. A natural prerequirement for a successful determination is the proper application of the STPF concept and usually also Zeeman-effect BC [6073]. Even disregarding considerations of principle (see Section 8.2.4.1), the

STPF concept is the only workable solution since the high time requirement prohibits the analyte addition technique for geochemical analyses. Publications on the direct determination of selected elements in digestion solutions using the STPF concept are compiled in Table 10-77.

Procedures in which *chemical vapor generation* is applied to separate the analyte from the bulk of the matrix (i.e. HG AAS and CV AAS) are advantageous for direct determinations due to the high sensitivity that is attained for those elements determinable by the respective technique. Nevertheless, it is well known that transition elements in Group VIII and Subgroup I can cause substantial interferences in the determination. This requires suitable buffers or complexing agents to be added. Additionally, the magnitude of the interferences depends on the type of system used, and is significantly lower in FI systems than in batch systems [1179]. Selected publications on the determination of hydride-forming elements directly from solutions of digested geological samples are cited in Table 10-78. The preconcentration of the hydride in a graphite tube with subsequent atomization by GF AAS is of particular interest due to an even further increase in the sensitivity and lower interferences [5771, 6487, 6540].

Practically only CV AAS is suitable for the determination of mercury in geochemical samples [1730]. Interferences due to transition metals are lower when tin(II) chloride is used as reductant compared to sodium tetrahydroborate [3078]. In strongly alkaline medium, noble metals such as gold or platinum and hydride-forming elements such as selenium or tellurium do not cause interferences in the determination of mercury [403]. FLANAGAN *et al.* [1919] vaporized mercury by heating larger quantities of sample and preconcentrating it by amalgamation on a gold foil prior to a determination by CV AAS.

Table 10-77. Direct determination of elements in digestion solutions of geological samples by GF AAS applying the STPF concept.

Analyte	Modifier	Background correction	References
Ag	thiourea–$NH_4H_2PO_4$	ZBC	[5301]
Ag, Cd	$(NH_4)_2HPO_4$	ZBC	[6566]
Ag, In, Tl	Pd–EDTA	deuterium	[6569]
As	Pd	deuterium	[5265]
As, Co, Cr, Ni, Se	Ni or Mg	ZBC	[4726]
Au	Pd	deuterium	[6457]
Au	V	Smith–Hieftje	[2067]
Be	Pd	deuterium	[6456]
Cd	$NH_4H_2PO_4$	deuterium	[3672]
Cr	EDTA	ZBC	[6565]
Cs, Rb	H_2SO_4	deuterium	[2228]
In	Pd–EDTA	ZBC	[6568]
Mn	Ni	ZBC	[3259]
Pt, Pd, Rh, Ir	NiS cupellation	deuterium	[4571, 6531]
Sb	Pd	deuterium	[4369]
Sn	ammonia	ZBC	[3665]
V	$Mg(NO_3)_2$	ZBC	[3864]

Table 10-78. Determination of hydride-forming elements in digestion solutions of geological samples by HG AAS.

Analyte	Buffer / complexing agent	Atomizer	References
As	0.7 mol/L HNO$_3$–0.02% L-cysteine	QTA	[1119]
As, Sb	6 mol/L HCl–KI–AlCl$_3$–ascorbic acid	QTA	[5797, 5799]
Bi	2 mol/L HCl–thiourea–AlCl$_3$–ascorbic acid	QTA	[5798]
Bi	thiosemicarbazide–1,10-phenanthroline	QTA	[1180]
Bi, Ge, Te	none	GF / Pd modifier	[6540]
Ge	citrate	QTA	[2287]
Ge	1.5–3 mol/L HCl	GF / Pd modifier	[5771]
Sb	KI	QTA	[1177]
Se	1,10-phenanthroline	QTA	[1178]
Se	(FI)	QTA	[1179]
Se	6 mol/L HCl–FeCl$_3$	QTA	[2762]
Se	9 mol/L HCl	QTA	[5096]
Te	none	GF	[6487]

10.4.3 Extraction and Preconcentration Procedures

Up to the middle of the 1980s virtually all papers on the preconcentration and separation of the analyte from digested geochemical samples referred to the subsequent determination by F AAS. In the meantime GF AAS is used as the detector in the majority of papers published, thus bringing a number of advantages. Firstly, much lower analyte contents can be detected and far lower enrichment factors are required than for F AAS, so that smaller quantities of sample are sufficient, both for the initial volume of sample solution and the volume of measurement solution. In other words, the procedures can be miniaturized. Since the same or similar preconcentration procedures are mostly used for both F AAS and GF AAS, we shall not treat them separately. Without doubt the most frequently used separation and preconcentration procedure in geochemical prospecting is the extraction of gold as the chloroaurate [527, 5800] or the bromoaurate in IBMK [774, 1205, 2876], pentyl acetate [1439, 6045], or ethyl acetate [5390]. In a similar manner silver [5803] and the platinum metals [862, 5802] can be extracted as the iodo complexes in IBMK. Another frequently used preconcentration procedure for gold and other precious metals is *coprecipitation* with mercury [513, 4370, 5808, 5809] or with selenium or tellurium after the addition of tin(II) chloride [4835]; in combination with GF AAS, this is a procedure suitable for extreme trace analysis [167, 1722, 5240]. Additionally, the preconcentration of precious metals on *ion exchangers* [775, 3292, 4880], *polyurethane foam* [6149], *dithiocarbamate-chitin* [5795] and other solid phases [3321] has been described. Instead of elution, a slurry of the solid phase can also be determined directly by GF AAS [3321]. We should also particularly like to mention the preconcentration of palladium, platinum, rhodium [550], and lead [5443] on *immobilized crown ethers*, which is highly selective compared to the solid phase extractions mentioned above.

Very similar preconcentration procedures have also been described for numerous other elements. For example, thallium can be extracted as the *chloro* [2760, 5801] or *iodo complex* [2288] with IBMK. This latter technique has been extended to a large number of other elements [1296, 1297, 6022]; back extraction into the aqueous phase is preferred on occasions prior to the determination by GF AAS. A number of authors added trioctylphosphine oxide (TOPO) to the IBMK to improve the extraction of cadmium, molybdenum and silver [1115], bismuth [4976], antimony [4975], and thallium [1115, 4977]. RUBEŠKA *et al.* [4992] have drawn to attention an interference to the extraction by nitrose gases if these are not removed following a digestion in *aqua regia*. Extraction with TOPO-IBMK following a lithium metaborate fusion has also been applied for beryllium [6046], molybdenum [6203], and tin [1731, 1732, 6046, 6203].

CLARK and VIETS [1295] described the *extraction* of 18 elements in a mixture of Alamin-336, Aliquat-336 and hexane with IBMK. Similar procedures, even if for a lower number of elements, have been described for the preconcentration of trace elements [4979, 6072]. DONALDSON [1575, 1577–1579] extracted a large number of elements as the xanthate complex with cyclohexane or chloroform. An alternative to this is sorbent extraction, for example on polyurethane foam [1237, 6152], which is also applicable to a large number of elements. The preconcentration of trace elements has been described for cobalt [187, 6066], gallium [186, 6139], and the rare-earth elements [2669] among others.

Very high enrichment factors can be attained by *coprecipitation*; the procedure described above for the coprecipitation of precious metals with mercury as the collector and tin(II) chloride as the reductant is also suitable for selenium and tellurium [4370]. Coprecipitation with hydrated iron(III) oxide has been described for antimony [4975], tellurium [1578], scandium, yttrium, lanthanum and the rare-earth elements [2007, 5234, 5236]. The major problem of coprecipitation reactions is that they are rather time consuming and require a good deal of effort, and naturally the risk of contamination also increases.

The technique of *continuous coprecipitation in a closed system*, proposed by VALCÁRCEL and co-workers [4052, 5087, 5088], opens completely new dimensions, for example for cobalt and nickel with 1-nitroso-2-naphthol. The precipitate is collected on a filter, dissolved in ethanol, and analyzed on-line by F AAS. The procedure was modified by PEI and FANG [4591], who dispensed with the filter, collected silver by on-line coprecipitation with Fe(II)-DDTC in the presence of 1,10-phenanthroline in a knotted reactor, dissolved in IBMK, and determined the silver by F AAS. Similar on-line preconcentration procedures using FI systems and packed columns for *sorbent extraction* have also been described for gold [4741, 6412] (refer to Chapter 6). The drastically reduced risk of contamination, the miniaturization with respect to sample and reagent consumption, and the high level of automation that can be attained by the coupling with FI certainly go a long way to meeting the requirements of the geochemical laboratory.

10.4.4 The Direct Analysis of Solid Samples

Since digestions must normally be performed in large series in the geochemical laboratory and a level of automation is applied, the analysis of the digestion solutions is mostly
quicker and more economical than the direct analysis of solids, particularly when a large
number of analytes are to be determined. Nevertheless, the direct analysis of geological
materials by GF AAS was investigated at an early stage (refer to Section 1.9) and proved
to be successful in a number of cases. LANGMYHR [3436] has published an overview on
the early investigations in which he also gives hints on possible problems with this technique. As we have repeatedly emphasized, the application of the STPF concept and
Zeeman-effect BC to eliminate possible interferences is particularly important in direct
solids analysis [2229, 3083]. NAKAMURA *et al.* [4276, 4277] found that 'dilution' of the
sample with graphite powder was of advantage. Selected publications on the analysis of
solid samples using the STPF concept are compiled in Table 10-79.

Dilution of the sample and its handling in general are greatly simplified by using the
slurry technique [1568]. The element most frequently determined by this technique is
lead [1568, 2854, 3610], followed by other easily atomized elements such as silver
[1246] or cadmium [1568]. Even the determination of difficultly volatilized elements
such as chromium and copper [3610] has been reported; the pyrolysis and heating out
steps were omitted to increase the sample frequency. JERROW *et al.* [2906] found that the
surface of geological samples took on ion exchanger properties during grinding, which

Table 10-79. Selected publications on the direct analysis of solid geological materials by GF AAS.

Analyte	Sample	Remarks	References
Ag	rocks, minerals	continuum source BC	[3673, 6550]
Ag, As, Cd, Pb, Sn, Zn	geological materials	ZBC, calibration with reference samples	[5192]
As, Cd, Cu, Pb, V	rocks, soils	ZBC	[5857]
As, Cd, Hg, Pb, Sb	rocks, soils, sediments	ZBC, 1+1 dilution with graphite powder	[3083]
As, Cd, Pb, Se	geological materials	ZBC	[4660]
Be, Bi, Co, Cs, Cu, Li, Ni, Pb, Rb	silicate rocks	ZBC, 1+1 dilution with graphite powder	[4277]
Cd, Pb	geological materials	comparison with slurry tech- nique	[1568]
Cu	carbonate rocks	ZBC, 1+1 dilution with graphite powder	[4276]
Pb	rocks	ZBC, modifier: phosphate	[2229]
Rb	granite	special tool for sample intro- duction	[6537]
Se	minerals	ZBC, modifier: Ni	[1776]

leads to an accumulation of the analyte on the particles when these, for example, are added to the solution during make up. This welcome behavior is perturbed by the addition of large quantities of buffer or modifier, but nevertheless the sorption of these substances on the particles can also be of advantage. It has even been possible using FI techniques to automatically introduce slurries of samples with high silica contents to F AAS [3611]. In order to avoid the preparation of highly diluted slurries, the slurries for the determination of calcium, magnesium, and iron were automatically diluted on-line.

As well as the papers cited in this section, a number of the procedures for the analysis of soils, sediments, ashes, and dust mentioned in Sections 10.3.2 and 10.3.3 can be applied to the geological materials discussed in this section.

10.5 Crude Oil and Petroleum Products

This section deals with the analysis of raw materials for refineries (crude oils, residual oils), low-boiling refinery intermediate and end products (dry gases, liquid gases, test gasoline), fuels for Otto motors, diesel engines, and jet aircraft engines, light and heavy fuels, and bunker oils for ships. The analysis of the base oil and the specific additives and fillers, and also of fresh oils and used oils (wear metals), is important for lubricants. We also deal with the analysis of bitumen and asphalt products, and rubber and plastics based on petroleum.

We have not arranged this section in general according to the individual oil products since analytically there is little difference between the determination of, say, nickel in light fuel oils and asphalt products. Major differences, both with respect to the concentration of the analyte and the form (compound) in which it is present, depend on whether the analyte derives from the formation of the crude oil, or whether it is an additive, or whether it enters the oil as a wear metal. For this reason we have arranged this section based on *origin-specific elements*, including contamination from the environment, refining and storage, *additives* in fresh oils and greases, and *wear metals* in used oils. Table 10-80 provides an overview on the typical origin of elements in crude oils and

Table 10-80. Origin and approximate concentration range of important elements in crude oil and petroleum products.

Product, cause	Elements	Concentration range (mg/kg)
Crude oil	Ni, V	200–2
– origin-specific	As, Cd, Co, Cr, Cu, Hg, Mn, Mo, Pb, Zn	1–0.001
– contamination	Na, Fe	200–2
Environment, refining, storage	Al, Cd, Cu, Fe, Pb, Si, Ti, Zn	20–0.001
Additives in fresh oils	Ba, Ca, Mg, Mo, Na, Pb, Zn	20 000–200
Soaps, thickeners in greases	Al, Ca, K, Li, Mg, Na, Si	20 000–200
Wear metals in used oils	Ag, Al, Co, Cr, Cu, Fe, Mg, Mn, Ni, Pb, Sn, Ti, V, W, Zn	200–0.1

petroleum products, as well as their approximate concentration ranges [2988, 2989]. Solely the analysis of *catalysts* and *plastics* is treated in a separate section since this is markedly different from crude oil analysis, particularly with respect to sample preparation. Comprehensive monographs and introductory articles on oil analysis have been published by EBDON and FISHER [1660], KÄGLER [2987], MARSHALL [3905], and NADKARNI [4262, 4263].

10.5.1 Origin-specific Elements and Contamination in Crude Oil

In crude oil analysis we must make a clear distinction between the elements iron, nickel, sodium, and vanadium, which are present in the mg/kg range (i.e. in the 'upper trace range'), and elements such as As, Cd, Co, Cr, Cu, Hg, Mn, Mo, Pb, Sb, Se, Sn, and Zn, which occur in the µg/kg range (i.e. in concentrations that are around three orders of magnitude lower). *Nickel* and *vanadium* derive largely from the formation of crude oil. The concentrations and ratio of these elements provide information as to whether a crude oil comes from the Middle East, the North Sea, or Venezuela. Additionally, nickel is a serious catalyst poison and vanadium leads to corrosion problems. *Iron* frequently enters oil as a result of corrosion, i.e. through chemical reaction with the oil pipeline or the storage tank, while the major source of *sodium* (as NaCl) is mainly from sea-water. These major elements, as well as numerous trace elements, are transported proportionally through crude oil refining procedures so that they are present in intermediate refining products, even if in markedly reduced concentrations, in proportions specific to a given oil well. The products with the most similar element concentrations to the original crude oil are heavy fuel oils and bitumen.

Flame AAS is the technique of choice for the concentration range of around 0.1–1 mg/kg and above since a direct determination of the analyte in the hydrocarbon matrix is possible after corresponding dilution. Provided that no additives have been added, the concentrations of the various elements are so low that mutual influences are hardly to be expected. Nevertheless, direct determinations are not completely free of problems since the solvent, the compounds used for calibration, and the container material have a substantial influence.

For the determination of sodium, nickel, and vanadium in crude oils it is important to use a *solvent mixture* which contains a component to completely dissolve asphalts and resins, which only have limited solubility in non-polar solvents, and a water-soluble component. The latter is required since sodium is not genuinely dissolved in crude oil and fuel oil, but is present in the form of agglomerates (NaCl or NaCl/H_2O islands), i.e. as dispersive colloidal particles [2988]. Mixtures of 80% xylene or toluene and 20% ethanol or propan-1-ol have proven to be satisfactory for the determination of sodium, nickel, and vanadium in crude oils, heavy fuel oils, and bitumen products [799, 2986, 2988].

KÄGLER [2986, 2988] made a detailed report on the complex relationship between solvent, container material, and sodium compound on the stability of the measurement solution and on the resulting measurement errors. A solution of sodium cyclohexanebutyrate in toluene was relatively stable in a PE container, but lost around 70% of the sodium within ten hours in a glass container due to wall adsorption. A solution of the same reagent in decalin became cloudy within a short time and lost around 50% of the sodium within 20 hours due to sedimentation, regardless of the container material. In the most unfavorable case this behavior can lead to *additive measurement errors*, for example when the sodium concentration in the calibration solution (cyclohexanebutyrate) decreases during storage due to wall adsorption or sedimentation, while the sodium concentration in the test sample solution increases due to the dissolution of NaCl islands. This explains the good repeatability often observed for each individual laboratory during an interlaboratory trial, which is in contrast to the very unfavorable comparability obtained for the measurement results [2978]. This problem cannot be solved by the analyte addition technique either, since the analyte in the added form (cyclohexanebutyrate) behaves in a different manner to that in the test sample which is present as NaCl [2979] (refer also to Section 5.2.2.3).

For a direct element determination in the diluted crude oil sample complete agreement between the element compound in the test sample solution and the calibration solution would be ideal. This is not possible because both nickel and vanadium are present in a number of various compounds in crude oil [1913] and the composition is *a priori* not known in advance. A number of reports have been published on differing sensitivities of the various organic compounds of iron, nickel, or vanadium by F AAS [2781, 3416, 3417, 5220]. By far the best results have been obtained using porphyrin complexes [2988, 4494], which should always be used as the standards for direct determinations.

As an alternative the use of *emulsions* instead of genuine solutions has been proposed; these allow the use of aqueous solutions of inorganic salts as calibration standards [254, 2249, 2254]. The treatment of oils and greases with formic acid in an ultrasonic bath and the subsequent dilution with ethyl acetate [5972] leads much more to the formation of an emulsion rather than a solution. This procedure is particularly elegant when the emulsion is produced on-line in a flow system under the addition of a surfactant in an ultrasonic bath and then immediately analyzed. In this case the stability of the emulsion does not play a role. Nevertheless even for this technique the analyte must be present in a suitable chemical form in the test sample so that the same sensitivity can be attained for test sample and calibration solutions. A prerequisite for this is that the analyte does not have a metal-carbon bond in the sample [2254]. We shall also discuss the analysis of emulsions in more detail in connection with additives (see Section 10.5.2).

The hydraulic high pressure nebulizer developed by BERNDT *et al.* [600] opens up completely new possibilities (refer to Section 4.1.3.4). Even undiluted oils can be directly nebulized, leading to a significant improvement in the sensitivity. The authors reported the determination of numerous trace elements in both fresh and used lubricating oils.

The most reliable procedure for the determination of nickel, sodium, and vanadium in crude oil and petroleum products is *ashing* of the sample and analysis of the residue after taking up in hydrochloric acid [1495, 2989, 4670]. The advantage of determining the 'true' concentrations overweighs the disadvantage of the long sample preparation time if it cannot be guaranteed for the direct analysis of the diluted sample that the 'correct' organic element standard is used to generate the calibration curve. Nevertheless care must be taken not to select an ashing temperature that is too high ($\leq 550\ °C$) to prevent losses of sodium [2988]. Next to dry ashing by combustion, wet ashing with sulfuric acid

[5311, 5949], ammonium nitrate and sulfuric acid [4981], or p-xylenesulfonic acid [5950] have been proposed. We shall discuss suitable digestion procedures together with the determination of trace elements.

An alternative to digestion, at least for the determination of nickel and vanadium, is GF AAS since the problem of differing sensitivity for various compounds does not appear to exist. The results from a simple dilution with xylene and ashing are practically identical [562]. Crude oil and high boiling fractions can be so strongly diluted that the hydrocarbon matrix no longer causes interferences [2172], while the sensitivity of GF AAS is of great advantage for the low boiling fractions with very low concentrations of nickel and vanadium [898, 1245]. BETTINELLI and TITTARELLI [621] described an interlaboratory trial in which F AAS and GF AAS were compared with other common procedures for the determination of nickel and vanadium in crude oil samples.

The *trace elements* are mostly present in crude oil in the concentration range 1–10 µg/kg and below; in low boiling oil fractions their concentrations can be orders of magnitude lower. They occur in various bonding forms as metalloporphyrins (Ni, V), other metal complexes (Co, Cr, Fe, Ni, V), organometallic compounds (Hg, As, Sb), salts of carboxylic acids of the polar functional groups of crude oil resins (Ge, Mo, Zn), and as colloidally distributed minerals (Al, Cd, Na, Si) [5985]. Preconcentration steps (ashing, extraction) are frequently required for their determination. Direct determinations are only possible by GF AAS; HG AAS and CV AAS mostly have the necessary sensitivity but require a digestion to mineralize the analyte.

The *mini-ash digestion procedure* is particularly suitable for the trace determination in petroleum products; this procedure was developed by ROBBINS and WALKER [4889] originally for the determination of cadmium and has subsequently been improved [2983, 2984]. The procedure is based on the maxim: *as much reagent as necessary* (to guarantee reliable ashing without element losses), *but as little reagent as possible* (to keep introduction of the analyte via the reagent to well below the limit of quantitation). Thus, for example, 10–20 drops of concentrated sulfuric acid are added to 0.5–2 g of sample in a platinum crucible and very carefully heated on a hot-plate. Initially the major source of heat to burn down the sample is provided by an infrared lamp; later the residual sulfuric acid is fumed off by slowly increasing the heating. Finally the crucible is heated to 550 °C to completely ash the sample and, after cooling, the residue is taken up in hydrochloric acid [2988].

Table 10-81 provides an overview on selected publications on the determination of trace elements in petroleum products. Earlier papers are cited in a review article on F AAS by GUARDIA and SALVADOR [2250], which discusses the literature up to 1984. Due to their toxicity the elements arsenic, cadmium, lead, and mercury are of especial significance since during combustion they largely reach the environment. The most sensitive technique for the determination of mercury is CV AAS and it is regularly used for the analysis of crude oil [2477, 3156, 5779, 5931, 6354] and natural gas [1978, 3382, 3383]. Nevertheless, the digestion essential for this technique can cause difficulties [3156], so that attempts have repeatedly been made to determine this element by GF AAS [111, 2477, 3338, 3493]. This situation is similar for HG AAS; it offers high sensitivity [4618] but requires a prior complete digestion, which is not always without problems [1048, 4288, 6133], so that GF AAS is also frequently preferred [1735, 1789].

Table 10-81. Selected publications on the determination of trace elements in petroleum products.

Analyte	Sample	Technique	References
Al	fuel oil	Standard method IP 363-83, dry ashing, F AAS	[2775]
Al, Si	fuel oil	Standard method IP 377-88, dry ashing, F AAS	[2776]
As, Se	petroleum products	ashing in an oxygen bomb, HG AAS	[4288]
Ca, Cu, Fe, Mg, Na, Ni, Zn	crude oil	F AAS	[5950]
Cd	crude oil, fuel oil, gasoline	GF AAS, STPF conditions, ZBC	[4294]
Cd	petroleum products	'mini-ash' procedure with H_2SO_4 and IR lamp, GF AAS	[2983, 2984]
Cd	petroleum products	oxygen ashing in the graphite tube, STPF	[4292]
Cd, Na	petroleum products	various, sources of error	[2985, 2986]
Co, Cu, Fe, Mo, Ni, V	crude oil	dilution with organic solvents, GF AAS	[4056]
Cu, Fe, Na, Ni, V	heavy oil fractions	various sample preparation procedures for F AAS	[6514]
Cu, Fe, Ni, Pb, V	petroleum fractions	GF AAS	[2172]
Fe, Mg, Na, Ni, V	crude oil	dry ashing, F AAS	[4670]
Fe, Ni	petroleum products	emulsion technique, F AAS	[5972]
Fe, Ni, V	FCC fractions (naphtha)	direct determination by GF AAS	[5360]
Hg	crude oil	amalgam technique	[6354]
Na, Ni, V	crude oil, fuel oil	dilution with organic solvents, F AAS, Standard methods IP 288/74 and BS-2000/288	[799]
Ni	petroleum products	various	[898, 2989]
Ni, V	fuel oil	various	[621]
Sb	crude oil	hydride generation following digestion and F AAS	[3913]
V	petroleum products	dry ashing with ashing aid, F AAS	[1301]

The major advantage of GF AAS is that numerous trace elements can be determined directly in petroleum products when the STPF concept and Zeeman-effect BC are applied [4294] and the sample is ashed in oxygen in the graphite tube [4292]; this leads to lower blank values and thus better limits of quantitation. The determination of lead in 'unleaded' gasoline is also best performed by GF AAS; the addition of iodine to stabilize volatile lead compounds has proven to be advantageous [210, 1748, 2988].

From the refining of crude oil, from cracking processes for example, very fine grain particles of the catalyst support material in the form of *cat-fines* [2988, 4881] can pass via the residual oils proportionally into the heavy bunker C oils used in shipping. These are *particles of aluminosilicates* in the magnitude of 1 μm, which due to their great hardness lead to abrasion in ships' diesel engines. In principle it should be possible to determine the aluminium directly from slurries, but nothing has been published on this to date. A digestion is usual in which either aluminium alone or both aluminium and silicon are determined. In the first case, following the burn down of the oil matrix, the test sample is fumed off with hydrofluoric acid and then fused with potassium hydrogen sulfate [801, 1496, 2988]. In the second case ashing is followed by a fusion with sodium carbonate-sodium tetraborate [1497, 2988].

Next to the determination of the total analyte concentration in petroleum products, a number of analysts have investigated *speciation analysis* using AAS as the detector. The coupling of HPLC with GF AAS was applied almost exclusively and a number of various vanadyl [1912–1914], nickel [1913, 1914], arsenic [1914, 2797], and phosphorus [5855] compounds were determined, and also indirectly various forms of sulfur [2599].

10.5.2 Additives to Fuels, Lubricating Oils, and Greases

One of the most well-known applications of additives for quality improvement of crude oil products is the *addition of lead alkyls to gasoline*; nevertheless, a practice that is nowadays severely restricted. As early as 1961 ROBINSON [4893] reported the determination of lead in gasoline by AAS and found it to be fast, free of interferences, and selective; he diluted the gasoline with isooctane to obtain the most favorable measurement range. Later, a number of authors observed differences in the absorbance of tetraethyllead and tetramethyl-lead; the solvent used, the length of the aspiration capillary, and the aspiration time played a role [352, 4207, 5890]. These differences can be eliminated by adding iodine, so that lead can be determined by F AAS, either directly after dilution with IBMK [2988, 3037] or after extraction [999, 1987, 2651]. Additionally, the analysis of emulsions against aqueous lead standards [1063, 1244], the determination of lead by GF AAS [3039], and the generation of the hydride directly from the gasoline sample diluted with DMF [328] have been described.

KOLB et al. [3224] in 1966 were the first to describe the coupling of GC with F AAS for the separation and element-specific detection of various lead alkyls. A number of working groups later applied a continuously heated graphite furnace as the GC detector mainly because of the better sensitivity [989, 1315, 4898, 5222]. In more recent work these have been replaced by heated quartz tubes [357, 905, 2403], which are more sensitive, but essentially much simpler (refer to Section 7.2.1). The combinations of HPLC-F AAS [731] and HPLC-GF AAS [2949] have also been described, as likewise a non-chromatographic procedure based on FI that operates with emulsions and exploits the differing behavior of tetraethyl-lead and tetramethyl-lead in the absence of iodine [716].

MOORE *et al.* [4182] determined copper and nickel, and MOSTYN and CUNNINGHAM [4207] determined zinc in gasoline by aspirating it undiluted. BARTELS and WILSON [399] investigated jet aviation fuel for manganese, which is added as methyl-cyclopentadienylmanganese-tricarbonyl to eliminate the typical vapor trails. If the fuel is aspirated without dilution, AAS can be used directly for the concentration range 0.02–0.3 weight percent of the additive. Higher concentrations can be determined after dilution. DE LA GUARDIA and coworkers investigated various ways of introducing the sample for the determination of manganese in gasoline by F AAS, either in gaseous form after prior vaporization [2251] or as an emulsion [2248].

The additives mixed with various *lubricating agents* to improve their properties are listed in Table 10-80 along with their concentration ranges. The most common additive elements for lubricating oils are barium, calcium, magnesium, and zinc, while greases contain in addition to the additive elements soaps of lithium, sodium, and calcium as thickeners. Since these alloyed lubricating oils are mostly *homogenous solutions*, their analysis is much more simple than that of crude oils (refer to Section 10.5.1) and used lubricating oils (refer to Section 10.5.3). Due to the high concentrations in which the elements are added, F AAS should be used for their determination, except for certain special cases. In addition, a *direct procedure* is preferred, not only because of the rapid determination, but also because the formation of difficultly soluble compounds such as calcium phosphate or barium sulfate during the ashing of highly alloyed lubricating oils can lead to erroneous measurements [2988]. To avoid or minimize interferences, the nitrous oxide-acetylene flame is to be preferred, except for zinc.

Table 10-82 provides an overview of selected publications on the determination of additive elements in lubricating agents by F AAS. The oil is usually diluted with a suitable solvent such as cyclohexane [2648], toluene [2988], isooctane [637], xylene [4624], or a mixture of solvents [2647, 2648, 2649]. *Oil-soluble metal salts*, such as cyclohex-anebutyrates, are used to establish the calibration curve. Errors due to varying sensitivities can nevertheless occur when different organometallic compounds are present in the test sample and calibration solutions [3653, 5220]. Proposals to eliminate these influences include among others chemical pretreatment with iodine [3038] and the use of *solvent mixtures* with HCl or SrCl$_2$ as the buffer [2648]. When a mixture of 70% toluene and 30% acetic acid (V/V) is used, solutions of calcium chloride, barium chloride, and zinc acetate in dimethylsulfoxide, for example, can be used to prepare calibration solutions instead of organometallic standards [2297]. HON *et al.* [2654] showed that aqueous, inorganic calibration solutions can be used for the direct determination of a number of elements in lubricating oil when isobutanoic acid is used for dilution instead of xylene. In addition the measurements were much more stable under these conditions.

HOPP [2662] developed a procedure in 1974 for the determination of barium, calcium, magnesium, and zinc in lubricating oil in which the test sample is not diluted with an organic solvent but is *emulsified in water* under the addition of surfactants. This procedure was used successfully by a number of working groups for the analysis of oil by F AAS [667, 715, 2523, 4681]. DE LA GUARDIA and SALVADOR [2254] have made a detailed study on sample introduction procedures and especially on the possibilities of automation using FI techniques.

Table 10-82. Determination of additive elements in lubricating oils and greases by F AAS.

Analyte	Solvent	Standards	References
Ba	mixture: cyclohexanone–1-butanol–ethanol–HCl–water	aqueous standards	[2649]
Ba	dry ashing at 550 °C, standard method IP 110/82	aqueous standards	[2779]
Ba, Ca, Mg, Zn	turpentine substitute, standard method IP 308/85	–	[2780]
Ba, Ca, Zn	petroleum or xylene–ethanol	oil-soluble compounds in 2-ethylhexanoic acid	[1498]
Ba, Ca, Zn	xylene	oil-soluble compounds	[4624]
Ba, Ca, Zn	isooctane	oil-soluble compounds	[637]
Ba, Ca, Zn	mixture: toluene–acetic acid	inorganic compounds in DMSO	[2297]
Ba, Ca, Cu, Fe, Zn	isobutanoic acid	aqueous standards	[2654]
Ca	Nemol–water (emulsion)	aqueous standards	[715]
Ca	dry ashing at 550 °C, standard method IP 111/82	aqueous standards	[2778]
Ca	mixture: toluene–acetic acid	inorganic compounds in DMSO	[1535]
Ca, Zn	mixture: cyclohexane–butanol	aqueous standards	[2648]
Ca, Mg, Zn	mixture: toluene–acetic acid	aqueous standards	[6365]
Ca, Mg, Zn	dry ashing	aqueous standards	[5951]
Fe	IBMK	oil-soluble compounds	[5881]
Li	toluene–methanol	oil-soluble compounds	[4727]
Li, Na, Zn	mixture: acetic anhydride–petroleum	inorganic compounds in acetic anhydride–petroleum	[2874]
Mo	water (emulsion)	aqueous standards	[667]
Mo, Zn	n-heptane	oil-soluble compounds	[4207]
Pb	water (emulsion)	aqueous standards	[2523]
Pb	2-ethylhexanoic acid–IBMK	oil-soluble compounds	[5703]
Sb	2-ethylhexanoic acid–IBMK	oil-soluble compounds	[5702]
Zn	n-heptane	oil-soluble compounds	[4207]
Zn	dry ashing at 550 °C, Standard Method IP 117/82	aqueous standards	[2777]

We have already mentioned that interferences during the determination of the alkaline-earth elements are best avoided by using the nitrous oxide-acetylene flame [637, 1498, 4624, 5413]. The addition of a 'petroleum-soluble' alkali salt, such as the sodium salt of petroleum sulfonic acid, is essential to suppress ionization [637, 1498]. Instead of this, DITS [1535] employed potassium hydroxide in 2-ethylhexanoic acid as the ionization buffer. KÄGLER [2988] reported an interference on the determination of barium by aluminium, which is present as the stearate in package additives. This interference could only be eliminated by the analyte addition technique.

As we mentioned at the beginning of this section, GF AAS is only applied in exceptional cases for the determination of additive elements in lubricating oils [1049, 2650, 3905, 5722]. One such case is the determination of phosphorus, which does not exhibit sufficient sensitivity by F AAS [5013, 5855]. The determination of iron has also been reported [5881]; better results are obtained by GF AAS than by F AAS. RENSHAW [4852]also determined barium by GF AAS.

Flame AAS is the technique of choice for the determination of calcium, lithium, and sodium in greases. KÄGLER [2978, 2979, 2988] described a 'quasi direct' method in which the grease is initially emulsified in cyclohexane by intense stirring and then the analyte is extracted in hydrochloric acid by forceful shaking. The determination can be performed directly on the aqueous layer after appropriate dilution [1510]. This procedure is simpler and more reliable than digestion procedures or dilution with organic solvents [4388, 4727]. Nevertheless an ashing step at 500 °C in a quartz crucible appears to be indispensable for the determination of lead [1511] and molybdenum [4207] in greases.

10.5.3   Wear Metals in Lubricating Oils

The determination of elements in used oils is largely performed for the following reasons:

i)   To indicate contamination of the oil.
ii)   To determine whether the oil can still be used.
iii) To monitor the operating condition of an engine or gears.

Oil as a source of information is the keyword when we require early warning of faults and damage in operating plant that is lubricated by oil. Figure 10-5 depicts schematically the course of damage development with the three areas *normal* (normal operating condition), *early warning*, and *damage*. Early warning is essential to effectively restrict damage, so that the sensitivity of the analytical technique applied plays a major role. An added difficulty for the analysis of used oils is that the wear metals must be determined in the presence of the additive elements (refer to Table 10-80), which are present in 1000- to 10 000-fold higher concentrations [2988].

Sampling is of major significance for the analysis of used oils since it must often be performed in very dirty surroundings. Under no circumstances should sampling implements and sample containers come into contact with dirty machine parts. In addition, sampling should be performed shortly after the machine is turned off while the oil is still

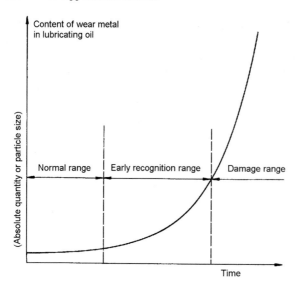

Figure 10-5. Course of damage development during the analysis of wear metals in lubricating oils [2988].

warm to prevent the sedimentation of particles as much as possible. If sampling is incorrect, the most sensitive element determination is useless, even if the repeatability is good.

The *comparability* of spectrometric techniques with respect to the results was investigated at an early stage of used oil analysis [971, 1693, 2980, 2981, 2997]; with appropriate sample preparation, AAS exhibited no systematic errors. In general the susceptibility of AAS to interferences due to matrix effects is markedly lower than for other techniques [2988]. The disadvantage of lower sample throughput in comparison to multielement techniques when a large number of elements is to be determined can be eliminated by restricting the determinations to a limited number of elements that provide clear information for a judgment of the operating condition. For example, the chromium concentration provides information on the condition of the piston rings of internal combustion engines; aluminium is a constituent of carburetor and piston materials; lead, silver and copper, and their concentration ratio, are indicators for the condition of the bearings; and general indications on the wear performance of machine parts can be derived from the concentrations of iron and chromium [2851, 2988].

In the earlier publications on the analysis of used lubricating oils the oil was mostly diluted with an organic solvent such as IBMK and *aspirated directly*; as with crude oil or fresh lubricating oil, calibration was performed against oil-soluble standards [25, 971, 4035, 5402, 5531]. Detailed investigations by BARTELS and co-workers [398, 3290, 3291], EISENTRAUT and co-workers [1693, 5028], and KÄGLER and JANTZEN [2982] however showed that at the best only particles < 1 μm could be completely analyzed by this technique. As can clearly be seen in Figure 10-6, the recovery decreases drastically with increasing *particle size*, so that we can no longer speak of a quantitative analysis. Complete ashing of the oil sample followed by acid treatment of the residue leads to the correct results, regardless of the particle size [25, 381, 4255], but is work intensive and time consuming.

KRISS and BARTELS [3290, 3291] developed a procedure in which the oil is diluted with IBMK and treated with a mixture of methanol, water and hydrochloric acid. The recovery after 1 h standing was 100% for particles of 1–2 μm, while the standing time was 16 h for particles of 3 μm. EISENTRAUT and co-workers [881, 1693, 5026] developed this procedure of acid treatment further so that nowadays it can be considered as mature [3051]; nevertheless the procedure never came into general use in used oil analysis. Table 10-83 presents a number of selected publications on this and derived procedures; the application of ultrasonics reduces the analysis time substantially [881, 1241].

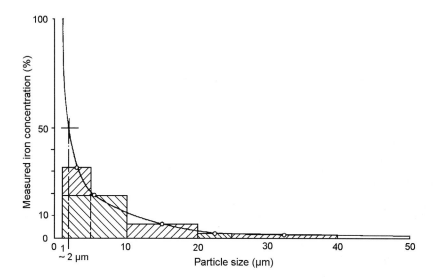

Figure 10-6. Dependence of the iron concentration determined by F AAS in gear oils on the particle size (from [2988]).

Table 10-83. Determination of wear metals in used oils by F AAS after acid treatment.

Analyte	Acid mixture	Treatment	References
Fe	HCl–H$_2$O	stand for 1–16 h	[3290]
Ti	HCl–HF	shake for 10 s	[5026]
Cr, Cu, Fe, Mn, Mo, V	HF–HCl–HNO$_3$	shake for 15 min	[5929]
Cr, Cu, Fe, Mn, Zn	HCl–H$_2$O$_2$	15–20 min ultrasonic bath	[369]
Al, Cu, Fe, Mg, Mo, Ni, Sn, Tl	HF–HCl–HNO$_3$	65 °C, ultrasonic bath	[881]
Al, Cr, Cu, Fe, Pb	HCl–HNO$_3$	60 °C, ultrasonic bath	[1241]
Cr, Cu, Fe, Pb	HCl–HNO$_3$	20 min hot-plate	[2521]
Mo	HF–HNO$_3$	shake for 2 min	[5027]
Pb	HCl–Aliquat 336	shake	[4522]
Fe	HF–HNO$_3$	shake for 5 min	[5068]

In contrast to F AAS, wear particles of up to about 20 μm can be quantitatively analyzed by GF AAS using platform atomization and the STPF concept [1693, 2988, 5029], so that digestion or dissolution of the wear particles is mostly not required for this technique. In addition, GF AAS is the technique of choice for the trace range, i.e. for the early recognition of wear metals which are present in alloys in relatively low concentrations but which can serve as typical indicators. SABA et al. [5029] determined 12 elements in used aviation lubricants by GF AAS. Due to the high performance of this technique a special multielement spectrometer was built for this application [4371].

10.5.4 Rubber and Plastics

For the analysis of polymers we must make a distinction between procedures in which the plastic material is dissolved in an organic solvent and those in which an oxidative wet or dry digestion is applied. When GF AAS is employed we have the direct analysis of the solid as an additional variant, which has found increasing use in recent times. Selected publications on the analysis of polymers after *dissolution in organic solvents* are compiled in Table 10-84. N,N-dimethylformamide (DMF) has proven to be a very universal solvent since it is miscible with both other organic solvents and water without the dissolved polymers from precipitating [4466]. For the subsequent analysis by GF AAS it is mostly sufficient to dilute the solution with acetone [490, 491], while for HG AAS the addition of hydrochloric acid [327] or bromine [4772] is required to separate the analyte from the organic bond and to mineralize it.

Selected publications on the analysis of polymers after an *oxidative wet digestion* are compiled in Table 10-85. The preferred digestion acid is quite clearly sulfuric acid, to which hydrogen peroxide is added drop by drop at the boiling temperature. After cooling the digestion solution is frequently neutralized with ammonia solution under the addition of EDTA (to prevent the formation of hydroxides) [485, 486, 488, 4589]. Sam- ples that contain a large amount of carbon must if necessary be further treated with

Table 10-84. Analysis of polymers after dissolution in an organic solvent.

Analyte	Polymer	Solvent *	Technique	References
Ca, Cu, Fe, K, Na, Si	PI	dioxin–cyclohexanone N-methylpyrrolidone	GF AAS	[3303]
Cd	PVC	DMSO–HNO$_3$	GF AAS	[2873]
Cd, Pb	PVC	DMF	F AAS	[489]
Cd, Pb	PVC	acetone and others	GF AAS	[491]
Sb	various	DMF–HCl	HG AAS	[327]
Sn	PVC	DMF	F AAS / GF AAS	[490]
Sn	PVC	DMF–Br$_2$	HG AAS	[4772]
Sn	PVC	THF	GF AAS	[5393]
Zn	rubber	ethylacetonate	F AAS	[2804]

* DMSO: dimethylsulfoxide; DMF: N,N-dimethylformamide; THF: tetrahydrofuran

Table 10-85. Analysis of polymers after oxidative wet digestion.

Analyte	Polymer	Digestion mixture	Technique	References
Al, Ca, Sb	PVC	$H_2SO_4-H_2O_2$ / Kjeldahl	F AAS	[485]
Al, Pb	PVC	$H_2SO_4-H_2O_2$ / Kjeldahl	F AAS	[488]
As, Ba, Hg	various	HNO_3, 8 h, 240 °C / pressure	various	[4948]
Cd	various	$HNO_3-H_2SO_4$ / pressure	F AAS	[863]
Cd, Pb, Sn, Zn	PVC	$H_2SO_4-H_2O_2$	F AAS	[4589]
Cd, Sb, Sn	PVC	$H_2SO_4-H_2O_2$ / Kjeldahl	F AAS	[487]
Mg, Pb	PVC	$H_2SO_4-H_2O_2$ / Kjeldahl	F AAS	[486]
Sb	PVC	$HCl-HNO_3-HClO_4-H_2SO_4$	HG AAS	[5090]
Sn	PVC	$H_2SO_4-H_2O_2$	GF AAS	[5393]
Sn	various	HNO_3-HCl / high pressure	GF AAS	[5230]

fuming nitric acid [487]. As well as complete digestion, investigations on leaching have been undertaken, for example tin from PVC [4750, 5393], or zinc from synthetic rubber, neoprene or Tygon [1784]. PUACZ [4728] described an interesting indirect procedure for the determination of elemental sulfur in rubber in which the test sample is shaken mechanically with DMF for 25 h at 20 °C. The extract is treated with a solution of 20 g/L sodium tetrahydroborate in a mixture of DMF, water and ammonium chloride and heated at 40 °C for 15 minutes. Residual sodium tetrahydroborate is then destroyed and a known quantity of mercury(II) chloride is added. The solution is stirred with tin(II) chloride in hydrochloric acid solution for 90 s and the released mercury is determined by CV AAS.

As an alternative to wet digestion, *dry ashing* at 250–300 °C followed by a *fusion* with potassium hydrogen sulfate has been investigated for the determination of aluminium and titanium in polypropylene [4453]. HEINRICHS *et al.* [2481] determined 37 elements in plastic waste after *ashing in an oxygen plasma*, followed by an acid digestion in an autoclave. VOLLRATH *et al.* [6102] compared ashing in oxygen with classical wet chemical digestions for cadmium and lead. NARASAKI applied oxygen ashing in a closed system for the subsequent determination of arsenic and selenium by HG AAS [4288] and mercury by CV AAS [4286]. TAKENAKA *et al.* [5757] determined Na, K, Ca, Cu, and Fe in PTFE by GF AAS after combustion in a stream of oxygen. With few exceptions dry ashing is required for the analysis of rubber, followed by dissolution of the residue in an acid mixture that is suitable for the analyte and the matrix. Selected publications on this subject are compiled in Table 10-86.

Table 10-87 provides an overview on a number of publications on the direct *solids analysis* of polymers by GF AAS. This technique is mainly applied because of its simplicity and speed, since it allows for example the direct investigation of pieces punched out of foils [1075, 3055] or textile fibers [6097, 6098]. A further reason for direct solids analysis is the high sensitivity, which can be of considerable advantage for the trace analysis of high purity materials [5115]. One investigation that is practically only possible by direct solids analysis is checking the homogeneity of polymers [2888, 4577]; this is of significance for reference materials and for quality control.

Table 10-86. Analysis of rubber by F AAS after dry ashing.

Analyte	Ashing temperature	Treatment of the ash	References
Cr	600 °C	solution in HNO_3	[6092]
Cu	550 °C	extraction with HCl, treatment with H_2SO_4–HF	[827]
Fe	550 °C	heated with HCl	[837]
Mn	550 °C	heated with HCl, treatment with H_2SO_4–HF	[826]
Pd	650 °C	treatment with formic acid, solution in *aqua regia*	[6091]
Zn	550 °C	solution in H_2SO_4–HNO_3	[844]

Table 10-87. Direct solids analysis of polymers by GF AAS.

Analyte	Polymer	References
Cd	various	[494, 3401, 4577, 4580, 4581, 5003]
Cd, Cr, Pb	foils, PVC	[1075]
Cd, Cu, Mn, Rb	various	[6098]
Co, Cr, Cu, Fe, Mn, Na, Ni	polycarbonate	[5115]
Cr, Cu, Ni, Pb	polyethylene	[2888]
Cu, Mn, Rb	polyester, nylon, perlon	[6097]
Mg	PVC	[493]
Pb	PVC	[492]

10.6 Metallurgical Samples

The analysis of metallurgical samples includes the following areas among others: iron, steel, ferrous and non-ferrous alloys, pure and ultrapure metals, refractory and noble metals, electrolytic baths, and slags. Analytically we make a distinction between the determination of the main and minor constituents (e.g. constituents of alloys) and trace determinations. Atomic absorption spectrometry is indispensable in particular for the latter since it offers not only high sensitivity, but also exceptional selectivity and low susceptibility to interferences in the presence of high matrix concentrations. It thus offers the best prerequisites for the accuracy of the analyses, the results of which are very frequently the basis of financial transactions. Various general review articles on the analysis of metallurgical samples have been published by BELCHER *et al.* [502, 1595], OHLS [4420] and KIPSCH [3119].

10.6.1 Iron, Steel and Ferrous Alloys

The application of AAS in the iron and steel laboratory essentially covers three areas: the control of raw materials, production control, and the analysis of final products. Production control usually demands a very rapid analysis which is normally performed almost exclusively by automatic multielement instruments, such as spark emission and X-ray spectrometers. The accuracy of the analyses must nevertheless be permanently checked against reference materials, and these must be analyzed for their trace element contents by independent techniques. The replacement of wet chemical procedures by AAS has led to a marked simplification of the analytical tasks of the steel laboratory [4424]. AAS can be applied directly for slow production processes or for when sample preparation can be performed rapidly. Examples here are the determination of magnesium in cast iron or of calcium and acid-soluble aluminium in steel. The major application of AAS is in the analysis of final products (quality control), which is almost always performed with simultaneous spectrometers, but which must be verified by an independent, standardized procedure. KOCH [3187, 4422] has published several review articles on the analysis of iron and steel.

10.6.1.1 Digestion Procedures

The determination of more than 25 elements in iron and steel by AAS has be reported [4424]; around 12 are determined on a routine basis [6030]. This means that for the multitude of samples presented to the steel laboratory there is a strong requirement for the simplest possible methods. This applies in particular to digestion procedures, since a method functions more rationally the greater is the number of elements that can be determined from a single weighing and a single sample solution. A very simple digestion procedure is with semiconcentrated hydrochloric acid to which concentrated nitric acid is added dropwise under boiling until the test sample is completely dissolved [622]. This digestion procedure can also be applied for the determination of acid-soluble aluminium in steel [722, 4615]. A digestion procedure utilizing hydrochloric acid, nitric acid, and perchloric acid is very universal for the majority of steels and for the determination of around 15 elements [4424, 5825]. If a residue of SiO_2 or Al_2O_3 remains, this must be fused with, for example, sodium tetraborate and added to the acidic digestion solution [4424]. It is possible to forgo a fusion for the determination of calcium in steels when hydrofluoric acid is added to the digestion mixture [2774]. If tungsten is also to be determined, the addition of phosphoric acid is recommended [501]. In recent years microwave-assisted digestions of steels and other ferrous alloys have been increasingly investigated [552, 616, 619, 719, 1598, 2169, 2339, 2596, 3964, 4176]. Metallic samples can absorb microwave energy directly and are thereby heated, thus additionally accelerating the reaction with the digestion acids. Ferrous alloys in particular tend to react violently, which can lead to damage of the magnetron and also possibly to ignition of the hydrogen released by reaction with the acid [3112]. For reasons of safety an open system should be used or the digestion vessel should only be sealed under the exclusion of oxygen when the first violent reaction has subsided [2169]. MATTHES [3963, 3964] has published information on the development of microwave-assisted digestions for metallurgy.

Digestion in hydrochloric acid under the addition of hydrofluoric acid, nitric acid, and sulfuric acid is suitable for oxide materials such as slags. After dilution the analysis is not noticeably different from that of steel. For the determination of higher tin contents in steel and also of ferrous alloys such as ferrosilicon, ferromanganese, or ferrochromium, a fusion with sodium or potassium carbonate, sodium carbonate-sodium peroxide, or sodium carbonate-sodium tetraborate is recommended [973, 4424]. Since the very high total salt concentration can cause difficulties, particularly for trace analyses, attempts have also been made to fume off ferrous alloys with hydrofluoric acid [974] or hydrofluoric acid-nitric acid [1412, 2973] and then to dissolve the residue in nitric acid-perchloric acid [1412].

10.6.1.2 Direct Determinations by F AAS

In F AAS interferences due to matrix constituents, particularly the main constituent iron, hardly play a role. Due to the complex and often relatively concentrated matrix, particularly for trace analysis, it is not usual to use pure aqueous calibration solutions in the steel laboratory but rather *reference materials* whose composition is similar to that of the test samples. These reference materials are digested in the same manner as the test samples so that if there is a matrix effect, it is the same for all measurement solutions [2733, 3168, 3238, 4424]. Table 10-88 presents a selection of publications, in particular standardized procedures, on the direct determination of a number of elements in iron, steel, and ferrous alloys.

Only few interferences are observed, particularly if the conditions stated in the literature are adhered to, and these can be attributed either to the use of an unsuitable flame, such as the air-acetylene flame for the determination of chromium and molybdenum [4525, 6493]—the interference disappears in the nitrous oxide-acetylene flame [5824]—or to known effects such as the enhancement of the titanium signal by alumin-

Table 10-88. Selected publications on the direct determination of trace elements in iron and steel by F AAS.

Analyte	Digestion *	Flame (oxidant)	References
Ag, Bi, Pb, Sb	$HCl–HNO_3$	air	[6575]
Al, As, Co, Pb, Sb, Sn, V	*aqua regia*	air or nitrous oxide	[421]
Al, Co, Cr, Cu, Mn, Mo, Ni, Pb, Sn, Ti, V	$HCl–HNO_3–HClO_4$	air or nitrous oxide	[5825]
Al	$HCl–HNO_3–H_2O_2$ / R: $H_2SO_4–HF$ / $H_3BO_3–K_2CO_3$	nitrous oxide	[824]
Ca	$HCl–HNO_3$	nitrous oxide	[820]
Cr	$HCl–HNO_3$ / R: $H_2SO_4–HF$ / $KHSO_4$	nitrous oxide	[821]
Cu	$HCl–HNO_3–HClO_4$	air	[813]
Mo	$HCl–H_3PO_4–HNO_3–HF–HClO_4$	nitrous oxide	[5550]
Ni	$HNO_3–HClO_4$	air	[814]
V	$HCl–HNO_3–HClO_4$	nitrous oxide (+Al)	[825]

* R = digestion of the residue

ium [4527], which can be controlled by adding aluminium to all measurement solutions, or the ionization of numerous elements in the nitrous oxide-acetylene flame, which can easily be eliminated by adding an ionization buffer (refer also to Section 8.1.2.5).

Flame AAS is suitable not only for the determination of trace elements but also for the accurate determination of alloy constituents in highly alloyed steels or ferrous alloys. In order to obtain the required accuracy, it is often advantageous to select a less sensitive analytical line, to use preanalyzed reference materials to prepare calibration solutions, and to calibrate using the bracketing technique [17]. When the parameters are appropriately optimized it is possible to perform the accurate determination of, for example, chromium and nickel in Cr-Ni-steel [17], the main constituents in Fe-Si and Fe-Si-Zr alloys [972], and also aluminium [802], iron, manganese [847, 974], and silicon [973] in ferrosilicon.

10.6.1.3 Separation and Preconcentration Procedures

Numerous extraction and preconcentration procedures to increase the sensitivity of F AAS for trace analyses have been described in the literature; a selection is compiled in Table 10-89. The major disadvantage of all these procedures is that they are mostly only applicable for one or a few analytes and are thus unsuitable for routine operation. Procedures that are less specific and thus allow the simultaneous preconcentration of a multitude of trace elements are more suitable, such as preconcentration on activated carbon [587, 589] or coprecipitation through partial precipitation of the matrix [1687, 2863]. All of these procedures, like those in Table 10-89, have the disadvantage that they are

Table 10-89. Selected publications on preconcentration procedures for trace element determinations in iron and steel by F AAS.

Analyte	Preconcentration procedure	References
Al	extraction of the Al acetylacetonate complex	[1573]
As	preconcentration as $AsCl_3$ in a cold trap	[5805]
As, Bi, Ga, In, Sb, Sn	extraction of the iodide in IBMK	[6022]
Co	preconcentration as the β-diketonate, vaporization	[1114]
Cr	preconcentration as the acetylacetonate, vaporization	[1109]
Cr	extraction with tribenzylamine in chloroform	[1572]
Mo	extraction with α-benzoinoxime in IBMK	[1100, 2723]
Mo	extraction with 5,5-methylene-disalicylhydroxamic acid	[5078]
Mo	extraction with caffeinic acid–Aliquat 336	[5257]
Ni	precipitation with glyoxime at pH 7.3	[3317]
Sc, Y	extraction with TOPO in IBMK	[2883]
Sn	extraction as the iodide in toluene	[1571]
Ti	extraction with caffeinic acid–Aliquat 336	[5258]
Ti	extraction with Amberlite in benzene solution	[4315]

time-consuming and require great care and experience. In the routine analytical laboratory they thus constantly conceal the risk of false measurements due to analyte losses or
contamination.

From the analytical point of view it is thus always much more sensible to *separate
the matrix* rather than to preconcentrate the analyte. Iron can be extracted from 7 mol/L
HCl as the chloro complex in IBMK [1236, 6028] or with 2-hexylpyridine in benzene
[4770] and thus be separated from numerous trace elements. The remaining solution can
be substantially reduced in volume so that the sensitivity is enhanced by around a power
of ten. A further improvement can be obtained if this extraction procedure is coupled
with the *injection technique* [421, 3855, 5219] or, better, with the *FI technique*. In this
case even smaller volumes and more highly concentrated solutions can be used [6148,
6511, 6575], so that even more elements can be determined with higher sensitivity. In
addition the FI technique offers an elegant possibility of automating the procedure,
which meets the requirements of routine analysis.

10.6.1.4 The Application of GF AAS

A further increase in sensitivity can be attained through the application of GF AAS; as to
be expected for this technique, earlier publications were largely devoted to interferences
and their elimination. In steel analysis it was fairly quickly established, however, that
these interferences could to the greater part be no longer observed when the STPF concept and Zeeman-effect BC were applied. Typical examples are the determination of
antimony, lead and tin in steel. Originally, strong interferences due to chloride [1957,
1959, 1960, 2053, 4820] and concomitant elements in steel had been reported, but these
could all be ascribed to unsuitable instruments or conditions, such as uncoated or poorly
coated graphite tubes, insufficient rates of heating, the ingress of oxygen, or the absence
of a modifier. These determinations are free of interferences when optimum instruments
and programs are used [1957, 1958, 2241, 3671, 4177, 5419].

Particularly in the far UV, for example in the determination of antimony [6426], arsenic [6422], and selenium [1892], spectral interferences due to the iron matrix and alloy
constituents of steels have been observed. These interferences are not due, however, to
direct line overlap but are mostly caused by the fact that iron exhibits a multitude of lines
and can absorb radiation from the continuum source within the spectral bandpass, thus
leading to overcorrection (refer to Sections 3.4.1 and 8.2.4.2). When Zeeman-effect BC
is applied, which is in principle always recommended in this spectral range and in the
presence of complex matrices, interferences are hardly observed [5994, 5995]. An exception is the interference by iron on the determination of bismuth at the 223.1 nm line,
which is best avoided by changing to the 306.8 nm line with about half the sensitivity
[5894].

A determination that only became possible under the application of STPF conditions
is that of phosphorus in iron and steel. On the basis of detailed theoretical and practical
investigations, PERSSON and FRECH [4619] showed that only under isothermal atomization conditions was the excited energy level from which absorption at the 213.6 nm line
emanates sufficiently populated to be able to perform an adequate determination of
phosphorus. The application of Zeeman-effect BC is also essential since iron causes a

spectral interference at this wavelength when a continuum source background corrector is used. When the STPF concept and Zeeman-effect BC are applied it is possible to employ the standard calibration technique for calibration and a limit of quantitation of 0.002% P in iron or steel can be attained. A similar application is the determination of acid-soluble aluminium using magnesium sulfate as the modifier [4268]. For the determination of boron in steel and ferrous alloys, magnesium and strontium [2916] and also nickel and zirconium [3556] have been proposed as modifiers. Nickel has also been used as a modifier for the determination of tellurium in cast iron [5769], and ammonium vanadate has been added for the determination of nickel in steel [2061].

Next to the direct determinations described above, various analyte-matrix separations prior to a determination by GF AAS have been described. For the determination of aluminium in iron, HIRAIDE *et al.* [2572] separated the matrix by electrolytic deposition on a mercury cathode. FRANKENBERGER *et al.* [1955] used the familiar extraction of iron as the chloro complex in IBMK for the determination of vanadium in steel. KAUKE *et al.* [3021] used the opposite approach and extracted arsenic from steel as the arsenomolybdic acid. ASHINO *et al.* [284, 287] described a reducing coprecipitation with palladium under the addition of ascorbic acid for the separation of Ge, Sb, Se, Sn, and Te from high purity iron. This procedure is particularly elegant since palladium as the carrier element simultaneously serves as an effective modifier for the determination by GF AAS.

10.6.1.5 The Application of HG AAS

The last-named elements along with the remaining hydride-forming elements are also determined preferentially by HG AAS. One of the reasons, as established by FLEMING and IDE [1926] in 1976, is that the interference of nickel in steel on the hydride-forming elements is far less pronounced than in other matrices or even in aqueous solutions. Detailed investigations later showed that this is due to the buffer action of iron [702, 704, 6212, 6222] (refer also to Section 8.3.3.2), so that for an interference-free determination it is merely necessary to ensure that an adequately high concentration of Fe(III) is present [1926, 6212]. An even greater freedom from interferences can be attained by adding a buffer or complexing agent such as sodium citrate [6161], 8-hydroxyquinoline [6420], or L-cysteine [6269]. The latter reagent can even replace potassium iodide which is usually added to reduce pentavalent antimony and arsenic to the trivalent oxidation state [6210].

Flow-injection HG AAS is particularly suited for routine applications since it combines high sensitivity with low sample consumption, greater freedom from interferences than batch HG AAS, and can easily be automated [6256, 6269]. BETTINELLI *et al.* [616, 620] compared FI-HG AAS with GF AAS for the determination of antimony, arsenic, bismuth, and selenium in steel and found that although both techniques gave good results, the former is better for routine applications. This derives in the first instance from the simple operation and the high sample throughput of the FI technique. Moreover FI-HG AAS exhibited better repeatability and provided limits of quantitation that were an order of magnitude better than those for GF AAS. Table 10-90 provides an overview of selected publications on the application of HG AAS for the analysis of iron and steel.

Table 10-90. Selected publications on the determination of hydride-forming elements in iron and steel by HG AAS.

Analyte	Remarks	System	References
As, Bi, Sb, Se, Sn, Te	acid concentration, Fe content	batch, QTA	[702, 704]
As, Bi, Sb, Se, Sn, Te	acid concentration	batch, QTA	[6212]
Sb	oxidation state	batch, QTA	[6210]
Se	sodium citrate buffer	batch, GF	[6161]
As, Bi, Sb	hardly any interferences	batch, QTA	[5994–5996]
As. Bi, Sb, Se	comparison with GF AAS	FI, QTA	[616, 620]
As, Sb	L-cysteine as buffer	FI, QTA	[6269]
Sn	acid and tetrahydroborate concentration	FI, QTA	[4020, 5193]

10.6.1.6 Direct Analysis of Solid Samples

The direct analysis of solid samples in steel analysis gained a relatively important role even at an early stage and a number of specially designed atomizers were built for this purpose (refer to Section 1.9). We merely need to mention the high-frequency induction furnaces of LANGMYHR and THOMASSEN [3247] and HEADRIDGE and co-workers [206, 207, 324, 978, 2457–2459], as well as the furnace designed by LUNDBERG and FRECH [3661] in which the test sample (e.g. metal turnings) is cast into a preheated atomizer. WALSH and co-workers [2085, 2182, 5011] employed a completely different approach with their *cathodic sputtering chamber* (see Figure 1-52) in which atoms are vaporized from a polished metal surface by glow discharge and determined by AAS. In contrast to inductively-heated furnaces, cathodic sputtering received further interest [2186, 2360, 6490] and was commercially developed [1152, 2735, 4407], but nevertheless up to the present this technique has not come into general use.

The direct analysis of solid samples has also been investigated in conventional graphite furnaces, for example the determination of antimony [339, 3660], bismuth [339, 1963, 5746], copper, manganese [5747], lead [339, 1963, 3660, 3661, 5746, 5747], silver [339, 5746, 5747], and zinc [339, 5746] in steel. The difficulties experienced with handling the test samples and the principal problems of direct solids analysis [2589] presumably were the cause that this technique did not gain general acceptance. As has been shown in numerous other application areas, direct solids analysis only has a chance of being accepted for routine use when the test samples can be handled in the form of slurries. Since the test samples must be very finely powdered for this technique, there are obviously principal difficulties for metallurgical samples. Solely a system proposed by KHARLAMOV and KARYAKIN [3092] for the production of colloidally dispersed 'solutions' by spark erosion would appear to offer a possibility in this direction [525, 3734]; detailed investigations on this technique have nevertheless still to be performed.

10.6.2 Non-ferrous Metals and Alloys

While the analysis of high and ultra-high purity metals involves essentially the determination of trace impurities and demands the highest sensitivity, the analysis of alloys is as much concerned with the determination of the major and minor constituents. This places very high requirements on the accuracy, which can often only be met by the choice of alternate lines, optimization of instrument parameters, and suitable calibration procedures, such as the bracketing technique [17]. In the following section we shall treat the individual types of metals and alloys separately according to their major constituents since these determine not only the digestion procedure required, but also frequently the specific interferences that influence the analytical method. Review articles on the analysis of non-ferrous alloys and pure metals have been published by NORTH [4386] and GIJBELS [2107].

10.6.2.1 Light Metals and Alloys

Aluminium alloys are mostly dissolved in semiconcentrated hydrochloric acid and some hydrogen peroxide is added toward the end of the reaction [4386]. The addition of hydrogen peroxide can be omitted for the analysis of high-purity aluminium [857]. If silicon should be determined as well, a little hydrofluoric acid is added to the digestion solution [4386]. Aluminium alloys with a high silicon content require digestion in hydrofluoric acid, hydrochloric acid, and nitric acid; the digestion solution is then fumed off with sulfuric acid to remove the silicon [3325]. YUAN *et al.* [6495] developed an FI system in which aluminium alloys can be dissolved electrolytically on-line and analyzed by F AAS.

The most important alloying element in aluminium alloys is magnesium. Aluminium interferes with the determination of magnesium in the air-acetylene flame and the addition of 5 g/L La is necessary to eliminate the suppressive influence of 0.5 g/L Al [4526]. Strontium [4848] and a mixture of strontium and 8-hydroxyquinoline [6578] have been proposed as alternatives. The use of the nitrous oxide-acetylene flame is nevertheless more suitable [17, 4386] since aluminium does not interfere in this flame. ABELE *et al.* [17] obtained the best accuracy for the determination of magnesium at the secondary line at 202.5 nm in the nitrous oxide-acetylene flame without additives. A further advantage mentioned by these authors was that Cr, Cu, Mg, Mn, Fe, and Zn could be determined from the same solution without further dilution. For the direct F AAS analysis of aluminium alloys, the hot, reducing flame is of advantage not only for 'nitrous oxide elements' such as silicon [3044], but also for chromium [17], indium [576], and strontium [145]. The air-acetylene flame appears to be more advantageous merely for iron [17, 4848], copper [17, 4848, 6495], lithium [514], manganese and zinc [17].

Flame AAS is mostly not sufficiently sensitive for the determination of trace contents in aluminium alloys and particularly in high-purity metals. Next to special preconcentration procedures, such as those for bismuth [2802], cobalt [4141, 4142], indium [144], or phosphorus [3325], procedures in which numerous trace elements can be separated and preconcentrated simultaneously are of particular interest. These include solvent extraction [272, 6022] or chelating ion exchangers [930]. APDC is a very universal complexing agent that finds application not only in connection with solvent extraction. The pre-concentration of trace elements and separation from the matrix can also be performed by membrane filtration of the APDC complexes [584] or by sorption on C-18 [595]. The preconcentration procedure on activated carbon proposed by JACKWERTH et al. [2856] has also be used for this purpose [586]. HÖHN and JACKWERTH [2627] described trace preconcentration from *high-purity aluminium* by partial dissolution of the matrix in the presence of mercury as an interceptor agent for the traces. JACKWERTH and MITTELSTÄDT [2864] determined 13 elements in *cast zinc alloys* after coprecipitation with only 1% of the matrix with hexahydroazepin-1-carbodithioate. JANSSEN et al. [2889] determined cadmium in zinc and zinc alloys after extraction of the dithizone complex in carbon tetrachloride, while ASHINO and HIROKAWA [286] applied coprecipi-tation with iron(III) hydroxide.

Reports on the application of GF AAS and HG AAS for the analysis of aluminium alloys are noticeably sparse. YOSHIMOTO et al. [6488] used sulfuric acid as the modifier for the direct determination of manganese, and SHIJO et al. determined indium [5308], silver [5310], and copper [3059] in high-purity aluminium by GF AAS after solvent extraction and back extraction with nitric acid. ASHINO and TAKADA [285] determined selenium and tellurium in a number of metals and alloys after coprecipitation with palla-dium, and HIRAIDE et al. [2581] determined copper in aluminium after extraction with ammoniacal EDTA solution. The direct analysis of *solid samples* was investigated for the determination of lead [315], gallium [316], and copper [5751]; various interferences were reported. AZNÁREZ et al. determined antimony [329] and tin [326] in aluminium alloys after extraction by hydride generation from non-aqueous medium.

10.6.2.2 Lead, Tin, and Lead-Tin Alloys

The digestion of lead and tin alloys causes a number of difficulties, since among other things tin, when present in high concentrations, precipitates as metastannic acid from nitric acid solutions. Silver, on the other hand, precipitates readily as the chloride when hydrochloric acid is used. According to the present author's own investigations [6208] the only usable procedure to dissolve lead-tin alloys is the digestion proposed by HWANG and SANDONATO [2737] using fluoroboric acid-nitric acid. When using this procedure it is important to strictly maintain the mixing ratio of 3 parts 65% nitric acid, 2 parts 35% fluoroboric acid, and 5 parts deionized water, since otherwise digestion of the sample will not be complete or metastannic acid will precipitate upon dilution. The Australian standard procedure for the analysis of solders [5546] is based on a similar digestion pro-cedure. If tin is only present in traces, lead alloys can be dissolved directly in nitric acid.

Using this digestion procedure with fluoroboric acid-nitric acid, it is possible by se-lecting suitable analytical lines to determine the main constituents lead (at 368.4 nm) and tin (at 266.1 nm) and also the trace elements Ag, As, Bi, Cu, Ni, and Sb from the same solution of *soft solders* without dilution [6208]. The nitrous oxide-acetylene flame is essential for the determination of tin. The air-acetylene flame, on the other hand, is suit-able for the determination of the trace elements antimony [5546, 5547, 6208], bismuth

[5546, 6208], cadmium [5546], copper [2266, 5546, 6208], gold [2266], iron [5118, 5546], silver [5546, 6208], and zinc [5546].

The sensitivity of F AAS is nevertheless inadequate in some cases for trace analysis. A number of precipitation and coprecipitation procedures have thus been described to preconcentrate trace elements from lead and lead alloys. JACKWERTH and co-workers investigated the preconcentration of traces by partial precipitation of the matrix with sodium tetrahydroborate [1421, 2862] and by dissociation of the bromo complexes after dissolution in bromine-bromic acid [1420]. CHOW [1270] determined bismuth and silver in lead-tin solders after coprecipitation with lead sulfate, and also Cd, Cu, Fe, Ni, and Zn from the solution after filtration. KOJIMA and TAKAYANAGI [3218] described the extraction of silver from lead solutions, and BABU et al. [335] extracted antimony from zinc-lead concentrates.

Although numerous trace elements can be determined free of problems by GF AAS directly from the digestion solution [6208] or by direct solids analysis [5748, 5749], nevertheless in the majority of cases separation and preconcentration procedures have been applied. HIRAIDE et al. [2580, 2582] determined Cd, Co, Cu, Fe, Ni, and Pb by GF AAS after separating the major quantity of tin by precipitation. ZHANG et al. [6555] determined aluminium in high-purity tin directly with good accuracy. To determine impurities in *high-purity lead*, HIRAIDE et al. [2578] fumed off the nitric acid solution to dryness and extracted the traces from the residue with ethanol. FOX [1951] separated selenium and tellurium from lead alloys by reducing them to the element in the presence of arsenic as a collector.

Investigations have also been made on the determination of trace elements in lead alloys and lead-tin alloys by HG AAS, despite the interferences that must be expected from the matrix which also reacts with sodium tetrahydroborate. BERNDT et al. [583] succeeded in determining arsenic in lead and lead alloys free of interferences by increasing the hydrochloric acid concentration to 6 mol/L. INUI et al. [2794] determined antimony in solder alloys using 1,10-phenanthroline as the buffer and atomizing in a graphite tube heated to 2300 °C. MA [3761] determined selenium in lead-antimony alloys by GF AAS after trapping the hydride in a graphite tube heated to 600 °C.

10.6.2.3 Copper and Copper Alloys

High-purity copper dissolves readily and completely in semiconcentrated hydrochloric acid [828, 829]. With copper alloys there are similar problems as with lead-tin alloys, namely that tin precipitates as metastannic acid when nitric acid is used and lead precipitates as lead chloride when hydrochloric acid is employed. For this reason a similar digestion procedure as for lead-tin alloys is prescribed in the majority of standard methods, i.e. with boric acid-hydrofluoric acid-nitric acid [816, 822, 5544]. If the digestion solution is not allowed to stand for too long, a digestion in hydrochloric acid-nitric acid is also possible, but care must be taken when diluting that a high hydrochloric acid concentration is maintained [4386]. BEDROSSIAN [464] has drawn attention to the risk of selenium losses in the form of a volatile chloride when hydrochloric acid is used for digestion. More recently microwave-assisted digestion procedures have been described [719, 1547] which have the advantage of greater speed.

ABELE *et al.* [17] optimized the instrument parameters and the experimental conditions for the determination of the main constituents copper and zinc in *brass* by determining copper at 244.2 nm and zinc at 307.6 nm in the air-acetylene flame. The use of a less sensitive line provides higher accuracy than dilution or turning the burner head in the radiation beam. Nowadays most standard methods are so compounded that they are applicable for the entire concentration range expected in copper alloys, for example 0.001–66% Zn [816], 0.01–5% Pb [822], or 0.003–2% Cr [815].Trace elements such as cadmium [5544], gold [5367], selenium, tellurium [5366], and silver [5548] can be determined directly in copper alloys by F AAS. The determination of Ag, Cd, Co, Cr, Fe, Mn, Ni, and Zn in *high-purity cathode copper* by F AAS is described in two British standard methods [828, 829]; the copper concentration is matched in the calibration solutions.

When analyzing copper and copper alloys, care must be taken not to use too highly concentrated solutions since otherwise the risk exists that highly explosive copper acetylide will be formed in the spray chamber or the burner. This risk can be substantially reduced by using smaller volumes of measurement solution, such as is possible with the injection [829] or flow injection [6497] techniques. Next to the advantage of automation, the latter technique also offers additional safety since the system is rinsed with carrier solution between the measurement solutions, so that solutions of higher concentration can still be used.

To increase the sensitivity for trace elements even further, BURNS *et al.* [966, 968, 969] introduced an 'atom trap' which comprised a slotted quartz tube located in the flame. Since the sensitivity could only be enhanced by a factor of 1.5–3, separation and preconcentration of the analyte is preferred in the majority of publications. Table 10-91 provides an overview on a few of the proposed procedures for solvent extraction. In the majority of cases the procedures are very specific and selective. Since copper itself also forms very stable complexes which can be readily extracted over a wide pH range, under given circumstances it can be more advantageous to remove the main alloy constituent by extraction [3388] or by ion exchange [43] rather than the analyte.

Next to solvent extraction, sorbent extraction has also been described for aluminium [626] and indium [3035]. A further interesting possibility is the preconcentration of cobalt as a β-diketonate, dithiocarbamine, or 8-hydroxyquinoline complex and introducing the sample into the flame by vaporizing the chelate complex [1114]. Coprecipitations are also particularly interesting since they can be applied for numerous elements and allow high enrichment factors. Worth mentioning in this respect is coprecipitation with lanthanum hydroxide for the determination of As, Bi, Cr, Fe, Pb, Sb, Se, and Te [5198], a procedure that has found acceptance in various standard methods [832]. As early as 1976 JACKWERTH and WILLMER [2859] proposed a preconcentration procedure for Cd, Co, Fe, In, Mn, Ni, Pb, Tl, and Zn in copper by coprecipitation of the matrix with thioacetamide as the coprecipitating agent. FRIGGE and JACKWERTH [1989] then later succeeded in preconcentrating 17 elements by coprecipitation with hexamethyleneamine-hexamethylenedithiocarbamate. Precipitations and coprecipitations can also be performed by flotation, usually under the addition of a wetting agent, and have been applied for example for the separation of lead as the sulfide or chromate [2103].

Table 10-91. Selected publications on the preconcentration of trace elements from copper and copper alloys by solvent extraction for determination by F AAS.

Analyte	Extraction procedure	References
Ag	tribenzylaminesilver bromide	[1574]
	4-heptanone-oxime / chloroform	[453]
	tri-N-octylmethylammoniumsilver bromide	[5922]
	1,2-bis(ethylthio)ethane-picrate / chloroform	[3218]
Al	Al-acetylacetonate	[1573]
As, Bi, Ga, In, Sb, Sn	extraction of the iodides in IBMK	[6022]
As, P	dioctyltindinitrate / chloroform-butanol	[5012]
Au	tri-N-octylmethylammoniumgold chloride	[5923]
	HCl / IBMK	[275, 916]
Cr	tribenzylamine / chloroform	[1572]
Cu, Mn, Ni	tetramethylammoniumbromide / thenoyltrifluoroacetone / IBMK	[4979]
P	phosphoantimonylmolybdate / IBMK	[5101]
Pd	tri-N-octylmethylammoniumtetrabromopalladate	[5921]
Se	methylmethacrylate	[2027]
Sn	tri-N-octylmethylammoniumhexachlorostannate / extraction of the iodide in IBMK	[5925]
Te	tri-N-octylmethylammoniumtellurium bromide	[5924]

It is an advantage for the direct determination of trace elements in high-purity copper by GF AAS that copper acts as a modifier for numerous elements, even if not as strongly as palladium. It was thus possible to determine various elements, such as Ag, As, Bi, Cd, Co, Cr, Fe, Mn, Ni, Pb, Sb, Se, Sn, Te, and Zn, under STPF conditions without interferences but without the addition of a modifier [1547, 4087, 4575, 5244, 6486]. The determination of As, Bi, Pb, Sb, Se, Sn, and Te in high-purity copper by GF AAS has already been accepted into national standards [831]. For the determination of platinum group elements, BEL'SKII and NEBOL'SINA [510] volatilized the copper matrix as the chloride at 1200 °C prior to the determination. Although background correction with a continuum source appears adequate for a large number of elements [1547], the majority of authors nevertheless prefer Zeeman-effect BC for this application [4087, 4575, 5244].

In contrast to GF AAS, where copper exercises a positive influence as a modifier, with HG AAS it causes considerable interferences (refer to Section 8.3.3.2). Various ways have been used to eliminate or to get round this problem. BYE [995] determined selenium in copper after removing the copper electrolytically. WICKSTROM et al. [6311] determined selenium after precipitating copper as the hydroxide, and NARASAKI [4289] separated copper by ion exchange but had to add an interceptor reagent in order to determine arsenic and selenium free of interferences. OFFLEY et al. [4417] developed a more elegant technique for the determination of selenium in copper by using an FI system and continuously removing the copper on a column packed with a strongly acidic cation exchanger. A further method for the interference-free determination of As, Bi, Sb,

Se, Sn, and Te by HG AAS is coprecipitation with lanthanum hydroxide [830]; this technique has been accepted as a national standard. Much more simple than the separation of the analyte and the matrix (unless this is performed automatically on-line in an FI system) is the elimination of the interference by an interceptor substance. This is an area that LINDSJÖ [3532] investigated at an early stage. It was thus possible to determine bismuth free of interferences in copper by adding thiosemicarbazide [5752]. One of the most successful additives, at least for the determination of antimony and arsenic, is L-cysteine; concentrations of 0.5–1 g/L Cu do not cause interferences in its presence [6269].

10.6.2.4 Nickel, Chromium, Cobalt, Manganese and their Alloys

The use of nickel and cobalt alloys has increased markedly in recent years, especially in the aerospace industry where there is a high demand for *high temperature alloys* that retain their metallurgical properties even at high temperatures. Since these alloys are strongly corrosion resistant, an aggressive digestion procedure is required which nevertheless can depend very much on the individual compositions. Pure nickel dissolves in semiconcentrated nitric acid [817]. A digestion in hydrochloric acid-nitric acid is suitable for Ni-Fe-Co alloys [1082], followed by fuming off with perchloric acid when required to precipitate SiO_2 [818]. Nevertheless fuming off with hydrofluoric acid is frequently required following a digestion in hydrochloric acid-nitric acid to take high temperature alloys completely into solution [4386, 6200, 6201]. Microwave-assisted pressure digestions of nickel alloys for the subsequent determination of boron, chromium, iron and silicon [6526] and also arsenic [2354, 4858], thallium [5098] and other metals [619, 3963, 6524] have been described.

As early as 1970 a number of papers were published dealing with the high precision determination of major alloy constituents by F AAS. CARPER [1082] determined the three major constituents in NiFe17Co3 alloys with standard deviations of 0.21%, 0.28% and 1.5%, respectively, in excellent agreement with the much more complicated wet chemical procedures. For this procedure the acid concentration and the contents of the major alloy constituents were matched in all solutions. WELCHER and KRIEGE [6201] determined the major constituents in a high temperature cobalt alloy using a similar principle and also developed a procedure for the highly accurate determination of Al, Co, Cr, Fe, Mo, Nb, Ta, Ti, V, and W in high temperature nickel alloys [6200] with the frequent use of reference materials. Nowadays there are a number of standard procedures for the determination of Ag, Bi, Cd, Co, Cu, Fe, Mn, Pb, and Zn in nickel [817], and of cobalt [839], chromium [840], copper [841], iron [842], and manganese [843] in nickel alloys, and of cobalt in ferronickel [818]. The determination of lead [3503] and magnesium [205, 1636] in nickel alloys can also be performed by F AAS free of interferences and with high sensitivity.

Preconcentration procedures for F AAS have been described sporadically. BURKE [959] determined thallium in nickel alloys after extraction with TOPO in IBMK, and also antimony, bismuth, lead, and tin in high-purity nickel after coprecipitation with manganese(IV) oxide [957]. BERNDT et al. [587] employed preconcentration on activated carbon to determine Bi, Cd, Cu, In, and Tl in high purity chromium and manganese. A trace-matrix separation using ion exchangers has been described for Ag, Bi, Cd, Pb, and Zn in nickel alloys [3126] as well as for gold and palladium in manganese and nickel [1715], and for Al, Be, Ca, K, Li, Mg, Na, Ni, Ti [1800], Bi, Cd, Co, Cu, Fe, Ga, In, Pb, U, and Zn [1801] in manganese.

One reason for the relatively seldom application of preconcentration procedures for trace determinations in nickel and nickel alloys is without doubt that the use of GF AAS was described in the mid 1970s [6202] and soon thereafter was proposed as a standard technique for the determination of Bi, Pb, Se, Te, and Tl [3438]. At the same time the determination of these elements in complex nickel alloys by direct solids analysis and calibration against preanalyzed reference materials was also described [3892].

A large part of the success of these early methods, even without the application of STPF conditions, can definitely be ascribed to the fact that nickel is a very suitable modifier for a large number of elements. This is reflected in a number of recent papers on the determination of bismuth [724, 5900], chromium, copper, iron, manganese [3726], gallium [3183], indium [3184], lead [5900], tellurium [3183, 5899], and thallium [5899] where STPF conditions are employed but no modifier is added. The British standard method for Ag, As, Bi, Cd, Pb, Sb, Se, Sn, Te, and Tl in nickel by GF AAS is also based on this principle [819]. Nevertheless in recent years the number of publications has increased in which more attention is paid to a modifier, such as the addition of potassium iodide for the determination of arsenic [5898], the addition of nickel and zirconium for the determination of boron [29], or of ammonium dihydrogenphosphate for the determination of lead [4085] and silver [4086], as well as the addition of hafnium and tartaric acid for bismuth [4086]. For the determination of Bi, Pb, Se, Te, and Tl in high temperature alloys, REICHARDT [4839] employed the Pd-Mg modifier and worked with Zeeman-effect BC, which is essential for the determination of at least selenium to avoid spectral interferences by nickel and iron [1895]. Likewise a spectral interference by iron in the determination of gallium has been observed but which can be countered by using the line at 294.4 nm. GONG et al. [2166] used a Pd-Mg modifier for the determination of selenium and tellurium in nickel and nickel-iron alloys.

Next to the numerous direct procedures a number of preconcentration procedures have been described for GF AAS, for example a solvent extraction for the determination of cadmium [5765], the separation of cobalt by ion exchange [5788], or a coprecipitation with arsenic for the determination of selenium [285, 3330] and tellurium [285, 3329] in high temperature alloys. A further possibility to increase the sensitivity is direct solids analysis by GF AAS [2800], which has been investigated for antimony, silver, zinc [339], bismuth [339, 3670], and lead [339, 3661] in nickel alloys.

Like copper, nickel also causes substantial interferences in HG AAS, so that either the analyte must be separated from the matrix or an interceptor reagent must be added. Included in the latter are EDTA [1612] or 8-hydroxyquinoline [6420] for the determination of bismuth, citrate for the determination of selenium [994, 6161], or L-cysteine for the determination of antimony and arsenic [6269]. The addition of tellurium for the determination of selenium [3133] or of iron for the determination of arsenic [6211] or selenium [996] can also markedly suppress the interferences due to nickel. Nevertheless, even more effective is the separation of the nickel matrix by precipitation as the hydroxide, where selenium remains quantitatively in solution [6218, 6311], or by ion exchange [4289]. The separation of the matrix is particularly elegant when it is performed continuously and on-line in an FI system [4858, 5943]. In a comparison between GF AAS and FI-HG AAS for the determination of antimony, arsenic, bismuth, and selenium in nickel alloys, BETTINELLI et al. [616, 620] found that FI-HG AAS was easier to use for routine applications and exhibited better repeatability and in particular higher sensitivity by around an order of magnitude.

10.6.2.5 Hard and Refractory Metals

Included in this group are metals such as molybdenum, niobium, tantalum and tungsten; alloys such as WTi; and silicides such as $MoSi_2$, $TaSi_2$ and $TiSi_2$. ORTNER et al. [4481, 4484, 4485] have published a number of review articles on the analysis of hard and refractory metals. A mixture of hydrofluoric acid and nitric acid is mostly used to digest these materials [811, 5467]; the addition of ammonium citrate prevents tungsten from precipitating out [4510]. Occasionally there have been reports on digestion procedures using ammonium fluoride-sulfuric acid [4073] or perchloric acid [846], followed by an oxidation with potassium persulfate and hydrogen peroxide as well as the addition of ammonium citrate. MIGNARDI and MASI [4079] described an alkaline fusion for wolframite. Microwave-assisted digestion procedures have been described for tantalum [5025].

Although F AAS does not exhibit sufficient sensitivity for trace analysis in the ng/g range in high-purity metals, nevertheless due to its simplicity and the almost total absence of line overlap (which can lead to substantial difficulties for these materials in OES) it is very suitable for determinations in the upper µg/g through % range. A number of standard procedures are thus based on a direct determination by F AAS; it is merely necessary to add cesium chloride and ammonium fluoride as buffers to the solutions. Examples are the determination of calcium, magnesium, potassium and sodium in the 10–200 µg/g range in hard metals [808], the determination of cobalt, iron, manganese and nickel in the range up to $w = 2\%$ [809, 811], of molybdenum, titanium and vanadium in the range $w = 0.01–0.5\%$ [810] or $w = 0.5–2\%$ [811], or the determination of chromium in the range $w = 0.01–2\%$ [846]. Further applications include the determination of copper ($w = 2.5\%$), nickel ($w = 7.5\%$) and iron in tungsten alloys [4510], the major constituents chromium, silicon and tungsten in Cr-Si-W materials [4223], or of tantalum ($w > 0.01\%$) in ferroniobium [2882].

Since a relatively large mass (a typical weighing is 1 g/100 mL) of a refractory, high melting metal reaches the flame with these direct determinations, the use of the nitrous oxide-acetylene flame is recommended; it is also recommended for less refractory elements. Possible interferences caused by the formation of compounds, such as can be observed in the cooler and less reducing air-acetylene flame, are substantially reduced. The use of the injection technique or the FI technique is advantageous for the more highly concentrated solutions since the volume of measurement solution is smaller and the system is regularly rinsed, thus effectively preventing clogging of the burner [420].

The sensitivity of F AAS can be considerably enhanced when the determination is preceded by a preconcentration and separation step. STRELOW [5601] determined 15 trace elements in the µg/g to ng/g range in molybdenum after separating them by ion chromatography. Tantalum was determined in titanium alloys after extraction of the fluoride in IBMK [4476], copper and nickel in titanium [4243], and Cu, Fe, Mn, Ni, Pb, and Zn in pure tungsten [4074] following extraction with APDC in DIBK or IBMK.

The pioneering work done on the application of GF AAS for hard metal analysis was performed by ORTNER and co-workers [4480, 4481]; they initially saturated the graphite tube with a solution of sodium tungstate to reduce matrix influences and to increase the lifetime of the graphite tube. This type of tube pretreatment is nowadays no longer necessary due to the greatly improved PG coating of the tubes. The direct determinations of cobalt, lead and manganese in molybdenum and zirconium [2418], sodium and potassium in high-purity tantalum [2417], chromium in pure tungsten [4073], and cadmium in zirconium and zirconium alloys [5755] have, among others, been reported.

For ultratrace analysis in the ng/g range by GF AAS, separation and preconcentration of the trace elements often cannot be avoided [672, 4484]. BLÖDORN et al. [671] described a procedure within a quality control program for the determination of 20 trace elements in high-purity chromium, molybdenum and tungsten via precipitation and adsorption of the traces on cellulose collectors. KRIVAN and co-workers [1557, 3302] have nevertheless shown that the risk of contamination and thus the risk of systematic errors in refractory metals analysis is very high due to the frequently difficult digestion procedures. If the digestion is followed by a preconcentration step, next to the increased risk of contamination there is an additional risk due to analyte losses. Analytical procedures must therefore be performed with the greatest of care if true results are to be obtained. For this reason these authors preferred direct solids analysis by GF AAS rather than a digestion procedure [1557, 1558]. From investigations using radiotracers, DOČEKAL et al. [1556] established that at temperatures of 1900–2700 °C more than 98% of the alkali and alkaline-earth elements and also copper and zinc were released from high-purity molybdenum or molybdenum oxide. DOČEKAL and KRIVAN [1554] conducted chloroform or carbon tetrachloride through the graphite tube at high temperature to remove the molybdenum matrix more rapidly. FRIESE et al. [1985] determined Cu, Fe, Mn, Na, and Zn in high-purity tantalum by direct solids analysis using GF AAS against aqueous standards.

10.6.2.6 Ultrahigh-purity Metals

In this section we shall discuss the analysis of the alkali metals, arsenic, cadmium, gallium, indium, mercury, selenium, tellurium, thallium, and zinc. Due to the high purity of the metals being investigated, the application of F AAS always requires a separation and preconcentration step. Very specific extraction procedures have been described, for example for silver from selenium and zinc as the tri-N-octylmethylammonium-silver bromide complex [5922], for indium from zinc with 1-phenyl-3-methyl-4-benzyl-5-pyrazolone [79], or for thallium as the iodo complex from zinc and cadmium [4251]. On the other hand, systems suitable for the preconcentration of as many elements as possible have also been described, such as the extraction of 14 elements from high-purity bismuth with 2-(2-benzoxazoyl)-cyanoacetaldehyde and -malonaldehyde [3030]. Numerous elements can be separated from gram quantities of thallium using cation exchangers [5600]. JACKWERTH and MESSERSCHMIDT [2858] applied the technique of partial dissolution of the matrix to preconcentrate trace elements in high-purity gallium. LUO et al. [3676]

determined impurities in high-purity mercury after largely separating the matrix by sub-boiling distillation.

For the metals and metalloids under consideration, GF AAS has the advantage that the relatively volatile matrix can be separated *in situ* prior to atomization of the trace elements. Applying STPF conditions and Zeeman-effect BC, around 15 elements can thus be determined directly in high-purity arsenic [5246]. Similarly, arsenic, tellurium and thallium can be determined in high-purity cadmium, and arsenic and thallium in high-purity tellurium using the Pd-Mg modifier [483]. Numerous trace elements can also be determined directly in metallic sodium [2063, 2064, 3258] or lithium [2089], where required under the addition of a modifier such as nickel or magnesium nitrate, following *in situ* separation of the matrix. A number of preconcentration procedures have also been described for GF AAS, such as precipitation with lanthanum hydroxide for the determination of chromium, iron and nickel in lithium [538] or preconcentration of cobalt from zinc on immobilized 1-nitroso-2-naphthol [2579].

10.6.2.7 Noble Metal Analysis

Pure noble metals or noble metal alloys are usually dissolved in *aqua regia* or, depending on the composition, in nitric acid with a small addition of hydrochloric acid. Microwave assistance accelerates these digestions substantially [2564, 3980].

Direct analysis by F AAS is only possible for higher concentrations, such as the determination of gold in jewelry alloys [3980] or for the analysis of intermediate products during smelting [41, 42, 3014]. The mutual influences of the noble metals in the flame are problematical, but these can be controlled by suitable additions such as 5 g/L Cu and 5 g/L Cd [5232], 5–10 g/L Li [2456], or 10 g/L Cu and 10 g/L Na [2886]. Lanthanum or uranium also appear to be suitable buffers [2468, 2469]. Performing the determinations from cyanidic solution is an alternative since the majority of the interelemental effects observed in acidic solution do not occur in the presence of 20 g/L KCN [41, 42, 3014]. A mathematical correction of the interferences in conjunction with the injection technique has also been proposed [213].

Preconcentration procedures for trace elements are relatively widely applied in noble metal analysis, and a number of papers are cited in Table 10-92. For F AAS, next to preconcentration the elimination of interelemental interferences also plays a major role [212]. As well as the extraction of the chloro complexes [211, 212, 2832], ion exchange chromatography in particular has been applied for separation and preconcentration [762–765, 4915]. An interesting procedure for trace determinations in silver via anodic dissolution of the sample with simultaneous electrolytic removal of the matrix was described by TANAKA *et al.* [5764]. Graphite furnace AAS exhibits adequate sensitivity for the determination of trace elements in noble metals, so that separation of the matrix by extraction [214, 2832] or ion exchange [766, 767] plays the more important role. Since interelemental interferences can hardly be observed in GF AAS [40], particularly when STPF conditions are applied, a separation step can usually be omitted. HINDS [2563] found that for the determination of traces in fine silver the matrix itself acted as a modifier so that he merely added 1 g/L Ag to the calibration solutions. Similarly the determi-

Table 10-92. Preconcentration procedures for trace elements in noble metals for determination by F AAS and GF AAS.

Analyte	Matrix	Separation and preconcentration procedure	Technique	References
Al, Ca, Cr, Cu, Mg, Mn, Ni, Pb	Pt	extraction as the chloro complex	F AAS	[211]
Au, Fe, Mo, Sb, Sn	Pt	extraction as the chloro complex	F AAS	[212]
20 elements	Au	matrix extracted in ethanol / chloroform	F AAS, GF AAS	[2832]
Ir, Rh, Ru	Pt	matrix extracted in isopentanol / IBMK	GF AAS	[214]
Pd	Pt, Ir	glycine, anion exchanger	F AAS	[762]
Traces	Rh, Ir	cellulose / tributylphosphate–toluene	F AAS	[4915]
Au	Pd, Pt	cellulose exchanger / amines	GF AAS	[766, 767]
Fe, Cu, Pb, Bi	Ag	matrix precipitation with HCl, precipitate extracted with HNO_3	F AAS	[5764]
Pb, Cd	Ag	electrolytic preconcentration	F AAS	[5764]

nation of trace elements in platinum [2564] and of silicon in fine gold [2566] were largely free of interferences when the calibration solutions were matched to the matrix content. To avoid interferences in the determination of Ag, Cd, Co, Fe, Ni, Mn, and Pb in palladium and platinum, the concentration of the platinum metals in the solution should not exceed 2 g/L [273]. The determination of gold, palladium, and platinum in high-purity silver [2565] and silicon in fine gold [880, 2566, 2567] by direct solids analysis has also been described.

Except for the investigation of interferences (refer to Section 8.3.3.2), there is little in the literature on the application of HG AAS in noble metal analysis. Graphite furnace AAS appears to be clearly the better technique for the determination of the hydride-forming elements. For the determination of mercury, on the other hand, only CV AAS offers sufficient sensitivity. The separation of the analyte from the matrix as the tetra-bromo complex has been proposed for the determination of mercury in silver [6301], while the determination of mercury in unrefined gold can be performed directly by CV AAS [5690].

10.6.2.8 Electrolytic Baths

Back in 1964 SHAFTO [5256] and WHITTINGTON and WILLIS [6304] reported in detail on the determination of numerous trace elements in plating baths. Since these determinations are mostly relatively simple, there are very few applications described in the modern literature [1473]. Plating baths can frequently be analyzed directly by F AAS after suitable dilution. Only in limited cases, such as when the plating bath contains ethylene-diamine, is it considered necessary to perform a digestion with sulfuric acid prior to the determination [917], or even to separate the analytes by extraction [4892].

The application of FI techniques in particular for the analysis of electrolytic baths has been described in the recent literature. DAVEY [1425] reported the determination of zinc in electrolytic prewash solutions by F AAS. DEBRAH *et al.* [1451] precipitated silver on-line from silver plating baths and separated it in a rotating reactor to prevent the formation of silver acetylide in the spray chamber. MAIER and NEUBAUER [3808] employed an FI-HG AAS system to automatically monitor and control the bismuth concentration in acid baths used to clean turbine blades. Table 10-93 presents a number of selected publications on the analysis of plating baths by F AAS.

Table 10-93. Selected publications on the analysis of electrolytic baths by F AAS.

Analyte	Bath	References
Ag	various	[2770]
Ag, Au, Cd, Cu, Fe, Pb, Pd, Sn, Zn	Cr, Ni	[5813]
Ag, Co, Cu, Fe, Ni, Pb	Au	[3234]
Ag, Cu, Fe	Cr, Ni, Sn	[3268]
Ag, Cu, Pd	Pb-Sn	[797]
Cd, Cu	various	[917]
Co	Au	[3028, 3268]
Cd, Cr, Cu, Fe, Ni, Pb, Zn	various	[1437]
Cu, Fe, Pb, Zn	Ni	[5256]

10.7 Miscellaneous Industrial Products

10.7.1 The Analysis of Glass

In this section we shall discuss the analysis of various *industrial and technical glasses*, from the *raw materials for the manufacture of glass*, such as quartz sand [4623], lead oxide [5118], potassium fluoride and ammonium fluoride [116, 2872], hafnium fluoride [2870] and zirconium fluoride [3249, 4365], to the analysis of finished glasses. The goal of an analysis can be *quality control*, both with respect to the composition, i.e. the major and minor constituents, and the trace impurities, which can have a major influence on the properties in particular of optical glasses. The *composition of glasses* can provide important information on their origin, which can be of particular interest in archeology [6584] and forensic science [975] (we shall discuss these topics in more detail in Sections 10.8.1 and 10.8.2). Particularly for pharmaceutical glass containers, such as ampoules for injection solutions, the quantity *released from the glass into the solution* is of interest [2244, 4151]. The release of mercury from glass packaging into foodstuffs is also discussed [2246].

The majority of glasses are so homogeneous that they do not have to be ground very finely to obtain a representative test sample. The most commonly employed *digestion procedure* in glass analysis is fuming off with hydrofluoric acid-perchloric acid, with the possible addition of boric acid in case insoluble fluorides are formed. The residue is taken up in hydrochloric acid, or for the determination of lead in nitric acid [4622, 6363]. Using this acidic digestion procedure, titanium oxide, zirconium oxide and α-aluminium oxide remain in the residue and silicon is vaporized as the tetrafluoride. If it is necessary to determine these elements a *fusion* is required instead of, or parallel to, an acid digestion. Lithium metaborate or a mixture of sodium or potassium carbonate and sodium tetraborate is the most suitable fusion agent [935, 6363]. Additionally a large number of special digestion procedures have been investigated, such as fusion with sodium peroxide [365] or digestion of borosilicate glass in hydrofluoric acid alone [3100]; a number of microwave-assisted digestion procedures have also been described [19, 1316, 3113]. BEARY *et al.* [451] have published an overview on sample preparation with particular reference to GF AAS.

For the analysis of glass we make a distinction between the major glass-forming constituents silicon and boron, the alkali elements, which mainly lower the melting point, the stabilizers, which include the alkaline-earth elements, cadmium and zinc (chemical resistance), aluminium (resistance to thermal shock) and lead (refractive index), as well as colorants such as iron and manganese and various trace elements. The application of F AAS for the analysis of glass was thoroughly investigated in the mid 1960s [33, 4560, 4561]. It covers essentially the determination of the elements silicon [935, 4822], boron [204, 3100], the alkali elements [915, 2918, 4622, 6538] and the alkaline-earth elements [33, 915, 2918, 6538], as well as aluminium [6362], iron, cobalt, nickel, copper [1708] and lead [681, 4705]. To avoid interferences, an ionization buffer is required for the alkali elements and the addition of EDTA and lanthanum for the alkaline-earth elements [33]. The determination of copper in the range 0.01–200 g/kg [4561, 5658], iron in the range 1–200 g/kg, and zinc in the range 10–100 g/kg [4560, 4562] is free of interferences. GROSSMANN [2237] investigated the interferences on the determination of titanium in ceramic glass in the nitrous oxide-acetylene flame. KOREČKOVÁ and WEISS [3249] determined the accompanying elements in chemicals for glasses on the basis of zirconium tetrafluoride and SAULS *et al.* [5118] determined chromium and iron in lead and lead oxide used for the production of glass. The release of potassium, sodium [4151] and ten further elements [4704] from glass into injection solutions has also been investigated by F AAS.

The determination of trace elements in glass is definitely the domain of GF AAS. Solely BRÜGGERHOFF and JACKERTH [896] reported a multielement preconcentration procedure for trace elements in optical glasses for F AAS using coprecipitation with HMA-HMDTC. Trace analysis could nevertheless acquire a new dimension by the application of on-line preconcentration employing the FI technique, as has been described for example for cobalt in glass [5980]. Extraction procedures have also been described for the subsequent determination of trace elements by GF AAS [2017, 2018, 2245], but simple volatilization of the matrix appears to be adequate for the interference-free determination of trace elements even in high-purity glasses, such as light fibers, by GF AAS [338, 5503].

Once the major constituent silicon has been almost completely removed by fuming off the glass with hydrofluoric acid, elements such as cobalt, copper, iron, and nickel can be determined free of interferences even in special glasses by GF AAS under application of the STPF concept [2869, 2870, 4336]. The same is true for the determination of iridium, platinum, and rhodium in the range 1–100 mg/kg [40], as well as numerous other elements in the µg/kg range [451]. The leaching behavior of aluminium from containers for pharmaceutical products has also been investigated by GF AAS [2244].

The direct solids analysis of glass for the trace elements bismuth, lead, and silver in an inductively heated furnace was reported in 1984 [2459]. Glass also appears to be particularly suitable for *slurry analysis*. The successful determinations of lead [116], cobalt, chromium, manganese [521], iron, nickel [116, 521], and copper [116, 521, 4198] have been reported. The determinations were performed either without a modifier since the matrix has a similar effect, or ammonium fluoride was added [4198]. The addition of hydrofluoric acid to the slurry additionally aided in removing part of the silicon prior to atomization [552]. HAUPTKORN and KRIVAN [2444] determined Al, Cr, Cu, Fe, K, Li, Mg, Mn, and Na in slurries of high-purity quartz powder against aqueous calibration solutions (except for aluminium) and found good agreement with the results obtained by GF AAS after prior digestion.

Hydride-generation AAS only finds very limited application in the analysis of glass. HERMANN [2520] determined selenium directly from the digestion solution, while ROHR and MECKEL [4936] determined arsenic after extraction with toluene and back extraction into the aqueous phase. The low interest in these elements is due to their volatility so that they are only present in glasses in very low concentrations. This is also almost certainly true for mercury, which has only recently received attention [2246].

10.7.2 Ceramic Materials, Superconductors

Ceramics are among the oldest products of human culture and in the form of commodities and ornaments have found the most varied applications in all periods of history. In the last few years the development of ceramics has gained an enormous impetus and modern 'hi-tech' ceramics have virtually nothing to do with the products of our ancestors. Highly developed ceramic materials based on pure or alloyed oxides, nitrides, carbides, or borides are synthesized from high-purity raw materials of clearly defined structure under strictly controlled conditions. The possible applications of these ceramic materials and their properties depend to a very large extent on the trace impurities that derive from the raw materials and the process of manufacture rather than on the raw materials themselves. The interested reader is referred to review articles for details [856, 3904, 5251, 6363].

In *archeology* the analysis of ceramic products of prehistoric [5598] and historical (e.g. Roman) [281, 668, 4136, 5927, 6584] origin is of considerable significance. The task in this case is mostly to determine the origin of the finds from the distribution pattern of the major and minor constituents or the traces. We shall discuss this application in somewhat more detail in Section 10.8.2.

A totally different analytical application that we shall mention here briefly is the *leachability of toxic heavy metals* from ceramic products that come into contact with foodstuffs. Nowadays the generally used and standardized procedure to determine the permeability of glazed ceramic surfaces to cadmium and lead [2549] was described by KRINITZ and FRANCO [3288] in 1973 and tested in an interlaboratory trial. The vessel being checked is allowed to stand at 22 ± 2 °C for 24 h with 4% acetic acid followed by the determination of the leached analyte either directly or after extraction [3217] by F AAS. To determine the actual risk in real situations it is possible to leach with drinks such as fruit juices, cola, or wine instead of using the standardized diluted acetic acid [2090].

The *digestion of ceramic materials* is naturally governed by their composition, so that we can only provide a general overview. Aluminium oxide ceramics are often digested in hydrochloric acid, sulfuric acid, or a mixture of both [2207] at high pressure and temperatures up to 240 °C, usually in PTFE containers. Phosphoric acid [2447] or a mixture of phosphoric acid and sulfuric acid [6140] has also been used. The alternative is a fusion with lithium carbonate and boric acid [2808]. Aluminium nitride dissolves in hydrochloric acid [2484] or in a mixture of hydrochloric acid and nitric acid [2207] under high pressure and at high temperature. Silicon carbide can be digested in hydrofluoric acid, nitric acid and fuming sulfuric acid in a high-pressure autoclave [1550, 2207]. Fusions are far less time consuming [1601], but far more susceptible to the risk of contamination. Silicon nitride is best dissolved in hydrofluoric acid-nitric acid under pressure [1954, 2207], molybdenum oxide also dissolves in hydrofluoric acid-nitric acid [3302], and zirconium oxide dissolves in hydrofluoric acid-sulfuric acid [4240]. As an alternative, BELYAEV et al. [512] proposed the digestion of ceramic materials by chlorination under pressure, for example with copper chloride and carbon tetrachloride at 600–800 °C or with hydrochloric acid and potassium permanganate at 160 °C. Superconductors of the type $YBa_xCu_yO_z$ are digested in semiconcentrated nitric acid [6586] or mixtures of hydrochloric and nitric acids [5254], while for vanadium silicide superconductors a fusion with sodium carbonate at 1100 °C is employed [508]. Microwave-assisted digestions can lead to a substantial saving in time for ceramic materials [1902, 3974].

Flame AAS is ideally suited for the determination of additive and doping elements in ceramic materials, whereas the sensitivity is inadequate for the determination of the majority of trace elements. ADELHELM and HIRSCHFELD [36] described an interlaboratory trial for the determination of Al, Ca, Fe, Mg, and Na in silicon nitride. Further typical applications are the determination of aluminium, chromium, iron, and manganese in silicon carbide and aluminium oxide-silicon carbide ceramics [4040], titanium in glass ceramics [2237], beryllium [2447] or magnesium [6140] in aluminium oxide ceramics, yttrium in hafnium oxide and zirconium oxide [4240], or silicon and vanadium in vanadium silicide superconductors [508]. BELAYA et al. determined iron [498] and magnesium [497] in a multitude of refractory carbides, borides, and nitrides. SHABUROVA et al. determined barium in piezoelectric ceramic materials [5253], the major constituents barium, copper and yttrium in ceramic superconductors [5254], and the doping additive thallium in ceramic high-temperature superconductors [5255]. YUDELEVICH et al. [6501] determined 15 elements in a number of high temperature superconducting materials. DE LA GUARDIA and co-workers [1061, 2255] described an FI-F AAS system with on-line dilution and buffer addition (lanthanum) for the determination of Al, Ca, Fe, K, Mg, and Na in ceramic materials.

The requirement for highly sensitive determination methods has grown in recent years to match the markedly increased purity requirements for ceramic materials. It has become quite obvious that it is not the instrumental detection limits of the determination method that are decisive but rather blank values and contamination during the sample digestion procedure [1550, 1984, 3302]. For this reason the relatively rapid fusion procedures, which are attractive for routine operation, must *a priori* be excluded for extreme trace analysis [1550]. Pressure digestions are more suitable, but purification of the digestion acids is mostly required, such as by subboiling distillation, and even at high pressure and temperature such digestions can often take many hours. There is therefore a substantial requirement for a simple analytical method, and the *direct analysis of solids or slurries* by GF AAS appears to offer a very attractive solution.

As early as 1981 SLOVÁK and DOČEKAL [5445] found that powdered aluminium oxide could be analyzed very well as an aqueous suspension by GF AAS for numerous trace impurities in the ng/g range. Nevertheless it took nearly a decade before the idea was taken up again and further investigated, particularly by Krivan and co-workers. Materials analyzed include silicon carbide [1550], silicon nitride [1984, 2238], boron nitride [2442, 3304], molybdenum oxide [1557, 3302], and zirconium dioxide [5174]. For the analysis of silicon carbide, for example, there were difficulties due to spectral interferences even under the application of Zeeman-effect BC, but these could nevertheless be eliminated by changing to alternate analytical lines [1550]. For the analysis of silicon nitride under STPF conditions with ammonium phosphate-magnesium nitrate as the modifier, it was possible to atomize numerous elements at 2200 °C before the matrix became sufficiently volatile to cause interferences [1984]; the majority of elements could thus be measured against aqueous calibration solutions. In numerous comparative measurements these authors found very good agreement of the results from GF AAS with the slurry technique and RNAA [3302, 5174] or INAA [1550] that were noticeably better than by solution analysis. Due to the much lower risk of contamination in the analysis of slurries compared to digestions, the former exhibit limits of detection for the test samples that are up to an order of magnitude better. The limits of detection for Ca, Cd, K, Mg, Na, and Zn in silicon nitride or zirconium dioxide were in the order of 1–20 ng/g, for Al, Co, Cr, Cu, Fe, Li, Mg, and Ni in the range 20–200 ng/g, and for silicon 2 µg/g [1550, 1984, 5174].

Next to slurries, solid powders have also been used directly, especially when the high density of the material being investigated precludes the formation of a stable slurry [1558].

10.7.3 Cement and its Raw Materials

The first papers on the determination of the major and minor constituents in cement were published in the early 1960s [1058, 5528, 5758] and cement chemists were among the first to recognize the significance of the, then, new technique and to utilize it [1367]. The results reported by the cement industry with respect to precision and trueness are perhaps the best that were ever obtained by AAS for the determination of high element concentrations [1369]. The reasons for this are most likely that the elements of interest in cement can be determined readily by AAS, the digestions are relatively easy to perform, and high accuracy can be attained by measuring against reference materials with a similar composition [1367] using the bracketing technique. The trueness of the measurements is typically 0.03% (0.3 g/kg) absolute or 0.5% relative and the repeatability is 0.02% (0.2 g/kg) absolute or 0.25% relative [1369].

The selection of the most suitable digestion procedure depends on which elements are to be determined. If silicon is also to be determined, dissolution in warm, dilute hydrochloric acid is recommended; less than 1% of non-silicate material remains undissolved in the residue [1367]. Chromium, manganese, titanium, and zinc can be determined directly from the undiluted digestion solution (0.5 g cement in 100 mL), while the solution must be diluted by a factor of five for the determination of Al, Fe, K, Na, and Sr, while for the determination of calcium and magnesium it must be further diluted by a factor of 100 under the addition of lanthanum chloride. The burner head is turned through 90° for the determination of calcium and the bracketing technique is employed. If the silicon content is of interest, a fusion with lithium metaborate is recommended which is then dissolved in nitric acid [1368]. The analyses are then performed in the same manner as for an acidic digestion; silicon is determined from the undiluted solution using the bracketing technique.

Next to these 'classical' digestion and determination procedures for cement analysis, LANGMYHR and PAUS [3423] proposed a digestion in hydrochloric, hydrofluoric and boric acids for cement, clinker, and raw materials. CHOI et al. [1264] used a fusion with sodium carbonate and lithium tetraborate (3 + 1) at 925 °C that only took 10 minutes. After dissolving the test sample in hydrochloric acid, they determined Al, Ca, Fe, Mg, and Si with an accuracy comparable to X-ray fluorescence analysis, and the entire procedure, including digestion and determinations, took only an hour. LAW et al. [3468] used equal quantities of oxalic acid, lithium carbonate, and lithium tetraborate under the same conditions and determined Al, Ca, Fe, K, Mg, Na, Si, and Ti. Further rapid procedures are compiled in a monograph by GEBHARDT [1980].

Back in 1975 FALINOWER [1804, 1805] described a fully automatic procedure for the analysis of cement and raw materials that encompassed a fusion and determinations by F AAS. The application of FI techniques has opened up new possibilities of automation that permit the application of slurries rather than solutions and also automatic dilution [431, 3915, 5070].

To determine environmentally-relevant trace elements such as As, Cd, Cr, Hg, Pb, Tl, and Zn in cement and related materials, ESSER [1774, 1775] applied GF AAS with direct solids analysis, while LÓPEZ-GARCIA et al. [3619] used slurries to determine cadmium, lead, and thallium. RECHENBERG and SPRUNG [4833] used both F AAS and GF AAS to check the leachability of environmentally-relevant trace elements such as cadmium, chromium, and thallium from cement-containing rubble to be reused in the construction of roads. The most sensitive procedure for the determination of mercury is CV AAS, which is also suitable for the analysis of raw materials used in the manufacture of cement [337, 2493]. BACHMANN and RECHENBERG [337] have pointed to the risk of mercury losses during grinding of the sample and adsorption on aged containers during digestion. As an alternative these authors proposed atomizing directly from the solid sample in a nickel tube heated to 1000 °C.

10.7.4 Chemicals

The majority of applications dealing with the investigation of chemicals are involved with the determination of trace impurities and thus with the production and quality control of products of guaranteed purity. Next to this the determination of major and minor constituents plays a role, such as in organometallic compounds or doping concentrations in crystals; F AAS is particularly suited due to its selectivity and accuracy. Examples are the determination of potassium and rubidium in potassium-rubidium dihydrogenphosphate [1482], ruthenium in hydrated ruthenium chloride [2395], or the doping concentration of indium in rubidium chloride crystals [2162]. The determination of the metal content in *organometallic compounds* has been described for aluminium [1468], molybdenum [230], rhenium [5873], ruthenium [753], selenium, tellurium [5704], and titanium [2395]. The organometallic compound is usually dissolved in an organic solvent and aspirated directly into the flame. AL-ABACHI and SALIH [98] described the indirect determination of bromine and chlorine in organic compounds after combustion in a Schöninger flask and treatment of the solution with silver chromate.

The analysis of *phosphors* also mostly involves the determination of higher quantities, so that the sensitivity of F AAS is generally adequate. Phosphors of the calcium halogenphosphate type, which are used in fluorescent lamps, dissolve in concentrated hydrochloric acid and can be analyzed directly by F AAS for impurities such as sodium [4612]. SCOTT [5216] determined europium in yttrium phosphors, which are used in color televisions, after a fusion with potassium carbonate and dissolution of the melt in hydrochloric acid. AMETANI *et al.* [164] dissolved zinc sulfide-silver phosphors in hot hydrochloric acid and determined magnesium, silver, sodium, and zinc after appropriate dilution.

The presence or absence of a number of metallic elements in determined concentrations in *photographic materials* such as films or photographic papers can be of decisive significance for their quality. The concentrations in photochemicals are often so high that the sensitivity of F AAS is adequate for determinations following simple dissolution and dilution of the sample. Care must nevertheless be taken to ensure that the concentration of silver is not too high and that the formation of explosive silver acetylide in the spray chamber is avoided. ANDRONOV *et al.* [208] avoided this problem by using the air-propane flame. For the analysis of photographic materials it is the sensitive layer that is usually of interest and not the support material (paper, film), so that it is more sensible to dissolve this layer rather than to ash the entire material. This can be done with potassium cyanide solution for the determination of silver [3350] or cadmium [3351], or even by enzymatic extraction [163, 1537]. The latter technique was also used in the determination of gold [1537] and palladium [163] traces in photographic emulsions by GF AAS.

Flame AAS is only suitable for the direct determination of trace impurities when high-purity chemicals are not involved. ABRAHAM *et al.* [22] thus determined siloxanes in used sulfuric acid after dilution with isopropanol. Further examples are the determination of copper in cobalt and nickel salts [2878], potassium and sodium in barium hydroxide [18], barium in strontium nitrate [3062], as well as calcium and magnesium in lithium hydroxide [6153] or sodium borates [812]. The application of injection or, better, FI techniques greatly facilitates the handling of solutions with high salt contents, such as are frequently encountered in the analysis of chemicals. KMETOV and FUTEKOV [3150] determined copper, iron, lead, and manganese in nitric acid after neutralization in ammonia. FU et al. [1998] determined potassium in high-purity $Na[Sb(OH)_6]$ and FANG et al. [1822] determined copper in a solution containing 300 g/L sodium chloride by FI-F AAS.

Since the materials being investigated in most cases contain compounds of known composition and high purity, it is fairly easy to develop systems for trace-matrix separation, a fact that is reflected in the literature. An elegant procedure, which can also be applied to other chemicals, was introduced by HEININGER et al. [2473]; they distilled off hydrofluoric acid at a temperature below the boiling point and determined numerous trace elements in the residue. JACKWERTH and WILLMER determined a multitude of trace elements in copper salts [2859], and in barium and strontium salts [2860] after precipitation of the matrix.

However, preconcentration procedures for trace elements using solvent extraction are to be found much more frequently in the literature; Table 10-94 presents a selection. In addition, preconcentration procedures employing ion exchange, sorption, or coprecipitation have also been proposed. Using a cation exchanger, KOHRI et al. [3192] separated around 20 trace elements from the vanadium matrix which was present as an anionic hydrogen peroxide complex. Lead can be preconcentrated on xanthate wool from sodium chloride [1024] and numerous other elements can be preconcentrated with HMDTC on polyurethane foam [121]. JACKWERTH and co-workers investigated the preconcentration of heavy metal traces in alkali and alkaline-earth salts [2857], in silver and thallium salts after complexing with xylenol orange [2855], and in chromium(III) salts after complexing with HMDTC [580] by sorption on activated carbon. VIRCAVS et al. [6079] determined Bi, Cd, Co, Cr, Cu, Fe, Ni, Pb, Ti, V, and Zn in high-purity aluminium chloride by coprecipitation with quinoline-5,8-dithiol, and HEININGER and HERION determined elements such as Co, Cu, Fe, Mn, Ni, Pb, and Zn in ammonium fluoride and hydrofluoric acid after coprecipitation with calcium fluoride [2474] and in phosphoric acid after coprecipitation with bismuth phosphate [2505]. VASSILEVA and HADJIIVANOV [6003] used a titanium dioxide column as a sorbent for the preconcentration of Cd, Co, Cu, Fe, Mn, Ni, and Pb from alkali salts of analytical reagent grade.

Because of the time and effort required to perform extractions or coprecipitations, and also because of the risk of contamination or loss of analyte, investigations on the direct determination of trace elements by GF AAS were performed at an early stage. In 1976 LANGMYHR and HÅKEDAL [3435] determined a number of trace elements in analytical grade acids with limits of detection of 1 μg/L. At the beginning of the 1980s the successful determinations of chromium, iron, nickel [2064], calcium and magnesium [5469] in saturated NaCl solutions were reported; ammonium hydroxide was used as the modifier. As in other areas, GF AAS only became accepted as the method of choice after the introduction of the STPF concept, effective modifiers [2866–2868], and in particular Zeeman-effect BC [736, 904, 1253, 2867, 2868, 2870, 6316], which permitted the use of concentrated and aggressive solutions such as 100 g/L HfF_4 in 24% HF [2870] or the direct analysis of solid samples [2871] or slurries [116, 448, 5126]. A further decisive contribution to the acceptance of GF AAS was the introduction of simultaneous spectrometers for this technique, which provide a sample throughput similar to that of F AAS but with a sensitivity that is three orders of magnitude higher [904]. The most important publications on the direct determination of trace elements in high-purity chemicals are summarized in Table 10-95.

Table 10-94. Preconcentration of trace elements in chemicals by solvent extraction for determination by F AAS.

Analyte	Matrix	Extraction system	References
Ag	Cu salts	heptan-4-one-oxime / chloroform	[453]
Ag, Bi, Cd, Co, Cu, Fe Mn, Ni, Pb, Zn	H_2SO_4, $(NH_4)_2SO_4$	DDTC / IBMK	[2829]
Ag, Bi, Cd, Cu, Fe, Mo, Pb, Sb, Zn	Co salts	TOMA–APDC / IBMK	[6001]
Ag, Bi, Cd, Fe, Mo, Pb, Sb, Zn	Cu salts	TOMA / IBMK	[274]
Ag, Cd, Cu, Fe, Zn	Ni salts	TOMA–APDC / IBMK	[6002]
Ba	Na salts	TTFA / crown ether	[635]
Bi, Cd, Cu, Pb	H_3PO_4	DDTC / butyl acetate	[4367]
Bi, Co, Cu, Fe, In, Mn, Ni, Pb, Sb, Zn	H_3PO_4, $(NH_4)_3PO_4$	APDC–PMBP / IBMK	[2450]
Cd	NaCl	APDC / chloroform	[833]
Cd, Co, Cr, Cu, Fe, Mn, Ni, Pb, Zn	H_3BO_3	DDTC / petroleum	[269]
Cd, Co, Cu, Fe, Ni, Pb	Zn salts	APDC / IBMK	[3029]
Cd, Co, Cu, Fe, Ni, Pb, Zn	Alkali and alkaline-earth salts	DDTC / IBMK	[266]
Cd, Co, Cu, Fe, Ni, Zn	$PbCl_2$, $PbSO_4$	DDTC / IBMK	[270]
Cd, Cu	NaCl	DDTC / IBMK	[3220]
Cd, Cu, Pb	NaCl	APDC / 4-methylpentan-2-one	[1777]
Cd, Cu, Pb, Zn	NaCl	DDTC / t-butanol	[3820]
Co, Cu, Fe, Mn, Ni, Zn	Alkali salts	PMBP / IBMK	[2828]
Co, Cu, Fe, Mn, Ni, Zn	Mo(VI) compounds	DDTC / IBMK	[2504]
Cr, Cu, Ni	Na_2CO_3, K_2CO_3 NH_4ReO_4	DDTC–cupferron / IBMK	[3262] [2952]
Cu, Ni, Zn	Pd salts	APDC / IBMK	[4688]
Hg, Pd	Cd, Cu, Pb salts	benzothiazolone derivatives / IBMK	[3944]
Ni	NaCl	5-nitrosalicylaldehyde-4-phenyl-3-thiosemicarbazone	[5955]
Pb	NaCl	APDC / chloroform	[834]
Pb	H_3PO_4	DDTC	[798]
17 elements	NH_4ReO_4, ultrapure	dithiocarbamate complexes / $CHCl_3$–CCl_4	[2952]

Table 10-95. Direct determination of trace elements in high-purity chemicals by GF AAS.

Analyte	Matrix	Remarks	References
Al	$(NH_4)_2WO_4$	modifier: Cu–Ni	[2455]
	Al alkoxides, and carboxylates	PC coating	[964]
Al, Ca, Cr, Cu, Fe, K, Mg, Na, Ni, Pb, Si, Sn, Zn	graphite	slurry	[5126]
As	H_3PO_4	modifier: Pd	[5085]
Au	$AgNO_3$	–	[6395]
Bi, Co, Cu, Fe, In, Mn, Ni, Pb, Sb, Zn	H_3PO_4, $(NH_4)_3PO_4$	ZBC	[2449]
Bi, Cu, Fe, Pb	NaCl, KBr, KI	–	[1242]
Ca, Mg	NaCl	–	[5469]
Cd	H_3PO_4	ZBC	[6316]
Cd, Co, Cr, Cu, Mn, Ni, Pb, Se, Sr	graphite	open microwave-assisted digestion	[3548]
Cd, Cr, Cu, Mn, Ni	$AgNO_3$	–	[6396]
Cd, Cu, Fe, Mn, Pb, Zn	HF, H_2SO_4, NH_4OH	–	[3435]
Cd, Mn, Pb	$ZnSO_4$	modifier: H_3PO_4	[2718]
Co, Cu, Fe, Ni	AlF_3	ZBC	[2872]
	HfF_4 in HF	ZBC	[2870]
	LaF_3	solid sample, ZBC	[2871]
	$La(NO_3)_3$	modifier: Pd–Mg	[2866]
Co, Fe	ZrF_4	modifier: Pd–HNO_3, ZBC	[2868]
Cr, Fe, Ni	NaCl	modifier: NH_4OH	[2064]
Cu, Fe, Ni	HF, HCl, HNO_3	simultaneous spectrometer, ZBC	[904]
Cu, Fe, Ni, Pb	KF, NH_4F	slurry	[116]
	metal fluorides	slurry	[448]
Cu, Ni	ZrF_4	modifier: Pd–HNO_3, ZBC	[2867]
Fe	ferrocene	ethanolic solution	[864]
Fe	phosphoryl chloride	direct determination	[3278]
Ge	organo-Ge compounds	modifier: $Ca(NO_3)_2$	[6461]
Hg	KI, NaI	atomization at 900 °C	[2431]
K, Na	acetone	–	[3526]
Pb	Pd salts	matrix functions as modifier	[4687]

In addition a number of separation and preconcentration procedures in combination with GF AAS have also been described. NÈVE *et al.* [4329] determined selenium(IV) after selective extraction with an aromatic o-diamine in toluene. HIRSCHFELDER and THIELE [2590] determined cadmium in organogallium compounds after extraction of the tetrachloro gallium complexes with diisopropyl ether. SAAB *et al.* [5023] extracted tungsten from high-purity tungsten hexafluoride with trioctylamine in xylene prior to determining 13 trace elements by GF AAS. HOPPSTOCK *et al.* [2664, 2665] investigated a trace preconcentration from ammonium fluoride by electrolytic deposition directly in the graphite tube. Procedures for sorbent extraction have been described for gold in palladium chloride [767], cobalt, copper, gold and lead in aluminium chloride [598], tin in iron(III) chloride [6530], and cadmium and lead in various high-purity salts [6266]. In contrast to the off-line preconcentration procedures described in the earlier literature, which demand considerable experience and the greatest of care particularly in the trace range, on-line preconcentration procedures using the FI technique offer in comparison a reliable and easy way of working [6266]. FANG *et al.* [1234] determined molybdenum in ultrapure chemicals using on-line preconcentration and GF AAS.

Reports on the application of HG AAS for trace determinations in chemicals are noticeably seldom. BYE and HOLEN [991] observed interferences in the determination of selenium in technical sulfuric acid, while the determination proceeded satisfactorily in the high-purity acid. ÅSTRÖM [291] determined bismuth in various acids and bases using FI-HG AAS. ERBER *et al.* [1754] determined antimony, arsenic, and selenium in organic chemicals after combustion in a Wickbold apparatus. For mercury, CV AAS is the method of choice due to its high sensitivity. BURNS *et al.* [962] determined mercury in organomercury compounds containing silicon after a pressure digestion in nitric acid-sulfuric acid. For the determination of mercury in sodium chloride prior heating with hydrochloric acid and sodium chlorate has been recommended [835]. WHITE and MURPHY [6301] determined mercury in silver salts after separation as the bromo complex. For the determination of mercury in mineral acids by GF AAS, HLADKY *et al.* [2593] preconcentrated the mercury on gold-coated graphite tubes after reduction with tin(II) chloride or sodium tetrahydroborate. DOBROWOLSKI and MIERWA [1548] compared CV AAS after a digestion with direct solids analysis by GF AAS for the determination of mercury in phosphors for fluorescent lamps.

The use of AAS for the determination of impurities in reactive process gases has also been sporadically reported [223, 331, 332, 5180].

10.7.5 Pharmaceuticals and Cosmetics

For the analysis of pharmaceutical and cosmetic products we make a distinction between the determination of metallic active ingredients or impurities and the indirect determination of organic active ingredients, which is used widely in the pharmaceutical industry.

10.7.5.1 Determination of Metallic Active Ingredients and Impurities

The procedure of *sample preparation* used for pharmaceutical and cosmetics depends very much on the nature of the substances being investigated. Numerous pharmaceuticals are already present as solutions or can be easily dissolved, so that hardly any sample preparation is required. Frequently wet ashing is performed with nitric acid in Kjeldahl flasks [86, 87], under pressure [2969], or with microwave assistance [651]. Medicinal herbs and herbal medicines can be digested by dry ashing at 300–550 °C [2875, 2969, 5283, 5829] and taking up the ash in hydrochloric acid or nitric acid. As well as complete digestion, occasionally an extract is examined to check the *bioavailability* of an element such as aluminium [3034, 3994] or lithium [3135]. PENNINCKX *et al.* [4597] introduced a software system using selected decision criteria that permits the selection of the most suitable dissolution procedure for subsequent AAS analysis.

Flame AAS can mostly be used for the determination of metallic or metal-containing *active ingredients* since they are usually present in relatively high concentrations; GF AAS is used almost exclusively for the determination of *trace impurities*. In many respects *aluminium* assumes a double role: on the one hand it is administered in the form of aluminium hydroxide for the treatment of minor stomach complaints and can be determined by F AAS in such preparations[3034, 3994], while on the other hand it must be detected often in very low concentrations due to its toxicity and this can only be done reliably by GF AAS [182, 4855]. In the latter respect infusion solutions [2452] and ingredients for dialysis solutions [2452, 4328] are of importance; dialysis solutions occasionally require a preconcentration step prior to the GF AAS determination [3563, 4602]. Since the determination of aluminium in the ultratrace range is exceedingly susceptible to contamination, the greatest of care is required [1764, 4328]; direct determinations by GF AAS offer the best prerequisites for correct values [4328].

Mercury also assumes a similar double role since it is contained as the active ingredient in the form of organomercury compounds, for example in antiseptic preparations and eye drops, and can then be determined by F AAS [1132, 2634]. Graphite furnace AAS has also been used for this purpose [2133], but the method of choice is CV AAS, both for the determination of higher mercury concentrations [2634, 2635] and trace impurities [3465, 5849, 5850]. Mixtures of hydrochloric and nitric acids [2634] or nitric and perchloric acids [5849, 5850] have been used for digestion. The determinations of further trace elements, such as cadmium, lead and thallium, in cosmetics [2482] and pharmaceuticals [86, 87] have sporadically been reported, but due to the usually very low dose that is ingested there should be no risk of intoxication [88]. Table 10-96 provides an overview of selected publications on the determination of metallic active ingredients and impurities in pharmaceuticals and cosmetics.

In recent years medicinal herbs [2875, 5829, 6560], ginseng [372, 6462], and traditional Chinese herbal medicines [3525, 5283, 6560, 6570] have been increasingly investigated, often for a multitude of elements. These investigations are aimed at finding active ingredients containing metals such as germanium [372, 3525] or molybdenum [5283], and also toxic heavy metals such as cadmium [2875, 5829], lead [5829], and mercury [2875].

A further area of application for pharmaceutical products is the analysis of injection and infusion solutions, mostly to determine the transfer of metal traces into the solution by leaching from the glass ampoule. Elements investigated in this respect include the alkali metals potassium and sodium [4151], copper and lead [4460], and a number of other trace elements [3058, 4704].

Table 10-96. Selected publications on the determination of metallic active ingredients and impurities in pharmaceuticals and cosmetics.

Element	Sample material	Technique	References
As	herbal medicines	GF AAS	[88, 2243, 5849]
Au	sodium aurothiomalate	F AAS	[5447]
Bi	medicines	GF AAS	[6418]
	eye shadow	GF AAS	[2140]
Ca	pharmaceutical preparations	F AAS	[1411]
	pain-killing tablets	F AAS	[4768]
Cd	pharmaceutical constituents and preparations	GF AAS	[86–88, 2969]
Co	vitamin B_{12}	GF AAS	[83, 4587]
Cr	vitamin tablets	GF AAS	[651]
Cu	pharmaceutical products	F AAS	[3831]
Cd, Co, Cr, Cu, Fe, Mn, Ni, Pb, Sb, Sn, V	pharmaceutical products	GF AAS	[276]
Fe	mineral tablets	F AAS	[4659]
	iron preparations	F AAS	[5778]
Ge	medicinal herbs	GF AAS	[4129]
	ginseng	GF AAS	[372]
	herbal medicines	GF AAS	[3525]
Li	lithium carbonate products	F AAS	[3135]
Mn	pharmaceutical preparations	F AAS	[5694]
Mo	herbal medicines	GF AAS	[5283]
Ni	heroin, cocaine	GF AAS	[567]
Pb	herbal medicines, pharmaceutical preparations	GF AAS	[86–88, 2243, 5849]
	toothpaste	GF AAS	[3090]
Pt	DNA platinum complexes	F AAS	[3772]
Sb	organo-Sb compounds	F AAS	[3894]
Si	hand lotion	F AAS	[3886]
Ti	sun blocker	F AAS	[3927]
	soap	GF AAS	[3495]
Tl	pharmaceutical preparations	GF AAS	[86, 88]
	radiopharmaceuticals	GF AAS	[6549]
Zn	pharmaceutical preparations	F AAS	[231, 3831, 4181]
	insulin	F AAS	[5525]

10.7.5.2 Indirect Determination of Organic Active Ingredients

The indirect determination of organic active ingredients in pharmaceuticals is mostly not specific for a given compound but rather for functional groups. This fact is normally of no major significance in practice since the ingredient itself and the composition of the sample are known, it is merely the *content* of the ingredient that must be determined. Table 10-97 presents a selection of such indirect procedures which are mostly based on the extraction of a metal complex or ion pair with the organic ingredient. The metal in the organic extract is usually determined by F AAS; GF AAS is only applied for the element determination in individual cases [1202, 2923].

Flow injection techniques have opened up completely new dimensions for the determination of organic active ingredients; on the one hand FI permits automated on-line operation of the procedure with F AAS determination, and on the other hand it enables completely new reactions and techniques to be accessed. Techniques described for FI-F AAS include continuous liquid-liquid extraction [4175, 5089, 5977], continuous on-line precipitations [1696, 3449, 3916, 4170, 4171, 5977], and sorbent extractions [5977]. The flow-through reactors specially developed for FI-F AAS are particularly interesting; these comprise a microcolumn packed with metallic cadmium or zinc on which NO groups are selectively reduced and simultaneously an equivalent quantity of metal ions is released [4172–4174]. Using this procedure sample throughputs of 150/h have been attained with sample volumes of less than 100 μL [4172, 4173]. MATÍNEZ CALATAYUD and co-workers prepared solid phase reactors by embedding $CuCO_3$ [2068], MnO_2 [3391] or PbO_2 [3392] in the polyester resin during polymerization. During reaction with the organic ingredient an equivalent quantity of metal ions is released which can be measured on-line by F AAS.

10.7.6 Catalysts

The effectivity of numerous catalysts depends on relatively low concentrations of elements, some of which are exceedingly unusual, while even traces of other elements such as lead can reduce the effectivity substantially. There is thus a broad field of application for AAS. Catalysts are used widely in the oil and chemical industries, and since the 1970s they find increasing use in the automobile industry.

For the analysis of catalysts we make a distinction between the determination of the *catalyst constituents* and the determination of trace elements that act as *catalyst poisons*. These poisons can be deposited on the catalyst surface during use and derive from the crude oil, transport and storage contamination, and possible residues from additives. The major elements in catalyst constituents are essentially silicon and aluminium from the support materials or cobalt, molybdenum, nickel or the noble metals gold, palladium or platinum from the surface coating. The determination of interfering trace elements largely involves Ni, V, Fe, Cr, Pb, and Ca [2988]; the catalyst becomes inactive when these 'poisoning elements' are above an application-specific value.

In principle the analysis of catalysts is similar to the analysis of metallurgical samples; we thus refer our readers to Section 10.6. A difference lies in the large excess of SiO_2 and Al_2O_3 present from the support material that can lead to substantial interferences. Fusions with potassium hydrogen sulfate are occasionally used, such as for the determination of iron and nickel in nickel oxide-aluminium oxide catalysts [5391]. Digestions with hydrofluoric acid [6529], hydrofluoric acid-sulfuric acid [1791, 1792, 6292], or hydrofluoric acid-nitric acid-sulfuric acid [4668] are however mostly used.

Table 10-97. Indirect determination of organic active ingredients in pharmaceuticals.

Element	Procedure	Active ingredients	References
Bi	extraction of the ion pair with [BiI$_4$]$^-$	alkaloids: sparteine, papaverine, bromhexine, amylocaine, avacine pyrrolizidine-alkaloids	[4321] [4323]
	precipitation of the ion pair with [BiI$_4$]$^-$	alkaloids: papaverine, cocaine, strychnine	[1696]
Cd	precipitation with metal ion	sulfonamide	[317]
	reduction on Cd metal, determination of the released metal ions	chloramphenicol	[4172]
		chlorodiazepoxide	[4174]
		methadone	[4173]
Co	extraction of the ion pair with [Co(SCN)$_4$]$^{2-}$	alkaloids	[4320]
		cocaine	[3507]
		aminochinoline antimalaria	[2420]
	extraction of the ion pair with [Co(NO$_3$)$_6$]$^{3-}$	cationic surfactants	[1202]
	extraction of the Co complex	bromhexine	[4319]
		salicylic acid	[1203]
	precipitation with metal ion	sulfonamide	[317]
		local anesthetics	[4171]
Cr	extraction with Reinecke-salt	amylocaine, bromhexine	[1697]
	precipitation as the Reineckate	cocaine, papaverine	[1698]
Cu	extraction of the Cu complex	amphetamine	[4175]
		ethambutol	[2421]
		lignocaine	[1710]
		azepine series	[103]
		bromazepam	[5089]
Cu	precipitation with the metal ion	lincomycin	[1709]
		sulfonamide	[317, 4170]
		chlorohexidine	[3916]
		tannin	[6479]
	reaction with CuCO$_3$	glycine	[2068]
	preconcentration on Nafion 117	arginine	[2923]

Table 10-97 continued

Element	Procedure	Active ingredients	References
Fe	extraction of the ion pair with $[Fe(SCN)_6]^{3-}$	amylocaine, avacine, bromhexine, papaverine, diphenylhydramine	[4322]
Hg	precipitation with HgI_2 / KI	levamisol	[3449]
		pilocarpin	[319]
		cinchona-alkaloids	[318]
Mn	oxidation with MnO_2	isoniazide	[3391]
Mo	extraction of the phosphomolybdate complex	alkaloids	[5371]
		caffeine	[449]
Pb	removal of sulfur with Pb(II), measure residual Pb(II)	vitamin B_1	[2419]
	reaction with PbO_2, measure resulting Pb(II)	ondansetrone	[3392]
Zn	reduction on metallic Zn, measure released Zn(II)	chloramphenicol	[4172]
		chlorodiazepoxide	[4174]

The silicon can be fumed off, possibly under the addition of perchloric acid, to such an extent that interferences can no longer be observed for example in the determination of vanadium in the nitrous oxide-acetylene flame [2988]. As an alternative, mostly for the analysis of automobile catalysts, digestions in hydrochloric acid-sulfuric acid [4701] or nitric acid-sulfuric acid [6189] have also been proposed. The application of microwave-assisted digestions leads to a substantial saving of time and allows the use of smaller quantities of reagents [2218, 4669]. A microwave-assisted digestion with nitric acid lasting only three minutes is perfectly adequate for the determination of cobalt, iron, molybdenum, and nickel in petrochemical catalysts [4190]. To avoid a digestion, suspensions have also been analyzed directly, both by F AAS [3036] and GF AAS [1259, 4274].

In the presence of higher aluminium contents, which cannot be removed so easily, it is necessary for the determination of cobalt, molybdenum, nickel, and vanadium to match the aluminium concentration in the calibration solutions by the addition of aluminium acetate [2988, 3378]. A number of authors thus added 1–2 g/L Al [1790, 3378] or a mixture of 1 g/L each of Al and La [1791] to all solutions to eliminate this influence. For GF AAS, on the other hand, aluminium does not appear to cause any interferences [2218, 4274]. For the determination of noble metals on catalysts by F AAS and the elimination of interferences by the addition of cadmium and copper [3677, 3829] we refer our readers to the information in Section 10.6.2.7. ROTTER described the determination of palladium [4699], platinum [4699, 4701], and rhodium [4700] in platinum-rhodium-coated automobile catalysts by GF AAS after dissolution of the sample in hydrochloric acid-sulfuric acid.

Various publications on the analysis of catalysts are compiled in Table 10-98. The majority are devoted to the determination of palladium, platinum, and a number of further platinum-group metals, which is hardly surprising since automobile and numerous other catalysts are prepared from these metals. KALLMANN and BLUMBERG [3015] have published an overview of papers covering this area up to 1980.

Table 10-98. Selected publications on the analysis of petrochemical catalysts.

Analyte	Catalyst	Digestion / Remarks	Technique	References
Al, Co, Fe, Mo, Ni, Si, Ti, V	Al-Si oxide	HNO_3–HF, microwave	F AAS	[4669]
Al, Fe, Mo, Na, Ni, Si, Ti, V	Al-Si oxide	HNO_3–HF–H_2SO_4	F AAS	[4668]
Al, Fe, Na, Ni, Sb, Sn, Ti, V	alumina	H_2SO_4–HF	F AAS	[1792]
As, Pb	Al-Si oxide and alumina	HF–H_3BO_3, microwave	GF AAS	[2218]
Cd, Cu, Pb	alumina	sorption on sulfhydryl cotton	F AAS	[3550]
Co, Fe, Mo, Ni	Laterite, Ketjen, Catal	HNO_3, microwave	F AAS	[4190]
Co, Mo	alumina	suspension	F AAS	[3036]
Co, Mo, Ni	Al-Si oxide	H_2SO_4–HF	F AAS	[3378]

Table 10-98 continued

Analyte	Catalyst	Digestion / Remarks	Technique	References
Cs, K, Li, Na, Rb, V	V_2O_5	H_2SO_4–HF	F AAS	[6292]
Fe	alumina	suspension	GF AAS	[4274]
K, Na, V	V_2O_5 on Ti-Si oxide	H_2SO_4–HF	F AAS	[6292]
Mo	impregnated clay	$HClO_4$–HF–HCl, extraction with IBMK	F AAS	[1084]
Pd	Pd on activated carbon	extraction with DMF	F AAS	[24]
Pd	powder	electrolytic preconcentration	F AAS	[6189]
Pd	polymer	suspension	GF AAS	[1259]
Pt	alumina	high pressure ashing, extraction with dithizone / IBMK, minimization of interferences	GF AAS	[3235]
Pt	alumina	H_2SO_4–HCl; matrix matching	GF AAS	[4701]
Pt	various	extraction with dithizone / IBMK	GF AAS	[3016]
Pt	Cordierite	interlaboratory trial	GF AAS	[6188]
Pt, Re	alumina	HF	F AAS	[6529]
Pt, Rh	Co oxide	*aqua regia*, extraction with TPP / IBMK	F AAS	[4980]
Pt, Rh	alumina	HF	F AAS	[6529]
Pt, Rh	alumina	comparison of methods	GF AAS	[4588]
Rh	titanium	fusion KOH–KNO_3	F AAS	[2396]
Rh	alumina	H_2SO_4–HCl	GF AAS	[4700]
Ru	alumina, Al-Si oxide	elimination of interferences	F AAS	[1790]
Ti	Al-Si oxide	H_2SO_4–HF–HCl	F AAS	[1791]
Zr	Zr–Ni/C and Zr–Pt/C	*aqua regia*–HF	F AAS	[5960]

10.7.7 Semiconductors

Semiconductor materials include essentially the following: silicon, germanium, germanium arsenide, indium antimonite, and the sulfides, selenides and tellurides of zinc, cadmium, and mercury. In these materials we must determine the doping elements, for which F AAS is occasionally adequate, and the trace impurities in the high-purity substances, for which GF AAS, HG AAS, or CV AAS offer the required sensitivity. Most materials can be dissolved relatively easily in nitric acid [6504], hydrochloric acid [6585], or mixtures of these [5736, 5737]. A mixture of hydrochloric acid and bromine is suitable for gallium arsenide [477, 478]. Hydrofluoric acid, or occasionally a mixture of nitric acid and hydrogen peroxide, is required solely for the digestion of silicon [3377,

3394]. An overview of various digestion procedures has been published by ZOLOTOV and GRASSERBAUER [6585]. To avoid contamination during the digestion, only high-purity or purified acids should be used and the digestion should be performed in a closed system. To reduce the risk of contamination even further for the digestion of silicon, PHELAN and POWELL [4635] proposed purification of the reagents in combination with the digestion. KOJIMA et al. [3215] performed a pressure digestion in the vapor phase in a PTFE autoclave and MAKSIMOV et al. [3823] heated silicon with xenon fluoride in an autoclave at 170 °C and dissolved the residue in water.

An alternative to digestions is direct solids analysis by GF AAS; this has been described for high-purity gallium [2554] and semiconductor silicon [2460], but does not appear to have found further application. The sputtering of semiconductor materials with argon ions has also been described; the sputtered material is collected directly on a L'vov platform and analyzed [907, 6278]. In this procedure the material was removed layer by layer, so that rather than a total analysis a surface analysis and a depth profile were possible. Layer-by-layer ablation combined with a depth profile of the distribution of trace elements had also been investigated in detail earlier by YUDELEVICH and co-workers [482, 4630, 6499, 6502, 6503] for various materials. They dissolved the layers with nitric acid and hydrofluoric acid or hydrobromic acid, or peeled them mechanically with a microtome [6499].

Because the concentrations of the trace elements to be determined in the high-purity materials are so low, but also to eliminate matrix effects, numerous separation and preconcentration procedures have been described. These include the preconcentration of traces from high-purity selenium on activated carbon for subsequent determination by F AAS [473, 2830] and separations by liquid-liquid extraction [3346, 5765] or ion exchange [2831, 3192] for subsequent GF AAS determination. Since these procedures require considerable effort and are highly susceptible to contamination, we shall not discuss them further. A procedure that is much more interesting for the current application is separation of the matrix by selective volatilization; this is in any case a procedure that is used routinely in GF AAS. As early as 1978 DITTRICH et al. [1538] investigated the vaporization behavior of various semiconductor materials for this purpose. BUSHEINA et al. [980] determined traces in gallium arsenide by first volatilizing arsenic at 1150 °C in a stream of argon and then gallium and residual arsenic in a chlorine-argon mixture at 250 °C. BEINROHR [475] used bromine vapor for the same purpose, a procedure that can also be performed directly in the graphite tube [474]. Germanium can also be easily volatilized and separated as the chloride prior to the analysis [5247].

The majority of all analyses of semiconductor materials are nevertheless carried out directly on the digestion solution by GF AAS. Table 10-99 presents a number of recent papers on this subject. An overview of work published prior to 1987 using GF AAS has been prepared by YUDELEVICH et al. [6500]. The determination of traces in semiconductor materials appears to be largely free of interferences when the STPF concept and, where necessary, Zeeman-effect BC are applied. In a direct comparison between GF AAS and ICP-MS for the determination of tin in indium phosphide, better accuracy was obtained with the former [5740].

Due to the matrix interferences that must be expected, HG AAS has been relatively seldom applied for the determination of trace elements in semiconductor materials. DITTRICH et al. [1540] compared GF AAS and HG AAS for the determination of traces of selenium and tellurium in A(III)B(V) semiconductors. KULDVERE [3336] determined arsenic in high-purity selenium after oxidizing Se(IV) to Se(VI) with potassium permanganate, and SAHAYAM et al. [5047] determined tin in high-purity gallium by collecting the tin hydride in silver nitrate solution and determining by GF AAS. SZMYD and BARANOWSKA [5727] reported the determination of mercury in elemental selenium by CV AAS. ZOLOTOV et al. [6585] have published a good overview on the determination of trace elements in semiconductor materials using various techniques covering the period prior to 1985.

Table 10-99. Selected publications on the determination of doping and trace elements in semiconductor materials by GF AAS.

Matrix	Analyte	Remarks	References
Cd	As, Te, Tl	modifier: Pd–Mg	[483]
CdHgTe	Ag, Cu, In	platform	[2273]
CdSe	Ag	platform	[3094]
	Ag, Zn		[3095]
Ga	Bi, Cd, Hg, Pb, Tl	solid sampling	[2554]
	Sn	modifier: Mo	[5046]
	Sn		[6504]
GaAs	Cr		[2931, 5734]
	Ge	modifier: $Ni(NO_3)_2$ or $Ba(NO_3)_2$	[1542]
	In	platform, matrix matching	[5738]
	Mg	platform	[5736]
	Pb	modifier: Cr(III)	[477]
	Si	platform, modifier: Ca	[5735]
	Si	indirect, determination of Mo	[5739]
	Te		[4284]
	Zn	platform, modifier: H_3PO_4	[5737]
In	21 elements	ZBC	[5245]
InAs	Ag, Au, Bi, Cd, Sn, Tl	platform, modifier: Ni	[1541]
InP	16 elements	platform, ZBC	[4088]
	Sn	platform, modifier: $Mg–PO_4$	[5740]
	Ag, Cd, Fe, Mn, Pb, Sb, Zn	solid sampling	[2460]
Si	Al	platform	[5733]
	Al, As, P	ZBC, modifier: Mg–Ni	[3394]
	Al, Ca, Cr, Fe, K, Na, Zn		[5599]
	Al, Cu, Fe, Ga, Li, In, Ni		[3823]
	B	indirect, determination of Cd	[3546]
Te	As, Tl	modifier: Pd–Mg	[483]

10.7.8 Nuclear Materials

Work with nuclear materials places special requirements on the laboratory and demands special levels of safety, a discussion of which is beyond the scope of this monograph. Introductory articles on the problems have been published by BAUDIN [426] and ACHE [28]. The minimum requirement is the installation of the analytical system in a glovebox [4677]; it is obviously only necessary to install the autosampler and the atomizer unit, i.e. the burner system [385, 5502] or the graphite furnace [353, 415, 2192, 2193, 2500, 4969, 5165], in the 'hot area', but not the spectrometer.

The purity requirements placed on all materials used in the nuclear area are very high. Even small contents of impurities in fuel rods can act as neutron poisons, for example, and markedly impair the effectiveness of the material. Corrosion products in the coolant provide information on incipient damage, but can also be deposited on the heat transfer surfaces and themselves become active.

For the majority of applications F AAS does not exhibit adequate sensitivity; moreover the air-acetylene flame is not hot enough to atomize a uranium or plutonium matrix so that considerable interferences must be expected. Also there is the added problem with radioactive samples that F AAS requires a relatively high amount of measurement solution and thus generates a great deal of waste, and the flame gases are a particular problem. The publications on direct determinations by F AAS are thus restricted to a few applications, such as the determination of potassium and sodium in uranium concentrates [5693] or ruthenium in synthetic waste solutions [2472]; interferences could largely be eliminated by the addition of a lanthanum buffer.

The sensitivity of the procedure can be increased when separation and preconcentration steps are applied and the waste problem can be reduced when the radioactive constituents are separated. GLEISBERG [2141] determined 14 trace elements in uranium after selectively separating it by extraction with tributylphosphate. BATISTONI and SMICHOWSKI [419] separated titanium and vanadium from uranium concentrates with tributylphosphate on a column packed with Kel-F, while BURBA and WILLMER [932] preconcentrated numerous trace elements in high-purity uranium on cellulose collectors. BANGIA et al. [368] extracted cadmium, cobalt, copper and nickel from high-purity uranium with dioctylsulfoxide in xylene, and YOUNG and BALDWIN [6491] extracted ruthenium with APDC in pentyl acetate. Both working groups used GF AAS for the determinations, as did RAJE et al. [4791] who separated and preconcentrated numerous trace elements from high-purity uranium and thorium by ion exchange. Likewise MAINKA and WEIS [3809] used an ion exchanger to separate cesium from highly concentrated salt solutions (331 g/kg salt content); to eliminate remaining interferences they applied Zeeman-effect BC. BRANDT et al. [780] separated solid corrosion products from the primary circuit of a pressurized water reactor by pressure filtration and determined the dissolved contents of chromium, cobalt, iron, and nickel after substantially reducing the volume of the filtrate in a rotary vaporizer. MATHEWS [3845] determined Ca, Co, Cr, Cu, Mg, Mn, and Pb in the sodium coolant of a fast breeder after distilling off most of the sodium.

Compared to F AAS for the analysis of radioactive materials, GF AAS has not only the advantage of much higher sensitivity but also requires considerably less sample. Additionally far fewer gases are released so that the waste problem is reduced by orders of magnitude. For this reason GF AAS has been relatively frequently applied for the direct determination of trace elements in nuclear materials, as the overview in Table 10-100 indicates. Next to the analysis of solutions, the direct analysis of solids and the analysis of slurries by GF AAS have been described. PAGE *et al.* [4517] determined a number of trace elements in uranium oxide powder after 'diluting' it with graphite, while SCHMIEDEL *et al.* [5165] investigated the determination of molybdenum, palladium, rhodium, and ruthenium in a simulated, non-active waste by analysis of slurries. These authors likewise diluted the test samples with graphite powder and applied Zeeman-effect BC.

Table 10-100. Selected publications on the analysis of nuclear materials by GF AAS.

Analyte	Test sample	Remarks	References
12 elements	radioactive waste		[4969]
13 elements	uranium, plutonium		[2192]
18 metals + 5 rare-earths	uranium, thorium, waste		[2097]
34 elements	uranium, plutonium		[2500]
Ag, Be, Ca, Fe, Pb, Sn	uranium		[2191]
Ag, Be, Cd, Li, Na, Sn, Zn	uranium, plutonium		[2193]
Ag, Be, Cd, Li, Pb	thorium		[5845]
Ag, Ca, K, Li, Mg, Na, Pb, Sn, Zn	uranium oxide	solid sampling	[4517]
Co, Cr, Fe, Ni	cooling water		[780]
Cs	radioactive waste		[1198, 3809]
Cu, K, Li, Na	uranium		[4564]
Fe	cooling water	influence of boric acid	[5475]
Mo, Pd, Rh, Ru	simulated waste	slurry sampling	[5165]

10.7.9 Fertilizers, Fungicides and Pesticides

The analysis of fertilizers is one of the oldest applications of AAS; DAVID [1434] has published an overview of papers up to 1978 covering this area. The complete digestion of fertilizers and the minerals used for their production is often only possible with a fusion [2731]. However, in practice it is the *bioavailable fraction* that is of interest, so that in individual cases only the water-soluble portion [3175, 4045] of an element or the portion that can be extracted with ammonium citrate [5844] is determined. For the determination of micronutrients such as copper, iron, manganese, or zinc the sample is mostly fumed off with hydrochloric acid and the residue is taken up in dilute acid [2347]. For manganese at least, a digestion in nitric acid-sulfuric acid provides a value that better reflects the bioavailable fraction than a digestion in hydrochloric acid [5843]. For the determination of elements of toxicological interest, such as cadmium, lead, or mercury, a digestion in *aqua regia* similar to the analysis of soils and sludges is mostly used [3881, 6109] since these elements are more completely extracted than with hydrochloric acid alone.

Since the analytes of interest in fertilizers are mostly present in relatively high concentrations, their determination is exclusively the domain of F AAS. Table 10-101 provides an overview of publications on the direct determination of a number of important elements by F AAS. In addition, a number of procedures for preconcentration and matrix separation have also been described, such as the extraction of boron with 2-ethyl-1,3-hexandiol in IBMK [4045] or chloroform [6183, 6184], or molybdenum with 5,5'-methylenedisalicylhyroxamic acid [5078]. A number of investigations on speciation are also worth mentioning, such as the determination of chelated iron [5362] or the separation of various metal chelates by gel permeation chromatography and subsequent F AAS determination [1446]. The separate determination of Cr(III) and Cr(VI) was achieved by coprecipitating the Cr(III) with aluminium hydroxide and determining Cr(VI) in the filtrate [4236].

Elements of toxicological interest, such as cadmium and lead, are nowadays preferentially determined directly by GF AAS [3881, 6109, 6376], but extraction procedures with tri-n-octylamine and potassium iodide in IBMK [1472] for the subsequent determination by F AAS have also been described. While manual extraction procedures for trace elements are used less and less due to the effort and the risk of errors, the importance of automatic on-line preconcentration coupled with FI-F AAS can increase dramatically in the future [6558].

Next to the determination of elements, several *indirect procedures* for fertilizer analysis have been described. ANWAR and UL HAQUE [226] determined nitrate via the formation of the diphenylsilver complex and the determination of the excess silver. CAMPBELL and TIOH [1046] determined sulfate in fertilizers by precipitation as barium sulfate and dissolution of the precipitate in EDTA. WOODIS et al. [6375] determined biuret in mixed fertilizers and urea via the addition of copper in strong alkaline solution;

Table 10-101. Selected publications on the direct determination of important elements in fertilizers by F AAS.

Analyte	Method / Test sample / Remarks	References
Al	aqueous extract	[3175]
B	liquid fertilizer	[4944]
Ca, Mg	SrCl$_2$ and EDTA as buffers	[5659]
Cu	liquid fertilizer	[4945]
Cu, Fe, Mg, Mn, Zn	interlaboratory trial	[4003-4006]
Fe, Mg, Zn	comparison with ICP OES	[300]
K		[2658, 4011, 5844]
Mn	interlaboratory trial	[5843]
Mo	interlaboratory trial	[2657, 3201, 3202]
Mo, Zn	interlaboratory trial	[4946]
Na	interlaboratory trial	[70, 1335]
P	direct determination	[2625]
P	indirect via Mg	[6390]

a biuret-copper complex is formed while the excess copper is precipitated. After filtration the dissolved copper, which corresponds to the concentration of the biuret, is determined

The determination of *fungicides and pesticides* and their residues on foodstuffs or wood can be performed by F AAS since the toxic agent in these substances is mostly an organometallic compound. An extraction with an organic solvent, occasionally followed by a back extraction [662], is often performed prior to the determination [663, 3626], or even a complete digestion [5576]. The procedure described for the determination of the fungicide Ziram comprising continuous extraction of the copper complex and on-line determination by FI-F AAS could be an indication of future developments in this direction [663]. Further examples are the determinations of organometallic fungicides on thiocarbamic acid basis via their contents of iron, manganese or zinc [2258], copper-8-hydroxyquinolate bleach on wood surfaces [5576], or the tin-containing pesticide Cyhexatin on apples [3626].

10.7.10 Paints, Varnishes, Pigments

The determination of *lead in paint* occupies by far the greatest space in the literature. The reason for this exceptional interest is the risk of long term exposure to this type of paint via abrasion and dust in interior rooms, to which small children in particular are exposed. Due to the relatively high analyte content the determination is mostly performed by F AAS and papers on the subject are largely devoted to digestion procedures. These include dry ashing at 300–600 °C [2445, 3773, 5218, 5987]; standard procedures nowadays recommend 475 °C under the addition of magnesium carbonate [803]. An alternative is digestion in nitric acid [2633, 4691], nitric acid-sulfuric acid or sulfuric acid-hydrogen peroxide [803]. Microwave-assisted digestions provide the same results as conventional digestions but are markedly quicker [638, 1326]. For the determination of lead and numerous other elements in fresh paints and additives the samples can be simply diluted with IBMK and analyzed by F AAS [1688]. Yet a further possibility is to mineralize with nitric acid and then to emulsify with Triton X-100 [2066].

Next to the total content of a potentially toxic element determined via a digestion, the 'soluble fraction' is of considerable interest for obvious reasons and this can be determined through suitable extraction. Corresponding procedures have been described for antimony [805], cadmium [806], lead [804], and mercury [807]. The *leachability* of antifouling paints is also of considerable interest, such as are used to paint the bottoms of ships. Corresponding procedures have been developed for copper [256], mercury [2877], and tin [3048].

The analysis of pigments such as Cr_2O_3, Fe_2O_3, or TiO_2 frequently requires a fusion with potassium hydrogen sulfate [4058] or lithium tetraborate [2446]. Dry ashing followed by a pressure digestion with sulfuric acid-potassium permanganate [4375] or with hydrofluoric acid [2016] has been proposed as an alternative. The *direct analysis of slurries* has proven to be particularly elegant and simple; this procedure was described by FULLER [2019] as early as 1976 for the determination of copper, iron, lead, and manganese in TiO_2 pigments using both F AAS and GF AAS. An advantage for this procedure is that pigments exhibit a very consistent particle size of 10 ± 0.3 µm. The injection technique has been applied for determinations in the flame to prevent the nebulizer and burner from becoming clogged. The results were in good agreement with other procedures; the direct procedure provides the greater speed and simplicity and the lowest blank values. This procedure was later extended by LÓPEZ GARCÍA and co-workers to the determination of numerous elements in Fe_2O_3 pigments and to all AAS techniques. They determined chromium, copper, manganese, and zinc directly in aqueous suspensions of pigments by FI-F AAS [3606]; arsenic under the addition of Triton X-100 and nickel as the modifier [3605], and lead with the ammonium phosphate modifier [3604] by GF AAS. The determination of lead from slurries by HG AAS under the addition of ammonium peroxodisulfate has also been described [3609]. The procedure has also been applied for the determination of mercury by CV AAS under the addition of hexametaphosphate and sodium tetrahydroborate as the reductant [3607].

10.7.11 Paper, Textiles, Leather

For the analysis of *paper, cardboard and raw materials for the production of paper*, next to the metals of technological importance such as Cr, Cu, Fe, Ti, and Zn, metals of toxicological interest such as arsenic, cadmium, lead, and mercury are also of significance [3163, 3166, 3345]. For the determination of relatively non-volatile elements such as copper, iron [3163] or titanium [838], dry ashing can be applied for the digestion procedure; this has a number of advantages compared to wet digestion and provides reproducible values. This type of digestion is less suitable for toxicologically relevant elements since there is a risk of uncontrollable losses, particularly for arsenic and mercury. For the determination of these elements a digestion in nitric acid [4495, 6169], frequently in an autoclave under pressure [2219, 3161, 3162, 3167, 4495], is thus mostly performed. An alternative to digestion is direct solids analysis in the graphite furnace [3081, 3165, 3432, 5370], which can be performed easily with punched pieces of paper and provides reproducible values [3165, 3345].

The elements present in higher concentrations, such as copper, iron [3163], or titanium [838], can be easily determined by F AAS, nevertheless like cadmium and lead they are frequently determined by GF AAS [3161, 3345]. Table 10-102 presents a number of publications on the determination of trace elements in paper and cardboard.

Table 10-102. Selected publications on the determination of trace elements in paper, cardboard, and raw materials for the production of paper.

Analyte	Remarks	References
Ag, Cd, Cr, Co, Cu, Fe, Mg, Mn, Sb	solid sampling	[5370]
Cd	pressure digestion	[2219, 3167]
	interlaboratory trial	[4495]
Cd, Cr, Cu, Pb, Zn	pressure digestion	[3161]
Cd, Cu, Pb, Mn	solid sampling	[3432]
Cu, Fe, Mn, Si	solid sampling	[3081]

For the determination of arsenic in paper HG AAS is normally employed following a digestion [3162], while for mercury practically only CV AAS comes into consideration [3164].

For wrapping paper used for foodstuffs, next to the total content of toxic heavy metals, the soluble or leachable fraction is of particular interest. Oil, water, and dilute acetic acid have been investigated as extracting agents; only the latter leaches out noticeable quantities of arsenic, cadmium, lead, or mercury [3166]. These elements were leached out to varying proportions from the materials investigated but by no means quantitatively, so that a risk to the consumer is hardly to be expected [3166]. PARALUSZ [4534] determined traces of organic silicon in paper products by extracting them with IBMK.

The contents of trace elements are of interest in natural fibers such as *wool* or *cotton*, but more especially in *synthetic fibers* which can contain residues of catalysts or stabilizers, flame retardants, or antibacterial additives (e.g. organotin compounds). The papers published by TONINI [5876, 5878] provide an overview on applications in this area. A number of elements can be determined after simple extraction with acid or after dry ashing and extraction of the residue in nitric acid or hydrochloric acid [5877]. Certain synthetic fibers also dissolve in organic solvents, acids or bases and can be measured directly by F AAS. Cellulose acetate, for example, dissolves in IBMK, polyacrylonitrile in DMF, nylon in formic acid, wool in sodium hydroxide, and cotton and cellulose in sulfuric acid [4465, 4466].

The sensitivity of F AAS is mostly adequate for the determination of trace elements in textile fibers. HARTLEY and INGLIS determined aluminium [2409], Ba, Cr, Cu, Hg, Sn, Sr, and Zn [2410] in wool; SIMONIAN [5372] determined copper in textiles after extraction in hot hydrochloric acid; and DUNK *et al.* [1629] described an indirect determination for sulfate. PRICE [4713] determined Co, Cu, Fe, Mn, Pb, Sb, Sn, and Zn in raw materials for synthetic fibers such as rayon and nylon directly by F AAS after dissolving them in sulfuric acid-hydrogen peroxide. EVANS [1779] determined silicon in wool by F AAS after aminolysis and extraction of the silicones with IBMK. KORENAGA [3250] determined traces of arsenic in acrylic fibers containing antimony by HG AAS following a digestion in nitric acid, perchloric acid and sulfuric acid, extraction into benzene, and back extraction into the aqueous phase. VÖLLKOPF *et al.* [6097, 6098] used direct solids sampling to determine cadmium, copper, manganese, and rubidium in textile fibers such as polyester and polyamide by GF AAS. The results were in good agreement with values obtained by F AAS after digestion and extraction; calibration was against aqueous solutions.

The use of chromium salts in *leather* tanning was a reason for a number of authors to investigate this element more closely. DELLA MONICA and MCDOWELL [4164] did not observe any interferences in the determination of chromium by F AAS due to glycine or hydrolyzed leather powder. On the other hand, KNECHT [3158] reported that the signal for Cr(III) was around 6% higher than for Cr(VI) in both the air-acetylene flame and the nitrous oxide-acetylene flame (refer to Section 9.12). MENDEN and RUTLAND [4050] examined the leachability of chromium from wastes discharged from tanning works. MENDEN *et al.* [4051] later found when extracting Cr(VI) from tanning waste with APDC/4-methyl-2-pentanone that organic Cr(III) compounds were coextracted, thus leading to higher values. MILAČIČ *et al.* [4083] described a sequential extraction procedure for water-soluble, exchangeable, organically-bound Cr(VI), as well as the carbonate, crystalline, hydroxide, and sulfide fractions. CHEN and HIN [1238] determined cadmium in leather after treatment with sulfuric acid-hydrogen peroxide.

10.8 Multidisciplinary Applications

In this section we shall discuss two further areas of application of AAS. The materials analyzed are the same as those treated in earlier sections, but the goal of the analyses is different so that different methodology can be required. We are referring to forensic and archeological analysis in which the goal is to prove the source or identity of the exhibit or find by means of a characteristic distribution pattern of elements. A common requirement in both disciplines is that the object being investigated be left as intact as possible, i.e. it is necessary to work with the smallest possible analytical samples.

10.8.1 Forensic Analysis

Forensic analysis is involved with a large number of materials, each of which is discussed in detail in the preceding sections, so that cross-reference to this information is adequate in most cases. One of the few applications specific to forensic analysis is the examination of residues remaining on the skin of a gunman after firing a handgun. DALE [1409] provides a good overview on the application of AAS in forensic analysis with a number of historical aspects on the development of the science.

The analysis of body fluids and tissues assumes a major role in forensic analysis. If there is suspicion that the victim was poisoned with arsenic or thallium, for example, then next to the analysis of blood the analysis of tissues such as liver, kidneys or stomach is of major interest since this can lead to conclusions on the time point of poisoning. In addition the analysis of hair and nails can provide valuable information on events that occurred several weeks previously, i.e. whether acute poisoning or long-term exposure is involved. KOWAL et al. [3266] made detailed investigations on the determination of lead in tissue, SOLOMONS and WALL [5491] were concerned with the analysis of arsenic from the viewpoint of forensics, and FELBY [1872] described the determination of organosilicon polymers in tissues. In general, however, we refer our readers to Section 10.1.3 (Toxicology) and Section 10.1.5 (Analysis of Tissues).

A typical forensic analysis that has been performed for decades is the examination of gunshot residues on the skin. Independent of the design of the weapon, residues remain on the hands and face of the gunman upon firing a shot. The igniter material, which comprises a mixture of lead styphnate, barium nitrate, antimony sulfide, and a number of other substances specific to the manufacturer, is of particular significance. Usually the affected areas on the skin of a suspect are wiped with a cotton swab wetted with dilute nitric acid (1 mol/L HNO_3). The swab is leached with 1 mol/L HNO_3 and examined for antimony, barium, and lead. The presence of these elements in a given ratio on the skin is not only evidence that the suspect had fired a weapon but also provides information on the type of weapon and the ammunition used. For this determination GF AAS offers adequate sensitivity and was thus employed for this application soon after its introduction [2158, 4851]. Recent work on this subject has been published by DAHL and LOTT [1406, 1407], KOONS et al. [3240, 3241], LICHTENBERG [3517], NEWTON [4338], and SWOVICK [5720].

The investigation of gunpowder [374], pistol bullets [903], shotgun pellets [3770], cartridge cases [2542], or metal traces remaining in bullet holes [4823] has a certain similarity to the analysis of gunshot residues. In all these investigations it is the goal to identify the source of the projectile or cartridge by means of the element distribution and thus to obtain information about the culprit. For the same reason a large number of other materials are investigated for their 'fingerprints', such as glass splinters [975, 2459], paint flakes, fibers, or even gasoline [1182], just to name a few. For the analytical procedures we refer our readers to the information provided for the individual materials; i.e. to Section 10.6.2.2 (analysis of lead alloys) for the projectiles, Section 10.6.2.3 (analysis of copper alloys) for cartridge cases, Section 10.7.1 for the analysis of glass, Section 10.7.10 for paint, and Section 10.7.11 for paper and textiles. Although frequently solely the major and minor constituents are used to determine the distribution pattern, GF AAS is used preferentially since in most cases only very small amounts of sample are available and in addition the 'evidence' should remain as intact as possible for the future trial.

10.8.2 Archeological Finds

The analysis of archeological finds (archeometry) serves primarily to answer important questions posed by archeologists:

i)   What is the object made from.
ii)   Where does the material come from.
iii) Where does the object come from.
iv)   Where was the object produced.
v)   When and by whom was the object produced.
vi)   Which extras (paint etc.) were used.

Answers to these questions provide among other things information on old trade routes, the techniques and technologies employed, and enable forgeries to be recognized. As early as 1976 HUGHES *et al.* [2709] published a review article on the application of AAS in archeometry. Recent review articles have been published by HUGHES [2710], EBDON and FISHER [1660], and ZIMMER [6584].

Although non-destructive techniques are clearly preferred in archeometry, AAS—and GF AAS in particular—is nevertheless a method of choice due to its high sensitivity and low sample requirement. This is clear when we consider for example that X-ray fluorescence analysis requires a smooth, clean surface, and that removing a layer of corrosion or a patina of several square millimeters from the surface can cause far more damage than drilling a hole with a diameter of less than 1 mm. SZONNTAGH [5728] developed a microdrilling technique for antique coins with which far more reliable results can be obtained than by surface analysis and only leaves a barely visible hole on the cylindrical surface while the face of the coin remains undamaged.

Objects examined include coins of gold, silver or bronze; figures, jewelry, and utensils made of noble metals, copper or bronze; axes, swords and other objects made of iron; and pottery, ceramics, tiles, and glass objects. For procedures of sample preparation we refer our readers to the respective sections: iron, steel and ferrous alloys (Section 10.6.1), copper and copper alloys (Section 10.6.2.3), noble metals (Section 10.6.2.7), glass (Section 10.7.1), and ceramic materials (Section 10.7.2). Table 10-103 provides an overview of selected papers on the application of AAS in archeometry.

Table 10-103. Selected publications on the analysis of archeological finds by AAS.

Object	Analyte	Remarks	References
Flint, neolithic	Al, Fe, K, Mg, Na	F AAS	[5348]
	Al, Ca, Fe, Mg, Mn	F AAS	[5349]
Marble	minor and trace elements		[4152]
Ceramic, neolithic	Ca, Cu, Fe, K, Mg, Mn, Na, Zn	comparison of digestion procedures	[5598]
Terra sigillata	Al, Ba, Ca, Fe, K, Mn, Na	fusion, F AAS	[5927]
	Al, Ba, Ca, Cr, Cu, Fe, Mn, T	GF AAS, pattern recognition	[4136]
	Al, Ca, Fe, K, Mg, Mn, Ti	GF AAS, pattern recognition	[281]
	Al, Ca, Fe, K, Mg, Mn, Ti	GF AAS, chemometrics	[4686]
Roman tiles	Ca, Fe, K, Mg, Na	F AAS	[668]
Glass	Co, S	F AAS	[4822]
	Ag, Bi, Pb	GF AAS, solid sampling	[2459]
Skeleton bones	Ag, Cd, Pb, Zn	GF AAS	[1762]
Arrow heads (iron)	Ag, Bi, Cd, Co, Cr, Cu, Mn, Se, Te, Zn	GF AAS	[320]
Copper alloys	Cu	F AAS	[6000]
	Sn	solvent extraction, F AAS	[3091]
	Ag, As, Bi, Co, Cu, Fe, Ni, Pb, Sb, Sn, Zn	GF AAS	[2137]
Coins	Bi, Cu, Pb	F AAS, GF AAS	[2096]
	Ag, Au, Cu, Fe, Pb, Sb, Sn, Zn	F AAS	[5728]
Gold jewelry, bronze age	Ag, Au, Cu	F AAS	[760]

Bibliography

[1]   Aadland, E.; J. Asseth, B. Radziuk, K. Saeed, Y. Thomassen, *Fresenius Z. Anal. Chem. 328/4-5*, 362–366 (1987).
[2]   Aadland, E.; J. Aaseth, K. Dahl, B. Radziuk, Y. Thomassen, *J. Trace Elem. Electrol. Health Dis. 4/4*, 233–236 (1990).
[3]   Abbas, M.; R. Bruns, I. Scarminio, *Environ. Pollut. 79*, 225 (1992).
[4]   Abbasi, S.A.; *Int. J. Environ. Anal. Chem. 33/2*, 149–160 (1988).
[5]   Abbasi, S.A.; *Int. J. Environ. Anal. Chem. 33/1*, 43–57 (1988).
[6]   Abbasi, S.A.; *Anal. Lett. 21/4*, 653–665 (1988).
[7]   Abbasi, S.A.; *Anal. Lett. 21/9*, 1705–1721 (1988).
[8]   Abbasi, S.A.; *Anal. Lett. 21*, 491–505 (1988).
[9]   Abbasi, S.A.; *Int. J. Environ. Anal. Chem. 35/3*, 139–147 (1989).
[10]   Abbasi, S.A.; *Anal. Lett. 22/1*, 237–255 (1989).
[11]   Abbasi, S.A.; *Int. J. Environ. Anal. Chem. 36*, 163–172 (1989).
[12]   Abbasi, S.A.; *Rev. Roum. Chim. 36*, 901 (1991).
[13]   Abbey, S.; *Geostand. Newsl. 15*, 191–194 (1991).
[14]   Abdallah, A.M.; M.A. Kabil, M.A. Mostafa, *Chem. Anal. (Warsaw) 31/1*, 3–13 (1986).
[15]   Abdallah, A.M.; M.A. Kabil, *Chem. Anal. (Warsaw) 33/1*, 75–84 (1988).
[16]   Abe, K. ; S. Terashima, *Bull. Geol. Surv. Jpn. 37*, 335–345 (1986).
[17]   Abele, C.; G. Weichbrodt, K.-H. Wichmann, *Fresenius Z. Anal. Chem. 322/1*, 11–16 (1985).
[18]   Abollino, O.; C. Sarzanini, V. Porta, E. Mentasti, *Ann. Chim. (Rom) 81/1-2*, 23–37 (1991).
[19]   Abollino, O.; E. Mentasti, C. Sarzanini, E. Modone, M. Braglia, *Fresenius J. Anal. Chem. 343/6*, 482–487 (1992).
[20]   Abollino, O.; M. Aceto, G. Sacchero, C. Sarzanini, E. Mentasti, *Anal. Chim. Acta 305*, 200–206 (1995).
[21]   Abraham, H.-J.; C. Schlums, *Staub. Reinhalt. Luft 52/9*, 353–360 (1992).
[22]   Abraham, V.J.; J.D. Hatheway, D.A. Anderson, E.M. Skelly Frame, *At. Spectrosc. 13/3*, 105–107 (1992).
[23]   Abrams, M.M.; R.G. Burau, *Commun. Soil. Sci. Plant Anal. 20/3-4*, 221–237 (1989).
[24]   Absalan, G.; A. Safavi, A. Massoumi, *Microchem. J. 37/2*, 212–215 (1988).
[25]   Abu-Elgheit, M.A.; *Indian J. Technol. 21/3*, 128–129 (1983).
[26]   Accomasso, G.M.; V. Zelano, P.G. Daniele, D. Gastaldi, M. Ginepro, G. Ostacoli, *Spectrochim. Acta, Part A 49/9*, 1205–1212 (1993).
[27]   Accominotti, M.; Y. Pegon, J.-J. Vallon, *Clin. Chim. Acta 173/2*, 99–106 (1988).
[28]   Ache, H.J.; Fresenius J. Anal. Chem. 343, 852–862 (1992).
[29]   Achterberg, E.P.; C.M.G. van den Berg, *Anal. Chim. Acta 291/3*, 213–232 (1994).
[30]   Adams, F.; *Pure Appl. Chem. 55/12*, 1925–1942 (1983).
[31]   Adams, M.J.; G.F. Kirkbright, *Can. J. Spectrosc. 21/5*, 127–133 (1976).
[32]   Adams, M.J.; G.W. Ewen, C.A. Shanda, *J. Autom. Chem. 10/3*, 130–134 (1988).
[33]   Adams, P.B.; W.O. Passmore, *Anal. Chem. 38/4*, 630–633 (1966).
[34]   Adams, P.B.; in: M. Zief, R. Speights (Editors), *Ultrapurity*, Marcel Dekker, New York, 1972, 293–351.
[35]   Adelantado, J.V.G. ; V. Peris Martínez, A. Pastor Garcia, F. Bosch Reig, *Talanta 38/9*, 959–963 (1991).
[36]   Adelhelm, C.; D. Hirschfeld, *Fresenius J. Anal. Chem. 342/1-2*, 125–127 (1992).
[37]   Adeloju, S.B.; A.M. Bond, *Anal. Chim. Acta 164*, 181 (1984).
[38]   Adeloju, S.B.; H.S. Dhindsa, R.K. Tandon, *Anal. Chim. Acta 285/3*, 359–364 (1994).
[39]   Adler, C.R.; W.R. Marshall, *Chem. Eng. Progress 47*, 515–522 (1951).
[40]   Adriaenssens, E.; P. Knoop, *Anal. Chim. Acta 68*, 37–48 (1973).
[41]   Adriaenssens, E.; F. Verbeek, *At. Absorpt. Newsletter 12/3*, 57–59 (1973).
[42]   Adriaenssens, E.; F. Verbeek, *At. Absorpt. Newsletter 13*, 41–44 (1974).
[43]   Adsul, J.S.; C.C. Dias, S.G. Iyer, C. Venkateswarlu, *Talanta 34/5*, 503–504 (1987).
[44]   Agarwal, R.P.; R.I. Henkin, *Biol. Trace Elem. Res. 4*, 117–124 (1982).
[45]   Agarwal, R.P.; R.I. Henkin, *Biol. Trace Elem. Res. 7/4*, 199–208 (1985).
[46]   Agazzi, E.J.; *Anal. Chem. 37/3*, 364–366 (1965).
[47]   Agemian, H.; K.I. Aspila, A.S.Y. Chau, *Anal. Chem. 47*, 1038–1041 (1975).
[48]   Agemian, H.; A.S.Y. Chau, *Anal. Chem. 50*, 13–16 (1978).

[49]   Agemian, H.; V. Cheam, *Anal. Chim. Acta 101*, 193–197 (1978).
[50]   Agemian, H.; J.A. DaSilva, *Anal. Chim. Acta 104*, 285–291 (1979).
[51]   Agemian, H.; D.P. Sturtevant, K.D. Austen, *Analyst (London) 105*, 125–130 (1980).
[52]   Agemian, H.; E. Bedek, *Anal. Chim. Acta 119*, 323–330 (1980).
[53]   Aggarwal, S.K.; M. Kinter, J. Nicholson, D.A. Herold, *Anal. Chem. 66/8*, 1316–1322 (1994).
[54]   Aggett, J.; A.C. Aspell, *Analyst (London) 101*, 341–347 (1976).
[55]   Aggett, J.; G.A. O'Brian, *Analyst (London) 106*, 497–505 (1981).
[56]   Aggett, J.; M.R. Kriegman, *Analyst (London) 112*, 153–157 (1987).
[57]   Agterdenbos, J.; *Anal. Chim. Acta 108*, 315–323 (1979).
[58]   Agterdenbos, J.; F.J.M.J. Maessen, J. Balke, *Anal. Chim. Acta 132*, 127–137 (1981).
[59]   Agterdenbos, J.; J.P.M. Van Noort, F.F. Peters, D. Bax, J.P. Ter Heege, *Spectrochim. Acta, Part B 40/3*, 501–515 (1985).
[60]   Agterdenbos, J.; J.P.M. van Noort, F.F. Peters, D. Bax, *Spectrochim. Acta, Part B 41/3*, 283–290 (1986).
[61]   Agterdenbos, J.; J.T. van Elteren, D. Bax, J.P. ter Heege, *Spectrochim. Acta, Part B 41/4*, 303–316 (1986).
[62]   Agterdenbos, J.; D. Bax, *Fresenius Z. Anal. Chem. 323/7*, 783–787 (1986).
[63]   Agterdenbos, J.; D. Bax, *Anal. Chim. Acta 188*, 127–135 (1986).
[64]   Agterdenbos, J.; R.W. Bussink, D. Bax, *Anal. Chim. Acta 232/2*, 405–407 (1990).
[65]   Ahlgrén, M.; A. Kontkanen, K. Vattulainen, H. Vehviläinen, *Analyst (London) 113/2*, 285–287 (1988).
[66]   Ahmed, R.; M. Stoeppler, *Analyst (London) 111/12*, 1371–1374 (1986).
[67]   Ahmed, R.; K. May, M. Stoeppler, *Fresenius Z. Anal. Chem. 326*, 510–516 (1987).
[68]   Ahmed, R.; M. Stoeppler, *Anal. Chim. Acta 192*, 109–113 (1987).
[69]   Aidorov, T.K.; R.S. Sadykov, *Zh. Prikl. Spektrosk. 16*, 379 (1972).
[70]   Aihara, M.; M. Kiboku, *Kinki Daigaku Kogakubu Kenkyu Hohoku 22*, 47 (1988).
[71]   Aitio, A.; J. Järvisalo, *Pure Appl. Chem. 56/4*, 549–566 (1984).
[72]   Aitio, A.; J. Järvisalo, *Ann. Clin. Lab. Sci. 15/2*, 121 (1985).
[73]   Aitio, A.; J. Järvisalo, M. Stoeppler, in: R.F.M. Herber, M. Stoeppler (Editors), *Trace Element Analysis in Biological Specimens*, Elsevier, Amsterdam, 1994, 3–19.
[74]   Aizawa, Y.; T. Takata, M. Kurihara, M. Tominaga, Y. Inoue, *Appl. Organomet. Chem. 2*, 395–398 (1988).
[75]   Aizpún-Fernández, B.; M.L. Fernández, E. Blanco, A. Sanz-Medel, *J. Anal. At. Spectrom. 9/11*, 1279–1284 (1994).
[76]   Ajayi, O.O.; D. Littlejohn, C.B. Boss, *Talanta 36/8*, 805–810 (1989).
[77]   Ajlec, R.; M. Cop, J. Štupar, *Analyst (London) 113/4*, 585–590 (1988).
[78]   Ajlec, R.; J. Štupar, *Analyst (London) 114/2*, 137–142 (1989).
[79]   Akama, Y.; A. Tong, S. Ishima, M. Kajitani, *Anal. Sci. 8/1*, 41–44 (1992).
[80]   Akatsuka, K.; I. Atsuya, *Anal. Chim. Acta 202*, 223–230 (1987).
[81]   Akatsuka, K.; N. Nobuyama, I. Atsuya, *Anal. Sci. 4/3*, 281–285 (1988).
[82]   Akatsuka, K.; N. Nobuyama, I. Atsuya, *Anal. Sci. 5/4*, 475–479 (1989).
[83]   Akatsuka, K.; I. Atsuya, *Fresenius Z. Anal. Chem. 335/2*, 200–204 (1989).
[84]   Akatsuka, K.; T. Katoh, N. Nobuyama, T. Okanaka, M. Okumura, S. Hoshi, *Anal. Sci. 12/2*, 209–213 (1996).
[85]   Akesson, B.; B. Martensson, *Int. J. Vitam. Nutr. Res. 61*, 72–76 (1991).
[86]   Akgün, E.; U. Pindur, *Pharm. Ind. 44/9*, 930–931 (1982).
[87]   Akgün, E.; U. Pindur, *Pharm. Acta Helv. 58/5-6*, 130–132 (1983).
[88]   Akgün, E.; U. Pindur, *Dtsch. Apoth.-Ztg. 123*, 2322–2323 (1983).
[89]   Akker, A.H. van den; H. van den Heuvel, *At. Spectrosc. 13/2*, 72–74 (1992).
[90]   Akman, S.; Ö. Genç, A.R. Özdural, T. Balkis, *Spectrochim. Acta, Part B 35/6*, 373–378 (1980).
[91]   Akman, S.; Ö. Genç, T. Balkis, *Spectrochim. Acta, Part B 36/11*, 1121–1130 (1981).
[92]   Akman, S.; S. Bektas, Ö. Genç, *Spectrochim. Acta, Part B 43/6-7*, 763–772 (1988).
[93]   Akman, S.; Ö. Genç, S. Bektas, *Spectrochim. Acta, Part B 46/14*, 1829–1839 (1991).
[94]   Akman, S.; H. Ince, Ü. Köklü, *J. Anal. At. Spectrom. 7/2*, 187–189 (1992).
[95]   Akman, S.; G. Döner, *Spectrochim. Acta, Part B 49/7*, 665–675 (1994).
[96]   Akman, S.; G. Döner, *Spectrochim. Acta, Part B 50/9*, 975–984 (1995).
[97]   Akman, S.; G. Döner, *Spectrochim. Acta, Part B 51/9-10*, 1163–1167 (1996).
[98]   Al-Abachi, M.Q.; E.S. Salih, *Microchem. J. 37/3*, 293–298 (1988).
[99]   Al-Sibaai, A A.; A.G. Fogg, *Analyst (London) 98*, 732–738 (1973).
[100] Alaejos, M.S.; C.D. Romero, *Chem. Rev.95/1*, 227–257 (1995).

[101]   Alarcón, O.M.; J.L. Burguera, M. Burguera, N. Suárez, J. Reinoza, *J. Trace Elem. Electrol. Health Dis. 3*, 203–208 (1989).
[102]   Alarcón, O.M.; J.L. Burguera, M. Burguera, *Trace Elem. Med. 8*, 19–22 (1991).
[103]   Alary, J.; A. Villet, A. Coeur, *Ann. Pharm. France 34*, 419 (1976).
[104]   Alary, J.; P. Bourbon, J. Vandaele, *Sci. Total Environ. 46*, 181–190 (1985).
[105]   Alary, J.; J. Vandaele, C. Escrieut, R. Haran, *Talanta 33/9*, 748–750 (1986).
[106]   Alberts, J.J.; J.E. Schindler, R.W. Miller, *Science 184*, 895–896 (1974).
[107]   Alcock, N.W.; *Biol. Trace Elem. Res. 13*, 363–370 (1987).
[108]   Alder, J.F.; D. Alger, A.J. Samuel, T.S. West, *Anal. Chim. Acta 87*, 301–311 (1976).
[109]   Alder, J.F.; B.C. Das, *Anal. Chim. Acta 94*, 193–194 (1977).
[110]   Alder, J.F.; B.C. Das, *Analyst (London) 102*, 564–568 (1977).
[111]   Alder, J.F.; D.A. Hickman, *Anal. Chem. 49/2*, 336–339 (1977).
[112]   Alder, J.F.; B.C. Das, *At. Absorpt. Newsletter 17/3*, 63–64 (1978).
[113]   Aldous, K.M.; R.F. Browner, R.M. Dagnall, T.S. West, *Anal. Chem. 42*, 939–941 (1970).
[114]   Aldous, K.M.; B.W. Bailey, J.M. Rankin, *Anal. Chem. 44/1*, 191–194 (1972).
[115]   Alegria, A.; R. Barberá, R. Farré, *J. Micronutr. Anal. 4*, 229 (1988).
[116]   Alemasova, A.S.; A.Ya. Makhno, I.A. Shevchuk, *Vysokochistye Veshchestva 5*, 176 (1991).
[117]   Aleshko-Ozhevskii, Yu.P.; N.N. Makhova, L.V. Shevyakova, *Zh. Anal. Khim. 40*, 100 (1985).
[118]   Alexander, J.; K. Saeed, Y. Thomassen, *Anal. Chim. Acta 120*, 377–382 (1980).
[119]   Alexander, N.M.; in: H.G. Seiler, A. Sigel, H. Sigel (Editors*), Handbook on Metals in Clinical and Analytical Chemistry*, Marcel Dekker, New York, 1994, 411–421.
[120]   Alexandrov, V.V.; A.I. Bezlepkin, A.S. Khomyak, *Spectrochim. Acta, Part B 36/12*, 1163–1172 (1981).
[121]   Alexandrova, A.; S. Arpadjan, *Analyst (London) 118/10*, 1309–1312 (1993).
[122]   Alfthan, G.; *Clin. Chem. (Winston-Salem) 31/3*, 500 (1985).
[123]   Alimonti, A.; F. Petrucci, N. Violante, S. Caroli, *Ann. Ist. Super. Sanita 19/4*, 665–668 (1983).
[124]   Alimonti, A.; F. Petrucci, N. Violante, S. Caroli, *Ann. Ist. Super. Sanita 19/4*, 66–67 (1985).
[125]   Alkemade, C.Th.J.; J.M.W. Milatz, *Appl. Sci. Res., Sect. B 4*, 289–299 (1955).
[126]   Alkemade, C.Th.J.; J.M.W. Milatz, *J. Opt. Soc. Am. 45*, 583–584 (1955).
[127]   Alkemade, C.Th.J.; M.H. Voorhuis, *Fresenius Z. Anal. Chem. 163*, 91 (1958).
[128]   Alkemade, C.Th.J.; *Anal. Chem. 38*, 1252–1253 (1966).
[129]   Alkemade, C.Th.J.; H.P. Hooymayers, *Combust. Flame 10*, 306–308 (1966).
[130]   Alkemade, C.Th.J.; in: J.A. Dean, T.C. Rains (Editors), *Flame Emission and Atomic Absorption Spectrometry, Vol. 1 Theory*, Marcel Dekker, New York - London, 1969, 101–150.
[131]   Alkemade, C.Th.J.; in: R. Mavrodineanu (Editor), *Analytical Flame Spectroscopy, Selected Topics*, MacMillan, London, 1970, 1–46.
[132]   Alkemade, C.Th.J.; R. Herrmann, *Fundamentals of Analytical Flame Spectroscopy*, Adam Hilger (John Wiley & Sons), Bristol (New York), 1979.
[133]   Allain, P.; Y. Mauras, *Anal. Chim. Acta 165/11*, 141–147 (1984).
[134]   Allain, P.; Y. Mauras, C. Grangeray, *Ann. Clin. Biochem. 24/5*, 518–519 (1987).
[135]   Allan, J.E.; *Analyst (London) 83*, 466–471 (1958).
[136]   Allan, J.E.; *Spectrochim. Acta 10*, 800–806 (1959).
[137]   Allan, J.E.; *Nature (London) 187*, 1110 (1960).
[138]   Allan, J.E.; *Analyst (London) 86*, 530–534 (1961).
[139]   Allan, J.E.; *Spectrochim. Acta 17*, 467–473 (1961).
[140]   Allan, J.E.; *4th Australian Spectrosc. Conf.*, 1963.
[141]   Allan, J.E.; *Spectrochim. Acta, Part B 24/1*, 13–18 (1969).
[142]   Allen, A.G.; M. Radojevic, R.M. Harrison, *Environ. Sci. Technol. 22/5*, 517–522 (1988).
[143]   Allen, S.; *Anal. Biochem. 138*, 346 (1984).
[144]   Aller, A.J.; *Anal. Chim. Acta 134*, 293–300 (1982).
[145]   Aller, A.J.; *Aluminium (Düsseldorf) 58*, 660–663 (1982).
[146]   Aller, A.J.; D. Bonilla, *Anal. Sci. 6/2*, 309–311 (1990).
[147]   Aller, A.J.; C. García-Olalla, *J. Anal. At. Spectrom. 7/5*, 753–760 (1992).
[148]   Aller, A.J.; J.M. Lumbreras, L.C. Robles, G.M. Fernández, *Anal. Proc. (London) 32/12*, 511–514 (1995).
[149]   Aller, A.J.; *Anal. Sci. 12/6*, 977–980 (1996).
[150]   Almeida, M.C.; W.R. Seitz, *Appl. Spectrosc. 40/1*, 4–8 (1986).
[151]   Alt, F.; in: P. Brätter, P. Schramel (Editors), *Trace Element Analytical Chemistry in Medicine and Biology*, Walter de Gruyter, Berlin - New York, 1988, 279–296.
[152]   Alt, F.; U. Jerono, J. Messerschmidt, G. Tölg, *Mikrochim. Acta (Vienna) 3*, 299–304 (1988).

[153]   Alvarado, J.; L.E. León, F. López, C. Lima, *J. Anal. At. Spectrom. 3/1*, 135–138 (1988).
[154]   Alvarado, J.; A. Petrola, *J. Anal. At. Spectrom. 4/5*, 411–414 (1989).
[155]   Alvarado, J.; R.T. Moreno, A.R. Cristiano, *J. Trace Elem. Electrol. Health Dis. 5/3*, 173–180 (1991).
[156]   Alvarado, J.; A.R. Cristiano, *J. Anal. At. Spectrom. 8/2*, 253–259 (1993).
[157]   Alvarez, G.H.; S.G. Capar, *Anal. Chem. 59/3*, 530–533 (1987).
[158]   Alvarez, G.H.; S.G. Capar, *Anal. Lett. 24*, 1695–1710 (1991).
[159]   Alvarez, M.A.; N. Carrión, H. Gutiérrez, *Spectrochim. Acta, Part B 50/13*, 1581–1594 (1995).
[160]   Alvin, J.F.; F.R. Gardiner, *Analyst (London) 111/8*, 897–899 (1986).
[161]   Alzieu, C.; P. Michel, J. Sanjuan, B. Averty, *Appl. Organomet. Chem. 4*, 55–61 (1990).
[162]   Ambe, M.; K. Suwabe, *Anal. Chim. Acta 92*, 55–60 (1977).
[163]   Ambrosetti, P.; F. Librici, *At. Absorpt. Newsletter 18/1*, 38 (1979).
[164]   Ametani, K.; E. Shima, T. Oka, *Microchem. J. 28/1*, 37–54 (1983).
[165]   Amos, M.D.; P.E. Thomas, *Anal. Chim. Acta 32*, 139–147 (1965).
[166]   Amos, M.D.; J.B. Willis, *Spectrochim. Acta 22*, 1325–1343 (1966).
[167]   Amossé, J.; W. Fischer, M. Allibert, M. Piboule, *Analusis 14/1*, 26–31 (1986).
[168]   Amran, B.; F. Lagarde, M.J.F. Leroy, A. Lamotte, C. Demesmay, M. Ollé, M. Albert, G. Rauret, J.F. López-Sánchez, in: Ph. Quevauviller, E.A. Maier, B. Griepink (Editors), *Quality Assurance for Environmental Analysis*, Elsevier, Amsterdam, 1995, 285–304.
[169]   An, Y.; S.N. Willie, R.E. Sturgeon, *Fresenius J. Anal. Chem. 344/1-2*, 64–65 (1992).
[170]   An, Y.; S.N. Willie, R.E. Sturgeon, *Spectrochim. Acta, Part B 47/12*, 1403–1410 (1992).
[171]   Analytical Methods Committee, *Analyst (London) 84*, 214–216 (1959).
[172]   Analytical Methods Committee, *Analyst (London) 102*, 769–776 (1977).
[173]   Analytical Methods Committee, *Analyst (London) 112*, 199–204 (1987).
[174]   Analytical Methods Committee, *Analyst (London) 112*, 679–686 (1987).
[175]   Analytical Methods Committee, *Analyst (London) 113*, 1469–1471 (1988).
[176]   Analytical Methods Committee, *Analyst (London) 114*, 1497–1503 (1989).
[177]   Analytical Methods Committee, *Analyst (London) 119/11*, 2363–2366 (1994).
[178]   Analytical Methods Committee, *Analyst (London) 120/9*, 2303–2308 (1995).
[179]   Analytical Quality Control (Harmonised Monitoring) Committee, *Analyst (London) 110/3*, 247–252 (1985).
[180]   Andersen, I.A.C. Høgetveit, *Fresenius Z. Anal. Chem. 318*, 41–44 (1984).
[181]   Andersen, J.R.; *Analyst (London) 110/3*, 315–316 (1985).
[182]   Andersen, J.R.; P. Helboe, *J. Pharm. Biomed. Anal. 4/1*, 111–114 (1986).
[183]   Andersen, J.R.; B. Gammelgaard, S. Reimert, *Analyst (London) 111/6*, 721–722 (1986).
[184]   Andersen, K.-J.; A. Wikshaland, A. Utheim, K. Julshamn, H. Vik, *Clin. Biochem. (Ottawa) 19/3*, 166–170 (1986).
[185]   Anderson, J.; *At. Absorpt. Newsletter 11*, 88–89 (1972).
[186]   Anderson, J.; T.N. Van Der Walt, F.W.E. Strelow, *Geostand. Newsl. 9*, 17–18 (1985).
[187]   Anderson, J.; A.H. Victor, *Geostand. Newsl. 10*, 27–28 (1986).
[188]   Anderson, K.A.; P.A. Talcott, in: H.G. Seiler, A. Sigel, H. Sigel (Editors), *Handbook on Metals in Clinical and Analytical Chemistry*, Marcel Dekker, New York, 1994, 453–466.
[189]   Anderson, P.; C.M. Davidson, D. Littlejohn, A.M. Ure, C.A. Shand, M.V. Cheshire, *Anal. Chim. Acta 327/3*, 53–60 (1996).
[190]   Anderson, R.G.; I.S. Maines, T.S. West, *Int. Sympos. Microchem.; Vol. B*, 1970.
[191]   Anderson, R.K.; M. Thompson, E. Culbard, *Analyst (London) 111/10*, 1143–1152 (1986).
[192]   Ando, A.; K. Fuwa, B.L. Vallee, *Anal. Chem. 42*, 818 (1970).
[193]   Andrade, J.C. de; C. Pasquini, N. Baccan, J.C. van Loon, *Spectrochim. Acta, Part B 38/10*, 1329–1338 (1983).
[194]   Andrade, J.C. de; F.C. Strong III, N.J. Martin, *Talanta 37/7*, 711–718 (1990).
[195]   Andreae, M.O.; *Anal. Chem. 49/6*, 820–823 (1977).
[196]   Andreae, M.O.; *Limnol. Oceanogr. 24/3*, 440–452 (1979).
[197]   Andreae, M.O.; P.N. Froelich, *Anal. Chem. 53/2*, 287–291 (1981).
[198]   Andreae, M.O.; J.-F. Asmodé, P. Foster, L. Van't Dack, *Anal. Chem. 53/12*, 1766–1771 (1981).
[199]   Andreae, M.O.; in: K. Grasshoff, M. Ehrhardt, K. Kremling (Editors), *Methods of seawater analysis*, Verlag Chemie, Weinheim, 1983, 218–236.
[200]   Andreae, M.O.; J.T. Byrd, *Anal. Chim. Acta 156*, 147–157 (1984).
[201]   Andreae, M.O.; P.N. Froelich, *Tellus 36B*, 101–117 (1984).
[202]   Andreae, M.O.; *Anal. Chem. 56/12*, 2064–2066 (1984).

[203] Andreae, M.O.; in: M. Bernhard, F.E. Brinckman, P.J. Sadler (Editors), *The Importance of Chemical 'Speciation' in Environmental Processes*, Springer, New York, 1986, 301–335.

[204] Andrew, B.E.; *Ceramic Bull. 55/6*, 583–584 (1976).

[205] Andrew, T.R.; P.N.R. Nichols, *Analyst (London) 87*, 25 (1962).

[206] Andrews, D.G.; J.B. Headridge, *Analyst (London) 102*, 436–445 (1977).

[207] Andrews, D.G.; A.M. Aziz-Alrahman, J.B. Headridge, *Analyst (London) 103*, 909–915 (1978).

[208] Andronov, Y.G.; V.K. Efimov, S.I. Polyakov, V.A. Cherkasov, *Zh. Nauchn., Priklk, Fotogr. Kinematogr. 27*, 129 (1982).

[209] Aneva, Z.; *Fresenius Z. Anal. Chem. 321/7*, 680–681 (1985).

[210] Aneva, Z.; M. Iancheva, *Anal. Chim. Acta 167*, 371–374 (1985).

[211] Aneva, Z.; M. Pankova, S. Arpadjan, St. Alexandrov, *Fresenius Z. Anal. Chem. 326*, 561–562 (1987).

[212] Aneva, Z.; S. Arpadjan, *J. Anal. At. Spectrom. 3/4*, 587–590 (1988).

[213] Aneva, Z.; *Anal. Chim. Acta 211*, 311–316 (1988).

[214] Aneva, Z.; S. Arpadjan, I. Kalaidjieva, *Anal. Chim. Acta 236*, 385–389 (1990).

[215] Angerer, J.; K.H. Schaller, in: Deutsche Forschungsgemeinschaft, *Analytische Methoden, Band 2, Analysen in biologischem Material*, Bonn, 1983, D1–D3.

[216] Angerer, J.; K.H. Schaller, *Analyses of Hazardous Substances in Biological Materials, Vol. 2*, DFG, Weinheim,1985.

[217] Angerer, J.; R. Heinrich-Ramm, G. Lehnert, *Int. J. Environ. Anal. Chem. 35*, 81–88 (1989).

[218] Angerer, J.; R. Heinrich-Ramm, in: J. Angerer, K.H. Schaller (Editors), *Analyses of Hazardous Substances in Biological Materials*, VCH Verlagsgesellschaft, Weinheim, 1991, 193–205.

[219] Angerer, J.; R. Heinrich-Ramm, J. Begerova, in: H. Günzler, R. Borsdorf, W. Fresenius, W. Huber, H. Kelker, I. Lüderwald (Editors), *Analytiker Taschenbuch, Band 10*, Springer, Berlin - Heidelberg - New York, 1991, 398–433.

[220] Anke, M.; M. Glei, in: H.G. Seiler, A. Sigel, H. Sigel (Editors), *Handbook on Metals in Clinical and Analytical Chemistry*, Marcel Dekker, New York, 1994, 495–501.

[221] Anke, M.; L. Angelow, *Fresenius J. Anal. Chem. 352/1-2*, 236–239 (1995).

[222] Anonymous, *LaborPraxis 17/5*, 98–99 (1993).

[223] Antonov, V.N.; S.L. Rusakov, L.N. Bludova, V.Z. Krasil'shchik, *Vysokochistye Veshchestva 1*, 142–146 (1993).

[224] Antonovich, V.P.; I.V. Bezlutskaya, Yu.V. Zelyukova, M.M. Novoselova, *Soviet J. of Water Chemistry and Technology (New York) 13/3*, 48–51(1991).

[225] Anwar, J.; I.L. Marr, *Anal. Proc. (London) 19/6*, 302–305 (1982).

[226] Anwar, J.; T. ul Haque, *J. Chem. Soc. Pak. 7*, 209 (1985).

[227] Anwar, J.; I.L. Marr, *J. Chem. Soc. Pak. 9*, 597 (1987).

[228] Anwar, J.; I.L. Marr, *J. Chem. Soc. Pak. 9*, 603 (1987).

[229] Anwar, J.; I.L. Marr, *J. Chem. Soc. Pak. 9*, 493 (1987).

[230] Anwar, J.; I.L. Marr, *Anal. Chim. Acta 207*, 259–268 (1988).

[231] Anwar, J.; A. Bokhari, K.H. Haider, R. Mahmood, *J. Chem. Soc. Pak. 12*, 157 (1990).

[232] Anwari, M.A.; G. Tuncel, O.Y. Ataman, *Int. J. Environ. Anal. Chem. 47*, 227–237 (1992).

[233] Anwari, M.A.; H.U. Abbasi, M. Volkan, O.Y. Ataman, *Fresenius J. Anal. Chem. 355/3-4*, 284–288 (1996).

[234] Anzano, J.M.; M. Paz Martínez-Garbayo, M.A. Belarra, J.R. Castillo, *J. Anal. At. Spectrom. 9/2*, 125–128 (1994).

[235] Aoki, H.; H. Murakami, M. Chikuma, *Bunseki Kagaku 42/11*, T147–T153 (1993).

[236] Apostoli, P.; L. Alessio, M. Dal Farra, P.L.Fabbri, *J. Anal. At. Spectrom. 3/3*, 471–474 (1988).

[237] Apostoli, P.; C. Minoia, S. Porru, A. Ronchi, in: C. Minoia, S. Caroli (Editors), *Appl. Zeeman Graphite Furnace At. Abs. Spectrom. Chem. Lab. Toxicology*, Pergamon, Oxford, 1992, 409–443.

[238] Apostoli, P.; C. Minoia, M.E. Gilberti, A. Ronchi, in: C. Minoia, S. Caroli (Editors), Appl. Zeeman Graphite Furnace At. Abs. Spectrom. Chem. Lab. Toxicology, Pergamon, Oxford, 1992, 495–516.

[239] Apte, S.C.; A.G. Howard, *J. Anal. At. Spectrom. 1/3*, 221–225 (1986).

[240] Apte, S.C.; A.G. Howard, *J. Anal. At. Spectrom. 1/5*, 379–382 (1986).

[241] Apte, S.C.; M.J. Gardner, *Talanta 35/7*, 539–544 (1988).

[242] Apte, S.C.; in: R.M. Harrison, S. Rapsomanikis (Editors), *Environ. Analysis using Chromatography Interfaced with Atomic Spectrometry*, Ellis Horwood, Chichester, 1989, 258–298.

776   Bibliography

[243]   Apte, S.C.; S.D.W. Comber, M.J. Gardner, A.M. Gunn, *J. Anal. At. Spectrom. 6/2*, 169–172 (1991).
[244]   Arafat, N.M.; W.A. Glooschenko, *Analyst (London) 106*, 1174–1178 (1981).
[245]   Araki, T.; J.P. Walters, S. Minami, *Appl. Spectrosc. 34/1*, 33–39 (1980).
[246]   Araujo, P.W.; M.J. Gómez, Z.A. de Benzo, C. Castillo, *Chemom. Intell. Lab. Sys. 16/3*, 203–211 (1992).
[247]   Araújo, A.N.; J.L.F.C. Lima, *J. Trace Elem. Electrol. Health Dis. 3/2*, 97–101 (1989).
[248]   Araújo, A.N.; J.L.F.C. Lima, *Quim. Anal. (Barcelona) 11/1*, 55–63 (1992).
[249]   Arbouine, M.W.; N.J. Smith, *At. Spectrosc. 12/2*, 54–58 (1991).
[250]   Archundia, C.; P.S. Bonato, J.F. Lugo Rivera, L.C. Mascioli, K.E.Collins, C.H. Collins, *Sci. Total Environ. 130-131*, 231–236 (1993).
[251]   Arenas, V.; M. Stoeppler, H. Müller, in: B. Welz (Editor), *5. Colloquium Atomspektrometrische Spurenanalytik*, Bodenseewerk Perkin-Elmer, Überlingen, 1989, 483–490.
[252]   Arenas, V.; *Report Juel 2512*, Korschungszentrum Jülich, 1991, 186.
[253]   Arendonk, M.D. Van; R.K. Skogerboe, C.L. Grant, *Anal. Chem. 53/14*, 2349–2350 (1981).
[254]   Arfelli, W.; *J. Test. Eval. 12/3*, 152–154 (1984).
[255]   Ari, Ü.; M. Volkan, N.K. Aras, *J. Agri. Food Chem. 39/12*, 2180–2183 (1991).
[256]   Arias, E.; P. Suau, *Pint. Acabados Ind. 35/202*, 53–58 (1993).
[257]   Arik, N.; A.R. Türker, *Fresenius J. Anal. Chem. 339/12*, 874–876 (1991).
[258]   Armannsson, H.; P.J. Ovenden, *Int. J. Environ. Anal. Chem. 8/2*, 127–136 (1980).
[259]   Arnaud, J.; A. Favier, J. Alary, *J. Anal. At. Spectrom. 6/8*, 647–652 (1991).
[260]   Arnaud, J.; A. Favier, J. Alary, *Lait (Paris) 71/1*, 87–97 (1991).
[261]   Arnaud, J.; A. Favier, *Analyst (London) 117/10*, 1593–1598 (1992).
[262]   Arnaud, J.; A. Favier, *Sci. Total Environ. 159*, 9–1519 (1995).
[263]   Aronson, S.; S. Lemont, J. Weiner, *Inorg. Chem. 10*, 1296 (1971).
[264]   Aroza, I.; M. Bonilla, Y. Madrid, C. Cámara-Rica, *J. Anal. At. Spectrom. 4/2*, 163–166 (1989).
[265]   Arpadjan, S.; D. Stojanova, *Fresenius Z. Anal. Chem. 302/3*, 206. (1980).
[266]   Arpadjan, S.; I. Karadjova, S. Alexandrov, D. Tsalev, *Fresenius Z. Anal. Chem. 320/6*, 581–583 (1985).
[267]   Arpadjan, S.; K. Chadjiivanov, D. Tsalev, *Spectrochim. Acta, Part B 40B/4*, 697–700 (1985).
[268]   Arpadjan, S.; V. Krivan, *Anal. Chem. 58/13*, 2611–2614 (1986).
[269]   Arpadjan, S.; M. Novkirishka, S. Alexandrov, P. Petkov, *Z. Chem. 26/12*, 430–433 (1986).
[270]   Arpadjan, S.; A. Kojnarska, R. Djingova, *Fresenius Z. Anal. Chem. 325*, 278–280 (1986).
[271]   Arpadjan, S.; V. Krivan, *Fresenius Z. Anal. Chem. 329/7*, 745–749 (1988).
[272]   Arpadjan, S.; M. Granda, O. Duque, *Dokl. Bolg. Akad. Nauk. 43/10*, 57–60 (1990).
[273]   Arpadjan, S.; I. Karadjova, E. Tserovsky, Z. Aneva, *J. Anal. At. Spectrom. 5/3*, 195–198 (1990).
[274]   Arpadjan, S.; E. Vassileva, S. Momchilova, *Analyst (London) 117/12*, 1933–1937 (1992).
[275]   Arpadjan, S.; L. Jordanova, I. Karadjova, *Fresenius J. Anal. Chem. 347/12*, 480–482 (1993).
[276]   Arpadjan, S.; A. Alexandrova, *J. Anal. At. Spectrom. 10/10*, 799–802 (1995).
[277]   Arruda, M.A.Z.; M. Gallego, M. Valcárcel, *Anal. Chem. 65/22*, 3331–3335 (1993).
[278]   Arruda, M.A.Z.; M.J. Quintela, M. Gallego, M. Valcárcel, *Analyst (London) 119/8*, 1695–1699 (1994).
[279]   Arruda, M.A.Z.; M. Gallego, M. Valcárcel, *J. Anal. At. Spectrom. 10/7*, 501–504 (1995).
[280]   Arruda, M.A.Z.; M. Gallego, M. Valcárcel, *J. Anal. At. Spectrom. 11/2*, 169–173 (1996).
[281]   Aruga, R.; P. Mirti, A. Casoli, *Anal. Chim. Acta 276/1*, 197–204 (1993).
[282]   Asami, T.; C. Mizui, T. Shimada, M. Kubota, *Fresenius J. Anal. Chem. 356/5*, 348–351 (1996).
[283]   Ashby, J.R.; P.J. Craig, *Sci. Total Environ. 78*, 219–232 (1989).
[284]   Ashino, T.; K. Takada, K. Hirokawa, *Anal. Chim. Acta 297/3*, 443–451 (1994).
[285]   Ashino, T.; K. Takada, *Anal. Chim. Acta 312/2*, 157–163 (1995).
[286]   Ashino, T.; K. Hirokawa, *Anal. Sci. 11/4*, 703–706 (1995).
[287]   Ashino, T.; K. Takada, *J. Anal. At. Spectrom. 11/8*, 577–583 (1996).
[288]   Assarian, G.S.; D. Oberleas, *Clin. Chem. (Winston-Salem) 23*, 1771–1772 (1977).
[289]   ASTM Committee E-11 on Statistical Methods, in: American Society for Testing and Materials, *Methods for Emission Spectrochemical Analysis*, ASTM, Easton, MD, USA, 1982, 125–135.
[290]   Aston, S.R.; B. Oregioni, A. Veglia, *At. Spectrosc. 8/5*, 155–158 (1987).
[291]   Åström, O.; *Dissertation*, University of Umea, Sweden, 1983.

[292] Åström, O.; *Anal. Chem. 54/2*, 190–193 (1982).
[293] Astruc, A.; R. Lavigne, V. Desauziers, R. Pinel, M. Astruc, *Appl. Organomet. Chem. 3*, 267–271 (1989).
[294] Astruc, A.; R. Pinel, M. Astruc, *Anal. Chim. Acta 228/1*, 129–137 (1990).
[295] Astruc, M.; R. Pinel, A. Astruc, *Mikrochim. Acta (Vienna) 109/1-4*, 73–77 (1992).
[296] Astruc, A.; M. Astruc, R. Pinel, M. Potin-Gautier, *Appl. Organomet. Chem. 6*, 39–47 (1992).
[297] Astruc, M.; A. Astruc, R. Pinel, *Mikrochim. Acta (Vienna) 109/1-4*, 83–86 (1992).
[298] Astruc, A.; X. Dauchy, F. Pannier, M. Potin-Gautier, M. Astruc, *Analusis 22*, 257 (1994).
[299] Atallah, R.H.; D.A. Kalman, *Talanta 38/2*, 167–173 (1991).
[300] Aten, C.F.; J.B. Bourke, J.C. Walton, *J. Assoc. Off. Anal. Chem. 66/3*, 766–768 (1983).
[301] Atkinson, R.J.; G.D. Chapman, L. Krause, *J. Opt. Soc. Am. 55/10*, 1269–1274 (1965).
[302] Atsuya, I.; K. Itoh, *Spectrochim. Acta, Part B 38/9*, 1259–1264 (1983).
[303] Atsuya, I.; K. Koch, K. Akatsuka, *Fresenius Z. Anal. Chem. 328/4-5*, 338–341 (1987).
[304] Atsuya, I.; K. Itoh, K. Akatsuka, K. Jin, *Fresenius Z. Anal. Chem. 326/1*, 53–56 (1987).
[305] Atsuya, I.; K. Itoh, *Fresenius Z. Anal. Chem. 329/7*, 750–755 (1988).
[306] Atsuya, I.; K. Akatsuka, K. Itoh, *Fresenius J. Anal. Chem. 337/3*, 294–298 (1990).
[307] Atsuya, I.; K. Aryu, Q.-b. Zhang, *Anal. Sci. 8/3*, 433–436 (1992).
[308] Atsuya, I.; H. Minami, Q.-b. Zhang, *Fresenius J. Anal. Chem. 346/12*, 1054–1057 (1993).
[309] Attiyat, A.S.; G.D. Christian, *Clin. Chim. Acta 137*, 151–157 (1984).
[310] Attiyat, A.S.; *Microchem. J. 36/2*, 228–234 (1987).
[311] Atwell, M.G.; J.Y. Hebert, *Appl. Spectrosc. 23*, 480–482 (1969).
[312] Austenfeld, F.A.; R.L. Berghoff, *Plant Soil 64/2*, 267–271 (1982).
[313] Avila, A.K.; A.J. Curtius, *J. Anal. At. Spectrom. 9/4*, 543–546 (1994).
[314] Avotins, P.; E.A. Jenne, *J. Environ. Qual. 4/4*, 515–519 (1975).
[315] Awad, N.A.N.; F. Jasim, *Microchem. J. 40/3*, 352–359 (1989).
[316] Awad, N.A.N.; F. Jasim, *Microchem. J. 40/2*, 187–196 (1989).
[317] Ayad, M.M.; L. Abd El Aziz, A. El Kheir, *Anal. Lett. 16, B16*, 1335–1342 (1983).
[318] Ayad, M.M.; S.E. Khayyal, N.M. Farag, *Spectrochim. Acta, Part B 40B/9*, 1205–1209 (1985).
[319] Ayad, M.M.; S.E. Khayyal, N.M. Farag, *Microchem. J. 33*, 371–375 (1986).
[320] Aycik, G.A.; E. Edgüer, *J. Radioanal. Nucl. Chem. 82*, 319–328 (1984).
[321] Aysola, P.; P.D. Anderson, C.H. Langford, *Anal. Chem. 59/11*, 1582–1583 (1987).
[322] Aysola, P.; P.D. Anderson, C.H. Langford, *Anal. Lett. 21/11*, 2003–2010 (1988).
[323] Azeredo, L.C.; R.E. Sturgeon, A.J. Curtius, *Spectrochim. Acta, Part B 48/1*, 91–98 (1993).
[324] Aziz-Alrahman, A.M.; J.B. Headridge, *Talanta 25*, 413–415 (1978).
[325] Aziz-Alrahman, A.M.; *Int. J. Environ. Anal. Chem. 22/3-4*, 251–257 (1985).
[326] Aznárez, J.; J.M. Rabadán, A. Ferrer, P. Cipres, *Talanta 33/5*, 458–460 (1986).
[327] Aznárez, J.; J.C. Vidal, J.M. Gascón, *At. Spectrosc. 7/2*, 59–60 (1986).
[328] Aznárez, J.; J.C. Vidal, R. Carnicer, *J. Anal. At. Spectrom. 2/1*, 55–58 (1987).
[329] Aznárez-Alduan, J.; F. Palacios, M.S. Ortega, J.C. Vidal. *Analyst (London) 109/2*, 123–125 (1984).
[330] Azzaria, L.M.; A. Aftabi, *Water, Air, Soil Pollut. 56*, 203–217 (1991).
[331] Baaske, B.; A. Golloch, U. Telgheder, *J. Anal. At. Spectrom. 9/8*, 867–870 (1994).
[332] Baaske, B.; U. Telgheder, *J. Anal. At. Spectrom. 10/12*, 1077–1080 (1995).
[333] Baasner, J.; F. Portala, D. Weber, in: B. Welz (Editor), *6. Colloquium Atomspektrometrische Spurenanalytik*, Bodenseewerk Perkin-Elmer, 1991, 281–288.
[334] Bablok, W.; H. Passing, *J. Autom. Chem. 7/2*, 74–79 (1985).
[335] Babu, R.R.; S.C.S. Rajan, L.S.A. Dikshitulu, *Talanta 42/12*, 2017–2020 (1995).
[336] Bachmann, G.; W. Rechenberg, in: B. Welz (Editor), *5. Colloquium Atomspektrometrische Spurenanalytik*, Bodenseewerk Perkin-Elmer, Überlingen, 1989, 573–591.
[337] Bachmann, G.; W. Rechenberg, in: B. Welz (Editor), *6. Colloquium Atomspektrometrische Spurenanalytik*, Bodenseewerk Perkin-Elmer, Überlingen, 1991, 699–706.
[338] Bächmann, K.; C. Spachidis, A. Weitz, *Fresenius Z. Anal. Chem. 301/1*, 3–6 (1980).
[339] Bäckman, S.; R.W. Karlsson, *Analyst (London) 104*, 1017–1029 (1979).
[340] Bäckström, K.; L.-G. Danielsson, L. Nord, *Analyst (London) 109/3*, 323–325 (1984).
[341] Bäckström, K.; L.-G. Danielsson, *Anal. Chem. 60/13*, 1354–1357 (1988).
[342] Bäckström, K.; L.-G. Danielsson, *Mar. Chem. 29/1*, 33–46 (1990).
[343] Bäckström, K.; L.-G. Danielsson, *Anal. Chim. Acta 232/2*, 301–315 (1990).
[344] Bader, R.; A.M. Reichlmayr-Lais, M. Kirchgessner, *Agribiol. Res. 44*, 212–218 (1991).
[345] Baedecker, P.A.(Editor); *U.S. Geol. Surv. Bull., No 1770*, 1987.
[346] Baeyens, W.: *Trends Anal. Chem. (Pers. Ed.) 11/7*, 245–254 (1992).
[347] Baez, M.E.; C. Gonzales, L. Bravo, *Analusis 15*, 424–425 (1987).

[348]   Baffi, F.; R. Frache, A. Dadone, *Ann. Chim. (Rom) 74*, 385–397 (1984).
[349]   Bagdi, G.; J. Lakatos, I. Lakatos, *Acta Chim. Hung. 125*, 543–554 (1988).
[350]   Bagdi, G.; J. Lakatos, I. Lakatos, *Magy. Kem. Foly. 95*, 275 (1989).
[351]   Bagdi, G.; J. Lakatos, I. Lakatos, *Acta Chim. Hung. 128/3*, 319–332 (1991).
[352]   Bagdi, G.; J. Lakatos, I. Lakatos, *J. Anal. At. Spectrom. 7/5*, 769–773 (1992).
[353]   Bagliano, G.; F. Benischek, I. Huber, *At. Absorpt. Newsletter 14/3*, 45–48 (1975).
[354]   Bagschik, U.; D. Quack, M. Stoeppler, *Fresenius J. Anal. Chem. 338/4*, 386–389 (1990).
[355]   Bahreyni-Toosi, M.-H.; J.B. Dawson, D.J. Ellis, R.J. Duffield, *Analyst (London) 109/12*, 1607–1612 (1984).
[356]   Bahreyni-Toosi, M.-H.; J.B. Dawson, R.J. Duffield, *Analyst (London) 109/7*, 943–946 (1984).
[357]   Bai, W.-m.; R. Feng, H.-q. Wong, *Guangpuxue Yu Guangpu Fenxi 7/2*, 46 (1987).
[358]   Baird, R.B.; S.M. Gabrielian, *Appl. Spectrosc. 28*, 273–274 (1974).
[359]   Baker, J.B. Headridge, R.A. Nicholson, *Anal. Chim. Acta 113*, 47–53 (1980).
[360]   Baker, A.A.; J.B. Headridge, *Anal. Chim. Acta 125*, 93–99 (1981).
[361]   Baker, R.A.; D.J. Hartshorne, A.G. Wilshire, *At. Absorpt. Newsletter 8/2*, 44–45 (1969).
[362]   Baldwin, S.; M. Deaker, W. Maher, *Analyst (London) 119/8*, 1701–1704 (1994).
[363]   Ballin, U.; R. Kruse, H.-A. Rüssel, *Fresenius J. Anal. Chem. 350/1-2*, 54–61 (1994).
[364]   Balls, P.W.; *Anal. Chim. Acta 197*, 309–313 (1987).
[365]   Banba, T.; H. Hagiya, Y. Tamura, C. Yonezawa, *Bunseki Kagaku 42/5*, 317–323 (1993).
[366]   Bandinelli, R.; *G. Ital. Chim. Clin. 17/6*, 455–461 (1992).
[367]   Bandyopadhyay, S.; A.K. Das, *Indian J. Environ. Prot. 12/3*, 166–170 (1992).
[368]   Bangia, T.R.; K.N.K. Kartha, M. Varghese, B.A. Dhawale, B.D. Joshi, *Fresenius Z. Anal. Chem. 310/5*, 410–412 (1982).
[369]   Bangroo, P.N.; C.R. Jagga, H.C. Arora, G.N. Rao, *At. Spectrosc. 16/3*, 118–120 (1995).
[370]   Bank, H.L.; J. Robson, J.B. Bigelow, J. Morrison, L.H. Spell, R. Kantor, *Clin. Chim. Acta 116*, 179–190 (1981).
[371]   Bannon, D.I.; C. Marashchik, C.R. Zapf, M.R. Farfel, J.J. Chisolm, *Clin. Chem. (Winston-Salem) 40/9*, 1730–1734 (1994).
[372]   Bao, C.-l.; X.-l. Cheng, Y. Li, Y.-d. Wei, Z. An, *Guangpuxue Yu Guangpu Fenxi 11/5*, 56 (1991).
[373]   Bao, C.-l.; X.-l. Cheng, C.-h. Liu, Y.-d. Wei, *Fenxi Huaxue 20*, 429 (1992).
[374]   Baqir, S.J.; F. Jasim, *Propellants, Explos., Pyrotech. 19*, 202 (1994).
[375]   Baralkiewicz, D.; H. Gramomwska, J. Zerbe, J. Siepak, *Chem. Anal. (Warsaw) 37/6*, 641–649 (1992).
[376]   Barbaro, M.; B. Passariello, R. Sbrilli, P. Millozzi, *At. Spectrosc. 4/4*, 155–156 (1983).
[377]   Barber, T.E.; P.E. Walters, J.D. Winefordner, N. Omenetto, *Appl. Spectrosc. 45/4*, 524–526 (1991).
[378]   Barber, T.E.; P.E. Walters, M.W. Wensing, J.D. Winefordner, *Spectrochim. Acta, Part B 46/6-7*, 1009–1014 (1991).
[379]   Barberá, R.; R. Farré, R. Montoro, *J. Assoc. Off. Anal. Chem. 68/3*, 511–513 (1985).
[380]   Barberá, R.; R. Farré, *At. Spectrosc. 9/1*, 6–8 (1988).
[381]   Barbooti, M.M.; N.S. Zaki, S.S. Baha-Uddin, E.B. Hassan, *Analyst (London) 115/8*, 1059–1061 (1990).
[382]   Barefoot, R.R.; J.C. Van Loon, *Determination of the Precious Metals: Selected Instrumental Methods,* John Wiley & Sons, Chichester, 1991.
[383]   Barefoot, R.R.; *Environ. Sci. Technol. 31/2*, 309–314 (1997).
[384]   Barghigiani, C.; T. Ristori, M. Cortopassi, *Environ. Technol. 12/10*, 935–941 (1991).
[385]   Barker, S.A.; S.G. Johnson, G.C. Knighton, M.T. Sayer, B.M. Candee, V.D. Dimick, *Appl. Spectrosc. 50/6*, 816–819 (1996).
[386]   Barnes, L.; *Anal. Chem. 38/8*, 1083–1085 (1966).
[387]   Barnett, N.W.; L. Ebdon, E. Hywel Evans, P. Ollivier, *Microchem. J. 44*, 168–178 (1991).
[388]   Barnett, W.B.; *Anal. Chem. 44*, 695–698 (1972).
[389]   Barnett, W.B.; *At. Absorpt. Newsletter 12*, 142–146 (1973).
[390]   Barnett, W.B.; J.W. Vollmer, S.M. DeNuzzo, *At. Absorpt. Newsletter 15/2*, 33–37 (1976).
[391]   Barnett, W.B.; M.M. Cooksey, *At. Absorpt. Newsletter 18/3*, 61–65 (1979).
[392]   Barnett, W.B.; W. Bohler, G.R. Carnrick, W. Slavin, *Spectrochim. Acta, Part B 40/10*, 1689–1703 (1985).
[393]   Barnett, W.B.; P. Seferovic, *At. Spectrosc. 17/5*, 190–195 (1996).
[394]   Bermejo Barrera, P.; R. Dominguez González, A. Bermejo Barrera, *Fresenius J. Anal. Chem. 357/4*, 457–461 (1997).
[395]   Barringer A.R.; *Trans. Inst. Min. Metal. (Bull. 714) 75B*, 8120 (1966).

[396] Barron, D.C.; B.W. Haynes, *Analyst (London) 111/1*, 19–21 (1986).
[397] Barros, J.S.; *Analyst (London) 114/3*, 369–373 (1989).
[398] Bartels, H.; *At. Absorpt. Newsletter 6*, 132 (1967).
[399] Bartels, T.T.; C.E. Wilson, *At. Absorpt. Newsletter 8/1*, 3–5 (1969).
[400] Bartenfelder, D.C.; A.D. Karathanasis, *Commun. Soil. Sci. Plant Anal. 19*, 471–492 (1988).
[401] Barto, R.; A.J.A.M. Sips, W.J.F. van der Vijgh, J.C. Netelenbos, *Clin. Chem. (Winston-Salem) 41/8*, 1159–1163 (1995).
[402] Barth, P.; V. Krivan, R. Hausbeck, *Anal. Chim. Acta 263/1*, 111–118 (1992).
[403] Bartha, A.; K. Ikrényi, *Anal. Chim. Acta 139*, 329–332 (1982).
[404] Barthel, W.F.; A.L. Smrek, G.P. Angel, J.A. Liddle, P.J. Landrigan, *J. Assoc. Off. Anal. Chem. 56/5*, 1252–1256 (1973).
[405] Bartlett, R.J.; J.M. Kimble, *J. Environ. Qual. 5/4*, 383–386 (1976).
[406] Baruthio, F.; O. Guillard, J. Arnaud, F. Pierre, R. Zawislak, *Clin. Chem. (Winston-Salem) 34/2*, 227–234 (1988).
[407] Baruthio, F.; F. Pierre, *Biol. Trace Elem. Res. 39*, 21–31 (1993).
[408] Barzev, A.; D. Dobreva, L. Futekov, V. Rusev, G. Bekjarov, G. Toneva, *Fresenius Z. Anal. Chem. 325*, 255–257 (1986).
[409] Basargin, N.N.; Z.S. Svanidze, Yu.G. Rozovskii, *Zavodsk. Lab. 59/2*, 8–9 (1993).
[410] Bascombe, K.N.; *10th. Symp. Combust.*, 1965.
[411] Bass, D.A.; J.A. Holcombe, *Anal. Chem. 59/7*, 974–980 (1987).
[412] Bass, D.A.; J.A. Holcombe, *Anal. Chem. 60/6*, 578–582 (1988).
[413] Bass, D.A.; J.A. Holcombe, *Spectrochim. Acta, Part B 43/12*, 1473–1483 (1988).
[414] Bass, D.A.; *Am. Lab. (Fairfield, CT) 21/12*, 24–28 (1989).
[415] Bass, D.A.; L.B. TenKate, A.M. Wroblewski, *At. Spectrosc. 17/2*, 92–97 (1996).
[416] Bastiaans, G.J.; G.M. Hieftje, *Anal. Chem. 46*, 901–910 (1974).
[417] Batho, A.; C. Phillips, *Lab. Pract. 38/3*, 70 (1989).
[418] Batifol, F.M.; C.F. Boutron, *Atmos. Environ. 18/11*, 2507–2515 (1984).
[419] Batistoni, D.A.; P.N. Smichowski, *Appl. Spectrosc. 39/2*, 222–226 (1985).
[420] Batistoni, D.A.; L.H. Erlijman, M.I. Fuertes, *Talanta 32/8A*, 641–644 (1985).
[421] Batistoni, D.A.; M.I. Fuertes, P.N. Smichowski, *At. Spectrosc. 10/1*, 12–16 (1989).
[422] Batley, G.E.; in: G.E. Batley (Editor), *Trace Element Speciation: Analytical Methods and Problems*, CRC Press, Boca Raton, FL, 1989, 1–24.
[423] Batley, G.E; *Trace Element Speciation: Analytical Methods and Problems*, CRC Press, Boca Raton, FL, 1989.
[424] Baucells, M.; G. Lacort, M. Roura, *Analyst (London) 110/12*, 1423–1429 (1985).
[425] Baudin, G.; R. Bonne, M. Chaput, L. Feve, *3rd CISAFA, Paris, 1971*, Adam Hilger, London, 1973, 853–870.
[426] Baudin, G.; *Prog. Anal. Spectrosc. 3/1*, 1–63 (1980).
[427] Bauer, H.-D.; D. Dahmann, H.-H. Fricke, S. Herwald, B. Neidhart, *Staub. Reinhalt. Luft 51*, 125–128 (1991).
[428] Baumgardt, B.; E. Jackwerth, H. Otto, G. Tölg, *Fresenius Z. Anal. Chem. 323/5*, 481–486 (1986).
[429] Bauslaugh, J.; B. Radziuk, K. Saeed, Y. Thomassen, *Anal. Chim. Acta 165/11*, 149–157 (1984).
[430] Bautista, M.A.; C. Pérez Sirvent, I. López García, M. Hernández Córdoba, *Fresenius J. Anal. Chem. 350/6*, 359–364 (1994).
[431] Bax, D.; F.F. Peters, J.P.M. van Noort, J. Agterdenbos, *Spectrochim. Acta, Part B 41/3*, 275–282 (1986).
[432] Bax, D.; J.T. van Elteren, J. Agterdenbos, *Spectrochim. Acta, Part B 41/9*, 1007–1013 (1986).
[433] Bax, D.; J. Agterdenbos, E. Worrell, J. Beneken Kolmer, *Spectrochim. Acta, Part B 43/9-11*, 1349–1354 (1988).
[434] Baxter, D.C.; W. Frech, *Spectrochim. Acta, Part B 42*, 1005–1010 (1987).
[435] Baxter, D.C.; W. Frech, *Anal. Chim. Acta 225/1*, 175–183 (1989).
[436] Baxter, D.C.; W. Frech, *Anal. Chim. Acta 236/2*, 377–384 (1990).
[437] Baxter, D.C.; J. Öhman, *Spectrochim. Acta, Part B 45/4-5*, 481–491 (1990).
[438] Baxter, D.C.; W. Frech, *Fresenius J. Anal. Chem. 337/3*, 253–263 (1990).
[439] Baxter, D.C.; W. Frech, I. Berglund, *J. Anal. At. Spectrom. 6/2*, 109–113 (1991).
[440] Baxter, D.C.; R. Nichol, D. Littlejohn, *Spectrochim. Acta, Part B 47/10*, 1155–1163 (1992).
[441] Baxter, D.C.; W. Frech, *Analyst (London) 118/5*, 495–504 (1993).
[442] Baxter, D.C.; W. Frech, *Pure Appl. Chem. 67/4*, 615–648 (1995).

[443] Baxter, M.J.; J.A. Burrell, H.M. Crews, R.C. Massey, D.J. McWeeny, *Food Addit. Contam. 6/3*, 341–349 (1989).
[444] Baxter, M.J.; J.A. Burrell, R.C. Massey, *Food Addit. Contam. 7*, 101–107 (1990).
[445] Bayer, S.; J.A. McHard, J.D. Winefordner, *J. Agri. Food Chem. 28*, 1307–1307 (1980).
[446] Bayer, W.; A. Zechmeister, *Appl. atom. spectrom. (Perkin-Elmer) 6.2./DE*, 1–2 (1988).
[447] Bayunov, P.A.; A.S. Savin, B.V. L'vov, *At. Spectrosc. 3/6*, 161–164 (1982).
[448] Bayunov, P.A.; Yu.P. Denisov, B.V. L'vov, Yu.I. Yarmak, *Zavodsk. Lab. 56*, 41–42 (1990).
[449] Bazzi, A.; J. Montgomery, G. Alent, *Analyst (London) 113/1*, 121–124 (1988).
[450] Bea-Barredo, F.; L. Polo-Díez, *Talanta 27/1*, 69–70 (1980).
[451] Beary, E.S.; P.J. Paulsen, T.C. Rains, K.J. Ewing, J. Jaganathan, I. Aggarwal, *L. Cryst. Growth 106*, 51 (1990).
[452] Beaty, M.; W.B. Barnett, Z. Grobenski, *At. Spectrosc. 1/3*, 72–77 (1980).
[453] Beaupré, P.W.; W.J. Holland, L.T. Dupuis, *Mikrochim. Acta (Vienna) III/1-2*, 41–44 (1983).
[454] Beccaluva, L.; G. Venturelli, *At. Absorpt. Newsletter 10/2*, 50–52 (1971).
[455] Beceiro-González, E.; J. Barciela-Garcia, P. Bermejo-Barrera, A. Bermejo-Barrera, *Fresenius J. Anal. Chem. 344/7-8*, 301–305 (1992).
[456] Beceiro-González, E.; P. Bermejo-Barrera, A. Bermejo-Barrera, J. Barciela-Garcia, C. Barciela-Alonso, *J. Anal. At. Spectrom. 8/4*, 649–653 (1993).
[457] Becker, D.A.; in: R. Zeisler, V.P. Guinn (Editors), *Nuclear Analytical Methods in the Life Science*, (Biol. Trace. Elem. Res.), Humana Press, Clifton, NJ, 1990, 571–577.
[458] Becker-Ross, H.; S. Florek, in: B. Welz (Editor), *CANAS'95 Colloquium Analytische Atomspektroskopie*, Bodenseewerk Perkin-Elmer, Überlingen, 1996, 1–15.
[459] Becker-Ross, H.; S. Florek, K.P. Schmidt, in: B. Welz (Editor), *6. Colloquium Atomspektrometrische Spurenanalytik*, Bodenseewerk Perkin-Elmer, Überlingen, 1991, 497–504.
[460] Becker-Ross, H.; S. Florek, R. Weiße, U. Heitmann, in: B. Welz (Editor), *CANAS'95. Colloquium Analytische Atomspektroskopie*, Bodenseewerk Perkin-Elmer, Überlingen, 1996, 23–29.
[461] Becker-Ross, H.; S. Florek, U. Heitmann, R. Weiße, *Fresenius J. Anal. Chem. 355/3-4*, 300–303 (1996).
[462] Bédard, M.; J.D. Kerbyson, *Anal. Chem. 47/8*, 1441–1444 (1975).
[463] Bédard, M.; J.D. Kerbyson, *Can. J. Spectrosc. 21*, 64–68 (1976).
[464] Bedrossian, M.; *Anal. Chem. 56/2*, 311–312 (1984).
[465] Beek, H. Van; A.J. Baars, *J. Chromatogr. 442*, 345–352 (1988).
[466] Beek, H. Van; H.C.A. Greefkes, A.J. Baars, *Talanta 34/6*, 580–582 (1987).
[467] Beer, C. de ; C. Alsen-Hinrichs, H. Kruse, G. Noack, H. Seibert, in: B. Welz (Editor), *5. Colloquium Atomspektrometrische Spurenanalytik*, Bodenseewerk Perkin-Elmer, Überlingen, 1989, 887–897.
[468] Begerow, J.; *Anal. Chim. Acta 340/1-3*, 277–283 (1997).
[469] Behmenburg, W.; *J. Quant. Spectrosc. Radiat. Transfer 4*, 177–193 (1964).
[470] Behmenburg, W.; *Z. Astrophys. 69*, 368 (1968).
[471] Behne, D.; P. Brätter, H. Gesser, G. Hube, W. Mertz, U. Rösick, *Fresenius Z. Anal. Chem. 278*, 269–272 (1976).
[472] Behne, D.; *J. Clin. Chem. Clin. Biochem. 19*, 115–120 (1981).
[473] Beinrohr, E.; H. Berndt, *Mikrochim. Acta (Vienna) I*, 199–208 (1985).
[474] Beinrohr, E.; S. Gergely, J. Izák, *Fresenius Z. Anal. Chem. 332/1*, 28–33 (1988).
[475] Beinrohr, E.; Mikrochim. *Acta (Vienna) I*, 121–128 (1989).
[476] Beinrohr, E.; *Spectrochim. Acta, Part B 45/1-2*, 131–137 (1990).
[477] Beinrohr, E.; M. Rapta, M. Taddia, V. Poluzzi, *J. Anal. At. Spectrom. 6/1*, 33–36 (1991).
[478] Beinrohr, E.; B. Siles, J. Stefanec, V. Rattay, *Chem. Papers 45/1*, 61–67 (1991).
[479] Beinrohr, E.; J. Mocak, X. Svobodova, O. L'Osova, *Chem. Listy 85/2*, 131–141 (1991).
[480] Beinrohr, E.; M.-l. Lee, P. Tschöpel, G. Tölg, *Fresenius J. Anal. Chem. 346/6-9*, 689–692 (1993).
[481] Beinrohr, E.; A. Manová, J. Dzurov, *Fresenius J. Anal. Chem. 355/5-6*, 528–531 (1996).
[482] Beizel, N.F.; G.I. Heinrich, G. Emrich, I.G. Yudelevich, *Vysokochistye Veshchestva 5*, 165 (1991).
[483] Beizel, N.F.; T.M. Korda, I.G. Yudelevich, *Vysokochistye Veshchestva 3*, 134–137 (1992).
[484] Bekjarov, G.L.; L. Futekov, G.V. Andreev, *Fresenius Z. Anal. Chem. 322/6*, 563–566 (1985).
[485] Belarra, M.A.; F. Gallarta, J.M. Anzano, J.R. Castillo-Suarez, *J. Anal. At. Spectrom. 1/2*, 141–144 (1986).

[486] Belarra, M.A.; J.M. Anzano, F. Gallarta, J.R. Castillo-Suarez, *J. Anal. At. Spectrom. 2/1*, 77–79 (1987).
[487] Belarra, M.A.; M.C. Azofra, J.M. Anzano, J.R.Castillo-Suarez, *J. Anal. At. Spectrom. 3/4*, 591–593 (1988).
[488] Belarra, M.A.; M.C. Azofra, J.M. Anzano, J.R.Castillo-Suarez, *J. Anal. At. Spectrom. 4/1*, 101–103 (1989).
[489] Belarra, M.A.; J.M. Anzano, J.R. Castillo-Suarez, *Fresenius Z. Anal. Chem. 334/2*, 118–121 (1989).
[490] Belarra, M.A.; J.M. Anzano, I. Lavilla, J.R. Castillo-Suarez, *Microchem. J. 41*, 377–383 (1990).
[491] Belarra, M.A.; J.M. Anzano, J.R. Castillo-Suarez, *Analyst (London) 115/7*, 955–957 (1990).
[492] Belarra, M.A.; I. Lavilla, J.M. Anzano, J.R. Castillo-Suarez, *J. Anal. At. Spectrom. 7/7*, 1075–1078 (1992).
[493] Belarra, M.A.; I. Lavilla, J.R. Castillo-Suarez, *Anal. Sci. 11/4*, 651–656 (1995).
[494] Belarra, M.A.; I. Lavilla, J.R. Castillo-Suarez, *Analyst (London) 120/12*, 2813–2917 (1995).
[495] Belarra, M.A.; M.P. Martínez-Garbayo, J.M. Anazano, J.R. Castillo, *Anal. Sci. 12/3*, 483–488 (1996).
[496] Belarra, M.A.; M. Resano, J.R. Castillo, *Spectrochim. Acta, Part B 51/7*, 697–705 (1996).
[497] Belaya, K.P.; L.V. Kustova, *Zavodsk. Lab. 52/9*, 40 (1986).
[498] Belaya, K.P.; L.V. Kustova, T.M. Kichina, *Zavodsk. Lab. 53/6*, 33 (1987).
[499] Belazi, A.U.; C.M. Davidson, G.E. Keating, D. Littlejohn, M. McCartney, *J. Anal. At. Spectrom. 10/3*, 233–240 (1995).
[500] Belcher, C.B.; K. Kinson, *Anal. Chim. Acta 30*, 483–487 (1964).
[501] Belcher, C.B.; R.M. Dagnall, T.S. West, *Talanta 11*, 1257 (1964).
[502] Belcher, C.B.; *Prog. Anal. Spectrosc. 1*, 299–346 (1978).
[503] Bell, W.E.; A.L. Bloom, J. Lynch, *Rev. Sci. Instrum. 32*, 688 (1961).
[504] Bell, Z.G.; H.B. Lovejoy, T.R. Vizena, *J. Occup. Med. 15*, 501–508 (1973).
[505] Bellinger, D.; *Environ. Health Perspect. 89*, 5 (1990).
[506] Belliveau, J.F.; J.H. Friedman, G.P. O'Leary, D. Guarrera, in: Ph. Collery, L.A. Poirier, M. Manfait, J.-C. Etienne (Editors), *Metal Ions in Biology and Medicine*, John Libbey Eurotext, London - Paris, 1990, 89–91.
[507] Belloni, J.; M. Haissinsky, H.N. Salama, *J. Phys. Chem. 53*, 881 (1959).
[508] Bel'skii, N.K.; L.I. Ochertyanova, *Zavodsk. Lab. 58/11*, 19–20 (1992).
[509] Bel'skii, N.K.; E.L. Timashuk, L.I. Ochertyanova, A.V. Garmash, *Zh. Anal. Khim. 49*, 825 (1994).
[510] Bel'skii, N.K.; L.A. Nebol'sina, *Zh. Anal. Khim. 50/9*, 942–944 (1995).
[511] Beltramo, J.-L.; P. Chomard, N. Autissier, *Ann. Falsif. Expert. Chim. Toxicol. 85*, 29–35 (1992).
[512] Belyaev, V.N.; I.N. Vladimirskaya, L.N. Kolonina, G.G. Kovalev, L.B. Kuznetsov, O.A. Shiryaeva, *Zh. Anal. Khim. 40*, 135 (1985).
[513] Bencini, A.; C. Ciurli, C. Verrucchi, *Mineral. Petrogr. Acta 31*, 207–215 (1988).
[514] Bencze, K.; in: B. Welz (Editor), *Fortschritte in der atomspektrometrischen Spurenanalytik*, VCH Verlagsgesellschaft, Weinheim, 1986, 207–216.
[515] Bencze, K.; in: B. Welz (Editor), *5. Colloquium Atomspektrometrische Spurenanalytik*, Bodenseewerk Perkin-Elmer, Überlingen, 1989, 863–877.
[516] Bencze, K.; Ch. Pelikan, A. Kronseder, *Aerztl. Lab. 35/5*, 102–106 (1989).
[517] Bencze, K.; *Fresenius J. Anal. Chem. 337/8*, 867–876 (1990).
[518] Bencze, K.; *Fresenius J. Anal. Chem. 338/1*, 58–61 (1990).
[519] Bencze, K.; in: H.G. Seiler, A. Sigel, H. Sigel (Editors), *Handbook on Metals in Clinical and Analytical Chemistry*, Marcel Dekker, New York, 1994, 201–216.
[520] Bendicho, C.; M.T.C. de Loos-Vollebregt, *Spectrochim. Acta, Part B 45/6*, 547–559 (1990).
[521] Bendicho, C.; M.T.C. de Loos-Vollebregt, *Spectrochim. Acta, Part B 45/7*, 679–693 (1990).
[522] Bendicho, C.; M.T.C. de Loos-Vollebregt, *Spectrochim. Acta, Part B 45/7*, 695–710 (1990).
[523] Bendicho, C.; M.T.C. de Loos-Vollebregt, *J. Anal. At. Spectrom. 6/5*, 353–377 (1991).
[524] Bendicho, C.; A. Sancho, *At. Spectrosc. 14/6*, 187–190 (1993).
[525] Bendicho, C.; *Fresenius J. Anal. Chem. 348/5-6*, 353–355 (1994).
[526] Bendicho, C.; *Anal. Chem. 66/23*, 4375–4381 (1994).
[527] Benedetti, M.F.; A.M. Kersabiec, J. Boulègue, *Geostand. Newsl. 11/1*, 127–129 (1987).
[528] Beneš, P.; *Radiochim. Acta 3*, 159 (1964).
[529] Beneš, P.; A. Garbe, *Radiochim. Acta 5*, 99 (1966).
[530] Beneš, P.; J. Smetana, *Radiochim. Acta 6*, 196 (1966).

[531] Beneš, P.; J. Smetana, V. Majer, *Collect. Czech. Chem. Commun. 33*, 3410 (1968).
[532] Beneš, P.; J. Smetana, *Collect. Czech. Chem. Commun. 34*, 1360–1374 (1969).
[533] Beneš, P.; I. Rajman, *Collect. Czech. Chem. Commun. 34*, 1375–1386 (1969).
[534] Beneš, P.; *Collect. Czech. Chem. Commun. 35*, 1349 (1970).
[535] Beneš, P.; J. Ku≠era, *Collect. Czech. Chem. Commun. 37*, 523 (1972).
[536] Beneš, P.; V. Jiránek, *Radiochim. Acta 21*, 49–53 (1974).
[537] Beneš, P.; E. Steines, *Water Research 9*, 741 (1975).
[538] Benischek-Huber, I.; F. Benischek, *Anal. Chim. Acta 140/1*, 205–212 (1982).
[539] Benjamin, M.M.; E.A. Jenne, *At. Absorpt. Newsletter 15/2*, 53–54 (1976).
[540] Benson, J.M.; A.F. Eidson, R.L. Hanson, R.F. Henderson, C.H. Hobbs, *J. Appl. Toxicol. 9/4*, 219–222 (1989).
[541] Benson, W.H.; P.C. Francis, W.J. Birge, J.A. Black, *At. Spectrosc. 4/6*, 212–213 (1983).
[542] Benzo, Z.A. De; R. Fraile, C. Gomez, N. Carrión, *Clin. Chim. Acta 197/2*, 141–148 (1991).
[543] Benzo, Z.A. De; H. Schorin, M. Velosa, *J. Food Sci. 51/1*, 222–224 (1986).
[544] Benzo, Z.A. De; R. Fraile, N. Carrión, *Anal. Chim. Acta 231/2*, 283–288 (1990).
[545] Berberan-Santos, M.N.; *J. Chem. Educ. 67*, 757–759 (1990).
[546] Berdén, M.; N. Clarke, L.-G. Danielsson, A. Sparén, *Water, Air, Soil Pollut. 72*, 213–233 (1994).
[547] Bergamin, H.; F.J. Krug, E.A.G. Zagatto, E.C. Arruda, C.A. Coutinho, *Anal. Chim. Acta 190*, 177–184 (1986).
[548] Bergdahl, I.A.; A. Schütz, G.-Å. Hansson, *Analyst (London) 120/4*, 1205–1209 (1995).
[549] Berger, H.; F. Meyberg, W. Dannecker, in: B. Welz (Editor), *Fortschritte in der atomspektrometrischen Spurenanalytik*, VCH Verlagsgesellschaft, Weinheim, 1986, 607–617.
[550] Bergeron, M.; M. Beaumier, A. Hébert, *Analyst (London) 116/10*, 1019–1024 (1991).
[551] Berghoff, R.L.; D.J. von Willert, *Fresenius Z. Anal. Chem. 331/1*, 42–45 (1988).
[552] Berglund, B.; C. Wickardt, *Anal. Chim. Acta 236*, 399 (1990).
[553] Berglund, M.; W. Frech, D.C. Baxter, *Spectrochim. Acta, Part B 46/13*, 1767–1777 (1991).
[554] Bergmann, K.; B. Neidhart, *Fresenius J. Anal. Chem. 356/1*, 57–61 (1996).
[555] Berman, E.; *4th ICAS, Toronto*, 1973.
[556] Berman, S.S.; P.A. Yeats, *CRC Crit. Rev. Anal. Chem. 16/1*, 1–14 (1985).
[557] Bermejo-Barrera, P.; E. Beceiro-González, A. Bermejo-Barrera, F. Bermejo-Martínez, *Microchem. J. 40*, 103–108 (1989).
[558] Bermejo-Barrera, P.; E. Beceiro-González, A. Bermejo-Barrera, *Anal. Chim. Acta 236*, 475–477 (1990).
[559] Bermejo-Barrera, P.; C. Pita-Calvo, J.A. Cocho de Juan, *Anal. Chim. Acta 231/2*, 321–324 (1990).
[560] Bermejo-Barrera, P.; C. Pita-Calvo, F. Bermejo-Martínez, *Analyst (London) 115/5*, 549–551 (1990).
[561] Bermejo-Barrera, P.; C. Pita-Calvo, A. Bermejo-Barrera, F. Bermejo-Martínez, *Fresenius J. Anal. Chem. 340/4*, 265–268 (1991).
[562] Bermejo-Barrera, P.; C. Pita-Calvo, F. Bermejo-Martínez, *Anal. Lett. 24/3*, 447–458 (1991).
[563] Bermejo-Barrera, P.; M. Aboal-Somoza, R.M. Soto-Ferreiro, R. Domínguez-González, *Analyst (London) 118/6*, 665–668 (1993).
[564] Bermejo-Barrera, P.; C. Barciela-Alonso, M. Aboal-Somoza, A. Bermejo-Barrera, *J. Anal. At. Spectrom. 9/3*, 469–475 (1994).
[565] Bermejo-Barrera, P.; J. Moreda-Piñeiro, A. Moreda-Piñeiro, A. Bermejo-Barrera, *Anal. Chim. Acta 296/2*, 181–193 (1994).
[566] Bermejo-Barrera, P.; M.C. Barciela-Alonso, M. Ferrón-Novais, A. Bermejo-Barrera, *J. Anal. At. Spectrom. 10/3*, 247–252 (1995).
[567] Bermejo-Barrera, P.; A. Moreda-Piñeiro, J. Moreda-Piñeiro, A. Bermejo-Barrera, *J. Anal. At. Spectrom. 10/11*, 1011–1017 (1995).
[568] Bermejo-Barrera, P.; R. Domínguez-González, R. Soto-Ferreiro, A. Bermejo-Barrera, *Analusis 23*, 135–136 (1995).
[569] Bermejo-Barrera, P.; A. Moreda-Piñeiro, J. Moreda-Piñeiro, A. Bermejo-Barrera, *Quim. Anal. (Barcelona) 14*, 201 (1995).
[570] Bermejo-Barrera, P.; J. Moreda-Piñeiro, A. Moreda-Piñeiro, A. Bermejo-Barrera, *Talanta 43/1*, 35–44 (1996).
[571] Bermejo-Barrera, P.; J. Moreda-Piñeiro, A. Moreda-Piñeiro, A. Bermejo-Barrera, *Mikrochim. Acta (Vienna) 124*, 111–122 (1996).

[572] Bermejo-Barrera, P.; M.C. Barciela-Alonso, J. Moreda-Piñeiro, C. González-Sixto, A. Bermejo-Barrera, *Spectrochim. Acta, Part B 51/9-10*, 1235–1244 (1996).
[573] Bermejo-Barrera, P.; A. Moreda-Piñeiro, T. Romero-Barbeito, J. Moreda-Piñeiro, A. Bermejo-Barrera, Talanta 43/7, 1099–1107 (1996).
[574] Bermejo-Barrera, P.; M.C. Barciela-Alonso, C. González-Sixto, A. Bermejo-Barrera, *Fresenius J. Anal. Chem. 357/3*, 274–278 (1997).
[575] Bermejo-Barrera, P.; R. Dominguez González, A. Bermejo-Barrera, *Fresenius J. Anal. Chem. 357/4*, 457–461 (1997).
[576] Bernal, J.L.; M.J. Del Nozal, L. Deban, A.J. Aller, *Talanta 29/12*, 1113–1116 (1982).
[577] Berndt, H.; E. Jackwerth, *Spectrochim. Acta, Part B 30*, 169–177 (1975).
[578] Berndt, H.; E. Jackwerth, *At. Absorpt. Newsletter 15/5*, 109–113 (1976).
[579] Berndt, H.; W. Slavin, *At. Absorpt. Newsletter 17/5*, 109–112 (1978).
[580] Berndt, H.; E. Jackwerth, *Fresenius Z. Anal. Chem. 290*, 369–371 (1978).
[581] Berndt, H.; E. Jackwerth, *Fresenius Z. Anal. Chem. 290*, 105–106 (1978).
[582] Berndt, H.; E. Jackwerth, *J. Clin. Chem. Clin. Biochem. 17*, 71–76 (1979).
[583] Berndt, H.; P.G. Willmer, E. Jackwerth, *Fresenius Z. Anal. Chem. 296*, 377–379 (1979).
[584] Berndt, H.; J. Messerschmidt, *Fresenius Z. Anal. Chem. 299*, 28–32 (1979).
[585] Berndt, H.; E. Jackwerth, *J. Clin. Chem. Clin. Biochem. 17*, 489–494 (1979).
[586] Berndt, H.; J. Messerschmidt, *Fresenius Z. Anal. Chem. 308/2*, 104–111 (1981).
[587] Berndt, H.; J. Messerschmidt, E. Reiter, *Fresenius Z. Anal. Chem. 310/3-4*, 230–233 (1982).
[588] Berndt, H.; J. Messerschmidt, *Fresenius Z. Anal. Chem. 316/2*, 2 (1983).
[589] Berndt, H.; *Arch. Eisenhüttenw. 54/12*, 503–505 (1983).
[590] Berndt, H.; D. Sopczak, *Fresenius Z. Anal. Chem. 329/1*, 18–26 (1987).
[591] Berndt, H.; G. Schaldach, R. Klockenkämper, *Anal. Chim. Acta 200/1*, 573–579 (1987).
[592] Berndt, H.; *Fresenius Z. Anal. Chem. 331/3-4*, 321–323 (1988).
[593] Berndt, H.; A. Müller, *Fresenius J. Anal. Chem. 345/1*, 18–24 (1993).
[594] Berndt, H.; G. Schaldach, B.Y. Spivakov, V.M. Shkinev, *Fresenius J. Anal. Chem. 345/6*, 428–431 (1993).
[595] Berndt, H.; A. Müller, G. Schaldach, *Fresenius J. Anal. Chem. 346*, 711–716 (1993).
[596] Berndt, H.; S.K. Luo, A. Müller, G. Schaldach, in: K. Dittrich, B. Welz (Editors), *CANAS'93, Colloq. Analytische Atomspektroskopie*, Universität Leipzig/UFZ Leipzig-Halle, Leipzig, 1993, 91–100.
[597] Berndt, H.; J. Posta, G. Schaldach, S.K. Luo, in: K. Dittrich, B. Welz (Editors), *CANAS'93, Colloq. Analytische Atomspektroskopie*, Universität Leipzig/UFZ Leipzig-Halle, Leipzig, 1993, 349–356.
[598] Berndt, H.; G. Schaldach, *J. Anal. At. Spectrom. 9/1*, 39–44 (1994).
[599] Berndt, H.; *GIT Fachz. Lab. 39/2*, 100–106 (1995).
[600] Berndt, H.; G. Schaldach, S.H. Kägler, *Fresenius J. Anal. Chem. 355/1*, 37–42 (1996).
[601] Berndt, H.; J. Yáñez, *J. Anal. At. Spectrom. 11*, 703–712 (1996).
[602] Berndt, H.; J. Yáñez, *Fresenius J. Anal. Chem. 355/5-6*, 555–558 (1996).
[603] Bernth, N.; K. Vendelbo, *Analyst (London) 109/3*, 309–311 (1984).
[604] Berrow, M.L.; W.M. Stein, *Analyst (London) 108*, 277–285 (1983).
[605] Berrow, M.L.; Anal. *Proc. (London) 25/4*, 116–118 (1988).
[606] Bettencourt, A.M.M. de; M.H. Florêncio, M.F.N. Duarte, M.L.R. Gomes, L.F.C. Vilas Boas, *Appl. Organomet. Chem. 8*, 43–56 (1994).
[607] Bettencourt, A.M.M. de; M.H.F.S. Florêncio, L.F. Vilas-Boas, *Mikrochim. Acta (Vienna) 109/1-4*, 53–59 (1992).
[608] Betti, M.; P. Papoff, *CRC Crit. Rev. Anal. Chem. 19/4*, 271–322 (1988).
[609] Bettinelli, M.; *At. Spectrosc. 4/1*, 5–9 (1983).
[610] Bettinelli, M.; *Anal. Chim. Acta 148*, 193–201 (1983).
[611] Bettinelli, M.; N. Pastorelli, U. Baroni, *At. Spectrosc. 7/1*, 45–48 (1986).
[612] Bettinelli, M.; U. Baroni, N. Pastorelli, *Analyst (London) 112/1*, 23–26 (1987).
[613] Bettinelli, M.; U. Baroni, N. Pastorelli, *J. Anal. At. Spectrom. 3/7*, 1005–1011 (1988).
[614] Bettinelli, M.; U. Baroni, N. Pastorelli, *Anal. Chim. Acta 225/1*, 159–174 (1989).
[615] Bettinelli, M.; U. Baroni, *Int. J. Environ. Anal. Chem. 43/1*, 33–40 (1990).
[616] Bettinelli, M.; S. Spezia, G. Bizzarri, *At. Spectrosc. 13/2*, 75–80 (1992).
[617] Bettinelli, M.; U. Baroni, N. Pastorelli, in: C. Minoia, S. Caroli (Editors), *Appl. Zeeman Graphite Furnace At. Abs. Spectrom. Chem. Lab. Toxicology*, Pergamon, Oxford, 1992, 47–77.

[618] Bettinelli, M.; U. Baroni, F. Fontana, P. Posetti, in: C. Minoia, S. Caroli (Editors), *Appl. Zeeman Graphite Furnace At. Abs. Spectrom. Chem. Lab. Toxicology*, Pergamon, Oxford, 1992, 445–458.

[619] Bettinelli, M.; U. Baroni, G. Bizzarri, in: R. Nauche (Editor), *Progress of Analytical Chemistry in the Iron and Steel Industry*, CEC, Report EUR, Brussels, 1992, 566–575.

[620] Bettinelli, M.; S. Spezia, U. Baroni, G. Bizzarri, *At. Spectrosc. 15/3*, 115–121 (1994).

[621] Bettinelli, M.; P. Tittarelli, *J. Anal. At. Spectrom. 9/7*, 805–812 (1994).

[622] Beyer, M.; *At. Absorpt. Newsletter 4/3*, 212–223 (1965).

[623] Bezúr, L.; *J. Anal. At. Spectrom. 3/1*, 217–225 (1988).

[624] Bhattacharya, S.K.; J.C. Williams, *Anal. Lett. 12/B4*, 397–414 (1979).

[625] Bhattacharyya, S.S.; A.K. Das, *At. Spectrosc. 9/2*, 68–70 (1988).

[626] Bhattacharyya, S.S.; A.K. Das, *At. Spectrosc. 9/2*, 70–72 (1988).

[627] Bhattacharyya, S.S.; A.K. Das, *At. Spectrosc. 10/1*, 9–11 (1989).

[628] Bhattacharyya, S.S.; S.R. Das (Biswas), A.K. Das, *Indian J. Environ. Prot. 10/8*, 619–621 (1990).

[629] Biester, H.; A. Hess, G. Müller, in: R. Reuther, *Geochemical Approaches to Environmental Enineering of Metals*, Springer, Heidelberg, 1996, 33–43.

[630] Biester, H.; C. Scholz, *Environ. Sci. Technol. 31/1*, 233–239 (1997).

[631] Bihan, A. Le; J.Y. Cabon, C. Elleouet, *Analusis 20*, 601–604 (1992).

[632] Bihan, A. Le; J.Y. Cabon, *J. Anal. At. Spectrom. 10*, 993–997 (1995).

[633] Bilhorn, R.B.; J.V. Sweedler, P.M. Epperson, M.B. Denton, *Appl. Spectrosc. 41/7*, 1114–1125 (1987).

[634] Billah, M.; T. Honjo, K. Terada, *Anal. Sci. 9/2*, 251–254 (1993).

[635] Billah, M.; T. Honjo, K. Terada, *Fresenius J. Anal. Chem. 347/3-4*, 107–110 (1993).

[636] Billings, G.K.; *At. Absorpt. Newsletter 4*, 357–362 (1965).

[637] Binding, U.; H. Gawlick, *Analysentechnische Berichte (Bodenseewerk Perkin-Elmer); ATB-18*, 1969.

[638] Binstock, D.A.; D.L. Hardison, P.M. Grohse, F. Gutknecht, *EPA Report 600/8-91/213*, 1991.

[639] Birch, J.; R.M. Harrison, D.P.H. Laxen, *Sci. Total Environ. 14*, 31–42 (1980).

[640] Birch, N.J.; C. Padgham, M.S. Hughes, in: H.G. Seiler, A. Sigel, H. Sigel (Editors), *Handbook on Metals in Clinical and Analytical Chemistry*, Marcel Dekker, New York, 1994, 441–450.

[641] Birch, N.J.; C. Padgham, in: H.G. Seiler, A. Sigel, H. Sigel (Editors), *Handbook on Metals in Clinical and Analytical Chemistry*, Marcel Dekker, New York, 1994, 531–534.

[642] Birnie, S.E.; *J. Autom. Chem. 10/3*, 140–143 (1988).

[643] Bishop, J.K.B.; *Anal. Chem. 62/6*, 553–557 (1990).

[644] Biswas, S.R. Dan; A.K. Das, *Fresenius Z. Anal. Chem. 329/7*, 794–795 (1988).

[645] Biswas, S.R. Dan; A.K. Das, *At. Spectrosc. 9/6*, 207–208 (1988).

[646] Biswas, S.R. Dan; A.K. Das, *Fresenius Z. Anal. Chem. 331*, 826–827 (1988).

[647] Bitron, N.D.; *Ind. Eng. Chem. 47*, 23–28 (1955).

[648] Bittó, A.; J. Posta, L. Tomcsányi, *Acta Chim. Hung. 127/4*, 407–415 (1990).

[649] Bjoerkman, L.; K. Mottet, M. Nylander, M. Vahter, B. Lind, L. Friberg, *Arch. Toxicol. 69*, 228 (1995).

[650] Black, S.S.; M.R. Riddle, J.A. Holcombe, *Appl. Spectrosc. 40/7*, 925–933 (1986).

[651] Black, S.S.; J.M. Babo, P.A. Stear, in: H.M. Kingston, Lois B. Jassie (Editors), *Introduction to Microwave Sample Preparation*, American Chemical Society, Washington, DC, 1988, 79–92.

[652] Blain, S.; P. Appriou, H. Handel, *Analyst (London) 116/8*, 815–820 (1991).

[653] Blain, S.; P. Appriou, H. Handel, *Anal. Chim. Acta 272/1*, 91–97 (1993).

[654] Blais, J.-S.; W.D. Marshall, *J. Environ. Qual. 15/3*, 255–260 (1986).

[655] Blais, J.-S.; W.D. Marshall, *J. Anal. At. Spectrom. 4/3*, 271–277 (1989).

[656] Blais, J.-S.; G.-M. Momplaisir, W.D. Marshall, *Anal. Chem. 62/11*, 1161–1166 (1990).

[657] Blais, J.-S.; A. Huyghues-Despointes, G.M. Momplaisir, W.D. Marshall, *J. Anal. At. Spectrom. 6/3*, 225–232 (1991).

[658] Blake, S.; *Tech. Rep.-Water Res. Cent. TR229*, 1985.

[659] Blanco, M.J.; R. Ribó, J. Obiols, *Anal. Proc. (London) 31/12*, 353–356 (1994).

[660] Blankley, M.; A. Henson, K.C. Thompson, in: S.J. Haswell (Editor), *Atomic Absorption Spectrometry - Theory, Design and Application*, Elsevier, Amsterdam, 1991, 79–123.

[661] Blanusa, M.; N. Ivicic, V. Simeon, *Bull. Environ. Contam. Toxicol. 45*, 478–485 (1990).

[662] Blas, O. Jiménez. de; J.L. Pereda de Paz, J. Hernández Méndez, *Analyst (London) 114/12*, 1675–1677 (1989).

[663]   Blas, O. Jiménez. de; J.L. Pereda de Paz, J. Hernández Méndez, *J. Anal. At. Spectrom. 5/8*, 693–696 (1990).
[664]   Blas, O. Jiménez. de; J.L. Pereda de Paz, J. Hernández Méndez, *Talanta 38/8*, 857–861 (1991).
[665]   Blas, O. Jiménez. de; J.L. Pereda de Paz, J. Hernández Méndez, *Quim. Anal. (Barcelona) 11*, 173 (1992).
[666]   Blas, O. Jiménez. de; R. Seisdedos Rodriguez, J. Hernández Méndez, J.A. Sanchez Tomero, B. de Leon Gomez, S.V. Gonzalez, *J. AOAC International 77*, 722–727 (1994).
[667]   Blasco, J.B. Beferull; M. de la Guardia-Cirugeda, A.S. Carreno, *Anal. Chim. Acta 174*, 353–357 (1985).
[668]   Blasius, E.; H. Wagner, H. Braun, R. Krumbholz, B. Thimmel, *Fresenius Z. Anal. Chem. 310*, 98–107 (1982).
[669]   Blasius, E.; R. Huth, W. Neumann, *Fresenius Z. Anal. Chem. 331/3-4*, 310–315 (1988).
[670]   Bloch, L.; E. Bloch, *Zeeman Verh. 1935*, 18 (1935).
[671]   Blödorn, W.; R. Krismer, H.M. Ortner, J. Stummeyer, P. Wilhartitz, G. Wünsch, *Mikrochim. Acta (Vienna) III/3-6*, 423–432 (1989).
[672]   Blödorn, W.; G. Wünsch, H.M. Ortner, P. Wilhartitz, in: B. Welz (Editor), *5. Colloquium Atomspektrometrische Spurenanalytik*, Bodenseewerk Perkin-Elmer, Überlingen, 1989, 475–481.
[673]   Bloom, N.S.; *At. Spectrosc. 4/6*, 204–207 (1983).
[674]   Bloom, N.S.; E.A. Crecelius, *Mar. Chem. 14/1*, 49–59 (1983).
[675]   Bloom, N.S.; E.A. Crecelius, *Anal. Chim. Acta 156*, 139–145 (1984).
[676]   Bloom, N.S.; M. Horvat, C.J. Watras, *Water, Air, Soil Pollut. 80*, 1257 (1995).
[677]   Blotcky, A.J.; W.C. Duckworth, F.G. Hamel, E.P. Rack, in: H.G. Seiler, A. Sigel, H. Sigel (Editors), *Handbook on Metals in Clinical and Analytical Chemistry*, Marcel Dekker, New York, 1994, 651–663.
[678]   Blunden, S.J.; P.A. Cusack, R. Hill, (Editors), *The industrial uses of tin chemicals*, The Royal Society of Chemistry, London, 1985.
[679]   Blunden, S.J.; A. Chapman, in: P.J. Craig (Editor), *Organometallic Compounds in the Environment: Principles and Reactions*, Longman Group, Harlow, Essex, 1986, 111–159.
[680]   Blust, R.; A. Van der Linden, E. Verheyen, W. Decleir, *J. Anal. At. Spectrom. 3/2*, 387–393 (1988).
[681]   Bober, A.; A.L. Mills, *Appl. Spectrosc. 22*, 62 (1968).
[682]   Bode, H.; H. Fabian, *Fresenius Z. Anal. Chem. 162*, 328 (1958).
[683]   Bode, H.; H. Fabian, *Fresenius Z. Anal. Chem. 163*, 187 (1958).
[684]   Bodenseewerk Perkin-Elmer, *Analytical Techniques for GFAAS*, Binder B332, 1984.
[685]   Boek, K.; *Z. Lebensm. Unters. Forsch. 155/4*, 209–215 (1974).
[686]   Boer, P.; R. Fransen, W.H. Boer, H.A. Koomans, *J. Anal. At. Spectrom. 8/4*, 611–614 (1993).
[687]   Boer, P.; B. Braam, R. Fransen, W.H. Boer, H.A. Koomans, *Kidney Int. 45*, 1211–1214 (1994).
[688]   Boer, P.; R. Fransen, W.H. Boer, H.A. Koomans, *Am. J. Physiol. 268/6,2*, F1229 (1995).
[689]   Bogacz, W.; *Chem. Anal. (Warsaw) 37/5*, 635–637 (1992).
[690]   Böhm, L.; *Lebensmittelchemie 48/3*, 53 (1994).
[691]   Böhmer, R.G.; J.J. Nieuwenhuis, J.J. Theron, *S. Afr. J. Chem. 36/1*, 27–31 (1983).
[692]   Boisvert, R.; M. Bergeron, J. Turcotte, *Anal. Chim. Acta 246/2*, 365–373 (1991).
[693]   Boiteau, H.L.; A. Pineau, in: H.A. McKenzie, L.E. Smythe (Editors), *Quantitative trace analysis of biological materials*, Elsevier, Amsterdam - New York - Oxford, 1988, 543–551.
[694]   Boiteau, H.L.; A. Pineau, in: H.A. McKenzie, L.E. Smythe (Editors), *Quantitative trace analysis of biological materials*, Elsevier, Amsterdam - New York - Oxford, 1988, 553–560.
[695]   Bokowski, D.L.; *At. Absorpt. Newsletter 6/5*, 97–100 (1967).
[696]   Bokros, J.C.; *Chemistry and Physics of Carbon, Vol. 9*, Marcel Dekker, New York, 1972.
[697]   Boline, D.R.; W.G. Schrenk, *J. Assoc. Off. Anal. Chem. 60/5*, 1170–1174 (1977).
[698]   Boling, A.E.; *Spectrochim. Acta 22/3*, 425–431 (1966).
[699]   Boling, A.E.; *Spectrochim. Acta, Part B 23/7*, 495–496 (1968).
[700]   Bombach, G.; K. Bombach, W. Klemm, *Fresenius J. Anal. Chem. 350/1-2*, 18–20 (1994).
[701]   Bombach, G.; A. Pierra, W. Klemm, *Fresenius J. Anal. Chem. 350/1-2*, 49–53 (1994).
[702]   Bombach, H.; B. Luft, E. Weinhold, F. Mohr, *Neue Hütte 29*, 233–236 (1984).
[703]   Bombach, H.; E. Weinhold, *Mikrochim. Acta (Vienna) I*, 229–236 (1989).
[704]   Bombach, H.; E. Weinhold, in: B. Welz (Editor), *6. Colloquium Atomspektrometrische Spurenanalytik*, Bodenseewerk Perkin-Elmer, Überlingen, 1991, 639–646.

[705] Bona, M.A.; M. Castellano, L. Plaza, A. Fernandez, *Human Experimental Toxicology 11*, 311–314 (1992).
[706] Bond, A.M.; T.A. O'Donnell, *Anal. Chem. 40/3*, 560–563 (1968).
[707] Bond, A.M.; J.B. Willis, *Anal. Chem. 40*, 2087 (1968).
[708] Bond, A.M.; *Anal. Chem. 42/8*, 932–935 (1970).
[709] Boniforti, R.; R. Ferraroli, P. Frigieri, D. Heltai, G, Queirazza,. *Anal. Chim. Acta 162*, 33–46 (1984).
[710] Bonilla, M.; L. Rodrigues, C. Cámara-Rica, *J. Anal. At. Spectrom. 2/2*, 157–161 (1987).
[711] Bookbinder, M.J.; K.J. Panosian, *Clin. Chem. (Winston-Salem) 33/7*, 1170–1176 (1987).
[712] Borg, H.; A. Edin, K. Holm, E. Skold, *Water Research 15/11*, 1291–1295 (1981).
[713] Borggaard, O.K.; H.E.M. Christensen, Soeren P. Lund, *Analyst (London) 109/9*, 1179–1182 (1984).
[714] Borggaard, O.K.; H.E.M. Christensen, C. Ilsoee, *Milchwissenschaft 39*,12, 725–727 (1984).
[715] Borja, R.M.; A. Salvador, M. de la Guardia, J.L. Burguera, M. Burguera, *Quim. Anal. (Barcelona) 8/2*, 241–246 (1989).
[716] Borja, R.M.; M. de la Guardia, A. Salvador, J.L. Burguera, M. Burguera, *Fresenius J. Anal. Chem. 338/1*, 9–15 (1990).
[717] Borsier, M.; M. Garcia, *Spectrochim. Acta, Part B 38/1-2*, 123–127 (1983).
[718] Borsier, M.; *Spectrochimica Acta Rev. 14/1-2*, 79–94 (1991).
[719] Borszéki, J.; P. Halmos, E. Gegus, P. Kárpáti, *Talanta 41/7*, 1089–1093 (1994).
[720] Bortlisz, J.; *Gewässerschutz, Wasser, Abwasser 57*, 213–247 (1982).
[721] Bortoli, A.; M. Gerotto, M. Marchiori, H. Muntau, A. Rehnert, *Mikrochim. Acta (Vienna) 119/3-4*, 305–310 (1995).
[722] Bosch, H.; E. Büchel, K. Lohau, *Analysentechnische Berichte (Bodenseewerk Perkin-Elmer) 20*, 1970.
[723] Bosch, J.C. van den; in: S. Flügge (Editor), *Handbuch der Physik, Bd. 28, Spektroskopie II*, Springer, Berlin, 1957, 296–332.
[724] Bosnak, C.P.; G.R. Carnrick, W. Slavin, *At. Spectrosc. 7/5*, 148–150 (1986).
[725] Bosnak, C.P.; D. Bradshaw, R. Hergenreder, K. Kingston, *At. Spectrosc. 14/3*, 80–82 (1993).
[726] Bosnak, C.P.; Z.A. Grosser, *At. Spectrosc. 17/6*, 211–214 (1996).
[727] Bosnak, C.P.; Z.A. Grosser, *At. Spectrosc. 17/6*, 218–224 (1996).
[728] Boss, C.B.; G.M. Hieftje, *Anal. Chem. 49*, 2112–2114 (1977).
[729] Botelho, G.M.A.; A.J. Curtius, R.C. Campos, *J. Anal. At. Spectrom. 9/11*, 1263–1267 (1994).
[730] Bothner, M.H.; D.E. Robertson, *Anal. Chem. 47/3*, 592–595 (1975).
[731] Botre, C.; F. Cacare, R. Cozzani, *Anal. Lett. 9/9*, 825–830 (1976).
[732] Boulding, J.R.; *Description and Sampling of Contaminated Soils, A Field Guide*, Lewis, Boca Raton, FL, 1994.
[733] Bouman, A.A.; A.J. Platenkamp, F.D. Posma, *Ann. Clin. Biochem. 23/3*, 346–350 (1986).
[734] Boumans, P.W.J.M.; *Spectrochim. Acta, Part B 33*, 625–634 (1978).
[735] Boumans, P.W.J.M.; J.J.A.M. Vrakking, *Spectrochim. Acta, Part B 40B/10*, 1423–1435 (1985).
[736] Bourgoin, B.P.; D. Boomer, M.J. Powell, S. Willie, D. Edgar, D. Evans, *Analyst (London) 117/1*, 19–22 (1992).
[737] Bourret, E.; I. Moynier, L. Bardet, M. Fussellier, *Anal. Chim. Acta 172*, 157–166 (1985).
[738] Bowen, H.J.M.; A. Peggs, *J. Sci. Food Agric. 35*, 1225 (1984).
[739] Bowen, H.J.M.; in: H.A. McKenzie, L.E. Smythe (Editors), *Quantitative trace analysis of biological materials*, Elsevier, Amsterdam - New York - Oxford, 1988, 607–622.
[740] Bower, N.W.; J.D. Ingle, *Anal. Chem. 48/4*, 686–692 (1976).
[741] Bower, N.W.; J.D. Ingle, *Anal. Chem. 49/4*, 574–579 (1977).
[742] Bower, N.W.; J.D. Ingle, *Anal. Chem. 50*, 544 (1978).
[743] Bower, N.W.; J.D. Ingle, *Anal. Chem. 51/1*, 72–76 (1979).
[744] Bower, N.W.; J.D. Ingle, *Appl. Spectrosc. 35/3*, 317–324 (1981).
[745] Bowman, J.A.; J.B. Willis, *Anal. Chem. 39/11*, 1210–1216 (1967).
[746] Bowman, J.A.; *Anal. Chim. Acta 37*, 465–471 (1967).
[747] Boyle, E.A.; J.M. Edmond, *Anal. Chim. Acta 91*, 189–197 (1977).
[748] Bozsai, G.; Z. Kárpáti, *Acta Chim. Hung. 126/3*, 377–383 (1989).
[749] Bozsai, G.; B. Welz, B. Radziuk, M. Sperling, in: B. Welz (Editor), *5. Colloquium Atomspektrometrische Spurenanalytik*, Bodenseewerk Perkin-Elmer, Überlingen, 1989, 235–246.
[750] Bozsai, G.; Z. Kárpáti, *XXVII CSI, Pre-symposium*, Lofthus, Norway, 1991, P-40.
[751] Bozsai, G.; *Microchem. J. 46/2*, 159–166 (1992).
[752] Bozsai, G.; M. Melegh, *Microchem. J. 51/1-2*, 39–45 (1995).

[753] Braca, G.; R. Cioni, G. Sbrana, G. Scandiffo, *At. Absorpt. Newsletter 14/2*, 39–40 (1975).
[754] Bradbury, M.W.B.; C.R. Kleeman, H. Bagdoyan, A. Berberian, *J. Clin. Lab. Med. 71*, 884 (1968).
[755] Bradley, C.; F. Ying Leung, *Clin. Chem. (Winston-Salem) 40/3*, 431–434 (1994).
[756] Bradshaw, D.K.; W. Slavin, *Spectrochim. Acta, Part B 44/12*, 1245–1256 (1989).
[757] Brady, D.V.; J.G. Montalvo, G. Glowacki, A. Pisciotta, *Anal. Chim. Acta 70*, 448–452 (1974).
[758] Brady, D.V.; J.G. Montalvo, J. Jung, R.A. Curran, *At. Absorpt. Newsletter 13*, 118 (1974).
[759] Brätter, V.E. Negretti de; P. Brätter, A. Tomiak, *J. Trace Elem. Electrol. Health Dis. 4*, 41–48 (1990).
[760] Brailsford, J.; J.E. Stapley, *Proc. Prehistoric Soc. 38*, 219 (1972).
[761] Braithwaite, R.A.; in: R.F.M. Herber, M. Stoeppler (Editors), *Trace Element Analysis in Biological Specimens*, Elsevier, Amsterdam, 1994.
[762] Brajter, K.; K. Slonawska, *Talanta 30/7*, 471–474 (1983).
[763] Brajter, K.; U. Ciborowska, I. Miazek, *Fresenius Z. Anal. Chem. 315/2*, 126–131 (1983).
[764] Brajter, K.; K. Klejny, *Talanta 32/7*, 521–524 (1985).
[765] Brajter, K.; K. Slonawska, *Fresenius Z. Anal. Chem. 323/2*, 145–147 (1986).
[766] Brajter, K.; K. Slonawska, *J. Anal. At. Spectrom. 2/2*, 167–170 (1987).
[767] Brajter, K.; K. Slonawska, *Anal. Lett. 21/2*, 311–318 (1988).
[768] Brajter, K.; K. Slonawska, *Water Research 22/11*, 1413–1416 (1988).
[769] Brajter, K.; K. Slonawska, *Mikrochim. Acta (Vienna) I*, 137–143 (1989).
[770] Bramall, N.; K.C. Thompson, *Analyst (London) 107*, 922–926 (1982).
[771] Braman, R.S.; L.L. Justen, C.C. Foreback, *Anal. Chem. 44/13*, 2195–2199 (1972).
[772] Braman, R.S.; in: E.A. Woolson (Editor), *Arsenical Pesticides*, ACS Symposium Series No. 7, ACS, 1975, 108–123.
[773] Branagh, W.; E.D. Salin, *Spectroscopy (Eugene, OR) 10/7*, 20 (1995).
[774] Branch, C.H.; D. Hutchison, *Analyst (London) 111/2*, 231–233 (1986).
[775] Branch, C.H.; D. Hutchison, *J. Anal. At. Spectrom. 1/6*, 433–436 (1986).
[776] Brandenberger, H.; H. Bader, *Helv. Chim. Acta 50/5*, 1409–1415 (1967).
[777] Brandenberger, H.; H. Bader, *At. Absorpt. Newsletter 6/5*, 101–103 (1967).
[778] Brandenberger, H.; H. Bader, *Chimia 21*, 597 (1967).
[779] Brandenberger, H.; H. Bader, *At. Absorpt. Newsletter 7/3*, 53–54 (1968).
[780] Brandt, F.; K. Thomas, R. Trost, in: B. Welz (Editor), *Fortschritte in der atomspektrometrischen Spurenanalytik*, Verlag Chemie, Weinheim, 1984, 625–636.
[781] Brandt, J.; B. Hitzmann, *Anal. Chim. Acta 291/1-2*, 29–40 (1994).
[782] Brandvold, D.K.; P. Martínez, C. Matlock, *Anal. Instr. (New York) 21/1-2*, 63–67 (1993).
[783] Braun, T.; S. Zsindely, *Anal. Proc. (London) 28*, 283 (1991).
[784] Breene, R.G.; *The Shift and Shape of Spectral Lines*, Pergamon Press, Oxford, New York, 1961.
[785] Breene, R.G.; *Theories of Spectral Line Shape*, John Wiley & Sons, New York - Chichester, 1981.
[786] Brennfleck, U.; G. Müller-Vogt, W. Wendl, *Spectrochim. Acta, Part B 51*, 1139–1145 (1996).
[787] Brescianini, C.; A. Mazzucotelli, F. Valerio, R. Frache, G. Scarponi, *Fresenius Z. Anal. Chem. 332/1*, 34–36 (1988).
[788] Bredthauer, U.; J. Agger, W. Dannecker, in: B. Welz (Editor), *CANAS'95 Colloquium Analytische Atomspektroskopie*, Bodenseewerk Perkin-Elmer, Überlingen, 1996, 299–305.
[789] Brewer, P.G.; D.W. Spencer, C.L. Smith, *ASTM Spec. Tech. Publ. 443*, 70 (1969).
[790] Brimhall, W.H.; *Anal. Chem. 41*, 1349 (1969).
[791] Brimmer, S.P.; W.R. Fawcett, K.A. Kulhavy, *Anal. Chem. 59/10*, 1470–1471 (1987).
[792] Brinckman, F.E.; W.R. Blair, K.L. Jewett, W.P. Iverson, *J. Chromatogr. Sci. 15/11*, 493–503 (1977).
[793] Brinckman, F.E.; K.L. Jewett, W.P. Iverson, K.J. Irgolic, K.C. Ehrhardt, R. Stockton, *J. Chromatogr. 191*, 31–46 (1980).
[794] Brindle, I.D.; X.-c. Le, X.-f. Li, *J. Anal. At. Spectrom. 4/2*, 227–232 (1989).
[795] Brindle, I.D.; H. Alarabi, S. Karshman, X.-c. Le, S.-g. Zheng, *Analyst (London) 117/3*, 407–411 (1992).
[796] Brindle, I.D.; S.-g. Zheng, *Spectrochim. Acta, Part B 41/14*, 1777–1780 (1996).
[797] Briska, M.; *Fresenius Z. Anal. Chem. 273*, 283 (1975).
[798] British Standards Institution, *British Standard BS-4258-11*, 1982.
[799] British Standards Institution, *British Standard BS-2000-288*, 1983.
[800] British Standards Institution, *British Standard BS-6068/2.5*, 1984.
[801] British Standards Institution, *British Standard BS-2000/363*, 1985.

[802]   British Standards Institution, *British Standard BS-6200/3.1.5*, 1985.
[803]   British Standards Institution, *British Standard BS-3900/B4*, 1986.
[804]   British Standards Institution, *British Standard BS-3900/B6*, 1986.
[805]   British Standards Institution, *British Standard BS-3900/B7*, 1986.
[806]   British Standards Institution, *British Standard BS-3900/B9*, 1986.
[807]   British Standards Institution, *British Standard BS-3900/B12*, 1986.
[808]   British Standards Institution, *British Standard BS-5600/4.17.2*, 1986.
[809]   British Standards Institution, *British Standard BS-5600/4.17.3*, 1986.
[810]   British Standards Institution, *British Standard BS-5600/4.17.4*, 1986.
[811]   British Standards Institution, *British Standard BS-5600/4.17.5*, 1986.
[812]   British Standards Institution, *British Standard BS-5688/28*, 1986.
[813]   British Standards Institution, *British Standard BS-6200/3.12.3*, 1986.
[814]   British Standards Institution, *British Standard BS-6200/3.20.4*, 1986.
[815]   British Standards Institution, *British Standard BS-6721/4*, 1986.
[816]   British Standards Institution, *British Standard BS-6721/5*, 1986.
[817]   British Standards Institution, *British Standard BS-6783/1*, 1986.
[818]   British Standards Institution, *British Standard BS-6783/3*, 1986.
[819]   British Standards Institution, *British Standard BS-6783/4*, 1986.
[820]   British Standards Institution, *British Standard BS-6200/3.7.1*, 1987.
[821]   British Standards Institution, *British Standard BS-6200/3.10.2*, 1989.
[822]   British Standards Institution, *British Standard BS-6721/9*, 1989.
[823]   British Standards Institution, *British Standard BS-6068/2.38*, 1990.
[824]   British Standards Institution, *British Standard BS-6200/3.1.4*, 1990.
[825]   British Standards Institution, *British Standard BS-6200/3.34.3*, 1990.
[826]   British Standards Institution, *British Standard BS-7164/26*, 1990.
[827]   British Standards Institution, *British Standard BS-7164/28.1*, 1990.
[828]   British Standards Institution, *British Standard BS-7317/1*, 1990.
[829]   British Standards Institution, *British Standard BS-7317/2*, 1990.
[830]   British Standards Institution, *British Standard BS-7317/3*, 1990.
[831]   British Standards Institution, *British Standard BS-7317/4*, 1990.
[832]   British Standards Institution, *British Standard BS-7317/7*, 1990.
[833]   British Standards Institution, *British Standard BS-7319/6*, 1990.
[834]   British Standards Institution, *British Standard BS-7319/8*, 1990.
[835]   British Standards Institution, *British Standard BS-7319/9*, 1990.
[836]   British Standards Institution, *British Standard BS-5666/3*, 1991.
[837]   British Standards Institution, *British Standard BS-7164/27*, 1991.
[838]   British Standards Institution, *British Standard BS-7427*, 1991.
[839]   British Standards Institution, *British Standard BS-7455/2*, 1991.
[840]   British Standards Institution, *British Standard BS-7455/3*, 1991.
[841]   British Standards Institution, *British Standard BS-7455/4*, 1991.
[842]   British Standards Institution, *British Standard BS-7455/5*, 1991.
[843]   British Standards Institution, *British Standard BS-7455/6*, 1991.
[844]   British Standards Institution, *British Standard BS-7164/29.1*, 1992.
[845]   British Standards Institution, *British Standard BS-6068/6.8*, 1993.
[846]   British Standards Institution, *British Standard BS-EN 27627-6*, 1993.
[847]   British Standards Institution, *British Standard BS-EN ISO 10700*, 1995.
[848]   Brockbank, C.I.; G.E. Batley, G.K.C. Low, *Environ. Technol. Lett. 9*, 1361–1366 (1988).
[849]   Brockmann, A.; H.-M. Kuß, A. Golloch, in: B. Welz (Editor), *6. Colloquium Atom-spektrometrische Spurenanalytik*, Bodenseewerk Perkin-Elmer, Überlingen, 1991, 439–447.
[850]   Brockmann, A.; Ch. Nonn, A. Golloch, *J. Anal. At. Spectrom. 8/3*, 397–401 (1993).
[851]   Broderick, B.; W. Cofino, R. Cornelis, K. Heydorn, W. Horwitz, D.Hunt, R. Hutton, H. Kingston, H.Muntau, R. Baudo, D. Rossi, J. van Raaphorst, *Mikrochim. Acta (Vienna) II*, 523–542 (1991).
[852]   Brodie, K.G.; J.P. Matoušek, *Anal. Chim. Acta 69*, 200–202 (1974).
[853]   Broe, S.; P.J. Joergensen, J.M. Christensen, M. Hoerder, *J. Trace Elem. Electrol. Health Dis. 2/1*, 31–35 (1988).
[854]   Broeckx, R.L.; *Anal. Chem. 58/2*, 275A-288A (1986).
[855]   Broek, W.M.G.T. van den; L. de Galan, *Anal. Chem. 49/14*, 2176–2186 (1977).
[856]   Broekaert, J.A.C.; T. Graule, H. Jenett, G. Tölg, P. Tschöpel, *Fresenius Z. Anal. Chem. 332/7*, 825–838 (1989).
[857]   Broekaert, J.A.C.; *Chem. Anal. (Warsaw) 35/1-3*, 5–16 (1990).

[858]   Brooke, P.J.; W.H. Evans, *Analyst (London) 106*, 514–520 (1981).
[859]   Brooks, I.B.; G.A. Luster, D.G. Easterly, *At. Absorpt. Newsletter 9/4*, 93–94 (1970).
[860]   Brooks, R.R.; A.K. Chatterjee, D.E. Ryan, *Chem. Geol. 33/1-2*, 163–169 (1981).
[861]   Brooks, R.R.; J.A. Willis, J.R. Liddle, *J. Assoc. Off. Anal. Chem. 66/1*, 130–134 (1983).
[862]   Brooks, R.R.; Bee-Saik Lee, *Anal. Chim. Acta 204*, 333–337 (1988).
[863]   Broske, P.; *LaborPraxis 9/3*, 152 (1985).
[864]   Brossier, P.; C. Moise, *Analusis 12/4*, 223–224 (1984).
[865]   Brovko, I.A.; A.S. Tursunov, M.A. Rish, A.D. Davirov, *Zh. Anal. Khim. 39*, 1768 (1984).
[866]   Brovko, I.A.; A. Davirov, G.M. Tolmacheva, *Uzb. Khim. Zh. (1)*, 25 (1986).
[867]   Brovko, I.A.; *Zh. Anal. Khim. 42/9*, 1627–1630 (1987).
[868]   Brovko, I.A.; *Zh. Anal. Khim. 43/11*, 1987–1993 (1988).
[869]   Browett, W.R.; M.J. Stillman, *Prog. Anal. Spectrosc. 12/1*, 73–110 (1989).
[870]   Brown, A.A.; P.J. Whiteside, *Int. Labmate 9/3*, 33–36 (1984).
[871]   Brown, A.A.; A. Taylor, *Analyst (London) 109/11*, 1455–1459 (1984).
[872]   Brown, A.A.; B.A. Milner, A. Taylor, *Analyst (London) 110/5*, 501–505 (1985).
[873]   Brown, A.A.; A. Taylor, *Analyst (London) 110/6*, 579–582 (1985).
[874]   Brown, A.A.; T.C. Dymott, *Int. Labmate 11/5*, 13–18 (1986).
[875]   Brown, A.A.; M. Lee, *Fresenius Z. Anal. Chem. 323/7*, 697–702 (1986).
[876]   Brown, A.A.; M. Lee, G. Küllmer, A. Rosopulo, *Fresenius Z. Anal. Chem. 328/4-5*, 354–358 (1987).
[877]   Brown, A.A.; *J. Anal. At. Spectrom. 3/1*, 67–71 (1988).
[878]   Brown, D.F.G.; A.M. Mackay, A. Turek, *Anal. Chem. 41,* 2091 (1969).
[879]   Brown, G.N.; D.L. Styris, *J. Anal. At. Spectrom. 8/2*, 211–216 (1993).
[880]   Brown, G.N.; D.L. Styris, M.W. Hinds, *J. Anal. At. Spectrom. 10/8*, 527–531 (1995).
[881]   Brown, J.J.; C.S. Saba, W.E. Rhine, K.J. Eisentraut, *Anal. Chem. 52/14*, 2365–2370 (1980).
[882]   Brown, R.J.; M.L. Parsons, *J. Quant. Spectrosc. Radiat. Transfer 21,* 553–561 (1979).
[883]   Brown, R.M.; R.C. Fry, J.L. Moyers, S.J. Northway, M.B. Denton, G.S. Wilson, *Anal. Chem. 53/11*, 1560–1566 (1981).
[884]   Brown, R.M.; C.J. Pickford, W.L. Davison, *Int. J. Environ. Anal. Chem. 18,* 135–141 (1984).
[885]   Brown, T.F.; L.K. Zeringue, *Commun. Soil. Sci. Plant Anal. 19/2,* 167–181 (1988).
[886]   Brown, T.F.; L.K. Zeringue, *Commun. Soil. Sci. Plant Anal. 22*, 399–407 (1991).
[887]   Browner, R.F.; R.M. Dagnall, T.S. West, *Talanta 16,* 75–81 (1969).
[888]   Browner, R.F.; R.M. Dagnall, T.S. West, *Anal. Chim. Acta 45,* 163–170 (1969).
[889]   Browner, R.F.; A.W. Boorn, D.D. Smith, *Anal. Chem. 54/8,* 1411–1419 (1982).
[890]   Browner, R.F.; A.W. Boorn, *Anal. Chem. 56/7*, 786A-798A (1984).
[891]   Browner, R.F.; *Microchem. J. 40/1*, 4–29 (1989).
[892]   Bruce, C.F.; P. Hannaford, *Spectrochim. Acta, Part B 26*, 207–235 (1971).
[893]   Brückner, H.P.; G. Drews, K. Kritsotakis, H.J. Tobschall, *Chem. Erde 45/1-2*, 53–56 (1986).
[894]   Brückner, H.P.; G. Drews, K. Kritsotakis, H.J. Tobschall, *Chem. Erde 45/1-2*, 57–74 (1986).
[895]   Brueggemeyer, T.W.; J.A. Caruso, *Anal. Chem. 54/6*, 872–875 (1982).
[896]   Brüggerhoff, S.; E. Jackwerth, *Fresenius Z. Anal. Chem. 326*, 528–535 (1987).
[897]   Bruening, R.L.; B.J. Tarbet, K.E. Krakowiak, M.L. Bruening, R.M. Izatt, *Anal. Chem. 63/10*, 1014–1017 (1991).
[898]   Bruhn, C.G.; V. Cabalín G., *Anal. Chim. Acta 147*, 193–203 (1983).
[899]   Bruland, K.W.; R.P. Franks, G.A. Knauer, J.H. Martin, *Anal. Chim. Acta 105*, 233–245 (1979).
[900]   Bruland, K.W.; K.H. Coale, L. Mart, *Mar. Chem. 17/4*, 285–300 (1985).
[901]   Brumbaugh, W.G.; M.J. Walther, *J. Assoc. Off. Anal. Chem. 72*, 484–486 (1989).
[902]   Brumbaugh, W.G.; M.J. Walther, *J. Assoc. Off. Anal. Chem. 74/3*, 570–571 (1991).
[903]   Brunelle, R.L.; C.M. Hoffman, K.B. Snow, *J. Assoc. Off. Anal. Chem. 53,* 470–474 (1970).
[904]   Brunetti, M.; A. Nicolotti, M. Feuerstein, G. Schlemmer, *At. Spectrosc. 15/5*, 209–212 (1994).
[905]   Brunetto, M.R.; J.L. Burguera, M. Burguera, D. Chakraborti, *At. Spectrosc. 13/4,* 123–126 (1992).
[906]   Brunker, R.L.; T.L. Bott, *Appl. Microbiol. 27*, 870–873 (1974).
[907]   Brunner, G.; F. Korneck, G. Müller-Vogt, W. Wendl, W. Send, *Spectrochim. Acta, Part B 47/9*, 1097–1105 (1992).
[908]   Bruno, S.N.F.; R.C. Campos, A.J. Curtius, *J. Anal. At. Spectrom. 9/3*, 341–344 (1994).
[909]   Bryan, G.W.; P.E. Gibbs, L.G. Hummerstone, G.R. Burt, *J. Mar. Biol. Ass. U.K. 66*, 611–640 (1986).
[910]   Brzezinska, A.; J.C. Van Loon, D. Williams, K. Oguma, K. Fuwa, I.H. Haraguchi, *Spectrochim. Acta, Part B 38/10*, 1339–1346 (1983).

[911] Brzezinska-Paudyn, A.; J.C. Van Loon, R. Haneock, *At. Spectrosc. 7/3*, 72–75 (1986).
[912] Brzezinska-Paudyn, A.; J.C. Van Loon, *Fresenius Z. Anal. Chem. 331/7*, 707–712 (1988).
[913] Bubert, H.; R. Klockenkämper, *Spectrochim. Acta, Part B 38/8*, 1087–1098 (1983).
[914] Buchanan, J.R.; T.T. Muraoka, *At. Absorpt. Newsletter 3*, 79–83 (1964).
[915] Buchmayer, P.; W. Hoffmann, L. Meckel, *Glastechn Ber. 60/2*, 47–54 (1987).
[916] Budesinsky, B.W.; *Microchem. J. 29/1*, 1–6 (1984).
[917] Budniok, A.; *Microchem. J. 25/4*, 531–534 (1980).
[918] Bulewicz, E.M.; C.G. James, T.M. Sugden, *Proc. Roy. Soc. (London), Ser. A 235*, 89–106 (1956).
[919] Bulska, E.; Z. Grobenski, G. Schlemmer, *J. Anal. At. Spectrom. 5/3*, 203–204 (1990).
[920] Bulska, E.; D.C. Baxter, W. Frech, *Anal. Chim. Acta 249/2*, 545–554 (1991).
[921] Bulska, E.; B. Godlewska, K. Wróbel, A. Hulanicki, *Can. J. Appl. Spectroscopy 36/4*, 89–93 (1991).
[922] Bulska, E.; K. Wróbel, A. Hulanicki, *Fresenius J. Anal. Chem. 342/9*, 740–743 (1992).
[923] Bulska, E.; W. Kandler, P. Paslawski, A. Hulanicki, *Mikrochim. Acta (Vienna) 119/1-2*, 137–146 (1995).
[924] Bulska, E.; W. Kandler, A. Hulanicki, *Spectrochim. Acta, Part B 51/9-10*, 1263–1270 (1996).
[925] Bulska, E.; K. Pyrzynska, *Fresenius J. Anal. Chem. 355/5-6*, 672–675 (1996).
[926] Bulska, E.; G. Chelmecki, A. Hulanicki, *Can. J. Appl. Spectrosc. 41/1*, 5 (1996).
[927] Bundesministerium des Inneren, *Technische Anleitung zur Reinhaltung der Luft (TA Luft)*, Heider-Verlag GM Bl. S.426, 1974.
[928] Bunker, V.W.; H.T. Delves, *Anal. Chim. Acta 201*, 331–334 (1987).
[929] Burba, P.; P.G. Willmer, M. Betz, F. Fuchs, *Int. J. Environ. Anal. Chem. 13/3*, 177–191 (1983).
[930] Burba, P.; P.G. Willmer, *Fresenius Z. Anal. Chem. 322/3*, 266–271 (1985).
[931] Burba, P.; P.G. Willmer, *Vom Wasser 66*, 33–47 (1986).
[932] Burba, P.; P.G. Willmer, *Fresenius Z. Anal. Chem. 323*, 811–817 (1986).
[933] Burba, P.; P.G. Willmer, R. Klockenkämper, *Vom Wasser 71*, 179–194 (1988).
[934] Burba, P.; *Fresenius J. Anal. Chem. 348/4*, 301–311 (1994).
[935] Burdo, R.A.; W.M. Wise, *Anal. Chem. 47/14*, 2360–2364 (1975).
[936] Burguera, J.L.; M. Burguera, M. Gallignani, O.M. Alarcón, *Clin. Chem. (Winston-Salem) 29/3*, 568–570 (1983).
[937] Burguera, J.L.; M. Burguera, M. Gallignani, *An. Acad. Bras. Cienc. 55*, 209–211 (1983).
[938] Burguera, J.L.; M. Burguera, O.M. Alarcón, *J. Anal. At. Spectrom. 1*, 79–83 (1986).
[939] Burguera, J.L.; M. Burguera, O.M. Alarcón, *Trace Elem. Med. 3*, 117–120 (1986).
[940] Burguera, J.L.; M. Burguera, C. Rivas, O.M. Alarcón, B. Ibarra de Díaz, *Acta Cient. Venez. 39*, 222–229 (1988).
[941] Burguera, J.L.; M. Burguera, O.M. Alarcón, M. Cañada de Zunzunegui, H.A. Carrasco, D. Dávila, J. Reinoza, *J. Trace Elem. Electrol. Health Dis. 2/4*, 215–219 (1988).
[942] Burguera, J.L.; M. Burguera, C. Rivas, O.M. Alarcón, B. Ibarra de Díaz, *Trace Elem. Med. 6*, 1–3 (1989).
[943] Burguera, J.L.; M. Burguera, C. Rivas, M. de la Guardia, A. Salvador, V. Carbonell, *J. Flow Injection Anal. 7/1*, 11–18 (1990).
[944] Burguera, J.L.; M. Burguera, G. Becerra, *Water, Air, Soil Pollut. 57-58*, 489–493 (1991).
[945] Burguera, J.L.; M. Burguera, *Lab. Robot. Autom. 3*, 119–124 (1991).
[946] Burguera, J.L.; M. Burguera, A. Matousek de Abel de la Cruz, N. Anez, O.M. Alarcon, *At. Spectrosc. 13/2*, 67–71 (1992).
[947] Burguera, J.L.; M. Burguera, *J. Anal. At. Spectrom. 8/2*, 235–241 (1993).
[948] Burguera, J.L.; M. Burguera, M.R. Brunetto, *At. Spectrosc. 14/4*, 90–94 (1993).
[949] Burguera, J.L.; M. Burguera, *J. Trace Elem. Electrol. Health Dis. 7/1*, 9–18 (1993).
[950] Burguera, J.L.; M. Burguera, P. Carrero, C. Rivas, M. Gallignani, M. Brunetto, *Anal. Chim. Acta 308/1-3*, 349–356 (1995).
[951] Burguera, M.; J.L. Burguera, O.M. Alarcón, *Anal. Chim. Acta 179*, 351–357 (1986).
[952] Burguera, M.; J.L. Burguera, P.C. Rivas, O.M. Alarcón, *At. Spectrosc. 7/3*, 79–81 (1986).
[953] Burguera, M.; J.L. Burguera, A.M. Garaboto, O.M. Alarcón, *Trace Elem. Med. 5/2*, 60–63 (1988).
[954] Burguera, M.; J.L. Burguera, O.M. Alarcón, *Anal. Chim. Acta 214/1-2*, 421–427 (1988).
[955] Burguera, M.; J.L. Burguera, M. Gallignani, *J. Flow Injection Anal. 9*, 13–19 (1992).
[956] Burguera, M.; J.L. Burguera, *J. Anal. At. Spectrom. 8/2*, 229–233 (1993).
[957] Burke, K.E.; *Anal. Chem. 42/13*, 1536–1540 (1970).
[958] Burke, K.E.; C.H. Albright, *J. Assoc. Off. Anal. Chem. 53/3*, 531–533 (1970).

[959] Burke, K.E.; *Appl. Spectrosc. 28/3*, 234–237 (1974).
[960] Burkhard, E.G.; G. Mehmood, L. Husain, *J. Radioanal. Nucl. Chem. 161/1*, 101–112 (1992).
[961] Burnett, R.W.; *Clin. Chem. (Winston-Salem) 26/5*, 644–646 (1980).
[962] Burns, D.T.; F. Glocking, V.B. Mahale, W.J. Swindall, *Analyst (London) 103*, 985 (1978).
[963] Burns, D.T.; F. Glocking, M. Harriott, *Analyst (London) 106*, 921–930 (1981).
[964] Burns, D.T.; D. Dadgar, M. Harriott, K. McBride, W.J. Swindall, *Analyst (London) 109/12*, 1613–1614 (1984).
[965] Burns, D.T.; M. Harriot, F. Glocking, *Fresenius Z. Anal. Chem. 327/7*, 701–703 (1987).
[966] Burns, D.T.; G.D. Atkinson, N. Chimpalee, M. Harriott, *Fresenius Z. Anal. Chem. 331*, 814–817 (1988).
[967] Burns, D.T.; N. Chimpalee, M. Harriott, *Fresenius J. Anal. Chem. 344/7-8*, 357–359 (1992).
[968] Burns, D.T.; N. Chimpalee, M. Harriott, *Fresenius J. Anal. Chem. 348/3*, 248–249 (1994).
[969] Burns, D.T.; N. Chimpalee, M. Harriott, *Fresenius J. Anal. Chem. 349/7*, 527–529 (1994).
[970] Burrell, D.C.; G.Güner Wood, *Anal. Chim. Acta 48*, 45–49 (1969).
[971] Burrows, J.A.; J.C. Heerdt, J.B. Willis, *Anal. Chem. 37*, 579–582 (1965).
[972] Burylin, M.Y.; Z.A. Temerdashev, A.P. Bayanov, L.V. Saprykin, *Zh. Anal. Khim. 41*, 2160 (1986).
[973] Burylin, M.Y.; Z.A. Temerdashev, A.P. Bayanov, *Zh. Anal. Khim. 41*, 675 (1986).
[974] Burylin, M.Y.; Z.A. Temerdashev, A.P. Bayanov, *Zavodsk. Lab. 53/4*, 32 (1987).
[975] Buscaglia, J.A.; *Anal. Chim. Acta 288/1-2*, 17–24 (1994).
[976] Busch, K.W.; G.H. Morrison, *Anal. Chem. 45/8*, 712A-722A (1973).
[977] Busch, K.W.; M.A. Busch, *Multielement Detection Systems for Spectrochemical Analysis* (Chemical Analysis Vol. 107), John Wiley & Sons, Chichester, 1990
[978] Busheina, I.S.; J.B. Headridge, *Anal. Chim. Acta 142*, 197–205 (1982).
[979] Busheina, I.S.; J.B. Headridge, *Anal. Chim. Acta 174*, 339–341 (1985).
[980] Busheina, I.S.; J.B. Headridge, D. Johnson, K.W. Jackson, C.W. McLeod, J.A. Roberts, *Anal. Chim. Acta 197*, 87–95 (1987).
[981] Butcher, D.J.; R.L. Irwin, J. Takahashi, G.-z. Su, G. Wei, R.G. Michel, *Appl. Spectrosc. 44/9*, 1521–1533 (1990).
[982] Butler, L.R.P.; A. Strasheim, *Spectrochim. Acta 21*, 1207–1216 (1965).
[983] Butler, L.R.P.; *At. Absorpt. Newsletter 5*, 99–101 (1966).
[984] Butler, L.R.P.; J.A. Brink, in: J.A. Dean, T.C. Rains (Editors), *Flame Emission and Atomic Absorption Spectrometry, Vol. 2*, Marcel Dekker, New York, 1971, 21–56.
[985] Butler, L.R.P.; *Spectrochim. Acta, Part B 38/5-6*, 913–919 (1983).
[986] Butler, L.R.P.; K. Laqua, A. Strasheim, *Pure Appl. Chem. 57/10*, 1453–1490 (1985).
[987] Butler, L.R.P.; K. Laqua, A. Strasheim, *Spectrochim. Acta, Part B 41/5*, 507–544 (1986).
[988] Butler, L.R.P.; K. Laqua, *Pure Appl. Chem. 67/10*, 1725–1744 (1995).
[989] Bye, R.; P.E. Paus, R. Solberg, Y. Thomassen, *At. Absorpt. Newsletter 17/6*, 131–134 (1978).
[990] Bye, R.; P.E. Paus, *Anal. Chim. Acta 107*, 169–175 (1979).
[991] Bye, R.; B. Holen, *Anal. Chim. Acta 144*, 235–238 (1982).
[992] Bye, R.; L. Engvik, W. Lund, *Anal. Chem. 55/14*, 2457–2458 (1983).
[993] Bye, R.; *Talanta 30/12*, 993–996 (1983).
[994] Bye, R.; *Analyst (London) 110/1*, 85–86 (1985).
[995] Bye, R.; *Anal. Chem. 57/7*, 1481–1482 (1985).
[996] Bye, R.; *Analyst (London) 111/1*, 111–113 (1986).
[997] Bye, R.; *Talanta 33/8*, 705–706 (1986).
[998] Bye, R.; *Anal. Chim. Acta 192*, 115–117 (1987).
[999] Bye, R.; *J. Chem. Educ. 64*, 188 (1987).
[1000] Bye, R.; W. Lund, *Fresenius Z. Anal. Chem. 332/3*, 242–244 (1988).
[1001] Bye, R.; *Talanta 37/10*, 1029–1030 (1990).
[1002] Bye, R.; *Anal. Proc. (London) 31/3*, 89–90 (1994).
[1003] Byrd, E.D.; D.J. Butcher, *Spectrosc. Lett. 26/9*, 1613–1624 (1993).
[1004] Byrne, A.R.; J. Versieck, in: R. Zeisler, V.P. Guinn (Editors), *Nuclear Analytical Methods in the Life Sciences*, Humana Press, Clifton, NJ, 1990, 529–540.
[1005] Byrne, J.P.; C.L. Chakrabarti, S.B. Chang, C.K. Tan, A.H. Delgado, *Fresenius Z. Anal. Chem. 324/5*, 448–455 (1986).
[1006] Byrne, J.P.; C.L. Chakrabarti, D.C. Grégoire, M.M. Lamoureux, Tam Ly, *J. Anal. At. Spectrom. 7/2*, 371–381 (1992).
[1007] Byrne, J.P.; C.L. Chakrabarti, G.F.R. Gilchrist, M.M. Lamoureux, P. Bertels, *Anal. Chem. 65/9*, 1267–1272 (1993).

[1008]   Byrne, J.P.; M.M. Lamoureux, C.L. Chakrabarti, Tam Ly, D.C. Grégoire, *J. Anal. At. Spectrom. 8/4*, 599–609 (1993).

[1008b] Byrne, J.P.; D.C. Grégoire, D.M. Goltz, C.L. Chakrabarti, *Spectrochim. Acta, Part B 49/5*, 433–443 (1994).

[1009]   Bysouth, S.R.; J.F. Tyson, *J. Anal. At. Spectrom. 1*, 85–87 (1986).

[1010]   Bysouth, S.R.; J.F. Tyson, *Anal. Chim. Acta 179*, 481–486 (1986).

[1011]   Cabanis, M.T.; G. Cassanas, J.C. Cabanis, S. Brun, *J. Assoc. Off. Anal. Chem. 71/5*, 1033–1037 (1988).

[1012]   Cabon, J.-Y.; A. Le Bihan, *Anal. Chim. Acta 198*, 87–101 (1987).

[1013]   Cabon, J.-Y.; A. Le Bihan, *Anal. Chim. Acta 198*, 103–111 (1987).

[1014]   Cabon, J.-Y.; A. Le Bihan, *Environ. Technol. 12*, 769–776 (1991).

[1015]   Cabon, J.-Y.; A. Le Bihan, *J. Anal. At. Spectrom. 9/3*, 477–481 (1994).

[1016]   Cabon, J.-Y.; A. Le Bihan, *Spectrochim. Acta, Part B 50/13*, 1703–1716 (1995).

[1017]   Cabon, J.-Y.; A. Le Bihan, *Spectrochim. Acta, Part B 51/6*, 619–631 (1996).

[1018]   Cabon, J.-Y.; A. Le Bihan, *Spectrochim. Acta, Part B 51/9-10*, 1245–1251 (1996).

[1019]   Cabrera, C.; M.L. Lorenzo, C. Gallego, M.C. López, *Anal. Chim. Acta 246/2*, 375–378 (1991).

[1020]   Cabrera, C.; M.L. Lorenzo, C. Gallego, M.C. López, E. Lillo, *J. Agri. Food Chem. 40/9*, 1631–1633 (1992).

[1021]   Cabrera, C.; Y. Madrid, C. Camara, *J. Anal. At. Spectrom. 9/12*, 1423–1426 (1994).

[1022]   Cabrera, C.; M.L. Lorenzo, M.C. Lopez, *J. AOAC International 78*, 1061 (1995).

[1023]   Cacho-Palomar, J.; I. Beltrán, C. Nerín de la Puerta, *J. Anal. At. Spectrom. 4/7*, 661–663 (1989).

[1024]   Cai, S.; *Yankuang Ceshi 10*, 304 (1991).

[1025]   Cai, Y.; S. Rapsomanikis, M.O. Andreae, *Mikrochim. Acta (Vienna) 109/1-4*, 67–71 (1992).

[1026]   Cai, Y.; S. Rapsomanikis, M.O. Andreae, *J. Anal. At. Spectrom. 8/1*, 119–125 (1993).

[1027]   Cai, Y.; S. Rapsomanikis, M.O. Andreae, *Anal. Chim. Acta 274/2*, 243–252 (1993).

[1028]   Cai, Y.; S. Rapsomanikis, M.O. Andreae, *Talanta 41/4*, 589–594 (1994).

[1029]   Caldas, E.D.; M.F. Gine-Rosias, J.G. Dorea, *Anal. Chim. Acta 254/1-2*, 113–118 (1991).

[1030]   Cali, J.P.; G.N. Bowers, D.S. Young, *Clin. Chem. (Winston-Salem) 19/10*, 1208–1213 (1973).

[1031]   Cali, J.P.; K.N. Marsh, *Pure Appl. Chem. 55*, 908–930 (1983).

[1032]   Calle-Guntiñas, M.B. de la; Y. Madrid, C. Cámara-Rica, *Analyst (London) 116/10*, 1029–1035 (1991).

[1033]   Calle-Guntiñas, M.B. de la; Y. Madrid, C. Cámara-Rica, *Anal. Chim. Acta 252/1-2*, 161–166 (1991).

[1034]   Calle-Guntiñas, M.B. de la; R. Torralba, Y. Madrid, M.A. Palacios, M. Bonilla, C. Cámara, *Spectrochim. Acta, Part B 47/10*, 1165–1172 (1992).

[1035]   Calle-Guntiñas, M.B. de la; Y. Madrid, C. Cámara-Rica, *Fresenius J. Anal. Chem. 344/1-2*, 27–29 (1992).

[1036]   Calle-Guntiñas, M.B. de la; Y. Madrid, C. Cámara-Rica, *Mikrochim. Acta (Vienna) 109/1-4*, 149–155 (1992).

[1037]   Calle-Guntiñas, M.B. de la; Y. Madrid, C. Cámara-Rica, *J. Anal. At. Spectrom. 8/5*, 745–748 (1993).

[1038]   Calle-Guntiñas, M.B. de la; Y. Madrid, C. Cámara-Rica, in: Ph. Quevauviller, E.A. Maier, B. Griepink (Editors), *Quality Assurance for Environmental Analysis*, Elsevier, Amsterdam, 1995, 263–283.

[1039]   Calmano, W.; K.H. Lieser, *Fresenius Z. Anal. Chem. 307*, 356–361 (1981).

[1040]   Calmano, W.; U. Förstner, *Sci. Total Environ. 28*, 77–90 (1983).

[1041]   Calmano, W.; W. Ahlf, T. Schilling, *Fresenius Z. Anal. Chem. 323/7*, 865–868 (1986).

[1042]   Camail, M.; B. Loiseau, A. Margaillan, J.L. Vernet, *Analusis 11*, 358–359 (1983).

[1043]   Cámara, C.; M.G. Cobo, M.A. Palacios, R. Muñoz, O.F.X. Donard, in: Ph. Quevauviller, E.A. Maier, B. Griepink (Editors), *Quality Assurance for Environmental Analysis*, Elsevier, Amsterdam, 1995, 235–262.

[1044]   Cammann, K.; J.T. Andersson, *Fresenius Z. Anal. Chem. 310/1*, 45–50 (1982).

[1045]   Campanella, L.; D. D'Orazio, B.M. Petronio, E. Pietrantonio, *Anal. Chim. Acta 309/1-3*, 387–393 (1995).

[1046]   Campbell, A.D.; N.H. Tioh, *Anal. Chim. Acta 100*, 451–455 (1978).

[1047]   Campbell, A.D.; *Pure Appl. Chem. 64/2*, 227–244 (1992).

[1048]   Campbell, M.B.; G.A. Kanert, *Analyst (London) 117/2*, 121–124 (1992).

[1049]   Campbell, W.C.; in: J.E. Cantle (Editor), *Atomic Absorption Spectrometry*, Elsevier, Amsterdam - Oxford - New York, 1982, 285–306.

[1050] Campos, R.C.; S.S. Moraes, *At. Spectrosc. 14/3*, 71–75 (1993).
[1051] Canals, A.; V. Hernandis, R.F. Browner, *J. Anal. At. Spectrom. 5/1*, 61–66 (1990).
[1052] Candelone, J.-P.; S.-m. Hong, C.F. Boutron, *Anal. Chim. Acta 299/1*, 9–16 (1994).
[1053] Candler, C.; *Atomic Spectra*, D. Van Nostrand Company, New York, 1964.
[1054] Capacho-Delgado, L.; D.C. Manning, *At. Absorpt. Newsletter 4*, 317–318 (1965).
[1055] Capacho-Delgado, L.; S. Sprague, *At. Absorpt. Newsletter 4*, 363–464 (1965).
[1056] Capacho-Delgado, L.; D.C. Manning, *Spectrochim. Acta 22*, 1505–1513 (1966).
[1057] Capacho-Delgado, L.; D.C. Manning, *At. Absorpt. Newsletter 5/1*, 1–3 (1966).
[1058] Capacho-Delgado, L.; D.C. Manning, *Analyst (London) 92*, 553–557 (1967).
[1059] Capar, S.G.; *J. Assoc. Off. Anal. Chem. 60/6*, 1400–1407 (1977).
[1060] Carbonell, V.; M. de la Guardia, A. Salvador, J.L. Burguera, M. Burguera, *Anal. Chim. Acta 238/2*, 417–421 (1990).
[1061] Carbonell, V.; A. Sanz, A. Salvador, M. de la Guardia, *J. Anal. At. Spectrom. 6/3*, 233–238 (1991).
[1062] Carbonell, V.; A. Morales-Rubio, A. Salvador, M. de la Guardia, José L. Burguera, *J. Anal. At. Spectrom. 7/7*, 1085–1089 (1992).
[1063] Cardarelli, E.; M. Cifani, M. Mecozzi, G. Sechi, *Talanta 33*, 279 (1986).
[1064] Cardellicchio, N.; S. Geraci, C. Marra, P. Paterno, *Appl. Organomet. Chem. 6*, 241–246 (1992).
[1065] Carelli, G.; A. Bergamaschi, M.C. Altavista, *At. Spectrosc. 5/2*, 46–50 (1984).
[1066] Caricchia, A.M.; S. Chiavarini, C. Cremisini, R. Morabito, R. Scerbo, *Anal. Chim. Acta 286/3*, 329–334 (1994).
[1067] Carillo-Camarero, F.; L. Polo-Díez, C. Cámara-Rica, *Analyst (London) 109/9*, 1171–1173 (1984).
[1068] Carillo-Camarero, F.; M. Bonilla, C. Cámara-Rica, *Microchem. J. 33/1*, 2–8 (1986).
[1069] Carlosa, A.; D. Prada, J.M. Andrade, P. López, S. Muniategui, *Fresenius J. Anal. Chem. 355/3-4*, 289–291 (1996).
[1070] Carman, R.J.; A. Maitland, *J. Phys. D.: Appl. Phys. 20*, 1021–1030 (1987).
[1071] Carneiro, M.C.; R.C. Campos, A.J. Curtius, *Talanta 40/12*, 1815–1822 (1993).
[1072] Carnrick, G.R.; W. Slavin, D.C. Manning, *Anal. Chem. 53/12*, 1866–1872 (1981).
[1073] Carnrick, G.R.; D.C. Manning, W. Slavin, *Analyst (London) 108*, 1297–1312 (1983).
[1074] Carnrick, G.R.; W.B. Barnett, *At. Spectrosc. 5/5*, 213–214 (1984).
[1075] Carnrick, G.R.; B.K. Lumas, W.B. Barnett, *J. Anal. At. Spectrom. 1/6*, 443–447 (1986).
[1076] Carnrick, G.R.; W.B. Barnett, W. Slavin, *Spectrochim. Acta, Part B 41/9*, 991–997 (1986).
[1077] Carnrick, G.R.; G. Daley, A. Fotinopoulos, *At. Spectrosc. 10/6*, 170–174 (1989).
[1078] Caroli, S.; (Editor), *Improved Hollow Cathode Lamps for Atomic Spectroscopy*, John Wiley & Sons, New York - Chichester, 1985.
[1079] Caroli, S.; A. Alimonti, F. Petrucci, in: S. Caroli (Editor), *Improved Hollow Cathode Lamps for Atomic Spectroscopy*, John Wiley & Sons, New York - Chichester, 1985, 13–34.
[1080] Caroli, S.; O. Senofonte, N. Violante, L. di Simone, *Appl. Spectrosc. 41/4*, 579–583 (1987).
[1081] Caroli, S.; *Microchem. J. 45/3*, 257–271 (1992).
[1082] Carper, J.L.; *At. Absorpt. Newsletter 9/2*, 48–49 (1970).
[1083] Carr, K.E.; *Carbon 8*, 155 (1976).
[1084] Carrión, N.; A. Llanos, Z.A. de Benzo, R. Fraile, *At. Spectrosc. 7/2*, 52–55 (1986).
[1085] Carrión, N.; Z.A. de Benzo, E.J. Eljuri, F. Ippditi, D. Flores, *J. Anal. At. Spectrom. 2/8*, 813–817 (1987).
[1086] Carrión, N.; Z.A. de Benzo, B. Moreno, A. Fernández, E.J. Eljuri, D. Flores, *J. Anal. At. Spectrom. 3/3*, 479–483 (1988).
[1087] Carroll, J.; J. Marshall, D. Littlejohn, J.M. Ottaway, *Fresenius Z. Anal. Chem. 322/2*, 145–150 (1985).
[1088] Carroll, J.; J. Marshall, D. Durie, D. Littlejohn, J.M. Ottaway, *Spectrochim. Acta, Part B 41/7*, 751–759 (1986).
[1089] Carron, J.; H. Agemian, *Anal. Chim. Acta 92*, 61–70 (1977).
[1090] Carrondo, M.J.T.; R. Perry, J.N. Lester, *Anal. Chim. Acta 106*, 309–317 (1979).
[1091] Carter, R.J.; V.S. Rajendram, *Sci. Total Environ. 125/1*, 33–38 (1992).
[1092] Cartwright, J.S.; C. Sebens, D.C. Manning, *At. Absorpt. Newsletter 5/4*, 91–96 (1966).
[1093] Cartwright, J.S.; D.C. Manning, *At. Absorpt. Newsletter 5/5*, 114–115 (1966).
[1094] Cartwright, J.S.; C. Sebens, W. Slavin, *At. Absorpt. Newsletter 5/2*, 22–27 (1966).
[1095] Cary, E.E.; O.E. Olson, *J. Assoc. Off. Anal. Chem. 58*, 433–435 (1975).
[1096] Cary, E.E.; *J. Assoc. Off. Anal. Chem. 68/3*, 495–498 (1985).

[1097]   Casetta, B.; F. Aldrighetti, in: C. Minoia, S. Caroli (Editors), *Appl. Zeeman Graphite Furnace At. Abs. Spectrom. Chem. Lab. Toxicology*, Pergamon, Oxford, 1992, 257–278.

[1098]   Castellano Giron, H.; *At. Absorpt. Newsletter 12/1*, 28–29 (1973).

[1099]   Castillo-Suarez, J.R.; J. Lanaja, M.A. Belarra, J. Aznárez-Alduan, *At. Spectrosc. 2/5*, 159–160 (1981).

[1100]   Castillo-Suarez, J.R.; M.A. Belarra, J. Aznárez-Alduan, *At. Spectrosc. 3/2*, 58–60 (1982).

[1101]   Castillo-Suarez, J.R.; J.M. Mir, J. Val, M.P. Colón, C. Martínez, *Analyst (London) 110/8*, 1219–1221 (1985).

[1102]   Castillo-Suarez, J.R.; J.M. Mir, C. Bendicho, C. Martínez, *At. Spectrosc. 6/6*, 152–155 (1985).

[1103]   Castillo-Suarez, J.R.; J.M. Mir, C. Martínez, J. Val, M.P. Colón, *Mikrochim. Acta (Vienna) I*, 253–263 (1985).

[1104]   Castillo-Suarez, J.R.; J.M. Mir, M.C. Martínez, T. Gómez, *At. Spectrosc. 9/1*, 9–12 (1988).

[1105]   Castillo-Suarez, J.R.; A. Fernandez, M.A. Bona, *At. Spectrosc. 9/6*, 200–203 (1988).

[1106]   Castillo-Suarez, J.R.; J.M. Mir, C. Bendicho, *Spectrochim. Acta, Part B 43/3*, 263–271 (1988).

[1107]   Castillo-Suarez, J.R.; J.M. Mir, C. Bendicho, F. Laborda, *Fresenius Z. Anal. Chem. 332/1*, 37–40 (1988).

[1108]   Castillo-Suarez, J.R.; J.M. Mir, C. Bendicho, *Fresenius Z. Anal. Chem. 332/7*, 783–786 (1988).

[1109]   Castillo-Suarez, J.R.; J.M. Mir, C. Bendicho, *J. Anal. At. Spectrom. 4/1*, 105–107 (1989).

[1110]   Castillo-Suarez, J.R.; A. Fernandez, *Microchem. J. 39/2*, 224–228 (1989).

[1111]   Castillo-Suarez, J.R.; J.M. Mir, *Microchem. J. 39*, 119–125 (1989).

[1112]   Castillo-Suarez, J.R.; J.M. Mir, M.E. Garcia-Ruiz, C. Bendicho, *Fresenius J. Anal. Chem. 338/6*, 721–725 (1990).

[1113]   Castillo-Suarez, J.R.; E. García, J. Delfa, J.M. Mir, C. Bendicho, *Microchem. J. 42/1*, 103–109 (1990).

[1114]   Castillo-Suarez, J.R.; J. Delfa, J.M. Mir, C. Bendicho, M. de la Guardia, A.R. Mauri, C. Mongay, E. Martínez, *J. Anal. At. Spectrom. 5/4*, 325–330 (1990).

[1115]   Castro, M.A. de; R. Bugagao, B. Ebarvia, N. Roque, I. Rubeška, *Geostand. Newsl. 12/1*, 47–51 (1988).

[1116]   Cathum, S.J.; C.L. Chakrabarti, J.C. Hutton, *Spectrochim. Acta, Part B 46/1*, 35–44 (1991).

[1117]   Cedergren, A.; I. Lindberg, E. Lundberg, D.C. Baxter, W. Frech, *Anal. Chim. Acta 180*, 373–388 (1986).

[1118]   Cedergren, A.; W. Frech, *Pure Appl. Chem. 59/2*, 221–228 (1987).

[1119]   Celková, J. Kubová, V. Stresko, *Fresenius J. Anal. Chem. 355/2*, 150–153 (1996).

[1120]   Cellier, K.M.; H.C.T. Stace, *Appl. Spectrosc. 20/1*, 26 (1966).

[1121]   Cernik, A.A.; M.H. Sayers, *Br. J. Ind. Med. 28*, 392–398 (1971).

[1122]   Cernik, A.A.; *4th ICAS, Toronto*, 1973.

[1123]   Cernik, A.A.; *At. Absorpt. Newsletter 12/2*, 42–44 (1973).

[1124]   Cernik, A.A.; *At. Absorpt. Newsletter 12*, 163–164 (1973).

[1125]   Černohorský, T.; S. Kotrlý, *J. Anal. At. Spectrom. 10/2*, 155–160 (1995).

[1126]   Černohorský, T.; *Spectrochim. Acta, Part B 50/13*, 1613–1620 (1995).

[1127]   Cervera, M.L.; A. Navarro, R. Montoro, R. Catalá, N. Ybáñez, *J. Assoc. Off. Anal. Chem. 72/2*, 282–285 (1989).

[1128]   Cervera, M.L.; A. Navarro, R. Montoro, R. Catalá, *At. Spectrosc. 10/5*, 154–159 (1989).

[1129]   Cervera, M.L.; R. Montoro, *Fresenius J. Anal. Chem. 348/5-6*, 331–340 (1994).

[1130]   Cervera, M.L.; J.C. Lopez, R. Montoro, *Microchem. J. 49/1*, 20–26 (1994).

[1131]   Cespón-Romero, R.M.; M.C. Yebra-Biurrun, M.P. Bermejo-Barrera, *Anal. Chim. Acta 327/3*, 37–45 (1996).

[1132]   Cha, K.-W.; S.G. Ha, *Anal. Sci. Technol. 5*, 235 (1992).

[1133]   Chakrabarti, C.L.; G.R. Lyles, F.B. Dowling, *Anal. Chim. Acta 29*, 489–499 (1963).

[1134]   Chakrabarti, C.L.; J.W. Robinson, P.W. West, *Anal. Chim. Acta 34*, 269 (1966).

[1135]   Chakrabarti, C.L.; M. Katyal, D.E. Willis, *Spectrochim. Acta, Part B 25*, 629–645 (1970).

[1136]   Chakrabarti, C.L.; *Can. J. Spectrosc. 23/4*, 134–145 (1978).

[1137]   Chakrabarti, C.L.; C.C. Wan, W.C. Li, *Spectrochim. Acta, Part B 35/2*, 93–105 (1980).

[1138]   Chakrabarti, C.L.; C.C. Wan, W.C. Li, *Spectrochim. Acta, Part B 35/9*, 547–560 (1980).

[1139]   Chakrabarti, C.L.; H.A. Hamed, C.C. Wan, W.C. Li, P.C. Bertels, D.C. Gregoire, S. Lee, *Anal. Chem. 52/1*, 167–176 (1980).

[1140]   Chakrabarti, C.L.; C.C. Wan, H.A. Hamed, P.C. Bertels, *Can. Res. 13/3*, 31–34 (1980).

[1141]   Chakrabarti, C.L.; C.C. Wan, H.A. Hamed, P.C. Bertels, *Anal. Chem. 53/3*, 444–450 (1981).

[1142] Chakrabarti, C.L.; S.B. Chang, T.J. Huston, P.C. Bertels, J.T. Rogers, R. Dick, *Anal. Chim. Acta 176*, 17–32 (1985).
[1143] Chakrabarti, C.L.; S.-l. Wu, R. Karwowska, J.T. Rogers, R. Dick, *Spectrochim. Acta, Part B 40/10*, 1663–1676 (1985).
[1144] Chakrabarti, C.L.; in: L.R.P. Butler (Editor), *Analytical Chemistry in the Exploration, Mining and Processing of Materials*, Blackwell Scientific Publications, Oxford - London, 1986, 57–66.
[1145] Chakrabarti, C.L.; A.H. Delgado, S.B. Chang, H. Falk, T.J. Huton, G. Runde, V. Sychra, J. Doležal, *Spectrochim. Acta, Part B 41/10*, 1075–1087 (1986).
[1146] Chakrabarti, C.L.; S.-l. Wu, F. Marcantonio, K.L. Headrick, *Fresenius Z. Anal. Chem. 323/7*, 730–736 (1986).
[1147] Chakrabarti, C.L.; R. Karwowska, B.R. Hollebone, P.M. Johnson, *Spectrochim. Acta, Part B 42/11*, 1217–1225 (1987).
[1148] Chakrabarti, C.L.; X.-r. He, S.-l. Wu, W.H. Schroeder, *Spectrochim. Acta, Part B 42*, 1227–1233 (1987).
[1149] Chakrabarti, C.L.; K.L. Headrick, P.C. Bertels, M.H. Back, *J. Anal. At. Spectrom. 3/5*, 713–723 (1988).
[1150] Chakrabarti, C.L.; A.H. Delgado, S.B. Chang, H. Falk, V. Sychra, J. Doležal, *Spectrochim. Acta, Part B 44/2*, 209–217 (1989).
[1151] Chakrabarti, C.L.; K.L. Headrick, J.C. Hutton, B. Marchand, M.H. Back, *Spectrochim. Acta, Part B 44/4*, 385–394 (1989).
[1152] Chakrabarti, C.L.; K.L. Headrick, J.C. Hutton, B.-c. Zhang, P.C. Bertels, M.H. Back, *Anal. Chem. 62/6*, 574–586 (1990).
[1153] Chakrabarti, C.L.; S.J. Cathum, *Talanta 37/12*, 1111–1117 (1990).
[1154] Chakrabarti, C.L.; K.L. Headrick, J.C. Hutton, P.C. Bertels, *Spectrochim. Acta, Part B 46/2*, 183–192 (1991).
[1155] Chakrabarti, C.L.; S.J. Cathum, *Talanta 38/2*, 157–166 (1991).
[1156] Chakrabarti, C.L.; A.Kh. Gilmutdinov, J.C. Hutton, *Anal. Chem. 65/6*, 716–723 (1993).
[1157] Chakrabarti, C.L.; Y.-J. Lu, J. Cheng, M.H. Back, W.H. Schroeder, *Anal. Chim. Acta 276*, 47–64 (1993).
[1158] Chakrabarti, C.L.; J.-g. Chen, M. Grenier, *Spectrochim. Acta, Part B 51/11*, 1335–1343 (1996).
[1159] Chakraborti, D.; W.R.A. de Jonghe, F.C. Adams, *Anal. Chim. Acta 120*, 121–127 (1980).
[1160] Chakraborti, D.; D.C.J. Hillman, K.J. Irgolic, R.A. Zingaro, *J. Chromatogr. 249/1*, 81–92 (1982).
[1161] Chakraborti, D.; W.R.A. de Jonghe, W.E. van Mol, R.J.A. van Cleuvenbergen, F.C. Adams, *Anal. Chem. 56/14*, 2692–2697 (1984).
[1162] Chakraborti, D.; K.J. Irgolic, in: T.D. Lekkas (Editor), *Heavy Metals in the Environment*, Proc. Int. Conf., 5th, Athens, Sept. 1985, CEP Consultants,., Edinburgh, 1985, 484–486.
[1163] Chakraborti, D.; R. Van Cleuvenbergen, F. Adams, in: J.N. Lester, R. Perry, R.M. Steritt (Editors), *Chemicals in the environment*, Proc. Int. Conf., Lisbon, Selper, London, 1986, 298–304.
[1164] Chakraborti, D.; F.C. Adams, K.J. Irgolic, *Fresenius Z. Anal. Chem. 323/4*, 340–342 (1986).
[1165] Chakraborti, D.; R.J.A. Van Cleuvenbergen, F.C. Adams, *Int. J. Environ. Anal. Chem. 30*, 233–242 (1987).
[1166] Chakraborti, D.; F.C. Adams, W. van Mol, K.J. Irgolic, *Anal. Chim. Acta 196*, 23–31 (1987).
[1167] Chakraborti, D.; W. Dirkx, R. J.A. Van Cleuvenbergen, F.C. Adams, *Sci. Total Environ. 84*, 249–257 (1989).
[1168] Chakraborti, D.; M. Burguera, J.L. Burguera, *Fresenius J. Anal. Chem. 347/6-7*, 233–237 (1993).
[1169] Chakraborty, D.; A.K. Das, *At. Spectrosc. 9/4*, 115–118 (1988).
[1170] Chakraborty, D.; A.K. Das, *At. Spectrosc. 9/6*, 189–190 (1988).
[1171] Chakraborty, D.; A.K. Das, *Analyst (London) 114/1*, 67–69 (1989).
[1172] Chakraborty, D.; A.K. Das, *Fresenius J. Anal. Chem. 349/10*, 774–775 (1994).
[1173] Chakraborty, R.; A.K. Das, M.L. Cervera, M. de la Guardia, *J. Anal. At. Spectrom. 10/5*, 353–358 (1995).
[1174] Chakraborty, R.; A.K. Das, M.L. Cervera, M. de la Guardia, *Fresenius J. Anal. Chem. 355/1*, 43–47 (1996).
[1175] Chakraborty, R.; A.K. Das, M.L. Cervera, M. de la Guardia, *Fresenius J. Anal. Chem. 355/2*, 99–111 (1996).
[1176] Chamsaz, M.; I.M. Khasawneh, J.D. Winefordner, *Talanta 35/7*, 519–523 (1988).

[1177] Chan, C.C.Y.; P.N. Vijan, *Anal. Chim. Acta 101*, 33–43 (1978).
[1178] Chan, C.C.Y.; M.W.A. Baig, *Anal. Lett. 17/A2*, 143–155 (1984).
[1179] Chan, C.C.Y.; *Anal. Chem. 57/7*, 1482–1485 (1985).
[1180] Chan, C.C.Y.; M.W.A. Baig, P.A. Lichti, *Anal. Lett. 23/12*, 2259–2272 (1990).
[1181] Chan, C.C.Y.; R.S. Sadana, *Anal. Chim. Acta 270/1*, 231–238 (1992).
[1182] Chan, L.; *Forensic Sci. Int. 18*, 57 (1981).
[1183] Chan, W.-F.; P.-K. Hon, *Analyst (London) 115/5*, 567–569 (1990).
[1184] Chan, W.H.; F. Tomassini, B. Loescher, *Atmos. Environ. 17/9*, 1779–1785 (1983).
[1185] Chana, B.S.; N.J. Smith, *Anal. Chim. Acta 197*, 177–186 (1987).
[1186] Chandler, H.A.; M. Scott, in: H.A. McKenzie, L.E. Smythe (Editors), *Quantitative trace analysis of biological materials*, Elsevier, Amsterdam - New York - Oxford, 1988, 561–571.
[1187] Chang, C.M.; H.J. Huang, *J. Chin. Chem. Soc. (Taipei) 40/5*, 425 (1993).
[1188] Chang, S.B.; C.L. Chakrabarti, T.J. Huston, P.C. Bertels, J.T. Rogers, R. Dick, *Anal. Chim. Acta 176*, 1–16 (1985).
[1189] Chao, T.T.; E.A. Jenne, L.M. Heppting, *U.S. Geol. Surv. Prof. Paper 66-D*, D13–D15, 1968.
[1190] Chao, T.T.; E.A. Jenne, L.M. Heppting, *U.S. Geol. Surv. Prof. Paper 66-D*, D16–D19, 1968.
[1191] Chao, T.T.; R.F. Sanzolone, *J. Geochem. Explor. 44*, 65–106 (1992).
[1192] Chapman, B.N.; *Glow Discharge Processes*, John Wiley & Sons, New York, 1980.
[1193] Chapman, J.F.; L.S. Dale, *Anal. Chim. Acta 87*, 91–95 (1976).
[1194] Chapman, J.F.; L.S. Dale, *Anal. Chim. Acta 89*, 363–368 (1977).
[1195] Chapman, J.F.; L.S. Dale, *Anal. Chim. Acta 111*, 137–144 (1979).
[1196] Chapman, J.F.; L.S. Dale, H.J. Fraser, *Anal. Chim. Acta 116/2*, 427–431 (1980).
[1197] Chapman, J.F.; L.S. Dale, *Anal. Chim. Acta 134*, 379–382 (1982).
[1198] Chapman, J.F.; L.S. Dale, S.A. Topham, *Anal. Chim. Acta 187*, 307–311 (1986).
[1199] Chappuis, P.; L. Duhaux, F. Paolaggi, M.C. De Vernejoul, *Clin. Chem. (Winston-Salem) 34/11*, 2253–2255 (1988).
[1200] Chappuis, P.; J. Poupon, J.F. Deschamps, P.J. Guillausseau, F. Rousselet, *Biol. Trace Elem. Res. 32*, 85–91 (1992).
[1201] Chattaraj, S.; A.K. Das, *Analyst (London) 116/7*, 739–741 (1991).
[1202] Chattaraj, S.; A.K. Das, *Anal. Lett. 25*, 2355–2366 (1992).
[1203] Chattaraj, S.; A.K. Das, *Indian J. Chem. (Section A) 32*, 1009–1011 (1993).
[1204] Chatterjee, A.; S. Basu, *Fresenius J. Anal. Chem. 340/1*, 61–62 (1991).
[1205] Chattopadhyay, P.; B.N. Sahoo, *Analyst (London) 117/9*, 1481–1484 (1992).
[1206] Chattopadhyay, P.; B.N. Sahoo, *Talanta 40/5*, 701–706 (1993).
[1207] Chattopadhyay, P.; M. Mistry, *Microchem. J. 50*, 78–87 (1994).
[1208] Chattopadhyay, P.; *Talanta 42/12*, 1965–1971 (1995).
[1209] Chattopadhyay, P.; *Chem. Anal. (Warsaw) 40/5*, 775 (1995).
[1210] Chau, Y.K.; P.T.S. Wong, P.D. Goulden, *Anal. Chem. 47*, 2279–2281 (1975).
[1211] Chau, Y.K.; P.T.S. Wong, H. Saitoh, *J. Chromatogr. Sci. 14*, 162–164 (1976).
[1212] Chau, Y.K.; P.T.S. Wong, P.D. Goulden, *Anal. Chim. Acta 85*, 421–424 (1976).
[1213] Chau, Y.K.; P.T.S. Wong, G.A. Bengert, O. Kramar, *Anal. Chem. 51*, 186–188 (1979).
[1214] Chau, Y.K.; P.T.S. Wong, O. Kramar, G.A. Bengert, R.A. Cruz, J.O. Kinrade, *Bull. Environ. Contam. Toxicol. 24*, 265–269 (1980).
[1215] Chau, Y.K.; P.T.S. Wong, G.A. Bengert, *Anal. Chem. 54/2*, 246–249 (1982).
[1216] Chau, Y.K.; P.T.S. Wong, O. Kramar, *Anal. Chim. Acta 146*, 211–217 (1983).
[1217] Chau, Y.K.; P.T.S. Wong, G.A. Bengert, J.L. Dunn, *Anal. Chem. 56/2*, 271–274 (1984).
[1218] Chau, Y.K.; in: P.J. Craig (Editor), *Organometallic compounds in the environment*, Longman Group, Harlow, 1986, 254–278.
[1219] Chau, Y.K.; S.-z. Zhang, R.J. Maguire, *Analyst (London) 117/7*, 1161–1164 (1992).
[1220] Chaudhry, M.M.; A.M. Ure, B.G. Cooksey, D. Littlejohn, D.J. Halls, *Anal. Proc. (London) 28/2*, 44–46 (1991).
[1221] Chaudhry, M.M.; D. Littlejohn, John E. Whitley, *J. Anal. At. Spectrom. 7/1*, 29–34 (1992).
[1222] Chaudhry, M.M.; D. Littlejohn, *Analyst (London) 117/4*, 713–715 (1992).
[1223] Chaudhry, M.M.; D. Mouillere, B.J. Ottaway, D. Littlejohn, J.E. Whitley, *J. Anal. At. Spectrom. 7/4*, 701–706 (1992).
[1224] Cheam, V.; H. Agemian, *Analyst (London) 105*, 737–743 (1980).
[1225] Cheam, V.; H. Agemian, *Anal. Chim. Acta 113*, 237–245 (1980).
[1226] Chen, D.-h.; Y.-e., Zeng, *Anal. Chim. Acta 235/2*, 337–342 (1990).
[1227] Chen, D.-h.; M.D. Luque de Castro, M. Valcárcel, *Analyst (London) 116/11*, 1095–1111 (1991).
[1228] Chen, F.-h.; H.-m. Wei, K.-l. Yang, S.-g. Dai, *Fenxi Huaxue 21/7*, 761–764 (1993).

[1229] Chen, G.-r.; K.W. Jackson, *Spectrochim. Acta, Part B 51/12*, 1505–1515 (1996).
[1230] Chen, H.-w.; I.D. Brindle, X.-c. Le, *Anal. Chem. 64/6*, 667–672 (1992).
[1231] Chen, H.-w.; I.D. Brindle, S.-g. Zheng, *Analyst (London) 117/10*, 1603–1608 (1992).
[1232] Chen, H.-w.; F.-l. Tang, C. Gu, I.D. Brindle, *Talanta 40/8*, 1147–1155 (1993).
[1233] Chen, H.-w.; S.-k. Xu, Z.-l. Fang, *Anal. Chim. Acta 298/2*, 167–173 (1994).
[1234] Chen, H.-w.; S.-k. Xu, Z.-l. Fang, *J. Anal. At. Spectrom. 10/8*, 533–537 (1995).
[1235] Chen, J.-J.; Y.-C. Lou, C.-W. Whang, *J. Chin. Chem. Soc. (Taipei) 39*, 461–464 (1992).
[1236] Chen, J.-s.; H. Berndt, G. Tölg, *Fresenius J. Anal. Chem. 344/12,* 526–534 (1992).
[1237] Chen, J.W.; F.G. Zeng, H.J. Shen, *Yankuang Ceshi 12/2*, 85–88,92 (1993).
[1238] Chen, R.; H. Xin, *Zhongguo Pige, (China) 19/5*, 42 (1990).
[1239] Chen, S.-c.; M.-y. Shiue, M.-h. Yang, *Fresenius J. Anal. Chem. 357/7*, 1192–1197 (1997).
[1240] Chen, S.-l.; S.R. Dzeng, M.-h. Yang, K.-h. Chiu, G.-m. Shieh, C.M. Wai, *Environ. Sci. Technol. 28*, 877–881 (1994).
[1241] Chen, S.-q.; *Lihua Jianyan, Huaxue Fence 29/5*, 268–269 (1993).
[1242] Chen, T.; C. Xu, *Fenxi Shiyanshi 10/6*, 38 (1991).
[1243] Chen, T.-j.; D. Littlejohn, *Analyst (London) 118/5*, 541–543 (1993).
[1244] Chen, X.; G. Jang, J. Chen, X. Chen, Z. Wen, K. Ge, *Biol. Trace Elem. Res. 2*, 91–107 (1980).
[1245] Chen, Y.; *Guangpuxue Yu Guangpu Fenxi 10/6*, 63 (1990).
[1246] Chen, Y.; *Yankuang Ceshi 10*, 291 (1991).
[1247] Chen, Y.-l.; W.-q. Qi, J.-s. Cao, M.-s. Chang, *J. Anal. At. Spectrom. 8/2*, 379–381 (1993).
[1248] Chen, Z.-s.; M. Hiraide, H. Kawaguchi, *Bunseki Kagaku 42/11*, 759–762 (1993).
[1249] Chen, Z.-s.; Y.-l. Chen, *Guangpuxue Yu Guangpu Fenxi 10/3*, 69 (1990).
[1250] Chen, Z.-s.; M. Hiraide, H. Kawaguchi, *Mikrochim. Acta (Vienna) 124*, 27–34 (1996).
[1251] Cheng, J.T.; W.F. Agnew, *At. Absorpt. Newsletter 13/5*, 123–124 (1974).
[1252] Cheng, J.-g.; C.L. Chakrabarti, M.H. Back, W.H. Schroeder, *Anal. Chim. Acta 288/3*, 141–156 (1994).
[1253] Cheng, Y.; Z. Chen, *Fenxi Ceshi Tongbao 10/4*, 46 (1991).
[1254] Chester, J.E.; R.M. Dagnall, M.R.G. Taylor, *Anal. Chim. Acta 51*, 95–107 (1970).
[1255] Chiasson, A.G.; *J. Chem. Ecology (New York) 16/8*, 2503–2510 (1990).
[1256] Chiavarini, S.; C. Cremisini, T. Ferri, R. Morabito, A. Perini, *Sci. Total Environ. 101*, 217–227 (1991).
[1257] Chiba, M.; A. Shinohara, Y. Inaba, *Microchem. J. 49/2-3*, 275–281 (1994).
[1258] Chikuma, M.; H. Aoki, H. Tanaka, *Anal. Sci. 7/Suppl*, 1131–1134 (1991).
[1259] Chiricosta, S.; G. Cum, R. Gallo, A. Spadaro, P. Vitarelli, *At. Spectrosc. 3/6*, 185–187 (1982).
[1260] Chiricosta, S.; G. Saija, R. Calapaj, E. Bruno, *At. Spectrosc. 10/6*, 183–187 (1989).
[1261] Chirkova, G.D.; N.V. Bondareva, E.S. Zolotovitskaya, V.G. Potapova, N.S. Granova, L.I. Pleskach, *Zh. Anal. Khim. 48/4*, 648–653 (1993).
[1262] Chiswell, B.; D. Johnson, in: H.G. Seiler, A. Sigel, H. Sigel (Editors), *Handbook on Metals in Clinical and Analytical Chemistry*, Marcel Dekker, New York, 1994, 467–478.
[1263] Chittleborough, G.; *Sci. Total Environ. 14/1*, 53–75 (1980).
[1264] Choi, K.-K.; L. Lam, S.-F. Luk, *Talanta 41/1*, 1–8 (1994).
[1265] Choi, S.-K.; H.-J. Kim, *Fresenius J. Anal. Chem. 355/3-4*, 308–311 (1996).
[1266] Chong, R.W.; D.F. Boltz, *Anal. Lett. 8/10*, 721–727 (1975).
[1267] Chork, C.Y.; *J. Geochem. Explor. 7*, 31–47 (1977).
[1268] Chormann, F.H.; M.J. Spencer, W.B. Lyons, P.A. Mayewski, *Chem. Geol. 53/1-2*, 25–30 (1985).
[1269] Chou, C.L.; J.F. Uthe, R.D. Guy, *J. AOAC International 76/4*, 794–798 (1993).
[1270] Chow, C.; *Talanta 33/1*, 91–94 (1986).
[1271] Christensen, J.M.; E. Holst, J.P. Bonde, L. Knudsen, *Sci. Total Environ. 132,* 11–25 (1993).
[1272] Christensen, J.M.; J. Kristiansen, in: H.G. Seiler, A. Sigel, H. Sigel (Editors), *Handbook on Metals in Clinical and Analytical Chemistry*, Marcel Dekker, New York, 1994, 425–440.
[1273] Christensen, S.; *At. Absorpt. Newsletter 9/6*, 126–128 (1970).
[1274] Christensen, S.; J.T.B. Anglov, J.M. Christensen, E. Olsen, O.M. Poulsen, *Fresenius J. Anal. Chem. 345*, 343–350 (1993).
[1275] Christensen, T.H.; *Int. J. Environ. Anal. Chem. 12/3-4*, 211–221 (1982).
[1276] Christensen, T.H.; L:I. Pedersen, J.C. Tjell, *Int. J. Environ. Anal. Chem. 12/1*, 41–50 (1982).
[1277] Christian, G.D.; F.J. Feldman, *Anal. Chim. Acta 40*, 173 (1968).
[1278] Christmann, D.R.; J.D. Ingle, *Anal. Chim. Acta 86*, 53–62 (1976).
[1279] Chu, R.C.; G.P. Barron, P.A.W. Baumgarner, *Anal. Chem. 44/8*, 1476–1479 (1972).
[1280] Chuang, F.S.; B.M. Patel, R.D. Reeves, M.T. Glenn, J.D. Winefordner, *Can. J. Spectrosc. 18/1*, 6–9 (1973).

[1281] Chuang, H.; S.-d. Huang, *Spectrochim. Acta, Part B 49/3*, 283–288 (1994).
[1282] Chung, C.-h.; E. Iwamoto, M. Yamamoto, Y. Yamamoto, *Spectrochim. Acta, Part B 39/2-3*, 459–466 (1984).
[1283] Chung, C.-h.; *Anal. Chem. 56/14*, 2714–2720 (1984).
[1284] Chung, H.K.; J.D. Ingle, *Anal. Chem. 62/23*, 2541–2547 (1990).
[1285] Ciappellano, S.; F. Brigheni, M. Porrini, G. Testolin, in: C. Minoia, S. Caroli (Editors), *Appl. Zeeman Graphite Furnace At. Abs. Spectrom. Chem. Lab. Toxicology*, Pergamon, Oxford, 1992, 279–304.
[1286] Ciappellano, S.; M. Porrini, G. Testolin, in: C. Minoia, S. Caroli (Editors), *Appl. Zeeman Graphite Furnace At. Abs. Spectrom. Chem. Lab. Toxicology*, Pergamon, Oxford, 1992, 593–612.
[1287] Ciceri, G.; R. Ferraroli, L. Guzzi, *Heavy Metals in the Hydrological Cycle*, Selper, London, 1988, 545–551.
[1288] Cichelli, A.; *Riv. Merceol. 24*, 7 (1985).
[1289] Cidu, R.; L. Fanfani, P. Shand, W.M. Edmunds, L. Van't Dack, R. Gijbels, *Anal. Chim. Acta 296/3*, 295–304 (1994).
[1290] Cioni, R.; R. Mazzucotelli, G. Ottonello, *Analyst (London) 101*, 956–960 (1976).
[1291] Ciopec, M.; *Chemom. Intell. Lab. Sys. 21/1*, 21–34 (1993).
[1292] Ciurea, I.C.; Y.F. Lipka, B.E. Humbert, *Mitt. Geb. Lebensmittelunters. Hyg. 77/4*, 509–519 (1986).
[1293] Clampitt, N.C.; G.M. Hieftje, *Anal. Chem. 44/7*, 1211–1219 (1972).
[1294] Clampitt, N.C.; G.M. Hieftje, *Anal. Chem. 46/3*, 382–386 (1974).
[1295] Clark, J.R.; J.G. Viets, *Anal. Chem. 53/1*, 61–65 (1981).
[1296] Clark, J.R.; J.G. Viets, *Anal. Chem. 53/1*, 65–70 (1981).
[1297] Clark, J.R.; *J. Anal. At. Spectrom. 1*, 301–308 (1986).
[1298] Clarke, N.; L.-G. Danielson, A. Sparén, *Int. J. Environ. Anal. Chem. 48*, 77–100 (1992).
[1299] Clarkson, T.W.; J.B. Hursh, P.R. Sager, T.L.M. Syversen, in: T.W. Clarkson, L. Friberg, G.F. Nordberg, P.R. Sager (Editors), *Biological monitoring of toxic metals*, Plenum Press, New York, 1988, 199–246.
[1300] Clavel, J.P.; P. Lavirotte, A. Galli, *Pathol. Biol. 31*, 851–854 (1983).
[1301] Clay, D.E.; L.J. Geiger, *Oil Gas J. 82*, 74 (1984).
[1302] Cleuvenbergen, R.J.A. Van; W.E. Van Mol, F.C. Adams, *J. Anal. At. Spectrom. 3/1*, 169–176 (1988).
[1303] Cleuvenbergen, R.J.A. Van; D. Chakraborti, F.C. Adams, *Anal. Chim. Acta 228/1*, 77–84 (1990).
[1304] Cleuvenbergen, R.J.A. Van; W. Dirkx, Ph. Quevauviller, F.C. Adams, *Int. J. Environ. Anal. Chem. 47*, 21–32 (1992).
[1305] Clyburn, S.A.; T. Kántor, C. Veillon, *Anal. Chem. 46/14*, 2214–2215 (1974).
[1306] Cobb, W.D.; T.S. Harrison, *Joint Symp. acc. Meth.Anal.Maj.Const.*, London, 1970.
[1307] Cobb, W.D.; W.W. Foster, T.S. Harrison, *Anal. Chim. Acta 78*, 293–298 (1975).
[1308] Cobb, W.D.; W.W. Foster, T.S. Harrison, *Analyst (London) 101*, 255–259 (1976).
[1309] Cobo-Fernández, M.G.; M.A. Palacios, C. Cámara, *Anal. Chim. Acta 283/1*, 386–392 (1993).
[1310] Cobo-Fernández, M.G.; M.A. Palacios, C. Cámara, F. Reis, Ph. Quevauviller, *Anal. Chim. Acta 286/3*, 371–379 (1994).
[1311] Cobo-Fernández, M.G.; M.A. Palacios, D. Chakraborti, Ph. Quevauviller, C. Cámara, *Fresenius J. Anal. Chem. 351/4-5*, 438–442 (1995).
[1312] Cocho, J.A.; J.R. Cervilla, M.L. Rey-Goldar, J.R. Fernandez-Lorenzo, J.M.Fraga, *Biol. Trace Elem. Res. 32*, 105–107 (1992).
[1313] Coe, M.; R. Cruz, J.C. van Loon, *Anal. Chim. Acta 120*, 171–176 (1980).
[1314] Coello, J.; L.-G. Danielsson, S. Hernandez-Cassou, *Anal. Chim. Acta 201*, 325–329 (1987).
[1315] Coker, D.T.; *Anal. Chem. 47*, 386–389 (1975).
[1316] Coleman, C.J.; N.E. Bibler, R.A. Dewberry, *Proc. Symp. Waste Management 2*, 651 (1990).
[1317] Collado-Gomez, G.; C. Bosch Ojeda, A. Garcia de Torres, J.M. Cano Pavon, *Analusis 23/5*, 224–227 (1995).
[1318] Collet, P.; L. Matter, B. Thomas, *Lebensmittelchemie 44/1*, 3–6 (1990).
[1319] Collett, D.L.; D.E. Fleming, G.A. Taylor, *Analyst (London) 105*, 897–901 (1980).
[1320] Comber, S.D.W.; A.G. Howard, *Anal. Proc. (London) 26/1*, 20–22 (1989).
[1321] Cook, N.J.; S.A. Wood, Y.-s. Zhang, *J. Geochem. Explor. 46*, 187–228 (1993).
[1322] Cooksey, M.M.; W.B. Barnett, *At. Absorpt. Newsletter 18/1*, 1–4 (1979).
[1323] Cooksey, M.M.; W.B. Barnett, *At. Absorpt. Newsletter 18/5*, 101–105 (1979).
[1324] Copeland, D.D.; M. Facer, R. Newton, P.J. Walker, *Analyst (London) 121/2*, 173–176 (1996).

[1325]   Corbin, D.R.; W.M. Barnard, *At. Absorpt. Newsletter 15/5*, 116–120 (1976).
[1326]   Corl, W.E.; *Spectroscopy (Eugene, OR) 6/8*, 40–43 (1991).
[1327]   Cornelis, R.; J. Versieck, L. Mees, *Biol. Trace Elem. Res. 3*, 257–263 (1981).
[1328]   Cornelis, R.; *Acta Pharmacol. Toxicol. 59/7*, 585–588 (1986).
[1329]   Cornelis, R.; in: J.A.C. Broekaert, S. Güçer, F. Adams (Editors), *Metal Speciation in the Environment*, NATO ASI Ser. G, Springer, Berlin - Heidelberg - New York, 1990, 169–194.
[1330]   Cornelis, R.; *Mikrochim. Acta (Vienna) III/1-3*, 37–44 (1991).
[1331]   Cornelis, R.; F. Borguet, S. Dyg, B. Griepink, *Mikrochim. Acta (Vienna) 109/1-4*, 145–148 (1992).
[1332]   Cornelis, R.; *Food Chem. 43*, 307–313 (1992).
[1333]   Cornelis, R.; J. de Kimpe, *J. Anal. At. Spectrom. 9/9*, 945–950 (1994).
[1334]   Cornelis, R.; B. Heinzow, R.F.M. Herber, J. Molin Christensen, O.M. Paulsen, E. Sabbioni, D.M. Templeton, Y. Thomassen, M. Vahter, O. Vesterberg, *Pure Appl. Chem. 67/8-9*, 1575–1608 (1995).
[1335]   Corominas, L.F.; R.A. Navarro, P. Rojas, *J. Assoc. Off. Anal. Chem. 66/5*, 1234–1241 (1983).
[1336]   Cortez, L.; Ph. Quevauviller, F. Martin, O.F.X. Donard, *Environ. Pollut. 82*, 57–62 (1993).
[1337]   Costantini, S.; R. Giordano, M. Rubbiani, *Microchem. J. 35*, 70–82 (1987).
[1338]   Coudert, M.A.; J.M. Vergnaud, *Anal. Chem. 42*, 1303 (1970).
[1339]   Cowan, R.D.; *Theory of Atomic Structure and Spectra*, University of California Press, Berkeley, 1981.
[1340]   Cowley, T.G.; V.A. Fassel, R.N. Kniseley, *Spectrochim. Acta, Part B 23/12*, 771–792 (1968).
[1341]   Cox, J.A.; S. Al-Shakshir, *Anal. Lett. 21/9*, 1757–1769 (1988).
[1342]   Cox, L.E.; *Anal. Chem. 47/8*, 1493–1494 (1975).
[1343]   Coyne, R.V.; J.A. Collins, *Anal. Chem. 44/6*, 1093–1096 (1972).
[1344]   Cragin, J.H.; M.M. Herron, *At. Absorpt. Newsletter 12*, 37–38 (1973).
[1345]   Cragin, J.H.; *Anal. Chim. Acta 110/2*, 313–319 (1979).
[1346]   Craig, P.J.; in: M. Bernhard, F.E. Brinckman, P.J. Sadler (Editors), *The Importance of Chemical 'Speciation' in Environmental Processes*, Springer, New York, 1986, 443–464.
[1347]   Craig, P.J.; in: P.J. Craig (Editor), *Organometallic Compounds in the Environment*, Longman Group, Harlow, Essex, 1986, 65–110.
[1348]   Craig, P.J.; R.J. Dewick, J.T. van Elteren, *Fresenius J. Anal. Chem. 351/4-5*, 467–470 (1995).
[1349]   Craney, C.L.; K. Swartout, F.W. Smith, C.D. West, *Anal. Chem. 58/3*, 656–658 (1986).
[1350]   Cresser, M.S.; R. Hargitt, *Anal. Chim. Acta 81*, 196–198 (1976).
[1351]   Cresser, M.S.; D.A. MacLeod, *Analyst (London) 101*, 86–90 (1976).
[1352]   Cresser, M.S.; R. Hargitt, *Talanta 23*, 153–154 (1976).
[1353]   Cresser, M.S.; R.F. Browner, *Anal. Chim. Acta 113/1*, 33–38 (1980).
[1354]   Cresser, M.S.; R.F. Browner, *Appl. Spectrosc. 34/3*, 364–368 (1980).
[1355]   Cresser, M.S.; *Prog. Anal. Spectrosc. 4/3*, 219–245 (1981).
[1356]   Cresser, M.S.; *Prog. Anal. Spectrosc. 5*, 35–62 (1982).
[1357]   Cresser, M.S.; C.E. O'Grady, I.L. Marr, *Prog. Anal. Spectrosc. 8*, 19–46 (1985).
[1358]   Cresser, M.S.; *Anal. Proc. (London) 22/3*, 65–66 (1985).
[1359]   Cresser, M.S.; *Anal. Proc. (London) 27/5*, 110–111 (1990).
[1360]   Cresser, M.S.; in: S.J. Haswell (Editor), *Atomic Absorption Spectrometry - Theory, Design and Applications*, Elsevier, Amsterdam - New York, 1991, 515–526.
[1361]   Cresser, M.S.; *J. Anal. At. Spectrom. 8/2*, 269–272 (1993).
[1362]   Crews, H.M.; J.A. Burrell, D.J. McWeeny, *Z. Lebensm. Unters. Forsch. 180/3*, 221–226 (1985).
[1363]   Criaud, A.; C. Foulliac, *Anal. Chim. Acta 167/1*, 257–267 (1985).
[1364]   Crick, J.; A.R. Flegal, *Clin. Chem. (Winston-Salem) 38/4*, 600–601 (1992).
[1365]   Crompton, T.R.; *Analysis of Seawater*, Butterworth, London, 1989.
[1366]   Crompton, T.R.; *Comprehensive Water Analysis. Volume I: Natural Waters*, Elsevier, London - New York, 1992.
[1367]   Crow, R.F.; W.G. Hime, J.D. Connolly, *J. Research and Development Lab. ; Bulletin 214*, 1967.
[1368]   Crow, R.F.; J.D. Connolly, *J. Test. Eval. 1/5*, 382–393 (1973).
[1369]   Crow, R.F.; J.D. Connolly, *Prog. Anal. Spectrosc. 1*, 347–352 (1978).
[1370]   Cruz, R.B.; C. Lorouso, S. George, Y. Thomassen, J.D. Kinrade, L.R.P. Butler, *Spectrochim. Acta, Part B 35/11-2*, 775–783 (1980).
[1371]   CSI XXVII-Pre-Symposium, *J. Anal. At. Spectrom. 7*, 471 (1992).
[1372]   Cui, J.-S.; J. Zhao, S.-G. Jiang, *Guangpuxue Yu Guangpu Fenxi 8*, 63 (1988).
[1373]   Cullen, W.R.; M. Dodd, *Appl. Organomet. Chem. 2*, 1–7 (1988).

[1374] Cullen, W.R.; K.J. Reimer, *Chem. Rev. 89*, 713–764 (1989).
[1375] Cunningham, A.F.; *At. Absorpt. Newsletter 8/3*, 70–71 (1969).
[1376] Cunningham, D.; *Anal. Chem. 60/7*, 471A-473A (1988).
[1377] Currey, N.A.; W.I. Benko, B.T. Yaru, R. Kabi, *Sci. Total Environ. 125/1*, 305–320 (1992).
[1378] Currie, L.A.; J.R. De Voe, in: J.R. De Voe (Editor), *Validation of the Measurement Process*, ACS Symp. Ser. Vol. 63, American Chemical Society, Washington, DC, 1977, 114–139.
[1379] Currie, L.A.; in: I.M. Kolthoff, P.J. Elving (Editors), *Treatise on Analytical Chemistry, Part 1. Theory and Practice*, J. Wiley & Sons, New York, 1978, 95–242.
[1380] Curtis, P.R.; J. Grusovin, *Commun. Soil Sci. Plant Anal. 16/12*, 1279–1291 (1985).
[1381] Curtius, A.J.; G. Schlemmer, B. Welz, *J. Anal. At. Spectrom. 1/6*, 421–427 (1986).
[1382] Curtius, A.J.; G. Schlemmer, B. Welz, *J. Anal. At. Spectrom. 2/1*, 115–124 (1987).
[1383] Curtius, A.J.; G. Schlemmer, B. Welz, *J. Anal. At. Spectrom. 2/3*, 311–315 (1987).
[1384] Cutter, G.A.; *Anal. Chim. Acta 98*, 59–66 (1978).
[1385] Cutter, G.A.; *Science 217*, 829–831 (1982).
[1386] Cutter, G.A.; *Anal. Chim. Acta 149*, 391–394 (1983).
[1387] Cutter, G.A.; *Anal. Chem. 57/14*, 2951–2955 (1985).
[1388] Cutter, L.S.; G.A. Cutter, M.L.C. San Diego-McGlove, *Anal. Chem. 63*, 1138–1142 (1991).
[1389] Czobik, E.J.; J.P. Matoušek, *Talanta 24*, 573–577 (1977).
[1390] Czobik, E.J.; J.P. Matoušek, *Spectrochim. Acta, Part B 35/11-2*, 741–751 (1980).
[1391] Dabeka, R.W.; *J. Assoc. Off. Anal. Chem. 65/4*, 1005–1009 (1982).
[1392] Dabeka, R.W.; *Analyst (London) 109*, 1259–1263 (1984).
[1393] Dabeka, R.W.; G.M.A. Lacroix, *Can. J. Spectrosc. 30/6*, 154–157 (1985).
[1394] Dabeka, R.W.; A.D. McKenzie, *Can. J. Spectrosc. 31/2*, 44–52 (1986).
[1395] Dabeka, R.W.; G.M.A. Lacroix, M.A. Gladys, *J. Assoc. Off. Anal. Chem. 70/5*, 866–870 (1987).
[1396] Dabeka, R.W.; A.D. McKenzie, *Can. J. Appl. Spectroscopy 36/4*, 123–126 (1991).
[1397] Dabeka, R.W.; A.D. McKenzie, G.M.A. Lacroix, C. Cleroux, S. Bowe, R.A. Graham, H.B.S. Conacher, P. Verdier, *J. AOAC International 76/1*, 14–25 (1993).
[1398] Dadfarnia, S.; K.C. Thompson, G. Hoult, *J. Anal. At. Spectrom. 9/1*, 7–9 (1994).
[1399] Dael, P. van; R. Van Cauwenbergh, H. Robberecht, H. Deelstra, *At. Spectrosc. 16/6*, 251–255 (1995).
[1400] Dagnall, R.M.; K.C. Thompson, T.S. West, *Talanta 14*, 551–555 (1967).
[1401] Dagnall, R.M.; K.C. Thompson, T.S. West, *Talanta 14*, 557–563 (1967).
[1402] Dagnall, R.M.; K.C. Thompson, T.S. West, *Talanta 14*, 1467–1475 (1967).
[1403] Dagnall, R.M.; K.C. Thompson, T.S. West, *Analyst (London) 92*, 506–512 (1967).
[1404] Dagnall, R.M.; K.C. Thompson, T.S. West, *At. Absorpt. Newsletter 6*, 117–120 (1967).
[1405] Dagnall, R.M.; T.S. West, *Appl. Opt. 7/7*, 1287–1294 (1968).
[1406] Dahl, D.B.; P.F. Lott, *Microchem. J. 35/3*, 347–359 (1987).
[1407] Dahl, D.B.; P.F. Lott, *J. Chem. Educ. 68*, 1025 (1991).
[1408] Dai, L.J.; Y.G. Xiao, J. Huang, *Lihua Jianyan, Huaxue Fence 30/1*, 15–17 (1994).
[1409] Dale, I.M.; in: S.J. Haswell (Editor), *Atomic Absorption Spectrometry - Theory, Design and Applications*, Elsevier, Amsterdam - New York, 1991, 439–462.
[1410] Dalen, H.P.J. van; L. de Galan, *Analyst (London) 106*, 695–701 (1981).
[1411] Dalrymple, B.A.; C.T. Kenner, *J. Pharm. Sci. 58/5*, 604–606 (1969).
[1412] Damiani, M.; M.G. Del Monte Tamba, F. Bianchi, *Analyst (London) 100*, 643–647 (1975).
[1413] Damkröger, G., M. Grote, E. Janßen, *Fresenius Z. Anal. Chem. 357/7*, 817–821 (1997).
[1414] Daniels, R.S.; D.C. Wigfield, *J. Anal. Toxicol. 13*, 214–217 (1989).
[1415] Daniels, R.S.; D.C. Wigfield, *Anal. Chim. Acta 248/2*, 575–577 (1991).
[1416] Daniels, R.S.; D.C. Wigfield, *J. Anal. Toxicol. 17/4*, 196–198 (1993).
[1417] Danielsson, L.-G.; A. Sparén, *Anal. Chim. Acta 306/2-3*, 173–181 (1995).
[1418] Danielsson, L.-G.; *Analyst (London) 120/10*, 2513–2519 (1995).
[1419] Dannecker, W.; in: B. Welz (Editor), *Atomspektrometrische Spurenanalytik*, Verlag Chemie, Weinheim, 1982, 187–211.
[1420] Danz, H.J.; E. Jackwerth, *325*, 157–162 (1986).
[1421] Danz, H.J.; E. Jackwerth, *Fresenius Z. Anal. Chem. 326*, 57–61 (1987).
[1422] Das, A.K.; S. Bandyopadhyay, *Indian J. Environ. Prot. 10/1*, 7–8 (1990).
[1423] Das, A.K.; R. Chakraborty, M.L. Cervera, M. de la Guardia, *Talanta 42/8*, 1007–1030 (1995).
[1424] Das, A.K.; R. Chakraborty, *Fresenius Z. Anal. Chem. 357/1*, 1–17 (1997).
[1425] Davey, D.E.; *Anal. Lett. 19/15*, 1573–1590 (1986).
[1426] David, D.J.; *Analyst (London) 83*, 655–661 (1958).
[1427] David, D.J.; *Analyst (London) 84*, 536–545 (1959).

[1428] David, D.J.; *Analyst (London) 85*, 459 (1960).
[1429] David, D.J.; *Nature (London) 187*, 1109 (1960).
[1430] David, D.J.; *Analyst (London) 86*, 730 (1961).
[1431] David, D.J.; *At. Absorpt. Newsletter 1*, 45–50 (1962).
[1432] David, D.J.; in: K. Paech, M.V. Tracey, H.F. Linskens (Editors), *Modern Methods of Plant Analysis*, Springer, Berlin - Göttingen - Heidelberg, 1962.
[1433] David, D.J.; *Analyst (London) 93*, 79 (1968).
[1434] David, D.J.; *Prog. Anal. Spectrosc. 1/3*, 225–254 (1978).
[1435] David, D.J.; *Commun. Soil. Sci. Plant Anal. 11/2*, 189–199 (1980).
[1436] Davidowski, L.; J.J. Werbicki, *At. Spectrosc. 4/3*, 104–107 (1983).
[1437] Davidowski, L.; Z.A. Grosser, *At. Spectrosc. 17/6*, 232–235 (1996).
[1438] Davidson, C.M.; R.P. Thomas, S.E. McVey, R. Perala, D. Littlejohn, A.M. Ure, *Anal. Chim. Acta 291/3*, 277–286 (1994).
[1439] Davidson, R.A.; *At. Spectrosc. 13/6*, 199–205 (1992).
[1440] Davies, I.M.; *Anal. Chim. Acta 102*, 189–194 (1978).
[1441] Davydov, Yu.P.; *Radiokhimiya 9*, 84 (1967).
[1442] Dawczynski, H.; E. Glatzel, E. Preu, A. Yersin, *Zentralbl. Pharm., Pharmakother. Lab. 125/3*, 149–161 (1986).
[1443] Dawson, J.B.; E. Grassam, D.J. Ellis, M.J. Keir, *Analyst (London) 101*, 315–316 (1976).
[1444] Dawson, J.B.; R.J. Duffield, P.R. King, M. Hajizadeh-Soffar, G.W. Fisher, *Spectrochim. Acta, Part B 43/9-11*, 1133–1140 (1988).
[1445] DeVine, J.C.; N.H. Suhr, *At. Absorpt. Newsletter 16/2*, 39–41 (1977).
[1446] Deacon, M.; M.R. Smyth, L.G.M.T. Tuinstra, *J. Chromatogr. A 657/1*, 69–76 (1993).
[1447] Dean, J.A.; W.J. Carnes, *Anal. Chem. 34/2*, 192–194 (1962).
[1448] Dean, Shaheb-ud ; P.J. Tscherwonyi, W.J. Riley, *Clin. Chem. (Winston-Salem) 38/1*, 119–122 (1992).
[1449] Deano, P.; J.W. Robinson, *Spectrosc. Lett. 19/1*, 11–19 (1986).
[1450] Debrah, E.; C.E. Adeeyinwo, S.R. Bysouth, J.F. Tyson, *Analyst (London) 115/12*, 1543–1547 (1990).
[1451] Debrah, E.; J.F. Tyson, M.W. Hinds, *Talanta 39/11*, 1525–1530 (1992).
[1452] Decker, C.F.; A. Aras, L.R. Decker, *Anal. Biochem. 8*, 344 (1964).
[1453] DØdina, J.; I. Rubeška, *Spectrochim. Acta, Part B 35/3*, 119–128 (1980).
[1454] Dědina, J.; *Anal. Chem. 54/12*, 2097–2102 (1982).
[1455] Dědina, J.; *Fresenius Z. Anal. Chem. 323/7*, 771–782 (1986).
[1456] Dědina, J.; W. Frech, I. Lindberg, E. Lundberg, A. Cedergren, *J. Anal. At. Spectrom. 2/3*, 287–291 (1987).
[1457] Dědina, J.; W. Frech, A. Cedergren, I. Lindberg, E. Lundberg, *J. Anal. At. Spectrom. 2*, 435–439 (1987).
[1458] Dědina, J.; *Prog. Anal. Spectrosc. 11/3-4*, 251–360 (1988).
[1459] Dědina, J.; W. Frech, E. Lundberg, A. Cedergren, *J. Anal. At. Spectrom. 4/2*, 143–148 (1989).
[1460] Dědina, J.; *Spectrochim. Acta, Part B 46/3*, 379–391 (1991).
[1461] Dědina, J.; B. Welz, *J. Anal. At. Spectrom. 7/2*, 307–314 (1992).
[1462] Dědina, J.; *Spectrochim. Acta, Part B 47/5*, 689–700 (1992).
[1463] Dědina, J.; B. Welz, *Spectrochim. Acta, Part B 48/3*, 301–314 (1993).
[1464] Dědina, J.; D.L. Tsalev, *Hydride Generation Atomic Absorption Spectrometry*, Wiley and Sons, Chichester, 1995.
[1465] Dědina, J.; T. Matoušek, W. Frech, *Spectrochim. Acta, Part B 51/9-10*, 1107–1119 (1996).
[1466] Dehairs, F.; M. De Bondt, W. Baeyens, P. van den Winkel, M. Hoenig, *Anal. Chim. Acta 196*, 33–40 (1987).
[1467] Deijck, W. van; A.M. Roelofsen, H.J. Pieters, R.F.M. Herber, *Spectrochim. Acta, Part B 38/5-6*, 791–797 (1983).
[1468] Deily, J.R.; *At. Absorpt. Newsletter 5/6*, 119–121 (1966).
[1469] Delft, W. van ; H.J. Horstman, H. Lammers, G. Vos, in: B. Welz (Editor), *5. Colloquium Atomspektrometrische Spurenanalytik*, Bodenseewerk Perkin-Elmer, Überlingen, 1989, 603–611.
[1470] Delijska, A.; T. Blazheva, I. Petkova, L. Dimov, *Fresenius Z. Anal. Chem. 332/4*, 362–365 (1988).
[1471] Dellar, D.; *Analyst (London) 108*, 759–763 (1983).
[1472] Dellien, I.; L. Persson, *Anal. Chim. Acta 160*, 217–225 (1984).
[1473] DeLon Hull, R.; J.C. Haartz, *Anal. Chim. Acta 121*, 187–196 (1980).
[1474] Delves, H.T.; *Analyst (London) 95*, 431–438 (1970).

802   Bibliography

[1475] Delves, H.T.; G. Shepherd, P. Vinter, *Analyst (London) 96*, 260–273 (1971).
[1476] Delves, H.T.; J. Woodward, *At. Spectrosc. 2/2*, 65–67 (1981).
[1477] Delves, H.T.; *Ann. Clin. Biochem. 24/6*, 529–551 (1987).
[1478] Delves, H.T.; in: H.A. McKenzie, L.E. Smythe (Editors), *Quantitative trace analysis of biological materials*, Elsevier, Amsterdam - New York - Oxford, 1988, 439–449.
[1479] Delves, H.T.; I.L. Shuttler, in: S.J. Haswell (Editor), *Atomic Absorption Spectrometry - Theory, Design and Applications*, Elsevier, Amsterdam - New York, 1991, 381–438.
[1480] Delves, H.T., M. Stoeppler, in: R.F.M. Herber, M. Stoeppler, *Trace Element Analysis in Biological Specimens*, Elsevier, Amsterdam - London - New York - Tokyo, 1994, 359–370.
[1481] Demange, M.; B. Hervé-Bazin, B. Carton, *Analusis 19*, 73–78 (1991).
[1482] Demarin, V.T.; V.N. Savinykh, *Fiz.-Khim. Metody Anal. 83* (1983).
[1483] Demidov, V.V.; Y.G. Kosovets, B.S. Kogar, *Nucleic Acids Res. 19*, 3155 (1991).
[1484] Demirata, B.; I. Torm, H. Filik, H. Afsar, *Fresenius J. Anal. chem. 356/6*, 375–377 (1996).
[1485] Dempster, W.S. ; H. de V. Heese, F.H. Pocock, H. Breuer, in: P. Brätter, P. Schramel (Editors), *Trace Element Analytical Chemistry in Medicine and Biology*, Walter de Gruyter, Berlin - New York, 1988, 90–95.
[1486] Denninger, R.; S. Ganz, W. Grimm, M. Gross, G. Hermann, A. Scharmann, *Prog. Anal. Spectrosc. 10*, 335–343 (1987).
[1487] Denton, M.B.; J.M. Freelin, T.R. Smith, in: J. Sneddon (Editor), *Sample Introduction in Atomic Spectroscopy (Anal. Spectrosc. Library Vol. 4)*, Elsevier, Amsterdam - Oxford - New York - Tokio, 1990, 73–106.
[1488] Department of the Environment (UK), Standing Committee of Analysts, *Li, Mg, Ca, Sr and Ba in waters and sewage effluents by AAS 1987*, Methods Exam. Waters Assoc. Mater., 1987.
[1489] Department of the Environment, (UK), *Methods for the determination of metals in soils, sediments and sewage sludge and plants by a hydrochloric - nitric acid digestion*, Methods Exam. Waters Assoc. Mater., 1987.
[1490] Derschau, H.A.M.; H. Prugger, *Fresenius Z. Anal. Chem. 247*, 8 (1969).
[1491] Desaulniers, J.A.H.; R.E. Sturgeon, S.S. Berman, *At. Spectrosc. 6/5*, 125–127 (1985).
[1492] Desauziers, V.; F. Leguille, R. Lavigne, M. Astruc, R. Pinel, *Appl. Organomet. Chem. 3*, 469–474 (1989).
[1493] Deschuytere, A.; K. Vermeylen, H. Deelstra, *Z. Lebensm. Unters. Forsch. 184*, 385–387 (1987).
[1494] Deshmukh, B.T.; A. Kumar, M.Z. Hasan, *Indian J. Chem. (Section A) 26*, 629–631 (1987).
[1495] Deutsches Institut für Normung e.V.; *Deutsche Norm DIN 51790-4*.
[1496] Deutsches Institut für Normung e.V.; *Deutsche Norm DIN 51416-1*.
[1497] Deutsches Institut für Normung e.V.; *Deutsche Norm DIN 51416-2*.
[1498] Deutsches Institut für Normung e.V.; *Deutsche Norm DIN 51391-1*.
[1499] Deutsches Institut für Normung e.V.; *Deutsche Norm DIN 51401-2*.
[1500] Deutsches Institut für Normung e.V.; Normenausschuß Wasserwesen (NAW), *Vom Wasser 77*, D13–D24 (1991).
[1501] Deutsches Institut für Normung e.V.; Normenausschuß Materialprüfung (NMP), *Deutsche Norm DIN 51401-1*, Beuth-Verlag, Berlin, 1992.
[1502] Deutsches Institut für Normung e.V.; Normenausschuß Wasserwesen, *Deutsche Norm DIN 38405-18*, Beuth-Verlag, Berlin, 1985.
[1503] Deutsches Institut für Normung e.V.; *Deutsche Norm DIN 38406-22*, Beuth-Verlag, Berlin, 1988.
[1504] Deutsches Institut für Normung e.V.; Normenausschuß Wasserwesen (NAW), *Deutsche Norm DIN 38406-E24*, Beuth-Verlag, Berlin, 1993.
[1505] Deutsches Institut für Normung e.V.; *Deutsche Norm DIN 38414-7*, Beuth-Verlag, Berlin, 1983.
[1506] Deutsches Institut für Normung e.V.; *Deutsche Norm DIN 38414-S1*, Beuth-Verlag, Berlin, 1986.
[1507] Deutsches Institut für Normung e.V.; *Deutsche Norm DIN 38414-11*, Beuth-Verlag, Berlin, 1987.
[1508] Deutsches Institut für Normung e.V, Normenausssschuß Wasserwesen (NAW) im DIN, *Deutsche Norm DIN 38405-23*, Beuth-Verlag, Berlin, 1992.
[1509] Deutsches Institut für Normung e.V.; *Deutsche Norm DIN 38405-32*, Beuth-Verlag, Berlin, 1996.
[1510] Deutsches Institut für Normung e.V.; *Deutsche Norm DIN 51815-1*.
[1511] Deutsches Institut für Normung e.V.; *Deutsche Norm DIN 51827*.

[1512] Deutsches Institut für Normung e.V.; Normenausschuß Wasserwesen (NAW), *Deutsche Norm, Entwurf DIN 38406-14*, Beuth-Verlag, Berlin, 1990.
[1513] Deutsches Institut für Normung e.V.; Normenausschuß Wasserwesen (NAW), *Deutsche Norm, Entwurf DIN 38406-13*, Beuth-Verlag, Berlin, 1990.
[1514] Deutsches Institut für Normung e.V.; Normenausschuß Wasserwesen (NAW), *Deutsche Norm DIN 38406-18*, Beuth-Verlag, Berlin, 1990.
[1515] Deutsches Institut für Normung e.V.; Normenausschuß Wasserwesen (NAW), *Deutsche Norm DIN 38406-7*, Beuth-Verlag, Berlin, 1991.
[1516] Deutsches Institut für Normung e.V.; Normenausschuß Wasserwesen (NAW), *Deutsche Norm DIN 38406-11*, Beuth-Verlag, Berlin, 1991.
[1517] Deutsches Institut für Normung e.V.; Normenausschuß Wasserwesen (NAW), *Deutsche Norm, Entwurf DIN 38406-6, Blei*, Beuth-Verlag, Berlin, 1994.
[1518] D'Haese, P.C.; L.V. Lamberts, L. Lian, F.L. Van de Vyver, M.E. De Broe, *Clin. Chem. (Winston-Salem) 37/9*, 1583–1588 (1991).
[1519] D'Haese, P.C.; G.F. Van Landeghem, L.V. Lamberts, M.E. De Broe, *Mikrochim. Acta (Vienna), 120/1-4*, 83 (1995).
[1520] D'Haese, P.C.; G.F. Van Landeghem, L.V. Lamberts, V.A. Bekaert, I. Schrooten, M.E. De Broe, *Clin. Chem. (Winston-Salem) 43/1*, 121–128 (1997).
[1521] Di, P.; D.E. Davey, *Talanta 42/8*, 1081–1088 (1995).
[1522] Díaz, J.M.; M. Caballero, J. Pérez-Bustamante, R. Cela, *Analyst (London) 115/9*, 1201–1205 (1990).
[1523] Dieijen-Visser, M.P. van ; G.M. Marell, J.L.L.M. Coenen, P.J. Brombacher, *Eur. J. Clin. Chem. Clin. Biochem. 29/6*, 381–384 (1991).
[1524] Dieke, G.H.; H.M. Crosswhite, *J. Opt. Soc. Am. 42*, 433 (1952).
[1525] Dietze, U.; J. Braun, H.-J. Peter, *Fresenius Z. Anal. Chem. 322/1*, 17–19 (1985).
[1526] Dilena, B.A.; L. Larsson, S. Öhman, in: H.G. Seiler, A. Sigel, H. Sigel (Editors), *Handbook on Metals in Clinical and Analytical Chemistry*, Marcel Dekker, New York, 1994, 299–310.
[1527] Ding, W.-w.; R.E. Sturgeon, *Spectrochim. Acta, Part B 51/11*, 1325–1334 (1996).
[1528] Ding, W.-w.; R.E. Sturgeon, *J. Anal. At. Spectrom. 11/3*, 225–230 (1996).
[1529] Ding, W.-w.; R.E. Sturgeon, *J. Anal. At. Spectrom. 11/6*, 421–425 (1996).
[1530] Dirkx, W.M.R.; F.C. Adams, *Mikrochim. Acta (Vienna) 109/1-4*, 79–81 (1992).
[1531] Dirkx, W.M.R.; R.J.A. Van Cleuvenbergen, F.C. Adams, *Mikrochim. Acta (Vienna) 109/1-4*, 133–135 (1992).
[1532] Dirkx, W.M.R.; R. Łobinski, F.C. Adams, *Anal. Chim. Acta 286/3*, 309–318 (1994).
[1533] Dirkx, W.M.R.; M.B. de la Calle, M. Ceulemans, F.C. Adams, *J. Chromatogr. A 683*, 51–58 (1994).
[1534] Dirscherl, C.; J. Klein, H. Schmidt, in: B. Welz (Editor), *4. Colloquium Atomspektrometrische Spurenanalytik*, Bodenseewerk Perkin-Elmer, Überlingen, 1987, 349–358.
[1535] Dits, J.S.; *Anal. Chim. Acta 130/2*, 395–400 (1981).
[1536] Dittfurth, C.; E. Ballesteros, M. Gallego, M. Valcárcel, *Spectrochim. Acta, Part B 51/14*, 1935–1941 (1996).
[1537] Dittrich, K.; W. Mothes, *Talanta 22*, 318 (1975).
[1538] Dittrich, K.; W. Mothes, P. Weber, *Spectrochim. Acta, Part B 33*, 325–336 (1978).
[1539] Dittrich, K.; S. Schneider, B.J. Spivakov, L.N. Suchowejewa, J.A. Zolotov, *Spectrochim. Acta, Part B 34/6*, 257–268 (1979).
[1540] Dittrich, K.; B. Vorberg, H. Wolters, *Talanta 26*, 747–754 (1979).
[1541] Dittrich, K.; W. Mothes, I.G. Yudelevich, T.S. Papina, *Talanta 32/3*, 195–201 (1985).
[1542] Dittrich, K.; R. Mandry, W. Mothes, I.G. Yudelevich, *Analyst (London) 110/2*, 169–175 (1985).
[1543] Dittrich, K.; R. Mandry, *Analyst (London) 111/3*, 277–280 (1986).
[1544] Dittrich, K.; R. Mandry, *Analyst (London) 111/3*, 269–275 (1986).
[1545] Dittrich, K.; R. Mandry, Ch. Udelnow, A. Udelnow, *Fresenius Z. Anal. Chem. 323/7*, 793–799 (1986).
[1546] Dobbener, A.; G. Schwedt, *LaborPraxis 18/11*, 70–77 (1994).
[1547] Dobreva, D.; A. Barzev, L. Futekov, V. Rusev, *Fresenius Z. Anal. Chem. 328/7*, 557–559 (1987).
[1548] Dobrowolski, R.; J. Mierzwa, *Analyst (London) 117/7*, 1165–1167 (1992).
[1549] Dobrowolski, R.; *Spectrochim. Acta, Part B 51/2*, 221–227 (1996).
[1550] Dočekal, B.; V. Krivan, *J. Anal. At. Spectrom. 7/3*, 521–528 (1992).
[1551] Dočekal, B.; V. Krivan, N. Pelz, *Fresenius J. Anal. Chem. 343*, 873–878 (1992).
[1552] Dočekal, B.; *J. Anal. At. Spectrom. 8/5*, 763–764 (1993).

[1553] Dočekal, B.; V. Krivan, *J. Anal. At. Spectrom. 8/4*, 637–641 (1993).
[1554] Dočekal, B.; V. Krivan, *Anal. Chim. Acta 279*, 253–260 (1993).
[1555] Dočekal, B.; V. Krivan, *Spectrochim. Acta, Part B 48/13*, 1645–1649 (1993).
[1556] Dočekal, B.; V. Krivan, M. Franek, *Spectrochim. Acta, Part B 49/6*, 577–582 (1994).
[1557] Dočekal, B.; V. Krivan, *Spectrochim. Acta, Part B*, im Druck.
[1558] Dočekal, B.; V. Krivan, *Spectrochim. Acta, Part B 50/4-7*, 517–526 (1995).
[1559] Dočekal, B.; J. Dědina, V. Krivan, *Spectrochim. Acta, Part B 52/6*, 787–794 (1997).
[1560] Dočekalová, H.; B. Dočekal, J. Komárek, I. Novotný, *J. Anal. At. Spectrom. 6/8*, 661–668 (1991).
[1561] Doidge, P.S.; B.T. Sturman, T.M. Rettberg, *J. Anal. At. Spectrom. 4/3*, 251–255 (1989).
[1562] Doidge, P.S.; *Spectrochim. Acta, Part B 46/14*, 1779–1787 (1991).
[1563] Doidge, P.S.; T. McAllister, *J. Anal. At. Spectrom. 8/3*, 403–408 (1993).
[1564] Doidge, P.S.; *Spectrochim. Acta, Part B 48/3*, 473–474 (1993).
[1565] Doidge, P.S.; *Spectrochim. Acta, Part B 50/3*, 209–263 (1995).
[1566] Dokiya, Y.; H. Ashikawa, S. Yamazaki, K. Fuwa, *Environ. Lett. 7/1*, 67–81 (1974).
[1567] Dokiya, Y.; H. Ashikawa, K. Fuwa, *Spectrosc. Lett. 7*, 551 (1974).
[1568] Dolinšek, F.; J. Štupar, V. Vrščaj, *J. Anal. At. Spectrom. 6/8*, 653–660 (1991).
[1569] Doll, R.; *Scand. J. Work Environ. Health 16*, 1–82 (1990).
[1570] Dolske, D.A.; J. Schneider, H. Sievering, *Atmos. Environ. 18/11*, 2557–2558 (1984).
[1571] Donaldson, E.M.; *Talanta 27/6*, 499–505 (1980).
[1572] Donaldson, E.M.; *Talanta 27/10*, 779–786 (1980).
[1573] Donaldson, E.M.; *Talanta 28/7A*, 461–467 (1981).
[1574] Donaldson, E.M.; *Talanta 29/12*, 1069–1075 (1982).
[1575] Donaldson, E.M.; *Talanta 35/1*, 47–53 (1988).
[1576] Donaldson, E.M.; M.E. Leaver, *Talanta 35/4*, 297–300 (1988).
[1577] Donaldson, E.M.; *Talanta 36/5*, 543–548 (1989).
[1578] Donaldson, E.M.; M.E. Leaver, *Talanta 37/2*, 173–183 (1990).
[1579] Donaldson, E.M.; *Talanta 37/10*, 955–964 (1990).
[1580] Donard, O.F.X.; Ph. Pedemay, *Anal. Chim. Acta 153*, 301–305 (1983).
[1581] Donard, O.F.X.; J.H. Weber, *Environ. Sci. Technol. 19*, 1104–1110 (1985).
[1582] Donard, O.F.X.; S. Rapsomanikis, J.H. Weber, *Anal. Chem. 58/4*, 772–777 (1986).
[1583] Donard, O.F.X.; L. Randall, S. Rapsomanikis, J.H. Weber, *Int. J. Environ. Anal. Chem. 27*, 55–67 (1986).
[1584] Donard, O.F.X.; in: B. Welz (Editor), *5. Colloquium Atomspektrometrische Spurenanalytik*, Bodenseewerk Perkin-Elmer, Überlingen, 1989, 395–417.
[1585] Donard, O.F.X.; F.M. Martin, *Trends Anal. Chem. (Pers. Ed.) 11/1*, 17–26 (1992).
[1586] Donard, O.F.X.; R. Ritsema, in: D. Barceló (Editor), *Environmental Analysis - Techniques, Applications and Quality Assurance*, Elsevier, Amsterdam, 1993, 549–606.
[1587] Donard, O.F.X.; Ph. Quevauviller, A. Bruchet, *Water Research 27/6*, 1085–1089 (1993).
[1588] Donat, J.R.; P.J. Statham, K.W. Bruland, *Mar. Chem. 18/1*, 85–99 (1986).
[1589] Döner, G.; S. Akman, *J. Anal. At. Spectrom. 9/3*, 333–336 (1994).
[1590] Döner, G.; S. Akman, *Spectrochim. Acta, Part B 51/1*, 181–187 (1996).
[1591] Dong, F.M.; A.E. Nevissi, B.A. Rasco, *J. Micronutr. Anal. 4*, 71–77 (1988).
[1592] Dong, L.-p.; Z.-l. Fang, *Fenxi Shiyanshi 11/6*, 5–9 (1992).
[1593] Dong, L.-p.; Z.-l. Fang, *Guangpuxue Yu Guangpu Fenxi 14/1*, 85–90,98 (1994).
[1594] Donhauser, S.; D. Wagner, F. Jacob, *Monatsschr. Brauwiss. 40/6*, 247–256 (1987).
[1595] Doolan, K.J.; C.B. Belcher, *Prog. Anal. Spectrosc. 3*, 125–179 (1980).
[1596] Doolan, K.J.; *Anal. Chim. Acta 140*, 187–195 (1982).
[1597] Dorado López, M.T.; M.T. Larrea Marín, A. Gómez-Coedo, *J. Anal. At. Spectrom. 3/3*, 447–452 (1988).
[1598] Dorado López, M.T.; M.A. Palacios Vida, A. Gómez-Coedo, *Congr. Nac. Cienc. Tecnol. Metal. 3*, 205 (1990).
[1599] Dornemann, A.; H. Kleist, *Fresenius Z. Anal. Chem. 300/3*, 197–199 (1980).
[1600] Dornemann, A.; H. Kleist, *Fresenius Z. Anal. Chem. 305/5*, 379–381 (1981).
[1601] Dornemann, A.; K.H. Kothen, D. Rudan, *Fresenius Z. Anal. Chem. 326*, 232 (1987).
[1602] Dougherty, J.P.; R.G. Michel, W. Slavin, *Anal. Lett. 18/A10*, 1231–1244 (1985).
[1603] Doughten, M.W.; J.R. Gillison, *Energy and Fuels 4/5*, 426–430 (1990).
[1604] Dowling, F.B.; C.L. Chakrabarti, G.R. Lyles, *Anal. Chim. Acta 28/4*, 392 (1963).
[1605] Drabæk, I.; Å. Iverfeldt, in: M. Stoeppler (Editor), *Hazardous Metals in the Environment*, Techn. Instr. Anal. Chem. Vol 12, Elsevier, Amsterdam, 1992.
[1606] Drasch, G.A.; L. von Meyer, G. Kauert, *Fresenius Z. Anal. Chem. 304/2-3*, 141–142 (1980).

[1607] Drasch, G.A.; I. Schupp, G. Riedl, *Dtsch. Zahnärztl. Ztsch. 47*, 490–496 (1992).
[1608] Drasch, G.A.; in: H.G. Seiler, A. Sigel, H. Sigel (Editors), *Handbook on Metals in Clinical and Analytical Chemistry*, Marcel Dekker, New York, 1994, 479–493.
[1609] Drasch, G.A.; I. Schupp, unveröffentlicht.
[1610] Drasch, G.A.; H.J. Gath, E. Heissler, I. Schupp, G. Roider, *J. Trace Elem. Med. Biol. 9/2*, 82–87 (1995).
[1611] Driehaus, W.; M. Jekel, *Fresenius J. Anal. Chem. 343/4*, 352–356 (1992).
[1612] Drinkwater, J.E.; *Analyst (London) 101*, 672–677 (1976).
[1613] Droessler, M.S.; J.A. Holcombe, *Spectrochim. Acta, Part B 42/8*, 981–994 (1987).
[1614] Droessler, M.S.; J.A. Holcombe, *J. Anal. At. Spectrom. 2/8*, 785–792 (1987).
[1615] Dube, P.; *At. Spectrosc. 9/2*, 55–58 (1988).
[1616] Dubois, J.P.; *J. Trace Microprobe Techniques 9*, 149–163 (1991).
[1617] Ducros, V.; D. Ruffieux, N. Belin, A. Favier, *Analyst (London) 119/8*, 1715–1717 (1994).
[1618] Duffy, S.J.; G.W. Hay, R.K. Micklethwaite, G.W. Vanloon, *Sci. Total Environ. 87*, 189–197 (1989).
[1619] D'Ulivo, A.; L. Lampugnani, I. Sfetsios, R. Zamboni, C. Forte, *Analyst (London) 119/4*, 633–640 (1994).
[1620] D'Ulivo, A.; J. Dědina, Spectrochim. Acta, *Part B 51/5*, 481–496 (1996).
[1621] Dulude, G.R.; J.J. Sotera, *Can. Res. 15/7*, 21–53 (1982).
[1622] Dulude, G.R.; J.J. Sotera, D.N. Peterson, *Spectroscopy Int. 1/4*, 44–49 (1989).
[1623] Dumarey, R.; P. Verbiest, R. Dams, *Bull. Soc. Chim. Belg. 94/5*, 351–357 (1985).
[1624] Dumarey, R.; R. Dams, in: T.D. Lekkas (Editor), *Heavy Metals in the Environment*, Proc. Int. Conf., 5th, Athens, CEP Consultants, Edinburgh, 1985, 501–503.
[1625] Dumarey, R.; M. van Ryckeghem, R. Dams, *J. Trace Microprobe Techniques 5/2-3*, 229–242 (1987).
[1626] Dunemann, L.; in: B. Welz (Editor), *5. Colloquium Atomspektrometrische Spurenanalytik*, Bodenseewerk Perkin-Elmer, Überlingen, 1989, 593–601.
[1627] Dunemann, L.; M. Meinerling, *Fresenius J. Anal. Chem. 342/9*, 714–718 (1992).
[1628] Dungs, K.W., B. Neidhart, in: B. Welz (Editor), *Fortschritte in der atomspektrometrischen Spurenanalytik, Band 2*, VCH Verlagsgesellschaft, Weinheim, 1986, 25–36.
[1629] Dunk, R. ; R.A. Mostyn, H.C. Hoare, *At. Absorpt. Newsletter 8/4*, 79–81 (1969).
[1630] Dunn, C.E.; G.E.M. Hall, E. Hoffman, *J. Geochem. Explor. 32*, 211–222 (1989).
[1631] Dürnberger, R.; P. Esser, A. Janßen, *Fresenius Z. Anal. Chem. 327/3-4*, 343–346 (1987).
[1632] Durst, R.A.; B.T. Duhart, *Anal. Chem. 42/9*, 1002–1004 (1970).
[1633] Dusci, L.J.; L.P. Hackett, *J. Assoc. Off. Anal. Chem. 59*, 1183–1185 (1976).
[1634] Duve, R.N.; J.P. Chandra, S.B. Singh, *J. Assoc. Off. Anal. Chem. 64/4*, 1027–1029 (1981).
[1635] Dybczynski, R.; K. Kulisa, M. Malusecka, M. Mandecka, H. Polkowska-Motrenko, S. Sterlinski, Z. Szopa, *Biol. Trace Elem. Res. 26*, 335 (1990).
[1636] Dyck, R.; *At. Absorpt. Newsletter 4*, 170–173 (1965).
[1637] Dyck, W.; *Anal. Chem. 40/2*, 454–455 (1968).
[1638] Dyg, S.; R. Cornelis, B. Griepink, P. Verbeek, in: J.A.C. Broekaert, S. Güçer, F. Adams (Editors), *Metal Speciation in the Environment*, NATO ASI Ser., Ser. G: Ecological Sci., Springer, Berlin - Heidelberg - New York, 1990, 361–376.
[1639] Dyg, S.; T. Anglov, J.M. Christensen, *Anal. Chim. Acta 286/3*, 273–282 (1994).
[1640] Dyg, S.; R. Cornelis, B. Griepink, Ph. Quevauviller, *Anal. Chim. Acta 286/3*, 297–308 (1994).
[1641] Dymott, T.C.; *CLB, Chem. Labor Betr. 32/12*, 580–584 (1981).
[1642] Dymott, T.C.; *Int. Labmate 10/3*, 25–28 (1985).
[1643] Dymott, T.C.; M.P. Wassall, P.J. Whiteside, *Analyst (London) 110/5*, 467–474 (1985).
[1644] Ebdon, L.; R.W. Ward, D.A. Leathard, *Analyst (London) 107*, 129–143 (1982).
[1645] Ebdon, L.; J.R. Wilkinson, K.W. Jackson, *Anal. Chim. Acta 136*, 191–199 (1982).
[1646] Ebdon, L.; A. Lechotycki, *Microchem. J. 34/3*, 340–348 (1986).
[1647] Ebdon, L.; A. Lechotycki, *Microchem. J. 36/2*, 207–215 (1987).
[1648] Ebdon, L.; E.H. Evans, *J. Anal. At. Spectrom. 2/3*, 317–320 (1987).
[1649] Ebdon, L.; H.G.M. Parry, *J. Anal. At. Spectrom. 2/1*, 131–134 (1987).
[1650] Ebdon, L.; J.R. Wilkinson, *Anal. Chim. Acta 194*, 177–187 (1987).
[1651] Ebdon, L.; A.P. Walton, G.E. Millward, M. Whitfield, *Appl. Organomet. Chem. 1*, 427–433 (1987).
[1652] Ebdon, L.; S.J. Hill, P. Jones, *J. Anal. At. Spectrom. 2/2*, 205–210 (1987).
[1653] Ebdon, L.; S.J. Hill, R.W. Ward, *Analyst (London) 112/1*, 1–16 (1987).
[1654] Ebdon, L.; S.J. Hill, P. Jones, *Analyst (London) 112*, 437–440 (1987).

[1655] Ebdon, L.; H.G.M. Parry, *J. Anal. At. Spectrom. 3/1*, 131–134 (1988).

[1656] Ebdon, L.; K. Evans, S.J. Hill, *Sci. Total Environ. 83*, 63–84 (1989).

[1657] Ebdon, L.; S.J. Hill, in: R.M. Harrison, S. Rapsomanikus (Editors), *Environ. Analysis using Chromatography Interfaced with Atomic Spectrometry*, Ellis Horwood, Chichester, 1989, 165–188.

[1658] Ebdon, L.; A.S. Fisher, H.G.M. Parry, A.A. Brown, *J. Anal. At. Spectrom. 5/4*, 321–324 (1990).

[1659] Ebdon, L.; S.J. Hill, P. Jones, *Talanta 38/6*, 607–611 (1991).

[1660] Ebdon, L.; A.S. Fisher, in: S.J. Haswell (Editor), *Atomic Absorption Spectrometry - Theory, Design and Application*, Elsevier, Amsterdam - Oxford - New York - Tokyo, 1991, 463–514.

[1661] Ebdon, L.; A.S. Fisher, S.J. Hill, *Anal. Chim. Acta 282*, 433–436 (1993).

[1662] Ebdon, L.; P. Goodall, S.J. Hill, P.B. Stockwell, K.C. Thompson, *J. Anal. At. Spectrom. 8/5*, 723–729 (1993).

[1663] Ebdon, L.; P. Goodall, S.J. Hill, P.B. Stockwell, K.C. Thompson, *J. Anal. At. Spectrom. 10/4*, 317–320 (1993).

[1664] Ebinghaus, R.; R.D. Wilken, *Appl. Organomet. Chem. 7*, 127–135 (1993).

[1665] Ebinghaus, R.; H. Hintelmann, R.D. Wilken, *Fresenius J. Anal. Chem. 350/1-2*, 21–29 (1994).

[1666] Echevarría, J.; M.T. Arcos, M.J. Ferrandez, J. Garrido Segovia, *Quim. Anal. (Barcelona) 11/1*, 25–33 (1992).

[1667] Eckelmans, V.; E. Graauwmans, S. De Jaegere, *Talanta 21*, 715 (1974).

[1668] Eckerlin, R.H.; D.W. Hourt, G.R. Carnrick, *At. Spectrosc. 8/1*, 64–66 (1987).

[1669] Edel, H.; L. Quick, K. Cammann, *Fresenius J. Anal. Chem. 351/6*, 479–483 (1995).

[1670] Edel, H.; L. Quick, K. Cammann, Anal. Chim. Acta 310/1, 181–187 (1995).

[1671] Edel, H.; D. Erber, R. Lehnert, W. Buscher, K. Cammann, *Fresenius J. Anal. Chem. 355/3-4*, 292–294 (1996).

[1672] Edgar, R.M.; *Anal. Chem. 48*, 1653 (1976).

[1673] Ediger, R.D.; R.L. Coleman, *At. Absorpt. Newsletter 11/2*, 33–36 (1972).

[1674] Ediger, R.D.; R.L. Coleman, *At. Absorpt. Newsletter 12/1*, 3–6 (1973).

[1675] Ediger, R.D.; *At. Absorpt. Newsletter 12*, 151–157 (1973).

[1676] Ediger, R.D.; G.E. Peterson, J.D. Kerber, *At. Absorpt. Newsletter 13/3*, 61–64 (1974).

[1677] Ediger, R.D.; *At. Absorpt. Newsletter 14/5*, 127–130 (1975).

[1678] Ediger, R.D.; *At. Absorpt. Newsletter 15/6*, 145 (1976).

[1679] Edmonds, J.S.; K.A. Francesconi, *Experientia 43*, 553–557 (1987).

[1680] Edmonstone, G.; J.C. Van Loon, *Spectrosc. Lett. 17/10*, 591–602 (1984).

[1681] Edwards, W.C.; T.A. Blackburn, *Vet. Hum. Toxicol. 28/1*, 12–13 (1986).

[1682] Egaas, E.; K. Julshamn, *At. Absorpt. Newsletter 17/6*, 135–138 (1978).

[1683] Egila, J.N.; D. Littlejohn, J.M. Ottaway, X.-q. Shan, *J. Anal. At. Spectrom. 2/3*, 293–298 (1987).

[1684] Eichholtz, G.G.; A.E. Nagel, R.B. Hughes, *Anal. Chem. 37*, 863–868 (1965).

[1685] Eickhoff, C.P.; B.J. Sykes, *J. Sci. Instr. 41*, 113 (1964).

[1686] Eidecker, R.; E. Jackwerth, *Fresenius Z. Anal. Chem. 328/6*, 469–474 (1987).

[1687] Eidecker, R.; E. Jackwerth, *Fresenius Z. Anal. Chem. 331/3-4*, 401–407 (1988).

[1688] Eider, N.G.; *Appl. Spectrosc. 25/3*, 313–316 (1971).

[1689] Einax, J.; K. Oswald, K. Danzer, *Fresenius J. Anal. Chem. 336/5*, 394–399 (1990).

[1690] Einax, J.; B. Machelett, S. Geiß, K. Danzer, *Fresenius J. Anal. Chem. 342/4-5*, 267–272 (1992).

[1691] Einhäuser, T.J.; M. Galanski, B.K. Keppler, *J. Anal. At. Spectrom. 11/9*, 747–750 (1996).

[1692] Eiras, S. de P.; P.G.P. Zamora, E.L. Reis, *Quim. Nova (Sao Paulo) 17*, 369 (1994).

[1693] Eisentraut, K.J.; R.W. Newman, C.S. Saba, *Anal. Chem. 56*, 1086A-94A (1984).

[1694] Eisman, M.; M. Gallego, M. Valcárcel, *J. Anal. At. Spectrom. 7/8*, 1295–1301 (1992).

[1695] Eisman, M.; M. Gallego, M. Valcárcel, *Anal. Chem. 64/14*, 1509–1512 (1992).

[1696] Eisman, M.; M. Gallego, M. Valcárcel, *J. Anal. At. Spectrom. 8/8*, 1117–1120 (1993).

[1697] Eisman, M.; M. Gallego, M. Valcárcel, *J. Pharm. Biomed. Anal. 11/4-5*, 301–305 (1993).

[1698] Eisman, M.; M. Gallego, M. Valcárcel, *J. Pharm. Biomed. Anal. 12/2*, 179–184 (1994).

[1699] Eklund, R.H.; J.A. Holcombe, *Talanta 26*, 1055–1057 (1979).

[1700] El-Awady, A.A.; R.B. Miller, M.J. Carter, *Anal. Chem. 48/1*, 110–116 (1976).

[1701] El-Azouzi, H.; M.Y. Pérez-Jordán, A. Salvador, M. de la Guardia, *Spectrochim. Acta, Part B 51/14*, 1747–1752 (1996).

[1702] El-Defrawy, M.M.; J. Posta, M.T. Beck, *Anal. Chim. Acta 102*, 185–188 (1978).

[1703] El-Defrawy, M.M.; A.M. Abdallah, M.A. Mostafa, M.A. Akl, *Analusis 14/6*, 306–309 (1986).

[1704]   El-Defrawy, M.M.; M.E. Khalifa, A.M. Abdallah, M.A. Akl, *J. Anal. At. Spectrom. 2*, 333–337 (1987).
[1705]   El-Defrawy, M.M.; *Analusis 19/9*, 292–296 (1991).
[1706]   El-Defrawy, M.M.; M.E. Khalifa, A.M. Abdallah, M.A.A. Mohamed, *Analusis 20/4*, 217–220 (1992).
[1707]   El-Defrawy, M.M.; *Analusis 21,* 91–93 (1993).
[1708]   El-Defrawy, M.M.; *Analusis 23/3*, 107–109 (1995).
[1709]   El-Ries, M.A.; *Anal. Lett. 27/8*, 1517–1531 (1994).
[1710]   El-Ries, M.A.; F.M. Abou Attia, F.M. Abdel-Gawad, S.M. Abu El-Wafa, *J. Pharm. Biomed. Anal. 12/9*, 1209–1213 (1994).
[1711]   El-Sayad, E.; M.S. Cresser, M. Abd El-Gawad, E.A. Khater, *Microchem. J. 38*, 307–312 (1988).
[1712]   El-Sayed, A.B.; N.E. Amine, S.H. Abd El-Haleem, M.F. El-Shahat, *Anal. Chim. Acta 165/11*, 113–119 (1984).
[1713]   El-Yazigi, A.; I. Al-Saleh, O. Al-Mefty, *Clin. Chem. (Winston-Salem) 30/8*, 1358–1360 (1984).
[1714]   Elçi, L.; M. Soylak, M. Dogan, *Fresenius J. Anal. Chem. 342/1-2*, 175–178 (1992).
[1715]   Elçi, L.; *Anal. Lett. 26/5*, 1025–1036 (1993).
[1716]   Elçi, L.; U. Sahin, Sibel Öztas, *Talanta 44/6*, 1017–1023.
[1717]   Elinder, C.G.; B. Lind, M. Piscator, K. Sundstedt, S. Åkerberg, *Bull. Environ. Contam. Toxicol. 27/6*, 810–815 (1981).
[1718]   Elkins, E.R.; A. Sulek, *J. Assoc. Off. Anal. Chem. 62/5*, 1050–1053 (1979).
[1719]   Ellen, G.; J.W. Van Loon, *Food Addit. Contam. 7/2*, 265–273 (1990).
[1720]   Ellend, N.; C. Rohrer, M. Grasserbauer, J.A.C. Broekaert, *Fresenius J. Anal. Chem. 356/1*, 99–101 (1996).
[1721]   Eller, P.M.; *Appl. Ind. Hyg. 1*, 115 (1986).
[1722]   Eller, R.; F. Alt, G. Tölg, H.J. Tobschall, *Fresenius Z. Anal. Chem. 334*, 723–739 (1989).
[1723]   Ellis, W.G.; V.F. Hodge, D.A. Darby, C.L. Jones, T.A. Hinners, *At. Spectrosc. 9/6*, 181–208 (1988).
[1724]   Elmahadi, H.A.M.; G.M. Greenway, *J. Anal. At. Spectrom. 6/8*, 643–646 (1991).
[1725]   Elmahadi, H.A.M.; G.M. Greenway, *J. Anal. At. Spectrom. 9/4*, 547–551 (1994).
[1726]   El-Moll, A.; R. Heimburger, F. Lagarde, M.J.F. Leroy, E. Maier, *Fresenius J. Anal. Chem. 354/5-6*, 550–556 (1996).
[1727]   Eloi, C.C.; J.D. Robertson, V. Majidi, *J. Anal. At. Spectrom. 8/2*, 217–222 (1993).
[1728]   Eloi, C.C.; J.D. Robertson, V. Majidi, *Anal. Chem. 67/2*, 335–340 (1995).
[1729]   Eloi, C.C.; J.D. Robertson, V. Majidi, *Appl. Spectrosc. 51/2*, 236–239 (1997).
[1730]   Elrick, K.A.; A.J. Horowitz, *Open-File Rep.-U.S. Geol. Surv. 86-529*, 1986.
[1731]   Elsheimer, H.N.; T.L. Fries, *Anal. Chim. Acta 239/1*, 145–149 (1990).
[1732]   Elsheimer, H.N.; *Anal. Sci. 9/5*, 681–685 (1993).
[1733]   Elsheimer, H.N.; *Mikrochim. Acta (Vienna) 112/5-6*, 189–196 (1994).
[1734]   Elsholz, O.; in: K. Dittrich, B. Welz (Editors), *CANAS'93, Colloq. Analytische Atomspektroskopie*, Universität Leipzig/UFZ Leipzig-Halle, Leipzig, 1993, 451–456.
[1735]   Elson, C.M.; E.M. Bem, R.G. Ackman, *J. Am. Oil Chem. Soc. 58,* 1024–1026 (1981).
[1736]   Elteren, J.T. Van; *Dissertation*, State University Utrecht, 1991.
[1737]   Elteren, J.T. Van; J. Hoegee, E.E. van der Hoek, H.A. Das, C.L. de Ligny, J. Agterdenbos, *J. Radioanal. Nucl. Chem. 154/5*, 343–355 (1991).
[1738]   Elton-Bott, R.R.; *Anal. Chim. Acta 86*, 281–284 (1976).
[1739]   Emancipator, K.; M.H. Kroll, *Clin. Chem. (Winston-Salem) 39/5*, 766–772 (1993).
[1740]   Emara, M.M.; M.M. Ali, Abdl El-Aziz Gharib, *Anal. Chim. Acta 102*, 181–184 (1978).
[1741]   Emerick, R.J.; *At. Spectrosc. 8/2*, 69–71 (1987).
[1742]   Emmerling, G.; K.H. Schaller, H. Valentin, *Zbl. Arbeitsmed. 36*, 258–265 (1986).
[1743]   Emmermann, R.; W. Luecke, *Fresenius Z. Anal. Chem. 248*, 325 (1969).
[1744]   Emteborg, H.; H.-W. Sinemus, B. Radziuk, D.C. Baxter, W. Frech, *Spectrochim. Acta, Part B 51/8*, 829–837 (1996).
[1745]   Englert, K.; G. Giebenhain, H.-J. Mosch, N. Müller, *GIT Fachz. Lab. 41/1*, 32–34 (1997).
[1746]   Enzweiler, J.; P.J. Potts, *Talanta 42/10*, 1411–1418 (1995).
[1747]   Epstein, M.S.; J.D. Winefordner, *Talanta 27/2*, 177–180 (1980).
[1748]   Epstein, M.S.; *At. Spectrosc. 4/2*, 62–63 (1983).
[1749]   Epstein, M.S.; G.R. Carnrick, W. Slavin, *Anal. Chem. 61/13*, 1414–1419 (1989).
[1750]   Epstein, M.S.; *Spectrochim. Acta, Part B 46/12*, 1583–1591 (1991).
[1751]   Epstein, M.S.; G.C. Turk, L.J. Yu, *Spectrochim. Acta, Part B 49/12-4*, 1681–1688 (1994).

[1752] Erard, M.; B. Zimmerli, *Mitt. Geb. Lebensmittelunters. Hyg. 80/4*, 452–466 (1989).
[1753] Erard, M.; M. Haldiman, B. Zimmerli, in: B. Welz (Editor), *5. Colloquium Atom-spektrometrische Spurenanalytik*, Bodenseewerk Perkin-Elmer, Überlingen, 1989, 789–798.
[1754] Erber, D.; L. Quick, J. Roth, K. Cammann, *Fresenius J. Anal. Chem. 346/4*, 420–425 (1993).
[1755] Erber, D.; L. Quick, F. Winter, J. Roth, K. Cammann, *Fresenius J. Anal. Chem. 349/7*, 502–509 (1994).
[1756] Erber, D.; J. Bettmer, K. Cammann, *GIT Fachz. Lab. 39/4*, 340–349 (1995).
[1757] Erber, D.; K. Cammann, *Analyst (London) 120/11*, 2699–2705 (1995).
[1758] Erber, D.; L. Quick, F. Winter, K. Cammann, *Talanta 42/7*, 927–936 (1995).
[1759] Erdman, J.A.; J.C. Olson, *J. Geochem. Explor. 24*, 281–304 (1985).
[1760] Ergenoglu, B.; A. Olcay, *Fuel Sci. Technol. Int. 8/7*, 743–752 (1990).
[1761] Ergenoglu, B.; A. Olcay, *Fuel 73/4*, 629–631 (1994).
[1762] Ericson, J.E ; D.R. Smith, A.R. Flegal, *Environ. Health Perspect. 93*, 217–223 (1991).
[1763] Ericson, S.P.; M.L. McHalsky, B.E. Rabinow, K.G. Kronholm, C.S. Arceo, J.A. Weltzner, S.W. Ayd, *Clin. Chem. (Winston-Salem) 32*, 1350–1356 (1986).
[1764] Ericson, S.P.; *At. Spectrosc. 13/6*, 208–212 (1992).
[1765] Erler, W.; R. Lehmann, U. Völlkopf, in: P. Brätter, P. Schramel (Editors), *Trace Element Analytical Chemistry in Medicine and Biology*, W. de Gruyter, Berlin - New York, 1987, 385–391.
[1766] Erler, W.; *Technical Summary, Atomic Absorption TSAA-41*, Bodenseewerk Perkin-Elmer, 1993.
[1767] Esmadi, F.T.; M.A. Kharoaf, A.S. Attiyat, *Talanta 37/12*, 1123–1128 (1990).
[1768] Esmadi, F.T.; M.A. Kharoaf, A.S. Attiyat, *Anal. Lett. 23/6*, 1069–1086 (1990).
[1769] Esmadi, F.T.; M.A. Kharoaf, A.S. Attiyat, *Analyst (London) 116/4*, 353–356 (1991).
[1770] Esmadi, F.T.; I.M. Khasawneh, M.A. Kharoaf, A.S. Attiyat, *Anal. Lett. 24/7*, 1231–1255 (1991).
[1771] Esmadi, F.T.; M.A. Kharoaf, A.S. Attiyat, *J. Flow Injection Anal. 10/1*, 33–47 (1993).
[1772] Esmadi, F.T.; A.S. Attiyat, *Anal. Sci. 10/4*, 687–690 (1994).
[1773] Espinosa Almendro, J.M.; C. Bosch Ojeda, A. García de Torres, J.M. Cano Pavón, *Talanta 40/11*, 1643–1648 (1993).
[1774] Esser, P.; *Fresenius Z. Anal. Chem. 322/7*, 677–680 (1985).
[1775] Esser, P.; in: B. Welz (Editor), *Fortschritte in der atomspektrometrischen Spurenanalytik*, VCH Verlagsgesellschaft, Weinheim, 1986, 307–316.
[1776] Esser, P.; R. Dürnberger, *Fresenius Z. Anal. Chem. 328/4-5*, 359–361 (1987).
[1777] Eulate, M.J.A. de; R. Montoro, N. Ybáñez, M. de la Guardia, *J. Assoc. Off. Anal. Chem. 69/5*, 871–873 (1986).
[1778] European Commission, Joint Research Centre, IRMM, *BCR Reference Materials*, Catalog, 1996.
[1779] Evans, D.J.; *Text. Res. 65/2*, 118–122 (1995).
[1780] Evans, W.H.; F.J. Jackson, D. Dellar, *Analyst (London) 104*, 16–34 (1979).
[1781] Evans, W.H.; P.J. Brooke, B.E. Lucas, *Anal. Chim. Acta 148*, 203–210 (1983).
[1782] Evenson, M.A.; B.L. Warren, *Clin. Chem. (Winston-Salem) 21/4*, 619–625 (1975).
[1783] Everett, K.; F.A. Graf, in: N.V. Steere (Editor), *Handbook of Laboratory Safety*, Chemical Rubber Co., Cleveland, OH, 1971, 265–276.
[1784] Everson, R.J.; *At. Absorpt. Newsletter 11/6*, 130 (1972).
[1785] Ewers, U.; A. Brockhaus, R. Dolgner, I. Freier, E. Jermann, A. Berbnard, RE. Siller-Winkler, R. Hahn, N. Manojlovic, *Int. Arch. Occup. Environ. Health 55*, 217 (1985).
[1786] Ewing, J.J.; R. Milstein, R.S. Berry, *J. Chem. Phys. 54*, 1752–1760 (1971).
[1787] Ezat, U.; *Analusis 16/3*, 168–172 (1988).
[1788] Ezell, J.B. *At. Absorpt. Newsletter 6*, 84–85 (1967).
[1789] Fabec, J.L; *Anal. Chem. 54/13*, 2170–2174 (1982).
[1790] Fabec, J.L; *At. Spectrosc. 4/4*, 46–48 (1983).
[1791] Fabec, J.L; M.L. Ruschak, *At. Spectrosc. 5/4*, 142–145 (1984).
[1792] Fabec, J.L; M.L. Ruschak, *At. Spectrosc. 6/4*, 81–87 (1985).
[1793] Fachgruppe Wasserchemie, Hauptausschuß I, *Vom Wasser 76*, (1991).
[1794] Fagioli, F.; S. Landi, G. Lucci, *Ann. Chim. (Rom) 72/1-2*, 63–71 (1982).
[1795] Fagioli, F.; S. Landi, C. Locatelli, C. Bighi, *Anal. Lett. 16/A4*, 275–288 (1983).
[1796] Fagioli, F.; S. Landi, C. Locatelli, C. Bighi, *At. Spectrosc. 7/2*, 49–51 (1986).
[1797] Fagioli, F.; C. Locatelli, R. Vechietti, G. Torsi, *J. Anal. At. Spectrom. 3/1*, 159–162 (1988).
[1798] Fagioli, F.; C. Locatelli, G. Torsi, *Ann. Chim. (Rom) 82/5-6*, 283–291 (1992).

[1799]   Fairman, B.; A. Sanz-Medel, M. Gallego, M.J. Quintela, P. Jones, R. Benson, *Anal. Chim. Acta 286/3*, 401–409 (1994).
[1800]   Faisca, A.M.M.M.; A.H. Victor, R.G. Böhmer, *Anal. Chim. Acta 215/1-2*, 111–118 (1988).
[1801]   Faisca, A.M.M.M.; A.H. Victor, R.G. Böhmer, *Anal. Chim. Acta 215/1-2*, 317–323 (1988).
[1802]   Falandysz, J.; *Z. Lebensm. Unters. Forsch. 181/2*, 117–120 (1985).
[1803]   Falbe, J.; M. Regitz, (Editors), *Römpp Chemie Lexikon Band 1: A - Cl*, G. Thieme, Stuttgart - New York, 1989.
[1804]   Falinower, C.; *At. Absorpt. Newsletter 14/6*, 145–148 (1975).
[1805]   Falinower, C.; *Analusis 4/5*, 227–235 (1976).
[1806]   Falk, H.; *Spectrochim. Acta, Part B 33*, 695–700 (1978).
[1807]   Falk, H.; *Prog. Anal. Spectrosc. 3*, 181–208 (1980).
[1808]   Falk, H.; *Prog. Anal. Spectrosc. 5/2-3*, 205–241 (1982).
[1809]   Falk, H.; A. Glismann, L. Bergann, G. Minkwitz, M. Schubert, J. Skole, *Spectrochim. Acta, Part B 40/4*, 533–542 (1985).
[1810]   Falk, H.; J. Tilch, *J. Anal. At. Spectrom. 2*, 527–531 (1987).
[1811]   Falk, H.; C. Schnürer, *Spectrochim. Acta, Part B 44/8*, 759–770 (1989).
[1812]   Falter, R.; H.F. Schöler, *Fresenius J. Anal. Chem. 348/3*, 253–254 (1994).
[1813]   Falter, R.; H.F. Schöler, *Chemosphere 29/6*, 1333–1338 (1994).
[1814]   Falter, R.; H.F. Schöler, *J. Chromatogr. A 675*, 253–256 (1994).
[1815]   Falter, R.; H.F. Schöler, *Fresenius J. Anal. Chem. 353/1*, 34–38 (1995).
[1816]   Falter, R.; H.F. Schöler, *Fresenius J. Anal. Chem. 354/4*, 492–493 (1996).
[1817]   Fang, Z.-l.; S.-k. Xu, S.-c. Zhang, *Anal. Chim. Acta 164*, 41–50 (1984).
[1818]   Fang, Z.-l.; *Anal. Chim. Acta 180*, 8–11 (1986).
[1819]   Fang, Z.-l.; S.-k. Xu, X. Wang, S.-c. Zhang, *Anal. Chim. Acta 179*, 325–340 (1986).
[1820]   Fang, Z.-l.; Z.-h. Zhu, S.-c. Zhang, S.-k. Xu, L. Guo, L.-j. Sun, *Anal. Chim. Acta 214/1-2*, 41–55 (1988).
[1821]   Fang, Z.-l.; B. Welz, *J. Anal. At. Spectrom. 4/1*, 83–89 (1989).
[1822]   Fang, Z.-l.; B. Welz, G. Schlemmer, *J. Anal. At. Spectrom. 4/1*, 91–95 (1989).
[1823]   Fang, Z.-l.; B. Welz, *J. Anal. At. Spectrom. 4/6*, 543–546 (1989).
[1824]   Fang, Z.-l.; M. Sperling, B. Welz, *J. Anal. At. Spectrom. 5/7*, 639–646 (1990).
[1825]   Fang, Z.-l.; B. Welz, M. Sperling, *J. Anal. At. Spectrom. 6/2*, 179–189 (1991).
[1826]   Fang, Z.-l.; T.-z. Guo, B. Welz, *Talanta 38/6*, 613–619 (1991).
[1827]   Fang, Z.-l.; M. Sperling, B. Welz, *J. Anal. At. Spectrom. 6/4*, 301–306 (1991).
[1828]   Fang, Z.-l.; *Spectrochimica Acta Rev. 14/3*, 235–259 (1991).
[1829]   Fang, Z.-l.; L.-p. Dong, S.-k. Xu, *J. Anal. At. Spectrom. 7/2*, 293–299 (1992).
[1830]   Fang, Z.-l.; L.-p. Dong, *J. Anal. At. Spectrom. 7/2*, 439–445 (1992).
[1831]   Fang, Z.-l.; *Microchem. J. 45*, 137–142 (1992).
[1832]   Fang, Z.-l.; M. Sperling, B. Welz, *Anal. Chim. Acta 269*, 9–19 (1992).
[1833]   Fang, Z.-l.; B. Welz, M. Sperling, *Anal. Chem. 65/13*, 1682–1688 (1993).
[1834]   Fang, Z.-l.; *Flow-injection Separation and Pre-concentration*, VCH Verlagsgesellschaft, Weinheim, 1993.
[1835]   Fang, Z.-l.; S.-k. Xu, L.-p. Dong, W.-q. Li, *Talanta 41*, 2165–2172 (1994).
[1836]   Fang, Z.-l.; *Flow Injection Atomic Absorption Spectrometry*, Wiley, Chichester - New York - Brisbane, 1995.
[1837]   Fang, Z.-l.; S.-k. Xu, X.-s. Bai, *Anal. Chim. Acta 326/1-3*, 49–55 (1996).
[1838]   Fang, Z.-l.; S.-k. Xu, G.-h. Tao, *J. Anal. At. Spectrom. 11/1*, 1–24 (1996).
[1839]   Fang, Z.-l.; G.-h. Tao, *Fresenius J. Anal. Chem. 355/5-6*, 576–580 (1996).
[1840]   Fang, Z.-l.; G.-h. Tao, M. Sperling, M. Leyrer, B. Welz, *J. Anal. At. Spectrom. 12*, submitted for publication (1997).
[1841]   Farah, K.S.; J. Sneddon, *Talanta 40/6*, 879–882 (1993).
[1842]   Farah, K.S.; J. Sneddon, *Appl. Spectrosc. Rev. 30/4*, 351–371 (1995).
[1843]   Farey, B.J.; L.A. Nelson, *Anal. Chem. 50*, 2147 (1978).
[1844]   Farey, B.J.; L.A. Nelson, M.G. Rolph, *Analyst (London) 103*, 656–660 (1978).
[1845]   Farey, B.J.; L.A. Nelson, in: J.E. Cantle (Editor), *Atomic Absorption Spectrometry*, Elsevier, Amsterdam - Oxford - New York, 1982, 67–94.
[1846]   Farino, J.; R.F. Browner, *Anal. Chem. 56/14*, 2709–2714 (1984).
[1847]   Farmer, J.G.; L.R. Johnson, *Br. J. Ind. Med. 47*, 342–348 (1990).
[1848]   Farnsworth, P.B.; J.P. Walters, in: S. Caroli (Editor), *Improved Hollow Cathode Lamps for Atomic Spectroscopy*, J. Wiley & Sons, New York - Chichester, 1985, 119–147.
[1849]   Farré, R.; M.J. Lagarda, R. Montoro, *J. Assoc. Off. Anal. Chem. 69/5*, 876–879 (1986).
[1850]   Fassel, V.A.; R.B. Myers, R.N. Kniseley, *Spectrochim. Acta 19*, 1187–1194 (1963).

[1851] Fassel, V.A.; V.G. Mossotti, W.E.L. Grossman, R.N. Kniseley, *Spectrochim. Acta 22*, 347–357 (1966).
[1852] Fassel, V.A.; J.O. Rasmuson, T.G. Cowley, *Spectrochim. Acta, Part B 23/9*, 579–586 (1968).
[1853] Fassel, V.A.; D.A. Becker, *Anal. Chem. 41/12*, 1522–1526 (1969).
[1854] Fassel, V.A.; J.O. Rasmuson, R.N. Kniseley, T.G. Cowley, *Spectrochim. Acta, Part B 25/10*, 559–576 (1970).
[1855] Faugeron, C.; R. Gill, T. Gerdei, *Spectra-2000 171*, 33–36 (1993).
[1856] Faure, H.; A. Favier, M. Tripier, J. Arnaud, *Biol. Trace Elem. Res. 24/1*, 25–37 (1990).
[1857] Favarato, M.; C.A. Mizzen, M.K. Sutherland, B. Krishnan, T.P.A. Kruck, D.R. Crapper McLachlan, *Clin. Chim. Acta 207*, 41 (1992).
[1858] Favretto, L.; G.P. Marletta, L. Favretto Gabrielli, *At. Absorpt. Newsletter 12/4*, 101–103 (1973).
[1859] Favretto, L.; G.P. Marletta, L. Favretto Gabrielli, *Mikrochim. Acta (Vienna) 1/5-6*, 387–398 (1981).
[1860] Favretto, L.; G.P. Marletta, L. Favretto Gabrielli, *At. Spectrosc. 5/2*, 51–54 (1984).
[1861] Favretto, L.; L. Favretto Gabrielli, L. Ceccon, D. Vojnovic, *Anal. Chim. Acta 248/1*, 51–58 (1991).
[1862] Favretto, L.; D. Vojnovic, B. Campisi, *Anal. Chim. Acta 293/3*, 295–300 (1994).
[1863] Fazakas, J.; *Anal. Lett. 15/A19*, 1523–1531 (1982).
[1864] Fazakas, J.; *Anal. Lett. 15/1*, 21–38 (1982).
[1865] Fazakas, J.; *Rev. Roum. Chim. 27/5*, 685–689 (1982).
[1866] Fazakas, J.; *Mikrochim. Acta (Vienna) II/3-4*, 217–224 (1983).
[1867] Fazakas, J.; *Spectrochim. Acta, Part B 38/3*, 455–459 (1983).
[1868] Fazakas, J.; *Mikrochim. Acta (Vienna) I/3-4*, 249–254 (1983).
[1869] Fazakas, J.; M. Hoenig, *Talanta 35/5*, 403–405 (1988).
[1870] Fazakas, J.; P.Gh. Zugravescu, *Spectrochim. Acta, Part B 43/8*, 897–900 (1988).
[1871] Fazakas, J.; P.Gh. Zugravescu, *Spectrosc. Lett. 24/4*, 525–538 (1991).
[1872] Felby, S.; *Forensic Sci. Int. 32/1*, 61–65 (1986).
[1873] Feldman, C.; *Anal. Chem. 33*, 1916–1920 (1961).
[1874] Feldman, C.; *Anal. Chem. 46/1*, 99–102 (1974).
[1875] Fell, G.S.; T.D.B. Lyon, in: R.F.M. Herber, M. Stoeppler (Editors), *Trace Element Analysis in Biological Specimens*, Elsevier, Amsterdam, 1994, 541–561.
[1876] Feng, B.; *Guangpuxue Yu Guangpu Fenxi 11/5*, 68 (1991).
[1877] Feng, Y.; K. Bencze, Ch. Pelikan; *Fresenius Z. Anal. Chem. 329/5*, 595–599 (1987).
[1878] Fergusson, J.E.; N.G. Purchase, *Environ. Pollut. 46/1*, 11–44 (1987).
[1879] Fernández C.; A.; R. Fernández, N. Carrión, D. Loreto, Z.A. de Benzo, R. Fraile, *At. Spectrosc. 12/4*, 111–118 (1991).
[1880] Fernandez, F.J.; D.C. Manning, *At. Absorpt. Newsletter 7/3*, 57–60 (1968).
[1881] Fernandez, F.J.; *At. Absorpt. Newsletter 8/4*, 90–91 (1969).
[1882] Fernandez, F.J.; D.C. Manning, J. Vollmer, *At. Absorpt. Newsletter 8/6*, 117–120 (1969).
[1883] Fernandez, F.J.; H.L. Kahn, *At. Absorpt. Newsletter 10/1*, 1–5 (1971).
[1884] Fernandez, F.J.; D.C. Manning, *At. Absorpt. Newsletter 10/4*, 86–88 (1971).
[1885] Fernandez, F.J.; *At. Absorpt. Newsletter 11/6*, 123–124 (1972).
[1886] Fernandez, F.J.; *At. Absorpt. Newsletter 12*, 70–72 (1973).
[1887] Fernandez, F.J.; *At. Absorpt. Newsletter 12*, 93–97 (1973).
[1888] Fernandez, F.J.; *Clin. Chem. (Winston-Salem) 21*, 558–561 (1975).
[1889] Fernandez, F.J.; J. Iannarone, *At. Absorpt. Newsletter 17/5*, 117–120 (1978).
[1890] Fernandez, F.J.; S.A. Myers, W. Slavin, *Anal. Chem. 52*, 741–746 (1980).
[1891] Fernandez, F.J.; M.M. Beaty, W.B. Barnett, *At. Spectrosc. 2/1*, 16–21 (1981).
[1892] Fernandez, F.J.; W. Bohler, M.M. Beaty, W.B. Barnett, *At. Spectrosc. 2/3*, 73–80 (1981).
[1893] Fernandez, F.J.; R. Giddings, *At. Spectrosc. 3/3*, 61–65 (1982).
[1894] Fernandez, F.J.; D. Hilligoss, *At. Spectrosc. 3/4*, 130–131 (1982).
[1895] Fernandez, F.J.; M.M. Beaty, *Spectrochim. Acta, Part B 39/2-3*, 519–523 (1984).
[1896] Fernández de la Campa, M.R.; E. Segovia García, M.C. Valdés-Hevia y Temprano, B. Aizpún Fernández, J.M. Marchante Gayón, A. Sanz-Medel, *Spectrochim. Acta, Part B 50/4-7*, 377–391 (1995).
[1897] Ferreira, A.M.R.; A.O.S.S. Rangel, J.L.F.C. Lima, *Commun. Soil. Sci. Plant Anal. 26/1-2*, 183–195 (1995).
[1898] Ferretti, M.; R. Udisti, E. Barbolani, *Fresenius J. Anal. Chem. 347/10*, 467–470 (1993).
[1899] Ferri, T.; E. Cardarelli, B.M. Petronio, *Talanta 36/4*, 513–517 (1989).
[1900] Ferris, A.P.; W.B. Jepson, R.C. Shapland, *Analyst (London) 95*, 574–578 (1970).

[1901]   Ferrús, R.; F. Torrades, *Anal. Chem. 60*, 1281–1285 (1988).
[1902]   Feuerbacher, H.; J. Böckler, in: B. Welz (Editor), *6. Colloquium Atomspektrometrische Spurenanalytik*, Bodenseewerk Perkin-Elmer, Überlingen, 1991, 647–653.
[1903]   Ficklin, W.H.; *Talanta 30/5*, 371–373 (1983).
[1904]   Ficklin, W.H.; *Talanta 37/8*, 831–834 (1990).
[1905]   Fidler, R.; A. Schöner, Analysis Europe 1/1, 39–42 (1994).
[1906]   Fieguth, P.; P. Hermann, *Staub. Reinhalt. Luft 43/9*, 362–366 (1983).
[1907]   Fio, J.L.; R. Fujii, *Soil Sci. Soc. Am. J. 54*, 363–369 (1990).
[1908]   Fiorino, J.A.; R.N. Kniseley, V.A. Fassel, *Spectrochim. Acta, Part B 23/6*, 413–425 (1968).
[1909]   Fiorino, J.A.; R.A. Moffitt, A.L. Woodson, R.J. Gajan, G.E. Huskey, R.G. Scholz, *J. Assoc. Off. Anal. Chem. 56/5*, 1246–1251 (1973).
[1910]   Fischer, R.; S. Rapsomanikis, M.O. Andreae, *Anal. Chem. 65/6*, 763–766 (1993).
[1911]   Fish, R.H.; F. Brinckman, K.L. Jewett, *Environ. Sci. Technol. 16/3*, 174–179 (1982).
[1912]   Fish, R.H.; J.J. Komlenic, *Anal. Chem. 56/3*, 510–517 (1984).
[1913]   Fish, R.H.; J.J. Komlenic, B.K. Wines, *Anal. Chem. 56*, 2452–2460 (1984).
[1914]   Fish, R.H.; J.G. Reynolds, *Trends Anal. Chem. (Pers. Ed.) 7/5*, 174–179 (1988).
[1915]   Fishbein, L.; *Int. J. Environ. Anal. Chem. 17*, 113–170 (1984).
[1916]   Fishman, M.J.; S.C. Downs, *US Geol. Surv. Water-Supply Papers 1540-C*, 1966.
[1917]   Fishman, M.J.; R. Spencer, *Anal. Chem. 49/11*, 1599–1602 (1977).
[1918]   Fitzgerald, W.F.; G.A. Gill, *Anal. Chem. 51/11*, 1714–1720 (1979).
[1919]   Flanagan, F.J.; R. Moore, P.J. Aruscavage, *Geostand. Newsl. 6*, 25–46 (1982).
[1920]   Flanjak, J.; *J. Sci. Food Agric. 33*, 579–583 (1982).
[1921]   Fleckenstein, J.; *Fresenius Z. Anal. Chem. 322/7*, 704–707 (1985).
[1922]   Fleet, B.; K.V. Liberty, T.S. West, *Talanta 17*, 203–210 (1970).
[1923]   Fleischer, M.; K.H. Schaller, R. Heinrich-Ramm, J. Angerer, in: J. Angerer, K.H. Schaller (Editors), *Analyses of Hazardous Substances in Biological Materials*, VCH Verlagsgesellschaft, Weinheim, 1988, 97–115.
[1924]   Fleischer, M.; K.H. Schaller, J. Angerer, in: B. Welz (Editor), *6. Colloquium Atomspektrometrische Spurenanalytik*, Bodenseewerk Perkin-Elmer, Überlingen, 1991, 851–860.
[1925]   Fleming, H.D.; *Spectrochim. Acta, Part B 23/3*, 207–212 (1967).
[1926]   Fleming, H.D.; E.G. Die, *Anal. Chim. Acta 83*, 67–82 (1976).
[1927]   Fleming, P.; *Appl. Spectrosc. 46/9*, 1400–1404 (1992).
[1928]   Fletcher, K.; S. Horsky, *J. Geochem. Explor. 30/1*, 29–34 (1988).
[1929]   Florence, T.M.; *Talanta 29/5*, 345–364 (1982).
[1930]   Florence, T.M.; G.M. Morrison, J.L. Stauber, *Sci. Total Environ. 125/1*, 1–13 (1992).
[1931]   Fonseca, R.W.; O.A. Güell, J.A. Holcombe, *Spectrochim. Acta, Part B 45/11*, 1257–1264 (1990).
[1932]   Fonseca, R.W.; O.A. Güell, J.A. Holcombe, *Spectrochim. Acta, Part B 47/4*, 573–576 (1992).
[1933]   Fonseca, R.W.; J. McNally, J.A. Holcombe, *Spectrochim. Acta, Part B 48/1*, 79–89 (1993).
[1933b]   Fonseca, R.W.; L.L. Pfefferkorn, J.A. Holcombe, *Spectrochim. Acta, Part B 49/12-14*, 1595–1608 (1994).
[1934]   Foote, J.W.; H.T. Delves, *Analyst (London) 107*, 121–124 (1982).
[1935]   Foote, J.W.; H.T. Delves, *Analyst (London) 108*, 492–504 (1983).
[1936]   Foote, J.W.; H.T. Delves, *Analyst (London) 109/6*, 709–711 (1984).
[1937]   Foote, J.W.; H.T. Delves, *Analyst (London) 113/6*, 911–915 (1988).
[1938]   Forehand, T.J.; A.E. Dupuy, H. Tai, *Anal. Chem. 48*, 999–1001 (1976).
[1939]   Forster, R.C.; A.G. Howard, *Anal. Proc. (London) 26/1*, 34–36 (1989).
[1940]   Förstner, U.; *Fresenius Z. Anal. Chem. 316/6*, 604–611 (1983).
[1941]   Förstner, U.; W. Calmano, *Vom Wasser 59*, 83–92 (1983).
[1942]   Forsyth, D.S.; *Sci. Total Environ. 89*, 299–304 (1989).
[1943]   Forsyth, D.S.; J.R. Iyengar, *Appl. Organomet. Chem. 3*, 211–218 (1989).
[1944]   Forsyth, D.S.; *Sci. Total Environ. 89*, 291–297 (1989).
[1945]   Forsyth, D.S.; R.W. Dabeka, C. Cléroux, *Appl. Organomet. Chem. 4*, 591–597 (1990).
[1946]   Forsyth, D.S.; R.W. Dabeka, C. Cléroux, *Food Addit. Contam. 8*, 477–484 (1991).
[1947]   Forsyth, D.S.; D. Weber, K. Dalglish, *J. AOAC International 75/6*, 964–973 (1992).
[1948]   Forsyth, D.S.; D. Weber, L. Barlow, *Appl. Organomet. Chem. 6*, 579–585 (1992).
[1949]   Foster, S.J.; R.J. Kraus, H.E. Ganther, *Arch. Biochem. Biophys. 251*, 77–86 (1986).
[1950]   Foster, W.H.; D.N. Hume, *Anal. Chem. 31/12*, 2033–2036 (1959).
[1951]   Fox, G.J.; *At. Spectrosc. 11/1*, 13–18 (1990).
[1952]   Fraile, R.; Z.A. de Benzo, M. Velosa, *Fresenius J. Anal. Chem. 343/3*, 319–323 (1992).

[1953] Frame, E.M.S.; J.A. King, D.A. Anderson, W.E. Balz, *Appl. Spectrosc. 47/8*, 1276–1282 (1993).
[1954] Franek, M.; V. Krivan, B. Gercken, J. Pavel, *Mikrochim. Acta (Vienna) 113*, 251–259 (1994).
[1955] Frankenberger, A.; R.R. Brooks, M. Hoashi, *Anal. Chim. Acta 246/2*, 359–363 (1991).
[1956] Frazer, A.; S.K. Secunda, J. Mendels, *Clin. Chim. Acta 36*, 499–509 (1972).
[1957] Frech, W.; *Talanta 21*, 565–571 (1974).
[1958] Frech, W.; G. Lundgren, S.-E. Lunner, *At. Absorpt. Newsletter 15/3*, 57–60 (1976).
[1959] Frech, W.; A. Cedergren, *Anal. Chim. Acta 82*, 83–92 (1976).
[1960] Frech, W.; A. Cedergren, *Anal. Chim. Acta 82*, 93–102 (1976).
[1961] Frech, W.; A. Cedergren, *Anal. Chim. Acta 88*, 57–67 (1977).
[1962] Frech, W.; J.-A. Persson, A. Cedergren, *Prog. Anal. Spectrosc. 3*, 279–297 (1980).
[1963] Frech, W.; E. Lundberg, M.M. Barbooti, *Anal. Chim. Acta 131/1*, 45–52 (1981).
[1964] Frech, W.; S. Jonsson, *Spectrochim. Acta, Part B 37*, 1021–1028 (1982).
[1965] Frech, W.; N.G. Zhou, E. Lundberg, *Spectrochim. Acta, Part B 37/8*, 691–702 (1982).
[1966] Frech, W.; A. Cedergren, E. Lundberg, D.D. Siemer, *Spectrochim. Acta, Part B 38/11*, 1435–1446 (1983).
[1967] Frech, W.; E. Lundberg, A. Cedergren, *Prog. Anal. Spectrosc. 8/3-4*, 257–370 (1985).
[1968] Frech, W.; J.M. Ottaway, L. Bezúr, J. Marshall, *Can. J. Spectrosc. 30/1*, 7–12 (1985).
[1969] Frech, W.; E. Lundberg, A. Cedergren, *Can. J. Spectrosc. 30/5*, 123–129 (1985).
[1970] Frech, W.; D.C. Baxter, B. Hütsch, *Anal. Chem. 58/9*, 1973–1977 (1986).
[1971] Frech, W.; D.C. Baxter, *Fresenius Z. Anal. Chem. 328/4-5*, 400–404 (1987).
[1972] Frech, W.; D.C. Baxter, E. Lundberg, *J. Anal. At. Spectrom. 3/1*, 21–25 (1988).
[1973] Frech, W.; M. Arshadi, D.C. Baxter, B. Hütsch, *J. Anal. At. Spectrom. 4/7*, 625–629 (1989).
[1974] Frech, W.; D.C. Baxter, *Spectrochim. Acta, Part B 45/8*, 867–886 (1990).
[1975] Frech, W.; B.V. L'vov, N.P. Romanova, *Spectrochim. Acta, Part B 47/13*, 1461–1469 (1992).
[1976] Frech, W.; A. Cedergren, in: M. Stoeppler (Editor), *Hazardous Metals in the Environment*, Techn. Instr. Anal. Chem. Vol 12, Elsevier, Amsterdam, 1992, 451–473.
[1977] Frech, W.; B.V. L'vov, *Spectrochim. Acta, Part B 48/11*, 1371–1379 (1993).
[1978] Frech, W.; D.C. Baxter, G. Dyvik, B. Dybdahl, *J. Anal. At. Spectrom. 10/10*, 769–775 (1995).
[1979] Frech, W.; *Fresenius J. Anal. Chem. 355/5-6*, 475–486 (1996).
[1980] Frechette, G.; J.C. Hebert, T.P. Thinh, R. Rousseau, *Anal. Chem. 51*, 957 (1979).
[1981] Freeman, G.H.C.; M. Outred, L.R. Morris, *Spectrochim. Acta, Part B 35/11-2*, 687–699 (1980).
[1982] Friberg, L.; C.G. Elinder, in: World Health Organization, *Environmental Health Criteria, Cadmium*, WHO, Geneva, 1992, 17.
[1983] Friend, M.T.; C.A. Smith, D. Wishart, *At. Absorpt. Newsletter 16/2*, 46–49 (1977).
[1984] Friese, K.C.; V. Krivan, *Anal. Chem. 67/2*, 354–359 (1995).
[1985] Friese, K.C.; V. Krivan, O. Schuierer, *Spectrochim. Acta, Part B 51/9-10*, 1223–1233 (1996).
[1986] Friese, K.H.; M. Roschig, G. Wuenscher, H. Matschiner, *Fresenius J. Anal. Chem. 337/8*, 860–866 (1990).
[1987] Frigerio, I.J.; M.J. McCormick, R.K. Symons, *Anal. Chim. Acta 143*, 261–264 (1982).
[1988] Frigge, Ch.; E. Jackwerth, *Spectrochim. Acta, Part B 47/6*, 787–791 (1992).
[1989] Frigge, Ch.; E. Jackwerth, *Anal. Chim. Acta 271/2*, 299–304 (1993).
[1990] Frigieri, P.; R. Trucco, *Analyst (London) 103*, 1089–1099 (1978).
[1991] Frigieri, P.; R. Trucco, I. Ciaccolini, G. Pampurini, *Analyst (London) 105*, 651–656 (1980).
[1992] Froment, D.H.; A.C. Alfrey, in: H.A. McKenzie, L.E. Smythe (Editors), *Quantitative trace analysis of biological materials*, Elsevier, Amsterdam - New York - Oxford, 1988, 633–647.
[1993] Froomes, P.R.A.; A.T. Wan, P.M. Harrison, A.J. McLean, *Clin. Chem. (Winston-Salem) 34/2*, 382–384 (1988).
[1994] Fry, R.C.; M.B. Denton, *Anal. Chem. 49*, 1413–1417 (1977).
[1995] Fry, R.C.; S.J. Northway, M.B. Denton, *Anal. Chem. 50/12*, 1719–1722 (1978).
[1996] Fry, R.C.; M.B. Denton, *Anal. Chem. 51/2*, 266–268 (1979).
[1997] Fry, R.C.; M.B. Denton, *Appl. Spectrosc. 33/4*, 393–399 (1979).
[1998] Fu, C.-g.; D.-j. Liu, G.-r. Zhang, *Guangpuxue Yu Guangpu Fenxi 13/6*, 87–90 (1993).
[1999] Fu, F.-F.; Z.-Z. Zhang, J.-D. Zheng, M.-W. Zhang, *Fenxi Huaxue 21/1*, 123 (1993).
[2000] Fuchs, C.; V.W. Armstrong, H. Hein, B. Kraft, *Nieren-Hochdruckkrankh. 12/5*, 164–169 (1983).
[2001] Fuente, M.A. de la ; M. Juárez, *Analyst (London) 120/1*, 107–111 (1995).
[2002] Fuente, M.A. de la ; G. Guerrero, M. Juárez, *At. Spectrosc. 16/5*, 219–223 (1995).
[2003] Fuente, M.A. de la ; G. Guerrero, M. Juárez, *J. Agri. Food Chem. 43/9*, 2406–2410 (1995).
[2004] Fuhrman, D.L.; *At. Absorpt. Newsletter 8/5*, 105 (1969).

[2005] Fujino, O.; M. Matsui, T. Nagahiro, Y. Nakaguchi, K. Hiraki, *Nippon Kagaku Kaishi (2)*, 153 (1988).
[2006] Fujino, O.; Y. Nishimura, Y. Koga, Y. Nakaguchi, T. Aono, K. Hiraki, *Kidorui 14*, 38 (1989).
[2007] Fujino, O.; S. Nishida, H. Togawa, K. Hiraki, *Anal. Sci. 7/6*, 889–892 (1991).
[2008] Fujita, K.; T. Takada, *Talanta 33/3*, 203–207 (1986).
[2009] Fujita, M.; E. Takabatake, Anal. Chem. 55/3, 454–457 (1983).
[2010] Fujiwara, K.; H. Haraguchi, K. Fuwa, *Anal. Chem. 47/9*, 1670–1673 (1975).
[2011] Fujiwara, K.; H. Haraguchi, K. Fuwa, *Bull. Chem. Soc. Jpn. 48/3*, 857–862 (1975).
[2012] Fukushima, M.; T. Ogata, K. Haraguchi, K. Nakagawa, S. Ito, M. Sumi, N. Asami, *J. Anal. At. Spectrom. 10/11*, 999–1002 (1995).
[2013] Fukushima, S.; *Mikrochim. Acta (Vienna) IV*, 596–618 (1959).
[2014] Fukuzaki, N.; Y. Ichikawa, *Bunseki Kagaku 33*, 178–182 (1984).
[2015] Fullana-Barceló, B.; F. Bosch-Serrat, R.M. Marin-Saez, A.R. Maurí-Aucejo, *Anal. Lett. 28/12*, 2247–2258 (1995).
[2016] Fuller, C.W.; *At. Absorpt. Newsletter 12/2*, 40–41 (1973).
[2017] Fuller, C.W.; J. Whitehead, *Anal. Chim. Acta 68*, 407–413 (1974).
[2018] Fuller, C.W.; *At. Absorpt. Newsletter 14/4*, 73–75 (1975).
[2019] Fuller, C.W.; *Analyst (London) 101*, 961–965 (1976).
[2020] Fuller, C.W.; I. Thompson, *Analyst (London) 102*, 141 (1977).
[2021] Fulton, A.; K.C. Thompson, T.S. West, *Anal. Chim. Acta 51*, 373–380 (1970).
[2022] Fung, Y.-S.; W.-C. Sham, *Analyst (London) 119/5*, 1029–1032 (1994).
[2023] Funk, W.; V. Dammann, C. Vonderheid, G. Oehlmann (Editors), *Statistische Methoden in der Wasseranalytik - Begriffe, Strategien, Anwendungen*, VCH Verlagsgesellschaft, Weinheim, 1985.
[2024] Funk, W.; V. Dammann, G. Donnevert, *Qualtätssicherung in der Analytischen Chemie*, VCH Verlagsgesellschaft Weinheim, 1992.
[2025] Furtmann, K.; D. Seifert, *Fresenius J. Anal. Chem. 338/1*, 73–74 (1990).
[2026] Furukawa, M.; E. Kamata, R. Nakashima, *Nagoya Kogyo Giyutsu Shikensho Hokoku 36/8-9*, 194 (1987).
[2027] Futekov, L.; R. Parichtkova, H. Specker, *Fresenius Z. Anal. Chem. 306/5*, 378–380 (1981).
[2028] Futekov, L.; G. Bekjarov, R. Paritschlova, *Fresenius Z. Anal. Chem. 315/1*, 12–19 (1983).
[2029] Fuwa, K.; B.L. Vallee, *Anal. Chem. 35*, 942–946 (1963).
[2030] Fuwa, K.; P. Pulido, R. McKay, B.L. Vallee, *Anal. Chem. 36*, 2407–2411 (1964).
[2031] Fuwa, K.; B.L. Vallee, *Spectrochim. Acta, Part B 35*, 657–661 (1980).
[2032] Fytianos, K.; V. Samanidou, *Sci. Total Environ. 92*, 265–268 (1990).
[2033] Favretto Gabrielli, L.; G.P. Marletta, L. Favretto, *At. Spectrosc. 1/1*, 35–37 (1980).
[2034] Favretto Gabrielli, L.; G.P. Marletta, L. Favretto, *Fresenius Z. Anal. Chem. 318*, 434–435 (1984).
[2035] Gailer, J.; K.J. Irgolic, *Appl. Organomet. Chem. 8*, 129–140 (1994).
[2036] Gaines, T.P.; J.W. West, J.F. McAllister, *J. Sci. Food Agric. 51*, 207–213 (1990).
[2037] Galan, L. de; R. Smith, J.D. Winefordner, *Spectrochim. Acta, Part B 23/8*, 521–525 (1968).
[2038] Galan, L. de; *Spectrochim. Acta, Part B 24/11*, 629–632 (1969).
[2039] Galan, L. de; G.F. Samaey, *Spectrochim. Acta, Part B 24/12*, 679–683 (1969).
[2040] Galan, L. de; G.F. Samaey, *Anal. Chim. Acta 50*, 39–50 (1970).
[2041] Galan, L. de; G.F. Samaey, *Spectrochim. Acta, Part B 25*, 245–259 (1970).
[2042] Galan, L. de; M.T.C. de Loos-Vollebregt, in: G.F. Kirkbright (Editor), *XXI CSI and 8th Conference on Atomic Spectroscopy*, Heyden & Sons, Cambridge, 1979, 64–85.
[2043] Galan, L. de; M.T.C. de Loos-Vollebregt, in: B. Welz (Editor), *Atomspektrometrische Spurenanalytik*, Verlag Chemie, Weinheim, 1982, 23–39.
[2044] Galan, L. de; M.T.C. de Loos-Vollebregt, R.A.M. Oosterling, *Analyst (London) 108*, 138–144 (1983).
[2045] Galan, L. de; M.T.C. de Loos-Vollebregt, *Spectrochim. Acta, Part B 39/8*, 1011–1019 (1984).
[2046] Galan, L. de; H.P.J. van Dalen, Guy R. Kornblum, *Analyst (London) 110/4*, 323–329 (1985).
[2047] Galgan, V.; A. Frank, in: P. Brätter, P. Schramel (Editors), *Trace Element Analytical Chemistry in Medicine and Biology*, W. de Gruyter, Berlin - New York, 1988, 84–89.
[2048] Gallego, M.; M. Valcárcel, *Anal. Chim. Acta 169*, 161–169 (1985).
[2049] Gallego, M.; M. Silva, M. Valcárcel, *Anal. Chem. 58/11*, 2265–2269 (1986).
[2050] Gallego, M.; M. Silva, M. Valcárcel, *Fresenius Z. Anal. Chem. 323/1*, 50–53 (1986).
[2051] Gallego, M.; M. Valcárcel, *Mikrochim. Acta (Vienna) III/4*, 163–168 (1991).
[2052] Gambrell, J.W.; *At. Absorpt. Newsletter 10/4*, 81–83 (1971).

[2053] Gamé, I.; L. Balabanoff, R. Valdebenito, L. Vivaldi, *Analyst (London) 111/10,* 1139–1141 (1986).
[2054] Gammelgaard, B.; O. Jøns, *J. Anal. At. Spectrom. 12/4,* 465–470 (1997).
[2055] Gandhi, M.N.; S.M. Khopkar, *Indian J. Chem. (Section A) 30A/8,* 706–710 (1991).
[2056] Gandhi, M.N.; S.M. Khopkar, *Chem. Anal. (Warsaw) 37/4,* 437–446 (1992).
[2057] Ganeyev, A.A.; S.E. Sholupov, *Spectrochim. Acta, Part B 47/11,* 1325–1338 (1992).
[2058] Ganje, T.J.; A.L. Page, *At. Absorpt. Newsletter 13/6,* 131–134 (1974).
[2059] Gao, S.B.; S.M. Gong, Z.T. Ji, *Lihua Jianyan, Huaxue Fence 29/4,* 214–215 (1993).
[2060] Gao, Y.-g.; B. Deng, *Spectrochim. Acta, Part B 51/9-10,* 1147–1153 (1996).
[2061] Gao, Z.-z.; Y.-q. Li, *Yejin Fenxi 12,* 18–21 (1992).
[2062] Gao, Z.-z.; Y.-q. Li, *Fenxi Shiyanshi 14/3,* 77 (1995).
[2063] Garbett, K.; G.I. Goodfellow, G.B. Marshall, *Anal. Chim. Acta 126,* 135–145 (1981).
[2064] Garbett, K.; G.I. Goodfellow, G.B. Marshall, *Anal. Chim. Acta 126,* 147–156 (1981).
[2065] Garbos, S.; M. Walcerz, E. Bulska, A. Hulanicki, *Spectrochim. Acta, Part B 50/13,* 1669–1677 (1995).
[2065b] Garbos, S.; E. Bulska, A. Hulanicki, N.I. Shcherbinia, E.M. Sedykh, *Anal. Chim. Acta 342,* 167-174 (1997).
[2066] Garcia-Anton, J.; J.L. Guinon, *Analusis 14/3,* 158–160 (1986).
[2067] García-Olalla, C.; A.J. Aller, *Anal. Chim. Acta 252/1-2,* 97–105 (1991).
[2068] Garcia Mateo, J.V.; J. Martínez Calatayud, *Anal. Chim. Acta 274/2,* 275–281 (1993).
[2069] Garcia Sanchez, E.; M. Hernandez Lopez, A. Arbaizar Ruiz de Dulanto, *Mikrochim. Acta (Vienna) 110,* 167–172 (1993).
[2070] Gardiner, J.; *Water Research 8,* 157 (1974).
[2071] Gardiner, P.E.; J.M. Ottaway, G.S. Fell, R.R. Burns, *Anal. Chim. Acta 124,* 281–294 (1981).
[2072] Gardiner, P.E.; E. Rösick, U. Rösick, P. Brätter, G. Kynast, *Clin. Chim. Acta 120,* 103–117 (1982).
[2073] Gardiner, P.E.; *Top. Curr. Chem. 141,* 145–174 (1987).
[2074] Gardiner, P.E.; M. Stoeppler, *J. Anal. At. Spectrom. 2/4,* 401–404 (1987).
[2075] Gardiner, P.E.; R. Schierl, K. Kreutzer, *Plant Soil 103,* 151 (1987).
[2076] Gardiner, P.E.; *J. Trace Elem. Electrol. Health Dis. 7,* 1–8 (1993).
[2077] Gardner, D.; *Mar. Pollut. Bull. 6,* 43 (1975).
[2078] Gardner, D.; *Anal. Chim. Acta 82,* 321–327 (1976).
[2079] Gardner, J.A.; S. Coleman, *Anal. Proc. (London) 30/7,* 292–295 (1993).
[2080] Gardner, M.J.; A.M. Gunn, *Fresenius Z. Anal. Chem. 330/2,* 103–106 (1988).
[2081] Gardner, M.J.; J.E. Ravenscroft, *Fresenius J. Anal. Chem. 354/5-6,* 602–605 (1996).
[2082] Garner, W.Y.; M.S. Barge, J.P. Ussary, *Good Laboratory Practice Standards. Application for Field and Laboratory Standards,* American Chemical Society, Washington, DC, 1992.
[2083] Garska, K.J.; *At. Absorpt. Newsletter 15/2,* 38–41 (1976).
[2084] Gasser, U.G.; W.J. Walker, R.A. Dahlgren, R.S. Borch, R.G. Burau, *Environ. Sci. Technol. 30/3,* 761–769 (1996).
[2085] Gatehouse, B.M.; A. Walsh, *Spectrochim. Acta 16/5,* 602–604 (1960).
[2086] Gatehouse, S.; D.W. Russell, J.C. Van Moort, *J. Geochem. Explor. 8,* 483–494 (1977).
[2087] Gaumer, M.W.; S. Sprague, W. Slavin, *At. Absorpt. Newsletter 5/3,* 58–61 (1966).
[2088] Gaydon, A.G.; H.G. Wolfhard, *Flames - Their Structure, Radiation and Temperature,* Chapman & Hall, London, 1970.
[2089] Geetha, R.; A. Thiruvengadasami, T.R. Mahalingam, *J. Anal. At. Spectrom. 4/5,* 447–450 (1989).
[2090] Gegiou, D.; M. Botsivali, *Analyst (London) 100,* 234–237 (1975).
[2091] Geiß, S.; J. Einax, K. Danzer, *Fresenius Z. Anal. Chem. 333/2,* 97–101 (1989).
[2092] Gelder, Z. van; *Spectrochim. Acta, Part B 25,* 669–681 (1970).
[2093] Gellert, G.; R. Wittassek, *Fresenius Z. Anal. Chem. 322/7,* 700–703 (1985).
[2094] Gelman, A.L.; *J. Sci. Food Agric. 23,* 299–305 (1972).
[2095] Genç, Ö.; S. Akman, A.R. Özdural, S. Ates, T. Balkis, *Spectrochim. Acta, Part B 36/3,* 163–168 (1981).
[2096] Gentner, W.; O. Müller, G.A. Wagner, N.H. Gale, *Naturwissenschaften 65/6,* 273–284 (1978).
[2097] Gerardi, M.; G.A. Pelliccia, *At. Spectrosc. 4/6,* 193–198 (1983).
[2098] Gerlach, W.; V. Krivan, K. Sprenger, in: B. Welz (Editor), *Fortschritte in der atom-spektrometrischen Spurenanalytik,* VCH Verlagsgesellschaft, Weinheim, 1986, 217–223.
[2099] Gerstenberger, S.L.; J. Pratt-Shelley, M.S. Beattie, J.A. Dellinger, *Bull. Environ. Contam. Toxicol. 50/4,* 612–617 (1993).
[2100] Gestring, W.D.; P.N. Soltanpour, *Commun. Soil. Sci. Plant Anal. 12/8,* 743–753 (1981).

[2101] Geugten, R.P. van der; *Fresenius Z. Anal. Chem. 306/1*, 13–14 (1981).
[2102] Ghazy, S.E.; M.A. Kabil, M. El-Defrawy, *Asian J. Chem. 6*, 1025 (1994).
[2103] Ghazy, S.E.; M.A. Kabil, *Analusis 23/3*, 117–119 (1995).
[2104] Giammarise, A.; *At. Absorpt. Newsletter 5/5*, 113–114 (1966).
[2105] Gibson, J.H.; W.E.L. Grossman, W.D. Cooke, *Anal. Chem. 35/3*, 266–277 (1963).
[2106] Giebenhain, G.; H.J. Rath, *GIT Fachz. Lab. 39/3*, 247–250 (1995).
[2107] Gijbels, R.; *ATB Metall. 30/4*, 91–98 (1990).
[2108] Gil, F.; M.L. Perez, A. Facio, E. Villaneuva, R. Tojo, A. Gil, *Clin. Chim. Acta 221/1-2*, 23–31 (1993).
[2109] Gilbert, T.R.; A.M. Clay, *Anal. Chim. Acta 67*, 289–295 (1973).
[2110] Gilchrist, G.F.R.; C.L. Chakrabarti, J.P. Byrne, M. Lamoureux, *J. Anal. At. Spectrom. 5/3*, 175–181 (1990).
[2111] Gill, G.A.; W.F. Fitzgerald, *Mar. Chem. 20*, 227–243 (1987).
[2112] Gillain, G.; in: H.A. McKenzie, L.E. Smythe (Editors*), Quantitative trace analysis of biological materials*, Elsevier, Amsterdam - New York - Oxford, 1988, 389–399.
[2113] Gilmutdinov, A.Kh.; I.S. Fishman, *Spectrochim. Acta, Part B 39/2-3*, 171–192 (1984).
[2114] Gilmutdinov, A.Kh.; Yu.A. Zakharov, V.P. Ivanov, *Zh. Anal. Khim. 43/7*, 1206–1213 (1988).
[2115] Gilmutdinov, A.Kh.; O.M. Yarkova, *Zh. Prikl. Spektrosk. 53/2*, 183–190 (1990); English translation see: *J. Appl. Spectros. 53*, 787–792 (1990).
[2116] Gilmutdinov, A.Kh.; Yu.A. Zakharov, A.V. Voloshin, V.P. Ivanov, *Zh. Prikl. Spektrosk. 53*, 359–364 (1990).
[2117] Gilmutdinov, A.Kh.; O.M. Shlyakhtina, *Spectrochim. Acta, Part B 46/8*, 1121–1141 (1991).
[2118] Gilmutdinov, A.Kh.; Yu.A. Zakharov, V.P. Ivanov, A.V. Voloshin, *J. Anal. At. Spectrom. 6/7*, 505–519 (1991).
[2119] Gilmutdinov, A.Kh.; Yu.A. Zakharov, V.P. Ivanov, A.V. Voloshin, K. Dittrich, *J. Anal. At. Spectrom. 7/4*, 675–683 (1992).
[2120] Gilmutdinov, A.Kh.; T.M. Abdullina, S.F. Gorbachev, V.L. Makarov, *Spectrochim. Acta, Part B 47/9*, 1075–1095 (1992).
[2121] Gilmutdinov, A.Kh.; C.L. Chakrabarti, J.C. Hutton, R.M. Mrasov, *J. Anal. At. Spectrom. 7/7*, 1047–1062 (1992).
[2122] Gilmutdinov, A.Kh.; Yu.A. Zakharov, A.V. Volosin, *J. Anal. At. Spectrom. 8/3*, 387–395 (1993).
[2123] Gilmutdinov, A.Kh.; R.M. Mrasov, A.R. Somov, C.L. Chakrabarti, J.C. Hutton, *Spectrochim. Acta, Part B 50/13*, 1637–1654 (1995).
[2124] Gilmutdinov, A.Kh.; B. Radziuk, M. Sperling, B. Welz, K.Yu. Nagulin, *Appl. Spectrosc. 49/4*, 413–424 (1995).
[2125] Gilmutdinov, A.Kh.; B. Radziuk, M. Sperling, B. Welz, K.Yu. Nagulin, *Appl. Spectrosc. 50/4*, 483–497 (1996).
[2126] Gilmutdinov, A.Kh.; B. Radziuk, M. Sperling, B. Welz, *Spectrochim. Acta, Part B 51*, 1023–1044 (1996).
[2127] Gilmutdinov, A.Kh.; M. Sperling, B. Welz, in: B. Welz (Editor), *CANAS'95 Colloquium Analytische Atomspektroskopie*, Bodenseewerk Perkin-Elmer, Überlingen, 1996, 51–60.
[2128] Gilmutdinov, A.Kh.; B. Radziuk, M. Sperling, B. Welz, K.Yu. Nagulin, *Spectrochim. Acta, Part B 51/9-10*, 931–940 (1996).
[2129] Giné, M.F., A.P. Packer, T. Blanco, B. Freire dos Reis, *Anal. Chim. Acta 323*, 47–53 (1996).
[2130] Giordano, R.; S. Costantini, I. Vernillo, A. Moramarco, V. Rasi, R. Giustolisi, C. Balacco Gabrieli, *Microchem. J. 46/2*, 184–190 (1992).
[2131] Giordano, R.; L. Ciarelli, G. Gattorta, M. Ciprotti, S. Costantini, *Microchem. J. 49/1*, 69–77 (1994).
[2132] Girard, L.; J. Hubert, *Talanta 43/11*, 1965–1974 (1996).
[2133] Girgis-Takla, P.; V. Valijanian, *Analyst (London) 107*, 378–384 (1982).
[2134] Girgis-Takla, P.; H.A. Mohammed, F. Fahmy, *Analyst (London) 112/12*, 1697–1699 (1987).
[2135] Giri, S.K.; D. Littlejohn, J.M. Ottaway, *Analyst (London) 107*, 1095–1098 (1982).
[2136] Gitelman, H.J.; F.R. Alderman, *J. Anal. At. Spectrom. 5/8*, 687–689 (1990).
[2137] Giumlia-Mair, A.R.; *Archaeometry 34/1*, 107–119 (1992).
[2138] Gladney, E.S.; K.E. Apt, *Anal. Chim. Acta 85*, 393–397 (1976).
[2139] Gladney, E.S.; W.E. Goode, *Anal. Chim. Acta 91*, 411–415 (1977).
[2140] Gladney, E.S.; *At. Absorpt. Newsletter 16/4*, 114–116 (1977).
[2141] Gleisberg, B.; *Isotopenpraxis 20/4*, 129–130 (1984).

816   Bibliography

[2142]   Gless, U.; Y. Schmitt, S. Ziegler, J.D. Kruse-Jarres, in: B. Welz (Editor), *6. Colloquium Atomspektrometrische Spurenanalytik*, Bodenseewerk Perkin-Elmer, Überlingen, 1991, 813–820.
[2143]   Glickstein, N.; *Mar. Pollut. Bull. 10*, 157 (1979).
[2144]   Gliksman, J.E.; J.E. Gibson, P.E. Kandetzki, *At. Spectrosc. 1/6*, 166–167 (1980).
[2145]   Gluodenis, T.J.; J.F. Tyson, *J. Anal. At. Spectrom. 7/2*, 301–306 (1992).
[2146]   Gluodenis, T.J.; J.F. Tyson, *J. Anal. At. Spectrom. 8/5*, 697–704 (1993).
[2147]   Goecke, R.F.; *Talanta 15*, 871 (1968).
[2148]   Goenaga Infante, H.; M.L. Fernández Sánchez, A. Sanz-Medel, *J. Anal. At. Spectrom. 11/8*, 571–575 (1996).
[2149]   Goguel, R.; *Spectrochim. Acta, Part B 26*, 313–330 (1971).
[2150]   Goguel, R.; *Anal. Chim. Acta 169*, 179–193 (1985).
[2151]   Gohda, S.; H. Yamazaki, T. Shigematsu, *Anal. Sci. 2/1*, 37–42 (1986).
[2152]   Goldberg, E.D.; *Pure Appl. Chem. 59/4*, 565–571 (1987).
[2153]   Goleb, J.A.; *Anal. Chem. 35*, 1978 (1963).
[2154]   Goleb, J.A.; J.K. Brody, *Anal. Chim. Acta 28/5*, 457–466 (1963).
[2155]   Goleb, J.A.; Y. Yokoyama, *Anal. Chim. Acta 30*, 213–222 (1964).
[2156]   Goleb, J.A.; *Anal. Chim. Acta 34*, 135–145 (1966).
[2157]   Goleb, J.A.; *Anal. Chim. Acta 36*, 130–131 (1966).
[2158]   Goleb, J.A.; C.R. Minkiff, *Appl. Spectrosc. 29/1*, 44–48 (1975).
[2159]   Goltz, D.M.; D.C. Grégoire, J.P. Byrne, C.L. Chakrabarti, *Spectrochim. Acta, Part B 50/8*, 803–814 (1995).
[2160]   Goltz, D.M.; C.L. Chakrabarti, R.E. Sturgeon, D.M. Hughes, D.C. Grégoire, *Appl. Spectrosc. 49/7*, 1006–1026 (1995).
[2161]   Gomez, A.; R. Lechber, P.L. Hermite, *Sampling Problems for the Chemical Analysis of Sludge, Soils and Plants*, Elsevier, London - New York, 1986.
[2162]   Gomez, M.T.; J.M. Mir, J.R. Castillo-Suarez, *At. Spectrosc. 9/1*, 46–47 (1988).
[2163]   Gómez Gómez, M.M.; M.M. Hidalgo García, M.A. Palacios Corvillo, *Analyst (London) 120/7*, 1911–1915 (1995).
[2164]   Gong, B.-l.; Y.-m. Liu, *At. Spectrosc. 11/6*, 229–232 (1990).
[2165]   Gong, B.-l.; H. Li, T. Ochiai, T.-z. Lin, K. Matsumoto, *Anal. Sci. 9/5*, 723–726 (1993).
[2166]   Gong, B.-l.; Y.-m. Liu, Z.-h. Li, T.-z. Lin, *Anal. Chim. Acta 304/1*, 115–120 (1995).
[2167]   Gong, B.-l.; Y.-m. Liu, Y.-l. Xu, Z.-h. Li, T.-z. Lin, *Talanta 42/10*, 1419–1423 (1995).
[2168]   Gong, B.-l.; T.-z. Lin, *Talanta 44/6*, 1003–1007 (1997).
[2169]   Gonon, L.; J.-M. Mermet, *Analusis 20/1*, 26M-29M (1992).
[2170]   González, E.B.; J.P. Parajón, J.I. García-Alonso, A. Sanz-Medel, *J. Anal. At. Spectrom. 4/2*, 175–179 (1989).
[2171]   Gonzalez, J.G.; R.T. Ross, *Anal. Lett. 5/10*, 683–694 (1972).
[2172]   González, M.C.; A.R. Rodriguez, V. Gonzalez, *Microchem. J. 35*, 94–106 (1987).
[2173]   Gonzalez Lafuente, J.M.; M.L. Fernandez Sanchez, A. Sanz-Medel, *J. Anal. At. Spectrom. 11*, 1163–1169 (1996).
[2174]   Goodfellow, G.I.; *Appl. Spectrosc. 21/7*, 39–42 (1967).
[2175]   Goodman, G.T.; T.M. Roberts, *Nature (London) 231*, 287–292 (1971).
[2176]   Goodnick, P.J.; R.R. Fieve, *Am. J. Psychiatry 142*, 761–762 (1985).
[2177]   Gosling, P.; *Ann. Clin. Biochem. 23*, 146–156 (1986).
[2178]   Goto, M.; T. Shibakawa, T. Arita, D. Ishii, *Anal. Chim. Acta 140*, 179–185 (1982).
[2179]   Goto, M.; S. Kumagai, D. Ishii, *Anal. Sci. 4/1*, 87–90 (1988).
[2180]   Goto, M.; E. Munaf, D. Ishii, *Fresenius Z. Anal. Chem. 332/6*, 745–749 (1988).
[2181]   Götz, R.; J.W. Elgersma, J.C. Kraak, H. Poppe, *Spectrochim. Acta, Part B 49/9*, 761–768 (1994).
[2182]   Gough, D.S.; P. Hannaford, A. Walsh, *Spectrochim. Acta, Part B 28*, 197–210 (1973).
[2183]   Gough, D.S.; *Anal. Chem. 48*, 1926–1931 (1976).
[2184]   Gough, D.S.; J.R. Meldrum, *Anal. Chem. 52/4*, 642–646 (1980).
[2185]   Gough, D.S.; J.V. Sullivan, *Anal. Chim. Acta 124*, 259–266 (1981).
[2186]   Gough, D.S.; P. Hannaford, R.M. Lowe, *Anal. Chem. 61/15*, 1652–1655 (1989).
[2187]   Goulden, P.D.; P. Brooksbank, *Anal. Chem. 46/11*, 1431–1436 (1974).
[2188]   Gouy, G.L.; *C.R. Acad. Sci. Paris 83*, 269–272 (1876).
[2189]   Govindaraju, K.; N. L'Homel, *At. Absorpt. Newsletter 11*, 115–117 (1972).
[2190]   Govindaraju, K.; G. Mevelle, C. Chouard, *Anal. Chem. 46/12*, 1672–1675 (1974).
[2191]   Goyal, N.; P.J. Purohit, A.R. Dhobale, B.M. Patel, A.G. Page, M.D. Sastry, *J. Anal. At. Spectrom. 2*, 459–461 (1987).

[2192] Goyal, N.; P.J. Purohit, A.R. Dhobale, A.G. Page, M.D. Sastry, *Fresenius Z. Anal. Chem. 330/2*, 114–115 (1988).

[2193] Goyal, N.; P.J. Purohit, A.G. Page, M.D. Sastry, *Fresenius J. Anal. Chem. 354/3*, 311–315 (1996).

[2194] Grabinski, A A.; *Anal. Chem. 53/7*, 966–968 (1981).

[2195] Granadillo, V.A.; R.A. Romero, J.A. Navarro, *Anal. Sci. 7/Suppl*, 1189–1192 (1991).

[2196] Granadillo, V.A.; J.A. Navarro, R.A. Romero, *Invest. Clin. 32/1*, 27–39 (1991).

[2197] Granadillo, V.A.; R.A. Romero, *Ciencia 1/1*, 55–63 (1993).

[2198] Granadillo, V.A.; R.A. Romero, in: M. Anke, D. Meissner, C.F. Mills (Editors), *Trace Elements in Man and Animals, TEMA-8*, Media Touristik, 1993, 86–89.

[2199] Granadillo, V.A.; H.S. Cubillán, J.M. Sánchez, J.E. Tahán, E.S. Márquez, R.A. Romero, *Anal. Chim. Acta 306/1*, 139–147 (1995).

[2200] Granadillo, V.A.; O. Salgado, L.Ch. Barrios, R.A. Romero, *Trace Elem. Electrolytes 12/2*, 76–80 (1995).

[2201] Gräser, K.; K. Staiger, *Z. gesamte Hyg. Ihre Grenzgeb. 29/12,* 734–737 (1983).

[2202] Gräser, K.; K. Staiger, *Z. gesamte Hyg. Ihre Grenzgeb. 29,12,* 737–739 (1983).

[2203] Grasserbauer, M.; S. Paleczek, J. Rendl, A. Kasper, H. Puxbaum, *Fresenius J. Anal. Chem. 350/7-8,* 431–439 (1994).

[2204] Grasshoff, K.; M. Ehrhardt, K. Kremling, *Methods of Seawater Analysis*, Verlag Chemie, Weinheim, 1983.

[2205] Grasso, G.; G. Bufalo, *At. Spectrosc. 7/4,* 93–95 (1986).

[2206] Grasso, G.; G. Bufalo, At. Spectrosc. 9/3, 84–86 (1988).

[2207] Graule, T.; A. von Bohlen, J.A.C. Broekaert, E. Grallath, R. Klockenkämper, P. Tschöpel, G. Tölg, *Fresenius Z. Anal. Chem. 335/7*, 637–642 (1989).

[2208] Greaves, M.C.; *Nature (London) 199*, 552 (1963).

[2209] Greeder, G.A.; J.A. Milner, *Science 209*, 825 (1980).

[2210] Green, H.C.; *Analyst (London) 100*, 640–642 (1975).

[2211] Greenberg, R.R.; H.M. Kingston, R.L. Watters, K.W. Pratt, *Fresenius J. Anal. Chem. 338/4*, 394–398 (1990).

[2212] Greenland, L.P.; E.Y. Campbell, *Anal. Chim. Acta 60*, 159 (1972).

[2213] Greenway, G.M.; S.N. Nelms, I. Skhosana, S.J.L. Dolman, *Spectrochim. Acta, Part B 51/14*, 1909–1915 (1996).

[2214] Grégoire, D.C.; C.L. Chakrabarti, *Anal. Chem. 49*, 2018–2023 (1977).

[2215] Grégoire, D.C.; C.L. Chakrabarti, *Spectrochim. Acta, Part B 37/7*, 611–623 (1982).

[2216] Grégoire, D.C.; M.M. Lamoureux, C.L. Chakrabarti, S. Al-Maawali, J.P. Byrne, *J. Anal. At. Spectrom. 7/4*, 579–585 (1992).

[2217] Gretzinger, K.; L. Kotz, P. Tschöpel, G. Tölg, *Talanta 29/11B*, 1011–1018 (1982).

[2218] Grey, P.; *Analyst (London) 115/2*, 159–165 (1990).

[2219] Griebenow, W.; B. Werthman, B. Schwarz, *Papier (Darmstadt) 39/3*, 105–109 (1985).

[2220] Griepink, B.; *Pure Appl. Chem. 56/10*, 1477–1498 (1984).

[2221] Griffin, A.C.; in: *Adv. Cancer Res. 29*, 419–441 (1979).

[2222] Griffin, H.R.; M.B. Hocking, D.G. Lowery, *Anal. Chem. 47*, 229–233 (1975).

[2223] Grimaldi, F.S.; M.M. Schnepfe, *Talanta 17*, 617–621 (1970).

[2224] Grinshtein, I.L.; D.A. Katskov, M.A. Khodorkovskii, *J. Appl. Spectrosc. (USSR) 44*, 439–444 (1986).

[2225] Grobecker, K.H.; B. Klüßendorf, *Fresenius Z. Anal. Chem. 322/7,* 673–676 (1985).

[2226] Grobenski, Z.; *Fresenius Z. Anal. Chem. 289*, 337–345 (1978).

[2227] Grobenski, Z.; R. Lehmann, R. Tamm, B. Welz, *Mikrochim. Acta (Vienna) 1/1-2*, 115–125 (1982).

[2228] Grobenski, Z.; D. Weber, B. Welz, J. Wolff, *Analyst (London) 108*, 925–932 (1983).

[2229] Grobenski, Z.; R. Lehmann, *At. Spectrosc. 4/3*, 111–112 (1983).

[2230] Grobenski, Z.; R. Lehmann, B. Radziuk, U. Völlkopf, *At. Spectrosc. 5/3*, 87–90 (1984).

[2231] Grognard, M.; *At. Spectrosc. 8/5*, 153–154 (1987).

[2232] Groll, H.; C. Schnürer-Patschan, A. Zybin, Y. Kuritzin, K, Niemax, *AIP Conf. Proc. 329*, 495–498 (1995).

[2233] Grosser, Z.A.; W.B. Barnett, *Spectroscopy (Eugene, OR) 1*, 58 (1986).

[2234] Grosser, Z.A.; *At. Spectrosc. 17/6*, 229–231 (1996).

[2235] Grossklaus, R.; I. Knoechel-Schiffer, in: J. Schaub (Editor), *Composition and Physiological Properties of Human Milk*, Elsevier, Amsterdam - New York - Oxford, 1985, 33–46.

[2236] Großmann, O.; E. Müller, *Fresenius Z. Anal. Chem. 308/4*, 327–331 (1981).

[2237] Großmann, O.; *Anal. Chim. Acta 203/1*, 55–66 (1987).

[2238]   Großmann, O.; E. Müller, in: B. Welz (Editor), *6. Colloquium Atomspektrometrische Spurenanalytik*, Bodenseewerk Perkin-Elmer, Überlingen, 1991, 691–697.

[2239]   Grotti, M.; E. Magi, R. Leardi, *Anal. Chim. Acta 327/3*, 47–51 (1996).

[2240]   Gruener, N.; O. Gozlan, T. Goldstein, J. Davis, T. Besner, T.C. Iancu, *Clin. Chem. (Winston-Salem) 37/2*, 263–265 (1991).

[2241]   Grüner, K.; *Neue Hütte 36/6*, 236–238 (1991).

[2242]   Grunbaum, B.W.; N. Pace, *Microchem. J. 15*, 666–672 (1970).

[2243]   Gu, X.; L.-H. Wang, *Zhongguo Yaoke Daxue Xuebao 23*, 373 (1992).

[2244]   Guadagnino, E.; M. Verita, C. Furlani, G. Poizonetti, *Glastechn Ber. 64/7*, 179–184 (1991).

[2245]   Guadagnino, E.; G.C. De Diana, B.M. Scalet, M.L. Scandellari, *Glass Technol. 33/6*, 209–213 (1992).

[2246]   Guadagnino, E.; G.C. De Diana, G. Rizzo, *Glastechn Ber. 66/4*, 100–104 (1993).

[2247]   Guardia, M. de La ; V. Carbonell, A. Morales-Rubio, A. Salvador, *Talanta 40/11*, 1609–1617 (1993).

[2248]   Guardia, M. de La ; M.J. Sanchez, *At. Spectrosc. 3/1*, 36–38 (1982).

[2249]   Guardia, M. de La ; M.J. Lizondo, *At. Spectrosc. 4/6*, 208–211 (1983).

[2250]   Guardia, M. de La ; A. Salvador, *At. Spectrosc. 5/4*, 150–155 (1984).

[2251]   Guardia, M. de La ; A.R. Mauri, C. Mongay, *J. Anal. At. Spectrom. 3/7*, 1035–1038 (1988).

[2252]   Guardia, M. de La ; A. Salvador, J.L. Burguera, M. Burguera, *J. Flow Injection Anal. 5/2*, 121–131 (1988).

[2253]   Guardia, M. de La ; V. Carbonell, A. Morales-Rubio, A. Salvador, *Fresenius Z. Anal. Chem. 335/8*, 975–979 (1989).

[2254]   Guardia, M. de La ; A. Salvador, *Analusis 19/6*, 52M-56M (1991).

[2255]   Guardia, M. de La ; A. Morales-Rubio, V. Carbonell, A. Salvador, J.L. Burguera, *Fresenius J. Anal. Chem. 345/8-9*, 579–584 (1993).

[2256]   Güçer, S.; H. Massmann, *XVII CSI Florence, Rep. 164*, 1973.

[2257]   Güçer, S.; M. Yaman, *J. Anal. At. Spectrom. 7/2*, 179–182 (1992).

[2258]   Gudzinowicz, B.J.; V.J. Luciano, *J. Assoc. Off. Anal. Chem. 49*, 1–8 (1966).

[2259]   Güell, O.A.; J.A. Holcombe, *Spectrochim. Acta, Part B 43/4-5*, 459–480 (1988).

[2260]   Güell, O.A.; J.A. Holcombe, *Spectrochim. Acta, Part B 44/2*, 185–196 (1989).

[2261]   Güell, O.A.; J.A. Holcombe, *Anal. Chem. 62/9*, 529A-542A (1990).

[2262]   Güell, O.A.; J.A. Holcombe, *Appl. Spectrosc. 45/7*, 1171–1176 (1991).

[2263]   Güell, O.A.; J.A. Holcombe, *J. Anal. At. Spectrom. 7/2*, 135–140 (1992).

[2264]   Güell, O.A.; J.A. Holcombe, *Spectrochim. Acta, Part B 47/14*, 1535–1544 (1992).

[2265]   Güell, O.A.; J.A. Holcombe, C.J. Rademeyer, *Anal. Chem. 65/6*, 748–751 (1993).

[2266]   Guerra, R.; *At. Spectrosc. 1/2*, 58–59 (1980).

[2267]   Guest, R.L.; H. Blutstein, *Anal. Chem. 53*, 727–731 (1981).

[2268]   Guevremont, R.; R.E. Sturgeon, S.S. Berman, *Anal. Chim. Acta 115*, 163–170 (1980).

[2269]   Guillard, O.; A. Piriou, P. Mura, D. Reiss, *Clin. Chem. (Winston-Salem) 28/7*, 1714–1715 (1982).

[2270]   Guillard, O.; K. Tiphaneau, D. Reiss, A. Piriou, *Anal. Lett. 17/B14*, 1593–1605 (1984).

[2271]   Gulmini, M.; G. Ostacoli, V. Zelano, *Analyst (London) 119/9*, 2075–2080 (1994).

[2272]   Gunshin, H.; M. Yoshikawa, T. Doudou, N. Kato, *Agric. Biol. Chem. 49/1*, 21–26 (1985).

[2273]   Günther, H.; T. Boeck, C. Franck, K. Jacobs, *Fresenius J. Anal. Chem. 343*, 756–759 (1992).

[2274]   Günther, K.; F. Umland, *Fresenius Z. Anal. Chem. 331/3-4*, 302–309 (1988).

[2275]   Günther, K.; H. Waldner, *Anal. Chim. Acta 259/1*, 165–173 (1992).

[2276]   Gunz, L.D.; E. Schnell, *Mikrochim. Acta (Vienna) III,5-6*, 409–416 (1984).

[2277]   Gunz, L.D.; E. Schnell, *Mikrochim. Acta (Vienna) III/1-2*, 125–133 (1983).

[2278]   Günzler, H.; (Editor), *Akkreditierung und Qualitätssicherung in der Analytischen Chemie*, Springer, Berlin, 1994.

[2279]   Guo, J.-h.; T. Fang, Z.-l. Xu, Y.-w. Chen, *Guangpuxue Yu Guangpu Fenxi 13/3*, 57–61 (1993).

[2280]   Guo, T.-z.; W. Erler, H. Schulze, S. McIntosh, *At. Spectrosc. 11/1*, 24–28 (1990).

[2281]   Guo, T.-z.; J. Baasner, *Anal. Chim. Acta 278/1*, 189–196 (1993).

[2282]   Guo, T.-z.; J. Baasner, *Talanta 40/12*, 1927–1936 (1993).

[2283]   Guo, T.-z.; J. Baasner, M. Gradl, A. Kistner, *Anal. Chim. Acta 320/2-3*, 171–176 (1996).

[2284]   Guo, T.-z.; J. Baasner, *J. Autom. Chem. 18/6*, 217–220 (1996).

[2285]   Guo, T.-z.; J. Baasner, *J. Autom. Chem. 18/6*, 221–223 (1996).

[2286]   Guo, T.-z.; J. Baasner, S. McIntosh, *Anal. Chim. Acta 331/3*, 263–270 (1996).

[2287]   Guo, X.; R.R. Brooks, *Anal. Chim. Acta 228/1*, 139–143 (1990).

[2288]   Guo, X.; M. Hoashi, R.R. Brooks, R.D. Reeves, *Anal. Chim. Acta 266/1*, 127–131 (1992).

[2289] Guo, X.-w.; X.-m. Guo, *Anal. Chim. Acta 330/2-3*, 237–243 (1996).
[2290] Gupta, H.K.L.; F.J. Amore, D.F. Boltz, *At. Absorpt. Newsletter 7*, 107–109 (1968).
[2291] Gustavsson, A.G.T.; *Spectrochim. Acta, Part B 39/1*, 85–94 (1984).
[2292] Gustavsson, A.G.T.; *Spectrochim. Acta, Part B 39/5*, 743–746 (1984).
[2293] Gustavsson, A.G.T.; *Anal. Chem. 56/7*, 815–817 (1984).
[2294] Gustavsson, A.G.T.; O. Nygren, *Spectrochim. Acta, Part B 42/6*, 883–888 (1987).
[2295] Guthrie, B.E.; W.R. Wolf, C. Veillon, *Anal. Chem. 50/13*, 1900–1902 (1978).
[2296] Gutiérrez, J.; H. Travieso, M.A. Pubillones, *Water, Air, Soil Pollut. 68*, 315–323 (1993).
[2297] Guttenberger, J.; M. Marold, *Fresenius Z. Anal. Chem. 262*, 102–103 (1972).
[2298] Gy, P.M.; *Sampling of Particulate Materials: Theory and Practice,* Elsevier, New York, 1982.
[2299] Gy, P.M.; *Analusis 11/9*, 413–440 (1983).
[2300] Gy, P.M.; *Anal. Chim. Acta 190,* 13–23 (1986).
[2301] Gy, P.M.; *Mikrochim. Acta (Vienna) II/1-6*, 457–466 (1991).
[2302] Gy, P.M.; *Sampling of Heterogeneous and Dynamic Material Systems*, Elsevier, Amsterdam, 1992.
[2303] Haan, K.E.C. De ; C.J. De Groot, H. Boxma, C.J.A. Van den Hamer, *Clin. Chim. Acta 170*, 111–112 (1987).
[2304] Haan, K.E.C. De ; U.D. Woroniecka, *Clin. Chem. (Winston-Salem) 35*, 888 (1989).
[2305] Habicht, J.; T. Prohaska, G. Friedbacher, M. Grasserbauer, H.M. Ortner, *Spectrochim. Acta, Part B 50/8*, 713–723 (1995).
[2306] Hadeishi, T.; R.D. McLaughlin, *Science 174*, 404–407 (1971).
[2307] Hadeishi, T.; R.D. McLaughlin, *Appl. Phys. Lett. 21*, 438–440 (1972).
[2308] Hadeishi, T.; D.A. Church, R.D. McLaughlin, B.D. Zak, M. Nakamura, B. Chang, *Science 187*, 348–349 (1975).
[2309] Hadeishi, T.; R.D. McLaughlin, *Anal. Chem. 48/7*, 1009–1011 (1976).
[2310] Hadeishi, T.; *U.S. Pat. 4,263,533,* 1981.
[2311] Hadeishi, T.; T. Le Vay, *Fresenius J. Anal. Chem. 337/3*, 264–270 (1990).
[2312] Hadgu, N.; A. Ohlsson, W. Frech, *Spectrochim. Acta, Part B 50/9*, 1077–1093 (1995).
[2313] Hadgu, N.; A. Ohlsson, W. Frech, *Spectrochim. Acta, Part B 51/9-10*, 1081–1092 (1996).
[2314] Haelen, P.; G. Cooper, C. Pampel, *At. Absorpt. Newsletter 13*, 1–3 (1974).
[2315] Hahn, E.; K. Hahn, H. Ellenberg, *Verh.-Ges. Ökol. 18*, 317 (1988).
[2316] Hahn, E.; K. Hahn, C. Mohl, M. Stoeppler, *Fresenius J. Anal. Chem. 337/3*, 306–309 (1990).
[2317] Hahn, L.; W. Wendl, G. Müller-Vogt, in: B. Welz (Editor), *5. Colloquium Atomspektrometrische Spurenanalytik*, Bodenseewerk Perkin-Elmer, Überlingen, 1989, 224–234.
[2318] Hahn, L.; G. Müller-Vogt, W. Wendl, *J. Anal. At. Spectrom. 8/2*, 223–227 (1993).
[2319] Hahn, M.H.; R.W. Kuennen, J. Caruso, F.L. Fricke, *J. Agri. Food Chem. 29/4*, 792–796 (1981).
[2320] Hahn, R.; M. Ikramuddin, *At. Spectrosc. 6/3*, 77–78 (1985).
[2321] Haj-Hussein, A.T.; G.D. Christian, J. Ruzicka, *Anal. Chem. 58/1*, 38–42 (1986).
[2322] Hakkala, E.; L. Pyy, *J. Anal. At. Spectrom. 7/2*, 191–196 (1992).
[2323] Haldimann, M.; T.Y. Venner, B. Zimmerli, *J. Trace Elem. Med. Biol. 10/1*, 31–45 (1996).
[2324] Hall, A.; M.C. Godinho, *Anal. Chim. Acta 113*, 369–373 (1980).
[2325] Hall, A.; *Chem. Geol. 30/1-2*, 135–142 (1980).
[2326] Hall, G.E.M.; J.E. Vaive, S.B. Ballantyne, *J. Geochem. Explor. 26*, 191–202 (1986).
[2327] Hall, G.E.M.; J.E. Vaive, J.A. Coope, E.F. Weiland, *J. Geochem. Explor. 34*, 157–171 (1989).
[2328] Hall, G.E.M.; J.-C. Pelchat, C.E. Dunn, *J. Geochem. Explor. 37/1*, 1–23 (1990).
[2329] Hall, G.E.M.; *Explore 68*, 18–20 (1990).
[2330] Hall, G.E.M.; G. Gauthier, J.-C. Pelchat, P. Pelchat, J.E. Vaive, *J. Anal. At. Spectrom. 11*, 787–796 (1996).
[2331] Hall, L.W.; *Mar. Pollut. Bull. 19*, 431 (1988).
[2332] Halls, D.J.; *Spectrochim. Acta, Part B 32/5-6*, 221–230 (1977).
[2333] Halls, D.J.; *Spectrochim. Acta, Part B 32/9-10*, 397–412 (1977).
[2334] Halls, D.J.; G.S. Fell, *Anal. Chim. Acta 129*, 205–211 (1981).
[2335] Halls, D.J.; *Analyst (London) 109/8*, 1081–1084 (1984).
[2336] Halls, D.J.; G.S. Fell, *J. Anal. At. Spectrom. 1/2*, 135–139 (1986).
[2337] Halls, D.J.; C. Mohl, M. Stoeppler, *Analyst (London) 112*, 185–189 (1987).
[2338] Halls, D.J.; M.M. Black, G.S. Fell, J.M. Ottaway, *J. Anal. At. Spectrom. 2/3*, 305–309 (1987).
[2339] Halmos, P.; E. Gegus, J. Borszeki, *Magy. Kem. Foly. 99*, 420 (1993).
[2340] Ham, N.S.; J.B. Willis, *Spectrochim. Acta, Part B 40/10-2*, 1607–1629 (1985).
[2341] Ham, N.S.; T. McAllister, *Spectrochim. Acta, Part B 43/6-7*, 789–797 (1988).

[2342] Hambly, A.N.; C.S. Rann, in: J.A. Dean, T.C. Rains (Editors), *Flame Emission and Atomic Absorption Spectrometry, Vol. 1 Theory*, Marcel Dekker, New York - London, 1969, 241–265.

[2343] Hambrick, G.A.; P.N. Froelich, M.O. Andreae, B.L. Lewis, *Anal. Chem. 56/3*, 421–424 (1984).

[2344] Hambsch, B.; B. Raue, H.-J. Brauch, *Acta Hydrochim. Hydrobiol. 23/4*, 166–172 (1995).

[2345] Hamid, H.A.; *At. Spectrosc. 10/1*, 16–18 (1989).

[2346] Hamlin, S.N.; *Water Resour. Bull. 25/2*, 255–262 (1989).

[2347] Hammar, H.E.; N.R. Page, *At. Absorpt. Newsletter 6/2*, 33–34 (1967).

[2348] Hams, G.A.; *Clin. Chem. (Winston-Salem) 33/5*, 719–720 (1987).

[2349] Hams, G.A.; J.K. Fabri, *Clin. Chem. (Winston-Salem) 34/6*, 1121–1123 (1988).

[2350] Han, H.-b.; G. Kaiser, G. Tölg, *Anal. Chim. Acta 128*, 9–21 (1981).

[2351] Han, H.-b.; Y.-b. Liu, S.-f. Mou, Z.-m. Ni, *J. Anal. At. Spectrom. 8/8*, 1085–1090 (1993).

[2352] Han, J.S.; J.H. Weber, *Anal. Chem. 60/4*, 316–319 (1988).

[2353] Han, W.-m.; *Huanjing Wuran Yu Fangzhi 9/6*, 26 (1987).

[2354] Hanna, C.P.; J.F. Tyson, S.G. Offley, *Spectrochim. Acta, Part B 47/9*, 1065–1073 (1992).

[2355] Hanna, C.P.; J.F. Tyson, S.A. McIntosh, *Anal. Chem. 65/5*, 653–656 (1993).

[2356] Hanna, C.P.; J.F. Tyson, S.A. McIntosh, *Clin. Chem. (Winston-Salem) 39/8*, 1662–1667 (1993).

[2357] Hanna, C.P.; G.R. Carnrick, S.A. McIntosh, L.C. Guyette, D.E. Bergemann, *At. Spectrosc. 16/2*, 82–85 (1995).

[2358] Hanna, C.P.; S.A. McIntosh, *At. Spectrosc. 16/3*, 106–114 (1995).

[2359] Hannaford, P.; R.M. Lowe, *Anal. Chem. 49*, 1852–1857 (1977).

[2360] Hannaford, P.; A. Walsh, *Spectrochim. Acta, Part B 43/9-11*, 1053–1068 (1988).

[2361] Hannaford, P.; *Spectrochim. Acta, Part B 49/12-4*, 1581–1593 (1994).

[2362] Hannaker, P.; Q.-l. Hou, *Talanta 31/12*, 1153–1157 (1984).

[2363] Hänni, E.; R.C. Daniel, *Mitt. Geb. Lebensmittelunters. Hyg. 73/1*, 94–105 (1982).

[2364] Hansen, E.H.; in: B. Welz (Editor), *5. Colloquium Atomspektrometrische Spurenanalytik*, Bodenseewerk Perkin-Elmer, Überlingen, 1989, 367–374.

[2365] Hansen, E.H.; *Anal. Chim. Acta 261/1-2*, 125–136 (1992).

[2366] Hansen, J.C.; N. Kromann, N.C. Wulf, K. Albøge, *Sci. Total Environ. 38,* 33–40 (1984).

[2367] Hansen, L.D.; G.L. Fisher, *Environ. Sci. Technol. 14*, 1111–1117 (1980).

[2368] Hansson, H.-C.; A.-K. Ekholm, H.B. Ross, *Environ. Sci. Technol. 22/5*, 527–531 (1988).

[2369] Hansson, L.; J. Pettersson, Å. Olin, *Talanta 34/10*, 829–833 (1987).

[2370] Hansson, L.; J. Pettersson, Å. Olin, *Analyst (London) 114/4*, 527–528 (1989).

[2371] Hansson, L.; J. Pettersson, L. Eriksson, Å. Olin, *Clin. Chem. (Winston-Salem) 35/4*, 537–540 (1989).

[2372] Haraguchi, H.; J. Takahashi, K. Tanabe, K. Fuwa, *Spectrochim. Acta, Part B 36/7*, 719–726 (1981).

[2373] Haraguchi, H.; T. Akagi, in: S.J. Haswell (Editor), *Atomic Absorption Spectrometry - Theory, Design and Applications*, Elsevier, Amsterdam - New York, 1991, 125–157.

[2374] Haraldsen, L.C.; M.A.B. Pougnet, *Analyst (London) 114/10*, 1331–1333 (1989).

[2375] Hardman, D.; A.A. Verbeek, *S. Afr. J. Chem. 38/3*, 145 (1985).

[2376] Hareland, W.A.; E.R. Ebersole, T.P. Ramachandran, *Anal. Chem. 44*, 520–523 (1972).

[2377] Harff, G.A.; W.T. Helversteijn, *Clin. Chem. (Winston-Salem) 33/7*, 1262 (1987).

[2378] Harms, U.; J. Kunze, *Z. Lebensm. Unters. Forsch. 164/3*, 204–207 (1977).

[2379] Harms, U.; *Z. Lebensm. Unters. Forsch. 172*, 118–122 (1981).

[2380] Harms, U.; B. Luckas, W. Lorenzen, A. Montag, *Fresenius Z. Anal. Chem. 316/6*, 600–603 (1983).

[2381] Harms, U.; B. Luckas, in: B. Welz (Editor), *Fortschritte in der atomspektrometrischen Spurenanalytik*, Verlag Chemie, Weinheim, 1984, 421–429.

[2382] Harms, U.; *Fresenius Z. Anal. Chem. 322/1*, 53–56 (1985).

[2383] Harms, U.; in: B. Welz (Editor), *Fortschritte in der atomspektrometrischen Spurenanalytik*, VCH Verlagsgesellschaft, Weinheim, 1986, 479–486.

[2384] Harms, U.; in: B. Welz (Editor), *5. Colloquium Atomspektrometrische Spurenanalytik*, Bodenseewerk Perkin-Elmer, Überlingen, 1989, 737–746.

[2385] Harms, U.; *Mikrochim. Acta (Vienna) 109/1-4,* 131–132 (1992).

[2386] Harnly, J.M.; T.C. O'Haver, *Anal. Chem. 53/8*, 1291–1298 (1981).

[2387] Harnly, J.M.; J.S. Kane, *Anal. Chem. 56/1*, 48–54 (1984).

[2388] Harnly, J.M.; J.A. Holcombe, *Anal. Chem. 57/9,* 1983–1986 (1985).

[2389] Harnly, J.M.; *Anal. Chem. 58/8*, 933A–943A (1986).

[2390] Harnly, J.M.; J.A. Holcombe, *J. Anal. At. Spectrom. 2/1*, 105–113 (1987).
[2391] Harnly, J.M.; *J. Anal. At. Spectrom.8/2*, 317–324 (1993).
[2392] Harnly, J.M.; B. Radziuk, *J. Anal. At. Spectrom. 10/3*, 197–206 (1995).
[2393] Harnly, J.M.; *Fresenius J. Anal. Chem. 355/5-6*, 501–509 (1996).
[2394] Harnly, J.M.; C.M.M. Smith, B. Radziuk, *Spectrochim. Acta, Part B 51/9-10*, 1055–1079 (1996).
[2395] Harrington, D.E.; *At. Absorpt. Newsletter 11/5*, 107–108 (1972).
[2396] Harrington, D.E.; W.R. Bramstedt, *At. Absorpt. Newsletter 15/6*, 125–128 (1976).
[2397] Harriott, M.; D.T. Burns, N. Chimpalee, *Anal. Proc. (London) 28/6*, 193–194 (1991).
[2398] Harriott, M.; D.T. Burns, C. Donaghy, *Anal. Proc. (London) 28/6*, 194–197 (1991).
[2399] Harrison, I.; D. Littlejohn, G.S. Fell, *J. Anal. At. Spectrom. 10/3*, 215–219 (1995).
[2400] Harrison, I.; D. Littlejohn, G.S. Fell, *Analyst (London) 121/2*, 189–194 (1996).
[2401] Harrison, R.M.; M. Radojevic, C.N. Hewitt, *Sci. Total Environ. 44*, 235–244 (1985).
[2402] Harrison, R.M.; M. Radojevic, *Environ. Technol. Lett. 6/3*, 129–136 (1985).
[2403] Harrison, R.M.; C.N. Hewitt, *Int. J. Environ. Anal. Chem. 21/1-2*, 89–104 (1985).
[2404] Harrison, R.M.; S. Rapsomanikis, *Heavy Met. Hydrol. Cycle*, Selper, London, 1988, 419.
[2405] Harrison, R.M.; S. Rapsomanikis, *Environmental Analysis Using Chromatography Interfaced With Atomic Spectroscopy*, Ellis Horwood, Chichester, W. Sussex, UK, 1989.
[2406] Harrison, W.W.; P.O. Juliano, *Anal. Chem. 41*, 1016 (1969).
[2407] Harrison, W.W.; W.H. Wadlin, *Anal. Chem. 41*, 374 (1969).
[2408] Harrison, W.W.; P.O. Juliano, *Anal. Chem. 43/2*, 248–252 (1971).
[2409] Hartley, F.R.; A.S. Inglis, *Analyst (London) 92*, 622 (1967).
[2410] Hartley, F.R.; A.S. Inglis, *Analyst (London) 93*, 394 (1968).
[2411] Hartmann, C.; J. Smeyers-Verbeke, D.L. Massart, *Analusis 21*, 125–132 (1993).
[2412] Hartmann, E.; *Z. Lebensm. Unters. Forsch. 180/2*, 87–95 (1985).
[2413] Hartstein, A.M.; R.W. Freedman, D.W. Platter, *Anal. Chem. 45/3*, 611–614 (1973).
[2414] Harzdorf, C.; *Spurenanalytik des Chroms*, G. Thieme Verlag, Stuttgart - New York, 1990.
[2415] Hase, A.; T. Kawabata, K. Terada, *Anal. Sci. 6/5*, 747–751 (1990).
[2416] Hasegawa, H.; Y. Sohrin, M. Matsui, M. Hojo, M. Kawashima, *Anal. Chem. 66/19*, 3247–3252 (1994).
[2417] Hasegawa, S.-i.; T. Kobayashi, K. Ide, R. Hasegawa, *Bunseki Kagaku 42/10*, 643 (1993).
[2418] Hasegawa, S.-i.; T. Kobayashi, K. Ide, H. Okochi, R. Hasegawa, *J. Japan Inst. Metals 58/1*, 23–29 (1994).
[2419] Hassan, S.S.M.; M.H. Eldesouki, *J. Assoc. Off. Anal. Chem. 62*, 315–319 (1979).
[2420] Hassan, S.S.M.; M.E.S. Metwally, A.A. Abou Ouf, *Analyst (London) 107*, 1235–1240 (1982).
[2421] Hassan, S.S.M.; A. Shalaby, *Mikrochim. Acta (Vienna) 109*, 193–199 (1992).
[2422] Hassell, D.C.; T.M. Rettberg, F.A. Fort, J.A. Holcombe, *Anal. Chem. 60/24*, 2680–2683 (1988).
[2423] Hassell, D.C.; V. Majidi, J.A. Holcombe, *J. Anal. At. Spectrom. 6/2*, 105–108 (1991).
[2424] Haswell, S.J.; P. O'Neill, K.C.C. Bancroft, *Talanta 32/1*, 69–72 (1985).
[2425] Haswell, S.J.; R.A. Stockton, K.C.C. Bancroft, P. O'Neill, A. Rahman, K. Irgolic, *J. Autom. Chem. 9/1*, 6–14 (1987).
[2426] Haswell, S.J.; J. Mendham, M.J. Butler, D.C. Smith, *J. Anal. At. Spectrom. 3/5*, 731–734 (1988).
[2427] Haswell, S.J.; in: S.J. Haswell (Editor), *Atomic Absorption Spectrometry - Theory, Design and Application*, Elsevier, Amsterdam - Oxford - New York - Tokyo, 1991, 21–49.
[2428] Haswell, S.J.; D. Barclay, *Analyst (London) 117/2*, 117–120 (1992).
[2429] Hatch, W.R.; W.L. Ott, *Anal. Chem. 40/14*, 2085–2087 (1968).
[2430] Hatfield, D.B.; *Anal. Chem. 59/14*, 1887–1888 (1987).
[2431] Hatterer, A.; V. Mougenel, S. Walter, *Analusis 15/9*, 486–489 (1987).
[2432] Haug, H.O.; C.-h. Ju, in: B. Welz (Editor), *5. Colloquium Atomspektrometrische Spurenanalytik*, Bodenseewerk Perkin-Elmer, Überlingen, 1989, 207–215.
[2433] Haug, H.O.; C.-h. Ju, *Bericht des Kernforschungszentrums Karlsruhe KfK 4661*, Karlsruhe, 1990.
[2434] Haug, H.O.; *Spectroscopy Int. 2/2*, 53 (1990).
[2435] Haug, H.O.; C.-h. Ju, *J. Anal. At. Spectrom. 5/3*, 215–223 (1990).
[2436] Haug, H.O.; *J. Anal. At. Spectrom. 7/2*, 451–455 (1992).
[2437] Haug, H.O.; Y.-p. Liao, *J. Anal. At. Spectrom. 10/12*, 1069–1076 (1995).
[2438] Haug, H.O.; Y.-p. Liao, *Spectrochim. Acta, Part B 50/11*, 1311–1324 (1995).
[2439] Haug, H.O.; Y.-p. Liao, in: B. Welz (Editor), *CANAS'95 Colloquium Analytische Atomspektroskopie*, Bodenseewerk Perkin-Elmer, Überlingen, 1996, 311–314.

[2440] Haug, H.O.; Y.-p. Liao, *Fresenius J. Anal. Chem. 356/7*, 435–444 (1996).
[2441] Haug, H.O.; *Spectrochim. Acta, Part B 51/11*, 1425–1433 (1996).
[2442] Hauptkorn, S.; V. Krivan, *Spectrochim. Acta, Part B 49/3*, 221–228 (1994).
[2443] Hauptkorn, S.; G. Schneider, V. Krivan, *J. Anal. At. Spectrom. 9/3*, 463–468 (1994).
[2444] Hauptkorn, S.; V. Krivan, *Spectrochim. Acta, Part B 51/9-10*, 1197–1210 (1996).
[2445] Hausknecht, K.A.; E.A. Ryan, L.P. Leonard, *At. Spectrosc. 3/2*, 53–55 (1982).
[2446] Hautbout, R.; G. Legrand, I.A. Voinovitch, *Analusis 14/3*, 139–147 (1986).
[2447] Havezov, L.; B. Tamnev, *Fresenius Z. Anal. Chem. 290*, 299–301 (1978).
[2448] Havezov, L.; E. Russeva, N. Jordanov, *Fresenius Z. Anal. Chem. 296/2-3*, 125–127 (1979).
[2449] Havezov, L.; G. Emrich, E. Wanova, P. Koehler, Nguyen van Ha, N. Jordanov, *Fresenius Z. Anal. Chem. 328/1-2*, 71–73 (1987).
[2450] Havezov, L.; E. Ivanova, P. Koehler, N. Jordanov, R. Matchat, K. Reiher, *Fresenius Z. Anal. Chem. 326*, 536–539 (1987).
[2451] Hayase, K.; K. Shitashima, H. Tsubota, *Talanta 33/9*, 754–756 (1986).
[2452] Hayes, P.; T.P. Martin, J. Pybus, *Aust. J. Hosp. Pharm. 22*, 353 (1990).
[2453] He, B.; Z.-m. Ni, *J. Anal. At. Spectrom. 11/2*, 165–168 (1996).
[2454] He, H.-w.; *Fenxi Huaxue 22/1*, 106 (1994).
[2455] He, J.-l.; Z.-j. Zheng, *Yankuang Ceshi 6*, 196 (1987).
[2456] Headridge, J.B.; M.A. Ashy, *4th ICAS Toronto*, 1973.
[2457] Headridge, J.B.; R. Thompson, *Anal. Chim. Acta 102*, 33–39 (1978).
[2458] Headridge, J.B.; I.M. Riddington, *Mikrochim. Acta (Vienna) II/5-6*, 457–467 (1982).
[2459] Headridge, J.B.; I.M. Riddington, *Analyst (London) 109/2*, 113–118 (1984).
[2460] Headridge, J.B.; D. Johnson, K.W. Jackson, J.A. Roberts, *Anal. Chim. Acta 201*, 311–315 (1987).
[2461] Heanes, D.L.; *Analyst (London) 106*, 172–181 (1981).
[2462] Heanes, D.L.; *Analyst (London) 106*, 182–187 (1981).
[2463] Heckmann, P.H.; E. Träbert, *Introduction to the Spectroscopy of Atoms*, Elsevier, Amsterdam, 1989.
[2464] Hedrich, M.; U. Rösick, P. Brätter, R.L. Bergmann, K.E. Bergmann, in: B. Welz (Editor), *5. Colloquium Atomspektrometrische Spurenanalytik*, Bodenseewerk Perkin-Elmer, Überlingen, 1989, 879–886.
[2465] Heiden, R.W.: D.A. Aikens, *Anal. Chem. 51/1*, 151–156 (1979).
[2466] Heiden, R.W.: D.A. Aikens, *Anal. Chem. 55/14*, 2327–2332 (1983).
[2467] Hein, H.; W. Kunze, *Umweltanalytik mit Spektrometrie und Chromatographie - Von der Laborgestaltung bis zur Dateninterpretation*, VCH Verlagsgesellschaft Weinheim, 1994.
[2468] Heinemann, W.; *Fresenius Z. Anal. Chem. 280*, 359–364 (1976).
[2469] Heinemann, W.; *Fresenius Z. Anal. Chem. 281*, 291–294 (1976).
[2470] Heinemann, W.; *Fresenius Z. Anal. Chem. 279*, 351–354 (1976).
[2471] Heinen Brown, J.; J.E. Vaz, Z.A. de Benzo, M. Velosa, *Analyst (London) 120/4*, 1215–1220 (1995).
[2472] Heinig, W.; K. Mauersberger, *Talanta 32/2*, 145–147 (1985).
[2473] Heininger, P.; V. Dünnbier, G. Henrion, *Z. Chem. 25/1*, 33 (1985).
[2474] Heininger, P.; G. Henrion, *Z. Chem. 25/2*, 73–74 (1985).
[2475] Heinrich, R.; J. Angerer, *Int. J. Environ. Anal. Chem. 16/4*, 305–314 (1984).
[2476] Heinrich, R.; J. Angerer, *Fresenius Z. Anal. Chem. 322/8*, 772–774 (1985).
[2477] Heinrichs, H.; *Fresenius Z. Anal. Chem. 273*, 197–201 (1975).
[2478] Heinrichs, H.; *LaborPraxis 13*, 1140–1146 (1989).
[2479] Heinrichs, H.; *LaborPraxis 14/1-2*, 20–25 (1990).
[2480] Heinrichs, H.; H.-J. Brumsack, W. Schultz, *LaborPraxis 15/9*, 709–715 (1991).
[2481] Heinrichs, H.; H.-J. Brumsack, W. Schultz, in: B. Welz (Editor), *6. Colloquium Atomspektrometrische Spurenanalytik*, Bodenseewerk Perkin-Elmer, Überlingen, 1991, 769–774.
[2482] Heisz, O.; *Seifen, Oele, Fette, Wachse 116/5*, 183–185 (1990).
[2483] Heitmann, U.; M. Schütz, H. Becker-Roß, S. Florek, *Spectrochim. Acta, Part B 51/9-10*, 1095–1105 (1996).
[2484] Hejduk, J.; J. Novak, *Fresenius Z. Anal. Chem. 234*, 327 (1968).
[2485] Hell, A.; W.F. Ulrich, N. Shifrin, J. Ramírez-Muñoz, *Appl. Opt. 7/7*, 1317–1323 (1968).
[2486] Hellmann, H.; in: H. Hulpke, H. Hartkamp, G. Tölg (Editors), *Analytische Chemie für die Praxis*, G. Thieme Verlag, Stuttgart, 1986, 1–237.
[2487] Helligsøe, J.E.T. Andersen, E.H. Hansen, *J. Anal. At. Spectrom. 12/5*, 585–588 (1997).
[2488] Helman, E.Z.; D.K. Wallick, I.M. Reingold, *Clin. Chem. (Winston-Salem) 17*, 61–62 (1971).
[2489] Helsby, C.A.; *Talanta 20*, 779–782 (1973).

[2490] Helsby, C.A.; *Talanta 24*, 46–48 (1977).
[2491] Heltai, G.; K. Percsich, *Talanta 41/7*, 1067–1072 (1994).
[2492] Hemmings, M.J.; E.A. Jones, *Talanta 38/2*, 151–155 (1991).
[2493] Hemptenmacher, P.; in: B. Welz (Editor), *4. Colloquium Atomspektrometrische Spurenanalytik*, Bodenseewerk Perkin-Elmer, Überlingen, 1987, 285–293.
[2494] Hendrikx-Jongerius, C.; L. de Galan, *Anal. Chim. Acta 87*, 259–271 (1976).
[2495] Hendzel, M.R.; B.W. Fallis, B.G.E. de March, *J. Assoc. Off. Anal. Chem. 69/5*, 863–868 (1986).
[2496] Heneage, P.; *At. Absorpt. Newsletter 5*, 67 (1966).
[2497] Héninger, I.; M. Potin-Gautier, I. de Gregori, H. Pinochet, *Fresenius J. Anal. Chem. 357/6*, 600–610 (1997).
[2498] Henn, E.L.; *At. Absorpt. Newsletter 12/5*, 109–111 (1973).
[2499] Henn, E.L.; *Anal. Chem. 47/3*, 428–432 (1975).
[2500] Henn, K.H.; R. Berg, L. Hörner, in: B. Welz (Editor), *Atomspektrometrische Spurenanalytik*, Verlag Chemie, Weinheim, 1982, 553–559.
[2501] Hennig, G.R.; *Progr. Inorg. Chem. 1*, 125–205 (1959).
[2502] Hennig, G.R.; *Proc. 5th Conf. on Carbon*, Pergamon Press, Oxford, 1962, 143.
[2503] Hennig, W.; H. Berndt, G. Schaldach, *Fresenius J. Anal. Chem. 337/3*, 275–279 (1990).
[2504] Henrion, G.; J. Gelbrecht, *Z. Chem. 21/12*, 453–454 (1981).
[2505] Henrion, G.; P. Heininger, *Z. Chem. 25*, 97 (1985).
[2506] Hensley, J.W.; A.O. Long, J.E. Willard, *Ind. Eng. Chem. 41*, 1415 (1949).
[2507] Henty, G.; *C.R. Acad. Sci. Paris C 264*, 376 (1967).
[2508] Henze, W.; F. Umland, *Anal. Sci. 3*, 225–227 (1987).
[2509] Herber, R.F.M.; A.M. Roelofsen, W. Hazelhoff-Roelfzema, J.H.J. Copius, *Fresenius Z. Anal. Chem. 322/7*, 743–746 (1985).
[2510] Herber, R.F.M.; H.J. Pieters, *Spectrochim. Acta, Part B 43/2*, 149–158 (1988).
[2511] Herber, R.F.M.; M. Stoeppler, D.B. Tonks, *Fresenius J. Anal. Chem. 338/3*, 269–278 (1990).
[2512] Herber, R.F.M.; *Pure Appl. Chem. 63/9*, 1213–1220 (1991).
[2513] Herber, R.F.M.; in: H.G. Seiler, A. Sigel, H. Sigel (Editors), *Handbook on Metals in Clinical and Analytical Chemistry*, Marcel Dekker, New York, 1994, 195–200.
[2514] Herber, R.F.M.; in: H.G. Seiler, A. Sigel, H. Sigel (Editors), *Handbook on Metals in Clinical and Analytical Chemistry*, Marcel Dekker, New York, 1994, 283–297.
[2515] Herber, R.F.M.; *Microchem. J. 51/1-2*, 46–52 (1995).
[2516] Herber, R.F.M.; K.-H. Grobecker, *Fresenius J. Anal. Chem. 351/6*, 577–582 (1995).
[2517] Herczynska, E.; I.G. Campbell, *Z. Phys. Chem. (Leipzig) 215*, 248 (1960).
[2518] Hergenröder, R.; K. Niemax, *Spectrochim. Acta, Part B 43/12*, 1443–1449 (1988).
[2519] Hergenröder, R.; K. Niemax, *Trends Anal. Chem. (Pers. Ed.) 8/8*, 333–335 (1989).
[2520] Hermann, R.; *At. Absorpt. Newsletter 16/2*, 44–45 (1977).
[2521] Hernandez-Artiga, M.P.; J.A. Muñoz-Leyva, R. Cozar-Sievert, *Analyst (London) 117/6*, 963–966 (1992).
[2522] Hernández-Córdoba, M.; I. López-García, *Talanta 38/11*, 1247–1251 (1991).
[2523] Hernández-Méndez, J.; L. Polo-Díez, A. Bernal-Melchor, *Anal. Chim. Acta 108*, 39–44 (1979).
[2524] Herold, D.A.; R.L. Fitzgerald, in: H.G. Seiler, A. Sigel, H. Sigel (Editors), *Handbook on Metals in Clinical and Analytical Chemistry*, Marcel Dekker, New York, 1994, 321–332.
[2525] Herrmann, R.; W. Lang, in: *Z. Ges. Exptl. Med. 134*, 268 (1961).
[2526] Herrmann, R.; W. Lang, D. Stamm, in: *Zeitschrift für Blutforschung 11*, 135 (1965).
[2527] Hershey, J.W.; P.N. Keliher, *Spectrochim. Acta, Part B 41/7*, 713–723 (1986).
[2528] Hershey, J.W.; P.N. Keliher, *Spectrochim. Acta, Part B 44/3*, 329–337 (1989).
[2529] Herwald, S.; U. Backes, B. Straka-Emden, B. Neidhart, in: B. Welz (Editor), *5. Colloquium Atomspektrometrische Spurenanalytik*, Bodenseewerk Perkin-Elmer, Überlingen, 1989, 419–428.
[2530] Herzberg, G.; *Atomspektren und Atomstruktur*, Steinkopff, Dresden, Leipzig, 1936.
[2531] Herzberg, G.; *Molecular Spectra and Molecular Structure I. Spectra of Diatomic Molecules*, Van Nostrand, New York, 1961.
[2532] Hessische Landesantalt für Umwelt.; *Verfahrensempfehlungen zur Probenahme von Boden, Abfall, Grundwasser, Sickerwasser für chemisch-physikalische Untersuchungen bei Altlasten.*
[2533] Hetherington, G.; L.W. Bell, in: M. Zief, R. Speights (Editors), *Ultrapurity*, Marcel Dekker, New York, 1972, 353–400.
[2534] Hetland, S.; Y. Thomassen, *Pure Appl. Chem. 65/12*, 2417–2422 (1993).
[2535] Hewitt, A.D.; C.M. Reynolds, *At. Spectrosc. 11/5*, 187–192 (1990).

[2536] Hewitt, C.N.; R.M. Harrison, *Anal. Chim. Acta 167/1*, 277–287 (1985).
[2537] Hewitt, C.N.; R.M. Harrison, M. Radojevic, *Anal. Chim. Acta 188*, 229–238 (1986).
[2538] Hey, H.; *Fresenius Z. Anal. Chem. 256*, 361–362 (1971).
[2539] Heydorn, K.; E. Damsgaard, *Talanta 29/11B*, 1019–1024 (1982).
[2540] Heydorn, K.; in: R. Zeisler, V.P. Guinn (Editors), *Nuclear Analytical Methods in the Life Sciences*, Humana Press, Clifton, NJ, 1990, 541.
[2541] Heydorn, K.; *Mikrochim. Acta (Vienna) III/1-3*, 1–10 (1991).
[2542] Heye, C.L.; J.I. Thornton, *Anal. Chim. Acta 288/1-2*, 83–96 (1994).
[2543] Hieftje, G.M.; H.V. Malmstadt, *Anal. Chem. 40*, 1860–1867 (1968).
[2544] Hieftje, G.M.; *Appl. Spectrosc. 25,6*, 653–659 (1971).
[2545] Hieftje, G.M.; *Anal. Chem. 44/6*, 81A-88A (1972).
[2546] Hieftje, G.M.; *Fresenius J. Anal. Chem. 337/5*, 528–537 (1990).
[2547] Higashiura, M.; H. Uchida, T. Uchida, H. Wada, *Anal. Chim. Acta 304/3*, 317–321 (1995).
[2548] High, K.A.; R. Azani, A.F. Fazekas, Z.A. Chee, J.S. Blais, *Anal. Chem. 64/24*, 3197–3201 (1992).
[2549] Hight, S.C.; *Ceram. Eng. Sci. Proc. 15*, 317 (1994).
[2550] Hilderbrand, D.C.; *At. Spectrosc. 4/4*, 164 (1983).
[2551] Hill, S.J.; in: M. Stoeppler (Editor), *Hazardous Metals in the Environment*, Techn. Instr. Anal. Chem. Vol 12, Elsevier, Amsterdam, 1992.
[2552] Hill, S.J.; L. Pitts, P. Worsfold, *J. Anal. At. Spectrom. 10/5*, 409–411 (1995).
[2553] Hillebrand, U.; *GIT Fachz. Lab. 35*, 1001–1007 (1991).
[2554] Hiltenkamp, E.; E. Jackwerth, *Fresenius Z. Anal. Chem. 332/2*, 134–139 (1988).
[2555] Hinderberger, E.J.; M.L. Kaiser, S.R. Koirtyohann, *At. Spectrosc. 2/1*, 1–7 (1981).
[2556] Hinds, M.W.; K.W. Jackson, A.P. Newman, *Analyst (London) 110/8*, 947–950 (1985).
[2557] Hinds, M.W.; K.W. Jackson, *J. Anal. At. Spectrom. 2*, 441–445 (1987).
[2558] Hinds, M.W.; K.W. Jackson, *J. Anal. At. Spectrom. 3/7*, 997–1003 (1988).
[2559] Hinds, M.W.; M. Katyal, K.W. Jackson, *J. Anal. At. Spectrom. 3/1*, 83–87 (1988).
[2560] Hinds, M.W.; K.W. Jackson, *J. Anal. At. Spectrom. 5/3*, 199–202 (1990).
[2561] Hinds, M.W.; K.W. Jackson, *At. Spectrosc. 12/4*, 109–110 (1991).
[2562] Hinds, M.W.; K.E. Latimer, K.W. Jackson, *J. Anal. At. Spectrom. 6/6*, 473–476 (1991).
[2563] Hinds, M.W.; *J. Anal. At. Spectrom. 7/4*, 685–688 (1992).
[2564] Hinds, M.W.; S. Littau, P. Moulinié, *Analyst (London) 117/9*, 1473–1475 (1992).
[2565] Hinds, M.W.; *Spectrochim. Acta, Part B 48/3*, 435–445 (1993).
[2566] Hinds, M.W.; V.V. Kogan, *J. Anal. At. Spectrom. 9/3*, 451–455 (1994).
[2567] Hinds, M.W.; G.N. Brown, D.L. Styris, J. Anal. At. Spectrom. 9/12, 1411–1416 (1994).
[2568] Hinkle, M.E.; R.E. Learned, *US Geol. Survey Prof. Paper 650-D*, 1969.
[2569] Hinners, T.A.; *Analyst (London) 105*, 751–755 (1980).
[2570] Hintelmann, H.; R.-D. Wilken, *Sci. Total Environ. 166*, 1–10 (1995).
[2571] Hiraide, M.; J. Mizutani, A. Mizuike, *Anal. Chim. Acta 151/2*, 329–337 (1983).
[2572] Hiraide, M.; P. Tschöpel, G. Tölg, *Anal. Chim. Acta 186*, 261–266 (1986).
[2573] Hiraide, M.; A. Mizuike, *Fresenius Z. Anal. Chem. 335/8*, 924–926 (1989).
[2574] Hiraide, M.; T. Ueda, A. Mizuike, *Anal. Chim. Acta 227/2*, 421–424 (1989).
[2575] Hiraide, M.; T. Usami, H. Kawaguchi, *Anal. Sci. 8/1*, 31–34 (1992).
[2576] Hiraide, M.; S.-h. Zhou, H. Kawaguchi, *Analyst (London) 118/11*, 1441–1443 (1993).
[2577] Hiraide, M.; S. Hiramatsu, H. Kawaguchi, *Fresenius J. Anal. Chem. 348/11*, 758–761 (1994).
[2578] Hiraide, M.; Y. Mikumi, H. Kawaguchi, *Analyst (London) 119/7*, 1451–1454 (1994).
[2579] Hiraide, M.; Y. Ohta, H. Kawaguchi, *Fresenius J. Anal. Chem. 350/11*, 648–650 (1994).
[2580] Hiraide, M.; Z.-s. Chen, K. Sugimoto, H. Kawaguchi, *Anal. Chim. Acta 302/1*, 103–107 (1995).
[2581] Hiraide, M.; Y. Mikumi, H. Kawaguchi, *Anal. Sci. 11/4*, 689–691 (1995).
[2582] Hiraide, M.; Z.-s. Chen, H. Nakamachi, H. Kawaguchi, *Anal. Sci. 11/6*, 1009–1011 (1995).
[2583] Hiraide, M.; Y. Mikumi, H. Kawaguchi, *Fresenius J. Anal. Chem. 354/2*, 212–215 (1996).
[2584] Hiraide, M.; Y. Mikumi, H. Kawaguchi, *Talanta, 43/7*, 1131–1136 (1996).
[2585] Hirano, Y.; Y. Nomura, K. Yasuda, K. Hirokawa, *Anal. Sci. 8/3*, 427–431 (1992).
[2586] Hirata, S.; Y. Umezaki, M. Ikeda, *J. Flow Injection Anal. 3/1*, 8–17 (1986).
[2587] Hirata, S.; K. Honda, T. Kumamaru, *Anal. Chim. Acta 221/1*, 65–76 (1989).
[2588] Hirner, A.V.; *Int. J. Environ. Anal. Chem. 46/1-3*, 77–85 (1992).
[2589] Hirokawa, K.; K. Yasuda, K. Takada, *Anal. Sci. 8/3*, 411–417 (1992).
[2590] Hirschfelder, D.; K.H. Thiele, *J. prakt. Chemie 333*, 165 (1991).
[2591] Histen, T.E.; O.A. Güell, I.A. Chavez, J.A. Holcombe, *Spectrochim. Acta, Part B 51*, 1279–1289 (1996).

[2592]   Hittorf, W.; *Ann. Phys. (Leipzig) 21*, 90–139 (1884).
[2593]   Hladky, Z.; J. Rísová, M. Fisera, *J. Anal. At. Spectrom. 5/8*, 691–692 (1990).
[2594]   Hlavay, J.; *Microchem. J. 46/1*, 121–129 (1992).
[2595]   Hlavay, J.; K. Polyák, I. Bódog, A. Molnár, E. Mészáros, *Fresenius. J. Anal. Chem. 354/2*, 227–232 (1996).
[2596]   Hlaváček, I.; I. Hlaváčková, *J. Anal. At. Spectrom. 6/7*, 535–540 (1991).
[2597]   Hoare, H.C. ; R.A. Mostyn, B.T.N. Newland, *Anal. Chim. Acta 40*, 181–186 (1968).
[2598]   Hobbs, R.S.; G.F. Kirkbright, M. Sargent, T.S. West, *Talanta 15*, 997–1007 (1968).
[2599]   Hocking, T.J.; W.M. Gulick, *Anal. Chim. Acta 151/1*, 195–202 (1983).
[2600]   Hocquaux, H.; *Zeeman Background Correction: Investigation of Analytical Performance for Complex Metallurgical Samples*, CSI Garmisch-Partenkirchen, 1985.
[2601]   Hocquellet, P.; M.-P. Candillier, *Analyst (London) 116/5*, 505–509 (1991).
[2602]   Hodge, V.F.; M. Stallard, M. Koide, E.D. Goldberg, *Anal. Chem. 58/3*, 616–620 (1986).
[2603]   Hoenig, M.; R. Vanderstappen, P. van Hoeyweghen, *Analusis 6/10*, 433–436 (1978).
[2604]   Hoenig, M.; P.O. Scokart, P. van Hoeyweghen, *Anal. Lett. 17/A17*, 1947–1962 (1984).
[2605]   Hoenig, M.; Y. van Elsen, R. van Cauter, *Anal. Chem. 58/4*, 777–780 (1986).
[2606]   Hoenig, M.; P. van Hoeyweghen, *Anal. Chem. 58/13*, 2614–2617 (1986).
[2607]   Hoenig, M.; *Varian instruments at work AA-61*, (1986).
[2608]   Hoenig, M.; P. van Hoeyweghen, *Varian instruments at work AA-67*, (1986).
[2609]   Hoenig, M.; P. van Hoeyweghen, *Varian instruments at work AA-66*, 1–6 (1986).
[2610]   Hoenig, M.; P. van Hoeyweghen, *Int. J. Environ. Anal. Chem. 24/3*, 193–202 (1986).
[2611]   Hoenig, M.; P. Regnier, A.M. de Kersabiec, *Analusis 18*, 420–425 (1990).
[2612]   Hoenig, M.; P. Regnier, L. Chou, *J. Anal. At. Spectrom. 6/4*, 273–275 (1991).
[2613]   Hoenig, M.; P. Regnier, L. Chou, *Analusis 19*, 163–166 (1991).
[2614]   Hoenig, M.; *Analusis 19/2*, 41–46 (1991).
[2615]   Hoenig, M.; E. Puskaric, P. Choisy, M. Wartel, *Analusis 19/9*, 285–291 (1991).
[2616]   Hoenig, M.; M.F. Guns, in: Ph. Quevauviller, E.A. Maier, B. Griepink (Editors), *Quality Assurance for Environmental Analysis*, Elsevier, Amsterdam, 1995, 63–88.
[2617]   Hoenig, M.; O. Dheere, *Mikrochim. Acta (Vienna) 119/3-4*, 259–264 (1995).
[2618]   Hoeyweghen, P. van ; M. Hoenig, *Analusis 13/6*, 275–278 (1985).
[2619]   Hoffmann, J.; in: B. Welz (Editor), *6. Colloquium Atomspektrometrische Spurenanalytik*, Bodenseewerk Perkin-Elmer, Überlingen, 1991, 113–117.
[2620]   Hoffmann, P.; K.H. Lieser, S. Abig, U. Stingl, N. Pilz, *Fresenius Z. Anal. Chem. 335/7*, 847–851 (1989).
[2621]   Hofmann, C.; J. Pauwels, C. Vandecasteele, *Fresenius J. Anal. Chem. 349/12*, 779–783 (1994).
[2622]   Hofmann, H.; P. Hoffmann, K.H. Lieser, *Fresenius J. Anal. Chem. 340/9*, 591–597 (1991).
[2623]   Hofsommer, H.-J.; H.-J. Bielig, *Z. Lebensm. Unters. Forsch. 173/3*, 213–218 (1981).
[2624]   Hofsommer, H.-J.; H. Gründing, H.-J. Bielig, in: B. Welz (Editor), *Fortschritte in der atomspektrometrischen Spurenanalytik*, Verlag Chemie, Weinheim, 1984, 441–455.
[2625]   Hoft, D.; J. Oxman, R.C. Gurira, *J. Agri. Food Chem. 27/1*, 145–147 (1979).
[2626]   Hogen, M.L.; *Cereal Chem. 60/5*, 403–405 (1983).
[2627]   Höhn, R.; E. Jackwerth, *Spectrochim. Acta, Part B 29*, 225–229 (1974).
[2628]   Höhn, R.; E. Jackwerth, *Anal. Chim. Acta 85*, 407–410 (1976).
[2629]   Hohnadel, D.C.; F.W. Sunderman, M.W. Nechay, M.D. McNeely, *Clin. Chem. (Winston-Salem) 19*, 1288–1292 (1973).
[2630]   Holak, W.; *Anal. Chem. 41*, 1712–1713 (1969).
[2631]   Holak, W.; B. Krinitz, J.C. Williams, *J. Assoc. Off. Anal. Chem. 55*, 741–742 (1972).
[2632]   Holak, W.; *At. Absorpt. Newsletter 12*, 63–65 (1973).
[2633]   Holak, W.; *Anal. Chim. Acta 74*, 216 (1975).
[2634]   Holak, W.; *J. Assoc. Off. Anal. Chem. 66/5*, 1203–1206 (1983).
[2635]   Holak, W.; *J. Liq. Chromatogr. 8/3*, 563–569 (1985).
[2636]   Holak, W.; J.J. Specchio, *At. Spectrosc. 12/4*, 105–108 (1991).
[2637]   Holcombe, J.A.; R.H. Eklund, K.E. Grice, *Anal. Chem. 50*, 2097 (1978).
[2638]   Holcombe, J.A.; R.H. Eklund, J.E. Smith, *Anal. Chem. 51/8*, 1205–1209 (1979).
[2639]   Holcombe, J.A.; G.D. Rayson, N. Akerlind, *Spectrochim. Acta, Part B 37/4*, 319–330 (1982).
[2640]   Holcombe, J.A.; *Spectrochim. Acta, Part B 38/4*, 609–615 (1983).
[2641]   Holcombe, J.A.; J.M. Harnly, *Anal. Chem. 58/13*, 2606–2611 (1986).
[2642]   Holcombe, J.A.; *Spectrochim. Acta, Part B 44/10*, 975–983 (1989).
[2643]   Holcombe, J.A.; D.L. Styris, J.D. Harris, *Spectrochim. Acta, Part B 46/5*, 629–639 (1991).
[2644]   Holcombe, J.A.; P.-x. Wang, *Fresenius J. Anal. Chem. 346/12*, 1047–1053 (1993).

[2645] Holcombe, J.A.; T.E. Histen, *Spectrochim. Acta, Part B 51/9-10*, 1045–1053 (1996).
[2646] Holden, A.J.; D. Littlejohn, G.S. Fell, *Anal. Proc. (London) 29/6*, 260–262 (1992).
[2647] Holding, S.T.; J. Noar, *Analyst (London) 95*, 1041 (1970).
[2648] Holding, S.T.; P.H.D. Matthews, *Analyst (London) 97*, 189–194 (1972).
[2649] Holding, S.T.; J.J. Rowson, *Analyst (London) 100*, 465–470 (1975).
[2650] Holding, S.T.; J.J. Rowson, in: G.B. Crump (Editor), *Petroanalysis 81: Advances in Analytical Chemistry in the Petroleum Industry*, J. Wiley & Sons, 1983.
[2651] Holding, S.T.; J.M. Palmer, *Analyst (London) 109/4*, 507–510 (1984).
[2652] Hollander, Tj.; B.J. Jansen, J.J. Plaat, Th.J. Alkemade, *J. Quant. Spectrosc. Radiat. Transfer 10*, 1301–1319 (1970).
[2653] Holtsmark, J.; *Z. Phys. 34*, 722–729 (1925).
[2654] Hon, P.-k.; O.-w. Lau, C.-s. Mok, *Analyst (London) 105*, 919–921 (1980).
[2655] Hon, P.-k.; O.-w. Lau, S.-k. Tsui, *J. Anal. At. Spectrom. 1/2*, 125–130 (1986).
[2656] Hongve, D.; S. Johansen, E. Andruchow, E. Bjertness, G. Becher, J. Alexander, *J. Trace Elem. Med. Biol. 10/1*, 6–11 (1996).
[2657] Hoover, W.L.; S.C. Duren, *J. Assoc. Off. Anal. Chem. 50*, 1269 (1967).
[2658] Hoover, W.L.; J.C. Reagor, *J. Assoc. Off. Anal. Chem. 51*, 211 (1968).
[2659] Hoover, W.L.; J.C. Reagor, J.C. Garner, *J. Assoc. Off. Anal. Chem. 52/4*, 708–714 (1969).
[2660] Hooymayers, H.P.; C.Th.J. Alkemade, *J. Quant. Spectrosc. Radiat. Transfer 6*, 501–526 (1966).
[2661] Hopfer, S.M.; L. Ziebka, F.W. Sunderman, J.R. Sporn, B.R. Greenberg, *Ann. Clin. Lab. Sci 19/6*, 389–396 (1989).
[2662] Hopp, H.U.; *Erdoel, Kohle, Erdgas, Petrochem. 27*, 435–436 (1974).
[2663] Hopps, H.C.; *J. Res. Natl. Bur. Stand. 91/2*, 47–50 (1986).
[2664] Hoppstock, K.; R.P.H. Garten, P. Tschöpel, *Fresenius J. Anal. Chem. 343*, 778–781 (1992).
[2665] Hoppstock, K.; R.P.H. Garten, P. Tschöpel, G. Tölg, *Anal. Chim. Acta 294/1*, 57–68 (1994).
[2666] Horner, E.; U. Kurfürst, *Fresenius Z. Anal. Chem. 328/4-5*, 386–387 (1987).
[2667] Hornig, U.; C. Krause, G. Sbieschni, *Acta Hydrochim. Hydrobiol. 19/3*, 285–293 (1991).
[2668] Horowitz, A.J.; *A Primer on Sediment Trace Element Chemistry*, Lewis, MI, 1991.
[2669] Horsky, S.J.; W.K. Fletcher, *Chem. Geol. 32*, 335–340 (1981).
[2670] Horvat, M.; K. May, M. Stoeppler, A.R. Byrne, *Appl. Organomet. Chem. 2*, 515–524 (1988).
[2671] Horvat, M.; V. Lupsina, B. Pihlar, *Anal. Chim. Acta 243/1*, 71–79 (1991).
[2672] Horvat, M.; *Water, Air, Soil Pollut. 56*, 95–102 (1991).
[2673] Horvat, M.; A.R. Byrne, *Analyst (London) 117/3*, 665–668 (1992).
[2674] Horváth, Z.; A. Lásztity, I. Varga, *Microchem. J. 46*, 130–135 (1992).
[2675] Horwitz, W. ; M. Perkany, *Pure Appl. Chem. 66*, 1903 (1994).
[2676] Hosking, J.W.; K.R. Oliver, B.T. Sturman, *Anal. Chem. 51/2*, 307–310 (1979).
[2677] Hou, X.-d.; P.-f. Xu, Z.-h. Sun, *Guangpuxue Yu Guangpu Fenxi 10/1*, 38 (1990).
[2678] Hou, Z.-h.; J. *Geochem. Explor. 27*, 323–328 (1987).
[2679] Howard, A.G.; M.H. Arbab-Zavar, *Analyst (London) 106*, 213–220 (1981).
[2680] Howard, A.G.; S.D.W. Comber, *Appl. Organomet. Chem. 3*, 509–514 (1989).
[2681] Howard, A.G.; S.D.W. Comber, *Mikrochim. Acta (Vienna) 109/1-4*, 27–33 (1992).
[2682] Howard, A.G.; L.E. Hunt, *Anal. Chem. 65/21*, 2995–2998 (1993).
[2683] Howard, A.G.; C. Salou, *Anal. Chim. Acta 333/1-2*, 89–96 (1996).
[2684] Howarth, R.J.; *Analyst (London) 120/7*, 1851–1873 (1995).
[2685] Howell, A.G.; S.R. Koirtyohann, *Appl. Spectrosc. 46/6*, 953–958 (1992).
[2686] Howlett, C.; A. Taylor, *Analyst (London) 103*, 916–920 (1978).
[2687] Hoyle, W.C.; A. Atkinson, *Appl. Spectrosc. 33*, 37–40 (1979).
[2688] Hsieh, C.-m.; S.C. Petrovic, H.L. Pardue, *Anal. Chem. 62/18*, 1983–1988 (1990).
[2689] Hsieh, C.-m.; H.L. Pardue, *Anal. Chem. 65/14*, 1809–1813 (1993).
[2690] Hsieh, L.L.; Y.-Y. Hung, C.-R. Lan, *J. Chin. Chem. Soc. (Taipei) 40*, 241–244 (1993).
[2691] Hsieh, T.P.; L.K. Liu, *Anal. Chim. Acta 282/1*, 221–225 (1993).
[2692] Hu, F.; X.-m. Chen, B. Xiao, L.-A. Shoa, *Spectrochim. Acta, Part B 46/13*, 1735–1744 (1991).
[2693] Hu, Y.-z.; J. Smeyers-Verbeke, D.L. Massart, *J. Anal. At. Spectrom. 4/7*, 605–611 (1989).
[2694] Huang, J.; D. Goltz, F. Smith, *Talanta 35/11*, 907–908 (1988).
[2695] Huang, S.-D.; K.-Y. Shih, *Spectrochim. Acta, Part B 48/12*, 1451–1460 (1993).
[2696] Huang, S.-D.; K.-Y. Shih, *Spectrochim. Acta, Part B 50/8*, 837–846 (1995).
[2697] Huang, S.-D.; W.-R. Lai, K.-Y. Shih, *Spectrochim. Acta, Part B 50*, 1237–1246 (1995).
[2698] Huang, S.-M.; X.-Y. Shong, X.-P. Ji, *Xiamen Daxue Xuebao, Ziran Kexueban 26/2*, 216 (1987).

[2699] Huang, Y.Q.; C.M. Wai, *Commun. Soil. Sci. Plant Anal. 17/2*, 125–133 (1986).
[2700] Huang, Z.-e.; *J. Anal. At. Spectrom. 9/1*, 11–15 (1994).
[2701] Huang, Z.-e.; *Spectrochim. Acta, Part B 50/11*, 1383–1393 (1995).
[2702] Huber, W.; *Fresenius Z. Anal. Chem. 319*, 379–383 (1984).
[2703] Hudnik, V.; M. Marolt-Gomišček, S. Gomišček, *Anal. Chim. Acta 157*, 303–311 (1984).
[2704] Hudnik, V.; M. Marolt-Gomišček, S. Gomišček, *Anal. Chim. Acta 157*, 183–186 (1984).
[2705] Hueber, D.M., J.D. Winefordner, *Anal. Chim. Acta 316/2*, 129–144 (1995).
[2706] Huettner, W.; C. Busche, *Fresenius Z. Anal. Chem. 323/7*, 674–680 (1986).
[2707] Hughes, D.M.; C.L. Chakrabarti, D.M. Goltz, R.E. Sturgeon, D.C. Grégoire, *Appl. Spectrosc. 50/6*, 715–731 (1996).
[2708] Hughes, H.; P.W. Hurley, *Analyst (London) 112*, 1445–1449 (1987).
[2709] Hughes, M.J.; M.R. Cowell, P.T. Craddock, *Archaeometry 18/1*, 19–37 (1976).
[2710] Hughes, M.J.; *Anal. Proc. (London) 22/3*, 75–76 (1985).
[2711] Hughes, R.O.; D.F. Wease, R.G. Troxler, *Clin. Chem. (Winston-Salem) 22*, 691–692 (1976).
[2712] Huie, C.W.; C.J. Curran, *Appl. Spectrosc. 42/7*, 1307–1311 (1988).
[2713] Huie, C.W.; C.J. Curran, *Appl. Spectrosc. 44/8*, 1329–1336 (1990).
[2714] Hulanicki, A.; E. Bulska, K. Wróbel, *Analyst (London) 110/8*, 1141–1145 (1985).
[2715] Hulanicki, A.; E. Bulska, *Analusis 16/9-10*, 12–16 Sup (1988).
[2716] Hulanicki, A.; E. Bulska, B. Godlewska, K. Wróbel, *Anal. Lett. 22/5*, 1341–1354 (1989).
[2717] Hulanicki, A.; *Anal. Proc. (London) 29/12*, 512–516 (1992).
[2718] Hulanicki, A.; E. Bulska, K. Wróbel, *Chem. Anal. (Warsaw) 37/1*, 93–97 (1992).
[2719] Hulanicki, A.; B. Godlewska, M. Brzóska, *Spectrochim. Acta, Part B 50/13*, 1717–1724 (1995).
[2720] Hull, D.A.; N. Muhammad, J.G. Lanese, S.D. Reich, T.T. Finkelstein, S. Fandrich, *J. Pharm. Sci. 70*, 500–502 (1981).
[2721] Human, H.G.C.; L.R.P. Butler, A. Strasheim, *Analyst (London) 94*, 81–88 (1969).
[2722] Human, H.G.C.; P.J.Th. Zeegers, J.A. van Elst, *Spectrochim. Acta, Part B 29*, 111–119 (1974).
[2723] Human, H.G.C.; R.H. Scott, *Spectrochim. Acta, Part B 31/8-9*, 459–473 (1976).
[2724] Human, H.G.C.; N.P. Ferreira, C.J. Rademeyer, P.K. Faure, *Spectrochim. Acta, Part B 37/7*, 593–602 (1982).
[2725] Human, H.G.C.; C.J. Rademeyer, *Fresenius Z. Anal. Chem. 323/7*, 754–758 (1986).
[2726] Hunt, D.T.E.; *Report Water Research Centre SCA/4.0/27/r.4*, 1981.
[2727] Hunt, D.T.E.; A.L. Wilson, *The Chemical Analysis of Water: General Principles and Techniques*, Royal Society of Chemistry, Cambridge, 1986.
[2728] Hunt, D.T.E.; A. Winnard, *Analyst (London) 111/7*, 785–789 (1986).
[2729] Hunt, S.M.; *At. Spectrosc. 9/4*, 100–106 (1988).
[2730] Hunter, J.S.; *J. Assoc. Off. Anal. Chem. 64/3*, 574–583 (1981).
[2731] Huo, G.; *Yejin Fenxi 13/1*, 49–50 (1993).
[2732] Hurlbut, J.A.; C.D. Chriswell, *Anal. Chem. 43/3, 465-466 (1971)*.
[2733] Husler, J.W.; *At. Absorpt. Newsletter 10/2*, 60–62 (1971).
[2734] Hutchinson, D.J.; F.J. Disinski, C.A. Nardelli, *J. Assoc. Off. Anal. Chem. 69/1*, 60–64 (1986).
[2735] Hutton, J.C.; C.L. Chakrabarti, P.C. Bertels, M.H. Back, *Spectrochim. Acta, Part B 46/2*, 193–202 (1991).
[2736] Huxtable, R.J.; *The Biochemistry of Selenium*, Plenum Press, New York, 1987.
[2737] Hwang, J.Y.; L.M. Sandonato, *Anal. Chem. 42/7*, 744–747 (1970).
[2738] Hyodoh, J.; K. Yokofujita, Y. Ishii, K. Takiyama, *Anal. Sci. 1/2*, 151–156 (1985).
[2739] Iadevaia, R.; N. Aharonson, E.A. Woolson, *J. Assoc. Off. Anal. Chem. 63/4*, 742–746 (1980).
[2740] IARC, *Monographs on the Evaluation of Carcinogenic Risks to Humans, Vol. 49: Chromium, Nickel and Welding Dust*, Lyon, 1990.
[2741] IARC, *Monographs on the Evaluation of Carcinogenic Risks to Humans, Vol. 52: Cobalt and Cobalt Compounds*, Lyon, 1991.
[2742] Ichikawa, F.; T. Sato, *Radiochim. Acta 12*, 89 (1969).
[2743] Ida, N.; H. Yoshikawa, Y. Ishibashi, N. Gunji, *Anal. Sci. 5/5*, 615–618 (1989).
[2744] Iffland, R.; in: H.G. Seiler, A. Sigel, H. Sigel (Editors), *Handbook on Metals in Clinical and Analytical Chemistry*, Marcel Dekker, New York, 1994, 237–253.
[2745] Ihnat, M.; H.J. Miller, *J. Assoc. Off. Anal. Chem. 60*, 813–824 (1977).
[2746] Ihnat, M.; *Can. J. Spectrosc. 23/4*, 112–125 (1978).
[2747] Ihnat, M.; *Commun. Soil. Sci. Plant Anal. 13*, 969–979 (1982).
[2748] Ihnat, M.; in: J.E. Cantle (Editor), *Atomic Absorption Spectrometry*, Elsevier, Amsterdam - Oxford - New York, 1982, 139–210.

[2749] Ihnat, M.; M.S. Wolynetz, Y. Thomassen, M. Verlinden, *Pure Appl. Chem. 58/7*, 1063–1076 (1986).
[2750] Ihnat, M.; *Fresenius Z. Anal. Chem. 326*, 739–741 (1987).
[2751] Ihnat, M.; *J. Res. Natl. Bur. Stand. 93/3*, 354–358 (1988).
[2752] Ihnat, M.; M. Stoeppler, *Fresenius J. Anal. Chem. 338/4*, 455–460 (1990).
[2753] Ihnat, M.; in: M. Stoeppler (Editor), *Hazardous Metals in the Environment*, Techn. Instr. Anal. Chem. Vol 12, Elsevier, Amsterdam, 1992.
[2754] Ihnat, M.; R.W. Dabeka, Mark S. Wolynetz, *Fresenius J. Anal. Chem. 348/7*, 445–451 (1994).
[2755] Ihnat, M.; M.S. Wolynetz, *Fresenius J. Anal. Chem. 348/7*, 452–458 (1994).
[2756] Ihnat, M.; *Fresenius J. Anal. Chem. 352/1-2*, 5–6 (1995).
[2757] Ihsanullah; *J. Radioanal. Nucl. Chem. 176/4*, 303–313 (1993).
[2758] Ikeda, M.; J. Nishibe, T. Nakahara, *Bunseki Kagaku 30/6*, 368–374 (1981).
[2759] Ikeuchi, I.; K. Daikatsu, I. Fujisaka, T. Amano, *Iyakuhin Kenkyu 21*, 1082 (1990).
[2760] Ikramuddin, M.; *At. Spectrosc. 4/3*, 101–103 (1983).
[2761] Ikrényi, K.; A. Bartha, *Anal. Chim. Acta 142*, 339–343 (1982).
[2762] Imai, N.; S. Terashima, A. Ando, *Geostand. Newsl. 8/1*, 39–41 (1984).
[2763] Imai, S.; Y. Hayashi, *Anal. Chem. 63/8*, 772–775 (1991).
[2764] Imai, S.; H. Ishikura, T. Tanaka, K. Saito, Y. Hayashi, *Eisei Kagaku 37*, 401 (1991).
[2765] Imai, S.; M. Ichinoseki, Y. Nishiyama, Y. Hayashi, *Bull. Chem. Soc. Jpn. 64*, 901–907 (1991).
[2766] Imai, S.; K. Okuhara, T. Tanaka, Y. Hayashi, K. Saito, *J. Anal. At. Spectrom. 10/1*, 37–41 (1995).
[2767] Imai, S.; N. Hasegawa, Y. Hayashi, K. Saito, *J. Anal. At. Spectrom. 11/7*, 515–520 (1996).
[2768] Imai, S.; N. Hasegawa, Y. Nishiyama, Y. Hayashi, K. Saito, *J. Anal. At. Spectrom. 11/8*, 601–606 (1996).
[2769] Imai, S.; M. Minezaki, Y. Hayashi, C. Jindoh, *Anal. Sci. 13/1*, 127–130 (1997).
[2770] Indyk, L.I.; V.V. Mel'nik, O.V. Knutova, *Zavodsk. Lab. 58/9*, 26–27 (1992).
[2771] Ingamelis, C.O.; F.F. Pitard, *Applied Geochemical Analysis*, Wiley Interscience, New York, 1986.
[2772] Ingle, J.D.; *Anal. Chem. 46/14*, 2161–2171 (1974).
[2773] Inglis, A.S.; P.W. Nicholls, *Mikrochim. Acta (Vienna) II*, 553–559 (1975).
[2774] Inokuma, Y.; J. Endo, *Bunseki Kagaku 37*, 493 (1988).
[2775] Institute of Petroleum, *Standard IP 363-83*, 1988.
[2776] Institute of Petroleum, *Standard IP 377-88*, 1988.
[2777] Institute of Petroleum, *Standard IP 117/82*, 1988.
[2778] Institute of Petroleum, *Standard IP 111/82*, 1988.
[2779] Institute of Petroleum, *Standard IP 110/82*, 1988.
[2780] Institute of Petroleum, *Standard IP 308/85*, 1988.
[2781] Institute of Petroleum, *Standard IP 362-83*, 1988.
[2782] International Organization for Standardization, *International Standard, Draft ISO/DIS 9174*, 1988.
[2783] International Organization for Standardization, *ISO Guide 35/4*, 1989.
[2784] International Organization for Standardization, *International Standard ISO 9964-1*, 1993.
[2785] International Organization for Standardization, *International Standard ISO 9964-2*, 1993.
[2786] International Organization for Standardization, Draft, *General guidelines for AAS - flame analysis, Part I- General subjects*, ISO/CD 13204/1, 1997.
[2787] International Organization for Standardization, Draft, *General guidelines for AAS - flame analysis, Part II- Method development and analysis*, ISO/CD 13204/2, 1997.
[2788] International Organization for Standardization, Draft, *General guidelines for AAS - graphite furnace analysis, Part I - General subjects*, ISO/CD 13812/1, 1997.
[2789] International Organization for Standardization, Draft, *General guidelines for AAS - graphite furnace analysis, Part II - Method development and analysis*, ISO/CD 13812/2, 1997.
[2790] Intonti, R.; A. Stacchini, *Spectrochim. Acta, Part B 23/7*, 437–442 (1968).
[2791] Inui, T.; S. Terada, H. Tamura, *Fresenius Z. Anal. Chem. 305/3*, 189–192 (1981).
[2792] Inui, T.; S. Terada, H. Tamura, N. Ichinose, *Fresenius Z. Anal. Chem. 311/5*, 492–495 (1982).
[2793] Inui, T.; S. Terada, H. Tamura, N. Ichinose, *Fresenius Z. Anal. Chem. 315/7*, 598–601 (1983).
[2794] Inui, T.; S. Terada, H. Tamura, N. Ichinose, *Fresenius Z. Anal. Chem. 318*, 502–504 (1984).
[2795] Ipach, R.; B. Altmayer, K.W. Eichhorn, *Fresenius Z. Anal. Chem. 314/2*, 157–158 (1983).
[2796] Iqbal, M.Z.; M.A. Qadir, *Int. J. Environ. Anal. Chem. 38/4*, 533–538 (1990).
[2797] Irgolic, K.J.; B.K. Puri, in: J.A.C. Broekaert, S. Güçer, F. Adams (Editors), *Metal Speciation in the Environment*, NATO ASI Ser., Ser. G: Ecological Sci., Springer, Berlin - Heidelberg - New York, 1990, 377–389.

[2798] Irgolic, K.J.; in: M. Stoeppler (Editor), *Hazardous Metals in the Environment*, Techn. Instr. Anal. Chem. Vol 12, Elsevier, Amsterdam, 1992, 288–340.
[2799] Irsch, B.; K. Schäfer, *Fresenius Z. Anal. Chem. 320/1*, 37–40 (1985).
[2800] Irwin, R.L.; A. Mikkelsen, R.G. Michel, J.P. Dougherty, F.R. Preli, *Spectrochim. Acta, Part B 45/8*, 903–915 (1990).
[2801] Isaac, R.A.; W.C. Johnson, *J. Assoc. Off. Anal. Chem. 58*, 436–440 (1975).
[2802] Ishida, K.; B.K. Puri, M. Satake, M.C. Mehra, *Talanta 32/3*, 207–208 (1985).
[2803] Ishida, K.; H. Orimo, in: H.G. Seiler, A. Sigel, H. Sigel (Editors), *Handbook on Metals in Clinical and Analytical Chemistry*, Marcel Dekker, New York, 1994, 387–397.
[2804] Ishii, T.; S. Musha, *Bunseki Kagaku 20*, 489 (1971).
[2805] Ishikawa, T.; E. Nakamura, *Anal. Chem. 62*, 2612–2616 (1990).
[2806] Ishinishi, N.; K. Tsuchiya, M. Vahter, B.A. Fowler, in: L. Friberg, G.F. Nordberg, V.B. Vouk (Editors), *Handbook on the toxicology of metals*, Elsevier, Amsterdam, 1986, 43–86.
[2807] Ishizuka, T.; Y. Uwamino, H. Sunahara, *Anal. Chem. 49*, 1340–1343 (1977).
[2808] Ishizuka, T.; Y. Uwamino, A. Tsuge, T. Kamiyanagi, *Anal. Chim. Acta 161*, 285–291 (1984).
[2809] Isozaki, A.; K. Kumagai, S. Utsumi, *Anal. Chim. Acta 153*, 15–22 (1983).
[2810] Issa Y.M.; H. Ibrahim, A.F. Shoukry, S.K. Mohamed, *Mikrochim. Acta 118/3-4*, 257–263 (1995).
[2811] Issaq, H.J.; W.L. Zielinski, *Anal. Chem. 46/9*, 1328–1329 (1974).
[2812] Issaq, H.J.; L.P. Morgenthaler, *Anal. Chem. 47*, 1661–1667 (1975).
[2813] Issaq, H.J.; L.P. Morgenthaler, *Anal. Chem. 47*, 1668–1669 (1975).
[2814] Issaq, H.J.; L.P. Morgenthaler, *Anal. Chem. 47*, 1748–1752 (1975).
[2815] Isshiki, K.; F. Tsuji, T. Kuwamoto, E. Nakayama, *Anal. Chem. 59/20*, 2491–2495 (1987).
[2816] Isshiki, K.; E. Nakayama, *Anal. Chem. 59/2*, 291–295 (1987).
[2817] Isshiki, K.; Y. Sohrin, H. Karatani, E. Nakayama, *Anal. Chim. Acta 224/1*, 55–64 (1989).
[2818] Itami, T.; M. Ema, H. Amano, H. Kawasaki, *J. Anal. Toxicol. 15/3*, 119–122 (1991).
[2819] Itoh, J.; M. Komata, N. Fijiyoshi, H. Tsushima, *Nippon Kagaku Kaishi (6)*, 715–718 (1993).
[2820] Itoh, K.; K. Akatsuya, I. Atsuya, *Bunseki Kagaku 33*, 301–305 (1984).
[2821] Itoh, K.; I. Atsuya, *Bunseki Kagaku 36*, 390 (1987).
[2822] Itoh, K.; M. Chikuma, H. Tanaka, *Fresenius Z. Anal. Chem. 330/7*, 600–604 (1988).
[2823] Iu, K.L.; I.D. Pulford, H.J. Duncan, *Anal. Chim. Acta 106*, 319–324 (1979).
[2824] IUPAC, *Pure Appl. Chem. 45*, 105–123 (1976).
[2825] IUPAC, Commission on Spectrochemical and Other Optical Procedures, *Pure Appl. Chem. 45*, 99–103 (1976).
[2826] IUPAC, *Spectrochim. Acta, Part B 33*, 242–245 (1978).
[2827] IUPAC, *Spectrochim. Acta, Part B 33*, 247–269 (1978).
[2828] Ivanova, E.; S. Mareva, N. Jordanov, *Fresenius Z. Anal. Chem. 303/5*, 378–380 (1980).
[2829] Ivanova, E.; I. Havezov, N. Vracheva, N. Jordanov, *Fresenius Z. Anal. Chem. 320/2*, 133–136 (1985).
[2830] Ivanova, E.; N. Vracheva, I. Havezov, N. Jordanov, *Fresenius Z. Anal. Chem. 323/5*, 477–480 (1986).
[2831] Ivanova, E.; I. Havezov, N. Jordanov, *Fresenius Z. Anal. Chem. 327/3-4*, 359–360 (1987).
[2832] Ivanova, E.; N. Jordanov, I. Havezov, M. Stoimenova, S. Kadieva, *Fresenius J. Anal. Chem. 336/6*, 501–502 (1990).
[2833] Ivanova, E.; G. Schaldach, H. Berndt, *Fresenius J. Anal. Chem. 342/1-2*, 47–50 (1992).
[2834] Ivanova, E.; G. Gentscheva, M. Stoimenonva, I. Havezov, *Anal. Lab. 4/1*, 14–17 (1995).
[2835] Iversen, B.S.; A. Panayi, J.P. Camblor, E. Sabbioni, *J. Anal. At. Spectrom. 11/8*, 591–594 (1996).
[2836] Ivicic, M.; M. Blanusa, *Fresenius Z. Anal. Chem. 330/7*, 643–644 (1988).
[2837] Iwamoto, E.; N. Miyazaki, S. Ohkubo, T. Kumamaru, *J. Anal. At. Spectrom. 4/5*, 433–437 (1989).
[2838] Iwamoto, E.; K.E.A. Ohlsson, D.C. Baxter, W. Frech, *J. Anal. At. Spectrom. 7/7*, 1063–1068 (1992).
[2839] Iwamoto, E.; H. Shimazu, K. Yokata, T. Kumamaru, *Anal. Chim. Acta 274/2*, 231–235 (1993).
[2840] Iyengar, G.V.; B. Sansoni, in: *Elemental Analysis of Biological Materials: Current Problems and Techn.*, Techn. Rep. Series No. 197, Int. Atomic Energy Agency, Vienna, 1980, 73–101.
[2841] Iyengar, G.V.; *Anal. Chem. 54*, 554A–558A (1982).
[2842] Iyengar, G.V.; in: H. Boström, N. Ljungstedt (Editors), *Trace Elements in Health and Disease*, Almqvist & Wiksell Int., Stockholm, 1985, 64–82.

[2843] Iyengar, G.V.; in: I.K. O'Neil, P. Schuller, L. Fishbein, *Environm. Carcinogens, selected methods of analysis*, Int. Agency Res. Cancer, Lyon, 1986, 141–158.
[2844] Iyengar, G.V.; *J. Res. Natl. Bur. Stand. 91/2*, 67–74 (1986).
[2845] Iyengar, G.V.; *Clin. Nutr. 6*, 147–153 (1987).
[2846] Iyengar, G.V.; V. Iyengar, in: H.A. McKenzie, L.E. Smythe (Editors), *Quantitative trace analysis of biological materials*, Elsevier, Amsterdam - New York - Oxford, 1988, 401–417.
[2847] Iyengar, G.V.; *Trace Elem. Med. 6*, 47 (1989).
[2848] Iyengar, G.V.; *Elemental Analyses of Biological Systems, Vol. 1 Biomedical, Environmental, Compositional, and Methodological Aspects of Trace Elements*, CRC Press, Boca Raton, FL, 1989.
[2849] Iyer, S.G.; C.K. Pillai, *Indian J. Technol. 28/12*, 713–714 (1990).
[2850] Izatt, R.M. ; R.L. Bruening, M.L. Bruening, B.J. Tarbet, K.E. Krakowiak, J.S. Bradshaw, J.J. Christensen, *Anal. Chem. 60*, 1825–1826 (1988).
[2851] Jackson, D.R.; C. Salama, R. Dunn, *Can. Spectroscopy 15*, 17–24 (1970).
[2852] Jackson, J.G.; R.W. Fonseca, J.A. Holcombe, *Spectrochim. Acta, Part B 50/14*, 1837–1846 (1995).
[2853] Jackson, K.W.; T.D. Fuller, D.G. Mitchell, K.M. Aldous, *At. Absorpt. Newsletter 14/5*, 121–123 (1975).
[2854] Jackson, K.W.; A.P. Newman, *Analyst (London) 108*, 261–264 (1983).
[2855] Jackwerth, E.; *Fresenius Z. Anal. Chem. 271*, 120–125 (1974).
[2856] Jackwerth, E.; J. Lohmar, G. Wittler, *Fresenius Z. Anal. Chem. 270*, 6 (1974).
[2857] Jackwerth, E.; H. Berndt, *Anal. Chim. Acta 74*, 299–307 (1975).
[2858] Jackwerth, E.; J. Messerschmidt, *Anal. Chim. Acta 87*, 341–351 (1976).
[2859] Jackwerth, E.; P.G. Willmer, *Fresenius Z. Anal. Chem. 279*, 23–27 (1976).
[2860] Jackwerth, E.; P.G. Willmer, *Spectrochim. Acta, Part B 33*, 343–348 (1978).
[2861] Jackwerth, E.; P.G. Willmer, R. Höhn, H. Berndt, *At. Absorpt. Newsletter 18/3*, 66–68 (1979).
[2862] Jackwerth, E.; R. Höhn, K. Musaick, *Fresenius Z. Anal. Chem. 299*, 362 (1979).
[2863] Jackwerth, E.; in: B. Welz (Editor), *Atomspektrometrische Spurenanalytik*, Verlag Chemie, Weinheim, 1982, 1–21.
[2864] Jackwerth, E.; H. Mittelstädt, *Mikrochim. Acta (Vienna) Suppl. 10*, 325–335 (1983).
[2865] Jacobson, B.E.; G. Lockitch, G. Quigley, *Clin. Chem. (Winston-Salem) 37/4*, 515–519 (1991).
[2866] Jaganathan, J.; K.J. Ewing, I.D. Aggarwal, *At. Spectrosc. 9/5*, 166–168 (1988).
[2867] Jaganathan, J.; K.J. Ewing, E.A. Buckley, L. Peitersen, I.D. Aggarwal, *Microchem. J. 41/1*, 106–112 (1990).
[2868] Jaganathan, J.; K.J. Ewing, L.E. Peiterson, E.A. O'Brien, I.D. Aggarwal, *Spectrochim. Acta, Part B 46/5*, 669–671 (1991).
[2869] Jaganathan, J.; K.J. Ewing, I. Aggarwal, *J. Anal. At. Spectrom. 7/8*, 1287–1290 (1992).
[2870] Jaganathan, J.; I. Aggarwal, *Appl. Spectrosc. 47/2*, 190–191 (1993).
[2871] Jaganathan, J.; I. Aggarwal, *Appl. Spectrosc. 47/8*, 1169–1170 (1993).
[2872] Jaganathan, J.; I. Aggarwal, *Microchem. J. 50*, 44–47 (1994).
[2873] Jäger, H.; *LaborPraxis 8/4*, 345–347 (1984).
[2874] Jain, M.C.; N.V.R. Appa Rao, *Res. Ind. 28/2*, 133–136 (1983).
[2875] Jain, N.; K.A. Magan, S.M. Sondhi, *Indian Drugs 30/5*, 190–194 (1993).
[2876] Jain, V.K.; P.M. Lall, J.S. Tiwari, *At. Spectrosc. 8/2*, 77–78 (1987).
[2877] James, A.D.; L.P. Judson, *Chem. N.Z. 46*, 60–61 (1982).
[2878] Jamro, G.H.; R.W. Frei, *Mikrochim. Acta (Vienna) 1970*, 429 (1970).
[2879] Jan, T.K.; D.R. Young, *Anal. Chem. 50/9*, 1250–1253 (1978).
[2880] Janauer, G.E.; F.E. Smith, J. Mangan, *At. Absorpt. Newsletter 6/1*, 3–5 (1967).
[2881] Janghorbani, M.; B.T.G. Ting, A. Nahapetian, V.R. Young, *Anal. Chem. 54/7*, 1188–1190 (1982).
[2882] Janousek, I.; *Fresenius Z. Anal. Chem. 315/3*, 248 (1983).
[2883] Janousek, I.; *Chem. Listy 78*, 1320 (1984).
[2884] Jansen, B.J.; Tj. Hollander, L.P.L. Franklin, *Spectrochim. Acta, Part B 29*, 37–49 (1974).
[2885] Jansen, B.J.; Tj. Hollander, C.Th.J. Alkemade, *J. Quant. Spectrosc. Radiat. Transfer 17*, 187–192 (1976).
[2886] Janssen, A.; F. Umland, *Fresenius Z. Anal. Chem. 251*, 101–107 (1970).
[2887] Janssen, J.H.; J.E. van den Enk, R. Bult, D.C. de Groot, *Anal. Chim. Acta 84*, 319–326 (1976).
[2888] Janßen, A.; B. Brückner, K.H. Grobecker, U. Kurfürst, *Fresenius Z. Anal. Chem. 322/7*, 713–716 (1985).
[2889] Janßen, A.; K.-H. Willmann, F.J. Simon, *Fresenius Z. Anal. Chem. 320/2*, 137–141 (1985).
[2890] Janßen, E.; *Landwirtsch. Forsch. 36*, 161–171 (1983).

[2891] Janßen, E.; *Fresenius Z. Anal. Chem. 325*, 381–386 (1986).
[2892] Janßen, E.; in: K. Dittrich, B. Welz (Editors), *CANAS'93, Colloq. Analytische Atomspektroskopie*, Proc. Conf. 1993, Oberhof, Universität Leipzig/UFZ Leipzig-Halle, Leipzig, 1993, 371–377.
[2893] Jantzen, E.; R.-D. Wilken, *Vom Wasser 76*, 1–11 (1991).
[2894] Jantzen, E.; *Dissertation,* Univerität Hamburg, 1992.
[2895] Jarosz, J.; J.-M. Mermet, J.P. Robin, *C.R. Acad. Sci. Paris 278B*, 885 (1974).
[2896] Jarvie, A.W.P.; R.N. Markall, H.R. Potter, *Environ. Res. (U.S.A) 25*, 241–249 (1981).
[2897] Jasim, F.; N.A.N. Awad, A.T. Al-Rawi, *Microchem. J. 38/3*, 337–342 (1988).
[2898] Jaudon, M.C.; B. Le Coz, J.-P. Clavel, *Clin. Chem. (Winston-Salem) 31/4*, 660 (1985).
[2899] Jaworowski, R.J.; R.P. Weberling, *At. Absorpt. Newsletter 5/6*, 125–126 (1966).
[2900] Jaworowski, R.J.; R.P. Weberling, D.J. Bracco, *Anal. Chim. Acta 37*, 284–294 (1967).
[2901] Jeffus, M.T.; J.S. Elkins, C.T. Kenner, *J. Assoc. Off. Anal. Chem. 53/6*, 1172–1175 (1970).
[2902] Jenkins, G.M.; K. Kawamura, *Polymeric Carbons*, Cambridge Univ. Press, Cambridge, UK, 1976.
[2903] Jenks, P.J.; *Fresenius J. Anal. Chem. 352/1-2*, 3–4 (1995).
[2904] Jenne, E.A.; *Adv. Chem. Ser. 73*, 337 (1968).
[2905] Jenne, E.A.; P. Avotins, *J. Environ. Qual. 4*, 427–431 (1975).
[2906] Jerrow, M.; I.L. Marr, M.S. Cresser, *J. Anal. At. Spectrom. 7/7*, 1117–1119 (1992).
[2907] Jewett, K.L.; F.E. Brinckman, *J. Chromatogr. Sci. 19/11*, 583–593 (1981).
[2908] Ji, G.; K. Günther, *Lebensmittelchemie 49/3*, 62–63 (1995).
[2909] Ji, X.-p.; *Fenxi Huaxue 21/4*, 492 (1993).
[2910] Jiang, G.-b.; Z.-m. Ni, S.-r. Wang, H.-b. Han, *J. Anal. At. Spectrom. 4/4*, 315–318 (1989).
[2911] Jiang, G.-b.; Z.-m. Ni, S.-r. Wang, H.-b. Han, *Fresenius Z. Anal. Chem. 334/1*, 27–30 (1989).
[2912] Jiang, G.-b.; Z.-m. Ni, Zhang Li, A. Li, H.-b. Han, X.-q. Shan, *J. Anal. At. Spectrom. 7/2*, 447–450 (1992).
[2913] Jiang, S.-g.; D. Chakraborti, W. de Jonghe, F.C. Adams, *Fresenius Z. Anal. Chem. 305/3*, 177–180 (1981).
[2914] Jiang, S.-g.; W. de Jonghe, F.C. Adams, *Anal. Chim. Acta 136*, 183–190 (1982).
[2915] Jiang, S.-g.; D. Chakraborti, F.C. Adams, *Anal. Chim. Acta 196*, 271–275 (1987).
[2916] Jiang, Y.-q.; J.-y. Yao, B.-l. Huang, *Fenxi Shiyanshi 7/12*, 21 (1988).
[2917] Jiang, Y.-q.; J.-y. Yao, B.-l. Huang, *Fenxi Huaxue 17/5*, 456 (1989).
[2918] Jiang, Z.-g.; *Boli Yu Tangci 17*, 27 (1989).
[2919] Jie, Z.; S.-x. Guo, *Analyst (London) 120/6*, 1661–1664 (1995).
[2920] Jiménez, M.S.; J.M. Mir, J.R. Castillo-Suarez, *J. Anal. At. Spectrom. 8/4*, 665–669 (1993).
[2921] Jiménez, M.S.; L. Martín, J.M. Mir, J.R. Castillo, *At. Spectrosc. 17/5*, 201–207 (1996).
[2922] Jin, K.; M. Taga, *Anal. Chim. Acta 143*, 229–236 (1982).
[2923] Jin, L.-t.; H. Zhu, T.-m. Xu, W. Tong, W.-l. Zhou, Y.-z. Fang, *Anal. Chim. Acta 268/1*, 159–162 (1992).
[2924] Jin, L.-z.; Z.-m. Ni, *Can. J. Spectrosc. 26/5*, 219–223 (1981).
[2925] Jin, L.-z.; *At. Spectrosc. 5/3*, 91–95 (1984).
[2926] Jin, L.-z.; D. Wu, Z.-m. Ni, *Huaxue Xuebao 45/3*, 808–812 (1987).
[2927] Jin, Z.; J. Shougui, C. Shikun, J. Desen, D. Chakraborti, *Fresenius J. Anal. Chem. 337/8*, 877–881 (1990).
[2928] Johannessen, J.-K.; B. Gammelgaard, O. Jøns, S.H. Hansen, *J. Anal. At. Spectrom. 8/7*, 999–1004 (1993).
[2929] Johansson, M.; D.C. Baxter, W. Frech, *J. Anal. At. Spectrom. 10/10*, 711–720 (1995).
[2930] Johansson, M.; D.C. Baxter, K.E.A. Ohlsson, W. Frech, *Spectrochim. Acta, Part B 52/5*, 643–656 (1997).
[2931] Johnson, D.; J.B. Headridge, C.W. McLeod, K.W. Jackson, J.A. Roberts, *Anal. Proc. (London) 23/1*, 8–9 (1986).
[2932] Johnson, H.N.; G.F. Kirkbright, R.J. Whitehouse, *Anal. Chem. 45/9*, 1603–1606 (1973).
[2933] Johnson, M.L.; *Anal. Biochem. 206*, 215–225 (1992).
[2934] Jones, A.H.; *At. Absorpt. Newsletter 9/1*, 1–5 (1970).
[2935] Jones, A.H.; B.W. Smith, J.D. Winefordner, *Anal. Chem. 61/15*, 1670–1674 (1989).
[2936] Jones, B.T.; M.A. Mignardi, B.W. Smith, J.D. Winefordner, *J. Anal. At. Spectrom. 4/7*, 647–651 (1989).
[2937] Jones, D.G.; *Anal. Chem. 57*, 1057A (1985).
[2938] Jones, D.G.; *Anal. Chem. 57*, 1207A (1985).
[2939] Jones, D.R.; H.C. Tung, S.E. Manahan, *Anal. Chem. 48/1*, 7–10 (1976).
[2940] Jones, D.R.; S.E. Manahan, *Anal. Chem. 48*, 502–505 (1976).

[2941] Jones, J.B.; V.W. Case, in: R.L. Westerman (Editor), *Soil Testing and Plant Analysis*, Soil Science Society of America, Madison, WI, 1990, 389–427.
[2942] Jones, J.B.; B. Wolf, H.A. Mills, *Plant Analysis Handbook - A Practical Sampling, Preparation, Analysis and Interpretation Guide*, Micro-Macro Publishing, Georgia, 1991.
[2943] Jones, J.W.; in: H.A. McKenzie, L.E. Smythe (Editors), *Quantitative trace analysis of biological materials*, Elsevier, Amsterdam - New York - Oxford, 1988, 353–365.
[2944] Jones, K.C.; P.J. Peterson, B.E. Davies, *Int. J. Environ. Anal. Chem. 20/3-4*, 247–254 (1985).
[2945] Jones, W.G.; A. Walsh, *Spectrochim. Acta 16*, 249–254 (1960).
[2946] Jong, G.J. de ; U.A.Th. Brinkman, *Anal. Chim. Acta 98*, 243–250 (1978).
[2947] Jonghe, W.R.A. de ; F.C. Adams, *Anal. Chim. Acta 108*, 21–30 (1979).
[2948] Jonghe, W.R.A. de ; D. Chakraborti, F.C. Adams, *Anal. Chem. 52/12*, 1974–1977 (1980).
[2949] Jonghe, W.R.A. de ; D. Chakraborti, F.C. Adams, *Anal. Chim. Acta 115*, 89–101 (1980).
[2950] Jonghe, W.R.A. de ; F.C. Adams, *Talanta 29/12*, 1057–1067 (1982).
[2951] Jordan, P.; J.M. Ives, G.R. Carnrick, W. Slavin, *At. Spectrosc. 10/6*, 165–169 (1989).
[2952] Jordanov, N.; I. Havezov, O. Bozhkov, *Fresenius Z. Anal. Chem. 335/8*, 910–913 (1989).
[2953] Jordanov, N.; Yo. Ivanova, *Fresenius J. Anal. Chem. 354/3*, 316–318 (1996).
[2954] Jorhem, L.; S. Slorach, *Food Addit. Contam. 4/3*, 309–316 (1987).
[2955] Jorhem, L.; B. Sundström, C. Åstrand, G. Haegglund, *Z. Lebensm. Unters. Forsch. 188*, 39–44 (1989).
[2956] Jorhem, L.; *J. AOAC International 76/4*, 798–813 (1993).
[2957] Jorhem, L.; *Mikrochim. Acta (Vienna) 119/3-4*, 211–218 (1995).
[2958] Joselow, M.M.; J.D. Bogden, *At. Absorpt. Newsletter 11/5*, 99–101 (1972).
[2959] Joselow, M.M.; J.D. Bogden, *At. Absorpt. Newsletter 11/6*, 127–128 (1972).
[2960] Joselow, M.M.; N.P. Singh, *At. Absorpt. Newsletter 12/5*, 128 (1973).
[2961] Joseph, K.T.; Parameswaran, S.D. Soman, *At. Absorpt. Newsletter 8/6*, 127–128 (1969).
[2962] Jotov, T.; *Bulg. J. Phys. 12/4*, 415–424 (1985).
[2963] Jowett, P.L.H.; M.I. Banton, *Anal. Lett. 19/11*, 1243–1258 (1986).
[2964] Julshamn, K.; O.R. Braekkan, *At. Absorpt. Newsletter 12*, 139–141 (1973).
[2965] Julshamn, K.; O.R. Braekkan, *At. Absorpt. Newsletter 14/3*, 49–52 (1975).
[2966] Julshamn, K.; *At. Absorpt. Newsletter 16/6*, 149–150 (1977).
[2967] Julshamn, K.; A. Maage, E.H. Larsen, *Fresenius J. Anal. Chem. 355/3-4*, 304–307 (1996).
[2968] Jung, P.D.; D. Clarke, *J. Assoc. Off. Anal. Chem. 57/2*, 379–381 (1974).
[2969] Jung, W.T.; J.Y. Shin, S.K. Kim, C.K. Yim, *At. Spectrosc. 15/3*, 122–125 (1994).
[2970] Jungreis, E.; Z. Anavi, *Anal. Chim. Acta 45*, 190 (1969).
[2971] Jungreis, E.; F. Ain, *Anal. Chim. Acta 88*, 191–192 (1977).
[2972] Jurasovic, J.; S. Telisman, *J. Anal. At. Spectrom. 8/3*, 419–425 (1993).
[2973] Jurczyk, J.; T. Glenc, I. Sheybal, K. Swiderska, *Chem. Anal. (Warsaw) 29/5*, 541–547 (1984).
[2974] Kabata-Pendias, A.; H. Pendias, *Trace Elements in Soils and Plants*, CRC Press, Boca Raton, FL, 1984.
[2975] Kabil, M.A.; M.A. Mostafa, *Bull. Soc. Chim. Belg. 94/4*, 253–260 (1985).
[2976] Kabil, M.A.; *J. Anal. At. Spectrom. 10/10*, 733–738 (1995).
[2977] Kacprzak, J.L.; *Int. J. Environ. Anal. Chem. 38/4*, 561–564 (1990).
[2978] Kägler, S.H.; *Erdoel, Kohle, Erdgas, Petrochem. 19*, 897–899 (1966).
[2979] Kägler, S.H.; *Erdoel, Kohle, Erdgas, Petrochem. 24*, 13–19 (1971).
[2980] Kägler, S.H.; *Schmiertechnik + Tribologie (Hannover) 25/2*, 46–51 (1978).
[2981] Kägler, S.H.; *Schmiertechnik + Tribologie (Hannover) 25/3*, 84–88 (1978).
[2982] Kägler, S.H.; E. Jantzen, *Fresenius Z. Anal. Chem. 310/5*, 401–405 (1982).
[2983] Kägler, S.H.; B. Dietzel, J. Frigge, W. Garbe, R. Kotzel, D. Kumar, *Deutsche Ges. Mineralöl. Kohlechemie ; Bericht 264*, 1983.
[2984] Kägler, S.H.; R. Kotzel, in: B. Welz (Editor), *Fortschritte in der atomspektrometrischen Spurenanalytik*, Verlag Chemie, Weinheim, 1984, 233–243.
[2985] Kägler, S.H.; *Erdoel, Kohle, Erdgas, Petrochem. 39/5*, 223–227 (1986).
[2986] Kägler, S.H.; *Erdoel, Kohle, Erdgas, Petrochem. 39/6*, 269–275 (1986).
[2987] Kägler, S.H.; *Neue Mineralölanalyse Bd. 1 Spektroskopie*, Dr. Alfred Hüthig Verlag, Heidelberg, 1987.
[2988] Kägler, S.H.; in: S.H. Kägler (Editor), *Neue Mineralölanalyse, Band 1: Spektroskopie*, Dr. Alfred Hüthig Verlag, Heidelberg, 1987, 253–452.
[2989] Kägler, S.H.; *Erdöl, Erdgas, Kohle 104/3*, 131–138 (1988).
[2990] Kahn, H.L.; *At. Absorpt. Newsletter 2*, 35–40 (1963).
[2991] Kahn, H.L.; D.C. Manning, *At. Absorpt. Newsletter 4/4*, 264–266 (1965).
[2992] Kahn, H.L.; *At. Absorpt. Newsletter 6/2*, 51–52 (1967).

[2993] Kahn, H.L.; J.E. Schallis, *At. Absorpt. Newsletter 7/1*, 5–9 (1968).
[2994] Kahn, H.L.; *At. Absorpt. Newsletter 7*, 40–43 (1968).
[2995] Kahn, H.L.; G.E. Peterson, J.E. Schallis, *At. Absorpt. Newsletter 7*, 35–39 (1968).
[2996] Kahn, H.L.; J.S. Sebestyen, *At. Absorpt. Newsletter 9/2*, 33–38 (1970).
[2997] Kahn, H.L.; G.E. Peterson, D.C. Manning, *At. Absorpt. Newsletter 9/3*, 79–80 (1970).
[2998] Kahn, H.L.; *At. Absorpt. Newsletter 10/2*, 58–59 (1971).
[2999] Kahn, H.L.; G. Dulude, Mary K. Conley, J.J. Sotera, *Can. Res. 14/1*, 27–31 (1981).
[3000] Kahn, H.L.; *Spectrochim. Acta, Part B 39/2-3*, 167 (1984).
[3001] Kaiser, G.; D. Götz, P. Schoch, G. Tölg, *Talanta 22*, 889–899 (1975).
[3002] Kaiser, G.; D. Götz, G. Tölg, G. Knapp, B. Maichin, H. Spitzy, *Fresenius Z. Anal. Chem. 291*, 278–291 (1978).
[3003] Kaiser, H.; *Fresenius Z. Anal. Chem. 216*, 80–93 (1966).
[3004] Kaiser, H.; A.C. Menzies, *The Limit of Detection of a Complete Analytical Procedure*, Adam Hilger, London, 1968
[3005] Kaiser, H.; *Spectrochim. Acta, Part B 33*, 551–576 (1978).
[3006] Kaiser, M.L.; S.R. Koirtyohann, E.J. Hindenberger, H.E. Taylor, *Spectrochim. Acta, Part B 36/8*, 773–783 (1981).
[3007] Kaiser, R.E.; *Fresenius Z. Anal. Chem. 256*, 1–6 (1971).
[3008] Kakulu, S.E.; O. Sibanjo, S.O. Ajayi, *Int. J. Environ. Anal. Chem. 30/3*, 209–217 (1987).
[3009] Kalac, P.; J. Burda, I. Stasková, *Sci. Total Environ. 105*, 109–119 (1991).
[3010] Kalaidjieva, *Anal. Lab. 4/4*, 235–241 (1995).
[3011] Kalantar, A.H.; *Analusis 21/2*, 119–122 (1993).
[3012] Kalb, G.W.; *At. Absorpt. Newsletter 9/4*, 84–87 (1970).
[3013] Kallmann, S.; E.W. Hobart, *Anal. Chim. Acta 51*, 120–124 (1970).
[3014] Kallmann, S.; E.W. Hobart, *Talanta 17*, 845–850 (1970).
[3015] Kallmann, S.; P. Blumberg, *Talanta 27*, 827–833 (1980).
[3016] Kalpalma, G.; V.J. Koshy, *Indian J. Chem. (Section A) 31A/1*, 39–42 (1992).
[3017] Kamel, H.; D.H. Brown, J.M. Ottaway, W.E. Smith, *Analyst (London) 101*, 790–797 (1976).
[3018] Kan, M.; T. Nasu, M. Taga, *Anal. Sci. 7/Suppl*, 1115–1116 (1991).
[3019] Kane, J.S.; *J. Anal. At. Spectrom. 3/7*, 1039–1045 (1988).
[3020] Kang, H.K.; J.L. Valentine, *Anal. Chem. 49/12*, 1829–1832 (1977).
[3021] Kanke, M.; T. Kumamaru, K. Sakai, Y. Yamamoto, *Anal. Chim. Acta 247/1*, 13–18 (1991).
[3022] Kántor, T.; P. Fodor, E. Pungor, *Anal. Chim. Acta 102*, 15–23 (1978).
[3023] Kántor, T.; L. Bezúr, E. Pungor, J.D. Winefordner, *Spectrochim. Acta, Part B 38/4*, 581–607 (1983).
[3024] Kántor, T.; *Spectrochim. Acta, Part B 38/11*, 1483–1495 (1983).
[3025] Kántor, T.; *Spectrochim. Acta, Part B 42/4*, 543–551 (1987).
[3026] Kántor, T.; B. Radziuk, B. Welz, *Spectrochim. Acta, Part B 49/9*, 875–891 (1994).
[3027] Kapeller, R.; F. Purtscheller, E. Schnell, *Mikrochim. Acta (Vienna) I*, 115–134 (1985).
[3028] Kapetan, J.P.; *ASTM Spec. Tech. Publ. 443*, 78 (1969).
[3029] Karadjova, I.; L. Jordanova, S. Arpadjan, *Z. Chem. 30/6*, 333 (1990).
[3030] Karadjova, I.; S. Arpadjan, T. DeliGeorgiev, *Anal. Chim. Acta 244/1*, 123–127 (1991).
[3031] Karadjova, I.; P. Mandjukov, S. Tsakovsky, V. Simeonov, J.A. Stratis, G.A. Zachariadis, *J. Aanal. At. Spectrom. 10/12*, 1065–1068 (1995).
[3032] Karanassios, V.; F.H. Li, B. Liu, E.D. Salin, *J. Anal. At. Spectrom. 6/6*, 457–463 (1991).
[3033] Karin, R.W.; J.A. Buono, J.L. Fasching, *Anal. Chem. 47/13*, 2296–2299 (1975).
[3034] Karkhanis, P.P.; J.R. Anfinsen, *J. Assoc. Off. Anal. Chem. 56/2*, 358–360 (1973).
[3035] Karve, M.A.; S.M. Khopkar, *Anal. Sci. 8/1*, 77–80 (1992).
[3036] Kashiki M; S. Oshima, *Anal. Chim. Acta 51*, 387–392 (1970).
[3037] Kashiki, M.; S. Yamazoe, S. Oshima, *Anal. Chim. Acta 53*, 95–100 (1971).
[3038] Kashiki, M.; S. Oshima, *Anal. Chim. Acta 55*, 436 (1971).
[3039] Kashiki, M.; S. Yamazoe, N. Ikeda, S. Oshima, *Anal. Lett. 7/1*, 53–64 (1974).
[3040] Kasimova, O.G.; N.I. Shcherbinina, E.M. Sedykh, L.I. Bol'shakova, G.V. Myasoedova, *Zh. Anal. Khim. 39*, 1823 (1984).
[3041] Kasper, H.U.; in: K. Dittrich, B. Welz (Editors), *CANAS'93, Colloq. Analytische Atomspektroskopie*, Universität Leipzig/UFZ Leipzig-Halle, Leipzig, 1993, 361–370.
[3042] Kaszermann, R.; K. Theurer, *At. Absorpt. Newsletter 15/6*, 129–133 (1976).
[3043] Kateman, G.; L. Buydens, *Quality Control in Analytical Chemistry*, J. Wiley & Sons., Chichester, 1993.
[3044] Kato, K.; *Fresenius Z. Anal. Chem. 326*, 524–527 (1987).

[3045] Katskov, D.A.; L.P. Kruglikova, B.V. L'vov, L.K. Polzik, *Zh. Prikl. Spektrosk. 20*, 739 (1974).
[3046] Katskov, D.A.; L.P. Kruglikova, B.V. L'vov, *Zh. Anal. Khim. 30/2*, 238–243 (1975).
[3047] Katskov, D.A.; A.M. Shtepan, I.L. Grinshtein, A.A. Pupyshev, *Spectrochim. Acta, Part B 47/8*, 1023–1041 (1992).
[3048] Katsura, T.; F. Kato, K. Matsumoto, *Anal. Chim. Acta 252/1-2*, 77–81 (1991).
[3049] Katz, A.; *Chem. Geol. 16*, 15–25 (1975).
[3050] Katz, S.A., *Int. Biotechnol. Lab. 3 (Juni)*, 10–16 (1985).
[3051] Kauffman, R.E.; C.S. Saba, W.E. Rhine, K.J. Eisentraut, *Anal. Chem. 54/6*, 975–979 (1982).
[3052] Kauppinen, M.; K. Smolander, *Anal. Chim. Acta 285/1-2*, 45–51 (1994).
[3053] Kauppinen, M.; K. Smolander, *Anal. Chim. Acta 296/2*, 195–203 (1994).
[3054] Kavlentis, E.; *Anal. Chim. Acta 208*, 212–316 (1988).
[3055] Kawai, H.; Y. Katayama, Y. Ninomiya, J. Okuda, *Bunseki Kagaku 43/12*, 1193 (1994).
[3056] Kawamoto, H.; R. Yokoyama, K. Tsunoda, H. Akaiwa, *Anal. Sci. 8/4*, 571– 573 (1992).
[3057] Kawamura, H.; G. Tanaka, N. Kurita, Y. Ohyagi, *Bull. Chem. Soc. Jpn. 43*, 970 (1970).
[3058] Kawamura, T.; Y. Sato, A. Yazawa, K. Sakurai, *Iyakuhin Kenkyu 23*, 76 (1992).
[3059] Kawamata, Y.; S. Sakurai, E. Yoshimoto, T. Kitamura, T. Shimizu, Y. Shijo, *Bunseki Kagaku 44/2*, 117–122 (1995).
[3060] Kaye, J.H.; N.E. Ballou, *Anal. Chem. 50*, 2076–2078 (1978).
[3061] Keating, A.D.; J.L. Keating, D.J. Halls, G.S. Fell, *Analyst (London) 112/10*, 1381–1385 (1987).
[3062] Kedziora, M.; A. Parczewski, *Chem. Anal. (Warsaw) 34/3-6*, 453–459 (1989).
[3063] Kedziora, M.; A. Parczewski, *Chem. Anal. (Warsaw) 35/6*, 781–787 (1990).
[3064] Kehoe, D.F.; D.M. Sullivan, R.L. Smith, *J. Assoc. Off. Anal. Chem. 71/6*, 1153–1155 (1988).
[3065] Keil, R.; *Fresenius Z. Anal. Chem. 297*, 44–48 (1979).
[3066] Keil, R.; *Fresenius Z. Anal. Chem. 319*, 391–394 (1984).
[3067] Keirsse, H.; J. Smeyers-Verbeke, D. Verbeelen, D.L. Massart, *Anal. Chim. Acta 196*, 103–114 (1987).
[3068] Keith, L.H.; W. Crummett, J. Deegan, R.A. Libby, J.K. Taylor, G. Wentler, *Anal. Chem. 55/14*, 2210–2218 (1983).
[3069] Keleman, J.; O. Szakács, A. Lásztity, *J. Anal. At. Spectrom. 5/5*, 377–384 (1990).
[3070] Keliher, P.N.; C.C. Wohlers, *Anal. Chem. 48/1*, 140–143 (1976).
[3071] Keliher, P.N.; C.C. Wohlers, *Anal. Chem. 48/3*, 333A-340A (1976).
[3072] Kelly, W.R.; C.B. Moore, *Anal. Chem. 45/7*, 1274–1275 (1973).
[3073] Kelner, M.J.; in: H.G. Seiler, A. Sigel, H. Sigel (Editors), *Handbook on Metals in Clinical and Analytical Chemistry*, Marcel Dekker, New York, 1994, 601–610.
[3074] Kemp, A.J.; C.J. Brown, *Analyst (London) 115/9*, 1197–1199 (1990).
[3075] Kemp, G.J.; *Clin. Chem. (Winston-Salem) 30/7*, 1168–1170 (1984).
[3076] Kenawy, I.M.M.; M.M. El-Defrawy, M.Sh. Khalil, K.S. El-Said, *Analusis 20*, 561–565 (1992).
[3077] Kendrick, M.J.; M.J. Plishka, K.D. Robinson, *Metals in Biological Systems* [Series in Inorganic Chemistry], Ellis Horwood, 1992.
[3078] Kennedy, K.R.; J.G. Crock, *Anal. Lett. 20/6*, 899–908 (1987).
[3079] Kepak, F.; *Chem. Rev. 71*, 357 (1971).
[3080] Kerber, J.D.; W.B. Barnett, *At. Absorpt. Newsletter 8/6*, 113–116 (1969).
[3081] Kerber, J.D.; A. Koch, G.E. Peterson, *At. Absorpt. Newsletter 12/4*, 104–105 (1973).
[3082] Kerbyson, J.D.; C. Ratzkowski, *Can. Spectroscopy 13/4*, 102–107 (1968).
[3083] Kersabiec, A.M. de ; M.F. Benedetti, *Fresenius Z. Anal. Chem. 328/4-5*, 342–345 (1987).
[3084] Kersabiec, A.M. de ; F. Vidot, J. Boulègue, I. Verhaeghe, *Analusis 18/3*, 214–216 (1990).
[3085] Kersten, M.; U. Förstner, *Water Sci. Technol. 18*, 121–130 (1986).
[3086] Kersten, M.; U. Förstner, in: A.M. Ure, C.M. Davidson (Editors), *Chemical Speciation in the Environment*, Blackie Academic Professional, Glasgow, 1995, 234–275.
[3087] Keshan Research Group, *Chin. Med. J. 92*, 471–476 (1979).
[3088] Khalid, N.; S. Rahman, R. Ahmed, I.H. Qureshi, *Int. J. Environ. Anal. Chem. 28/1-2*, 133–141 (1987).
[3089] Khammas, Z.A.; J. Marshall, D. Littlejohn, J.M. Ottaway, S.C. Stephen, *Mikrochim. Acta (Vienna) I*, 333–355 (1985).
[3090] Khammas, Z.A.-A.; M.H. Farhan, M.M. Barbooti, *Talanta 36/10*, 1027–1030 (1989).
[3091] Kharbade, B.; K. Agarwal, *At. Spectrosc. 14*, 13–15 (1993).
[3092] Kharlamov, I.P.; V.Yu. Karyakin, *Zavodsk. Lab. 55/8*, 36 (1989).
[3093] Kheboian, C.; C.F. Bauer, *Anal. Chem. 59/10*, 1417–1423 (1987).

[3094] Khozhainov, Y.M.; O.A. Tyurin, N.P. Deinikina, *Zavodsk. Lab. 51/4*, 30 (1985).
[3095] Khozhainov, Y.M.; N.V. Dolova, *Zavodsk. Lab. 53/4*, 28 (1987).
[3096] Kielkopf, J.F.; in: B. Wende (Editor), *Spectral Line Shapes*, W. de Gruyter, Berlin, New York, 1981, 665–688.
[3097] Kiilunen, M.; J. Jarvisalo, O. Makitie, A. Aitio, *Int. Arch. Occup. Environ. Health 59/1*, 43–50 (1987).
[3098] Kildahl, B.T.; W. Lund, *Fresenius J. Anal. Chem. 354*, 93–96 (1996).
[3099] Kilroe-Smith, T.A.; H.B. Röllin, *Microchem. J. 42*, 349–354 (1990).
[3100] Kilroy, W.P.; C. Moynihan, *Anal. Chim. Acta 83*, 389 (1976).
[3101] Kim, H.J.; E.H. Piepmeier, *Anal. Chem. 60/19*, 2040–2046 (1988).
[3102] Kim, N.D.; J.E. Fergusson, *Sci. Total Environ. 105*, 191–209 (1991).
[3103] Kim, N.D.; S.J. Hill, *Environ. Technol. 14/11*, 1015–1026 (1993).
[3104] Kimberly, M.M.; G.G. Bailey, D.C. Paschal, *Analyst (London) 112*, 287–290 (1987).
[3105] Kimbrough, D.E.; J. Wakakuma, *Analyst (London) 119/3*, 383–388 (1994).
[3106] King, A.S.; *Astrophys. J. 28*, 300 (1908).
[3107] King, E.E.; R.C. Fry, M. van Swaay, *Anal. Chem. 58/3*, 642–647 (1986).
[3108] King, J.N.; J.S. Fritz, *Anal. Chem. 57/6*, 1016–1020 (1985).
[3109] King, W.G.; J.M. Rodriguez, C.M. Wai, *Anal. Chem. 46*, 771–773 (1974).
[3110] King, W.H.; *Isotopic Shifts in Atomic Spectra*, Plenum Press, New York, 1984.
[3111] Kingston, H.M.; I.L Barnes, T.J. Brady, T.C. Rains, M.A. Champ, *Anal. Chem. 50/14*, 2064–2070 (1978).
[3112] Kingston, H.M.; L.B. Jassie, (Editors), *Introduction to Microwave Sample Preparation: Theory and Practice*, American Chemical Society, Washington, DC, 1988.
[3113] Kingston, H.M.; L.B. Jassie, *J. Res. Natl. Bur. Stand. 93/3*, 269–274 (1988).
[3114] Kingston, H.M.; P.J. Walter, *Spectroscopy (Eugene, OR) 7/9*, 20–27 (1992).
[3115] Kinoshita, M.; N. Yoshimura, H. Ogata, D. Tsujino, T. Takahashi, S. Takahashi, Y. Wada, K. Someya, T. Ohno, K. Masuhara, Y. Tanaka, *J. Chromatogr. 529*, 462 (1990).
[3116] Kinsella, B.; R.L. Willix, *Anal. Chem. 54/14*, 2614–2616 (1982).
[3117] Kinson, K.; C.B. Belcher, *Anal. Chim. Acta 30*, 64–67 (1964).
[3118] Kinson, K.; C.B. Belcher, *Anal. Chim. Acta 31*, 180–183 (1964).
[3119] Kipsch, D.; *Neue Hütte 36/1*, 27–32 (1991).
[3120] Kirchhof, H.; *Spectrochim. Acta, Part B 24/4*, 235–241 (1969).
[3121] Kirchhoff, G.; *Ann. Physik (Poggendorf's Annalen) 109*, 275 (1860).
[3122] Kirchhoff, G.; *Phil. Mag. 20*, 1 (1860).
[3123] Kirchhoff, G.; R. Bunsen, *Phil. Mag. 20*, 89 (1860).
[3124] Kirchhoff, G.; R. Bunsen, *Phil. Mag. 22*, 329–349 (1861).
[3125] Kirchner, C.H.; G.A. Eagle, H. Hennig, *Int. J. Environ. Anal. Chem. 32/1*, 9–21 (1988).
[3126] Kirk, M.; E.G. Perry, J.M. Arritt, *Anal. Chim. Acta 80*, 163–169 (1975).
[3127] Kirkbright, G.F.; T.S. West, P.J. Wilson, *At. Absorpt. Newsletter 11/3*, 53–56 (1972).
[3128] Kirkbright, G.F.; *Some studies in flame AA and high frequency plasma emission spectrometry at wavelengths less than 200 nm*, II CS. Conf. Flame Spectrosc., Zvikov, 1973.
[3129] Kirkbright, G.F.; P.J. Wilson, *Anal. Chem. 46/11*, 1414–1418 (1974).
[3130] Kirkbright, G.F.; P.J. Wilson, *At. Absorpt. Newsletter 13/6*, 140–143 (1974).
[3131] Kirkbright, G.F.; M. Sargent, *Atomic Absorption and Fluorescence Spectroscopy*, Academic Press, London, 1974.
[3132] Kirkbright, G.F.; M. Taddia, *Anal. Chim. Acta 100*, 145–150 (1978).
[3133] Kirkbright, G.F.; M. Taddia, *At. Absorpt. Newsletter 18/3*, 68–70 (1979).
[3134] Kirkbright, G.F.; X.-q. Shan, R.D. Snook, *At. Spectrosc. 1/4*, 85–89 (1980).
[3135] Kirkwood, C.; S.K. Wilson, P.E. Hayes, W.H. Barr, M.A. Sarkar, P.G. Ettigi, *Am. J. Hosp. Pharm. 51*, 486 (1994).
[3136] Kisfaludi, G.; M. Lenhof, *Anal. Chim. Acta 55*, 442 (1971).
[3137] Kiss, E.; *Anal. Chim. Acta 140*, 197–204 (1982).
[3138] Kitagawa, K.; T. Takeuchi, *Anal. Chim. Acta 68*, 212–216 (1974).
[3139] Kitagawa, K.; Y. Ide, T. Takeuchi, *Anal. Chim. Acta 113*, 21–32 (1980).
[3140] Klaassen, C.D.; L.D. Lehman-McKeeman, *Methods Emzymol. 205*, 190–198 (1991).
[3141] Klaessens, J.W.A.; G. Kateman, *Fresenius Z. Anal. Chem. 326*, 203–213 (1987).
[3142] Klaos, E.; V. Odinets, *Talanta 37/5*, 519–526 (1990).
[3143] Klein, B.; J.H. Kaufman, M. Oklander, *Clin. Chem. (Winston-Salem) 13*, 797 (1967).
[3144] Kleinpoppen, H.; A. Scharmann, in: W. Hanle, H. Kleinpoppen (Editors), *Progress in Atomic Spectroscopy*, Plenum Press, New York, 1978, 329–390.
[3145] Klemm, W.; G. Bombach, *Fresenius J. Anal. Chem. 353/1*, 12–15 (1995).

[3146] Klich, H.; R. Pradel, *Fresenius J. Anal. Chem. 352/1-2*, 23–27 (1995).
[3147] Klinkhammer, G.P.; *Anal. Chem. 52/1*, 117–120 (1980).
[3148] Klockow, D.; *Fresenius Z. Anal. Chem. 326*, 5–24 (1987).
[3149] Klüßendorf, B.; A. Rosopulo, W. Kreuzer, *Fresenius Z. Anal. Chem. 322/7*, 721–727 (1985).
[3150] Kmetov, V.; L. Futekov, *Fresenius J. Anal. Chem. 338/8*, 895–897 (1990).
[3151] Knapp, G.; *Fresenius Z. Anal. Chem. 274*, 271–273 (1975).
[3152] Knapp, G.; B. Sadjadi, H. Spitzy, *Fresenius Z. Anal. Chem. 274*, 275–278 (1975).
[3153] Knapp, G.; S.E. Raptis, B. Schreiber, in: P. Brätter, P. Schramel (Editors), *Trace Element Analytical Chemistry in Medicine and Biology*, W. de Gruyter, Berlin - New York, 1980, 523–529.
[3154] Knapp, G.; *Anal. Proc. (London) 27/5*, 112–116 (1990).
[3155] Knapp, G.; B. Maichin, F. Panholzer, in: B. Welz (Editor), *6. Colloquium Atomspektrometrische Spurenanalytik*, Bodenseewerk Perkin-Elmer, Überlingen, 1991, 571–580.
[3156] Knauer, H.E.; G.E. Milliman, *Anal. Chem. 47*, 1263–1268 (1975).
[3157] Knecht, J.; G. Stork, *Fresenius Z. Anal. Chem. 270*, 97–99 (1974).
[3158] Knecht, J.; *Fresenius Z. Anal. Chem. 316/4*, 409–412 (1983).
[3159] Knecht, U.; *Aerztl. Lab. 31*, 215 (1985).
[3160] Knechtel, J.R.; *Analyst (London) 105*, 826–829 (1980).
[3161] Knezevic, G.; *Papier (Darmstadt) 34/6*, 226–228 (1980).
[3162] Knezevic, G.; *Papier (Darmstadt) 36/11*, 534–536 (1982).
[3163] Knezevic, G.; *GIT Fachz. Lab. 28/3*, 178–181 (1984).
[3164] Knezevic, G.; *Papier (Darmstadt) 38/9*, 430–431 (1984).
[3165] Knezevic, G.; U. Kurfürst, *Fresenius Z. Anal. Chem. 322/7*, 717–718 (1985).
[3166] Knezevic, G.; *Verpack-Rundsch. 37*, 39–42 (1986).
[3167] Knezevic, G.; O. Toeppel, *Papier (Darmstadt) 45*, 285 (1991).
[3168] Knight, D.M.; M.K. Pyzyna, *At. Absorpt. Newsletter 8/6*, 129–130 (1969).
[3169] Kniseley, R.N.; A.P. D'Silva, V.A. Fassel, *Anal. Chem. 35*, 910–911 (1963).
[3170] Knowles, M.B.; K.G. Brodie, *J. Anal. At. Spectrom. 3/4*, 511–516 (1988).
[3171] Knowles, M.B.; *J. Anal. At. Spectrom. 4/3*, 257–260 (1989).
[3172] Knowles, M.B.; K.G. Brodie, *J. Anal. At. Spectrom. 4/3*, 305–306 (1989).
[3173] Knudson, E.J.; G.D. Christian, *Anal. Lett. 6/12*, 1039–1054 (1973).
[3174] Knudsen, E.; G. Wibetoe, I. Martinsen, *J. Anal. At. Spectrom. 10/10*, 757–761 (1995).
[3175] Knutson, A.; *J. Assoc. Off. Anal. Chem. 66/4*, 946–948 (1983).
[3176] Knutti, R.; in: B. Welz (Editor), *Fortschritte in der atomspektrometrischen Spurenanalytik*, VCH Verlagsgesellschaft, Weinheim, 1986, 397–408.
[3177] Kobayashi, K.; T. Akagi, H. Haraguchi, *Bull. Chem. Soc. Jpn. 63*, 554–558 (1990).
[3178] Kobayashi, R.; K. Imaizumi, *Anal. Sci. 6/1*, 83–90 (1990).
[3179] Kobayashi, R.; K. Imaizumi, *Biomed. Res. Trace Elem. 1*, 235 (1990).
[3180] Kobayashi, R.; K. Imaizumi, *Anal. Sci. 7/3*, 447–450 (1991).
[3181] Kobayashi, R.; K. Imaizumi, *Anal. Sci. 7/Suppl*, 841–844 (1991).
[3182] Kobayashi, R.; K. Imaizumi, M. Kudo, *Anal. Sci. 11/2*, 267–269 (1995).
[3183] Kobayashi, T.; F. Hirose, S. Hasegawa, H. Okochi, *J. Japan Inst. Metals 49*, 656 (1985).
[3184] Kobayashi, T.; H. Okochi, *J. Japan Inst. Metals 53*, 1123 (1989).
[3185] Koch, H.; H.J. Eichler, *J. Appl. Phys. 54/9*, 4939–4946 (1983).
[3186] Koch, K.-H.; CLB, *Chem. Labor Betr. 37*, 282 (1986).
[3187] Koch, K.-H.; *Mikrochim. Acta (Vienna) I*, 151–167 (1987).
[3188] Koch, K.R.; M.A.B. Pougnet, S. De Villiers, *Analyst (London) 114/8*, 911–913 (1989).
[3189] Kodama, M.; S. Shimizu, T. Tominaga, *Anal. Lett. 10*, 591–598 (1977).
[3190] Kodama, M.; S. Miyagawa, *Anal. Chem. 52/14*, 2358–2361 (1980).
[3191] Koeman, J.H.; W.H.M. Peeters, C.H.M. Kondstaal, P.S. Tjioe, J.J.M. DeGoed, *Nature (London) 245*, 385–386 (1973).
[3192] Kohri, M.; O. Kujirai, H. Okochi, *Anal. Sci. 7/5*, 767–771 (1991).
[3193] Koide, M.; V. Hodge, J.S. Yang, E.D. Goldberg, *Anal. Chem. 59/14*, 1802–1805 (1987).
[3194] Koide, M.; V. Hodge, E.D. Goldberg, K. Bertine, *Appl. Geochem. 3*, 237–241 (1988).
[3195] Koirtyohann, S.R.; C. Feldman, in: J.E. Forrette, E. Lanterman (Editors), *Developments in Applied Spectroscopy*, Plenum Press, New York, 1964, 180.
[3196] Koirtyohann, S.R.; E.E. Pickett, *Anal. Chem. 37/4*, 601–603 (1965).
[3197] Koirtyohann, S.R.; *At. Absorpt. Newsletter 6*, 77–84 (1967).
[3198] Koirtyohann, S.R.; E.E. Pickett, *Anal. Chem. 40/13*, 2068–2070 (1968).
[3199] Koirtyohann, S.R.; E.E. Pickett, *Spectrochim. Acta, Part B 23/10,* 673–685 (1968).

[3200] Koirtyohann, S.R.; in: J.A. Dean, T.C. Rains (Editors), *Flame Emission and Atomic Absorption Spectrometry, Vol. 1 Theory,* Marcel Dekker, New York - London, 1969, 295–315.
[3201] Koirtyohann, S.R.; M. Hamilton, *J. Assoc. Off. Anal. Chem. 54,* 787 (1971).
[3202] Koirtyohann, S.R.; *J. Assoc. Off. Anal. Chem. 55/5,* 989–990 (1972).
[3203] Koirtyohann, S.R.; *4th ICAS, Toronto 1973, Beitrag 27,* 1973.
[3204] Koirtyohann, S.R.; M. Khalil, *Anal. Chem. 48,* 136–139 (1976).
[3205] Koirtyohann, S.R.; *Spectrochim. Acta, Part B 35,* 663–670 (1980).
[3206] Koirtyohann, S.R.; E.D. Glass, F.E. Lichte, *Appl. Spectrosc. 35/1,* 22–26 (1981).
[3207] Koirtyohann, S.R.; M.L. Kaiser, E.J. Hindenberger, *J. Assoc. Off. Anal. Chem. 65/4,* 999–1004 (1982).
[3208] Koizumi, H.; K. Yasuda, *Anal. Chem. 47/9,* 1679–1682 (1975).
[3209] Koizumi, H.; K. Yasuda, *Anal. Chem. 48/8,* 1178–1182 (1976).
[3210] Koizumi, H.; K. Yasuda, *Spectrochim. Acta, Part B 31/5,* 237–255 (1976).
[3211] Koizumi, H.; K. Yasuda, *Spectrochim. Acta, Part B 31/10,* 523–535 (1976).
[3212] Koizumi, H.; K. Yasuda, M. Katayama, *Anal. Chem. 49/8,* 1106–1112 (1977).
[3213] Koizumi, H.; *Anal. Chem. 50/8,* 1101–1105 (1978).
[3214] Kojima, I.; C. Iida, *J. Anal. At. Spectrom. 2,* 463–467 (1987).
[3215] Kojima, I.; F. Jinno, Y. Noda, C. Iida, *Anal. Chim. Acta 245/1,* 35–41 (1991).
[3216] Kojima, I.; H. Fukumori, C. Iida, *Anal. Sci. 8/4,* 533–537 (1992).
[3217] Kojima, I.; A. Suzuki, *Bunseki Kagaku 42/7,* 435–437 (1993).
[3218] Kojima, I.; A. Takayanagi, *J. Anal. At. Spectrom. 11/8,* 607–610 (1996).
[3219] Kokkonen, P.; M. Palko, L.H.J. Lajunen, *At. Spectrosc. 8/3,* 98–100 (1987).
[3220] Köklü, Ü.; S. Akman, Acta Chim. Hung. 129/6, 825–829 (1992).
[3221] Kokot, M.L.; *At. Absorpt. Newsletter 15/4,* 105 (1976).
[3222] Kokot, S.; G. King, H.R. Keller, D.L. Massart, *Anal. Chim. Acta 259/2,* 267–279 (1992).
[3223] Kolb, A.; G. Müller-Vogt, W. Wendl, W. Stößel, *Spectrochim. Acta, Part B 42/8,* 951–957 (1987).
[3224] Kolb, B.; G. Kemmner, F.H. Schleser, E. Wiedeking, *Fresenius Z. Anal. Chem. 221,* 166–175 (1966).
[3225] Kölbl, G.; K. Kalcher, K.J. Irgolic, *Anal. Chim. Acta 284/2,* 301–310 (1993).
[3226] Kölbl, G.; K. Kalcher, K.J. Irgolic, *J. Autom. Chem. 15/2,* 37–45 (1993).
[3227] Kölbl, G.; J. Lintschinger, K. Kalcher, K.J. Irgolic, *Mikrochim. Acta (Vienna), 119/1-2,* 113–127 (1995).
[3228] Kölbl, G.; *Mar. Chem. 48,* 185–197 (1995).
[3229] Kölbl, G.; K. Kalcher, K.J. Irgolic, in: V.G. Kumar (Editor), *Main Group Elements and Their Compounds,* Narosa Publ., New Delhi, 1996, 161–172.
[3230] Kolesov, G.M.; V. Anikiev, S.K. Prasad, E.M. Sedykh, Chem. *Anal. (Warsaw) 38/5,* 625–637 (1993).
[3231] Kolosova, L.P.; N.V. Novatskaya, R.I. Ryzhova, A.E. Aladyshkina, *Zh. Anal. Khim. 39,* 1475 (1984).
[3232] Kolosova, L.P.; L.A. Ushinskaya, T.N. Kopylova, *Zh. Anal. Khim. 48/4,* 644–647 (1993).
[3233] Komárek, J.; M. Ganoczy, *Collect. Czech. Chem. Commun. 56/4,* 764–773 (1991).
[3234] Kometani, T.Y.; *Plating 56,* 1251 (1969).
[3235] König, H.P.; H. Kock, R.F. Hertel, in: B. Welz (Editor), *5. Colloquium Atom-spektrometrische Spurenanalytik,* Bodenseewerk Perkin-Elmer, Überlingen, 1989, 647–656.
[3236] König, K.-H.; P. Neumann, *Fresenius Z. Anal. Chem. 279,* 337–345 (1976).
[3237] König, K.-H.; M. Schuster, B. Steinbrech, G. Schneeweis, R. Schlodder, *Fresenius Z. Anal. Chem. 221/5,* 457 (1985).
[3238] König, P.; K.-H. Schmitz, E. Thiemann, *Arch. Eisenhüttenw. 40/1,* 53–56 (1969).
[3239] Konishi, T.; H. Takahashi, *Analyst (London) 108,* 827–834 (1983).
[3240] Koons, R.D.; D.G. Havekost, C.A. Peters, *J. Forensic. Sci. 32/4,* 846–865 (1987).
[3241] Koons, R.D.; D.G. Havekost, C.A. Peters, *J. Forensic. Sci. 34,* 218 (1989).
[3242] Koons, R.D.; C.A. Peters, *J. Anal. Toxicol. 18/1,* 36–40 (1994).
[3243] Koops, J.; C. de Graaf, D. Westerbeek, *Neth. Milk Dairy J. 38/4,* 223–239 (1984).
[3244] Koops, J.; H. Klomp, D. Westerbeek, *Neth. Milk Dairy J. 42/2,* 99–110 (1988).
[3245] Koops, J.; H. Klomp, D. Westerbeek, *Neth. Milk Dairy J. 43/2,* 185–198 (1989).
[3246] Kopp, J.F.; M.C. Longbottom, L.B. Lobring, *J. Am. Water Works Assoc. 64,* 20 (1972).
[3247] Korde, R.; J. Geist, *Appl. Opt. 26,* 5284 (1987).
[3248] Korečková, J.; W. Frech, E. Lundberg, J.-Å. Persson, *Anal. Chim. Acta 130/2,* 267–280 (1981).

[3249] Korečková, J.; D. Weiss, *Chem. Listy 84*, 431 (1990).
[3250] Korenaga, T.; *Analyst (London) 106*, 40–46 (1981).
[3251] Kornahrens, H.; K.D. Cook, D.W. Armstrong, *Anal. Chem. 54/8*, 1325–1329 (1982).
[3252] Kornblum, G.R.; L. de Galan, *Spectrochim. Acta, Part B 28*, 139–147 (1973).
[3253] Koropchak, J.A.; D.H. Winn, *Trends Anal. Chem. (Pers. Ed.) 6/7*, 171–175 (1987).
[3254] Korunová, V.; J. Dědina, *Analyst (London) 105*, 48–51 (1980).
[3255] Koscielniak, P.; A. Parczewski, *Fresenius Z. Anal. Chem. 321/6*, 572–574 (1985).
[3256] Koscielniak, P.; M. Sperling, B. Welz, *Chem. Anal. (Warsaw) 41/4*, 587 (1996).
[3257] Koshino, Y.; A. Narukawa, *Analyst (London) 117/6*, 967–969 (1992).
[3258] Koshino, Y.; A. Narukawa, *Talanta 40/6*, 799–803 (1993).
[3259] Koshino, Y.; A. Narukawa, *Analyst (London) 118/8*, 1027–1030 (1993).
[3260] Kostyniak, P.J.; *J. Anal. Toxicol. 7/1*, 20–23 (1983).
[3261] Kothny, E.L.; *Appl. Spectrosc. 41/4*, 700–702 (1987).
[3262] Kots, L.S.; L.Ya. Kheifets, V.F. Osyka, I.G. Yudelevich, *Vysokochistye Veshchestva 3*, 141–147 (1993).
[3263] Kotz, L.; G. Henze, G. Kaiser, S. Pahlke, M. Veber, G. Knapp, *Talanta 26*, 681–691 (1979).
[3264] Kovatsis, A.V.; M.A. Tsougas, *Bull. Environ. Contam. Toxicol. 15/4*, 412–420 (1975).
[3265] Kovatsis, A.V.; *At. Absorpt. Newsletter 17/5*, 104–106 (1978).
[3266] Kowal, W.A.; P.M. Krahn, O.B. Beattie, *Int. J. Environ. Anal. Chem. 35*, 119–126 (1989).
[3267] Krachler, M.; H. Radner, K.J. Irgolic, *Fresenius J. Anal. Chem. 355/2*, 120–128 (1996).
[3268] Kraft, E.A.; *Am. Lab. (Fairfield, CT) 1/Aug*, 8 (1969).
[3269] Kraft, G.; D. Lindenberger, H. Beck, *Fresenius Z. Anal. Chem. 282*, 119–121 (1976).
[3270] Kraft, J.; D. Truckenbrodt, J. Einax, *GIT Fachz. Lab. 40/1*, 33–36 (1996).
[3271] Kragten, J.; *At. Spectrosc. 2/4*, 135–136 (1981).
[3272] Krakovská, E.; *J. Anal. At. Spectrom. 5/3*, 205–207 (1990).
[3273] Krakovská, E.; *CLB, Chem. Labor Betr. 42/5*, 247–250 (1991).
[3274] Krakovská, E.; Zh. Rybárová, in: B. Welz (Editor), *6. Colloquium Atomspektrometrische Spurenanalytik*, Bodenseewerk Perkin-Elmer, Überlingen, 1991, 87–97.
[3275] Krakovská, E.; P. Puliš, in: K. Dittrich, B. Welz (Editors), *CANAS'93, Colloq. Analytische Atomspektroskopie*, Proc. Conf. 1993, Oberhof, Universität Leipzig/UFZ Leipzig-Halle, Leipzig, 1993, 357–390.
[3276] Krakovská, E.; N. Pliesovská, K. Flórián, *CLB, Chem. Labor Betr. 46/8*, 368–373 (1995).
[3277] Krakovská, E.; P. Puliš, *Spectrochim. Acta, Part B 51/9-10*, 1271–1275 (1996).
[3278] Král, M. Sybr, Z. Plzák, *Anal. Chim. Acta 304/2*, 237–241 (1995).
[3279] Kramer, J.R.; in: R.M. Harrison, R. Perry (Editors), *Handbook of Air Pollution Analysis*, Chapman and Hall, New York, 1986, 535–561.
[3280] Kranz, E.; *Emissionsspektroskopie*, Akademie Verlag, Berlin, 1964.
[3281] Kranz, E.; *Spectrochim. Acta, Part B 27/8*, 327–343 (1972).
[3282] Kratochvil, B.G.; J.K. Taylor, *NBS Tech. Note (U.S.) 1153*, 1982.
[3283] Kratochvil, B.G.; N. Motkosky, *Anal. Chem. 59/7*, 1064–1066 (1987).
[3284] Kratochvil, B.G.; S. Mamba, *Can. J. Chem. 68/2*, 360–362 (1990).
[3285] Kreppel, H.; F.X. Reichl, W. Forth, in: B. Welz (Editor), *4. Colloq. Atomspektrom. Spurenanalytik*, Bodenseewerk Perkin-Elmer, Überlingen, 1987, 587–592.
[3286] Kreppel, H.; F.X. Reichl, W. Forth, in: B. Welz (Editor), *5. Colloq. Atomspektrom. Spurenanalytik*, Bodenseewerk Perkin-Elmer, Überlingen, 1989, 439–448.
[3287] Krieg, M.; J. Einax, *Fresenius J. Anal. Chem. 348/8-9*, 490–495 (1994).
[3288] Krinitz, B.; V. Franco, *J. Assoc. Off. Anal. Chem. 56*, 869–875 (1973).
[3289] Krishnamurti, G.S.R.; P.M. Huang, K.C.J. Van Rees, L.M. Kozak, H.P.W. Rostad, *Commun. Soil. Sci. Plant Anal. 25/5-6*, 615–625 (1994).
[3290] Kriss, R.H.; T.T. Bartels, *At. Absorpt. Newsletter 9/3*, 78–79 (1970).
[3291] Kriss, R.H.; T.T. Bartels, *At. Absorpt. Newsletter 11/5*, 110–112 (1972).
[3292] Kritsotakis, K.; H.J. Tobschall, *Fresenius Z. Anal. Chem. 320/1*, 15–21 (1985).
[3293] Krivan, V.; K. Petrick, B. Welz, M. Melcher, *Anal. Chem. 57/8*, 1703–1706 (1985).
[3294] Krivan, V.; G. Schaldach, *Fresenius Z. Anal. Chem. 324*, 158–167 (1986).
[3295] Krivan, V.; *Sci. Total Environ. 64*, 21–40 (1987).
[3296] Krivan, V.; H.F. Haas, *Fresenius Z. Anal. Chem. 332/1*, 1–6 (1988).
[3297] Krivan, V.; S. Arpadjan, *Fresenius Z. Anal. Chem. 335/7*, 743–747 (1989).
[3298] Krivan, V.; K. Petrick, *Fresenius J. Anal. Chem. 336/6*, 480–483 (1990).
[3299] Krivan, V.; M. Kückenwaitz, *Fresenius J. Anal. Chem. 342/9*, 692–697 (1992).
[3300] Krivan, V.; *J. Anal. At. Spectrom. 7/2*, 155–164 (1992).
[3301] Krivan, V.; P. Barth, A. Feria Morales, *Mikrochim. Acta (Vienna) 110*, 217–236 (1993).

[3302] Krivan, V.; M. Franek, *Fresenius J. Anal. Chem. 351/1*, 117–124 (1995).
[3303] Krivan, V.; B. Koch, *Anal. Chem. 67/18*, 3148–3153 (1995).
[3304] Krivan, V.; T. Römmelt, *J. Anal. At. Spectrom. 12/2*, 137–141 (1997).
[3305] Kroll, M.H.; K. Emancipator, *Clin. Chem. (Winston-Salem) 39/3*, 405–413 (1993).
[3306] Kron, T.; C. Hansen, E. Werner, in: P. Brätter, P. Schramel (Editors), *Trace Elem. Anal. Chem. Med. Biol.*, Vol. 5, W. de Gruyter, Berlin - New York, 1988, 412–417.
[3307] Kron, T.; C. Hansen, E. Werner, in: B. Welz (Editor), *5. Colloquium Atomspektrometrische Spurenanalytik*, Bodenseewerk Perkin-Elmer, Überlingen, 1989, 809–814.
[3308] Kronen, R.C.; D.G. Ballinger, in: I.M. Kolthoff, P.J. Elving (Editors), *Treatise on Analytical Chemistry*, Part III, J. Wiley & Sons, New York, 1971, 343.
[3309] Kronholm, K.G.; R.K. Skogerboe, *Appl. Spectrosc. 40/8*, 1161–1166 (1986).
[3310] Krupa, R.J.; T.F. Culbreth, B.W. Smith, J.D. Winefordner, *Appl. Spectrosc. 40/6*, 729–733 (1986).
[3311] Kruse, R.; *Lebensmittelchem. Gerichtl. Chem. 40*, 103–104 (1986).
[3312] Kruse, R.; U. Ballin, H.A. Rüssel-Sinn, in: B. Welz (Editor), *6. Colloquium Atomspektrometrische Spurenanalytik*, Bodenseewerk Perkin-Elmer, Überlingen, 1991, 557–564.
[3313] Krushevska, A.; R.Barnes, C. Amarasiriwaradena, *Analyst (London) 118/9*, 1175–1181 (1993).
[3314] Krynitsky, A.J.; *Anal. Chem. 59/14*, 1884–1886 (1987).
[3315] Kubán, V.; J. Komárek, D. Cajková, *Chem. Listy 84*, 376 (1990).
[3316] Kubasik, N.P.; M.T. Volosin, *Clin. Chem. (Winston-Salem) 19*, 954–958 (1973).
[3317] Kubo, M.; *Bunseki Kagaku 37*, 159 (1988).
[3318] Kubota, T.; K. Uchida, T. Ueda, T. Okutani, *Anal. Chim. Acta 208*, 351–356 (1988).
[3319] Kubota, T.; K. Suzuki, T. Okutani, *Talanta 42/7*, 949–955 (1995).
[3320] Kubrakova, I.V.; T. Kudinova, A. Formanovsky, N. Kuz'min, G. Tsysin, Y. Zolotov, *Analyst (London) 119/11*, 2477–2480 (1994).
[3321] Kubrakova, I.V.; T. Kudinova, N. Kuz'min, I.A. Kovalev, G. Tsysin, Y. Zolotov, *Anal. Chim. Acta 334/1-2*, 167–175 (1996).
[3322] Kubán, V.; J. Komárek, D. Cajková, *Collect. Czech. Chem. Commun. 54/10*, 2683–2691 (1989).
[3323] Kubán, V.; J. Komárek, D. Cajková, Z. Zdráhal, *Chem. Papers 44/3*, 339–346 (1990).
[3324] Kubán, V.; *Fresenius J. Anal. Chem. 346*, 873–881 (1993).
[3325] Kudermann, G.; K.-H. Blaufuß, C. Lührs, *Aluminium (Düsseldorf) 64*, 625–629 (1988).
[3326] Kudo, A.; S. Miyahara, *Water Sci. Technol. 23*, 283–290 (1991).
[3327] Kuehner, E.C.; D.H. Freeman, in: M. Zief (Editor), *Purification of Inorganic and Organic Materials*, Marcel Dekker, New York, 1969, 297–306.
[3328] Kuhn, H.G.; *Atomic Spectra*, Longmans-Green (Academic), London (New York), 1969.
[3329] Kujirai, O.; T. Kobayashi, K. Ide, E. Sudo, *Talanta 29/1*, 27–30 (1982).
[3330] Kujirai, O.; T. Kobayashi, K. Ide, E. Sudo, *Talanta 30/1*, 9–14 (1983).
[3331] Kukier, U.; M.E. Summer, W.P. Miller, *Commun. Soil. Sci. Plant Anal. 25/7-8*, 1149–1159 (1994).
[3332] Kuldvere, A.; B.T. Andreassen, *At. Absorpt. Newsletter 18/5*, 106–110 (1979).
[3333] Kuldvere, A.; *At. Spectrosc. 1/5*, 138–142 (1980).
[3334] Kuldvere, A.; *Analyst (London) 107*, 179–184 (1982).
[3335] Kuldvere, A.; *Analyst (London) 107*, 1343–1349 (1982).
[3336] Kuldvere, A.; *Analyst (London) 113/2*, 277–280 (1988).
[3337] Kuldvere, A.; *Analyst (London) 115/5*, 559–562 (1990).
[3338] Küllmer, G.; S.F.N. Morton, *CLB, Chem. Labor Betr. 34/6*, 243–248 (1983).
[3339] Külpmann, W.R. ; R. Buchholz, C. Dyrssen, D. Ruschke, *J. Clin. Chem. Clin. Biochem. 27*, 631–637 (1989).
[3340] Kumamaru, T.; E. Tao, N. Okamoto, Y. Yamamoto, *Bull. Chem. Soc. Jpn. 38*, 2204. (1965).
[3341] Kumamaru, T.; Y. Hayashi, N. Okamoto, E. Tao, Y. Yamamoto, *Anal. Chim. Acta 35*, 524 (1966).
[3342] Kumamaru, T.; *Anal. Chim. Acta 43*, 19 (1968).
[3343] Kumamaru, T.; Y. Okamoto, S. Hara, H. Matsuo, M. Kiboku, *Anal. Chim. Acta 218/1*, 173–178 (1989).
[3344] Kumar, A.; M.Z. Hasan, B.T. Deshmukh, *Indian J. Pure Appl. Phys. 24*, 465 (1986).
[3345] Kumar, A.; M.Z. Hasan, B.T. Deshmukh, *Indian J. Pure Appl. Phys. 25/2*, 49 (1987).
[3346] Kumar, S.J.; S. Natarajan, J. Arunachalam, S. Gangadharan, *Fresenius J. Anal. Chem. 338/7*, 836–840 (1990).
[3347] Kumar, S.J.; S. Gangadharan, *J. Anal. At. Spectrom. 8/1*, 127–129 (1993).

[3348] Kumar, S.J.; N.N. Meeravali, *At. Spectrosc. 17/1*, 27–29 (1996).
[3349] Kumina, D.M.; E.N. Savinova, T.V. Shumskaya, M.D. Alybaeva, A.V. Karyakin, *Zh. Anal. Khim. 44*, 567 (1989).
[3350] Kumine, N.; H. Kawada, *Bunseki Kagaku 16*, 185 (1967).
[3351] Kumine, N.; H. Kawada, *Bunseki Kagaku 16*, 189 (1967).
[3352] Kumpulainen, J.T.; J. Lehto, P. Koivistoinen, M. Uusitupa, E. Vuori, *Sci. Total Environ. 31*, 71–80 (1983).
[3353] Kumpulainen, J.T.; in: H.A. McKenzie, L.E. Smythe (Editors), *Quantitative trace analysis of biological materials*, Elsevier, Amsterdam - New York - Oxford, 1988, 451–462.
[3354] Kunishi, M.; S. Ohno, *At. Absorpt. Newsletter 13*, 29–30 (1974).
[3355] Kunvar, U.K.; D. Littlejohn, D.J. Halls, *J. Anal. At. Spectrom. 4/2*, 153–156 (1989).
[3356] Kunvar, U.K.; D. Littlejohn, D.J. Halls, *Talanta 37/6*, 555–559 (1990).
[3357] Kuo, N.-W.; C.-R. Lan, Z.B. Alfassi, *J. Radioanal. Nucl. Chem. 172/1*, 117–123 (1993).
[3358] Kurfürst, U.; B. Rues, *Fresenius Z. Anal. Chem. 308/1*, 1–6 (1981).
[3359] Kurfürst, U.; *Fresenius Z. Anal. Chem. 315/5*, 303–320 (1983).
[3360] Kurfürst, U.; *Fresenius Z. Anal. Chem. 322/7*, 660–665 (1985).
[3361] Kurfürst, U.; *Fresenius Z. Anal. Chem. 328/4-5*, 316–318 (1987).
[3362] Kurfürst, U.; M. Kempeneer, M. Stoeppler, O. Schuierer, *Fresenius J. Anal. Chem. 337/3*, 248–252 (1990).
[3363] Kurfürst, U.; *Pure Appl. Chem. 63/9*, 1205–1211 (1991).
[3364] Kurfürst, U.; in: H. Günzler, R. Borsdorf, W. Fresenius, W. Huber, H. Kelker, I. Lüderwald, (Editors), *Analytiker Taschenbuch*, Springer, Berlin - Heidelberg - New York, 1991, 189–248.
[3365] Kurfürst, U.; J. Pauwels, K.-H. Grobecker, M. Stoeppler, H. Muntau, *Fresenius J. Anal. Chem. 345/2-4*, 112–120 (1993).
[3366] Kurfürst, U.; *Fresenius J. Anal. Chem. 346/6-9*, 556–559 (1993).
[3367] Kurfürst, U.; J. Pauwels, *J. Anal. At. Spectrom. 9/4*, 531–534 (1994).
[3368] Kurfürst, U.; A. Rehnert, H. Muntau, *Spectrochim. Acta, Part B 51/2*, 229–244 (1996).
[3369] Kurfürst, U.; *Solid Sample Analysis - Direct and Slurry Sampling using ETAAS and ETV-ICP*, Springer, Heidelberg, 1997.
[3370] Kuroda, R.; N. Matsumoto, K. Oguma, *Fresenius Z. Anal. Chem. 330/2*, 111–113 (1988).
[3371] Kuroda, R.; K. Oguma, Y. Mori, H. Okabe, T. Takizawa, *Chromatographia 32/11*, 583–585 (1991).
[3372] Kushida, I; A. Tanaka, Y. Kidani, J. Hirose, T. Sakai, Z. Fengze, M. Noji, *Chem. Pharm. Bull. 39/5*, 1315–1316 (1991).
[3373] Kuss, H.-M.; *Fresenius J. Anal. Chem. 343*, 788–793 (1992).
[3374] Kuusi, T.; K. Laaksovirta, H. Liukkonen-Lilja, M. Lodenius, Piepponen, *Z. Lebensm. Unters. Forsch. 173*, 261–267 (1981).
[3375] Kuzovlev, I.A.; Y.N. Kuznetsov, O.A. Sverdlina, *Zavodsk. Lab. 39/4*, 428–430 (1973).
[3376] Kožušniková, J.; *Chem. Listy 78*, 1209 (1984).
[3377] Kvaratskheli, Yu.K. ; Z.A. Nikitina, I.P. Zharkova, *Zh. Anal. Khim. 47/9*, 1721–1726 (1992).
[3378] LaBrecque, J.J.; *Appl. Spectrosc. 30/6*, 625–627 (1976).
[3379] LaBrecque, J.J.; *Chem. Geol. 26*, 321–329 (1979).
[3380] LaBrecque, J.J.; *Appl. Spectrosc. 33*, 389–393 (1979).
[3381] LaBrecque, J.J.; P. Rosales, G. Rada, R. Villalba, S. Yariv, E. Mendolovici, *Chem. Geol. 29*, 313–321 (1980).
[3382] LaVilla, F.; F. Queraud, *Rev. Inst. Francais Pétrole 32*, 413 (1977).
[3383] LaVilla, F.; Y. Pean, Rev. *Inst. Francais Pétrole 32*, 89–101 (1977).
[3384] Laborda, F.; D. Chakraborti, J.M. Mir, J.R. Castillo-Suarez, *J. Anal. At. Spectrom. 8/4*, 643–648 (1993).
[3385] Laborda, F.; M.V. Vicente, J.M. Mir, J.R. Castillo, *Fresenius J. Anal. Chem. 357/7*, 837–843 (1997).
[3386] Lacayo, M.L.; A. Cruz, S. Calero, J. Lacayo, I. Fomsgaard, *Bull. Environ. Contam. Toxicol. 49/3*, 463–470 (1992).
[3387] Lachica, M.; *Analusis 18/5*, 331–333 (1990).
[3388] Lachowicz, E.; A. Kaliszuk, *Fresenius Z. Anal. Chem. 323/1*, 54–56 (1986).
[3389] Lahiri, S.; M.J. Stillman, *Anal. Chem. 64/4*, 283A-291A (1992).
[3390] Lahl, H.; L. Schöllmann, in: B. Welz (Editor), *5. Colloquium Atomspektrometrische Spurenanalytik*, Bodenseewerk Perkin-Elmer, Überlingen, 1989, 697–701.
[3391] Lahuerta-Zamora, L.; J.V. Garcia Mateo, J. Martínez Calatayud, *Anal. Chim. Acta 265*, 81–86 (1992).

[3392] Lahuerta-Zamora, L.; J. Martínez Calatayud, *Anal. Chim. Acta 300/1-3*, 143–148 (1995).
[3393] Lajunen, L.H.J.; A. Kinnunen, E. Yrjänheikki, *At. Spectrosc. 6/2*, 49–52 (1985).
[3394] Lajunen, L.H.J.; P. Kokkonen, J. Karijoki, V. Porkka, E. Saari, *At. Spectrosc. 11/5*, 193–197 (1990).
[3395] Lajunen, L.H.J.; J. Piispanen, E. Saari, *At. Spectrosc. 13/4*, 127–131 (1992).
[3396] Lakatos, J.; L. Bezúr, I. Lakatos, *Fresenius J. Anal. Chem. 355/3-4*, 312–314 (1996).
[3397] Lakomaa, E.-L.; H. Mussalo-Rauhamaa, S. Salmela, *J. Trace Elem. Electrol. Health Dis. 2/1*, 37–41 (1988).
[3398] Lam, R.B.; T.L. Isenhour, *Anal. Chem. 52/7*, 1158–1161 (1980).
[3399] LamLeung, S.Y.; V.K.W. Cheng, Y.W. Lam, *Analyst (London) 116/9*, 957–959 (1991).
[3400] Lambert, M.J.; *J. Inst. Wood Sci. 27*, 27 (1969).
[3401] Lamberty, A.; P. De Bièvre, A. Götz, *Fresenius J. Anal. Chem. 345/2-4*, 310–313 (1993).
[3402] Lamble, S.J. Hill, *Anal. Chim. Acta 334/3*, 261–270 (1996).
[3403] Lamm, S.; B. Cole, K. Glynn, W. Ullman, *New England J. Med. 289*, 574–575 (1973).
[3404] Lamothe, P.J.; T.L. Fries, J.J. Consul, *Anal. Chem. 58/8*, 1881–1886 (1986).
[3405] Lamoureux, M.M.; C.L. Chakrabarti, J.C. Hutton, A.Kh. Gilmutdinov, Yu.A. Zakharov, D.C. Grégoire, *Spectrochim. Acta, Part B 50/14*, 1847–1867 (1995).
[3406] Lan, C.-R.; *Analyst (London) 118/2*, 189–192 (1993).
[3407] Lan, W.G.; M.K. Wong, Y.M. Sin, *Talanta 41/1*, 53–58 (1994).
[3408] Lan, W.G.; M.K. Wong, Y.M. Sin, *Talanta 41/2*, 195–200 (1994).
[3409] Lan, W.G.; M.K. Wong, N. Chen, Y.M. Sin, *Analyst (London) 119/8*, 1659–1667 (1994).
[3410] Lan, W.G.; M.K. Wong, N. Chen, Y.M. Sin, *Analyst (London) 119/8*, 1669–1675 (1994).
[3411] Landeghem, G.F. Van; P.C. D'Haese, L.V. Lamberts, M.E. De Broe, *Anal. Chem. 66/2*, 216–222 (1994).
[3412] Landi, S.; F. Fagioli, C. Locatelli, R. Vecchietti, *Analyst (London) 115/2*, 173–177 (1990).
[3413] Landi, S.; F. Fagioli, *Anal. Chim. Acta 298/3*, 363–374 (1994).
[3414] Landner, L.; *Speciation of Metals in Water, Sediment and Soil Systems*, Springer, Berlin, 1987.
[3415] Lane, W.R.; *Industrial and Engineering Chemistry 43/6*, 1312–1317 (1951).
[3416] Lang, I.; G. Šebor, V. Sychra, D. Kolihová, O. Weisser, *Anal. Chim. Acta 84*, 299–305 (1976).
[3417] Lang, I.; G. Šebor, O. Weisser, V. Sychra, *Anal. Chim. Acta 88*, 313–318 (1977).
[3418] Lang, W.; *Fresenius Z. Anal. Chem. 223*, 241 (1966).
[3419] Lange-Hesse, K.; L. Dunemann, G. Schwedt, *Fresenius J. Anal. Chem. 349/6*, 460–464 (1994).
[3420] Lange-Hesse, K.; *Fresenius J. Anal. Chem. 350/1-2*, 68–73 (1994).
[3421] Langmyhr, F.J.; P.E. Paus, *Anal. Chim. Acta 43*, 397–408 (1968).
[3422] Langmyhr, F.J.; P.E. Paus, *At. Absorpt. Newsletter 7*, 103–106 (1968).
[3423] Langmyhr, F.J.; P.E. Paus, *Anal. Chim. Acta 44*, 445–446 (1969).
[3424] Langmyhr, F.J.; P.E. Paus, *Anal. Chim. Acta 45*, 176–179 (1969).
[3425] Langmyhr, F.J.; P.E. Paus, *Anal. Chim. Acta 47*, 371–373 (1969).
[3426] Langmyhr, F.J.; P.E. Paus, *Anal. Chim. Acta 50*, 515 (1970).
[3427] Langmyhr, F.J.; Y. Thomassen, *Fresenius Z. Anal. Chem. 264*, 122–127 (1973).
[3428] Langmyhr, F.J.; S. Rasmussen, *Anal. Chim. Acta 72*, 79–84 (1974).
[3429] Langmyhr, F.J.; J.R. Stubergh, Y. Thomassen, J. Doležal, *Anal. Chim. Acta 71*, 35–42 (1974).
[3430] Langmyhr, F.J.; A. Sundli, J. Jonsen, *Anal. Chim. Acta 73*, 81–85 (1974).
[3431] Langmyhr, F.J.; R. Solberg, L.T. Wold, *Anal. Chim. Acta 69*, 267–273 (1974).
[3432] Langmyhr, F.J.; Y. Thomassen, A. Massoumi, *Anal. Chim. Acta 68*, 305–309 (1974).
[3433] Langmyhr, F.J.; T. Lind, J. Jonsen, *Anal. Chim. Acta 80*, 297–301 (1975).
[3434] Langmyhr, F.J.; J. Aamodt, *Anal. Chim. Acta 87*, 483–486 (1976).
[3435] Langmyhr, F.J.; J.T. Håkedal, *Anal. Chim. Acta 83*, 127–131 (1976).
[3436] Langmyhr, F.J.; *Talanta 24*, 277–282 (1977).
[3437] Langmyhr, F.J.; R. Solberg, Y. Thomassen, *Anal. Chim. Acta 92*, 105–109 (1977).
[3438] Langmyhr, F.J.; I. Kjuus, *Anal. Chim. Acta 100*, 139–144 (1978).
[3439] Langmyhr, F.J.; S. Orre, *Anal. Chim. Acta 118*, 307–311 (1980).
[3440] Langmyhr, F.J.; I.M. Dahl, *Anal. Chim. Acta 131*, 303–306 (1981).
[3441] Langmyhr, F.J.; *Fresenius Z. Anal. Chem. 322/7*, 654–656 (1985).
[3442] Langmyhr, F.J.; G. Wibetoe, *Prog. Anal. Spectrosc. 8/3-4*, 193–256 (1985).
[3443] Lansdown, R.; in: R. Lansdown, W. Yule (Editors), *Lead Toxicity: History and Environmental Impact*, Johns Hopkins University Press, Baltimore, 1986, 235.
[3444] Lansens, P.; C. Meuleman, W. Baeyens, *Anal. Chim. Acta 229/2*, 281–285 (1990).
[3445] Lansford, M.; E.M. McPherson, M.J. Fishman, *At. Absorpt. Newsletter 13/4*, 103–105 (1974).

[3446] Lapointe, S.; A. LeBlanc, *At. Spectrosc. 17/4,* 163–166 (1996).
[3447] Laqua, K.; B. Schrader, G.G. Hoffmann, D.S. Moore, T. Vo-Dinh, *Pure Appl. Chem.67/10,* 1745–1760 (1995).
[3448] Laqua, K.; B. Schrader, G.G. Hoffmann, D.S. Moore, T. Vo-Dinh, *Spectrochim. Acta, Part B 52/5,* 537–552 (1997).
[3449] Laredo-Ortiz, S.; J. Garcia Mateo, J. Martínez Calatayud, *Microchem. J. 48/1,* 112–117 (1993).
[3450] Larkins, P.L.; *Spectrochim. Acta, Part B 39/9-11,* 1365–1376 (1984).
[3451] Larkins, P.L.; *Spectrochim. Acta, Part B 40/10,* 1585–1598 (1985).
[3452] Larkins, P.L.; *Spectrochim. Acta, Part B 43/9-11,* 1175–1186 (1988).
[3453] Larkins, P.L.; *J. Anal. At. Spectrom. 7/2,* 265–272 (1992).
[3454] Larsen, E.H.; L. Rasmussen, *Z. Lebensm. Unters. Forsch. 192/2,* 136–141 (1991).
[3455] Larsen, E.H.; S.H. Hansen, *Mikrochim. Acta (Vienna) 109/1-4,* 47–51 (1992).
[3456] Larsen, E.H.; J.-S. Blais, *J. Anal. At. Spectrom. 8/4,* 659–664 (1993).
[3457] Larsen, I.L.; N.A. Hartmann, J.J. Wagner, *Anal. Chem. 45/8,* 1511–1513 (1973).
[3458] Latino, J.C.; Z.A. Grosser, *At. Spectrosc. 17/6,* 215–217 (1996).
[3459] Latino, J.C.; D.C. Sears, F. Portala, I.L. Shuttler, *At. Spectrosc. 16/3,* 121–126 (1995).
[3460] Latorre, M.J.; C. Garcia-Jares, B. Medina, C. Herrero, *J. Agri. Food Chem. 42,* 1451 (1994).
[3461] Lau, C.; A. Held, R. Stephens, *Can. J. Spectrosc. 21,* 100–104 (1976).
[3462] Lau, C.M.; A.M. Ure, T.S. West, *Anal. Proc. (London) 20/3,* 114–117 (1983).
[3463] Lau, O.-W.; K.L. Li, *Analyst (London) 100,* 430–437 (1975).
[3464] Lau, O.-W.; P.-K. Hon, C.-Y. Cheung, M.-C. Wong, *Analyst (London) 109/9,* 1175–1178 (1984).
[3465] Lau, O.-W.; P.-K. Hon, C.-Y. Cheung, M.-H. Chau, *Analyst (London) 110/5,* 483–485 (1985).
[3466] Lauder, L. (Editor); *Speciation of Metals in Water, Sediments and Soil Systems,* Springer, Berlin, 1987.
[3467] Lauwerys, R.R.; H.A. Roels, J.P. Buchet, A. Bernard, D. Stanescu, *Environ. Persp. 28,* 137 (1979).
[3468] Law, O.-W.; L. Lam, S.-F. Luk, *Talanta 42/9,* 1265–1271 (1995).
[3469] Lawrence, C.B.; M. Phillippo, *Anal. Chim. Acta 118,* 153–157 (1980).
[3470] Lawrenz, J.; K. Niemax, *Spectrochim. Acta, Part B 44/2,* 155–164 (1989).
[3471] Lawson, S.R.; J.A. Nichols, P. Viswanadham, R. Woodriff, *Appl. Spectrosc. 36/4,* 375–378 (1982).
[3472] Lázaro, F.; M.D. Luque de Castro, M. Valcárcel, *Anal. Chim. Acta 242/2,* 283–289 (1991).
[3473] Le, X.-c.; W.R. Cullen, K.J. Reimer, I.D. Brindle, *Anal. Chim. Acta 258/2,* 307–315 (1992).
[3474] Le, X.-c.; W.R. Cullen, K.J. Reimer, *Appl. Organomet. Chem. 6,* 161–171 (1992).
[3475] Le, X.-c.; W.R. Cullen, K.J. Reimer, *Talanta 40/2,* 185–193 (1993).
[3476] Le, X.-c.; W.R. Cullen, K.J. Reimer, *Anal. Chim. Acta 285/3,* 277–285 (1994).
[3477] Le, X.-c.; W.R. Cullen, K.J. Reimer, *Talanta 41/4,* 495–502 (1994).
[3478] Leach, C.N.; J.V. Linden, S.M. Hopfer, M.C. Crisostomo, F.W. Sunderman, *Clin. Chem. (Winston-Salem) 31,* 556–560 (1985).
[3479] Lechotycki, A.; *J. Anal. At. Spectrom. 5/1,* 25–28 (1990).
[3480] Lecompte, R.; P. Paradis, S. Monaro, M. Barette, G. Lamoureux, H.A. Menard, *Int. J. Nucl. Med. Biol. 6,* 207 (1979).
[3481] Lee, D.S.; *Anal. Chem. 54/7,* 1182–1184 (1982).
[3482] Lee, D.S.; *Anal. Chem. 54/11,* 1682–1686 (1982).
[3483] Lee, J.J. van der; E. Temminghoff, V.J.G. Houba, I. Novozamsky, *Appl. Spectrosc. 41/3,* 388–390 (1987).
[3484] Lee, M.L.; G. Tölg, E. Beinrohr, P. Tschöpel, *Anal. Chim. Acta 272/2,* 193–203 (1993).
[3485] Lee, Y.-K.; D.S. Lee, B.M. Yoon, H. Hwang, *Bull. Korean Chem. Soc. 12,* 290–295 (1991).
[3486] Leenheer, J.; in: *Bull. Int. Dairy Fed. 235,* 34 (1988).
[3487] Leermakers, M.; P. Lansens, W. Baeyens, *Fresenius J. Anal. Chem. 336/8,* 655–662 (1990).
[3488] Leeuwenkamp, O.R.; W.J.F. van der Vijgh, B.C.P. Huesken, P. Lips, J.C. Netelenbos, *Clin. Chem. (Winston-Salem) 35/9,* 1911–1914 (1989).
[3489] Legret, M.; L. Divet, *Anal. Chim. Acta 189,* 313–321 (1986).
[3490] Legret, M.; L. Divet, *Analusis 16/2,* 97–106 (1988).
[3491] Legret, M.; *Int. J. Environ. Anal. Chem. 51,* 161–165 (1993).
[3492] Leloux, M.S.; N.P. Lich, L.-R. Claude, *At. Spectrosc. 8/2,* 71–75 (1987).
[3493] Lendero, L.; V. Krivan, *Anal. Chem. 54/3,* 579–581 (1982).
[3494] Lepretre, A.; S. Martin, *Analusis 22/3,* M40–M43 (1994).
[3495] Lercari, C.; B. Sartorel, L. Sedea, G. Toninelli, *J. Am. Oil Chem. Soc. 60/4,* 856–857 (1983).

[3496] Lernhardt, U.; J. Kleiner, in: B. Welz (Editor), *CANAS'95 Colloquium Analytische Atomspektroskopie,* Bodenseewerk Perkin-Elmer, Überlingen, 1996, 231–236.
[3497] Letourneau, V.A.; B.M. Joshi, L.C. Butler, *At. Spectrosc. 8/5,* 145–149 (1987).
[3498] Leung, F.Y.; A.E. Niblock, C. Bradley, A.R. Henderson, *Sci. Total Environ. 71/1,* 49–56 (1988).
[3499] Levander, O.A.; in: W. Mertz (Editor), *Trace Elements in Human and Animal Nutrition,* Academic Press, New York, 1988, 209–279.
[3500] Levine, J.R.; S.G. Moore, S.L. Levine, *Anal. Chem. 42,* 412 (1970).
[3501] Lewalter, J.; C. Domik, H. Weidemann, in: J. Angerer, K.H. Schaller (Editors), *Analyses of Hazardous Substances in Biological Materials,* VCH Verlagsgesellschaft, Weinheim, 1991, 109–126.
[3502] Lewis, B.L.; H.P. Mayer, in: H. Sigel, A. Sigel (Editors), *Metal ions in biological systems,* Marcel Dekker, New York, 1993, 79–100.
[3503] Lewis, C.; W.L. Ott, N.M. Sine, *The Analysis of Nickel,* Pergamon Press, Oxford, 1966, 192.
[3504] Lewis, E.L.; *Physics Report 58,* 1–71 (1980).
[3505] Li, H.-t.; B.-j. Ma, *Guangpuxue Yu Guangpu Fenxi 13/1,* 127–130 (1993).
[3506] Li, Q.; *Yankuang Ceshi 5/1,* 29 (1986).
[3507] Li, X.-p.; *Lihua Jianyan, Huaxue Fence 28/6,* 366–367 (1992).
[3508] Lian, L.; *Spectrochim. Acta, Part B 47/2,* 239–244 (1992).
[3509] Liang, L.; P.C. D'Haese, L.V. Lamberts, M.E. De Broe, *Clin. Chem. (Winston-Salem) 37/3,* 461–466 (1991).
[3510] Liang, L.; P. Danilchik, Z. Huang, *At. Spectrosc. 15/4,* 151–155 (1994).
[3511] Liang, Y.-z., Z.-m. Ni, P.-y. Yang, *J. Anal. At. Spectrom. 10/10,* 699–702 (1995).
[3512] Liang, Y.-z., M. Li, Z. Rao, *Anal. Sci. 12/4,* 629–633 (1996).
[3513] Liang, Y.-z., M. Li, Z. Rao, *Fresenius J. Anal. Chem. 357/1,* 112–116 (1997).
[3514] Liao, Y.-p.; A.-m. Li, *J. Anal. At. Spectrom. 8/4,* 633–636 (1993).
[3515] Lichon, M.J. ; K.W. James, *J. Assoc. Off. Anal. Chem. 73/5,* 820–825 (1990).
[3516] Lichte, F.E.; R.K. Skogerboe, *Anal. Chem. 44,* 1480–1482 (1972).
[3517] Lichtenberg, W.; *Fresenius Z. Anal. Chem. 328/4-5,* 367–369 (1987).
[3518] Liddell, P.R.; P.C. Wildy, *Spectrochim. Acta, Part B 35/4,* 193–198 (1980).
[3519] Lieberman, K.W.; *Clin. Chim. Acta 46,* 217–221 (1973).
[3520] Lima, É.C.; F.J. Krug, J.A. Nóbrega, E.A.N. Fernandes, *J. Anal. At. Spectrom. 12/4,* 475–478 (1997).
[3521] Lima, R.; K.C. Leandro, R.E. Santelli, Talanta, 43/6, 977–983 (1996).
[3522] Lin, J.-L.; B.K. Puri, M. Satake, M. Endo, *Mikrochim. Acta (Vienna) 3,*1–2, 87–94 (1984).
[3523] Lin, S.-l.; H.-p. Hwang, *Talanta 40/7,* 1077–1083 (1993).
[3524] Lin, S.-l.; Q. Shuai, H.-o. Qiu, Z.-y. Tang, *Spectrochim. Acta, Part B, 51/14,* 1769–1775 (1996).
[3525] Lin, W.; Y. He, J. Luo, L. Huang, L. Kin, S. Wu, J. Chen, *Fenxi Shiyanshi 10,* 30 (1991).
[3526] Lin, Y.P.; S.L. Su, C.N. Chao, I.L. Juang, H.Y.W. Chang, *Huaxue 50/3,* 193 (1992).
[3527] Lin, Y.-h.; X.-r. Wang, D.-x. Yuan, P.-Y. Yang, B.-l. Huang, Z.-x. Zhuang, *J. Anal. At. Spectrom. 7/2,* 287–291 (1992).
[3528] Lind, B.; R. Body, L. Friberg, *Fresenius J. Anal. Chem. 345,* 314–317 (1993).
[3529] Lindahl, P.C.; A.M. Bischop, *Fuel 61,* 658–662 (1982).
[3530] Lindberg, I.; E. Lundberg, P. Arkhammar, P.-O. Berggren, *J. Anal. At. Spectrom. 3/4,* 497–501 (1988).
[3531] Linden, W.E. van der; *Pure Appl. Chem. 61/1,* 91–95 (1989).
[3532] Lindsjö, O.; in: B. Welz (Editor), *Atomspektrometrische Spurenanalytik,* Verlag Chemie, Weinheim, 1982, 437–455.
[3533] Ling, C.; *Anal. Chem. 39,* 798 (1967).
[3534] Liou, D.-C.; Y.-C. Lou, C.-W. Whang, *J. Chin. Chem. Soc. (Taipei) 39,* 217–222 (1992).
[3535] Liphard, K.G.; *Staub. Reinhalt. Luft 47/1-2,* M126 (1987).
[3536] Lisy, V.; L. Turna, *Czech. J. Phys. B33,* 480 (1983).
[3537] Litman, R.; H.L. Finston, E.T. Williams, *Anal. Chem. 47/14,* 2364–2369 (1975).
[3538] Little, P.; M.H. Martin, *Environ. Pollut. 6,* 1–9 (1974).
[3539] Littlejohn, D.; J.M. Ottaway, *Analyst (London) 103,* 662–665 (1978).
[3540] Littlejohn, D.; J.M. Ottaway, *Can. J. Spectrosc. 24/6,* 154–162 (1979).
[3541] Littlejohn, D.; J.M. Ottaway, *Analyst (London) 104,* 1138–1150 (1979).
[3542] Littlejohn, D.; J. Marshall, J. Caroll, W. Cormack, J.M. Ottaway, *Analyst (London) 108,* 893–896 (1983).
[3543] Littlejohn, D.; I.S. Duncan, J. Marshall, J.M. Ottaway, *Anal. Chim. Acta 157,* 291–302 (1984).

[3544] Littlejohn, D.; S. Cook, D. Durie, J.M. Ottaway, *Spectrochim. Acta, Part B 39/2-3*, 295–304 (1984).

[3545] Littlejohn, D.; I.S. Duncan, J.B.M. Hendry, J. Marshall, J.M.Ottaway, *Spectrochim. Acta, Part B 40B/10*, 1677–1687 (1985).

[3546] Liu, C.Y.; P.Y. Chen, H.M. Lin, M.H. Yang, *Fresenius Z. Anal. Chem. 320/1*, 22–28 (1985).

[3547] Liu, J.-h.; R.E. Sturgeon, S.N. Willie, *Analyst 120/7*, 1905–1909 (1995).

[3548] Liu, J.-h.; R.E. Sturgeon, *Ciencia 3/2*, 127–138 (1995).

[3549] Liu, J.-h.; R.E. Sturgeon, V.J. Boyko, S.N. Willie, *Fresenius J. Anal. Chem. 356*, 416–419 (1996).

[3550] Liu, L.-h.; S.-L. Zhang, C.-S. Zhao, *Shiyou Huagong 18/9*, 635 (1989).

[3551] Liu, X.-z.; S.-k. Xu, Z.-l. Fang, *At. Spectrosc. 15/6*, 229–233 (1994).

[3552] Liu, X.-z.; Z.-l. Fang, *Anal. Chim. Acta 316/3*, 329–335 (1995).

[3553] Liu, Y.; J. Zhao, *Fenxi Huaxue 16*, 94 (1988).

[3554] Liu, Y.; J.D. Ingle, *Anal. Chem. 61/6*, 520–524 (1989).

[3555] Liu, Y.; J.D. Ingle, *Anal. Chem. 61/6*, 525–529 (1989).

[3556] Liu, Y.-m.; B.-l. Gong, Y.-l. Xu, Z.-h. Li, T.-z. Lin, *Anal. Chim. Acta 292/3*, 325–328 (1994).

[3557] Liu, Y.-m.; B.-l. Gong, Z.-h. Li, Z.-h. Li, T.-z. Lin, *Talanta 43/7*, 985–989 (1996).

[3558] Liu, Z.; L. Dai, J.M. Qian, J. Peng, *Fenxi Shiyanshi 9/6, 60-61,71 (1990)*.

[3559] Liu, Z.-S.; S.-d. Huang, *Anal. Chim. Acta 267/1*, 31–37 (1992).

[3560] Livardjani, F.; R. Heimburger, M.J.F. Leroy, M. Dahlet, A. Jaeger, *Analusis 19/7*, 205–207 (1991).

[3561] Lizondo, F.; M.T. Vidal, M. De la Guardia, *Analusis 19/4*, 136–138 (1991).

[3562] Ljung, P.; O. Axner, *Spectrochim. Acta, Part B 52/3*, 305–319 (1997).

[3563] Ljunggren, L.; I. Altrell, L. Risinger, G. Johansson, *Anal. Chim. Acta 256/1*, 75–80 (1992).

[3564] Lo, D.B.; G.D. Christian, *Microchem. J. 23*, 481–487 (1978).

[3565] Lo, D.B.; R.L. Coleman, *At. Absorpt. Newsletter 18/1*, 10–12 (1979).

[3566] Lo, J.-M.; C.M. Wai, *Anal. Chem. 47/11*, 1869–1970 (1975).

[3567] Lo, J.-M.; J.C. Yu, F.I. Hutchison, C.M. Wai, *Anal. Chem. 54/14*, 2536–2539 (1982).

[3568] Łobinski, R.; W.M.R. Dirkx, J. Szpunar-Łobinska, F.C. Adams, *Anal. Chim. Acta 286/3*, 381–390 (1994).

[3569] Łobinski, R.; in: Ph. Quevauviller, E.A. Maier, B. Griepink (Editors), *Quality Assurance for Environmental Analysis*, Elsevier, Amsterdam, 1995, 319–356.

[3570] Lockyer, J.N.; *Studies in Spectrum Analysis*, Appleton, London, 1878.

[3571] Loenen, D.C. van ; C.A. Weers, in: B. Welz (Editor), *Fortschritte in der atom-spektrometrischen Spurenanalytik*, VCH Verlagsgesellschaft, Weinheim, 1986, 635–647.

[3572] Loenen-Imming, D.C. van ; C.A. Weers, in: B. Welz (Editor), *Fortschritte in der atomspektrometrischen Spurenanalytik*, Verlag Chemie, Weinheim, 1984, 571–586.

[3573] Lofty, D.; Lab. Equip. Dig. 30/10, 13–15 (1992).

[3574] Loken, H.F.; J.S. Teal, E. Eisenberg, *Anal. Chem. 35*, 875 (1963).

[3575] Long, G.L.; J.D. Winefordner, *Anal. Chem. 55/7*, 712A-724A (1983).

[3576] Longbottom, J.E.; *Anal. Chem. 44*, 1111 (1972).

[3577] Loon, J.C. Van; *At. Absorpt. Newsletter 7/1*, 3–4 (1968).

[3578] Loon, J.C. Van; C.M. Parissis, *Analyst (London) 94*, 1057–1062 (1969).

[3579] Loon, J.C. Van; *At. Absorpt. Newsletter 11/3*, 60–62 (1972).

[3580] Loon, J.C. Van; B. Radziuk, *Can. J. Spectrosc. 21/2*, 46–50 (1976).

[3581] Loon, J.C. Van; *Pure Appl. Chem. 49*, 1495–1505 (1977).

[3582] Loon, J.C. Van; *Spectrochim. Acta, Part B 38/11*, 1509–1524 (1983).

[3583] Loon, J.C. Van; *Fresenius Z. Anal. Chem. 246*, 122–126 (1969).

[3584] Loon, J.C. Van; *Anal. Chem. 51/11*, 1139A-50A (1979).

[3585] Loos-Vollebregt, M.T.C. de ; L. de Galan, *Spectrochim. Acta, Part B 33*, 495–511 (1978).

[3586] Loos-Vollebregt, M.T.C. de ; L. de Galan, *Appl. Spectrosc. 33*, 616–626 (1979).

[3587] Loos-Vollebregt, M.T.C. de ; L. de Galan, *Spectrochim. Acta, Part B 35/8*, 495–506 (1980).

[3588] Loos-Vollebregt, M.T.C. de ; L. de Galan, *Appl. Spectrosc. 34/4*, 464–472 (1980).

[3589] Loos-Vollebregt, M.T.C. de ; L. de Galan, *Spectrochim. Acta, Part B 37*, 659–672 (1982).

[3590] Loos-Vollebregt, M.T.C. de ; L. de Galan, J.W.M. van Uffelen, W. Slavin, D.C. Manning, *Spectrochim. Acta, Part B 38/5-6*, 799–807 (1983).

[3591] Loos-Vollebregt, M.T.C. de ; L. de Galan, *Spectrochim. Acta, Part B 39/2-3*, 449–458 (1984).

[3592] Loos-Vollebregt, M.T.C. de ; L. de Galan, *Appl. Spectrosc. 38/2*, 141–148 (1984).

[3593] Loos-Vollebregt, M.T.C. de ; L. de Galan, *Prog. Anal. Spectrosc. 8*, 47–81 (1985).

[3594] Loos-Vollebregt, M.T.C. de ; L. de Galan, J.W.M. van Uffelen, *Spectrochim. Acta, Part B 41/8*, 825–835 (1986).

[3595] Loos-Vollebregt, M.T.C. de ; L. de Galan, *Spectrochim. Acta, Part B 41/6*, 597–610 (1986).
[3596] Loos-Vollebregt, M.T.C. de ; L. de Galan, *J. Anal. At. Spectrom. 3/1*, 151–154 (1988).
[3597] Loos-Vollebregt, M.T.C. de ; J.P. Koot, J. Padmos, *J. Anal. At. Spectrom. 4/5*, 387–391 (1989).
[3598] Loos-Vollebregt, M.T.C. de ; P. Van Oosten, M.J. de Koning, J. Padmos, *Spectrochim. Acta, Part B 48/12*, 1505–1515 (1993).
[3599] López, A.; R. Torralba, M.A. Palacios-Corvillo, C. Cámara-Rica, *Talanta 39/10*, 1343–1348 (1992).
[3600] López, M.A.; M.M. Gómez, M.A. Palacios, C. Cámara, *Fresenius J. Anal. Chem. 346/6-9*, 643–647 (1993).
[3601] López-Fernández, J.M.; A. Ríos, M. Valcárcel, *Analyst (London) 120/9*, 2393–2400 (1995).
[3602] López-García, A.; E. Blanco González, A. Sanz-Medel, *Mikrochim. Acta (Vienna) 112/1-4*, 19–29 (1993).
[3603] López-García, I.; C. O'Grady, M. Cresser, *J. Anal. At. Spectrom. 2/2*, 221–225 (1987).
[3604] López-García, I.; M. Hernández-Córdoba, *J. Anal. At. Spectrom. 4/8*, 701–704 (1989).
[3605] López-García, I.; M. Hernández-Córdoba, *J. Anal. At. Spectrom. 5/7*, 647–650 (1990).
[3606] López-García, I.; F. Ortiz-Sobejano, M. Hernández-Córdoba, *Analyst (London) 116/5*, 517–520 (1991).
[3607] López-García, I.; M.J. Vizcaíno Martínez, M. Hernández-Córdoba, *J. Anal. At. Spectrom. 6/8*, 627–630 (1991).
[3608] López-García, I.; P. Viñas, M. Hernández Córdoba, *J. Anal. At. Spectrom. 7/3*, 529–532 (1992).
[3609] López-García, I.; I. Nuño Peñalver, M. Hernández Córdoba, *Mikrochim. Acta (Vienna) 109*, 211–219 (1992).
[3610] López-García, I.; J. Arroyo Cortéz, M. Hernández Córdoba *J. Anal. At. Spectrom. 8/1*, 103–108 (1993).
[3611] López-García, I.; J. Arroyo Cortéz, M. Hernández Córdoba, *Talanta 40/11*, 1677–1685 (1993).
[3612] López-García, I.; J. Arroyo Cortéz, M. Hernández Córdoba, *At. Spectrosc. 14/5*, 144–147 (1993).
[3613] López-García, I.; P. Viñas, N. Campillo, M. Hernández-Córdoba, *Anal. Chim. Acta 308/1-3*, 85–95 (1995).
[3614] López-García, I.; M. Sánchez-Merlos, M. Hernández-Córdoba, *At. Spectrosc. 17/3*, 107–111 (1996).
[3615] López-García, I.; M. Sánchez-Merlos, M. Hernández-Córdoba, *J. Anal. At. Spectrom.11/10*, 1003–1006 (1996).
[3616] López-García, I.; M. Sánchez-Merlos, P. Viñas, M. Hernández-Córdoba, *Spectrochim. Acta, Part B 51/14*, 1761–1768 (1996).
[3617] López-García, I.; P. Viñas, N. Campillo, M. Hernández-Córdoba, *J. Agri. Food Chem. 44*, 836–841 (1996).
[3618] López-García, I.; M. Sánchez-Merlos, M. Hernández-Córdoba, *Anal. Chim. Acta 328/1*, 19–25 (1996).
[3619] López-García, I.; E. Navarro, P. Viñas, M. Hernández-Córdoba, *Fresenius J. Anal. Chem. 357/6*, 642–646 (1997).
[3620] López-García, I.; M. Sánchez-Merlos, M. Hernández-Córdoba, *Spectrochim. Acta, Part B 52/4*, 437–443 (1997).
[3621] López-Sánchez, J.F.; R. Rubio, G. Rauret, *Int. J. Environ. Anal. Chem. 51/1-4*, 113–121 (1993).
[3622] Lord, D.A.; J.W. McLaren, R.C. Wheeler, *Anal. Chem. 49*, 257–261 (1977).
[3623] Lorentzen, E.M.L.; H.M.S. Kingston, *Anal. Chem. 68/24*, 4316–4320 (1996).
[3624] Lothian, G.F.; *Analyst (London) 88*, 678–685 (1963).
[3625] Louie, H.; *Analyst (London) 108*, 1313–1317 (1983).
[3626] Love, J.L.; J.E. Patterson, *J. Assoc. Off. Anal. Chem. 61*, 627–628 (1978).
[3627] Lovett, R.J. ; D.L. Welch, M.L. Parsons, *Appl. Spectrosc. 29/6*, 470–477 (1975).
[3628] Low, P.S. ; G.J. Hsu, *Fresenius J. Anal. Chem. 337/3*, 299–305 (1990).
[3629] Loya, E.W.; *At. Spectrosc. 10/2*, 61–65 (1989).
[3630] Lu, G.-h.; J.-y. Xu, T.-m. Xu, L.-t. Jin, Y.-z. Fang, *Talanta 39/1*, 51–53 (1992).
[3631] Lu, G.-h.; W. Xia, J.-l. Wan, F. Shong, X.-H. Ying, *Talanta 42/4*, 557–560 (1995).
[3632] Lu, X.-k. ; J. Li, S.-z. Chen, G.-s. Dai, *Haiyang Xuebao 11/4*, 444 (1989).
[3633] Luca, C.; I. Tanase, A.F. Danet, I. Ioneci, *Rev. Anal. Chem. 9/1*, 1–47 (1989).
[3634] Lücker, E.; A. Rosopulo, W. Kreuzer, *Fresenius Z. Anal. Chem. 328/4-5*, 370–377 (1987).

[3635] Lücker, E.; A. Rosopulo, S. Koberstein, W. Kreuzer, *Fresenius Z. Anal. Chem. 329/1*, 31–34 (1987).
[3636] Lücker, E.; A. Rosopulo, W. Kreuzer, *Fresenius J. Anal. Chem. 340/4*, 234–241 (1991).
[3637] Lücker, E.; H. König, W. Gabriel, A. Rosopulo, *Fresenius J. Anal. Chem. 342/12*, 941–949 (1992).
[3638] Lücker, E.; *Fresenius J. Anal. Chem. 343/4*, 386–390 (1992).
[3639] Lücker, E.; C. Gerbig, W. Kreuzer, *Fresenius J. Anal. Chem. 346/12*, 1062–1067 (1993).
[3640] Lücker, E.; J. Meuthen, W. Kreuzer, *Fresenius J. Anal. Chem. 346/12*, 1068–1071 (1993).
[3641] Lücker, E.; S. Thorius-Ehrler, *Fresenius J. Anal. Chem. 346/12*, 1072–1076 (1993).
[3642] Luecke, W.; *N. Jb. Mineral., Mh. 10*, 469–476 (1971).
[3643] Luecke, W.; R. Emmermann, *At. Absorpt. Newsletter 10/2*, 45–49 (1971).
[3644] Luecke, W.; H.-J. Zielke, *Fresenius Z. Anal. Chem. 253*, 20–23 (1971).
[3645] Luecke, W.; *Chem. Erde 38/1*, 1–39 (1979).
[3646] Luecke, W.; *J. Anal. At. Spectrom. 7/5*, 765–768 (1992).
[3647] Luecke, W.; *Fresenius J. Anal. Chem. 344/6*, 242–246 (1992).
[3648] Luecke, W.; *Chem. Geol. 98*, 323–326 (1992).
[3649] Luecke, W.; *J. Anal. At. Spectrom. 9/2*, 105–109 (1994).
[3650] Lugowski, S.J. ; D.C. Smith, A.D. McHugh, J.C. Van Loon, *J. Trace Elem. Electrol. Health Dis. 5/1*, 23–29 (1991).
[3651] Luguera, M.; Y. Madrid, C. Cámara-Rica, *J. Anal. At. Spectrom. 6/8*, 669–671 (1991).
[3652] Luipen, J. van; *At. Absorpt. Newsletter 17/6*, 144–145 (1978).
[3653] Lukasiewicz, R.J. ; B.E. Buell, *Anal. Chem. 47/9*, 1673–1676 (1975).
[3654] Lum, K.R.; D.G. Edgar, *Analyst (London) 108*, 918–924 (1983).
[3655] Lum, K.R.; D.G. Edgar, *Int. J. Environ. Anal. Chem. 15/4*, 241–248 (1983).
[3656] Lum, K.R.; M. Callaghan, *Anal. Chim. Acta 187*, 157–162 (1986).
[3657] Luna, J. ; M.R. Brunetto, J.M.L. Burguera, M. Burguera, M. Gallignani, *Fresenius J. Anal. Chem. 354/3*, 367–369 (1996).
[3658] Lundberg, E.; G. Johansson, *Anal. Chem. 48/13*, 1922–1926 (1976).
[3659] Lundberg, E.; *Appl. Spectrosc. 32*, 276–281 (1978).
[3660] Lundberg, E.; W. Frech, *Anal. Chim. Acta 104/1*, 67–74 (1979).
[3661] Lundberg, E.; W. Frech, *Anal. Chim. Acta 104/1*, 75–84 (1979).
[3662] Lundberg, E.; W. Frech, *Anal. Chem. 53/9*, 1437–1442 (1981).
[3663] Lundberg, E.; B. Bergmark, W. Frech, *Anal. Chim. Acta 142*, 129–142 (1982).
[3664] Lundberg, E.; W. Frech, Ingela Lindberg, *Anal. Chim. Acta 160*, 205–215 (1984).
[3665] Lundberg, E.; B. Bergmark, *Anal. Chim. Acta 188*, 111–118 (1986).
[3666] Lundberg, E.; W. Frech, J.M. Harnly, *J. Anal. At. Spectrom. 3/8*, 1115–1119 (1988).
[3667] Lundberg, E.; W. Frech, D.C. Baxter, A. Cedergren, *Spectrochim. Acta, Part B 43/4-5*, 451–457 (1988).
[3668] Lundegårdh, H.; *Die quantitative Spektralanalyse der Elemente Teil 1*, Gustav Fischer, Jena, 1929.
[3669] Lundgren, G.; L. Lundmark, G. Johansson, *Anal. Chem. 46*, 1028–1031 (1974).
[3670] Luo, C.-A.; S.-R. Zhao, Z.-M. Hong, *Yejin Fenxi 9/3*, 26 (1989).
[3671] Luo, F.; Y. Ding, *Fenxi Huaxue 15*, 285 (1987).
[3672] Luo, F.-r.; X.-d. Hou, *Yankuang Ceshi 6*, 255 (1987).
[3673] Luo, F.-r.; S.-K. Wu, *Fenxi Huaxue 15*, 739–742 (1987).
[3674] Luo, F.-r.; X.-y. Li, M.-h. Wang, *At. Spectrosc. 11/2*, 78–82 (1990).
[3675] Luo, F.-r.; X.-d. Hou, *At. Spectrosc. 15/5*, 216–219 (1994).
[3676] Luo, J. ; W.-Y. Wang, J.W. Zhou, J.-R. Hua, S.-D. Lin, *Huaxue Shijie 28*, 547 (1987).
[3677] Lush, J.F.P.; in: D.R. Hodges (Editor), *Recent Analytical Developments in the Petroleum Industry*, Applied Science Publ., Barking, Essex, 1974, 185–211.
[3678] Luterotti, S.; T. Zanic-Grubisic, D. Juretic, *Analyst (London) 117/2*, 141–143 (1992).
[3679] L'vov, B.V.; *Inzhener.-Fiz. Zhur., Akad. Nauk Belorus. SSR (J. Eng. Phys. [USSR]) 2/2*, 44–52 (1959).
[3680] L'vov, B.V.; *Spectrochim. Acta 17*, 761–770 (1961).
[3681] L'vov, B.V.; *Tr. Gos. Inst. Prikl. Khim. 49*, 256 (1962).
[3682] L'vov, B.V.; *Optik. i Spektrosk. 19*, 507–510 (1965).
[3683] L'vov, B.V.; *Atomic Absorption Spectrochemical Analysis*, Nauka, Moscow, 1966.
[3684] L'vov, B.V.; *Spectrochim. Acta, Part B 24/1*, 53–70 (1969).
[3685] L'vov, B.V.; A.D. Khartsyzov, *Zh. Prikl. Spektrosk. 11*, 413 (1969).
[3686] L'vov, B.V.; A.D. Khartsyzov, *Zh. Anal. Khim. 24/5*, 799–800 (1969).
[3687] L'vov, B.V.; A.D. Khartsyzov, *Zh. Prikl. Spektrosk. 11*, 9 (1969).

[3688] L'vov, B.V.; *Pure Appl. Chem. 23*, 11–34 (1970).
[3689] L'vov, B.V.; A.D. Khartsyzov, *Zh. Anal. Khim. 25/9*, 1824–1826 (1970).
[3690] L'vov, B.V.; *Atomic Absorption Spectrochemical Analysis*, Adam Hilger (Elsevier), London (New York), 1970.
[3691] L'vov, B.V.; *Zh. Anal. Khim. 26*, 510 (1971).
[3692] L'vov, B.V.; *Méthodes Physiques D'Analyse (GAMS) 8/1*, 3 (1972).
[3693] L'vov, B.V.; L.K. Polzik, D.A. Katskov, L.P. Kruglikova, *Zh. Prikl. Spektrosk. 22*, 787–793 (1975).
[3694] L'vov, B.V.; L.P. Kruglikova, L.K. Polzik, D.A. Katskov, *Zh. Anal. Khim. 30/4*, 645–651 (1975).
[3695] L'vov, B.V.; L.P. Kruglikova, L.K. Polzik, D.A. Katskov, *Zh. Anal. Khim. 30/4*, 652–658 (1975).
[3696] L'vov, B.V.; L.P, Kruglikova, L.K. Polzik, D.A. Katskov, *Zh. Anal. Khim. 30/5*, 846–853 (1975).
[3697] L'vov, B.V.; N.A. Orlov, *Zh. Anal. Khim. 30*, 1661 (1975).
[3698] L'vov, B.V.; L.P. Kruglikova, L.K. Polzik, D.A. Katskov, *Zh. Anal. Khim. 30/5*, 839–845 (1975).
[3699] L'vov, B.V.; *Talanta 23*, 109–118 (1976).
[3700] L'vov, B.V.; D.A. Katskov, L.P. Kruglikova, L. K. Polzik, *Spectrochim. Acta, Part B 31/2*, 49–80 (1976).
[3701] L'vov, B.V.; L.A. Pelieva, A.I. Sharnopolsky, *Zh. Prikl. Spektrosk. 27/5*, 395–399 (1977).
[3702] L'vov, B.V.; N.A. Orlov, L.K. Polzik, *Zh. Anal. Khim. 32*, 5 (1977).
[3703] L'vov, B.V.; L.A. Pelieva, *Can. J. Spectrosc. 23*, 1–4 (1978).
[3704] L'vov, B.V.; *Spectrochim. Acta, Part B 33*, 153–193 (1978).
[3705] L'vov, B.V.; L.A. Pelieva, *Zh. Anal. Khim. 33/9*, 1695–1704 (1978).
[3706] L'vov, B.V.; L.A. Pelieva, *Zh. Anal. Khim. 33*, 1572–1575 (1978).
[3707] L'vov, B.V.; L.A. Pelieva, *Zh. Anal. Khim. 34*, 1744–1755 (1979).
[3708] L'vov, B.V.; L.A. Pelieva, *Zh. Prikl. Spektrosk. 31/1*, 16–23 (1979).
[3709] L'vov, B.V.; *Keynote lectures, 8th ICAS, XXI CSI*, Heyden, London - Philadelphia - Rheine, 1979.
[3710] L'vov, B.V.; P.A. Bayunov, I.B. Patrov, T.B. Polobeiko, *Zh. Anal. Khim. 35*, 1877–1884 (1980).
[3711] L'vov, B.V.; L.K. Polzik, *Zh. Anal. Khim. 35/3*, 421–426 (1980).
[3712] L'vov, B.V.; G.N. Ryabchuk, *Zh. Prikl. Spektrosk. 33*, 1013 (1980).
[3713] L'vov, B.V.; L.A. Pelieva, *Prog. Anal. Spectrosc. 3/1*, 65–86 (1980).
[3714] L'vov, B.V.; P.A. Bayunov, G.N. Ryabchuk, *Spectrochim. Acta, Part B 36/5*, 397–425 (1981).
[3715] L'vov, B.V.; *Zh. Anal. Khim. 37/12*, 2116–2124 (1982).
[3716] L'vov, B.V.; G.N. Ryabchuk, *Spectrochim. Acta, Part B 37/8*, 673–684 (1982).
[3717] L'vov, B.V.; *Dokl. Akad. Nauk SSSR 271/1*, 119–121 (1983).
[3718] L'vov, B.V.; A.S. Savin, *Zh. Anal. Khim. 38*, 1925–1932 (1983).
[3719] L'vov, B.V.; L.F. Yatsenko, *Zh. Anal. Khim. 39/10*, 1773–1780 (1984).
[3720] L'vov, B.V.; *Spectrochim. Acta, Part B 39/2-3*, 149–157 (1984).
[3721] L'vov, B.V.; *Spectrochim. Acta, Part B 39/2-3*, 159–166 (1984).
[3722] L'vov, B.V.; *Zh. Anal. Khim. 39/11*, 1953–1960 (1984).
[3723] L'vov, B.V.; L.F. Yatsenko, *Zh. Anal. Khim. 40/4*, 626–629 (1985).
[3724] L'vov, B.V.; V.G. Nikolaev, E.A. Norman, L. K. Polzik, M. Mojica, *Spectrochim. Acta, Part B 41/10*, 1043–1053 (1986).
[3725] L'vov, B.V.; E.A. Norman, L.K. Polzik, *Zh. Prikl. Spektrosk. 47/5*, 711–715 (1987).
[3726] L'vov, B.V.; E.A. Norman, M. Mojica, *Zavodsk. Lab. 53/6*, 26 (1987).
[3727] L'vov, B.V.; *J. Anal. At. Spectrom. 2/1*, 95–104 (1987).
[3728] L'vov, B.V.; *Analyst (London) 112*, 355–364 (1987).
[3729] L'vov, B.V.; V.G. Nikolaev, E.A. Norman, *Zh. Anal. Khim. 43/1*, 46–52 (1988).
[3730] L'vov, B.V.; *J. Anal. At. Spectrom. 3/1*, 9–12 (1988).
[3731] L'vov, B.V.; V.G. Nikolaev, A.V. Novichikhin, L.K. Polzik, *Spectrochim. Acta, Part B 43/9-11*, 1141–1146 (1988).
[3732] L'vov, B.V.; *Anal. Proc. (London) 25/7*, 222–224 (1988).
[3733] L'vov, B.V.; *Spectrochim. Acta, Part B 44/12*, 1257–1271 (1989).
[3734] L'vov, B.V.; A.V. Novichikhin, *At. Spectrosc. 11/1*, 1–6 (1990).
[3735] L'vov, B.V.; *Spectrochim. Acta, Part B 45/7*, 633–655 (1990).

[3736] L'vov, B.V.; L.K. Polzik, N.P. Romanova, A.I. Yuzefovskii, *J. Anal. At. Spectrom. 5/3*, 163–169 (1990).

[3737] L'vov, B.V.; L.K. Polzik, N.P. Romanova, *Zh. Anal. Khim. 46*, 837 (1991).

[3738] L'vov, B.V.; A.V. Novichikhin, L.K. Polzik, *Spectrochim. Acta, Part B 47/2*, 289–296 (1992).

[3739] L'vov, B.V.; N.V. Kocharova, L. K. Polzik, N.P. Romanova, Yu.I. Yarmak, *Spectrochim. Acta, Part B 47/6*, 843–854 (1992).

[3740] L'vov, B.V.; L.K. Polzik, N.V. Kocharova, *Spectrochim. Acta, Part B 47/7*, 889–895 (1992).

[3741] L'vov, B.V.; L.K. Polzik, N.V. Kocharova, Yu.A. Nemets, A.V. Novichikhin, *Spectrochim. Acta, Part B 47/10*, 1187–1202 (1992).

[3742] L'vov, B.V.; L.K. Polzik, P.N. Fedorov, W. Slavin, *Spectrochim. Acta, Part B 47/12*, 1411–1420 (1992).

[3743] L'vov, B.V.; *Spectrochim. Acta, Part B 48/13*, 1633–1637 (1993).

[3744] L'vov, B.V.; L.K. Polzik, A.V. Novichikhin, P.N. Fedorov, A.V. Borodin, *Spectrochim. Acta, Part B 48/13*, 1625–1632 (1993).

[3745] L'vov, B.V.; L.K. Polzik, A.V. Borodin, P.N. Fedorov, A.V. Novichikhin, *Spectrochim. Acta, Part B 49/12-4*, 1609–1627 (1994).

[3746] L'vov, B.V.; L.K. Polzik, A.V. Borodin, A.V. Novichikhin, *J. Anal. At. Spectrom. 10/10*, 703–709 (1995).

[3747] L'vov, B.V.; L.K. Polzik, A.V. Novichikhin, A.V. Borodin, A.O. Dyakov, *Spectrochim. Acta, Part B 50/14*, 1757–1768 (1995).

[3748] L'vov, B.V.; L.K. Polzik, P.N. Fedorov, A.V. Novichikhin, A.V. Borodin, *Spectrochim. Acta, Part B 50/13*, 1621–1636 (1995).

[3749] L'vov, B.V.; *Spectrochim. Acta, Part B 51/5*, 533–541 (1996).

[3750] L'vov, B.V.; L.K. Polzik, A.V. Novichikin, A.V. Borodin, A.O. Dyakov, *Spectrochim. Acta, Part B 51/6*, 609–618 (1996).

[3751] L'vov, B.V.; *Fresenius J. Anal. Chem. 355/3-4*, 222–226 (1996).

[3752] Lydersen, E.; B. Salbu, A.B.S. Polèo, *Analyst (London) 117*, 613–617 (1992).

[3753] Lynch, S.; D. Littlejohn, *J. Anal. At. Spectrom. 4/2*, 157–161 (1989).

[3754] Lynch, S.; R.E. Sturgeon, V.T. Luong, D. Littlejohn, *J. Anal. At. Spectrom. 5/4*, 311–319 (1990).

[3755] Lynch, S.; D. Littlejohn, *Talanta 37/8*, 825–830 (1990).

[3756] Lyon, T.D.B. ; G.S. Fell, D.J. Halls, J. Clark, F. McKenna, *J. Trace Elem. Electrol. Health Dis. 3/2*, 109–118 (1989).

[3757] Ma, R.-l.; W. Van Mol, F. Adams, *Anal. Chim. Acta 285/1-2*, 33–43 (1994).

[3758] Ma, R.-l.; W. Van Mol, F. Adams, *Anal. Chim. Acta 293/3*, 251–260 (1994).

[3759] Ma, R.-l.; W. Van Mol, F. Adams, *Mikrochim. Acta (Vienna) 119/1-2*, 95–102 (1995).

[3760] Ma, Y.-p.; G.l. Zhan, Y.-h. Chen, *Fenxi Huaxue 20/10*, 1227 (1992).

[3761] Ma, Y.-p.; *Lihua Jianyan, Huaxue Fence 30/1*, 29–31 (1994).

[3762] Ma, Y.-z.; J. Bai, D.-j. Sun, *J. Anal. At. Spectrom. 7/1*, 35–42 (1992).

[3763] Ma, Y.-z.; J. Bai, J.-z. Wang, Z.-k. Li, L. Zhu, Y.-q. Li, H. Zheng, B. Li, *J. Anal. At. Spectrom. 7/2*, 425–431 (1992).

[3764] Ma, Y.-z.; B.-w. Li, Z.-k. Li, J.-z. Wang, Y.-q. Li, *Fenxi Huaxue 21/7*, 745–749 (1993).

[3765] Ma, Y.-z.; Z.-k. Li, X.-h. Wang, J.-z. Wang, Y.-q. Li, *J. Anal. At. Spectrom. 9/6*, 679–683 (1994).

[3766] Maage, A.; K. Julshamn, K.-J. Andersen, *J. Anal. At. Spectrom. 6/4*, 277–281 (1991).

[3767] MacPherson, H.B.; S.S. Berman, *Analyst (London) 108*, 639–641 (1983).

[3768] Macalalad, E.; R. Bayoran, B. Ebarvia, I. Rubeška, *J. Geochem. Explor. 30*, 167–177 (1988).

[3769] Machata, G.; R. Binder, *Z. Rechtsmed. 73/1*, 29–34 (1973).

[3770] Machata, G.; R. Binder, *Arch. Kriminologie 155/3-4*, 87–90 (1975).

[3771] Macpherson, A.K.; B. Sampson, A.T. Diplock, *Analyst (London) 113/2*, 281–283 (1988).

[3772] Macquet, J.P.; T. Theophanides, *Biochim. Biophys. Acta 442*, 142–146 (1976).

[3773] Madany, I.M.; S.M. Ali, M.S. Akhter, *Environ. Int. 13*, 331–333 (1987).

[3774] Mader, P.; J. Száková, E. Curdová, *Talanta 43/4*, 521–534 (1996).

[3775] Madrid, Y.; M. Bonilla, C. Cámara-Rica, *J. Anal. At. Spectrom. 3/8*, 1097–1100 (1988).

[3776] Madrid, Y.; M. Bonilla, C. Cámara-Rica, *J. Anal. At. Spectrom. 4/2*, 167–169 (1989).

[3777] Madrid, Y.; J. Meseguer, M. Bonilla, C. Cámara-Rica, *Anal. Chim. Acta 237/1*, 181–187 (1990).

[3778] Madrid, Y.; M. Bonilla, C. Cámara-Rica, *Analyst (London) 115/5*, 563–569 (1990).

[3779] Madrid, Y.; C. Cámara-Rica, *Analyst (London) 119/8*, 1647–1658 (1994).

[3780] Madrid, Y.; C. Cabrera, T. Perez-Corona, C. Cámara-Rica, *Anal. Chem. 67/4*, 750–754 (1995).

[3781] Maeda, T.; Y. Tanimoto, N. Okazaki, *Anal. Sci. 3/4*, 291–295 (1987).
[3782] Maeda, T.; M. Nakatani, Y. Tanimoto, *Bunseki Kagaku 38*, 734 (1989).
[3783] Maessen, F.J.M.J.; F.D. Posma, J. Balke, *Anal. Chem. 46/11*, 1445–1449 (1974).
[3784] Maessen, F.J.M.J.; F.D. Posma, *Anal. Chem. 46/11*, 1439–1444 (1974).
[3785] Magee, R.J. ; B.D. James, in: H.G. Seiler, A. Sigel, H. Sigel (Editors), *Handbook on Metals in Clinical and Analytical Chemistry*, Marcel Dekker, New York, 1994, 551–562.
[3786] Magill, W.A.; G. Svehla, *Fresenius Z. Anal. Chem. 268*, 180–184 (1974).
[3787] Magill, W.A.; G. Svehla, *Fresenius Z. Anal. Chem. 268*, 177–180 (1974).
[3788] Magonov, S.N.; M.-H. Whangbo, *Surface Analysis with STM and AFM*, VCH Verlagsgesellschaft Weinheim, 1996.
[3789] Magos, L.; *Analyst (London) 96*, 847–853 (1971).
[3790] Magos, L.; in: A. Vercruysse (Editor), *Hazardous Metals in Human Toxicology*, Elsevier, Amsterdam, 1984.
[3791] Maguire, R.J.; R.J. Tkacz, *J. Chromatogr. 268/1*, 99–101 (1983).
[3792] Maguire, R.J.; R.J. Tkacz, *J. Agri. Food Chem. 33/5*, 947–953 (1985).
[3793] Maguire, R.J.; R.J. Tkacz, Y.K. Chau, G.A. Bengert, P.T.S. Wong, *Chemosphere 15/3*, 253–274 (1986).
[3794] Magyar, B.; *Guide Lines to Planning Atomic Spectrometric Analysis*, Elsevier (Akadémiai Kiadò), Amsterdam (Budapest), 1982.
[3795] Magyar, B.; K. Ikrényi, E. Bertalan, *Spectrochim. Acta, Part B 45/10*, 1139–1150 (1990).
[3796] Mahan, C.A.; J.A. Holcombe, *Anal. Chem. 64/17*, 1933–1939 (1992).
[3797] Mahan, K.I.; S.E. Mahan, *Anal. Chem. 49/4*, 662–664 (1977).
[3798] Mahan, W.A.; *Anal. Chim. Acta 126*, 157–165 (1981).
[3799] Mahan, W.A.; *Talanta 29/5*, 532–534 (1982).
[3800] Mahan, W.A.; *Anal. Chim. Acta 138*, 365–370 (1982).
[3801] Mahan, W.A.; *Talanta 30/7*, 534–536 (1983).
[3802] Mahan, W.A.; *Anal. Lett. 19/3-4*, 295–305 (1986).
[3803] Mahan, W.A.; *Microchem. J. 35*, 125–129 (1987).
[3804] Mahan, W.A.; *Microchem. J. 40/1*, 132–135 (1989).
[3805] Mahmood T.M.; K.W. Jackson, *Spectrochim. Acta, Part B 51/9-10*, 1155–1162 (1996).
[3806] Mai, S.; C. de Beer, C. Alsen-Hinrichs, in: B. Welz (Editor), *6. Colloquium Atomspektrometrische Spurenanalytik*, Bodenseewerk Perkin-Elmer, Überlingen, 1991, 829–840.
[3807] Maier, D.; H.W. Sinemus, E. Wiedeking, *Fresenius Z. Anal. Chem. 296*, 114–124 (1979).
[3808] Maier, K.; J. Neubauer, *Fresenius J. Anal. Chem. 340/3*, 187–189 (1991).
[3809] Mainka, E.; S. Weis, in: B. Welz (Editor), *4. Colloquium Atomspektrometrische Spurenanalytik*, Bodenseewerk Perkin-Elmer, Überlingen, 1987, 247–253.
[3810] Maitani, T.; S. Uchiyama, Y. Saito, *J. Chromatogr. 391/1*, 161–168 (1987).
[3811] Maitra, A.M.; E. Patsalides, *Anal. Chim. Acta 193*, 179–191 (1987).
[3812] Majer, J.R.; S.E.A. Khalil, *Anal. Chim. Acta 126*, 175–183 (1981).
[3813] Majidi, V.; J.A. Holcombe, *Spectrochim. Acta, Part B 43/12*, 1423–1429 (1988).
[3814] Majidi, V.; J.A. Holcombe, *J. Anal. At. Spectrom. 4/5*, 439–442 (1989).
[3815] Majidi, V.; J.A. Holcombe, *Spectrochim. Acta, Part B 45/7*, 753–761 (1990).
[3816] Majidi, V.; J. Ratliff, M. Owens, *Appl. Spectrosc. 45/3*, 473–476 (1991).
[3817] Majidi, V.; J.D. Robertson, *Spectrochim. Acta, Part B 46/13*, 1723–1733 (1991).
[3818] Majidi, V.; R.G. Smith, R.E. Bossio, R.T. Pogue, M.W. McMahon, *Spectrochim. Acta, Part B 51/9-10*, 941–959 (1996).
[3819] Mak, Y.T. ; D.W.Y. Ho, C.W.K. Lam, *Med. Lab. Sci. 46/3*, 272–275 (1989).
[3820] Makhno, A.Ya.; A.S. Alemasova, *Zavodsk. Lab. 58/12*, 23–24 (1992).
[3821] Makino, T.; K. Takahara, *Clin. Chem. (Winston-Salem) 27/8*, 1445–1447 (1981).
[3822] Makino, T.; K. Takahara, *Clin. Chem. (Winston-Salem) 27/12*, 2073 (1981).
[3823] Maksimov, G.A.; V.G. Pimenov, D.A. Timonin, *Vysokochistye Veshchestva 3*, 127–134 (1993).
[3824] Makun, H.A.; *Trace Elem. Electrolytes 12/1*, 52–54 (1995).
[3825] Malaiyandi, M.; J.P. Barrette, *J. Assoc. Off. Anal. Chem. 55/5*, 951–959 (1972).
[3826] Malhotra, R.K.; D.S.R. Murty, G. Srinivasan, *At. Spectrosc. 8/6*, 161–163 (1987).
[3827] Malhotra, R.K.; K. Satyanarayana, G. Srinivasan, D.S.R. Murty, *At. Spectrosc. 8/6*, 164–166 (1987).
[3828] Malla, M.E.; M.B. Alvarez, N.L. Belloni, E. Roso, *Bol. Soc. Quim. Peru 59*, 253 (1993).
[3829] Mallett, R.C.; D.C.G. Pearton, E.J. Ring, T.W. Steele, *Talanta 19*, 181–195 (1972).
[3830] Mallett, R.C.; S.J. Royal, T.W. Steele, *Anal. Chem. 51/11*, 1617–1620 (1979).
[3831] Malvankar, P.L.; V.M. Shinde, *Analyst (London) 116/10*, 1081–1084 (1991).

[3832] Manahan, S.E.; D.R. Jones, *Anal. Lett. 6/8*, 745–753 (1973).
[3833] Manahan, S.E.; R. Kunkel, *Anal. Lett. 6/6*, 547–553 (1973).
[3834] Mancy, K.H.; W.J. Weber, in: I.M. Kolthoff, P.J. Elving (Editors), *Treatise on Analytical Chemistry, Part III*, J. Wiley & Sons, New York, 1971, 413.
[3835] Mandel, J.; *The Statistical Analysis of Experimental Data*, J. Wiley & Sons (Interscience), New York, 1964.
[3836] Maney, J.P.; V.J. Luciano, *Anal. Chim. Acta 125*, 183–186 (1981).
[3837] Manning, B.A.; D.A. Martens, *Environ. Sci. Technol. 31/1*, 171–177 (1997).
[3838] Manning, D.C.; W. Slavin, *At. Absorpt. Newsletter 1*, 39–43 (1962).
[3839] Manning, D.C.; *At. Absorpt. Newsletter 3*, 84–88 (1964).
[3840] Manning, D.C.; H.L. Kahn, *At. Absorpt. Newsletter 4/3*, 224–227 (1965).
[3841] Manning, D.C.; D.J. Trent, S. Sprague, W. Slavin, *At. Absorpt. Newsletter 4*, 255–263 (1965).
[3842] Manning, D.C.; D.J. Trent, J. Vollmer, *At. Absorpt. Newsletter 4*, 234–236 (1965).
[3843] Manning, D.C.; *At. Absorpt. Newsletter 5/3*, 63–64 (1966).
[3844] Manning, D.C.; *At. Absorpt. Newsletter 5/6*, 127–134 (1966).
[3845] Manning, D.C.; L. Capacho-Delgado, *Anal. Chim. Acta 36*, 312 (1966).
[3846] Manning, D.C.; *At. Absorpt. Newsletter 6/2*, 35–37 (1967).
[3847] Manning, D.C.; *At. Absorpt. Newsletter 6/3*, 75 (1967).
[3848] Manning, D.C.; F.J. Fernandez, *At. Absorpt. Newsletter 6*, 15–16 (1967).
[3849] Manning, D.C.; J. Vollmer, *At. Absorpt. Newsletter 6/2*, 38–41 (1967).
[3850] Manning, D.C.; J. Vollmer, F.J. Fernandez, *At. Absorpt. Newsletter 6*, 17–18 (1967).
[3851] Manning, D.C.; F.J. Fernandez, *At. Absorpt. Newsletter 7*, 24 (1968).
[3852] Manning, D.C.; S. Slavin, *At. Absorpt. Newsletter 8/6*, 132 (1969).
[3853] Manning, D.C.; F.J. Fernandez, *At. Absorpt. Newsletter 9/3*, 65–70 (1970).
[3854] Manning, D.C.; *At. Absorpt. Newsletter 10/6*, 123–124 (1971).
[3855] Manning, D.C.; *At. Absorpt. Newsletter 14/4*, 99–102 (1975).
[3856] Manning, D.C.; R.D. Ediger, *At. Absorpt. Newsletter 15/2*, 42–44 (1976).
[3857] Manning, D.C.; *At. Absorpt. Newsletter 17/5*, 107–108 (1978).
[3858] Manning, D.C.; W. Slavin, *Anal. Chem. 50*, 1234–1238 (1978).
[3859] Manning, D.C.; W. Slavin, *At. Absorpt. Newsletter 17/2*, 43–46 (1978).
[3860] Manning, D.C.; W. Slavin, S. Myers, *Anal. Chem. 51/14*, 2375–2378 (1979).
[3861] Manning, D.C.; W. Slavin, *Anal. Chim. Acta 118*, 301–306 (1980).
[3862] Manning, D.C.; W. Slavin, G.R. Carnrick, *Spectrochim. Acta, Part B 37/4*, 331–341 (1982).
[3863] Manning, D.C.; W. Slavin, *Appl. Spectrosc. 37/1*, 1–11 (1983).
[3864] Manning, D.C.; W. Slavin, *Spectrochim. Acta, Part B 40/3*, 461–473 (1985).
[3865] Manning, D.C.; W. Slavin, *At. Spectrosc. 7/4*, 123–126 (1986).
[3866] Manning, D.C.; W. Slavin, *Spectrochim. Acta, Part B 42*, 755–763 (1987).
[3867] Mansfield, J.M.; M.P. Bratzel, H.O. Norgordon, D.O. Knapp, K.E. Zacha, J.D. Winefordner, *Spectrochim. Acta, Part B 23/6*, 389–402 (1968).
[3868] Manthei, K.; *Patent DE 2,245,610*, Bodenseewerk Perkin-Elmer, Überlingen, 1974.
[3869] Manthey, G.; H. Berge, *Nahrung 24*, 413–421 (1980).
[3870] Manzoori, J.L.; A. Saleemi, *J. Anal. At. Spectrom. 9/3*, 337–339 (1994).
[3871] Maqueda, C.; E. Morillo, *Fresenius J. Anal. Chem. 338/3*, 253–254 (1990).
[3872] Maquieira, A.; H.A.M. Elmahadi, R. Puchades, *Anal. Chem. 66/9*, 1462–1467 (1994).
[3873] Maquieira, A.; H.A.M. Elmahadi, R. Puchades, *Anal. Chem. 66/21*, 3632–3638 (1994).
[3874] Marabini, A.M.; M. Barbaro, B. Passariello, *At. Spectrosc. 3/5*, 140–142 (1982).
[3875] March, J.G.; R. Forteza, F. Grases, *Microchem. J. 33/1*, 39–45 (1986).
[3876] Marchandise, H.; *Fresenius J. Anal. Chem. 345/2-4*, 82–86 (1993).
[3877] Marchante-Gayón, J.M.; J. Pérez Parajón, A. Sanz-Medel, C.S. Fellows, *J. Anal. At. Spectrom. 7/5*, 743–747 (1992).
[3878] Marchante-Gayón, J.M.; J.E. Sánchez-Uría, A. Sanz-Medel, *J. Anal. At. Spectrom. 8/5*, 731–736 (1993).
[3879] Marchante-Gayón, J.M.; J.M. González, M.L. Fernández, E. Blanco, A. Sanz-Medel, *Fresenius J. Anal. Chem. 355/5-6*, 615–622 (1996).
[3880] Marczenko, Z.; R. Łobinski, *Pure Appl. Chem. 63/11*, 1627–1636 (1991).
[3881] Marecek, J.; V. Synek, D. Tomkova, *Chem. Prum. 40*, 140 (1990).
[3882] Marella, M.; R. Milanino, *At. Spectrosc. 7/1*, 40–42 (1986).
[3883] Margel, S.; J. Hirsh, *Clin. Chem. (Winston-Salem) 30/2*, 243–245 (1984).
[3884] Marin, S.R.; S.G. Olave, O.E. Andonie, O.G. Arlegui, *Int. J. Environ. Anal. Chem. 52*, 127–136 (1993).
[3885] Marinkovic, M.; P.J. Slevin, T.J. Vickers, *Appl. Spectrosc. 25/3*, 372–374 (1971).

[3886] Mario, E.; R.E. Gerner, *J. Pharm. Sci. 57*, 1243 (1968).
[3887] Markert, B.; *Fresenius Z. Anal. Chem. 335/6*, 562–565 (1989).
[3888] Markert, B.; *Fresenius J. Anal. Chem. 342/4-5*, 409–412 (1992).
[3889] Markert, B.; (Editor), *Plants as Biomonitors; Indicators for Heavy Metals in the Terrestrial Environment*, VCH Verlagsgesellschaft Weinheim - New York - Basel -Cambridge, 1993.
[3890] Marks, J.Y.; G.G. Welcher, *Anal. Chem. 42/9*, 1033–1040 (1970).
[3891] Marks, J.Y.; R.J. Spellman, B. Wysocki, *Anal. Chem. 48/11*, 1474–1478 (1976).
[3892] Marks, J.Y.; G.G. Welcher, R.J. Spellman, *Appl. Spectrosc. 31/1*, 9–11 (1977).
[3893] Marple, V.A.; K. Willeke, in: B.Y.H. Liu (Editor), *Fine Particles: Aerosol Generation, Measurement, Sampling and Analysis*, Academic Press, New York, 1976, 411–446.
[3894] Marr, I.L.; J. Anwar, B.B. Sithole, *Analyst (London) 107*, 1212–1217 (1982).
[3895] Marr, I.L.; M.S. Cresser, L.J. Ottendorfer, *Umweltanalytik - Eine allgemeine Einführung*, G. Thieme, Stuttgart, 1988.
[3896] Marr, I.L.; P. Kluge, L. Main, V. Margerin, C. Lescop, *Mikrochim. Acta (Vienna) 119/3-4*, 219–232 (1995).
[3897] Marsh, H.; A.P. Warburton, *J. Appl. Chem. 20*, 113–142 (1970).
[3898] Marsh, H.; D. Crawford, D.W. Taylor, *Carbon 21*, 81–87 (1983).
[3899] Marshall, G.; D. Midgley, *Anal. Chem. 53/12*, 1760–1765 (1981).
[3900] Marshall, G.D.; *Mintek Report M276*, 11 (1986).
[3901] Marshall, G.D.; J.F. van Staden, *J. Anal. At. Spectrom. 5/8*, 681–686 (1990).
[3902] Marshall, J.; J. Carroll, D. Littlejohn, J.M. Ottaway, T.C. O'Haver, J.M. Harnly, *Anal. Proc. (London) 22/3*, 67–69 (1985).
[3903] Marshall, J.; B.J. Ottaway, J.M. Ottaway, D. Littlejohn, *Anal. Chim. Acta 180*, 357–371 (1986).
[3904] Marshall, J.; J. Franks, *Anal. Proc. (London) 27/9*, 240–241 (1990).
[3905] Marshall, J.; in: S.J. Haswell (Editor), *Atomic Absorption Spectrometry - Theory, Design and Applications*, Elsevier, Amsterdam - New York, 1991, 321–340.
[3906] Marshall, W.D.; J.S. Blais, F.C. Adams, in: J.A.C. Broekaert, S. Güçer, F. Adams (Editors), *Metal Speciation in the Environment*, Springer, Berlin - Heidelberg - New York, 1990, 253–273.
[3907] Martens, D.A.; D.L. Suarez, *Environ. Sci.Technol. 31/1*, 133–139 (1997).
[3908] Martin, C.K.; J.C. Williams, *J. Anal. At. Spectrom. 4/8*, 691–695 (1989).
[3909] Martin, F.M.; C.-m. Tseng, C. Belin, Ph. Quevauviller, O.F.X. Donard, *Anal. Chim. Acta 286/3*, 343–355 (1994).
[3910] Martin, I.; M. Lopez-Gonzalvez, M. Gomez, C. Camara, M. Palacios, *J. Chromatogr. 666*, 101 (1995).
[3911] Martin, R.F.; *Clin. Chem. (Winston-Salem) 26/10*, 1509–1510 (1980).
[3912] Martin, T.D.M.; J.K. Riley, *At. Spectrosc. 3/6*, 174–179 (1982).
[3913] Martínez, C.; J.R. Castillo-Suarez, *At. Spectrosc. 4/2*, 63–65 (1983).
[3914] Martínez-Avila, R.; V. Carbonell, M. de la Guardia, A. Salvador, *J. Assoc. Off. Anal. Chem. 73/3*, 389–393 (1990).
[3915] Martínez-Avila, R.; V. Carbonell, A. Salvador, M. de la Guardia, *Talanta 40/1*, 107–112 (1993).
[3916] Martínez-Calatayud, J.; J.V. Garcia-Mateo, *J. Pharm. Biomed. Anal. 7/12*, 1441–1445 (1989).
[3917] Martínez Gonzalez, P.; C. Camara-Rica, L. Polo-Díez, *J. Anal. At. Spectrom. 2/8*, 809–811 (1987).
[3918] Martínez-Jiménez, P.; M. Gallego, M. Valcárcel, *At. Spectrosc. 6/5*, 137–141 (1985).
[3919] Martínez-Jiménez, P.; M. Gallego, M. Valcárcel, *Anal. Chim. Acta 193*, 127–135 (1987).
[3920] Martínez-Jiménez, P.; M. Gallego, M. Valcárcel, *J. Anal. At. Spectrom. 2/2*, 211 (1987).
[3921] Martínez-Jiménez, P.; M. Gallego, M. Valcárcel, *Anal. Chem. 59/1*, 69–74 (1987).
[3922] Martínez-Jiménez, P.; M. Gallego, M. Valcárcel, *Anal. Chim. Acta 215/1-2*, 233–240 (1988).
[3923] Martínez-Soria, M.T.; J. Sanz Asencio, J. Galbán Bernal, *J. Anal. At. Spectrom. 10/11*, 975–980 (1995).
[3924] Martinsen, I.; F.J. Langmyhr, *Anal. Chim. Acta 135/1*, 137–143 (1982).
[3925] Martinsen, I.; B. Radziuk, Y. Thomassen, *J. Anal. At. Spectrom. 3/7*, 1013–1022 (1988).
[3926] Marts, R.W.; J.J. Blaha, *J. Assoc. Off. Anal. Chem. 66/6*, 1421–1423 (1983).
[3927] Mason, J.T.; *J. Pharm. Sci. 69/1*, 101–102 (1980).
[3928] Mason, R.P.; W.F. Fitzgerald, *Nature (London) 347*, 457–459 (1990).
[3929] Mason, R.W.; S.J. Hogg, I.R. Edwards, *Toxicology 38*, 219–226 (1986).
[3930] Massart, D.L.; *Trends Anal. Chem. (Pers. Ed.) 7/5*, 157–158 (1988).

[3931]   Massart, D.L.; *Patent GB 9008922 [20.04.90], EUR. Pat. Appl. 91200880.2*, N.V. Philips' Gloelampenfabrieken, Eindhoven, 1991.
[3932]   Massart, D.L.; F.J.M.J. Maessen, J.J.M. de Goeij, *Anal. Chim. Acta 127*, 181–193 (1981).
[3933]   Massey, R.C.; D. Taylor, *Aluminium in Food and the Environment*, Royal Society of Chemistry, Cambridge, 1989.
[3934]   Massmann, H.; *Z. Instrumentenkunde 71*, 225–229 (1963).
[3935]   Massmann, H.; in: G. Ehrlich (Editor), *2. Int. Symp. Reinststoffe in Wissenschaft und Technik*, Akademie Verlag, Berlin, 1966, 297–308.
[3936]   Massmann, H.; *Fresenius Z. Anal. Chem. 225/2*, 203–213 (1967).
[3937]   Massmann, H.; *Spectrochim. Acta, Part B 23/4*, 215–226 (1968).
[3938]   Massmann, H.; Z. El Gohary, S. Güçer, *Spectrochim. Acta, Part B 31/7*, 399–409 (1976).
[3939]   Massmann, H.; in: *Ullmanns Encyklopädie der technischen Chemie, Band 5*, Verlag Chemie, Weinheim, 1980, 423–440.
[3940]   Massmann, H.; Z. El Gohary, in: K.-H. Koch, H. Massmann (Editors), *13. Spektrometertagung*, W. de Gruyter, Berlin, 1981, 359–375.
[3941]   Massmann, H.; *Talanta 29/11B*, 1051–1055 (1982).
[3942]   Masters, R.; C.-m. Hsiech, H.L. Pardue, *Anal. Chim. Acta 199*, 253–257 (1987).
[3943]   Masters, R.; C.-m. Hsiech, H.L. Pardue, *Talanta 36/1-2*, 133–139 (1989).
[3944]   Mateeva, N.; S. Arpadjan, T. Delingeorgiev, M. Mitewa, *Analyst (London) 117/10*, 1599–1601 (1992).
[3945]   Mathews, C.K.; *Pure Appl. Chem. 54/4*, 807–818 (1982).
[3946]   Matni, G.; R. Azani, M.R. Van Calsteren, M.C. Bissonnette, J.S. Blais, *Analyst (London) 120/2*, 395–401 (1995).
[3947]   Matoušek, J.P.; *Spectrochim. Acta, Part B 39/2-3*, 205–212 (1984).
[3948]   Matoušek, J.P.; H.K.J. Powell, *Appl. Spectrosc. 42/1*, 166–168 (1988).
[3949]   Matsumoto, K.; T. Solin, K. Fuwa, *Spectrochim. Acta, Part B 39/2-3*, 481–483 (1984).
[3950]   Matsunaga, K.; S. Takahashi, *Anal. Chim. Acta 87*, 487–489 (1976).
[3951]   Matsunaga, K.; S. Konishi, M. Nishimura, *Environ. Sci. Technol. 13/1*, 63–65 (1979).
[3952]   Matsunaga, K.; M. Negushi, S. Fukase, *Geochim. Cosmochim. Acta 44*, 1615–1619 (1980).
[3953]   Matsunaga, K.; T. Yoshino, *Anal. Chim. Acta 157*, 193–197 (1984).
[3954]   Matsunaga, K.; T. Yoshino, Y. Yamamoto, *Anal. Chim. Acta 167/1*, 299–304 (1985).
[3955]   Matsunaga, K.; M. Izuchi, *Anal. Sci. 7/1*, 159–161 (1991).
[3956]   Matsusaki, K.; *Anal. Chim. Acta 248/1*, 251–255 (1991).
[3957]   Matsusaki, K.; T. Oishi, *Anal. Sci. 9/3*, 381–384 (1993).
[3958]   Matsusaki, K.; K. Okada, T. Oishi, T. Sata, *Anal. Sci. 10/2*, 281–285 (1994).
[3959]   Matsusaki, K.; T. Yamaguchi, Y. Yamamoto, *Anal. Sci. 12/2*, 301–305 (1996).
[3960]   Matsusaki, K.; T. Yamaguchi, T. Sata, *Anal. Sci. 12/5*, 797–800 (1996).
[3961]   Matter, L.; W. Schneider, *Lebensmittelchem. Gerichtl. Chem. 28*, 231–262 (1974).
[3962]   Matter, L.; in: M. Stoeppler (Editor), *Probennahme und Aufschluß*, Springer, Berlin - Heidelberg - New York, 1994, 101–109.
[3963]   Matthes, S.A.; R.F. Farrell, A.J. Mackie, *US Bureau of Mines, Technical Progress Report 120*, Washington, 1983.
[3964]   Matthes, S.A.; in: H.M. Kingston, L.B. Jassie (Editors), *Introduction to Microwave Sample Preparation*, American Chemical Society, Washington, DC, 1988, 33–51.
[3965]   Matthes, W.; R. Flucht, M. Stoeppler, *Fresenius Z. Anal. Chem. 291*, 20–26 (1978).
[3966]   Matthys, M.; V. Blaton, J. Spincemaille, *Biol. Trace Elem. Res. 32*, 349–354 (1992).
[3967]   Matusiewicz, H.; *Anal. Chim.Acta 136*, 215–223 (1982).
[3968]   Matusiewicz, H.; *Spectroscopy (Eugene, OR) 1/12*, 32–34 (1986).
[3969]   Matusiewicz, H.; R.E. Sturgeon, *Prog. Anal. Spectrosc. 12/1*, 21–39 (1989).
[3970]   Matusiewicz, H.; *J. Anal. At. Spectrom. 4/3*, 265–269 (1989).
[3971]   Matusiewicz, H.; R.E. Sturgeon, S.S. Berman, *J. Anal. At. Spectrom. 4/4*, 323–327 (1989).
[3972]   Matusiewicz, H.; *Spectroscopy Int. 3/2*, 22–27 (1991).
[3973]   Matusiewicz, H.; R.E. Sturgeon, S.S. Berman, *J. Anal. At. Spectrom. 6/4*, 283–287 (1991).
[3974]   Matusiewicz, H.; *Mikrochim. Acta (Vienna) 111/1-3*, 71–82 (1993).
[3975]   Matusiewicz, H.; R.E. Sturgeon, *Fresenius J. Anal. Chem. 349/6*, 428–433 (1994).
[3976]   Matusiewicz, H.; M. Kopras, A. Suszka, *Microchem. J. 52/3*, 282–289 (1995).
[3977]   Matusiewicz, H.; R.E. Sturgeon, *Spectrochim. Acta, Part B 51/4*, 377–397 (1996).
[3978]   Maurer, J.; *Z. Lebensm. Unters. Forsch. 165*, 1–4 (1977).
[3979]   Maurer, W.; G. Schaldach, W. Wagener, J. Peters, *Fresenius Z. Anal. Chem. 322/3*, 359–364 (1985).
[3980]   Maurí, A.R.; E. Huerta, M. de la Guardia, *Fresenius J. Anal. Chem. 338/6*, 699–702 (1990).

[3981] Mausbach, G.; *GIT Fachz. Lab. 23/10*, 898–907 (1979).
[3982] Mausbach, G.; *LaborPraxis 8/7-8*, 764–768 (1984).
[3983] Mavrodineanu, R.; H. Boiteeux, *Flame Spectroscopy*, J. Wiley & Sons, New York - London - Sydney, 1965.
[3984] Mavrodineanu, R.; R.C. Hughes, *Appl. Opt. 7/7*, 1281–1285 (1968).
[3985] Maxfield, R.E.; *At. Absorpt. Newsletter 18*, 100 (1979).
[3986] May, I.; L.P. Greenland, *Anal. Chem. 49*, 2376–2377 (1977).
[3987] May, K.; K. Reisinger, R. Flucht, M. Stoeppler, *Vom Wasser 55*, 63–76 (1980).
[3988] May, K.; M. Stoeppler, *Fresenius Z. Anal. Chem. 317/3-4*, 248–251 (1984).
[3989] May, K.; K. Reisinger, B. Torres, M. Stoeppler, *Fresenius Z. Anal. Chem. 320/7*, 646 (1985).
[3990] May, K.; M. Stoeppler, K. Reisinger, *Toxicol. Environm. Chem. 13/3-4*, 153–159 (1987).
[3991] May, T.W.; W.G. Brumbaugh, *Anal. Chem. 54/7*, 1032–1037 (1982).
[3992] May, T.W.; D.A. Kane, *Anal. Chim. Acta 161*, 387–391 (1984).
[3993] May, T.W.; J.L. Johnson, *At. Spectrosc. 6/1*, 9–15 (1985).
[3994] Maya, M.; J. Morais, J. Aranda da Silva, H. Rebelo, A. Godinho, C. Silveira, *Rev. Port. Farm. 40/3-4*, 9 (1990).
[3995] Mayer, D.R.; W. Kosmus, H. Pogglitsch, D. Mayer, W. Beyer, *Biol. Trace Elem. Res. 37/1*, 27–38 (1993).
[3996] Mazzucotelli, A.; B. Vivaldi, F. De Paz, E. Ambrogio, *Ann. Chim. (Rom) 82*, 107–120 (1992).
[3997] Mazzucotelli, A.; P. Rivaro, *Ann. Chim. (Rom) 83/3-4*, 105–115 (1993).
[3998] Mazzucotelli, A.; M. Grotti, *Spectrochim. Acta, Part B 50/14*, 1897–1804 (1995).
[3999] Mbofung, C.M.F.; T. Atinmo, A. Omololu, *Nutr. Rep. Int. 30*, 1137–1146 (1984).
[4000] McAlister, J.J.; K.R. Hirons, *Microchem. J. 30*, 79–91 (1984).
[4001] McAllister, T.; *J. Anal. At. Spectrom. 5/3*, 171–174 (1990).
[4002] McAuliffe, C.A.; H.L. Sharma, N.D. Tinker, in: F.R. Hartley (Editor), *Chemistry of the Platinum Group Metals*, Elsevier, Amsterdam, 1991.
[4003] McBride, C.H.; *At. Absorpt. Newsletter 3*, 144–159 (1964).
[4004] McBride, C.H.; *J. Assoc. Off. Anal. Chem. 48/2*, 406 (1965).
[4005] McBride, C.H.; *J. Assoc. Off. Anal. Chem. 48/6*, 1100 (1965).
[4006] McBride, C.H.; *J. Assoc. Off. Anal. Chem. 51*, 847 (1968).
[4007] McCaffrey, J.T.; R.G. Michel, *Appl. Spectrosc. 43/2*, 310–319 (1989).
[4008] McCaffrey, J.T.; R.G. Michel, *Appl. Spectrosc. 44/6*, 919–933 (1990).
[4009] McCall, S.; *Chem. N.Z. 55/Dec*, 10–11 (1991).
[4010] McCarthy, H.T.; P.C. Ellis, *J. Assoc. Off. Anal. Chem. 74*, 566–569 (1991).
[4011] McCrackan, M.L.; H.J. Webb, H.E. Hammar, C.B. Loadholt, *J. Assoc. Off. Anal. Chem. 50*, 5–7 (1967).
[4012] McDaniel, M.; A.D. Shendrikar, K.D. Reiszner, P.W. West, *Anal. Chem. 48*, 2240–2243 (1976).
[4013] McGahan, M.C.; L.Z. Bito, *Anal. Biochem. 135/1*, 186–192 (1983).
[4014] McGahan, M.C.; L.N. Fleisher, *Anal. Biochem. 156/2*, 397–402 (1986).
[4015] McGahan, M.C.; A.M. Grimes, *Ophthalmic Res. 23*, 45–50 (1991).
[4016] McGee, W.W.; J.D. Winefordner, *Anal. Chim. Acta 37*, 429–435 (1967).
[4017] McHugh, J.B.; J.H. Turner, *U.S. Geol. Surv. Prof. Pap. 1129-A-1*, 1980.
[4018] McHugh, J.B.; *At. Spectrosc. 4/2*, 66–68 (1983).
[4019] McHugh, J.B.; *J. Geochem. Explor. 20/3*, 303–310 (1984).
[4020] McIntosh, S.; Z. Li, G.R. Carnrick, W. Slavin, *Spectrochim. Acta, Part B 47/7*, 897–906 (1992).
[4021] McIntosh, S.; B. Welz, Perkin-Elmer, *Document ENV-12A*, 1992.
[4022] McIntosh, S.; *At. Spectrosc. 14/2*, 47–49 (1993).
[4023] McIntosh, S.; J. Baasner, Z. Grosser, C. Hanna, *At. Spectrosc. 15/4*, 161–163 (1994).
[4024] McKenzie, H.A.; L.E. Smythe, (Eds.), *Quantitative Trace Analysis of Biological Materials - Principles and Methods for Determination of Trace Elements and Trace Amounts of Some Macro Elements*, Elsevier, Amsterdam - New York - Oxford, 1988.
[4025] McKie, J.C.; *Anal. Chim. Acta 197*, 303–308 (1987).
[4026] McKinney, G.L.; *J. Res. Natl. Bur. Stand. 93/3*, 307–310 (1988).
[4027] McLaren, J.W.; R.C. Wheeler, *Analyst (London) 102*, 542–546 (1977).
[4028] McLaughlin, K.; D. Dadgar, M.R. Smyth, D. McMaster, *Analyst (London) 115/3*, 275–278 (1990).
[4029] McMahon, J.W.; A.E. Docherty, J.M.A. Judd, S.-R. Gentner, *Int. J. Environ. Anal. Chem. 24/4*, 297–303 (1986).
[4030] McNally, J.; J.A. Holcombe, *Anal. Chem. 59/8*, 1105–1112 (1987).

[4031] McNally, J.; J.A. Holcombe, *Anal. Chem. 63/18*, 1918–1926 (1991).
[4032] McPherson, G.L.; *At. Absorpt. Newsletter 4*, 186–191 (1965).
[4033] McWilliam, I.G.; H.C. Bolton, *Anal. Chem. 41/13*, 1755–1762 (1969).
[4034] McWilliam, I.G.; H.C. Bolton, *Anal. Chem. 41/13*, 1762–1770 (1969).
[4035] Means, E.A.; D. Ratcliff, *At. Absorpt. Newsletter 4*, 174–179 (1965).
[4036] Measures, C.I.; J.D. Burton, *Anal. Chim. Acta 120*, 177–186 (1980).
[4037] Medeiros, L.C.; R.P. Belden, E.S. Williams, *J. Food Sci. 58/4*, 731–733 (1993).
[4038] Medina Escriche, J.; F. Hernández Hernández, M. Conesa Casanova, A.P. García, *Analusis 15/1*, 47–53 (1987).
[4039] Medlin, J.H.; N.H. Suhr, J.B. Bodkin, *At. Absorpt. Newsletter 8/2*, 25–29 (1969).
[4040] Medvecky, L.; J. Mihalik, J. Dusza, *Chem. Listy 86/5*, 375–377 (1992).
[4041] Mei, N.-s.; B.-x. Xu, Y.-z. Fang, *Lihua Jianyan, Huaxue Fence 24/6*, 322 (1988).
[4042] Meier, A.L.; *Anal. Chem. 54/13*, 2158–2161 (1982).
[4043] Melcher, M.; B. Welz, *AA/AS Lab Notes (Perkin-Elmer) 20/E*, 9 (1978).
[4044] Melcher, R.G.; T.L. Peters, H.W. Emmel, *Top. Curr. Chem. 134*, 59–123 (1986).
[4045] Melton, J.R.; W.L. Hoover, P.A. Howard, *J. Assoc. Off. Anal. Chem. 52*, 950–953 (1969).
[4046] Melton, J.R.; W.L. Hoover, P.A. Howard, J.L. Ayers, *J. Assoc. Off. Anal. Chem. 53*, 682–685 (1970).
[4047] Memon, M.A.; Z.-x. Zhuang, Z.-l. Fang, *At. Spectrosc. 14/2*, 50–54 (1993).
[4048] Memon, M.A.; X.-r. Wang, B.-l. Huang, *At. Spectrosc. 14/4*, 99–102 (1993).
[4049] Menden, E.E.; D. Brockman, H. Choudhury, H.G. Petering, *Anal. Chem. 49/11*, 1644–1645 (1977).
[4050] Menden, E.E.; F.H. Rutland, *J. Am. Leather Chem. Assoc. 83*, 220–231 (1988).
[4051] Menden, E.E.; F.H. Rutland, W.E. Kallenberger, *J. Am. Leather Chem. Assoc. 85/10*, 363–375 (1990).
[4052] Mendes, P.C.S.; R.E. Santelli, M. Gallego, M. Valcárcel, *J. Anal. At. Spectrom. 9/5*, 663–666 (1994).
[4053] Menis, O.; T.C. Rains, in: R. Mavrodineanu (Editor), *Analytical Flame Spectroscopy, Selected Topics*, MacMillan, London, 1970, 47–77.
[4054] Menné, T.; H.I. Maibach, *Nickel and the Skin: Immunology and Toxicology*, CRC Press, Boca Raton, 1989.
[4055] Mentser, M.; S. Ergun, *Carbon 5*, 331 (1967).
[4056] Menz, D.; G. Conradi, *Erdoel, Kohle, Erdgas, Petrochem. 35/1*, 33 (1982).
[4057] Menzies, A.C.; *Anal. Chem. 32/8*, 898–904 (1960).
[4058] Méranger, J.C.; E. Somers, *Analyst (London) 93*, 799–801 (1968).
[4059] Méranger, J.C.; E. Somers, *Bull. Environ. Contam. Toxicol. 3/6*, 360–365 (1969).
[4060] Méranger, J.C.; K.S. Subramanian, C.H. Langford, *Rev. Anal. Chem. 5/1-2*, 29–51 (1980).
[4061] Méranger, J.C.; B.R. Hollebone, G.A. Blanchette, *J. Anal. Toxicol. 5/1*, 33–41 (1981).
[4062] Méranger, J.C.; K.S. Subramanian, R.F. McCurdy, *Sci. Total Environ. 39*, 49–55 (1984).
[4063] Meret, S.; R.I. Henkin, *Clin. Chem. (Winston-Salem) 17*, 369–373 (1971).
[4064] Merz, W.; R. Willinger, *Mikrochim. Acta (Vienna) III/1-3*, 11–16 (1991).
[4065] Metcalfe, P.J.; *Anal. Proc. (London) 26/4*, 134–136 (1989).
[4066] Methodenbuch, Band 1: *Die Untersuchung von Böden*, VDLUFA-Verlag, Darmstadt, 1991.
[4067] Meyer, A.; Ch. Hofer, G. Tölg, S. Raptis, G. Knapp, *Fresenius Z. Anal. Chem. 296*, 337–344 (1979).
[4068] Meyer, A.; L. Dunemann, in: B. Welz (Editor), *6. Colloquium Atomspektrometrische Spurenanalytik*, Bodenseewerk Perkin-Elmer, Überlingen, 1991, 315–321.
[4069] Meyer, A.; G. Schwedt, *LaborPraxis 17/4*, 44–46,48 (1993).
[4070] Meyer, M.L.; P.R. Bloom, *Plant Soil 153*, 281 (1993).
[4071] Michaelis, M.R.A.; W. Wegscheider, H.M. Ortner, *J. Anal. At. Spectrom. 3/4*, 503–509 (1988).
[4072] Michaelis, M.R.A.; W. Wegscheider, H.M. Ortner, *J. Res. Natl. Bur. Stand. 93/3*, 467–469 (1988).
[4073] Michailova, M.; V. Stoitschkova, S. Arpadjan, *Z. Chem. 28*, 149–150 (1988).
[4074] Michailova, M.; V. Stoitschkova, S. Arpadjan, *Z. Chem. 30*, 418–419 (1990).
[4075] Michalke, B.; *Fresenius J. Anal. Chem. 350/1-2*, 2–6 (1994).
[4076] Michel, P.; B. Averty, V. Colandini, *Mikrochim. Acta (Vienna) 109/1-4*, 35–38 (1992).
[4077] Michie, N.D.; E.J. Dixon, *J. Sci. Food Agric. 28*, 215–224 (1977).
[4078] Mignardi, M.A.; B.T. Jones, B.W. Smith, J.D. Winefordner, *Anal. Chim. Acta 227/2*, 331–342 (1989).
[4079] Mignardi, M.A.; U. Masi, *Talanta 42/12*, 2059–2061 (1995).

[4080] Mikac, N.; M. Branica, Y. Wang, R.M. Harrison, *Environ. Sci. Technol. 30*, 499–508 (1996).
[4081] Mikac, N.; Y. Wang, R.M. Harrison, *Anal. Chim. Acta 326/1*, 57–66 (1996).
[4082] Milačič, R.; J. Štupar, N. Kožuh, J. Korošin, *Analyst (London) 117/2*, 125–130 (1992).
[4083] Milačič, R.; J. Štupar, N. Kožuh, J. Korošin, I. Glazer, *J. Am. Leather Chem. Assoc. 87*, 221–234 (1992).
[4084] Milačič, R.; J. Štupar, *Analyst (London) 119/4*, 627–632 (1994).
[4085] Mile, B.; C.C. Rowlands, A.V. Jones, *J. Anal. At. Spectrom. 7/7*, 1069–1073 (1992).
[4086] Mile, B.; C.C. Rowlands, A.V. Jones, *J. Anal. At. Spectrom. 10/10*, 785–789 (1995).
[4087] Milella, E.; E. Sentimenti, G. Mazzetto, *At. Spectrosc. 14/1*, 1–3 (1993).
[4088] Milella, E.; E. Sentimenti, G. Mazzetto, L. Meregalli, M. Battagliarin, *Anal. Chim. Acta 272/1*, 99–103 (1993).
[4089] Miller, D.B.; *At. Spectrosc. 9/1*, 43–45 (1988).
[4090] Miller, D.T.; D.C. Paschal, E.W. Gunter, Ph.E. Stroud, J. D'Angelo, *Analyst (London) 112/12*, 1701–1704 (1987).
[4091] Miller, G.R.; *Spectroscopy (Eugene, OR) 9/8*, 36–38 (1994).
[4092] Miller, J.N.; *Spectroscopy Int. 3/4*, 41–43 (1991).
[4093] Miller, L.A.; K.W. Bruland, *Anal. Chim. Acta 284/3*, 573–586 (1994).
[4094] Miller, N.L.; J.A. Durr, A.C. Alfrey, *Anal. Biochem. 182/2*, 245–249 (1989).
[4095] Miller, W.P.; W.W. McFee, *J. Environ. Qual. 12*, 29–33 (1983).
[4096] Miller-Ihli, N.J.; T.C. O'Haver, J.M. Harnly, *Anal. Chem. 54*, 2590–2591 (1982).
[4097] Miller-Ihli, N.J.; T.C. O'Haver, J.M. Harnly, *Anal. Chem. 54/4*, 799–803 (1982).
[4098] Miller-Ihli, N.J.; T.C. O'Haver, J.M. Harnly, *Spectrochim. Acta, Part B 39/12*, 1603–1614 (1984).
[4099] Miller-Ihli, N.J.; T.C. O'Haver, J.M. Harnly, *Anal. Chem. 56/2*, 176–181 (1984).
[4100] Miller-Ihli, N.J.; *J. Anal. At. Spectrom. 3/1*, 73–81 (1988).
[4101] Miller-Ihli, N.J.; *J. Res. Natl. Bur. Stand. 93/3*, 350–354 (1988).
[4102] Miller-Ihli, N.J.; *J. Anal. At. Spectrom. 4/3*, 295–297 (1989).
[4103] Miller-Ihli, N.J.; *Spectrochim. Acta, Part B 44/12*, 1221–1227 (1989).
[4104] Miller-Ihli, N.J.; *At. Spectrosc. 13/1*, 1–6 (1992).
[4105] Miller-Ihli, N.J.; F.E. Greene, *J. AOAC International 75/2*, 354–359 (1992).
[4106] Miller-Ihli, N.J.; in: M. Stoeppler (Editor), *Hazardous Metals in the Environment*, Techn. Instr. Anal. Chem. Vol 12, Elsevier, Amsterdam, 1992.
[4107] Miller-Ihli, N.J.; *Fresenius J. Anal. Chem. 345/7*, 482–489 (1993).
[4108] Miller-Ihli, N.J.; *J. Anal. At. Spectrom. 9/10*, 1129–1134 (1994).
[4109] Miller-Ihli, N.J.; *Spectrochim. Acta, Part B 50/4-7*, 477–488 (1995).
[4110] Miller-Ihli, N.J.; *J. Agri. Food Chem. 43/4*, 923–927 (1995).
[4111] Miller-Ihli, N.J.; *J. Anal. At. Spectrom. 12/2*, 205–212 (1997).
[4112] Miller-Ihli, N.J.; *Spectrochim. Acta, Part B 52/4*, 431–436 (1997).
[4113] Mills, G.L.; J.G. Quinn, *Mar. Chem. 15*, 151–172 (1984).
[4114] Mills, J.C.; C.B. Belcher, *Prog. Anal. Spectrosc. 4*, 49–88 (1981).
[4115] Mills, J.W.; G.M. Hieftje, *Spectrochim. Acta, Part B 39/7*, 859–866 (1984).
[4116] Millward, C.G.; P.D. Kluckner, *J. Anal. At. Spectrom. 4/8*, 709–713 (1989).
[4117] Millward, C.G.; P.D. Kluckner, *J. Anal. At. Spectrom. 6/1*, 37–40 (1991).
[4118] Milne, D.B.; R.L. Sims, N.V.C. Ralston, *Clin. Chem. (Winston-Salem) 36/3*, 450–452 (1990).
[4119] Milner, B.A.; *Am. Lab. (Fairfield, CT) 15*, 74–76 (1983).
[4120] Milner, B.A.; E. Heiden, *CLB, Chem. Labor Betr. 34/5*, 200–204 (1983).
[4121] Milosavljevic, E.B.; L. Solujic, J.L. Hendrix, J.H. Nelson, *Analyst (London) 114/7*, 805–808 (1989).
[4122] Min, R.W.; E.H. Hansen, *Chem. Anal. (Warsaw) 40/3*, 361–371 (1995).
[4123] Minami, H.; T. Honjyo, I. Atsuya, *Spectrochim. Acta, Part B 51/2*, 211–220 (1996).
[4124] Minamisawa, H.; N. Arai, T. Okutani, *Bunseki Kagaku 42/11*, 767–771 (1993).
[4125] Mincey, D.W.; R.C. Williams, J.J. Giglio, G.A. Graves, A.J. Pacella, *Anal. Chim. Acta 264*, 97–100 (1992).
[4126] Mindel, B.D.; B. Karlberg, *Lab. Pract. 30/7*, 719–723 (1981).
[4127] Mingorance, M.D.; M. Lachica, *Anal. Lett. 18/A12*, 1519–1531 (1985).
[4128] Mingorance, M.D.; M.L. Pérez-Vazquez, M. Lachica, *J. Anal. At. Spectrom. 8/6*, 853–858 (1993).
[4129] Mino, Y.; N. Ota, S. Sakao, S. Shimomura, *Chem. Pharm. Bull. 28/9*, 2687–2691 (1980).
[4130] Minoia, C.; R. Pietra, E. Sabbioni, A. Ronchi, A. Gatti, A. Cavalleri, L. Manzo, *Sci. Total Environ. 120*, 63–79 (1992).

[4131] Minoia, C.; S. Canedoli, L. Vescovi, E. Rizzio, E. Sabbioni, R. Pietra, L. Manzo, in: C. Minoia, S. Caroli (Editors), *Appl. Zeeman Graphite Furnace At. Abs. Spectrom. Chem. Lab. Toxicology*, Pergamon, Oxford, 1992, 179–207.

[4132] Minoia, C.; A. Alimonti, E. Sabbioni, R. Pietra, S. Caroli, in: C. Minoia, S. Caroli (Editors), *Appl. Zeeman Graphite Furnace At. Abs. Spectrom. Chem. Lab. Toxicology*, Pergamon, Oxford, 1992, 475–493.

[4133] Minoia, C.; A. Ronchi, M. Bettinelli, G. Santagostino, L. Manzo, F. Candura, in: C. Minoia, S. Caroli (Editors), *Appl. Zeeman Graphite Furnace At. Abs. Spectrom. Chem. Lab. Toxicology*, Pergamon, Oxford, 1992, 517–537.

[4134] Minoia, C.; E. Sabbioni, R. Pietra, A. Ronchi, F. Poggio, A. Salvadeo, in: C. Minoia, S. Caroli (Editors), *Appl. Zeeman Graphite Furnace At. Abs. Spectrom. Chem. Lab. Toxicology*, Pergamon, Oxford, 1992, 627–646.

[4135] Mir, P.S.; C.M. Kalnin, S.A. Garvey, *J. Dairy Sci. 72*, 2549 (1989).

[4136] Mirti, P.; R. Aruga, V. Zelano, L. Appolonia, M. Aceto, *Fresenius J. Anal. Chem. 336/3*, 215–221 (1990).

[4137] Mitchell, A.C.G.; M.W. Zemansky, *Resonance Radiation and Excited Atoms*, Cambridge University Press, London - New York, 1971.

[4138] Mitchell, D.G.; K.W. Jackson, K.M. Aldous, *Anal. Chem. 45/14*, 1215A-23A (1973).

[4139] Mitchell, D.G.; K.M. Aldous, A.F. Ward, *At. Absorpt. Newsletter 13/5*, 121–122 (1974).

[4140] Mitra, S.; *Mercury in the ecosystem: its dispersion and pollution today*, Trans Tech. Publications, Aedermannsdorf, 1986.

[4141] Miura, J.; S. Arima, M. Satake, *Anal. Chim. Acta 237/1*, 201–206 (1990).

[4142] Miura, J.; N. Sugita, M. Satake, *Microchem. J. 42*, 306–313 (1990).

[4143] Miwa, T.; S. Tilleketatne, A. Mizuike, *Anal. Sci. 2/6*, 591–592 (1986).

[4144] Miwa, T.; M. Murakami, A. Mizuike, *Anal. Chim. Acta 219/1*, 1–8 (1989).

[4145] Miyazawa, M.; M. Pavan, M.F.M. Block, *Commun. Soil. Sci. Plant Anal. 15/2*, 141–147 (1984).

[4146] Mniszek, W.; *Chem. Anal. (Warsaw) 41/2*, 269 (1996).

[4147] Mo, S.C.; D.S. Choi, J.W. Robinson, *J. Environ. Sci. Health, Part A 23A/5*, 441–451 (1988).

[4148] Mo, S.-j.; J.A. Holcombe, *Anal. Chem. 62/18*, 1994–1997 (1990).

[4149] Mo, S.-j.; J.A. Holcombe, *Talanta 38/5*, 503–510 (1991).

[4150] Mocak, J.; M. Vanickova, J. Labuda, *Mikrochim. Acta (Vienna) 2/2*, 231–246 (1985).

[4151] Moenke-Blankenburg, L.; P. Decker, *Anal. Chim. Acta 173*, 327–330 (1985).

[4152] Moens, L.; P. Roose, J. De Rudder, J. Hoste, P. De Paepe, J. Van Hende, R. Marechal, M. Waelkens, *J. Radioanal. Nucl. Chem. 123*, 333–348 (1988).

[4153] Mohammad, B.; A.M. Ure, J. Reglinski, D. Littlejohn, *Chem. Speciation Bioavailability 3*, 117–122 (1990).

[4154] Mohammad, B.; A.M. Ure, D. Littlejohn, *J. Anal. At. Spectrom. 8/2*, 325–331 (1993).

[4155] Mohammed, D.A.; Analyst (London) 112, 209–211 (1987).

[4156] Mohay, J.; Zs. Végh, *Fresenius Z. Anal. Chem. 329/8*, 856–860 (1988).

[4157] Mohd, A.A.; J.R. Dean, W.R. Tomlinson, *Analyst (London) 117/11*, 1743–1748 (1992).

[4158] Mohl, C.; H.D. Narres, M. Stoeppler, in: B. Welz (Editor), *Fortschritte in der atomspektrometrischen Spurenanalytik*, VCH Verlagsgesellschaft, Weinheim, 1986, 439–446.

[4159] Mohl, C.; K.H. Grobecker, M. Stoeppler, *Fresenius Z. Anal. Chem. 328/4-5*, 413–418 (1987).

[4160] Moldan, B.; I. Rubeška, M. Miksovsky, M. Huka, *Anal. Chim. Acta 52*, 91–99 (1970).

[4161] Moldan, B.; in: R.M. Dagnall, G.F. Kirkbright (Editors), *Atomic Absorption Spectroscopy*, Butterworth, London, 1970, 127–143.

[4162] Momplaisir, G.M.; J.-S. Blais, M. Quinteiro, W.D. Marshall, *J. Agri. Food Chem. 39/8*, 1448–1451 (1991).

[4163] Monder, C.; N. Sells, *Anal. Biochem. 20*, 215 (1967).

[4164] Monica, E.S.Della; P.E. McDowell, *J. Am. Leather Chem. Assoc. 66*, 21–30 (1971).

[4165] Monien, H.; R. Bovenkerk, K.P. Kringe, D. Rath, *Fresenius Z. Anal. Chem. 300/5*, 363–371 (1980).

[4166] Monien, H.; R. Stangel, *Fresenius Z. Anal. Chem. 311/3*, 209–213 (1982).

[4167] Montaser, A.; S.R. Crouch, *Anal. Chem. 46/12*, 1817–1820 (1974).

[4168] Montaser, A.; S.R. Goode, S.R. Crouch, *Anal. Chem. 46/4*, 599–601 (1974).

[4169] Monteiro, M.I.C.; A.J. Curtius, *J. Anal. At. Spectrom. 10/4*, 329–334 (1995).

[4170] Montero, R.; M. Gallego, M. Valcárcel, *J. Anal. At. Spectrom. 3/5*, 725–729 (1988).

[4171] Montero, R.; M. Gallego, M. Valcárcel, *Anal. Chim. Acta 215/1-2*, 241–248 (1988).

[4172] Montero, R.; M. Gallego, M. Valcárcel, *Talanta 37/12*, 1129–1132 (1990).

[4173] Montero, R.; M. Gallego, M. Valcárcel, *Anal. Chim. Acta 234/2*, 433–437 (1990).

[4174] Montero, R.; M. Gallego, M. Valcárcel, *Analyst (London) 115*, 943 (1990).
[4175] Montero, R.; M. Gallego, M. Valcárcel, *Anal. Chim. Acta 252/1-2*, 83–88 (1991).
[4176] Monte Tamba, M.G. Del; R. Falciani, T. Dorado López, A. Gómez Coedo, *Analyst (London) 119/9*, 2081–2085 (1994).
[4177] Monte Tamba, M.G. Del; N. Luperi, *Analyst (London) 102*, 489–494 (1977).
[4178] Moody, J.R.; R.M. Lindstrom, *Anal. Chem. 49/14*, 2264–2267 (1977).
[4179] Moody, J.R.; *Trends Anal. Chem. (Pers. Ed.) 2/5*, 116–118 (1983).
[4180] Moody, J.R.; R.R. Greenberg, K.W. Pratt, T.C. Rains, *Anal. Chem. 60/21*, 1203A-18A (1988).
[4181] Moody, R.R.; R.B. Taylor, *J. Pharm. Pharmacol. 24*, 848–852 (1972).
[4182] Moore, E.J.; O.I. Milner, J.R. Glass, *Microchem. J. 10*, 148 (1966).
[4183] Moore, M.R.; P.A. Meredith, *Clin. Chim. Acta 75*, 167–170 (1977).
[4184] Mora, J.; A. Canals, V. Hernandis, *J. Anal. At. Spectrom. 6/2*, 139–143 (1991).
[4185] Mora, J.; V. Hernandis, A. Canals, *J. Anal. At. Spectrom. 6/7*, 573–579 (1991).
[4186] Mora, J.; A. Canals, V. Hernandis, *Spectrochim. Acta, Part B 51/12*, 1535–1549 (1996).
[4187] Mora, S.J. de; N.G. King, M.C. Miller, *Environ. Technol. Lett. 10*, 901 (1989).
[4188] Morabito, R.; *Fresenius J. Anal. Chem. 351/4-5*, 378–385 (1995).
[4189] Morales-Rubio, A.; F. Pomares, M. de la Guardia, A. Salvador, *J. Anal. At. Spectrom. 4/4*, 329–332 (1989).
[4190] Morales-Rubio, A.; A. Salvador Carreno, M. De la Guardia Cirugeda, *Anal. Chim. Acta 235/2*, 405–411 (1990).
[4191] Morales-Rubio, A.; A. Salvador, M. De la Guardia, R. Ros, *At. Spectrosc. 14/1*, 8–12 (1993).
[4192] Morgan, M.E.; *At. Absorpt. Newsletter 3*, 43–45 (1964).
[4193] Moreira, M.; A. Curtius, R. Calixto de Campos, *Analyst (London) 120/3*, 947 (1995).
[4194] Morisi, G.; M. Patriarca, G. Marano, *G. Ital. Chim. Clin. 13*, 193 (1988).
[4195] Morisi, G.; M. Patriarca, A. Menotti, *Clin. Chem. (Winston-Salem) 34/1*, 127–130 (1988).
[4196] Morita, H.; T. Mitsubashi, H. Sakurai, S. Shimomura, *Anal. Chim. Acta 153*, 351–355 (1983).
[4197] Morita, M.; J.S. Edmonds, *Pure Appl. Chem. 64/4*, 575–590 (1992).
[4198] Morita, Y.; H. Tsukada, A. Isozaki, *Bunseki Kagaku 42/9*, 551–556 (1993).
[4199] Morita, Y.; A. Isozaki, *Bunseki Kagaku 44/9*, 703 (1995).
[4200] Morrow, R.W.; R.J. McElhaney, *Appl. Spectrosc. 27*, 386–387 (1973).
[4201] Morrow, R.W.; R.J. McElhaney, *At. Absorpt. Newsletter 13*, 45–46 (1974).
[4202] Mortensen, G.; B. Pedersen, G. Pritzl, *Appl. Organomet. Chem. 9*, 65–73 (1995).
[4203] Mossotti, V.G.; V.A. Fassel, *Spectrochim. Acta 20*, 1117–1127 (1964).
[4204] Mossotti, V.G.; K. Laqua, W.D. Hagenah, *Spectrochim. Acta, Part B 23/3*, 197–206 (1967).
[4205] Mossotti, V.G.; M. Duggan, *Appl. Opt. 7/7*, 1325–1330 (1968).
[4206] Mostafa, M.A.; S.E. Ghazy, M.A. El-Tanbouly, *Analusis 14/10*, 543–548 (1986).
[4207] Mostyn, R.A.; A.F. Cunningham, *J. Inst. Petroleum 53*, 101–111 (1967).
[4208] Mostyn, R.A.; A.F. Cunningham, *Anal. Chem. 39*, 433–435 (1967).
[4209] Moulton, G.P.; T.C. O'Haver, J.M. Harnly, *J. Anal. At. Spectrom. 4/7*, 673–674 (1989).
[4210] Moulton, G.P.; T.C. O'Haver, J.M. Harnly, *J. Anal. At. Spectrom. 5/2*, 145–150 (1990).
[4211] Moura, M.J.M.P.; M.T.S.D. Vasconcelos, A.A.S.C. Machado, *J. Anal. At. Spectrom. 2*, 451–454 (1987).
[4212] Moynier, L.; E. Bourret, M. Fussellier, L. Bardet, *Analusis 15/6*, 306–310 (1987).
[4213] Mrozowski, S.; *Z. Phys. 112*, 223 (1939).
[4214] Muangnoicharoen, S.; K.-Y. Chiou, O.K. Manuel, *Anal. Chem. 58/13*, 2811–2813 (1986).
[4215] Muangnoicharoen, S.; K.-Y. Chiou, O.K. Manuel, *Talanta 35/9*, 679–683 (1988).
[4216] Mücke, G.; *Fresenius Z. Anal. Chem. 320/7*, 639–641 (1985).
[4217] Mukai, H.; Y. Ambe, *Anal. Chim. Acta 193*, 219–229 (1987).
[4218] Mulford, C.E.; *At. Absorpt. Newsletter 5*, 28–30 (1966).
[4219] Mulford, C.E.; *At. Absorpt. Newsletter 5/3*, 63 (1966).
[4220] Mulford, C.E.; *At. Absorpt. Newsletter 5/4*, 88–90 (1966).
[4221] Muller, F.L.L.; J.D. Burton, P.J. Stratham, *Anal. Chim. Acta 245/1*, 21–25 (1991).
[4222] Müller, C.; R. Eckard, in: M. Stoeppler (Editor), *Probenahme und Aufschluß*, Springer, Berlin, 1994, 1–7.
[4223] Müller, E.; R. Stahlberg, *Fresenius Z. Anal. Chem. 322/4*, 401–403 (1985).
[4224] Müller, G.; *Chem. Ztg. 109/7-8*, 245–250 (1985).
[4225] Müller, K.; P. Pringsheim, *Naturwissenschaften 18*, 364 (1930).
[4226] Müller, M; M. Anke, H. Illing, E. Hartmann, *GIT Fachz. Lab. 39/9*, 795–796 (1995).
[4227] Müller-Vogt, G.; W. Wendl, *Anal. Chem. 53/4*, 651–653 (1981).
[4228] Müller-Vogt, G.; W. Wendl, P. Pfundstein, *Fresenius Z. Anal. Chem. 314/7*, 638–641 (1983).

[4229]   Müller-Vogt, G.; L. Hahn, H. Müller, W. Wendl, D. Jacquiers-Roux, *J. Anal. At. Spectrom.*
         10/10, 777–783 (1995).
[4229b] Müller-Vogt, G.; F. Weigend, W. Wendl, *Spectrochim. Acta, Part B 51*, 1133–1137 (1996).
[4230]   Mullins, E.; *Analyst (London) 119/3*, 369–375 (1994).
[4231]   Mullins, T.L.; *Anal. Chim. Acta 165/11*, 97–103 (1984).
[4232]   Munaf, E.; H. Haraguchi, D. Ishii, T. Takeuchi, M. Goto, *Anal. Chim. Acta 235/2*, 399–404
         (1990).
[4233]   Munaf, E.; H. Haraguchi, D. Ishii, T. Takeuchi, M. Goto, *Sci. Total Environ. 99/1-2*,
         205–209 (1990).
[4234]   Munaf, E.; T. Takeuchi, D. Ishii, H. Haraguchi, *Anal. Sci. 7/4*, 605–609 (1991).
[4235]   Munaf, E.; T. Takeuchi, H. Haraguchi, *Fresenius J. Anal. Chem. 342/1-2*, 154–156 (1992).
[4236]   Munk, H.; *VDLUFA-Schriftenreihe 32*, 1990.
[4237]   Munns, R.K.; D.C. Holland, *J. Assoc. Off. Anal. Chem. 54/1*, 202–205 (1971).
[4238]   Muñoz-Olivas, R.; O.F.X. Donard, C. Cámara, Ph. Quevauviller, *Anal. Chim. Acta 286/3*,
         357–370 (1994).
[4239]   Muntau, H.; *Fresenius Z. Anal. Chem. 324/7*, 678–682 (1986).
[4240]   Muntz, J.H.; *At. Absorpt. Newsletter 10/1*, 9–11 (1971).
[4241]   Muny, R.P.; *At. Absorpt. Newsletter 3*, 129 (1964).
[4242]   Murakami, K.; Y. Okamoto, T. Kumamaru, *J. Flow Injection Anal. 9/2*, 195–203 (1992).
[4243]   Murakami, M.; T. Takada, *Talanta 39/10*, 1293–1298 (1992).
[4244]   Murdock, J.A.; F.W. Heaton, *Clin. Chim. Acta 28*, 505 (1970).
[4245]   Mürer, A.J.L.; A. Abildtrup, O.M. Poulsen, J.M. Christensen, *Talanta 39/5*, 469–474 (1992).
[4246]   Murie, R.A.; R.C. Bourke, *Appl. Spectrosc. 18/4*, 116 (1964).
[4247]   Murphy, G.F.; R. Stephens, *Talanta 25*, 223–225 (1978).
[4248]   Murphy, J.; *At. Absorpt. Newsletter 14/6*, 151–152 (1975).
[4249]   Murphy, J.; P. Jones, S.J. Hill, *Spectrochim. Acta, Part B 51/14*, 1867–1873 (1996).
[4250]   Murphy, V.A.; *Anal. Biochem. 161/1*, 144–151 (1987).
[4251]   Murti, S.S.; I.V. Sambasiva Rao, S.C.S. Rajan, J. Subramanyam, *Talanta 36/5*, 601–602
         (1989).
[4252]   Murty, D.S.R.; B.N. Tikoo, *At. Spectrosc. 8/2*, 79–80 (1987).
[4253]   Musaick, K.; E. Jackwerth, *Fresenius Z. Anal. Chem. 325*, 175–177 (1986).
[4254]   Muse, J.O.; M.B. Tudino, L. D'Huicque, O.E. Troccoli, C.N. Carducci, *Environ. Pollut. 58*,
         303–312 (1989).
[4255]   Muse, W.T.; *Rep., Order No. AD-A223070* (Avail NTIS) CRDEC-TR-167, 1990.
[4256]   Musil, J.; M. Nehasilová, *Talanta 23*, 729–731 (1976).
[4257]   Musil, J.; I. Rubeška, *Analyst (London) 107*, 588–590 (1982).
[4258]   Muzgin, V.N.; Yu.B. Atnashev, V.E. Korepanov, A.A. Pupyshev, *Talanta 34/1*, 197–200
         (1987).
[4259]   Myasoedova, G.V.; I.I. Anatolkol'skaya, S.B. Savvin, *Talanta, 32/12*, 1105–1112 (1985).
[4260]   Nadkarni, R.A.; R.I. Botto, S.E. Smith, *At. Spectrosc. 3/6*, 180–184 (1982).
[4261]   Nadkarni, R.A.; *Anal. Chim. Acta 135*, 363–368 (1982).
[4262]   Nadkarni, R.A.; *ASTM Spec. Tech. Publ. 1109*, 5 (1991).
[4263]   Nadkarni, R.A.; *ASTM Spec. Tech. Publ. 1109*, 19 (1991).
[4264]   Nagahiro, T.; K. Uesugi, M. Satake, *Analusis 16/2*, 120–123 (1988).
[4265]   Naganuma, A.; T. Okutani, *Anal. Sci. 6/1*, 77–81 (1990).
[4266]   Nagourney, S.J.; M. Heit, D.C. Bogen, *Talanta 34/5*, 465–472 (1987).
[4267]   Nagourney, S.J.; N.J. Tummillo, J. Birri, K. Peist, J.S. Kane, *Talanta 44/2*, 189–196 (1997).
[4268]   Naka, H.; H. Kurayasu, Y. Inokuma, *Bunseki Kagaku 39*, 171 (1990).
[4269]   Nakahara, T.; M. Munemori, S. Musha, *Anal. Chim. Acta 62*, 267–278 (1972).
[4270]   Nakahara, T.; M. Munemori, S. Musha, *Bull. Chem. Soc. Jpn. 46*, 1172–1177 (1973).
[4271]   Nakahara, T.; S. Musha, *Can. J. Spectrosc. 24/5*, 138–143 (1979).
[4272]   Nakahara, T.; C.L. Chakrabarti, *Anal. Chim. Acta 104*, 99–111 (1979).
[4273]   Nakamura, S.; *Anal. Chim. Acta 167*, 365–370 (1985).
[4274]   Nakamura, S.; H. Suzuki, M. Kubota, *Bunseki Kagaku 38*, 399 (1989).
[4275]   Nakamura, S.; M. Kubota, *Analyst (London) 115/3*, 283–286 (1990).
[4276]   Nakamura, T.; K. Okubo, J. Sato, *Anal. Chim. Acta 209*, 287–292 (1988).
[4277]   Nakamura, T.; H. Oka, H. Morikawa, J. Sato, *Analyst (London) 117/2*, 131–135 (1992).
[4278]   Nakamura, T.; H. Oka, M. Ishii, J. Sato, *Analyst (London) 119/6*, 1397–1401 (1994).
[4279]   Nakashima, S.; *Analyst (London) 104*, 172–173 (1979).
[4280]   Nakashima, S.; R.E. Sturgeon, S.N. Willie, S.S. Berman, *Fresenius Z. Anal. Chem. 330/7*,
         592–595 (1988).

[4281] Nakashima, S.; R.E. Sturgeon, S.N. Willie, S.S. Berman, *Anal. Chim. Acta 207*, 291–299 (1988).
[4282] Nakashima, S.; R.E. Sturgeon, S.N. Willie, S.S. Berman, *Analyst (London) 113/1*, 159–163 (1988).
[4283] Nakashima, S.; *Fresenius J. Anal. Chem. 343*, 614–615 (1992).
[4284] Nakayama, S.; M. Shibata, H. Mizusuna, S. Harada, *Bunseki Kagaku 36*, 499 (1987).
[4285] Nara, T.; K. Oguma, R. Kuroda, *Bunseki Kagaku 36*, 852–855 (1987).
[4286] Narasaki, H.; *Anal. Chim. Acta 125*, 187–191 (1981).
[4287] Narasaki, H.; M. Ikeda, *Anal. Chem. 56/12*, 2059–2063 (1984).
[4288] Narasaki, H.; *Anal. Chem. 57/13*, 2481–2486 (1985).
[4289] Narasaki, H.; *Anal. Sci. 2*, 141–144 (1986).
[4290] Narasaki, H.; Y. Kato, H. Kimura, *Anal. Sci. 8/6*, 893–896 (1992).
[4291] Narayanan, S.; F.C. Lin, H.J. Walder, *Acta Pharmacol. Toxicol. 59/7*, 498–560 (1986).
[4292] Narres, H.D.; C. Mohl, M. Stoeppler, *Int. J. Environ. Anal. Chem. 18/4*, 267–279 (1984).
[4293] Narres, H.D.; C. Mohl, M. Stoeppler, *Z. Lebensm. Unters. Forsch. 181/2*, 111–116 (1985).
[4294] Narres, H.D.; C. Mohl, M. Stoeppler, *Erdoel, Kohle, Erdgas, Petrochem. 39*, 193–194 (1986).
[4295] Narsito, J. Agterdenbos, *Anal. Chim. Acta 197*, 315–321 (1987).
[4296] Narsito, J. Agterdenbos, S.J. Santosa, *Anal. Chim. Acta 237/1*, 189–199 (1990).
[4297] Narsito, J. Agterdenbos, D. Bax, *Anal. Chim. Acta 244/1*, 129–134 (1991).
[4298] Nash, L.A.; L.N. Peterson, S.P. Nadler, D.Z. Levine, *Anal. Chem. 60/21*, 2413–2418 (1988).
[4299] Nasu, T.; M. Kan, *Analyst (London) 113/11*, 1683–1686 (1988).
[4300] Nater, E.A.; R.G. Burau, *Anal. Chim. Acta 220/1*, 83–92 (1989).
[4301] National Academy of Sciences / National Research Council, USA, *Medical and Biological Effects of Environmental Pollutants: Arsenic*, Washington, D.C., 1977.
[4302] National Institute for Occupational Safety and Health (NIOSH), *Manual of Analytical Methods*, 2nd Edition, Vol. 3 USDHEW No PB 276 838, Cincinnati, OH, 1977.
[4303] National Institute of Standards and Technology, *SRM 2709 San Joaquim Soil, Certificate of Analysis*, Gaithersburg, MD, 1993.
[4304] National Institute of Standards and Technology, *SRM 2711 Montana Soil, Certificate of Analysis*, Gaithersburg, MD, 1993.
[4305] National Research Council Canada, *Information Sheet NASS-1*, 1982.
[4306] Naumann, K.; J. Bergmann, W. Dannecker, in: B. Welz (Editor), *Fortschritte in der atomspektrometrischen Spurenanalytik*, Verlag Chemie, Weinheim, 1984, 543–555.
[4307] Navarro, J.A.; V.A. Granadillo, O.E. Parra, R.A. Romero, *J. Anal. At. Spectrom. 4/5*, 401–406 (1989).
[4308] Navarro, J.A.; V.A. Granadillo, O. Salgado, B. Rodríguez-Iturbe, R. García, G. Delling, R.A. Romero, *Clin. Chim. Acta 211*, 133–142 (1992).
[4309] Navarro, M.; M.C. López, H. López, M. Sánchez, *Anal. Chim. Acta 257/1*, 155–158 (1992).
[4310] Navarro, M.; H. López, M.C. López, M. Sánchez, *J. Anal. Toxicol. 16/3*, 169–171 (1992).
[4311] Navarro, M.; M.C. López, H. López, M. Sánchez, *J. AOAC International 75/6*, 1029–1031 (1992).
[4312] Navarro-Alarcón, M.; M.C. López-Martínez, M. Sánchez-Viñas, H. López-García de la Serrana, *J. Agri. Food Chem. 39/12*, 2223–2225 (1991).
[4313] Nazario, C.L.; E.E. Menden, *J. Am. Leather Chem. Assoc. 85*, 212–224 (1990).
[4314] Nägele, D.; W. Brunn, *Z. Pflanzenernähr. Bodenkd. 151/1*, 55–56 (1988).
[4315] Negishi, A.; C. Tsurumi, *Nippon Kagaku Kaishi (6)*, 683–686 (1992).
[4316] Neidhart, B.; Ch. Lippmann, *Fresenius Z. Anal. Chem. 306/4*, 259–263 (1981).
[4317] Neidhart, B.; C. Tausch, *Mikrochim. Acta (Vienna) 109/1-4*, 137–140 (1992).
[4318] Nelson, L.A.; *Anal. Chem. 51/13*, 2289–2290 (1979).
[4319] Nerín de la Puerta, C.; J. Cacho Palomar, A. Garnica, *Anal. Lett. 18/B15*, 1887–1896 (1985).
[4320] Nerín de la Puerta, C.; A. Garnica, J. Cacho Palomar, *Anal. Chem. 57/1*, 34–38 (1985).
[4321] Nerín de la Puerta, C.; A. Garnica, J. Cacho Palomar, *Anal. Chem. 58/13*, 2617–2621 (1986).
[4322] Nerín de la Puerta, C.; A. Garnica, J. Cacho Palomar, *Mikrochim. Acta (Vienna) 1986/III*, 117–126 (1987).
[4323] Nerín de la Puerta, C.; J. Cacho Palomar, A. Garnica, *Anal. Lett. 22/15*, 3041–3055 (1989).
[4324] Nerín de la Puerta, C.; S. Olavide, J. Cacho Palomar, A. Garnica, *Water, Air, Soil Pollut. 44*, 339–345 (1989).
[4325] Nerín de la Puerta, C.; R. Zufiaurre, J. Cacho Palomar, *Mikrochim. Acta (Vienna) 108/3-6*, 241–249 (1992).
[4326] Nesbitt, R.W.; *Anal. Chim. Acta 35*, 413–420 (1966).
[4327] Neubauer, J.; K. Maier, *At. Spectrosc. 13/6*, 206–212 (1992).

[4328] Neunteufel, R.; S. Meyer, in: B. Welz (Editor), *4. Colloquium Atomspektrometrische Spurenanalytik*, Bodenseewerk Perkin-Elmer, Überlingen, 1987, 509–516.

[4329] Nève, J.; M. Hanocq, L. Molle, *Fresenius Z. Anal. Chem. 308/5*, 448–451 (1981).

[4330] Nève, J.; M. Hanocq, L. Molle, G. Lefebvre, *Analyst (London) 107*, 934–941 (1982).

[4331] Nève, J.; M. Hanocq, L. Molle, in: P. Brätter, P. Schramel (Editors), *Trace Element Analytical Chemistry in Medicine and Biology, Vol. 2*, Walter de Gruyter, Berlin -New York, 1983, 859–876.

[4332] Nève, J.; L. Molle, *Acta Pharmacol. Toxicol. 59/S7*, 606–609 (1986).

[4333] Nève, J.; S. Chamart, L. Molle, in: P. Brätter, P. Schramel (Editors), *Trace Element Analytical Chemistry in Medicine and Biology*, Walter de Gruyter, Berlin - New York, 1987, 349–358.

[4334] Nève, J.; S. Chamart, P. Trigaux, F. Vertongen, *At. Spectrosc. 8/6*, 167–169 (1987).

[4335] Nève, J.; N. Leclercq, *Clin. Chem. (Winston-Salem) 37/5*, 723–728 (1991).

[4336] Newman, P.; A. Voelkel, D. MacFarlane, *J. Non-Cryst. Solids 184*, 324 (1995).

[4337] Newton, D.W.; R. Ellis, *J. Environ. Qual. 3/1*, 20–23 (1974).

[4338] Newton, J.T.; *J. Forensic. Sci. 26*, 302–312 (1981).

[4339] Ng, K.C.; A.H. Ali, T.E. Barber, J.D. Winefordner, *Anal. Chem. 62/17*, 1893–1895 (1990).

[4340] Ng, K.C.; A.H. Ali, T.E. Barber, J.D. Winefordner, *Appl. Spectrosc. 44/5*, 849–852 (1990).

[4341] Ng, K.C.; A.H. Ali, T.E. Barber, J.D. Winefordner, *Appl. Spectrosc. 44/6*, 1094–1096 (1990).

[4342] Ngim, C.-H.; S.-C. Foo, W.-O. Phoon, *J. Anal. Toxicol. 12/3*, 132–135 (1988).

[4343] Nham, T.T.; K.G. Brodie, *J. Anal. At. Spectrom. 4/8*, 697–700 (1989).

[4344] Ni, Z.-m.; H.-b. Hang, A. Li, B. He, F.-z. Xu, *J. Anal. At. Spectrom. 6/5*, 385–387 (1991).

[4345] Ni, Z.-m.; X.-q. Shan, L.-z. Jin, S. Luan, L. Zhang, K.S. Subramanian, in: K.S. Subramanian, G.V. Iyengar, K. Okamoto (Editors), *Biol. Trace Elem. Research*; ACS Symp. Series No. 445, ACS, Washington, D.C., 1991, 206–214.

[4346] Ni, Z.-m.; B. He, H.-b. Han, *Can. J. Appl. Spectroscopy 38/1*, 11–14 (1993).

[4347] Ni, Z.-m.; B. He, H.-b. Han, *J. Anal. At. Spectrom. 8/7*, 995–998 (1993).

[4348] Ni, Z.-m.; B. He, H.-b. Han, *Spectrochim. Acta, Part B 49/10*, 947–953 (1994).

[4349] Ni, Z.-m.; B. He, *J. Anal. At. Spectrom. 10/10*, 747–751 (1995).

[4350] Ni, Z.-m.; D.-q. Zhang, *Spectrochim. Acta, Part B 50/14*, 1779–1786 (1995).

[4351] Ni, Z.-m.; Z. Rao, M. Li, *Anal. Chim. Acta 334/1-2*, 177–182 (1996).

[4352] Niazi, S.B.; D. Littlejohn, D.J. Halls, *Analyst (London) 118/7*, 821–825 (1993).

[4353] Nichols, J.A.; R.D. Jones, R. Woodriff, *Anal. Chem. 50*, 2071–2076 (1978).

[4354] Nicholson, J.R.P.; M.G. Savory, J. Savory, M.R. Wills, *Clin. Chem. (Winston-Salem) 35/3*, 488–490 (1989).

[4355] Nickson, R.A.; S.J. Hill, P.J. Worsfold, *Anal. Proc. (London) 32/9*, 387–395 (1995).

[4356] Nieboer, E.; D.H.S. Richardson, in: S.J. Eisenreich (Editor), *Atmospheric Pollutants in Natural Waters*, Ann Arbor Science Publisher, Ann Arbor, 1981, 339–388.

[4357] Nieboer, E.; A.A. Jusys, in: Jerome O. Nriagu, Evert Nieboer (Editors), *Chromium in the Natural and Human Environment* (Adv. Environ. Sci. Techn. 20), J. Wiley & Sons, New York, 1988, 21–79.

[4358] Nieboer, E.; A.A. Jusys, *Pure Appl. Chem.* , (1988).

[4359] Nielsen, F.H.; in: K.T. Smith (Editor), *Trace Minerals in Foods*, Marcel Dekker, New York - Basel, 1988, 357–428.

[4360] Nielsen, S.; J.J. Sloth, E.H. Hansen, *Talanta 43/6*, 867–880 (1996).

[4361] Nielsen, S.; E.H. Hansen, *Anal. Chim. Acta 343/1-2*, 5–7 (1997).

[4362] Niemax, K.; A. Zybin, C. Schnürer-Patchan, H. Groll, *Anal. Chem. 68/11*, 351A-56A (1996).

[4363] Nieuwenhuize, J.; C.H. Poley-Vos, A. Goud, M.A. Hemminga, *At. Spectrosc. 9/6*, 204–206 (1988).

[4364] Nieuwenhuize, J.; C.H. Poley-Vos, A.H. van den Akker, W. van Delft, *Analyst (London) 116/4*, 347–351 (1991).

[4365] Nikitina, Z.A.; N.M. Kuznetsova, I.P. Zharkova, N.G. Monakhova, *J. Anal. Chem. [Engl. Transl.] 50/1*, 90–92 (1995).

[4366] Nikolaev, G.I.; V.B. Aleskovskii, *Zh. Anal. Khim. 18*, 816 (1963).

[4367] Nikolova, B.; N. Jordanov, *Talanta 29/10*, 861–866 (1982).

[4368] Nilsson, T.; P.-O. Berggren, *Anal. Chim. Acta 159*, 381–385 (1984).

[4369] Niskavaara, H.; J. Virtasalo, L.H.J. Lajunen, *Spectrochim. Acta, Part B 40B/9*, 1219–1225 (1985).

[4370] Niskavaara, H.; E. Kontas, *Anal. Chim. Acta 231/2*, 273–282 (1990).

[4371] Niu, W.; R. Haring, R. Newman, *Am. Lab. (Fairfield, CT) 19/11*, 40–47 (1987).

[4372] Nixon, D.E.; T.P. Moyer, D.P. Squillace, J.T. McCarthy, *Analyst (London) 114/12*, 1671–1674 (1989).

[4373] Nixon, D.E.; G.V. Mussmann, S.J. Eckdahl, T.P. Moyer, *Clin. Chem. (Winston-Salem) 37/9*, 1575–1579 (1991).
[4374] Nixon, D.E.; T.P. Moyer, *Clin. Chem. (Winston-Salem) 38/12*, 2479–2483 (1992).
[4375] Noga, R.J.; *Anal. Chem. 47/2*, 332–333 (1975).
[4376] Noller, B.N.; P.J. Cusbert, N.A. Currey, P.H. Bradley, M. Tuor, *Environ. Technol. Lett. 6/9*, 381–390 (1985).
[4377] Nomoto, S.; S.-i. Shoji, *Clin. Chem. (Winston-Salem) 33/11*, 2004–2007 (1987).
[4378] Nomura, T.; I. Karasawa, *Anal. Chim. Acta 126*, 241–245 (1981).
[4379] Nord, L.; B. Karlberg, *Anal. Chim. Acta 125*, 199–202 (1981).
[4380] Nord, L.; B. Karlberg, *Anal. Chim. Acta 145*, 151–158 (1983).
[4381] Nordahl, K.; B. Radziuk, Y. Thomassen, R. Weberg, *Fresenius J. Anal. Chem. 337/3*, 310–315 (1990).
[4382] Norin, H.; A. Christakopoulos, *Chemosphere 11/3*, 287–298 (1982).
[4383] Norin, H.; R. Ryhage, A. Christakopoulos, M. Sandstrom, *Chemosphere 12/3*, 299–315 (1983).
[4384] Norin, H.; M. Vahter, A. Christakopoulos, M. Sandstrom, *Chemosphere 14/3-4*, 325–334 (1985).
[4385] Norris, J.D.; T.S. West, *Anal. Chem. 46/11*, 1423–1425 (1974).
[4386] North, M.R.; in: S.J. Haswell (Editor), *Atomic Absorption Spectrometry - Theory, Design and Applications*, Elsevier, Amsterdam - New York, 1991, 275–287.
[4387] Norval, E.; H.G. Human, L.R. Butler, *Anal. Chem. 51*, 2045–2048 (1979).
[4388] Norwitz, G.; H. Gordon, *Talanta 20*, 905–906 (1973).
[4389] Novak, L.; M. Stoeppler, *Fresenius Z. Anal. Chem. 323/7*, 737–741 (1986).
[4390] Novozamsky, I.; R. Van Eck, V.J.G. Houba, *Commun. Soil. Sci. Plant Anal. 15/3*, 205–211 (1984).
[4391] Novozamsky, I.; V.J.G. Houba, H.J. Van der Lee, R. Van Eck, M.D. Mignorance, *Commun. Soil. Sci. Plant Anal. 24/19*, 2595–2605 (1993).
[4392] Novozamsky, I.; H.J. van der Lee,; V.J.G. Houba, *Mikrochim. Acta (Vienna) 119/3-4*, 183–189 (1995).
[4393] Nowka, R.; H. Müller, K. Venth, *GIT Fachz. Lab. 41/6*, 622–625 (1997).
[4394] Nukatsuka, I.; K. Sakai, R. Kudo, K. Ohzeki, *Analyst (London) 120/12*, 2819–2822 (1995).
[4395] Nukiyama, S.; Y. Tanasawa, *Trans. Soc.Mech. Eng. (Japan), Report 4, 5 und 6, 1938-1940* (translated by E. Hope), Defense Research Bord, 1950.
[4396] Nygren, O.; C.-A. Nilsson, A. Gustavsson, *Analyst (London) 113/4*, 591–594 (1988).
[4397] Nygren, O.; C.-A. Nilsson, W. Frech, *Anal. Chem. 60/20*, 2204–2208 (1988).
[4398] Nygren, O.; *Spectrochim. Acta, Part B 48/8*, 977–983 (1993).
[4399] Nygren, O.; *Appl. Organomet. Chem. 8*, 601–605 (1994).
[4400] Nyholm, N.E.I.; *Ann. Chim. (Rom) 85/7-8*, 343 (1995).
[4401] Nylander, M.; J. Weiner, *Br. J. Ind. Med. 48*, 729–734 (1991).
[4402] O'Grady, C.E.; I.L. Marr, M.S. Cresser, *Analyst (London) 109/8*, 1085–1089 (1984).
[4403] O'Grady, C.E.; I.L. Marr, M.S. Cresser, *Analyst (London) 109/9*, 1183–1185 (1984).
[4404] O'Grady, C.E.; I.L. Marr, M.S. Cresser, *Analyst (London) 110/6*, 729–731 (1985).
[4405] O'Grady, C.E.; I.L. Marr, M.S. Cresser, *Analyst (London) 110/5*, 435–438 (1985).
[4406] O'Grady, C.E.; I.L. Marr, M.S. Cresser, *J. Anal. At. Spectrom. 1/1*, 51–54 (1986).
[4407] O'Gram, S.J.; J.R. Dean, W.R. Tomlinson, J. Marshall, *J. Anal. At. Spectrom. 7/2*, 229–234 (1992).
[4408] O'Gram, S.J.; J.R. Dean, J. Marshall, *Anal. Proc. (London) 30/3*, 135–136 (1993).
[4409] O'Haver, T.C.; J.M. Harnly, A.T. Zander, *Anal. Chem. 50/8*, 1218–1221 (1978).
[4410] O'Haver, T.C.; J.D. Messman, *Prog. Anal. Spectrosc. 9*, 483–503 (1986).
[4411] O'Haver, T.C.; J. Carroll, R. Nichol, D. Littlejohn, *J. Anal. At. Spectrom. 3/1*, 155–157 (1988).
[4412] O'Leary, R.M.; J.G. Viets, *At. Spectrosc. 7/1*, 4–8 (1986).
[4413] Oatts, T.J.; L.G. Hamilton, N.P. Buddin, *Spectrosc. 12/1*, 4–10 (1991).
[4414] Oatts, T.J.; L.G. Hamilton, N.P. Buddin, *At. Spectrosc. 16/4*, 145–148 (1995).
[4415] Ochsenkühn-Petropulu, M.; J. Varsamis, G. Parissakis, *Anal. Chim. Acta 337/3*, 323–327 (1997).
[4416] Oda, C.E.; J.D. Ingle, *Anal. Chem. 53/14*, 2305–2309 (1981).
[4417] Offley, S.G.; N.J. Seare, J.F. Tyson, H.A.B. Kibble, *J. Anal. At. Spectrom. 6/2*, 133–138 (1991).
[4418] Ogata, K.; S. Tanabe, T. Imanari, *Chem. Pharm. Bull. 31/4*, 1419–1421 (1983).
[4419] Oguma, K.; T. Nara, R. Kuroda, *Bunseki Kagaku 35*, 690–693 (1986).

[4420] Ohls, K.D.; *Mikrochim. Acta (Vienna) Suppl.9*, 49–70 (1981).
[4421] Ohls, K.D.; D. Sommer, *Fresenius Z. Anal. Chem. 312/3*, 195–220 (1982).
[4422] Ohls, K.D.; *Eur. Spectros. News 61*, 10–20 (1985).
[4423] Ohls, K.D.; *Fresenius J. Anal. Chem. 337/3*, 280–283 (1990).
[4424] Ohls, K.D.; D. Sommer, in: S.J. Haswell (Editor), *Atomic Absorption Spectrometry - Theory, Design and Applications*, Elsevier; Amsterdam - New York, 1991, 227–274.
[4425] Ohlsson, K.E.A.; W. Frech, *J. Anal. At. Spectrom. 4/5*, 379–385 (1989).
[4426] Ohlsson, K.E.A.; *J. Anal. At. Spectrom. 7/2*, 357–363 (1992).
[4427] Ohlsson, K.E.A.; E. Iwamoto, W. Frech, A. Cedergren, *Spectrochim. Acta, Part B 47/12*, 1341–1352 (1992).
[4428] Ohta, K.; M. Suzuki, *Anal. Chim. Acta 85*, 83–88 (1976).
[4429] Ohta, K.; M. Suzuki, *Talanta 25*, 160–162 (1978).
[4430] Ohta, K.; M. Suzuki, *Anal. Chim. Acta 104*, 293–297 (1979).
[4431] Ohta, K.; M. Suzuki, *Talanta 26*, 207–210 (1979).
[4432] Ohta, K.; M. Suzuki, *Anal. Chim. Acta 110/1*, 49–54 (1979).
[4433] Ohta, K.; M. Suzuki, *Fresenius Z. Anal. Chem. 298*, 140–143 (1979).
[4434] Ohta, K.; M. Suzuki, *Anal. Chim. Acta 108*, 69–74 (1979).
[4435] Ohta, K.; M. Suzuki, *Anal. Chim. Acta 107*, 245–251 (1979).
[4436] Ohta, K.; M. Suzuki, *Fresenius Z. Anal. Chem. 302/3*, 177–180 (1980).
[4437] Ohta, K.; S. Su, *Anal. Chem. 59/3*, 539–540 (1987).
[4438] Ohta, K.; *Fresenius Z. Anal. Chem. 326/2*, 132–134 (1987).
[4439] Ohta, K.; T. Mizuno, *Microchem. J. 37/2*, 203–211 (1988).
[4440] Ohta, K.; W. Aoki, T. Mizuno, *J. Anal. At. Spectrom. 3/7*, 1027–1030 (1988).
[4441] Ohta, K.; W. Aoki, T. Mizuno, *Talanta 35/11*, 831–836 (1988).
[4442] Ohta, K.; T. Mizuno, *Anal. Chim. Acta 217/2*, 377–382 (1989).
[4443] Ohta, K.; W. Aoki, T. Mizuno, *Mikrochim. Acta (Vienna) 1/1-2*, 81–86 (1990).
[4444] Ohta, K.; S.-I. Itoh, T Mizuno, *Anal. Sci. 7/Supp*, 457–461 (1991).
[4445] Ohta, K.; S.-I. Itoh, S. Kaneco, T. Mizuno, *Anal. Sci. 8/3*, 423–426 (1992).
[4446] Ohta, K.; S. Kaneco, S.-I. Itoh, T. Mizuno, *Anal. Chim. Acta 267/1*, 131–136 (1992).
[4447] Ohta, K.; K. Mizutani, S.-I. Itoh, T. Mizuno, *Microchem. J. 48/2*, 184–191 (1993).
[4448] Ohta, K.; J. Ogawa, T. Mizuno, *Fresenius J. Anal. Chem. 357/7*, 995–997 (1997).
[4449] Ohzeki, K.; M. Minorikawa, F. Yokota, I. Nukatsuka, R. Ishida, *Analyst (London) 115/1*, 23–28 (1990).
[4450] Oilunkaniemi, R.; P. Perämäki, L.H.J. Lajunen, *At. Spectrosc. 15/3*, 126–130 (1994).
[4451] Oishi, K.; *Bunko Kenkyu, 39*, 88 (1990).
[4452] Oishi, K.; K. Yasuda, K. Hirokawa, *Anal. Sci. 7/6*, 883–887 (1991).
[4453] Oktavec, D.; J. Lehotay, *At. Spectrosc. 14/4*, 103–105 (1993).
[4454] Okuno, I.; R.A. Wilson, R.E. White, *J. Assoc. Off. Anal. Chem. 55/1*, 96–100 (1972).
[4455] Okuno, I.; J.A. Whitehead, R.E. White, *J. Assoc. Off. Anal. Chem. 61/3*, 664–667 (1978).
[4456] Okutani, T.; Y. Tsuruta, A. Sakuragawa, *Anal. Chem. 65/9*, 1273–1276 (1993).
[4457] Olade, M.; K. Fletcher, *J. Geochem. Explor. 3*, 337–344 (1974).
[4458] Olafsson, J.; in: C.S. Wong, E.A. Boyle, K.W. Bruland (Editors), *Trace Metals in Seawater*, Plenum Press, New York, 1983, 475–486.
[4459] Olaofe, O.; E.O. Oladeji, O.I. Ayodeji, *J. Sci. Food Agric. 41*, 241 (1987).
[4460] Olech, A.; I. Sokolowska, K. Usiekniewicz, in: *Farm. Pol. 43*, 265 (1987).
[4461] Oles, P.J.; S. Siggia, *Anal. Chem. 46/7*, 911–914 (1974).
[4462] Oles, P.J.; S. Siggia, *Anal. Chem. 46/14*, 2197–2200 (1974).
[4463] Oles, P.J.; W.M. Graham, *J. Assoc. Off. Anal. Chem. 74*, 812–814 (1991).
[4464] Oles, P.J.; *J. AOAC International 76/3*, 615–620 (1993).
[4465] Olivier, M.; *Fresenius Z. Anal. Chem. 248*, 145–148 (1969).
[4466] Olivier, M.; *Fresenius Z. Anal. Chem. 257*, 135 (1971).
[4467] Olsen, E.D.; P.I. Jatlow, F.J. Fernandez, H.L. Kahn, *Clin. Chem. (Winston-Salem) 19/3*, 326–329 (1973).
[4468] Olsen, S.; J. Ruzicka, E.H. Hansen, *Anal. Chim. Acta 136*, 101–112 (1982).
[4469] Olsen, S.; L.C.R. Pessenda, J. Ruzicka, E.H. Hansen, *Analyst (London) 108*, 905–917 (1983).
[4470] Olson, A.D.; W.B. Hamlin, *Clin. Chem. (Winston-Salem) 15/6*, 438–444 (1969).
[4471] Omang, S.H.; *Anal. Chim. Acta 53*, 415–419 (1971).
[4472] Omang, S.H.; O.D. Vellar, *Fresenius Z. Anal. Chem. 269*, 177–181 (1974).
[4473] Omenetto, N.; P. Cavalli, G. Rossi, *Rev. Anal. Chem. 5/3-4*, 185–205 (1981).
[4474] Ong, C.N.; L.H. Chua, B.L. Lee, H.Y. Ong, K.S. Chia, *J. Anal. Toxicol. 14*, 29–33 (1990).
[4475] Orheim, R.M.; H.H. Bovee, *Anal. Chem. 46/7*, 921–922 (1974).

[4476] Orlova, M.N.; O.L. Skorskaya, *Zavodsk. Lab. 59/8*, 23–24 (1993).
[4477] Örnemark, U.; J. Pettersson, Å. Olin, *Talanta 39/9*, 1089–1096 (1992).
[4478] Örnemark, U.; Å. Olin, *Talanta 41/1*, 67–74 (1994).
[4479] Orsini, L.; A. Bermond, Int. *J. Environ. Anal. Chem. 51/1-4*, 97–108 (1993).
[4480] Ortner, H.M.; E. Kantuscher, *Talanta 22*, 581–586 (1975).
[4481] Ortner, H.M.; E. Lassner, *Mikrochim. Acta (Vienna) Suppl.7*, 41–62 (1977).
[4482] Ortner, H.M.; G. Schlemmer, B. Welz, W. Wegscheider, *Spectrochim. Acta, Part B 40B/7*, 959–977 (1985).
[4483] Ortner, H.M.; W. Birzer, B. Welz, G. Schlemmer, J.A. Curtius, W. Wegscheider, V. Sychra, *Fresenius Z. Anal. Chem. 323/7*, 681–688 (1986).
[4484] Ortner, H.M.; W. Bloedorn, G. Friedbacher, M. Grasserbauer, V. Krivan, A. Virag, P. Wilhartitz, G. Wünsch, *Mikrochim. Acta (Vienna) I*, 233–260 (1987).
[4485] Ortner, H.M.; *GIT Fachz. Lab. 35/8*, 891–900 (1991).
[4486] Ortner, H.M.; B. Welz, G. Schlemmer, V. Sychra, in: K. Dittrich, B. Welz (Editors), *CANAS'93, Colloq Analytische Atomspektroskopie*, Universität Leipzig/UFZ Leipzig-Halle, Leipzig, 1993, 51–78.
[4487] Ortner, H.M.; U. Rohr, G. Schlemmer, B. Welz, R. Lehmann, P. Brückner, U. Völlkopf, *CSI XXXIX Post Symposium, Book of Abstracts*, Ulm, 1995.
[4488] Ortner, H.M.; U. Rohr, S. Weinbruch, G. Schlemmer, B. Welz, *2nd Eur. Furnace Meeting, Book of Abstracts OIV/2*, St. Petersburg, 1996.
[4489] Ortner, H.M.; H.H. Xu, J. Dahmen, K. Englert, H. Opfermann, W. Görtz, *Fresenius J. Anal. Chem. 355/5-6*, 657–664 (1996).
[4490] Osberghaus, U.; *Probennahme und Aufschluß*, Springer, Berlin - Heidelberg - New York, 1994, 41–53.
[4491] Osborn, K.R.; H.E. Gunning, *J. Opt. Soc. Am. 45*, 552–556 (1955).
[4492] Osheim, D.L.; P.F. Ross, *J. Assoc. Off. Anal. Chem. 66/5*, 1140–1142 (1983).
[4493] Osheim, D.L.; P.F. Ross, *J. Assoc. Off. Anal. Chem. 68/1*, 44–45 (1985).
[4494] Osibanjo, O.; S.E. Kakulu, S.O. Ajayi, *Analyst (London) 109/2*, 127–129 (1984).
[4495] Ostapczuk, P.; G. Knezevic, L. Matter, G. Steinle, *Fresenius J. Anal. Chem. 345/2-4*, 308–309 (1993).
[4496] Oster, O.; *J. Clin. Chem. Clin. Biochem. 19/7*, 471–477 (1981).
[4497] Oster, O.; W. Prellwitz, in: H. Günzler, R. Borsdorf, W. Fresenius, W. Huber, H. Kelker, I. Lüderwald (Editors), *Analytiker Taschenbuch*, Springer, Berlin - Heidelberg - New York, 1990, 399–414.
[4498] Oster, O.; in: B. Welz (Editor), *6 Colloquium Atomspektrometrische Spurenanalytik*, Bodenseewerk Perkin-Elmer, Überlingen, 1991, 861–866.
[4499] Otruba, V.; J. Jambor, J. Komárek, J. Horák, L. Sommer, *Anal. Chim. Acta 101*, 367–374 (1978).
[4500] Ottaviani, M.; P. Magnatti, *J. Anal. At. Spectrom. 1/3*, 243–245 (1986).
[4501] Ottaway, J.M.; D.T. Coker, W.B. Rowston, D.R. Bhattarai, *Analyst (London) 95*, 567–573 (1970).
[4502] Ottaway, J.M.; F. Shaw, *Appl. Spectrosc. 31*, 12–17 (1977).
[4503] Ottaway, J.M.; J. Carroll, S. Cook, S.P. Corr, D. Littlejohn, J. Marshall, *Fresenius Z. Anal. Chem. 323/7*, 742–747 (1986).
[4504] Ottaway, J.M.; G.S. Fell, *Pure Appl. Chem. 58/12*, 1707–1720 (1986).
[4505] Ouchi, G.L.; *Am. Lab. (Fairfield, CT) 19*, 82–95 (1987).
[4506] Owens, G.D.; R.J. Eckstein, *J. Lab. Rob. Autom. 1*, 141–155 (1989).
[4507] Owens, J.W.; E.S. Gladney, D. Knab, *Anal. Chim. Acta 135/1*, 169–172 (1982).
[4508] Oya, Y.; S. Otani, *Carbon 17*, 131 (1979).
[4509] Oyamada, N.; M. Ishizaki, *Anal. Sci. 2*, 365–369 (1986).
[4510] Pabalkar, M.A.; S.V. Naik, N.R. Sanjana, *Analyst (London) 106*, 47–53 (1981).
[4511] Pacer, R.A.; .J. Weber, *J. Radioanal. Nucl. Chem. 129/1*, 181–190 (1989).
[4512] Pacer, R.A.; *J. Radioanal. Nucl. Chem. 186/3*, 237–243 (1994).
[4513] Pacey, G.E.; J.A. Ford, *Talanta 28/12*, 935–938 (1981).
[4514] Pacey, G.E.; M.R. Straka, J.R. Gord, *Anal. Chem. 58/2*, 502–504 (1986).
[4515] Padberg, S.; M. Burow, K. May, M. Stoeppler, in: B. Welz (Editor), *6. Colloquium Atomspektrometrische Spurenanalytik*, Bodenseewerk Perkin-Elmer, Überlingen, 1991, 751–759.
[4516] Padberg, S.; M. Burow, M. Stoeppler, *Fresenius J. Anal. Chem. 346/6-9*, 686–688 (1993).
[4517] Page, A.G.; S.V. Godbole, M.J. Kulkarni, N.K. Porwal, S.S. Shelar, B.D. Joshi, *Talanta 30/10*, 783–786 (1983).
[4518] Pai, S.; F. Lin, C. Tseng, D. Sheu, *Int. J. Environ. Anal. Chem. 50/3*, 193–205 (1993).

[4519] Pakkanen, T.A.; R.E. Hillamo, W. Maenhaut, *J. Anal. At. Spectrom. 8/1*, 79–84 (1993).
[4520] Palacios, M.A.; *Anal. Chim. Acta 340/1-3*, 209–220 (1997).
[4521] Palm, R.; R. Sjöström, G. Hallmans, *Clin. Chem. (Winston-Salem) 29/3*, 486–491 (1983).
[4522] Palmer, J.M.; M.W. Rush, *Analyst (London) 107*, 994–999 (1982).
[4523] Palmisano, F.; P.G. Zambonin, N. Cardellicchio, *Fresenius J. Anal. Chem. 346/6-9*, 648–652 (1993).
[4524] Panday, V.K.; A.K. Ganguly, *Spectrosc. Lett. 9/2*, 73–80 (1976).
[4525] Pandey, L.P.; A. Ghose, P. Dasgupta, A.S. Rao, *Talanta 25*, 482–483 (1978).
[4526] Pandey, L.P.; K.K. Gupta, S.N. Iha, P. Dasgupta, N.N. Chatterjee, *Fresenius Z. Anal. Chem. 329/7*, 795 (1988).
[4527] Pandey, L.P.; K.K. Gupta, P. Dasgupta, B.K. Verma, S. Bhattacharjee, *Spectroscopy Int. 3/3*, 38–40 (1991).
[4528] Pang, F.Y.; Y.L. Ng, S.M. Phang, S.L. Tong, *Int. J. Environ. Anal. Chem. 53*, 53–61 (1993).
[4529] Pannetier, G.; P. Toffoli, *Bull. Soc. Chim. Fr. 10*, 3775 (1971).
[4530] Pannier, F.; A. Astruc, M. Astruc, *Anal. Chim. Acta 287/1-2*, 17–24 (1994).
[4531] Papaspyrou, M.; L.E. Feinendegen, C. Mohl, M.J. Schwuger, *J. Anal. At. Spectrom. 9/7*, 791–795 (1994).
[4532] Papp, C.S.E.; L.B. Fisher, *Analyst (London) 112*, 337–338 (1987).
[4533] Pappas, E.G.; L.A. Rosenberg, *J. Assoc. Off. Anal. Chem. 49*, 792–793 (1966).
[4534] Paralusz, C.M.; *Appl. Spectrosc. 22/5*, 520–526 (1968).
[4535] Parisis, N.E.; A. Heyndrickx, *Analyst (London) 111/3*, 281–284 (1986).
[4536] Parker, H.E.; *At. Absorpt. Newsletter 2*, 23–29 (1963).
[4537] Parks, E.J.; F.E. Brinckman, W.R. Blair, *J. Chromatogr. 185*, 563–572 (1979).
[4538] Parks, E.J.; W.R. Blair, F.E. Brinckman, *Talanta 32/8A*, 633–639 (1985).
[4539] Parks, E.J.; F.E. Brinckman, K.L. Jewett, W.R. Blair, C.S. Weiss, *Appl. Organomet. Chem. 2*, 441–450 (1988).
[4540] Parr, R.M.; *J. Res. Natl. Bur. Stand. 91/2*, 51–57 (1986).
[4541] Parsley, D.H.; *J. Anal. At. Spectrom. 6/4*, 289–293 (1991).
[4542] Parsons, M.L.; W.J. McCarthy, J.D. Winefordner, *Appl. Spectrosc. 20*, 223–230 (1966).
[4543] Parsons, M.L.; J.D. Winefordner, *Anal. Chem. 38/11*, 1593–1595 (1966).
[4544] Parsons, M.L.; J.D. Winefordner, *Appl. Spectrosc. 21*, 368 (1967).
[4545] Parsons, M.L.; B.W. Smith, P.M. McElfresh, *Appl. Spectrosc. 27/6*, 471–480 (1973).
[4546] Partha, N.; J. Buffle, *Anal. Chim. Acta 284/3*, 649–659 (1994).
[4547] Paschal, D.C.; M.M. Kimberly, *At. Spectrosc. 6/5*, 134–136 (1985).
[4548] Paschal, D.C.; G.G. Bailey, *At. Spectrosc. 7/1*, 1–3 (1986).
[4549] Paschal, D.C.; M.M. Kimberly, G.G. Bailey, *Anal. Chim. Acta 181*, 179–186 (1986).
[4550] Paschal, D.C.; G.G. Bailey, *At. Spectrosc. 8/5*, 150–152 (1987).
[4551] Paschal, D.C.; G.G. Bailey, *J. Res. Natl. Bur. Stand. 93/3*, 323–326 (1988).
[4552] Paschal, D.C.; G.G. Bailey, *Sci. Total Environ. 89*, 305–310 (1989).
[4553] Paschal, D.C.; G.G. Bailey, *At. Spectrosc. 11/2*, 65–69 (1990).
[4554] Paschal, D.C.; G.G. Bailey, *At. Spectrosc. 12/5*, 151–154 (1991).
[4555] Paschen, A.; *Ann. Phys. (Leipzig) 50/1*, 901–940 (1916).
[4556] Paschen, K.; *Deutsche Medizinische Wochenschrift 95/51*, 2570–2773 (1970).
[4557] Pascucci, P.R.; *Anal. Lett. 26/3*, 445–455 (1993).
[4558] Pascucci, P.R.; J. Sneddon, *J. Environ. Sci. Health, Part A 28/7*, 1483–1493 (1993).
[4559] Passey, R.B.; K.C. Maluf, R. Fuller, *Anal. Biochem. 151/2*, 462–465 (1985).
[4560] Passmore, W.O.; P.B. Adams, *At. Absorpt. Newsletter 4*, 237–242 (1965).
[4561] Passmore, W.O.; P.B. Adams, *At. Absorpt. Newsletter 5/4*, 77–83 (1966).
[4562] Passmore, W.O.; P.B. Adams, in: American Society of Testing and Materials, *Methods for Emission Spectrochemical Analysis*, ASTM, Easton, MD., 1982, 902–906.
[4563] Patassy, F.Z.; *Plant Soil 22*, 395 (1965).
[4564] Patel, B.M.; N. Gupta, P. Purohit, B.D. Joshi, *Anal. Chim. Acta 118*, 163–168 (1980).
[4565] Patriarca, M.; E. Petrozzi, G. Morisi, *G. Ital. Chim. Clin. 13*, 111 (1988).
[4566] Patterson, C.C.; *Arch. Environ. Health 11*, 344 (1965).
[4567] Patterson, J.W.; R. Passimo, (Editors), *Metals Speciation, Separation and Recovery*, Book two, CRC Press, Boca Raton, FL, 1990.
[4568] Patterson, K.Y.; C. Veillon, H.M. Kingston, in: H.M. Kingston, L.B. Jassie (Editors), *Introduction to Microwave Sample Preparation*, American Chemical Society, Washington, DC, 1988, 155–166.
[4569] Paudyn, A.M.; J.C. Van Loon, *Fresenius Z. Anal. Chem. 325*, 369–376 (1986).
[4570] Paudyn, A.M.; D.M. Templeton, A.D. Baines, *Sci. Total Environ. 89/3*, 342–352 (1989).

[4571] Paukert, T.; I. Rubeška, *Anal. Chim. Acta 278/1*, 125–136 (1993).
[4572] Paulson, A.J.; *Anal. Chem. 58/1*, 183–187 (1986).
[4573] Paus, P.E.; *At. Absorpt. Newsletter 11/6*, 129–130 (1972).
[4574] Pauwels, J.; G.N. Kramer, L. De Angelis, K.H. Grobecker, *Fresenius J. Anal. Chem. 338/4*, 515–519 (1990).
[4575] Pauwels, J.; L. De Angelis, F. Peetermans, C. Ingelbrecht, *Fresenius J. Anal. Chem. 337/3*, 290–293 (1990).
[4576] Pauwels, J.; C. Vandecasteele, *Fresenius J. Anal. Chem. 345/2-4*, 121–123 (1993).
[4577] Pauwels, J.; C. Hofmann, K.H. Grobecker, *Fresenius J. Anal. Chem. 345/7*, 475–477 (1993).
[4578] Pauwels, J.; U. Kurfürst, K.H. Grobecker, Ph. Quevauviller, *Fresenius J. Anal. Chem. 345/7*, 478–481 (1993).
[4579] Pauwels, J.; C. Hofmann, C. Vandecasteele, *Fresenius J. Anal. Chem. 348/7*, 411–417 (1994).
[4580] Pauwels, J.; C. Hofmann, C. Vandecasteele, *Fresenius J. Anal. Chem. 348/7*, 418–421 (1994).
[4581] Pauwels, J.; A. Lamberty, P. De Bièvre, K.-H. Grobecker, C.Bauspiess, *Fresenius J. Anal. Chem. 349/6*, 409–411 (1994).
[4582] Pavel, J.; J. Kliment, S. Stoerk, O. Suter, *Fresenius Z. Anal. Chem. 321/6*, 587–591 (1985).
[4583] Paveri-Fontana, S.L.; G. Tessari, G.C. Torsi, *Anal. Chem. 46/8*, 1032–1038 (1974).
[4584] Pavski, V.; A. Corsini, S. Landsberger, *Talanta 36/3*, 367–372 (1989).
[4585] Pavski, V.; C.L. Chakrabarti, J.C. Hutton, M.H. Back, *Spectrochim. Acta, Part B 48/3*, 413–423 (1993).
[4586] Peats, S.; *At. Absorpt. Newsletter 18/6*, 118–120 (1979).
[4587] Peck, E.; *Anal. Lett. 11B/2*, 103–117 (1978).
[4588] Peddy, R.V.C.; G. Kalpana, V.J. Koshy, N.V.R. Apparao, M.C. Jain, R.V. Patel, *Analyst (London) 116*, 847 (1991).
[4589] Peddy, R.V.C.; G. Kalpana, V.J. Koshy, *Analyst (London) 117/1*, 27–30 (1992).
[4590] Pedersen, B.; E. Thomsen, R.M. Stern, *Ann. Occup. Hyg. 31*, 325–338 (1987).
[4591] Pei, S.-q.; Z.-l. Fang, *Anal. Chim. Acta 294/2*, 185–193 (1994).
[4592] Peile, R.; R. Grey, R. Starek, *J. Anal. At. Spectrom. 4/5*, 407–410 (1989).
[4593] Pellar, T.G.; J.F. Tuckerman, A.R. Henderson, *J. Autom. Chem. 7/2*, 95–98 (1985).
[4594] Pellenbarg, R.E.; T.M. Church, *Anal. Chim. Acta 97*, 81–86 (1978).
[4595] Peña, Y.P. de; M. Gallego, M. Valcárcel, *J. Anal. At. Spectrom. 9/6*, 691–696 (1994).
[4596] Penninckx, W.; D.L. Massart, J. Smeyers-Verbeke, *Fresenius J. Anal. Chem. 343/6*, 526–531 (1992).
[4597] Penninckx, W.; J. Smeyers-Verbeke, D.L. Massart, L.G.C.W. Spanjers, F. Maris, *Chemom. Intell. Lab. Sys. 17*, 193–200 (1992).
[4598] Penninckx, W.; J. Smeyers-Verbeke, C. Hartmann, B. Bourguignon, D.L. Massart, *Chemom. Intell. Lab. Sys. 23*, 137–148 (1994).
[4599] Penninckx, W.; *Analusis 22*, M22–M23 (1994).
[4600] Penninckx, W.; C. Hartmann, D.L. Massart, J. Smeyers-Verbeke, *J. Anal. At. Spectrom. 11/4*, 237–246 (1996).
[4601] Percelay, L.; P. Appriou, H. Handel, R. Guglielmetti, *Anal. Chim. Acta 209*, 249–258 (1988).
[4602] Pereiro-García, M.R.; A. López-García, M.E. Díaz-García, A. Sanz-Medel, *J. Anal. At. Spectrom. 5/1*, 15–19 (1990).
[4603] Pérez-Cid, B.; I. Lavilla, C. Bendicho, *Analyst (London) 121/10*, 1479–1484 (1996).
[4604] Pérez-Parajón, J.M.; A. Sanz-Medel, *Quim. Anal. (Barcelona) 12/1*, 30–34 (1993).
[4605] Pérez-Parajón, J.M.; A. Sanz-Medel, *J. Anal. At. Spectrom. 9/2*, 111–116 (1994).
[4606] Pergantis, S.A.; W.R. Cullen, A.P. Wade, *Talanta 41/2*, 205–209 (1994).
[4607] Peris, M.; A. Maquieira, R. Puchades, V. Chirivella, R. Ors, J. Serrano, A. Bonastre, *Chemom. Intell. Lab. Sys. 21/2-3*, 243–247 (1993).
[4608] Perkampus, H.H.; *Encyclopedia of Spectroscopy*, VCH Verlagsgesellschaft Weinheim, 1995.
[4609] Perkin-Elmer, *Analytical Methods for Atomic Absorption Spectrometry*, Handbook 0303–0152, Überlingen, 1994.
[4610] Perkin-Elmer, *Analytical Methods using Mercury/Hydride Systems*, Handbook, Überlingen, 1979.
[4611] Perkin-Elmer, *Flow Injection Mercury/Hydride Analyses, Recommended Analytical Conditions and General Information*, Publication B3505.10, Überlingen, 1996.
[4612] Perkins, J.; *Analyst (London) 88*, 324 (1963).
[4613] Perring, M.A.; *J. Sci. Food Agric. 25*, 247–250 (1974).
[4614] Perry, D.F.; *J. Assoc. Off. Anal. Chem. 73/4*, 619–621 (1990).
[4615] Persson, J.-Å.; W. Frech, A. Cedergren, *Anal. Chim. Acta 89*, 119–126 (1977).
[4616] Persson, J.-Å.; W. Frech, A. Cedergren, *Anal. Chim. Acta 92*, 85–93 (1977).

[4617] Persson, J.-Å.; W. Frech, A. Cedergren, *Anal. Chim. Acta 92*, 95–104 (1977).
[4618] Persson, J.-Å.; W. Frech, G. Pohl, K. Lundgren, *Analyst (London) 105*, 1163–1170 (1980).
[4619] Persson, J.-Å.; W. Frech, *Anal. Chim. Acta 119*, 75–89 (1980).
[4620] Pesch, H.-J.; Th. Kraus, in: B. Welz (Editor), *6. Colloquium Atomspektrometrische Spurenanalytik*, Bodenseewerk Perkin-Elmer, Überlingen, 1991, 867–878.
[4621] Pesch, H.-J.; S. Bloß, J. Schubert, H. Seibold, *Fresenius J. Anal. Chem. 343/1*, 152–153 (1992).
[4622] Peters, A.; W.W. Fletcher, *Glastechn Ber. 60/5*, 163–173 (1987).
[4623] Peters, A.; *Glass Technol. 34*, 239 (1993).
[4624] Peterson, G.G.; H.L. Kahn, *At. Absorpt. Newsletter 9/3*, 71–74 (1970).
[4625] Petersson, B.A.; Z.-l. Fang, J. Ruzicka, E.H. Hansen, *Anal. Chim. Acta 184*, 165–172 (1986).
[4626] Petrick, K.; V. Krivan, *Anal. Chem. 59/20*, 2476–2479 (1987).
[4627] Petrick, K.; V. Krivan, *Fresenius Z. Anal. Chem. 327/3-4*, 338–342 (1987).
[4628] Petrov, I.I.; D.L. Tsalev, A.I. Barsev, *At. Spectrosc. 1/2*, 47–51 (1980).
[4629] Petrov, Y.M.; *Zh. Anal. Khim. 29*, 686 (1974).
[4630] Petrova, N.I.; I.G. Yudelevich, S.I. Chikichev, L.M. Buyanova, *Sib. Khim. Zh. 2*, 51 (1991).
[4631] Pettersson, J.; L. Hansson, Å. Olin, *Talanta 33/3*, 249–254 (1986).
[4632] Pettersson, J.; Å. Olin, *Talanta 38/4*, 413–417 (1991).
[4633] Pettine, M.; A. Liberatori, D. Mastroianni, *Metodi Anal. Acque 6/2*, 16 (1986).
[4634] Pharr, D.Y.; H.E. Selnau, E.A. Pickral, R.L. Gordon, *Analyst (London) 116/5*, 511–515 (1991).
[4635] Phelan, V.J.; R.J.W. Powell, *Analyst (London) 109/10*, 1269–1272 (1984).
[4636] Phillips, D.J.H.; *Environ. Pollut. 13*, 281–317 (1977).
[4637] Piccolino, S.P.; *J. Chem. Educ. 60/3*, 235. (1983).
[4638] Pickering, W.F.; P.E. Thomas, *Talanta 22*, 691 (1975).
[4639] Pickering, W.F.; *CRC Crit. Rev. Anal. Chem. 12*, 233–266 (1981).
[4640] Pickett, E.E.; S.R. Koirtyohann, *Spectrochim. Acta, Part B 23/4*, 235–244 (1968).
[4641] Pickett, E.E.; M. Aldreshaidat, S. Broadway, S.R. Koirtyohann, *Spectrochim. Acta, Part B 44/12*, 1273–1284 (1989).
[4642] Pickford, C.J.; G. Rossi, *Analyst (London) 97*, 647–652 (1972).
[4643] Pickford, C.J.; G. Rossi *At. Absorpt. Newsletter 14/4*, 78–80 (1975).
[4644] Pickup, J.F.; M.J. Jackson, E.M. Price, S.S. Brown, *Clin. Chem. (Winston-Salem) 20/10*, 1324–1330 (1974).
[4645] Pickup, J.F.; M.J. Jackson, E.M. Price, M.J.R. Healy, S.S. Brown, *Clin. Chem. (Winston-Salem) 21/10*, 1416–1421 (1975).
[4646] Pieczonka, K.; A. Rosopulo, *Fresenius Z. Anal. Chem. 322/7*, 697–699 (1985).
[4647] Piepmeier, E.H.; L. de Galan, *Spectrochim. Acta, Part B 31/4*, 163–177 (1976).
[4648] Piepmeier, E.H.; *Spectrochim. Acta, Part B 44/6*, 609–616 (1989).
[4649] Piepponen, S.; H. Liukkonen-Lilja, T. Kuusi, *Z. Lebensm. Unters. Forsch. 177/4*, 257–260 (1983).
[4650] Pierce, F.D.; H.R. Brown, R.S. Fraser, *Appl. Spectrosc. 29/6*, 489–493 (1975).
[4651] Pierce, F.D.; T.C. Lamoreaux, H.R. Brown, R.S. Fraser, *Appl. Spectrosc. 30*, 38–42 (1976).
[4652] Pierce, F.D.; H.R. Brown, *Anal. Chem. 48/4*, 693–695 (1976).
[4653] Pilate, A.; P. Geladi, F. Adams, *Talanta 24*, 512 (1977).
[4654] Pillow, M.E.; *Spectrochim. Acta, Part B 36/8*, 821–843 (1981).
[4655] Pineau, A.; O. Guillard, F. Chauvelon, J.-F. Risse, *Biol. Trace Elem. Res. 32*, 139–143 (1992).
[4656] Pineau, A.; O. Guillard, P. Chappuis, J. Arnaud, R. Zawislak, *CRC Crit. Rev. Clin. Lab. Sci. 30*, 203–222 (1993).
[4657] Pinel, R.; M.Z. Benabdallah, A. Astruc, M. Potier-Gautier, M. Astruc, *Analusis 12/7*, 344–349 (1984).
[4658] Pinna, K.; N. Klein Amy, J.R. Turnlund, *J. Micronutr. Anal. 5*, 127–138 (1989).
[4659] Pinnell, R.P.; A.W. Zanella, *J. Chem. Educ. 58*, 444 (1981).
[4660] Pinta, M.; A.M. de Kersabiec, M.L. Richard, *Analusis 10/5*, 207–215 (1982).
[4661] Pio, C.A.; in: R.M. Harrison, R. Perry (Editors), *Handbook of Air Pollution Analysis*, Chapman and Hall, New York, 1986, 1–93.
[4662] Pita Calvo, C.; P. Bermejo Barrera, A. Bermejo Barrera, *Quim. Anal. (Barcelona) 11/1*, 35–43 (1992).
[4663] Pita Calvo, C.; P. Bermejo Barrera, A. Bermejo Barrera, *Anal. Chim. Acta 310/1*, 189–198 (1995).
[4664] Pitts, A.E.; J.C. Van Loon, F.E. Beamish, *Anal. Chim. Acta 50*, 181–194 (1970).
[4665] Pizent, A.; S. Telisman, *At. Spectrosc. 17/2*, 88–91 (1996).

[4666] Place, J.F.; A. Truchaud, K. Ozawa, H. Pardue, P. Schnipelsky, *J. Autom. Chem. 17/1*, 115 (1995).
[4667] Platte, J.A.; V.M. Marcy, *At. Absorpt. Newsletter 4/6*, 289–292 (1965).
[4668] Platteau, O.; *Analyst (London) 119/2*, 339–348 (1994).
[4669] Platteau, O.; D. Casabiell, *Analyst (London) 119/8*, 1705–1713 (1994).
[4670] Platteau, O.; M. Carillo, *Fuel 74/5*, 761–767 (1995).
[4671] Plessi, M.; A. Monzani, *J. Assoc. Off. Anal. Chem. 73/5*, 798–800 (1990).
[4672] Plessi, M.; A. Monzani, *J. Food Compos. Anal. 8/1*, 21–26 (1995).
[4673] Pohl, B.; M. Knowles, A. Grund, *GIT Lab. Med. 10/6*, 269–272 (1987).
[4674] Pohl, B.; M. Knowles, A. Grund, *Fresenius Z. Anal. Chem. 327*, 19–20 (1987).
[4675] Pohl, B.; M. Lange, *Lab. med. 18*, 38–40 (1994).
[4676] Pohl, B.; G. Weichbrodt, S. Fraunhofer, *GIT Fachz. Lab. 39/12*, 1134–1135 (1995).
[4677] Pollard, P.M.; J.W. McMillan, T. McCullogh, J.L. Bowen, L.C. Buckley, *U.K. At. Energy Res. Establ., [Rep.] AERE R 11499*, 1986.
[4678] Pollman, R.M.; *J. Assoc. Off. Anal. Chem. 74/1*, 27–31 (1991).
[4679] Pollock, E.N.; S.J. West, *At. Absorpt. Newsletter 11/5*, 104–106 (1972).
[4680] Pollock, E.N.; S.J. West, *At. Absorpt. Newsletter 12/1*, 6–8 (1973).
[4681] Polo-Díez, L.; J. Hernández-Méndez, F. Pedraz-Penalva, *Analyst (London) 105*, 37–42 (1980).
[4682] Poluektov, N.S.; R.A. Vitkun, *Zh. Anal. Khim. 18*, 33 (1963).
[4683] Poluektov, N.S.; R.A. Vitkun, Y.A. Zelyukova, *Zh. Anal. Khim. 19*, 937 (1964).
[4684] Polyák, K.; I. Bódog, J. Hlavay, *Talanta 41/7*, 1151–1159 (1994).
[4685] Poole, S.; *Commun. Soil. Sci. Plant Anal. 19/15*, 1681–1691 (1988).
[4686] Pop, H.F.; D. Dumitrescu, C. Sârbu, *Anal. Chim. Acta 310/2*, 269–279 (1995).
[4687] Popova, S.A.; S.P. Bratinova, *J. Anal. At. Spectrom. 5/1*, 35–38 (1990).
[4688] Popova, S.A.; S.P. Bratinova, C.R. Ivanova, *Analyst (London) 116/5*, 525–528 (1991).
[4689] Porta, V.; O. Abollino, E. Mentasti, C. Sarzanini, *J. Anal. At. Spectrom. 6/2*, 119–122 (1991).
[4690] Porter, B.L.; V.A. Fuavao, J. Sneddon, *At. Spectrosc. 4/5*, 185–187 (1983).
[4691] Porter, W.K.; *J. Assoc. Off. Anal. Chem. 57*, 614–617 (1974).
[4692] Posselt, H.S.; W.J. Weber, *Tech. Report T-71-1, University of Michigan*, Ann Arbor, 1971.
[4693] Posta, J.; L. Szücs, *Acta Chim. Hung. 126*, 325–332 (1989).
[4694] Posta, J.; H. Berndt, *Spectrochim. Acta, Part B 47/8*, 993–999 (1992).
[4695] Posta, J.; H. Berndt, B. Dereckei, *Anal. Chim. Acta 262*, 261–267 (1992).
[4696] Posta, J.; B. Dereckei, *Microchem. J. 46/3*, 271–279 (1992).
[4697] Posta, J.; H. Berndt, S.-k. Luo, G. Schaldach, *Anal. Chem. 65/19*, 2590–2595 (1993).
[4698] Potter, A.L.; E.D. Ducay, R.M. McCready, *J. Assoc. Off. Anal. Chem. 51*, 748 (1968).
[4699] Potter, N.M.; *Anal. Chem. 48*, 531–534 (1976).
[4700] Potter, N.M.; *Anal. Chem. 50*, 769 (1978).
[4701] Potter, N.M.; R.A. Waldo, *Anal. Chim. Acta 110/1*, 29–34 (1979).
[4702] Potts, P.J.; *A Handbook of Silicate Rock Analysis*, Blackie & Sons, Glasgow, 1987.
[4703] Pougnet, M.A.B.; L.C. Haraldsen, *S. Afr. J. Chem. 45/2-3*, 50–53 (1992).
[4704] Pradeau, D.; J. Petiot, V. Bissery, M. Hamon, R. Grangeon, M. Lacaze, J. Chenebaux, *Int. Pharm. J. 2/6*, 209 (1988).
[4705] Prager, M.J.; D. Graves, *J. Assoc. Off. Anal. Chem. 60*, 609–612 (1977).
[4706] Pratt, L.K.; K.A. Elrick, *At. Spectrosc. 8/6*, 170–171 (1987).
[4707] Preer, J.R.; B.R. Stephens, C. Wilson Bland, *J. Assoc. Off. Anal. Chem. 65/4*, 1010–1015 (1982).
[4708] Prell, L.J.; D.L. Styris, D.A. Redfield, *J. Anal. At. Spectrom. 5/3*, 231–238 (1990).
[4709] Prell, L.J.; D.L. Styris, *Spectrochim. Acta, Part B 46/1*, 45–49 (1991).
[4710] Prell, L.J.; D.L. Styris, D.A. Redfield, *J. Anal. At. Spectrom. 6/1*, 25–32 (1991).
[4711] Preu, E.; H. Schroeter, K. Winnefeld, U. Schmidt, H. Müller, *Zentralbl. Pharm., Pharmakother. Lab. 127/6*, 422–424 (1988).
[4712] Prewett, W.; M. Promphutha, *Anal. Chim. Acta 339/3*, 297–302 (1997).
[4713] Price, J.P.; *At. Absorpt. Newsletter 11/1*, 1–4 (1972).
[4714] Price, W.J.; J.T.H. Roos, *Analyst (London) 93*, 709–714 (1968).
[4715] Price, W.J.; T.C. Dymott, P.J. Whiteside, *Spectrochim. Acta, Part B 35/1*, 3–10 (1980).
[4716] Price, W.J.; T.C. Dymott, M.P. Wassall, P.J. Whiteside, *Analusis 13/2*, 92–96 (1985).
[4717] Pritchard, M.W.; J. Lee, in: H.A. McKenzie, L.E. Smythe (Editors), *Quantitative trace analysis of biological materials*, Elsevier, Amsterdam - New York - Oxford, 1988, 367–387.
[4718] Pronin, V.A.; A.P. Galanova, A.K. Kudryavina, Z.N. Shastina, V.N. Apolitskii, *Zh. Anal. Khim. 28*, 2328 (1973).

[4719] Prudnikov, E.D.; H. Bradaczek, H. Labischinski, *Fresenius Z. Anal. Chem. 308/4*, 342–346 (1981).
[4720] Prugger, H.; *Optik 21/7*, 320 (1964).
[4721] Prugger, H.; R. Torge, *Patent, Ger. Offen. DE 1,964,469*, 1969.
[4722] Pruszkowska, E.; G.R. Carnrick, W. Slavin, *Anal. Chem. 55/2*, 182–186 (1983).
[4723] Pruszkowska, E.; G.R. Carnrick, W. Slavin, *At. Spectrosc. 4/2*, 59–61 (1983).
[4724] Pruszkowska, E.; D.C. Manning, G.R. Carnrick, W. Slavin, *At. Spectrosc. 4/3*, 87–93 (1983).
[4725] Pruszkowska, E.; G.R. Carnrick, W. Slavin, *Clin. Chem. (Winston-Salem) 29/3*, 477–480 (1983).
[4726] Pruszkowska, E.; P. Barrett, *Spectrochim. Acta, Part B 39/2-3*, 485–491 (1984).
[4727] Psenicnik, O.; *At. Spectrosc. 12/6*, 207–209 (1991).
[4728] Puacz, W.; *Acta Chim. Hung. 124/2*, 293–298 (1987).
[4729] Puchades, R.; A. Maquieira, M. Plantá, *Analyst (London) 114/11*, 1397–1399 (1989).
[4730] Puchelt, H.; T. Nöltner, *Fresenius Z. Anal. Chem. 331*, 216–219 (1988).
[4731] Puchyr, R.F.; R. Shapiro, *J. Assoc. Off. Anal. Chem. 69/5*, 868–870 (1986).
[4732] Puderbach, H.; *Labo 23/5*, 7–11 (1992).
[4733] Puk, R.; J.H. Weber, *Anal. Chim. Acta 292/1-2*, 175–183 (1994).
[4734] Purchase, N.G.; J.E. Ferguson, *Sci. Total Environ. 52*, 239–250 (1986).
[4735] Purushottam, A.; P.P. Naidu, S.S. Lal, *Talanta 20*, 631–637 (1973).
[4736] Püschel, P.; Z. Formánek, R. Hlaváč, D. Kolihová, V. Sychra, *Anal. Chim. Acta 127*, 109–120 (1981).
[4737] Puttemans, F.; D.L. Massart, *Anal. Chim. Acta 141*, 225–231 (1982).
[4738] Pyrzynska, K.; *Anal. Lett. 22/13*, 2847–2859 (1989).
[4739] Pyrzynska, K.; *Spectrochim. Acta, Part B 50/13*, 1595–1598 (1995).
[4740] Qi, W.-q.; J. Liu, L. Ying, *Zhongguo Huanjing Jiance 9/4*, 23 (1993).
[4741] Qi, W.-y.; X. Wu, C. Zhou, H.-z. Wu, Y.-g. Gao, *Anal. Chim. Acta 270*, 205–211 (1992).
[4742] Qian, S.; P. Yang, *Fenxi Huaxue 18/11*, 1064 (1990).
[4743] Qiao, H.-c.; K.W. Jackson, *Spectrochim. Acta, Part B 46/14*, 1841–1859 (1991).
[4744] Qiao, H.-c.; K.W. Jackson, *Spectrochim. Acta, Part B 47/11*, 1267–1276 (1992).
[4745] Qiao, H.-c.; T.M. Mahmood, K.W. Jackson, *Spectrochim. Acta, Part B 48/12*, 1495–1503 (1993).
[4746] Qiao, H.-c.; P.J. Parsons, W. Slavin, *Clin. Chem. (Winston-Salem) 41/10*, 1451–1454 (1995).
[4747] Quevauviller, Ph.; R. Lavigne, R. Pinel, M. Astruc, *Environ. Pollut. 57*, 149–166 (1989).
[4748] Quevauviller, Ph.; O.F.X. Donard, *Appl. Organomet. Chem. 4*, 353–367 (1990).
[4749] Quevauviller, Ph.; O.F.X. Donard, *Fresenius J. Anal. Chem. 339/1*, 6–14 (1991).
[4750] Quevauviller, Ph.; A. Bruchet, O.F.X. Donard, *Appl. Organomet. Chem. 5*, 125–129 (1991).
[4751] Quevauviller, Ph.; H. Etcheber, C. Raoux, O.F.X. Donard, *Oceanol. Acta SP11*, 247–255 (1991).
[4752] Quevauviller, Ph.; O.F.X. Donard, E.A. Maier, *Mikrochim. Acta (Vienna) 109/1-4*, 169–190 (1992).
[4753] Quevauviller, Ph.; O.F.X. Donard, J.C. Wasserman, F.M. Martin, J. Schneider, *Appl. Organomet. Chem. 6*, 221–228 (1992).
[4754] Quevauviller, Ph.; F.Martin, C. Belin, O. Donard, *Appl. Organomet. Chem. 7*, 149–157 (1993).
[4755] Quevauviller, Ph.; I. Drabaek, H. Muntau, B. Griepink, *Appl. Organomet. Chem. 7*, 413–420 (1993).
[4756] Quevauviller, Ph.; in: Ph. Quevauviller, E.A. Maier, B. Griepink (Editors), *Quality Assurance for Environmental Analysis*, Elsevier, Amsterdam, 1995, 1–25.
[4757] Quevauviller, Ph.; Y. Wang, A.B. Turnbull, W.M.R. Dirkx, R.M. Harrison, F.C. Adams, *Appl. Organomet. Chem. 9*, 89–93 (1995).
[4758] Quevauviller, Ph.; M.B. de la Calle-Guntiñas, E.A. Maier, C. Cámara, *Mikrochim. Acta (Vienna) 118*, 131–141 (1995).
[4759] Quevauviller, Ph.; H. Muntau, U. Fortunati, K. Vercoutere, *Report EUR 16890 EN*, EC, Brussels, 1996.
[4760] Quevauviller, Ph.; H. Muntau, U. Fortunati, K. Vercoutere, *Report EUR 16892 EN*, EC, Brussels, 1996.
[4761] Quevauviller, Ph.; K.J.M. Kramer, T. Vinhas, *Fresenius J. Anal. Chem. 354/4*, 397–404 (1996).
[4762] Quevauviller, Ph.; *Fresenius J. Anal. Chem. 354/5-6*, 515–520 (1996).
[4763] Quevauviller, Ph.; U. Fortunati, M. Fillipelli, F. Baldi, M. Bianchi, H. Muntau, *Appl. Organomet. Chem. 10*, 537–544 (1996).

[4764] Quevauviller, Ph.; K.J.M. Kramer, E.M. van der Vlies, W. Dorten, B. Griepink, *Fresenius J. Anal. Chem. 356/7*, 411–415 (1996).
[4765] Quevauviller, Ph.; G. Rauret, R. Rubio, J.-F. López-Sánchez, A. Ure, J. Bacon, H. Muntau, *Fresenius J. Anal. Chem. 357/6*, 611–618 (1997).
[4766] Quigley, M.N.; F. Vernon, *Anal. Proc. (London) 28/6*, 175–176 (1991).
[4767] Quigley, M.N.; F. Vernon, *Anal. Proc. (London) 29/1*, 1–3 (1992).
[4768] Quigley, M.N.; *J. Chem. Educ. 71*, 800 (1994).
[4769] Quiñonero, J.; C. Mongay, M. de la Guardia, *Ann. Chim. (Rom) 79/5-6*, 311–318 (1989).
[4770] Qureshi, M.A.; M. Farid, A. Aziz, M. Ejaz, *Talanta 26*, 166–168 (1979).
[4771] Raaphorst, J.G. van, H.M. Haremaker, P.A. Deurloo, B. Beemsterboer, *Anal. Chim. Acta 286/3*, 291–296 (1994).
[4772] Rabadán, J.M.; J. Galbán, J. Aznárez, *At. Spectrosc. 14/4*, 95–98 (1993).
[4773] Räde, H.S.; *At. Absorpt. Newsletter 13/4*, 81–83 (1974).
[4774] Rademeyer, C.J.; I. Vermaak, *J. Anal. At. Spectrom. 7/2*, 347–351 (1992).
[4775] Radojevic, M.; R.M. Harrison, *Environ. Technol. Lett. 7*, 519–524 (1986).
[4776] Radojevic, M.; R.M. Harrison, *Environ. Technol. Lett. 7*, 525–530 (1986).
[4777] Radojevic, M.; R.M. Harrison, *Atmos. Environ. 21/11*, 2403–2411 (1987).
[4778] Radunovic, A.; M.W.B. Bradbury, H.T. Delves, *Analyst (London) 118/5*, 533–536 (1993).
[4779] Radziuk, B.; J.C. van Loon, *Sci. Total Environ. 6*, 251–257 (1976).
[4780] Radziuk, B.; Y. Thomassen, J.C. Van Loon, Y.K. Chau, *Anal. Chim. Acta 105*, 255–262 (1979).
[4781] Radziuk, B.; I.L. Shuttler, Y. Thomassen, *J. Anal. At. Spectrom. 5*, 413 (1990).
[4782] Radziuk, B.; Y. Thomassen, *J. Anal. At. Spectrom. 7/2*, 397–403 (1992).
[4783] Radziuk, B.; G. Rödel, H. Stenz, H. Becker-Ross, S. Florek, *J. Anal. At. Spectrom. 10/2*, 127–136 (1995).
[4784] Radziuk, B.; G. Rödel, M. Zeiher, S. Mizuno, K. Yamamoto, *J. Anal. At. Spectrom. 10/6*, 415–422 (1995).
[4785] Raij, B. Van; J.A. Quaggio, N.M. Da Silva, *Commun. Soil. Sci. Plant Anal. 17/5*, 547–566 (1986).
[4786] Rains, T.C.; O. Menis, *J. Assoc. Off. Anal. Chem. 55/6*, 1339–1344 (1972).
[4787] Rains, T.C.; T.A. Rush, T.A. Butler, *J. Assoc. Off. Anal. Chem. 65/4*, 994–998 (1982).
[4788] Rains, T.C.; in: S.J. Haswell (Editor), *Atomic Absorption Spectrometry - Theory, Design and Application*, Elsevier, Amsterdam - Oxford - New York - Tokyo, 1991, 191–226.
[4789] Räisänen, M.L.; L. Hämäläinen, L.M. Westerberg, *Analyst (London) 117/3*, 623–627 (1992).
[4790] Raithel, H.J.; K.H. Schaller, *Fresenius J. Anal. Chem. 338/4*, 534–537 (1990).
[4791] Raje, N.; S. Kayasth, T.P.S. Sari, S. Gangadharan, *Anal. Chim. Acta 290/3*, 371–377 (1994).
[4792] Raksasataya, M.; A.G. Langdon, N.D. Kim, *Anal. Chim. Acta 332/1*, 1–14 (1996).
[4793] Ramakrishna, T.V.; J.W. Robinson, P.W. West, *Anal. Chim. Acta 37*, 20–26 (1967).
[4794] Ramakrishna, T.V.; J.W. Robinson, P.W. West, *Anal. Chim. Acta 39*, 81 (1967).
[4795] Ramelow, G.J.; L. Liu, C. Himel, D. Fralick, Y. Zhao, C. Tong, *Int. J. Environ. Anal. Chem. 53*, 219 (1993).
[4796] Ramírez-Muñoz, J.; *Microchem. J. 15*, 244 (1970).
[4797] Ramsey, J.M.; *Anal. Chem. 52/13*, 2141–2147 (1980).
[4798] Ramsey, M.H.; A. Argyraki, M. Thompson, *Analyst (London) 120/5*, 1353–1356 (1995).
[4799] Randall, L.; J.S. Han, J.H. Weber, *Environ. Technol. Lett. 7/11*, 571–576 (1986).
[4800] Randlesome, J.E.; S.R. Aston, *Environ. Technol. Lett. 1/1*, 3–8 (1980).
[4801] Rann, C.S.; A.N. Hambly, *Anal. Chem. 37/7*, 879–884 (1965).
[4802] Rann, C.S.; *Spectrochim. Acta, Part B 23/12*, 827–847 (1968).
[4803] Rann, C.S.; *Spectrochim. Acta, Part B 24/12*, 685–687 (1969).
[4804] Rantala, R.T.T.; D.H. Loring, *Anal. Chim. Acta 220/1*, 263–267 (1989).
[4805] Rao, C.R.M.; *Anal. Chim. Acta 318/1*, 113–116 (1995).
[4806] Rao, P.D.; *At. Absorpt. Newsletter 11*, 49–50 (1972).
[4807] Rao, V.M.; P.E. Koehler, *J. Food Sci. 44*, 977–981 (1979).
[4808] Rao, V.M.; M.N. Sastri, *J. Sci. Ind. Res. 41/10*, 607–615 (1982).
[4809] Raoot, S.; S.V. Athavale, T.H. Rao, *Analyst (London) 111/1*, 115–117 (1986).
[4810] Rapsomanikis, S.; O.F.X. Donard, J.H. Weber, *Anal. Chem. 58/1*, 35–38 (1986).
[4811] Rapsomanikis, S.; R.M. Harrison, *Appl. Organomet. Chem. 2*, 151–157 (1988).
[4812] Rapsomanikis, S.; M.O. Andreae, *Int. J. Environ. Anal. Chem. 49/1-2*, 43–48 (1993).
[4813] Rapsomanikis, S.; *Analyst (London) 119/7*, 1429–1439 (1994).
[4814] Raptis, S.E.; G. Knapp, A. Meyer, G. Tölg, *Fresenius Z. Anal. Chem. 300/1*, 18–21 (1980).
[4815] Raptis, S.E.; G. Kaiser, G. Tölg, *Fresenius Z. Anal. Chem. 316/2*, 105–123 (1983).

[4816] Rasberry, S.D.; T.E. Gills, *Spectrochim. Acta, Part B 46/12*, 1577–1582 (1991).
[4817] Rasmuson, J.O.; V.A. Fassel, R.N. Kniseley, *Spectrochim. Acta, Part B 28*, 365–406 (1973).
[4818] Rasmussen, L.; *Anal. Chim. Acta 125*, 117–130 (1981).
[4819] Rasmussen, P.E.; G. Mierle, J.O. Nriagu, *Water, Air, Soil Pollut. 56*, 379–390 (1991).
[4820] Ratcliffe, D.B.; C.S. Byford, P.B. Osman, *Anal. Chim. Acta 75*, 457–459 (1975).
[4821] Ratliff, J.; V. Majidi, *Anal. Chem. 64/22*, 2743–2750 (1992).
[4822] Rauret, G.; E. Casassas, M. Baucells, *Archaeometry 27*, 195–201 (1985).
[4823] Ravreby, M.; *J. Forensic. Sci. 27*, 92 (1982).
[4824] Rawat, J.P.; M. Alam, M.S. Rathi, *Microchem. J. 32/2*, 153–160 (1985).
[4825] Rawling, B.S.; C. Greaves, M.D. Amos, *Nature (London) 188*, 137 (1960).
[4826] Rawson, R.A.G.; in: *Le Groupement pour l'Avancement des Méthodes Physiques d'Analyse (GAMS)*, 3rd Int. Congress of Atomic Absorption and Atomic Fluorescence Spectrometry, J. Wiley & Sons, New York - Toronto, 1973, 165–173.
[4827] Ray, P.K. Nayar, A.K. Misra, N. Sethunathan, *Analyst (London) 105*, 984–986 (1980).
[4828] Rayson, G.D.; J.A. Holcombe, *Anal. Chim. Acta 136*, 249–260 (1982).
[4829] Rayson, G.D.; J.A. Holcombe, *Spectrochim. Acta, Part B 38/7*, 987–993 (1983).
[4830] Rayson, G.D.; D.A. Bass, *Appl. Spectrosc. 45/6*, 1049–1050 (1991).
[4831] Reamer, D.C.; C. Veillon, P.T. Tokousbalides, *Anal. Chem. 53/2*, 245–248 (1981).
[4832] Reamer, D.C.; C. Veillon, *Anal. Chem. 53/8*, 1192–1195 (1981).
[4833] Rechenberg, W.; S. Sprung, in: B. Welz (Editor), *4. Colloquium Atomspektrometrische Spurenanalytik*, Bodenseewerk Perkin-Elmer, Überlingen, 1987, 295–302.
[4834] Rechenberg, W.; *GIT Fachz. Lab. 35*, 208–217 (1991).
[4835] Reddi, G.S.; S. Ganesh, C.R.M. Rao, V. Ramanan, *Anal. Chim. Acta 260*, 131–134 (1992).
[4836] Reddy, M.M.; M.A. Benefiel, H.C. Claassen, *Mikrochim. Acta (Vienna) I*, 159–170 (1986).
[4837] Redfield, D.A.; W. Frech, *J. Anal. At. Spectrom. 4/8*, 685–690 (1989).
[4838] Reeve, R.N.; J.D. Barnes, *Environmental Analysis* [Analytical Chemistry by Open Learning, ACOL Series Vol. 32], Wiley, Chichester, 1994.
[4839] Reichardt, M.S.; *At. Spectrosc. 13/5*, 178–184 (1992).
[4840] Reichert, J.K.; H. Gruber, *Vom Wasser 51*, 191–197 (1979).
[4841] Reichert, J.K.; H. Gruber, *Vom Wasser 52*, 289 (1979).
[4842] Reif, I.; V.A. Fassel, R.N. Kniseley, *Spectrochim. Acta, Part B 28*, 105–123 (1973).
[4843] Reimann, C.; F. Wurzer, *Mikrochim. Acta (Vienna) II*, 31–42 (1986).
[4844] Reimold, E.W.; D.J. Besch, *Clin. Chem. (Winston-Salem) 24*, 675–680 (1978).
[4845] Reis, B.F. dos; A.O Jacintho, J. Mortatti, F.J. Krug, E.A.G. Zagatto, H. Bergamin F°, *Anal. Chim. Acta 123*, 221–228 (1981).
[4846] Reis, B.F. dos; M.A.Z. Arruda, E.A.G. Zagatto, *Anal. Chim. Acta 206*, 253–262 (1988).
[4847] Reitznerová, E.; M. Kopcáková, in: K. Dittrich, B. Welz (Editors), *CANAS'93, Colloq. Analytische Atomspektroskopie*, Universität Leipzig/UFZ Leipzig-Halle, Leipzig, 1993, 403–406.
[4848] Ren, Z.; M. Qu, *Yejin Fenxi 13/2*, 61–62 (1993).
[4849] Renoe, B.W.; C.E. Shideler, J. Savory, *Clin. Chem. (Winston-Salem) 27/9*, 1546–1550 (1981).
[4850] Renoe, B.W.; *J. Autom. Chem. 4/2*, 61–64 (1982).
[4851] Renshaw, G.D.; C.A. Pounds, E.F. Pearson, *At. Absorpt. Newsletter 12/2*, 55–56 (1973).
[4852] Renshaw, G.D.; *At. Absorpt. Newsletter 12*, 158–160 (1973).
[4853] Rettberg, T.M.; L.M. Beach, *J. Anal. At. Spectrom. 4/5*, 427–432 (1989).
[4854] Retzik, M.; D. Bass, *Int. Lab. 18/8*, 49–56 (1988).
[4855] Reust, J.B.; in: B. Welz (Editor), *5. Colloquium Atomspektrometrische Spurenanalytik*, Bodenseewerk Perkin-Elmer, Überlingen, 1989, 657–665.
[4856] Rezende, M. do Carmo R.; R.C. Campos, A.J. Curtius, *J. Anal. At. Spectrom. 8/2*, 247–251 (1993).
[4857] Riandey, C.; R. Gavinelli, M. Pinta, *Spectrochim. Acta, Part B 35/11-2*, 765–773 (1980).
[4858] Riby, P.G.; S.J. Haswell, R. Grzeskowiak, *J. Anal. At. Spectrom. 4/2*, 181–184 (1989).
[4859] Ricci, G.R.; L.S. Shepard, G. Colovos, N.E. Hester, *Anal. Chem. 53/4*, 610–613 (1981).
[4860] Richards, M.P.; *J. Chromatogr. 482*, 87–97 (1989).
[4861] Ridder, T.B.; T.A. Buishand, H.F.R. Reijnders, M.J. 't Hart, J. Slanina, *Atmos. Environ. 19*, 759–762 (1985).
[4862] Riddle, C.; (Editor), *Analysis of Geological Materials*, Marcel Dekker New York, 1993.
[4863] Riedel, U.; G. Küllmer, *CLB, Chem. Labor Betr. 38/6*, 298–308 (1987).
[4864] Rigin, V.I.; G.N. Verkhoturov, *Zh. Anal. Khim. 32/10*, 1965–1968 (1977).
[4865] Rigin, V.I.; *Zh. Anal. Khim. 33/10*, 1966–1971 (1978).
[4866] Rigin, V.I.; *Zh. Anal. Khim. 34/8*, 1569–1573 (1979).

[4867] Riley, C.M.; L.A. Sternson, A.J. Repta, *Anal. Biochem. 124*, 167–179 (1982).
[4868] Riley, C.M.; L.A. Sternson, A.J. Repta, R.W. Siegler, *J. Chromatogr. 229*, 373–386 (1982).
[4869] Riley, J.P.; S.A. Siddiqui, *Anal. Chim. Acta 181*, 117–123 (1986).
[4870] Riley, K.W.; *At. Spectrosc. 3/4*, 120–121 (1982).
[4871] Riley, K.W.; *Analyst (London) 109/2*, 181–182 (1984).
[4872] Riley, K.W.; *At. Spectrosc. 6/3*, 76–77 (1985).
[4873] Ringdal, O.; K. Julshamn, K.-J. Andersen, E. Svendsen, in: P. Brätter, P. Schramel (Editors), *Trace Element Analytical Chemistry in Medicine and Biology, Vol. 3*, W. de Gruyter, Berlin - New York, 1984, 189–199.
[4874] Ríos, Â.; M. Valcárcel, *Analyst (London) 119/1*, 109–112 (1994).
[4875] Ríos, C.; H. Valero, T. Sanchez, *At. Spectrosc. 8/1*, 67–68 (1987).
[4876] Ríos, C.; R. Guzmán-Méndez, *J. Pharmacological Methods (New York) 24*, 327–332 (1990).
[4877] Ritchie, G.S.P.; G. Sposito, in: A.M. Ure, C.M. Davidson (Editors), *Chemical Speciation in the Environment*, Blackie Academic Professional, Glasgow, 1995, 201–233.
[4878] Ritsema, R.; R.W.P.M. Laane, O.F.X. Donard, *Mar. Environ. Res. 32*, 243–260 (1991).
[4879] Rivas, G.A.; J. Martínez Calatayud, *Talanta 42/9*, 1285–1289 (1995).
[4880] Rivoldini, A.; T. Haile, *At. Spectrosc. 10/3*, 89–91 (1989).
[4881] Rixen, F.; W. Bölts, *Hansa-Schiffahrt-Schiffbau-Hafen 121*, 1804–1812 (1984).
[4882] Robards, K.; P.J. Worsfold, *Trends Anal. Chem. (Pers. Ed.) 9/7*, 228–231 (1990).
[4883] Robards, K.; P.J. Worsfold, *Analyst (London) 116/6*, 549–568 (1991).
[4884] Robards, K.; M. Antolovich, *Analyst (London) 120/1*, 1–28 (1995).
[4885] Robberecht, H.J.; R.E. Van Grieken, P.A. Van Den Bosch, H.A. Deelstra, D.A. Vanden Berghe, *Talanta 29/11*, 1025–1028 (1982).
[4886] Robberecht, H.J.; R.E. Van Grieken, *Talanta 29/10*, 823–844 (1982).
[4887] Robberecht, H.J.; H.A. Deelstra, *Clin. Chim. Acta 136*, 107–120 (1984).
[4888] Robberecht, H.J.; H.A. Deelstra, *Talanta 31/7*, 497–508 (1984).
[4889] Robbins, W.K.; H.H. Walker, *Anal. Chem. 47/8*, 1269–1275 (1975).
[4890] Robertson, D.E.; *Anal. Chim. Acta 42*, 533–536 (1968).
[4891] Robertson, D.E.; in: M. Zief, R. Speights (Editors), *Ultrapurity*, Marcel Dekker, New York, 1972, 207–253.
[4892] Robinson, J.L.; R.G. Barnekow, P.F. Lott, *At. Absorpt. Newsletter 8/3*, 60–64 (1969).
[4893] Robinson, J.W.; *Anal. Chim. Acta 24/5*, 451 (1961).
[4894] Robinson, J.W.; *Anal. Chem. 33*, 1067–1071 (1961).
[4895] Robinson, J.W.; L.E. Vidarreta, D.K. Wolcott, J.P. Goodbread, E. Kiesel, *Spectrosc. Lett. 7*, 491 (1974).
[4896] Robinson, J.W.; D.K. Wolcott, *Environ. Lett. 6/4*, 321–333 (1974).
[4897] Robinson, J.W.; L.E. Vidaurretta, D.K. Wolcott, J.P. Goodbread, E. Kiesel, *Spectrosc. Lett. 8/7*, 491–507 (1975).
[4898] Robinson, J.W.; E.L. Kiesel, J.P. Goodbread, R. Bliss, R. Marshall, *Anal. Chim. Acta 92*, 321–328 (1977).
[4899] Robinson, J.W.; E.L. Kiesel, *J. Environ. Sci. Health, Part A 12/8*, 411–422 (1977).
[4900] Robinson, J.W.; S. Weiss, *Spectrosc. Lett. 11/9*, 715–724 (1978).
[4901] Robinson, J.W.; E.L. Kiesel, I.A.L. Rhodes, *J. Environ. Sci. Health, Part A 14*, 65 (1979).
[4902] Robinson, J.W.; S. Weiss, *J. Environ. Sci. Health, Part A 15/6*, 635–662 (1980).
[4903] Robinson, J.W.; S. Weiss, *J. Environ. Sci. Health, Part A 15/6*, 663–697 (1980).
[4904] Robinson, J.W.; E.M. Skelly, *J. Environ. Sci. Health, Part A 17/3*, 391–425 (1982).
[4905] Robinson, J.W.; P.L.H. Jowett, *Spectrosc. Lett. 16/3*, 159–179 (1983).
[4906] Robinson, J.W.; T.A. Ekman, *Spectrosc. Lett. 17/10*, 615–632 (1984).
[4907] Robinson, J.W.; E.D. Boothe, *Spectrosc. Lett. 17/11*, 653–671 (1984).
[4908] Robinson, J.W.; J.C. Wu, *Spectrosc. Lett. 18/6*, 399–418 (1985).
[4909] Robinson, J.W.; D.S. Choi, *Spectrosc. Lett. 20/4*, 375–390 (1987).
[4910] Robinson, K.G.; M.S. Shuman, *Int. J. Environ. Anal. Chem. 36/2*, 111–123 (1989).
[4911] Robles, L.C.; C. García-Olalla, M.T. Alemany, A. Javier Aller, *Analyst (London) 116/7*, 735–737 (1991).
[4912] Robles, L.C.; C. Garcia-Olalla, A.J. Aller, *J. Anal. At. Spectrom. 8/7*, 1015–1022 (1993).
[4913] Robles, L.C.; A.J. Aller, *J. Anal. At. Spectrom. 9/8*, 871–879 (1994).
[4914] Robles, L.C.; A.J. Aller, *Talanta 42/11*, 1731–1744 (1995).
[4915] Robért, R.V.D.; S.M. Graham, *Mintek Report M395*, Randburg, (1989).
[4916] Rocks, B.F.; R.A. Sherwood, C. Riley, *Clin. Chem. (Winston-Salem) 28/3*, 440–443 (1982).
[4917] Rocks, B.F.; R.A. Sherwood, L.M. Bayford, C. Riley, *Ann. Clin. Biochem. 19/5*, 338–344 (1982).

[4918] Rocks, B.F.; R.A. Sherwood, Zoe J. Turner, C. Riley, *Ann. Clin. Biochem.* 20/2, 72–76 (1983).
[4919] Rocks, B.F.; R.A. Sherwood, C. Riley, *Ann. Clin. Biochem.* 21, 51–56 (1984).
[4920] Roden, D.R.; D.E. Tallman, *Anal. Chem.* 54/2, 307–309 (1982).
[4921] Rodriguez, L.C.; A.M. García Campaña, C. Jiménez Linares, M. Román Ceba, *Anal. Lett.* 26/6, 1243–1258 (1993).
[4922] Rodriguez, M.P.; A.M. Narizano, V. Demczylo, A. Cid, *At. Spectrosc.* 10/2, 68–70 (1989).
[4923] Rodríguez, M.C.; J.M. Sánchez, H.S. Cubillán, R.A. Romero, *Ciencia* 2/2, 103–112 (1994).
[4924] Rodriguez Pereiro, I.; V.O. Schmitt, J. Szupunar, O.F.X. Donard, R. Łobinski, *Anal. Chem.* 68/23, 4135–4140 (1996).
[4925] Rodríguez-R.; E.; C. Díaz-R., *J. Trace Elem. Med. Biol.* 9/4, 200–209 (1995).
[4926] Roe, D.A.; P.S. Miller, L. Lutwak, *Anal. Biochem.* 15, 313–322 (1966).
[4927] Roe, K.K.; P.N. Froelich, *Anal. Chem.* 56/14, 2724–2726 (1984).
[4928] Roelandts, I.; *Spectrochim. Acta, Part B* 46/9, 1299–1303 (1991).
[4929] Roelandts, I.; *Spectrochim. Acta, Part B* 48/10, 1291–1295 (1993).
[4930] Roelandts, I.; *Spectrochim. Acta, Part B* 51/1, 189–196 (1996).
[4931] Rogers, L.B.; *J. Chem. Educ.* 63/1, 3–6 (1986).
[4932] Rohbock, E.; H.W. Georgii, C. Perseke, L. Kins, in: B. Welz (Editor), *Atomspektrometrische Spurenanalytik*, Verlag Chemie, Weinheim, 1982, 295–304.
[4933] Röhl, R.; *Anal. Chem.* 58/13, 2891–2893 (1986).
[4934] Rohleder, H.A.; F. Dietl, B. Sansoni, *Spectrochim. Acta, Part B* 29, 19–31 (1974).
[4935] Rohling, O.; B. Neidhart, *Fresenius J. Anal. Chem.* 351/1, 33–40 (1995).
[4936] Rohr, U.; L. Meckel, *Fresenius J. Anal. Chem.* 342/4-5, 370–375 (1992).
[4937] Rohr, U.; *Dissertation,* Techn. Hochschule, Darmstadt, 1996.
[4938] Rohrer, C.; W. Wegscheider, *Spectrochim. Acta, Part B* 48/3, 315–329 (1993).
[4939] Rojas, D.; W. Olivares, *Spectrochim. Acta, Part B* 47/3, 387–397 (1992).
[4940] Rojas, D.; *Spectrochim. Acta, Part B* 47/13, 1423–1433 (1992).
[4941] Rojas, D.; W. Olivares, *Spectrochim. Acta, Part B* 50/9, 1011–1030 (1995).
[4942] Rojas, D.; *Spectrochim. Acta, Part B* 50/9, 1031–1044 (1995).
[4943] Rokkjaer, I.; B. Hoyer, N. Jensen, *Talanta* 40/5, 729–735 (1993).
[4944] Romanova, N.S.; T.N. Vladimirovskaya, *Khim. Sel'sk. Khoz.* (7), 74 (1986).
[4945] Romanova, N.S.; T.A. Gatina, *Khim. Sel'sk. Khoz. (12),* 54 (1989).
[4946] Romanova, N.S.; T.A. Gatina, *Khim. Sel'sk. Khoz. (12),* 57 (1989).
[4947] Rombach, N.; K. Kock, *Fresenius Z. Anal. Chem.* 292, 365–369 (1978).
[4948] Rombach, N.; R. Apel, F. Tschochner, *GIT Fachz. Lab.* 24/12, 1165–1168 (1980).
[4949] Romero, C.D.; *Alimentaria (Madrid)* 29/230, 29–31 (1992).
[4950] Rondón, C.E.; M. Burguera, J.L. Burguera, M.R. Brunetto, P. Carrero, *J. Trace Elem. Med. Biol.* 9/1, 49–54 (1995).
[4951] Rooney, R.C.; *Analyst (London)* 101, 749–752 (1976).
[4952] Roos, J.T.H.; *Spectrochim. Acta, Part B* 25, 539–544 (1970).
[4953] Roos, J.T.H.; W.J. Price, *Spectrochim. Acta, Part B* 26, 279–284 (1971).
[4954] Roos, J.T.H.; *Spectrochim. Acta, Part B* 26, 285–290 (1971).
[4955] Roos, J.T.H.; *Spectrochim. Acta, Part B* 27/11, 473–478 (1972).
[4956] Roos, J.T.H.; *Spectrochim. Acta, Part B* 28, 407–415 (1973).
[4957] Rorsman, P.; P.-O. Berggren, *Anal. Chim. Acta* 140, 325–329 (1982).
[4958] Rosain, R.M.; C.M. Wai, *Anal. Chim. Acta* 65/2, 279–284 (1973).
[4959] Rosales, A.T.; *At. Absorpt. Newsletter* 15/2, 51–52 (1976).
[4960] Rose, G.A.; E.G. Wilden, *Analyst (London)* 98, 243–245 (1973).
[4961] Rose, S.A.; D.F. Boltz, *Anal. Chim. Acta* 44, 239 (1969).
[4962] Rösick, U.; A. Parlow, G. Fürstenau, in: B. Welz (Editor), *Fortschritte in der atom-spektrometrischen Spurenanalytik*, VCH Verlagsgesellschaft, Weinheim, 1986, 225–239.
[4963] Rösick, U.; S. Recknagel, P. Brätter, in: P. Brätter, H.-J. Gramm (Editors), *Mineralstoffe und Spurenelemente in der Ernährung des Menschen*, Blackwell Wissenschaft, Berlin, 1991, 104–117.
[4964] Roso, E.; N. Belloni, M. Malla, M. Alvarez, *Bol. Soc. Quim. Peru* 56, 192 (1990).
[4965] Rosopulo, A.; K.H. Grobecker, U. Kurfürst, *Fresenius Z. Anal. Chem.* 319, 540–546 (1984).
[4966] Rosopulo, A.; W. Kreuzer, in: B. Welz (Editor), *Fortschritte in der atomspektrometrischen Spurenanalytik*, VCH Verlagsgesellschaft, Weinheim, 1986, 455–463.
[4967] Rosopulo, A.; E. Hornung, C. Busche, G. Küllmer, *LaborPraxis 11*, 1436–1444 (1987).
[4968] Rossi, G.; N. Omenetto, *Appl. Spectrosc.* 21, 329–331 (1967).

[4969] Rossi, G.; N. Omenetto, G. Pigozzi, R. Vivian, U. Mattiuz, F.Mousty, G.Crabi, *At. Spectrosc. 4/4*, 113–117 (1983).
[4970] Rotruck, J.T.; A.E. Pope, H.E. Ganther, A.B. Swanson, D.G. Hafeman, W.G. Hoekstra, *Science 179*, 588 (1973).
[4971] Routh, M.W.; *Appl. Spectrosc. 36/5*, 585–587 (1982).
[4972] Rowan, C.A.; O.T. Zajicek, E.J. Calabrese, *Anal. Chem. 54/1*, 149–151 (1982).
[4973] Rowston, W.B.; J.M. Ottaway, *Anal. Lett. 3*, 411–417 (1970).
[4974] Rowston, W.B.; J.M. Ottaway, *Analyst (London) 104*, 645–659 (1979).
[4975] Roy, N.K.; A.K. Das, *Talanta 35/5*, 406–408 (1988).
[4976] Roy, N.K.; A.K. Das, *J. Indian Chem. Soc. 66/5*, 353–355 (1989).
[4977] Roy, N.K.; A.K. Das, *J. Indian Chem. Soc. 67*, 976 (1990).
[4978] Royset, O.; Y. Thomassen, *Anal. Chim. Acta 188*, 247–255 (1986).
[4979] Rózanska, B.; E. Lachowicz, *Talanta 33/12*, 1027–1029 (1986).
[4980] Rózanska, B.; *Analyst (London) 120/2*, 407–411 (1995).
[4981] Ruana, J.F.; J.I. Candelas, J.M. Gonzalez, *An. Quim., Ser. B 76*, 135 (1980).
[4982] Rubeška, I.; V. Svoboda, *Anal. Chim. Acta 32*, 253–261 (1965).
[4983] Rubeška, I.; J. Štupar, *At. Absorpt. Newsletter 5/4*, 69–72 (1966).
[4984] Rubeška, I.; B. Moldan, *Anal. Chim. Acta 37*, 421–428 (1967).
[4985] Rubeška, I.; B. Moldan, *Appl. Opt. 7/7*, 1341–1344 (1968).
[4986] Rubeška, I.; *Anal. Chim. Acta 40*, 187–194 (1968).
[4987] Rubeška, I.; M. Miksovsky, *At. Absorpt. Newsletter 11/3*, 57–59 (1972).
[4988] Rubeška, I.; *Spectrochim. Acta, Part B 29*, 263–268 (1974).
[4989] Rubeška, I.; *Can. J. Spectrosc. 20/6*, 156–161 (1975).
[4990] Rubeška, I.; *Chem. Anal. (Warsaw) 22*, 403–411 (1977).
[4991] Rubeška, I.; J. Musil, *Prog. Anal. Spectrosc. 2*, 309–353 (1979).
[4992] Rubeška, I.; B. Ebarvia, E. Macalalad, D. Ravis, N. Roque, *Analyst (London) 112*, 27–29 (1987).
[4993] Rubeška, I.; in: T.S. West, H.W. Nürnberg, IUPAC, *The Determination of Trace Metals in Natural Waters*, Blackwell Scientific Publ., Oxford, 1988, 91–104.
[4994] Rubi, E.; M.S. Jiménez, F. Bauzá de Mirabó, R. Forteza, V. Cerdà, *Talanta 44*, 553–562 (1997).
[4995] Rubio, R.; A. Sahuquillo, G. Rauret, Ph. Quevauviller, *Int. J. Environ. Anal. Chem. 47*, 99–128 (1992).
[4996] Rubio, R.; A. Sahuquillo, G. Rauret, L.G. Beltran, Ph. Quevauviller, *Anal. Chim. Acta 283/1*, 207–212 (1993).
[4997] Rubio, R.; A.M. Ure, *Int. J. Environ. Anal. Chem. 51/1-4*, 205–217 (1993).
[4998] Rubio, R.; A. Padró, G. Rauret, *Fresenius J. Anal. Chem. 351/2-3*, 331–333 (1995).
[4999] Rubio, R.; J. Albertí, A. Padró, G. Rauret, *Trends Anal. Chem. (Pers. Ed.) 14/6*, 274–279 (1995).
[5000] Ruck, A. de; C. Vandecasteele, R. Dams, *Mikrochim. Acta (Vienna) 1987/II*, 187–193 (1988).
[5001] Rudnevskii, A.N.; V.N. Shishov, D.E. Maksimov, *Zh. Prikl. Spektrosk. 46*, 481–483 (1987).
[5002] Rüdorff, W.; *Adv. Inorg. Chem. Radiochem. 1*, 223–266 (1959).
[5003] Rühl, W.J.; *Fresenius Z. Anal. Chem. 322/7*, 710–712 (1985).
[5004] Ruig, W.G. de; *J. Assoc. Off. Anal. Chem. 69/6*, 1009–1013 (1986).
[5005] Ruiz, A.I.; A. Canals, V. Hernandis, *J. Anal. At. Spectrom. 8/1*, 109–113 (1993).
[5006] Ruiz, E.; F. Romero, G. Besga, *Toxicol. Environm. Chem. 33*, 1 (1991).
[5007] Rükgauer, M.; A.Zeyfang, K. Uhland, J.D. Kruse-Jarres, *J. Trace Elem. Med. Biol. 9/3*, 130–135 (1995).
[5008] Rump, H.H.; H. Krist, *Laboratory Manual for the Examination of Water, Waste Water and Soil*, VCH Publishers, New York, 1988.
[5009] Runnels, J.H ; R.N. Merryfield, H.B. Fisher, *Anal. Chem. 47/8*, 1258–1263 (1975).
[5010] Russell, B.J.; J.P. Shelton, A. Walsh, *Spectrochim. Acta 8*, 317–328 (1957).
[5011] Russell, B.J.; A. Walsh, *Spectrochim. Acta 15*, 883–885 (1959).
[5012] Russeva, E.; I. Havezov, B.Ya. Spivakov, V.M. Shkinev, *Fresenius Z. Anal. Chem. 315/6*, 499–501 (1983).
[5013] Russeva, E.; I. Havezov, *Bulgarian Acad. Sci./Commun. Departm. Chem. 19/3*, 422–425 (1986).
[5014] Russeva, E.; I. Havezov, A. Detcheva, *Fresenius J. Anal. Chem. 347/8-9*, 320–323 (1993).
[5015] Ruzicka, J.; E.H. Hansen, *Flow Injection Analysis*, J. Wiley & Sons, New York, 1981.
[5016] Ruzicka, J.; E.H. Hansen, *Flow Injection Analysis, 2nd Edition*, J. Wiley & Sons, New York, 1988.

[5017] Ruzicka, J.; A. Arndal, *Anal. Chim. Acta 216/1-2*, 243–255 (1989).
[5018] Ryan, N.; J.D. Glennon, D. Muller, *Anal. Chim. Acta 283/1*, 344–349 (1993).
[5019] Ryan, P.; F. O'Donoghue-Ryan, *Lab. Pract. 38/6*, 29–33 (1989).
[5020] Rygh, G.; K.W. Jackson, *J. Anal. At. Spectrom. 2/4*, 397–400 (1987).
[5021] Ryssen, J.B.J. Van; J.T. Deagan, M.A. Beilstein, P.D. Whanger, *J. Agri. Food Chem. 37*, 1358 (1989).
[5022] Ryzhinsky, M.V.; *Fresenius J. Anal. Chem. 345/2-4*, 166–168 (1993).
[5023] Saab, W.; A. Sarda, G. Cote, *Anal. Chim. Acta 248/1*, 235–239 (1991).
[5024] Saatçi, A.O.; M. Irkeç, H. Özgünes, *Ophthalmic Res. 23*, 31–32 (1991).
[5025] Saati, M.R.; *Adv. Lab. Autom. Rob. 6*, 521 (1990).
[5026] Saba, C.S.; K.J. Eisentraut, *Anal. Chem. 49/3*, 454–457 (1977).
[5027] Saba, C.S.; K.J. Eisentraut, *Anal. Chem. 51/12*, 1927–1930 (1979).
[5028] Saba, C.S.; W.E. Rhine, K.J. Eisentraut, *Anal. Chem. 53/7*, 1099–1103 (1981).
[5029] Saba, C.S.; W.E. Rhine, K.J. Eisentraut, *Appl. Spectrosc. 39/4*, 689–693 (1985).
[5030] Sabbatini, L.; G. Tessari, *Ann. Chim. (Rom) 74*, 779–793 (1984).
[5031] Sachde, Z.G.; J. Bundt, *Dtsche Lebensm. Rundsch. 85*, 108 (1989).
[5032] Sachdev, S.L.; J.W. Robinson, P.W. West, *Anal. Chim. Acta 37*, 12–19 (1967).
[5033] Sachsenberg, S.; Th. Klenke, W.E. Krumbein, E. Zeeck, *Fresenius J. Anal. Chem. 342/1-2*, 163–166 (1992).
[5034] Sachsenberg, S.; T. Klenke, W.E. Krumbein, H.J. Schellnhuber, E. Zeeck, *Anal. Chim. Acta 279*, 241–251 (1993).
[5035] Sadin, Y.; P. Deldime, *Anal. Lett. 12/B6*, 563–572 (1979).
[5036] Sadiq, M.; *Toxic Metals in Marine Environments*, Marcel Dekker, New York, 1992.
[5037] Saeed, K.; Y. Thomassen, F.J. Langmyhr, *Anal. Chim. Acta 110*, 285–289 (1979).
[5038] Saeed, K.; Y. Thomassen, *Anal. Chim. Acta 130/2*, 281–287 (1981).
[5039] Saeed, K.; Y. Thomassen, *At. Spectrosc. 4/4*, 163 (1983).
[5040] Saeki, S.; M. Kubota, T. Asami, *Water, Air, Soil Pollut. 83/3-4*, 253 (1995).
[5041] Sager, M.; *Mikrochim. Acta (Vienna) 3/3-4*, 129–139 (1986).
[5042] Sager, M.; *Mikrochim. Acta (Vienna) 106*, 241–251 (1992).
[5043] Sager, M.; in: M. Stoeppler (Editor), *Hazardous Metals in the Environment*, Techn. Instr. Anal. Chem. Vol 12, Elsevier, Amsterdam, 1992.
[5044] Sager, M.; *Fuel 72/9*, 1327–1330 (1993).
[5045] Sah, R.N.; R.O. Miller, *Anal. Chem. 64/2*, 230–233 (1992).
[5046] Sahayam, A.C.; S. Gangadharan, *At. Spectrosc. 14/3*, 83–84 (1993).
[5047] Sahayam, A.C.; S. Natarajan, S. Gangadharan, *Fresenius J. Anal. Chem. 346/10*, 961–963 (1993).
[5048] Sahayam, A.C.; A.K. Tyagi, S. Gangadharan, *Fresenius J. Anal. Chem. 347/10*, 461–462 (1993).
[5049] Sahuquillo, A.; R. Rubio, G. Rauret, in: Ph. Quevauviller, E.A. Maier, B. Griepink (Editors), *Quality Assurance for Environmental Analysis*, Elsevier, Amsterdam, 1995, 39–62.
[5050] Saito, K.; H. Freiser, *Anal. Sci. 5*, 583–586 (1989).
[5051] Saito, L.; H. Oshima, N. Kawamura, M. Yamada, *J. Assoc. Off. Anal. Chem. 71/4*, 829–832 (1988).
[5052] Sakae, S.; N. Teramae, T. Sawada, *Bunseki Kagaku 38*, 429 (1989).
[5053] Sakamoto, H.; T. Tomiyasu, N. Yonehara, *Anal. Sci. 8/1*, 35–39 (1992).
[5054] Salbu, B.; *Mikrochim. Acta (Vienna) 2*, 29–37 (1991).
[5055] Salim, R.; B.G. Cooksey, *J. Electroanal. Chem. 106*, 251 (1980).
[5056] Salim, R.; *J. Environ. Sci. Health, Part A 22A(2)*, 125 (1987).
[5057] Salinas, F.; J.G. March, *Int. J. Environ. Anal. Chem. 18/3*, 209–214 (1984).
[5058] Salinas, F.; M.C. Mahedero, M. Jimenez Arrabal, *Quim. Anal. (Barcelona) 11/1*, 11–16 (1992).
[5059] Salles, L.C.; A.J. Curtius, *Mikrochim. Acta (Vienna) 2/1-2*, 125–130 (1983).
[5060] Salmela, S.; E. Vuori, *Talanta 26*, 175–176 (1979).
[5061] Salmela, S.; E. Vuori, J.O. Kilpiö, *Anal. Chim. Acta 125*, 131–137 (1981).
[5062] Salmela, S.; E. Vuori, *At. Spectrosc. 5/4*, 146–149 (1984).
[5063] Salmon, S.G.; J.A. Holcombe, *Anal. Chem. 51/6*, 648–650 (1979).
[5064] Salmon, S.G.; R.H. Davis, J.A. Holcombe, *Anal. Chem. 53/2*, 324–330 (1981).
[5065] Salmon, S.G.; J.A. Holcombe, *Anal. Chem. 54/4*, 630–634 (1982).
[5066] Salomons, W.; U. Förstner, *Environ. Technol. Lett. 1*, 506–517 (1980).
[5067] Salonen, J.T.; G. Alfthan, J. Pikkarainen, J.K. Huttunen, P. Paska, *Lancet 2*, 175–179 (1982).

[5068] Salvador, A.; M. de la Guardia-Cirugeda, V. Berenguer-Navarro, *Talanta 30/12*, 986–988 (1983).
[5069] Salvador, A.; M. de la Guardia-Cirugeda, A. Mauri, *At. Spectrosc. 9/6*, 195–198 (1988).
[5070] Salvador, A.; R. Martínez-Avila, V. Carbonell, M. de la Guardia, *Fresenius J. Anal. Chem. 347/8-9*, 356–360 (1993).
[5071] Salvato, N.; G. Banfi, F. Taccani, A. Rottoli, G. Banderari, in: C. Minoia, S. Caroli (Editors), *Appl. Zeeman Graphite Furnace At. Abs. Spectrom. Chem. Lab. Toxicology*, Pergamon, Oxford, 1992, 305–323.
[5072] Samiullah, Y.; *J. Geochem. Explor. 23*, 193–202 (1985).
[5073] Sampson, B.; *J. Anal. At. Spectrom. 3/3*, 465–469 (1988).
[5074] Sampson, B.; *Anal. Proc. (London) 25/7*, 229–230 (1988).
[5075] Sampson, B.; *J. Anal. At. Spectrom. 6/2*, 115–118 (1991).
[5076] Sánchez, J.; E. Millán, *Quim. Anal. (Barcelona) 11/1*, 3–10 (1992).
[5077] Sánchez, J.M.; H.S. Cubillan, M. Hernandez, B.I. Semprun, V.A. Granadillo, *Quim. Anal. (Barcelona) 15/2*, 178 (1996).
[5078] Sanchez, M.; D. Gazquez, P. Garcia, *Talanta 38/7*, 747–752 (1991).
[5079] Sand, J.R.; J.H. Liu, C.O. Huber, *Anal. Chim. Acta 87*, 79–90 (1976).
[5080] Sandoval, L.; J.-C. Herraez, G. Steadman, K.I. Mahan, *Mikrochim. Acta (Vienna) 108/1-2*, 19–27 (1992).
[5081] Sanemasa, L.; *Bull. Chem. Soc. Jpn. 48*, 1795 (1975).
[5082] Sanemasa, L.; T. Deguchi, K. Urata, J. Tomooka, H. Nagai, *Anal. Chim. Acta 87*, 479–481 (1976).
[5083] Sanemasa, L.; T. Deguchi, K. Urata, J. Tomooka, H. Nagai, *Anal. Chim. Acta 94*, 421–427 (1977).
[5084] Sangsila, S.; G. Labinaz, J.S. Poland, G.W. van Loon, *J. Chem. Educ. 66/4*, 351–353 (1989).
[5085] Sansonetti, J.E.; J. Reader, C.J. Sansonetti, N. Acquista, *J. Res. Natl. Inst. Stand. Technol. 97/1*, 1–211 (1992).
[5086] Sansoni, B.; R.K. Iyer, R. Kurth, *Fresenius Z. Anal. Chem. 306*, 212–232 (1981).
[5087] Santelli, R.E.; M. Gallego, M. Valcárcel, *Anal. Chem. 61/13*, 1427–1430 (1989).
[5088] Santelli, R.E.; M. Gallego, M. Valcárcel, *J. Anal. At. Spectrom. 4/6*, 547–550 (1989).
[5089] Santelli, R.E.; M. Gallego, M. Valcárcel, *Talanta 38/11*, 1241–1245 (1991).
[5090] Sanz, J.; M.T. Martínez, J. Galbán, J.R. Castillo-Suarez, *At. Spectrosc. 9/2*, 63–67 (1988).
[5091] Sanz Alaejos, M; C. Díaz Romero, *Clin. Chem. (Winston-Salem) 39/10*, 2040–2052 (1993).
[5092] Sanz-Medel, A.; R. Rodriguez Roza, C. Perez-Conde, *Analyst (London) 108*, 204–212 (1983).
[5093] Sanz-Medel, A.; M.R. Fernández de la Campa, M.C. Valdés-Hevia y Temprano, B. Aizpún Fernandez, Y.M. Liu, *Talanta 40/11*, 1759–1768 (1993).
[5094] Sanz-Medel, A.; M.C. Valdés-Hevia y Temprano, N. Bordel García, M.R. Fernández de la Campa, Anal. Chem. 67/13, 2216–2223 (1995).
[5095] Sanz-Medel, A.; M.C. Valdés-Hevia y Temprano, N. Bordel García, M.R. Fernández de la Campa, Anal. Proc. (London). 32/2, 49–52 (1995).
[5096] Sanzolone, R.F.; T.T. Chao, *Geostand. Newsl. 11/1*, 81–85 (1987).
[5097] Saraswati, R.; R.L. Watters, *Talanta 41/10*, 1785–1790 (1994).
[5098] Saraswati, R.; N.R. Desikan, T.H. Rao, *J. Anal. At. Spectrom. 9/11*, 1289–1291 (1994).
[5099] Saraswati, R.; T.W. Vetter, R.L. Watters, *Mikrochim. Acta (Vienna) 118*, 163–175 (1995).
[5100] Sargent, M.; G. MacKay, *Guidelines for Achieving Quality in Trace Analysis*, Royal Society of Chemistry, Cambridge, 1995.
[5101] Sarkar, A.K.; D.C. Parashar, *At. Spectrosc. 12/1*, 19–20 (1991).
[5102] Sarkar, B.; in: H.G. Seiler, A. Sigel, H. Sigel (Editors), *Handbook on Metals in Clinical and Analytical Chemistry*, Marcel Dekker, New York, 1994, 339–347.
[5103] Sarudi, I.; *Fresenius Z. Anal. Chem. 303/3*, 197–199 (1980).
[5104] Sarx, B.; K. Bächmann, *Fresenius Z. Anal. Chem. 316/6*, 621–626 (1983).
[5105] Sarx, B.; K. Bächmann, in: B. Welz (Editor), *Fortschritte in der atomspektrometrischen Spurenanalytik*, VCH Verlagsgesellschaft, Weinheim, 1986, 619–626.
[5106] Sarzanini, C.; G. Sacchero, M. Aceto, O. Abollino, E. Mentasti, *J. Chromatogr. 626/1*, 151–159 (1992).
[5107] Sarzanini, C.; G. Sacchero, M. Aceto, O. Abollino, *Anal. Chim. Acta 284/3*, 661–667 (1994).
[5108] Sastri, V.S.; C.L. Chakrabarti, D.E. Willis, *Can. J. Chem. 47*, 587–596 (1969).
[5109] Sastri, V.S.; C.L. Chakrabarti, D.E. Willis, *Talanta 16*, 1093 (1969).
[5110] Satake, M.; K. Ishida, B.K. Puri, S. Usami, *Anal. Chem. 58/12*, 2502–2504 (1986).
[5111] Satake, M.; G. Kano, B.K. Puri, S. Usami, *Anal. Chim. Acta 199*, 209–214 (1987).
[5112] Satake, M.; N. Sugita, M. Katyal, *Ann. Chim. (Rome) 80/7-8*, 385–391 (1990).

[5113]   Satake, M.; T. Nagahiro, B.K. Puri, *Analyst (London) 118/1*, 85–88 (1993).
[5114]   Sato, C.; *Hokkaidoritsu Eisei Kenkyushoho (38)*, 66 (1988).
[5115]   Satoh, E.; Y. Yamamoto, T. Takyu, *Anal. Sci. 7/Suppl*, 1645–1648 (1991).
[5116]   Sattur, T.W.; *At. Absorpt. Newsletter 5/3*, 37–41 (1966).
[5117]   Sauerhoff, S.T.; Z.A. Grosser, G.R. Carnrick, *At. Spectrosc. 17/6*, 225–228 (1996).
[5118]   Sauls, F.C.; W.F. Berry, J.J. Natishan, D.P. Condo, *At. Spectrosc. 5/6*, 226–227 (1984).
[5119]   Savory, J.; J.W. Wiggins, M.G. Heintges, *Am. J. Clin. Pathol. 51*, 720–726 (1969).
[5120]   Savory, M.G.; J. Savory, in: F.W. Sunderman (Editor), *Manual of procedures for the seminar on laboratory diagnosis of disorders of the fetus, newborn and early childhood*, Inst. Clin. Sci., Philadelphia, 1990, 65–77.
[5121]   Savory, J.; M.R. Wills, *Clin. Chem. (Winston-Salem) 38*, 1565–1573 (1992).
[5122]   Savory, J.; R.L. Bertholf, S. Brown, M.R. Wills, in: R.F.M. Herber, M. Stoeppler, *Trace Element Analysis in Biological Specimens*, Elsevier, Amsterdam - London - New York - Tokyo, 1994, 273–289.
[5123]   Sawada, K.; S. Ohgake, M. Kobayashi, T. Suzuki, *Bunseki Kagaku 42/11*, 741–745 (1993).
[5124]   Saxena, R.; A.K. Singh, S.S. Sambi, *Anal. Chim. Acta 295/1-2*, 199–204 (1994).
[5125]   Schäfer, K.; D. Behne, in: B. Welz (Editor), *4. Colloquium Atomspektrometrische Spurenanalytik*, Bodenseewerk Perkin-Elmer, Überlingen, 1987, 375–384.
[5126]   Schäffer, U.; V. Krivan, *Spectrochim. Acta, Part B 51/9-10*, 1211–1222 (1996).
[5127]   Schaller, K.-H.; G. Manke, H.J. Raithel, G. Bühlmeyer, M. Schmidt, H. Valention, *Int. Arch. Occup. Environ. Health 47*, 223 (1980).
[5128]   Schaller, K.-H.; in: J. Angerer, K.H. Schaller (Editors), *Analyses of Hazardous Substances in Biological Materials, Vol. 3*, VCH Verlagsgesellschaft, Weinheim, 1991, 81–92.
[5129]   Schaller, K.-H.; S. Letzel, J. Angerer, in: H.G. Seiler, A. Sigel, H. Sigel (Editors), *Handbook on Metals in Clinical and Analytical Chemistry*, Marcel Dekker, New York, 1994, 217–226.
[5130]   Schaller, K.-H.; H.-J. Raithel, J. Angerer, in: H.G. Seiler, A. Sigel, H. Sigel (Editors), *Handbook on Metals in Clinical and Analytical Chemistry*, Marcel Dekker, New York, 1994, 505–518.
[5131]   Schallis, J.E.; H.L. Kahn, *At. Absorpt. Newsletter 7/4*, 75–79 (1968).
[5132]   Schauenburg, H.; P. Weigert, *Fresenius J. Anal. Chem. 342/12*, 950–956 (1992).
[5133]   Schaumlöffel, D.; B. Neidhart, *Fresenius J. Anal. Chem. 354/7-8*, 866–869 (1996).
[5134]   Schebek, L.; M.O. Andreae, H.J. Tobschall, *Environ. Sci. Technol. 25/5*, 871–878 (1991).
[5135]   Schebek, L.; M.O. Andreae, H.J. Tobschall, *Int. J. Environ. Anal. Chem. 45/4*, 257–273 (1991).
[5136]   Schelenz, R.; R.M. Parr, E. Zeiller, S.A. Clements, *Fresenius Z. Anal. Chem. 333/1*, 33–34 (1989).
[5137]   Schermaier, A.J.; L.H. O'Connor, K.H. Pearson, *Clin. Chim. Acta 152*, 123–134 (1985).
[5138]   Scheuermann, H.; H. Hartkamp, *Fresenius Z. Anal. Chem. 315/5*, 430–433 (1983).
[5139]   Scheuhammer, A.M.; D. Bond, *Biol. Trace Elem. Res. 31*, 119–139 (1991).
[5140]   Schierling, P.; K.H. Schaller, *Arbeitsmed., Sozialmed., Präventivmed. 3/3*, 57–61 (1981).
[5141]   Schindler, E.; *Dtsche Lebensm. Rundsch. 79/10*, 334–337 (1983).
[5142]   Schindler, E.; *Dtsche Lebensm. Rundsch. 81/8*, 250–252 (1985).
[5143]   Schinkel, H.; *Fresenius Z. Anal. Chem. 317/1*, 10–26 (1984).
[5144]   Schintu, M.; T. Kauri, A. Kudo, *Water Research 23/6*, 699–704 (1989).
[5145]   Schintu, M.; F. Jean-Caurant, J.-C. Amiard, *Ecotoxicol. Environ. Safety 24*, 95–101 (1992).
[5146]   Schladot, J.-D.; F. Backhaus, in: M. Stoeppler (Editor), *Probennahme und Aufschluß*, Springer, Berlin - Heidelberg - New York, 1994, 55–70.
[5147]   Schleich, C.; G. Henze, *Fresenius J. Anal. Chem. 338/2*, 140–144 (1990).
[5148]   Schlemmer, G.; B. Welz, *Perkin-Elmer, Application Study 691*, 1984.
[5149]   Schlemmer, G.; B. Welz, *Fresenius Z. Anal. Chem. 320/7*, 648–649 (1985).
[5150]   Schlemmer, G.; B. Welz, *Spectrochim. Acta, Part B 41/11*, 1157–1165 (1986).
[5151]   Schlemmer, G.; B. Welz, *Fresenius Z. Anal. Chem. 323/7*, 703–709 (1986).
[5152]   Schlemmer, G.; B. Welz, *Fresenius Z. Anal. Chem. 328/4-5*, 405–409 (1987).
[5153]   Schlemmer, G.; *GIT Fachz. Lab. 33*, 561–568 (1989).
[5154]   Schlemmer, G.; W. Schrader, H. Schulze, *LaborPraxis 14/10*, 822–828 (1990).
[5155]   Schlemmer, G.; M. Feuerstein, in: K. Dittrich, B. Welz (Editors), *CANAS'93, Colloq. Analytische Atomspektroskopie*, Universität Leipzig/UFZ Leipzig-Halle, Leipzig, 1993, 431–441.
[5156]   Schmid, W.; V. Krivan, *Anal. Chem. 57/1*, 30–34 (1985).
[5157]   Schmidt, D.; *Water, Air, Soil Pollut. 62*, 43–55 (1992).
[5158]   Schmidt, F.J.; J.L.Royer, *Anal. Lett. 6/1*, 17–23 (1973).

[5159] Schmidt, K.P.; H. Falk, *Spectrochim. Acta, Part B 42/3*, 431–443 (1987).
[5160] Schmidt, K.P.; H. Becker-Roß, S. Florek, *Spectrochim. Acta, Part B 45/11*, 1203–1210 (1990).
[5161] Schmidt, P.; P. Walzel, *Chem.-Ing.-Tech. 52/4*, 304–311 (1980).
[5162] Schmidt, P.; P. Walzel, *Physik in unserer Zeit 15/4*, 113–120 (1984).
[5163] Schmidt, W.F.; F. Dietl, *Fresenius Z. Anal. Chem. 295*, 110–115 (1979).
[5164] Schmidt, W.F.; F. Dietl, *Fresenius Z. Anal. Chem. 315/8*, 687–690 (1983).
[5165] Schmiedel, G.; E. Mainka, H.J. Ache, *Fresenius Z. Anal. Chem. 335/2*, 195–199 (1989).
[5166] Schmitt, Y.; *J. Trace Elem. Electrol. Health Dis. 1/2*, 107–114 (1987).
[5167] Schmitt, Y.; J.D. Kruse-Jarres, in: B. Welz (Editor), *5. Colloquium Atomspektrometrische Spurenanalytik*, Bodenseewerk Perkin-Elmer, Überlingen, 1989, 823–832.
[5168] Schmitt, Y.; A. Ott, J.D. Kruse-Jarres, in: B. Welz (Editor), *6. Colloquium Atomspektrometrische Spurenanalytik*, Bodenseewerk Perkin-Elmer, Überlingen, 1991, 821–828.
[5169] Schmitt, Y.; J.D. Kruse-Jarres, in: P. Brätter, H.J. Gramm (Editors), *Mineralstoffe und Spurenelemente in der Ernährung der Menschen*, Blackwell Wissenschaft, Berlin, 1992, 52–56.
[5170] Schmitt, Y.; V. Moser, J.D. Kruse-Jarres, *Fresenius J. Anal. Chem. 346*, 852–858 (1993).
[5171] Schnarz, K.; D.B. Milne, E. Vinyard, *Biochem. Biophys. Res. Commun. 40*, 22 (1970).
[5172] Schneider, C.A.; H. Schulze, J. Baasner, S. McIntosh, C. Hanna, *Am. Lab. (Fairfield, CT) 26/Feb*, 18 (1994).
[5173] Schneider, C.A.; H. Schulze, J. Baasner, S. McIntosh, C. Hanna, *Int. Lab. 24/4*, 13–15 (1994).
[5174] Schneider, G.; V. Krivan, *Spectrochim. Acta, Part B 50/13*, 1557–1571 (1995).
[5175] Schnepfe, M.M.; F.S. Grimaldi, *Talanta 16*, 591–595 (1969).
[5176] Schnitzer, G.; C. Pellerin, C. Clouet, *Lab. Pract. 37/1*, 63–65 (1988).
[5177] Schnitzer, G.; A. Soubelet, C. Testu, C. Chafey, *Mikrochim. Acta (Vienna) 119/3-4*, 199–209 (1995).
[5178] Schothorst, R.C.; H.M.A. Géron, D. Spitsbergen, R.F.M. Herber, *Fresenius Z. Anal. Chem. 328/4-5*, 393–395 (1987).
[5179] Schottel, V.; A. Mandal, D. Clark, S. Simon, R.W. Hedges, *Nature (London) 251*, 335–337 (1974).
[5180] Schram, J.; *Fresenius J. Anal. Chem. 343*, 727–732 (1992).
[5181] Schramel, P.; *Anal. Chim. Acta 72*, 414–418 (1974).
[5182] Schramel, P.; S. Hasse, J. Ovcar-Pavlu, *Biol. Trace Elem. Res. 15*, 111–138 (1988).
[5183] Schramel, P.; G. Lill, S. Hasse, B.J. Klose,. *Biol. Trace Elem Res. 16*, 67–75 (1988).
[5184] Schramel, P.; S. Hasse, *Fresenius J. Anal. Chem. 346*, 794–799 (1993).
[5185] Schramel, P.; *J. Radioanal. Nucl. Chem. 168/1*, 215–221 (1993).
[5186] Schrauzer, G.N.; D.A. White, C.J. Schneider, *Bioinorg. Chem. 7*, 23–24 (1977).
[5187] Schrauzer, G.N.; *Am. J. Clin. Nutr. 33*, 1892–1894 (1980).
[5188] Schrenk, W.G.; D.A. Lehman, L. Neufeld, *Appl. Spectrosc. 20*, 389–391 (1966).
[5189] Schrenk, W.G.; C.E. Meloan, C.W. Frank, *Spectrosc. Lett. 1*, 237 (1968).
[5190] Schroeder, W.H.; M.C. Hamilton, S.R. Stobart, *Rev. Anal. Chem. 8/3*, 179–209 (1985).
[5191] Schroeder, W.H.; G. Keeler, H. Kock, P. Roussel, D. Schneeberger, F. Schädlich, *Water, Air, Soil Pollut. 80*, 611 (1995).
[5192] Schrön, W.; B. Dressler, in: B. Welz (Editor), *CANAS'95 Colloquium Analytische Atomspektroskopie*, Bodenseewerk Perkin-Elmer, Überlingen, 1996, 121–130.
[5193] Schubert-Jacobs, M.; T. Guo, B. Welz, in: B. Welz (Editor), *6. Colloquium Atomspektrometrische Spurenanalytik*, Bodenseewerk Perkin-Elmer, Überlingen, 1991, 337–348.
[5194] Schuhmacher, M.; J.L. Domingo, J.M. Llobet, J. Corbella, *Bull. Environ. Contam. Toxicol. 50/4*, 514–521 (1993).
[5195] Schulz, W.; K. Böttger, B. Meder, E. Grallath, *J. Clin. Chem. Clin. Biochem. 19/10*, 1063–1066 (1981).
[5196] Schulz, W.; L. Kotz, *CLB, Chem. Labor Betr. 33/11*, 483–493 (1982).
[5197] Schulze, G.; O. Elsholz, R. Hielscher, A. Rauth, S. Recknagel, A. Thiele, *Fresenius Z. Anal. Chem. 334/1*, 9–12 (1989).
[5198] Schulze, G.; R. Mertens-Menzel, *Fresenius J. Anal. Chem. 346/6-9*, 663–666 (1993).
[5199] Schulze, G.; C. Lehmann, *Anal. Chim. Acta 288/3*, 215–220 (1994).
[5200] Schulze, H.; J. Baasner, *CAV, Chem. Anl. Verf. 27*, 92 (1994).
[5201] Schumann, H.; R. Bärwald, M. Ernst, in: B. Welz (Editor), *6. Colloquium Atomspektrometrische Spurenanalytik*, Bodenseewerk Perkin-Elmer, Überlingen, 1991, 681–689.
[5202] Schuster, M.; *Fresenius J. Anal. Chem. 342/10*, 791–794 (1992).

[5203]   Schuster, M.; M. Schwarzer, in: B. Welz (Editor), *CANAS'95 Colloquium Analytische Atomspektroskopie*, Bodenseewerk Perkin-Elmer, Überlingen, 1996, 289–297.

[5204]   Schuster, M.; M. Schwarzer, *Anal. Chim. Acta 328/1*, 1–11 (1996).

[5205]   Schuster, M.; S. Ringmann, R. Gärtner, X. Lin, J. Dahmen, *Fresenius J. Anal. Chem.357/3*, 258–265 (1997).

[5206]   Schutz, D.F.; K.K. Turekian, *Geochim. Cosmochim. Acta 29*, 259 (1965).

[5207]   Schwab, M.R.; N.H. Hembree, *At. Absorpt. Newsletter 10/1*, 15–16 (1971).

[5208]   Schwartz, R.S.; L.T. Hecking, *J. Anal. At. Spectrom. 6/8*, 637–642 (1991).

[5209]   Schwarz, K.; C.M. Foltz, *J. Am. Chem. Soc. 79*, 3292–3293 (1957).

[5210]   Schwediman, L.C.; A.K. Postma, L.F. Coleman, *L.C. 10*, 947–953 (1966).

[5211]   Schwedt, G.; K. Wedepohl, D. Zöltzer, *Fresenius Z. Anal. Chem. 316/4*, 390–396 (1983).

[5212]   Schwedt, G.; G. Jahns, *Fresenius Z. Anal. Chem. 328/1-2*, 85–88 (1987).

[5213]   Schwedt, G.; *LaborPraxis 17/5*, 44–46 (1993).

[5214]   Schwinger, R.; D.H. Antoni, W.G. Guder, *J. Trace Elem. Electrol. Health Dis. 1/2*, 89–98 (1987).

[5215]   Scoble, H.A.; R. Litman, *Anal. Lett. 11B/2*, 183–189 (1978).

[5216]   Scott, R.L.; *At. Absorpt. Newsletter 9/2*, 46–47 (1970).

[5217]   Scudlark, J.R.; T.M. Church, K.M. Conko, S.M. Moore, *Gen. Tech. Rep. NC (North Cent. For. Exp. Stn.) 150*, 57 (1992).

[5218]   Searle, B.; W. Chan, C. Jensen, B. Davidow, *At. Absorpt. Newsletter 8/6*, 126–127 (1969).

[5219]   Sebastiani, E.; K. Ohls, G. Riemer, *Fresenius Z. Anal. Chem. 264*, 105–109 (1973).

[5220]   Šebor, G.; I. Lang, D. Kolihová, O. Weisser, *Analyst (London) 107*, 1350–1355 (1982).

[5221]   Seeling, W.; A. Grünert, K.H. Kienle, R. Opferkuch, M. Swobodnik, *Fresenius Z. Anal. Chem. 299*, 368–374 (1979).

[5222]   Segar, D.A.; *Anal. Lett. 7/1*, 89–95 (1974).

[5223]   Segar, D.A.; A.Y. Cantillo, *Adv. Chem. Ser. 147*, 56 (1975).

[5224]   Segar, D.A.; A.Y. Cantillo, *Anal. Chem. 52/11*, 1766–1767 (1980).

[5225]   Seifert, D.; *Landwirtsch. Forsch. 35/3-4*, 214–219 (1982).

[5226]   Seiler, H.G.; in H.G. Seiler, H. Sigel (Editors), *Handbook on toxicity of inorganic compounds*, Marcel Dekker, New York, 1988, 39–49.

[5227]   Seimiya, T.; H. Kozai, T. Sasaki, *Bull. Chem. Soc. Jpn. 42*, 2797–2800 (1969).

[5228]   Sekine, T.; K. Inaba, T. Morimoto, H. Aikawa, *Bull. Chem. Soc. Jpn. 61*, 1131–1134 (1988).

[5229]   Seligman, P.F.; J.G. Grovhoug, A.O. Valkirs, P.M. Stang, R. Fransham, M.O. Stallard, B. Davidson, R.F. Lee, *Appl. Organomet. Chem. 3*, 31–47 (1989).

[5230]   Seltner, H.; G. Hermann, C. Heppler, in: B. Welz (Editor), *6. Colloquium Atomspektrometrische Spurenanalytik*, Bodenseewerk Perkin-Elmer, Überlingen, 1991, 609–616.

[5231]   Semu, E.; A.R. Selmer-Olsen, B.R. Singh, K. Steenberg, *Fresenius Z. Anal. Chem. 322/4*, 440. (1985).

[5232]   Sen Gupta, J.G.; *Anal. Chim. Acta 58*, 23–37 (1972).

[5233]   Sen Gupta, J.G.; *Miner. Sci. Engng. 5/3*, 207–218 (1973).

[5234]   Sen Gupta, J.G.; *Geostand. Newsl. 1/2*, 149–154 (1977).

[5235]   Sen Gupta, J.G.; *Talanta 28/1*, 31–36 (1981).

[5236]   Sen Gupta, J.G.; *Anal. Chim. Acta 138*, 295–302 (1982).

[5237]   Sen Gupta, J.G.; *Talanta 31/12*, 1053–1056 (1984).

[5238]   Sen Gupta, J.G.; *Talanta 32/1*, 1–6 (1985).

[5239]   Sen Gupta, J.G.; *Talanta 34/4*, 427–431 (1987).

[5240]   Sen Gupta, J.G.; *Talanta 36/6*, 651–656 (1989).

[5241]   Sen Gupta, J.G.; *J. Anal. At. Spectrom. 8/1*, 93–101 (1993).

[5242]   Sen Gupta, J.G.; *Talanta 40/6*, 791–797 (1993).

[5243]   Sen Gupta, J.G.; *Can. J. Appl. Spectroscopy 38/5*, 145–149 (1993).

[5244]   Sentimenti, E.; G. Mazzetto, *At. Spectrosc. 7/6*, 181–182 (1986).

[5245]   Sentimenti, E.; G. Mazzetto, G. Pannocchia, *Anal. Chim. Acta 234*, 425–431 (1990).

[5246]   Sentimenti, E.; G. Mazzetto, in: C. Minoia, S. Caroli (Editors), *Appl. Zeeman Graphite Furnace Atom. Absorption Spectrom. Chem. Lab. Toxicol.*, Pergamon Press, Oxford, 1992, 95–106.

[5247]   Sentimenti, E.; G. Mazzetto, E. Milella, *J. Anal. At. Spectrom. 8/1*, 89–92 (1993).

[5248]   Seow, F.; J. Chrisp, J.R. Lawrence, M.J. Field, *Kidney Int. 37/2*, 818–821 (1990).

[5249]   Septon, J.C.; in: S.J. Haswell (Editor), *Atomic Absorption Spectrometry - Theory, Design and Applications*, Elsevier, Amsterdam - New York, 1991, 159–190.

[5250]   Sesana, G.; A. Baj, in: C. Minoia, S. Caroli (Editors), *Appl. Zeeman Graphite Furnace At. Abs. Spectrom. Chem. Lab. Toxicology*, Pergamon, Oxford, 1992, 459–474.

[5251] Seymour, R.B.; in: I.M. Kolthoff, P.J. Elving (Editors), *Treatise on Analytical Chemistry, Part III*, J. Wiley & Sons, New York, 1976, 341.
[5252] Seymour, R.J.; C.B. Boss, *Anal. Chem. 54/7*, 1037–1042 (1982).
[5253] Shaburova, V.P.; E.D. Ruchkin, *Sib. Khim. Zh. (1)*, 97 (1985).
[5254] Shaburova, V.P.; I.G. Yudelevich, *Zh. Anal. Khim. 43*, 603 (1988).
[5255] Shaburova, V.P.; *Sib. Khim. Zh. (5)*, 30 (1992).
[5256] Shafto, R.G.; *At. Absorpt. Newsletter 3/9*, 115–117 (1964).
[5257] Shah, N.; S.K. Menon, M.N. Desai, Y.K. Agarwal, *Anal. Lett. 22*, 1807–1817 (1989).
[5258] Shah, N.; M.N. Deasi, Y.K. Agrawal, *Int. J. Environ. Anal. Chem. 42/1-4*, 53–60 (1990).
[5259] Shaikh, A.U.; D.E. Tallmann, *Anal. Chim. Acta 98*, 251–259 (1978).
[5260] Shalmi, M.; J. Kibble, J.P. Day, P. Christensen, J.C. Ahterton, *Am. J. Physiol. 267*, F695 (1994).
[5261] Shalmi, M.; *Am. J. Physiol. 268/6,2*, F1230 (1995).
[5262] Shamberger, R.J.; *J. Environm. Pathol. Toxicol. 4*, 305–311 (1980).
[5263] Shan, X.-q.; Z.-m. Ni, *Huaxue Xuebao 37/4*, 261–266 (1979).
[5264] Shan, X.-q.; Z.-m. Ni, *Huaxue Xuebao 39/6*, 575–578 (1981).
[5265] Shan, X.-q.; Z.-m. Ni, L. Zhang, *Anal. Chim. Acta 151/1*, 179–185 (1983).
[5266] Shan, X.-q.; Z.-m. Ni, L. Zhang, *Talanta 31/2*, 150–152 (1984).
[5267] Shan, X.-q.; Z.-m. Ni, L. Zhang, *At. Spectrosc. 5/1*, 1–4 (1984).
[5268] Shan, X.-q.; Z.-m. Ni, Z.-n. Yuan, *Anal. Chim. Acta 171*, 269–277 (1985).
[5269] Shan, X.-q.; D.-x. Wang, *Anal. Chim. Acta 173*, 315–319 (1985).
[5270] Shan, X.-q.; Z.-n. Yuan, Z.-m. Ni, *Anal. Chem. 57/4*, 857–861 (1985).
[5271] Shan, X.-q.; Z.-n. Yuan, Z.-m. Ni, *Can. J. Spectrosc. 31/2*, 35–39 (1986).
[5272] Shan, X.-q.; J. Egila, D. Littlejohn, J.M. Ottaway, *J. Anal. At. Spectrom. 2/3*, 299–303 (1987).
[5273] Shan, X.-q.; L. Shen, Z.-m. Ni, *J. Anal. At. Spectrom. 3/1*, 99–103 (1988).
[5274] Shan, X.-q.; Z. Yiang, Z.-m. Ni, *Anal. Chim. Acta 217/2*, 271–280 (1989).
[5275] Shan, X.-q.; B. Radziuk, B. Welz, in: B. Welz (Editor), *6. Colloquium Atomspektrometrische Spurenanalytik*, Bodenseewerk Perkin-Elmer, Überlingen, 1991, 99–111.
[5276] Shan, X.-q.; B. Radziuk, B. Welz, V. Sychra, *J. Anal. At. Spectrom. 7/2*, 389–396 (1992).
[5277] Shan, X.-q.; W. Wang, B. Wen, *J. Anal. At. Spectrom. 7/5*, 761–764 (1992).
[5278] Shan, X.-q.; B. Wen, *J. Anal. At. Spectrom. 10/10*, 791–798 (1995).
[5279] Shan, X.-q.; Z.-m. Ni, *Huanjing Kexue 1/2*, 24 (1980).
[5280] Shan, X.-q.; B. Radziuk, B. Welz, O. Vyskočilová, *J. Anal. At. Spectrom. 8/3*, 409–413 (1993).
[5281] Shand, C.A.; A.M. Ure, *J. Anal. At. Spectrom. 2/2*, 143–149 (1987).
[5282] Shang, S.; W. Hong, *Fresenius J. Anal. Chem. 357/7*, 997–999 (1997).
[5283] Shang, W.-x.; Y.-z. Gu, *Fenxi Shiyanshi 12/3*, 22–24 (1993).
[5284] Shao, E.Y.; J.D. Winefordner, *Microchem. J. 39*, 229–234 (1989).
[5285] Shao, G.D.; *Fenxi Huaxue 17*, 441 (1989).
[5286] Shapiro, J.; *J. Am. Water Works Assoc. 56*, 1062 (1964).
[5287] Sharma, R.P.; *Ther. Drug Monit. 4/2*, 219–224 (1982).
[5288] Sharma, R.P.; I.R. Edwards, *Ther. Drug Monit. 5/3*, 367–370 (1983).
[5289] Sharpe, F.S.; D.R. Williams, *J. Am. Soc. Brew. Chem. 53/2*, 85–91 (1995).
[5290] Shawky, S.; H. Emons, H.W. Dürbeck, *Anal. Commun. 33/3*, 107–110 (1996).
[5291] Shcherbakov, V.I.; Yu.I. Belyaev, B.F. Myasoedov, *Zh. Prikl. Spektrosk. 36*, 893 (1982).
[5292] Shekiro, J.M.; R.K. Skogerboe, H.E. Taylor, *Anal. Chem. 60/23*, 2578–2582 (1988).
[5293] Shelley, M.D.; R.G. Fish, A. Brewster, M. Adams, *Ann. Clin. Biochem. 24/S1*, 83–84 (1987).
[5294] Shen, L.; X.-q. Shan, Z.-m. Ni, *J. Anal. At. Spectrom. 3/7*, 989–995 (1988).
[5295] Shendrikar, A.D.; P.W. West, *Anal. Chim. Acta 72*, 91–96 (1974).
[5296] Shendrikar, A.D.; P.W. West, *Anal. Chim. Acta 74*, 189 (1975).
[5297] Shendrikar, A.D.; V. Dharmarajan, H. Walker-Merrick, P.W. West, *Anal. Chim. Acta 84*, 409–417 (1976).
[5298] Shepard, M.R.; C.E. Lee, R.S. Woosley, D.J. Butcher, *Mikrochem. J. 52/1*, 118–126 (1995).
[5299] Sherwood, R.A.; B.F. Rocks, C. Riley, *Analyst (London) 110/5*, 493–496 (1985).
[5300] Shewhart, W.A.; *Economic Control of Manufactured Products*, MacMillan, London, 1931.
[5301] Shi, Q.-g.; *Yankuang Ceshi 7/2*, 137 (1988).
[5302] Shi, Y.; J.-h. Shen, X.-z. Chi, *Fenxi Shiyanshi 12/6*, 58–61 (1993).
[5303] Shijo, Y.; Y. Kimura, T. Shimizu, K. Sakai, *Bunseki Kagaku 32*, E285–291 (1983).
[5304] Shijo, Y.; T. Shimizu, K. Sakai, *Anal. Sci. 1/5*, 479–480 (1985).
[5305] Shijo, Y.; T. Shimizu, T. Tsunoda, *Anal. Sci. 5/1*, 65–68 (1989).

[5306] Shijo, Y.; T. Shimizu, T. Tsunoda, S.-q. Tao, M. Suzuki, *Anal. Chim. Acta 242/2*, 209–213 (1991).
[5307] Shijo, Y.; M. Mitsuhashi, T. Shimizu, S. Sakurai, *Analyst (London) 117/12*, 1929–1931 (1992).
[5308] Shijo, Y.; E. Yoshimoto, S. Sakurai, T. Shimizu, Y. Kawamata, *Anal. Sci. 10/2*, 277–280 (1994).
[5309] Shijo, Y.; M. Suzuki, T. Shimizu, S. Aratake, N. Uehara, *Anal. Sci. 12/6*, 953–957 (1996).
[5310] Shijo, Y.; E. Yoshimoto, T. Kitamura, H. Ono, N. Uehara, T. Shimizu, *Anal. Sci. 12/6*, 959–962 (1996).
[5311] Shimizu, T.; Y. Shijo, K. Sakai, *Bunseki Kagaku 29*, 685 (1980).
[5312] Shimizu, T.; H. Koyanagi, Y. Shijo, K. Sakai, *Chem. Lett. (3)*, 319–322 (1986).
[5313] Shimomura, S.; *Anal. Sci. 5/6*, 633–639 (1989).
[5314] Shinogi, M.; T. Masaki, I. Mori, *J. Trace Elem. Electrol. Health Dis. 3/1*, 25–28 (1989).
[5315] Shiowatana, A. Sirioinyanond, *At. Spectrosc. 17/3*, 122–127 (1996).
[5316] Shirasaki, T.; A. Yonetani, K. Uchino, K. Sakai, *Bunseki Kagaku 40*, 163 (1991).
[5317] Shkinev, V.M.; V.N. Gomolitskii, B.Ya. Spivakov, K.E. Geckeler, E. Bayer, *Talanta 36/8*, 861–863 (1989).
[5318] Shkolnik, G.M.; R.F. Bevill, *At. Absorpt. Newsletter 12/5*, 112–114 (1973).
[5319] Shore, B.W.; D.H. Menzel, *Principles of Atomic Spectra*, J. Wiley & Sons, New York, 1968.
[5320] Short, J.W.; *Bull. Environ. Contam. Toxicol. 39*, 412–416 (1987).
[5321] Shrivastava, A.K.; S.G. Tandon, *Int. J. Environ. Anal. Chem. 11/3-4*, 221–226 (1982).
[5322] Shrivastava, A.K.; S.G. Tandon, *Toxicol. Environm. Chem. 5/3-4*, 311–329 (1982).
[5323] Shum, G.T.C.; H.C. Freeman, J.F. Uthe, *Anal. Chem. 51/3*, 414–416 (1979).
[5324] Shuttler, I.L.; H.T. Delves, *Analyst (London) 111/6*, 651–656 (1986).
[5325] Shuttler, I.L.; H.T. Delves, *J. Anal. At. Spectrom. 2/2*, 171–176 (1987).
[5326] Shuttler, I.L.; H.T. Delves, *J. Anal. At. Spectrom. 3/1*, 145–149 (1988).
[5327] Shuttler, I.L.; H.T. Delves, B. Hütsch, *J. Anal. At. Spectrom. 4/2*, 137–141 (1989).
[5328] Shuttler, I.L.; G. Schlemmer, G.R. Carnrick, W. Slavin, *Spectrochim. Acta, Part B 46/5*, 583–602 (1991).
[5329] Shuttler, I.L.; *Int. Labmate 17*, 15–19 (1992).
[5330] Shuttler, I.L.; *At. Spectrosc. 13/5*, 174–177 (1992).
[5331] Shuttler, I.L.; M. Feuerstein, G. Schlemmer, *J. Anal. At. Spectrom. 7/8*, 1299–1301 (1992).
[5332] Shuttler, I.L.; H. Schulze, *analysis europa 1/1*, 44–47 (1994).
[5333] Shuttler, I.L.; in: G. Subramanian (Editor), *Quality assurance in environmental monitoring*, VCH Verlagsgesellschaft Weinheim - New York - Basel - Tokyo, 1995, 55–93.
[5334] Shvoeva, O.P.; G.P. Kuchava, I.V. Kubrakova, G.V. Myasoedova, S.B. Savvin, *Zh. Anal. Khim. 41*, 2186 (1986).
[5335] Siddik, Z.H.; R.A. Newman, *Anal. Biochem. 172/1*, 190–196 (1988).
[5336] Sideman, L.; J.J. Murphy, D.T. Wilson, *Clin. Chem. (Winston-Salem) 16/7*, 597–601 (1970).
[5337] Siemer, D.D.; J.F. Lech, R. Woodriff, *Spectrochim. Acta, Part B 28*, 469–471 (1973).
[5338] Siemer, D.D.; R. Woodriff, *Spectrochim. Acta, Part B 29*, 269–276 (1974).
[5339] Siemer, D.D.; L. Hageman, *Anal. Lett. 8/5*, 323–337 (1975).
[5340] Siemer, D.D.; R.W. Stone, *Appl. Spectrosc. 29*, 240 (1975).
[5341] Siemer, D.D.; P. Koteel, V. Jariwala, *Anal. Chem. 48/6*, 836–840 (1976).
[5342] Siemer, D.D.; R.K. Vitek, P. Koteel, W.C. Houser, *Anal. Lett. 10*, 357–369 (1977).
[5343] Siemer, D.D.; J.M. Baldwin, *Anal. Chem. 52/2*, 295–300 (1980).
[5344] Siemer, D.D.; *Anal. Chem. 55/4*, 692–697 (1983).
[5345] Siemer, D.D.; *Appl. Spectrosc. 37/6*, 552–557 (1983).
[5346] Siemer, D.D.; W. Frech, *Spectrochim. Acta, Part B 39/2-3*, 261–269 (1984).
[5347] Siemer, D.D.; *Anal. Chem. 56/8*, 1517–1519 (1984).
[5348] Sieveking, G. de G.; P.T. Craddock, M.J. Hughes, P. Bush, J. Ferguson, *Nature (London) 228*, 251–254 (1970).
[5349] Sieveking, G. de G.; P. Bush, J. Ferguson, P.T. Craddock, M.J. Hughes, M.R. Cowell, *Archaeometry 14/2*, 151–176 (1972).
[5350] Sievers, E.; H.D. Oldigs, G. Schulz-Lell, K. Dörner, J. Schaub, in: J. Schaub (Editor), *Composition and Physiological Properties of Human Milk*, Elsevier, Amsterdam - New York - Oxford, 1985, 65–76.
[5351] Sigel, H.; A. Sigel, *Metal Ions in Biological Systems. Vol. 12 Properties of Copper*, Marcel Dekker, New York, 1981.
[5352] Sigel, H.; A. Sigel, *Metal Ions in Biological Systems. Vol. 13 Copper Proteins*, Marcel Dekker, New York, 1981.

[5353] Sigel, H.; A. Sigel, *Metal Ions in Biological Systems. Vol. 15 Zinc and Its Role in Biology and Nutrition*, Marcel Dekker, New York, 1983.

[5354] Sigel, H.; A. Sigel, *Metal Ions in Biological Systems. Vol. 17 Calcium and Its Role in Biology*, Marcel Dekker, New York, 1984.

[5355] Sigel, H.; A. Sigel, *Metal Ions in Biological Systems. Vol. 23 Nickel and Its Role in Biology*, Marcel Dekker, New York, 1988.

[5356] Sigel, H.; A. Sigel, *Metal Ions in Biological Systems. Vol. 26 Compendium on Magnesium and Its Role in Biology, Nutrition and Physiology*, Marcel Dekker, New York, 1990.

[5357] Sighinolfi, G.P.; C. Gorgoni, A.H. Mohamed, *Geostand. Newsl. 8*, 25–29 (1994).

[5358] Sighinolfi, G.P.; C. Gorgoni, O. Bonori, E. Cantoni, M. Martelli, L. Simonetti, *Mikrochim. Acta (Vienna) I*, 171–179 (1989).

[5359] Siitonen, P.H.; H.C. Thompson, *J. AOAC International 77*, 1299 (1994).

[5360] Sikora, Z.; Z. Kowalewska, *Nafta (Katowice.) 45/10-2*, 174 (1989).

[5361] Silberman, D.; W.R. Harris, *Int. J. Environ. Anal. Chem. 17/1*, 73–83 (1984).

[5362] Silkey, J.R.; *J. Assoc. Off. Anal. Chem. 66*, 952–955 (1983).

[5363] Silva, E.C.; M.C.U. Araújo, R.S. Honorato, J.L.F. Costa, *Anal. Chim. Acta 319/1-2*, 153–158 (1996).

[5364] Silva, M.; M. Valcárcel, *Analyst (London) 107*, 511–518 (1982).

[5365] Silva, M.; M. Gallego, M. Valcárcel, *Anal. Chim. Acta 179*, 341–349 (1986).

[5366] Simeonova, S.; G. Bekjarov, L. Futekov, *Fresenius Z. Anal. Chem. 325*, 478–479 (1986).

[5367] Simeonova, S.; G. Toneva, L. Futekov, *Fresenius Z. Anal. Chem. 332/4*, 367–368 (1988).

[5368] Simmons, W.J.; *Anal. Chem. 45/11*, 1947–1949 (1973).

[5369] Simmons, W.J.; J.F. Loneragan, *Anal. Chem. 47/3*, 566–568 (1975).

[5370] Simon, P.J.; B.C. Giessen, T.R. Copeland, *Anal. Chem. 49*, 2285–2288 (1977).

[5371] Simon, S.J.; D.F. Boltz, *Microchem. J. 20*, 468 (1975).

[5372] Simonian, J.V.; *At. Absorpt. Newsletter 7/3*, 63–64 (1968).

[5373] Simons, J.W.; R.E. McClean, R.C. Oldenborg, *J. Phys. Chem. 95/6*, 2361–2364 (1991).

[5374] Simpson, R.T.; *At. Spectrosc. 10/3*, 82–84 (1989).

[5375] Sinemus, H.-W.; M. Melcher, B. Welz, *At. Spectrosc. 2/3*, 81–86 (1981).

[5376] Sinemus, H.-W.; D. Maier, *Fresenius Z. Anal. Chem. 322/4*, 440 (1985).

[5377] Sinemus, H.-W.; D. Maier, M. Schubert-Jacobs, B. Welz, in: B. Welz (Editor), *Fortschritte in der atomspektrometrischen Spurenanalytik*, VCH Verlagsgesellschaft, Weinheim, 1986, 571–577.

[5378] Sinemus, H.-W.; D. Maier, B. Radziuk, in: B. Welz (Editor), *4. Colloquium Atomspektrometrische Spurenanalytik*, Bodenseewerk Perkin-Elmer, Überlingen, 1987, 607–615.

[5379] Sinemus, H.-W.; H.-H. Stabel, in: B. Welz (Editor), *5. Colloquium Atomspektrometrische Spurenanalytik*, Bodenseewerk Perkin-Elmer, Überlingen, 1989, 711–713.

[5380] Sinemus, H.-W.; J. Kleiner, H.-H. Stabel, B. Radziuk, *J. Anal. At. Spectrom. 7/2*, 433–437 (1992).

[5381] Sinemus, H.-W.; H.-H. Stabel, B. Radziuk, J. Kleiner, *Spectrochim. Acta, Part B 48/5*, 643–648 (1993).

[5382] Sinemus, H.-W.; H.-H. Stabel, *Spectrochim. Acta, Part B 48/14*, 1719–1722 (1993).

[5383] Sinemus, H.-W.; J. Kleiner, B. Radziuk, H.-H. Stabel, in: K. Dittrich, B. Welz (Editors), *CANAS'93, Colloq. Analytische Atomspektroskopie*, Universität Leipzig/UFZ Leipzig-Halle, Leipzig, 1993, 657–664.

[5384] Sinemus, H.-W.; J. Kleiner, B. Radziuk, H.-H. Stabel, in: B. Welz (Editor), *CANAS'95 Colloquium Analytische Atomspektroskopie*, Bodenseewerk Perkin-Elmer, Überlingen, 1996, 307–310.

[5385] Sinemus, H.-W.; J. Kleiner, *GIT Fachz. Lab. 41/5*, 475–479 (1997).

[5386] Singh, A.K.; S. Sharma, *Microchem. J. 35/3*, 365–368 (1987).

[5387] Singh, A.K.; T.G. Sampath Kumar, *Microchem. J. 40/2*, 197–200 (1989).

[5388] Singh, A.K.; P. Rita, *Microchem. J. 43/2*, 112–115 (1991).

[5389] Singh, A.K.; S.K. Dhingra, *Analyst (London) 117/5*, 889–891 (1992).

[5390] Singh, A.K.; *Talanta 43/11*, 1843–1846 (1996).

[5391] Sinha, R.C.H.; P.C. Gupta, *Fert. Technol. 17*, 181 (1980).

[5392] Sinner, T.; P. Hoffmann, H.M. Ortner, *Spectrochim. Acta, Part B 48/2*, 255–261 (1993).

[5393] Sipos, M.; T. Adamis, *Acta Chim. Hung. 128/4-5*, 869–876 (1991).

[5394] Skelly, E.M.; F.T. DiStefano, *Appl. Spectrosc. 42/7*, 1302–1306 (1988).

[5395] Skogerboe, R.K.; R.A. Woodriff, *Anal. Chem. 35/12*, 1977. (1963).

[5396] Skogerboe, R.K.; in: J.A. Dean, T.C. Rains (Editors), *Flame Emission and Atomic Absorption Spectrometry, Vol. 1 Theory*, Marcel Dekker, New York - London, 1969, 381–411.

[5397] Skogerboe, R.K.; D.L. Dick, D.A. Pavlica, F.E. Lichte, *Anal. Chem. 47/3*, 568–570 (1975).
[5398] Skogerboe, R.K.; S.A. Wilson, *Anal. Chem. 53/2*, 228–232 (1981).
[5399] Skogerboe, R.K.; W.A. Hanagan, H.E. Taylor, *Anal. Chem. 57/14*, 2815–2818 (1985).
[5400] Slaveykova, V.I.; S. Manev, Dobri Lazarov, *Spectrochim. Acta, Part B 50/13*, 1725–1732 (1995).
[5401] Slaveykova, V.I.; F. Rastgar, M.J.F. Leroy, *J. Anal. At. Spectrom. 11/10*, 997–1002 (1996).
[5402] Slavin, S.; W. Slavin, *At. Absorpt. Newsletter 5/5*, 106–112 (1966).
[5403] Slavin, S.; T.W. Sattur, *At. Absorpt. Newsletter 7*, 99 (1968).
[5404] Slavin, W.; *At. Absorpt. Newsletter 1*, (1962).
[5405] Slavin, W.; D.C. Manning, *Anal. Chem. 35*, 253–254 (1963).
[5406] Slavin, W.; *At. Absorpt. Newsletter 3*, 93–109 (1964).
[5407] Slavin, W.; S. Sprague, *At. Absorpt. Newsletter 3*, 1–6 (1964).
[5408] Slavin, W.; D.C. Manning, *Appl. Spectrosc. 19/3*, 65–68 (1965).
[5409] Slavin, W.; D.J. Trent, S. Sprague, *At. Absorpt. Newsletter 4*, 180–185 (1965).
[5410] Slavin, W.; *Appl. Spectrosc. 19*, 32 (1965).
[5411] Slavin, W.; A. Venghiattis, D.C. Manning, *At. Absorpt. Newsletter 5/4*, 84–88 (1966).
[5412] Slavin, W.; *At. Absorpt. Newsletter 6/1*, 9–13 (1967).
[5413] Slavin, W.; *Atomic Absorption Spectroscopy*, Interscience, New York - London - Sidney, 1968.
[5414] Slavin, W.; G.J. Schmidt, *J. Chromatogr. Sci. 17*, 610–613 (1979).
[5415] Slavin, W.; D.C. Manning, *Anal. Chem. 51/2*, 261–265 (1979).
[5416] Slavin, W.; D.C. Manning, *Spectrochim. Acta, Part B 35/11-2*, 701–714 (1980).
[5417] Slavin, W.; S.A. Myers, D.C. Manning, *Anal. Chim. Acta 117*, 267–273 (1980).
[5418] Slavin, W.; D.C. Manning, G.R. Carnrick, *Anal. Chem. 53/9*, 1504–1509 (1981).
[5419] Slavin, W.; D.C. Manning, G.R. Carnrick, *At. Spectrosc. 2/5*, 137–145 (1981).
[5420] Slavin, W.; *At. Spectrosc. 2/1*, 8–11 (1981).
[5421] Slavin, W.; G.R. Carnrick, D.C. Manning, *Anal. Chem. 54/4*, 621–624 (1982).
[5422] Slavin, W.; G.R. Carnrick, D.C. Manning, *Anal. Chim. Acta 138*, 103–110 (1982).
[5423] Slavin, W.; D.C. Manning, *Spectrochim. Acta, Part B 37/11*, 955–964 (1982).
[5424] Slavin, W.; *Fresenius Z. Anal. Chem. 316/3*, 319–320 (1983).
[5425] Slavin, W.; D.C. Manning, G.R. Carnrick, E. Pruszkowska, *Spectrochim. Acta, Part B 38/8*, 1157–1170 (1983).
[5426] Slavin, W.; G.R. Carnrick, D.C. Manning, E. Pruszkowska, *At. Spectrosc. 4/3*, 69–86 (1983).
[5427] Slavin, W.; G.R. Carnrick, *Spectrochim. Acta, Part B 39/2-3*, 271–282 (1984).
[5428] Slavin, W.; *Graphite Furnace AAS - A Source Book*, Perkin Elmer, Ridgefield, CT, 1984.
[5429] Slavin, W.; *Anal. Proc. (London) 21/2*, 59–61 (1984).
[5430] Slavin, W.; *J. Anal. At. Spectrom. 1*, 281–285 (1986).
[5431] Slavin, W.; G.R. Carnrick, *At. Spectrosc. 7/1*, 9–13 (1986).
[5432] Slavin, W.; *Spectrochim. Acta, Part B 42*, 933–935 (1987).
[5433] Slavin, W.; *Methods Emzymol. 158*, 117–145 (1988).
[5434] Slavin, W.; G.R. Carnrick, *CRC Crit. Rev. Anal. Chem. 19/2*, 95–134 (1988).
[5435] Slavin, W.; *Sci. Total Environ. 71/1*, 17–35 (1988).
[5436] Slavin, W.; D.C. Manning, G.R. Carnrick, *Talanta 36/1-2*, 171–178 (1989).
[5437] Slavin, W.; D.C. Manning, G.R. Carnrick, *Spectrochim. Acta, Part B 44/12*, 1237–1243 (1989).
[5438] Slavin, W.; N.J. Miller-Ihli, G.R. Carnrick, *Am. Lab. (Fairfield, CT) 22/Oct*, 80–92 (1990).
[5439] Slavin, W.; *Anal. Chem. 63/21*, 1033A-38A (1991).
[5440] Slikkerveer, A.; R.B. Helmich, P.M. Edelbroek, G.B. Van der Voet, F.A. De Wolff, *Clin. Chim. Acta 201/1-2*, 17–26 (1991).
[5441] Slikkerveer, A.; R.B. Helmich, F.A. De Wolff, *Clin. Chem. (Winston-Salem) 39/5*, 800–803 (1993).
[5442] Slonawska, K.; K. Brajter, *J. Anal. At. Spectrom. 4/7*, 653–655 (1989).
[5443] Sloof, J.E.; J.R.W. Woittiez, U. Woroniecka, *Fresenius J. Anal. Chem. 354/1*, 16–20 (1996).
[5444] Slovák, Z.; M. Smrz, B. Dočekal, S. Slovákova, *Anal. Chim. Acta 111*, 243–249 (1979).
[5445] Slovák, Z.; B. Dočekal, *Anal. Chim. Acta 129*, 263–267 (1981).
[5446] Smart, G.A.; J.C. Sherlock, *Food Addit. Contam. 2/2*, 139–147 (1985).
[5447] Smart, H.T.; D.J. Campbell, *Can. J. Pharm. Sci. 4*, 73 (1969).
[5448] Smets, B.; *Analyst (London) 105*, 482–490 (1980).
[5449] Smets, B.; *Spectrochim. Acta, Part B 35/1*, 33–42 (1980).
[5450] Smeyers-Verbeke, J.; M.R. Detaevernier, L. Denis, D.L. Massart, *Clin. Chim. Acta 113/3*, 329–333 (1981).

[5451] Smeyers-Verbeke, J.; Q. Yang, W. Penninckx, F. Vandervoort, *J. Anal. At. Spectrom. 5/5*, 393–398 (1990).
[5452] Smichowski, P.; M.B. de la Calle Guntiñas, C. Cámara, *Fresenius J. Anal. Chem. 348/5-6*, 380–384 (1994).
[5453] Smichowski, P.; M.B. de la Calle Guntiñas, Y. Madrid, M.G. Cobo, C. Cámara, *Spectrochim. Acta, Part B 49/11*, 1049–1055 (1994).
[5454] Smichowski, P.; Y. Madrid, M.B. de la Calle Guntiñas, C. Cámara, *J. Anal. At. Spectrom. 10/10*, 815–821 (1995).
[5455] Smith, A.E.; *Analyst (London) 98*, 209–212 (1973).
[5456] Smith, A.E.; *Analyst (London) 98*, 65–68 (1973).
[5457] Smith, A.E.; *Analyst (London) 100*, 300–306 (1975).
[5458] Smith, B.M.; in: H.A. McKenzie, L.E. Smythe (Editors), *Quantitative trace analysis of biological materials*, Elsevier, Amsterdam - New York - Oxford, 1988, 649–657.
[5459] Smith, C.I.; B.L. Carson, *Trace iMetals in the Environment, Vol. 1, Thallium*, Ann arbor Science, Ann Arbor, MI, 1977.
[5460] Smith, C.M.M.; R. Nichol, D. Littlejohn, *Anal. Proc. (London) 29/6*, 262–264 (1992).
[5461] Smith, C.M.M.; J.M. Harnly, *Spectrochim. Acta, Part B 49/4*, 387–398 (1994).
[5462] Smith, C.M.M.; J.M. Harnly, *J. Anal. At. Spectrom. 11/11*, 1055–1061 (1996).
[5463] Smith, D.D.; R.F. Browner, *Anal. Chem. 54/3*, 533–537 (1982).
[5464] Smith, D.D.; R.F. Browner, *Anal. Chem. 56/14*, 2702–2708 (1984).
[5465] Smith, F.; B. Cousins, J. Bozic, W. Flora, *Anal. Chim. Acta 177*, 243–245 (1985).
[5466] Smith, J.C.; J.T. Holbrook, D.E. Danford, *J. Am. Coll. Nutr. 4/6*, 627–638 (1985).
[5467] Smith, K.; J.R. Dean, *Analyst (London) 118/11*, 1445–1447 (1993).
[5468] Smith K.A. (Editor), *Soil Analysis: Modern Instrumental Techniques*, Marcel Dekker, New York, 1991.
[5469] Smith, M.R.; H.B. Cochran, *At. Spectrosc. 2/4*, 97–100 (1981).
[5470] Smith, R.; J.D. Winefordner, *Spectrosc. Lett. 1/4*, 157–165 (1968).
[5471] Smith, R.G.; J.C. Van Loon, J.R. Knechtel, J.L. Fraser, A.E. Pitts, A. Hodges, *Anal. Chim. Acta 93*, 61–67 (1977).
[5472] Smith, R.G.; *Talanta 25*, 173–175 (1978).
[5473] Smith, R.G.; H.L. Windom, *Anal. Chim. Acta 113*, 39–46 (1980).
[5474] Smith, S.B.; G.M. Hieftje, *Appl. Spectrosc. 37/5*, 419–424 (1983).
[5475] Smolander, K.; R. Järnström, A. Sajalinna, *Fresenius Z. Anal. Chem. 335/4*, 392–394 (1989).
[5476] Sneddon, J.; J.M. Ottaway, R.W. Rowston, *Analyst (London) 103*, 776–779 (1978).
[5477] Sneddon, J.; *Talanta 30/9*, 631–648 (1983).
[5478] Sneddon, J.; *Anal. Chem. 56*, 1982–1986 (1984).
[5479] Sneddon, J.; *Anal. Lett. 18/A10*, 1261–1280 (1985).
[5480] Sneddon, J.; R.F. Browner, P.N. Keliher, J.D. Winefordner, D.J. Butcher, R.G. Michel, *Prog. Anal. Spectrosc. 12/4*, 369–402 (1989).
[5481] Sneddon, J.; in: J. Sneddon (Editor), *Sample Introduction in Atomic Spectroscopy (Anal. Spectrosc. Library Vol. 4)*, Elsevier ; Amsterdam, 1990, 329–351.
[5482] Sneddon, J.; *Anal. Chim. Acta 245/2*, 203–206 (1991).
[5483] Sneddon, J.; *Determination of Metals in the Atmosphere by Atomic Spectrometry*, Ellis Horwood, Chichester, 1992.
[5484] Sneddon, J.; M.V. Smith, S. Indurthy, Y.-i. Lee, *Spectroscopy (Eugene, OR) 10/1*, 26–32 (1995).
[5485] Snell, J.P.; S. Sandberg, W. Frech, *J. Anal. At. Spectrom. 12/4*, 491–494 (1997).
[5486] Soares, M.E.; M.L. Bastos, M.A. Ferreira, *J. Anal. At. Spectrom. 8/4*, 655–657 (1993).
[5487] Soares, M.E.; M.L. Bastos, M.A. Ferreira, *At. Spectrosc. 16/6*, 256–260 (1995).
[5488] Sobelman, I.I.; *Introduction to the Theory of Atomic Spectra*, Pergamon Press, Oxford, 1972.
[5489] Sohrin, Y.; K. Isshiki, T. Kuwamoto, E. Nakayama, *Talanta 34/3*, 341–344 (1987).
[5490] Solchaga, M.; R. Montoro, M. de la Guardia, *J. Assoc. Off. Anal. Chem. 69/5*, 874–876 (1986).
[5491] Solomons, E.T.; H.C. Walls, *J. Anal. Toxicol. 7/5*, 220–222 (1983).
[5492] Sommer, K.; A.P. Thorne, R.C.M. Learner, *J. Phys. D.: Appl. Phys. 16/3*, 233–244 (1983).
[5493] Sommerfeld, A.; *Atombau und Spektrallinien Band I*, Vieweg, Braunschweig, Germany, 1960.
[5494] Sommerfeld, M.R.; T.D. Love, R.D. Olsen, *At. Absorpt. Newsletter 14/1*, 31–32 (1975).
[5495] Son, M. van; H. Muntau, *Fresenius Z. Anal. Chem. 328/4-5*, 390–392 (1987).
[5496] Sonnekalb, U.; H. Helms, H. Schrübbers, in: B. Welz (Editor), *6. Colloquium Atomspektrometrische Spurenanalytik*, Bodenseewerk Perkin-Elmer, Überlingen, 1991, 789–795.

[5497] Soriano, M.J.; M. de la Guardia, *Talanta 31/5*, 347–352 (1984).
[5498] Sotera, J.J.; G.R. Dulude, R.L. Stux, *Sci. Total Environ. 71/1*, 45–48 (1988).
[5499] Soto, E.G.; E. Alonso Rodriguez, D. Prada Rodriguez, P. López Mahía, S. Muniategui Lorenzo, *Sci. Total Environ. 141/1*, 87–91 (1994).
[5500] Soto-Ferreiro, R.M.; C. Casais Laiño, P. Bermejo-Barrera, *Anal. Lett. 24/12*, 2277–2292 (1991).
[5501] Soto-Ferreiro, R.M.; P. Bermejo-Barrera, *Analusis 21/4*, 197–199 (1993).
[5502] Sovereign, W.R.; E.R. Ebersole, R. Villarreal, W.A. Hareland, *Nucl. Appl. Tech. 9*, 416 (1970).
[5503] Spachidis, C.; A. Weitz, K. Bächmann, *Fresenius Z. Anal. Chem. 306/4*, 268–274 (1981).
[5504] Spellmaeker, M.; B. Ouddane, J.C. Fischer, M. Wartel, M. Hoenig, *Analusis 24/3*, 76 (1996).
[5505] Sperling, K.R.; *Fresenius Z. Anal. Chem. 299*, 103–107 (1979).
[5506] Sperling, K.R.; Bahr, B.; *Fresenius Z. Anal. Chem. 306/1*, 7–12 (1981).
[5507] Sperling, K.R.; *Fresenius Z. Anal. Chem. 311/7*, 656–664 (1982).
[5508] Sperling, M.; B. Welz, in: B. Welz (Editor), *5. Colloquium Atomspektrometrische Spurenanalytik*, Bodenseewerk Perkin-Elmer, Überlingen, 1989, 531–541.
[5509] Sperling, M.; Z.-l. Fang, B. Welz, *Anal. Chem. 63/2*, 151–159 (1991).
[5510] Sperling, M.; X.-f. Yin, B. Welz, *J. Anal. At. Spectrom. 6/4*, 295–300 (1991).
[5511] Sperling, M.; X.-f. Yin, S.-k. Xu, B. Welz, in: B. Welz (Editor), *6. Colloquium Atomspektrometrische Spurenanalytik*, Bodenseewerk Perkin-Elmer, Überlingen, 1991, 215–242.
[5512] Sperling, M.; P. Koscielniak, B. Welz, in: B. Welz (Editor), *6. Colloquium Atomspektrometrische Spurenanalytik*, Bodenseewerk Perkin-Elmer, Überlingen, 1991, 264–280.
[5513] Sperling, M.; X.-f. Yin, B. Welz, *J. Anal. At. Spectrom. 6/8*, 615–621 (1991).
[5514] Sperling, M.; X.-f. Yin, B. Welz, *Spectrochim. Acta, Part B 46/14*, 1789–1801 (1991).
[5515] Sperling, M.; X.-f. Yin, B. Welz, *Analyst (London) 117/3*, 629–635 (1992).
[5516] Sperling, M.; P. Koscielniak, B. Welz, *Anal. Chim. Acta 261/1-2*, 115–123 (1992).
[5517] Sperling, M.; X.-f. Yin, B. Welz, *Fresenius J. Anal. Chem. 343*, 754–755 (1992).
[5518] Sperling, M.; S.-k. Xu, B. Welz, *Anal. Chem. 64/24*, 3101–3108 (1992).
[5519] Sperling, M.; Y.-z. He, B. Welz, in: K. Dittrich, B. Welz (Editors), *CANAS'93, Colloq. Analytische Atomspektroskopie,* Universität Leipzig/UFZ Leipzig-Halle, Leipzig, 1993, 989–1004.
[5520] Sperling, M.; B. Welz, J. Hertzberg, C. Rieck, G. Marowsky, *Spectrochim. Acta, Part B 51/9-10*, 897–930 (1996).
[5521] Sperling, M.; X.-p. Yan, B. Welz, *Spectrochim. Acta, Part B 51/14*, 1891–1908 (1996).
[5522] Sperling, M.; X.-p. Yan, B. Welz, *Spectrochim. Acta, Part B 51/14*, 1875–1889 (1996).
[5523] Sperling, M.; X.-p. Yan, B. Welz, *J. Anal. At. Spectrom.*, submitted (1997).
[5524] Speváčková, V.; K. Kratzer, M. Cejchanova, *J. Anal. At. Spectrom. 6/8*, 673–674 (1991).
[5525] Spielholtz, G.; G.C. Toralballa, *Analyst (London) 94*, 1072–1074 (1969).
[5526] Spitz, J.; G. Uny, *Appl. Opt. 7/7*, 1345–1349 (1968).
[5527] Spitz, J.; G. Uny, M. Roux, J. Besson, *Spectrochim. Acta, Part B 24/7*, 399–403 (1969).
[5528] Sprague, S.; *At. Absorpt. Newsletter 2*, 9–10 (1963).
[5529] Sprague, S.; W. Slavin, *At. Absorpt. Newsletter 4/6*, 293–295 (1965).
[5530] Sprague, S.; W. Slavin, *At. Absorpt. Newsletter 4*, 228–233 (1965).
[5531] Sprague, S.; W. Slavin, *At. Absorpt. Newsletter 4*, 367–369 (1965).
[5532] Sprenger, F.J.; *Water Research 15*, 233–241 (1981).
[5533] Squib, S.K.; B.A. Fowler, in: B.A. Fowler (Editor), *Biological and Environmental Effects of Arsenic*, Elsevier/North Holland; Amsterdam, 1983, 233–270.
[5534] Staats, G.; *Fresenius Z. Anal. Chem. 315/1*, 1–5 (1983).
[5535] Stacchini, A.; M. Baldini, E. Coni, in: C. Minoia, S. Caroli (Editors), *Appl. Zeeman Graphite Furnace At. Abs. Spectrom. Chem. Lab. Toxicology*, Pergamon, Oxford, 1992, 325–348.
[5536] Staden, J.F. van; A. van Rensburg, *Analyst (London) 115/5*, 605–608 (1990).
[5537] Staden, J.F. van; A. van Rensburg, *Fresenius J. Anal. Chem. 337/4*, 393–397 (1990).
[5538] Stadtman, T.C.; *Ann. Rev. Biochem. 59*, (1990).
[5539] Stafford, J.D.; J.A. Holcombe, *J. Anal. At. Spectrom. 3/1*, 35–42 (1988).
[5540] Stalikas, C.D.; G.A. Pilidis, M.I. Karayannis, *J. Anal. At. Spectrom. 11/1*, 595–599 (1996).
[5541] Stallard, M.O.; V. Hodge, E.D. Goldberg, *Environ. Monit. Assess. 9*, 195–220 (1987).
[5542] Stallard, M.O.; S.Y. Cola, C.A. Dooley, *Appl. Organomet. Chem. 3*, 105–114 (1989).
[5543] Standards Association of Australia, *Australian Standard ; AS-1038*, 1986.
[5544] Standards Association of Australia, *Australian Standard ; AS-1515.5*, 1987.
[5545] Standards Association of Australia, *Australian Standard ; AS-3503*, 1987.
[5546] Standards Association of Australia, *Australian Standard ; AS-2292*, 1987.

[5547] Standards Association of Australia, *Australian Standard ; AS-1671*, 1987.
[5548] Standards Association of Australia, *Australian Standard ; AS-1515.3*, 1989.
[5549] Standards Association of Australia, *Australian Standard ; AS-3550.6*, 1990.
[5550] Standards Association of Australia, *Australian Standard ; AS-1050.23*, 1990.
[5551] Standards Australia, *Australian Standard AS-4090*, 1993.
[5552] Standards Australia, *Australian Standard AS 1038.10.5*, 1993.
[5553] Standards Australia, *Australian Standard AS 4205.2*, 1994.
[5554] Starn, T.K.; G.M. Hieftje, *Spectrochim. Acta, Part B 49*, 533 (1994).
[5555] Stauss, P.; *Dissertation, Universität Konstanz*, 1993.
[5556] Steglich, F.; R. Stahlberg, *Fresenius Z. Anal. Chem. 314/4*, 402–406 (1983).
[5557] Steglich, F.; R. Stahlberg, *Fresenius Z. Anal. Chem. 315/4*, 321–328 (1983).
[5558] Steglich, F.; R. Stahlberg, *Fresenius Z. Anal. Chem. 315/5*, 329–333 (1983).
[5559] Steglich, F.; R. Stahlberg, M. Lufter, H. Pawlik, *Fresenius Z. Anal. Chem. 322/6*, 555–562 (1985).
[5560] Stein, K.; G. Schwedt, *Fresenius J. Anal. Chem. 350/1-2*, 38–43 (1994).
[5561] Steiner, J.W.; D.C. Moy, H.L. Kramer, *Analyst (London) 112/8*, 1113–1115 (1987).
[5562] Steiner, L.; M.L. Xing, B.Y. Pu, S. Hartland, *Anal. Chim. Acta 246/2*, 347–349 (1991).
[5563] Stelz, A.; *Lebensmittelchem. Gerichtl. Chem. 41/5*, 110–112 (1987).
[5564] Stendal, H.; *Chem. Erde 39/3*, 276–282 (1980).
[5565] Stephen, S.C.; D. Littlejohn, J.M. Ottaway, *Analyst (London) 110/8*, 1147–1151 (1985).
[5566] Stephen, S.C.; J.M. Ottaway, D. Littlejohn, *Fresenius Z. Anal. Chem. 328/4-5*, 346–353 (1987).
[5567] Stephens, R.; D.E. Ryan, *Talanta 22*, 655–658 (1975).
[5568] Stephens, R.; D.E. Ryan, *Talanta 22*, 659–662 (1975).
[5569] Stephens, R.; *Talanta 24*, 233–239 (1977).
[5570] Stephens, R.; *Talanta 25*, 435–440 (1978).
[5571] Stephens, R.; G.F. Murphy, *Talanta 25*, 441–445 (1978).
[5572] Stephens, R.; *Talanta 26*, 57–59 (1979).
[5573] Stephens, R.; *Spectrochim. Acta, Part B 43/9-11*, 1247–1254 (1988).
[5574] Stephenson, M.D.; D.R. Smith, *Anal. Chem. 60/7*, 696–698 (1988).
[5575] Stobbaerts, R.F.J.; M. Ieven, H. Deelstra, I. De Leeuw, *Z. Ernährungswiss. 31*, 138–146 (1992).
[5576] Stock, D.J.; *J. Anal. At. Spectrom. 5/7*, 631–634 (1990).
[5577] Stockinger, K.; V. Muntean, *Lab. med. 16*, 419–422 (1992).
[5578] Stockton, R.A.; Kurt J. Irgolic, *Int. J. Environ. Anal. Chem. 6/4*, 313–319 (1979).
[5579] Stoeppler, M.; M. Kampel, B. Welz, *Fresenius Z. Anal. Chem. 282*, 369–378 (1976).
[5580] Stoeppler, M.; M. Kampel, *Report Jül-1360*, 1976.
[5581] Stoeppler, M.; W. Matthes, *Anal. Chim. Acta 98*, 389–392 (1978).
[5582] Stoeppler, M.; U. Bagschik, K. May, *Fresenius Z. Anal. Chem. 301/2*, 106–107 (1980).
[5583] Stoeppler, M.; U. Kurfürst, K.-H. Grobecker, *Fresenius Z. Anal. Chem. 322/7*, 687–691 (1985).
[5584] Stoeppler, M.; in: P. Brätter, P. Schramel, *Trace Element Analytical Chemistry in Medicine and Biology, Vol. 2*, 1983, 909–928.
[5585] Stoeppler, M.; in: W. Fresenius, H. Günzler, W. Huber, I. Lüderwald, G. Tölg, H. Wisser (Editors), *Analytiker Taschenbuch,* Springer, Berlin - Heidelberg - New York, 1985, 199–216.
[5586] Stoeppler, M.; P. Ostapczuk, *Nickel und Cobalt*, in: M. Stoeppler (Editor), *Hazardous Metals in the Environment, Techn. Instr. Anal. Chem. Vol 12*, Elsevier, Amsterdam, 1992.
[5587] Stoeppler, M.; *Cadmium*, in: M. Stoeppler (Editor), *Hazardous Metals in the Environment, Techn. Instr. Anal. Chem. Vol 12*, Elsevier, Amsterdam, 1992.
[5588] Stoeppler, M.; U. Kurfürst, in: U. Kurfürst (Editor), *Solid Sample Analysis - Direct and Slurry Sampling using ETAAS and ETV-ICP*, Springer, Heidelberg, 1997.
[5589] Stojanovic, D.D.J.; J. Bradshaw, J.D. Winefordner, *Anal. Chim. Acta 96*, 45 (1978).
[5590] Stojanovic, D.D.J.; V.J. Vajgand, *Spectrochim. Acta, Part B 39/6*, 767–775 (1984).
[5591] Stollenwerk, K.G.; D.B. Grove, *J. Environ. Qual. 14/3*, 396–399 (1985).
[5592] Stone, S.F.; M.C. Freitas, R.M. Parr, R. Zeissler, *Fresenius J. Anal. Chem. 352/1-2*, 227–231 (1995).
[5593] Strasheim, A.; L.R.P. Butler, *Appl. Spectrosc. 16/3*, 109 (1962).
[5594] Strasheim, A.; *Nature (London) 196*, 1194 (1962).
[5595] Strasheim, A.; G.J. Wessels, *Appl. Spectrosc. 17*, 65–70 (1963).
[5596] Strasheim, A.; H.G.C. Human, *Spectrochim. Acta, Part B 23/4,* 265–275 (1968).

[5597] Stratham, P.J.; *Anal. Chim. Acta 169*, 149–159 (1985).

[5598] Stratis, J.A.; G.A. Zachariadis, E.A. Dimitrakoudi, V. Simeonov, *Fresenius Z. Anal. Chem. 331/7*, 725–729 (1988).

[5599] Streckfuß, N.; L. Frey, G. Zielonka, F. Kroninger, C. Ryzlewicz, H. Ryssel, *Fresenius J. Anal. Chem. 343*, 765–768 (1992).

[5600] Strelow, F.W.E.; T.N. van der Walt, *Anal. Chim. Acta 136*, 429–434 (1982).

[5601] Strelow, F.W.E.; *S. Afr. J. Chem. 40/1*, 1–6 (1987).

[5602] Stripp, R.A.; D.C. Bogen, *J. Anal. Toxicol. 13/1*, 57–59 (1989).

[5603] Strübel, G.; V. Rzepka-Glinder, K.H. Grobecker, *Fresenius Z. Anal. Chem. 328/4-5*, 382–385 (1987).

[5604] Strübel, G.; V. Rzepka-Glinder, in: B. Welz (Editor), *5. Colloquium Atomspektrometrische Spurenanalytik*, Bodenseewerk Perkin-Elmer, Überlingen, 1989, 747–754.

[5605] Strübel, G.; V. Rzepka-Glinder, K.H. Grobecker, K. Jarrar, *Fresenius J. Anal. Chem. 337/3*, 316–319 (1990).

[5606] Struempler, A.W.; *Anal. Chem. 45*, 2251–2254 (1973).

[5607] Stuart, D.C.; *Anal. Chim. Acta 101*, 429–432 (1978).

[5608] Stuart, D.C.; *Anal. Chim. Acta 106*, 411–415 (1979).

[5609] Stummeyer, J.; B. Harazim, T. Wippermann, *Fresenius J. Anal. Chem. 354/3*, 344–351 (1996).

[5610] Stummeyer, J.; B. Harazim, T. Wippermann, in: B. Welz, *CANAS'95 Colloquium Analytische Atomspektroskopie*, Bodenseewerk Perkin-Elmer, Überlingen, 1996, 425–428.

[5611] Stünzi, H.; in: B. Welz (Editor), *Fortschritte in der atomspektrometrischen Spurenanalytik*, VCH Verlagsgesellschaft, Weinheim, 1986, 487–492.

[5612] Stünzi, H.; *Mitt. Geb. Lebensmittelunters. Hyg. 79/1*, 99–111 (1988).

[5613] Štupar, J.; B. Podobnik, J. Korošin, *Croat. Chem. Acta. 27*, 141 (1965).

[5614] Štupar, J.; *Mikrochim. Acta (Vienna) 4-5*, 722–736 (1966).

[5615] Štupar, J.; J.B. Dawson, *Appl. Opt. 7/7*, 1351–1358 (1968).

[5616] Štupar, J.; W. Frech, *J. Chromatogr. 541*, 243–255 (1991).

[5617] Sturgeon, R.E.; C.L. Chakrabarti, I.S. Maines, P.C. Bertels, *Anal. Chem. 47/8*, 1240–1249 (1975).

[5618] Sturgeon, R.E.; C.L. Chakrabarti, *Anal. Chem. 48/4*, 677–686 (1976).

[5619] Sturgeon, R.E.; C.L. Chakrabarti, C.H. Langford, *Anal. Chem. 48/12*, 1792–1807 (1976).

[5620] Sturgeon, R.E.; C.L. Chakrabarti, P.C. Bertels, *Spectrochim. Acta, Part B 32/5-6*, 257–277 (1977).

[5621] Sturgeon, R.E.; C.L. Chakrabarti, *Anal. Chem. 49/8*, 1100–1106 (1977).

[5622] Sturgeon, R.E.; C.L. Chakrabarti, *Prog. Anal. Spectrosc. 1/1-2*, 5–199 (1978).

[5623] Sturgeon, R.E.; S.S. Berman, S. Kaschyop, *Anal. Chem. 52/7*, 1049–1053 (1980).

[5624] Sturgeon, R.E.; S.S. Berman, S.N. Willie, J.A.H. Desaulniers, *Anal. Chem. 53/14*, 2337–2340 (1981).

[5625] Sturgeon, R.E.; S.S. Berman, S.N. Willie, *Talanta 29/3*, 167–171 (1982).

[5626] Sturgeon, R.E.; D.F. Mitchell, S.S. Berman, *Anal. Chem. 55/7*, 1059–1064 (1983).

[5627] Sturgeon, R.E.; S.N. Willie, S.S. Berman, *Anal. Chem. 57/1*, 6–9 (1985).

[5628] Sturgeon, R.E.; S.S. Berman, *Anal. Chem. 57/7*, 1268–1275 (1985).

[5629] Sturgeon, R.E.; S.N. Willie, S.S. Berman, *Anal. Chem. 57/12*, 2311–2314 (1985).

[5630] Sturgeon, R.E.; J.S. Arlow, *J. Anal. At. Spectrom. 1/5*, 359–363 (1986).

[5631] Sturgeon, R.E.; S.N. Willie, S.S. Berman, *J. Anal. At. Spectrom. 1/2*, 115–118 (1986).

[5632] Sturgeon, R.E.; S.N. Willie, S.S. Berman, *Fresenius Z. Anal. Chem. 323/7*, 788–792 (1986).

[5633] Sturgeon, R.E.; *Fresenius Z. Anal. Chem. 324/8*, 807–818 (1986).

[5634] Sturgeon, R.E.; J.W. McLaren, S.N. Willie, D. Beauchemin, S.S. Berman, *Can. J. Chem. 65/5*, 961–964 (1987).

[5635] Sturgeon, R.E.; S.N. Willie, S.S. Berman, *Anal. Chem. 59/19*, 2441–2444 (1987).

[5636] Sturgeon, R.E.; S.N. Willie, G.I.Sproule, S.S.Berman, *J. Anal. At. Spectrom. 2/7*, 719–722 (1987).

[5637] Sturgeon, R.E.; *Can. J. Spectrosc. 32/4*, 79–85 (1987).

[5638] Sturgeon, R.E.; S.S. Berman, *CRC Crit. Rev. Anal. Chem. 18/3*, 209–244 (1987).

[5639] Sturgeon, R.E.; S.N. Willie, S.S. Berman, *J. Anal. At. Spectrom. 4/5*, 443–446 (1989).

[5640] Sturgeon, R.E.; S.N. Willie, G.I. Sproule, P.T. Robinson, S.S. Berman, *Spectrochim. Acta, Part B 44/7*, 667–682 (1989).

[5641] Sturgeon, R.E.; S.N. Willie, S.S. Berman, *Anal. Chem. 61/17*, 1867–1869 (1989).

[5642] Sturgeon, R.E.; K.W.M. Siu, S.N. Willie, S.S. Berman, *Analyst (London) 114/11*, 1393–1396 (1989).

[5643] Sturgeon, R.E.; *Spectrochim. Acta, Part B 44/12*, 1209–1220 (1989).
[5644] Sturgeon, R.E.; *Fresenius J. Anal. Chem. 355/5-6*, 425–432 (1996).
[5645] Sturgeon, R.E.; J. Liu, J. Boyko, V.T. Luong, *Anal. Chem. 68*, 1883–1887 (1996).
[5646] Sturman, B.T.; *Appl. Spectrosc. 39/1*, 48–56 (1985).
[5647] Sturman, B.T.; *J. Anal. At. Spectrom. 1/1*, 55–58 (1986).
[5648] Stuyfzand, P.J.; *Trends Anal. Chem. (Pers. Ed.) 6/2*, 50–54 (1987).
[5649] Styris, D.L.; J.H. Kaye, *Spectrochim. Acta, Part B 36/1*, 41–47 (1981).
[5650] Styris, D.L.; J.H. Kaye, *Anal. Chem. 54/6*, 864–869 (1982).
[5651] Styris, D.L.; *Anal. Chem. 56/7*, 1070–1076 (1984).
[5652] Styris, D.L.; *Fresenius Z. Anal. Chem. 323/7*, 710–715 (1986). ·
[5653] Styris, D.L.; D.A. Redfield, *Anal. Chem. 59/24*, 2891–2897 (1987).
[5654] Styris, D.L.; D.A. Redfield, *Anal. Chem. 59/24*, 2897–2903 (1987).
[5655] Styris, D.L.; L.J. Prell, D.A. Redfield, *Anal. Chem. 63/5*, 503–507 (1991).
[5656] Styris, D.L.; L.J. Prell, D.A. Redfield, J.A. Holcombe, D.A. Bass, V. Majidi, *Anal. Chem. 63/5*, 508–517 (1991).
[5657] Styris, D.L.; D.A. Redfield, *Spectrochimica Acta Rev. 15/2*, 71–123 (1993).
[5658] Su, Y.-s.; R.A. Burdo, W.R. Strzegowski, *Mikrochim. Acta (Vienna) I/5-6*, 321–331 (1984).
[5659] Su, Z.-g.; Y.-d. Cao, J.-h. Qiao, J.-q. Fang, *Guangpuxue Yu Guangpu Fenxi 7/4*, 58 (1987).
[5660] Subramanian, K.S.; C.L. Chakrabarti, J.E. Sueiras, I.S. Maines, *Anal. Chem. 50/3*, 444–448 (1978).
[5661] Subramanian, K.S.; C.L. Chakrabarti, *Prog. Anal. Spectrosc. 2*, 287–308 (1979).
[5662] Subramanian, K.S.; J.-C. Méranger, *Anal. Chim. Acta 124/1*, 131–142 (1981).
[5663] Subramanian, K.S.; *Fresenius Z. Anal. Chem. 305/5*, 382–386 (1981).
[5664] Subramanian, K.S.; J.-C. Méranger, J.E. MacKeen, *Anal. Chem. 55/7*, 1064–1067 (1983).
[5665] Subramanian, K.S.; J.-C. Méranger, J. Connor, *J. Anal. Toxicol. 7/1*, 15–19 (1983).
[5666] Subramanian, K.S.; J.-C. Méranger, J. Connor, *Talanta 32/5*, 435–438 (1985).
[5667] Subramanian, K.S.; J.-C. Méranger, *Anal. Chem. 57/13*, 2478–2481 (1985).
[5668] Subramanian, K.S.; *Prog. Anal. Spectrosc. 9/2*, 237–334 (1986).
[5669] Subramanian, K.S.; in: H.A. McKenzie, L.E. Smythe (Editors), *Quantitative trace analysis of biological materials*, Elsevier, Amsterdam - New York - Oxford, 1988, 589–603.
[5670] Subramanian, K.S.; *Anal. Chem. 60/1*, 11–15 (1988).
[5671] Subramanian, K.S.; *Sci. Total Environ. 89/3*, 311–315 (1989).
[5672] Subramanian, K.S.; in: K.S. Subramanian, G.V. Iyengar, K. Okamoto (Editors), *Biological Trace Element Research*, ACS Symposium Ser. 445, ACS, Washington, D.C., 1991, 130–157.
[5673] Šucman, E.; H. Šucmanová, O. Čelechovská, S. Zima, in: B. Welz (Editor), *6. Colloquium Atomspektrometrische Spurenanalytik*, Bodenseewerk Perkin-Elmer, Überlingen, 1991, 617–624.
[5674] Suddendorf, R.F.; M.B. Denton, *Appl. Spectrosc. 28/1*, 8–13 (1974).
[5675] Suddendorf, R.F.; J.O. Watts, K. Boyer, *J. Assoc. Off. Anal. Chem. 64/5*, 1105–1110 (1981).
[5676] Suetomi, K.; H. Takahashi, *Analyst (London) 116/3*, 261–264 (1991).
[5677] Sugimoto, F.; K. Yoshikawa, Y. Maeda, *Chem. Express 6/7*, 467–470 (1991).
[5678] Sugiyama, M.; O. Fujino, M. Matsui, *Bunseki Kagaku 33*, E123 (1984).
[5679] Suhr, N.H.; C.O. Ingamells, *Anal. Chem. 38*, 730–734 (1966).
[5680] Sule, P.A.; J.D. Ingle, *Anal. Chim. Acta 326*, 85–93 (1996).
[5681] Sullivan, J.V.; A. Walsh, *Spectrochim. Acta 21*, 721–726 (1965).
[5682] Sullivan, J.V.; A. Walsh, *Spectrochim. Acta 21*, 727–730 (1965).
[5683] Sullivan, J.V.; A. Walsh, *Spectrochim. Acta 22*, 1843–1852 (1966).
[5684] Sullivan, J.V.; A. Walsh, *VI Aust. Spectrosc. Conf., Brisbane*, 1967.
[5685] Sullivan, J.V.; A. Walsh, *XIII CSI, Ottawa*, 1967.
[5686] Sullivan, J.V.; A. Walsh, *Appl. Opt. 7/7*, 1271–1280 (1968).
[5687] Sullivan, J.V.; *Prog. Anal. Spectrosc. 4/4*, 311–340 (1981).
[5688] Sun, F.-s.; K. Julshamn, *Spectrochim. Acta, Part B 42/6*, 889–894 (1987).
[5689] Sun, H.-w.; X.-q. Shan, Z.-m. Ni, *Talanta 29/7*, 589–593 (1982).
[5690] Sun, L.-g.; *Fenxi Shiyanshi 12/3*, 58–59 (1993).
[5691] Sun, P.; X. Shan, Y. Zheng, L. Jin, W. Xu, *J. Chromatogr. 110*, 73–84 (1991).
[5692] Sun, R.; J. Luo, Z. Shen, *Fenxi Huaxue 18/10*, 982 (1990).
[5693] Sun, S.-q.; X. Song, W.-d. Kang, *Youkuangye 10*, 46 (1991).
[5694] Sundaramurthi, N.M.; V.M. Shinde, *Analyst (London) 116/5*, 541–544 (1991).
[5695] Sundberg, L.L.; *Anal. Chem. 45/8*, 1460–1464 (1973).
[5696] Sunderman, F.W.; in: A. Vercruysse (Editor), *Hazardous Metals in Human Toxicology*, Vol. 4, Elsevier, Amsterdam, 1984, 279–306.

[5697] Sunderman, F.W.; M.C. Crisostomo, M.C. Reid, S.M. Hopfer, S. Nomoto, *Ann. Clin. Lab. Sci* *14/3*, 232–241 (1984).
[5698] Sunderman, F.W.; S.M. Hopfer, M.C. Crisostomo, M. Stoeppler, *Ann. Clin. Lab. Sci 16/3*, 219–230 (1986).
[5699] Sunderman, F.W.; S.M. Hopfer, M.C. Crisostomo, *Methods Emzymol. 158*, 382 (1988).
[5700] Sundin, N.G.; J.F. Tyson, C.P. Hanna, S.A. McIntosh, *Spectrochim. Acta, Part B 50/4-7*, 369–375 (1995).
[5701] Sung, Y.-h.; Z.-s. Liu, S.-d. Huang, *Spectrochim. Acta, Part B 52/7*, 755–764 (1997).
[5702] Supp, G.R.; *At. Absorpt. Newsletter 11/6*, 122–123 (1972).
[5703] Supp, G.R.; I. Gibbs, M. Juszli, *At. Absorpt. Newsletter 12*, 66–67 (1973).
[5704] Supp, G.R.; I. Gibbs, *At. Absorpt. Newsletter 13/3*, 71–73 (1974).
[5705] Sussman, I.P.; L.C. MacGregor, B.R. Masters, F.M. Matachinsky, *J. Histochem. Cytochem. 36*, 237 (1988).
[5706] Sutarno, R.; H.F. Steger, *Talanta 32/11*, 1088–1091 (1985).
[5707] Sutcliffe, P.; *Analyst (London) 101*, 949–955 (1976).
[5708] Suttie, E.D.; E.W. Wolff, *Anal. Chim. Acta 258/2*, 229–236 (1992).
[5709] Suzuki, K.T.; H. Sunaga, T. Yajima, *J. Chromatogr. 303*, 131–136 (1984).
[5710] Suzuki, M.; K. Ohta, T. Yamakita, *Anal. Chem. 53/1*, 9–13 (1981).
[5711] Suzuki, M.; K. Ohta, *Talanta 28/3*, 177–181 (1981).
[5712] Suzuki, M.; K. Ohta, *Anal. Chim. Acta 151*, 401–407 (1983).
[5713] Suzuki, M.; K. Ohta, *Prog. Anal. Spectrosc. 6/1-2*, 49–162 (1983).
[5714] Suzuki, M.; K. Ohta, *Fresenius Z. Anal. Chem. 322/5*, 480–485 (1985).
[5715] Suzuki, T.; M. Sensui, *Anal. Chim. Acta 245/1*, 43–48 (1991).
[5716] Svensmark, O.; *Fresenius Z. Anal. Chem. 309/3*, 189–192 (1981).
[5717] Sweedler, J.V.; R.B. Bilhorn, P.M. Epperson, G.R. Sims, M.B. Denton, *Anal. Chem. 60/4*, 282A–291A (1988).
[5718] Sweileh, J.A.; F.F. Cantwell, *Can. J. Chem. 63/9*, 2559–2563 (1985).
[5719] Sweileh, J.A.; D. Lucyk, B. Kratochvil, F.F. Cantwell, *Anal. Chem. 59/4*, 586–592 (1987).
[5720] Swovick, M.; *At. Spectrosc. 6/3*, 79–90 (1985).
[5721] Sychra, V.; D. Kolihová, O. Vyskočilová, R. Hlaváč, *Anal. Chim. Acta 105*, 263–270 (1979).
[5722] Sychra, V.; I. Lang, G. Šebor, *Prog. Anal. Spectrosc. 4*, 341–426 (1981).
[5723] Sychra, V.; J. Doležal, R. Hlaváč, L. Petroš, O. Vyskočilová, *J. Anal. At. Spectrom. 6/7*, 521–526 (1991).
[5724] Szaplonczay, A.M.; *Analyst (London) 97*, 29–35 (1972).
[5725] Szeto, M.S.; J.H.L. Opdebeeck, J.C. Van Loon, J. Kreins, *At. Spectrosc. 5/4*, 186–188 (1984).
[5726] Szivós, K.; L. Pólos, E. Pungor, *Spectrochim. Acta, Part B 31/5*, 289–294 (1976).
[5727] Szmyd, E.; I. Baranowska, *Chem. Anal. (Warsaw) 39*, 61–66 (1994).
[5728] Szonntagh, E.L.; *Mikrochim. Acta (Vienna) I/3-4*, 191–205 (1981).
[5729] Szpunar, J.; M. Ceulemans, V.O. Schmitt, F.C. Adams, R. Łobinski, *Anal. Chim. Acta 332/2-3*, 225–232 (1996).
[5730] Szpunar, J.; V.O. Schmitt, O.F.X. Donard, R. Łobinski, *Trends Anal. Chem. (Pers. Ed.) 15/4*, 181 (1996).
[5731] Sztraka, A.; D. Bischoff, in: B. Welz (Editor), *6. Colloquium Atomspektrometrische Spurenanalytik*, Bodenseewerk Perkin-Elmer, Überlingen, 1991, 625–632.
[5732] Tabeling, R.W.; J. Devaney, in: W.D. Ashby (Editor), *Developments in Applied Spectroscopy*, Plenum Press, New York, 1962, 175.
[5733] Taddia, M.; *Anal. Chim. Acta 158/1*, 131–136 (1984).
[5734] Taddia, M.; P. Lanza, *Anal. Chim. Acta 159*, 375–380 (1984).
[5735] Taddia, M.; *J. Anal. At. Spectrom. 1/6*, 437–441 (1986).
[5736] Taddia, M.; O. Filippini, *Fresenius Z. Anal. Chem. 328/1-2*, 64–67 (1987).
[5737] Taddia, M.; O. Filippini, *Fresenius Z. Anal. Chem. 330/6*, 506–509 (1988).
[5738] Taddia, M.; G. Clauser, *Fresenius Z. Anal. Chem. 334/2*, 148–153 (1989).
[5739] Taddia, M.; G. Clauser, *Fresenius J. Anal. Chem. 345/8-9*, 575–578 (1993).
[5740] Taddia, M.; M. Bosi, V. Poluzzi, *J. Anal. At. Spectrom. 8/5*, 755–758 (1993).
[5741] Taga, M.; M. Kan, *Talanta 36/9*, 955–956 (1989).
[5742] Tahán, J.E.; V.A. Granadillo, J.M. Sánchez, H.S. Cubillán, R.A. Romero, *J. Anal. At. Spectrom. 8/7*, 1005–1010 (1993).
[5743] Tahán, J.E.; V.A. Granadillo, R.A. Romero, *Anal. Chim. Acta 295/1-2*, 187–197 (1994).
[5744] Tahvonen, R.; J. Kumpulainen, *Fresenius J. Anal. Chem. 340/4*, 242–244 (1991).
[5745] Tahvonen, R.; *J. Food Compos. Anal. 6*, 75–86 (1993).
[5746] Takada, K.; K. Hirokawa, *Talanta 29/10*, 849–855 (1982).

[5747] Takada, K.; K. Hirokawa, *Fresenius Z. Anal. Chem. 312/2*, 109–113 (1982).
[5748] Takada, K.; T. Shoji, *Fresenius Z. Anal. Chem. 315/1*, 34–37 (1983).
[5749] Takada, K.; K. Hirokawa, *Talanta 30/5*, 329–332 (1983).
[5750] Takada, K.; *Talanta 32/9*, 921–925 (1985).
[5751] Takada, K.; *Anal. Sci. 3/3*, 221–224 (1987).
[5752] Takada, K.; K. Fujita, *Talanta 32/7*, 571–573 (1985).
[5753] Takahashi, M.; K. Takiyama, *Bunseki Kagaku 37*, 674 (1988).
[5754] Takamatsu, T.; H. Aoki, T. Yoshida, *Soil Science 133*, 239–246 (1982).
[5755] Takashima, K.; Y. Toida, *Bunseki Kagaku 43/5*, 489 (1994).
[5756] Takenaka, M.; S. Kozuka, Y. Hashimoto, *Bunseki Kagaku 42/5*, T71–T75 (1993).
[5757] Takenaka, M.; Y. Yamada, M. Hayashi, H. Endo, *Anal. Chim. Acta 336/1-3*, 151–156 (1996).
[5758] Takeuchi, T.; M. Suzuki, *Talanta 11/10*, 1391–1397 (1964).
[5759] Takiyama, K.; Y. Ishii, *Anal. Sci. 8/3*, 419–421 (1992).
[5760] Tallman, D.E.; A.U. Shaikh, *Anal. Chem. 52*, 196–199 (1980).
[5761] Tamm, R.; in: B. Welz (Editor), *Fortschritte in der atomspektrometrischen Spurenanalytik*, VCH Verlagsgesellschaft, Weinheim, 1986, 275–283.
[5762] Tamura, H.; T. Arai, M. Nagase, N. Ichinose, *Bunseki Kagaku 41*, T13 (1992).
[5763] Tan, Y.-x.; W.D. Marshall, *Analyst (London) 122/1*, 13–18 (1997).
[5764] Tanaka, T.; Y. Maki, Y. Kobayashi, A. Mizuike, *Anal. Chim. Acta 252/1-2*, 211–213 (1991).
[5765] Tanaka, T.; N. Makino, A. Mizuike, *Mikrochim. Acta (Vienna) 106*, 253–259 (1992).
[5766] Tanaka, T.; Y. Aoki, K. Tamase, F. Umoto, H. Ohbayashi, M. Imou, M. Sasaki, A. Okayama, *Shokuhin Eiseigaku Zasshi 33/4*, 359–364 (1992).
[5767] Tang, S.; P.J. Parsons, W. Slavin, *J. Anal. At. Spectrom. 10/8*, 521–526 (1995).
[5768] Tang, S.; P.J. Parsons, W. Slavin, *Analyst (London) 121/2*, 195–200 (1996).
[5769] Tang, S.-j.; Z.-q. Huang, G.-h. Zhang, *Yejin Fenxi 9/6*, 31 (1989).
[5770] Tang, Y.-w.; B.-g. Chen, S.-j. Mo, *Talanta 43/5*, 761–765 (1996).
[5771] Tao, G.-h.; Z.-l. Fang, *J. Anal. At. Spectrom. 8/4*, 577–584 (1993).
[5772] Tao, G.-h.; E.H. Hansen, *Analyst (London) 119/2*, 333–337 (1994).
[5773] Tao, G.-h.; Z.-l. Fang, *Talanta 42/3*, 375–383 (1995).
[5774] Tao, G.-h.; Z.-l. Fang, *Spectrochim. Acta, Part B 50/14*, 1747–1755 (1995).
[5775] Tao, G.-h.; Z.-l. Fang, *At. Spectrosc. 17/1*, 22–26 (1996).
[5776] Tao, H.; A. Miyazaki, K. Bansho, *Bunseki Kagaku 34*, 188–192 (1985).
[5777] Tao, S.-q.; Y. Shijo, L. Wu, L. Lin, *Analyst (London) 119/7*, 1455–1458 (1994).
[5778] Tarlin, H.I.; M. Batchelder, *J. Pharm. Sci. 59*, 1328–1330 (1970).
[5779] Tarui, T.; H. Tokairin, *Bunseki Kagaku 30*, T76 (1981).
[5780] Tatro, M.E.; W.L. Raynolds, F.M. Costa, *At. Absorpt. Newsletter 16/6*, 143–145 (1977).
[5781] Taulli, T.A.; E.F. Kaeble, *At. Absorpt. Newsletter 9*, 100–102 (1970).
[5782] Tavares, H.M.F.; M.T.S.D. Vasconcelos, A.A.S.C. Machado, P.A.P. Silva, *Analyst (London) 118/11*, 1433–1439 (1993).
[5783] Tavenier, P.; H.B.A. Hellendoorn, *Clin. Chim. Acta 23*, 47–52 (1969).
[5784] Tawali, A.B.; G. Schwedt, *Fresenius J. Anal. Chem. 357/1*, 50–55 (1997).
[5785] Taylor, A.; A.W. Walker, *Ann. Clin. Biochem. 29*, 377–389 (1992).
[5786] Taylor, J.H.; *At. Absorpt. Newsletter 8/4*, 95–96 (1969).
[5787] Taylor, J.K.; *Quality Assurance of Chemical Measurements*, Lewis Publishers, Chelsea, MI, 1987.
[5788] Taylor, P.D.P.; B. Desmet, R. Dams, *Anal. Lett. 18/A19*, 2477–2487 (1985).
[5789] Teague-Nishimura, J.E.; T. Tominaga, T. Katsura, K. Matsumoto, *Anal. Chem. 59/13*, 1647–1651 (1987).
[5790] Teissèdre, P.-L.; M.-T. Cabanis, J.-C. Cabanis, *Analusis 21/5*, 249–254 (1993).
[5791] Tekgül, H.I.; S. Akman, *Spectrochim. Acta, Part B 52/5*, 621–631 (1997).
[5792] Temminghoff, E.J.M.; *J. Anal. At. Spectrom. 5/3*, 273 (1990).
[5793] Temminghoff, E.J.M.; I. Novozamsky, *Analyst (London) 117/1*, 23–26 (1992).
[5794] Templeton, D.M.; F.W. Sunderman, R.F.M. Herber, *Sci. Total Environ. 148*, 243–251 (1994).
[5795] Terada, K.; H. Kawamura, *Anal. Sci. 7/Supp*, 71–74 (1991).
[5796] Terashima, S.; *Bunseki Kagaku 18*, 1259 (1969).
[5797] Terashima, S.; *Bunseki Kagaku 33*, 561–563 (1984).
[5798] Terashima, S.; *Geostand. Newsl. 8/2*, 155–158 (1984).
[5799] Terashima, S.; *Geostand. Newsl. 10/2*, 127–130 (1986).
[5800] Terashima, S.; *Geostand. Newsl. 12/1*, 57–60 (1988).
[5801] Terashima, S.; *Geostand. Newsl. 13/2*, 277–281 (1989).
[5802] Terashima, S.; *Geostand. Newsl. 15/1*, 125–128 (1991).

[5803] Terashima, S.; *Geostand. Newsl. 15/2*, 195–198 (1991).
[5804] Tesfalidet, S.; K. Irgum, *Anal. Chem. 60/19*, 2031–2035 (1988).
[5805] Tesfalidet, S.; K. Irgum, *Fresenius J. Anal. Chem. 338/6*, 741–744 (1990).
[5806] Tessari, G.; G.C. Torsi, *Anal. Chem. 47/6*, 842–849 (1975).
[5807] Tessier, A.; P.G.C. Campbell, M. Bisson, *J. Geochem. Explor. 16*, 77–104 (1982).
[5808] Tewari, R.K.; V.K. Tarsekar, M.B. Lokhande, *At. Spectrosc. 11/4*, 125–127 (1990).
[5809] Tewari, R.K.; J.R. Gupta, *At. Spectrosc. 14/5*, 152–153 (1993).
[5810] Thayer, J.S.; *Appl. Organomet. Chem. 1*, 227–234 (1987).
[5811] Thayer, J.S.; *Met. Ions Biol. Sys. 29*, 1–36 (1993).
[5812] Thibaud, Y.; D. Cossa, *Appl. Organomet. Chem. 3*, 257–266 (1989).
[5813] Thiem, T.L.; Y.-I. Lee, J. Sneddon, *Microchem. J. 48/1*, 65–71 (1993).
[5814] Thiemann, W.; D. Spuziak-Salzenberg, *VGB Kraftwerkstech. 69/3*, 266–267 (1989).
[5815] Thiemann, W.; D. Spuziak-Salzenberg, *VGB Kraftwerkstech. 69/3*, 299–301 (1989).
[5816] Thiers, R.E.; in: D. Glick (Editor), *Methods of Biochemical Analysis*, Interscience, New York, 1957, 273–309.
[5817] Thomaidis, N.S.; E.A. Piperaki, C.E. Efstathiou, *J. Anal. At. Spectrom. 10/3*, 221–226 (1995).
[5818] Thomaidis, N.S.; E.A. Piperaki, P.A. Siskos, *Mikrochim. Acta (Vienna) 119/3-4*, 233–241 (1995).
[5819] Thomaidis, N.S.; E.A. Piperaki, C.K. Polydorou, C.E. Efstathiou, *J. Anal. At. Spectrom. 11/1*, 31–36 (1996).
[5820] Thomaidis, N.S.; E.A. Piperaki, *Analyst (London) 121/2*, 111–117 (1996).
[5821] Thomas, B.; J.A. Roughan, E.D. Watters, *J. Sci. Food Agric. 25*, 771–776 (1974).
[5822] Thomas, J.M.; *Chemistry and Physics of Carbon Vol. 1*, Marcel Dekker, New York, 1965.
[5823] Thomas, R.P.; A.M. Ure, C.M. Davidson, D. Littlejohn, G. Rauret, R. Rubio, J.F. López-Sánchez, *Anal. Chim. Acta 286/3*, 423–429 (1994).
[5824] Thomerson, D.R.; W.J. Price, *Analyst (London) 96*, 321–329 (1971).
[5825] Thomerson, D.R.; W.J. Price, *Analyst (London) 96*, 825 (1971).
[5826] Thomerson, D.R.; W.J. Price, *Anal. Chim. Acta 72*, 188–193 (1974).
[5827] Thompson, A.J.; P.A. Thoresby, *Analyst (London) 102*, 9–16 (1977).
[5828] Thompson, D.D.; R.J. Allen, *At. Spectrosc. 2/2*, 53–58 (1981).
[5829] Thompson, J.; N.I. Ward, *J. Micronutr. Anal. 6*, 85–96 (1989).
[5830] Thompson, K.C.; D.R. Thomerson, *Analyst (London) 99*, 595–601 (1974).
[5831] Thompson, K.C.; R.G. Godden, D.R. Thomerson, *Anal. Chim. Acta 74*, 289–297 (1975).
[5832] Thompson, K.C.; K. Wagstaff, K.C. Wheatstone, *Analyst (London) 102*, 310 (1977).
[5833] Thompson, K.C.; K. Wagstaff, *Analyst (London) 105*, 641–650 (1980).
[5834] Thompson, K.C.; P.M. Cummings, *Analyst (London) 109/4*, 511–514 (1984).
[5835] Thompson, M.; B. Pahlavanpour, S.J. Walton, G.F. Kirkbright, *Analyst (London) 103*, 705–713 (1978).
[5836] Thompson, M.; *Analyst (London) 107*, 1169–1180 (1982).
[5837] Thompson, M.; *Anal. Proc. (London) 27/6*, 142–144 (1990).
[5838] Thompson, M.; E.K. Banerjee, in: S.J. Haswell (Editor), *Atomic Absorption Spectrometry - Theory, Design and Applications*, Elsevier, Amsterdam - New York, 1991, 289–320.
[5839] Thompson, M.; M. Maguire, *Analyst (London) 118/9*, 1107–1110 (1993).
[5840] Thompson, M.; H. Ramsey, *Analyst (London) 120/2*, 261–270 (1995).
[5841] Thorat, D.D.; P.N. Bhat, T.N. Mahadevan, S. Narayan, G. Subramanian, *Anal. Lett. 28/11*, 1947–1958 (1995).
[5842] Thorpe, V.A.; *J. Assoc. Off. Anal. Chem. 54/1*, 206–210 (1971).
[5843] Thorpe, V.A.; *J. Assoc. Off. Anal. Chem. 55/4*, 695–698 (1972).
[5844] Thorpe, V.A.; *J. Assoc. Off. Anal. Chem. 56/1*, 147–150 (1973).
[5845] Thulasidas, S.K.; M.J. Kulkarni, N.K. Porwal, A.G. Page, M.D. Sastry, *Anal. Lett. 21/2*, 265–278 (1988).
[5846] Thunus, L.; J.-F. Dauphin, *Anal. Chim. Acta 235/2*, 393–398 (1990).
[5847] Thunus, L.; R. Lejeune, in: H.G. Seiler, A. Sigel, H. Sigel (Editors), *Handbook on Metals in Clinical and Analytical Chemistry*, Marcel Dekker, New York, 1994, 333–338.
[5848] Thunus, L.; R. Lejeune, in: H.G. Seiler, A. Sigel, H. Sigel (Editors), *Handbook on Metals in Clinical and Analytical Chemistry*, Marcel Dekker, New York, 1994, 667–674.
[5849] Tian, J.; S. Shi, *Yaowu Fenxi Zazhi 13/4*, 260–262 (1993).
[5850] Tian, J.Q.; S.M. Shi, *Fenxi Shiyanshi 12/4*, 81–82 (1993).
[5851] Tian, L.-C.; F. Peng, X.-r. Wang, *Anal. Sci. 7/Suppl*, 1155–1158 (1991).
[5852] Tindall, F.M.; *At. Absorpt. Newsletter 5/6*, 140 (1966).

[5853] Ting, B.T.G.; A. Nahapetian, V.R. Young, M. Janghorbani, *Analyst (London) 107*, 1495–1498 (1982).
[5854] Tinggi, U.; C. Reilly, S. Hahn, M.J. Capra, *Sci. Total Environ. 125*, 15–23 (1992).
[5855] Tittarelli, P.; A. Mascherpa, *Anal. Chem. 53/9*, 1466–1469 (1981).
[5856] Tittarelli, P.; C. Biffi, *J. Anal. At. Spectrom. 7/2*, 409–415 (1992).
[5857] Tittarelli, P.; C. Biffi, in: C. Minoia, S. Caroli (Editors), *Appl. Zeeman Graphite Furnace Atom. Absorption Spectrom. Chem. Lab. Toxicol.*, Pergamon Press, Oxford, 1992, 79–94.
[5858] Tittarelli, P.; A. Anselmi, C. Biffi, V. Kmetov, *Spectrochim. Acta, Part B 50/13*, 1687–1701 (1995).
[5859] Tobies, K.H.; W. Großmann, in: B. Welz (Editor), *5. Colloquium Atomspektrometrische Spurenanalytik*, Bodenseewerk Perkin-Elmer, Überlingen, 1989, 513–521.
[5860] Toda, W.; J. Lux, J.C. van Loon, *Anal. Lett. 13/B13*, 1105–1113 (1980).
[5861] Toffaletti, J.; J. Savory, *Anal. Chem. 47/13*, 2092–2095 (1975).
[5862] Toffoli, P.; G. Pannetier, *3rd CISAFA, Paris, 1971*, Adam Hilger, London, 1973, 707–712.
[5863] Tohda, K.; in: H.G. Seiler, A. Sigel, H. Sigel (Editors), *Handbook on Metals in Clinical and Analytical Chemistry*, Marcel Dekker, New York, 1994, 571–576.
[5864] Tölg, G.; *Fresenius Z. Anal. Chem. 283*, 257–267 (1977).
[5865] Tölg, G.; in: B. Welz (Editor), *Fortschritte in der atomspektrometrischen Spurenanalytik*, Verlag Chemie, Weinheim, 1984, 5–28.
[5866] Tölg, G.; P. Tschöpel, *Anal. Sci. 3/3*, 199–208 (1987).
[5867] Tölg, G.; A. Mizuike, Yu.A. Zolotov, M. Hiraide, N.M. Kuz'min, *Pure Appl. Chem. 60/9*, 1417–1424 (1988).
[5868] Tölg, G.; *CLB, Chem. Labor Betr. 44/6*, 271–281 (1993).
[5869] Tominaga, M.; Y. Umezaki, *Anal. Chim. Acta 139*, 279–285 (1982).
[5870] Tominaga, M.; Y. Umezaki, *Anal. Chim. Acta 148*, 285–291 (1983).
[5871] Tominaga, M.; K. Bansho, Y. Umezaki, *Anal. Chim. Acta 169*, 171–177 (1985).
[5872] Tomiyasu, T.; A. Nagano, H. Sakamoto, N. Yonehara, *Anal. Sci. 12/3*, 477–481 (1996).
[5873] Tomkins, D.F.; C.W. Frank, *Appl. Spectrosc. 25/5*, 539–541 (1971).
[5874] Tong, A.; Y. Akama, *Chem. Express 5/10*, 705–708 (1990).
[5875] Tong, S.-l.; *Anal. Chem. 50/3*, 412–414 (1978).
[5876] Tonini, C.; *Tinctoria 75*, 160 (1978).
[5877] Tonini, C.; *Tinctoria 76*, 117 (1979).
[5878] Tonini, C.; *Tinctoria 77*, 358 (1980).
[5879] Topping, G.; *Sci. Total Environ. 49*, 9–25 (1986).
[5880] Toribara, T.Y.; C.P. Shields, L. Korval, *Talanta 17*, 1025–1028 (1970).
[5881] Torre, M.; M.C. González, O. Jiménez, A.R. Rodríguez, *Anal. Lett. 23/8*, 1519–1536 (1990).
[5882] Torsi, G.; G. Tessari, *Anal. Chem. 47/6*, 839–842 (1975).
[5883] Torsi, G.; F. Palmisano, *Spectrochim. Acta, Part B 41/3*, 257–264 (1986).
[5884] Torsi, G.; P. Reschiglian, F. Fagioli, C. Locatelli, *Spectrochim. Acta, Part B 48/5*, 681–689 (1993).
[5885] Torsi, G.; *Spectrochim. Acta, Part B 50/8*, 707–712 (1995).
[5886] Trahey, N.M.; (Editor), *NIST Standard Reference Materials Catalog 1995-1996*, NIST Special Publication 260, U.S. Gov. Printing Off., Washingthon, 1995.
[5887] Tramblay, G.; F.J. Vastola, P.L. Walker, *Carbon 16*, 35 (1978).
[5888] Trapp, G.A.; *Anal. Biochem. 148/1*, 127–132 (1985).
[5889] Trent, D.J.; W. Slavin, *At. Absorpt. Newsletter 3*, 53–64 (1964).
[5890] Trent, D.J.; *At. Absorpt. Newsletter 4*, 348–350 (1965).
[5891] Trent, D.J.; D.C. Manning, W. Slavin, *At. Absorpt. Newsletter 4*, 335–337 (1965).
[5892] Trigt, C. van ; Tj. Hollander, C.Th.J. Alkemade, *J. Quant. Spectrosc. Radiat. Transfer 5*, 813–833 (1965).
[5893] Trojanowicz, M.; K. Pyrzynska, *Anal. Chim. Acta 287/3*, 247–252 (1994).
[5894] Trostle, D.; T. Beals, R. Kuczenski, M. Shaver, *At. Spectrosc. 12/2*, 64–68 (1991).
[5895] Trüb, U.-H.; *Phoenix International 6*, 36–39 (1986).
[5896] Truckenbrodt, D.; J. Einax, *Fresenius J. Anal. Chem. 352/5*, 437–443 (1995).
[5897] Trudeau, D.L.; E.F. Freier, *Clin. Chem. (Winston-Salem) 13*, 101–114 (1967).
[5898] Tsai, S.-J.J.; Y.-L. Bae, *Analyst (London) 118/3*, 297–300 (1993).
[5899] Tsai, S.-J.J.; C.-C. Jan, *Analyst (London) 118/9*, 1183–1191 (1993).
[5900] Tsai, S.-J.J.; C.-C. Jan, L.-L. Chang, *Spectrochim. Acta, Part B 49/8*, 773–785 (1994).
[5901] Tsakovski, S.; E. Ivanova, I. Havezov, *Talanta 41/5*, 721–724 (1994).
[5902] Tsalev, D.L.; I. Petrov, *Anal. Chim. Acta 111*, 155–162 (1979).
[5903] Tsalev, D.L.; I. Petrov, *Dokl. Bolg. Akad. Nauk. 34/10*, 1413–1416 (1981).

[5904] Tsalev, D.L.; V.I. Slaveykova, P.B. Mandjukov, *Spectrochimica Acta Rev. 13/3*, 225–274 (1990).
[5905] Tsalev, D.L.; T.A. Dimitrov, P.B. Mandjukov, *J. Anal. At. Spectrom. 5/3*, 189–194 (1990).
[5906] Tsalev, D.L.; M. Sperling, B. Welz, *Analyst (London) 117/11*, 1729–1733 (1992).
[5907] Tsalev, D.L.; M. Sperling, B. Welz, *Analyst (London) 117/11*, 1735–1741 (1992).
[5908] Tsalev, D.L.; V.I. Sleveykova, *Spectrosc. Lett. 25/2*, 221–238 (1992).
[5909] Tsalev, D.L.; E.I. Tserovsky, A.G. Raitcheva, A.I. Barzev, R.G. Georgieva, Z.K. Zaprianov, *Spectrosc. Lett. 26/2*, 331–346 (1993).
[5910] Tsalev, D.L.; A. D'Ulivo, L. Lampugnani, M. Di Marco, R. Zamboni, *J. Anal. At. Spectrom. 10/11*, 1003–1009 (1995).
[5911] Tsalev, D.L.; *Atomic Absorption Spectrometry in Occupational and Environmental Health Practice. Vol. III: Progress in Analytical Methodology*, CRC Press, Boca Raton, FL, 1995.
[5912] Tsalev, D.L.; V.I. Sleveykova, R.B. Georgieva, *Anal. Lett. 29/1*, 73–88 (1996).
[5913] Tsalev, D.L.; A. D'Ulivo, L. Lampugnani, M. Di Marco, R. Zamboni, *J. Anal. At. Spectrom. 11/10*, 979–988 (1996).
[5914] Tsalev, D.L.; A. D'Ulivo, L. Lampugnani, M. Di Marco, R. Zamboni, *J. Anal. At. Spectrom. 11/10*, 989–995 (1996).
[5915] Tsalev, D.L.; V.I. Sleveykova,.Chemical modification in electrothermal atomic absorption spectrometry, in: J. Sneddon, *Advances in Atomic Spectroscopy*, Volume IV, JAI Press, Greenwich, 1997.
[5916] Tschöpel, P.; L. Kotz, W. Schulz, M. Veber, G. Tölg, *Fresenius Z. Anal. Chem. 302/1*, 1–14 (1980).
[5917] Tschöpel, P.; G. Tölg, *J. Trace Microprobe Techniques 1*, 1–77 (1982).
[5918] Tschöpel, P.; *Labo 19/4*, 9–23 (1988).
[5919] Tsuchiya, S.; *Bull. Chem. Soc. Jpn. 37/6*, 828–842 (1964).
[5920] Tsuda, T.; M. Wada, S. Aoki, Y. Matsui, *J. Assoc. Off. Anal. Chem. 71/2*, 373–374 (1988).
[5921] Tsukahara, I.; M. Tanaka, *Anal. Chim. Acta 116*, 383–389 (1980).
[5922] Tsukahara, I.; M. Tanaka, *Talanta 27/3*, 237–241 (1980).
[5923] Tsukahara, I.; M. Tanaka, *Talanta 27/8,* 655–658 (1980).
[5924] Tsukahara, I.; T. Yamamoto, *Talanta 28/8*, 585–589 (1981).
[5925] Tsukahara, I.; T. Yamamoto, *Anal. Chim. Acta 135*, 235–241 (1982).
[5926] Tsunoda, K.-i.; H. Haraguchi, K. Fuwa, *Spectrochim. Acta, Part B 35/11-2*, 715–729 (1980).
[5927] Tubb, A.; G. Nickless, *Anal. Proc. (London) 19*, 334–336 (1982).
[5928] Tubino, M.; J.R. de Oliveira Torres, *Anal. Lett. 23/12*, 2339–2349 (1990).
[5929] Tuell, T.M.; A.H. Ullman, B.D. Pollard, A. Massoumi, J.D. Bradshaw, J.N. Bower, J.D. Winefordner, *Anal. Chim. Acta 108*, 351–356 (1979).
[5930] Türkan, I.; E. Henden, Ü. Celik, S. Kivilcim, *Sci. Total Environ. 166*, 61–67 (1995).
[5931] Turunen, M.; S. Peräniemi, M. Ahlgrén, H. Westerholm, *Anal. Chim. Acta 311/1*, 85–91 (1995).
[5932] Twardowska, I.; J. Kyziol, *Fresenius J. Anal. Chem. 354/5-6*, 580–586 (1996).
[5933] Tye, C.T.; S.J. Haswell, P. O'Neill, K.C.C. Bancroft, *Anal. Chim. Acta 169*, 195–200 (1985).
[5934] Tyndall, J.; *Six Lectures on Light*, D. Appleton, New York, 1898.
[5935] Tyson, J.F.; *Anal. Proc. (London) 18/12*, 542–545 (1981).
[5936] Tyson, J.F.; *Analyst (London) 109/3*, 313–317 (1984).
[5937] Tyson, J.F.; J.M.H. Appleton, *Talanta 31/1*, 9–14 (1984).
[5938] Tyson, J.F.; C.E. Adeeyinwo, J.M.H. Appleton, S.R. Bysouth, A.B. Idris, *Analyst (London) 110/5*, 487–492 (1985).
[5939] Tyson, J.F.; *Analyst (London) 112*, 523–526 (1987).
[5940] Tyson, J.F.; S.R. Bysouth, *J. Anal. At. Spectrom. 3/1*, 211–215 (1988).
[5941] Tyson, J.F.; *Fresenius Z. Anal. Chem. 329/6*, 663–667 (1988).
[5942] Tyson, J.F.; *Spectrochimica Acta Rev. 14/3*, 169–233 (1991).
[5943] Tyson, J.F.; S.G. Offley, N.J. Seare, H.A.B. Kibble, C.S. Fellows, *J. Anal. At. Spectrom. 7/2*, 315–322 (1992).
[5944] Uchida, T.; I. Kojima, C. Iida, *Anal. Chim. Acta 116*, 205–210 (1980).
[5945] Uchida, T.; B.L. Vallee, *Anal. Sci. 2/3*, 243–248 (1986).
[5946] Uchida, Y.; S. Hattori, *Oyo Butsuri 44*, 852 (1975).
[5947] Uchino, E.; T. Kosuga, S. Konishi, M. Nishimura, *Environ. Sci. Technol. 21/9*, 920–922 (1987).
[5948] Udelnow, C.; R. Rautschke, *Z. Chem. 26/8*, 297–298 (1986).
[5949] Udoh, A.P.; S.A. Thomas, E.J. Ekanem, *Bull. Chem. Soc. Ethiop. 4/1*, 13–18 (1990).
[5950] Udoh, A.P.; S.A. Thomas, E.J. Ekanem, *Talanta 39/12*, 1591–1595 (1992).

[5951] Udoh, A.P.; *Talanta 42/12*, 1827–1831 (1995).
[5952] Ueda, J.; C. Mizui, *Anal. Sci. 4/4*, 417–421 (1988).
[5953] Ueda, J.; M. Takagi, *Bull. Chem. Soc. Jpn. 63/2*, 544–547 (1990).
[5954] Ueda, K.; K. Kubo, O. Yoshimura, Y. Yamamoto, *Bull. Chem. Soc. Jpn. 61/8*, 2791–2795 (1988).
[5955] Uesugi, K.; S. Suehiro, H. Nishioka, T. Kumagai, T. Nagahiro, *Nippon Kaisui Gakkaishi 46*, 22 (1992).
[5956] Uggerud, H.; W. Lund, *J. Anal. At. Spectrom. 10/5*, 405–408 (1995).
[5957] Ulberth, F.; S. Blineder, *Fat. Sci. Technol. 5*, 185–188 (1989).
[5958] Ulfvarsons, U.; *Acta Chem. Scand. 21*, 641 (1967).
[5959] Urasa, I.T.; W.J. Mavura, *Int. J. Environ. Anal. Chem. 48/3*, 229–240 (1992).
[5960] Urbain, H.; A. Chambosse, *At. Spectrosc. 3/5*, 143–145 (1982).
[5961] Ure, A.M.; L.R.P. Butler, R.G. Scott, R. Jenkins, *Pure Appl. Chem. 60/9*, 1461–1472 (1988).
[5962] Ure, A.M.; *Fresenius J. Anal. Chem. 337/5*, 577–581 (1990).
[5963] Ure, A.M.; R. Thomas, D. Littlejohn, *Int. J. Environ. Anal. Chem. 51*, 65–84 (1993).
[5964] Ure, A.M.; Ph. Quevauviller, H. Munttau, B. Griepink, *CEC Report EUR 14763*, Brussels, 1993.
[5965] Ure, A.M.; L.R.P. Butler, R.G. Scott, R. Jenkins, *Spectrochim. Acta, Part B 52/4*, 409–420 (1997).
[5966] Usami, S.; S. Yamada, B.K. Puri, M. Satake, *Mikrochim. Acta (Vienna) I*, 263–270 (1989).
[5967] Usenko, S.I.; M.M. Prorok, *Zavodsk. Lab. 58/6*, 6–7 (1992).
[5968] Uthe, J.F.; F.A.J. Armstrong, M.P. Stainton, *J. Fish Res. Bd. Canada 27*, 805–811 (1970).
[5969] Uthus, E.O.; M.E. Collings, W.E. Cornatzer, F.H. Nielsen, *Anal. Chem. 53/14*, 2221–2224 (1981).
[5970] Vaessen, H.A.M.G.; C.G. Van de Kamp, *Pure Appl. Chem. 61/1*, 113–120 (1989).
[5971] Vaessen, H.A.M.G.; C.G. Van de Kamp, *Z. Lebensm. Unters. Forsch. 190*, 199–204 (1990).
[5972] Vaeth, E.; E. Holz, *Fette, Seifen, Anstrichm. 87/3*, 97–99 (1985).
[5973] Vajda, F.; *Anal. Chim. Acta 128*, 31–43 (1981).
[5974] Valcárcel, M.; M. Gallego, P. Martínez-Jiménez, *Anal. Proc. (London) 23*, 233 (1986).
[5975] Valcárcel, M.; M.D. Luque de Castro, *Automatic Methods of Analysis*, Elsevier, Amsterdam, 1988.
[5976] Valcárcel, M.; M. Gallego, *Trends Anal. Chem. (Pers. Ed.) 8/1*, 34–40 (1989).
[5977] Valcárcel, M.; M. Gallego, R. Montero, *J. Pharm. Biomed. Anal. 8/8-12*, 655–661 (1990).
[5978] Valcárcel, M.; M.D. Luque de Castro, *Non-Chromatographic Continuous Separation Techniques*, Royal Society of Chemistry, Cambridge, 1991.
[5979] Valcárcel, M.; M.D. Luque de Castro, *Microchem. J. 45/2*, 189–209 (1992).
[5980] Valdés-Hevia y Temprano, M.C.; J. Pérez Parajón, M.E. Díaz-García, A. Sanz-Medel, *Analyst (London) 116/11*, 1141–1144 (1991).
[5981] Valdés-Hevia y Temprano, M.C.; M.R. Fernández de la Campa, A. Sanz-Medel, *J. Anal. At. Spectrom. 8/6*, 847–852 (1993).
[5982] Valdés-Hevia y Temprano, M.C.; M.R. Fernández de la Campa, A. Sanz-Medel, *J. Anal. At. Spectrom. 9/3*, 231–236 (1994).
[5983] Valkirs, A.O.; P.F. Seligman, P.M. Stang, V. Homer, S.H. Lieberman, G. Vafa, C.A. Dooley, *Mar. Pollut. Bull. 17*, 319–324 (1986).
[5984] Valkirs, A.O.; M.O. Stallard, P.M. Stang, S. Frank, *Analyst (London) 115/10*, 1327–1328 (1990).
[5985] Valkovic, V.; *Trace Elements in Petroleum*, The Petroleum Publishing Company, Tusla, Oklahoma, USA, 74101, 1978.
[5986] Vallee, B.L.; D.S. Auld, *Biochemistry 29*, 5647–5659 (1990).
[5987] Vandeberg, J.T.; H.D. Swafford, R.W. Scott, *J. Paint Technol. 47*, 84–91 (1975).
[5988] Vandecasteele, C.; G. Windels, B. Desmet, A. de Ruck, R. Dams, *Analyst (London) 113/11*, 1691–1694 (1988).
[5989] Vanhaecke, F.; H. Vanhoe, L Moens, R. Dams, *Bull. Soc. Chim. Belg. 104/11*, 653–661 (1995).
[5990] Vanhoe, H.; C. Vandecasteele, B. Desmet, R. Dams, *J. Anal. At. Spectrom. 3/5*, 703–707 (1988).
[5991] Vankeerberghen, P.; J. Smeyers-Verbeke, A. Thielemans, D.L. Massart, *Analusis 20/3*, 103–109 (1992).
[5992] Vankeerberghen, P.; J. Smeyers-Verbeke, *Chemom. Intell. Lab. Sys. 15/2-3*, 195–202 (1992).
[5993] Vankeerberghen, P.; J. Smeyers-Verbeke, D.L. Massart, *J. Anal. At. Spectromn. 11/2*, 149–158 (1996).

[5994] Vanloo, B.; R. Dams, J. Hoste, *Anal. Chim. Acta 151/2*, 391–400 (1983).
[5995] Vanloo, B.; R. Dams, J. Hoste, *Anal. Chim. Acta 160*, 329–333 (1984).
[5996] Vanloo, B.; R. Dams, J. Hoste, *Anal. Chim. Acta 175*, 325–328 (1985).
[5997] Vargas, M.C. de; R.A. Romero, *At. Spectrosc. 10/5*, 160–164 (1989).
[5998] Vargas, M.C. de; R.A. Romero, *Analyst (London) 117/3*, 645–647 (1992).
[5999] Varley, J.A.; P.Y. Chin, *Analyst (London) 95*, 592–595 (1970).
[6000] Varma, A.; *Talanta 28/10*, 785–787 (1981).
[6001] Vassileva, E.; S. Arpadjan, *Analyst (London) 115/4*, 399–403 (1990).
[6002] Vassileva, E.; S. Arpadjan, R. Stefanova, V. Vassileva, *Fresenius J. Anal. Chem. 336/7*, 582–585 (1990).
[6003] Vassileva, E.; K. Hadjiivanov, *Fresenius J. Anal. Chem. 357/7*, 881–885 (1997).
[6004] Vaughn, W.W.; J.H. McCarthy, *U.S. Geological Survey Prof. Paper 501-D*, 1964.
[6005] Vazquez-Gonzalez, J.-F.; P. Bermejo-Barrera, F. Bermejo-Martínez, *Anal. Chim. Acta 219/1*, 79–87 (1989).
[6006] Veber, M.; K. Čujes, S. Gomišček, *J. Anal. At. Spectrom. 9/3*, 285–290 (1994).
[6007] Vecchietti, R.; F. Fagioli, C. Locatelli, G. Torsi, *Talanta 36/7*, 743–748 (1989).
[6008] Veen, N.G. van der ; H.J. Keukens, G. Vos, *Anal. Chim. Acta 171*, 285–291 (1985).
[6009] Veillon, C.; J.M. Mansfield, M.L. Parsons, J.D. Winefordner, *Anal. Chem. 38/2*, 204–208 (1966).
[6010] Veillon, C.; in: J.A. Dean, T.C. Rains (Editors), *Flame Emission and Atomic Absorption Spectrometry, Vol. 2*, Marcel Dekker, New York, 1971, 177–195.
[6011] Veillon, C.; B.E. Guthrie, W.R. Wolf, *Anal. Chem. 52/3*, 457–459 (1980).
[6012] Veillon, C.; *Anal. Chem. 58/8*, 851A–866A (1986).
[6013] Veillon, C.; *Methods Emzymol. 158*, 334–343 (1988).
[6014] Veillon, C.; *Sci. Total Environ. 86*, 65–68 (1989).
[6015] Veillon, C.; K.Y. Patterson, *Analyst (London) 121/7*, 983–985 (1996).
[6016] Vélez, D.; N. Ybáñez, R. Montoro, *J. Anal. At. Spectrom. 11/4*, 271–277 (1996).
[6017] Vélez, D.; N. Ybáñez, R. Montoro, *J. Anal. At. Spectrom. 12/1*, 91–96 (1997).
[6018] Velghe, N.; A. Campe, A. Claeys, *At. Absorpt. Newsletter 17/2*, 37–40 (1978).
[6019] Velghe, N.; A. Campe, A. Claeys, *At. Absorpt. Newsletter 17/6*, 139–143 (1978).
[6020] Venghiattis, A.A.; *Spectrochim. Acta, Part B 23/2*, 67–78 (1967).
[6021] Venghiattis, A.A.; *Appl. Opt. 7/7*, 1313–1316 (1968).
[6022] Venkaji, K.; P.P. Naidu, T.J.P. Rao, *Talanta 41/8*, 1281–1290 (1994).
[6023] Verbeek, A.A.; M.C. Mitchell, A.M. Ure, *Anal. Chim. Acta 135*, 215–228 (1982).
[6024] Vercammen, M.; W. Goedhuys, A. Broeyckens, R. De Roy, J. Sennesael, F. Gorus, *Clin. Chem. (Winston-Salem) 36/10*, 1812–1815 (1990).
[6025] Verebey, K.; J.F. Rosen, D.J. Schonfeld, D. Carriero, Y.M. Eng, J. Deutsch, S. Reimer, J. Hogan, *Clin. Chem. (Winston-Salem) 41/3*, 469–470 (1995).
[6026] Vereda Alonso, E.; A. García de Torres, J.M. Cano Pavón, *Mikrochim. Acta (Vienna) 110/1-3*, 41–45 (1993).
[6027] Vereda Alonso, E.; J.M. Cano Pavon, A. García de Torres, M.T. Siles Cordero, *Anal. Chim. Acta 283/1*, 224–229 (1993).
[6028] Verein Deutscher Eisenhüttenleute, *Handbuch für das Eisenhüttenlaboratorium Band 5*, Verlag Stahleisen, Düsseldorf, 1977.
[6029] Verein Deutscher Ingenieure e.V., *Richtlinie VDI-2268/1*, 1987.
[6030] Verein Deutscher Eisenhüttenleute, *Handbuch für das Eisenhüttenlaboratorium Band 2*, Verlag Stahleisen, Düsseldorf, 1987.
[6031] Verein Deutscher Ingenieure e.V., *Vorentwurf VDI 3868/Bl.2*, 1987.
[6032] Verein Deutscher Ingenieure e.V., *VDI-Richtlinie 2268/2*, 1990.
[6033] Verein Deutscher Ingenieure e.V., *VDI-Richtlinie 2268/4*, 1990.
[6034] Verein Deutscher Ingenieure e.V.; *VDI-Richtlinie 2310, Atembare Stäube, VDI-Handbuch Reinhaltung der Luft*, Beuth Verlag, Berlin.
[6035] Verein Deutscher Ingenieure e.V.; *VDI-Richtlinie 2066, Blatt 2, Emissionsmessungen, VDI-Handbuch Reinhaltung der Luft*, Beuth Verlag, Berlin
[6036] Verlinden, M.; H. Deelstra, *Fresenius Z. Anal. Chem. 296*, 253–258 (1979).
[6037] Verlinden, M.; W. Cooreman, H. Deelstra, in: P. Brätter, P. Schramel (Editors), *Trace Element Analytical Chemistry in Medicine and Biology*, W. de Gruyter, Berlin - New York, 1980, 513–522.
[6038] Verlinden, M.; J. Baart, H. Deelstra, *Talanta 27/8*, 633–639 (1980).
[6039] Verlinden, M.; *Anal. Chim. Acta 140*, 229–235 (1982).

[6040] Verlinden, M.; *Talanta 29/10*, 875–882 (1982).
[6041] Verlinden, M.; W. Cooreman, H. Deelstra, in: P. Brätter, P. Schramel (Editors), *Trace Element Analytical Chemistry in Medicine and Biology, Vol. 2*, W. de Gruyter, Berlin - New York, 1983, 513–522.
[6042] Vermeir, G.; C. Vandecasteele, E. Temmerman, R. Dams, J. Versieck, *Mikrochim. Acta (Vienna) 1988/III*, 305–313 (1988).
[6043] Vermeir, G.; C. Vandecasteele, R. Dams, *Anal. Chim. Acta 220/1*, 257–261 (1989).
[6044] Vermette, S.J.; M.E. Peden, S. Hamdy, T.C. Willoughby, L. Schroeder, S.E. Lindberg, J.G. Owens, A.D. Weiss, *Gen. Tech. Rep. NC 150*, North Centre for Exp. Stn., 1992.
[6045] Vermeulen, J.M.; *J. Anal. At. Spectrom. 4/1*, 77–82 (1989).
[6046] Vernet, M.; L. Marin, S. Boulmier, J. Lhomme, *Analusis 15/2*, 87–94 (1987).
[6047] Vernon, F.; C.D. Wani, *Anal. Proc. (London) 30/11*, 442–445 (1993).
[6048] Verschoor, M.A.; R. Herber, J. van Hemmen, A. Wibowo, R. Zielhuis, *Scand. J. Work Environ. Health 13*, 232 (1987).
[6049] Versieck, J.; F. Barbier, R. Cornelis, J. Hoste, *Talanta 29/11B*, 973–984 (1982).
[6050] Versieck, J.; *Trends Anal. Chem. (Pers. Ed.) 2/5*, 110–113 (1983).
[6051] Versieck, J.; *Trace Elem. Med. 1/1*, 2–12 (1984).
[6052] Versieck, J.; *CRC Crit. Rev. Clin. Lab. Sci. 22/2*, 97–184 (1985).
[6053] Versieck, J.; *J. Res. Natl. Bur. Stand. 91/2*, 87–92 (1986).
[6054] Versieck, J.; *Anal. Chim. Acta 204*, 299 (1987).
[6055] Versieck, J.; L. Vanballenberghe, *Biol. Trace Elem. Res. 12*, 45–54 (1987).
[6056] Versieck, J.; L. Vanballenberghe, A. De Kesel, J. Hoste, B. Wallaeys, J. Vandenhaute, N. Baeck, F.W. Sunderman, *Anal. Chim. Acta 204/1-2*, 63–75 (1988).
[6057] Versieck, J.; L. Vanballenberghe, A. De Kesel, *Clin. Chem. (Winston-Salem) 34/8*, 1659–1660 (1988).
[6058] Versieck, J.; R. Cornelis, *Trace Elements in Human Plasma or Serum*, CRC Press, Boca Raton, FL, 1989.
[6059] Versieck, J.; L. Vanballenberghe, *Anal. Chem. 63*, 1143–1146 (1991).
[6060] Versieck, J.; L. Vanballenberghe, in: H.G. Seiler, A. Sigel, H. Sigel (Editors*), Handbook on Metals in Clinical and Analytical Chemistry*, Marcel Dekker, New York, 1994, 31–44.
[6061] Vertacnik, A.; J. Biscan, *J. Radioanal. Nucl. Chem. 175/5*, 401–413 (1993).
[6062] Vest, P.; M. Schuster, K.-H. König, *Fresenius Z. Anal. Chem. 335/7*, 759–763 (1989).
[6063] Vest, P.; M. Schuster, K.-H. König, *Fresenius J. Anal. Chem. 339/3*, 142–144 (1991).
[6064] Vevey, E. De ; G. Bitton, D. Rossel, L.D. Ramos, L. Mungia Guerrero, J. Tarradellas, *Bull. Environ. Contam. Toxicol. 50/2*, 253–259 (1993).
[6065] Vickrey, T.M.; H.E. Howell, M.T. Paradise, *Anal. Chem. 51*, 1880–1883 (1979).
[6066] Victor, A.H.; *Geostand. Newsl. 10*, 13–18 (1986).
[6067] Vidal, J.C.; J.M. Sanz, J.R. Castillo-Suarez, *Fresenius J. Anal. Chem. 344/6*, 234–242 (1992).
[6068] Vidal, M.T.; M. de la Guardia, *Talanta 34/10*, 892–894 (1987).
[6069] Vidal, M.T.; F. Lizondo, M. de la Guardia, *At. Spectrosc. 11/2*, 75–77 (1990).
[6070] Vidale, G.L.; *Report R60SD330, General Electric T.I.S.*, 1961.
[6071] Vidale, G.L.; *Report R60SD331, General Electric T.I.S.*, 1962.
[6072] Viets, J.G.; R.M. O'Leary, J.R. Clark, *Analyst (London) 109/12*, 1589–1592 (1984).
[6073] Viets, J.G.; R.M. O'Leary, *J. Geochem. Explor. 44/1-3*, 107–138 (1992).
[6074] Vijan, P.N.; C.Y. Chan, *Anal. Chem. 48*, 1788–1792 (1976).
[6075] Vijan, P.N.; A.C. Rayner, D. Sturgis, G.R. Wood, *Anal. Chim. Acta 82*, 329–336 (1976).
[6076] Vijan, P.N.; G.R. Wood, *Talanta 23*, 89–94 (1976).
[6077] Vijan, P.N.; *At. Spectrosc. 1/5*, 143–144 (1980).
[6078] Villalba, A.; M.A. Elorza, R. Coll, *Clin. Chim. Acta 214/2*, 243–244 (1993).
[6079] Vircavs, M.; V. Rone, A.M. Garbar, V.G. Pimenov, D. Vircava, M.B. Kozlova, I. Mednis, *Vysokochistye Veshchestva 3*, 138–144 (1992).
[6080] Viñas, P.; N. Campillo, I. López García, V.C. Hernandez, *Food Chem. 46*, 307–311 (1993).
[6081] Viñas, P.; N. Campillo, I. López García, M.H. Córdoba, *J. Agri. Food Chem. 41*, 2024–2027 (1993).
[6082] Viñas, P.; N. Campillo, I. López García, M.H. Córdoba, *Anal. Chim. Acta 283/1*, 393–400 (1993).
[6083] Viñas, P.; N. Campillo, I. López García, M.H. Córdoba, *Fresenius J. Anal. Chem. 349/4*, 306–310 (1994).
[6084] Viñas, P.; N. Campillo, I. López García, M.H. Córdoba, *Food Chem. 50*, 317–321 (1994).
[6085] Viñas, P.; N. Campillo, I. López García, M.H. Córdoba, *Talanta 42/4*, 527–533 (1995).
[6086] Viñas, P.; N. Campillo, I. López García, M.H. Córdoba, *At. Spectrosc. 16/2*, 86–89 (1995).

896   Bibliography

[6087] Viñas, P.; N. Campillo, I. López García, M.H. Córdoba, *Fresenius J. Anal. Chem. 351/7*, 695–696 (1995).
[6088] Vogliotti, F.L.; *At. Absorpt. Newsletter 9/6*, 123–125 (1970).
[6089] Voigtman, E.; *Appl. Spectrosc. 45/2*, 237–241 (1991).
[6090] Voigtman, E.; *Anal. Instr. (New York) 21/1-2*, 43–62 (1993).
[6091] Volchikhina, E.M.; A.P. Garshin, I.P. Zornikov, *Zavodsk. Lab. 56/2*, 33 (1990).
[6092] Volchikhina, E.M.; A.P. Garshin, V.V. Krivoshein, *Zavodsk. Lab. 57/12*, 22–23 (1991).
[6093] Volland, G.; G. Kölblin, P. Tschöpel, G. Tölg, *Fresenius Z. Anal. Chem. 284*, 1–12 (1977).
[6094] Völlkopf, U.; Z. Grobenski, B. Welz, *At. Spectrosc. 4/5*, 165–170 (1983).
[6095] Völlkopf, U.; Z. Grobenski, *At. Spectrosc. 5/3*, 115–122 (1984).
[6096] Völlkopf, U.; Z. Grobenski, R. Tamm, B. Welz, *Analyst (London) 110/6*, 573–577 (1985).
[6097] Völlkopf, U.; R. Lehmann, D. Weber, *LaborPraxis 9*, 990 (1985).
[6098] Völlkopf, U.; R. Lehmann, D. Weber, *J. Anal. At. Spectrom. 2*, 455–458 (1987).
[6099] Vollmer, J.; C. Sebens, W. Slavin, *At. Absorpt. Newsletter 4/6*, 306–308 (1965).
[6100] Vollmer, J.; *At. Absorpt. Newsletter 5*, 12 (1966).
[6101] Vollmer, J.; *At. Absorpt. Newsletter 5*, 35 (1966).
[6102] Vollrath, A.; T. Otz, C. Hohl, H.G. Seiler, *Fresenius J. Anal. Chem. 344/6*, 269–274 (1992).
[6103] Volynskii, A.B.; B.Y. Spivakov, Y.A. Zolotov, *Talanta 31/6*, 449–458 (1984).
[6104] Volynskii, A.B.; E.M. Sedykh, B.Y. Spivakov, I. Havezov, *Anal. Chim. Acta 174*, 173–182 (1985).
[6105] Volynskii, A.B.; V. Krivan, *J. Anal. At. Spectrom. 11*, 159–164 (1996).
[6106] Volynskii, A.B.; V. Krivan, S.V. Tikhomirov, *Spectrochim. Acta, Part B 51/9-10*, 1253–1261 (1996).
[6107] Vos, G.; *Fresenius Z. Anal. Chem. 320/6*, 556–561 (1985).
[6108] Voth-Beach, L.M.; D.E. Shrader, *J. Anal. At. Spectrom. 2/1*, 45–50 (1987).
[6109] Voulgaropoulos, A.; M. Paneli, E. Papaefstathiou, S. Stavroulias, *Fresenius J. Anal. Chem. 341/9*, 568–569 (1991).
[6110] Vrána, A.; E. Petrzelová, Keyzlarová, in: B. Welz (Editor), *CANAS'95 Colloquium Analytische Atomspektroskopie*, Bodenseewerk Perkin-Elmer, Überlingen, 1996, 747–751.
[6111] Vulfson, E.K.; A.V. Karyakin, A.I. Shidlovsky, *Zh. Anal. Khim. 28*, 1253–1256 (1973).
[6112] Vyskočilová, O.; V. Sychra, D. Kolihová, *Anal. Chim. Acta 105*, 271–279 (1979).
[6113] Waal, W.A.J. de; F.J.M.J. Maessen, J.C. Kraak, *J. Pharm. Biomed. Anal. 8/1*, 1–30 (1990).
[6114] Wabner, C.L.; D.C. Sears, *At. Spectrosc. 17/3*, 119–121 (1996).
[6115] Wagemann, R.; *Water Research 12*, 139 (1978).
[6116] Wagenaar, H.C.; L. de Galan, *Spectrochim. Acta, Part B 28*, 157–177 (1973).
[6117] Wagenaar, H.C.; C.J. Pickford, L. de Galan, *Spectrochim. Acta, Part B 29*, 211–224 (1974).
[6118] Wagenaar, H.C.; I. Novotný, L. de Galan, *Spectrochim. Acta, Part B 29*, 301–317 (1974).
[6119] Wagenaar, H.C.; L. de Galan, *Spectrochim. Acta, Part B 30*, 361–381 (1975).
[6120] Wagenaar, H.C.; *Dissertation, Techn. Hochschule* Delft, 1976.
[6121] Wagenen, S. Van; D.E. Carter, A.G. Ragheb, Q. Fernando, *Anal. Chem. 59/6*, 891–896 (1987).
[6122] Wagley, D.; G. Schmiedel, E. Mainka, H.J. Ache, *At. Spectrosc. 10/4*, 106–111 (1989).
[6123] Wagner, G.; in: H. Lieth, B. Markert (Editors), *Element Concentration Cadasters in Ecosystems (ECCE)*, Verlag Chemie, Weinheim, 1990, 41–54.
[6124] Wagner, G.; in: M. Stoeppler (Editor), *Probennahme und Aufschluß*, Springer, Berlin - Heidelberg - New York, 1994, 71–83.
[6125] Wagner, H.P.; M.J. McGarrity, *J. Am. Soc. Brew. Chem. 49*, 151 (1991).
[6126] Wagner, H.P.; *J. Am. Soc. Brew. Chem. 53/3*, 141–144 (1995).
[6127] Wahab, H.S.; C.L. Chakrabarti, *Spectrochim. Acta, Part B 36/5*, 463–474 (1981).
[6128] Wahab, H.S.; C.L. Chakrabarti, *Spectrochim. Acta, Part B 36/5*, 475–481 (1981).
[6129] Wai, C.M.; L.M. Tsay, J.C. Yu, *Mikrochim. Acta (Vienna) 2*, 73–78 (1987).
[6130] Walcerz, M.; E. Bulska, A. Hulanicki, *Fresenius J. Anal. Chem. 346/6-9*, 622–626 (1993).
[6131] Walcerz, M.; S. Garbos, E. Bulska, A. Hulanicki, *Fresenius J. Anal. Chem. 350/12*, 662–666 (1994).
[6132] Walder, H.; F.C. Lin, S. Narayanan, *Ann. Clin. Biochem. 24/S2*, 268 (1987).
[6133] Walker, H.H.; J.H. Runnels, R.N. Merryfield, *Anal. Chem. 48/14*, 2056–2060 (1976).
[6134] Wallace, G.F.; B.K. Lumas, F.J. Fernandez, W.B. Barnett, *At. Spectrosc. 2/4*, 130–133 (1981).
[6135] Walsh, A.; *Spectrochim. Acta 7*, 108–117 (1955).
[6136] Walsh, A.; *X CSI, Maryland*, 1962.
[6137] Walsh, A.; *Anal. Chem. 46*, 698A (1974).
[6138] Walsh, A.; *Spectrochim. Acta, Part B 35*, 643–652 (1980).

[6139]   Walt, T.N. van der; F.W.E. Strelow, *Geostand. Newsl. 8/2*, 163–167 (1984).
[6140]   Walt, T.N. van der; F.W.E. Strelow, *Anal. Chem. 57/14*, 2889–2891 (1985).
[6141]   Walters, P.E.; J. Lanauze, J.D. Winefordner, *Spectrochim. Acta, Part B 39/1*, 125–129 (1984).
[6142]   Walton, A.P.; L. Ebdon, G.E. Millward, *Anal. Proc. (London) 23*, 422–423 (1986).
[6143]   Walton, W.H.; W.C. Prewett, *Proc. Phys. Soc. B62*, 341 (1949).
[6144]   Wan, A.T.; P. Froomes, *At. Spectrosc. 12/3*, 77–80 (1991).
[6145]   Wan, A.T.; R.A.J. Conyers, Chris J. Coombs, J.P. Masterton, *Clin. Chem. (Winston-Salem) 37/10*, 1683–1687 (1991).
[6146]   Wang, C.-f.; J.-y. Yang, C.-h. Ke, *Anal. Chim. Acta 320/2-3*, 207–216 (1996).
[6147]   Wang, H.; *Lihua Jianyan, Huaxue Fence 27*, 175 (1991).
[6148]   Wang, H.-n.; *Yejin Fenxi 6/5*, 53 (1986).
[6149]   Wang, J.; J. Zhao, *Fenxi Ceshi Tongbao 10/4*, 56 (1991).
[6150]   Wang, J.; W.D. Marshall, *Anal. Chem. 66/22*, 3900–3907 (1994).
[6151]   Wang, J.-p.; B. Deng, *Spectrochim. Acta, Part B 47/5*, 711–718 (1992).
[6152]   Wang, J.-s.; H.-y. Zhou, *Fenxi Huaxue 17*, 1000 (1989).
[6153]   Wang, M.; Y. Feng, *Guangpuxue Yu Guangpu Fenxi 5/3*, 59 (1985).
[6154]   Wang, M.-h.; A.I. Yuzefovskii, R.G. Michel, *Microchem. J. 48*, 326–342 (1993).
[6155]   Wang, P.-x.; W.-m. Zhang, G.-x. Xiong, T.-z. Lin, *Spectrochim. Acta, Part B 43/2*, 141–147 (1988).
[6156]   Wang, P.-x.; V. Majidi, J.A. Holcombe, *Anal. Chem. 61/23*, 2652–2658 (1989).
[6157]   Wang, P.-x.; J.A. Holcombe, *Spectrochim. Acta, Part B 47/11*, 1277–1286 (1992).
[6158]   Wang, S.T.; F. Peter, *J. Anal. Toxicol. 9*, 85–88 (1985).
[6159]   Wang, S.T.; S. Pizzolato, H.P. Demshar, *J. Anal. Toxicol. 15/2*, 66–70 (1991).
[6160]   Wang, S.T.; H.P. Demshar, *Clin. Chem. (Winston-Salem) 39/9*, 1907–1910 (1993).
[6161]   Wang, W.; *Lihua Jianyan, Huaxue Fence 23*, 167 (1987).
[6162]   Wang, W.-j.; S. Hanamura, J.D. Winefordner, *Anal. Chim. Acta 184*, 213–218 (1986).
[6163]   Wang, W.-q.; J.-z. Wei, *Guangpuxue Yu Guangpu Fenxi 15*, 71 (1995).
[6164]   Wang, Y.; C. Lu, Z.-f. Xiao, G.-j. Wang, *J. Agri. Food Chem. 39*, 724–726 (1991).
[6165]   Wang, Z.; S. Luo, Z. Gong, *Fenxi Huaxue 18/9*, 859 (1990).
[6166]   Ward, D.A.; D.G. Biechler, *At. Absorpt. Newsletter 14*, 29 (1975).
[6167]   Ward, J.; J. Cooper, E.W. Smith, *J. Quant. Spectrosc. Radiat. Transfer 14*, 555–590 (1974).
[6168]   Warmerdam, L.J.C.; O. van Tellingen, R.A.A. Maes, J.H. Beijnen, *Fresenius J. Anal. Chem. 351/8*, 777–781 (1995).
[6169]   Watkins, D.; T. Corbyons, J. Bradshaw, J.D. Winefordner, *Anal. Chim. Acta 85*, 403–406 (1976).
[6170]   Watling, R.J.; *Anal. Chim. Acta 94*, 181–186 (1977).
[6171]   Watling, R.J.; *Anal. Chim. Acta 97*, 395–398 (1978).
[6172]   Watson, C.; *Trends Anal. Chem. (Pers. Ed.) 3/1*, 25–28 (1984).
[6173]   Watson, C.A.; *J. Anal. At. Spectrom. 3/2*, 407–414 (1988).
[6174]   Webb, D.P.; E.D. Salin, *Intell. Instr. Comput. 5*, 185–193 (1992).
[6175]   Webb, D.P.; J. Hamier, E.D. Salin, *Trends Anal. Chem. (Pers. Ed.) 13/2*, 44–53 (1994).
[6176]   Webb, D.R.; D.E. Carter, *J. Anal. Toxicol. 8/3*, 118–123 (1984).
[6177]   Weber, G.; *Fresenius Z. Anal. Chem. 321/3*, 217–224 (1985).
[6178]   Weber, G.; *J. Trace Elem. Electrol. Health Dis. 2*, 61–65 (1988).
[6179]   Weber, G.; *Fresenius J. Anal. Chem. 340/3*, 161–165 (1991).
[6180]   Weber, G.; H. Berndt, *Chromatographia 29/5-6*, 254–258 (1990).
[6181]   Weber, G.; H. Berndt, *Int. J. Environ. Anal. Chem. 52*, 195–202 (1993).
[6182]   Wedepohl, K.; G. Schwedt, *Anal. Chim. Acta 203*, 23–34 (1987).
[6183]   Weger, S.J.; L.R. Hossner, L.W. Ferrara, *J. Agri. Food Chem. 17/6*, 1276–1278 (1969).
[6184]   Weger, S.J.; L.R. Hossner, L.W. Ferrara, *At. Absorpt. Newsletter 9/3*, 58–60 (1970).
[6185]   Wegscheider, W.; M. Michaelius, H.M. Ortner, in: B. Welz (Editor), *4. Colloquium Atomspektrometrische Spurenanalytik*, Bodenseewerk Perkin-Elmer, Überlingen, 1987, 125–136.
[6186]   Wegscheider, W.; C. Rohrer, H.M. Ortner, in: B. Welz (Editor), *5. Colloquium Atomspektrometrische Spurenanalytik*, Bodenseewerk Perkin-Elmer, Überlingen, 1989, 289–298.
[6187]   Wegscheider, W.; L. Jancar, Thai Phe Mai, M.R. Michaelis, H.M. Ortner, *Chemom. Intell. Lab. Sys. 7*, 281–293 (1990).
[6188]   Wegscheider, W.; M. Zischka, *Fresenius J. Anal. Chem. 346/6-9*, 525–529 (1993).
[6189]   Wei, J.Z.; H.W. Sun, *Lihua Jianyan, Huaxue Fence 29/3*, 140–170 (1993).
[6190]   Wei, J.-s.; Q. Liu, T. Okutani, *Anal. Sci. 10/3*, 465–468 (1994).
[6191]   Weibust, G.; F.J. Langmyhr, Y. Thomassen, *Anal. Chim. Acta 128*, 23–29 (1981).
[6192]   Weinstock, N.; *J. Clin. Chem. Clin. Biochem. 18*, 712–713 (1980).

[6193] Weinstock, N.; M. Uhlemann, *Clin. Chem. (Winston-Salem) 27/8*, 1438–1440 (1981).
[6194] Weir, D.R.; R.P. Kofluk, *At. Absorpt. Newsletter 6/2*, 24–27 (1967).
[6195] Weisel, C.P.; R.A. Duce, J.L. Fasching, *Anal. Chem. 56*, 1050–1052 (1984).
[6196] Weiss, H.V.; W.H. Shipman, M.A. Guttman, *Anal. Chim. Acta 81/1*, 211–217 (1976).
[6197] Weissberg, N.; B.A. Brooks, G. Schwartz, U. Eylath, A.S. Abraham, *Clin. Chim. Acta 187*, 281–288 (1990).
[6198] Welch, M.W.; D.W. Hamar, M.J. Fettman, *Clin. Chem. (Winston-Salem) 36/2*, 351–354 (1990).
[6199] Welch, S.; W. Berry Lyons, C.-A. Kling, *Environ. Technol. 11/2*, 141–144 (1990).
[6200] Welcher, G.G.; O.H. Kriege, *At. Absorpt. Newsletter 8/5*, 97–101 (1969).
[6201] Welcher, G.G.; O.H. Kriege, *At. Absorpt. Newsletter 9/3*, 61–64 (1970).
[6202] Welcher, G.G.; O.H. Kriege, J.Y. Marks, *Anal. Chem. 46*, 1227–1231 (1974).
[6203] Welsch, E.P.; *Talanta 32/10*, 996–998 (1985).
[6204] Welz, B.; *Int. At. Abs. Spectrosc. Conf. G6, Sheffield*, 1969.
[6205] Welz, B.; *XVII CSI 128, Florence*, 1973.
[6206] Welz, B.; *Fresenius Z. Anal. Chem. 279/2*, 103–104 (1976).
[6207] Welz, B.; *Proceedings Symp. III on electrothermal atomization in AAS*, Chlum u Třeboně, 1977, 126.
[6208] Welz, B.; Z. Grobenski, *Appl. atom. spectrom. (Perkin-Elmer) 10*, (1978).
[6209] Welz, B.; M. Melcher, *At. Absorpt. Newsletter 18/6*, 121–122 (1979).
[6210] Welz, B.; M. Melcher, *At. Spectrosc. 1/5*, 145–147 (1980).
[6211] Welz, B.; Z. Grobenski, M. Melcher, in: K.H. Koch, H. Massmann (Editors), *13 Spektrometertagung*, W. de Gruyter, Berlin, 1981, 337–347.
[6212] Welz, B.; M. Melcher, *Spectrochim. Acta, Part B 36/5*, 439–462 (1981).
[6213] Welz, B.; M. Melcher, *Anal. Chim. Acta 131/1*, 17–25 (1981).
[6214] Welz, B.; U. Völlkopf, Z. Grobenski, *Anal. Chim. Acta 136*, 201–214 (1982).
[6215] Welz, B.; M. Melcher, *Vom Wasser 59*, 407–416 (1982).
[6216] Welz, B.; M. Melcher, *Analyst (London) 108*, 213–224 (1983).
[6217] Welz, B.; M. Melcher, G. Schlemmer, *Fresenius Z. Anal. Chem. 316/2*, 271–276 (1983).
[6218] Welz, B.; M. Melcher, *Anal. Chim. Acta 153*, 297–300 (1983).
[6219] Welz, B.; G. Schlemmer, U. Völlkopf, *Spectrochim. Acta, Part B 39/2-3*, 501–510 (1984).
[6220] Welz, B.; M. Melcher, *Analyst (London) 109/5*, 569–572 (1984).
[6221] Welz, B.; M. Melcher, *Analyst (London) 109/5*, 573–575 (1984).
[6222] Welz, B.; M. Melcher, *Analyst (London) 109/5*, 577–579 (1984).
[6223] Welz, B.; M. Melcher, H.W. Sinemus, D. Maier, *At. Spectrosc. 5/2*, 37–42 (1984).
[6224] Welz, B.; M. Melcher, *At. Spectrosc. 5/2*, 59–61 (1984).
[6225] Welz, B.; M. Melcher, J. Nève, *Anal. Chim. Acta 165/11*, 131–140 (1984).
[6226] Welz, B.; M. Melcher, *Vom Wasser 62*, 137–148 (1984).
[6227] Welz, B.; S. Akman, G. Schlemmer, *Analyst (London) 110/5*, 459–465 (1985).
[6228] Welz, B.; M. Melcher, *Anal. Chem. 57/2*, 427–431 (1985).
[6229] Welz, B.; A.J. Curtius, G. Schlemmer, H.M. Ortner, W. Birzer, *Spectrochim. Acta, Part B 41/11*, 1175–1201 (1986).
[6230] Welz, B.; G. Schlemmer, H. M. Ortner, *Spectrochim. Acta, Part B 41/6*, 567–589 (1986).
[6231] Welz, B.; G. Schlemmer, *J. Anal. At. Spectrom. 1*, 119–124 (1986).
[6232] Welz, B.; M. Schubert-Jacobs, *J. Anal. At. Spectrom. 1/1*, 23–27 (1986).
[6233] Welz, B.; *Fresenius Z. Anal. Chem. 325*, 95–101 (1986).
[6234] Welz, B.; M. Schubert-Jacobs, *Fresenius Z. Anal. Chem. 324/8*, 832–838 (1986).
[6235] Welz, B.; M. Verlinden, *Acta Pharmacol. Toxicol. 59*, 577–580 (1986).
[6236] Welz, B.; G. Schlemmer, U. Völlkopf, *Acta Pharmacol. Toxicol. 59/7*, 589–592 (1986).
[6237] Welz, B.; *Instrumentation Research 2/1*, 46–51 (1986).
[6238] Welz, B.; G. Schlemmer, *Fresenius Z. Anal. Chem. 327/2*, 246–252 (1987).
[6239] Welz, B.; M.S. Wolynetz, M. Verlinden, *Pure Appl. Chem. 59/7*, 927–936 (1987).
[6240] Welz, B.; S. Akman, G. Schlemmer, *J. Anal. At. Spectrom. 2/8*, 793–799 (1987).
[6241] Welz, B.; G. Schlemmer, J.R. Mudakavi, *Anal. Chem. 60/23*, 2567–2572 (1988).
[6242] Welz, B.; G. Schlemmer, *At. Spectrosc. 9/3*, 76–80 (1988).
[6243] Welz, B.; G. Schlemmer, *At. Spectrosc. 9/3*, 81–83 (1988).
[6244] Welz, B.; G. Schlemmer, J.R. Mudakavi, *J. Anal. At. Spectrom. 3/1*, 93–97 (1988).
[6245] Welz, B.; G. Schlemmer, J.R. Mudakavi, *J. Anal. At. Spectrom. 3/5*, 695–701 (1988).
[6246] Welz, B.; B. Radziuk, G. Schlemmer, *Spectrochim. Acta, Part B 43/6-7*, 749–762 (1988).
[6247] Welz, B.; M. Sperling, G. Schlemmer, N. Wenzel, G. Marowsky, *Spectrochim. Acta, Part B 43/9-11*, 1187–1207 (1988).

[6248] Welz, B.; B. Radziuk, G. Schlemmer, *Anal. Proc. (London) 25/7*, 220–222 (1988).
[6249] Welz, B.; M. Schubert-Jacobs, *Fresenius Z. Anal. Chem. 331/3-4*, 324–329 (1988).
[6250] Welz, B.; M. Schubert-Jacobs, G. Schlemmer, M. Sperling, in: P. Brätter, P. Schramel (Editors), *Trace Element Analytical Chemistry in Medicine and Biology*, W. de Gruyter, Berlin - New York, 1988, 25–41.
[6251] Welz, B.; G. Schlemmer, Hugo M. Ortner, W. Wegscheider, *Prog. Anal. Spectrosc. 12/2*, 111–245 (1989).
[6252] Welz, B.; G. Schlemmer, H.M. Ortner, W. Birzer, *Spectrochim. Acta, Part B 44/11*, 1125–1161 (1989).
[6253] Welz, B.; M. Schubert-Jacobs, in: B. Welz (Editor), *5. Colloquium Atomspektrometrische Spurenanalytik*, Bodenseewerk Perkin-Elmer, Überlingen, 1989, 327–345.
[6254] Welz, B.; M. Schubert-Jacobs, M. Sperling, D.L. Styris, D.A. Redfield, *Spectrochim. Acta, Part B 45/11*, 1235–1256 (1990).
[6255] Welz, B.; G. Schlemmer, H.M. Ortner, W. Birzer, *Spectrochim. Acta, Part B 45/4-5*, 367–407 (1990).
[6256] Welz, B.; M. Schubert-Jacobs, *At. Spectrosc. 12/4*, 91–104 (1991).
[6257] Welz, B.; S.-k. Xu, M. Sperling, *Appl. Spectrosc. 45/9*, 1433–1443 (1991).
[6258] Welz, B.; *Microchem. J. 45*, 163–177 (1992).
[6259] Welz, B.; T.-z. Guo, *Spectrochim. Acta, Part B 47/5*, 645–658 (1992).
[6260] Welz, B.; G. Schlemmer, J.R. Mudakavi, *J. Anal. At. Spectrom. 7/3*, 499–503 (1992).
[6261] Welz, B.; G. Bozsai, M. Sperling, B. Radziuk, *J. Anal. At. Spectrom. 7/3*, 505–509 (1992).
[6262] Welz, B.; X.-f. Yin, M. Sperling, *Anal. Chim. Acta 261/1-2*, 477–487 (1992).
[6263] Welz, B.; D.L. Tsalev, M. Sperling, *Anal. Chim. Acta 261/1-2*, 91–103 (1992).
[6264] Welz, B.; *Spectrochim. Acta, Part B 47/8*, 1043–1044 (1992).
[6265] Welz, B.; G. Schlemmer, J.R. Mudakavi, *J. Anal. At. Spectrom. 7/8*, 1257–1271 (1992).
[6266] Welz, B.; M. Sperling, X.-j. Sun, *Fresenius J. Anal. Chem. 346/6-9*, 550–555 (1993).
[6267] Welz, B.; P. Stauss, *Spectrochim. Acta, Part B 48/8*, 951–976 (1993).
[6268] Welz, B.; M. Šucmanová, *Analyst (London) 118/11*, 1417–1423 (1993).
[6269] Welz, B.; M. Šucmanová, *Analyst (London) 118/11*, 1425–1432 (1993).
[6270] Welz, B.; Y.-z. He, M. Sperling, *Talanta 40/12*, 1917–1926 (1993).
[6271] Welz, B.; W. Luecke, *Spectrochim. Acta, Part B 48/14*, 1703–1711 (1993).
[6272] Wendl, W.; G. Müller-Vogt, A. Bienhaus, in: B. Welz (Editor), *Fortschritte in der atomspektrometrischen Spurenanalytik*, Verlag Chemie, Weinheim, 1984, 125–133.
[6273] Wendl, W.; G. Müller-Vogt, *Spectrochim. Acta, Part B 39/2-3*, 237–242 (1984).
[6274] Wendl, W.; G. Müller-Vogt, *Spectrochim. Acta, Part B 40/4*, 527–531 (1985).
[6275] Wendl, W.; A. Kolb, G. Müller-Vogt, in: B. Welz (Editor), *Fortschritte in der atomspektrometrischen Spurenanalytik*, VCH Verlagsgesellschaft, Weinheim, 1986, 15–24.
[6276] Wendl, W.; G. Müller-Vogt, *J. Anal. At. Spectrom. 3/1*, 63–66 (1988).
[6277] Wendl, W.; M. Gieb, G. Müller-Vogt, in: B. Welz (Editor), *5. Colloquium Atomspektrometrische Spurenanalytik*, Bodenseewerk Perkin-Elmer, Überlingen, 1989, 217–224.
[6278] Wendl, W.; F. Korneck, G. Müller-Vogt, W. Send, in: B. Welz (Editor), *6. Colloquium Atomspektrometrische Spurenanalytik*, Bodenseewerk Perkin-Elmer, Überlingen, 1991, 459–468.
[6279] Wennrich, R.; W. Frech, E. Lundberg, *Spectrochim. Acta, Part B 44/3*, 239–246 (1989).
[6280] Werner, M.; in: Mario W. (Editor), *Microtechniques for the Clinical Laboratory, Concepts and Applications*, J. Wiley & Sons, New York, 1976, 37–54.
[6281] West, A.C.; V.A. Fassel, R.N. Kniseley, *Anal. Chem. 45/4*, 815–816 (1973).
[6282] West, A.C.; V.A. Fassel, R.N. Kniseley, *Anal. Chem. 45/9*, 1586–1594 (1973).
[6283] West, A.C.; V.A. Fassel, R.N. Kniseley, *Anal. Chem. 45*, 2420–2422 (1973).
[6284] West, C.D.; D.N. Hume, *Anal. Chem. 36*, 412–415 (1964).
[6285] West, F.K.; P.W. West, F.A. Iddings, *Anal. Chem. 38/11*, 1566–1570 (1966).
[6286] West, F.K.; P.W. West, T.V. Ramakrishna, *Environ. Sci. Technol. 1*, 717 (1967).
[6287] West, F.K.; P.W. West, F.A. Iddings, *Anal. Chim. Acta 37*, 112–121 (1967).
[6288] West, T.S.; X.K. Williams, *Anal. Chim. Acta 45*, 27–41 (1969).
[6289] West, T.S.; H.W. Nürnberg, *The Determination of Trace Metals in Natural Waters*, Blackwell Scientific Publications, Oxford, 1988.
[6290] West, T.S.; *Anal. Proc. (London) 25/7*, 240–244 (1988).
[6291] Westerlund-Helmerson, U.; *At. Absorpt. Newsletter 5*, 97 (1966).
[6292] Westerman, D.W.B.; I.E. Ruffio, M.S. Wainwright, N.R. Foster, *Anal. Chim. Acta 117*, 285–291 (1980).
[6293] Westgard, J.O.; *J. Res. Natl. Bur. Stand. 93/3*, 218–221 (1988).

[6294] Wetherill, G.B.; *Sampling Inspection and Quality Control*, Chapman and Hall, London, 1977.
[6295] Wheat, J.A.; *Appl. Spectrosc. 25*, 328–330 (1971).
[6296] Wheeling, K.; G.D. Christian, *Anal. Lett. 17/B3*, 217–227 (1984).
[6297] White, A.D.; *J. Appl. Phys. 30*, 711–719 (1959).
[6298] White, J.A.; W.L. Harper, A.P. Friedman, V.E. Banas, *Appl. Spectrosc. 28*, 192 (1974).
[6299] White, M.A.; A.M. Boran, *At. Spectrosc. 10/5*, 141–143 (1989).
[6300] White, M.A.; S.A. O'Hagan, A.L. Wright, H.K. Wilson, *J. Exp. Anal. Environ. Epidemiol. 2/2*, 195–206 (1992).
[6301] White, W.W.; P.J. Murphy, *Anal. Chem. 49*, 255–256 (1977).
[6302] Whitehouse, R.C.; A.S. Prasad, Z.T. Cossack, *Clin. Chem. (Winston-Salem) 29/11*, 1974–1977 (1983).
[6303] Whitman, D.A.; G.D. Christian, *Talanta 36/1-2*, 205–211 (1989).
[6304] Whittington, C.M.; J.B. Willis, *Plating Magazine 51/Aug.*, 767 (1964).
[6305] Wibetoe, G.; F.J. Langmyhr, *Anal. Chim. Acta 165/11*, 87–96 (1984).
[6306] Wibetoe, G.; F.J. Langmyhr, *Anal. Chim. Acta 176*, 33–40 (1985).
[6307] Wibetoe, G.; F.J. Langmyhr, *Anal. Chim. Acta 186*, 155–162 (1986).
[6308] Wibetoe, G.; F.J. Langmyhr, *Anal. Chim. Acta 198*, 81–86 (1987).
[6309] Wichems, D.N.; C.P. Calloway, R. Fernando, B.T. Jones, M.J. Morykwas, *Appl. Spectrosc. 47/10*, 1577–1579 (1993).
[6310] Wickstrøm, T.; W. Lund, R. Bye, *Anal. Chim. Acta 208*, 347–350 (1988).
[6311] Wickstrøm, T.; W. Lund, R. Bye, *J. Anal. At. Spectrom. 6/5*, 389–391 (1991).
[6312] Wienke, D.; T. Vijn, L. Buydens, *Anal. Chem. 66*, 841–849 (1994).
[6313] Wiese Simonsen, K.; B. Nielsen, A. Jensen, J.R. Andersen, *J. Anal. At. Spectrom. 1/6*, 453–456 (1986).
[6314] Wieteska, E.; A. Ziótek, A. Drzewińska, *Anal. Chim. Acta 330/2-3*, 251–257 (1996).
[6315] Wifladt, A.-M.; W. Lund, R. Bye, *Talanta 36/3*, 395–399 (1989).
[6316] Wifladt, A.-M.; R. Bye, W. Lund, *Fresenius J. Anal. Chem. 344/12*, 541–544 (1992).
[6317] Wigfield, D.C.; S.L. Perkins, *Anal. Chim. Acta 167*, 419–424 (1985).
[6318] Wigfield, D.C.; S.A. Eatock, *J. Anal. Toxicol. 11/4*, 137–139 (1987).
[6319] Wigfield, D.C.; R.S. Daniels, *J. Anal. Toxicol. 12/2*, 94–97 (1988).
[6320] Wigfield, D.C.; R.S. Daniels, *J. Anal. Toxicol. 13*, 191–192 (1989).
[6321] Wilber, C.G.; *Clin. Toxicol. 17/2*, 171–230 (1980).
[6322] Wildhagen, D.; V. Krivan, B. Gercken, J. Pavel, *J. Anal. At. Spectrom. 11/5*, 371–377 (1996).
[6323] Wilhelm, M.; F.K. Ohnesorge, *J. Anal. Toxicol. 14/4*, 206–210 (1990).
[6324] Wilken, R.-D.; H. Hintelmann, in: J.A.C. Broekaert, S. Güçer, F. Adams (Editors), *Metal Speciation in the Environment*, NATO ASI Ser., Ser. G: Ecological Sci., Springer, Berlin - Heidelberg - New York, 1990, 339–359.
[6325] Wilken, R.-D.; *Fresenius J. Anal. Chem. 342/10*, 795–801 (1992).
[6326] Wilken, R.-D.; J. Kuballa, E. Jantzen, *Fresenius J. Anal. Chem. 350/1-2*, 77–84 (1994).
[6327] Willet, W.C.; M.J. Stampfer, D. Hunter, G.A. Colditz, in: A. Aitio, A. Aro, J. Järvisalo, H. Vainio (Editors), *Trace Elements in Health and Disease*, The Royal Society of Chemistry, Cambridge, 1991, 141–155.
[6328] Williams, A.; *Anal. Proc. (London) 30/6*, 248–250 (1993).
[6329] Williams, D.M.; *Clin. Chim. Acta 99*, 23 (1979).
[6330] Williams, R.R.; G.N. Coleman, *Spectrochim. Acta, Part B 38/8*, 1171–1178 (1983).
[6331] Willie, S.N.; R.E. Sturgeon, S.S. Berman, *Anal. Chim. Acta 149*, 59–66 (1983).
[6332] Willie, S.N.; R.E. Sturgeon, S.S. Berman, *Anal. Chem. 58/6*, 1140–1143 (1986).
[6333] Willie, S.N.; *At. Spectrosc. 15/5*, 205–208 (1994).
[6334] Willie, S.N.; *Spectrochim. Acta, Part B 51/14*, 1781–1790 (1996).
[6335] Willis, J.B.; *Spectrochim. Acta 16*, 259–272 (1960).
[6336] Willis, J.B.; *Spectrochim. Acta 16*, 551–558 (1960).
[6337] Willis, J.B.; *Anal. Chem. 33/4*, 556–559 (1961).
[6338] Willis, J.B.; *Anal. Chem. 34/6*, 614–617 (1962).
[6339] Willis, J.B.; in: D. Glick (Editor), *Methods of Biochemical Analysis*, Interscience, New York - London, Sydney, 1963, 1–67.
[6340] Willis, J.B.; *Nature (London) 207*, 715–716 (1965).
[6341] Willis, J.B.; *Spectrochim. Acta 23A*, 811–830 (1967).
[6342] Willis, J.B.; J.O. Rasmuson, R.N. Kniseley, V.A. Fassel, *Spectrochim. Acta, Part B 23/11*, 725–730 (1968).
[6343] Willis, J.B.; *Anal. Chem. 47*, 1753–1758 (1975).
[6344] Willis, J.B.; B.T. Sturman, B.D. Frary, *J. Anal. At. Spectrom. 5/5*, 399–405 (1990).

[6345] Willis, J.B.; B.T. Sturman, *Appl. Spectrosc. 46/8*, 1231–1234 (1992).
[6346] Willis, J.B.; *Clin. Chem. (Winston-Salem) 39/1*, 155–160 (1993).
[6347] Willis, J.B.; *Spectrochim. Acta, Part B 52/6*, 667–674 (1997).
[6348] Wills, M.R.; C.S. Brown, R.L. Bertholf, R. Ross, J. Savory, *Clin. Chim. Acta 145*, 193 (1985).
[6349] Wilson, A.L.; *Talanta 17*, 21–29 (1970).
[6350] Wilson, A.L.; *Talanta 21*, 1109–1121 (1974).
[6351] Wilson, D.O.; *Commun. Soil. Sci. Plant Anal. 10/10*, 1319–1330 (1979).
[6352] Wilson, L.; *Anal. Chim. Acta 30*, 377–383 (1964).
[6353] Wiltshire, G.A.; D.T. Bolland, D. Littlejohn, *J. Anal. At. Spectrom. 9/11*, 1255–1262 (1994).
[6354] Wimberley, J.W.; *Anal. Lett. 19/15*, 1561–1572 (1986).
[6355] Winchester, M.R.; R.K. Marcus, *J. Anal. At. Spectrom. 5/1*, 9–14 (1990).
[6356] Winchester, M.R.; S.M. Hayes, R.K. Marcus, *Spectrochim. Acta, Part B 46/5*, 615–627 (1991).
[6357] Winefordner, J.D.; T.J. Vickers, *Anal. Chem. 36/10*, 1947–1954 (1964).
[6358] Wingo, C.S.; G.B. Bixler, C.H. Park, S.G. Straub, *Kidney Int. 31*, 1225–1228 (1987).
[6359] Winkler, P.; S. Jobst, C. Harder, *BPT-Bericht 1/89,* GSF München, 1989.
[6360] Winkler, P.; M. Schulz, W. Dannecker, *Fresenius J. Anal. Chem. 340/9*, 575–579 (1991).
[6361] Wirz, P.; U. Kurfürst, K.-H. Grobecker, *LaborPraxis 4/5*, 16 (1980).
[6362] Wise, W.M.; S.D. Solsky, *Anal. Lett. 10*, 273–282 (1977).
[6363] Wise, W.M.; R.A. Burdo, D.E. Goforth, in: S.J. Haswell (Editor), *Atomic Absorption Spectrometry - Theory, Design and Applications*, Elsevier, Amsterdam - New York, 1991, 341–357.
[6364] Wittassek, R.; in: B. Welz (Editor), *4. Colloquium Atomspektrometrische Spurenanalytik*, Bodenseewerk Perkin-Elmer, Überlingen, 1987, 393–402.
[6365] Wittmann, Z.; *Analyst (London) 104*, 156–160 (1979).
[6366] Woidich, H.; W. Pfannhauser, *Fresenius Z. Anal. Chem. 276/1*, 61–66 (1975).
[6367] Woidich, H.; W. Pfannhauser, *Nahrung 24/4-5*, 367–371 (1980).
[6368] Woitke, P.; A. Thiesies, G. Henrion, G. Steppuhn, *Z. Chem. 29/5*, 180–181 (1989).
[6369] Wolcott, J.F.; C. Butler Sobel, *Appl. Spectrosc. 32/6*, 591–593 (1978).
[6370] Wolcott, J.F.; C. Butler Sobel, *Appl. Spectrosc. 36/6*, 685–686 (1982).
[6371] Wolf, W.R.; K.K. Stewart, *Anal. Chem. 51/8*, 1201–1205 (1979).
[6372] Wong, P.T.S.; Y.K. Chau, Y. Yaromich, P. Hodson, M. Whittle, *Appl. Organomet. Chem. 3*, 59–70 (1989).
[6373] Woo, I.H.; K. Watanabe, Y. Hashimoto, Y.K. Lee, *Anal. Sci. 3/1*, 49–52 (1987).
[6374] Wood, S.A.; D. Vlassopoulos, A. Mucci, *Anal. Chim. Acta 229/2*, 227–238 (1990).
[6375] Woodis, T.C.; G.B. Hunter, F.J. Johnson, *J. Assoc. Off. Anal. Chem. 59*, 22–25 (1976).
[6376] Woodis, T.C.; G.B. Hunter, F.J. Johnson, *Anal. Chim. Acta 90*, 127–136 (1977).
[6377] Woodriff, R.; G. Ramelow, *Spectrochim. Acta, Part B 23/10*, 665–671 (1968).
[6378] Woodriff, R.; R.W. Stone, A.M. Held, *Appl. Spectrosc. 22/5*, 408–411 (1968).
[6379] Woodriff, R.; R.W. Stone, *Appl. Opt. 7/7*, 1337–1339 (1968).
[6380] Woodson, T.T.; *Rev. Sci. Instrum. 10*, 308–311 (1939).
[6381] Woodward, C.; *At. Absorpt. Newsletter 8/6*, 121–123 (1969).
[6382] Woolson, E.A.; N. Aharonson, *J. Assoc. Off. Anal. Chem. 63*, 523–528 (1980).
[6383] Woolson, E.A.; N. Aharonson, R. Iadevaia, *J. Agri. Food Chem. 30*, 580–584 (1982).
[6384] Workman, S.M.; P.N. Soltanpour, *Soil Sci. Soc. Am. J. 44/6*, 1331–1333 (1980).
[6385] World Health Organization; *Environmental Health Criteria No. 101*, 1990.
[6386] World Health Organization; *Environmental Health Criteria No. 118*, 1991.
[6387] Wright, F.C.; J.C. Riner, *J. Assoc. Off. Anal. Chem. 55/3*, 662–663 (1972).
[6388] Wróbel, K.; E. Blanco González, A. Sanz-Medel, *J. Anal. At. Spectrom. 9/3*, 281–284 (1994).
[6389] Wróbel, K.; E. Blanco González, A. Sanz-Medel, *Analyst (London) 120/3*, 809–815 (1995).
[6390] Wu, D.; J. Wu, *Zheijiang Gongxueyuan Xuebao 35*, 69 (1987).
[6391] Wu, J.C.G.; *Spectrosc. Lett. 24/5*, 681–697 (1991).
[6392] Wu, M.-x.; *Yankuang Ceshi 7*, 251 (1988).
[6393] Wu, S.-l.; C.L. Chakrabarti, F. Marcantonio, K.L. Headrick, *Spectrochim. Acta, Part B 41/7*, 651–667 (1986).
[6394] Wu, X.-a.; K.-H. Bellgardt, *Anal. Chim. Acta 313*, 161–176 (1995).-
[6395] Wu, X.-h.; S.-h. Lu, Y. Peng, G.-n. Cui, *Ganguang Kexue Yu Kuang Huaxue 9*, 233 (1991).
[6396] Wu, X.-h.; G.-c. Cui, Y. Peng, *Fenxi Huaxue 19*, 1352 (1991).
[6397] Wu, X.-j.; S.-c. Liang, *Fresenius J. Anal. Chem. 336/2*, 120–123 (1990).
[6398] Wu, X.-w.; *Fenxi Huaxue 12*, 613–615 (1984).
[6399] Wu, Z.; Y. Ma, *Fenxi Huaxue 11/6*, 423–427 (1983).

[6400] Wu, Z.-a.; J.-y. Yao, B.-W. Wang, *Fenxi Huaxue 17*, 1068 (1989).
[6401] Wunderlich, E.; W. Hädeler, *Fresenius Z. Anal. Chem. 281*, 300 (1976).
[6402] Würfels, M.; E. Jackwerth, M. Stoeppler, *Anal. Chim. Acta 226/1*, 1–16 (1989).
[6403] Würfels, M.; E. Jackwerth, M. Stoeppler, *Anal. Chim. Acta 226/1*, 17–30 (1989).
[6404] Wycislik, A.; in: R. Nauche (Editor), *Progress of Analytical Chemistry in the Iron and Steel Industry*, CEC, Report EUR 14113, Brussels, 1992, 542–546.
[6405] Wyck, D.B. Van; R.B. Schifman, J.C. Stivelman, J. Ruiz, D. Martin, *Clin. Chem. (Winston-Salem) 34/6*, 1128–1130 (1988).
[6406] Xu, B.-x.; T.-m. Xu, M.-n. Shen, Y.-z. Fang, *Talanta 32/3*, 215–217 (1985).
[6407] Xu, D.-q.; G.-p. Guo, H.-w. Sun, *J. Anal. At. Spectrom. 10/10*, 753–755 (1995).
[6408] Xu, L.-q.; J.-f. Zhu, L.-j. Zhang, *Fenxi Huaxue 15*, 1053 (1987).
[6409] Xu, L.-q.; J.-f. Zhu, L.-j. Zhang, C. Qiu, *Lihua Jianyan, Huaxue Fence 24*, 278 (1988).
[6410] Xu, N.; V. Majidi, W.D. Ehmann, W.R. Markesbery, *J. Anal. At. Spectrom. 7/5*, 749–751 (1992).
[6411] Xu, N.; V. Majidi, W.R. Markesbery, W.D. Ehmann, *Neurotoxicol. 13*, 735 (1992).
[6412] Xu, S.-k.; L.-j. Sun, Z.-l. Fang, *Anal. Chim. Acta 245/1*, 7–11 (1991).
[6413] Xu, S.-k.; L.-j. Sun, Z.-l. Fang, *Talanta 39/6*, 581–587 (1992).
[6414] Xu, S.-k.; M. Sperling, B. Welz, *Fresenius J. Anal. Chem. 344/12*, 535–540 (1992).
[6415] Xu, S.-k.; Z.-l. Fang, *Chin. Chem. Lett. 3/11*, 915–918 (1992).
[6416] Xu, S.-k.; Z.-l. Fang, *Fenxi Shiyanshi 13/2*, 20–22 (1994).
[6417] Xu, S.-k.; Z.-l. Fang, *Microchem. J. 51*, 360–366 (1995).
[6418] Xu, T.-m.; S.-r. Zhao, H. Zhu, B.-x. Xu, *Huaxue Shijie 33*, 220 (1992).
[6419] Xu, Y.-p., Y.-z. Liang, *J. Anal. At. Spectrom. 12/4*, 471–474 (1997).
[6420] Xuan, W.-k.; M.-b. Chen, *Yejin Fenxi 9*, 36 (1989).
[6421] Xuan, W.-k.; J.-g. Li, *Spectrochim. Acta, Part B 45/7*, 669–677 (1990).
[6422] Xuan, W.-k.; *Yejin Fenxi 11/3*, 55 (1991).
[6423] Xuan, W.-k.; *Spectrochim. Acta, Part B 47/4*, 545–551 (1992).
[6424] Ya, J.-y.; Y.-q. Jiang, *Fenxi Huaxue 17*, 399 (1989).
[6425] Yamada, E.; T. Yamada, M. Sato, *Anal. Sci. 8/6*, 863–868 (1992).
[6426] Yamada, H.; K. Uchino, H. Koizumi, T. Noda, K. Yasuda, *Anal. Lett. 11A/10*, 855–868 (1978).
[6427] Yamamoto, L.; Y. Kaneda, Y. Hikasa, *Int. J. Environ. Anal. Chem. 16/1*, 1–16 (1983).
[6428] Yamamoto, M.; K. Urata, Y. Yamamoto, *Anal. Lett. 14/A1*, 21–26 (1981).
[6429] Yamamoto, M.; K. Urata, K. Murashige, Y. Yamamoto, *Spectrochim. Acta, Part B 36/7*, 671–677 (1981).
[6430] Yamamoto, M.; Y. Yamamoto, T. Yamashige, *Analyst (London) 109/11*, 1461–1463 (1984).
[6431] Yamamoto, M.; K. Fujishige, H. Tsubota, Y. Yamamoto, *Anal. Sci. 1/1*, 47–50 (1985).
[6432] Yamamoto, M.; M. Yasuda, Y. Yamamoto, *Anal. Chem. 57/7*, 1382–1385 (1985).
[6433] Yamamoto, M.; K. Takada, T. Kumamaru, M. Yasuda, S. Yokoyama, Y. Yamamoto, *Anal. Chem. 59/19*, 2446–2448 (1987).
[6434] Yamamoto, Y.; T. Kumamaru, Y. Hayashi, *Talanta 14*, 611 (1967).
[6435] Yamamoto, Y.; T. Kumamaru, *Fresenius Z. Anal. Chem. 282*, 139. (1976).
[6436] Yamamoto, Y.; T. Kumamaru, A. Shiraki, *Fresenius Z. Anal. Chem. 292*, 273–277 (1978).
[6437] Yaman, M.; S. Güçer, *Analyst (London) 120/1*, 101–105 (1995).
[6438] Yaman, M.; S. Güçer, *Analusis 23/4*, 168–171 (1995).
[6439] Yan, X.-p.; Z.-m. Ni, *J. Anal. At. Spectrom. 6/6*, 483–486 (1991).
[6440] Yan, X.-p.; Z.-m. Ni, Q.-l. Guo, *Anal. Chim. Acta 272/1*, 105–114 (1993).
[6441] Yan, X.-p.; Z.-m. Ni, X.-t. Yang, G.-q. Hong, *Spectrochim. Acta, Part B 48/4*, 605–624 (1993).
[6442] Yan, X.-p.; Z.-m. Ni, *Spectrochim. Acta, Part B 48/11*, 1315–1323 (1993).
[6443] Yan, X.-p.; W. Van Mol, F. Adams, *Analyst (London) 121/8*, 1061–1067 (1996).
[6444] Yan, X.-p.; F. Adams, *J. Anal. At. Spectrom. 12/4*, 459–464 (1997).
[6445] Yan, Z.-y.; W.-m. Zhang, *J. Anal. At. Spectrom. 4/8*, 797–799 (1989).
[6446] Yanagi, K.; *Bunseki Kagaku 29/3*, 194–199 (1980).
[6447] Yanagisawa, M.; M. Suzuki, T. Takeuchi, *Anal. Chim. Acta 52*, 386–389 (1970).
[6448] Yanagisawa, M.; M. Suzuki, T. Takeuchi, *Mikrochim. Acta (Vienna) I*, 475–480 (1973).
[6449] Yanagisawa, M.; T. Takeuchi, *Anal. Chim. Acta 123*, 364 (1974).
[6450] Yang, G.-y.; D.-x. Xu, R.-x. Jin, *Analyst (London) 120/6*, 1657–1659 (1995).
[6451] Yang, J.-j.; R.-f. Su, *Lihua Jianyan, Huaxue Fence 26*, 36 (1990).
[6452] Yang, J.-f.; L.-j. Dai, *Yankuang Ceshi 9/1*, 44–47 (1990).
[6453] Yang, P.; *Guangpuxue Yu Guangpu Fenxi 12/1*, 116–118 (1992).

[6454] Yang, P.-y.; Z.-m. Ni, Z.-x. Zhuang, F.-c. Xu, A.-b. Jiang, *J. Anal. At. Spectrom. 7/3*, 515–519 (1992).
[6455] Yang, Q.; J. Smeyers-Verbeke, *Clin. Chim. Acta 204/1-3*, 23–36 (1991).
[6456] Yang, W.-m.; Z.-m. Ni, *Spectrochim. Acta, Part B 49/11*, 1067–1079 (1994).
[6457] Yang, W.-m.; Z.-m. Ni, *Spectrochim. Acta, Part B 51/1*, 65–73 (1996).
[6458] Yang, X.-h.; G.-k. Li, B.-y Tang, Z.-x. Zhang, *Redai Haiyang 10/3*, 101–105 (1991).
[6459] Yao, J.; Z. Wu, G. Zhi, *Fenxi Huaxue 19*, 594 (1991).
[6460] Yao, J.; Y. Ji, *Guangpuxue Yu Guangpu Fenxi 12/3*, 93–96,82 (1992).
[6461] Yao, J.-y.; Y.-q. Jiang, *Fenxi Ceshi Tongbao 9/3*, 35 (1990).
[6462] Yao, J.-y.; J.-y. Song, J.-r. Zhou, Y.-d. Wei, *Fenxi Shiyanshi 9*, 16–21 (1990).
[6463] Yao, J.-y.; R.-y. Xiang, *Huaxue Tongbao (5)*, 43–44,42 (1993).
[6464] Yao, Y.; Y. Jiang, *Guangpuxue Yu Guangpu Fenxi 11/3*, 45–50 (1991).
[6465] Yasuda, K.; H. Koizumi, K. Ohishi, T. Noda, *Prog. Anal. Spectrosc. 3/4*, 299–368 (1980).
[6466] Yasuda, K.; Y. Hirano, T. Kamino, K. Hirokawa, *Anal. Sci. 10/4*, 623–631 (1994).
[6467] Yasuda, K.; Y. Hirano, T. Kamino, T. Yaguchi, K. Hirokawa, *Anal. Sci. 11/3*, 437–440 (1995).
[6468] Yasuda, K.; Y. Hirano, T. Kamino, T. Yaguchi, K. Hirokawa, *Anal. Sci. 12/4*, 659–663 (1996).
[6469] Yasuda, M.; S. Murayama, *Spectrochim. Acta, Part B 36/7*, 641–647 (1981).
[6470] Ybáñez, N.; M.L. Cervera, R. Montoro, R. Catala, *Rev. Agroquim. Tecnol. Aliment. 27*, 590 (1987).
[6471] Ybáñez, N.; M.L. Cervera, Rosa Montoro, M. de la Guardia, *J. Anal. At. Spectrom. 6/5*, 379–384 (1991).
[6472] Yebra, M.C.; M. Gallego, M. Valcárcel, *Anal. Chim. Acta 308/1-3*, 275–280 (1995).
[6473] Yebra-Biurrun, M.C.; A. Bermejo-Barrera, M.L. Mella Louzao, M.P. Bermejo Barrera, *Quim. Anal. (Barcelona) 10/1*, 59–64 (1991).
[6474] Yebra-Biurrun, M.C.; A. Bermejo-Barrera, M.P. Bermejo-Barrera, *Alimentaria (Madrid) 28/222*, 53–55 (1991).
[6475] Yebra-Biurrun, M.C.; M.C. García-Dopazo, A. Bermejo-Barrera, M.P. Bermejo-Barrera, *Talanta 39/6*, 671–674 (1992).
[6476] Yebra-Biurrun, M.C.; A. Bermejo-Barrera, M.P. Bermejo-Barrera, *Bull. Soc. Chim. Belg. 101/6*, 473 (1992).
[6477] Yebra-Biurrun, M.C.; A. Bermejo-Barrera, M.P. Bermejo-Barrera, *Mikrochim. Acta (Vienna) 109*, 243–251 (1992).
[6478] Yebra-Biurrun, M.C.; A. Bermejo-Barrera, M.P. Bermejo-Barrera, M.C. Barciela-Alonso, *Anal. Chim. Acta 303/2-3*, 341–345 (1995).
[6479] Yebra-Biurrun, M.C.; M. Gallego, M. Valcárcel, *Anal. Chim. Acta 308/1-3*, 357–363 (1995).
[6480] Yeh, Y.-y.; P. Zee, *Clin. Chem. (Winston-Salem) 20/3*, 360–364 (1974).
[6481] Yin, X.-f.; G. Schlemmer, B. Welz, *Anal. Chem. 59/10*, 1462–1466 (1987).
[6482] Yen, C.-c.; W.-k. Chen, C.-c. Hu, B.-l. Wei, C. Chung, *At. Spectrosc. 18/2*, 64–69 (1997).
[6483] Yokoi, K.; M. Kimura, K. Sekine, Y. Itokawa, *Trace Nutrients Research 4*, 133–138 (1988).
[6484] Yokoyama, T.; Y. Takahashi, T. Tarutani, *Chem. Geol. 103*, 103–111 (1993).
[6485] Yokoyama, T.; T. Watarai, T. Uehara, K.-i. Mizuoka, K. Kohara, M. Kido, M. Zenki, *Fresenius J. Anal. Chem. 357/7*, 860–863 (1997).
[6486] Yoo, K.S.; J.D. Kyung, J.G. Kwon, J.J. Lee, *J. Korean Che. Soc. 39/4*, 252 (1995).
[6487] Yoon, B.M.; S.C. Shim, H.C. Pyun, D.S. Lee, *Anal. Sci. 6/4*, 561–566 (1990).
[6488] Yoshimoto, E.; T. Kitamura, M. Hasegawa, *Bunseki Kagaku 36*, 490 (1987).
[6489] Yoshimura, C.; N. Shinya, *Nippon Kagaku Kaishi 8*, 1545 (1987).
[6490] You, J.-l.; Z.-s. Wang, G.-s. Zhang, J.-s. Ren, *Anal. Lett. 26/3*, 541–556 (1993).
[6491] Young, C.M.; J.M. Baldwin, *Microchem. J. 23*, 265–277 (1978).
[6492] Yu, K.-l.; *Fenxi Huaxue 22/1*, 108 (1994).
[6493] Yu, S.; M. Zou, G. Ye, *Guangpuxue Yu Guangpu Fenxi 4/4*, 56 (1984).
[6494] Yu, Z.-q.; C. Vandecasteele, B. Desmet, R. Dams, *Mikrochim. Acta (Vienna) I/1-2*, 41–48 (1990).
[6495] Yuan, D.-x.; X.-r. Wang, P.-y. Yang, B.-l. Huang, *Anal. Chim. Acta 243/1*, 65–69 (1991).
[6496] Yuan, D.-x.; I.L. Shuttler, *Anal. Chim. Acta 316/3*, 313–322 (1995).
[6497] Yuan, Y.; X.-w. Guo, *Fenxi Shiyanshi 12/4*, 19–22 (1993).
[6498] Yuan, Z.-n.; *Guangpuxue Yu Guangpu Fenxi 15*, 89 (1995).
[6499] Yudelevich, I.G.; N.F. Beisel, T.S. Papina, K. Dittrich, *Spectrochim. Acta, Part B 39/2-3*, 467–472 (1984).
[6500] Yudelevich, I.G.; L.V. Zelentsova, N.F. Beisel, *Talanta 34/1*, 147–151 (1987).

[6501] Yudelevich, I.G.; B.I. Zaksas, V.P. Shaburova, A.S. Cherevko, *At. Spectrosc. 13/3*, 108–111 (1992).
[6502] Yudelevich, I.G.; N.I. Petrova, L.M. Buyanova, *Vysokochistye Veshchestva (4)*, 128–131 (1992).
[6503] Yudelevich, I.G.; N.I. Petrova, L.M. Buyanova, N.F. Beisel, *Vysokochistye Veshchestva (5-6)*, 211 (1992).
[6504] Yudelevich, I.G.; N.I. Petrova, L.M. Buyanova, *Vysokochistye Veshchestva (2)*, 83–86 (1993).
[6505] Yule, W.; M. Rutter, in: K.R. Mahaffey (Editor), *Dietary and Environmental Lead: Human Health Effects*, Elsevier, Amsterdam, 1985, 211.
[6506] Yuzefovskii, A.I.; E.G. Su, R.G. Michel, W. Slavin, J.T. McCaffrey, *Spectrochim. Acta, Part B 49/12-14*, 1643–1656 (1994).
[6507] Zacha, K.E.; J.D. Winefordner, *Anal. Chem. 38/11*, 1537–1539 (1966).
[6508] Zachariadis, G.G.; J.A. Stratis, *J. Anal. At. Spectrom. 6/3*, 239–245 (1991).
[6509] Zagatto, E.A.G.; F.J. Krug, H. Bergamin, S.S. Jørgensen, B.F. Reis, *Anal. Chim. Acta 104*, 279–284 (1979).
[6510] Zagatto, E.A.G.; M.F. Giné, E.A.N. Fernandez, B.F. Reis, F.J. Krug, *Anal. Chim. Acta 173*, 289–297 (1985).
[6511] Zagatto, E.A.G.; O. Bahia, M.F. Giné, H. Bergamin Fº, *Anal. Chim. Acta 181*, 265–270 (1986).
[6512] Zaidel, A.N.; E.P. Korennoi, *Opt. Spectrosc. [Engl. Transl.] 10*, 299 (1961).
[6513] Zaino, E.C.; *At. Absorpt. Newsletter 6*, 93–94 (1967).
[6514] Zaki, N.S.; M.M. Barbooti, S.S. Baha-Uddin, E.B. Hassan, *Appl. Spectrosc. 43/7*, 1257–1259 (1989).
[6515] Zander, A.T.; T.C. O'Haver, P.N. Keliher, *Anal. Chem. 48/8*, 1166–1175 (1976).
[6516] Zander, A.T.; T.C. O'Haver, P.N. Keliher, *Anal. Chem. 49/6*, 838–842 (1977).
[6517] Zantopoulos, N.; V. Antoniou, H. Tsoukali-Papadopoulou, *J. Environ. Sci. Health, Part A 27A/6*, 1453–1458 (1992).
[6518] Zaranyika, M.F.; P. Makuhunga, *Fresenius J. Anal. Chem. 357/3*, 249–257 (1997).
[6519] Zasoski, R.J.; R.G. Burau, *Commun. Soil. Sci. Plant Anal. 8/5*, 425–436 (1977).
[6520] Zatka, V.J.; *Anal. Chem. 50/3*, 538–541 (1978).
[6521] Zayed, M.A.; A.F. Nour El-Dien, K.A. Rabie, *Microchem. J. 49*, 27–35 (1994).
[6522] Zeegers, P.J.Th.; R. Smith, J.D. Winefordner, *Anal. Chem. 40/13*, 26A-47A (1968).
[6523] Zeeman, P.; *Phil. Mag. 5*, 226 (1897).
[6524] Zehr, B.D.; M.A. Fedorchak, *Am. Lab. (Fairfield, CT) 23/Febr*, 40–45 (1991).
[6525] Zehr, B.D.; M.A. Maryott, *Spectrochim. Acta, Part B 48/10*, 1275–1280 (1993).
[6526] Zehr, B.D.; *At. Spectrosc. 15/5*, 213–215 (1994).
[6527] Zeisler, R.; *J. Res. Natl. Bur. Stand. 91/2*, 75–85 (1986).
[6528] Zeisler, R.; D.A. Becker, T.E. Gills, *Fresenius J. Anal. Chem. 352/1-2*, 111–115 (1995).
[6529] Zelentsova, L.V.; V.A. Shchipachev, L.M. Levchenko, *Zh. Anal. Khim. 47/10-1*, 1889–1992 (1992).
[6530] Zeng, Z.; *Lihua Jianyan, Huaxue Fence 28/6*, 367–368 (1992).
[6531] Zereini, F.; H. Urban, H.M. Lueschow, *Erzmetall 47/1*, 45–52 (1994).
[6532] Zettner, A.; D. Seligson, *Clin. Chem. (Winston-Salem) 10*, 869–890 (1964).
[6533] Zettner, A.; L.C. Sylvia, L. Capacho-Delgado, *Am. J. Clin. Pathol. 45*, 533–540 (1966).
[6534] Zettner, A.; A.H. Mensch, *Am. J. Clin. Pathol. 48*, 225–228 (1967).
[6535] Zettner, A.; A.H. Mensch, *Am. J. Clin. Pathol. 49*, 196–199 (1968).
[6536] Zhai, G.-x.; *Guangpuxue Yu Guangpu Fenxi 12/1*, 119–122 (1992).
[6537] Zhang, G.-c.; H. Brian, *Fenxi Huaxue 15*, 330 (1987).
[6538] Zhang, J.-k.; Y.-l. Zhou, X.-y. Luo, P. Jin, *Lihua Jianyan, Huaxue Fence 30*, 144 (1994).
[6539] Zhang, L.; Z.-m. Ni, X.-q. Shan, *Spectrochim. Acta, Part B 44/3*, 339–346 (1989).
[6540] Zhang, L.; Z.-m. Ni, X.-q. Shan, *Spectrochim. Acta, Part B 44/8*, 751–758 (1989).
[6541] Zhang, L.; G.R. Carnrick, J. Schickli, S. McIntosh, W. Slavin, *At. Spectrosc. 11/6*, 216–221 (1990).
[6542] Zhang, L.; Z.-m. Ni, X.-q. Shan, *Can. J. Appl. Spectroscopy 36/2*, 47–52 (1991).
[6543] Zhang, L.; S. McIntosh, G.R. Carnrick, W. Slavin, *Spectrochim. Acta, Part B 47/5*, 701–709 (1992).
[6544] Zhang, L.; G.R. Carnrick, W. Slavin, *Spectrochim. Acta, Part B 48/11*, 1435–1443 (1993).
[6545] Zhang, Q.-b.; H. Minami, I. Atsuya, *Bunseki Kagaku 43/1*, 39–43 (1994).
[6546] Zhang, S.-c.; L.-j. Sun, H.-c. Jiang, Z.-l. Fang, *Guangpuxue Yu Guangpu Fenxi 4/3*, 42 (1984).
[6547] Zhang, S.-c.; S.-k. Xu, Z.-l. Fang, *Quim. Anal. (Barcelona) 8/2*, 191–199 (1989).
[6548] Zhang, S.-z.; Y.-k. Chau, W.-c. Li, A.S.Y. Chau, *Appl. Organomet. Chem. 5*, 431–434 (1991).

[6549] Zhang, W.-w.; S.-y. He, *Hejishu 14*, 490 (1991).
[6550] Zhang, X.; B.-x. Li, D.-c. Zhou, G.-z. Yu, X.-m. Zhan, *Guangpuxue Yu Guangpu Fenxi 12/5*, 83–89 (1992).
[6551] Zhang, X.-r.; R. Cornelis, J. de Kimpe, L. Mees, *Anal. Chim. Acta 319/1-2*, 177–185 (1996).
[6552] Zhang, X.-r.; R. Cornelis, J. de Kimpe, L. Mees, *J. Anal. At. Spectrom. 11*, 1075–1079 (1996).
[6553] Zhang, X.-r.; R. Cornelis, J. de Kimpe, L. Mees, V. Vanderbiesen, A. de Cubber, R. Vanholder, *Clin. Chem. (Winston-Salem, NC) 42/8*, 1231–1237 (1996).
[6554] Zhang, Y.-a.; P. Riby, A.G. Cox, C.W. McLeod, A.R. Date, Y.Y. Cheung, *Analyst (London) 113/1*, 125–128 (1988).
[6555] Zhang, Y.-f.; K. Zhang, Z. Fang, Y.-z. Wang, *J. Anal. At. Spectrom. 10/5*, 359–362 (1995).
[6556] Zhang, Z.; J. Zhang, *Yejin Fenxi 13/1*, 53–54 (1993).
[6557] Zhang, Z.-z.; J.-d. Zheng, F.-f. Fu, *Fenxi Shiyanshi 13/2*, 72–74 (1994).
[6558] Zhao, H.-z.; L.G. Danielsson, F. Ingman, *Zhongguo Kexue Jishu Daxue Xuebao 20/1*, 61–66 (1990).
[6559] Zhao, Y.-b.; A. Li, *Fenxi Huaxue 17*, 673–676 (1989).
[6560] Zhao, Z.; H. Chen, *Fenxi Shiyanshi 10/5*, 55 (1991).
[6561] Zheng, Y.; X.-q. Shan, P. Sun, L.-z. Jin, *Chem. Speciation Bioavailability 3/1*, 30–36 (1991).
[6562] Zheng, Y.-s.; R. Woodriff, J.A. Nichols, *Anal. Chem. 56/8*, 1388–1391 (1984).
[6563] Zheng, Y.-s.; D.-l. Zhang, *Guangpuxue Yu Guangpu Fenxi 9/6*, 54 (1989).
[6564] Zheng, Y.-s.; D.-l. Zhang, *Anal. Chem. 64/15*, 1656–1659 (1992).
[6565] Zheng, Y.-s.; X.-g. Su, *Talanta 40/3*, 347–350 (1993).
[6566] Zheng, Y.-s.; X.-g. Su, *Can. J. Appl. Spectroscopy 38/4*, 109–113 (1993).
[6567] Zheng, Y.-s.; Y. Wang, *Spectrosc. Lett. 27*, 817 (1994).
[6568] Zheng, Y.-s.; X.-g. Su, *Mikrochim. Acta (Vienna) 112/5-6*, 237–243 (1994).
[6569] Zheng, Y.-s.; Y.-q. Wang, *Talanta 42/3*, 361–364 (1995).
[6570] Zhi, Z.; L. Wang, S. Dong, *Zhongguo Yaoke Daxue Xuebao 22/1*, 33 (1991).
[6571] Zhou, C.Y.; M.K. Wong, L.L. Koh, Y.C. Wee, *Anal. Sci. 12/3*, 471–476 (1996).
[6572] Zhou, C.Y.; M.K. Wong, L.L. Koh, Y.C. Wee, *Talanta 43/7*, 1061–1068 (1996).
[6573] Zhou, C.Y.; M.K. Wong, L.L. Koh, Y.C. Wee, *J. Anal. At. Spectrom. 11/8*, 585–590 (1996).
[6574] Zhou, H.-g.; X.-y. Peng, R. Tao, *Guangpuxue Yu Guangpu Fenxi 12/3*, 97–100 (1992).
[6575] Zhou, N.-g.; W. Frech, E. Lundberg, *Anal. Chim. Acta 153*, 23–31 (1983).
[6576] Zhou, N.-g.; W. Frech, L. de Galan, *Spectrochim. Acta, Part B 39/2-3*, 225–235 (1984).
[6577] Zhou, Z.-f.; W.-b. Xu, S.-g. Ye, *Huanjing Wuran Yu Fangzhi 11/4*, 43 (1989).
[6578] Zhu, X.-m.; G. Duc, R. Faure, *Analusis 20/7*, 397–399 (1992).
[6579] Zhuang, Z.-x.; P.-y. Yang, J. Luo, X.-r. Wang, B.-l. Huang, *Can. J. Appl. Spectroscopy 36*, 9–14 (1991).
[6580] Zhuang, Z.-x.; P.-y. Yang, X.-r. Wang, Z.-w. Deng, B.-l. Huang, *J. Anal. At. Spectrom. 8/8*, 1109–1111 (1993).
[6581] Zhuang, Z.-x.; C.-l. Yang, X.-r. Wang, P.-y. Yang, B.-l. Huang, *Fresenius J. Anal. Chem. 355/3-4*, 277–280 (1996).
[6582] Zielke, H.J.; W. Luecke, *Fresenius Z. Anal. Chem. 271*, 29–30 (1974).
[6583] Zima, S.; M. Šucmanová, O. Čelechovská, I. Řehůřková, E. Šucman, in: B. Welz (Editor), *5. Colloquium Atomspektrometrische Spurenanalytik*, Bodenseewerk Perkin-Elmer, Überlingen, 1989, 769–778.
[6584] Zimmer, K.; *Fresenius Z. Anal. Chem. 324*, 875–885 (1986).
[6585] Zolotov, Yu.A.; M. Grasserbauer, G.H. Morrison, Yu.A. Karpov, *Pure Appl. Chem. 57/8*, 1133–1152 (1985).
[6586] Zolotovitskaya, E.S.; V.G. Potapova, T.V. Druzenko, *Zh. Anal. Khim. 47/10-1*, 1918–1921 (1992).
[6587] Zong, Y.Y.; P.J. Parsons, W. Slavin, *J. Anal. At. Spectrom. 11/1*, 25–30 (1996).
[6588] Zorro, J.; M. Gallego, M. Valcárcel, *Microchem. J. 39/1*, 71–75 (1989).
[6589] Zou, H.-f.; S.-k. Xu, Z.-l. Fang, *At. Spectrosc. 17/3*, 112–118 (1996).
[6590] Zsákó, J.; *Anal. Chem. 50/8*, 1105–1107 (1978).
[6591] Zucchetti, S.; G. Contarini, *At. Spectrosc. 14/2*, 60–64 (1993).
[6592] Zunk, B.; *Anal. Chim. Acta 236/1*, 337–343 (1990).
[6593] Zurera-Cosano, G.; R. Moreno-Rojas, F. Rincón-León, M.A. Amaro-Lopez, *J. Sci. Food Agric. 57*, 565 (1991).
[6594] Zurera-Cosano, G.; R. Moreno-Rojas, M. Amaro-Lopez, *Food Chem. 51*, 75–78 (1994).
[6595] Zytner, R.G.; N. Biswas, J.K. Bewtra, *Int. J. Environ. Stud. 33/4*, 299–306 (1989).

Index

A

absolute analysis 235
–, using GF AAS 432
absolute calibration 91
absolute sensitivity, dependence on graphite tube dimensions 176
absorbance 91, 143
–, integrated 30, 36, 147, 170, 364
–, negative 370
absorbance coefficient, molar 143
–, specific 143
absorbance measurements, spatially resolved 96
absorbance noise 258
absorbing species, distribution in the absorption volume 92
absorptance 91, 143
absorption, by molecules 15
–, non-specific 15
–, of radiation 2, 64
absorption cells, for CV AAS 218
absorption coefficient 86, 89
absorption cross-section 89
absorption factor, spectral 91, 143
absorption line, intensity 68
–, total width 275
absorption profile, broadening 342
–, –, at higher temperatures 365
absorption profiles, of spectral lines 82
absorption signals, time-resolved, for recognizing interferences 284
absorption spectrum 65
absorption volume 67, 149
–, influence on spatial-distribution interferences 345
–, temporally non-isothermal 34
absorption zone, non-isothermal 34
accuracy 221
acid concentration, influence on hydride generation 460
–, influence on interferences, in HG AAS 457
acid digestion, microwave-assisted 316
acid digestions, for sample pretreatment 227
activation energies, for the atomization of copper 395

activation energy, calculation, from atomization signals 363, 383
A/D conversion 145
addition calibration technique 239
additive interferences 240
adsorption, of the analyte, on graphite surfaces 383
adsorption-desorption mechanism, for atomization, in GF AAS 363, 387, 394
adsorption volume 69
aerosol, conditioning 163
–, enhanced exploitation 169
–, formation 156
–, primary 157
–, secondary 157
–, tertiary 157, 163, 165
–, transport 156
–, transport through the spray chamber 163
aerosol formation, efficiency 165
aerosol particles, rate of volatilization, in flames 346
air, analysis 694
–, speciation analysis 702
air-acetylene flame 150
–, fuel rich 151
–, –, influence on interferences 356
air-hydrogen flame 23, 153
aldehydes, indirect determination 535
algae, as sorbent, for trace elements 640
alkali metals, atomization mechanism, in GF AAS 387
alkaline-earth elements, atomization mechanism, in GF AAS 387
–, flame requirements 150
–, half-width of resonance line 75
–, interference of aluminium 320
–, solute-volatilization interferences 350
alternate lines 68
aluminium, acid-soluble, determination in steel 725, 729
–, analyte atom distribution, during atomization in GF AAS 376
–, atomization mechanism, in GF AAS 395
–, bioavailability, in pharmaceuticals 753
–, degree of atomization 339
–, determination 479